Principles of Organic Synthesis

D0643209

Principles of
Organic Synthesis

R. O. C. NORMAN FRS
Ministry of Defence

SECOND EDITION

LONDON   NEW YORK
CHAPMAN AND HALL

HOUSTON PUBLIC LIBRARY

RO137865407
bstca

First published 1968
by Methuen & Co Ltd
Second edition 1978 published
by Chapman and Hall Ltd
11 New Fetter Lane, London EC4P 4EE
Reprinted 1981, 1986
Published in the USA by Chapman and Hall
29 West 35th Street, New York NY 10001

© *1968, 1978 R. O. C. Norman*

Printed in Great Britain by
Richard Clay (The Chaucer Press) Ltd, Bungay, Suffolk

ISBN 0 412 15520 6

All rights reserved. No part of this book may be reprinted, or repro-
duced or utilized in any form or by any electronic, mechanical or
other means, now known or hereafter invented, including photocopy-
ing and recording, or in any information storage and retrieval system,
without permission in writing from the Publisher.

This paperback edition is sold subject to the condition that it shall
not, by way of trade or otherwise, be lent, re-sold, hired out, or
otherwise circulated without the publisher's prior consent in any form
of binding or cover other than that in which it is published and without
a similar condition including this condition being imposed on the sub-
sequent purchaser.

To R. P. B.
and W. A. W.

Preface to the First Edition

The last thirty years have witnessed a profound increase in our understanding of the ways in which organic compounds react together—their mechanisms of reaction. This has, on the one hand, become a large, discrete branch of organic chemistry; but it has also, on the other, had a considerable impact on our approach to devising methods for the synthesis of organic compounds. To the student, reaction mechanism can have a two-fold appeal: it is, in its own right, an intellectually stimulating subject in its rationalization and unification of complex processes; and it also provides a relatively simple superstructure on which the vast array of the facts of organic chemistry can be hung. In a paradoxical way, the amount to be usefully *learned* in a subject to which an array of facts is being added daily remains, as our *understanding* grows, almost unchanged.

The purpose of this book is to show how an understanding of these mechanistic principles can usefully be applied in thinking about and planning the construction of organic compounds. It is designed for those who have had a brief introduction to organic chemistry; an elementary knowledge of the nomenclature and structures of organic compounds is assumed. The text is divided into two parts. In the first five Chapters, mechanism is set in its wider context of the basic principles and concepts underlying chemical reactions: chemical thermodynamics, structural theory, theories of rates of reaction, mechanism itself, and stereochemistry. In the remaining fourteen Chapters, these principles and concepts are applied to the problems involved in putting together particular types of bonds, groupings, and compounds. The account is not intended to be exhaustive; for example, the vast body of evidence on which mechanisms are based has been omitted, nor are experimental details included. The object has been to convey a broad understanding rather than to produce a reference text.

I should like to acknowledge the help I have received from many of my former colleagues at the Dyson Perrins Laboratory, Oxford, my present colleagues in the University of York, and in particular Professor A. W. Johnson, F.R.S., who read the whole of the manuscript and made many helpful suggestions, and two of my former pupils, Messrs. A. J. Hart-Davis and J. C. MacDougall, who helped with the preparation of several Chapters.

R. O. C. NORMAN

Preface to the Second Edition

The ten years that have passed since this book was first published have served to emphasize the value to the organic chemist of approaching problems in synthesis with the aid of a thorough understanding of reaction mechanism. This is apparent in the design both of new synthetic methods and of the multi-stage synthesis of complex target molecules.

In this new edition, then, the basic Chapters which comprise Part I remain essentially unchanged while the Chapters, in Part II, which develop these basic ideas and show their operation in practice, have been brought up to date. In particular, there have been important advances in our understanding of the course of pericyclic reactions—concerted processes which occur within a cyclic array of the participating atomic centres—and this has led to the complete re-drafting of Chapter 9; valuable new methods which make use of reagents containing phosphorus, sulphur, or boron have been introduced (Chapter 15); and the use of photochemical methods in synthesis has advanced to the stage where a separate Chapter is justified. In addition, numerous new reagents have been included, especially in the Chapters on oxidation and reduction. Finally, the chance has been taken to transfer to S.I. units.

I am indebted to my colleagues Dr B. C. Gilbert and Dr J. M. Vernon for their advice in producing this edition, and especially to my colleague Dr Peter Hanson for his help in the preparation of the new material.

R. O. C. NORMAN

Contents

Part I

Introduction to Part I

The five Chapters which comprise Part I are concerned with the principles which govern organic reactions.

The first Chapter examines the implications of the laws of thermodynamics. Reactions can "go"—that is, have equilibrium constants greater than unity—only if the products have a lower free-energy content than the reactants. The free energy of a species is related to its enthalpy, which is determined essentially by the strengths of the bonds it contains, and to its entropy, which is a measure of its degree of disorder; a low free energy corresponds to a system's having strong bonding forces and a high degree of disorder. From thermodynamic considerations there follows, for example, an understanding of why it is possible to reduce acetylene to ethylene at room temperature whereas ethylene is successfully dehydrogenated to give acetylene at a temperature of about 1,000°C.

The second Chapter considers the current theories of bonding in organic molecules and relates these to the strengths of the bonds in typical chemical groupings. Thus, the different chemical properties of benzene and ethylene are then seen to be related to the very considerable stabilization energy of the benzene ring. Other properties of organic compounds which are of importance in synthesis, such as the acidities of C—H bonds in various environments, also follow from structural theory.

That there should be a negative free-energy change is in practice a necessary but not a sufficient condition for a reaction to occur, for the *rate* at which it takes place may be negligible. Thermodynamic considerations alone indicate that hydrocarbons should not coexist with air, for the free-energy change involved in their oxidation to carbon dioxide and water is significantly negative; in practice, however, their rates of combustion at ordinary temperatures are negligible. The third Chapter sets out the theories of reaction kinetics and the effects of temperature on rate, and then introduces correlations of the rates of specific types of reaction with structure.

The planning of syntheses is helped considerably by an understanding of the mechanisms by which reactions occur. It must be emphasized that mechanisms are *theories* and not *facts* of the subject; they have been deduced from experimental observations and in some instances they transpire to be incorrect or at least in need of refinement; one should say, rather, 'the mechanism is thought to be,' than 'the mechanism is.' Nonetheless, the current mechanistic theories, which are surveyed in the fourth Chapter, not only provide an intellectually

3

satisfying and unifying picture of the complexity of reactions, but also enable predictions to be made, with increasing assurance as the degree of rationalization of the subject increases, of the effects which structural modifications will have on the course of a reaction.

Stereochemistry—the study of the spatial relationships of atoms and bonds— would in the past have been a natural adjunct of the study of molecular structure. It is now as important to considerations of chemical dynamics as to those of chemical statics, and follows naturally, in the last Chapter, the study of kinetics and mechanism. Indeed, it is closely intertwined with mechanism; many naturally occurring compounds have a complex and highly specific stereochemistry, and it has only been through an understanding of the stereoelectronic principles of reactions that their syntheses have been successfully planned and executed.

1. Chemical Thermodynamics

1.1 Equilibrium

All chemical reactions are in principle reversible: reactants and products eventually reach *equilibrium*. In some cases, such as the esterification of an acid by an alcohol,

$$\text{RCO}_2\text{H} + \text{R}'\text{OH} \rightleftharpoons \text{RCO}_2\text{R}' + \text{H}_2\text{O}$$

the equilibrium situation is quite closely balanced between reactants and products, whereas in other cases the equilibrium constant is either very high or very low, so that the reaction goes essentially to completion in one direction or the other (given the appropriate conditions). From the point of view of devising an organic synthesis it is necessary to know whether the position of equilibrium will favour the desired product. The factors which determine the equilibrium constant of a reaction and its variation with changes in conditions follow from two of the most firmly established natural laws: the First and Second Laws of Thermodynamics.

1.2 The First Law

The First Law of Thermodynamics is commonly expressed as the Law of Conservation of Energy: energy can be neither created nor destroyed. Consider a system into which an amount of heat q is introduced. The absorption of this heat may bring about both an increase in the energy of the system (manifested, for example, by a rise in the temperature) and also the performance of work, w, by the system (as, for example, the pushing back of a piston by a heated gas*). Then it follows from the First Law that, for a change from state A to state B,

$$q = \Delta E + w$$

where ΔE is the change in energy of the system. It is convenient to define a function E as the *internal energy* of the system; ΔE is then equal to $E_B - E_A$, where E_A and E_B are the energies of the initial and final states.

If the process is carried out at constant volume, no mechanical work is done by the system and $q = \Delta E$. On the other hand, if the process is carried out at

---

*There are other forms of work, such as electrical work, but we shall not be concerned with these.

constant pressure the work done is $P\Delta V$, where ΔV is the change in volume
of the system. Then

$$q = \Delta E + P\Delta V$$
$$= E_B - E_A + PV_B - PV_A$$
$$= (E_B + PV_B) - (E_A + PV_A)$$

Since E, P, and V are functions of the state of the system, $(E + PV)$ is also
such a function. This property is termed the *heat content* or *enthalpy*, H, and
proves to be important in chemical systems where we are usually concerned
with changes taking place at constant (*e.g.*, atmospheric) pressure.

1.3 The Second Law

If a box-full of red balls and a box-full of blue balls are poured into a container and shaken up, we shall expect to find when the balls are poured back
into the two boxes that each box contains approximately equal numbers of red
and blue balls. On the other hand, if we start with boxes containing mixtures
of the balls we shall not expect to find, after the mixing process, that all the red
balls end in one box and the blue balls in the other. Again, if a poker is made
red-hot at one end, the heat gradually diffuses along the poker until eventually
its temperature is uniform along its whole length. We do not, however, observe
the reverse of this process: a poker at ambient temperature never becomes
hotter at one end and colder at the other, even though this would not necessarily
contravene the First Law.

Observations of this type led to the enunciation of the Second Law of Thermodynamics, the classical form of which is that 'heat does not flow spontaneously
from a colder to a hotter body.' The relevance of this Law to the example of
the poker is obvious, but its applicability to the problem of mixing balls of
different colours is not immediately apparent. It becomes so, however, when it
is realized that the Second Law is concerned with *probabilities*: it is extremely
improbable that a mixture of red and blue balls will, by a shuffling process, end
in the ordered condition of separate groups of red balls and blue balls. Given
only six red and six blue balls, it is 200 times as probable that the first six balls
poured from the mixing container into one box will consist of three red and three
blue than that they will be all red or all blue. Moreover, the ratio of the number
of ways in which a system can be arranged in a 'random' manner to
those corresponding to a particular 'ordered' arrangement increases rapidly
with the number of species contained in the system, so that in the case of chemical
molecules, which in any particular system under observation are numbered in
many powers of ten, the likelihood of an 'ordered' system emerging spontaneously from a 'disordered' one is negligible. We should not expect that the
random collisions between the molecules in the poker would result in all the

faster moving (*i.e.*, hotter) molecules accumulating at one end of the poker and all the slower moving (*i.e.*, colder) molecules at the other, but, given this ordered situation (by heating the poker at one end), we should expect the faster and slower moving molecules to attain a random arrangement, as a result of molecular movements and collisions, corresponding to a uniform temperature along the poker.

There is thus a tendency for ordered systems to become disordered. It is convenient to have a measure of the *degree of disorder* of a system, and this is defined as its *entropy*, S, where $S = \mathbf{k} \ln W$, W being the number of ways in which the system may be arranged and $\mathbf{k}$ being Boltzmann's constant. In the two *irreversible* processes described above (the mixing of differently coloured balls and the achievement of uniform temperature in the poker) there is an increase in the entropy of the system, and an alternative formulation of the Second Law of Thermodynamics is that 'the entropy of an isolated system tends to increase.'

1.4 Free Energy

We assumed, in considering the mixing of red and blue balls, that no forces were operative. Suppose now that the red balls exert strong attractive forces on balls of their own kind and repulsive forces on blue balls. There will then be a tendency for the red balls to stay together so as to decrease the potential energy of the system, and this will oppose the tendency for the entropy of the system to increase. It transpires, from thermodynamical arguments, that the resulting compromise of these opposed trends is determined, for a system at constant pressure and temperature, by the value of the function $(H - TS)$: this tends to decrease, and the compromise situation (i.e., equilibrium) corresponds to its minimum value. It is convenient to define a new function, G, the *Gibbs free energy*, as $G = H - TS$. A process will occur spontaneously if, as a result, G decreases. It will continue until G reaches a minimum, and this point corresponds to the equilibrium situation; forward and reverse reactions continue, but at equal rates; that is, the equilibrium is a dynamic one. Consider a reaction occurring at temperature T in the gas phase. Since

$$G = H - TS = E + PV - TS,$$

$$dG = dE + PdV + VdP - TdS$$

For an equilibrium situation it can be shown that $TdS = dq$, and since (from 1·2) $dq = dE + PdV$

$$dG = VdP$$

Since, for 1 mole of a perfect gas at a constant temperature, $PV = RT$, i.e.,

$V = RT/P$, integration gives,

$$[G]_0^1 = \int_0^1 V dP = \int_0^1 (RT/P) dP = [RT \ln P]_0^1$$

for a change in the system between the two states for which $G = G^0$ and G^1, respectively,

$$\textit{i.e., } G^1 - G^0 = RT \ln (P_1/P_0)$$

It is convenient to take G^0 as the free energy of the gas at temperature T in its *standard state* (i.e., for 1 mole at 1 atmosphere pressure). Then $G^1 - G^0 = RT \ln P_1$.

Consider the equilibrium,

$$A + B \rightleftharpoons C + D$$

The free-energy change in the reaction is the sum of the free energies of the products minus the sum of the free energies of the reactants, i.e., $\Delta G = G(C) + G(D) - G(A) - G(B)$. Thus,

$$\Delta G^1 - \Delta G^0 = RT \ln (p_1^C p_1^D / p_1^A p_1^B)$$

Now, if the set of pressures, p_1, are the *equilibrium* pressures in the gas mixture, then, since at equilibrium G must be a minimum with respect to any displacement of the system, $\Delta G^1 = 0$, and it follows that

$$- \Delta G^0 = RT \ln (p_1^C p_1^D / p_1^A p_1^B)$$

Since the G^0 values are dependent only on temperature,

$$- \Delta G^0 = RT \ln K_p$$

where K_p is also dependent only on temperature.

For a reaction in solution it may readily be shown that

$$- \Delta G^0 = RT \ln (c^C c^D / c^A c^B) = RT \ln K_c$$

where c^i is the concentration of the ith reactant at equilibrium. (Strictly, c^i should be replaced by the *activity*, a^i.)

Thus, equilibrium constants are related to the standard free energies of the reactants and products, and a knowledge of these enables us to predict whether a given set of reactants is likely to yield a desired set of products. If ΔG^0 is negative, the equilibrium constant (greater than 1) is favourable to the formation of the products, whereas if ΔG^0 is positive it is correspondingly unfavourable.*

Standard free energies have been measured for a large number of organic and inorganic compounds. For example, those of ethylene, ethane, and hydrogen in the gaseous state at 25°C are 68, -33, and 0 kJ mol^{-1}, respectively. (By convention, the most stable allotrope of an *element* in its standard state at 25°C is assigned a free-energy value of zero. The scale of free energies is therefore an arbitrary one, but this is immaterial since we are always concerned with *differences* in free energies.) Therefore the reaction

$$C_2H_6 \rightleftharpoons C_2H_4 + H_2$$

has $\Delta G^0 = +101$ kJ mol^{-1}, so that we cannot obtain ethylene from ethane in significant amount under these conditions although we may expect to obtain ethane from ethylene. Notice, however, that even though a free-energy change may be favourable, the *rate* of formation of the product may be too slow for the reaction to be practicable, or the reaction may take a different course (e.g., ethylene might react with two molecules of hydrogen to give two molecules of methane, which is also a thermodynamically favourable process). Fulfilment of the thermodynamic criterion is therefore a *necessary* but not a *sufficient* condition for a reaction to occur. The factors which control rates of reaction and the pathways which reactions take are discussed in subsequent chapters.

Thermodynamic data also give no information about the *reasons* why the free energies of compounds have particular values. Ethylene, for example, has a positive standard free energy of formation, whereas that for ethane is negative; that is, ethylene is unstable with respect to carbon and hydrogen and ethane is stable. These are empirical *facts*; the reasons underlying them lie in the realm of *theory*, namely, the theory of chemical structure and bonding, which is discussed in the next chapter.

1.5 The Effect of Temperature on Equilibrium

Since $RT \ln K_p = -\Delta G^0 = -\Delta H^0 + T\Delta S^0$,

$$d \ln K_p/dT = \Delta H^0/RT^2$$

This expression, the van't Hoff isochore, shows how the equilibrium constant varies with temperature.

The isochore not only enables ΔH^0 to be evaluated by measurement of the equilibrium constant at a series of temperatures, but also indicates that the equilibrium constant in an exothermic reaction (negative ΔH^0) decreases with

---

*An 'uphill' reaction may, however, occur if the product is *removed* from equilibrium by, for example, precipitation or involvement in a subsequent reaction.

rise in temperature whereas that of an endothermic reaction (positive ΔH^0) increases. In effect, the isochore is a mathematical formulation of the application of Le Chatelier's principle to temperature changes.

An example of the application of the equation occurs in the commercial production of Buna S rubber, a synthetic rubber obtained by copolymerizing styrene and butadiene. These components are made from benzene and ethylene, and from 1-butene, respectively:

$$C_6H_6 + C_2H_4 \rightleftharpoons C_6H_5\text{---}C_2H_5$$

$$C_6H_5\text{---}C_2H_5 \rightleftharpoons C_6H_5\text{---}CH\text{==}CH_2 + H_2$$
Styrene

$$CH_3\text{---}CH_2\text{---}CH\text{==}CH_2 \rightleftharpoons CH_2\text{==}CH\text{---}CH\text{==}CH_2 + H_2$$
Butadiene

The first reaction is exothermic and the others endothermic. For the formation of ethylbenzene, $\ln K_p$ is calculated from thermodynamic data to be about $+5$ at 200°C and zero at 800°C, so that the yield of ethylbenzene is increased by operating at as low a temperature as possible, consistent with the occurrence of the reaction at a practicable rate. Conversely, the yields of styrene and butadiene are improved by carrying out the reactions at the highest temperatures compatible with the non-occurrence of side-reactions.

1.6 Bond Energies

Most reactions result in the absorption or liberation of heat, e.g.,

$$CH_4(g) + 2\,O_2(g) \rightarrow CO_2(g) + 2\,H_2O(l) \qquad \Delta H^0 = -890 \text{ kJ}$$

Heats of reactions may be employed in an additive manner. For example, the heats of the following three reactions have been measured (the first two by a spectroscopic and the third by a calorimetric procedure):

(1) $H_2 \rightarrow 2\,H$ $\Delta H^0 = 435$ kJ

(2) $O_2 \rightarrow 2\,O$ $\Delta H^0 = 494$ kJ

(3) $H_2 + \frac{1}{2}\,O_2 \rightarrow H_2O$ $\Delta H^0 = -242$ kJ

It follows that, for the reaction

(4) $2\,H + O \rightarrow H_2O$

$$\Delta H(4) = \Delta H(3) - \Delta H(1) - \tfrac{1}{2}\Delta H(2) = -924 \text{ kJ}$$

for if $\Delta H(4)$ had any other value, it would be possible to carry out the four reactions in such a sequence that the starting materials could be recovered in their original state and energy could be gained at the same time, contrary to the First Law of Thermodynamics.

Measurement of the heats of certain reactions enables the heats of others to be calculated. Moreover, such calculations are facilitated by constructing a table of *bond energies* from the observed data, taking as the reference point the heat of formation of molecules from their constituent atoms. For instance, from the heat of combustion of methane quoted above together with the heats of formation of carbon dioxide and water, the heat of the reaction,

$$CH_4(g) \rightarrow C(g) + 4H(g)$$

is calculated to be 1655 kJ,* and since four C—H bonds are broken in the process, the bond energy of the C—H bond is defined as one-quarter of this, 414 kJ. The O—H bond energy is one-half the heat of formation of water, *i.e.*, 462 kJ. To a close approximation, bond energies are constant for a particular bond in different structural environments, so that, given the bond energies 414, 357, and 462 kJ for C—H, C—O, and O—H bonds respectively, the total bond energy of methanol, which contains three C—H, one C—O, and one O—H bond, is calculated to be 2061 kJ, in fair agreement with the experimental value (1985 kJ) derived from the heat of combustion of methanol.

Bond energies for a number of commonly occurring structural units are given in Table 1.1.

Table 1.1   Bond energies (kJ mol^{-1}) at 25°C

H—H	435	C—C	347	C—O	357
H—F	560	C=C	610	C=O^a	694
H—Cl	428	C≡C	836	C=O^b	736
H—Br	364	N—N	163	C=O^c	748
H—I	297	N=N	418	H—C	414
F—F	150	N≡N	940	H—N	389
Cl—Cl	238	C—N	305	H—O	462
Br—Br	188	C=N	614	O—O	155
I—I	150	C≡N	890	S—S	251

^{a}In formaldehyde. ^{b}In other aldehydes. ^{c}In ketones.

*This is the sum of the heats of the following five reactions:

$$CH_4(g) + 2\ O_2(g) \rightarrow CO_2(g) + 2\ H_2O$$
$$CO_2(g) \rightarrow C(s) + O_2(g)$$
$$2\ H_2O(l) \rightarrow 2\ H_2(g) + O_2(g)$$
$$2\ H_2(g) \rightarrow 4\ H(g)$$
$$C(s) \rightarrow C(g)$$

The heat of the last reaction, the atomization of carbon, is necessary because the heat of formation of carbon dioxide is measured for solid carbon whereas bond-energy data refer to atoms, *i.e.*, 'gaseous' carbon. It has been difficult to measure the heat of atomization of carbon, but the value now generally accepted is 718 kJ mol^{-1}.

Bond energies quoted above are *average* bond energies, and it is often more useful to know the energy required to break a particular bond in a compound, *i.e.*, the *bond dissociation energy*. Although the O—H bond energy is, by definition, one-half the heat of formation of water, in fact a greater quantity of energy is required to break the first O—H bond in water ($H_2O \rightarrow HO + H$; $\Delta H = 491$ kJ) than the second ($HO \rightarrow O + H$; $\Delta H = 433$ kJ). The bond dissociation energy is therefore greater than the bond energy, and this is commonly the case, except for diatomic molecules for which the values are necessarily identical. Some typical bond dissociation energies are set out in Table 1.2.

Table 1.2   Bond dissociation energies (kJ mol^{-1})

$H-CH_3$	426	$H-OH$	491	CH_3-F	447
$H-CH_2CH_3$	401	$H-NH_2$	426	CH_3-Cl	339
$H-CH(CH_3)_2$	385	CH_3-OH	376	CH_3-Br	280
$H-C(CH_3)_3$	372	CH_3-NH_2	334	CH_3-I	226
$H-CH_2Ph$	322	CH_3-CH_3	347	CH_3-NO_2	238

The premise on which the use of Table 1.1 is based, that bond energies are constant for a particular bond in different environments, requires closer examination. If it were strictly true, for example, the heats of combustion of the three isomeric pentanes, each of which contains four C—C and twelve C—H bonds, would be the same. In fact, they are different: n-pentane, 3533; isopentane, 3525; neopentane, 3513 kJ mol^{-1}. Nevertheless, the percentage differences are so small in this and many similar cases that bond-energy data are of considerable use. However, there are certain structural environments in which appreciable differences between observed and predicted bond energies occur. One example, in Table 1.1, concerns the carbonyl group, whose bond energy is nearly 10% greater in ketones than in formaldehyde, and is greater still (804 kJ) in carbon dioxide.

A more general deviation occurs with *conjugated* compounds, that is, compounds possessing alternating single and double bonds. For example, consider the following hydrogenations:

$CH_2=CH-CH=CH_2 + 2 H_2 \rightarrow CH_3CH_2CH_2CH_3$ $\qquad \Delta H^\circ = -238 \cdot 7$ kJ

$CH_2=CH-CH_2-CH=CH_2 + 2 H_2 \rightarrow CH_3CH_2CH_2CH_2CH_3$ $\qquad \Delta H^\circ = -254 \cdot 2$ kJ

If bond energies were strictly additive, the heats of hydrogenation in the above reactions should be the same, since in each, two C=C double bonds are converted into single bonds, two H—H bonds are broken, and four C—H bonds are formed. The lower value for butadiene shows that this is a more stable compound than predicted by examination of 1,4-pentadiene, and this proves to be so for other conjugated systems. We shall refer to the difference between predicted and observed values of heats of hydrogenation as the *stabilization energy*.

The attachment of alkyl groups to olefinic carbon also leads to the introduction of stabilization energy. For example, whereas the heat of hydrogenation of ethylene is 137 kJ mol^{-1}, that of propylene ($CH_3CH=CH_2$; one alkyl group) is 126 kJ mol^{-1} and that of *trans*-2-butene ($CH_3CH=CHCH_3$; two alkyl groups) is 115 kJ mol^{-1}; i.e. the stabilization energies are 11 and 22 kJ mol^{-1}, respectively.

For those cyclic, conjugated compounds which are aromatic (2.6b) heats of hydrogenation are in many cases very considerably less than the predicted values. The stabilization energies of benzene, naphthalene, and anthracene are respectively, 150, 255, and 349 kJ mol^{-1}.

An understanding of the origin of stabilization energies follows from current theories of molecular structure (Chapter 2). It is here necessary to emphasize that bond-energy data must be used with care but that nevertheless, conjugated and aromatic systems excepted, they provide a useful working basis for predictive purposes.

1.7 Entropy

In the problem of the mixing of red and blue balls, the entropy of the ordered system (red and blue balls each in their initial containers) is zero since $S = k \ln W$ and there is only one way in which the system can be arranged ($W = 1$). If, however, the red and blue balls were atoms or molecules the entropy of the initial (unmixed) system would not be zero, except for perfect crystalline solids at 0 K, for as a result of the motions which atoms and molecules undergo at all temperatures above absolute zero, there is a number of ways in which an aggregate of such species may be arranged.

The energy possessed by a molecule is manifested as translational, rotational, and vibrational energy. In each case, the energy levels are *quantized*: that is, only certain values occur (see 2.2). Consider any one form of motion, for which the energy levels correspond to energies ϵ_1, $\epsilon_2 \ldots \epsilon_j \ldots$ Of the total of N molecules, if N_1 have energy ϵ_1, N_2 have energy ϵ_2, and in general N_j have energy ϵ_j, the number of ways in which the system can be arranged is given by

$$W = \frac{N!}{N_1! \, N_2! \, \ldots \, N_j! \, \ldots}$$

Further, statistical treatments lead to a relation between the number of molecules in the *j*th energy level and the energy of that level:

$$N_j/N = g_j e^{-\epsilon_j/kT} \Big/ \sum_j g_j e^{-\epsilon_j/kT}$$

(The term g_j is introduced to take account of the fact that a particular energy level may be degenerate; that is, there may be more than one state with this energy.)

This result shows that the molecules are distributed throughout the energy states in a manner dependent only on the energies of those states, the absolute temperature, and the statistical factor, g_j. Since the entropy is related, through the probability W, to N and N_j, entropies.can be evaluated by determining the values of ϵ_j for translational, rotational, and vibrational motion. It is a general property of the relationship that, as the difference in energies between successive states is decreased, the entropy increases.

The translational energy levels are very closely spaced so that the higher levels are well populated and the entropy due to translational motion is large. S_{tr}^0 for an ideal gas at 1 atmosphere pressure is given by the expression, $\frac{3}{2}R \ln M + \frac{5}{2}R \ln T - 9{\cdot}6$ J K^{-1}, where M is the molecular weight. Typical values at 25°C are 153 J K^{-1} for hydrogen chloride, and 144 J K^{-1} for methane.

The rotational energy levels are less closely spaced and the resulting contribution to the total entropy is correspondingly smaller. Typical values of $S_{rot.}^0$ at 25°C are 34 J K^{-1} for hydrogen chloride and 32 J K^{-1} for methane.

There are two forms of vibrational motion: stretching and bending vibrations. * The energy levels for stretching vibrations are usually widely spaced, so that the vibrational contribution to the entropy is negligible. The energy levels for bending vibrations are closer together, and the contribution to the entropy, particularly for a molecule which has a large number of bending modes, is significant.

Many polyatomic molecules possess entropy by virtue of their undergoing *internal rotations*. For example, the methyl groups in ethane, CH_3—CH_3, rotate with respect to each other, so that the molecule can adopt any of an infinite number of possible *conformations*. In one of the two shown in the projection diagram below, the hydrogen atoms on adjacent carbons eclipse each other, and in the other they are fully staggered (p. 46). Two conformations of n-butane, showing possible arrangements of the four carbon atoms, are also shown.

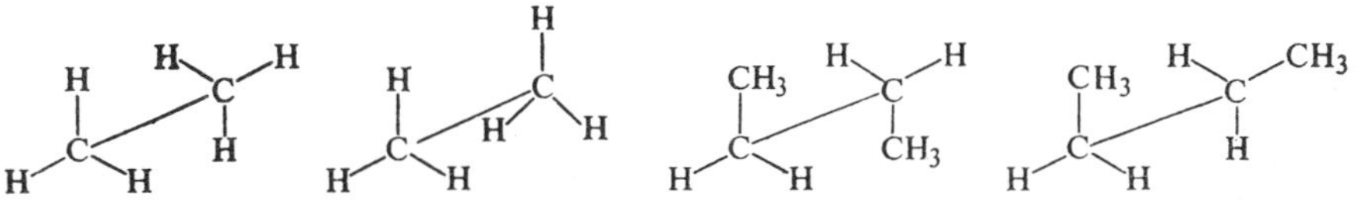

*A non-linear molecule containing n atoms ($n > 2$) has $3n - 6$ vibrational modes of which $n - 1$ are stretching modes and $2n - 5$ are bending modes, together with 3 translational and 3 rotational degrees of freedom.

The conformations are not necessarily equally energetic; in n-butane, for example, the first of the above two structures is of lower energy-content than the second, largely because of the repulsive forces between the two methyl groups in the latter (p. 46). Nevertheless, the energy differences are normally small, so that the higher-energy conformations are quite heavily populated. As a consequence, there is a contribution to the entropy of the molecule which can be thought of as arising from a particularly loose form of bending vibration. The magnitude of conformational entropy increases with the number of available conformations and hence with the length of the chain of atoms.

The most significant aspects of the above discussion are, first, the importance of the translational entropy in determining the total entropy and, secondly, the significance of the conformational entropy.

(*i*) In a reaction which results in an increase in the number of species, such as the fragmentation $A \rightarrow B + C$, there is a considerable increase in entropy because of the gain of three degrees of translational freedom. (There are also changes in the rotational and vibrational contributions to the entropy, but these are normally much smaller and can be neglected for the present argument.) On the other hand, such reactions may result in a decrease in the number of bonds and, associated with this, a decrease in the enthalpy of the system; the simplest example of such a reaction is the dissociation of the hydrogen molecule, $H_2 \rightarrow 2 H\cdot$ ($\Delta H = +435$ kJ mol^{-1}). Even when there is no overall change in the number of bonds, there may still be a decrease in the enthalpy of the system; an example is the dehydrogenation of ethylene, $CH_2{=}CH_2 \rightarrow CH{\equiv}CH + H_2$ ($\Delta H = +167$ kJ mol^{-1}). In these cases, the enthalpy and entropy terms are opposed and whether ΔG is negative or positive (and hence whether K is greater or less than unity) depends on whether the gain in translational entropy is larger or smaller than the decrease in enthalpy.

Now, $\Delta S_{tr} = \frac{3}{2}R \ln (M_B M_C / M_A) + \frac{5}{2}R \ln T - 9{\cdot}6$ J K^{-1} mol^{-1} (p. 14). Thus, at 25°C, for the dissociation of the hydrogen molecule, $\Delta S_{tr} \sim 100$ J K^{-1} mol^{-1}, and for the dehydrogenation of ethylene, $\Delta S_{tr} \sim 117$ J K^{-1} mol^{-1}. The contributions from the increase in translational freedom to the decrease in free energy during the forward reaction ($= T\Delta S$) at 25°C are therefore ~ 30 and 35 kJ mol^{-1}, respectively. In each case, these values are far smaller than the values for the decrease in enthalpy, so that ΔG is large and positive and equilibrium lies essentially completely to the left. In general, this is true of such fragmentations unless ΔH is less than about 40 kJ mol^{-1}.

At much higher temperatures, however, this situation may be reversed. For example, for the dehydrogenation of ethylene at 1,000°C, $\Delta S_{tr} \sim 150$ J K^{-1} mol^{-1}, so that $T\Delta S \sim 190$ kJ mol^{-1}. The free-energy change ($= \Delta H - T\Delta S$) is now significantly negative and equilibrium lies almost completely to the right. Thus, whereas it is possible to reduce acetylene at or near room temperature to obtain ethylene, it is also possible to dehydrogenate ethylene at very high temperatures in order to produce acetylene. This is in fact of considerable industrial importance; the older method for making acetylene, from calcium carbide,* has now been largely superseded by the method from ethylene since this material is so cheaply obtained ($\sim$£200 or \$350 per tonne) from the cracking of petroleum.

In contrast to fragmentations, the entropy changes which accompany reactions in which there is no change in the number of species, such as A + B $\rightarrow$ C + D, are usually small and the effect of temperature on ΔS is also small; for a typical case, the esterification of acetic acid by methanol, ΔS_{tr} is $-4\cdot6$ J K^{-1} mol^{-1} at 25°C, equivalent to a contribution of only 1·4 kJ mol^{-1} to the free-energy change at this temperature. The equilibrium constant in such reactions tends, therefore, to be governed over a wide temperature range by the enthalpy change. However, exceptions to this generalization occur when, particularly in equilibria involving ions, one or more of the species is *solvated*; entropy changes associated with the 'freezing' of solvent molecules become important (p. 21).

(*ii*) Consider the equilibrium between 1-hexene and cyclohexane,

$$CH_3{-}CH_2{-}CH_2{-}CH_2{-}CH{=}CH_2 \quad \rightleftharpoons \quad \begin{matrix} & CH_2 & \\ CH_2 & & CH_2 \\ | & & | \\ CH_2 & & CH_2 \\ & CH_2 & \end{matrix}$$

for which $K = 6 \times 10^9$ at 25°C. The enthalpy change is favourable: ΔH from bond-energy data is -84 kJ mol^{-1}, in close agreement with the experimental value of -82 kJ mol^{-1}. If this factor alone were involved, K would be $\sim 10^{14}$. The entropy change is, however, unfavourable, because the number of conformations of 1-hexene (which arise from rotations about C$-$C bonds) is much greater than that for cyclohexane, where rotations are prevented by the ring structure. It is this factor which accounts for almost the entire entropy change of $-86\cdot5$ J K^{-1} mol^{-1}

For the formation of the six-membered ring above, the entropy change, though markedly negative, does not offset the favourable enthalpy change; ΔG^0 is -52 kJ mol^{-1}. Smaller rings, however, are *strained*, because the bond angles are distorted from their natural values (2.6a) so that the ΔH term is less favourable. Although less internal freedom is lost on ring-closure (for the smaller acyclic compounds have fewer possible conformations), the free-energy change is less favourable. This is also true of the formation of larger

---

*This method involves heating calcium oxide with carbon in an electric furnace at about 2,000°C and decomposing the resulting calcium carbide with water (CaC$_2$ + 2H$_2$O $\rightarrow$ HC$\equiv$CH + Ca(OH)$_2$). The electric-power requirement is 9,000 kWh per ton of acetylene.

rings, for although the ring-strain is very small (for 7- to 11-membered rings) or zero (for larger rings) (2.6a), the loss of internal freedom increases with increase in ring-size.

1.8 Further Applications of Thermodynamic Principles

(a) TAUTOMERISM

In principle, it should be possible to establish equilibrium between any two isomeric substances, e.g.

$$CH_3-CH_2-CH_2-CH_3 \rightleftharpoons \underset{CH_3}{\overset{CH_3}{\diagdown}} CH-CH_3$$

n–Butane   Isobutane

In practice, it is common to find that conditions are not available for attainment of the equilibrium; each isomer may be isolated and is stable with respect to the other indefinitely. There are, however, many such systems where equilibrium is attained, more or less rapidly, by migration of a group from one position to another. The isomers are then known as *tautomers* and the phenomenon as tautomerism.

The most frequently encountered tautomeric systems are those in which the tautomers differ in the position of a hydrogen atom. For example, ethyl aceto-acetate consists of a mixture of keto and enol forms:

$$CH_3-\underset{\underset{O}{\|}}{C}-CH_2-CO_2Et \rightleftharpoons CH_3-\underset{\underset{OH}{|}}{C}=CH-CO_2Et$$

keto form   *enol* form

The pure ester contains 8% of the enol and 92% of the keto tautomer and exhibits the reactions typical of both the carbonyl group, C—O, and the enol group, C=C—OH. On the other hand, acetone, whose tautomeric equilibrium is also between keto and enol structures,

$$CH_3-\underset{\underset{O}{\|}}{C}-CH_3 \rightleftharpoons CH_3-\underset{\underset{OH}{|}}{C}=CH_2$$

contains less than $10^{-4}\%$ of the enol form.

The reason for the difference lies essentially in the enthalpy terms for these equilibria. Although no precise data are available, the following crude calculation is illustrative. The conversion of the keto into the enol form of acetone involves the replacement of one C=O, one C—C, and one C—H bond by one C- -O, one C=C, and one O—H bond, and inspection of bond-energy data (Table 1.1) shows that ΔH should be approximately $+80$ kJ mol^{-1}. Since the entropy change should be small, the ΔG^0 value should be significantly positive; the experimental value for K is $ca.$ 10^{-6}, corresponding to $\Delta G^0 = +33$ kJ mol^{-1}.

For ethyl acetoacetate, however, two factors increase the bonding (i.e. enthalpy) of the enol form relative to the keto form. First, the carbon-carbon double bond is conjugated to the carbonyl double bond of the ester group, corresponding to an increase in bonding estimated to be about 17 kJ mol^{-1}. Secondly, the hydroxylic hydrogen is *hydrogen-bonded* to the carboxylic oxygen (p. 54):

$$CH_3-C{\overset{CH}{<}}{\underset{O}{}}C-OEt$$

The strength of this bond is about 25 kJ mol^{-1}. A rough estimate for the enthalpy change in the keto-enol transformation is therefore $80 - 42 \simeq 38$ kJ mol^{-1}, and it is understandable that ethyl acetoacetate is much more highly enolized than acetone. *

For simple phenolic compounds, on the other hand, equilibrium favours the enolic over the ketonic form, e.g.

Phenol does not show any of the properties of a ketone. Compared with the situation for acetone, the enolic form of phenol possesses the stabilization energy of the aromatic ring to which oxygen is conjugated ($ca.$ 150 kJ mol^{-1}; see p. 54) whereas the ketonic form, which is conjugated but not aromatic, has a very much smaller stabilization energy ($ca.$ 20 kJ mol^{-1}). The enthalpy change on enolization is therefore approximately $80 + 20 - 150 = -50$ kJ mol^{-1}, that is, markedly in favour of the enol form.

The situation is different when a second phenolic group is introduced, *meta* to the first, as in resorcinol,

----

*These calculations are necessarily very approximate because ΔH is a small difference between two large quantities, the individual contributions to which are themselves not known accurately. Nevertheless, they may be used satisfactorily to obtain an indication of whether one tautomer is likely to predominate in a particular equilibrium or, as in the example above, to predict the effect of a structural variation.

Simple calculation leads to a ΔH value close to zero, essentially because two strongly bonded carbonyl groups are present in the keto structure to offset the aromatic stabilization energy of the enol. Consistent with this, resorcinol has properties characteristic of both a phenol and a ketone: for example, it undergoes the rapid electrophilic substitutions such as bromination which are characteristic of phenols (p. 411) and it is reduced by sodium amalgam to 1,3-cyclohexanedione in the manner characteristic of $\alpha\beta$-unsaturated carbonyl compounds (p. 617). An extension of the argument rationalizes the behaviour of phloroglucinol (1,3,5-trihydroxybenzene), which is more fully ketonized.

Finally, β-naphthol, unlike phenol, has certain ketonic properties:

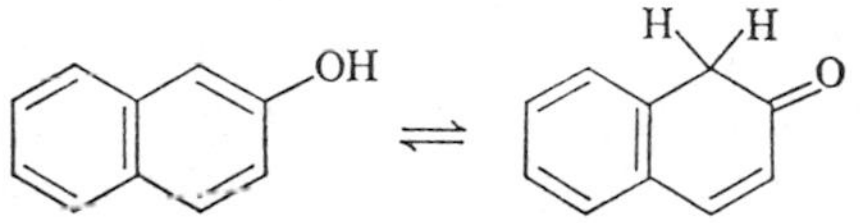

The calculation for phenol is here modified because both tautomers have aromatic stabilization energy. The loss of this energy on ketonization is approximately the difference in stabilization energies of naphthalene and benzene (105 kJ mol^{-1}), so that, compared with the ketonization of phenol for which the corresponding loss is 150 kJ mol^{-1}, the ketonization of β-naphthol is more favourable by about 45 kJ mol^{-1}.

Many other systems of general structure X=Y—Z—H are tautomeric: X=Y—Z—H $\rightleftharpoons$ H—X—Y=Z. Those most frequently met are the following:

Carbon triad
$$\underset{H}{C}-C{=}C \rightleftharpoons C{=}C-\underset{H}{C}$$

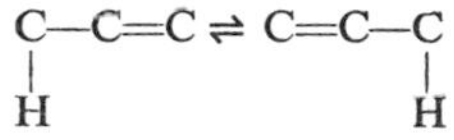

e.g.* PhCH$_2$CH=CH$_2$ $\rightleftharpoons$ PhCH=CHCH$_3$

Azomethine
$$\underset{H}{C}-N{=}C \rightleftharpoons C{=}N-\underset{H}{C}$$

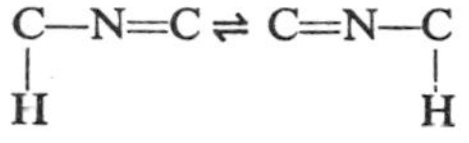

e.g. ArCH$_2$N=CHAr' $\rightleftharpoons$ ArCH=NCH$_2$Ar'

Nitro—aci-nitro 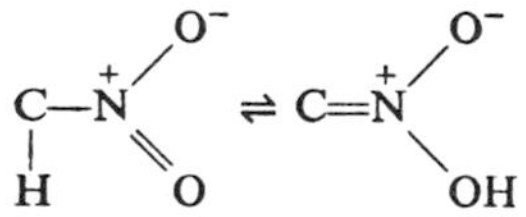
$$\underset{H}{C}-\overset{+}{N}\overset{O^-}{\underset{O}{\diagdown}} \rightleftharpoons C{=}\overset{+}{N}\overset{O^-}{\underset{OH}{\diagdown}}$$

e.g. CH$_3$—NO$_2$ $\rightleftharpoons$ CH$_2$=$\overset{+}{N}\overset{O^-}{\underset{OH}{\diagdown}}$

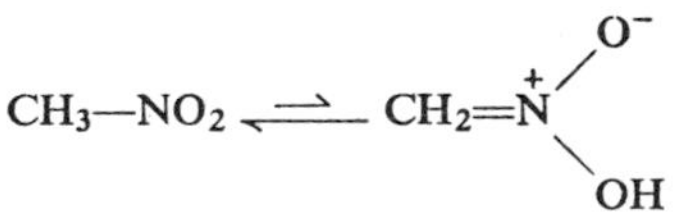

*Equilibrium, attained by heating with a strong base, lies to the right because the product possesses stabilization energy owing to the conjugation between the olefinic bond and the aromatic ring.

Nitroso—oxime $C-N=O \rightleftharpoons C=N-OH$
$\qquad\qquad\qquad\qquad\qquad |$
$\qquad\qquad\qquad\qquad\qquad H$

e.g. $CH_3-NO \rightleftharpoons CH_2=N-OH$

Diazoamino $N-N=N \rightleftharpoons N=N-N$
$\qquad\qquad\qquad\quad |\qquad\qquad\qquad\quad |$
$\qquad\qquad\qquad\quad H\qquad\qquad\qquad\quad H$

e.g. $ArNHN=NAr' \rightleftharpoons ArN=NNHAr'$

Tautomerism in which the tautomers differ in the position of a hydrogen atom is referred to as *prototropy*, for the mechanism of interconversion usually involves the elimination and re-addition of a proton. Tautomerism involving compounds which differ in the position of a group which can (at least in principle) migrate as an anion is referred to as *anionotropy*. For example, 1-methylallyl alcohol and crotyl alcohol may be equilibrated by treating either compound with dilute sulphuric acid at about 100° for 5 hours; the equilibrium mixture contains about 70% of the primary alcohol:

$$CH_3-CH-CH=CH_2 \quad \rightleftharpoons \quad CH_3-CH=CH-CH_2$$
$$\qquad\quad | \qquad\qquad\qquad\qquad\qquad\qquad\qquad\qquad |$$
$$\qquad\quad OH \qquad\qquad\qquad\qquad\qquad\qquad\qquad OH$$

1–Methylallyl alcohol   Crotyl alcohol
30%   70%

As with prototropic systems, the equilibrium constant in anionotropic systems is strongly dependent upon the structures of the tautomers. For example, although in the anionotropic system above the equilibrium is relatively balanced between the tautomers, in that between 1-phenylallyl alcohol and cinnamyl alcohol,

$$Ph-CH-CH=CH_2 \quad \rightleftharpoons \quad Ph-CH=CH-CH_2$$
$$\qquad\quad | \qquad\qquad\qquad\qquad\qquad\qquad\qquad\qquad |$$
$$\qquad\quad OH \qquad\qquad\qquad\qquad\qquad\qquad\qquad OH$$

1–Phenylallyl alcohol   Cinnamyl alcohol

equilibrium is so strongly in favour of the latter tautomer that the former is not detectable in the equilibrated mixture.

The position of equilibrium is dictated mainly by the enthalpy term, for the entropy change in a reaction $A \rightleftharpoons B$ is small. The more highly conjugated of two tautomers is therefore the predominant one at equilibrium, for it possesses the larger stabilization energy; thus, the conjugated cinnamyl alcohol has a lower enthalpy than the non-conjugated 1-phenylallyl alcohol. Neither 1-methylallyl nor crotyl alcohol is conjugated, but in the latter the olefinic bond is attached to two alkyl substituents whereas in the former it is attached to one; this leads to a *small* enthalpy difference (p. 53). (The difference in free energies corresponding to an equilibrium of 70% A and 30% B is only 2 kJ mol^{-1} at 25°C.)

The rates of interconversion in anionotropic systems vary widely, tending to be greater under given conditions when more highly conjugated systems are involved. This can be important in synthesis: e.g. an attempt to convert 1-phenylallyl alcohol into an ester in acid-catalyzed conditions would give the cinnamyl ester:

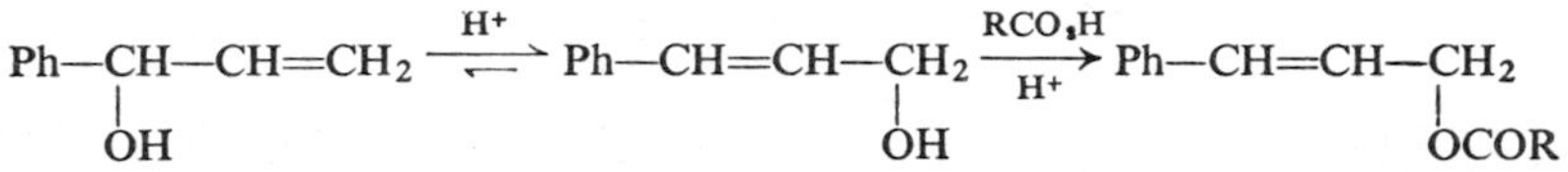

$$Ph-\underset{\underset{OH}{|}}{CH}-CH=CH_2 \underset{}{\overset{H^+}{\rightleftharpoons}} Ph-CH=CH-\underset{\underset{OH}{|}}{CH_2} \xrightarrow[H^+]{RCO_*H} Ph-CH=CH-\underset{\underset{OCOR}{|}}{CH_2}$$

(b) ACIDITY AND BASICITY

For the ionization of acetic acid in water at 25°C,

$$CH_3CO_2H + H_2O \rightleftharpoons CH_3CO_2^- + H_3O^+$$

$\Delta G^0 = 27$ kJ, $\Delta H^0 = \sim -0.4$ kJ, and $\Delta S^0 = -92$ J K^{-1}. From the earlier discussion, so large a (negative) entropy of ionization would not be expected, for there are two particles on each side of the equilibrium. The result is due to the fact that both the acetate ion and the hydronium ion are surrounded by sheaths of solvent molecules appropriately oriented; thus, the hydronium ion has three molecules of water hydrogen-bonded to it:

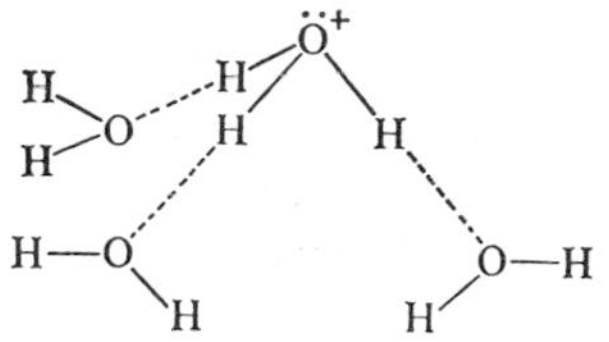

The ions are said to be *solvated*. A particular ion is surrounded by a number of solvent molecules which are constantly changing places with other solvent molecules, so that the process of solvation corresponds to the establishment of a dynamic equilibrium. The solvating bonds contribute to a decrease in both the enthalpy and the entropy (the latter owing to the orientation of the solvent; i.e. translational and rotational freedom are lost).

Thermodynamic data for the protonation of neutral bases (i.e. the reverse of the ionization of cationic acids) are in marked contrast to those for the ionization of (neutral) acids. For example, for

$$NH_3 + H_3O^+ \rightleftharpoons NH_4^+ + H_2O$$

$\Delta G^0 = -52.7$ kJ, $\Delta H^0 = -51.8$ kJ, and $\Delta S^0 = +2.9$ J K^{-1}. Here, the entropy change is negligible because there is one ion on each side of the equilibrium and the solvation entropies tend to cancel out.

Acidity and basicity are of considerable importance in organic synthesis because of the variety of reactions which are acid- or base-catalyzed. The strength
of the acid or base is frequently important. Consider a hypothetical pathway for
the ionization of acetic acid in the gas phase:

$$CH_3CO_2\text{---}H \longrightarrow CH_3CO_2\cdot + H\cdot$$
$$H\cdot \longrightarrow H^+ + electron$$
$$CH_3CO_2\cdot + electron \longrightarrow CH_3CO_2^-$$
$$\overline{CH_3CO_2\text{---}H \longrightarrow CH_3CO_2^- + H^+}$$

The sum of the enthalpy changes for these three steps must equal that of the
'direct' ionization (First Law); that is,

$$\Delta H = D_{OH} + I_H + EA_{CH_3CO_2}.$$

where D_{OH} is the bond dissociation energy of the O—H bond, I_H is the ionization
potential of the hydrogen atom, and $EA_{CH_3CO_2}.$ is the electron affinity of the
acetate radical.

Now, for the dissociation of the related acid, CCl_3CO_2H, D_{OH} is approximately the same and I_H is common to both, so that

$$\Delta H \text{ (acetic)} - \Delta H \text{ (trichloroacetic)} \sim EA_{CH_3CO_2}. - EA_{CCl_3CO_2}.$$

The electron affinity of the trichloroacetate radical is greater than that of the
acetate radical, for the negative charge in the trichloroacetate anion can be
absorbed to some extent into the electronegative chlorine atoms (2.6c), so that
we would expect that the enthalpy term would favour the ionization of trichloroacetic acid, making this the stronger acid. However, while it is true that trichloroacetic acid $(K \sim 1)$ is a far stronger acid than acetic acid $(K \sim 10^{-5})$, the
enthalpy term actually favours the ionization of acetic acid. Approximate data,
for aqueous solutions, are as follows:

	ΔH^0 (kJ)	ΔS^0 (J K^{-1})
$CH_3CO_2H \rightleftharpoons CH_3CO_2^- + H^+$	$-0\cdot4$	-92
$CCl_3CO_2H \rightleftharpoons CCl_3CO_2^- + H^+$	$+4$	$+8$

This is again the result of solvation. Of the two anions, the trichloroacetate
ion, whose negative charge is more uniformly spread throughout the species, is
correspondingly less strongly solvated: the ΔH term for solvation is less favourable than that for acetate ion, but as a result of the lower degree of ordering of
solvent molecules, less entropy is lost. On balance, the free energy of ionization
is less positive than that for acetic acid and trichloroacetic acid is much the
stronger acid.

It is not always the case that the stronger of two acids has the more favourable
entropy of ionization and the less favourable energy of ionization, but it is a

general fact that the introduction of an electron-attracting group into an acid causes changes in both the enthalpy and entropy terms such that the free energy of ionization becomes more negative (or less positive) and the dissociation constant is increased.

(c) RING-CLOSURE

Organic reactions frequently lead to the formation of cyclic compounds from open-chain (acyclic) compounds. We shall consider two general classes of ring-closure reactions: (i) A → B, and (ii) A → B + C.

(i) The comparison of a ring-closure of type A → B with an analogous inter-molecular reaction is instructive. For example, acetaldehyde exists in aqueous solution in equilibrium with its hydrate:

$$CH_3CH{=}O + H_2O \rightleftharpoons CH_3CH(OH)_2$$

The equilibrium lies to the left, despite a favourable decrease in enthalpy, because of the unfavourable entropy change ($-68 \cdot 5$ J K^{-1} mol^{-1}, equivalent to a contribution of $20 \cdot 4$ kJ mol^{-1} to ΔG at 25°C) which results mainly from the loss of translational freedom. For the analogous equilibrium between the acyclic and cyclic forms of ω-hydroxyvaleraldehyde, however, the equilibrium lies well to the right ($K \sim 16$):

$$HO{-}(CH_2)_4{-}CH{=}O \rightleftharpoons \begin{array}{c} CH_2 \\ CH_2 \quad O \\ CH_2 \quad CH{-}OH \\ CH_2 \end{array}$$

This is because ΔS_{tr} is zero, and the principal source of the entropy change is the loss of internal freedom which, for six-membered cyclization, is usually less than the loss of translational freedom for the corresponding intermolecular reaction.

The equilibrium between ω-hydroxybutyraldehyde, $HO{-}(CH_2)_3{-}CHO$, and its cyclic form is slightly less favourable ($K = 8$) because, although the entropy factor is more favourable than for formation of the six-membered ring, the five-membered ring is slightly strained (p. 48) so that the enthalpy factor is less favourable. The equilibrium constants for the corresponding three- and four-membered rings are negligible because ring-strain is considerable,* and those

---

*α-Hydroxy-aldehydes form solid *dimers* (six-membered ring) which revert to the monomers in aqueous solution:

$$2\ HO{-}CH_2{-}CHO \rightleftharpoons \begin{array}{c} CH_2 \\ HO{-}CH \quad O \\ O \quad CH{-}OH \\ CH_2 \end{array}$$

for rings with more than six atoms decrease rapidly because the entropy factor becomes increasingly unfavourable (and, for rings containing 7–11 members, there is also some internal strain). Thus, $K = 0.2$ and 0.1, respectively, for the seven- and eight-membered rings, so that conditions favour the intermolecular reaction of the aldehyde and water as compared with the intramolecular cyclization.

(*ii*) Consider the lactonization of ω-hydroxybutyric acid in comparison with the esterification of acetic acid by methanol:

$$\begin{array}{ccc}
\text{CH}_2\text{—OH} & & \text{CH}_2\text{—O} \\
\diagup & & \diagup \quad\quad | \\
\text{CH}_2 & \rightleftharpoons & \text{CH}_2 \quad\quad | \quad + \text{H}_2\text{O} \\
\diagdown & & \diagdown \quad\quad | \\
\text{CH}_2\text{—CO}_2\text{H} & & \text{CH}_2\text{—CO}
\end{array}$$

$$\text{CH}_3\text{CO}_2\text{H} + \text{CH}_3\text{OH} \rightleftharpoons \text{CH}_3\text{CO—OCH}_3 + \text{H}_2\text{O}$$

In each case, the bonds formed (C—O and O—H) correspond to those broken, so ΔH is likely to be very small. (This would not be true if there were significant strain in the lactone ring.) In the esterification, the changes in both translational entropy $[= 3/2 \; RT \ln(M_C M_D/M_A M_B) = 1.2 \; \text{J} \; \text{K}^{-1} \; \text{mol}^{-1}]$ and rotational entropy are negligible since there is no change in the number of particles, and the change in internal freedom (ΔS_{vib}) is also likely to be negligible. Hence both ΔH and ΔS are not far from zero, so that $\Delta G \sim 0$ and K is close to unity. For the lactonization, however, there is an increase in the number of particles and hence in S_{tr} (143 J K^{-1} mol^{-1} at 25°C, corresponding to a contribution of -43 kJ to ΔG) and in S_{rot}. There is a corresponding loss in internal freedom, equivalent to $\Delta S_{vib} \sim -84$ J K^{-1} mol^{-1}, but the overall change in entropy is positive. Consequently ΔG is markedly negative and lactonization is essentially complete.

The same considerations apply to the formation of six-membered lactone rings and to the formation of other five- and six-membered cyclic systems by reactions involving the generation of two species from one. For smaller rings, however, the enthalpy change is less favourable because of the strain in the ring, and for larger rings, the entropy term becomes increasingly less favourable as the size of the ring is increased.

(d) UNSTABLE COMPOUNDS

The use of the words *stability* and *instability* often gives rise to confusion. The thermodynamic definitions are clear: the stability of a compound refers to the standard free energy of its formation; a compound may be stable or unstable with respect to its elements. This use must be distinguished from that relating to other reactions: for example, a compound may be unstable to heat (e.g. a peroxide), to water (e.g. an acetal in the presence of acid) or to air (e.g. a drying-oil). We shall always refer to the conditions in which a compound is unstable, and here draw attention to the thermal stability of organic compounds.

Most well-known organic compounds are stable to heat to temperatures of over 200°C, and since reactions are usually carried out below this temperature the decomposition of one of the reactants does not usually present a problem. Thermal instability within this temperature range is, however, associated with certain structural groups. The extent of decomposition at a particular temperature and within a given time does not depend directly on the thermodynamic properties of the material, but is determined by the *rate* of decomposition. Nevertheless, the rate is usually related to the thermodynamic properties (see 3.7), and in general weaker bonds undergo faster bond-cleavage than stronger bonds. For example, the extent of decomposition of methane ($D_{C-H} = 426$ kJ mol^{-1}) is negligible at 100°, whereas that of dibenzoyl peroxide (C_6H_5COO- $OCOC_6H_5 \rightarrow 2C_6H_5CO_2$; $D_{O-O} = 130$ kJ mol^{-1}) is about 50% after 30 minutes.

Such instability may be usefully applied. For example, certain organic reactions may be initiated by the introduction of a reactive species such as the benzoyloxy radical (17.1) and these reactions can be brought about by introducing the peroxide and heating to a temperature at which the rate of generation of benzoyloxy radicals brings about the initiated reaction at a practicable rate.

Other commonly occurring weak bonds are those of the halogens. Many compounds containing C—H bonds may be chlorinated or brominated* by being heated with the halogen, reaction occurring through the mediation of halogen atoms, e.g.

$$CH_4 + Cl_2 \xrightarrow{\text{via } Cl\cdot} CH_3Cl + HCl$$

Again, however the thermodynamic criterion for a reaction to occur ultimately dictates the issue. In the chlorination of methane, ΔS is very small and ΔH (-100 kJ mol^{-1}; cf. Tables 1.1 and 1.2) is dominant in the free-energy change. For the iodination of methane, however, $\Delta H = +53$ kJ mol^{-1} and the free-energy change is unfavourable. Thus, despite the ease with which suitable conditions for iodination can be established (i.e. the generation of iodine atoms), reaction does not occur.

Finally, certain compounds are unstable to heat not because they contain any intrinsically weak bonds but because decomposition can lead to the formation of a strongly bonded molecule. For example, azobisisobutyronitrile decomposes fairly rapidly below 100°C because of the favourable enthalpy change in the process owing mainly to the formation of the strongly bonded molecular nitrogen:

---

*Fluorine is not normally introduced in this way since the reactions are so strongly exothermic (because of the very weak bond energy of F—F as compared with H—F and C—F) as to be violent.

$$(CH_3)_2C-N=N-C(CH_3)_2 \rightarrow 2 \ (CH_3)_2C \cdot \ + N_2$$
$$\qquad | \qquad\qquad\quad | \qquad\qquad\qquad\quad |$$
$$\qquad CN \qquad\quad CN \qquad\qquad\qquad CN$$

Such compounds are also useful as initiators of free-radical reactions (17.1).

Further Reading

SMITH, E. B., *Basic Chemical Thermodynamics*, Clarendon Press (Oxford 1973).
WARN, J. R. W., *Concise Chemical Thermodynamics*, Van Nostrand Reinhold (London 1969).

Problems

1. Isopropanol is dehydrogenated when it is passed, in the gas phase, over a heated catalyst:

$$(CH_3)_2CHOH \rightleftharpoons (CH_3)_2CO + H_2$$

   (*i*) Formulate the equilibrium constant, K, in terms of the partial pressures of the reactant and the products.

   (*ii*) If, at equilibrium, the degree of dissociation of isopropanol is α and the total pressure is P, show that

$$K = \frac{\alpha^2 P}{(1 - \alpha^2)}$$

   (*iii*) Given that, at 450 K, $\alpha = 0.56$ and $P = 0.95$ atmospheres, calculate ΔG^0.

   (*iv*) Calculate $\Delta G_{450 \ K}$ under the following conditions:
   and comment on the values of ΔG^0 and ΔG at 450 K.

$$(CH_3)_2CHOH \ (g; P = 1 \ \text{atm.}) \rightarrow (CH_3)_2CO \ (g; P = 0.1 \ \text{atm.}) + H_2 \ (g; P = 0.1 \ \text{atm.})$$

   and comment on the values of ΔG^0 and ΔG at 450 K.

2. The bond energies (25°C) of C—C, C=C, C—H, and H—H are, respectively, 347, 610, 414, and 435 kJ mol^{-1}. Calculate the enthalpy of the reaction,

$$CH_2=CH_2 + H_2 \rightarrow CH_3-CH_3$$

   In practice, the reduction of ethylene to ethane can be carried out readily at room temperature. Under what conditions might it be possible to carry out the reverse reaction?

3. Estimate ΔH (from the data in Tables 1.1 and 1.2) for the following reactions:

(a) $CH_4 + Cl_2 \rightleftharpoons CH_3Cl + HCl$

(b) $C_2H_5OH \rightleftharpoons C_2H_4 + H_2O$

(c) $CH_3CHO + H_2O \rightleftharpoons CH_3CH(OH)_2$

In which cases would you expect ΔS to contribute significantly to ΔG? Which are likely to have favourable equilibrium constants for the forward reaction at room temperature? What would be the effect on the equilibrium constants of increasing the temperature?

4. The entropy changes for the formation of ethyl chloride by (a) the chlorination of ethane ($C_2H_6 + Cl_2 \rightarrow C_2H_5Cl + HCl$), and (b) the addition of hydrogen chloride to ethylene ($CH_2{=}CH_2 + HCl \rightarrow C_2H_5Cl$), are (a) $+2$ and (b) -130 J K^{-1} mol^{-1}. Comment.

5. What product would you expect from the addition of one molecule of hydrogen to anthracene? Why would you expect this reaction to be more exothermic than the addition of a molecule of hydrogen to benzene?

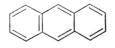

Anthracene

2. Molecular Structure

2.1 Bonding

The principles of thermodynamics relate the concentrations of chemical species in equilibrium to the enthalpies and entropies of those species. Bond energies, closely related to enthalpies, have precise values which may be measured, but thermodynamic principles give no information about the *origin* of these bond energies. It is the purpose of this chapter to outline the current theories of molecular structure, with especial reference to the strengths of bonds and other physical properties of organic compounds.

2.2 Quantum Theory

After the discovery of the electron in 1897, the 'planetary' theory of atomic structure evolved during the first two decades of the present century. The atom was then thought to consist of minute, nearly weightless, negatively charged particles (electrons) surrounding a much heavier, positively charged nucleus. The theory had, however, one obvious failing: if an electron were stationary with respect to the nucleus, it should fall into the nucleus as a result of electrostatic attraction, whereas if it were moving round the nucleus, electromagnetic radiation should be continually emitted and the electron should move gradually nearer the nucleus, reducing the potential energy of the system to compensate for the radiated energy, and should finally collapse into the nucleus. To obviate this and other difficulties, Bohr postulated that electrons exist in *stationary states* around the nucleus, each corresponding to a discrete energy determined by the electrostatic attraction between nucleus and electron, and that, although an electron may move from one such state to another, its translation to an intermediate position does not occur. This theory accounted not only for the fact that radiation is not continuously emitted but also for the observed spectroscopic properties of atoms, namely, that an atom absorbs (or, when excited, emits) only particular frequencies of radiation. Thus, absorption corresponds to the excitation of an electron from a lower-energy state of energy E_1 to a higher-energy state of energy E_2, emission corresponds to the reverse process, and the frequency associated with these electronic changes is given by $h\nu = E_2 - E_1$.

However, the theory of stationary states involves an arbitrary postulate for which there is no basis in classical theory. A more satisfactory picture of atomic structure, which in particular embraces the concept of stationary states, emerged as a result of de Broglie's suggestion in 1924 that electrons have wave properties described by the equation $\lambda = h/mv$, where λ is the wavelength of the electron-wave and m and v are the mass and velocity of the electron. The suggestion was confirmed experimentally two years later when it was found that electrons, like light waves, may be diffracted and that the wavelength derived from the diffraction experiments is that predicted by de Broglie's relationship.

This theory was then applied by Schrödinger to the problem of atomic structure. Assuming that the electron may be described as a plane wave,

$$\psi = A \sin 2\pi \, (x/\lambda)$$

where ψ describes the wave motion and is conveniently taken to measure the amplitude of the wave, A is the maximum value of ψ, and x is the space coordinate. Then:

$$\frac{d^2\psi}{dx^2} = -\frac{4\pi^2}{\lambda^2} A \sin 2\pi \, (x/\lambda) = -\frac{4\pi^2}{\lambda^2} \psi$$

The kinetic energy, T, of the electron is $\frac{1}{2}mv^2$, or, from the de Broglie relationship, $(1/2m)(h^2/\lambda^2)$. Thus,

$$T = -\frac{h^2}{8\pi^2 m} \cdot \frac{1}{\psi} \cdot \frac{d^2\psi}{dx^2}$$

i.e.
$$\frac{d^2\psi}{dx^2} + \frac{8\pi^2 m}{h^2} \cdot T \cdot \psi = 0$$

We have so far assumed that the electron is in field-free space (i.e. the potential energy, V, is constant). For most systems, such as an electron moving in the field of a positive nucleus, this is not so; the potential energy can vary too. The total energy, E, is given by $E = T + V$, so that we postulate, by analogy with the equation for T,

$$\frac{d^2\psi}{dx^2} + \frac{8\pi^2 m}{h^2} (E - V)\psi = 0$$

For three-dimensional space, the Schrödinger equation,

$$\nabla^2\psi + \frac{8\pi^2 m}{h^2} (E - V)\psi = 0$$

(where ∇^2, the Laplacian operator, is given by $\nabla^2 = \partial^2/\partial x^2 + \partial^2/\partial y^2 + \partial^2/\partial z^2$) can be constructed similarly. There is no proof of the validity of the equation, but in every case for which solutions can be obtained there is close agreement between the predicted and the observed data. It is customary to interpret ψ^2 as measuring the *probability density* of the electron, just as, for electromagnetic radiation, ψ^2 measures the radiation density; that is, $\psi^2 d\tau$ measures the probability that the electron will be found in a small volume $d\tau$.

The arbitrary restrictive condition of Bohr's theory (that only certain stationary states occur) is a natural consequence of the wave theory. Consider a particle in a one-dimensional box, so that it is constrained to move only along the x-axis. If the particle is prevented from escaping from the box, the potential energy, V, may be taken as infinitely great outside the box, while if no force acts on the particle within the box its potential energy here is constant and may be arbitrarily chosen as zero. Then

$$\frac{d^2\psi}{dx^2} = -\frac{8\pi^2 mE}{h^2}\psi$$

This equation is satisfied by $\psi = A \sin kx + B \cos kx$, where $k^2 = 8\pi^2 mE/h^2$ and A and B are constants. Outside the box, where $V = \infty$, ψ must be zero, for there is no chance of finding the particle in this region, so ψ is also zero at the walls of the box for otherwise there would be a discontinuous change in ψ at this point. Therefore, since $\psi = 0$ at $x = 0$, the solution simplifies to $\psi = A \sin kx$, and since, for a box of length L, ψ is also 0 at $x = L$, it follows that $kL = n\pi$ where n is any integer. Hence

$$\psi = A \sin n\pi(x/L)$$

and
$$E = n^2 h^2 / 8mL^2$$

Thus, the energy of the particle can only have certain discrete values, dependent on the integral value of n, which is termed a *quantum number*. The appearance of quantum restrictions follows from the application of the Schrödinger equation not only in this case but also for other forms of motion in atomic and molecular systems (e.g. the rotational motion of a molecule).

Consider next the hydrogen atom. One electron moves in the coulombic field of a proton, and since the mass of the electron is negligible compared with that of the proton, the latter may be considered at rest. The potential energy of the electron is given by $V = -e^2/r$, where the charges on the electron and the proton are $-e$ and $+e$, respectively, and r is the distance between the species. Introduction of this value for V into the Schrödinger equation leads to a general value for the total energy, E, given by

$$E = -\frac{2\pi^2 me^4}{n^2 h^2}$$

Again, therefore, a quantum restriction appears, but in this instance solution of the equation leads to the appearance of two further quantum numbers, l and *m*. For a given value of *n*, *l* may have any of the values 0, 1, ... (*n* − 1), and for a given value of *l*, *m* can have any of the values −*l*, (−*l* + 1) ... (*l* − 1), *l*. The three quantum numbers relate to the three polar co-ordinates of the electron, *n* being the principal quantum number, *l* determining the total angular momentum of the electron, and *m* determining the component of angular momentum about a particular axis. The electron may therefore be in one of a number of states, of which the five of lowest energy are:

Quantum numbers			Energy
n	*l*	*m*	
1	0	0	E_1
2	0	0	E_2
2	1	0	E_2
2	1	−1	E_2
2	1	+1	E_2

where $E_2 = \frac{1}{4}E_1$, from the relationship between E and n. Thus, for the state defined by $n = 1$ the electron is more tightly bound than in the case $n = 2$, and the energy of the system is lower in the former state.

Detailed analysis shows that, for the energy state defined by $n = 1$, $l = 0$, $m = 0$, ψ varies with r according to Fig. 2.1 for any direction from the nucleus. Since the probability of finding an electron in a shell of thickness dr and at a distance r from the nucleus is proportional to ψ^2 times the volume of the shell $(4\pi r^2 dr)$, a graph of $(4\pi r^2 \psi^2)$ against r describes the relative probability of finding the electron in a unit volume at a distance r from the nucleus (Fig. 2.2). The probability falls off sharply with distance after the distance corresponding to maximum probability, and it is possible to draw a boundary line about the nucleus within which there is a particular probability, say 99%, or 99·9% if the contour is extended, of the electron's being found.

The solutions of the wave equation for the various electronic states of an atom are termed *atomic orbitals*, and an electron is described as occupying a

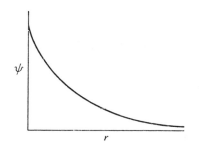

Fig. 2.1   Variation of ψ with r.

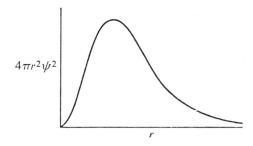

Fig. 2.2   Variation of $4\pi r^2\psi^2$ with r.

particular atomic orbital. The orbitals are conveniently represented by contour diagrams. The representation for the present case ($n = 1, l = 0, m = 0$) would be by the construction of a sphere about the nucleus as centre; a two-dimensional representation is shown in Fig. 2.3a. Thus, the orbital has zero angular momentum and spherical symmetry and is defined as an s orbital, and this applies to other orbitals for which $l = 0$. Hence there is a series of s orbitals corresponding to $n = 1, 2, 3 \ldots$ and described as $1s, 2s, 3s \ldots$ orbitals.

The wave function of an orbital defined by $l = 1$ is not spherically symmetrical about the nucleus. For $l = 1, m = 0$, the contour lines are as in Fig. 2.3b; the orbitals have two spherical lobes which differ in sign and are separated by a plane though the nucleus in which $\psi = 0$ (nodal plane). The orbitals corresponding to $l = 1, m = +1$ or -1 are similar, except that they are directed along the y- and z-axes, respectively. Each of the three orbitals for which $l = 1$ (termed p orbitals) has angular momentum, the p_y and p_z orbitals, unlike the p_x orbital, having positive values for the component of angular momentum about the x-axis. Orbitals of other types (e.g. d orbitals, which are defined by $l = 2$, $m = 0, \pm 1, \pm 2$) are not usually of direct concern in organic structures.

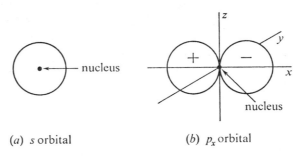

(a) s orbital   (b) p_x orbital

Fig. 2.3   Distribution of s and p_x orbitals.

2.3 Electronic Structure of Atoms

The electronic structure of the hydrogen atom follows immediately from the principles outlined above. In the unexcited state of the atom the one electron occupies the lowest-energy state, which is the $1s$ orbital. For other atoms, the same general principle applies: the electronic structure is built up by feeding electrons into the atomic orbitals in order of increasing energy. Two further principles need to be explained, however.

First, although for the hydrogen atom all orbitals having the same principal quantum number n have the same energy (p. 31), this is not true for atoms having more than one electron. This is because new energy terms appear which relate to the inter-electronic repulsive forces. Since the probability density of an s orbital is greater near the nucleus than that of a p orbital, an electron in a p

orbital is more effectively screened from the nucleus by electrons in lower-lying orbitals and is consequently less tightly bound. The order of increasing orbital energies becomes:

$$1s < 2s < 2p_x = 2p_y = 2p_z < 3s < 3p_x = 3p_y = 3p_z.$$

Secondly, only two electrons can occupy any one orbital. This principle, known after Pauli, has wide implications in bonding theory and can be understood as follows. An electron possesses a magnetic moment, and in a magnetic field it may align itself in one of two senses with respect to the field. Two 'states' are therefore possible, and these are conveniently described as corresponding to the electron's having one of two *spins*, characterized by a spin quantum number m_s which can have only two values. In effect, Pauli's principle states that only one electron with a particular spin may exist in an atomic orbital; the orbital may itself contain two electrons, one of each spin, which are then said to be 'paired'.

Consider now the carbon atom, which has six electrons. The first two electrons are fed into the lowest energy orbital, $1s$, and the next two into the $2s$ orbital. The remaining two electrons occupy the $2p$ orbitals, but since there are three of these of equal energy, there are a number of possible dispositions. Since electrons repel each other, it is understandable that the electrons should occupy *different* $2p$ orbitals, and in fact the lowest-energy situation corresponds to these two electrons having the same spin (Hund's rule). The resulting electronic configuration for carbon is shown in Fig. 2.4a, ↑ and ↓ denoting electrons in the two spin states. The configurations for other commonly occurring elements are given in Fig. 2.4, b, c, d.

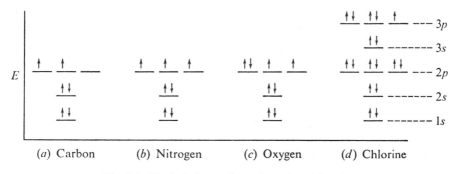

| (a) Carbon | (b) Nitrogen | (c) Oxygen | (d) Chlorine |

Fig. 2.4   Typical electronic configurations of atoms.

It should be noted that the vertical axis does not represent energy differences between atomic orbitals *quantitatively*. As the nuclear charge is increased, the energies of the orbitals are lowered as a result of the increased electrostatic attraction between electron and nucleus.

2.4 The Covalent Bond

The treatment of molecules by wave mechanics is in principle similar to the treatment of atoms, save that the electrons move in the field of more than one nucleus. The principal problem which arises in this treatment is that the wave equation cannot be solved exactly for molecules containing more than one electron, and with the exception, therefore, of the hydrogen molecule-ion, H_2^+, approximate methods of solution must be used. The two most commonly employed are the molecular-orbital (MO) and valence-bond (VB) methods.

(a) THE MOLECULAR-ORBITAL METHOD

In this treatment the electrons are fed into orbitals in the order of their increasing energies, just as in the building up of the electronic structures of atoms, except that the orbitals are polycentric since more than one nucleus is involved. As usual, two electrons, one of each spin, may occupy each orbital.

The problem of calculating the energies of electrons in these molecular orbitals (MOs) is usually approached by means of the LCAO (linear combination of atomic orbitals) approximation. Consider a diatomic molecule A—B. It is assumed that when an electron is in the vicinity of one of the two nuclei, the forces acting on it are due mainly to that nucleus, so that in the neighbourhood of the nucleus of A the MO resembles the AO ψ_A, and similarly in the neighbourhood of the nucleus of B the MO resembles the AO ψ_B. Thus the MO has characteristics of both ψ_A and ψ_B and it is logical to write the complete MO, ψ, as

$$\psi = N(\psi_A + \lambda\psi_B)$$

The parameter λ is introduced to take account of the relative importance of the two orbitals which combine to form the MO, ψ_A and ψ_B, for if the electronegativities of the atoms differ there will be a greater probability of the electron's being associated with one nucleus than with the other ($\lambda = 1$ for a homonuclear diatomic molecule). N is a normalizing factor chosen to ensure that the total probability of finding the electron somewhere in space is unity. A second solution is also possible, namely,

$$\psi = N(\psi_A - \lambda\psi_B)$$

The first of these two MOs corresponds to the system A—B having a lower energy than the separated atoms, whereas the second corresponds to its having a higher energy. The former is therefore described as a *bonding orbital* and the latter as an *antibonding orbital*. From the point of view of chemical structure the crucial question is whether the formation of a molecular orbital from atomic orbitals yields a significant change in the energy of the system. It transpires that this is so only when three conditions are fulfilled: (*i*) the energies of ψ_A and ψ_B are comparable; (*ii*) the charge clouds of ψ_A and ψ_B overlap to a significant extent; and (*iii*) ψ_A and ψ_B have the same symmetry. Hence the problem of computing MOs is simplified by neglecting possible combinations of AOs which do not fulfil these conditions.

Consider the interaction of two hydrogen atoms. Each has one electron in an orbital of the same energy ($1s$); the charge clouds can overlap, and the symmetry condition is fulfilled. The energy levels of the two MOs which are established are shown in Fig. 2.5. Both $1s$ electrons can occupy the lower-lying MO, so the total energy is lower in H_2 than in the atomic systems, i.e. bonding occurs. The interaction of two helium atoms, on the other hand, each of which has two $1s$ electrons, would lead to the establishment of two analogous MOs both of which would be occupied, so that there would be no net binding energy. Consistent with this, helium is monatomic.

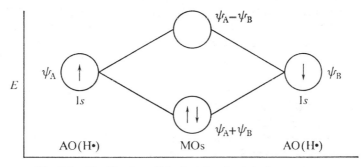

Fig. 2.5   Combination of the $1s$ AOs of two hydrogen atoms.

An orbital representation of the MOs formed by hydrogen atoms is shown in Fig. 2.6. The contour lines show that in the lower-energy MO, termed the $\sigma 1s$ bonding orbital (σ being used to denote that the orbital has zero angular momentum about the molecular axis), there is an accumulation of charge in the region between the nuclei, whereas the higher-energy MO (the σ^*1s antibonding orbital) has a nodal plane which bisects the H—H axis where the electron density is zero.

Consider next the interaction of two fluorine atoms. From condition (i) above, significant interaction of the AOs occurs only between the $1s$ orbitals of each, the $2s$ orbitals of each, and the $2p$ orbitals of each. Moreover, from condition (ii), the $2p_x$ orbitals interact more strongly than the $2p_y$ or $2p_z$ orbitals (Fig. 2.7)

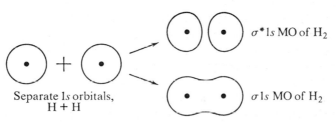

Fig. 2.6   Molecular orbitals for H_2.

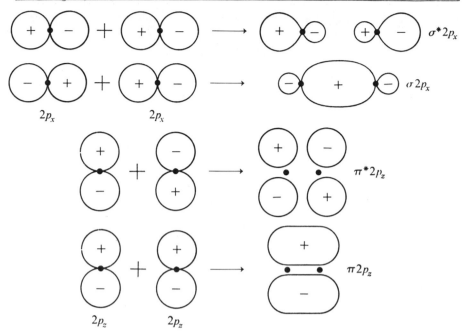

Fig. 2.7   Molecular orbitals formed by $2p$ atomic orbitals.*

since the degree of overlap is greater, while there can be no interaction between those $2p$ orbitals which are directed along different axes. The MOs for F_2 may therefore be written down as in Fig. 2.8. The resultant interaction is due to the pair of electrons in the $\sigma 2p_x$ MO, all other interactions nullifying each other, so F_2 may be regarded as having a single σ-bond. (Note that MOs formed from $2p_y$ and $2p_z$ AOs are termed π-orbitals, having a nodal plane through the molecular axis and angular momentum about this axis.)

It can now readily be appreciated that the interaction of AOs in which there are pairs of electrons does not lead to any resultant bonding, and that bonding is associated with the *pairing* of electrons from AOs of the appropriate types.

Two further factors are introduced in considering heteronuclear diatomic molecules. First, the appropriate AOs for strong interaction are not necessarily those of the same type for each atom because the energies of these AOs are different for each atom (see condition (*i*)). For example, in HCl the bond is formed by the $1s$ electron of hydrogen and a $2p$ electron of chlorine (in fact, the $2p_x$ electron, from condition (*ii*)). Secondly, λ (p. 34) is no longer unity, for an electron has a higher probability of being found in the vicinity of the more electronegative nucleus (i.e. chlorine, in HCl). The centres of action of the

*Overlap of the p-orbital lobes of like sign gives the lower-energy, bonding MO, whereas overlap of lobes of opposite sign gives the higher-energy, antibonding MO. For simplicity, the signs will be omitted from now on except in discussion of pericyclic reactions, where the nature of the antibonding orbitals is important (Chapter 9).

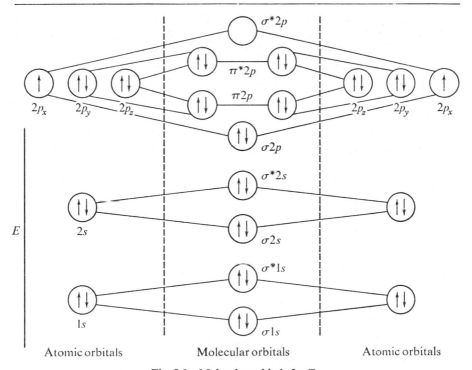

Fig. 2.8   Molecular orbitals for F_2.

positive and negative charges in the molecule will not coincide and there will be a dipole moment ($\mu \simeq 1$ D for HCl), so λ is related to μ.

With polyatomic molecules, a new factor, molecular shape, is introduced. A simple example is provided by H_2O. Oxygen has two unpaired $2p$ electrons, which can be regarded as being in the $2p_x$ and $2p_y$ orbitals. Maximum overlap with the hydrogen $1s$ orbitals requires the hydrogen nuclei to be on the x- and y-axes, so the two σ-bonds should be at 90° to each other:

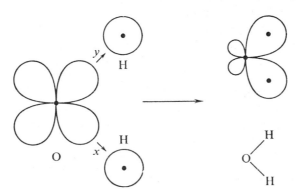

The observed angle ($104\frac{1}{2}°$) is somewhat larger than this, in part because the hydrogen atoms are the positive ends of dipoles and consequently repel each other and in part because the pair of $2s$ electrons on oxygen combines with the $2p$ electrons to form hybridized bonds (see below), so that the bonds formed are not derived from 'pure' p orbitals but have some s character. Similarly, we should expect the three $2p$ electrons of nitrogen to form three bonds with hydrogen atoms which were mutually perpendicular, whereas the observed angle in NH_3 is 108°; the bonds formed by nitrogen have some s character.

Since carbon has two unpaired $2p$ electrons, the principles outlined so far might lead us to expect that it should be divalent and form bonds which were approximately mutually perpendicular. In fact, carbon is tetravalent in almost all stable organic compounds; the few stable tervalent carbon compounds (free radicals) owe their stability to unusual structural features (17.1), and divalent carbon species exist only as short-lived intermediates in certain reactions (see, e.g. p. 125).

The tetravalence of carbon may be understood in terms of the concept of *hybridization*. The difference in energies of the $2s$ and $2p$ orbitals of the carbon atom is approximately 400 kJ mol^{-1}, so for the expenditure of this amount of energy a $2s$ electron could be promoted to the empty $2p$ orbital, giving four unpaired electrons and hence four covalent bonds. This proves to be thermodynamically practicable, for the formation of two extra covalent bonds more than offsets the excitation energy. However, an atom which formed bonds by using three unpaired p electrons and one unpaired s electron would yield a molecule lacking spherical symmetry because three of the bonds would be different from the fourth, whereas the saturated compound, methane, possesses spherical symmetry, the four bonds being directed towards the corners of a regular tetrahedron of which carbon is the centre. The only way in which four such equivalent bonds can be formed is by the 'mixing' of the three $2p$ orbitals and the one $2s$ orbital; the new orbitals are combinations of these and are described as sp^3-hybridized orbitals.

Moreover, as well as providing equivalent bonds, the hybridized orbitals are strongly directional, as shown in Fig. 2.9. The resulting overlap with the AOs of four atoms which are suitably placed is greater than that which could be achieved by the three directional p orbitals and the non-directional s orbital.

There are two other types of hybridization of particular importance. In one, the $2s$ orbital is mixed with two of the $2p$ orbitals, giving three sp^2-hybridized orbitals which are coplanar and at 120° to each other. The remaining $2p$ orbital

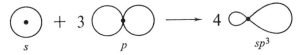

Fig. 2.9 The formation of an sp^3-hybrid atomic orbital.

is perpendicular to this plane. In this state, two carbon atoms are able to form a double bond, one part of which arises from the overlap of two sp^2 orbitals along the internuclear axis (σ-bond) and the other from the lateral overlap of the two p orbitals (π-bond), as shown for ethylene in Fig. 2.10.

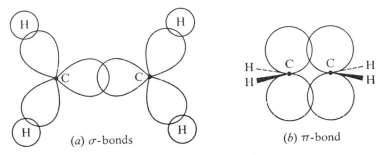

(a) σ-bonds   *(b)* π-bond

Fig. 2.10   Orbital structure of ethylene,
(a) viewed perpendicularly to the plane of the sp^2 orbitals,
(b) viewed along the plane of the sp^2 orbitals.

A number of properties of compounds containing carbon-carbon double bonds follow from this description. First, because the bond strength is related to the extent of overlap of the AOs, there is a tendency for the lateral interaction of the p orbitals to bring the nuclei closer together than in singly bonded molecules, the extra bonding offsetting the extra internuclear repulsion; e.g. the C=C bond-length in ethylene is 1·34 Å whereas the C—C bond-length in ethane is 1·54 Å. Secondly, since lateral p-orbital overlap is not so effective as the sp^2 overlap along the internuclear axis, the π-bond is weaker than the σ-bond. Consequently, the bond-energy of C=C is less than twice that of C—C (p. 11). Thirdly, p-orbital overlap is reduced by rotation of one carbon atom relative to the other around the internuclear axis, falling to zero when the angle of rotation is 90°. Since this results in reduction of the bond-energy, there is a strong tendency for π-bonded systems to resist rotation. This gives rise to the existence of geometrical isomers, for the energy required to convert, e.g. *cis*-2-butene (1) into *trans*-2-butene (2) is far greater than is available at ordinary temperatures. Similarly, in a bridgehead olefin such as (3) the fixed geometry of the ring system would not allow effective p orbital interaction and the thermodynamic stability of the compound would be correspondingly low; in fact, this compound cannot be obtained.

Nitrogen, like carbon, can form double bonds by promotion of one $2s$ electron followed by sp^2-hybridization. Four electrons then occupy the three hybridized orbitals, one of the orbitals containing a pair of electrons and the other two orbitals each containing a single electron; the fifth electron occupies a p orbital perpendicular to the plane of the sp^2 hybrids. σ-Bonds may then be formed by the singly occupied sp^2 orbitals and a π-bond may be formed by the p orbital, so that, e.g. acetoxime has the structure (4).

$$
\begin{array}{c}
CH_3 \\
\quad\diagdown \\
\qquad C{=}N \\
\quad\diagup \qquad \diagdown \\
CH_3 \qquad\quad OH
\end{array}
$$

(4)

As in doubly bound carbon, the geometrical requirement for the lateral overlap of p orbitals leads to coplanar systems around the double bond and resistance to rotation. Geometrical isomerism, as in unsymmetrical oximes, (5) and (6), and azo-compounds, (7) and (8), therefore occurs.

$$
\begin{array}{cccc}
\text{R}\diagdown\qquad & \text{R}\diagdown\quad\diagup\text{OH} & \text{R}\diagdown & \text{R}\diagdown\quad\diagup\text{R} \\
\quad C{=}N & \quad C{=}N & \quad N{=}N & \quad N{=}N \\
\text{R}'\diagup\quad\diagdown\text{OH} & \text{R}'\diagup & \quad\diagdown\text{R} & \\
(5) & (6) & (7) & (8)
\end{array}
$$

The third important type of hybridized orbital is obtained by the mixing of one $2s$ and one $2p$ orbital. Two sp-hybridized orbitals are formed which point in opposite directions, and the remaining $2p$ orbitals are each singly occupied and are available for forming two π-bonds. The simplest example of a compound formed in this way is acetylene, which is constructed as follows:

$$
\text{H}\underset{sp-s}{\overset{s-sp}{\text{—}}}\text{C}\underset{sp-sp}{\text{—}}\text{C}\text{—H}
$$

$p_y \qquad p_y$

$p_z \qquad p_z$

(b) THE VALENCE-BOND METHOD

Whereas in the molecular-orbital method the electronic configuration is established by first bringing the nuclei into position and then allotting the electrons to polycentric MOs, in the valence-bond method the complete atoms are brought together and then allowed to interact.

Consider the formation of H_2 from two hydrogen atoms. There is a wave function corresponding to electron 1 being near nucleus A and electron 2 being near nucleus B, given by $\psi = \psi_A(1)\,\psi_B(2)$. Since electrons are indistinguishable, the solution $\psi = \psi_A(2)\,\psi_B(1)$ is equally likely, so more appropriate solutions are obtained by the linear combinations of these functions,

$$\psi_s = \psi_A(1)\,\psi_B(2) + \psi_A(2)\,\psi_B(1)$$

$$\psi_a = \psi_A(1)\,\psi_B(2) - \psi_A(2)\,\psi_B(1)$$

The function ψ_s is *symmetric* in the co-ordinates of the electrons whereas ψ_a is *antisymmetric*, and it is the symmetric function, combined with the appropriate antisymmetric spin function (which must be introduced to allow for the spin properties of the electrons), which corresponds to the hydrogen molecule in its unexcited state.

We have so far considered only the possibility that each nucleus is associated with one electron. There is also the possibility that both electrons should simultaneously be near the same nucleus. For both electrons in the vicinity of nucleus A, $\psi = \psi_A(1)\,\psi_A(2)$, and for both in the vicinity of nucleus B, $\psi = \psi_B(1)\,\psi_B(2)$. Since these situations must be equally probable for a homonuclear di-atomic molecule, the combination

$$\psi_{\text{ionic}} = \psi_A(1)\,\psi_A(2) + \psi_B(1)\,\psi_B(2)$$

is appropriate. The wave function ψ_{ionic} corresponds to the likelihood that the hydrogen molecule is H^+H^-, that is, purely ionic, and this must now be combined with ψ_s (above), which represents the bond in H_2 as purely covalent, to give the complete molecular wave function,

$$\psi = \psi_{\text{covalent}} + \lambda\psi_{\text{ionic}}$$

where λ represents the relative importance of the ionic wave function.

In other words, the valence-bond description treats a molecule as a 'mixture' of covalent and ionic bonds. A molecule is commonly described as a *resonance hybrid* of various *canonical structures*; the canonical structures for H_2 are H—H, H^+H^-, and H^-H^+, and the resonance hybrid is represented as

$$\text{H—H} \leftrightarrow H^+H^- \leftrightarrow H^-H^+$$

For a heteronuclear diatomic molecule such as HCl, the contribution of the ionic structure H^+Cl^- is relatively more important (larger λ) than that of H^+H^- in H_2 because of the greater electronegativity (due to the larger nuclear charge) of chlorine than hydrogen. Conversely, the contribution of H^-Cl^+ is negligible, and the molecule is represented as the resonance hybrid

$$H—Cl \leftrightarrow H^+Cl^-$$

It is helpful to compare the treatments of molecules by the valence-bond and molecular-orbital methods. Both describe bonding as resulting from the pairing of electrons derived from different atoms, corresponding to there being an increase in the charge-density between the nuclei. Both also embrace the phenomenon of polarity in a bond, but in the first approximation they lead to different predictions of the magnitude of the polarity. For example, the VB wave function for the heteronuclear diatomic A—B is

$$\psi_{VB} = \psi_A(1)\,\psi_B(2) + \psi_B(1)\,\psi_A(2)$$

which may be combined with ionic wave functions (p. 41). The MO wave function for the same system is the product of the wave functions for each of the two electrons, that is,

$$\psi_{MO} = \psi_1\psi_2 = [\psi_A(1) + \lambda\psi_B(1)]\,[\psi_A(2) + \lambda\psi_B(2)]$$

$$= \psi_A(1)\,\psi_A(2) + \lambda^2\psi_B(1)\,\psi_B(2) + \lambda[\psi_A(1)\,\psi_B(2) + \psi_B(1)\,\psi_A(2)]$$

Comparison shows that the MO method gives a larger weight to configurations in which both electrons are on the same nucleus, for such configurations in the VB method are either neglected or else introduced (with an appropriate factor) to yield better solutions (i.e. a lower value for the energy, corresponding to a stronger bond). The difference arises because the MO method fails to allow for electron-correlation, that is, the fact that the electrons do not move independently of each other but are mutually repulsive. As a result the MO wave function overestimates the importance of the ionic configurations A^+B^- and A^-B^+, whereas in its simplest form the VB method underestimates these configurations by neglecting them.

2.5 Delocalization

In both the MO and VB treatments of simple organic compounds, the bonds are regarded as *localized*. Thus, in MO language, they are formed by the pairing of electrons derived from two AOs by the appropriate overlapping of these AOs, the characteristics of a particular bond—its length and its strength— being determined solely by the natures of these AOs and being independent of other electronic features of the molecule. This is justified, theoretically, by the existence of the conditions which lead to significant bonding (p. 34) and, experimentally, by the fact that a bond of given type has the same properties in a wide variety of structural environments. For example, a bond formed by two sp^3-hybridized carbon atoms is always 1·54 Å long and about 347 kJ mol⁻¹ strong.

The situation is different in *conjugated* molecules such as butadiene, $CH_2{=}CH{-}CH{=}CH_2$. First, the heat of hydrogenation of butadiene (238·7 kJ mol⁻¹) is less than twice that of 1-butene (126·7 kJ mol⁻¹), so that the compound is more stable by 14·7 kJ mol⁻¹ than expected from the principle of additivity of bond energies. Secondly, the central C—C bond is 1·46 Å long, compared with 1·54 Å for the C—C bond length in saturated compounds. The constancy of bond properties, which holds so well for saturated and simple unsaturated compounds and for which there is wave-mechanical justification, breaks down for conjugated systems.

A new concept, again justified wave-mechanically, is introduced: namely, delocalization. In the MO method, butadiene is treated in the usual way insofar as its σ-bond electrons are concerned, but the π-bond electrons, derived from the $2p$ orbitals of the sp^2-hybridized carbons, occupy two delocalized MOs which extend over the whole molecule. The representation in Fig. 2.11 shows that this arises from the overlap of the p orbitals of the central carbon atoms with the p orbitals both of their outer neighbours and of each other. The extra overlap leads to an increase in the total bonding and to the contraction of the central C—C bond which now has some double-bond character.

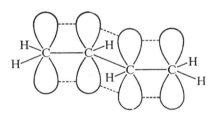

Fig. 2.11   Formation of delocalized π-orbitals in butadiene.

Valence-bond theory approaches the problem by developing the concept of resonance. The ionic structures (b) and (c) and the non-polar structure (d) (in which the single electrons on the terminal carbons are formally paired) contribute to the resonance hybrid:

$$CH_2{=}CH{-}CH{=}CH_2 \quad \overset{+}{C}H_2{-}CH{=}CH{-}\overset{-}{C}H_2 \quad \overset{-}{C}H_2{-}CH{=}CH{-}\overset{+}{C}H_2 \quad \overset{\cdot}{C}H_2{-}CH{=}CH{-}\overset{\cdot}{C}H_2$$
$$\quad\;\; (a) \qquad\qquad\qquad (b) \qquad\qquad\qquad (c) \qquad\qquad\qquad (d)$$

The complete wave function is

$$\psi = c_a\psi_a + c_b\psi_b + c_c\psi_c + c_d\psi_d$$

where the coefficients c_a, c_b, c_c, and c_d indicate the relative importance of each structure. * Since the structures (b), (c), and (d) contain one fewer covalent bond than (a), and the formation of (b) and (c) requires the expenditure of energy to separate unlike charges, the energies of (b), (c), and (d) are higher than that of (a) and the structures are of less importance in the hybrid (i.e. $c_a > c_b$, c_c, c_d). However, because each of the three describes the central C—C bond as a double bond, the complete wave function corresponds to there being some double-bond character in this bond, and its shorter length than that of C—C in a saturated molecule is accounted for. Moreover, inclusion of the wave functions ψ_b, ψ_c, and ψ_d leads to a lower energy than that corresponding to ψ_a alone. The value of the decrease in energy achieved by inclusion of the canonical structures (b—d) is defined as the *resonance energy*, or, perhaps more satisfactorily, the *stabilization energy* of the molecule.†

While both the MO and the VB theories encompass the properties of conjugated systems such as butadiene, there is one important aspect in which the two treatments differ. Looked at from a simple viewpoint, the overlap of the p orbitals of the central carbons of butadiene (Fig. 2.11) should be just as effective as the overlap of each of these p orbitals with its outside neighbour. That this cannot be so is indicated by the fact that the central C—C bond is only shorter than a single C—C bond by 0·08 Å, whereas the terminal bonds are shorter by 0·2 Å and are the same length as the double bond in ethylene. The MO treatment overestimates the importance of p-orbital interaction at the centre of the molecule because it ignores electron correlation. The four p electrons which occupy the two delocalized MOs stay apart as a result of their mutual repulsion and are most probably aligned as pairs at opposite ends of the molecule, corresponding to the classical structure (a). The VB method does not overestimate the importance of structures (b–d); appropriate coefficients are introduced to allow for the relative importance of each canonical structure.

From this it would seem that the VB method is more suitable for the consideration of conjugated systems, but its qualitative applications are easily abused and care must be exercized in its use. The following principles provide general guidance for students of the theory.

*Contrast the weighting of *structures* in the VB treatment with the weighting of *atomic orbitals* in the MO treatment.

†It is not certain that the extra stability of conjugated compounds compared with nonconjugated compounds, as measured, for example, by heats of reduction, is entirely attributable to delocalization. We would expect that the strength of a C—C bond would be dependent on the state of hybridization of each carbon atom. Then, since in the reduction of 1-butene an sp^2–sp^3 C—C bond is converted into an sp^3–sp^3 C—C bond whereas in the reduction of butadiene an sp^2–sp^2 C—C bond is converted into an sp^3–sp^3 C—C bond, there could be a difference in the heats of hydrogenation of butadiene as compared with that of two molecules of 1-butene irrespective of there being a special stabilization associated with the π-orbital system. Likewise, the smaller length of the single C—C bond in butadiene than that in saturated compounds could, at least in part, result from the fact that sp^2-hybridized carbon has a smaller covalent radius than sp^3-hybridized carbon.

(*i*) Resonance is only significant when the canonical structures are of comparable energy. This is true, for example, of the two Kekulé structures for benzene, (9) and (10), which must have the same energy; the resulting stabilization energy is considerable (p. 55). It is not true, however, of the structures (11) and (12) for ethylene for which the high-energy structure (12) may be neglected; that is, inclusion of the wave function for (12) has a negligible effect on the calculated energy.

$$CH_2{=}CH_2 \qquad \overset{+}{C}H_2{-}\overset{-}{C}H_2$$

(9)   (10)   (11)   (12)

(*ii*) The nuclei in each of the canonical structures must be in the same relative positions. For example, the structure (14) does not contribute to the structure of isobutylene (13) but is instead the isomeric compound, *trans*-2-butene.

$$\overset{\cdot}{C}H_2{-}CH{=}CH{-}\overset{\cdot}{C}H_2$$

(13)   (14)   (15)

(*iii*) Electron spin must be conserved. Thus, the structure (15) does not contribute to the structure of butadiene but is one representation of an (electronically) excited state of this molecule in which the complete pairing of the ground state has been lost as the result of a spin-inversion.

Finally, despite the criticism of MO theory mentioned above, it is conceptually the more realistic approach to bonding, giving a clearer idea of the important structural features of molecules. The VB theory may be described as artificial in that the canonical structures whose superimposition is used to describe a molecule have no real existence. Nevertheless, VB descriptions are usually easier to follow in discussing organic reactions, and we shall make considerable use of the method throughout this book.

2.6 Applications of Structural Theory

(*a*) MOLECULAR STRUCTURE AND SHAPE

The origin of directed valencies has already been briefly discussed in terms of the shape of atomic orbitals and hybridized orbitals. We shall now describe the bonding and the geometry of the more important groups and systems in organic chemistry in the same way.

(*i*) *Saturated compounds.* Methane has a tetrahedral structure because the four sp^3-hybridized orbitals of carbon are directed towards the corners of a regular tetrahedron. Substitution of one hydrogen atom by, say, a chlorine atom to give methyl chloride, CH_3Cl, leads to a slight distortion of the tetrahedron because the repulsive forces between pairs of atoms, and pairs of electron-pairs, are no longer the same. This holds also for methylene chloride, CH_2Cl_2, and chloroform, $CHCl_3$, but with carbon tetrachloride, CCl_4, the regular tetrahedral structure is again established.

Substitution of one hydrogen in methane by a *group* of atoms leads to a different effect. In ethane, for example, not only is there very slight distortion from tetrahedral angles, but there is also an infinite number of *conformations* which are made possible by the axial symmetry of the C—C σ-bond. Two of these, the *staggered* conformation (16) and the *eclipsed* conformation (17) are shown; they are perhaps more easily visualized by reference to the *Newman projections* (18) and (19), respectively, which correspond to viewing the molecule from one end along the C—C bond.

(16)   (17)   (18)   (19)

As a result of the interactions between pairs of protons, and pairs of electron-pairs, on the two carbon atoms, the conformations differ in energy. The staggered conformation is the most stable of all the possible conformations and the eclipsed conformation is the least stable; the difference in energy between these two extremes is about 12 kJ mol^{-1}. Application of the Boltzmann distribution relationship shows that at any instant there are about 100 times as many molecules in the staggered as in the eclipsed conformation. The difference in energies of these two conformations provides a barrier to rotation about the single bond, but this is so small that, at ordinary temperatures, rotation is extremely rapid.

For more complex molecules a larger number of eclipsed and staggered conformations is possible. Three of those for n-butane, which differ in the arrangement about the central C—C bond, are shown. The *anti* conformation is about 4 kJ mol^{-1} more stable than the *gauche* conformation and about 25 kJ mol^{-1} more stable than the *eclipsed* conformation so that the *anti* conformation is the most heavily populated. However, rotation about the C—C

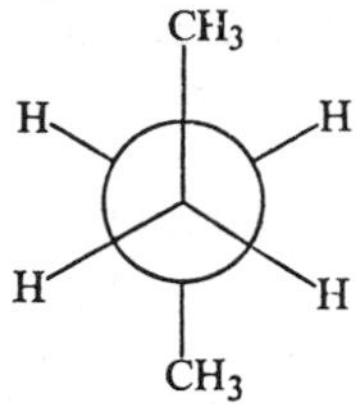

CH₃ structures (Newman projections):

anti
(20)

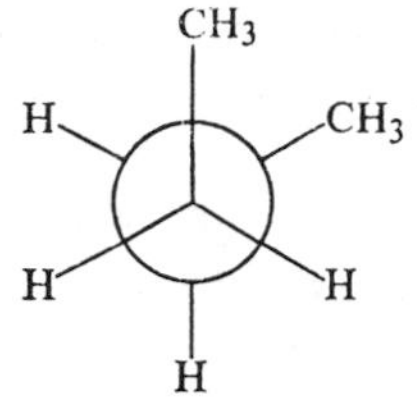

gauche
(21)

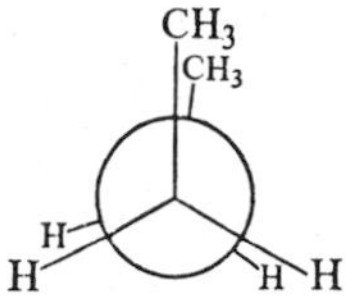

eclipsed
(22)

bond is rapid so that it is not possible to isolate the *anti* and *gauche* conformations as discrete compounds.

Alicyclic compounds with three members are necessarily coplanar and therefore possess not only angular strain (the CCC angle being 60° rather than the preferred tetrahedral angle) but also conformational strain, for the hydrogen atoms on adjacent carbons are eclipsed. For cyclobutane and cyclopentane and their derivatives, the tendency for the CCC angles to be as near to tetrahedral as possible would lead to the carbon skeletons being coplanar, but this situation corresponds to the complete eclipsing of the substituent hydrogen atoms and it proves to be thermodynamically more favourable for the rings to be slightly buckled. For cyclohexane, both angular strain and conformational strain are eliminated by the ring's adopting the *chair* conformation (23) which is shown also in the Newman projection (24) (the molecule being viewed along two parallel C—C bonds). Other conformations possess no angular strain but all have conformational strain: e.g. the *boat* conformation (25) has four eclipsing

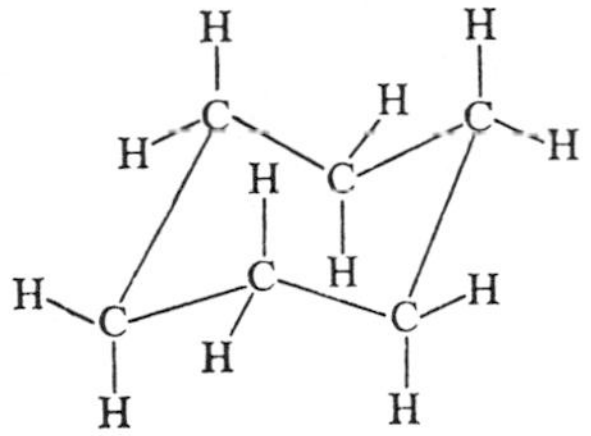

(23)

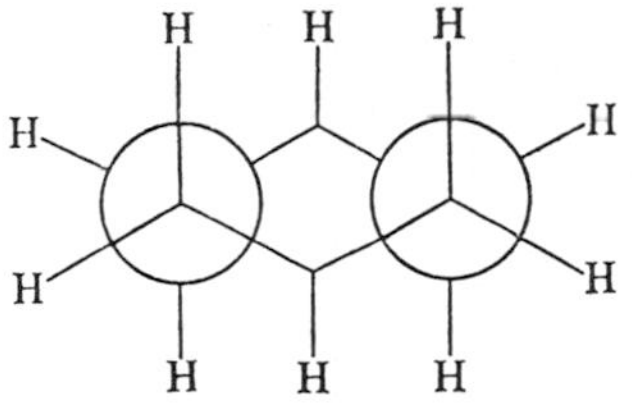

(24)

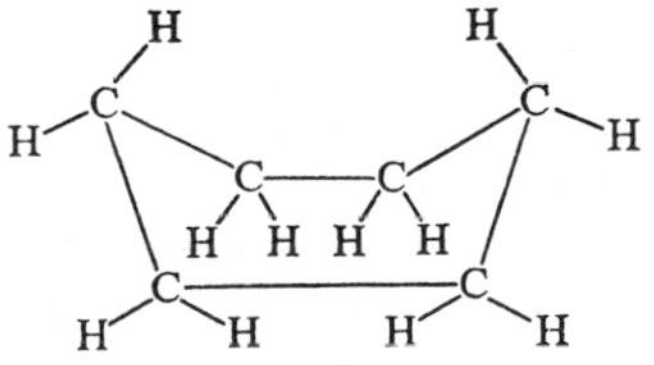

(25)

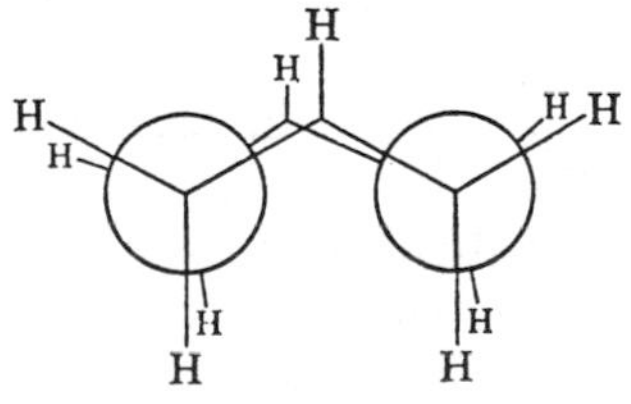

(26)

interactions (see (26)) as well as a strong repulsive force between the hydrogen atoms at the opposite ends of the boat. Consequently cyclohexane exists predominantly in the chair form.

Seven-membered rings and above can all be constructed with strain-free CCC angles, but for C_7 to C_{11} rings all conformations possess some eclipsing strain. For rings larger than eleven-membered, strain-free conformations again occur.

A measure of the strain in alicyclic rings is given by the heat of combustion of the molecule per methylene group less the value for the strainless cyclohexane. Data for rings possessing three to eleven members are as follows:

No. of carbons in the ring	Heat of combustion per methylene group (kJ mol^{-1})	Strain energy (kJ mol^{-1})
3	696	38
4	685	27
5	663	5
6	658	0
7	662	4
8	663	5
9	664	6
10	663	5
11	662	4

(ii) *Dipolar bonds: the inductive effect.* Whenever two dissimilar atoms are bonded, the bonding pair is not symmetrically placed with respect to the two nuclei because these have different charges. Theoretically, bond-polarity is accommodated by the inclusion in the wave function of a parameter which measures the relative probabilities that the bonding electrons are in the vicinity of each nucleus. Experimentally, this asymmetry of bonds is manifested in the occurrence of dipole moments. These are measured by the distance between the centres of positive and negative charge in the molecule multiplied by the size of the charge and have values in the region 10^{-30} coulomb metre. The Debye unit (D) provides a useful scale of measurement: $1D = 3.3 \times 10^{-30}$ C m.

It is convenient to refer also to *bond moments*, for, like bond length and bond energy, the bond moment of a particular bond varies little with the structural environment except when conjugative influences are introduced. The vector sum of the individual bond moments in a molecule gives the resultant dipole moment.

Heteronuclear diatomic molecules invariably possess dipole moments. That of hydrogen chloride is 1·03D, the negative end of the dipole being the chlorine atom, i.e. $\overset{+}{H}$—$\overset{-}{Cl}$. Symmetrical polyatomic molecules have no dipole moment, even though the individual bonds may have large moments, for the vector sum of the bond moments is zero. Thus methane and carbon tetrachloride have zero resultant moments, although both C—H and C—Cl bonds have moments, whereas methyl chloride, methylene chloride, and chloroform have significant moments. Again, *p*-dichlorobenzene, unlike its *ortho* and *meta* isomers, has a zero moment, although each nuclear carbon is positively polarized, relative to those in benzene, by the electronegative chlorine substituents.

The concept of bond moments can be usefully extended to establish a scale of *inductive* effects. Since methane has no dipole but methyl chloride has a dipole whose negative end is the chlorine atom, chlorine is described as having an electron-withdrawing inductive effect $(-I)$. Groups of atoms may also have inductive effects; e.g. in nitromethane carbon is attached to a positively charged nitrogen atom (p. 50), so that the $-NO_2$ group has a strong $-I$ effect. A procedure has been established for deriving a quantitative scale of inductive effects relative to hydrogen, and the order for some commonly occurring atoms and groups is:

$$(CH_3)_3C < CH_3 < (H) < OCH_3 < CF_3 < Br < Cl < F < CN < NO_2 < \overset{+}{N}Me_3.$$

The inductive effect of an atom or group acts most powerfully on the atom to which it is bonded but atoms further away are affected to some degree. For example, in n-butyl chloride the electron-pair in the C—Cl bond is displaced towards chlorine. The nucleus of the carbon atom is less strongly shielded than that of the corresponding carbon in butane, so that it is effectively a more electronegative atom. The electron-pair of the adjacent C—C bond is accordingly displaced towards the carbon to which chlorine is attached, and in turn the second carbon is slightly affected. In this way, the replacement of hydrogen by a more electronegative atom results in electron displacements throughout the molecule, the effect on successive atoms falling off with their increasing distance from the electronegative centre. The effect may be represented for n-butyl chloride as in (27).

$$\overset{\delta\delta\delta\delta+}{CH_3}\!-\!\!-\!\!-\!\overset{\delta\delta\delta+}{CH_2}\!-\!\!-\!\!-\!\overset{\delta\delta+}{CH_2}\!-\!\!-\!\!-\!\overset{\delta+}{CH_2}\!-\!\!-\!\!-\!\overset{\delta-}{Cl}$$

$$(27)$$

(*iii*) *Unsaturated compounds*. The elements most commonly involved in unsaturated systems are carbon, oxygen, and nitrogen.

In double-bonded groupings both carbon and nitrogen are sp^2-hybridized. Carbon forms three σ-bonds, by means of its three singly occupied sp^2 orbitals, which are coplanar and at (or close to) 120° to each other and a π-bond by means of its singly occupied p orbital which is perpendicular to this plane. Nitrogen systems are similar save that one of the sp^2 orbitals contains two electrons and constitutes a non-bonding pair. In heteropolar double bonds it is appropriate to include ionic structures in the valence-bond descriptions; thus, as a result of the increasing order of electronegativities, carbon < nitrogen < oxygen, canonical structures of the types (28)–(30) contribute to the total wave functions of carbonyl compounds, imines, and nitroso compounds, respectively, and these bonds have significant polarity and slightly *less* double-bond character than the homopolar double bonds, $C{=}C$ and $N{=}N$.

$$\ce{>C=O} \leftrightarrow \ce{>\overset{+}{C}-\overset{-}{O}} \qquad \ce{>C=N<} \leftrightarrow \ce{>\overset{+}{C}-\overset{-}{N}<} \qquad \ce{N=O} \leftrightarrow \ce{\overset{+}{N}-\overset{-}{O}}$$

$$(28) \qquad\qquad\qquad (29) \qquad\qquad\qquad (30)$$

The nitro group also contains sp^2-hybridized carbon and may be regarded as derived from the nitroso group by the transference of one electron to oxygen from nitrogen's non-bonding pair followed by the formation of a covalent bond:

$$\ce{R-\underset{\cdot\cdot}{N}\backslash\!\!=\!\!O} \; + \; \ce{\overset{\cdot\cdot}{\underset{\cdot\cdot}{O}}:} \;\longrightarrow\; \ce{R-\overset{+}{N}\backslash\!\!=\!\!O} \; + \; \ce{\overset{\cdot\cdot}{\underset{\cdot\cdot}{O}}:^{-}} \;\longrightarrow\; \ce{R-\overset{+}{N}{<}^{\overset{-}{O}}_{O}}$$

It should be noted that the oxygen atoms now become equivalent, so that a nitro compound is described as a resonance hybrid of the structures (31) and (32).

$$\ce{R-\overset{+}{N}{<}^{\overset{-}{O}}_{O}} \leftrightarrow \ce{R-\overset{+}{N}{<}^{O}_{\overset{-}{O}}}$$

$$(31) \qquad\qquad (32)$$

In the molecular-orbital description, nitrogen forms three coplanar σ-bonds by means of singly occupied sp^2 orbitals, and its p orbital, which contains a pair of electrons, interacts with the p orbitals of each of the two oxygen atoms, as shown in (33).

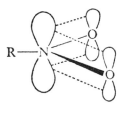

$$(33)$$

In triple-bonded groupings carbon and nitrogen are sp-hybridized. Carbon forms two collinear σ-bonds by means of its singly occupied sp orbitals and two π-bonds by means of its two singly occupied and mutually perpendicular p

orbitals. For nitrogen, one of the *sp* orbitals contains a non-bonding electron pair. In the valence-bond description of the $C \equiv N$ bond it is appropriate to include the ionic structure $-\overset{+}{C}=\overset{-}{N}$, so that the bond is significantly polar.

(*iv*) *Conjugated compounds*. Special properties are associated with systems in which a π-bond is conjugated either with a second π-bond or with an atom which possesses a pair of electrons in a *p* orbital. These are: stabilization energy, single-bond lengths which are shorter than those in non-conjugated compounds, and (in some cases) the modification of dipolar properties.

1. π-*Bond*–π-*bond conjugation*. The simplest example is butadiene, $CH_2 = CH-CH=CH_2$, which has been described (p. 43). This is a symmetrical molecule, and conjugation does not lead to the appearance of a dipole; in VB language, contributions from the ionic structures $\overset{+}{C}H_2-CH=CH-\overset{-}{C}H_2$ and $\overset{-}{C}H_2-CH=CH-\overset{+}{C}H_2$ are necessarily equal and their dipoles therefore nullify each other. This is not true, however, when π-bonds of different types are in conjugation. For example, in an $\alpha\beta$-unsaturated carbonyl compound such as crotonaldehyde the contribution of the canonical structure (34) is greater than that of (35) because oxygen is significantly more electronegative than carbon; the carbonyl group thus polarizes the olefinic bond.

$$CH_3-\overset{+}{C}H-CH=CH-\overset{-}{O} \qquad\qquad CH_3-\overset{-}{C}H-CH=CH-\overset{+}{O}$$
$$\text{(34)} \qquad\qquad\qquad\qquad \text{(35)}$$

A group such as carbonyl which withdraws electrons from an adjacent group *via* the π-bonding framework is described as having a $-M$ effect, the negative sign indicating electron-withdrawal and M standing for *mesomeric*. Other groups of $-M$ type include carboalkoxyl (36), nitrile (37), and nitro (38); the curved arrows in the representations denote the direction of the mesomeric effect.

$$\text{(36)} \qquad\qquad\qquad \text{(37)} \qquad\qquad\qquad \text{(38)}$$

2. π-*Bond*–*p*-*orbital conjugation*. In vinyl chloride the *p* orbital on the carbon which is attached to chlorine can overlap with both the *p* orbital on the second carbon atom and one of the *p* orbitals on chlorine, as shown in (39). Since the *p* orbital of chlorine is initially filled, its participation in a delocalized π-system leads to the partial removal of electrons from chlorine and the appearance of a

dipole moment directed from chlorine towards carbon. This is opposed to the dipole established in the C—Cl σ-bond as a result of the $-I$ effect of chlorine, so that the overall result is that the dipole moment of vinyl chloride (1·44 D) is considerably smaller than that of ethyl chloride (2·0 D) in which only the $-I$ effect is operative. The capacity of chlorine for donating electrons into a molecular π-system is described as a $+M$ effect.

$$CH_2{=}CH{-}Cl \leftrightarrow \bar{C}H_2{-}CH{=}\overset{+}{C}l$$

$$(40) \qquad\qquad (41)$$

(39)

Alternatively, vinyl chloride may be described in valence-bond terminology as a hybrid of the structures (40) and (41), the latter symbolizing the $+M$ effect. Both descriptions also indicate that the C—Cl bond should be shorter than that in a saturated alkyl halide, as is found.

Other elements with unshared p-electrons which take part in forming delocalized π-systems include the other halogens, oxygen, and nitrogen, e.g.

$$CH_2{=}CH{-}\overset{..}{O}CH_3 \qquad\qquad CH_2{=}CH{-}\overset{..}{N}(CH_3)_2$$

$$(42) \qquad\qquad\qquad\qquad (43)$$

In each case the substituent has a $+M$ effect.

Two further points should be noted. First, bond-strength is dependent on the extent of the overlap of the combining atomic orbitals (p. 34), so that in these conjugated systems the more nearly equal in size the p orbitals are, the more effective is the π-orbital overlap. Hence fluorine is more effective than chlorine in conjugating with carbon, and oxygen is more effective than sulphur. Secondly, as the nuclear charge in an atom is increased, so also is the hold of the nucleus on the surrounding electrons, so that, for comparably sized atoms, the ability to conjugate decreases as the atomic number increases. Hence the order of $+M$ effects is —NR$_2$ > —OR > —F.

3. *Hyperconjugation.* There is some evidence that a C—H bond which is adjacent to a π-bond can take part in a delocalized π-system. For example, propylene can be described as a hybrid of the structures (44) and (45); alternatively, the MO description is that the C—H bonding orbital interacts with the

adjacent *p*-orbital on carbon to form a delocalized π-molecular orbital. This type of conjugation is termed *hyperconjugation*.

$$\underset{(44)}{\overset{H}{\diagdown}CH_2-CH\!\!=\!\!CH_2} \quad \longleftrightarrow \quad \underset{(45)}{\overset{H^+}{}CH_2\!\!=\!\!CH\!-\!\bar{C}H_2}$$

   The evidence for hyperconjugation in this case is that the C—C single bond is slightly shorter than that in ethane and that the heat of hydrogenation (126 kJ mol^{-1}) is less than that of ethylene (137 kJ mol^{-1}). However, it may be that part, if not all, of the measured stabilization energy of 11 kJ mol^{-1} is due to the fact that an sp^2–sp^3 carbon-carbon bond is stronger than an sp^3–sp^3 carbon-carbon bond, while the shortened bond length may result from sp^2-carbon having a smaller covalent radius than sp^3-carbon (cf. footnote, p. 44). Nevertheless, whatever the origin of the stabilization energy, it is a general fact that, amongst a given group of isomeric olefins, the most stable (unless strain factors are involved) is the one in which the olefinic bond is conjugated to the largest number of alkyl groups. This fact has many consequences in organic synthesis, as, for example, in the direction taken in elimination reactions (p. 120).

   (*v*) *Hydrogen-bonding and chelation.* A hydrogen atom which is bonded to an electronegative atom can form a *hydrogen-bond* to a second electronegative atom. The binding results from the polarization of the electron cloud of the second electronegative atom by the small, positively polarized hydrogen atom. Hydrogen-bonds are longer than covalent bonds and are much weaker: their strengths lie in the range 10–40 kJ mol^{-1}, which is of the order of one-tenth the strength of covalent bonds. Conventionally, the bonds are represented by dotted lines, e.g.

$$\overset{R}{\diagdown}O\!-\!H\,-\!-\!-\!-\!-\,\overset{\diagup R}{O}\underset{\diagdown H}{}$$

   The only elements concerned in hydrogen-bonding are nitrogen, oxygen, fluorine, and, to a lesser extent, chlorine. The bond may be formed both between molecules of the same type, as in water, alcohols, and carboxylic acids, and between molecules of different types, as in the interaction between the proton of an amino group and the oxygen of a carbonyl group:

$$\overset{\diagdown}{\underset{\diagup}{N}}\!-\!H\,-\!-\!-\!-\!-\,O\!\!=\!\!\overset{\diagup}{\underset{\diagdown}{C}}$$

   Hydrogen-bonding leads to an increase in intermolecular 'aggregation' forces and is manifested particularly in the boiling-point and solubility of the organic compound. The boiling-point is raised because energy is required to separate the hydrogen-bonded molecules in their translation to the gaseous state; e.g. ethanol boils more than 100°C higher than dimethyl ether, which has the same molecular weight, and alcohols boil at higher temperatures than the corresponding thiols. Two compounds which can form hydrogen-bonds tend to be more soluble in each other than in those which cannot; alcohols, and to a lesser extent ethers, are more soluble in water than are hydrocarbons of comparable molecular weight.

Intramolecular hydrogen-bonds may also be formed, and are of particular significance when the resulting ring is five- or six-membered. The phenomenon is then described as *chelation* and is illustrated for the enolic form of acetoacetic

(46)   (47)

ester (46) and for *o*-nitrophenol (47). Since chelation does not give rise to intermolecular aggregation forces, chelated compounds have 'normal' boiling points. For example, *o*-nitrophenol is much more volatile than its *para*-isomer, for only the latter forms intermolecular hydrogen-bonds.

(b) AROMATIC CHARACTER

Many cyclic, conjugated compounds possess markedly different physical and chemical properties from those expected by comparison of their structures with acyclic analogues. The simplest example is benzene, which may be regarded as the parent compound of the aromatic series. The only possible classical representations are (48) and (49), but neither alone is adequate, for there is not an alternation of single and double bonds: each C—C bond is the same length (1·39 Å), rather closer to a normal double bond (1·34 Å) than to a normal single bond (1·54 Å). Moreover, whereas the heat of hydrogenation of the double bond in cyclohexene is 119 kJ mol^{-1}, that for the three double bonds in benzene is 207 kJ mol^{-1}; the latter value is 150 kJ mol^{-1} less than three times the former, so that benzene is more stable by this amount than predicted on the basis of its formal similarity to an alicyclic compound. Finally, benzene does not exhibit the ready addition reactions characteristic of olefins. These facts are accommodated by both the valence-bond and the molecular-orbital theories. According to the former, benzene is described as a hybrid of the two canonical structures (48) and (49) (which are known as Kekulé structures), and since these are of equal energy content, the resulting stabilization energy is considerable.

(48)   (49)   (50)   (51)   (52)

The more energetic Dewar structures (50–52) are less important, but the inclusion of wave functions corresponding to them results in a still greater reduction in the total energy. The total stabilization energy is 150 kJ mol^{-1}.

The molecular-orbital treatment describes benzene as containing six sp^2-hybridized carbon atoms, each of which forms σ-bonds with two carbon atoms and one hydrogen atom, the six remaining electrons being in p orbitals each of which overlaps with two neighbours, as shown in (53). The planarity of the ring, which corresponds to strainless CCC angles (120°), allows maximum overlap of these p orbitals. Six delocalized π-orbitals are thereby established, of which the three of lowest energy (i.e. bonding orbitals) are occupied. The relative energies of these are shown in Fig. 2.12.

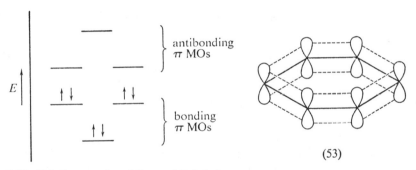

(53)

Fig. 2.12   Relative energies of the π-orbitals in benzene.

It should be noted that, unlike the situation in butadiene, in which the freedom of movement of the π-electrons in the delocalized MOs is opposed by electron repulsion, the π-electrons in benzene can circulate round the ring in a synchronized manner without increasing their mutually repulsive forces. Consequently the delocalization energy in benzene is far larger than in butadiene.

In addition to accounting for the stability of benzene, both the treatments outlined above also embrace the equivalence of the bonds and account for the inertness of benzene to addition, for the addition of, e.g. one molecule of bromine would lead to the simple conjugated system (54) and the loss of over 125 kJ of stabilization energy. This results in the reaction being endothermic by about 40 kJ mol^{-1}, whereas the addition of bromine to ethylene is exothermic.

Polycyclic hydrocarbons composed of fused benzene rings are also aromatic. The simplest examples are naphthalene, anthracene, and phenanthrene, which

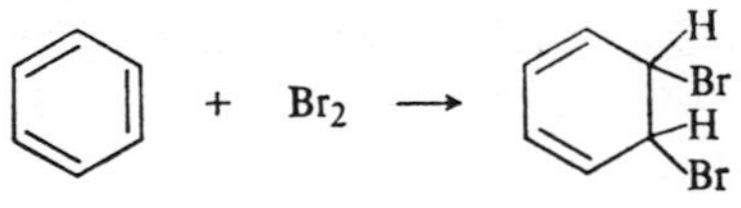

(54)

can be represented as the hybrids (55), (56), and (57), respectively. In each, delocalization is achieved by extensive *p*-orbital overlap, and the stabilization energies are greater than that of benzene. In general, for compounds containing equal numbers of benzene rings, that for which the greatest number of Kekulé structures can be drawn has the largest stabilization energy (e.g. phenanthrene > anthracene).

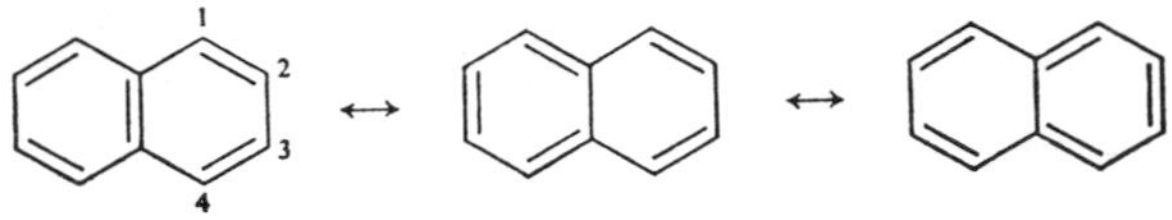

(55) Stabilization energy: 255 kJ mol^{-1}.

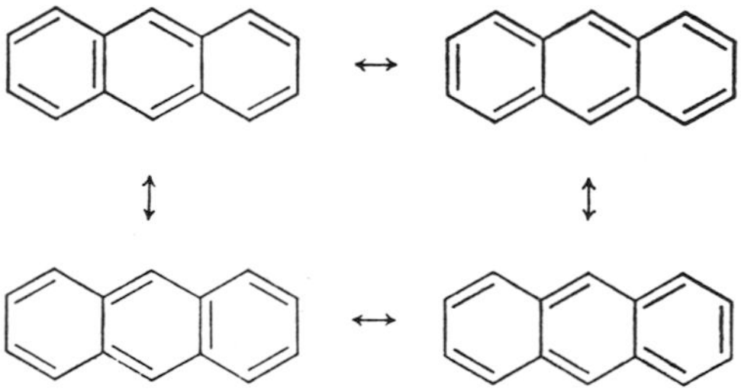

(56) Stabilization energy: 349 kJ mol^{-1}.

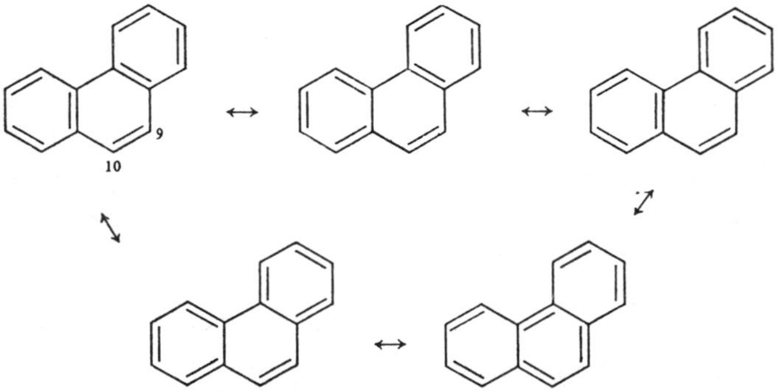

(57) Stabilization energy: 380 kJ mol^{-1}.

The bond lengths are not equal in these systems. For instance, the 1,2-bond in naphthalene (1·36 Å) is shorter than the 2,3-bond (1·42 Å). A rough estimate of the relative lengths of the bonds can be obtained by examining the Kekulé structures: for naphthalene, the 1,2-bond is represented as a double bond in two of these structures and as a single bond in the third, while the opposite is the case for the 2,3-bond. Inspection of phenanthrene shows that the 9,10-bond is a double bond in four of the five structures; consistently with this, its length is close to that in an olefin.

The stabilization energy of a polycyclic hydrocarbon is less than that of the sum of its constituent benzenoid rings and this fact has important consequences in reactivity. For example, addition across the central ring in anthracene gives a compound containing two benzene rings, and therefore results in the loss of only $(349 - 2 \times 150)$, i.e. 49 kJ mol^{-1} of stabilization energy, compared with the loss of nearly 150 kJ mol^{-1} on addition to benzene. In practice, anthracene undergoes many addition reactions at the 9,10-positions: e.g. bromine gives 9,10-dibromo-9,10-dihydroanthracene (58) and maleic anhydride gives a Diels-Alder adduct (9.2).

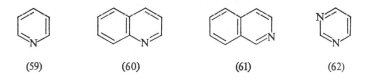

(58)

The properties which are associated with the carbocyclic, six-membered rings of benzene and the polycyclic hydrocarbons, and which result from the extensive *p*-orbital interaction in these planar, conjugated systems, are summarized in the phrase *aromatic character*. To display aromatic properties, however, compounds do not necessarily need to be either entirely carbocyclic or six-membered, and aromatic compounds of different types will now be introduced.

(1) sp^2-Hybridized nitrogen may replace a CH group. The nitrogen atom forms two σ-bonds with the adjacent carbon atoms by using two singly occupied sp^2 orbitals, possesses an unshared electron-pair in its third sp^2 orbital, and contributes one *p* electron to the delocalized π orbitals. The most important examples of such compounds are pyridine (59) and its benzo-derivatives, quinoline (60) and isoquinoline (61), and pyrimidine (62).

(59)   (60)   (61)   (62)

In valence-bond terms, pyridine may be described as a hybrid of the canonical structures (63) and (64), and because nitrogen is more electronegative than carbon, it is appropriate to include also, though with reduced weighting, the ionic structures (65–67). As a result, the electron-densities at the 2-, 4-, and 6-positions are less than at the carbon atoms in benzene, and this is of importance in connection with the chemical properties of the compound (pp. 382, 424). The same considerations hold for the related systems (60–62).

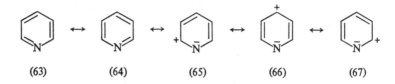

(63)   (64)   (65)   (66)   (67)

(2) Two CH groups, each of which provides one p electron for the delocalized π-system, may be replaced by one atom which supplies two p electrons. For example, furan (68) is aromatic because it contains a conjugated, cyclic, and planar system and six π-electrons in delocalized MOs, four of which are provided by carbon atoms and the remaining two by oxygen. The valence-bond representation is in terms of the canonical structures (68–72).

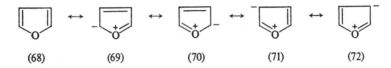

(68)   (69)   (70)   (71)   (72)

The contribution of the ionic structures (69–72) corresponds to there being a negative charge on each of the nuclear carbons, although this is small because the dipolar structures are of higher energy, and are therefore less important contributors, than the 'classical' structure (68). Nevertheless, this is of importance in determining the reactivity of the carbon atoms. In addition, again because of the relatively high energy of the ionic structures, the stabilization energy of furan (ca. 85 kJ mol^{-1}) is much less than that of benzene. As a result, addition reactions occur more easily, for there is less stabilization energy to be lost (e.g. the Diels-Alder reaction; 9.2).

Pyrrole (73) and thiophen (74) and their benzo-derivatives such as indole (75) are aromatic for the same reason as furan. For thiophen and its derivatives, a new characteristic is introduced: since sulphur has relatively low-lying and unfilled $3d$ orbitals, it may accept π-electrons, so that canonical structures such as (76) contribute to the hybrid. As a result, thiophen is more strongly stabilized than furan and its carbon atoms are less strongly negatively polarized, and these characteristics are reflected in its chemistry (e.g. thiophen, unlike furan, does not undergo the Diels-Alder reaction).

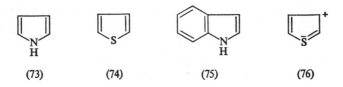

(73)   (74)   (75)   (76)

The cyclopentadienyl anion (78) fulfils the criteria for aromaticity (it is iso-electronic with pyrrole) and proves to be a relatively stable species; thus, the ion is formed from cyclopentadiene (77) on treatment with base in conditions in which acyclic olefins such as 1,4-pentadiene are unreactive.

(77)   (78)

(3) An atom which contributes an empty p orbital may be introduced into six-membered aromatic systems. For example, cycloheptatrienyl bromide (79) is a salt, unlike aliphatic bromides, and this may be ascribed to the fact that the cycloheptatrienyl (tropylium) cation is aromatic and consequently strongly resonance-stabilized. Again, tropone (80) has markedly different properties from aliphatic ketones, being miscible with water, very high-boiling, strongly polar ($\mu = 4\cdot3$ D), and lacking ketonic properties. It is evident that dipolar structures such as (81) and (82), which constitute the aromatic tropylium system, provide a better representation of the compound.

(79)   (80)   (81)   (82)

It is necessary to introduce here a final criterion for the occurrence of aromatic character. So far, each of the aromatic systems we have described has contained six π-electrons. Insofar as the simple application of resonance theory is concerned, there is no reason why cyclopentadienyl anions and cycloheptatrienyl cations should be comparatively stable whereas cyclopentadienyl cations and cycloheptatrienyl anions should not be; as many equivalent canonical structures can be drawn for $C_5H_5^+$ as for $C_5H_5^-$ and for $C_7H_7^-$ as for $C_7H_7^+$. Yet cyclopentadienyl bromide, unlike tropylium bromide, is not ionic, and cyclo-heptatriene, unlike cyclopentadiene, does not form an anion in basic solution. The explanation is that only systems which contain ($4n + 2$) π-electrons (where *n* is an integer) can be expected to be aromatic, and the reason for this lies in the relative energies of the π MOs. For example, cyclobutadiene might exist as a square structure in which the four *p* AOs combined to give four de-localized π MOs whose relative energies are shown in Fig. 2.13*a*. Since four *p* electrons are available, two would occupy the lowest MO and the other two would be distributed one each in the degenerate MOs and would have the same spin: two further electrons would be required to complete these MOs in order to provide significant binding energy. A more stable disposition is in fact available: the molecule adopts a rectangular shape (83) in which the four *p* orbitals form two *localized* π bonds. Even this is a relatively high-energy structure because of the strain in the CCC angles; it has only a fleeting existence, rapidly dimerizing to a mixture of (84) and (85).

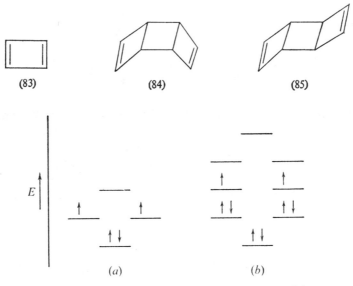

(83)   (84)   (85)

E

(*a*)   (*b*)

Fig. 2.13   Molecular orbitals formed by systems containing
(*a*) four *p* AOs and
(*b*) eight *p* AOs.

For a compound such as cyclo-octatetraene in which eight *p* AOs were able to overlap, the energy levels of the eight delocalized π MOs would be as in Fig. 2.13*b*; there would again be incomplete pairing in the MOs. Instead of its adopting the planar structure (86) necessary for the complete overlapping of its *p* AOs, cyclo-octatetraene exists as the tub-shaped structure (87) in which each *p* orbital can overlap with only one other, and this gives it the properties of an aliphatic olefin (e.g. ready addition of bromine).

It might be argued that cyclo-octatetraene is not aromatic because the angular

(86)   (87)

strain in the necessary planar structure would more than offset the resulting
stabilization energy. This thesis is, however, at variance with other observations,
such as the comparative stability of the cyclo-octatetraene di-anion (88) and the
cyclopropenyl cation (89); both these systems should be highly strained and
their stabilities are evidently associated with their possessing 10 and 2 electrons,
respectively, in delocalized MOs which form complete-shell configurations.

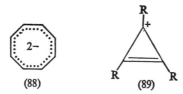

(88)   (89)

Finally, attempts to make pentalene (90) and heptalene (91) have failed,
whereas azulene (92) is an aromatic compound with a stabilization energy of
about 170 kJ mol^{-1}. Pentalene and heptalene do not contain $(4n + 2)$
π-electrons, but azulene does, and it is especially interesting that azulene has a
dipole moment, suggesting that the seven-membered ring in some degree donates
one π-electron to the five-membered ring so that each ring approximates to a
six π-electron system. Thus, in valence-bond terms, ionic structures such as (93)
are important contributors to the hybrid.

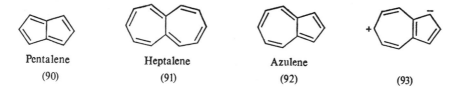

Pentalene   Heptalene   Azulene
(90)   (91)   (92)   (93)

The following conclusions may be noted. (1) Cyclic, conjugated, planar compounds which possess $(4n + 2)$ π-electrons tend to be strongly stabilized and
are described as aromatic. (2) Such compounds do not have the properties
typical of their aliphatic analogues; in particular, addition reactions are rare
except when the stabilization energy is relatively small or the product of addition
is itself strongly resonance-stabilized. (3) Because of their stability, aromatic
compounds are often readily formed both from aliphatic compounds by ring-

closure and from alicyclic compounds by dehydrogenation or elimination of other groups. Illustrations of conclusions (2) and (3) are numerous throughout this book.

(c) ACIDITY AND BASICITY

The customary definitions of acids and bases, due to Brönsted, are that an acid is a species having a tendency to lose a proton and a base is a species having a tendency to add a proton. An alternative definition of each is due to Lewis who defined an acid as a species capable of accepting an electron-pair and a base as a species capable of donating an electron-pair. Since a base can accept a proton only by virtue of being able to donate an electron-pair, all Lewis bases are also Brönsted bases and *vice versa*. However, Lewis's definition of acids embraces compounds which do not have a tendency to lose a proton: e.g. boron trifluoride is a Lewis acid, as illustrated by its reaction with ammonia:

$$BF_3 + :NH_3 \rightarrow F_3\bar{B}-\overset{+}{N}H_3$$

We shall use the term acid according to the Brönsted definition and shall refer to the compounds embraced only by Lewis's definition as Lewis acids.

Acids and bases are necessarily conjugate entities: that is, if the species HA can donate a proton to the species B, then since the reaction is reversible, the species BH^+ can donate a proton to A^-:

$$HA + B \rightleftharpoons A^- + BH^+$$

The species A^- is referred to as the *conjugate base* of the acid HA, and BH^+ is termed the *conjugate acid* of the base B. Clearly, if HA is a strong acid, A^- must be a weak base. It is, however, more convenient to discuss acidity and basicity separately.

It must also be remembered that the acidity or basicity of a species depends not only on the structure of that species but also on the nature of the solvent. Aniline is a weak base in water (which is a weak proton-donor) but a strong base in sulphuric acid; amide ion is a far stronger base in water than in liquid ammonia (for water is a stronger proton-donor than ammonia). Unless specified, the reference solvent is taken to be water.

(i) *Acids.* The most important elements in organic systems from which protons are donated are oxygen, sulphur, nitrogen, and carbon. The acidities of the groups —O—H, —S—H, $\geq$N—H, and $\geq$C—H vary widely with the structure of the remainder of the molecule, but one principle is of pre-eminent importance in determining acidity: any factor which stabilizes the anion of an acid relative to the acid itself increases the strength of the acid. This follows immediately from the thermodynamic principles governing an equilibrium (1.8) and is implicit throughout this discussion.

1. *Acidity of O—H groups.* Ethanol (pK 16) is a very weak acid compared with acetic acid (pK 4·8). Two factors are responsible. First, the carbonyl group is electron-attracting compared with the methylene group, so that the negative charge is better accommodated in CH_3—CO—O^- than in CH_3—CH_2—O^-. Secondly, both acetic acid and its anion are resonance-stabilized, being hybrids of (94) and (95), and (96) and (97), respectively. However, the stabilization energy of the ion, in which the canonical structures are of equal energy, is greater than that of the acid, for which the dipolar structure (95) is of high energy compared with (94). Consequently, there is an increase in stabilization energy when acetic acid ionizes whereas there is none in the ionization of ethanol. These two factors are responsible for a difference of about 60 kJ mol^{-1} in the free energies of ionization. (This difference is reflected partly in the enthalpy terms and partly in the entropy terms; see p. 22).

$$
CH_3—C\underset{OH}{\overset{O}{<}} \leftrightarrow CH_3—C\underset{\overset{+}{OH}}{\overset{\bar{O}}{<}} \qquad CH_3—C\underset{O}{\overset{O}{<}} \leftrightarrow CH_3—C\underset{O}{\overset{\bar{O}}{<}}
$$

$$\quad (94) \qquad\qquad (95) \qquad\qquad (96) \qquad\qquad (97)$$

The discussion of acetic acid has revealed two of the principal factors which govern acidity: the inductive and mesomeric effects. Further illustration of the operation of these effects is given in the following examples.

(1) The strengths of the mono-halogen-substituted acetic acids increase with the increasing $-I$ effect of the halogen (I < Br ~ Cl < F):

X—CH_2—CO_2H X=	H	F	Cl	Br	I
pK	4·75	2·66	2·86	2·86	3·12

(2) Increase in the number of $-I$ substituents further promotes acidity:

	$ClCH_2$—CO_2H	Cl_2CH—CO_2H	Cl_3C—CO_2H
pK	2·86	1·30	0·65

(3) The acid-strengthening influence of a group of $-I$ type is reduced as the group is moved further from the acid centre:

	$ClCH_2$—CO_2H	$ClCH_2$—CH_2—CO_2H	$ClCH_2$—CH_2—CH_2—CO_2H
pK	2·86	4·08	4·52

(4) Alkyl groups are acid-weakening to an extent depending on their $+I$ effects: $(CH_3)_3C > CH_3$—$CH_2 > CH_3 >$ H:

	H—CO_2H	CH_3—CO_2H	CH_3—CH_2—CO_2H	$(CH_3)_3C$—CO_2H
pK	3·77	4·75	4·88	5·05

(5) Phenol (pK 10) is more acidic than ethanol because the negative charge in the phenoxide ion is delocalized over the aromatic ring, as symbolized by the contributions of the canonical structures (100–102):

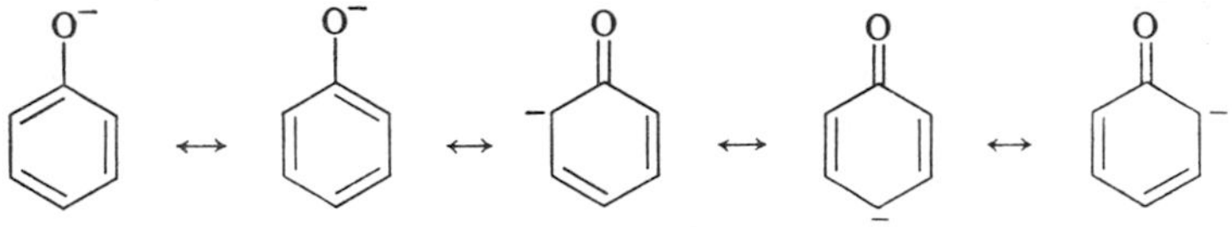

(98)   (99)   (100)   (101)   (102)

(6) The introduction of a substituent of electron-attracting type into the nucleus of phenol has an acid-strengthening effect and is greater when the sub-stituent is *ortho* or *para* to the phenolic group than when it is *meta*, for in the former case it interacts directly with a negatively polarized carbon atom (see structures (100) to (102)). For example, the greater acidity of *p*-nitrophenol (pK 7·1) than phenol may be attributed to the large contribution to the anion of the structure (103); the effect of the nitro group in *m*-nitrophenol (pK 8·35) is not directly transmitted to the oxyanion (see structure (104)). Groups of electron-releasing type reduce the acidity of phenol in the same way.

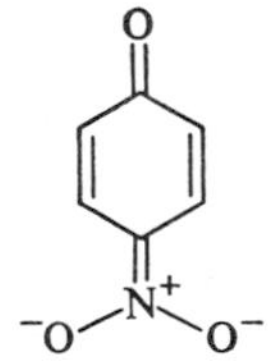

(103) 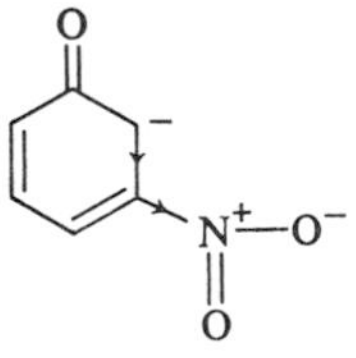   (104)

(7) In a conjugated acid, both the position of a substituent and the relative importance of its inductive and mesomeric effects are important in determining acidity. Consider, for example, the acid-strengths of the following benzoic acids:

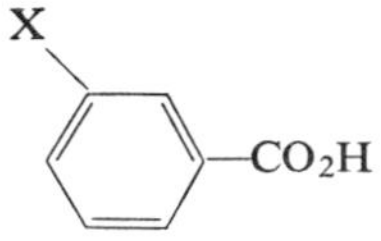

X	pK
H	4·2
Cl	3·8
OCH₃	4·1

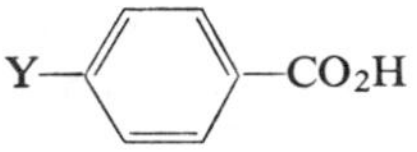

Y	pK
H	4·2
Cl	4·0
OCH₃	4·5

Both *m*-Cl and *m*-OCH₃ have an acid-strengthening influence because of the electron-withdrawing inductive effects of the groups, that of chlorine being greater than that of methoxyl. In the *para*-substituted acids, the $-I$ effects are opposed by the acid-weakening $+M$ effects of the substituents, as symbolized by the contributions of the canonical structures (105) and (106). For chlorine, the $-I$ effect is more powerful than the $+M$ effect so that, whereas p-chloro-benzoic acid is weaker than its *meta*-isomer, it is stronger than benzoic acid. For methoxyl, on the other hand, the $-I$ effect is less powerful than the $+M$ effect so that p-methoxybenzoic acid is a weaker acid than both its *meta*-isomer and benzoic acid. *

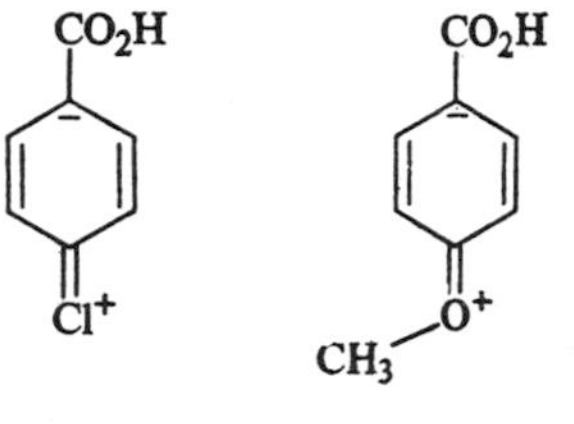

(105)   (106)

Similar considerations serve to rationalize the acid strengths (and hence the Hammett σ-values; p. 87) of other substituted benzoic acids.

(8) Sulphonic acids are very much stronger than carboxylic acids because of the strong $-I$ effect of the sulphone group ($-SO_2-$) and the greater delocalization of charge in the sulphonate anion (107–110) than in the carboxylate anion. Benzenesulphonic acid and its derivatives are useful as strong acid catalysts in organic synthesis as alternatives to sulphuric acid for, while being strong acids, they do not bring about the side reactions (e.g. oxidation and sulphonation) characteristic of sulphuric acid.

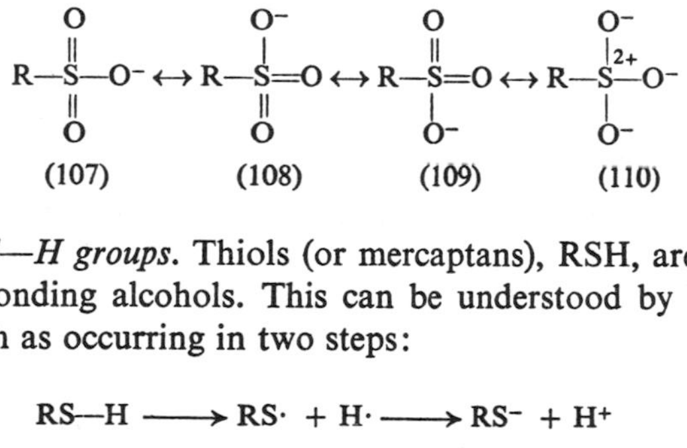

(107)   (108)   (109)   (110)

2. *Acidity of S—H groups.* Thiols (or mercaptans), RSH, are stronger acids than the corresponding alcohols. This can be understood by considering the ionization of each as occurring in two steps:

$$RS-H \longrightarrow RS\cdot + H\cdot \longrightarrow RS^- + H^+$$
$$RO-H \longrightarrow RO\cdot + H\cdot \longrightarrow RO^- + H^+$$

Although the second step is energetically more favourable for RO· than RS· because oxygen is more electronegative than sulphur, the first step is more favourable for the thiol for the S—H bond (340 kJ mol^{-1}) is considerably

*The differences in *pK*-values quoted above are numerically small, but it should be remembered that the scale is logarithmic: a difference in *pK*s of 0·3 corresponds to a factor of 2 in the dissociation constants.

weaker than the O—H bond (462 kJ mol⁻¹) and it is the latter factor which
dominates in the values for the free energies of ionization.

3. *Acidity of N—H groups.* The N—H bond is intrinsically less acidic than
the O—H bond because nitrogen is less electronegative than oxygen. Thus
carboxamides are very feebly acidic compared with carboxylic acids; e.g. they
are not dissolved by caustic soda solution. However, just as sulphonic acids are
much stronger than carboxylic acids, so sulphonamides, RSO_2NH_2 and
RSO_2NHR', are much more strongly acidic than carboxamides, giving delocalized anions represented by structures (111–113). Thus, although they are not
acidic enough to liberate carbon dioxide from sodium bicarbonate, they are
dissolved by caustic soda solution. Use may be made of this in separating primary and secondary amines: the mixture is treated with a sulphonyl chloride to
give a mixture of the sulphonamides, $R'SO_2NHR$ and $R'SO_2NR_2$, of which
only the former has an acidic proton and can be extracted into caustic soda
solution.

$$
\begin{array}{ccc}
\underset{\overset{\displaystyle\|}{O}}{\overset{\displaystyle O}{R-S-\bar{N}R'}} & \longleftrightarrow & \underset{\overset{\displaystyle\|}{O}}{\overset{\displaystyle O^-}{R-S=NR'}} & \longleftrightarrow & \underset{\overset{\displaystyle|}{O^-}}{\overset{\displaystyle O}{R-S=NR'}} \\
(111) & & (112) & & (113)
\end{array}
$$

Two carbonyl groups bonded to NH yield an effect comparable with that of
one sulphone group: imides, such as succinimide (114), do not react with sodium
bicarbonate but are soluble in caustic soda solution. The delocalized structure of

$$O=C-\bar{N}-C=O \longleftrightarrow \bar{O}-C=N-C=O \longleftrightarrow O=C-N=C-\bar{O}$$

(114)

(115)   (116)   (117)

the anion from an imide is represented as (115–117). Use is made of the acidity
of imides in a method for forming C—N bonds (p. 326).

4. *Acidity of C—H groups.* The sp^3C—H bond is less acidic than the N—H
bond because carbon is less electronegative than nitrogen. As would be expected
from the earlier discussion, the acidity is increased when the CH group is
attached to an increasing number of carbonyl groups, for the negative charge of
the conjugate base is then increasingly effectively delocalized, as shown for the
ion from triacetylmethane:

$$CH_3\overset{O}{\overset{\|}{C}}-\overset{-}{\underset{\underset{\underset{CH_3}{|}}{\underset{C=O}{|}}}{C}}-\overset{O}{\overset{\|}{C}}CH_3 \leftrightarrow CH_3\overset{O^-}{\overset{|}{C}}=\underset{\underset{\underset{CH_3}{|}}{\underset{C=O}{|}}}{C}-\overset{O}{\overset{\|}{C}}CH_3 \leftrightarrow CH_3\overset{O}{\overset{\|}{C}}-\underset{\underset{\underset{CH_3}{|}}{\underset{C=O}{|}}}{C}=\overset{O^-}{\overset{|}{C}}CH_3 \leftrightarrow CH_3\overset{O}{\overset{\|}{C}}-\overset{O}{\overset{\|}{C}}-\underset{\underset{\underset{CH_3}{|}}{\underset{C-O^-}{|}}}{C}CH_3$$

(118)   (119)   (120)   (121)

Thus, the pK-values of CH_3COCH_2—H, $(CH_3CO)_2CH$—H, and $(CH_3CO)_3C$—H are respectively 20, 9, and 6.

Carbonyl is one of a number of groups of $-M$ type which have a marked acid-promoting influence; the more important, in decreasing order of effectiveness, are: $-NO_2$, —CHO, —COR, —CN, and $-CO_2R$. For example, the acidity of nitromethane (pK 10·2) derives from the delocalization of the charge on the anion onto the oxygen atoms of the nitro group:

$$\overset{-}{C}H_2-\overset{+}{N}\overset{\nearrow O}{\underset{\searrow \overset{-}{O}}{}} \leftrightarrow CH_2=\overset{+}{N}\overset{\nearrow \overset{-}{O}}{\underset{\searrow \overset{-}{O}}{}}$$

(122)   (123)

The acidity of C—H bonds is of great significance in organic chemistry for the anions formed by ionization of the bond are intermediates in many synthetic processes (Chapter 7). The following structural situations are of particular importance.

(1) The C—H bond is adjacent to *one* group of $-M$ type; pK-values lie in the range 10–20. Reactions involving anions derived from compounds containing these structures are the aldol condensation ($>$CH—CHO and $>$CH—COR) (p. 227), the Claisen condensation ($>$CH—CO$_2$R) (p. 238), and the Thorpe reaction ($>$CH—CN) (p. 243).

(2) The C—H bond is adjacent to *two* groups of $-M$ type. Typical examples are ethyl acetoacetate, $CH_3COCH_2CO_2Et$, and diethyl malonate, $CH_2(CO_2Et)_2$, whose uses are described in 7.4. The pK-values of such compounds are in the range 4–12.

(3) The C—H bond is part of cyclopentadiene or a derivative. The compounds are acidic because the derived anion, having six π-electrons and fulfilling the criteria for aromaticity, is strongly resonance-stabilized (p. 59).

(4) The C—H bond in the haloforms. This is acidic because the conjugate base is stabilized both by the inductive effects of three halogen atoms and by charge-delocalization (see, e.g. structures (124–127)), for the halogens (other than fluorine) have unfilled and relatively low-lying d orbitals. These anions, by loss of

a halide ion, form carbenes, which are intermediates in many reactions (e.g. p. 125).

(124)   (125)   (126)   (127)

(5) The C—H bond in acetylene itself and mono-substituted acetylenes, $RC{\equiv}CH$. This is very much more acidic than that in a paraffin or olefin; e.g. the acetylene is ionized when treated with amide ion in liquid ammonia. The synthetic applications of this property in the formation of C—C bonds are described later (7.6). The reason is that an *s* electron is on average nearer the nucleus than a *p* electron (p. 32), so that it is more strongly held by that nucleus; thus the electrons in a bond formed by *sp*-hybridized carbon are more strongly held by the carbon nucleus than those in bonds formed by carbon in the sp^2 or sp^3 states because of the greater proportion of *s* character of the electrons in the first case. *sp*-Hybridized carbon is therefore more electronegative than carbon in other hybridized states and the charge on the acetylene anion is more readily accommodated.

(*ii*) *Bases*. The common bases include both anions, e.g. NH_2^- and EtO^-, and neutral molecules containing at least one unshared pair of electrons, e.g. NH_3 and EtOH. For the former group, the basicity is weakened by any factor which stabilizes the negative charge of the anion, while for the latter, the basicity is increased by any factor which stabilizes the positive charge on the conjugate acid of the base. Anionic bases are far stronger than their neutral analogues (e.g. $NH_2^- \gg NH_3$).

Amongst anionic bases, the charge is normally associated with oxygen, nitrogen or carbon. Since the electronegativities of these elements fall in the order O > N > C, the order of basicities is $R_3C^- > R_2N^- > RO^-$. For example, amide ion is a far stronger base than hydroxide ion, and methide ion (CH_3^-) is of such high energy that it does not exist in organic media. However, Ph_3C^-, in which the negative charge is delocalized by the aromatic rings, is more stable and is used (as sodium triphenylmethyl) in certain reactions which require a particularly powerful base (see, e.g. p. 239). The order of basicities of oxy-anions, $(CH_3)_3CO^- > CH_3O^- > PhO^- > CH_3CO_2^-$, follows immediately from the principles discussed above which govern acidity: t-butoxide ion is a stronger base than methoxide ion because the three electron-releasing methyl groups in the former destabilize the negative charge; phenoxide ion is a weaker base than an alkoxide ion because the charge is delocalized over the aromatic ring; and acetate ion is still weaker because the charge delocalization by oxygen is even more effective.

Nitrogen is the most important basic element in uncharged bases. Typical alkylamines have pK-values in the region 9–11 (where pK refers to the conjugate acid formed by the base; e.g. $CH_3—NH_3^+$ has pK 10·6). The conjugation of the amino group to a carbon-carbon double bond or an aromatic ring reduces the basicity considerably because the resonance-stabilization in the base (represented, in the case of aniline (pK 4·6), by the contribution of the ionic structures such as (128)) is lost on protonation. The protonation of pyrrole and related compounds, in which the unshared electron-pair on nitrogen completes the aromatic sextet, results in the complete loss of the aromatic stabilization energy, and these compounds are essentially non-basic.

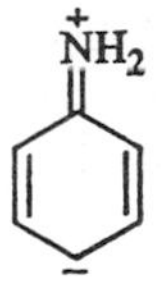

(128) 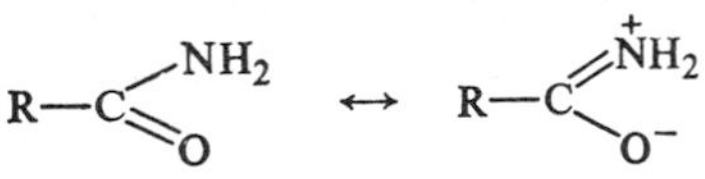   (129)   (130)

Amides are much less basic than amines because the resonance energy of the delocalized system (129–130) is lost on protonation of the nitrogen atom.

Nitriles are only very weakly basic. The explanation corresponds to that given for the greater acidity of acetylenes than paraffins (p. 68): namely, since the sp-hybridized nitrogen in a nitrile is a more electronegative species than the nitrogen in an amine, $RC{\equiv}NH^+$ is more acidic than RNH_3^+. Imines and heterocyclic aromatic compounds such as pyridine (pK 5·2) occupy an intermediate position. Amidines, however, are much stronger bases than both imines and amines. This is because, although both the base (131) and its conjugate acid (132) are resonance-stabilized, the stabilization energy of the latter, whose principal canonical structures are equivalent, is greater than that of the former, in which one of the corresponding structures is dipolar and of high energy. The stability of the conjugate acid formed by guanidine (133) serves to make this as strong a base as the alkali-metal hydroxides.

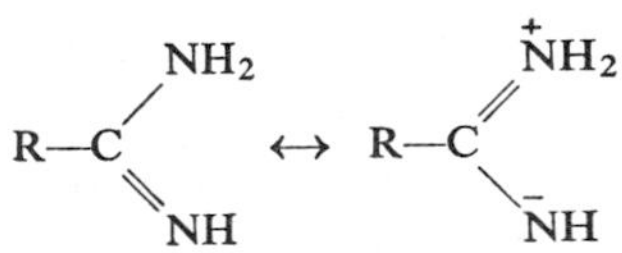

(131) 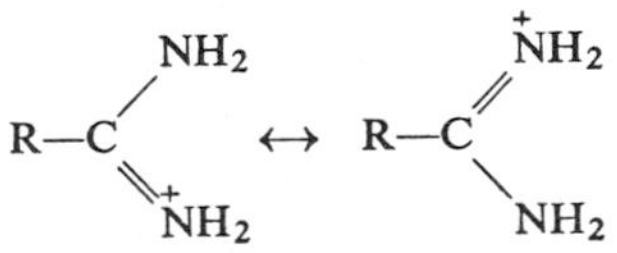   (132)

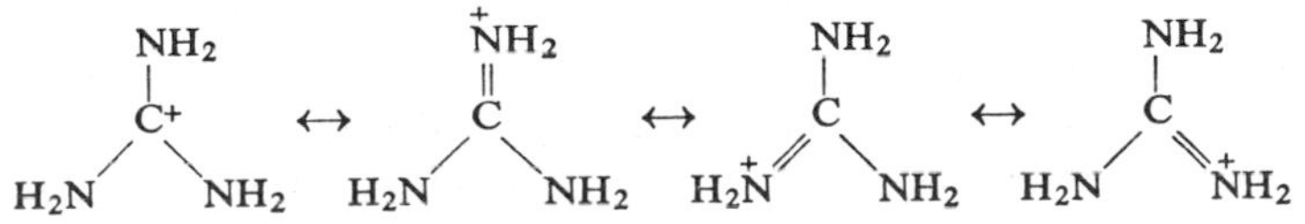

(133)

The oxygen atom in neutral molecules is only weakly basic. Nevertheless, many reactions of alcohols, aldehydes, ketones, and esters occur *via* the conjugate acids formed by these compounds, and the solubility in sulphuric acid of oxygen-containing compounds such as esters, ethers, and nitro compounds is due to the formation of oxonium ion salts in this medium (e.g. $R_2OH^+\ HSO_4^-$). Both olefinic and aromatic C=C bonds are also weakly basic: e.g. benzenoid compounds form salts such as (134) with hydrogen chloride in the presence of aluminium trichloride. The basicity of polycyclic aromatic hydrocarbons increases with increase in the number of carbon atoms which can delocalize the positive charge of the conjugate acid.

(134)

Further Reading

ALLINGER, N. L., and ALLINGER, J., *Structures of Organic Molecules*, Prentice-Hall (New Jersey 1965).

COMPANION, A. L., *Chemical Bonding*, McGraw-Hill (New York 1964).

COULSON, C. A., *Valence*, 2nd Ed., Oxford University Press (London 1961).

MURRELL, J. N., KETTLE, S. F. A., and TEDDER, J. M., *Valence Theory*, 2nd Ed., Wiley (New York and London 1970).

Problems

1. Draw orbital representations of the following compounds: ethylene; allene (CH_2=C=CH_2); butadiene; nitromethane; acrylonitrile (CH_2=CHCN); hydrazine (NH_2NH_2).

2. The stabilization energies of conjugated compounds are usually obtained by comparing either the heats of hydrogenation or the heats of combustion of these compounds with those of appropriate non-conjugated compounds. Why does the former method give the more reliable values?

   Would you expect biphenyl (Ph—Ph) to have a greater stabilization energy than twice that of benzene? If so, by approximately how much?

   Given that the heat of hydrogenation of styrene (PhCH=CH_2) is -326 kJ mol^{-1}, calculate the stabilization energy of styrene (see Table 1.1 for bond-energy values).

3 Arrange each of the following groups in decreasing order of acid-strength:

   (*i*) HCO_2H, CH_3CO_2H, $ClCH_2CO_2H$, FCH_2CO_2H.

   (*ii*) Phenol, *m*- and *p*-chlorophenol, *m*- and *p*-cresol.

  (*iii*) Benzoic acid, *m*- and *p*-nitrobenzoic acid, *m*- and *p*-methoxybenzoic acid.

  (*iv*) Penta-1,4-diene and cyclopentadiene.

4. Arrange each of the following groups in decreasing order of base-strength:

  (*i*) Ammonia, aniline, *m*- and *p*-nitroaniline.

  (*ii*) Ethoxide ion, t-butoxide ion, acetate ion, and phenoxide ion.

  (*iii*) Pyrrole and pyrrolidine (tetrahydropyrrole).

  (*iv*) Pyridine and piperidine (hexahydropyridine).

5. How can you account for the following:

  (*i*) The dipole moment of ethylene dichloride (CH_2Cl—CH_2Cl) increases as the temperature is raised.

  (*ii*) The dipole moment of *p*-nitroaniline (6·2 D) is larger than the sum of the values for nitrobenzene (3·98 D) and aniline (1·53 D).

  (*iii*) The dipole moment of acrolein ($CH_2{=}CHCHO$; 3·04 D) is greater than that of propionaldehyde (CH_3CH_2CHO; 2·73 D).

  (*iv*) Picric acid (2,4,6-trinitrophenol) liberates carbon dioxide from aqueous sodium carbonate, but phenol does not.

  (*v*) *NN*-Dimethylation triples the basicity of aniline but increases the basicity of 2,4,6-trinitroaniline by 40,000-fold.

  (*vi*) The bond dissociation energy of the $PhCH_2$—H bond (322 kJ mol^{-1}) is considerably smaller than that of the CH_3—H bond (426 kJ mol^{-1}).

  (*vii*) The boiling point of ethanol is very much higher than that of its isomer, dimethyl ether.

  (*viii*) Boron trifluoride and aluminium trichloride are Lewis acids.

3. Chemical Kinetics

3.1 Rates of Reaction

It is possible to deduce from thermodynamic data whether or not a particular reaction can *in principle* yield a particular set of products in significant amounts. Thus, if the reaction A → B is accompanied by a decrease in free energy, it is possible in principle to convert A largely into B. However, thermodynamic data do not give information about the *rates* of reactions, and it is quite possible that the percentage conversion of A into B will be negligible even after many years. For example, the reaction between hydrogen and oxygen to give water is accompanied by a large negative free energy change, but nevertheless the rate at which water is formed is insignificant at room temperature, although it is very rapid indeed at high temperatures. Again, the reaction between acetaldehyde and hydrogen cyanide to give the cyanohydrin,

$$CH_3{-}CH{=}O + HCN \longrightarrow CH_3{-}CH\begin{smallmatrix} \diagup OH \\ \diagdown CN \end{smallmatrix}$$

is accompanied by a favourable free-energy change but is very slow unless a small quantity of a base is added.

In effect, the thermodynamic criterion for a reaction to proceed is a necessary but not a sufficient one; metastable equilibria are common, as in the case of the hydrogen-oxygen reaction. It is necessary that other conditions should be met which result in equilibrium being attained at a practicable rate. It is the purpose of this Chapter to set out the principles which govern the rates of reactions.

3.2 The Orders of Reactions

In many reactions, measurement of the rate of reaction at a given instant shows that the rate is proportional to the concentration of one or more of the reactants present at that time. For example, for certain reactions of the type A → B, the rate of loss of A can be expressed as

$$-d[A]/dt = k[A]$$

where [A] is the concentration of A at time t and k is a constant defined as the *velocity constant* or *rate constant* of the reaction. If the initial concentration of

A is $[A_0]$ and its concentration at time t is $[A]$,

$$-[\ln[A]]_{[A_0]}^{[A]} = [kt]_0^t$$

i.e.
$$\ln[A_0] - \ln[A] = kt$$

For such a reaction, a plot of log $[A]$ against time is linear and of gradient $-k$, and the reaction is described as being of *first order* with respect to A.

The rate can also be expressed in terms of the increase in the concentration of B: $d[B]/dt = k[A]$, where $[B] = [A_0] - [A]$.

For some reactions of the simple type $A \rightarrow B$, more complex kinetics are observed. One obvious complication is introduced when the reverse reaction, $B \rightarrow A$, occurs at a significant rate. If k_1 and k_{-1} are the velocity constants for the forward and reverse reactions,

$$-d[A]/dt = k_1[A] - k_{-1}([A_0] - [A])$$

The maximum value of $-d[A]/dt$ occurs when $[A] = [A_0]$, i.e. at the beginning of the reaction, and here the rate is first-order with respect to A. As reaction proceeds and $[A]$ decreases, the term in $([A_0] - [A])$ becomes increasingly important until eventually $k_1[A] = k_{-1}([A_0] - [A])$ and $-d[A]/dt = 0$. Reaction stops, in that the concentrations of A and B do not thereafter change, although the two are in dynamic equilibrium. In these reactions, then, as reaction proceeds the loss of A becomes slower than the simple first-order equation predicts, the extent of deviation from the first-order plot depending on the magnitude of k_{-1} compared with k_1.

The next most simple kinetic description applies in reactions of the type $A + B \rightarrow C$, where the rates of loss of both A and B are proportional to their concentrations and the rate of formation of C is proportional to the products of the concentrations of A and B:

$$d[C]/dt = k[A][B]$$

If, initially, the concentrations of A and B are equal $([A_0])$ and the concentration of C is zero, and after time t the concentrations of A and B are each $[A_0] - [A]$ and the concentration of C is therefore $[A]$,

$$d[A]/dt = k([A_0] - [A])^2$$

i.e.
$$[A]/[A_0]([A_0] - [A]) = kt$$

Such reactions are described as *second order*, being first-order in each of A and B. It should be noted that if either A or B is in considerable excess of the other so that its concentration remains sensibly constant during the reaction, the reaction follows first-order kinetics.

Reactions which are of non-integral order with respect to one or more of the reactants are well known in organic chemistry. They are normally associated with processes which occur by complex mechanisms, particularly chain reactions (p. 102). An example is the decomposition of acetaldehyde at high temperatures:

$$CH_3CHO \xrightarrow{k_1} \cdot CH_3 + \cdot CHO$$

$$CH_3CHO + \cdot CH_3 \xrightarrow{k_2} CH_3CO + CH_4$$

$$CH_3CO \xrightarrow{k_3} \cdot CH_3 + CO$$

$$2 \cdot CH_3 \xrightarrow{k_4} C_2H_6$$

Reaction is initiated by the fragmentation of a molecule of the aldehyde. The methyl radical which is formed abstracts hydrogen from a second molecule of the aldehyde, giving methane and an acetyl radical; the latter rapidly gives carbon monoxide and a new methyl radical which reacts with a third molecule of aldehyde. The chain continues to be propagated in this way until two methyl radicals meet and dimerize. Since the concentration of methyl radicals is low, the probability of occurrence of the dimerization step compared with the chain-propagating step is small, so that a relatively small number of radicals from the initiation step can yield a large quantity of methane. It can be shown that the kinetics should then be given by

$$d[CH_4]/dt = k_2 \left[\frac{k_1}{k_4}\right]^{1/2} [CH_3CHO]^{3/2}$$

and this is indeed observed.

Reactions of order higher than second are also well known and again usually correspond to processes which involve two or more steps. They will be referred to as appropriate.

One further complicating feature arises in kinetic observations: namely, the effect of the competition of two species for a reactive intermediate.

For example, in the solvolysis of t-butyl chloride in excess of water,

$$(CH_3)_3C-Cl + H_2O \longrightarrow (CH_3)_3C-OH + HCl$$

first-order kinetics are obeyed at the beginning of the reaction, i.e.

$$d[(CH_3)_3COH]/dt = k[(CH_3)_3CCl]$$

but at later stages the kinetics are more accurately described by

$$d[(CH_3)_3COH]/dt = k[(CH_3)_3CCl]/(k' + k''[Cl^-])$$

The slow step in the reaction is the cleavage of t-butyl chloride into two ions,

$$(CH_3)_3C—Cl \xrightarrow{k_1} (CH_3)_3C^+ + Cl^-$$

and this is followed by a rapid reaction of the t-butyl ion with water,

$$(CH_3)_3C^+ + H_2O \xrightarrow{k_2} (CH_3)_3C—OH + H^+$$

In addition, the t-butyl ion can react with chloride ion by the reverse of the reaction producing it,

$$(CH_3)_3C^+ + Cl^- \xrightarrow{k_{-1}} (CH_3)_3C—Cl$$

Now, since the reactions which destroy the t-butyl ion have much higher rate constants than that which forms it, this ion can never accumulate in a high concentration. It will build up until its rate of destruction is equal to its rate of formation, i.e.

$$k_2[(CH_3)_3C^+][H_2O] + k_{-1}[(CH_3)_3C^+][Cl^-] = k_1[(CH_3)_3CCl]$$

at which point its concentration is given by

$$[(CH_3)_3C^+] = k_1[(CH_3)_3CCl]/(k_2[H_2O] + k_{-1}[Cl^-])$$

which is described as its *steady-state* concentration. Since the rate of formation of the product is given by

$$d[(CH_3)_3COH)]/dt = k_2[(CH_3)_3C^+][H_2O]$$

substitution gives

$$d[(CH_3)_3COH]/dt = k_1k_2[H_2O][(CH_3)_3CCl]/(k_2[H_2O] + k_{-1}[Cl^-])$$

This is equated with the observed kinetic form if $k_1k_2[H_2O] = k$, $k_2[H_2O] = k'$, and $k_{-1} = k''$, and all these are permissible equalities since the concentration of water, being considerably in excess of that of t-butyl chloride, remains effectively constant during the reaction.

In this example we have assumed the mechanism of the reaction in order to show how it leads to the observed kinetics. In practice, it is usual to adopt the opposite approach; the mechanism is deduced from the observed kinetics and other experimental observations. This can be illustrated by the acid-catalyzed bromination of acetone,

$$CH_3—CO—CH_3 + Br_2 \xrightarrow{H^+} CH_3—CO—CH_2Br + HBr$$

The rate of consumption of bromine is independent of the concentration of bromine until this is very low, whereas it is dependent on the concentration both of acetone and of the protons in solution, i.e.

$$- d[Br_2]/dt = k[CH_3COCH_3][H^+]$$

It follows that the rate of the reaction is determined by a step in which bromine plays no part. This step might be the acid-catalyzed enolization of acetone,

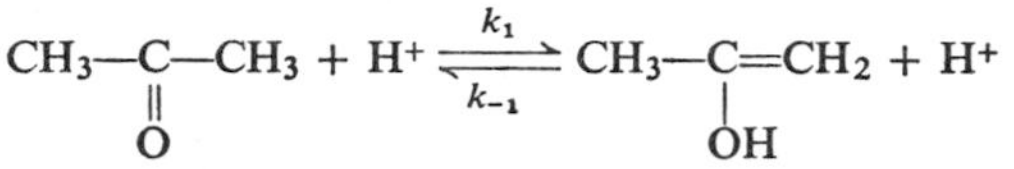

$$CH_3\!-\!\underset{\underset{O}{\|}}{C}\!-\!CH_3 + H^+ \underset{k_{-1}}{\overset{k_1}{\rightleftharpoons}} CH_3\!-\!\underset{\underset{OH}{|}}{C}\!=\!CH_2 + H^+$$

the enol then reacting with bromine as follows:

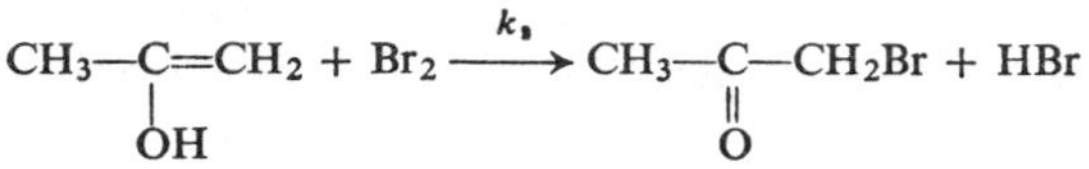

$$CH_3\!-\!\underset{\underset{OH}{|}}{C}\!=\!CH_2 + Br_2 \xrightarrow{k_2} CH_3\!-\!\underset{\underset{O}{\|}}{C}\!-\!CH_2Br + HBr$$

Now, if the rate of formation of the enol is slow compared with the rates of both reactions which destroy it, the enol can be present in only very small concentrations, and the steady-state treatment gives

$$d[enol]/dt = k_1[CH_3COCH_3] [H^+] - k_{-1}[enol] [H^+] - k_2[enol][Br_2] = 0$$

i.e. $$[enol] = k_1[CH_3COCH_3] [H^+]/(k_{-1}[H^+] + k_2[Br_2])$$

Hence, $-d[Br_2]/dt = k_2[enol] [Br_2]$

$$= k_1k_2[CH_3COCH_3] [H^+] [Br_2]/(k_{-1}[H^+] + k_2[Br_2])$$

$$= k_1[CH_3COCH_3] [H^+], \text{ providing}$$

that $k_2 \gg k_{-1}$ and $[Br_2]$ does not fall to too low a value. Thus, the observed kinetics have enabled a mechanism to be deduced, and independent evidence is in agreement with this mechanism.

Two other examples are illustrative. Acetaldehyde condenses with itself to give acetaldol, and the reaction is catalyzed by bases such as hydroxide ion,

$$2\ CH_3\!-\!CHO \xrightarrow{OH^-} CH_3\!-\!CH(OH)\!-\!CH_2\!-\!CHO$$

Although the complete kinetics are complex, they are given approximately by

$$- d[CH_3CHO]/dt = k[CH_3CHO][OH^-]$$

except at very low aldehyde concentrations; i.e. the reaction is first-order in acetaldehyde although two molecules are involved in the reaction. This leads to the conclusion, which is supported by other evidence, that the slow step is the formation of the carbanion, $(CH_2CHO)^-$, which then reacts rapidly with a second molecule of acetaldehyde:

$$CH_3—CHO + OH^- \underset{k_{-1}}{\overset{k_1}{\rightleftharpoons}} \bar{C}H_2—CHO + H_2O$$

$$\bar{C}H_2—CHO + CH_3—CHO \overset{k_2}{\longrightarrow} CH_3—CH(O^-)—CH_2—CHO$$

$$CH_3—CH(O^-)—CH_2—CHO + H_2O \overset{k_3}{\longrightarrow} CH_3—CH(OH)—CH_2—CHO + OH^-$$
Acetaldol

The observed kinetics can be shown to result, analogously to those for the acid-catalyzed bromination of acetone, provided that $k_2 \gg k_{-1}$.

For the analogous condensation of acetone to give diacetone alcohol, however, the kinetics are second-order in the ketone. These kinetics can be shown to follow providing that the reverse of the formation of the carbanion from acetone is fast compared with the reaction of the carbanion with a second molecule of acetone, i.e. $k_2 \ll k_{-1}$:

$$CH_3—CO—CH_3 + OH^- \underset{k_{-1}}{\overset{k_1}{\rightleftharpoons}} CH_3—CO—\bar{C}H_2 + H_2O$$

$$CH_3—CO—\bar{C}H_2 + CH_3 \quad CO \quad CH_3 \overset{k_2}{\longrightarrow} (CH_3)_2C(O^-)—CH_2—CO—CH_3$$

$$(CH_3)_2C(O^-)—CH_2—CO—CH_3 + H_2O \overset{k_3}{\longrightarrow} (CH_3)_2C(OH)—CH_2—CO—CH_3 + OH^-$$
Diacetone alcohol

The kinetic examination of a reaction frequently provides useful evidence about the mechanism of the process, but supplementary evidence is also required to *prove* that the suggested mechanism is correct. Kinetic data give information about the slowest step of the reaction (referred to as the *rate-determining step*), but not necessarily about the structures of the intermediates involved. Thus, in the above examples, it is possible to conclude only that the kinetics *are consistent with* the view that the enolic tautomer of acetone and the carbanion from acetaldehyde are involved in the bromination of acetone and the aldol condensation of acetaldehyde, respectively.

3.3 Molecularity

The molecularity of a reaction describes the number of species that take part in the formation of the transition state (p. 81). For example, in the hydrolysis of ethyl bromide,

$$C_2H_5Br + OH^- \rightarrow C_2H_5OH + Br^-$$

both reactants come together to form the transition state (4.5), so that the reaction is said to be bimolecular. Since the reaction is second order, this is a case where the order corresponds to the molecularity, but in reactions in which two or more steps are involved, order and molecularity usually differ. For example, in the solvolysis of t-butyl chloride, the slow step is the cleavage of the C—Cl bond to give two ions; this is a unimolecular step, but the solvolysis is only first order at the beginning of the reaction (p. 74).

3.4 The Effect of Temperature on Reaction Rates

For many reactions a plot of the logarithm of the rate constant against the reciprocal of the absolute temperature is approximately linear with negative gradient (i.e. the reaction is faster at higher temperatures): $\ln k = B - C/T$, where B and C are constants. Differentiation gives d $(\ln k)/dT = C/T^2$ which is of the same form as the van't Hoff isochore (1.5), d $(\ln K)/dT = \Delta H/RT^2$, which describes the effect of temperature on the equilibrium constant of a reaction. Moreover, the equilibrium constant of a reaction is equal to the ratio of the rate constants for the forward and reverse steps of the process, and Arrhenius (1889) therefore suggested that the constant C above should be replaced by $\Delta E/R$, where ΔE is an energy term in the expression for the rate of the reaction analogous to the enthalpy term ΔH in the isochore. The expression d $(\ln k)/dT = \Delta E/RT^2$ gives, on integration, the *Arrhenius* equation,

$$k = Ae^{-\Delta E/RT}$$

where A, like ΔE, is a constant for the reaction concerned. ΔE represents a critical energy which the molecules must possess in order for reaction to occur; for a given value of A, the rate constant decreases as ΔE becomes larger.

The significance of A and ΔE are discussed in the two succeeding sections. For the moment it should be emphasized that the fact that a reaction may have a favourable free energy change does not imply that reaction will occur at other than an infinitesimal rate; it is necessary also that the values of A and $\Delta E/RT$ should be such that k is in the range of what may be termed practicable rate constants.

3.5 Collision Theory

The simplest interpretation of the Arrhenius equation for a bimolecular reaction is that, for reaction to occur, the two reactants must collide and the total energy possessed by them must be at least equal to ΔE. Thus, the rate constant should be given by the frequency of collisions times the fraction of the collisions which involve suitably activated molecules, $e^{-\Delta E/RT}$. It ought then to be possible to equate A with Z, the collision number (i.e. the number of collisions per second when there is only one molecule of reactant per unit volume*). We shall first examine the validity of this proposal.

Values of Z lie in the region of 10^{11} dm^3 mol^{-1} s^{-1}, and for many simple gas-phase reactions and for some reactions in solution the experimental value of A is of this order. For these cases, equating A with Z is justified. There are, however, many other reactions for which A is considerably smaller than Z; examples are the dimerization of cyclopentadiene, both in the gas-phase and in solution, and the reaction between butadiene and acrolein, for which $A \sim 10^6$ dm^3 mol^{-1} s^{-1}:

$$
\begin{array}{c}
\text{CH} \overset{\text{CH}_2}{\underset{\text{CH}}{\big|}} \\
\text{CH}_2
\end{array}
\quad + \quad
\begin{array}{c}
\text{CH} \overset{\text{CHO}}{\underset{\text{CH}_2}{\parallel}}
\end{array}
\quad \longrightarrow \quad
\begin{array}{c}
\text{CH}\overset{\text{CH}_2}{\text{CH}}\overset{\text{CHO}}{}\\
\text{CH}\underset{\text{CH}_2}{}\text{CH}_2
\end{array}
$$

In general, collision theory fails to account for values of A which are substantially less than Z, and this is because it does not allow for the fact that two molecules possessing the necessary activation may, on collision, be unsuitably oriented with respect to each other for reaction to occur. A factor P, representing the *probability* that a collision between suitably activated molecules will result in reaction, is therefore introduced:

$$k = PZ\, e^{-\Delta E/RT}$$

It is not immediately obvious how collision theory can be applied to first-order reactions, such as the solvolysis of t-butyl chloride whose kinetics were outlined above. The problem was solved by Lindemann, who suggested that the necessary activation energy is acquired by the reactant molecule as the result of its colliding with other molecules. The activated molecule can then either undergo reaction or dissipate its extra energy in further collisions. Then, for a reaction $A \rightarrow B$,

$$A + A \underset{k_{-1}}{\overset{k_1}{\rightleftharpoons}} A + A^*$$

$$A^* \overset{k_2}{\longrightarrow} B$$

where A* represents A in the suitably activated state. A small concentration of A* is thereby maintained, and the steady-state treatment gives

$$d[B]/dt = k_1 k_2 [A]^2 / (k_{-1}[A] + k_2)$$

*The expression for Z contains a temperature term: e.g. for collisions between molecules of the same type which have molecular weight M and diameter σ, $Z = 4\sigma^2(\pi RT/M)^{\frac{1}{2}}$. However, the effect of changes in T on the value of Z is negligible compared with the effect on the value of the exponential term, unless ΔE is very small.

Hence, if $k_{-1} \gg k_2$, $d[B]/dt = k_1k_2[A]/k_{-1}$ except for low values of [A]; i.e. first-order kinetics should be followed. At very low values of [A], $d[B]/dt = k_1[A]^2$; i.e. second-order kinetics should be followed. At intermediate values of [A] the kinetics should not be described by either the first- or the second-order equations. Where these predictions have been tested, they have been found to hold.

As with bimolecular reactions, it is sometimes difficult to interpret the value of A in unimolecular reactions. In many, A is of the order of $10^{11}\,\mathrm{dm^3\,mol^{-1}\,s^{-1}}$ and may be equated with the collision frequency, but for others it is less than Z by several powers of ten. The most satisfactory explanation is that when a complex molecule possesses the critical activation energy, the chances may be small that this energy is in that area of the molecule at which reaction is to occur. Again, therefore, a probability factor, P, is necessary.

The interpretation of E in collision theory is more satisfactory than that of A. Consider the reaction between a hydrogen molecule and a hydrogen atom, $H_2 + H \rightarrow H + H_2$. As the atom approaches the molecule, the existing H—H bond begins to break and the new H—H bond begins to form. This corresponds to a total energy for the system which is greater than that of the original atom and molecule in isolation, and the total energy continues to rise until a point is reached at which the hydrogen atom which is being transferred is symmetrically placed between the other two hydrogens, as represent by H·····H·····H. (Thus, the bonding energy of the two *partial* bonds in H·····H·····H is less than that of one *full* bond.) Thereafter, the energy of the system falls again until the new system of hydrogen atom and hydrogen molecule is obtained. The energy-

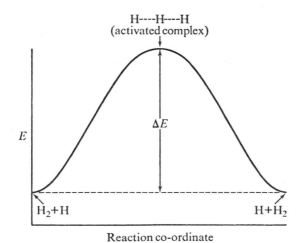

Fig. 3.1   Energy profile for the reaction,

$$H_2 + H\cdot \rightarrow H\cdot + H_2.$$

changes during the reaction are shown in the *energy profile*, Fig. 3.1, as a function of the progress made in the formation of the new H—H bond.

There is in effect an energy barrier between the two pairs of reactants, and ΔE is the energy necessary for the reactants to surmount the barrier, being described as the *activation energy* of the reaction. The structure corresponding to the highest point on the energy profile is termed the *activated complex*, and in this case corresponds to the linear arrangement of three hydrogen atoms which are joined by two partial bonds. For this particular reaction, the difference in energies of the activated complex and the reactants has been calculated theoretically and the value agrees closely with the experimental result (36 kJ mol^{-1}), providing justification for this interpretation of ΔE in the Arrhenius equation. It should also be noted that calculations show that the energy of the activated complex in this reaction is minimal when the complex is linear; that is, the hydrogen atom most easily reacts by approaching the hydrogen molecule along the molecular axis.

It can be appreciated from this discussion why the hydrogen atom-molecule reaction consists of a one-step bimolecular process rather than of the alternative two-step process in which the hydrogen molecule first decomposes to two hydrogen atoms one of which then combines with the third hydrogen atom. The latter path would have an activation energy of at least 435 kJ mol^{-1} (the dissociation energy of H_2), whereas in the bimolecular path the energy evolved as the new H—H bond is formed in effect helps to bring about the dissociation of the original H—H bond.

3.6 Transition-state Theory

An alternative approach to the collision theory is based on the application of thermodynamic principles to the activated complex. Consider a reaction A + B → C in which the activated complex is represented as AB* and whose energy profile is shown in Fig. 3.2. Transition-state theory treats AB* as a normal chemical species one of whose vibrations is replaced by a translational degree of freedom; a loose bond between A and B, instead of undergoing a stretching-and-contracting vibrational motion, flies apart either to give A and B or to give C. It can be shown that the frequency, v, with which this happens is given by $v = kT/h$ where k is the Boltzmann constant and h is Planck's constant. The activated complex, or transition state, is regarded as being in equilibrium with the reactants, i.e.

$$\frac{[AB^*]}{[A][B]} = K^{\ddagger}$$

where the symbol $\ddagger$ signifies that this is not a conventional stable equilibrium. The rate at which AB* is transformed into C is given by the product of the concentration of AB* and the frequency with which AB* breaks down to products, so that the rate of loss of A is given by

$$-d[A]/dt = d[C]/dt = [AB*]kT/h$$

Since $-d[A]/dt = k_2[A][B]$, where k_2 is the bimolecular rate constant,

$$k_2 = kT/h \frac{[AB*]}{[A][B]} = \frac{kT}{h} K^{\ddagger}$$

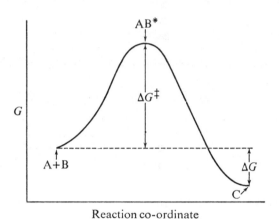

Fig. 3.2 Energy profile for A + B → C.

Hence the rate constant* is related to the equilibrium constant, $K^{\ddagger}$. Now, $-RT \ln K^{\ddagger} = \Delta G^{\ddagger} = \Delta H^{\ddagger} - T\Delta S^{\ddagger}$, where $\Delta G^{\ddagger}$, $\Delta H^{\ddagger}$, and $\Delta S^{\ddagger}$ are the differences in free energies, enthalpies, and entropies, respectively, of AB* and A + B. It follows that

$$k_2 = \frac{kT}{h} e^{-\Delta G^{\ddagger}/RT} = \frac{kT}{h} e^{-\Delta H^{\ddagger}/RT} e^{\Delta S^{\ddagger}/R}$$

The term $\Delta H^{\ddagger}$, the *enthalpy of activation*, corresponds closely to the activation energy term, ΔE, in the Arrhenius equation (for liquids and solids, $\Delta E = \Delta H^{\ddagger} + RT$). The term $(kT/h)e^{\Delta S^{\ddagger}/R}$ corresponds to the A factor; $\Delta S^{\ddagger}$ is known as the *entropy of activation*. One important merit of transition-state theory is that it rationalizes the probability factor, P. A reaction in which the transition state is

---

Strictly, k_2 should be multiplied by a factor (the transmission coefficient) which is the probability that AB will dissociate into products instead of back into the reactants. In most cases this factor is close to unity.

highly organized will have a large, negative $\Delta S^{\ddagger}$, corresponding to a small value for P. Examples of the dependence of $\Delta S^{\ddagger}$ on the reaction type are given in the following section.

3.7 Applications of Kinetic Principles

(a) STRUCTURE, RATES AND EQUILIBRIA

The addition of hydrogen chloride to propylene could in principle give two products:

$$CH_3-CH=CH_2 + HCl \longrightarrow CH_3-CH_2-\underset{\underset{Cl}{|}}{CH_2} \text{ and } CH_3-\underset{\underset{Cl}{|}}{CH}-CH_3$$

In practice the latter product predominates, and in general the orientation in the addition of a compound HX to an unsymmetrical olefin is described by *Markovnikov's rule* (1870) which states that 'the hydrogen atom of HX adds to that carbon which bears the greater number of hydrogens.' The basis of the rule is as follows.

The rate-determining step in the reaction is the addition of a proton to one of the carbon atoms of the double bond, giving a *carbonium ion* which then reacts rapidly with the anion X^- from HX. One of two such carbonium ions might be formed, e.g.

$$CH_3-CH=CH_2 + H^+ \begin{cases} CH_3-CH_2-\overset{+}{C}H_2 \xrightarrow{Cl^-} CH_3-CH_2-CH_2Cl \\ \\ CH_3-\overset{+}{C}H-CH_3 \xrightarrow{Cl^-} CH_3-CHCl-CH_3 \end{cases}$$

Of the two intermediate carbonium ions, the secondary ion, $CH_3-\overset{+}{C}H-CH_3$, is thermodynamically the more stable, for the concentration of positive charge is reduced by the two electron-releasing methyl groups attached to the positive carbon, whereas in the primary ion only one electron-releasing group (ethyl) is attached to the positive carbon.* Now, the rate of formation of these two ions is not immediately dependent on their stabilities, but rather on the relative free energies of the *transition states* which precede the ions. The detailed structures

---

*Relatively to a hydrogen atom substituent, an alkyl group stabilizes positively charged carbon by virtue of its inductive effect, just as it destabilizes an anion (cf. the acid strengths of formic and acetic acids; p. 63). There is evidence that, in addition, an alkyl group stabilizes an adjacent charged carbon by delocalization (hyperconjugation; p. 52); that is, structures of the type

$$\overset{H^+}{\underset{CH_2=CH_2}{}}$$

contribute to the stability of the ion $CH_3-\overset{+}{C}H_2$. Both the inductive and hyperconjugative influences increase with the number of alkyl groups attached to the cationic centre, so that the order of stability of simple alkyl cations is $(CH_3)_3C^+ > (CH_3)_2CH^+ > CH_3CH_2^+ > CH_3^+$.

of the transition states are not known; the new C—H bond is partly formed at the transition state, but to an unknown degree. However, it is known that the carbonium ions are of relatively high free energies compared with the reactants, as represented in Fig. 3.3, and there is therefore a smaller change in energy when the transition states are transformed into the intermediate carbonium ions

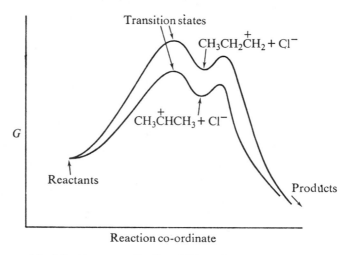

Fig. 3.3　Energy profiles for addition of HCl to propylene.

than when they revert to the reactants. Consequently, it is reasonable to conclude that there is a smaller reorganization of the molecular geometry in passage from transition state to intermediate than in passage from transition state to reactants; that is, the transition states somewhat resemble the carbonium ions to which they give rise. Thus, the two transition states can be represented as

$$CH_3—CH{\cdots\cdots}\overset{\delta+}{CH_2} \quad \text{and} \quad CH_3—\overset{\delta+}{CH}{\cdots\cdots}CH_2$$
$$| \qquad\qquad\qquad\qquad |$$
$$H \qquad\qquad\qquad\qquad H$$
$$| \qquad\qquad\qquad\qquad |$$
$$Cl^{\delta-} \qquad\qquad\qquad\quad Cl^{\delta-}$$

corresponding to there being a proportion of the unit positive charge which is ultimately possessed by the carbonium ions on the appropriate carbon atoms of the transition states. It follows that the factors which determine the relative stabilities of the carbonium ions are also effective in determining the relative stabilities of the transition states, so that the more stable of the two ions is formed the faster and corresponds to a lower maximum on the energy profile (Fig. 3.3).

This discussion is generalized by *Hammond's postulate* which states that if a reaction step, $A + B \rightarrow C$, is strongly endothermic, the transition state resembles the product C, whereas if it is exothermic, the transition state resembles the reactants.

Hammond's postulate provides a useful working principle, for many reactions occur *via* intermediates whose structures are known and which are of relatively high energy content. In such cases, it is helpful to regard the intermediate as a model for the transition state, for conclusions can then be drawn about the effect which a change in the structure of a reactant is likely to have on the rate of the reaction. Some examples are illustrative.

1. Olefins substituted with electron-withdrawing groups undergo addition of HX in the opposite manner to that predicted by Markovnikov's rule, e.g.

$$CH_2=CH-CO_2Et + HCl \rightarrow CH_2Cl-CH_2-CO_2Et$$

This is because the intermediate $\overset{+}{C}H_2-CH_2-CO_2Et$ is of lower energy than $CH_3-\overset{+}{C}H-CO_2Et$, in which the positive charge is *adjacent* to the strongly electron-withdrawing carboethoxy group.

2. The rates of addition of HX to three typical olefins decrease in the order, $(CH_3)_2C=CH_2 > CH_3-CH=CH_2 > CH_2=CH_2$. This is because the relative stabilities of the carbonium ions formed in the rate-determining step are,

$$\underset{CH_3}{\overset{CH_3}{\diagdown}}\overset{+}{C}-CH_3 > CH_3-\overset{+}{C}H-CH_3 > \overset{+}{C}H_2-CH_3$$

which in turn follows from the fact that the stability is increased by the electron-releasing methyl group $(+I)$, three such groups being more effective than two, and two more effective than one.

3. Compounds in which a C—H bond is adjacent to one or more groups of $-M$ type are weakly acidic (p. 66). Typical pK-values are:

$CH_3-CO-\underset{H}{\overset{|}{C}}H_2$	$EtO_2C-\underset{H}{\overset{|}{C}}H-CO_2Et$	$CH_3-CO-\underset{H}{\overset{|}{C}}H-CO_2Et$	$CH_3-CO-\underset{H}{\overset{|}{C}}H-CO-CH_3$
pK 20	13·3	10·7	9

These compounds undergo a number of base-catalyzed reactions, such as bromination (p. 102), for which typical first-order rate constants are:

CH_3COCH_3	$EtO_2CCH_2CO_2Et$	$CH_3COCH_2CO_2Et$	$CH_3COCH_2COCH_3$
k (s^{-1}) 5×10^{-10}	2×10^{-5}	10^{-3}	2×10^{-2}

Thus, the more acidic the C—H group, the faster is the rate at which bromination occurs. This is because bromination, and the other base-catalyzed reactions,
occur *via* the corresponding *carbanions*, e.g. CH_3—CO—$\bar{C}H_2$ from acetone,
and it is the formation of these anions which constitutes the rate-determining
step:

$$CH_3\text{—CO—}CH_3 + OH^- \xrightarrow{\text{slow}} CH_3\text{—CO—}\bar{C}H_2 + H_2O$$

$$CH_3\text{—CO—}\bar{C}H_2 + Br_2 \xrightarrow[-Br^-]{\text{fast}} CH_3\text{—CO—}CH_2Br \rightarrow \text{(further bromination)}$$

Hence any factor which stabilizes the carbanion, and thereby increases the
acidity of the C—H bond, stabilizes the transition state which precedes the
anion and enhances the rate of reaction.

Some exceptions to this generalization are known; e.g. the *pK* of nitromethane is 10·2, so that it is a slightly stronger acid than ethyl acetoacetate (*pK*
10·7), but its rate of ionization (5×10^{-8} per second) is much less. It must be
emphasized, therefore, that this rate-equilibrium correlation is not quantitative
and should be used only as a helpful guide.

4. Quantitative correlations of rate and equilibria do, however, occur. The
second-order rate constants for the alkaline hydrolysis of the benzoic esters
R—C_6H_4—CO_2Et (in aqueous acetone at 25°C), where R is a *meta-* or *para-*
substituent on the benzene nucleus, together with the *pK*-values for the corresponding acids (in water at 25°C) are:

R	*p*-CH₃	*m*-CH₃	H	*p*-Cl	*m*-Cl	*m*-NO₂	*p*-NO₂
$10^3 k$ (dm³ mol⁻¹ s⁻¹)	2·3	3·5	4·9	21·2	36·3	310	510
pK	4·37	4·27	4·20	3·98	3·83	3·49	3·42

A plot of log k against *pK* is linear:

$$\log k = -\rho pK + A$$

where ρ and A are constants. Since the point for the unsubstituted compound is
fitted by the straight line, $\log k_0 = -\rho pK_0 + A$, so that

$$\log (k/k_0) = \rho(pK_0 - pK)$$

Since log $k \propto \Delta G^{\ddagger}$ and *pK* $\propto \Delta G$, where $\Delta G^{\ddagger}$ and ΔG are the free energies of
activation and ionization, respectively, there is a *linear free energy relationship*
between the hydrolysis rates and the acid strengths.

This correlation has the following basis. The rate-determining step of the hydrolysis is the addition of hydroxide ion to the carbonyl group of the ester:

$$
\text{Ar—C—OEt} + \text{OH}^- \longrightarrow \text{Ar—}\overset{\displaystyle \text{OH}}{\underset{\displaystyle \text{O}^-}{\text{C}}}\text{—OEt}
$$
$$
\underset{\displaystyle \text{O}}{\overset{\|}{}}
$$

The transition state has some of the character of the resulting intermediate and can be represented as

$$
\text{Ar—}\overset{\displaystyle \text{OH}^{\delta-}}{\underset{\displaystyle \text{O}^{\delta-}}{\overset{\|}{\text{C}}}}\text{—OEt}
$$

In the passage to the transition state, therefore, the electron-density in the vicinity of the carboethoxy group is increased, just as it is in the ionization of the corresponding acid $(ArCO_2H \rightarrow ArCO_2^-)$. A substituent which withdraws electrons from the nuclear carbon adjacent to the functional centre (e.g. m-NO_2) stabilizes both the transition state of the hydrolysis and the product of the ionization, resulting in a reduction of the free energy of activation of the former (increasing k relative to k_0) and in the free energy of ionization of the latter (decreasing pK relative to pK_0). An electron-releasing group (e.g. p-CH_3) has the opposite effect.

Analogous correlations are obtained between log k for reactions of other derivatives of benzoic acid, such as the acid-catalyzed hydrolysis of benzamides, and the pK-values of the corresponding benzoic acids. The magnitude of ρ varies with the reaction, and its sign (positive or negative) depends on whether the reaction rate is increased or decreased, respectively, by the withdrawal of electron density from the functional centre. It is convenient to define $(pK_0 - pK)$ as σ, which is a constant for a given substituent. Then

$$
\log (k/k_0) = \sigma\rho
$$

which is known as the *Hammett equation*. Typical σ-values are:

Substituent		CH₃	F	Cl	NO₂	OCH₃
σ-value	*meta*-position	−0·07	0·34	0·37	0·71	0·11
	para-position	−0·17	0·06	0·23	0·78	−0·27

The order of effects measured by σ-values follows qualitatively from the principles which govern the polar properties of substituents and the strengths of acids. As a typical example, the σ-value of m-NO_2 is less than that of p-NO_2

because the $-I$ effect of this group (p. 49) is reinforced, when the substituent is in the *para*-position, by its $-M$ effect (p. 51).

Linear free-energy relationships do not hold for *ortho*-substituents or in aliphatic systems because steric effects become important; a substituent can physically impede the approach of the reagent to the functional centre. This can be of considerable significance; e.g. ethyl 2,6-dimethylbenzoate is effectively inert to base-catalyzed hydrolysis. Many other examples of *steric hindrance* will be described in later chapters.

(*b*) THE EFFECT OF SOLVENT

The rate of a reaction in solution is almost always dependent, and often very strongly dependent, on the nature of the solvent.

Consider again the solvolysis of t-butyl chloride, the rate-determining step of which is the formation of the t-butyl cation (p. 75):

$$(CH_3)_3C\!\!-\!\!Cl \xrightarrow{\text{slow}} (CH_3)_3C^+ + Cl^-$$

At the transition state, the carbon-chlorine bond is partially broken and the covalent bonding-pair, which is ultimately associated completely with chlorine, has been considerably displaced towards chlorine:

$$(CH_3)_3C^{\delta+} \text{-----} Cl^{\delta-}$$

Thus, there is a separation of unlike charges in passage from reactant to transition state.

Two characteristics of the solvent play a part in determining the relative free energies of reactant and transition state and therefore the rate of reaction. First, energy is needed to separate the unlike charges, and the amount of energy decreases as the dielectric constant of the solvent increases.* Consequently the reaction rate increases with the dielectric constant.

If this were the only factor responsible, a solvent with a very high dielectric constant such as hydrogen cyanide ($\epsilon = 118$) would give rise to more rapid rates for such reactions than are observed. A second characteristic, the *solvating power* of the solvent, is also important. The transition state, being more polar than the reactant, is the more effectively solvated of the two, so that a powerful solvating solvent stabilizes the transition state relative to the reactant and increases the rate. Either or both of the developing ions may be solvated: for example, in an alcoholic solvent the developing carbonium ion is solvated by the electron-rich hydroxylic oxygen and the developing chloride is solvated by the electron-deficient hydroxylic proton:

*The dielectric constant, ϵ, of a medium is equal to the attractive force between opposite charges in a vacuum relative to that in the medium. Values for commonly used solvents are: water, 80; methanol, 34; acetone, 21; acetic acid, 6·1; ether, 4·3; benzene, 2·3.

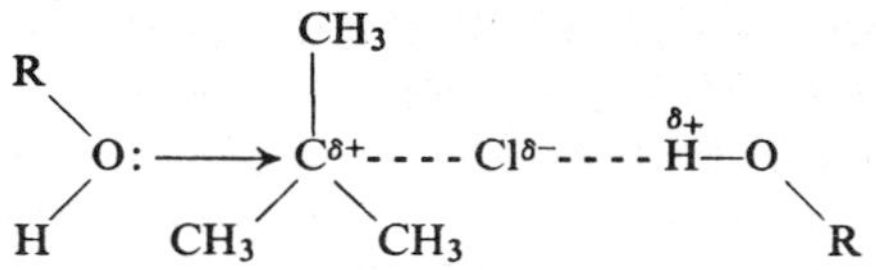

In this example, unlike charges are separated during passage to the transition state. In others, e.g.

$$HO^- + (CH_3)_3S^+ \longrightarrow HO-CH_3 + (CH_3)_2S$$

there is a partial neutralization of unlike charges in passage to the transition state,

$$\overset{\delta-}{HO} \cdots CH_3 \cdots \overset{\delta+}{S(CH_3)_2}$$
transition state

and these reactions occur more slowly as the dielectric constant and solvating power of the solvent are increased.

In a third class of reactions there is a slight dispersal of charge in the rate-determining step, e.g. the alkaline hydrolysis of methyl iodide:

$$HO^- + CH_3-I \longrightarrow \left[\overset{\delta-}{HO} \cdots CH_3 \cdots \overset{\delta-}{I} \right] \longrightarrow HO-CH_3 + I^-$$
transition state

The reactant hydroxide ion, in which the charge is more concentrated, is more effectively solvated and stabilized than the transition state, so that the reaction occurs more slowly as the anion-solvating power of the solvent is increased.

An extreme example of this type is provided by the reaction of bromobenzene with an alkoxide ion:

$$Ph-Br + OR^- \longrightarrow Ph-OR + Br^-$$

The formation of the transition state,

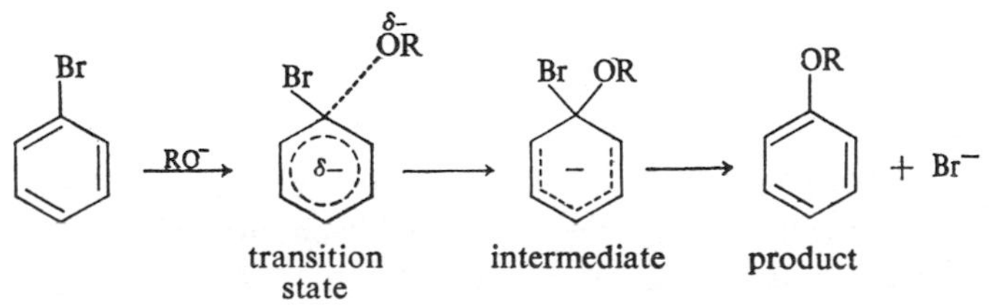

transition state   intermediate   product

involves the dispersal of charge, so that the reaction is slower in a solvent which solvates anions strongly, such as an alcohol, than in a non-solvating medium. In practice, reaction with t-butoxide ion occurs about nine powers of ten faster in dimethyl sulphoxide, which has little ability to stabilize $(CH_3)_3C\!-\!O^-$, than in t-butanol, where the rate is negligible.

In general, reactions where the activation process involves the separation of unlike charges occur more rapidly as the polarity of the medium is increased, whereas those involving the partial neutralization of charge or the dispersal of charge occur more slowly. Most synthetic processes are of ionic type, as will be seen in the course of the text; free-radical reactions, in which the effect of the solvent on the rate is not usually significant, are relatively rarely employed. It is necessary, therefore, to pay careful attention to the choice of the solvent in planning an organic synthesis.

(c) RING-CLOSURE

Whereas the esterification of acetic acid by methanol is very slow in the absence of an acid catalyst, the lactonization of γ-hydroxybutyric acid,

$$\begin{array}{c}
CH_2\!-\!CH_2 \\
| \qquad\qquad\searrow CO_2H \\
CH_2\!-\!OH
\end{array}
\longrightarrow
\begin{array}{c}
CH_2\!-\!CH_2 \\
| \qquad\qquad\searrow CO \;+\; H_2O \\
CH_2\!-\!\!-\!O
\end{array}$$

occurs essentially spontaneously. The reason for the difference becomes apparent when the transition states for the processes are considered. In each case, the rate-determining step involves the formation of the new C—O bond, and at the transition states this bond is partially formed:

$$\begin{array}{c}
\quad\; O \\
\quad\; \| \\
CH_3\!-\!C\!-\!OH \\
\\
CH_3OH
\end{array}
\longrightarrow
\begin{array}{c}
\quad\; O^{\delta-} \\
\quad\; \| \\
CH_3\!-\!C\!-\!OH \\
\quad\; | \\
\quad\; O^{\delta+} \\
\nearrow \quad\; \nwarrow \\
CH_3 \qquad H
\end{array}$$

$$\begin{array}{c}
CH_2\!-\!CH_2 \qquad O \\
| \qquad\qquad\searrow C \nearrow \\
CH_2\!-\!OH \qquad \searrow OH
\end{array}
\longrightarrow
\begin{array}{c}
CH_2\!-\!CH_2 \qquad O^{\delta-} \\
| \qquad\qquad\searrow C \\
CH_2\!-\!\!-\!O \quad\;\; \searrow OH \\
\quad\qquad {}^{\delta+}\;\;\; \\
\qquad\qquad\qquad H
\end{array}$$

In the esterification, passage to the transition state necessitates two molecules coming together to form one species, so that translational freedom is lost and $\Delta S^{\ddagger}_{tr}$ is correspondingly large and negative. On the other hand, in the lactonization only internal, or vibrational, freedom is lost, and the $\Delta S^{\ddagger}_{vib.}$ term is much smaller than the $\Delta S^{\ddagger}_{tr.}$ term for the intermolecular analogue, though still negative. The $\Delta H^{\ddagger}$ terms are similar in magnitude, for the same types of bond-forming and bond-breaking processes occur in each case and there is comparatively little strain in the lactone ring which is being formed. Consequently $\Delta G^{\ddagger}$ is smaller for lactonization and this process occurs faster than esterification.

This discussion both reveals the significance of the PZ factor of the Arrhenius equation in these reactions and shows that the factors which determine the relative rates of the two reactions are essentially the same as those which determine the relative equilibrium constants (cf. p. 24).

This similarity between the factors controlling rates and those controlling equilibria can be taken further. For example, in an intramolecular reaction leading to a four-membered ring, $\Delta S^{\ddagger}$ is less negative than in one leading to a five-membered ring for the acyclic reactant in the former case has fewer conformations than that in the latter, so that there is less loss of internal freedom $(\Delta S_{\text{vib.}}^{\ddagger})$ in the formation of the transition state. However, $\Delta H^{\ddagger}$ is larger for the four-membered ring because of the strain in the cyclic transition state, and this factor outweighs the $T\Delta S^{\ddagger}$ term. Hence $\Delta G^{\ddagger}$ is larger for the formation of a four-membered ring, and ring-closure occurs more slowly (e.g. β-hydroxy-acids, unlike γ-hydroxy-acids, do not lactonize spontaneously).

At first sight surprisingly, three-membered rings are usually formed faster than four-membered rings. This is because the greater strain (leading to a higher $\Delta H^{\ddagger}$) in the transition state for the former is more than offset by the greater probability (less negative $\Delta S^{\ddagger}$) that the ends of the chain shall come together for bond-formation.

The entropy term is less favourable for the formation of a six-membered than a five-membered ring and, unless there is a significant strain factor involved in the formation of the five-membered ring, this is formed the faster. For larger sized rings the entropy term becomes increasingly unfavourable and there is in addition a small amount of conformational strain in the cyclic transition states for seven- to twelve-membered rings.

The combined effects of the enthalpy and entropy factors are illustrated for the cyclization of some ω-bromo-amines:

$$H_2N \overset{(CH_2)_{n-2}}{\diagup\diagdown} CH_2Br \longrightarrow HN \overset{(CH_2)_{n-2}}{\diagup\diagdown} CH_2$$

n	3	4	5	6	7	10	15
k_{relative}	0·12	0·002	100	1·7	0·03	10^{-8}	10^{-4}

Cyclizations of bifunctional compounds compete with the corresponding intermolecular reaction between two molecules of the compound. In typical reaction conditions, the formation of five- and six-membered rings is strongly favoured over the intermolecular process, and indeed such cyclizations often occur in milder conditions than analogous intermolecular reactions, as in the lactonization of γ-hydroxybutyric acid described above. The formation of three-, four-, and seven-membered rings competes less successfully and yields of the cyclic products are usually low; and rings with more than seven members are usually not formed in significant amounts. It should, however, be remembered that the competition between the intra- and inter-molecular reactions depends not only on the ring-size but also on the concentration of the reactant; since two molecules are necessary for the intermolecular process and only one for cyclization, high dilutions favour cyclization, and this fact has been employed in a technique for synthesizing large rings (p. 243).

Ring-closures involving the participation of two molecules often have very low PZ factors (i.e. large negative $\Delta S^{\ddagger}$); e.g. that for the Diels-Alder reaction,

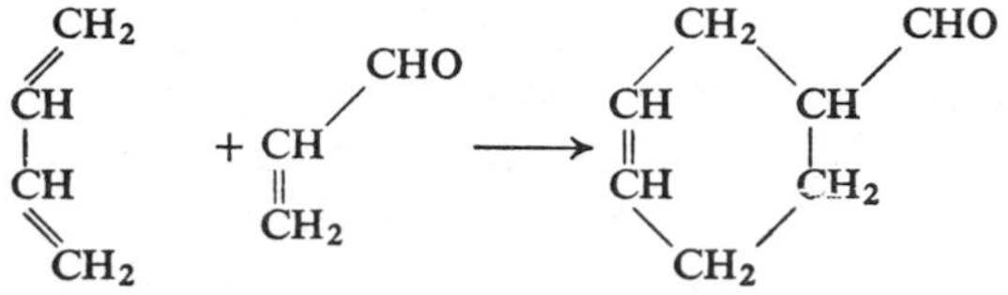

is about 10^6. This is because the transition state is highly ordered, the four carbon atoms which take part in the formation of the two new carbon-carbon bonds having to be aligned in the appropriate positions.

(d) THERMODYNAMIC VERSUS KINETIC CONTROL

In most reactions which can proceed by two or more pathways each of which gives a different set of products, the products isolated are those derived from the pathway of lowest free energy of activation, regardless of whether this path results in the greatest decrease in the free energy of the system. These reactions are described as being *kinetically controlled*.

If, however, the reaction conditions are suitable for equilibrium to be established between the reactants and the kinetically controlled products, a different set of products, formed more slowly but corresponding to a lower free energy for the system, can in some instances be isolated. Such reactions are described as being *thermodynamically controlled*.

Consider the reaction of naphthalene with concentrated sulphuric acid. Two monosulphonated products are in principle obtainable:

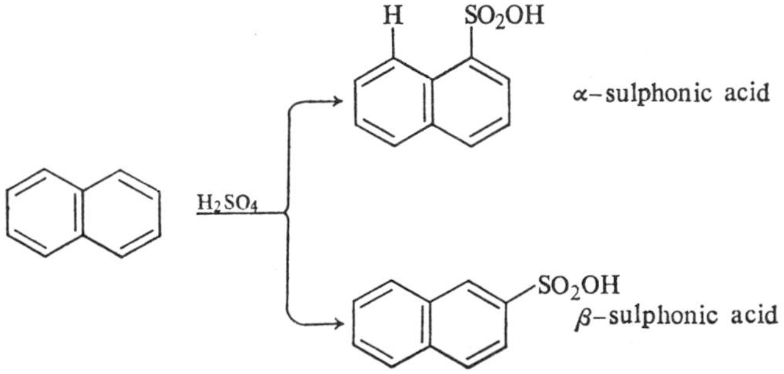

The α-derivative is formed the faster of the two but the β-derivative is thermo-dynamically the more stable (at least in part because of the repulsive forces between the sulphonic acid group and the *peri*-hydrogen, shown, in the α-isomer). The situation can be represented schematically as in Fig. 3.4.

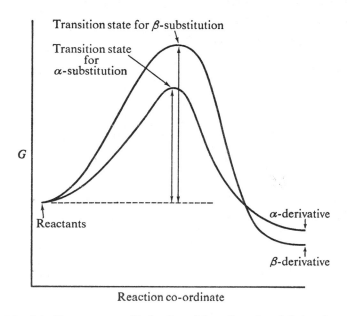

Fig. 3.4   Free-energy profile for the sulphonation of naphthalene.*

At low temperatures (*ca.* 80°C), sulphonation at the α-position occurs fairly rapidly whereas that at the β-position is very slow. The free energy of activation for the desulphonation of the α-sulphonic acid is such that in these conditions this product is essentially inert and is therefore isolated. At higher temperatures (*ca.* 160°C), desulphonation of the α-sulphonic acid becomes important and equilibrium is fairly rapidly established between this product and the reactants. The rate of formation of the β-sulphonic acid is now also greater, so that gradu-ally most of the naphthalene is converted into the β-derivative and this becomes the major product.

Further Reading

LATHAM, J. L., and BURGESS, A. E., *Elementary Reaction Kinetics*, 3rd Ed., Butterworths (London 1977).

PILLING, M. J., *Reaction Kinetics*, Clarendon Press (Oxford 1975).

*For simplicity, intermediates are not shown.

Problems

1. It is said in some elementary text-books that 'reaction rates double when the temperature rises by 10°C.' What is the activation energy of such reactions? (Assume that the statement refers to reactions occurring at or near room temperature.)

2. In a particular first-order reaction, 50% of the reactant has been consumed after 60 minutes. What is the rate constant?

3. The second-order rate constant for the alkaline hydrolysis of ethyl phenyl-acetate ($PhCH_2CO_2Et$) in aqueous acetone varies with temperature as follows:

T (°C)	10	25	40	55
10^2k (dm^3 mol^{-1} s^{-1})	1·74	4·40	10·4	23·6

Calculate the activation energy, ΔE, and the pre-exponential factor, PZ, for the reaction.

4. Two reactions have, respectively, $A = 10^9$ dm^3 mol^{-1} s^{-1}, $\Delta E = 60$ kJ mol^{-1}, and $A = 10^{10}$ dm^3 mol^{-1} s^{-1}, $\Delta E = 70$ kJ mol^{-1}. At what temperature would the rate constants be equal?

5. The kinetics for the solvolysis of benzhydryl chloride (Ph_2CHCl) in aqueous acetone are given by

$$-d[Ph_2CHCl]/dt = k[Ph_2CHCl]/(1 + k'[Cl^-])$$

Suggest an explanation.

6. Phenylnitromethane, $PhCH_2NO_2$, is a liquid which dissolves in sodium hydroxide solution. Acidification of the solution precipitates the tautomer,

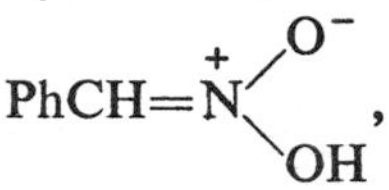

$$PhCH=\overset{+}{N}\overset{O^-}{\underset{OH}{\diagup}}$$
, as a solid, but this then slowly reverts to the original liquid.

Explain the chemistry of the formation of the tautomer and draw an energy profile consistent with the observations.

7. The Cannizzaro reaction on benzaldehyde is thought to have the following mechanism:

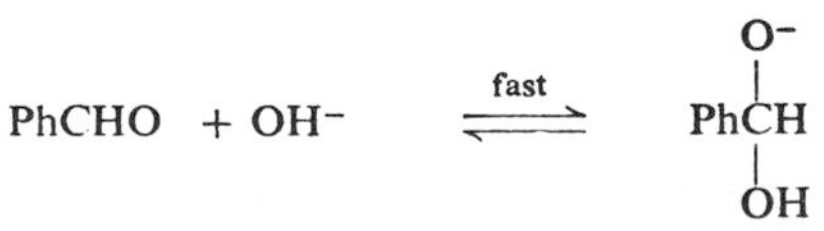

$$PhCHO + OH^- \underset{fast}{\rightleftharpoons} PhC\overset{O^-}{\underset{OH}{H}}$$

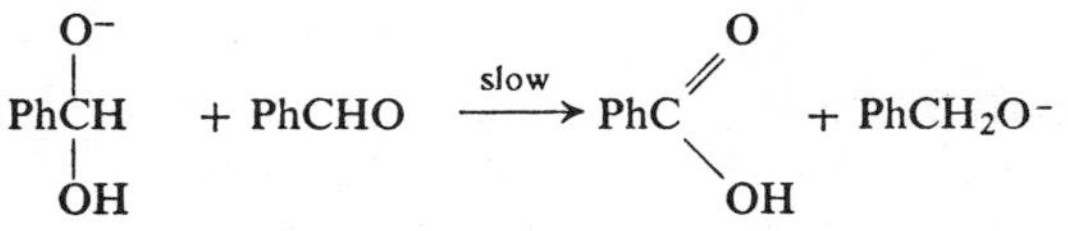

$$\underset{\underset{\text{OH}}{|}}{\overset{\overset{\text{O}^-}{|}}{\text{PhCH}}} + \text{PhCHO} \xrightarrow{\text{slow}} \underset{\text{OH}}{\overset{\overset{\text{O}}{\parallel}}{\text{PhC}}} + \text{PhCH}_2\text{O}^-$$

$$\text{PhCO}_2\text{H} + \text{PhCH}_2\text{O}^- \xrightarrow{\text{fast}} \text{PhCO}_2^- + \text{PhCH}_2\text{OH}$$

What kinetics would you expect the reaction to follow?

8. Derive the kinetic expression given on p. 74 for the decarbonylation of acetaldehyde.

9. The decomposition $(CH_3)_2O \rightarrow CO + H_2 + CH_4$ was followed manometrically. The following results were obtained:

(a) At 504°C

Time (s)	0	390	665	916	1625	2643	∞
Pressure (mm Hg)	312	408	468	512	·624	749	931

(b) At 551°C

Time (s)	0	114	182	261	464	564	∞
Pressure (mm Hg)	420	743	891	1031	1163	1197	1258

Find the order of reaction, the rate constant at the two temperatures, and the activation energy.

10. The following mechanistic scheme has been postulated for the reaction between iodine and an organic compound, HA, in the presence of hydroxide ions in aqueous solution:

$$\text{HA} + \text{OH}^- \underset{k_2}{\overset{k_1}{\rightleftharpoons}} \text{A}^- + \text{H}_2\text{O}$$

$$\text{A}^- + \text{I}_2 \xrightarrow{k_3} \text{IA} + \text{I}^-$$

Assuming that the concentration of A^- is at all times very much smaller than that of HA and IA, derive an equation for the rate of change of [HA] in terms of the rate constants and the concentrations of HA, OH^-, and I_2 (OH^- and I_2 being present in excess).

4. Mechanism

4.1 Unit Processes in Organic Reactions

Many organic reactions seem at first glance to be highly complex, involving extensive reorganization of the bonds of the reactants. An example is the Skraup synthesis in which quinoline is obtained in yields of up to 90% by heating a mixture of aniline, glycerol, nitrobenzene, and concentrated sulphuric acid:

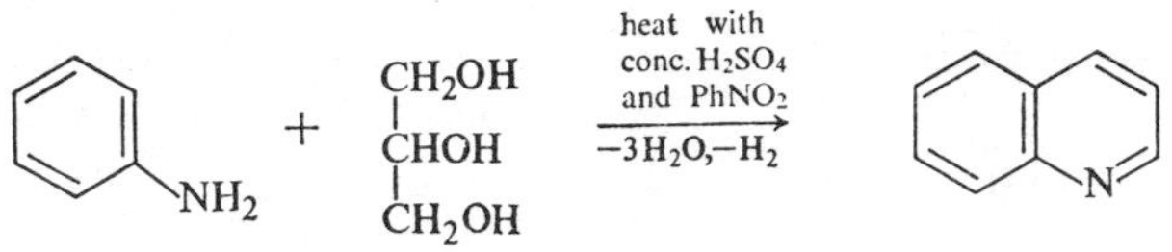

Such complex reactions usually consist of a number of simple steps in each of which a comparatively small reorganization of bonds takes place and which lead to intermediates which undergo further reaction until the final product is reached. The Skraup synthesis, for example, consists of the following steps:

$$CH_2OH-CH(OH)-CH_2OH \xrightarrow[-H_2O]{(H_2SO_4)} CH_2OH-CH=CHOH$$

$$\rightleftharpoons CH_2OH-CH_2-CH=O \xrightarrow[-H_2O]{(H_2SO_4)} CH_2=CH-CHO$$

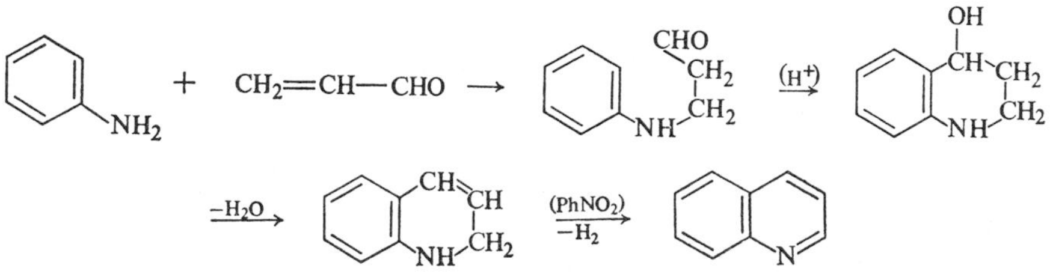

Even reactions which may appear simpler often consist of a number of steps. For instance, the nitration of benzene by a mixture of concentrated nitric and sulphuric acids, whose stoicheiometry is represented by

$$PhH + HNO_3 \xrightarrow{(H_2SO_4)} PhNO_2 + H_2O$$

occurs through the following stages:

$$HO-NO_2 + H_2SO_4 \rightleftharpoons H_2\overset{+}{O}-NO_2 + HSO_4{}^-$$

$$H_2\overset{+}{O}-NO_2 \rightleftharpoons H_2O + NO_2{}^+$$

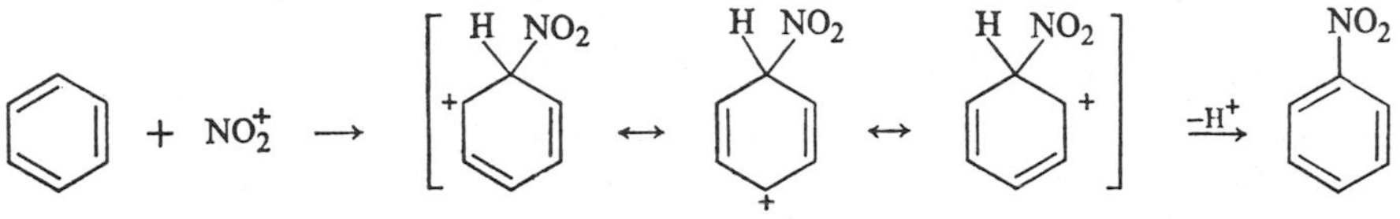

There is, indeed, a comparatively small number of basic processes, and each reaction is a combination of these. Five such processes may be recognized: (a) bond-breaking, (b) bond-forming, (c) synchronous bond-breakage and bond-formation, (d) intramolecular migration, and .(e) electron-transfer.

(a) BOND-BREAKING

A covalent bond A—B may break in one of two ways: either *homolytically*, to give A· + B·, or *heterolytically*, to give $A^+ + B^-$ or $A^- + B^+$.

(i) *Homolysis*, or homolytic fission, occurs on heating to moderate temperatures those compounds which either contain an intrinsically weak bond, such as O—O, or which, on dissociation, liberate a particularly strongly bonded molecule, such as N_2. Peroxides illustrate the former class, e.g.

$$PhCO-O-O-COPh \longrightarrow 2\ PhCO-O\cdot$$
$$(CH_3)_3C-O-O-C(CH_3)_3 \longrightarrow 2\ (CH_3)_3C-O\cdot$$

and certain azo compounds illustrate the latter, e.g.

$$(CH_3)_2\underset{\overset{|}{CN}}{C}-N=N-\underset{\overset{|}{CN}}{C}(CH_3)_2 \longrightarrow 2\ (CH_3)_2\underset{\overset{|}{CN}}{C}\cdot + N_2$$

The activation energy for homolysis is usually barely greater than the energy of the bond which breaks, for the activation energy for the reverse reaction is negligible. In turn, the dissociation energy for a bond is affected by the presence of substituents which can stabilize the radicals which are formed by delocalization of the unpaired electron in each. Consequently, the rate of homolysis of a particular type of bond is related to the delocalization energy in the resulting radicals; for example, the faster rate of dissociation of dibenzoyl compared with di-t-butyl peroxide can be related to the delocalization of the unpaired electron in the benzoyloxy radical,

$$Ph-C\begin{matrix} O \\ \backslash \\ O\cdot \end{matrix} \longleftrightarrow Ph-C\begin{matrix} O\cdot \\ \diagup \\ O \end{matrix}$$

which has no counterpart in the t-butoxy radical.

Homolysis can also be induced by ultraviolet (or sometimes visible) light, providing that the molecule possesses an appropriate light-absorbing group (chromophore). The principles are discussed in detail in Chapter 16.

(ii) *Heterolysis*, or heterolytic fission, occurs when the species formed are relatively stable. For example, whereas methyl chloride is relatively inert to water, t-butyl chloride is fairly rapidly hydrolyzed, reaction occurring through the t-butyl carbonium ion:

$$(CH_3)_3C-Cl \rightleftharpoons (CH_3)_3C^+ + Cl^-$$

$$(CH_3)_3C^+ + H_2O \longrightarrow (CH_3)_3C-\overset{+}{O}H_2 \xrightarrow{-H^+} (CH_3)_3C-OH$$

The difference in behaviour is rationalized as follows. The transition state for the heterolysis involves the partial breakage of the C—Cl bond, the bonding-pair departing with chlorine, and may be represented as $(CH_3)_3C^{\delta+} ---- Cl^{\delta-}$. The partially positively polarized carbon species is stabilized, relatively to that involved in the heterolysis of methyl chloride $(CH_3{}^{\delta+} ---- Cl^{\delta-})$, by the three electron-releasing methyl substituents, just as the t-butyl cation is stabilized relative to the methyl cation. Consequently the activation energy for the heterolysis of t-butyl chloride is less than that for methyl chloride.

The stability of the departing anion is also important. Thus, the order of ease of heterolysis of some t-butyl compounds is

$$(CH_3)_3C-OH < (CH_3)_3C-OAc < (CH_3)_3C-Cl,$$

just as the acid strengths lie in the order,

$$H-OH < H-OAc < H-Cl.$$

Chloride ion is described as a better *leaving group* than acetate, and acetate as better than hydroxide. The capacity of hydroxyl as a leaving group is increased by protonation (H_3O^+ is a far stronger acid than H_2O), and this has synthetic utility: for example, t-butanol is unaffected by chloride ion but is converted into t-butyl chloride by hydrogen chloride:*

---

*This is simply the reverse of the hydrolysis of t-butyl chloride. The direction which the reaction takes depends on the relative concentrations of water and chloride ion.

$$(CH_3)_3C—OH \xrightleftharpoons{HCl} (CH_3)_3C—\overset{+}{O}H_2 + Cl^-$$

$$(CH_3)_3C—\overset{+}{O}H_2 \rightleftharpoons (CH_3)_3C^+ + H_2O$$

$$(CH_3)_3C^+ + Cl^- \longrightarrow (CH_3)_3C—Cl$$

(b) BOND-FORMING
(i) Two radicals (or atoms) may combine, e.g.

$$2 \cdot CH_3 \longrightarrow CH_3—CH_3$$

$$\cdot CH_3 + \cdot Cl \longrightarrow CH_3—Cl$$

$$2 \cdot Cl \longrightarrow Cl_2$$

These reactions are the reverse of homolysis. They normally have very low activation energies and occur rapidly, although some radicals, particularly those which are sterically hindered (17.1), are stable in certain conditions.

Many reactions involve free-radical intermediates; the radicals react with neutral molecules, giving new radicals which perpetuate a chain reaction. Combinations of radicals such as those above lead to the termination of the chain and are disadvantageous. However, despite the large rate-constants for combination, long chains can be propagated because the concentration of radicals is usually small so that their rate of combining (which is proportional to the product of the concentrations of two radicals) does not necessarily compete effectively with the chain-propagating steps in which a radical reacts with a neutral molecule which may be present in far higher concentration.

(ii) Oppositely charged ions may combine, e.g.

$$(CH_3)_3C^+ + Cl^- \longrightarrow (CH_3)_3C—Cl$$

This is the reverse of heterolysis. When the positive ion is a carbonium ion, reaction is usually rapid and rather unselective with respect to different anions. It should be noted that positive ions which are electronically saturated do not combine with anions: e.g. quaternary ammonium ions such as $(CH_3)_4N^+$ form stable salts in solution, comparable with those formed by alkali-metal cations, since the quaternary nitrogen, unlike the electron-deficient carbon in a carbonium ion, does not possess a low-lying orbital which can accept two more electrons.

(iii) An ion may add to a neutral molecule, e.g.

$$(CH_3)_3C^+ + H_2O \longrightarrow (CH_3)_3C—\overset{+}{O}H_2$$

$$Cl^- + AlCl_3 \longrightarrow AlCl_4{}^-$$

For addition to a *cation*, the neutral molecule must possess at least one unshared pair of electrons (i.e. it must be a base, the positive ion acting as a Lewis acid), whereas for addition to an *anion* it must possess the ability to accept an electron-pair (i.e. it must be a Lewis acid). It is convenient to define new descriptive terms for these and other species. Cations and electron-deficient molecules which have a tendency to form a bond by accepting an electron-pair from another species are termed *electrophilic* ('electron-seeking'), and anions and molecules with unshared pairs of electrons which have a tendency to form a bond by donating an electron-pair to another species are termed *nucleophilic* ('nucleus-seeking'). These terms are most commonly applied to species reacting at carbon; e.g. in the reaction,

$$(CH_3)_3C^+ + OAc^- \longrightarrow (CH_3)_3C—OAc$$

the acetate ion is acting as a nucleophile.

(c) SYNCHRONOUS BOND-BREAKAGE AND BOND-FORMATION
In the transition state of these reactions one or more bonds are partially broken and one or more are partially formed. The energy required for bond-breakage is partly supplied by the energy evolved in bond-formation, so that the overall activation energy is less than that for the bond-breaking reactions alone.

There are two major categories of synchronous reactions: *(i)* a single bond is broken and a single bond is formed; *(ii)* a double bond is converted into a single bond (or a triple bond into a double bond) while a single bond is formed. Examples of the two are:

(i) $CN^- + CH_3—I \longrightarrow CH_3—CN + I^-$

(ii) $CN^- + CH_3—CH{=}O \longrightarrow CH_3—\underset{\underset{CN}{|}}{C}H—O^-$

These two reactions clearly have the same basis: one bond is cleaved heterolytically while a second bond is formed by nucleophilic attack on carbon. Despite this similarity, it is convenient to discuss the two types of reaction separately.

(i) The examples below are classified with respect to the element which undergoes synchronous substitution.

Carbon is substituted by nucleophiles when it is attached to an electronegative group which can depart as a relatively stable anion. Thus whereas displacement on a paraffin, e.g.

$$HO^- + CH_3—H \longrightarrow HO—CH_3 + H^-$$

does not take place because of the high energy of hydride ion, the reactions of alkyl halides and alkyl sulphates and sulphonates, such as

$$H_3N + CH_3—I \longrightarrow H_3\overset{+}{N}—CH_3 + I^-$$

$$RO^- + CH_3—OSO_2OCH_3 \longrightarrow RO—CH_3 + CH_3OSO_2O^-$$

$$RO^- + CH_3—OSO_2Ar \longrightarrow RO—CH_3 + ArSO_2O^-$$

occur readily. The transition states are represented as, e.g.

$$
RO \overset{\delta-}{-----} \underset{\underset{H}{\diagup} \underset{H}{\diagdown}}{\overset{\overset{H}{|}}{C}} \overset{\delta-}{-----} OSO_2OCH_3
$$

and a common symbolism for this reaction type, which indicates the electronic movements, is:

$$
RO^- \qquad CH_3\!-\!OSO_2OCH_3 \longrightarrow RO\!-\!CH_3 + CH_3OSO_2O^-
$$

Note that groups which form less stable anions, such as —OH and —NH$_2$, are not displaced from carbon; e.g. it is not possible to form an ether from an amine,

$$
RO^- + CH_3\!-\!NH_2 \nrightarrow RO\!-\!CH_3 + NH_2^-
$$

Carbon is substituted by electrophiles when it is attached to a strongly electropositive group, *viz.* a metal, as in the reaction of a Grignard reagent with an acid (p. 206):

$$
H^+ + CH_3\!-\!MgBr \longrightarrow CH_4 + Mg^{2+} + Br^-
$$

Carbon does not undergo synchronous substitutions with radicals or atoms.

Hydrogen undergoes attack by nucleophiles when it is attached to an electronegative group, e.g.

$$
H_2O + H\!-\!Cl \longrightarrow H_3O^+ + Cl^-
$$

$$
HO^- + H\!-\!CH_2CHO \longrightarrow H_2O + \bar{C}H_2CHO
$$

These reactions are reversible and are, of course, better described as acid-base equilibria; the principles which govern the position of the equilibrium have been described (p. 21). It should be noted that, whereas reaction is normally very fast when hydrogen is attached to oxygen, nitrogen, or a halogen, abstraction from carbon is often very slow; for example, the first-order rate constants for the ionization of acetic acid and nitromethane in aqueous solution at 25°C are, respectively, about 10^6 and 10^{-8} per second.

Hydrogen is attacked by electrophiles when it is bonded to a strongly electropositive element, as in certain metal hydrides such as lithium aluminium hydride, LiAlH$_4$, which contains the AlH$_4{}^-$ ion. For example, the electrophilic carbon of the carbonyl group reacts to form a C—H bond (p. 146):

$$
H_3\bar{Al}\!-\!H \quad \overset{|}{\underset{|}{C}}\!=\!O \longrightarrow H\!-\!\overset{|}{\underset{|}{C}}\!-\!O^- + AlH_3 \longrightarrow H\!-\!\overset{|}{\underset{|}{C}}\!-\!O\!-\!\bar{Al}H_3
$$

Hydrogen is attacked by radicals and atoms in a number of different structural environments, e.g.

$$\Delta H$$
$$(kJ\ mol^{-1})$$

$$\cdot Cl + H - CH_3 \rightarrow HCl + \cdot CH_3 \qquad\qquad -2$$

$$\cdot Br + H - CH_3 \rightarrow HBr + \cdot CH_3 \qquad\qquad +62$$

$$CH_3CH_2 \cdot + H - Br \rightarrow CH_3CH_3 + \cdot Br \qquad\qquad -37$$

The chief characteristic of these reactions is that they inevitably lead to the formation of a new radical which can itself abstract an atom from another molecule. Thus, chain reactions are propagated. Those abstractions which are exothermic, such as that of the ethyl radical with hydrogen bromide, usually occur rapidly (a rough guide is that the activation energy is about 5% of the dissociation energy of the bond which is broken), whereas those which are endothermic are relatively slow (the activation energy is necessarily at least as great as the reaction enthalpy). Nevertheless, processes which are only moderately endothermic take part in chain reactions: for example, the reaction of the bromine atom with methane constitutes one step in the radical-catalyzed bromination of methane (p. 542). Conversely, reactions of the reverse type, such as that of the ethyl radical with hydrogen bromide, also occur in chain processes such as the radical-catalyzed addition of hydrogen bromide to olefins (p. 545).

Oxygen and *nitrogen* undergo synchronous substitution only rarely; both elements, as, e.g. hydroxide ion and ammonia, instead react primarily as nucleophiles, adding to unsaturated electrophilic centres and substituting at saturated electrophilic centres. Some useful examples are known, however, such as the formation of *N*-oxides, e.g.

and the formation of amines from Grignard reagents and *O*-methylhydroxylamine (p. 215),

$$BrMg - R \cap NH_2 - OCH_3 \longrightarrow R - NH_2 + MgBr(OCH_3)$$

The *halogens*, chlorine, bromine, and iodine, undergo displacement by nucleophiles as in the halogenation of carbanions, e.g.

$$CH_3COCH_2^- \cap Br - Br \longrightarrow CH_3COCH_2Br + Br^-$$

The success of the reaction is related to the ability of a halogen atom to depart as an anion.

Halogen molecules also react with radicals, e.g.

$$\cdot CH_3 + Cl_2 \longrightarrow CH_3Cl + \cdot Cl$$

as in the radical-catalyzed halogenation of hydrocarbons (p. 541).

The fluorine molecule reacts so violently with most organic compounds that it has little use as a synthetic reagent. The vigour of reaction is a result of the small bond dissociation energy of F—F together with the very large bond energies of H—F and C—F (p. 11).

(*ii*) Nucleophiles, electrophiles, radicals, and atoms may add to unsaturated bonds, of which those most frequently encountered are:

$$\diagup C = C \diagdown \quad \text{in olefins and aromatic compounds}$$

—C≡C— in acetylenes

$$\diagdown C = O \quad \text{in aldehydes, ketones, amides, acids, esters, anhydrides, and acid halides}$$

—C≡N in nitriles (cyanides)

Other unsaturated groupings include C=N (in imines), N=O (in nitroso and nitro compounds), N=N (in azo compounds), and C=S (in thio analogues of carbonyl-containing compounds).

There are marked differences in the ease with which addition occurs. Thus, simple olefins react readily with electrophiles and radicals, e.g.

$$CH_2{=}CH_2 + H^+ \longrightarrow CH_3{-}CH_2^+$$

$$CH_2{=}CH_2 + \cdot Br \longrightarrow CH_2Br{-}CH_2\cdot$$

but do not react with nucleophiles. On the other hand, the carbonyl group reacts with nucleophiles, e.g.

$$R_2C{=}O + CN^- \longrightarrow R_2C{-}O^-$$
$$\underset{\displaystyle CN}{|}$$

this difference from olefins resulting from the fact that oxygen is better able than carbon to accommodate the negative charge in the adduct. Likewise, the cyano group reacts with nucleophiles to give an adduct in which the charge is accommodated by nitrogen, e.g.

$$-C{\equiv}N + OH^- \longrightarrow -C{=}N^-$$
$$\underset{\displaystyle OH}{|}$$

The acetylenic bond reacts with electrophiles less readily, and with nucleophiles more readily, than olefins. For example, acetylene itself, unlike ethylene, reacts with alkoxide ions in alcoholic solution at high temperatures and pressures:

$$RO^- + HC{\equiv}CH \longrightarrow RO{-}CH{=}\bar{C}H$$

The difference is ascribed to the fact that unsaturated carbon is more electronegative than saturated carbon (cf. the greater acidity of acetylene then ethylene).

An aromatic double bond in benzene reacts less readily than olefins with all reagents as a result of the loss of aromatic stabilization energy which accompanies addition, e.g.

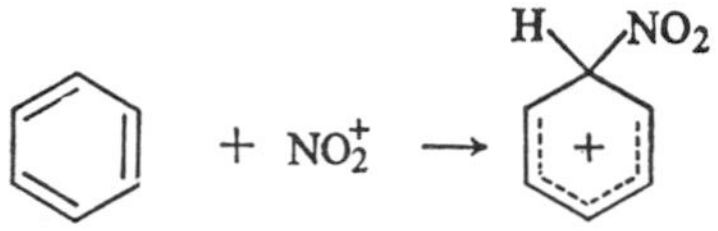

However, the presence of substituents which can stabilize the resulting adduct increases the rate of addition.

(*d*) INTRAMOLECULAR MIGRATION
In certain structural situations a group migrates from one atom to another in the same species without becoming 'free' of that species. Three classes of migration may be distinguished.

(1) The group migrates together with the bonding-pair by which it was attached in the original species. For example, the neopentyl cation rearranges to the t-amyl cation by the migration of a methyl group, represented as follows:

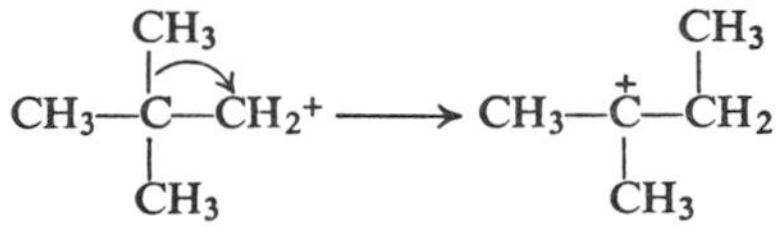

The thermodynamic driving force for the reaction lies in the greater stability of a tertiary than a primary carbonium ion (p. 83).

(2) The group migrates with one electron of the original bonding-pair, e.g.

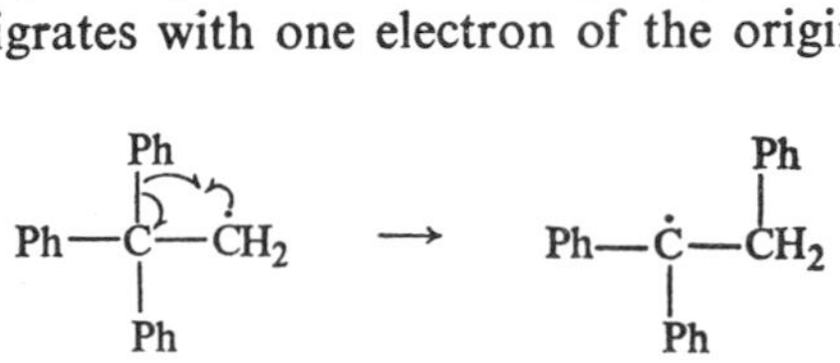

The driving force is the greater stability of the tertiary radical, in which the unpaired electron is delocalized over two aromatic rings, than the primary radical.

(3) The group migrates with neither of the original bonding electrons, e.g.

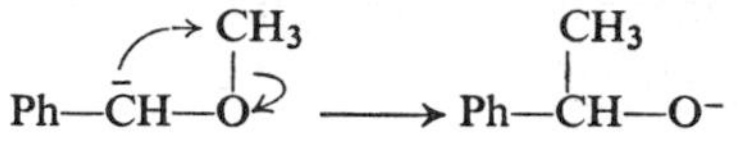

$$\overset{\curvearrowright CH_3}{Ph-\overset{-}{C}H-\overset{\curvearrowright}{O}} \longrightarrow \overset{CH_3}{Ph-\overset{|}{C}H-O^-}$$

The driving force lies in the greater capacity of oxygen than of carbon to be anionic.

These, and related, migrations occur as unit steps in intramolecular rearrangements (Chapter 14).

(e) ELECTRON-TRANSFER

Species which have a strong tendency to give up one electron (low ionization potential) can react with species with a strong tendency for electron-acceptance (high electron affinity) by electron-transfer. Thus the well known mode of formation of inorganic salts from electropositive metals and electronegative elements or groups has as its organic analogues such reactions as

$$Na\cdot + Ph_2C{=}O \longrightarrow Na^+ + Ph_2\overset{\cdot}{C}{-}O^-$$

$$Fe^{2+} + RO{-}OH \longrightarrow Fe^{3+} + RO\cdot + OH^-$$

Organic compounds may be electron-acceptors, as in the above examples, or donors, as in

$$Fe^{3+} + PhO{-}H \longrightarrow Fe^{2+} + PhO\cdot + H^+$$

Donation or acceptance by a neutral organic compound gives rise to a radical which normally undergoes further reaction. For example, the acceptance of an electron by acetone from magnesium gives $Me_2\overset{\cdot}{C}{-}O^-$ which dimerizes to

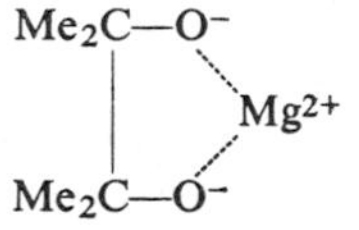

$$\begin{array}{c} Me_2C{-}O^- \\ | \quad\diagdown \\ \quad\quad Mg^{2+} \\ | \quad\diagup \\ Me_2C{-}O^- \end{array}$$

Treatment with acid leads to pinacol, $Me_2C(OH){-}CMe_2(OH)$ (pinacol reduction, p. 637).

It should be noted that the uptake of electrons by an organic compound constitutes reduction (Chapter 19) and the loss of electrons constitutes oxidation (Chapter 18).

4.2 Types of Reaction

Combinations of the unit processes described in the previous section give reactions of various types. For example, when ethyl acetate is refluxed in ethanol containing sodium ethoxide, the sodium derivative of ethyl acetoacetate is formed:

$$2\ CH_3CO_2Et + Na^+OEt^- \longrightarrow [CH_3COCHCO_2Et]^-Na^+ + 2\ EtOH$$

The unit processes which combine in the overall reaction are: an acid-base equilibrium (1), bond-formation (2), heterolytic bond-breakage (3), and a second acid-base equilibrium which lies well to the right (4):

$$CH_3CO_2Et + EtO^- \underset{}{\overset{(1)}{\rightleftharpoons}} \bar{C}H_2CO_2Et + EtOH$$

$$CH_3-\underset{\underset{\bar{C}H_2CO_2Et}{\uparrow}}{\overset{\overset{O}{\parallel}}{C}OEt} \quad \underset{}{\overset{(2)}{\rightleftharpoons}} \quad CH_3-\underset{CH_2CO_2Et}{\overset{\overset{O}{\parallel}}{C}-OEt} \quad \underset{}{\overset{(3)}{\rightleftharpoons}} \quad CH_3-\overset{\overset{O}{\parallel}}{C}-CH_2CO_2Et + EtO^-$$

$$CH_3COCH_2CO_2Et + EtO^- \underset{}{\overset{(4)}{\rightleftharpoons}} CH_3CO\bar{C}HCO_2Et + EtOH$$

It is convenient to employ descriptive terms to overall processes such as this, and the reactions described hereafter are classified as addition, elimination, substitution, condensation, rearrangement, multi-centre reactions, and oxidation-reduction. The example above is a condensation (4.6).

Subdivision of these reactions is helpful, and in many cases the reaction can be classified with respect to the nature of the reagent which reacts with a given class of organic compound. For example, additions to unsaturated bonds are divided into those in which the reagent is, respectively, an electrophile, a nucleophile, and a radical or atom. The following are the more important reagents.

(i) Electrophiles

Hydrogen, in the form of a proton-donor such as H_3O^+ or a carboxylic acid
Halogens: chlorine, bromine, and iodine
Nitrogen, as the nitronium ion (NO_2^+), the nitrosonium ion (NO^+), 'carriers' of these ions from which the ions are readily abstracted (e.g. alkyl nitrites, RO—NO), and aromatic diazonium ions (ArN_2^+)
Sulphur, as sulphur trioxide, usually derived from sulphuric acid
Oxygen, as hydrogen peroxide and ozone
Carbon, as a carbonium ion (e.g. $(CH_3)_3C^+$)

(ii) Nucleophiles

Oxygen, in H_2O, ROH, RCO_2H, RSO_3H, and their conjugate bases (OH^-, RO^-, etc.)
Sulphur, in H_2S and RSH and their conjugate bases, and bisulphite ion (HSO_3^-)
Nitrogen, in NH_3 and NH_2^-, amines, and amine derivatives (e.g. NH_2OH)

Carbon, in cyanide ion (CN^-), acetylide ions ($HC{\equiv}C^-$ and $RC{\equiv}C^-$), carbanions (e.g. $CH_3COCH_2{}^-$), and organometallic compounds (e.g. $CH_3{-}MgI$)

Halide ions (F^-, Cl^-, Br^-, I^-)

Hydrogen, as hydride ion and 'carriers' of this ion (e.g. $LiAlH_4$).

(*iii*) *Radicals and atoms*

Halogen atoms ($\cdot Cl$, $\cdot Br$)

Alkyl and aryl radicals (e.g. $\cdot CH_3$, $Ph\cdot$)

4.3 Addition

(*a*) ELECTROPHILIC ADDITION

(*i*) *Olefins.* The principal features of electrophilic addition to olefins are as follows.

(1) Addition occurs in two stages, with the intermediation of a carbonium ion, e.g.

$$RCH{=}CH_2 + HCl \longrightarrow R\overset{+}{C}H{-}CH_3 + Cl^-$$

$$R\overset{+}{C}H{-}CH_3 + Cl^- \longrightarrow RCHCl{-}CH_3$$

The first step is rate-determining, and this has the following consequences.

First, since the formation of the transition state in the addition of a neutral electrophile to an olefin involves the separation of unlike charges, e.g.

$$CH_2{=}CH_2 + H{-}Cl \longrightarrow \left[\overset{\delta+}{C}H_2 = CH_2 - H - \overset{\delta-}{Cl}\right] \longrightarrow \overset{+}{C}H_2{-}CH_3 + Cl^-$$

transition state

reaction occurs faster in a polar than in a non-polar solvent.

Secondly, since the organic reactant acquires positive charge in passage to the transition state, electron-releasing substituents increase the rate of addition and electron-attracting substituents decrease it.

Thirdly, groups of $+I$ type (alkyl groups) cause the electrophile to add so as to give predominantly the more highly alkyl-substituted of the two possible carbonium ions (Markovnikov's rule), e.g.

$$CH_3{-}CH{=}CH_2 + HCl \longrightarrow CH_3{-}\overset{+}{C}H{-}CH_3 + Cl^- \longrightarrow CH_3{-}CHCl{-}CH_3$$

$$(CH_3)_2C{=}CH{-}CH_3 + HCl \longrightarrow (CH_3)_2\overset{+}{C}{-}CH_2{-}CH_3 + Cl^- \longrightarrow (CH_3)_2CCl{-}CH_2{-}CH_3$$

The underlying principle is that the electron-releasing alkyl groups stabilize carbonium ions more effectively when they are bound directly to the positively charged carbon than when they are further removed from it, so that, for example,

$CH_3—\overset{+}{C}H—CH_3$ (two stabilizing methyl groups) is more stable than $CH_3—$

$\overset{+}{C}H_2—CH_2$ (one stabilizing ethyl group) and is formed the faster by the addition of a proton to propylene.

Conversely, groups of electron-attracting type $(-I$ and/or $-M)$ destabilize carbonium ions more effectively when they are bound directly to the positively charged carbon than when they are further removed from it. Consequently, these groups cause the electrophile to add so as to give the less highly substituted of the two possible carbonium ions, e.g.

$$F_3C—CH=CH_2 + HCl \longrightarrow F_3C—CH_2—\overset{+}{C}H_2 + Cl^- \longrightarrow F_3C—CH_2—CH_2Cl$$

$$O_2N—CH=CH_2 + HCl \longrightarrow O_2N—CH_2—\overset{+}{C}H_2 + {}^{\cdot}Cl^- \longrightarrow O_2N—CH_2—CH_2Cl$$

Halogen substituents appear at first sight to behave anomalously: they retard addition, acting in this sense as electron-attracting groups, but orient in the same way as the electron-releasing alkyl substituents. The reason is that the halogens have opposed inductive $(-I)$ and mesomeric $(+M)$ effects. Consider the addition of a proton to ethylene, giving $CH_3—\overset{+}{C}H_2$, and to vinyl chloride, giving $CH_3—\overset{+}{C}HCl$ or $\overset{+}{C}H_2—CH_2Cl$. The third of these ions, $\overset{+}{C}H_2—CH_2Cl$, is less stable than the ethyl cation because of the $-I$ effect of chlorine. The second, $CH_3—\overset{+}{C}HCl$, is more strongly destabilized by chlorine's $-I$ effect, since the halogen is nearer the positively charged centre, but is stabilized by the mesomeric effect,

$$CH_3—\overset{+}{C}H—\overset{..}{\underset{}{C}}l \leftrightarrow CH_3—CH=\overset{+}{C}l$$

and this factor serves to make $CH_3—\overset{+}{C}HCl$ more stable than $\overset{+}{C}H_2—CH_2Cl$ although it is less stable than $CH_3—\overset{+}{C}H_2$ (i.e. the $-I$ effect of chlorine outweighs its $+M$ effect). Consequently, the ease of formation of the three ions decreases in the order $CH_3—\overset{+}{C}H_2 > CH_3—\overset{+}{C}HCl > \overset{+}{C}H_2—CH_2Cl$, so that vinyl chloride reacts less rapidly than ethylene but orients in the same manner as propylene.

In the other common substituents of $-I, +M$ type, alkoxyl and amino, the $+M$ effect is of greater consequence than the $-I$ effect so that not only is the direction of addition in accordance with Markovnikov's rule but also the rate of addition is enhanced relative to the unsubstituted compound. For example, methyl vinyl ether reacts rapidly with hydrogen chloride:

$$CH_3O—CH=CH_2 + HCl \longrightarrow CH_3O—CHCl—CH_3$$

(2) Addition is stereochemically specific: the *trans* product is formed; e.g. *trans*-2-butene and bromine give *meso*-2,3-dibromobutane. This is apparently because the intermediate carbonium ion is not able to undergo free rotation about the new C—C single bond but is held rigid by interaction with the electrophile which has been added. In the addition of bromine, for example, the carbonium ion is thought to have bromonium-ion character, as represented by the contribution of the canonical structure (*b*):

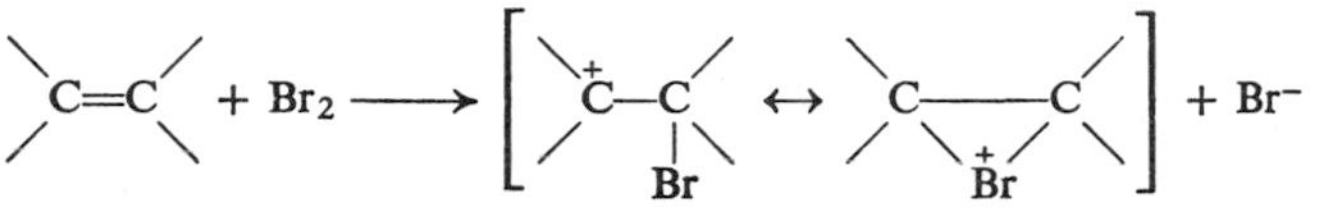

This ion is then attacked by bromide ion from the opposite side to the bromine atom already present (cf. S_N2 reactions, p. 126, and the opening of epoxide rings, p. 571), giving the *trans* product:

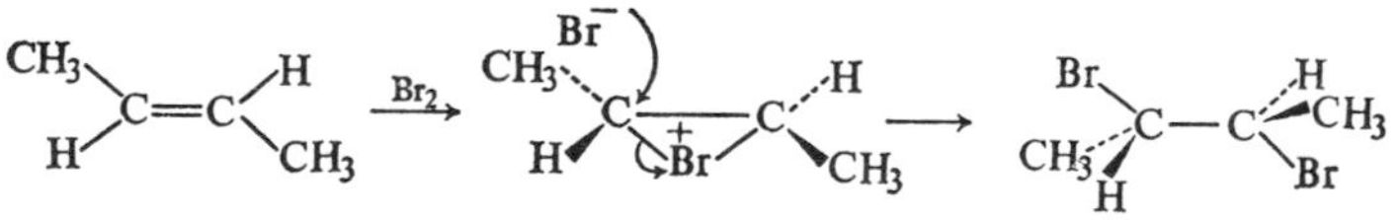

meso-2, 3-Dibromobutane

The stereochemistry of addition is discussed further in a later Chapter (p. 179). For convenience in the present context, however, the carbonium ion is represented as such.

(3) The intermediate carbonium ion may be attacked by any nucleophile which is present. For example, the addition of bromine to ethylene in aqueous solution gives both the dibromide and the bromohydrin, since water is nucleophilic:

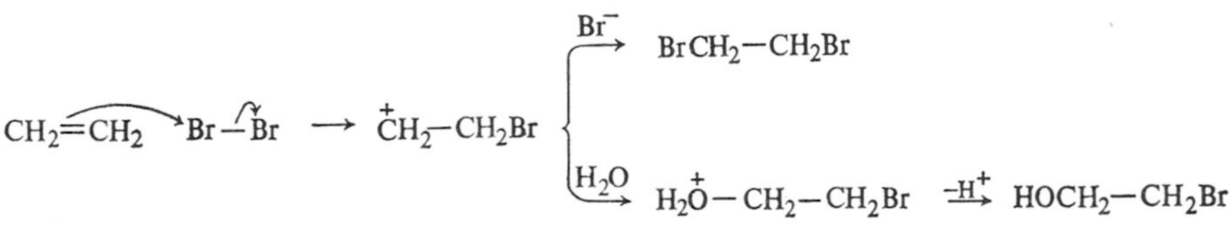

Water can be made the principal nucleophilic species by employing as the electrophile a reagent such as sulphuric acid whose 'nucleophilic half', bisulphate ion, is itself very weakly nucleophilic. This makes it possible to hydrate an olefin with dilute sulphuric acid, e.g.

$$(CH_3)_2C{=}CH_2 + H^+(H_2SO_4) \longrightarrow (CH_3)_3C^+ \xrightarrow{H_2O} (CH_3)_3C{-}\overset{+}{O}H_2 \xrightarrow{-H^+} (CH_3)_3C{-}OH$$

When the concentration of water is sufficiently low, a second molecule of the olefin can act as the nucleophile. The resulting adduct then loses a proton to the water (acting here as a base) to give a mixture of olefins:

$$(CH_3)_2C{=}CH_2 + H^+ \longrightarrow (CH_3)_3C^+ \xrightarrow{\ (CH_3)_2C=CH_2\ }$$

$$(CH_3)_3C{-}CH_2{-}\overset{+}{C}(CH_3)_2 \xrightarrow{\ -H^+\ } (CH_3)_3C{-}CH_2{-}C(CH_3){=}CH_2 + (CH_3)_3C{-}CH{=}C(CH_3)_2$$

When still less water is present, the dimeric carbonium ion adds to a further molecule of the olefin and a trimeric ion is formed. Further such additions lead to a chain polymer (p. 266).

The course of the reaction of vinyl ethers with acids is also changed by the presence of water: thus, whereas methyl vinyl ether reacts with hydrogen chloride to give the adduct, $CH_3{-}CHCl{-}OCH_3$ (p. 108), reaction with hydrochloric acid gives acetaldehyde and methanol. Here, the intermediate carbonium ion reacts with water to give a hemi-acetal which is further hydrolyzed (cf. the hydrolysis of acetals; p. 113).

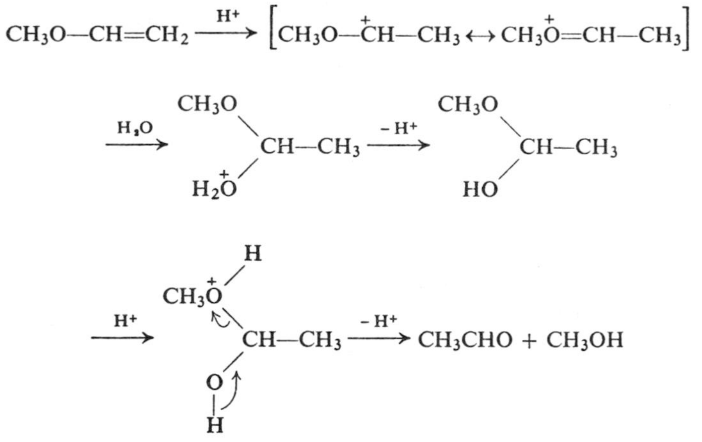

$$CH_3O{-}CH{=}CH_2 \xrightarrow{\ H^+\ } \left[CH_3O{-}\overset{+}{C}H{-}CH_3 \leftrightarrow CH_3\overset{+}{O}{=}CH{-}CH_3 \right]$$

(4) The example of the addition to isobutylene above demonstrates another property of the intermediate carbonium ions: the elimination of a proton competes with the addition of a nucleophile. Elimination is favoured both when addition is sterically hindered, as in the reaction

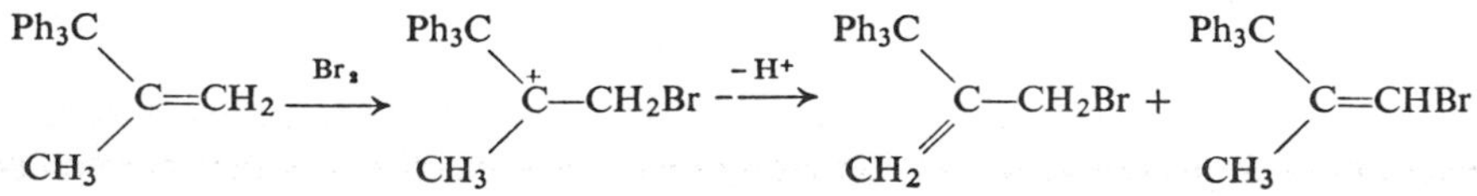

and in addition to enols, e.g.

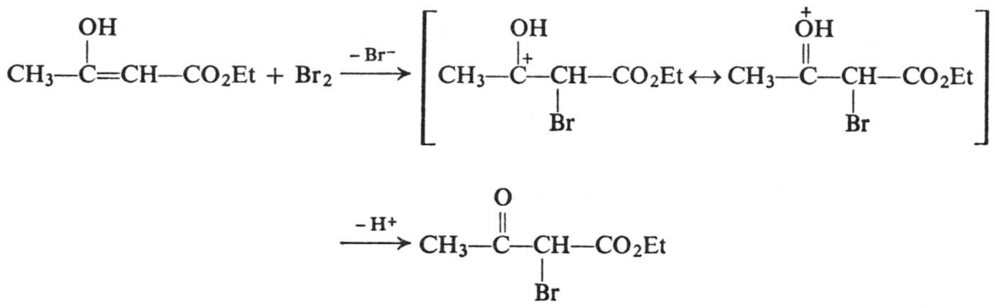

$$CH_3-\underset{\underset{\textstyle |}{OH}}{C}=CH-CO_2Et + Br_2 \xrightarrow{-Br^-} \left[CH_3-\underset{\underset{\textstyle |}{\underset{\textstyle Br}{}}}{\overset{\overset{\textstyle OH}{|}}{C}}-CH-CO_2Et \leftrightarrow CH_3-\underset{\underset{\textstyle |}{\underset{\textstyle Br}{}}}{\overset{\overset{\textstyle \overset{+}{O}H}{\|}}{C}}-CH-CO_2Et \right]$$

$$\xrightarrow{-H^+} CH_3-\underset{\underset{\textstyle |}{\underset{\textstyle Br}{}}}{\overset{\overset{\textstyle O}{\|}}{C}}-CH-CO_2Et$$

(5) Addition to a system containing two or more *conjugated* double bonds gives mixtures of products because the charge in the intermediate carbonium ion is delocalized over two or more carbon atoms each of which can be attacked in the second step, e.g.

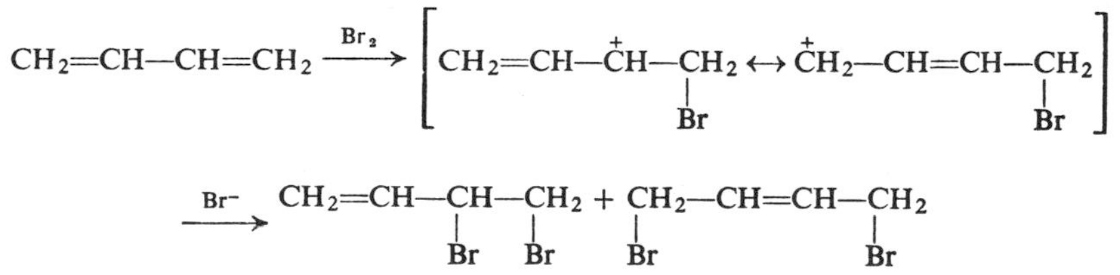

$$CH_2=CH-CH=CH_2 \xrightarrow{Br_2} \left[CH_2=CH-\overset{+}{C}H-\underset{\underset{\textstyle Br}{|}}{C}H_2 \leftrightarrow \overset{+}{C}H_2-CH=CH-\underset{\underset{\textstyle Br}{|}}{C}H_2 \right]$$

$$\xrightarrow{Br^-} \underset{\underset{\textstyle Br}{|}}{C}H_2=CH-\underset{\underset{\textstyle Br}{|}}{C}H-\underset{\underset{\textstyle Br}{|}}{C}H_2 + \underset{\underset{\textstyle Br}{|}}{C}H_2-CH=CH-\underset{\underset{\textstyle Br}{|}}{C}H_2$$

(*ii*) *Acetylenes* are less susceptible than olefins to electrophilic addition (p. 104) but share the same general characteristics: reaction occurs in two stages, involving a carbonium ion intermediate, and gives the *trans* adduct.

(*iii*) *Aromatic* carbon-carbon double bonds normally react with electrophiles by substitution rather than addition because addition would result in the loss of the aromatic stabilization energy. If, however, the product of addition is itself strongly stabilized, addition can compete with substitution. For example, anthracene and bromine give the 9,10-dibromo-adduct which possesses the aromatic stabilization energy of two benzene rings:

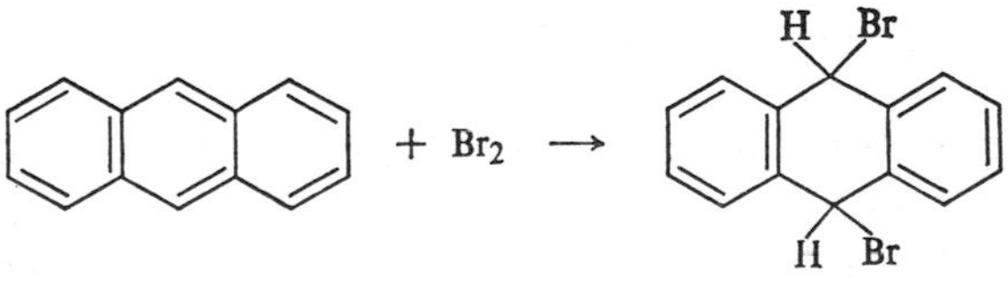

(*iv*) *Carbonyl* groups do not react with electrophiles such as bromine, and their reactions with proton acids are so readily reversible that the products cannot be isolated, e.g.

$$R_2C=O + HCl \rightleftharpoons R_2C \overset{OH}{\underset{Cl}{<}}$$

However, the addition of an acid to the oxygen atom of a carbonyl group promotes the ease of addition of a nucleophile to the carbon atom. These reactions are more conveniently classified as nucleophilic additions (see below).

(b) NUCLEOPHILIC ADDITION

(i) *Carbonyl groups.* The main features of the addition of nucleophiles to the carbonyl groups in aldehydes and ketones are as follows.

(1) The rate-determining step is the addition of the nucleophile, e.g.

$$R_2C=O + CN^- \longrightarrow R_2\overset{\displaystyle |}{\underset{\displaystyle CN}{C}}-O^-$$

$$R_2C=O \longrightarrow R_2\overset{\displaystyle |}{\underset{\displaystyle O=S-OH}{C}}-O^-$$

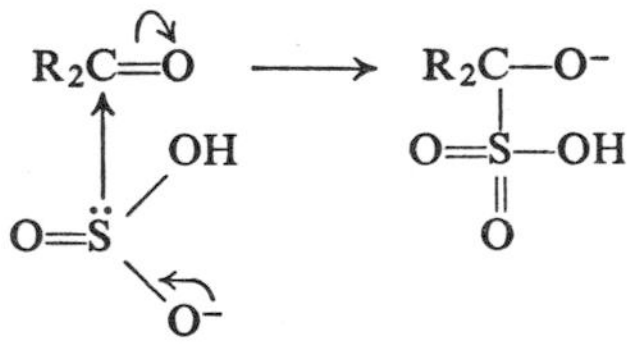

bisulphite ion

Since the carbonyl group acquires a partial negative charge in the transition state of this step, electron-withdrawing groups enhance the rate and electron-releasing groups retard it.

(2) Addition is completed by the uptake of a proton from the solvent. The complete reaction scheme for the addition of hydrogen cyanide to a ketone in aqueous solution is therefore:

$$R_2C=O + CN^- \longrightarrow R_2\overset{\displaystyle |}{\underset{\displaystyle CN}{C}}-O^- \overset{H_2O}{\longrightarrow} R_2\overset{\displaystyle |}{\underset{\displaystyle CN}{C}}-OH + OH^-$$

It is now apparent why this reaction is base-catalyzed: hydrogen cyanide is a rather weak acid and the function of the base is to increase the concentration of the active entity, cyanide ion, the base being regenerated in the second step.

(3) Bulky groups in the vicinity of the carbonyl group retard the addition: e.g. acetone reacts much faster than diethyl ketone with sodium bisulphite, and gives a much higher yield of the adduct (56% and 2%, respectively). The car-

bonyl group in 2,4,6-trimethylbenzaldehyde is so strongly hindered that the compound fails to react with bisulphite.

(4) Aromatic aldehydes and ketones react less rapidly than their aliphatic analogues. This is a result of the fact that, in passage to the transition state, the stabilization due to the conjugation between the carbonyl double bond and the aromatic ring is destroyed. Electron-attracting substituents on the aromatic ring facilitate addition and electron-releasing groups retard it.

(5) Aliphatic $\alpha\beta$-unsaturated carbonyl compounds are similarly less reactive than their saturated analogues. Addition usually occurs instead at the olefinic double bond (p. 253).

(6) Whereas reactive nucleophiles such as cyanide ion add readily to carbonyl, less reactive nucleophiles such as water often react slowly unless the activity of the carbonyl group is increased by hydrogen-bonding between the oxygen atom and an acid, HA, e.g.

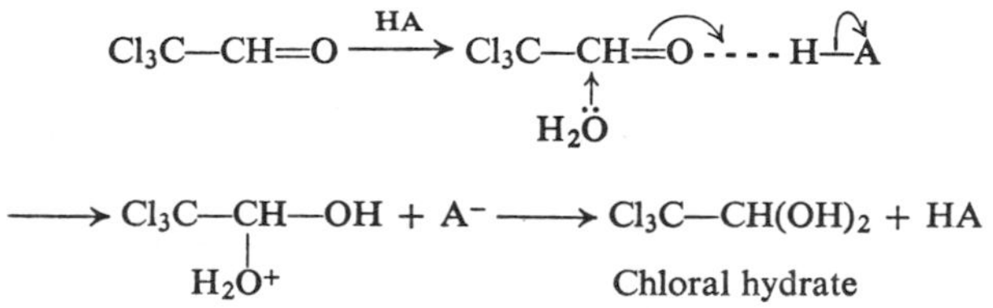

$$Cl_3C-CH=O \xrightarrow{HA} Cl_3C-CH=\overset{\uparrow}{O}\cdots H-A$$
$$\underset{H_2\overset{..}{O}}{}$$

$$\longrightarrow Cl_3C-\underset{\underset{H_2O^+}{|}}{CH}-OH + A^- \longrightarrow Cl_3C-CH(OH)_2 + HA$$
$$\text{Chloral hydrate}$$

Hydrate-formation, as illustrated above for chloral, is readily reversible, and hydrates can only be isolated from compounds which contain strongly electron-withdrawing substituents at the carbonyl group, such as $-CCl_3$ or a second carbonyl group. However, aldehydes react with alcohols in the presence of acid to give isolable acetals, e.g.

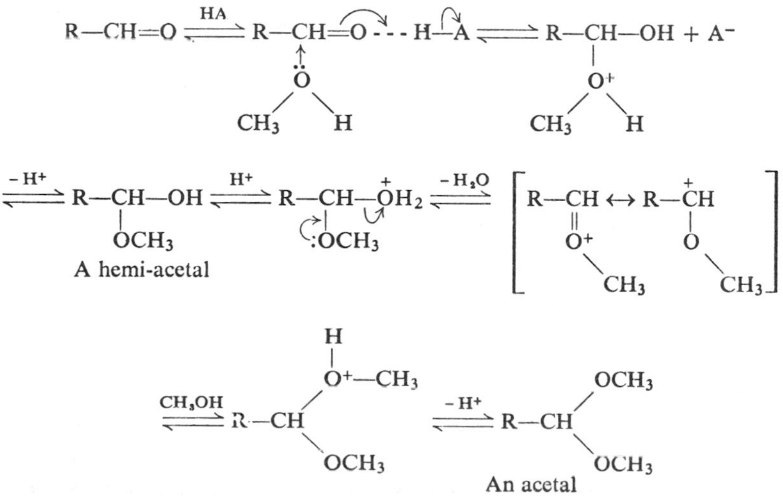

$$R-CH=O \underset{HA}{\rightleftharpoons} R-CH=\overset{\uparrow}{O}\cdots H-A \rightleftharpoons R-\underset{\underset{O^+}{|}}{CH}-OH + A^-$$
$$\underset{\underset{CH_3\quad H}{\diagup\diagdown}}{\overset{..}{O}} \qquad \underset{\underset{CH_3\quad H}{\diagup\diagdown}}{O^+}$$

$$\underset{-H^+}{\rightleftharpoons} R-\underset{\underset{OCH_3}{|}}{CH}-OH \underset{H^+}{\rightleftharpoons} R-\underset{\underset{:OCH_3}{|}}{CH}-\overset{+}{O}H_2 \underset{-H_2O}{\rightleftharpoons} \left[R-\underset{\underset{O^+}{\|}}{CH} \leftrightarrow R-\overset{+}{\underset{\underset{O}{|}}{CH}} \right]$$
$$\text{A hemi-acetal} \qquad\qquad\qquad\qquad \underset{CH_3}{\diagdown}\qquad\underset{CH_3}{\diagdown}$$

$$\underset{CH_3OH}{\rightleftharpoons} R-CH\underset{\underset{OCH_3}{\diagdown}}{\overset{\overset{\overset{H}{|}}{O^+-CH_3}}{\diagup}} \underset{-H^+}{\rightleftharpoons} R-CH\underset{\underset{OCH_3}{\diagdown}}{\overset{OCH_3}{\diagup}}$$
$$\text{An acetal}$$

The intermediate hemi-acetal is not isolable* but undergoes acid-catalyzed elimination of water to give a new cation; here, water acts as a good leaving-group and cation-formation is aided by the electron-releasing ability of the alkoxy-substituent (cf. the ready S_N1 solvolysis of α-halo-ethers; p. 132). It should be noted that the reactions involved in the formation of an acetal are all readily reversible, and the position of the resulting equilibrium depends on the relative concentrations of the alcohol and water. Thus, whereas acetals are formed from aldehydes by treatment with an alcohol in the presence of an *anhydrous* acid, they are readily hydrolyzed to the aldehyde by treatment with *aqueous* acid.

Ketones do not react with monohydric alcohols to give ketals but do so with 1,2-diols to give cyclic ketals, e.g.

$$(CH_3)_2C{=}O + \begin{matrix} HO{-}CH_2 \\ | \\ HO{-}CH_2 \end{matrix} \xrightarrow{\text{H}^+} (CH_3)_2C \begin{matrix} O{-}CH_2 \\ | \\ O{-}CH_2 \end{matrix} + H_2O$$

This difference represents a further example of the importance of the entropy factor in determining the positions of equilibria: whereas the formation of a ketal from a ketone with two molecules of a monohydric alcohol involves the loss of three degrees of translational freedom, reaction of a ketone with a di-hydric alcohol involves no change in the number of degrees of translational freedom.

Both aldehydes and ketones give the thio-analogues of acetals and ketals when treated with mercaptans in the presence of acid:

$$R_2C{=}O + 2\ R'SH \xrightarrow{\text{H}^+} R_2C \begin{matrix} SR' \\ \\ SR' \end{matrix} + H_2O$$

(*ii*) *Other carbonyl groups.* The carbonyl group in derivatives of acids (acid halides, anhydrides, esters, and amides) is also attacked by nucleophiles, but reaction is completed by the departure of an electronegative group and not by the addition of a proton, e.g.

$$R{-}\underset{\underset{Cl}{|}}{C}{=}O + OH^- \longrightarrow R{-}\underset{\underset{Cl}{|}}{\overset{\overset{OH}{|}}{C}}{-}O^- \xrightarrow{-\,Cl^-} R{-}CO_2H \xrightarrow{OH^-} R{-}CO_2^-$$

These reactions are therefore classified as substitutions (cf. 4.5).

---

* See, however, the discussion of ω-hydroxy-aldehydes, which exist predominantly as hemi-acetals in certain cases, p. 23.

(*iii*) *Nitriles.* The carbon-nitrogen triple bond in nitriles is attacked by powerful nucleophiles such as hydroxide ion in water:

$$R-C\equiv N + OH^- \longrightarrow \underset{\underset{OH}{|}}{R-C=N^-} \xrightarrow{H_2O} \underset{\underset{OH}{|}}{R-\ C=NH} \xrightarrow{tautomerizes} \underset{\underset{O}{\|}}{R-C-NH_2}$$

and by weaker nucleophiles such as water and alcohols in the presence of acid:

$$R-C\equiv N + H^+ \;\rightleftharpoons\; \left[R-C\equiv \overset{+}{N}H \leftrightarrow R-\overset{+}{C}=NH \right]$$

(a) $+ H_2O \longrightarrow \underset{\underset{+OH_2}{|}}{R-C=NH} \xrightarrow[\text{(2) tautomerizes}]{\text{(1) }-H^+} \underset{\underset{O}{\|}}{R-C-NH_2}$

(b) $+ R'OH \longrightarrow \underset{\underset{O^+}{|}}{R-C=NH} \longrightarrow \underset{\underset{OR'}{|}}{R-C=\overset{+}{N}H_2}$

                            R'   H   the conjugate
                                            acid of an
                                            imino-ester

(*iv*) *Olefins* are not attacked by nucleophiles unless the carbon-carbon double bond is conjugated to a group of $-M$ type (p. 51). The more powerful nucleophiles such as carbanions then add to the double bond as follows (Michael addition; 7.5):

$$\underset{\underset{R'}{\overset{O}{\|}}}{R-CH=CH-C} \xrightarrow{\ \overline{C}H(CO_2Et)_2\ } \left[\underset{\underset{CH(CO_2Et)_2}{|}}{\underset{\overset{O}{\|}}{R-CH-\overset{-}{C}H-C-R'}} \leftrightarrow \underset{\underset{CH(CO_2Et)_2}{|}}{\underset{\overset{O^-}{|}}{R-CH-CH=C-R'}} \right]$$

$$\xrightarrow{H^+} \underset{\underset{CH(CO_2Et)_2}{|}}{\underset{\overset{O}{\|}}{R-CH-CH_2-C-R'}}$$

$$\underset{\overset{O}{\|}}{PhCH=CH-C-Ph} \xrightarrow{\ KCN-HOAc/EtOH\ } \underset{\underset{CN}{|}}{\underset{\overset{O}{\|}}{PhCH-CH_2-C-Ph}}$$

It is apparent that the increased ease of addition compared with that of an unconjugated olefin arises from the stability given to the intermediate anion, and therefore to the preceding transition state, by the delocalization of the charge onto the electronegative oxygen. Reaction is completed by the uptake of a proton from the solvent, which is usually an alcohol.

Less powerful nucleophiles such as alcohols add to these systems only in the presence of an acid, e.g.

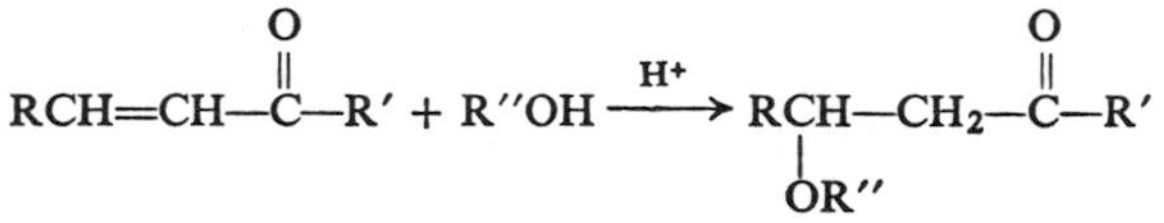

$$\underset{\substack{\parallel \\ O}}{RCH\!=\!CH\!-\!C\!-\!R'} + R''OH \xrightarrow{\ H^+\ } \underset{\substack{| \\ OR''}}{RCH\!-\!CH_2\!-\!\underset{\substack{\parallel \\ O}}{C}\!-\!R'}$$

(*v*) *Acetylenes* react with powerful nucleophiles such as alkoxide ion in an alcoholic solvent:

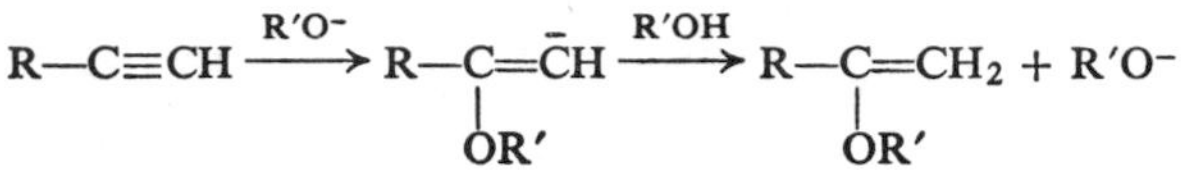

$$R\!-\!C\!\equiv\!CH \xrightarrow{\ R'O^-\ } \underset{\substack{| \\ OR'}}{R\!-\!C\!=\!\bar{C}H} \xrightarrow{\ R'OH\ } \underset{\substack{| \\ OR'}}{R\!-\!C\!=\!CH_2} + R'O^-$$

Less powerful nucleophiles require catalysis, and mercury(II) ion is frequently used because of its tendency to complex with, and draw electrons from, the triple bond. For example, water reacts in the presence of mercury(II) sulphate and dilute sulphuric acid, giving a vinyl alcohol which tautomerizes to a ketone (in the case of acetylene itself, to acetaldehyde):

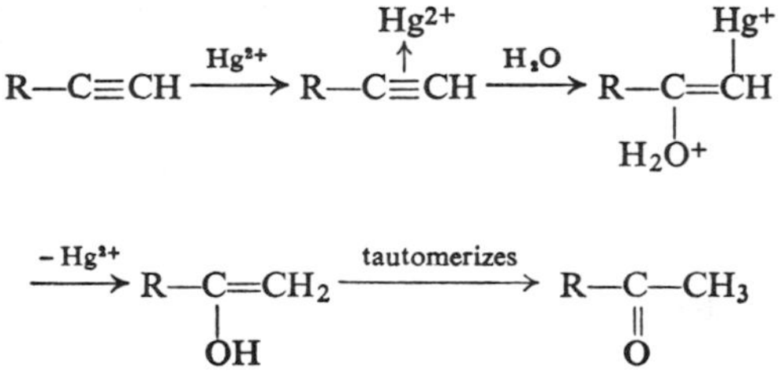

$$R\!-\!C\!\equiv\!CH \xrightarrow{\ Hg^{2+}\ } \overset{Hg^{2+}}{\overset{\uparrow}{R\!-\!C\!\equiv\!CH}} \xrightarrow{\ H_2O\ } \underset{\substack{| \\ H_2O^+}}{\overset{Hg^+}{R\!-\!C\!=\!CH}}$$

$$\xrightarrow{\ -Hg^{2+}\ } \underset{\substack{| \\ OH}}{R\!-\!C\!=\!CH_2} \xrightarrow{\ tautomerizes\ } \underset{\substack{\parallel \\ O}}{R\!-\!C\!-\!CH_3}$$

4.4 Elimination

Elimination reactions are classified under two general headings: (*a*) β-eliminations, in which groups on adjacent atoms are eliminated with the formation of an unsaturated bond,

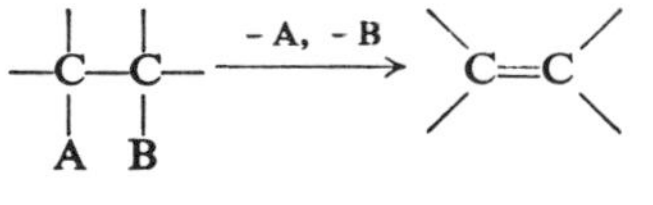

$$\underset{\substack{| \ | \\ A\ B}}{-\!C\!-\!C\!-} \xrightarrow{\ -A,\ -B\ } \overset{\diagdown}{\underset{\diagup}{C}}\!=\!\overset{\diagup}{\underset{\diagdown}{C}}$$

e.g. $CH_3CHBrCH_3 \xrightarrow[-H^+,\ -Br^-]{\ NaOEt\ } CH_3CH\!=\!CH_2$

and (b) α-eliminations, in which two groups are eliminated from the same atom,

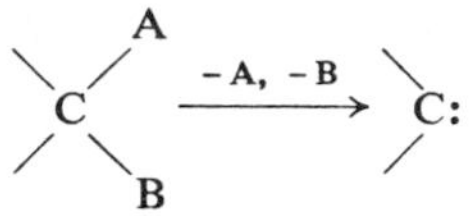

$$\diagdown \!\! \underset{\diagup}{\overset{\overset{\displaystyle A}{|}}{C}} \!\! \diagdown_{B} \xrightarrow{\ -A,\ -B\ } \diagdown C: \diagup$$

$$\text{e.g. } CHCl_3 \xrightarrow[-H^+]{OH^-} CCl_3^- \xrightarrow{\ -Cl^-\ } :CCl_2$$

In the latter case, unstable species are formed which undergo further reactions.

(a) β-ELIMINATION

The unsaturated bond formed by elimination may be any one of a number of types ($C=C$, $C\equiv C$, $C=O$, $C\equiv N$, etc.). We shall consider in detail only those eliminations leading to $C=C$ unsaturation, and in particular the case in which one of the two groups eliminated is hydrogen, e.g.

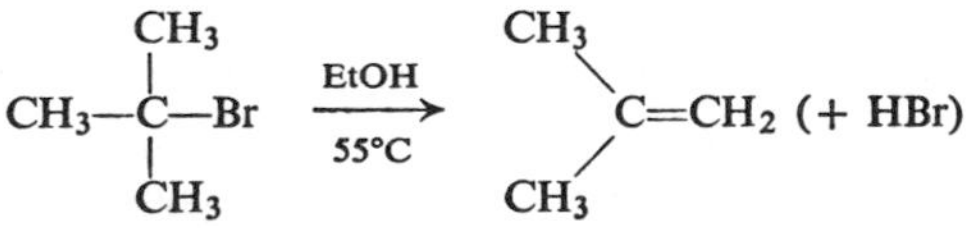

$$CH_3-\underset{\underset{\displaystyle CH_3}{|}}{\overset{\overset{\displaystyle CH_3}{|}}{C}}-Br \xrightarrow[55°C]{EtOH} \underset{CH_3}{\overset{CH_3}{\diagdown}}C=CH_2 \ (+ \ HBr)$$

This group of reactions may be further subdivided into those which occur by a bimolecular mechanism (E2) and those which occur by a unimolecular mechanism (E1).

(i) *The bimolecular process.* Here, elimination is facilitated by attack by a base on the hydrogen atom which is to be removed, e.g.

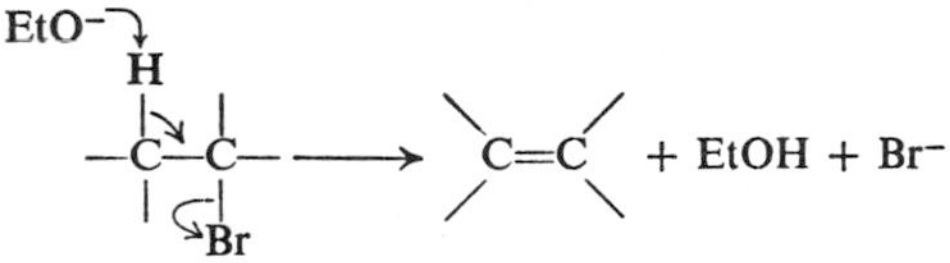

$$EtO^- \!\!\diagdown \quad \underset{\underset{\displaystyle Br}{|}}{\overset{\overset{\displaystyle H}{|}}{-C-C-}} \longrightarrow \diagdown C=C \diagup + EtOH + Br^-$$

The other group eliminated must be one capable of departing as a neutral molecule or a relatively stable anion. In the transition state, the new double bond is partially formed and the groups being eliminated are partially removed:

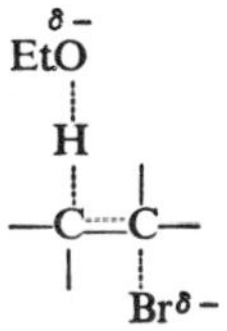

$$\overset{\delta -}{EtO} \\ \vdots \\ H \\ \vdots \\ -\underset{|}{C}\cdots\underset{|}{C}- \\ \quad\ Br^{\delta -}$$

The following are the principal features of the reaction.

1. The reaction rate increases with increasing strength of the base, e.g.

$$CH_3CO_2^- < HO^- < EtO^- < Me_3CO^- < NH_2^-$$

2. The rate increases with increase in the capacity of the second eliminated group to depart with the covalent bonding-pair. For example, an alcohol is stable to elimination by base because of the weak capacity of hydroxyl to be eliminated as hydroxide ion, whereas a sulphonate undergoes elimination readily,

because of the stability of the sulphonate anion (the conjugate base of a strong acid).

There is therefore a general tendency for the ease of departure of a group to be related to the stability of the group as an anion. However, when different elements are being compared, a second factor, the dissociation energy of the C—X bond which is broken on the departure of X, is also concerned. Thus, the ease of the E2 reaction on alkyl halides is —I > —Br > —Cl > —F, for, although the electronegativities of the groups lie in the opposite order, the bond strengths are in the order C—I < C—Br < C—Cl < C—F, and this factor is here dominant, so that iodide is the best leaving-group of the series and fluoride is the worst.

(3) Amongst alkyl groups, the order of reactivity is tertiary > secondary > primary, e.g.

$$(CH_3)_3C\text{—}Br \xrightarrow{\;B:\;} (CH_3)_2C\text{=}CH_2 \text{ faster than}$$

$$(CH_3)_2CH\text{—}Br \xrightarrow{\;B:\;} CH_3CH\text{=}CH_2 \text{ faster than}$$

$$CH_3CH_2\text{—}Br \xrightarrow{\;B:\;} CH_2\text{=}CH_2$$

The reason is that the thermodynamic stability of an olefinic system increases as the C=C double bond is conjugated to an increasing number of alkyl groups (p. 53) and the stabilizing influence of this conjugation in the product of the eliminations is reflected in some degree in the preceding transition state in which the π-bond is partly formed.

(4) Elimination occurs more readily when the new double bond comes into conjugation with existing unsaturated bonds, because the stabilization energy due to the conjugation in the products is partly developed at the transition state.

For example, elimination of hydrogen bromide occurs more readily from $CH_2=CH—CH_2—CH_2Br$, to give $CH_2=CH—CH=CH_2$, than from n-butyl bromide.

An existing unsaturated group of $-M$ type facilitates elimination more strongly when the proton to be eliminated is attached to the α-carbon atom than when it is attached to the β-carbon; for example, base-catalyzed elimination occurs faster from β-bromopropionaldehyde than from α-bromopropionaldehyde:

$$CH_2{-}\overset{H}{\underset{Br}{C}}H—CH=O \longrightarrow CH_2=CH—CH=O \quad \text{faster than}$$

$$\overset{H}{\underset{Br}{C}}H_2{-}CH—CH=O \longrightarrow CH_2=CH—CH=O$$

This is because the $-M$ group stabilizes structures such as;

$$\overset{H—\overset{+}{B}}{\underset{\underset{Br}{|}}{CH_2—\overset{-}{C}H—CH=O}}$$

which contribute to the transition state (B = the base catalyst), thereby lowering the activation energy. The activating effect is sufficient to promote the elimination of water from β-hydroxy carbonyl, nitrile, or nitro compounds, whereas simple alcohols are inert to base-catalyzed elimination. This has synthetic consequences, for several general reactions lead to such β-hydroxy-substituted compounds from which the corresponding $\alpha\beta$-unsaturated compounds may therefore readily be obtained (see especially 7.2). *

(5) The reaction occurs fastest when the two eliminated groups are *trans* to each other and they and the connecting carbon atoms are coplanar. For example, whereas the compound (1) readily eliminates toluene-p-sulphonic acid, its

---

*β-Hydroxy-acids eliminate water on being heated to give $\alpha\beta$-unsaturated acids (e.g., $CH_2OH—CH_2—CO_2H \rightarrow CH_2=CH—CO_2H + H_2O$), whereas the α-hydroxy isomers preferentially form *lactides*, e.g.,

$$2\ CH_3CH\overset{\displaystyle OH}{\underset{\displaystyle CO_2H}{\big\langle}} \xrightarrow{-2H_2O} CH_3—CH\overset{\displaystyle O—CO}{\underset{\displaystyle CO—O}{\big\langle \quad \big\rangle}}CH—CH_3$$

The γ- and δ-hydroxy isomers form *lactones* (p. 90).

isomer (2), in which there is no β-hydrogen *trans* to the toluene-*p*-sulphonate group, is inert even to strong bases (see also p. 189).

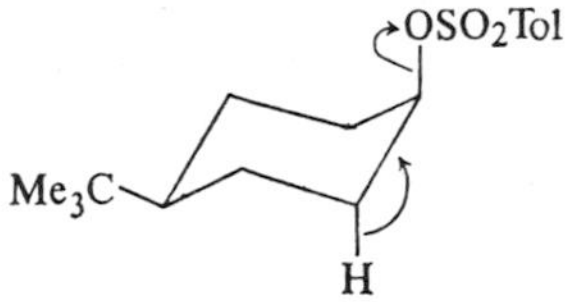

(1)    (2)

(6) A compound which can eliminate in each of two ways to give different olefins normally gives as the major product the more highly conjugated olefin, e.g.

$$CH_3—CH_2—\underset{\underset{Br}{|}}{CH}—CH_3 \xrightarrow[-HBr]{EtO^-/EtOH} CH_3—CH=CH—CH_3 + CH_3—CH_2—CH=CH_2$$

                                                    4 parts   1 part

This generalization is known as the *Saytzeff rule* and has the same underlying principle as described in (3), namely, that the more conjugated of the two olefins is the more stable and this difference is reflected in the rate-determining transition states.

Exceptions to the Saytzeff rule occur in two circumstances. First, when the proton to be removed for Saytzeff elimination is in the sterically more hindered environment, the use of a base whose basic centre is also sterically hindered can lead to the predominance of the less highly conjugated olefin. The following data are illustrative.

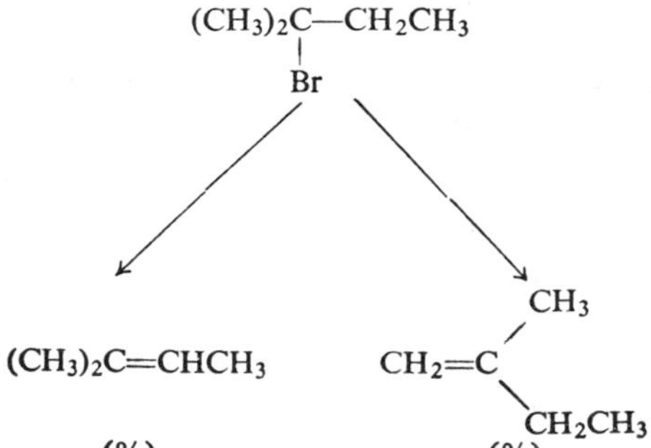

$$(CH_3)_2\underset{\underset{Br}{|}}{C}—CH_2CH_3$$

$(CH_3)_2C=CHCH_3$ $CH_2=C\overset{\diagup CH_3}{\diagdown CH_2CH_3}$

            (%)   (%)

Base	$(CH_3)_2C=CHCH_3$ (%)	$CH_2=C(CH_3)CH_2CH_3$ (%)
EtO$^-$	70	30
Me$_3$CO$^-$	28	72
Et$_3$CO$^-$	12	88

Secondly, elimination from quaternary ammonium ions (Hofmann elimination) usually gives the less highly conjugated olefin (*Hofmann rule*), e.g.

$$CH_3-CH-CH_2-CH_3 \xrightarrow{OH^-} CH_2{=}CH-CH_2-CH_3 + Me_3N + H_2O$$
$$|_{\overset{+}{N}Me_3}$$

The difference between these eliminations and those which follow the Saytzeff rule probably stems from the large size of the quaternary ammonium substituent. Thus, for elimination from the quaternary ion above, the required *trans* coplanar transition states for the Saytzeff direction of elimination are related to the conformations (3) and (4). Because of the large size of the nitrogen substituent, repulsive forces between it and the methyl groups on the adjacent carbon are considerable (the majority of molecules adopt the conformation (5)) and these transition states are of higher energy than those for the Hofmann elimination which are related to the conformation (6). On the other hand, the repulsive forces between halogen substituents and alkyl groups on the adjacent carbon in conformations corresponding to (3) and (4) are much smaller; the steric factor is then subordinated to the stabilizing factor referred to above which favours the development of the transition state leading to the more highly alkylated olefin.

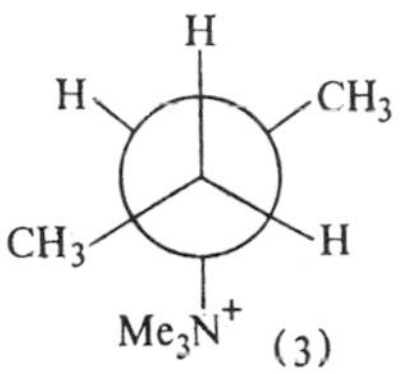

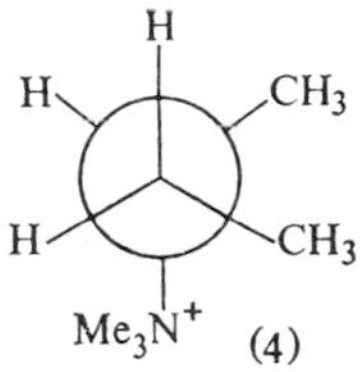

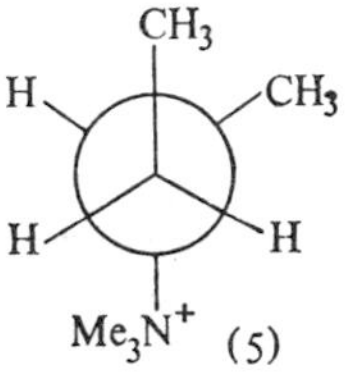

The stereochemistry of E2 eliminations is considered further in Chapter 5.

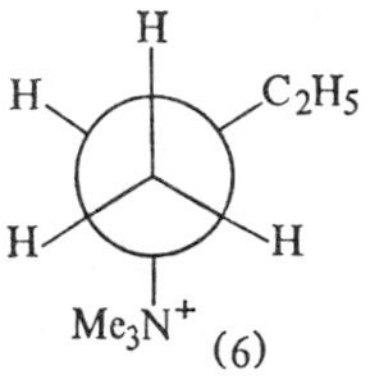

(7) The S_N2 reaction always competes with the E2 reaction. The proportion of elimination is determined by the natures of the base and the alkyl group; the underlying theory is discussed below (4.5).

(8) Elimination does not occur to give 'bridgehead' olefins in which the geometry of the ring-system precludes the close attainment to coplanarity required for *p*-orbital overlap in the olefin (*Bredt's rule*). For example,

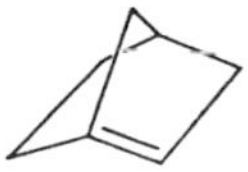

has never been obtained.

(*ii*) *The unimolecular process.* Elimination can also occur without the participation of a base. The reaction occurs in two stages, the first step, unimolecular heterolysis, being rate-determining:

$$
-\overset{\underset{\displaystyle |}{H}}{\underset{\displaystyle |}{C}}-\overset{\underset{\displaystyle |}{}}{\underset{\displaystyle |}{C}}-X \xrightarrow{\text{slow}} -\overset{\underset{\displaystyle |}{H}}{\underset{\displaystyle |}{C}}-\overset{+}{C}\diagup + X^-
$$

$$
-\overset{\underset{\displaystyle |}{H}}{\underset{\displaystyle |}{C}}-\overset{+}{C}\diagup \xrightarrow{\text{fast}} \diagdown C{=}C\diagup + H^+
$$

The characteristics of this process (E1) are as follows.

(1) The order of reactivity of alkyl groups is tertiary > secondary > primary. This is because the rate-determining step is the formation of a carbonium ion, and the stabilities of these ions (and the transition states preceding them) increase in the order primary < secondary < tertiary (p. 83). The rate of elimination from primary alkyl groups is usually negligible.

(2) The effects on the rate of the nature of the leaving group are the same as in the E2 process.

(3) The direction of elimination from the carbonium ion usually follows the Saytzeff rule (p. 120), e.g.

$$
\text{CH}_3-\overset{\underset{\displaystyle \text{CH}_2\text{CH}_3}{|}}{\overset{\displaystyle \overset{\text{CH}_3}{|}}{\text{C}}}-\text{Cl} \longrightarrow \text{CH}_3-\overset{+}{\underset{\diagdown \text{CH}_2\text{CH}_3}{\overset{\diagup \text{CH}_3}{\text{C}}}} \longrightarrow (\text{CH}_3)_2\text{C}{=}\text{CHCH}_3 + \text{CH}_2{=}\underset{\diagdown \text{CH}_2\text{CH}_3}{\overset{\diagup \text{CH}_3}{\text{C}}}
$$

$$
\text{4 parts} \qquad\qquad \text{1 part}
$$

However, when the product of Saytzeff elimination is the more sterically compressed, the less highly conjugated olefin may be formed, e.g.

$$
(\text{CH}_3)_3\text{C}-\text{CH}_2-\overset{+}{\underset{\diagdown \text{CH}_3}{\overset{\diagup \text{CH}_3}{\text{C}}}} \longrightarrow (\text{CH}_3)_3\text{C}-\text{CH}_2-\underset{\diagdown \text{CH}_2}{\overset{\diagup \text{CH}_3}{\text{C}}}
$$

because the alternative product,

$$
\underset{\text{CH}_3 \diagup \quad \diagdown \text{C}=\text{C} \diagdown}{\overset{\text{CH}_3 \diagdown \quad \diagup \text{CH}_3}{\text{C}}}\ \underset{\overset{\diagup}{\text{H}} \quad \diagdown \text{CH}_3}{\overset{\diagdown \text{CH}_3 \diagup \text{CH}_3}{}}
$$

possesses significant non-bonding repulsive forces between the hydrogens on the t-butyl group and those on the methyl group *cis* to it.

(4) The carbonium ion may not only eliminate but may also add a nucleophilic species to give a substitution product (S_N1 reaction). The E1 and S_N1 reactions are therefore competitive (4.5). Further, the carbonium ion may rearrange (4.7).

(*iii*) *Elimination to give C≡C bonds.* Base-catalyzed elimination from olefinic derivatives,

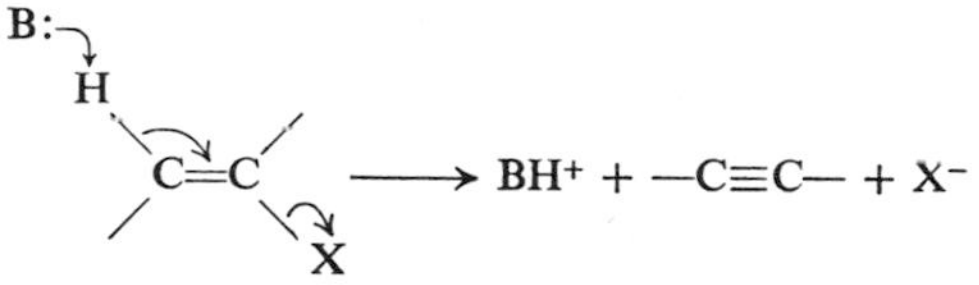

$$\longrightarrow BH^+ + -C{\equiv}C- + X^-$$

is similar in principle to the E2 reaction for the formation of olefins. Strong bases are necessary, and amide ion is commonly used.

The product of elimination is usually a terminal acetylene, e.g.

$$RC{=}CHCH_3 \xrightarrow{NH_2^-/NH_3} RCH_2C{\equiv}C^- \xrightarrow{H_2O} RCH_2C{\equiv}CH$$
$$\underset{Br}{|}$$

This is because the first product formed undergoes a series of prototropic shifts, catalyzed by the strong base:

$$RC{=}CHCH_3 \xrightarrow{-HBr} RC{\equiv}C-CH_3 \overset{NH_2^-}{\rightleftharpoons} [RC{\equiv}C-\bar{C}H_2 \leftrightarrow R\bar{C}{=}C{=}CH_2]$$
$$\underset{Br}{|}$$

$$\overset{NH_3}{\rightleftharpoons} RCH{=}C{=}CH_2 \overset{NH_2^-}{\rightleftharpoons} [RCH{=}C{=}\bar{C}H \leftrightarrow R\bar{C}H-C{\equiv}CH]$$

$$\overset{NH_3}{\rightleftharpoons} RCH_2-C{\equiv}CH \overset{NH_2^-}{\rightleftharpoons} RCH_2-C{\equiv}C^-$$

The thermodynamic driving force is the stability of the acetylenic anion (p. 68) which, on the addition of water, gives the terminal acetylene.

(*iv*) *Elimination to give C=O bonds.* Base-catalyzed elimination can give carbonyl compounds, as in the case of nitrate esters:

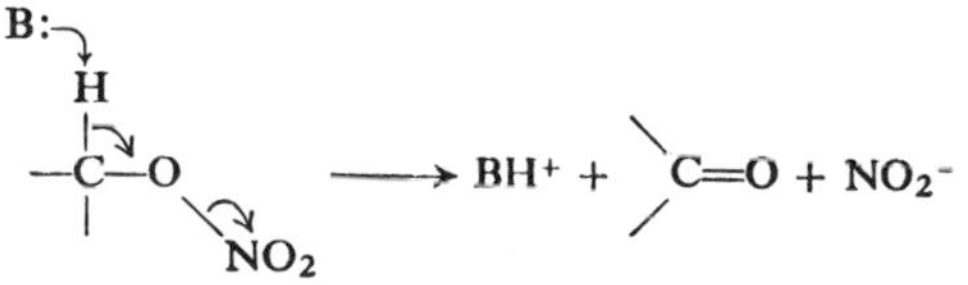

$$\longrightarrow BH^+ + {>}C{=}O + NO_2^-$$

(*v*) *Elimination to give C≡N bonds.* Derivatives of aldoximes undergo base-catalyzed elimination to give nitriles:

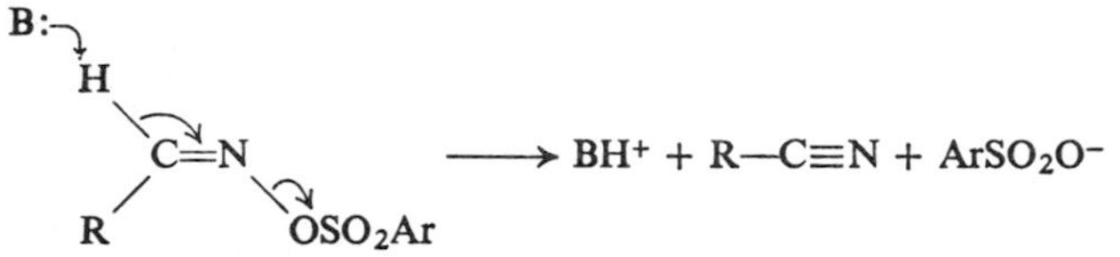

$$\text{B:} \quad \overset{H}{\underset{R}{\underset{|}{C}}} \text{=N} \quad \overset{}{\underset{OSO_2Ar}{}} \longrightarrow BH^+ + R—C≡N + ArSO_2O^-$$

As in the E2 reaction, the rate depends on the capacity of the leaving group to depart with the bonding-pair (toluene-*p*-sulphonates are therefore suitable derivatives for elimination) and a *trans* geometry is required.

(*b*) α-ELIMINATION

The haloforms undergo base-catalyzed α-elimination with the formation of *carbenes*. The reaction normally occurs in two stages, the second of which is rate-determining, e.g.

$$CHCl_3 + HO^- \rightleftharpoons CCl_3{}^- + H_2O$$

$$CCl_3{}^- \longrightarrow :CCl_2 + Cl^-$$

The carbanion intermediates owe their stability in part to the ability of all the halogens except fluorine to accommodate the negative charge in *d* orbitals, e.g.

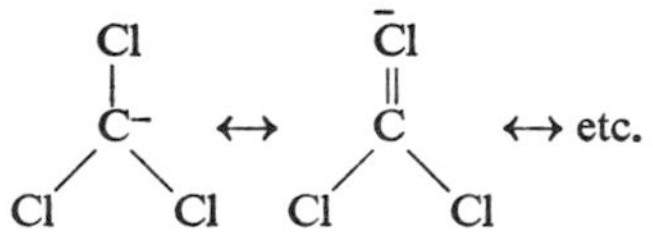

$$\underset{Cl}{\overset{Cl}{\underset{|}{\underset{Cl}{C^-}}}} \leftrightarrow \underset{Cl}{\overset{\bar{C}l}{\underset{||}{\underset{Cl}{C}}}} \leftrightarrow \text{etc.}$$

The resulting order of reactivity in the first step is $CHI_3 > CHBr_3 > CHCl_3 \gg CHF_3$.

The stability of the carbenes is determined by the ability of the halogens to supply *p*-electrons to the electron-deficient carbon, e.g.

$$F—C—F \leftrightarrow F—\bar{C}=\overset{+}{F} \leftrightarrow \overset{+}{F}=\bar{C}—F$$

Since the $+M$ effects of the halogens lie in the order $F > Cl > Br > I$, the ease of the second step in carbene-formation is greatest when the haloform contains fluorine atoms. Indeed, the three mixed haloforms, CHF_2Cl, CHF_2Br, and CHF_2I, undergo *concerted* elimination to carbenes,

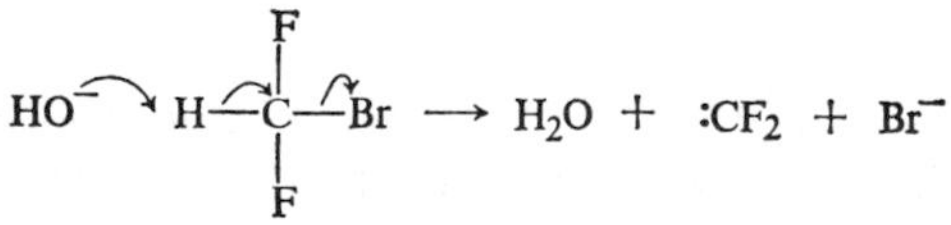

$$HO^- \quad H—\overset{F}{\underset{F}{\underset{|}{C}}}—Br \longrightarrow H_2O + :CF_2 + Br^-$$

Dihalocarbenes are reactive species which cannot be isolated. In the normal conditions for their formation (using hydroxide ion in a hydroxylic medium) they are hydrolyzed to acids,

$$HO^- + :CCl_2 \longrightarrow HO-\bar{C}Cl_2 \xrightarrow{H_2O} HO-CHCl_2 \xrightarrow{hydrolysis} HCO_2H \xrightarrow{OH^-} HCO_2^-$$

The addition to the reaction medium of a more strongly nucleophilic species than hydroxide ion (e.g. RS^-) diverts the carbene from hydrolysis. The addition of an olefin leads to cyclopropane derivatives whose formation occurs in a stereo-specific manner (p. 181), e.g.

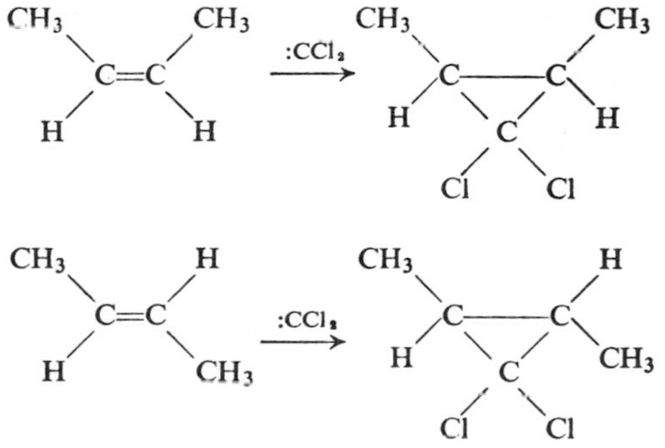

Carbenes also react with phenoxide ions (Reimer-Tiemann reaction; p. 398) and with primary amines (carbylamine reaction; p. 328).

Methylene itself, $:CH_2$, is of higher energy-content than the dihalocarbenes and is correspondingly more difficult to obtain, more reactive, and less selective. It may be prepared by ultraviolet-irradiation of diazomethane:

$$CH_2N_2 \xrightarrow{h\nu} :CH_2 + N_2$$

Its reactivity is illustrated by its reactions with cyclohexene, which include both addition to the double bond and insertion in the saturated and unsaturated C—H bonds,

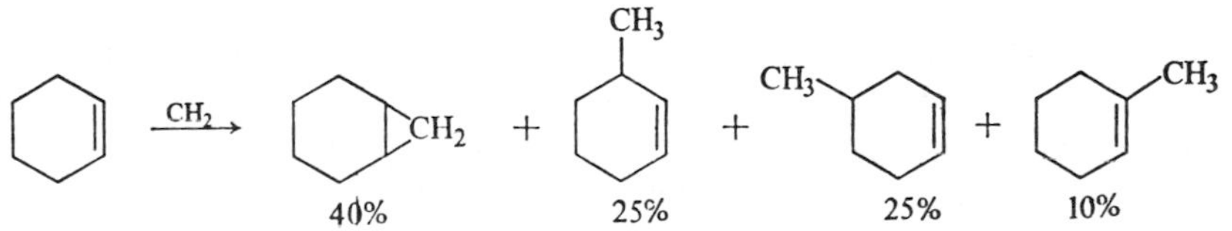

4.5 Substitution

One group attached to carbon may be replaced by another group in one of three ways: (a) by synchronous substitution, (b) by elimination followed by addition, and (c) by addition followed by elimination. The last applies only to unsaturated carbon.

(a) SYNCHRONOUS SUBSTITUTION

The reagent may be a nucleophile (S_N2 reaction) or an electrophile (S_E2 reaction). Atoms and radicals do not substitute directly at carbon.

(i) *The S_N2 reaction.* The reaction may be represented in general by

$$\text{Nu:} \quad \overset{\diagdown}{\underset{\diagup}{\text{C}}}\text{--Le} \longrightarrow \text{Nu--}\overset{\diagup}{\underset{\diagdown}{\text{C}}}\text{--} + \text{Le}^-$$

where Nu: is a nucleophile and Le is a leaving group. The following are the principal characteristics of the process.

(1) The relative reactivities of different leaving groups are the same as in the E2 reaction (p. 118) (e.g. $I > Br > Cl > F$). As in that process, hydroxy, alkoxy, and amino groups are not displaced as the corresponding anions, since these are of too high an energy-content, so that alcohols, ethers, and amines are inert to nucleophiles (but see electrophilic catalysis, below). Sulphates and sulphonates are particularly reactive since the leaving group is in each case the anion of a strong acid. This makes dimethyl sulphate a useful methylating agent for alcohols in basic solution:

$$\text{ROH} + \text{OH}^- \rightleftharpoons \text{RO}^- + \text{H}_2\text{O}$$

$$\text{RO}^- + \text{CH}_3\text{--OSO}_2\text{OCH}_3 \longrightarrow \text{RO--CH}_3 + \bar{\text{O}}\text{SO}_2\text{OCH}_3$$

(2) The carbon atom at which substitution occurs undergoes inversion of its configuration (Walden inversion; p. 175) because the nucleophile approaches along a line diametrically opposite the bond to the leaving group, e.g.

$$\text{HO}^- \text{-----} \overset{\diagdown}{\underset{\diagup}{\text{C}}}\text{--Br} \longrightarrow \text{HO--}\overset{\diagup}{\underset{\diagdown}{\text{C}}}\text{<} + \text{Br}^-$$

(3) The nucleophiles may be any of those listed on p. 106. Typical examples of reactions on alkyl halides, RX, are:

Nucleophile	Product
OH^-	Alcohols, R—OH
$R'O^-$	Ethers, R—OR'
$R'S^-$	Thioethers (sulphides), R—SR'
$R'CO_2^-$	Esters, R—OCOR'

Nucleophile	Product
$R'C\equiv C^-$	Acetylenes, $R-C\equiv CR'$
CN^-	Nitriles, $R-CN$
NH_3	Amines, $R-NH_2$
R'_3N	Quaternary ammonium salts, $R'_3RN^+X^-$

A nucleophile is also a base, so that elimination (E2) competes with substitution. Just as in the E2 reaction the rate of elimination increases with the strength of the base, so in the S_N2 reaction the rate increases with the power of the nucleophile. Since a species is both base and nucleophile by virtue of its possessing the same characteristic—an unshared electron-pair—it might be expected that nucleophilic power should parallel basic strength. This is true, however, only in two general circumstances: first, when a common nucleophilic atom is involved (e.g. anions such as HO^- and RO^- are both stronger bases and more powerful nucleophiles than their conjugate acids, H_2O and ROH); and secondly, when elements in the same *period* are involved (e.g. ammonia is both a stronger base and a more powerful nucleophile than water). It is not necessarily true in other circumstances, particularly when elements in the same *group* are compared: for example, a thiol anion, RS^-, although less basic than its oxygen analogue, RO^-, is more strongly nucleophilic, so that the S_N2/E2 ratio is higher for reactions with thiol anions than for those with alkoxide anions.

The main reason for this reversal of basicity and nucleophilicity can be understood by reference to the factors which determine the relative strengths of acids in a given medium, namely, the dissociation energy of the bond which breaks on ionization, and the electron affinity of the group from which the proton is lost. For thiols and alcohols, these factors are in opposition (S—H bonds (340 kJ mol^{-1}) are weaker than O—H bonds (462 kJ mol^{-1}), but RS· is less electronegative than RO·). However, the very large difference in bond energies dominates the situation, with the result that thiols are stronger acids than the analogous alcohols (i.e. thiol anions are weaker bases than alkoxide anions). Nucleophilicity is also determined, in part,[*] by the strength of the bond formed by the nucleophile, in this case with a carbon atom, and by the ease with which the nucleophile can release an electron to form a bonding pair. For thiols and alcohols, these factors are again in opposition, but the bond-strength factor (C—S, 272; C—O, 357 kJ mol^{-1}) is in this instance of less importance than the electronegativity factor, with the result that RS^- is more strongly nucleophilic than RO^-

At least one other factor is of importance in determining nucleophilicity. A

---

[*]It must be remembered that nucleophilicity refers to a *kinetic* phenomenon (the rate at which a species reacts at carbon) whereas basicity refers to an *equilibrium*. In particular, the processes discussed here have not occurred completely at the transition state for the substitution but are completed in the equilibrium.

species whose nucleophilic centre is attached to an atom possessing an unshared pair of electrons (e.g. NH_2—NH_2) is more strongly nucleophilic than predicted by consideration of its basicity. The reason is not clear.

(4) The order of reactivity of alkyl groups is primary > secondary > tertiary, at least in part because of the increased steric hindrance to the approach of the reagent as the carbon atom is more heavily substituted. As an example, the relative reactivities of the following alkyl bromides towards iodide ion in acetone are:

CH_3—Br	CH_3CH_2—Br	$(CH_3)_2CH$—Br	$(CH_3)_3C$—Br
145	1	8×10^{-3}	5×10^{-4}

The order of reactivities in the E2 reaction is the opposite of this (p. 118), so that it follows that the S_N2/E2 ratio is greatest for a primary halide and least for a tertiary halide. The following results for the reactions of alkyl bromides with ethoxide ion in ethanol at 55°C are typical:

$$CH_3—CH_2—Br \longrightarrow CH_3—CH_2—OEt + CH_2{=}CH_2$$
$$ 90\% 10\%$$

$$\begin{matrix} CH_3 \\ \diagdown \\ CH—Br \longrightarrow (CH_3)_2CH—OEt + CH_3—CH{=}CH_2 \\ \diagup \\ CH_3 21\% 79\% \end{matrix}$$

$$\begin{matrix} CH_3 \\ | \\ CH_3—C—Br \longrightarrow (CH_3)_2C{=}CH_2 \\ | \\ CH_3 100\% \end{matrix}$$

It should be noted that tertiary halides rarely give significant yields in S_N2 reactions: e.g. t-butyl cyanide cannot be obtained from t-butyl chloride and cyanide ion; the product is isobutylene.

(5) Steric hindrance is also pronounced when the β-carbon atom is increasingly heavily substituted by alkyl groups. For the reaction with iodide ion in acetone, relative reactivities of alkyl bromides are:

CH_3—CH_2—Br	CH_3—CH_2—CH_2—Br	$(CH_3)_2CH$—CH_2—Br	$(CH_3)_3C$—CH_2—Br
1	0·8	4×10^{-2}	10^{-5}

Inspection of models shows that, for neopentyl bromide, approach of the nucleophile along the line of the C—Br bond is inevitably impeded by a methyl group, whatever geometry is established by rotation about the single bonds.

$$
\begin{array}{c}
CH_3 \\
CH_3\!\!\diagdown\!\!\underset{\displaystyle |}{C} \\
CH_3\!\!\diagup \\
HO\text{-----------}\overset{\displaystyle}{C}\!-\!Br \\
\underset{\displaystyle H}{H}
\end{array}
$$

(6) The reaction rate can be increased by the presence of electron-rich substituents which are stereochemically suited to interact with the carbon atom undergoing substitution. For example, the β-chloro-sulphide, $C_2H_5SCH_2CH_2Cl$, is hydrolyzed in aqueous dioxan 10,000-times faster than its ether analogue, $C_2H_5OCH_2CH_2Cl$. This has been ascribed to the participation of the sulphur atom:

$$
\begin{array}{ccc}
\begin{array}{c}
C_2H_5\diagdown \\
\quad S\!:\!\diagdown \\
\diagup \\
CH_2\!\!-\!\!CH_2 \\
\qquad \diagdown \\
\qquad Cl
\end{array}
&\longrightarrow&
\begin{array}{c}
C_2H_5\diagdown \\
\quad S^{+} \\
\diagup \\
CH_2\!\!-\!\!CH_2 \\
\qquad \diagdown \\
\qquad :OH_2
\end{array}
\xrightarrow{-H^+}
\begin{array}{c}
C_2H_5\diagdown \\
\quad S \\
\diagup \\
CH_2\!\!-\!\!CH_2OH
\end{array}
\end{array}
$$

Instead of there being a relatively slow one-step reaction, two relatively rapid reactions occur. In the first, sulphur acts as an internal nucleophile. This step is favoured as compared with the intermolecular reaction with water because of the much more favourable entropy of activation for the intramolecular process (p. 90); and it is favoured as compared with the corresponding reaction of the oxygen analogue because sulphur is a more powerful nucleophile than oxygen (p. 127). In the second step, the three-membered ring is opened by attack by water.

Other atoms and groups which enhance S_N2 rates in this way include nitrogen in amines, oxygen in carboxylate and alkoxide ions, and aromatic rings (p. 140). Participation is only effective when the interaction involves three-, five-, and six-membered rings (cf. p. 91).

Participation by the oxygen in carboxylates and alkoxides and by nitrogen in primary or secondary amines can lead to cyclic compounds if these are stable in the reaction conditions. For example, ethylene chlorohydrin with alkali gives ethylene oxide,

$$
\begin{array}{ccc}
\begin{array}{c}
CH_2\!\!-\!\!CH_2Cl \\
| \\
OH
\end{array}
\underset{\displaystyle}{\overset{\displaystyle OH^-}{\rightleftharpoons}}
&
\begin{array}{c}
\quad\quad Cl \\
CH_2\!\!-\!\!CH_2 \\
| \\
O^-
\end{array}
\xrightarrow{-Cl^-}
&
\begin{array}{c}
CH_2\!\!-\!\!CH_2 \\
\diagdown\!\!O\!\!\diagup
\end{array}
\end{array}
$$

and pentamethylenediamine hydrochloride gives piperidine on being heated,

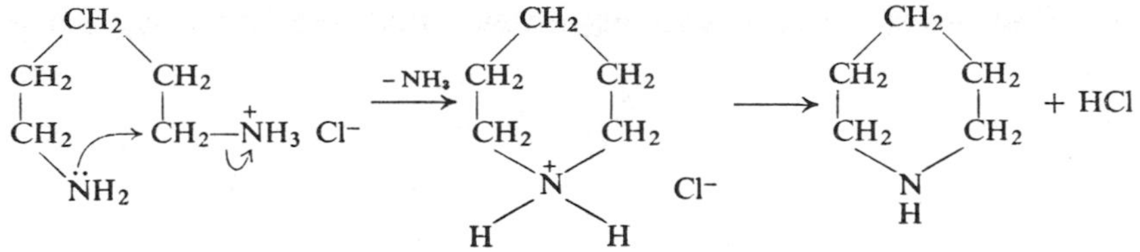

Neighbouring-group participation can also lead to rearrangement (Chapter 14).

(7) Reaction rates are increased by electrophilic catalysis (*cf.* addition to carbonyl groups, p. 113). For example, although alcohols are inert to chloride ion, they react with hydrogen chloride,

$$R-OH \underset{}{\overset{H^+}{\rightleftharpoons}} R-\overset{+}{O}H_2 \xrightarrow{Cl^-} R-Cl + H_2O$$

since water is a far better leaving group than hydroxide ion. Ethers may be cleaved by hydrogen iodide, reaction occurring on the conjugate acid of the ether,

Hydrogen iodide is more effective in this respect than the other halogen acids because iodide is the most strongly nucleophilic of the halide ions.

The reactivity of alkyl halides is increased by Lewis acids and cations such as Ag^+ and Hg^{2+} with which the halide ions form strong bonds.

(8) Conjugated (allylic) systems can undergo substitution with rearrangement (S_N2' reaction) in addition to the S_N2 reaction:

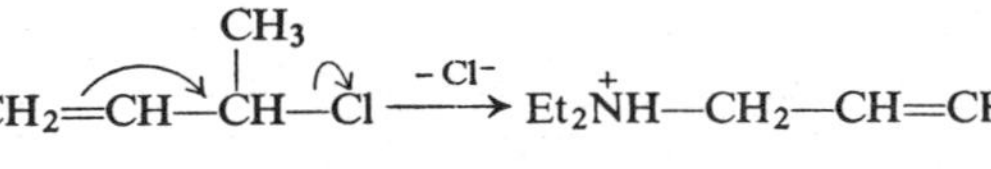

The S_N2' reaction normally occurs only when the S_N2 process is sterically hindered. For example, α-methylallyl chloride reacts with diethylamine to give a crotylamine derivative:

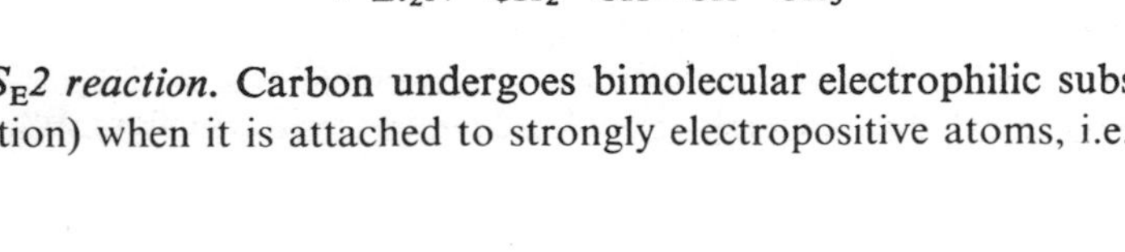

(*ii*) *The S_E2 reaction.* Carbon undergoes bimolecular electrophilic substitution (S_E2 reaction) when it is attached to strongly electropositive atoms, i.e. metals.

A typical reaction is that of an organo-mercurial with bromine,

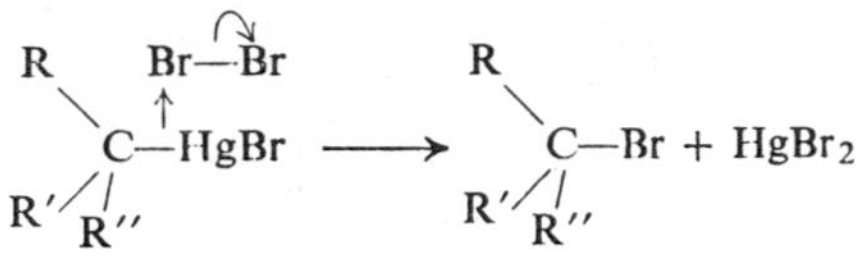

$$
\begin{array}{c}
R \\
\diagdown \\
C\text{---}HgBr \\
\diagup\diagdown \\
R'\ R''
\end{array}
\quad
\begin{array}{c}
Br\text{---}\overset{\frown}{Br} \\
\uparrow
\end{array}
\quad\longrightarrow\quad
\begin{array}{c}
R \\
\diagdown \\
C\text{---}Br \;+\; HgBr_2 \\
\diagup\diagdown \\
R'\ R''
\end{array}
$$

The reaction occurs with *retention* of configuration at carbon, as shown above, in contrast to the *inversion* which is characteristic of the S_N2 reaction.

(*iii*) *The S_Ni reaction.* Alcohols which possess an aromatic substituent react with thionyl chloride to give the corresponding chlorides with retention of the configuration of the hydroxyl-bearing carbon atom, e.g.

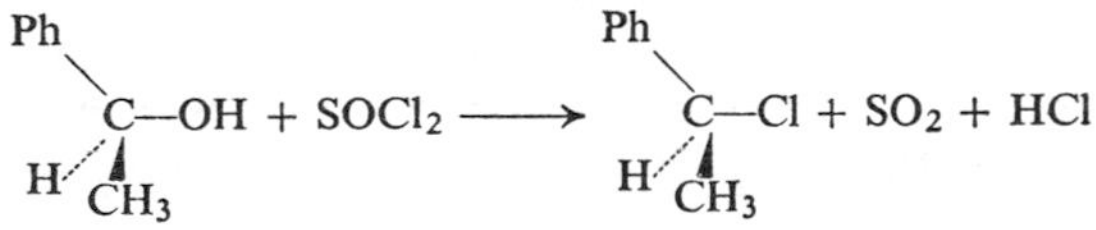

$$
\begin{array}{c}
Ph \\
\diagdown \\
C\text{---}OH \;+\; SOCl_2 \\
\diagup\diagdown \\
H\ CH_3
\end{array}
\quad\longrightarrow\quad
\begin{array}{c}
Ph \\
\diagdown \\
C\text{---}Cl \;+\; SO_2 \;+\; HCl \\
\diagup\diagdown \\
H\ CH_3
\end{array}
$$

This is described as an S_Ni reaction (substitution, nucleophilic, internal). Reaction occurs through the chlorosulphite which collapses, with the elimination of sulphur dioxide, to give an ion-pair and thence the chloride:

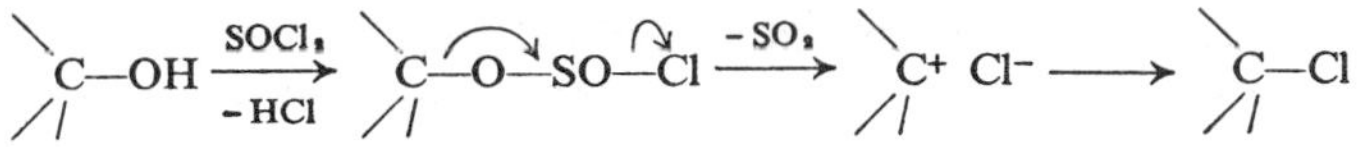

$$
\begin{array}{c}
\diagdown \\
C\text{---}OH \\
\diagup
\end{array}
\xrightarrow[\text{--HCl}]{SOCl_2}
\begin{array}{c}
\diagdown \\
C\text{---}O\text{---}SO\text{---}Cl \\
\diagup
\end{array}
\xrightarrow{-SO_2}
\begin{array}{c}
\diagdown \\
C^+\ Cl^- \\
\diagup
\end{array}
\longrightarrow
\begin{array}{c}
\diagdown \\
C\text{---}Cl \\
\diagup
\end{array}
$$

The role of the aromatic substituent is probably to stabilize the cationic part of the ion-pair by delocalizing the positive charge.

In the presence of pyridine, however, inversion of configuration occurs, probably because the pyridine removes a proton from the hydrogen chloride generated in the first step and the resulting chloride then reacts with the chlorosulphite in the S_N2 manner:

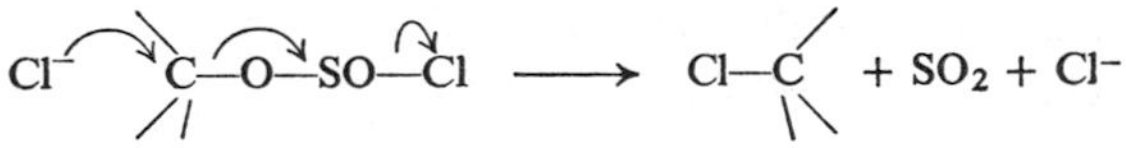

$$
Cl^- \overset{\frown}{}
\begin{array}{c}
\diagdown \\
C\text{---}O\text{---}SO\text{---}Cl \\
\diagup
\end{array}
\longrightarrow
\begin{array}{c}
\diagup \\
Cl\text{---}C \\
\diagdown
\end{array}
\;+\; SO_2 \;+\; Cl^-
$$

The other reagents commonly used for converting alcohols into chlorides (phosphorus trichloride, oxychloride, and pentachloride, and hydrogen chloride) lead to exclusive (S_N2) or predominant (S_N1) inversion of configuration.

(*b*) ELIMINATION FOLLOWED BY ADDITION
When a carbon atom is attached to a group which has a strong capacity for departure with the bonding electron-pair, a unimolecular solvolysis (S_N1

reaction) can occur, e.g.

$$(CH_3)_3C\text{—}Cl \longrightarrow (CH_3)_3C^+ + Cl^-$$

$$(CH_3)_3C^+ + H_2O \longrightarrow (CH_3)_3C\text{—}\overset{+}{O}H_2 \xrightarrow{-H^+} (CH_3)_3C\text{—}OH$$

Thus, the S_N1 process is related to the S_N2 reaction in the same way as E1 elimination is related to E2 elimination. The following are the characteristics of the reaction.

(1) Reaction is facilitated by substituents which stabilize the carbonium ion, i.e. groups of $+I$ and/or $+M$ type. Thus, among alkyl halides the order of reactivity is tertiary > secondary > primary, which is the opposite of that obtaining in the S_N2 reaction. Ether groups typify those of $+M$ type: e.g. methyl chloromethyl ether is rapidly hydrolyzed in water, reaction occurring through a delocalized carbonium-oxonium ion (note that the first product, a hemi-acetal, is further hydrolyzed, p. 110):

$$CH_3O\text{—}CH_2\text{—}Cl \xrightarrow{-Cl^-} [CH_3O\text{—}\overset{+}{C}H_2 \leftrightarrow CH_3\overset{+}{O}\text{=}CH_2]$$

$$\xrightarrow{H_2O} CH_3O\text{—}CH_2\text{—}\overset{+}{O}H_2 \xrightarrow{-H^+} CH_3O\text{—}CH_2\text{—}OH \longrightarrow CH_2\text{=}O + CH_3OH$$

(2) Reaction does not occur at unsaturated carbon centres; thus, ethylenic, acetylenic, and aromatic halides do not react *via* the S_N1 mechanism.

(3) The carbon atom at which substitution occurs does not maintain its configuration. This is a result of the fact that a carbonium ion forms three coplanar bonds and can be attacked from either side, so that an optically active compound gives a mixture of products:

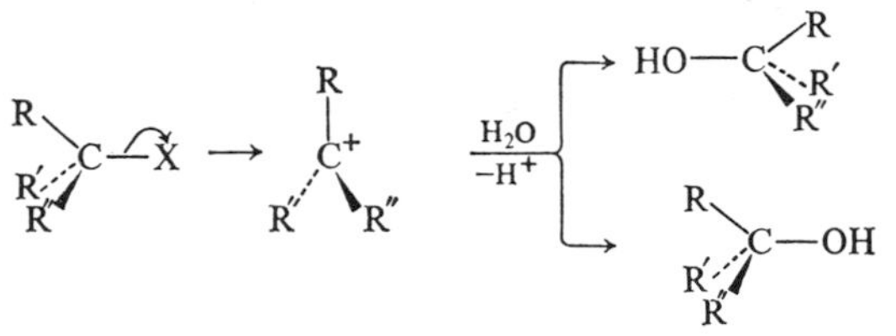

Attack would appear to be equally probable from both sides of the ion, but the two products are not normally formed in equal amounts: the major product has the opposite configuration to that of the reactant (i.e. *inversion* predominates over *retention*). A possible explanation is that the departing group shields the side of the carbon atom at which it was attached so that the incoming nucleophile more easily approaches from the other side.

(4) E1 elimination competes with S_N1 solvolysis, e.g.

$$(CH_3)_3C\text{—}Cl \xrightarrow{H_2O/EtOH} (CH_3)_3C\text{—}OH + (CH_3)_2C\text{=}CH_2$$
$$\phantom{(CH_3)_3C\text{—}Cl \xrightarrow{H_2O/EtOH} (CH_3)_3C} 83\% 17\%$$

Elimination relative to substitution is favoured by more highly alkylated compounds since the olefinic product is then more strongly conjugated to alkyl groups (p. 118). The $E1/S_N1$ ratio is, however, independent of the nature of the leaving group because competition between the two processes occurs only after the carbonium ion has been formed.

(5) Allylic systems give mixtures of products: e.g. the solvolysis of crotyl chloride in aqueous acetone gives both crotyl alcohol and α-methylallyl alcohol:

$$CH_3\text{—}CH\text{=}CH\text{—}CH_2\text{—}Cl \xrightarrow[-HCl]{H_2O} CH_3\text{—}CH\text{=}CH\text{—}CH_2 + CH_3\text{—}CH\text{—}CH\text{=}CH_2$$
$$ \underset{OH}{|} \underset{OH}{|}$$

This is a result of the fact that the intermediate carbonium ion is delocalized and can react with the nucleophile (water) at each of two carbon atoms:

$$CH_3\text{—}CH\text{=}CH\text{—}\overset{+}{C}H_2 \longleftrightarrow CH_3\text{—}\overset{+}{C}H\text{—}CH\text{=}CH_2$$
$$\phantom{CH_3\text{—}CH\text{=}CH\text{—}} \uparrow \phantom{\longleftrightarrow CH_3\text{—}\overset{+}{C}H} \uparrow$$
$$\phantom{CH_3\text{—}CH\text{=}CH\text{—}} H_2\overset{..}{O} \phantom{\longleftrightarrow CH_3\text{—}} H_2\overset{..}{O}$$

(*c*) ADDITION FOLLOWED BY ELIMINATION
Unsaturated carbon in suitable environments undergoes substitution *via* a two-step process consisting of addition followed by elimination. The more important classes of reaction are nucleophilic substitution at carbonyl groups, and nucleophilic, electrophilic, and free-radical substitution at aromatic carbon.

(*i*) *Nucleophilic substitution at carbonyl.* Derivatives of carboxylic acids, in which the carbonyl group is adjacent to an electronegative substituent, are susceptible to substitution *via* the addition-elimination mechanism, e.g.

$$R\text{—}\overset{\overset{O}{\|}}{C}\text{—}Cl + OH^- \xrightarrow{slow} R\text{—}\underset{\underset{OH}{|}}{\overset{\overset{O^-}{|}}{C}}\text{—}Cl \xrightarrow[-Cl^-]{fast} R\text{—}\underset{\underset{OH}{|}}{\overset{\overset{O}{\|}}{C}} \xrightarrow{OH^-} RCO_2^-$$

It should be noted that the carbonyl group in these environments differs from that in aldehydes and ketones only by virtue of the fact that the anionic intermediate derived by addition can eliminate a group as an anion (Cl^-, in the above example) whereas the corresponding anionic adducts derived from aldehydes and ketones cannot (H^- and R^- being of too high an energy-content); the latter intermediates react instead with a proton (p. 112). Further, the reaction of an acid halide with a nucleophile differs from the S_N2 reaction of an alkyl halide only in that the former gives an intermediate whereas the latter cannot. The factors governing the ease of reaction (nature of the group R, the leaving group, and the nucleophile) are essentially the same in the two types of reaction, although acyl derivatives are in general more reactive than the corresponding alkyl derivatives. The following points are notable.

(1) Amongst the commoner acyl derivatives, the order of reactivity is acid halide > anhydride > ester > amide. Acid halides react rapidly with the weak nucleophile, water; anhydrides react slowly; and esters and amides are essentially inert (but see below). The order is the result of two electronic factors which operate in the same sense: first, the $-I$ effects of the halogens, oxygen, and nitrogen decrease in that order, so that the charge on the intermediate

$$\begin{array}{c} O^- \\ | \\ R-C-X \\ | \\ Nu \end{array}$$

is best accommodated when X = halogen and least well when X = NH_2; and secondly, the $+M$ effects of these elements, which stabilize the reactant,

$$\begin{array}{ccc} O & & O^- \\ \| & & | \\ R-C-X & \leftrightarrow & R-C=\overset{+}{X} \end{array}$$

fall in the order N > O > halogen (p. 52), so that the greatest amount of resonance energy is lost in passage to the transition state when X is nitrogen and the least when X is halogen.

(2) Since passage to the transition state involves the uptake of negative charge, the rates of reaction are increased by the presence of electron-attracting substituents and decreased by electron-releasing substituents. For example, ethyl *m*-nitrobenzoate is hydrolyzed by hydroxide ion more than ten times as rapidly as ethyl benzoate (see also p. 87). The polar effect of the carboethoxyl group in diethyl oxalate (EtO_2C-CO_2Et) is great enough to cause this ester to be hydrolyzed even by water, towards which ethyl acetate is inert.

(3) The reactions are reversible. The final position of equilibrium in many cases lies well to one side; for example, in the hydrolysis of an ester by hydroxide ion,

$$R-CO_2Et + OH^- \rightleftharpoons R-CO_2H + EtO^-$$

the carboxylic acid is essentially all removed by ionization in the basic conditions so that reaction goes to completion. In other cases, the equilibrium is fairly closely balanced between reactants and products, as in the reaction of an ester with an alkoxide ion (transesterification):

$$RCO_2R' + R''O^- \rightleftharpoons RCO_2R'' + R'O^-$$

Nevertheless, it is usually possible to force such reactions in the desired direction. For example, in the above transesterification a high yield of the ester RCO_2R'' could be obtained by carrying out the reaction with the alkoxide $R''O^-$ in a considerable excess of the corresponding alcohol, $R''OH$, so that $R'O^-$ is removed through establishment of the equilibrium,

$$R'O^- + R''OH \text{ (excess)} \rightleftharpoons R'OH + R''O^-$$

(4) The reactions are subject to electrophilic catalysis. For example, although carboxylic acids are unreactive to alcohols alone, they can usually be esterified in the presence of a small quantity of concentrated sulphuric acid or about 3% by weight (of the alcohol) of hydrogen chloride:

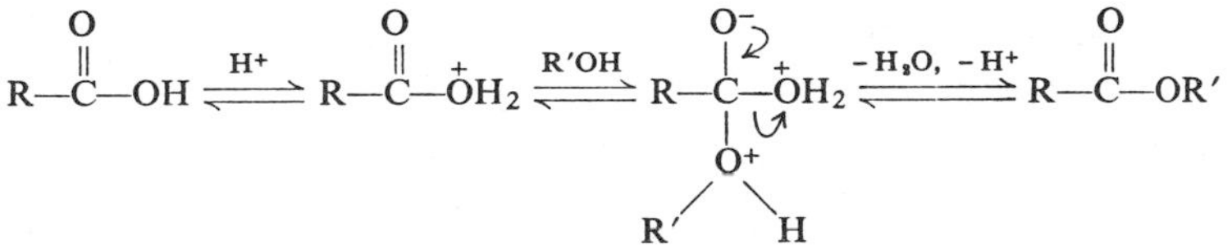

Conversely, esters can be hydrolyzed by aqueous acid. However, this procedure, leading to an equilibrium, is not so efficient as the base-catalyzed method of hydrolysis which goes essentially to completion and it is chosen only when the ester contains another functional group which is susceptible to an unwanted reaction with base.

(5) Carboxylic acids have one property which places them in a separate class from their derivatives: nucleophiles which are of moderate basic power convert them into their conjugate bases, RCO_2^-, which are inert to all but the most powerful nucleophiles. For example, whereas esters react with hydrazine to give acid hydrazides, acids form only salts. Reactions may be effected with nucleophiles such as alcohols by employing an acid catalyst as described above, but nucleophiles which are stronger bases, e.g. ammonia, do not react in acidic conditions because the nucleophile is itself more or less fully protonated and therefore deactivated. The carboxylate ion is attacked by the AlH_4^- ion (p. 639) and the exceptionally powerful nucleophilic organic moiety in organo-lithium compounds (p. 218):

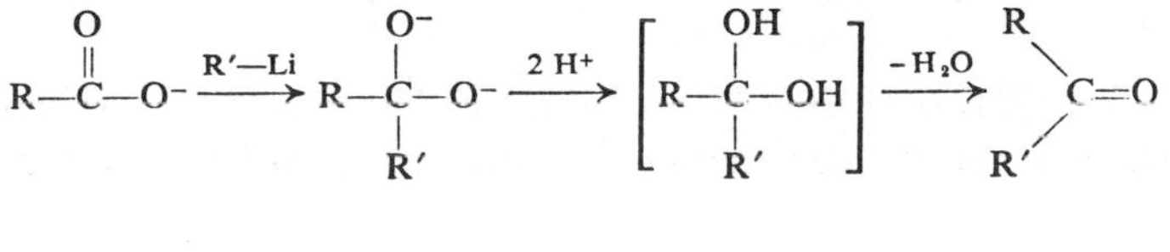

(6) Substitution at carbonyl groups is very markedly subject to steric hindrance. For instance, tertiary acids such as $(CH_3)_3C—CO_2H$, unlike primary and secondary acids, cannot be esterified by an alcohol in the presence of acid, nor can their esters be hydrolyzed by treatment with hydroxide ion or aqueous acid.

(*ii*) *Nucleophilic substitution at aromatic carbon.* Benzene and its halogen-derivatives, like ethylene and vinyl halides, are inert to nucleophiles in normal laboratory conditions. More vigorous conditions lead to substitution, e.g.

$$PhCl \xrightarrow[\text{350°C (pressure)}]{\text{NaOH}} PhOH$$

However, the introduction of groups of $-M$ type, *ortho* or *para* with respect to the leaving group, causes significant increases in rate. For example, *p*-nitro-chlorobenzene is hydrolyzed on boiling with caustic soda, and 2,4,6-trinitro-chlorobenzene (picryl chloride) is hydrolyzed even by water.

The reason is that these reactions occur by addition of the nucleophile to the aromatic ring, giving an anionic intermediate from which elimination occurs:

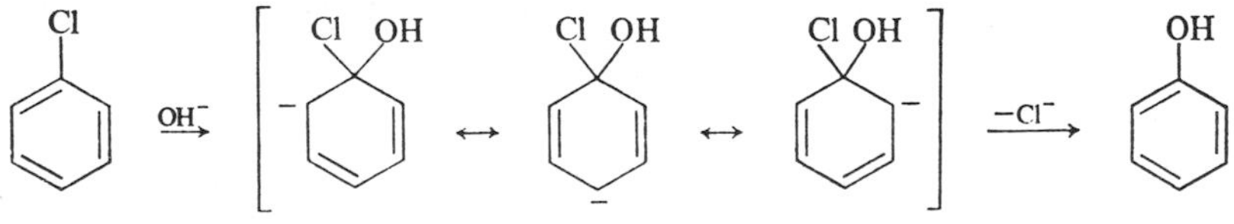

The intermediate is of high energy-content relative to the reactants, for the loss of aromatic stabilization energy is far greater than the delocalization energy of the anion, but the activation energy is reduced by the attachment of $-M$ groups at *ortho* or *para* positions because the charge is then further delocalized over electronegative atoms, e.g.

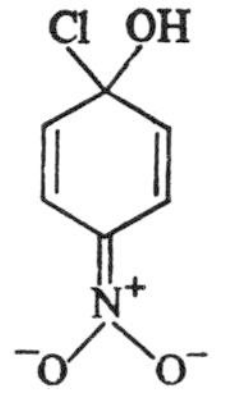

Note that a *meta* substituent can act only through its inductive effect, so that *m*-nitrochlorobenzene is less reactive than its *para* isomer.

An entirely different mechanism for nucleophilic aromatic substitution applies to reactions of the halobenzenes with amide ion in liquid ammonia. Here the power of amide ion as a base take precedence over its nucleophilic property and elimination occurs to give an unstable *benzyne* (which may be regarded as a strained acetylene). The benzyne then reacts with amide ion.

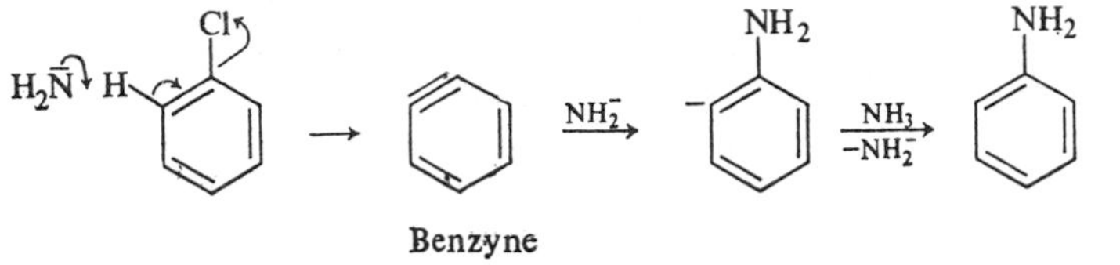

Benzyne

These reactions are discussed in Chapter 12.

(iii) Electrophilic substitution at aromatic carbon. Aromatic compounds react with electrophiles by the addition-elimination mechanism illustrated by the nitration of benzene:

$$HNO_3 + 2\ H_2SO_4 \rightleftharpoons NO_2^+ + H_3O^+ + 2\ HSO_4^-$$

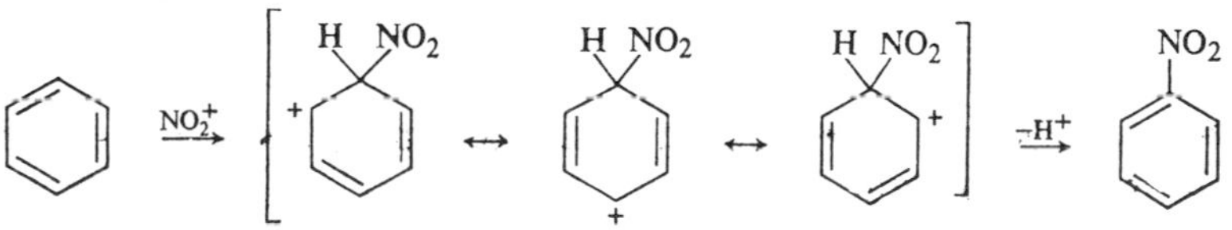

The activation energy is lowered by the presence of an electron-releasing substituent on the benzene ring, the effect being greater when the substituent is *ortho* or *para* to the entering reagent (cf. nucleophilic aromatic substitution, above). Hence substituents of $+I$ and/or $+M$ type direct reagents to the *ortho* and *para* positions and activate these positions relative to those in benzene, e.g.

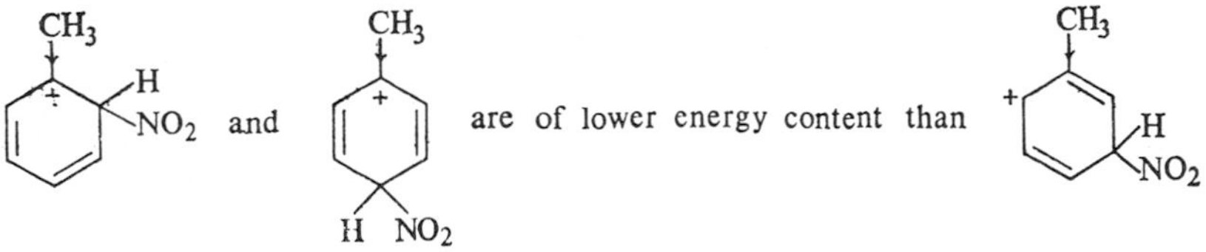

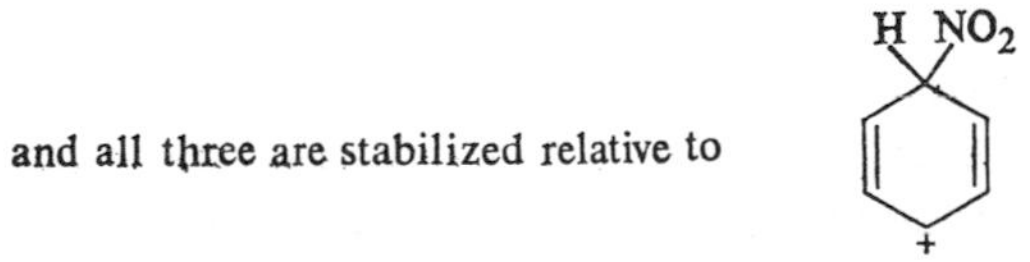

Conversely, electron-attracting substituents, which destabilize the intermediate cation with respect to the reactant (more strongly when the reagent enters the *ortho* or *para* positions than the *meta* position), deactivate the molecule and cause substitution to occur predominantly at the *meta* position, e.g.

$$PhNO_2 \xrightarrow{\text{nitration}} \textit{m}\text{–dinitrobenzene (90\%)} + \text{small proportions of the } \textit{o}\text{– and } \textit{p}\text{–isomers}$$

These reactions are discussed fully in Chapter 11.

(*iv*) *Free-radical substitution at aromatic carbon.* Like electrophiles and nucleo- philes, free radicals and atoms react with aromatic compounds by the addition- elimination mechanism. An example is the formation of biphenyl by the thermal decomposition of dibenzoyl peroxide in benzene:

$$PhCOO\text{—}OCOPh \longrightarrow 2 \; PhCO_2\cdot$$

$$PhCO_2\cdot \longrightarrow Ph\cdot + CO_2$$

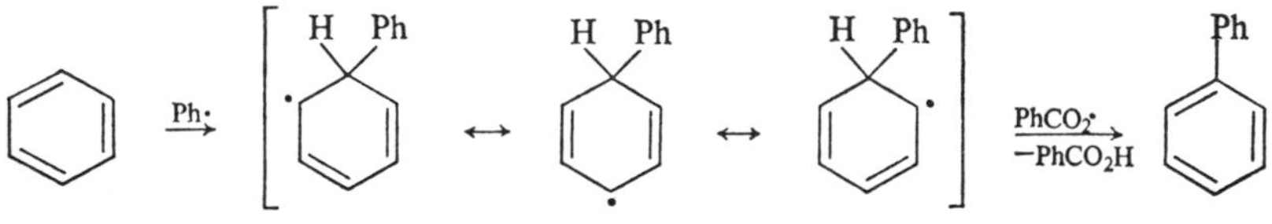

Substituents in the *ortho* or *para* positions are able to increase the delocalization of the intermediate radical, e.g.

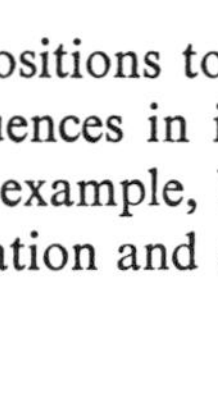

and consequently activate these positions to substitution. However, compared with the powerful stabilizing influences in ionic reactions, the effects in free- radical substitution are small: for example, benzene is about one-million times as reactive as nitrobenzene in nitration and about one-third as reactive in free- radical phenylation.

4.6 Condensation

The term condensation was originally applied to those reactions in which a small molecule such as water or an alcohol is eliminated between two reactants, as in the *Claisen condensation* (p. 238):

$$2 \; CH_3CO_2Et \xrightarrow{(EtO^-)} CH_3COCH_2CO_2Et + EtOH$$

However, since there are many closely related reactions in which a molecule is not necessarily eliminated, such as the *aldol condensation* (p. 227),

$$2 \ CH_3CHO \xrightarrow{\text{(HO}^-)} CH_3CH(OH)CH_2CHO$$

it is convenient to define condensation more widely, and it will here be used to describe reactions in which new carbon-carbon bonds are formed.

It should be emphasized that a condensation is not a reaction of a special mechanistic type but consists of a combination of the reaction types so far described. A key step in the Claisen condensation, for example, is a nucleophilic substitution by a carbanion at the carbonyl group of an ester,

$$\underset{\displaystyle CH_3\overset{\textstyle O}{\overset{\|}{C}}-OEt}{} + \bar{C}H_2CO_2Et \longrightarrow CH_3\overset{\textstyle O}{\overset{\|}{C}}-CH_2CO_2Et + OEt^-$$

and the corresponding step in the aldol condensation is a nucleophilic addition to the carbonyl group of an aldehyde or ketone,

$$CH_3-\overset{\textstyle O}{\overset{\|}{C}}-H + \bar{C}H_2CHO \longrightarrow CH_3-\underset{\displaystyle CH_2CHO}{\overset{\textstyle O^-}{\overset{|}{\underset{|}{C}}}}-H \xrightarrow[-OH^-]{H_2O} CH_3-\underset{}{\overset{\textstyle OH}{\overset{|}{CH}}}-CH_2-CHO$$

4.7 Rearrangement

Rearrangement reactions fall into two categories: (*a*) those in which the migrating group is never fully detached from the system in which it migrates (intramolecular); and (*b*) those in which it becomes completely detached and is later re-attached (intermolecular).

(*a*) INTRAMOLECULAR REARRANGEMENTS

Some of the systems in which migration occurs have been described (p. 104). The principles involved will be discussed with reference to the rearrangements of carbonium ions, for these are of wide synthetic application (Chapter 14).

The simplest example is the solvolysis of neopentyl bromide. In a polar solvent such as ethanol, S_N1-heterolysis occurs to give the neopentyl carbonium ion, a methyl group migrates to give the t-amyl cation, and this undergoes part-elimination and part-substitution:

$$CH_3-\underset{\displaystyle CH_3}{\overset{\textstyle CH_3}{\overset{|}{\underset{|}{C}}}}-CH_2-Br \longrightarrow CH_3-\underset{\displaystyle CH_3}{\overset{\textstyle CH_3}{\overset{|}{\underset{|}{C}}}}-CH_2^+ \longrightarrow CH_3-\underset{\displaystyle CH_3}{\overset{\textstyle CH_3}{\overset{|}{\underset{|}{\overset{+}{C}}}}}-CH_2$$

$$\xrightarrow{\text{EtOH}} (CH_3)_2C{=}CHCH_3 + (CH_3)_2C\underset{\displaystyle OEt}{\overset{}{\overset{|}{-}}}CH_2CH_3$$

The chief characteristics are the following.

(1) The thermodynamic driving-force for the migration is the increased stability of the tertiary carbonium ion compared with the primary carbonium ion. This step is rapid, for it competes successfully with the expected rapid attack of the solvent on the initially formed carbonium ion.

(2) Other reactions leading to carbonium ions which can rearrange to give more stable ions also result in rearrangement. For example, the addition of hydrogen iodide to t-butylethylene gives mainly a rearranged product:

$$
\begin{array}{ccc}
\underset{\substack{| \\ CH_3}}{\overset{\substack{CH_3 \\ |}}{CH_3-C-CH=CH_2}} & \xrightarrow[-I^-]{HI} & \underset{\substack{| \\ CH_3}}{\overset{\substack{CH_3 \\ |}}{CH_3-C-\overset{+}{C}H-CH_3}} & \longrightarrow & \underset{\substack{| \\ CH_3}}{\overset{\substack{CH_3 \\ |}}{CH_3-\overset{+}{C}-CH-CH_3}}
\end{array}
$$

$$
\xrightarrow{I^-} \underset{\substack{| \\ I}}{(CH_3)_2C-CH(CH_3)_2}
$$

Thus, the possibility that rearrangement will occur must always be considered when S_N1 and E1 reactions and electrophilic addition to double bonds are concerned.

(3) In related compounds in which two or three different groups are attached to the β-carbon, that group migrates which is best able to supply electrons to the carbonium ion; e.g. phenyl migrates in preference to methyl (see below).

(4) Aryl groups on the β-carbon not only have a much stronger tendency than alkyl groups to migrate, but also speed reaction by participating in the rate-determining step (cf. p. 129). Thus $PhC(CH_3)_2-CH_2Cl$ undergoes solvolytic re-arrangement thousands of times faster than neopentyl chloride because the rate-determining step in the former case involves the formation not of the high-energy primary carbonium ion, $PhC(CH_3)_2-CH_2^+$ but of the delocalized ('bridged') phenonium ion:

$$
\underset{\text{Neophyl chloride}}{\underset{\substack{| \\ CH_3}}{CH_3-C-CH_2-Cl}} \longrightarrow \underset{\text{phenonium ion}}{\underset{\substack{| \\ CH_3}}{CH_3-C-CH_2}} \longrightarrow (CH_3)_2\overset{+}{C}-CH_2
$$

As expected for a reaction involving the formation of an ion in which positive charge is delocalized over an aromatic system, the rate of reaction is increased by the presence of electron-releasing groups in the aromatic ring and retarded by electron-withdrawing groups.

(b) INTERMOLECULAR REARRANGEMENTS

These reactions are strictly not representative of a new mechanistic type because they are combinations of the processes described above. For example, the rearrangement of N-chloroacetanilide to o- and p-chloroacetanilide, catalyzed by hydrochloric acid, consists of the formation of chlorine by a displacement process followed by the electrophilic substitution of acetanilide by chlorine:

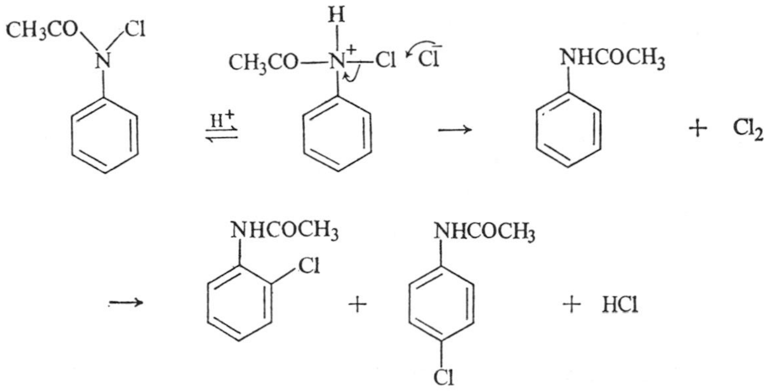

Prototropic (p. 17) and anionotropic (p. 20) shifts are also intermolecular rearrangements (see also the S_N1 solvolysis of allylic halides; p. 133).

4.8 Pericyclic Reactions

A pericyclic reaction is one involving a concerted bond reorganization in which the essential bonding changes occur within a cyclic array of the participating atomic centres. An example is the reaction between butadiene and acrolein, which occurs on heating:

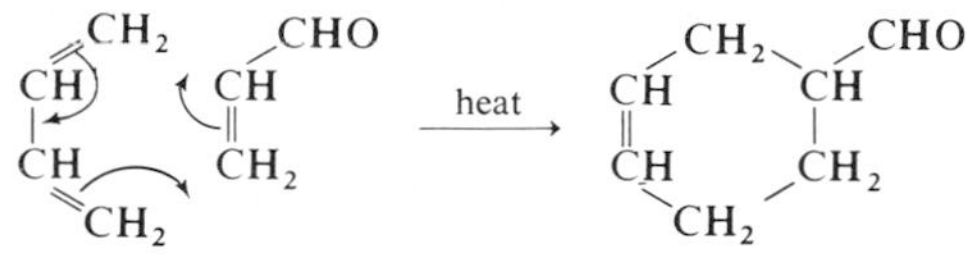

It must be emphasized that it is not possible to determine the exact manner in which electron reorganization takes place; in this example the arrows could equally well have been directed in the opposite sense. The important feature is that neither ions nor radicals are formed as intermediates. The theory of the course of the reactions is discussed in Chapter 9.

Pericyclic reactions can be subdivided as follows:

(a) CYCLOADDITIONS

The reaction above is an example of a cycloaddition.

(b) ELECTROCYCLIC REACTIONS

These are intramolecular pericyclic reactions which involve either the formation of a ring, with the generation of one new σ-bond and the consumption of one π-bond, or the converse. An example is the thermal interconversion of hexa-1,3,5-triene and 1,2-dihydrobenzene:

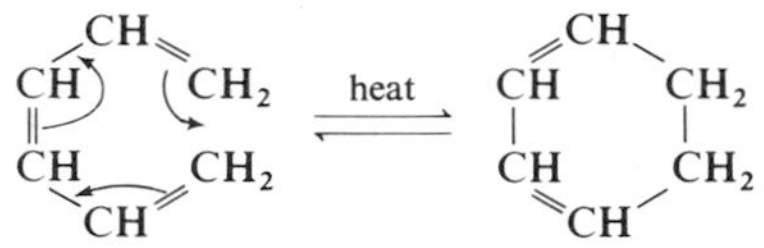

(c) CHELETROPIC REACTIONS

These are processes in which two σ-bonds which terminate at a single atom are made or broken in a concerted reaction, for example:

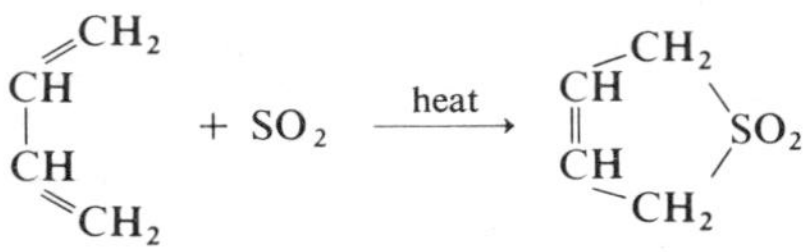

(d) SIGMATROPIC REARRANGEMENTS

In these, an atom or a group migrates within a π-electron system without change in the number of σ- or π-bonds. An example is the *Claisen rearrangement* in which allyl aryl ethers rearrange to *o*-allylphenols at about 200°C:

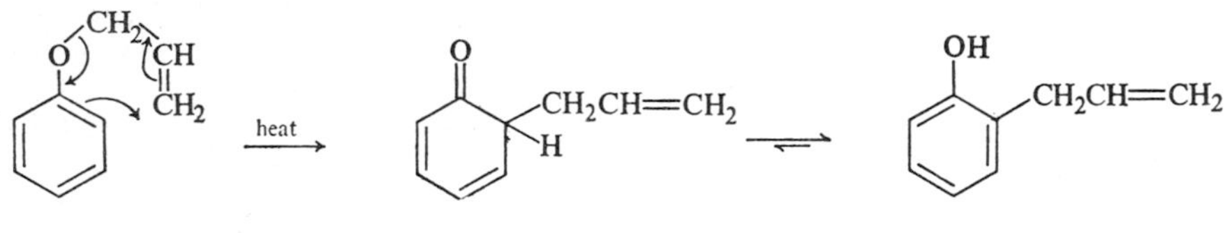

(e) THE ENE-REACTION

This is the reaction of an allylic compound with an olefin:

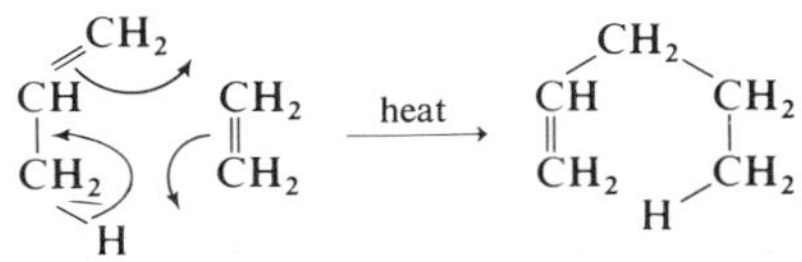

4.9 Oxidation-Reduction

As strictly defined, a compound or group is described as undergoing *oxidation* when electrons are wholly or partly removed from it. For example, the methyl group is oxidized when methane is converted by bromine into methyl bromide, because the electron-pair in the C—Br bond is less under the control of the carbon atom than the pair in the original C—H bond since bromine is more electronegative than hydrogen. It is more convenient, however, to employ a somewhat narrower definition: oxidation throughout this book is used to cover reactions in which an electron is completely lost from an organic compound, e.g.

$$PhO^- \xrightarrow{Ce^{4+}} PhO\cdot$$

oxygen is gained, e.g.

$$RCHO \xrightarrow{[O]} RCO_2H$$

or hydrogen is lost, e.g.

$$RCH_2OH \xrightarrow{-[2H]} RCHO$$

Reduction is used to describe the converse reactions.

Oxidation and reduction are complementary in that in any system in which one species is oxidized, another is reduced. The term used is customarily that appropriate to the reaction undergone by the organic compound concerned: e.g. the reaction of an alcohol with dichromate is described as an oxidation, although of course the dichromate ion is reduced. In some cases, however, this usage is unsatisfactory, as in the *Cannizzaro reaction* in which the disproportionation of two molecules of certain aldehydes leads to the oxidation of one and the reduction of the other:

$$2 \, PhCHO \xrightarrow{OH^-} PhCO_2H + PhCH_2OH$$

(*a*) OXIDATION
Oxidation is normally brought about in one of the following ways:

(1) By removal of an electron, as in the oxidation of phenols by ferricyanide (p. 578), e.g.

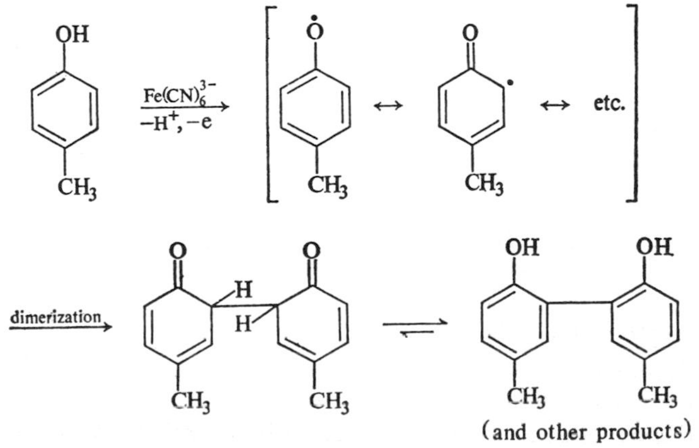

(and other products)

The requirement for the oxidizing agent is that it should be capable of one-electron reduction (here, Fe(III) to Fe(II)) characterized by a suitable redox potential. The requirement for the organic compound is that it should give a relatively stable radical on oxidation (achieved in this case by the delocalization of the unpaired electron over the benzene ring).

(2) By removal of a hydrogen atom, as in the radical-catalyzed autoxidation of aldehydes (p. 563),

$$R-CHO \xrightarrow{R'\cdot} R-\overset{\cdot}{C}=O + R'-H$$

$$R-\overset{\cdot}{C}=O \xrightarrow{O_2} R-\overset{\overset{\displaystyle O\cdot}{|}}{\underset{|}{C}}=O \xrightarrow{RCHO} R-\overset{\overset{\displaystyle OH}{|}}{\underset{|}{C}}=O + R-\overset{\cdot}{C}O$$

$$R-\overset{\overset{\displaystyle O_2H}{|}}{C}=O + RCHO \longrightarrow 2\ RCO_2H$$

(3) By removal of hydride ion, as in the *Cannizzaro* reaction (p. 634), e.g.

$$PhCHO \xrightarrow{OH^-} Ph-\overset{\overset{\displaystyle OH}{|}}{\underset{\underset{\displaystyle O^-}{|}}{C}}-H$$

$$Ph-\overset{\overset{\displaystyle OH}{|}}{\underset{\underset{\displaystyle O^-}{|}}{C}}-H \quad \overset{Ph}{\underset{\underset{\displaystyle H}{|}}{C}}=O \longrightarrow Ph-C\overset{OH}{\underset{O}{\diagdown}} + H-\overset{\overset{\displaystyle Ph}{|}}{\underset{\underset{\displaystyle H}{|}}{C}}-O^-$$

(4) By the insertion of oxygen, as in the epoxidation of an olefin by a peracid (p. 568):

$$\overset{\diagdown}{\diagup}C=C\overset{\diagup}{\diagdown} + R-\overset{\overset{\displaystyle}{\underset{\underset{\displaystyle O}{\|}}{C}}}{}-O-OH \longrightarrow \overset{\diagdown}{\diagup}C\underset{O}{\diagdown\diagup}C\overset{\diagup}{\diagdown} + RCO_2H$$

(5) By a concerted reaction in which the oxidizing agent undergoes a two-electron reduction, as in the oxidation of glycols by lead tetra-acetate (p. 595),

$$\begin{matrix} -\overset{|}{C}-OH \\ | \\ -\overset{|}{C}-OH \end{matrix} + Pb(OAc)_4 \xrightarrow{-2HOAc} \begin{matrix} -\overset{|}{C}-O \\ \diagdown Pb \overset{OAc}{\underset{OAc}{\diagup}} \\ -\overset{|}{C}-O \diagup \end{matrix}$$

$$\longrightarrow \begin{matrix} -\overset{|}{C}=O \\ + \\ -\overset{|}{C}=O \end{matrix} + Pb(OAc)_2$$

(6) By catalytic dehydrogenation, as in the palladium-catalyzed conversion of cyclohexane into benzene (p. 585).

(*b*) REDUCTION

The commoner mechanisms of reduction are the following:

(1) By addition of an electron, as in the formation of pinacols,

$$R_2CO \xrightarrow{\ e\ } R_2\overset{\cdot}{C}{-}O^-$$

$$2\ R_2\overset{\cdot}{C}{-}O^- \longrightarrow \underset{\underset{O^-\ O^-}{|\quad|}}{R_2C{-}CR_2} \xrightarrow{2H^+} \underset{\underset{OH\ OH}{|\quad|}}{R_2C{-}CR_2}$$

Two electrons may be transferred, as in the *trans*-reduction of acetylenes by sodium in liquid ammonia:

$$-C{\equiv}C- \xrightarrow{2e} {}^-\overset{\diagdown}{C}{=}\overset{\diagup}{C}{}^- \xrightarrow{2\ H^+} \overset{\diagdown}{\underset{H}{C}}{=}\overset{H}{\underset{\diagdown}{C}}$$

The requirement of the reducing agent is that it should have a strong tendency to denote an electron: e.g. electropositive metals such as sodium, and transition-metal ions in low valency states such as Cr(II) and Ti(III).

(2) **By addition of hydride ion, usually from a complex metal hydride (p. 633),** e.g.

$$H_3\overline{Al}{-}H \quad \overset{\diagdown}{C}{=}O \longrightarrow H{-}\overset{|}{\underset{|}{C}}{-}O^- + AlH_3$$

Reaction may occur *via* a cyclic transition state, as in the *Meerwein-Ponndorf-Verley* reduction of aldehydes and ketones (p. 635):

$$\begin{array}{c} R_2C{=}\overset{+}{O} \quad OPr^i \\ \diagdown \quad \diagup \\ H\diagdown \quad Al \\ C{-}O \quad OPr^i \\ CH_3 \diagup \diagdown CH_3 \end{array} \longrightarrow \begin{array}{c} R_2C{-}O \quad OPr^i \\ |\diagdown\diagup \\ H \quad Al \\ \diagdown OPr^i \\ + \\ (CH_3)_2C{=}O \end{array}$$

(3) **By catalytic hydrogenation, as in the reduction of olefins over Raney nickel (a form of nickel in a very finely divided state):**

$$\overset{\diagdown}{\underset{\diagup}{C}}{=}\overset{\diagup}{\underset{\diagdown}{C}} \xrightarrow{H_2\text{-}Ni} \underset{\underset{H\ H}{|\quad|}}{-C{-}C{-}}$$

These reactions occur in a stereospecific manner, giving the *cis*-dihydro adduct.

Further Reading

ALDER, R. W., BAKER, R., and BROWN, J. M., *Mechanism in Organic Chemistry*, Wiley-Interscience (London 1971).

BRESLOW, R., *Organic Reaction Mechanisms: An Introduction*, 2nd Ed., W. A. Benjamin (New York 1969).

GOULD, E. S., *Mechanism and Structure in Organic Chemistry*, Holt, Rinehart, and Winston (New York 1959).

HINE, J., *Physical Organic Chemistry*, 2nd Ed., McGraw-Hill (New York 1962).

INGOLD, C. K., *Structure and Mechanism in Organic Chemistry*, 2nd Ed., Bell (London 1970).

SYKES, P., *A Guidebook to Mechanism in Organic Chemistry*, 4th Ed., Longmans (London 1975).

SYKES, P., *The Search for Organic Reaction Pathways*, Longmans (London 1972).

Problems

1. Arrange the following in decreasing order of their reactivity towards an electrophilic reagent (e.g. H^+):

$$CH_2{=}CH{-}\overset{+}{N}Me_3 \qquad CH_2{=}CH_2 \qquad CH_2{=}CH{-}CH_3$$

$$CH_2{=}CH{-}OCH_3 \qquad CH_2{=}CH{-}Br \qquad CH_2{=}CH{-}NO_2$$

2. What reactivity differences would you expect between the compounds in each of the following pairs:

    (*i*) $CH_3CH{=}CHCO_2Et$ and $CH_2{=}CHCH_2CO_2Et$
    (*ii*) $C_2H_5OC_2H_5$ and $CH_2{=}CHOC_2H_5$
    (*iii*) $PhOCH_3$ and $PhNO_2$
    (*iv*) CH_3COCl and CH_3CH_2Cl
    (*v*) CH_3COCH_3 and $(CH_3)_3C{-}CO{-}C(CH_3)_3$
    (*vi*) $CH_3{-}CH_2Br$ and $(CH_3)_3C{-}CH_2Br$
    (*vii*) $CH_3SCH_2CH_2Cl$ and $CH_3SCH_2CH_2CH_2Cl$

3. Account for the following:

    (*i*) Whereas t-butyl chloride almost instantly gives a precipitate with alcoholic silver nitrate, apocamphyl chloride is inert, even on prolonged boiling.

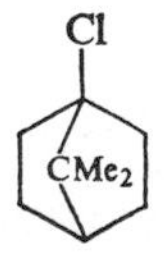

Apocamphyl chloride

(*ii*) Acetals are stable to bases but are readily hydrolyzed by acids.

(*iii*) When the solvent polarity is increased, the rate of the S_N2 reaction,

$$HO^- + CH_3OSO_2Ph \rightarrow CH_3OH + PhSO_2O^-$$

is slightly reduced but that of the S_N2 reaction,

$$Et_3N + EtI \rightarrow Et_4N^+ I^-$$

is greatly increased.

(*iv*) The alkaline hydrolysis of ethyl bromide is catalyzed by iodide ion.

(*v*) The dehydration of methyl-t-butylcarbinol, CH_3—$CH(OH)$—$C(CH_3)_3$, with concentrated sulphuric acid gives tetramethylethylene.

(*vi*) Simple β-keto-acids, $RCOCH_2CO_2H$, readily decarboxylate on being heated, but the compound (I) is stable.

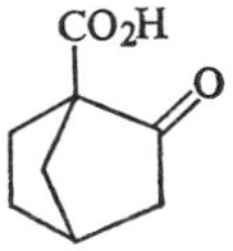

(I)

4. What products would you expect from the following reactions:

(*i*) $PhCH{=}CHCH_3 + HCl \rightarrow$

(*ii*) $(CH_3)_3C{-}Cl + KCN \rightarrow$

(*iii*)

$$\underset{CH_3}{\overset{CH_3CH_2}{>}}CH{-}\overset{+}{N}Me_3 \ OH^- \xrightarrow{\text{heat}}$$

(*iv*) $CH_3CH_2Br + C_2H_5ONa \rightarrow$

(*v*) $CH_3CHClCH_2CH_3 + C_2H_5ONa \rightarrow$

(*vi*) $CH_2{=}CH{-}CH{=}CH_2 + Br_2 \rightarrow$

(*vii*) $CH_3C{\equiv}CH + H_2O \xrightarrow{HgSO_4}$

(*viii*)

$$\underset{H}{\overset{CH_3}{>}}C{=}C\underset{H}{\overset{CH_3}{<}} \quad + Br_2 \rightarrow$$

(*ix*) $(CH_3)_3C{-}CH{=}CH_2 + HCl \rightarrow$

(x) $CH_2(OH)-CH_2Cl + NaOH \rightarrow$

(xi) $CH_3CH=CHCH_2CO_2H \xrightarrow{\text{heat}}$

(xii) $CH_2=CH-O-CH_2CH=CH_2 \xrightarrow{\text{heat}}$

5. Indicate schematically the mechanisms of the following reactions:

(i) The acid-catalyzed hydrolysis of an amide, $RCONH_2$.

(ii) The base-catalyzed hydrolysis of an ester, RCO_2CH_3.

(iii) The acid-catalyzed bromination of acetone.

(iv) The base-catalyzed self-condensation of ethyl acetate.

5. Stereochemistry

5.1 Optical Isomerism

(*a*) ENANTIOMERS

A compound whose structure is such that it is not superimposable on its mirror image exists in two forms, known as *enantiomers* or *enantiomorphs*. Lactic acid is an example; the two forms may be represented two-dimensionally as (1) and (2).

$$
\begin{array}{cc}
CO_2H & CO_2H \\
| & | \\
C & C \\
H \quad OH & HO \quad H \\
CH_3 & CH_3 \\
(1) & (2)
\end{array}
$$

Other compounds of the general type *Cabcd* also exist in enantiomeric forms; the central carbon atom is described as *chiral*.

The physical and chemical properties of enantiomers are identical in a symmetrical environment but differ in a dissymmetric environment. In particular, enantiomers rotate the plane of plane-polarized light in opposite directions, although to the same extent per mole in the same conditions; they are described as being *optically active*. For example, lactic acid (1) is laevorotatory in aqueous solution and is described as (−)-lactic acid; the enantiomer (2) is dextrorotatory and is described as (+)-lactic acid.

A dissymmetric environment can also be provided by one enantiomer of another enantiomeric pair: thus, (+) and (−) enantiomers usually react at different rates with another enantiomer (see, e.g. p. 165) and give different products (diastereoisomers; p. 153).

The phenomenon of the existence of enantiomers is known as optical isomerism and its importance stems from the fact that many naturally occurring compounds are optically active. Lactic acid, for example, occurs in sour milk as the dextrorotatory enantiomer and in muscle tissue as the laevorotatory enantiomer.

The criterion of non-superimposability of a structure and its mirror-image is a necessary and sufficient one for the existence of enantiomers. An alternative approach for deciding whether a particular structure is capable of existing in optically active forms is to examine the symmetry of the molecule: a structure lacking an alternating axis of symmetry is not superimposable on its mirror image and therefore represents an optically active species. * It is simpler in practice to recognize a centre or plane of symmetry: for example, trans-1,3-dichloro-cyclobutane (3) cannot exist in optically active forms because it has both a plane and a centre of symmetry; and the cis isomer (4) cannot because it has two planes of symmetry.

(3)   (4)   (5)

Caution must be exercised if this approach is adopted; for example, the spiro-compound (5) possesses neither a centre nor a plane of symmetry, but it cannot exist in enantiomeric forms because it possesses a four-fold alternating axis of symmetry. (The reader should show that (5) is superimposable on its mirror image.)

(b) PROJECTION DIAGRAMS

Since stereochemistry is a three-dimensional science which requires two-dimensional representation on paper, a number of conventions for representing structures has been developed. The earliest was Fischer's projection formula, illustrated for compounds of the type $Cabcd$, such as lactic acid. The molecule is first oriented so that the atom C is in the plane of the paper; two substituents, e.g. a and d, are arranged at the top and bottom relative to C, each below the plane of the paper and inclined equally to it; and the other two substituents, b and c, are arranged to the left and right of C, each above the plane of the paper and inclined equally to it. The resulting structure (6) is then projected onto the plane to give (7). Thus, a molecule shown in Fischer projection as (7) has the structure (6).

(6)   (7)   (8)

*An object possesses an n-fold alternating axis of symmetry if rotation of $360°/n$ about an axis followed by reflection in a plane perpendicular to that axis brings the object into a position indistinguishable from its original one. A plane of symmetry is equivalent to a one-fold alternating axis of symmetry, and a centre of symmetry is equivalent to a two-fold alternating axis of symmetry.

The Fischer projection has certain disadvantages. It is easily misused; thus, it must be remembered that (7) and (8) are not representations of the same compound but represent enantiomers. Further, the representation of compounds containing more than one chiral carbon atom is inadequate: e.g. the projection (9) for (−)-threose implies an eclipsing relationship of the groups attached to the two chiral atoms, whereas staggered conformations are more stable (cf. the conformational preferences of n-butane; p. 46). Moreover, the realistic presentation of many of the reactions of such compounds necessitates the representation of staggered structures (5.3).

$$
\begin{array}{c}
\text{CHO} \\
| \\
\text{HO—C—H} \\
| \\
\text{H—C—OH} \\
| \\
\text{CH}_2\text{OH}
\end{array}
$$

(9)

There are two conventions which avoid these difficulties. In one, the 'sawhorse' representation, the bond between the chiral carbon atoms is drawn diagonally, implying that it runs downwards through the plane of the paper, and is slightly elongated for clarity. The substituents on each of the carbons are then projected on to the plane of the paper and can be represented as staggered or eclipsed, e.g. for (−)-threose,

Two of the staggered conformations
of (−)-threose

An eclipsed conformation
of (−)-threose

In the second, the Newman projection, the molecule is viewed along the bond joining the chiral carbon atoms, and these atoms are represented as superimposed circles, only one circle being drawn. The remaining bonds and the substituents are then projected into the plane of the paper, the bonds to the nearer carbon being drawn to the centre of the circle and those to the further carbon being drawn only to the perimeter. The projections for (−)-threose which correspond to those above in the sawhorse representation are (10)–(12).

(c) STRUCTURAL SITUATIONS WHICH GIVE RISE TO OPTICAL ISOMERISM
In addition to compounds of the type Cabcd which possess one chiral carbon atom and fulfil the conditions necessary for the occurrence of enantiomeric pairs, there are several other structural situations which lead to the occurrence

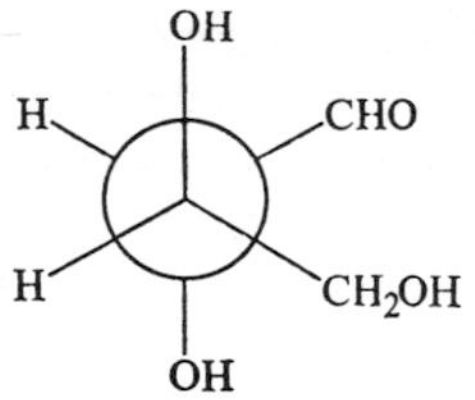

(10) 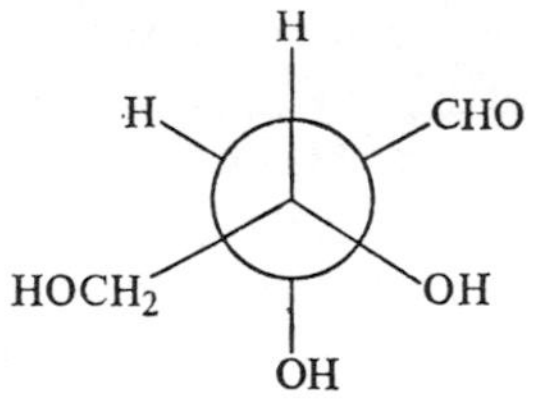   (11) 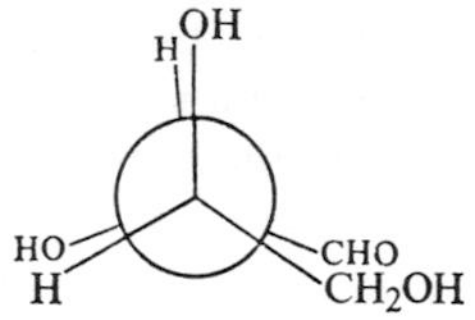   (12)

of optical isomerism. Some of these structures contain one or more chiral atoms and others owe their optical activity to the presence of other dissymmetric features.

(i) *Compounds possessing a chiral atom other than carbon.* Since the valencies of nitrogen in an amine are directed approximately towards three corners of a tetrahedron with the lone-pair being directed towards the fourth corner, it might be expected that an amine of the type $Nabc$ would exist in enantiomeric forms: the mirror images (13) and (14) are not superimposable. However, the separation of such amines into optically active isomers has never been achieved

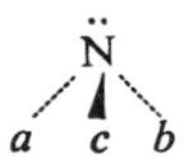

(13)    (14)

because the isomers are too rapidly interconvertible by a flapping mechanism, i.e.

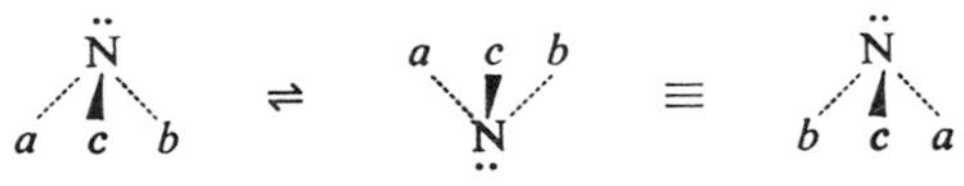

i.e. (13) ⇌ (14)

If flapping is prevented by the joining together of the groups *abc*, optical isomers can be obtained. A historically interesting example is Tröger's base whose enantiomeric forms have the structures (15) and (16).

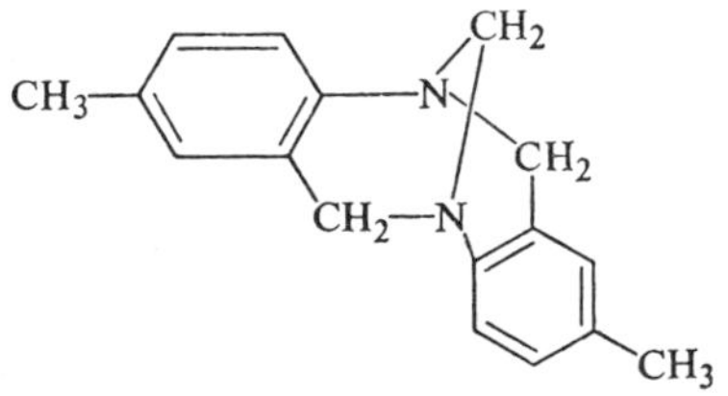

(15) 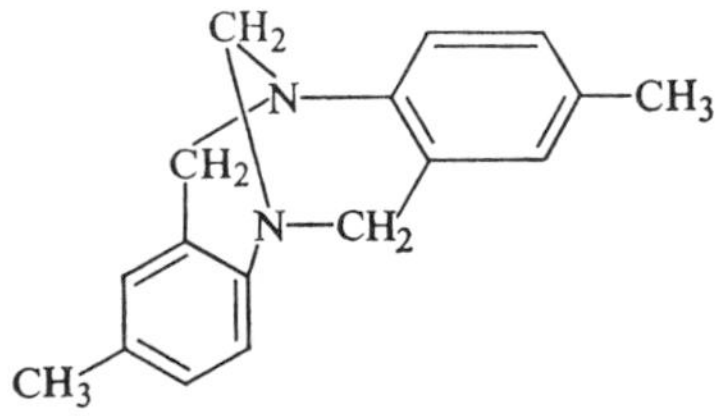   (16)

Ternary phosphorus, arsenic, and antimony compounds are configurationally more stable than their nitrogen analogues and several have been resolved, e.g. PMeEtPh, whose enantiomeric forms have the structures (17) and (18). Since quaternary ammonium ions cannot flap in the manner of amines, those of the type $Nabcd^+$ exist as enantiomeric pairs some of which have been separated, e.g. (19) and (20). Phosphonium and arsonium salts have also been resolved.

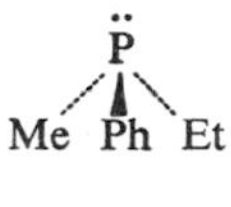

$$\ddot{P}$$
Me Ph Et

(17)

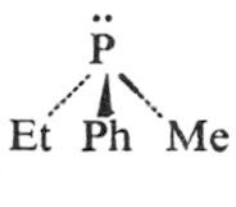

$$\ddot{P}$$
Et Ph Me

(18)

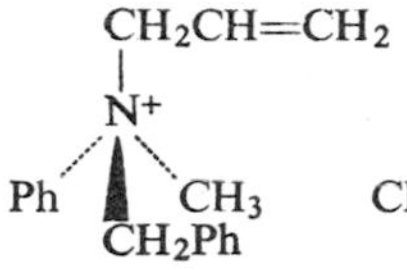

$$CH_2CH{=}CH_2$$
$$N^+$$
Ph CH₃
CH₂Ph

(19)

$$CH_2CH{=}CH_2$$
$$N^+$$
CH₃ Ph
CH₂Ph

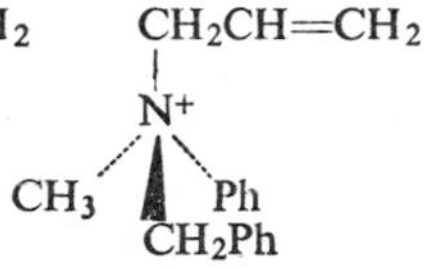

(20)

Optically active sulphonium salts, sulphoxides and silicon and germanium analogues of hydrocarbons have also been obtained, e.g.

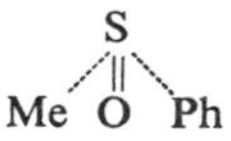

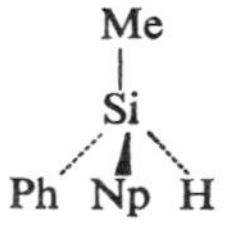

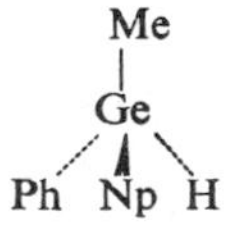

$$S^+$$
Me Et
CH₂COPh

$$S$$
Me O Ph

Me
Si
Ph Np H

Me
Ge
Ph Np H

$$(Np = \alpha\text{-naphthyl})$$

(*ii*) *Compounds possessing two or more chiral carbon atoms.* Compounds with two chiral carbon atoms in which at least one substituent is not common to both carbons occur in four optically isomeric forms; e.g. the four isomers of structure $CH(OH)(CH_2OH){-}CH(OH)(CHO)$ are:

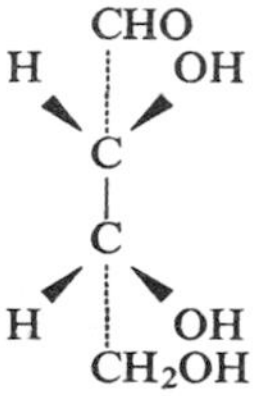

CHO
H OH
C
C
H OH
CH₂OH

(−)-Erythrose

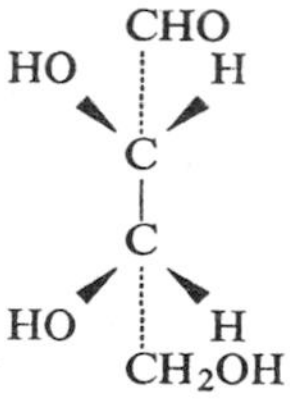

CHO
HO H
C
C
HO H
CH₂OH

(+)-Erythrose

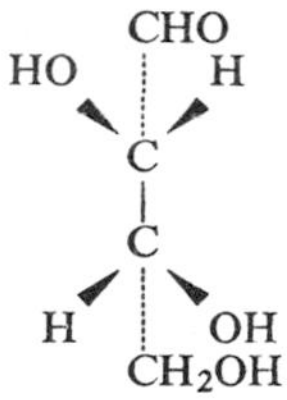

CHO
HO H
C
C
H OH
CH₂OH

(−)-Threose

CHO
H OH
C
C
HO H
CH₂OH

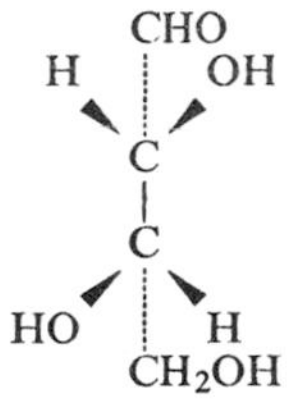

(+)-Threose

In general, a compound possessing n distinct chiral carbon atoms exists in 2^n optically active forms.

It should be noted that, whereas (+)- and (−)-erythrose, and (+)- and (−)-threose, are mirror-image pairs and therefore have identical properties in a symmetric environment, neither of the erythroses bears a mirror-image relationship to either of the threoses. These and other stereoisomers which are not enantiomers are described as *diastereoisomers*, and, unlike enantiomers, they

differ in physical and chemical properties. This is because, whereas all intramolecular distances between corresponding groups are the same in enantiomers,
they are different in diastereoisomers. For example, the diol (22) reacts with
acetone in the presence of anhydrous acid, to give the cyclic ketal, faster than
its diastereoisomer (24). This is because it is necessary for the hydroxyl groups
to be brought into a coplanar *cis*-configuration (21) for reaction to occur, and
the appropriate conformation (23) for the diol (22) is of lower energy than the
corresponding conformation (25) for the diol (24) since the steric repulsion
forces between the alkyl groups are smaller.

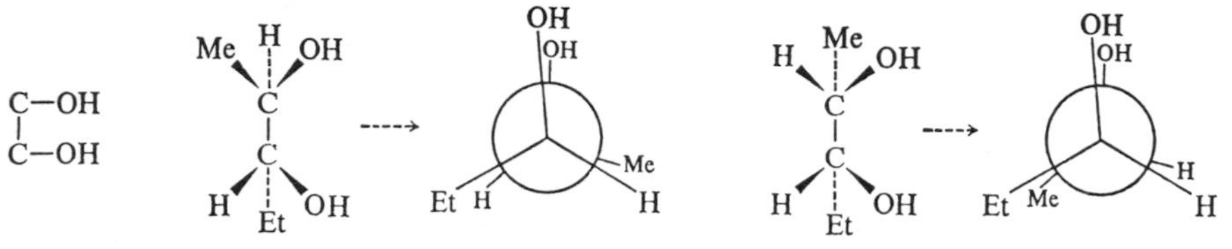

Compounds which contain two chiral carbon atoms and are of the type
C*abc*–C*abc* exist in only three isomeric forms. Two of these are non-superimposable mirror images of each other and are optically active and the third, a
diastereoisomer of the first two, contains a plane of symmetry, is superimposable
on its mirror image, and is not optically active:

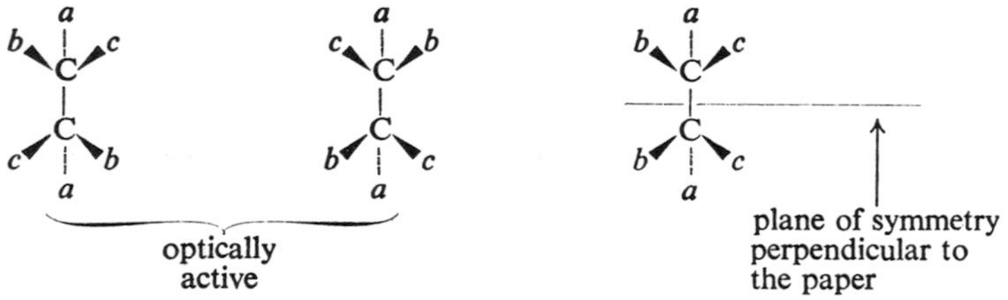

The inactive diastereoisomer is described as a *meso* form. As with other examples
of diastereoisomers, the properties of *meso* forms are different from those of the
isomeric mirror-image pairs; e.g. of the tartaric acids, $CH(OH)(CO_2H)$—
$CH(OH)(CO_2H)$, *meso*-tartaric acid melts at a lower temperature (140°C) than
the (+)- and (−)-isomers (170°C), and is less dense, less soluble in water, and a
weaker acid.

It is simple to extend this discussion to compounds which contain more than
two chiral carbon atoms. For example, the aldohexoses, CH_2OH—$(CHOH)_4$—
CHO, possess four chiral carbon atoms, each of which is distinct from the
remainder, and there are therefore $2^4 = 16$ stereoisomers all of which are
optically active. (One of these is glucose, whose (−)-form, in Fischer projection,

is (26)). On the other hand, some of the derived saccharic acids, $CO_2H—(CHOH)_4—CO_2H$, have symmetry properties which make them optically inactive; 10 stereoisomers exist of which 8 are optically active 4 (($\pm$)-pairs) and 2 are *meso* forms.

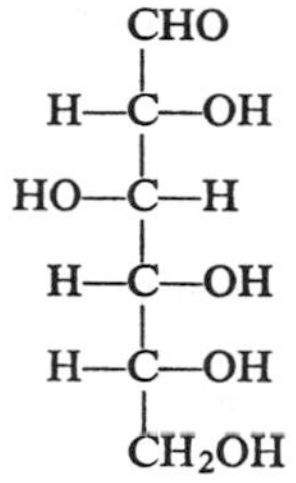

(26) 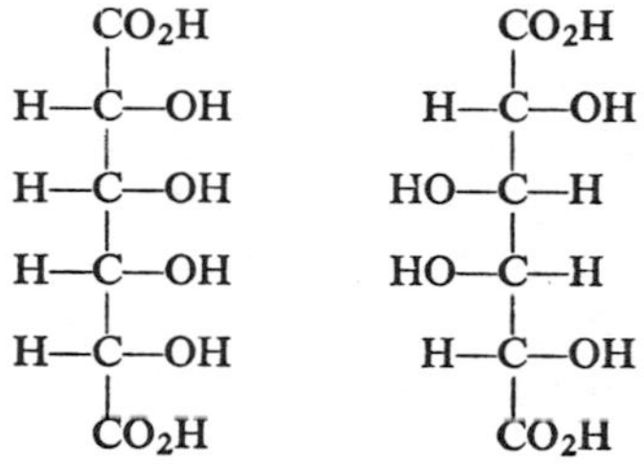   *meso* forms of saccharic acids

(*iii*) *Allenes.* An *sp*-hybridized carbon atom possesses one electron in each of two mutually perpendicular *p* orbitals. When it is joined to two *sp²*-hybridized carbon atoms, as in an allene, two mutually perpendicular π-bonds are formed and consequently the π-bonds to the *sp²*-carbons are in perpendicular planes (27). Allenes of the type $abC{=}C{=}Cab$ ($a \neq b$) are therefore not superimposable on their mirror images and, despite the absence of any chiral atoms, exist as enantiomers (28) and (29). Several optically active compounds have been obtained (e.g. a = phenyl, b = α-naphthyl).

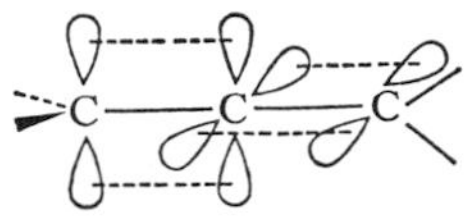

(27) 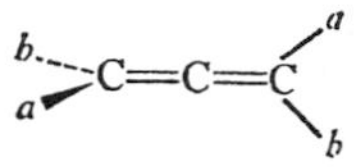   (28) 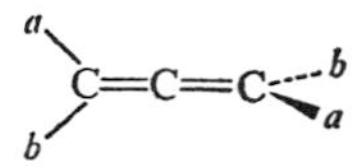   (29)

(*iv*) *Alkylidenecycloalkanes.* The replacement of one double bond in an allene by a ring does not alter the basic geometry of the system and appropriately substituted compounds exist in optically active forms, e.g. 4-methylcyclohexylidene-acetic acid (30). Related compounds in which *sp²*-carbon is replaced by nitrogen, e.g. (31), have also been obtained as optical isomers.

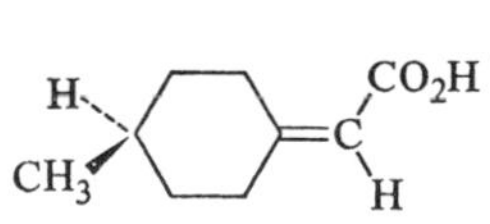

(30) 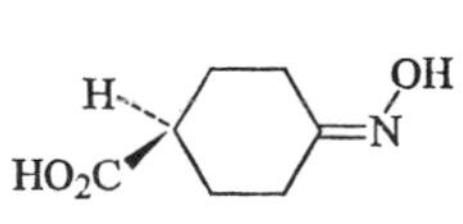   (31) 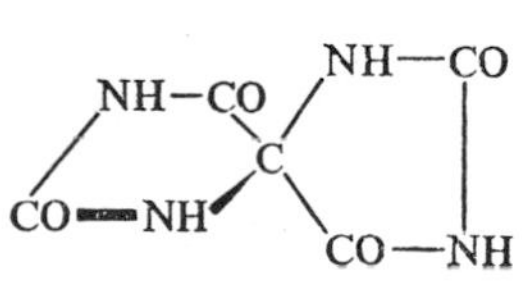   (32)

(*v*) *Spirans.* The replacement of both double bonds in an allene by ring systems gives a spiran; appropriately substituted compounds have been obtained in optically active forms, e.g. (32).

(*vi*) *Biphenyls.* The two conformations, (33) and (34), of *meso*-2,3-dichlorobutane are non-superimposable mirror-images of each other. However, it is not possible to isolate optically active forms because rotation about the central C—C bond occurs rapidly and results in the interconversion of the two structures. On the other hand, when the barrier to rotation about a C—C bond exceeds about 80 kJ mol^{-1} in suitably substituted compounds, rotation is slow enough at room temperature for the isolation of optically active isomers to become practicable. This situation holds for certain biphenyls of the general structure (35).

          (33)   (34)   (35)

     Optical isomerism in biphenyls is possible because the conformation in which the benzenoid rings are coplanar and which possesses a plane of symmetry is strained with respect to non-coplanar conformations such as (35) as a result of the steric repulsions between pairs of *ortho*-substituents on the adjacent rings. The substituents X and Y require to be fairly bulky for the occurrence of the necessary barrier to rotation; and their presence also ensures that the molecule has the necessary lack of symmetry for optical activity. Examples of compounds of this type which have been obtained in optically active form are biphenyl-2,2′-disulphonic acid (X=Y=SO$_2$OH) and the diamine X=Y=NMe$_2$.

     A given energy barrier between a biphenyl and its enantiomer is surmounted more rapidly at higher temperatures. Hence one enantiomer is converted, with first-order kinetics, into a mixture of the enantiomers at a rate which increases with temperature, and the optical activity of the species falls eventually to zero when equilibrium is reached. This process—racemization (p. 160)—occurs less rapidly as the sizes of X and Y are increased. The introduction of one or two more *ortho*-substituents in place of hydrogen atoms therefore reduces the rate of racemization to an extent depending on the sizes of the groups: e.g. racemization of the compounds (36) occurs at 118°C with a half-life of 91 minutes (Z = OCH$_3$), 125 minutes (Z = NO$_2$), and 179 minutes (Z = CH$_3$).

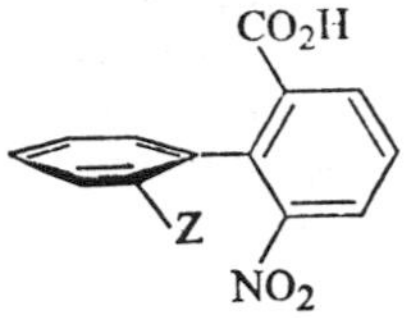

(36)

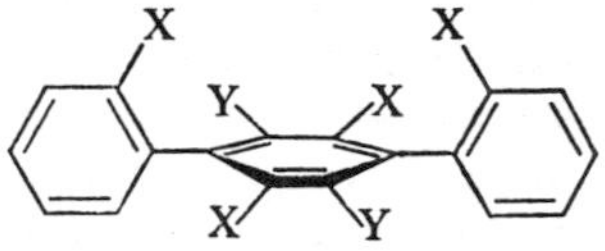

(37)

Optical isomerism occurs also in suitably substituted polyphenyls, but the stereochemistry is more complex since both *meso* forms and geometrical isomerism are also possible. For example, compounds of the type (37) exist in three stereoisomeric forms: there are enantiomeric *cis* isomers, one of which is (37), and a *trans* isomer (in which the X groups on the end rings are on opposite sides) which, having a centre of symmetry, is an optically inactive (*meso*) form.

There are other compounds which are optically active due to restricted rotation about single bonds, of which the following are examples:

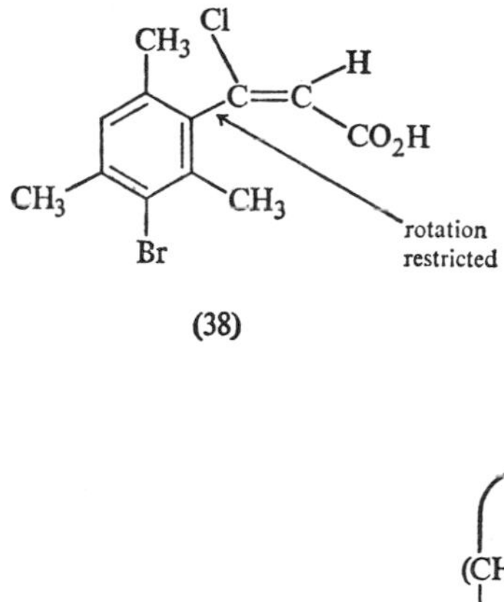

(38)

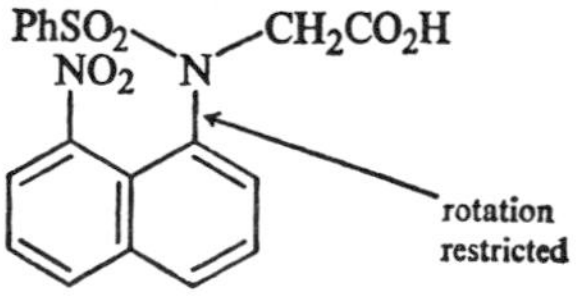

(39)

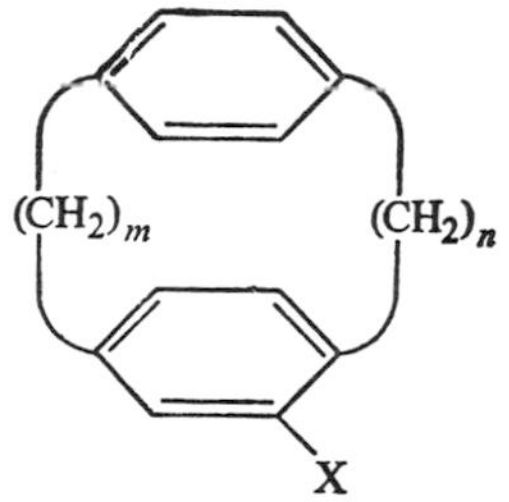

(40)

Compounds of the type (40) ('paracyclophanes') have been resolved when m and n are fairly small (e.g. when $m = 3$, $n = 4$, and $X = CO_2H$), for then the benzenoid rings cannot rotate at a significant rate. When the connecting methylene chains are longer, however, rotation is so fast that the enantiomers rapidly equilibrate.

(d) ABSOLUTE CONFIGURATION

The complete structure of a chiral molecule is not elucidated until the absolute configuration of the compound—the actual arrangement of the atoms in space—is known. The first determination of absolute configuration, by an X-ray method (1951), showed that sodium rubidium (+)-tartrate has the absolute configuration (41).

$$
\begin{array}{c}
\text{CO}_2\text{Rb} \\
\text{H} \diagdown \mid \diagup \text{OH} \\
\text{C} \\
\mid \\
\text{C} \\
\text{HO} \diagup \mid \diagdown \text{H} \\
\text{CO}_2\text{Na}
\end{array}
\qquad (41)
$$

The technique is limited to suitably crystalline solids and has not been widely applied, but fortunately the absolute configurations of other optically active compounds can be obtained from that of the (+)-tartrate by correlative methods. The simplest of these is to convert one optically active compound into another without breaking any bond to the relevant chiral centre. The absolute configurations of reactant and product are then necessarily the same, so that if one is known the other follows. For example, the known sodium rubidium (+)-tartrate can be converted in this way successively into (+)-malic acid, (+)-isoserine, (−)-bromolactic acid, and (−)-lactic acid, all of whose absolute configurations are thereby established:

$$
\begin{array}{ccccc}
\text{CO}_2\text{H} & & \text{CO}_2\text{H} & & \text{CO}_2\text{H} \\
\text{H} \diagdown \mid \diagup \text{OH} & & \text{H} \diagdown \mid \diagup \text{OH} & & \text{H} \diagdown \mid \diagup \text{OH} \\
\text{C} & & \text{C} & & \text{C} \\
\mid & \longrightarrow & \mid & \longrightarrow & \mid \\
\text{C} & & \text{CH}_2\text{CO}_2\text{H} & & \text{CH}_2\text{NH}_2 \\
\text{HO} \diagup \mid \diagdown \text{H} & & \text{(+)-Malic acid} & & \text{(+)-Isoserine} \\
\text{CO}_2\text{H} & & & &
\end{array}
$$

$$
\begin{array}{ccccc}
& \text{CO}_2\text{H} & & \text{CO}_2\text{H} & \\
& \text{H} \diagdown \mid \diagup \text{OH} & & \text{H} \diagdown \mid \diagup \text{OH} & [\equiv (1),\ \text{p. 149}] \\
\longrightarrow & \text{C} & \longrightarrow & \text{C} & \\
& \mid & & \mid & \\
& \text{CH}_2\text{Br} & & \text{CH}_3 & \\
& \text{(−)-Bromolactic acid} & & \text{(−)-Lactic acid} &
\end{array}
$$

It must be emphasized that there is no simple relationship between absolute configuration and the sign of rotation of an optically active compound; a small structural alteration in which the absolute configuration of the chiral centre is maintained can result in a change of the direction of rotation, as in the above conversion of (+)-isoserine into (−)-bromolactic acid. It is sometimes useful to specify the absolute configuration of a structure. One convention employs the prefix D for (+)-glyceraldehyde and compounds having the same absolute configuration, and L for compounds having the mirror-image configuration. Since both D-glyceraldehyde and (+)-isoserine are converted into (−)-glyceric acid by reactions which do not involve change of configuration, (+)-tartaric acid and all the structures in the inter-related scheme above are members of the D-series.

An alternative and increasingly widely used convention is applied as follows. First, the atoms attached to the chiral carbon atom are listed in a priority order, defined as that of decreasing atomic weight, a, b, c, d (if two or more are the same, the atoms attached to these are considered; and if an atom forms a double or triple bond to another atom X, the first atom is regarded as attached to 2 or 3 atoms of X). Then the chiral carbon atom is viewed from the side opposite the atom or group of lowest priority, d. If the other groups then appear in the order a, b, c in a clockwise direction, the compound is given the prefix R (Latin *rectus* = right); if anticlockwise, S (Latin *sinister* = left). For example, for glyceraldehyde, the priority order is $OH > CHO > CH_2OH > H$; viewing from the direction opposite the hydrogen atom gives

OH   OH

HOCH₂   CHO   OCH   CH₂OH

R-glyceraldehyde S-glyceraldehyde

Note that R-glyceraldehyde is also D-glyceraldehyde, but it must be emphasized that R and D do not necessarily correspond.

This order of precedence is also applied in a convention for naming geometrical isomers. If the two substituents on the same side of a C=C bond are of higher precedence, the prefix Z (German *zusammen* = together) is used; if they are on opposite sides, E (German *entgegen* = across), e.g.

CH₃   OCH₃   H   OCH₃

    C=C   C=C

H   CH₂OH   CH₃   CH₂OH

Z E

This has much wider generality than *cis,trans* nomenclature, which is applicable only if at least one pair of substituents on adjacent olefinic carbon atoms are the same.

(e) RACEMIC MODIFICATIONS AND RACEMIZATION

The optical activities of equal numbers of molecules of enantiomers nullify each other. Such an assembly is known as a *racemic modification* and is symbolized by *dl* or ($\pm$).

In the liquid and the gaseous states the properties of a racemic modification are usually identical with those of the individual enantiomers, for the mixture is normally essentially an ideal one. In the solid state, however, the properties of the racemic modification and the enantiomers can differ greatly. Three types of racemic modification occur. First, each enantiomer may have a greater affinity for molecules of the same kind than for those of the other enantiomer; the enantiomers then tend to crystallize separately, giving a *racemic mixture* of two types of crystals. Secondly, each enantiomer may have a greater affinity for molecules of the other enantiomer; the crystal then grows by the laying down of (+) and (−) molecules alternately, giving a crystal of different type, and hence different properties, from the individual enantiomers (a *racemic compound*). Thirdly, there may be little difference between the affinities of one enantiomer for molecules of its own type and for those of the other enantiomer (i.e. nearly ideal behaviour); the arrangement of molecules in the solid is then random and a *racemic solid solution* is obtained, identical in properties with the individual enantiomers.

Racemic modifications may be obtained in three ways: by mixing equal amounts of the (+) and (−) isomers; by synthesis; and by racemization.

Synthesis. The synthesis of a chiral compound from achiral reactants necessarily produces a racemic modification. Consider, for example, the addition of hydrogen cyanide to acetaldehyde. If the approach of cyanide ion to the aldehyde in the conformation (42) is as shown, the product is D-lactonitrile, but it is statistically as probable that the approach will be to the mirror-image conformation (43), giving L-lactonitrile. It is apparent that any particular pathway to the D-product corresponds to an equally probable pathway to the L-product, so that a racemic modification is produced.

Racemization. Racemization is defined as the production of a racemic modification from one enantiomer. It is normally an undesirable process: that is, when

(42)

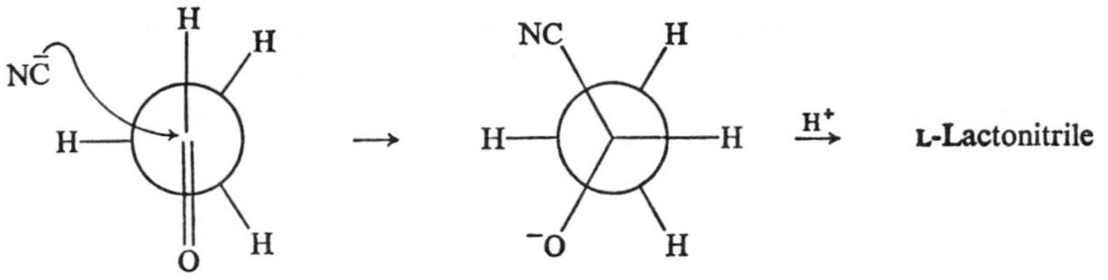

(43)

it is necessary to synthesize an optically active compound, it is necessary to avoid steps which result in racemization.

Racemization can occur in a number of ways, one or more of which may apply to a particular optically active compound. The commoner mechanisms are as follows.

(1) *By rotation about a single bond.* The biphenyls and related compounds, in which optical activity results from the restriction of rotation about a single bond, racemize when enough thermal energy is present for the energy barrier between the enantiomers to be surmounted at a practicable rate (p. 156).

(2) *Via an anion.* Since carbanions $RR'R''C^-$, like amines, are not optically stable, an enantiomer of the type $RR'R''C—X$ racemizes in an environment in which X is reversibly removed without its bonding-pair of electrons. The group X is most commonly hydrogen, and if one or more of the other substituents is of the type which stabilizes a carbanion, racemization can occur in basic conditions. For example, enantiomers containing a carbonyl group adjacent to the chiral carbon racemize in this way:

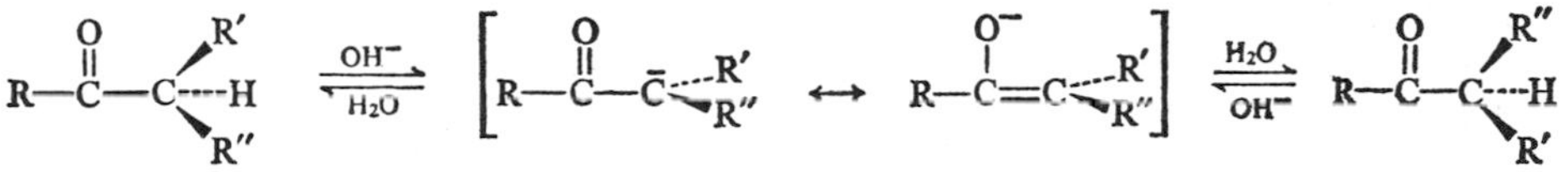

Other groups of $-M$ type, such as nitro and carboalkoxyl (p. 51), have the same effect as carbonyl.

(3) *Via a cation.* Carbonium ions are coplanar at the tervalent carbon atom, so that if the ion $RR'R''C^+$ is formed from an enantiomer $RR'R''C—X$ by the reversible removal of X with its bonding-pair, racemization occurs. The group X is usually a halogen, and its removal can be aided by a Lewis acid, e.g.

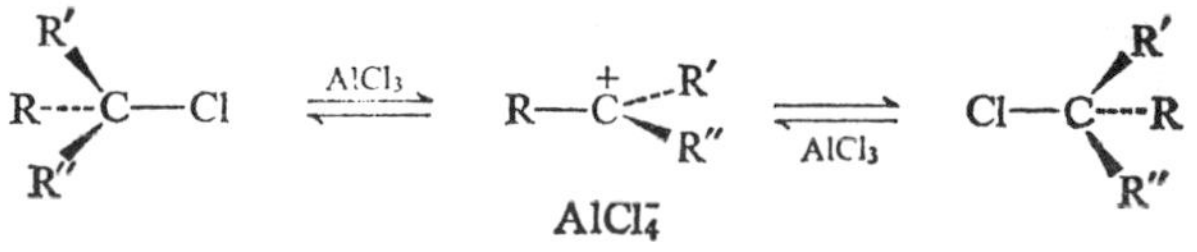

(4) *Via reversibly formed, inactive intermediates.* This mechanism is similar to methods 2 and 3 save that the intermediate is a relatively stable compound as compared with the transient carbanions and carbonium ions. For example, optically active α-phenethyl chloride racemizes in polar solvents such as formic acid through optically inactive styrene:

$$
\underset{\substack{Ph}}{\overset{\substack{H}}{CH_3\text{---}C\text{---}Cl}} \quad \underset{}{\overset{-HCl}{\rightleftharpoons}} \quad \underset{\substack{Ph}}{\overset{\substack{H}}{\text{---}C}}{=}CH_2 \quad \underset{}{\overset{+HCl}{\rightleftharpoons}} \quad \underset{\substack{Ph}}{\overset{\substack{H}}{Cl\text{---}C\text{---}CH_3}}
$$

(5) *By S_N2-reaction.* The racemization of optically active halides, particularly iodides, in which the halogen is attached to the chiral carbon, occurs in the presence of the corresponding halide ion. This is a result of the fact that the S_N2-reaction causes inversion of configuration (p. 175).

$$
I^- \quad \underset{\substack{R' \\ R''}}{\overset{R}{C\text{---}I}} \quad \rightleftharpoons \quad I\text{---}C\underset{\substack{R' \\ R''}}{\overset{R}{}}\ ^-I
$$

(*f*) EPIMERIZATION

Epimerization is said to occur when there is a change in configuration at one chiral centre in a compound which possesses more than one such centre. It results in the formation of a diastereoisomer, not the enantiomer, of the starting material. The mechanisms of epimerization parallel those of racemization.

(*g*) RESOLUTION

The resolution of a racemic modification consists of separating the enantiomers and isolating them in a pure state. (In synthesis, it is common practice to attempt to obtain only the required enantiomer in pure form.) There are two widely used approaches to resolution: *via* diastereoisomers, and by biochemical methods.

(*i*) *via Diastereoisomers.* The reaction of each of a pair of enantiomers with an optically active compound gives two diastereoisomers. Since diastereoisomers do not have identical properties, it is usually possible to separate them; crystallization is most commonly employed. Provided that the reaction leading to the diastereoisomers can be reversed, resolution can be achieved.

   The criteria for a satisfactory resolution by this method are that the diastereoisomers should be easily formed, well crystalline, and easily broken up. These conditions are met by employing salts: if the required enantiomers are acids or bases, or can be converted readily into them, treatment with an optically active base or acid, respectively, gives the diastereoisomeric salts and, after separation, the required enantiomers can be obtained by treating each diastereoisomer with mineral acid, to isolate an organic acid, or with base, to isolate an organic base. Many optically active acids and bases which are suitable as resolving agents occur naturally, e.g. (+)-tartaric acid and the alkaloids quinine (p. 729) and brucine.

Enantiomers which are neither acids nor bases are usually converted into acidic derivatives for resolution. For example, alcohols can be treated with phthalic anhydride to give the acid phthalate esters (44) which, after separation, are reconverted into the enantiomeric alcohols by hydrolysis with sodium hydroxide or reduction with lithium aluminium hydride (p. 639). Carbonyl compounds can be resolved by treatment with the semicarbazide (45), resolution of the resulting enantiomeric acids (46) with an optically active base, and reconversion by hydrolysis. Amino-acids, which are amphoteric, are usually first converted into their N-formyl derivatives which are resolved as acids and reconverted by hydrolysis.

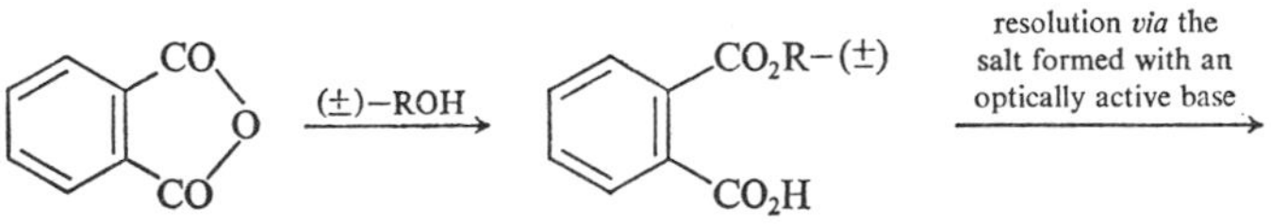

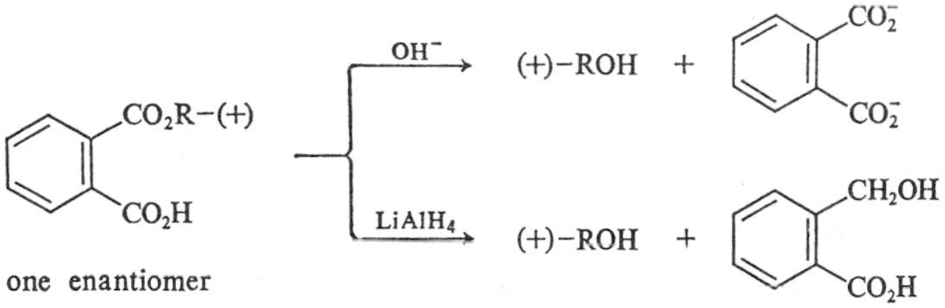

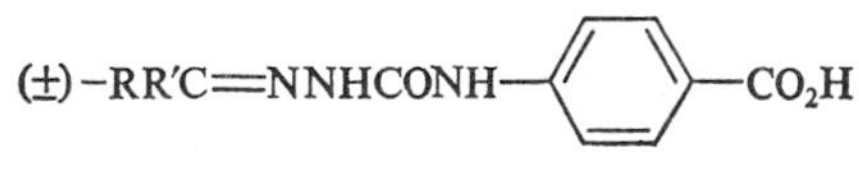

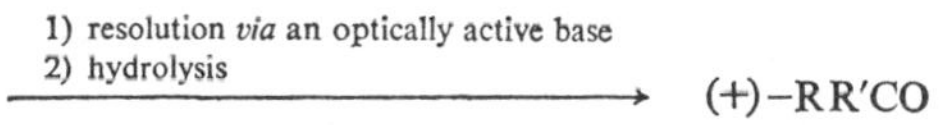

Compounds which cannot be converted readily into acidic or basic derivatives are usually obtained in optically active form by synthesis from an optically active precursor. There are, however, some special, rather than general, methods of resolution *via* diastereoisomers.

First, the diastereoisomeric molecular complexes formed with a chiral reagent may be employed. For example, the optically active fluorenone derivative (47), itself obtained by resolution as an acid, can be used to resolve derivatives of aromatic compounds such as naphthalenes which complex with it;* the separated diastereoisomers are usually reconverted readily into the reactants by heating or chromatography. A related method employs the inclusion complexes (clathrates) formed by certain chiral compounds.

(47)

Secondly, in some cases enantiomers can be separated by chromatography on an optically active support; equilibration occurs with diastereoisomeric adsorbates which have different stabilities, so that the enantiomer forming the less stable adsorbate is eluted first.

(ii) By biochemical methods. These methods differ from those in which diastereoisomers are formed and separated in employing the differences in *rates* of reactions of each of a pair of enantiomers with, or in the presence of, a chiral material of natural origin. The chiral compounds used include both enzymes and living organisms such as moulds. Their use is special rather than general, but where they are successful they are usually very efficient. The following are examples.

(1) Whereas the laboratory reaction between acetaldehyde and hydrogen cyanide gives optically inactive lactonitrile (p. 160) the presence of the enzyme emulsin gives a nearly optically pure product.

(2) The mould *Penicillium glaucum* preferentially destroys the (+)-isomer of racemic ammonium tartrate, leaving the pure laevorotatory salt.

(3) The reaction of racemic acetylphenylalanine with *p*-toluidine catalyzed by the enzyme papain gives the *p*-toluidide of acetyl-L-phenylalanine and leaves unchanged D-phenylalanine:

---

*These complexes are of the same type as the picrates formed by picric acid with aromatic compounds such as anthracene.

$$(\pm)-\text{PhCH}_2\text{CH}\overset{\text{NHAc}}{\underset{\text{CO}_2\text{H}}{<}} \quad + \quad \text{H}_2\text{N}-\!\!\left\langle\bigcirc\right\rangle\!\!-\text{CH}_3 \quad \xrightarrow{\text{papain}}$$

$$\text{PhCH}_2\text{CH}\overset{\text{NHAc}}{\underset{\text{CO}-\text{NH}-\!\!\left\langle\bigcirc\right\rangle\!\!-\text{CH}_3}{<}} \quad + \quad \text{PhCH}_2\text{CH}\overset{\text{NHAc}}{\underset{\text{CO}_2\text{H}}{<}}$$

L-isomer   D-isomer

These three examples are typical of methods which are described respectively as asymmetric synthesis, asymmetric destruction, and asymmetric kinetic resolution.

(*iii*) *Asymmetric transformations.* Each of the biochemical methods has a counterpart in the use of chiral compounds which are laboratory chemicals rather than living organisms (or materials such as enzymes which are derived from living organisms).

Asymmetric synthesis—the introduction of a chiral centre in the course of a reaction, with at least partial configurational specificity—is the most important of these methods. Two classes of reaction are recognized: in one, the reactant is not chiral but the reagent or catalyst is, as in the production of optically active lactonitrile described above; and in the other, the reactant contains at least one chiral centre but the reagent is not necessarily chiral. The principle of the methods is that the energies of the transition states leading to two diastereoisomers are not usually the same and hence there can be differences in the rates at which the diastereoisomers are formed.

Consider the reaction of a nucleophile Nu^- with an optically active aldehyde in which the chiral carbon is attached to a small group, S, a medium-sized group, M, and a large group, L. The preferential direction of approach, for steric reasons, is as shown in (48), so that the product contains more of the product (49) than its diastereoisomer (50); i.e. the new chiral centre is formed in a partially selective manner. Understandably, the degree of selectivity decreases as the chiral centre in the reactant is moved further from the functional group at which the new chiral centre is being created.

(48)   (49)   (50)

(*iv*) *Other methods.*, A variety of less general methods has been used for resolution. The earliest ever employed was the hand-picking of the chiral crystals of the enantiomeric sodium ammonium tartrates, but this method is limited because it is applicable only to racemic mixtures (p. 160), it requires that the crystals be easily distinguishable by eye, and it is excessively tedious.

Another method is to seed a saturated solution of the enantiomers with a crystal of one form which can induce the crystallization of the same enantiomer. If a crystal of one form is not available, it is sometimes possible to induce selective crystallization by seeding with a crystal of an optically active form of another molecule. In other cases one enantiomer may crystallize spontaneously from a supersaturated solution.

Finally, resolution has been achieved in one instance by inducing a photochemical reaction with circularly polarized light: irradiation of racemic $CH_3CH(N_3)CONMe_2$ preferentially destroys one enantiomer, depending on the direction of polarization of the light, and leaves some of the reactant enriched in the other enantiomer.

5.2 The Stereochemistry of Cyclic Compounds

The shapes of cycloalkanes containing 3 to 6 carbon atoms were discussed briefly in Chapter 2 (p. 47) and these compounds are here considered in the wider context of their stereochemical properties.

(*a*) TYPES OF STEREOISOMERISM IN CYCLOALKANES AND THEIR DERIVATIVES

(*i*) *Cyclopropanes.* The carbon atoms in cyclopropanes must inevitably lie in a plane. A monosubstituted (51) and a 1,1-disubstituted (52) cyclopropane exist in only one form, for the structures possess a plane of symmetry. A 1,2-disubstituted compound exists in four forms: a pair of *cis* enantiomers, (53) and (54), and a

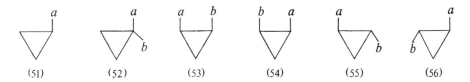

pair of *trans* enantiomers, (55) and (56). If the substituents are the same ($a = b$) the *cis* form possesses a plane of symmetry and is therefore a *meso* type.

(*ii*) *Cyclobutanes.* The CCC-angles would be least strained if cyclobutanes adopted a coplanar-ring structure, but this structure corresponds to maximum steric repulsion between substituents on adjacent carbon atoms. As a result of these opposing factors, the ring buckles slightly from the plane.

As with cyclopropanes, monosubstituted and 1,1-disubstituted cyclobutanes exist in only one form whereas 1,2-disubstituted cyclobutanes exist as diastereo-isomeric pairs of enantiomers (57, 58) (or a pair of enantiomers and a *meso* form if *a* = *b*). However, both the *cis* and the *trans* 1,3-disubstituted derivatives (59) and (60) possess a plane of symmetry, so that there are only two isomers and neither is optically active. (This is true of any disubstituted cycloalkane containing an even number of carbon atoms where the substituents are on the opposite sides of the ring.)

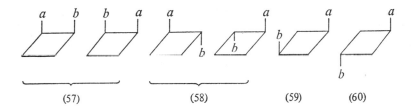

(57)   (58)   (59)   (60)

(*iii*) *Cyclopentanes.* The opposing angular and eclipsing strain factors which apply to cyclobutanes apply also to cyclopentanes, and these compounds adopt an 'envelope' structure. The 1,2- and 1,3-disubstituted compounds exist in four forms, there being two diastereoisomeric (*cis* and *trans*) pairs of enantiomers (cf. cyclopropanes).

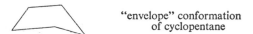

"envelope" conformation
of cyclopentane

(*iv*) *Cyclohexanes.* The cyclohexane system is by far the most commonly occurring of the cycloalkanes in natural products, doubtless because of its stability and the (related) ease of its formation (p. 47).

Unlike the 3- to 5-membered rings, cyclohexanes can adopt a conformation which is free both of angular strain and of conformational strain: this is the *chair* conformation (61) shown also in Newman projection (63). By co-ordinated rotations about its single bonds, the molecule can flip into the mirror-image chair conformation (62), and since the energy barrier between these forms is only 42 kJ mol^{-1}, flipping occurs rapidly at room temperature. In between the chair structures there are two other notable conformations: the skew-boat conformation (64), which is 23 kJ mol^{-1} less stable than the chair, and the boat conformation (65) [Newman projection (66)], which is 30 kJ mol^{-1} less stable. It should be noted that the latter is free of angular strain but possesses considerable conformational strain as a result of four eclipsing interactions (*xx*, etc.) and repulsion between the two groups *y* (the so-called bowsprit-flagpole interaction). Although the skew-boat form is not strainless, the conformational strain forces are considerably less than those in the boat form and the structure is (7 kJ mol^{-1}) more stable. The complete energy profile is shown in Fig. 5.1; the chair conformations constitute over 99·9% of the mixture of isomers.

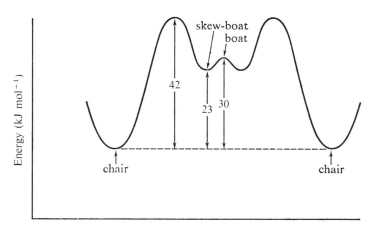

Fig. 5.1   Energy profile and conformations of cyclohexane.

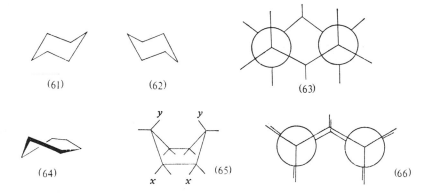

Axial and equatorial bonds. The twelve C—H bonds in the chair structure of cyclohexane are of two types (67). Six of the bonds are parallel to the three-fold axis of symmetry of the molecule, i.e. they are represented by vertical lines in the plane of the paper, and are described as *axial*; and the remaining six bonds are inclined at an angle of 109° 28′ to the three-fold axis and are described as *equatorial*. In derivatives of cyclohexane, the substituents which are in one or other of these two situations are described as axial and equatorial substituents, respectively.

(67)   (68)   (69)

Monosubstituted cyclohexanes. The substituent in a monosubstituted cyclohexane may be either axial (68) or equatorial (69). These structures are isomeric,
but it is not possible to isolate two isomers—*conformational isomers*—because
the rate of interconversion of the two, through ring-flipping, is too rapid, and an
equilibrium is maintained. The two isomers normally have different energies, so
that the equilibrium constant is not unity. The difference arises because there are
repulsive forces between the axial substituent and the hydrogen atoms on C_3
and C_5 [shown in (68)] which are not present when the substituent is equatorial
and which destabilize the axially substituted compound [*cf.* (68) and (69)]. The
difference in stability increases with the size of the substituent; e.g. for methyl,
phenyl, and t-butyl, ΔG is *ca.* 7·1, 11·3, and 23·4 kJ mol^{-1}, so that the equilibrium constants are about 20, 100, and 10^4, respectively, in favour of the
equatorially substituted conformational isomer.

Disubstituted cyclohexanes. There are three geometrically different relationships between the substituents, X, in a 1,2-disubstituted cyclohexane of the
formula $C_6H_{10}X_2$: each substituent may be axial (70), each may be equatorial
(72), or one may be axial and the other equatorial (74). The three structures are
shown together with their mirror images.

None of the structures is immediately superimposable upon its mirror image,
and three pairs of enantiomers might be expected. In practice, however, there are
fewer stereoisomers as a result of the rapid interconversions of chair conformations. Thus, (70) is in equilibrium with (73) (as can be seen by converting (70) into

(70)   (71)   (72)   (73)   (74)   (75)

the alternative chair conformation and rotating this through 120° about the
axis of symmetry), and likewise (71) is in equilibrium with (72). These four
structures therefore constitute one pair of enantiomers. The remaining pair are
superimposable by chair-flapping, and therefore constitute an optically inactive
($\pm$)-pair which are unresolvable because they interconvert too rapidly.

Largely for historical reasons* the four structures which constitute the enantiomeric pair are described as *trans* and the two which constitute the ($\pm$)-pair
are described as *cis*. Clearly, the relationship between the X groups in (70) and
(71) is a *trans* one, but it is not so in the conformational isomers (73) and (72).
It is useful to retain the terminology, however, and *trans* substituents in cyclohexanes (and other rings) are defined as substituents which are on opposite sides
of the ring when the ring is moulded into a planar form; *cis* substituents are those
which are correspondingly on the same side of the ring. The *trans* isomer is
usually more stable when the substituents are both equatorial (*cf.* monosubstituted cyclohexanes); e.g. for $X = CH_3$, ΔG is about 11·3 kJ mol^{-1}, so
that only about one molecule in a hundred is present in the diaxial form. The
trans isomer is more stable than the *cis* isomer, in which one substituent is
necessarily in the unfavourable axial situation; e.g. for $X = CH_3$, ΔG is about
7·5 kJ mol^{-1}

This discussion of the 1,2-disubstituted cyclohexanes can readily be extended
to the 1,3- and 1,4-analogues. The 1,3-compounds exist in three discrete stereoisomeric forms: an enantiomeric *trans* pair in which one substituent is axial and
the other equatorial, (76) and (77); and a *cis* isomer which has a plane of symmetry (not resolvable) and which consists of interconvertible conformational
isomers having both substituents respectively equatorial (78) and axial (79), the

(76)   (77)   (78)   (79)

former predominating. The *cis* isomer is in this case the more stable, e.g. for
$X = CH_3$, by 7·5 kJ mol^{-1}.

The 1,4-compounds exist in only two forms: a *trans* compound, with both
substituents either equatorial (80) or axial (81); and a *cis* compound, with one

---

*It was originally thought that cyclohexanes have planar rings. The relationship between
the isomers above was therefore imagined to be similar to that of the geometrical isomers
of olefins, i.e.

The non-resolvable *cis*-isomer is now seen to be the conformational isomer with one axial
and one equatorial substituent, and the resolvable *trans*-isomer to be the conformational
isomer with both substituents either axial or equatorial.

axial and one equatorial substituent (82). Both *cis* and *trans* compounds possess a plane of symmetry and are optically inactive, and the *trans* isomer is the more stable.

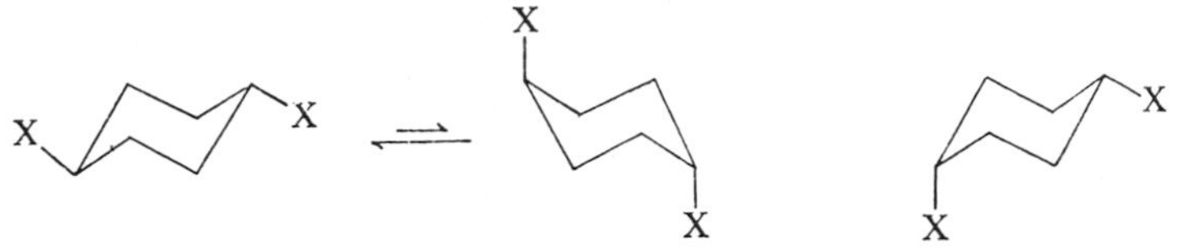

(80) (81) (82)

(*v*) *Cyclohexene.* Since the four carbon atoms in the fragment C—C=C—C are necessarily coplanar, cyclohexene cannot adopt either a chair or a boat form. Instead, its preferred conformation is the half-chair (83).

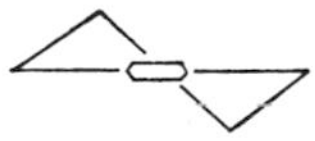

(83) 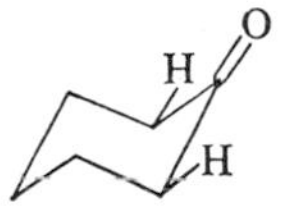 (84)

(*vi*) *Cyclohexanone.* Compared with cyclohexane, the ring in cyclohexanone is buckled very slightly from the chair structure in order to accommodate the trigonal carbon atom whose optimal CCC angle is 120°. As shown in (84), the equatorial hydrogen atoms on the α-carbon atom are then nearly eclipsed by the carbonyl oxygen, so that equatorial substituents are somewhat destabilized by steric repulsions. In some cases this results in the greater stability of the axially substituted conformational isomer (e.g. 2-bromocyclohexanone) and in others, of the boat-shaped conformation [e.g. *cis*-2,4-di-t-butylcyclohexanone (85)].

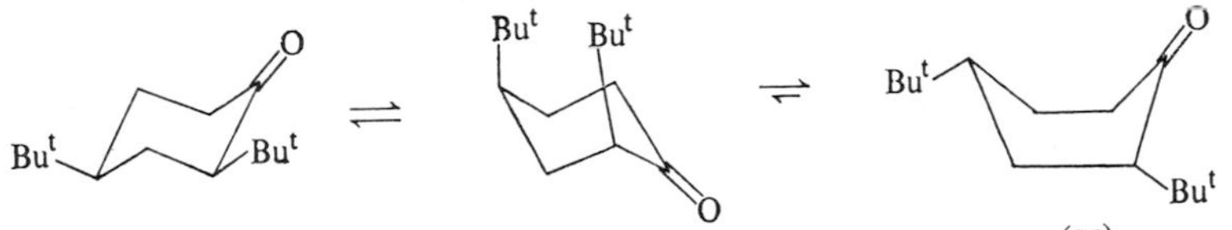

(85)

(*vii*) *Larger rings.* Heat of combustion data show that cycloalkanes containing 7- to 11-membered rings are all slightly strained (p. 48). This is because the conformations in which the CCC angles are tetrahedral do not correspond to the optimal staggered arrangement of hydrogen atoms on adjacent carbons. Indeed, to reduce the strain due to this partial eclipsing, the rings deform slightly and an optimum situation is achieved in which the total strain (angular + eclipsing) is minimized. Rings containing 12 or more carbon atoms can, however, adopt conformations in which there is neither angular nor conformational strain.

Detailed information about the preferred conformations of cycloalkanes containing 7- or more members is not complete. It is, however, known that cyclo-heptanone preferentially adopts the somewhat deformed chair structure (86).

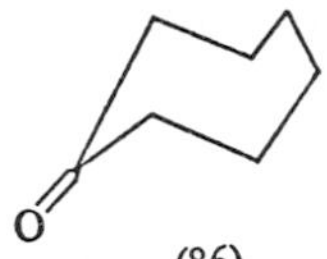

(86)

(b) FUSED RINGS

A fused ring-system is one in which adjacent rings have two atoms in common. By far the most important is the decalin system, of which there are two isomers; the preferred conformations of each are shown.

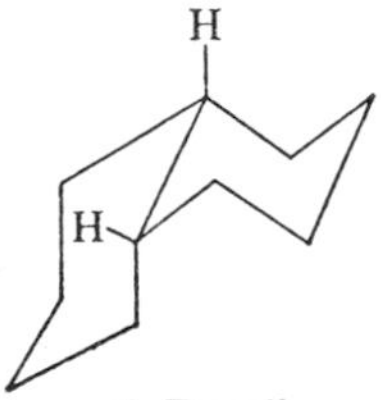

cis-Decalin
(87)

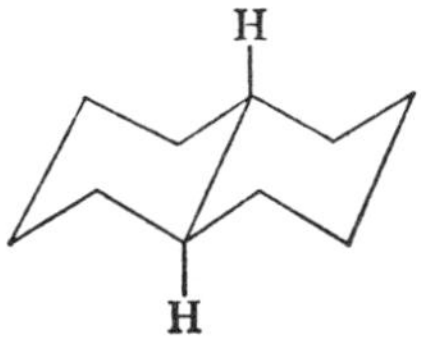

trans-Decalin
(88)

In the *trans* isomer each ring is fused to the other by equatorial bonds, whereas in the *cis* isomer one ring is joined to the other by an axial bond. Consequently, in line with the greater stability of *trans*- than *cis*-1,2-dimethylcyclohexane (p. 170), the *trans* isomer is the more stable (by 11 kJ mol^{-1}).

trans-Decalin is incapable of undergoing ring-flipping, whereas the *cis* compound interconverts rapidly with the conformational isomer in which the rings shown in (87) are in the alternative chair forms. The *trans* compound possesses a centre of symmetry and is therefore optically inactive. The *cis* compound is chiral in both conformations, which are non-superimposable mirror images of each other, but since ring-flipping between them is rapid the compound is a non-resolvable ($\pm$)-pair.

The 9-methyldecalin system occurs in many natural products (e.g. cholesterol; p. 717). It differs from the decalin system in that the difference in stabilities between the *cis* and *trans* forms is smaller, as a result of the fact that in the *trans* compound (89) the methyl group is sterically compressed by four hydrogen atoms and in the *cis* compound (90) by only two. Depending on the presence of other groups in substituted decalins, either the *cis* or the *trans* isomer may be the more stable.

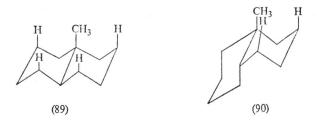

(89) (90)

Polycyclic systems composed of fused six-membered rings also occur commonly, particularly the perhydrophenanthrene system. Perhydrophenanthrene itself exists in ten stereoisomeric forms (four enantiomeric pairs and two *meso* isomers); the most stable isomer (*trans-anti-trans*) and the least stable (*trans-syn-trans*, in which the central ring is in the boat conformation) are shown.*

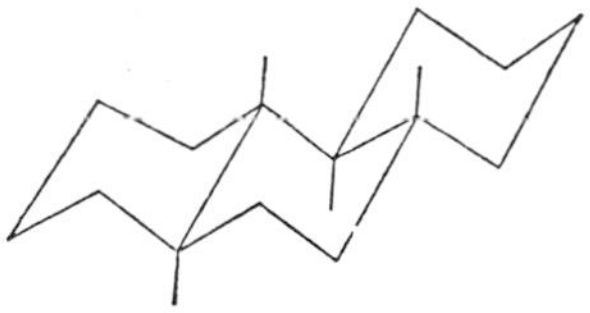

trans-anti-trans
(91)

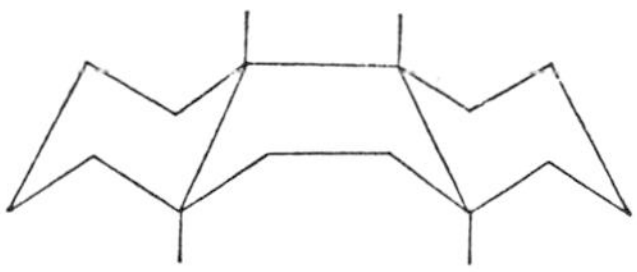

trans-syn-trans
(92)

(*c*) BRIDGED RINGS
A bridged ring-system is one in which adjacent rings have more than two atoms in common. Many such systems occur naturally: examples are derivatives of bornane (93) in terpenes, and tropane (94), in alkaloids.

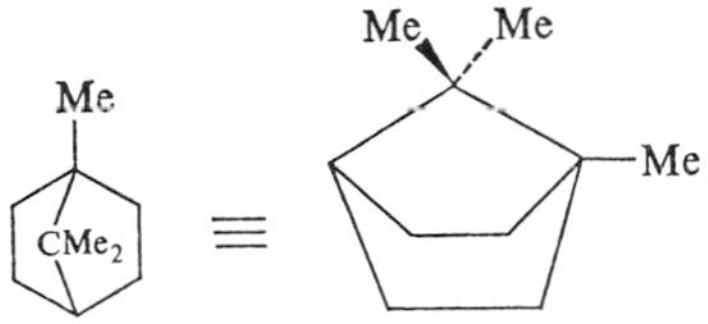

(93) 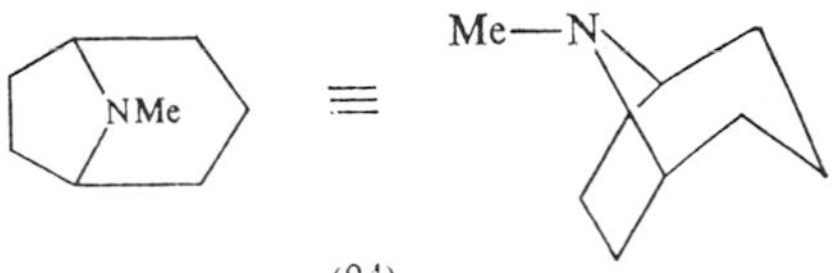   (94)

Many other bridged-ring systems have been synthesized, among them, norbornadiene, nortricyclane, 'Dewar benzene' (bicyclo[2.2.0]hexadiene), barrelene (bicyclo[2,2,2]octatriene), cubane, and adamantane. With the exception of the last, each is strained in some degree and 'Dewar benzene' is particularly unstable, but in adamantane the four rings each adopt the strainless chair con-

---

*The prefixes *cis* and *trans* are used to denote the stereochemistry at the bonds of fusion between rings, and *syn* and *anti* refer to the orientation of the terminal rings with respect to each other.

formation and the compound is the most stable isomer of molecular formula $C_{10}H_{16}$. Urotropin (hexamethylenetetramine) (p. 330) is a heterocyclic analogue of adamantane in which nitrogen atoms replace CH at the four bridgeheads.

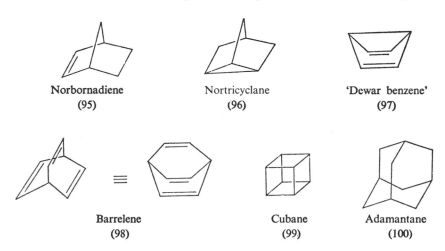

Norbornadiene   Nortricyclane   'Dewar benzene'
(95)   (96)   (97)

Barrelene   Cubane   Adamantane
(98)   (99)   (100)

5.3 Stereochemistry and Reactivity

Reactivity is related to stereochemistry in two ways. First, many reactions have specific requirements for the relative conformations of the groups involved; for example, in E2 eliminations the eliminated groups are *trans* to each other (p. 119). Secondly, the conformational requirements of the non-reacting groups help to determine reactivity; for example, S_N2 displacements on neopentyl bromide are far slower than on ethyl bromide because the three β-methyl groups hinder the approach of the nucleophile (p. 128). These characteristics are termed respectively *stereoelectronic* and *steric* factors.

These factors play an important part in synthesis. Thus, in building up molecules containing several chiral centres such as the steroids and some of the alkaloids (Chapter 21), it is desirable to be able to carry out reactions which give wholly, or at least predominantly, the one desired stereoisomer when two or more exist. Fortunately, knowledge of the mechanisms of reactions and of the inter-relationship of mechanism and stereochemistry is now at a stage where such control is possible and can be widely applied.

Two levels of stereochemical control can be recognized. A reaction may be *stereospecific*: that is, stereochemically different starting materials give stereochemically different products; or it may be *stereoselective*: that is, of two or more possible stereoisomeric products, one is formed in predominance.

The stereochemical factors controlling reactivity and the stereospecificity and stereoselectivity of reactions will be discussed with respect first to acyclic and then to cyclic compounds.

(a) ACYCLIC COMPOUNDS

(i) S_N2 Reactions. S_N2 Processes are stereospecific, involving inversion of configuration at the reaction centre:

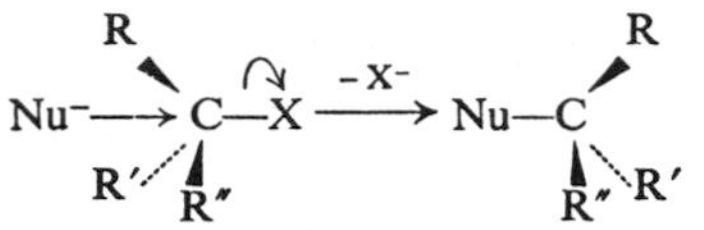

(101)

Thus, if the reaction centre is a chiral carbon having the L configuration, the product has the D configuration.

If inversion of configuration is impossible, S_N2 displacements do not occur: e.g. apocamphyl chloride (102) is inert to hydroxide ion.

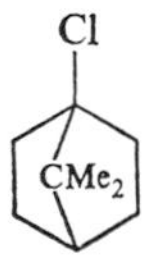

(102)

S_N2 reactions are also susceptible to steric hindrance (p. 128).

(ii) S_N1 Reactions. Unlike their bimolecular analogues, S_N1 reactions are not stereospecific. This is because the intermediate carbonium ion is coplanar and can be attacked from both sides to give a mixture of the two possible products:

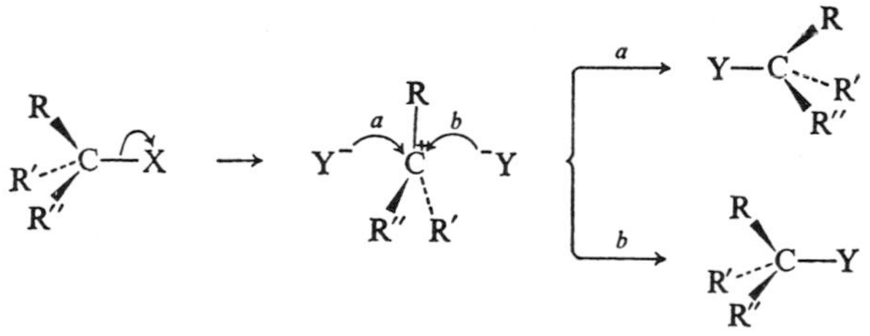

In practice, the product is not always the racemic modification indicated by the above formulation but contains more of that enantiomer in which the original configuration has been inverted. A simple view of this is that the departing group, X, shields the carbonium ion from one side and so favours attack from the other.

That carbonium ions are coplanar results in the fact that a compound which cannot achieve coplanarity of the bonds in such an ion does not undergo the

S_N1 reaction. For example, apocamphyl chloride (102; p. 175) is inert to displacement of the chlorine, although its acyclic analogues, tertiary chlorides, are very reactive.

(*iii*) *E2 Eliminations.* For reaction to occur, the two eliminated groups must be able to attain the *anti* conformation:

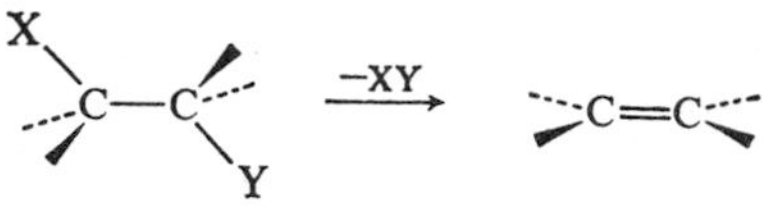

Consider the elimination of hydrogen bromide, induced by hydroxide ion, from the diastereoisomers of PhCHBr—CHPhCH$_3$. Reaction occurs through the *anti* conformations (103) and (104) respectively, with the result that the former diastereoisomer gives specifically *trans*-α-methylstilbene and the latter gives specifically *cis*-α-methylstilbene:

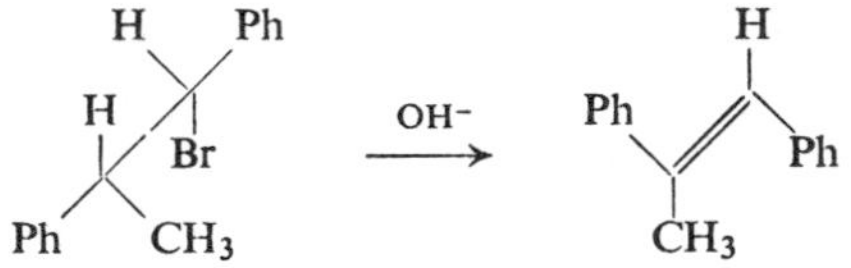

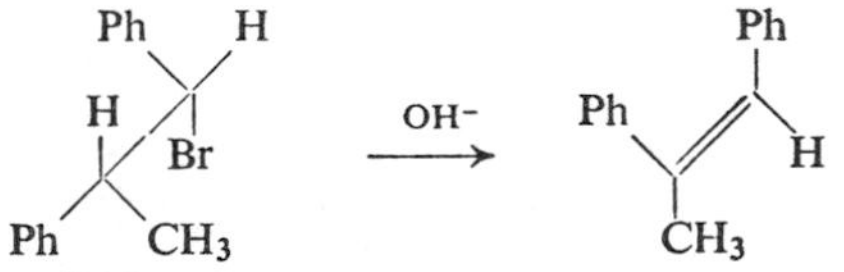

Similarly, the elimination of bromine, induced by iodide ion, gives specifically *trans*-2-butene from *meso*-2,3-dibromobutane and specifically *cis*-2-butene from both (+)- and (−)-2,3-dibromobutane:

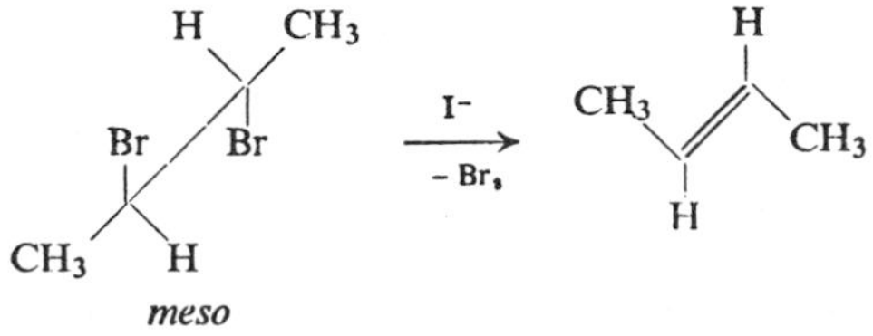

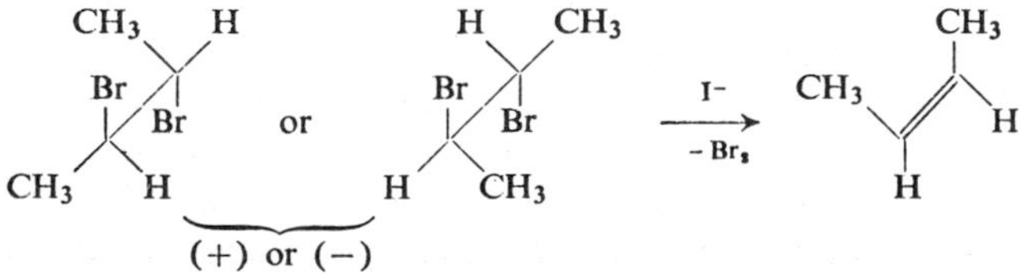

In addition to the fact that each of the above reactions is stereospecific, the rate of formation of the *trans* product is greater in each case than that of the *cis* product. This is because the repulsive forces between the nearby aromatic rings or methyl groups which are present in the *cis* isomers, and serve to make them less stable than the *trans* isomers, are present in some degree in the transition states in which the *cis* products are being developed.

These are examples of stereospecificity in elimination reactions; there are also situations in which a particular compound reacts stereoselectively to give a predominance of one of two or more products. For example, the base-induced elimination of hydrogen chloride from 1,2-diphenyl-1-chloroethane can occur through each of two *anti* conformations to give *cis*- and *trans*-stilbene respectively, the latter in predominance:

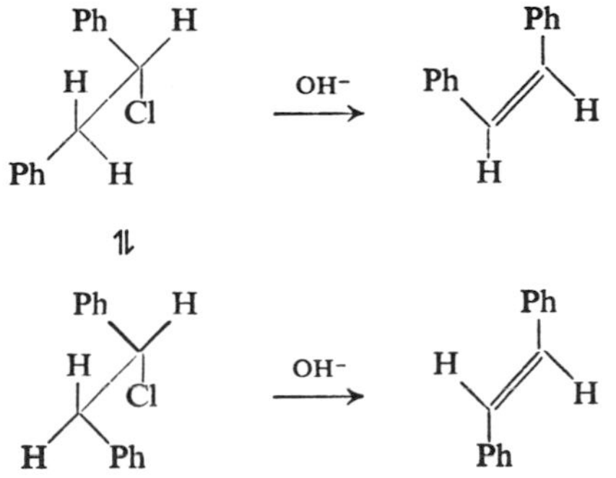

A priori, either or both of two factors might be responsible for the preponderance of the *trans*-isomer: the conformation of the reactant leading to the major product might be more heavily populated than that leading to the minor product; and the transition state for the formation of the major product might be of lower energy than that leading to the minor product. Now, the activation energy of the elimination is large compared with the rotational energy barrier which separates the two significant conformations, and in these conditions a general proposition—the Curtin-Hammett principle—applies: the relative proportions of the products are not determined by the relative populations of the ground state conformations but depend only on the relative energies of the transition states of the various processes. Thus, in the present example, *trans*-stilbene predominates over *cis*-stilbene because the transition state which leads to it is of lower energy; this in turn follows because of the repulsive forces between the two aromatic rings in the transition state leading to *cis*-stilbene (see above).

The basis of the Curtin-Hammett principle is as follows. Suppose the two conformers, A and B, have free energies G_1 and G_2 and the transition states for the formation of the products from each have free energies $G_1^{\ddagger}$ and $G_2^{\ddagger}$ (Fig. 5.2). The ratio, R, of the products, P_1 and P_2, is equal to the ratio of their

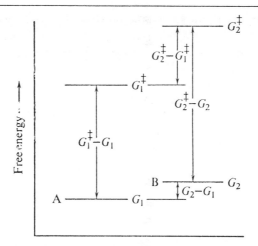

Fig. 5.2   Energies of reactants and transition states.

rates   for   formation,   i.e. $R = (dP_1/dt)/(dP_2/dt) = k_1[A][X]/k_2[B][X] = k_1[A]/k_2[B]$ (where $[X]$ symbolizes the concentration of all other species in the rate equation). From transition-state theory (p. 81),

$$k_1 = (kT/h)e^{-\Delta G_1^{\ddagger}/RT} \text{ and } k_2 = (kT/h)e^{-\Delta G_2^{\ddagger}/RT}$$

where $\Delta G_1^{\ddagger} = G_1^{\ddagger} - G_1$ and $\Delta G_2^{\ddagger} = G_2^{\ddagger} - G_2$. In addition, the conformers A and B are in equilibrium, i.e. $[B]/[A] = K = e^{-\Delta G/RT}$ where $\Delta G = G_2 - G_1$. It follows that

$$R = \frac{e^{-\Delta G_1^{\ddagger}/RT}}{e^{-\Delta G_2^{\ddagger}/RT}} \cdot e^{\Delta G/RT} = e^{(\Delta G + \Delta G_2^{\ddagger} - \Delta G_1^{\ddagger})/RT}$$

$$= e^{(G_2^{\ddagger} - G_1^{\ddagger})/RT}$$

which is dependent only on the difference in free energies of the two transition states.

(*iv*) *Pyrolytic eliminations.* Certain eliminations which occur on heating, and without the need for an added reagent, take place through cyclic transition states in which the eliminated groups are *cis* to each other (p. 318), e.g. xanthate pyrolysis:

Consequently one or other of the geometrically isomeric products is formed from a particular reactant, although reaction is not exclusively stereospecific and a small proportion of the alternative isomer is usually produced.

(*v*) *E1 Elimination.* Unlike E2 eliminations, E1 eliminations are not stereospecific. Essentially, this is because the intermediate carbonium ion can undergo rotation about the bonds to the tervalent carbon atom before a proton is eliminated (however, more subtle factors are involved in determining the precise ratios of the possible products), e.g.

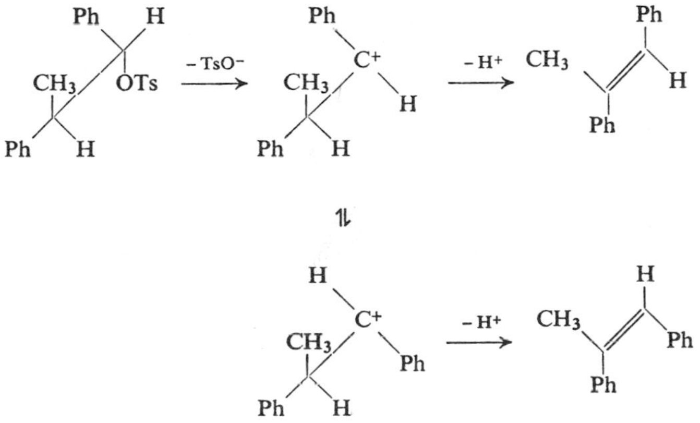

(*vi*) *Electrophilic addition to olefins.* In most cases where the stereochemistry of electrophilic addition to olefins has been studied, reaction has been found to occur in a stereospecifically *trans* manner; an exception, which will not be considered here, is the addition of hydrogen bromide to certain olefins.

The addition of bromine to *cis*- and *trans*-2-butene is illustrative: the former gives ($\pm$)-2,3-dibromobutane and the latter gives *meso*-2,3-dibromobutane (cf. the iodide-catalyzed elimination of bromine from each of these dibromobutanes; p. 176). This stereospecificity indicates that the intermediate carbonium ion cannot undergo rotation about the bonds to the tervalent carbon before reaction is completed, and to account for this it has been suggested that the carbonium ion's stereochemistry is held by interaction between the ion and the electrophile which has been added, e.g. for bromination:

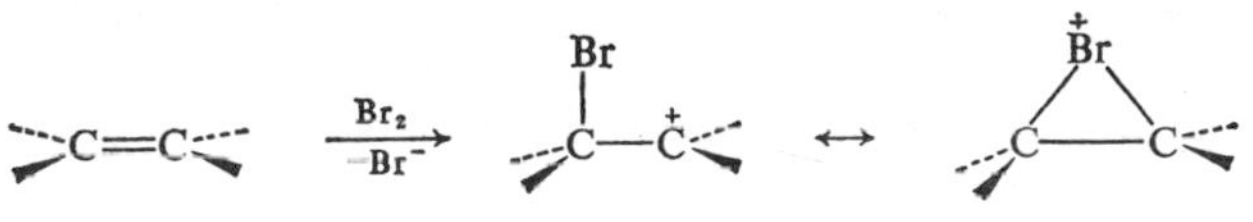

The anion then attacks the intermediate cation in the S_N2 manner, i.e. from the side opposite the bromine atom already present:

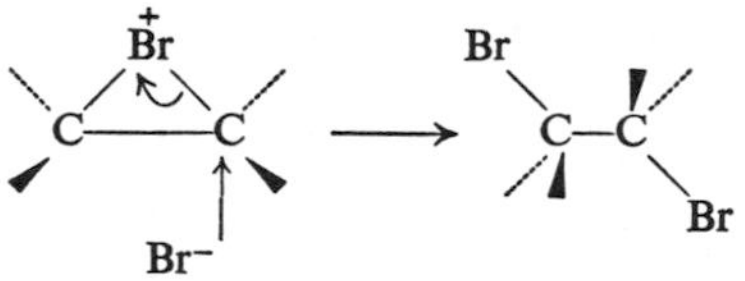

The full course of the bromination of the isomeric 2-butenes is therefore as follows:

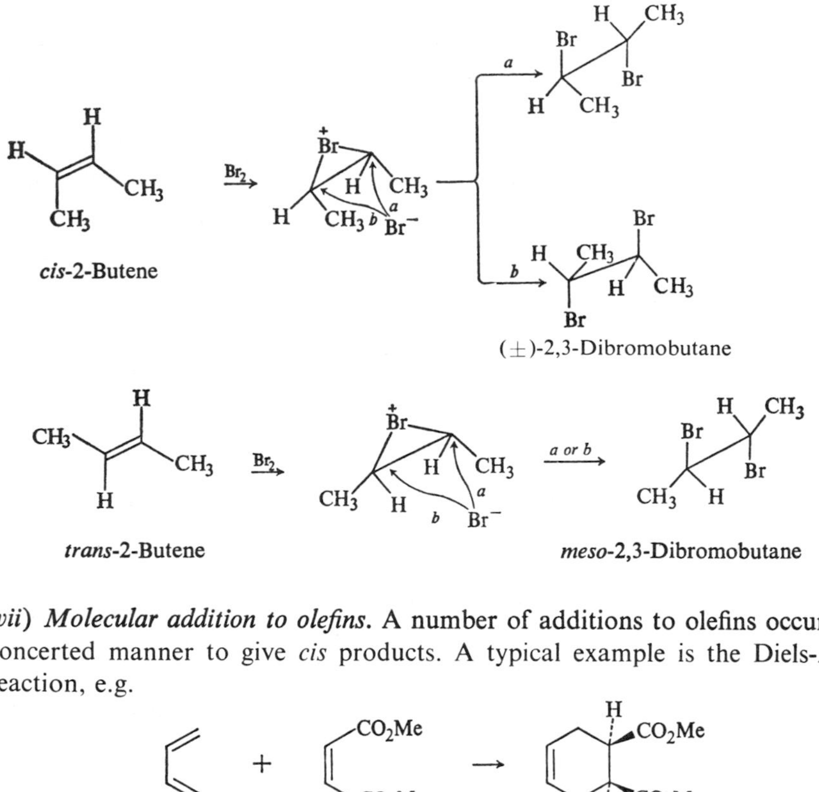

cis-2-Butene

($\pm$)-2,3-Dibromobutane

trans-2-Butene

meso-2,3-Dibromobutane

(*vii*) *Molecular addition to olefins.* A number of additions to olefins occur in a concerted manner to give *cis* products. A typical example is the Diels-Alder reaction, e.g.

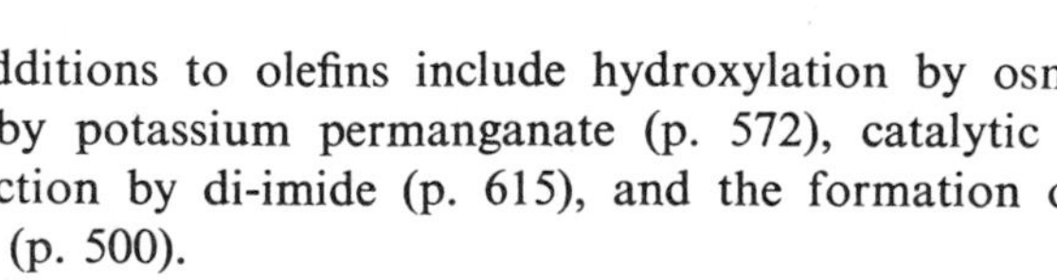

The full stereochemical details of Diels-Alder addition are described later (p. 292).

Other *cis* additions to olefins include hydroxylation by osmium tetroxide (p. 571) and by potassium permanganate (p. 572), catalytic hydrogenation (p. 611), reduction by di-imide (p. 615), and the formation of alcohols *via* hydroboration (p. 500).

Carbenes (divalent carbon species) in which the two non-bonding electrons on carbon have paired spins add to olefins in a concerted and stereospecifically *cis* manner, e.g.

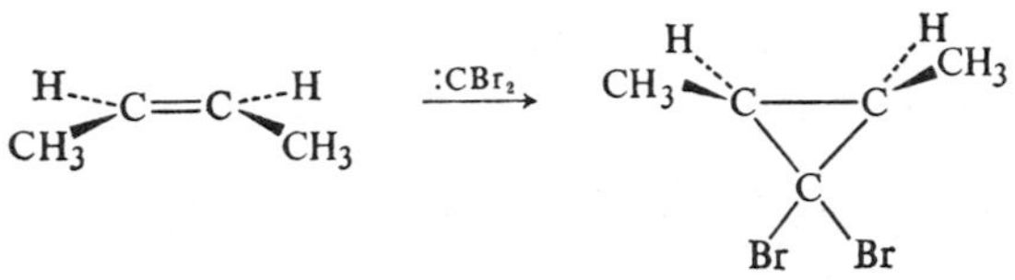

Carbenes in which the non-bonding electrons have parallel spins add non-stereospecifically, because of the requirement that one electron must invert its spin before the reaction can be completed. The two new bonds are formed with a time-lag between them while spin-inversion occurs, and during this time the intermediate can undergo rotation about a single bond so that a mixture of *cis* and *trans* products results, e.g.

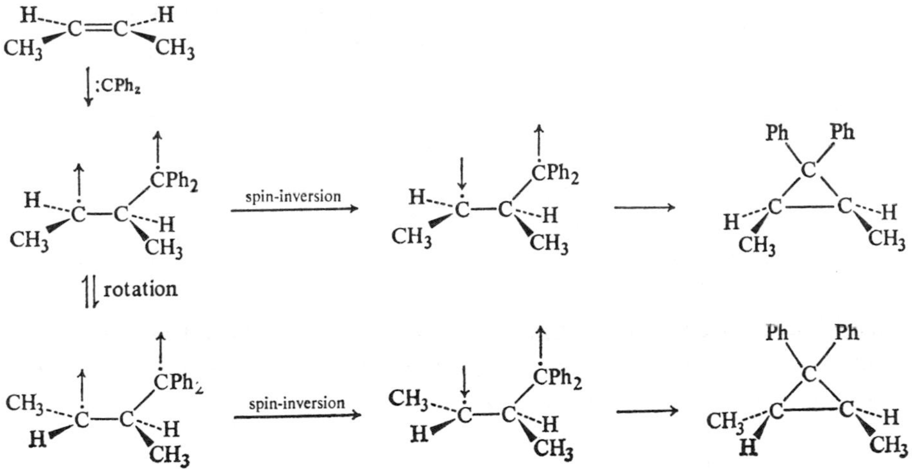

(*viii*) *Addition to acetylenes.* Electrophilic addition gives mainly, but by no means exclusively, the *trans* product: e.g.

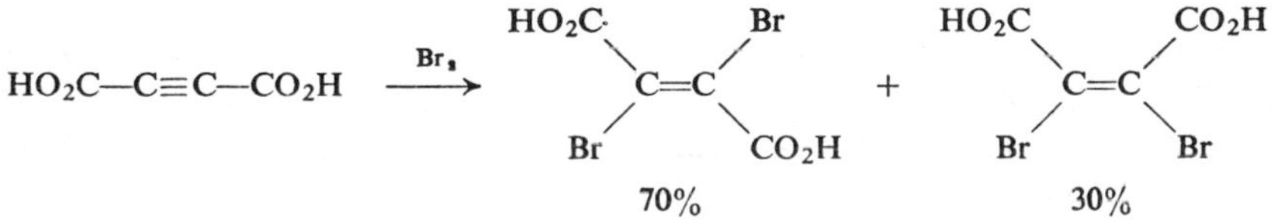

Nucleophilic addition occurs in the *trans* manner, e.g.

$$Ph-C\equiv C-H \xrightarrow{\text{CH}_3\text{O}^-/\text{CH}_3\text{OH}} \underset{\substack{\diagup \\ H}}{\overset{\substack{Ph \\ \diagdown}}{C}} = \underset{\substack{\diagdown \\ H}}{\overset{\substack{OCH_3 \\ \diagup}}{C}}$$

Catalytic hydrogenation gives specifically the *cis* olefin (p. 613), whereas chemical reduction with sodium in liquid ammonia gives mainly the *trans* olefin (p. 620), as does the reduction of α-acetylenic alcohols by lithium aluminium hydride (p. 620).

(*ix*) *Addition to carbonyl groups.* Addition of an achiral reagent to the carbonyl group of an optically inactive aldehyde or ketone necessarily occurs with neither stereospecificity nor stereoselectivity (p. 160). If, however, the carbonyl group is part of a chiral system, addition can occur with some degree of selectivity to give a predominance of one diastereoisomer (p. 165).

(*x*) *Intramolecular rearrangements.* Rearrangements of electrophilic type (14.1) display, in appropriate cases, three stereochemical properties: the migrating group retains its configuration; there is inversion of configuration at the migration origin; and there is also inversion of configuration at the migration terminus.

The first property is shown in those reactions in which a chiral alkyl group migrates, as in the following Hofmann rearrangement (p. 468):

$$\underset{\substack{| \\ C_2H_5}}{\overset{\substack{H \\ |}}{CH_3\text{-----}C-CONH_2}} \xrightarrow{\text{Br}_2\text{-OH}^-} \underset{\substack{| \\ C_2H_5}}{\overset{\substack{H \\ |}}{CH_3\text{-----}C-NCO}} \xrightarrow{\text{H}_2\text{O}} \underset{\substack{| \\ C_2H_5}}{\overset{\substack{H \\ |}}{CH_3\text{-----}C-NH_2}}$$

Manifestation of the second property is rare; an example is the conversion of camphene hydrochloride into isobornyl chloride which occurs on standing and which may be represented as follows:

The third property is shown in reactions of the pinacol type (p. 461), e.g.

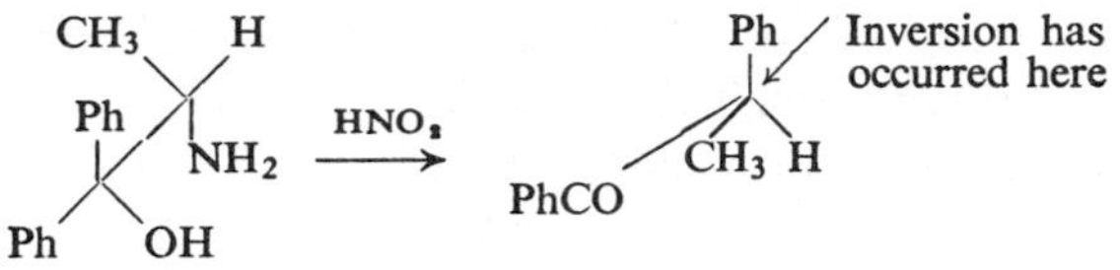

The last reaction has a further stereochemical characteristic. In a compound in which either of two aryl groups might migrate, such as $PhArC(OH)$—$CH(NH_2)CH_3$ (Ar $=p$-methoxyphenyl), the phenyl group migrates through conformation (105) and the p-methoxyphenyl group through conformation (106). The former conformation is the more stable, and therefore the more heavily populated, because the two largest groups, Ph and Ar, are gauche to the smallest group (hydrogen) on the adjacent carbon atom rather than to the methyl group, as in the latter conformation. On the other hand, p-methoxyphenyl, having a greater capacity than phenyl for electron-supply to electron-deficient transition states, has the greater intrinsic migratory aptitude. Were the Curtin-Hammett principle to apply, the relative populations of the appropriate conformational states would be irrelevant and p-methoxyphenyl would migrate preferentially. However, in this case the principle does not apply because the activation energy for the loss of nitrogen from the intermediate cation which leads to migration is of the same order of magnitude as the barrier to internal rotation. Consequently the relative proportions of the initial conformational states become significant, and in fact the predominant product is that in which phenyl has migrated.

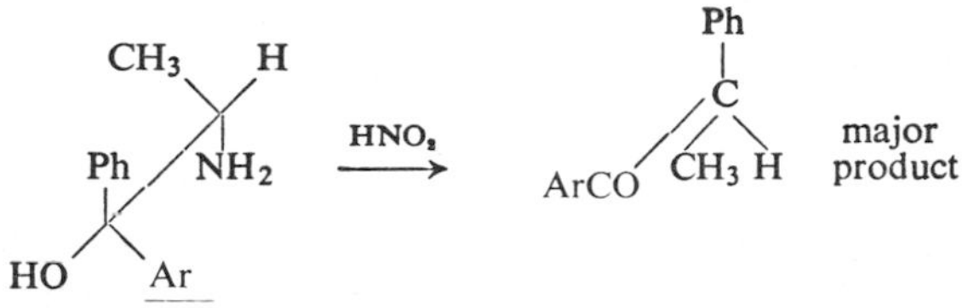

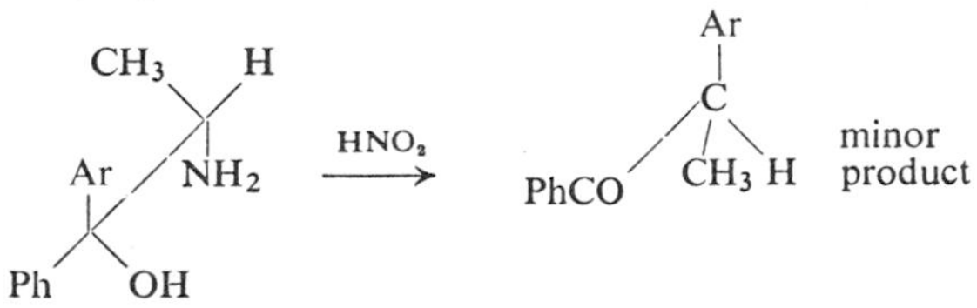

It should be noted that, in the other diastereoisomer of the above compound, the more stable of the two critical conformations corresponds to the *anti* arrangement of p-methoxyphenyl and amino, and p-methoxyphenyl migrates essentially exclusively.

(xi) *Neighbouring-group participation.* Participation by neighbouring groups can result not only in an enhancement of the rate of a reaction (p. 129) but also in the alteration of the normal stereochemical course. This is particularly significant in substitutions; the following are examples.

(1) The optically active 3-bromo-2-butanol (107) reacts with hydrogen bromide to give ($\pm$)-2,3-dibromobutane, whereas the diastereoisomeric (108) gives *meso*-2,3-dibromobutane. If reaction occurred through 'open' carbonium ions, each compound would give a mixture of ($\pm$) and *meso* products; however, the neighbouring bromine atom fixes the stereochemistry of the intermediate ion, as shown, in a way similar to that in which the *trans* addition of bromine to olefins is controlled (p. 179).

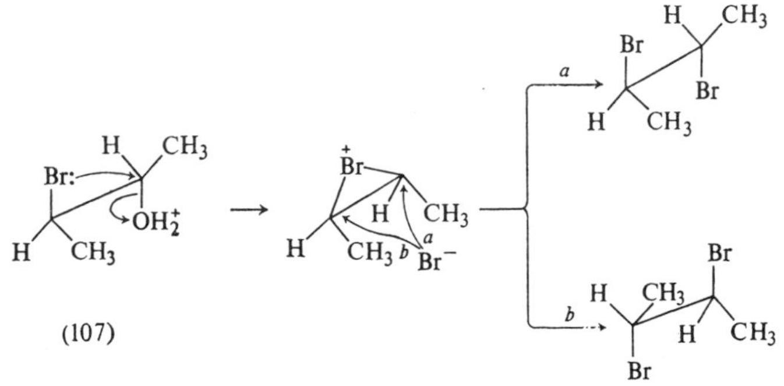

(107)

($\pm$)-Product

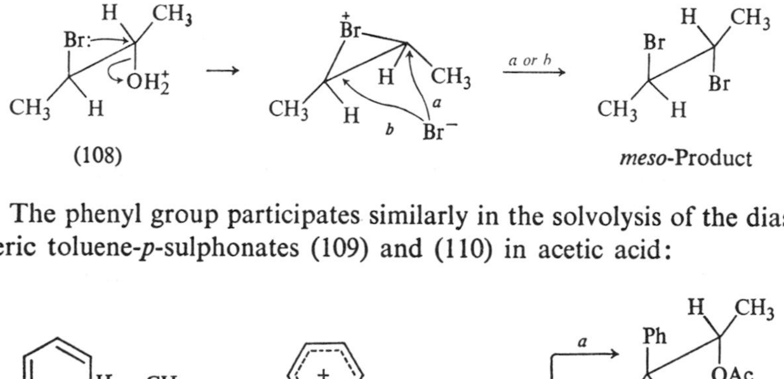

(108)

meso-Product

(2) The phenyl group participates similarly in the solvolysis of the diastereoisomeric toluene-*p*-sulphonates (109) and (110) in acetic acid:

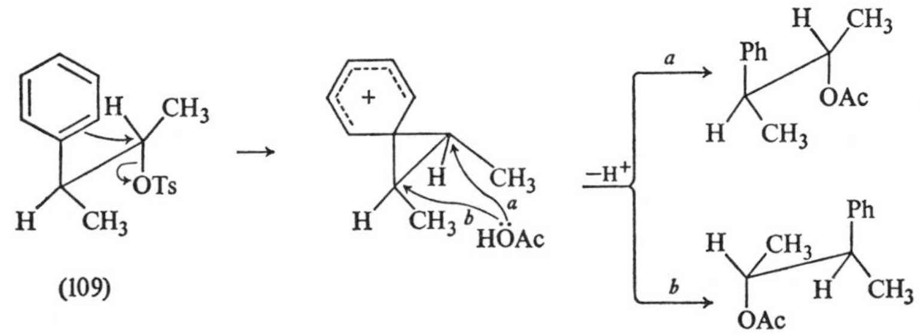

(109)

One pair of
enantiomeric products

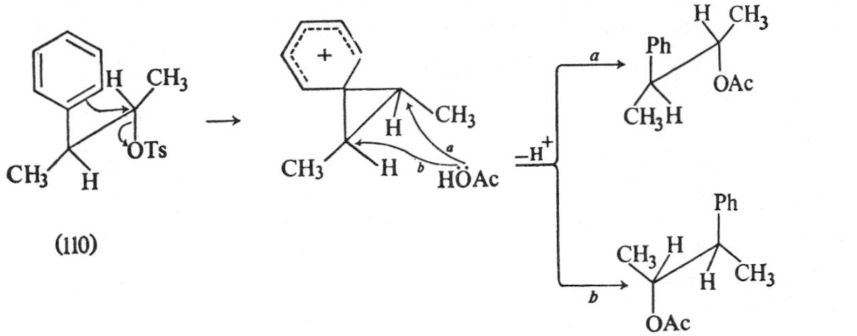

A diastereoisomer of the
enantiomers from (109) (paths
a and *b* giving the *same* product)

(3) The carboxylate group in the α-bromopropionate ion (111) directs the stereochemistry of the unimolecular solvolysis below by preventing the attack of solvent from the side opposite to the departing bromide ion, with the result that the configuration of the chiral carbon atom is retained:

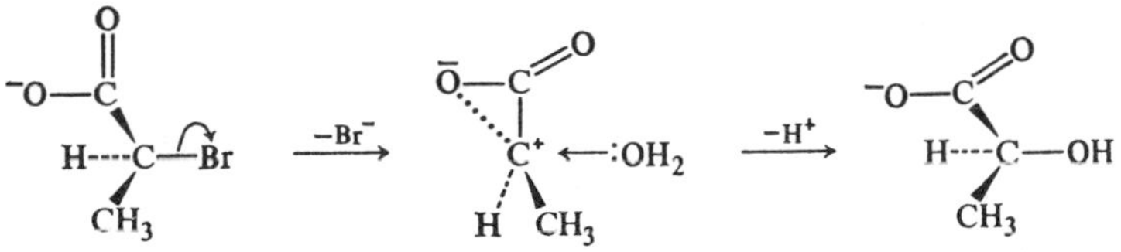

(b) CYCLIC COMPOUNDS

With the exception of the final section, the following discussion concerns only cyclohexane systems. This is because the six-membered ring system is the most widely occurring in natural products and has been investigated the most extensively. It should, however, be clear that many of the propositions apply in a general sense to other ring systems.

(i) *Equilibria.* Monosubstituted cyclohexanes exist as equilibrium systems in which axially and equatorially substituted compounds interconvert rapidly through the flipping of the ring, although one or other of the conformational isomers usually predominates (p. 169). However, disubstituted and more complex cyclohexanes can exist in stereoisomeric forms which are not normally interconvertible, whether or not the ring is able to flip. For example, *cis*- and *trans*-menthane exist as separate compounds even though ring-flipping occurs in each; and *cis*- and *trans*-decalin exist as separate compounds because ring-flipping cannot occur to interconvert the two (p. 172).

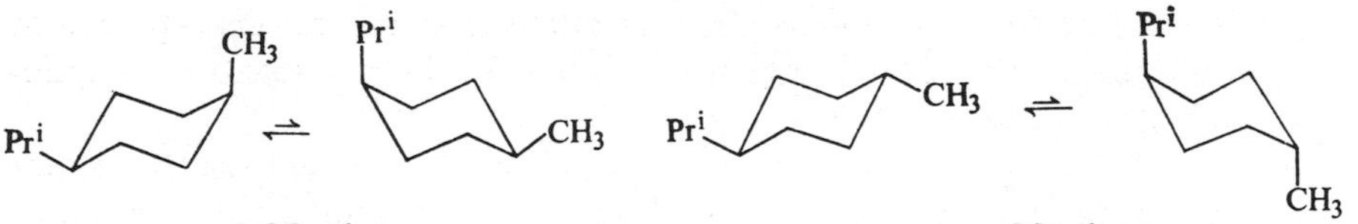

cis-Menthane   trans-Menthane

There are, however, structural situations which do allow interconversion of such stereoisomers in certain conditions. For example, the incorporation of a carbonyl group in menthane gives *cis* and *trans* isomers which are interconvertible in both acidic and basic conditions; the *trans* isomer, having both alkyl substituents equatorial, predominates.

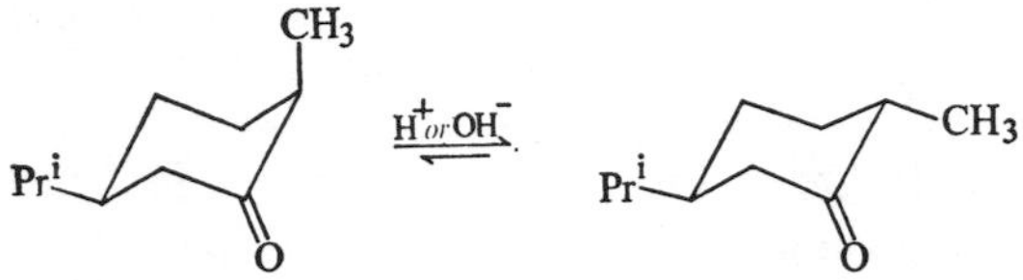

Isocarvomenthone   Carvomenthone
20%   80%

Interconversion occurs, in acid conditions, through the enolic tautomer of the ketone, and, in basic conditions, through the enolate anion (carbanion):

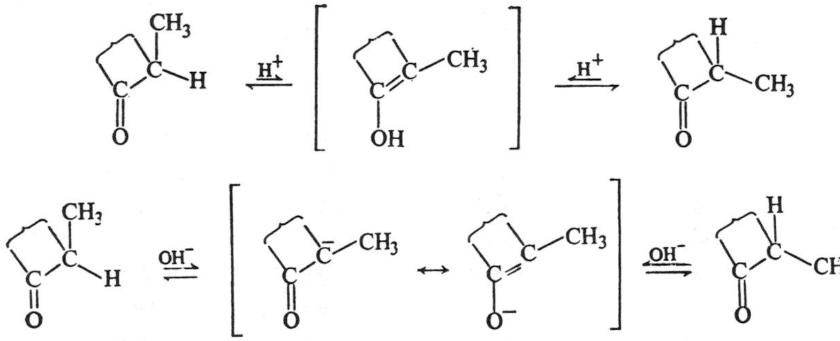

A second example is the interconversion of the decalin derivatives (112) and (113) by the action of ethoxide ion; a carbanion is again the critical intermediate. Equilibrium lies to the right because (113) has its carbethoxyl group in the equatorial position.

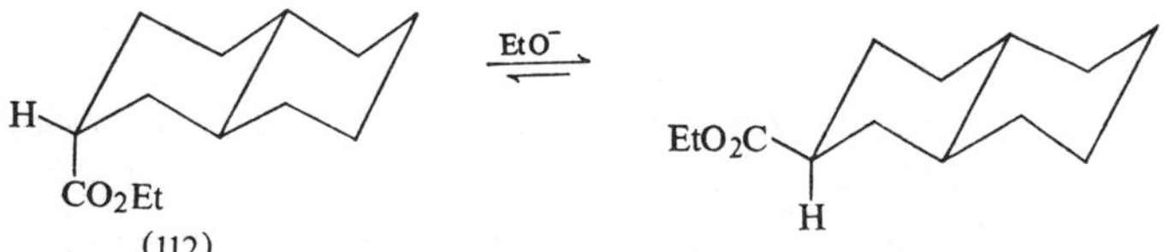

(112)   (113)

It will be clear from these examples that interconversion of this type can in general be brought about by the methods suitable for the racemization of optically active compounds (p. 160).

Interconversion can also be brought about if the ring system readily undergoes ring-opening and ring-closure. The best known example is glucose: α-D-glucose and β-D-glucose are interconverted fairly rapidly in solution through the open-chain tautomer (114); the β form predominates (β:α ~ 2:1) because all the substituents are in equatorial positions.

| α-D-Glucose | (114) | β-D-Glucose |

Thus, when crystalline α-D-glucose is dissolved in water, the initial rotation of the solution, corresponding to $[\alpha]_D^{20} = 111°$, falls gradually to an equilibrium value of 52·5° (the β form has $[\alpha]_D^{20} = 19°$). The phenomenon is known as *mutarotation*.

(*ii*) *Substitutions.* Equatorially and axially substituted conformational isomers normally react at different rates. Which isomer reacts the faster depends on whether the steric requirements of the transition state are less than, greater than, or comparable with, those of the reactants.

The first case is illustrated by the S_N1 solvolysis of *cis*- and *trans*-4-t-butyl-cyclohexyl toluene-*p*-sulphonate.* The *cis* isomer is the less stable, having the ester group in the axial conformation, but the transition states differ less in stability because the substituent has become somewhat removed from the ring system. The situation is as in Fig. 5.3, from which it is clear that the activation energy should be smaller for the *cis* compound. In practice, solvolysis of the *cis* isomer is faster by a factor of about 3 and is described as being *sterically assisted*.

---

*The large size of the t-butyl group results in the conformational isomer in which the group is axial being of such high energy-content relative to that in which it is equatorial that the latter isomer constitutes essentially 100% of the material. The discussion would, of course, apply to other systems of this sort or ones in which the conformation was rigidly held, such as the *trans*-decalin system.

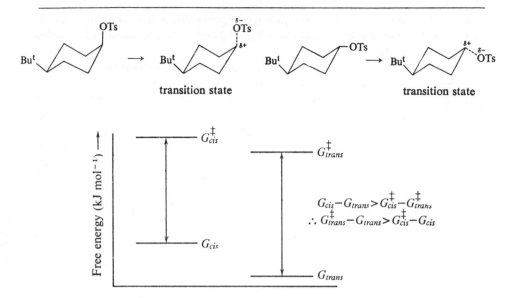

Fig. 5.3   Energies of reactants and transition states.

The second case is illustrated by the S_N2 hydrolysis of the *cis* and *trans* esters (115) and (116). Here, the *cis* compound is again the less stable, but the difference in stabilities of the transition states is greater than that of the reactants because the addition of hydroxide ion further increases the molecular crowding. Consequently the *trans* isomer reacts the more rapidly, by a factor of about 20.

These two cases are examples of the more general proposition that, if passage to the transition state is associated with a decrease in steric crowding, the axial

substituent will react faster, whereas if it is associated with an increase in crowding, the equatorial substituent will react faster.

(*iii*) *E2 Eliminations*. The stereoelectronic requirement is the availability of the *anti* conformation for the groups to be eliminated. This requirement can affect both rates and products.

For example, neomenthyl chloride undergoes base-catalyzed dehydro-chlorination readily, giving the expected 3-menthene (Saytzeff rule; p. 120); reaction occurs through the favoured conformation in which both alkyl groups are equatorial. The diastereoisomeric menthyl chloride, on the other hand, reacts at only one-twohundredth the rate and gives 2-menthene; elimination occurs through the unfavourable conformation in which all three substituents are axial and in which the proton for Saytzeff-elimination is not in the necessary *anti* position:

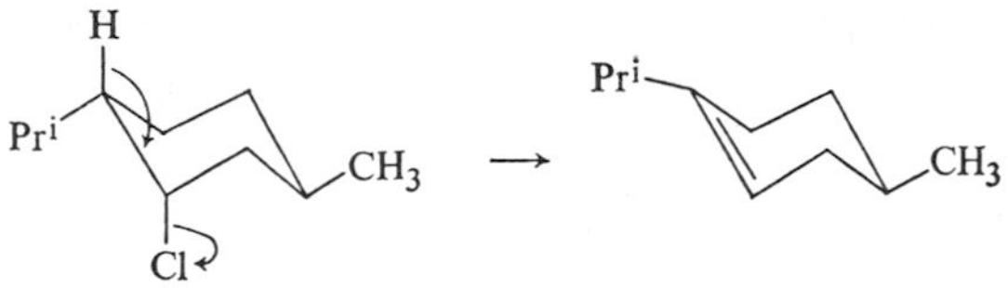

Neomenthyl chloride   3-Menthene

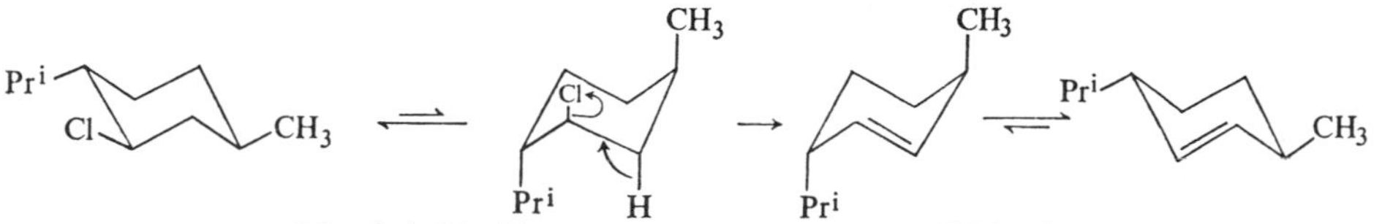

Menthyl chloride   2-Menthene

It can be readily appreciated why β-hexachlorocyclohexane undergoes de-hydrochlorination more slowly (by several powers of ten) than any of its diastereoisomers.

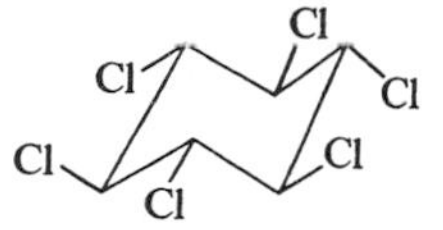

β-Hexachlorocyclohexane

When the achievement of the *anti* conformation is impossible, rearrangement may precede elimination, e.g.

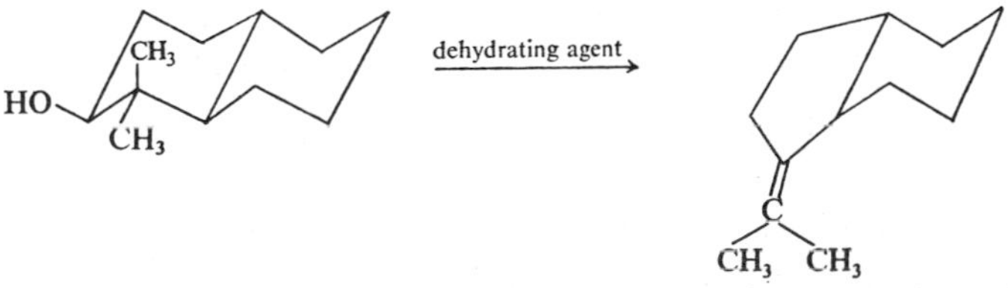

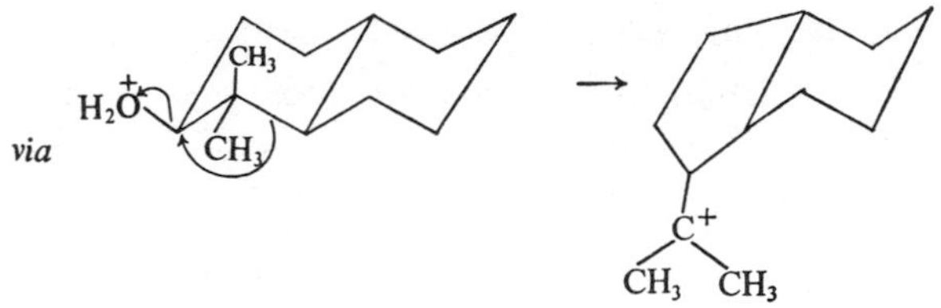

(Note that the requirement of such eliminations—inversion at the migration terminus—is met; p. 182).

(*iv*) *Fragmentations*. 1,3-Disubstituted cyclohexanes which possess a hydroxyl substituent and a good leaving group such as toluene-*p*-sulphonate undergo base-induced fragmentation, providing that the leaving group can attain the equatorial position:

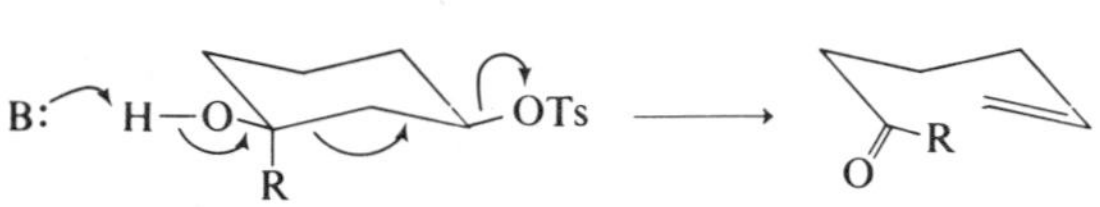

That is, the stereoelectronic requirement is that the leaving group and the C—C bond which breaks should be in the *anti* conformation.

Use can be made of this reaction in the formation of *trans*-cyclodecane derivatives:

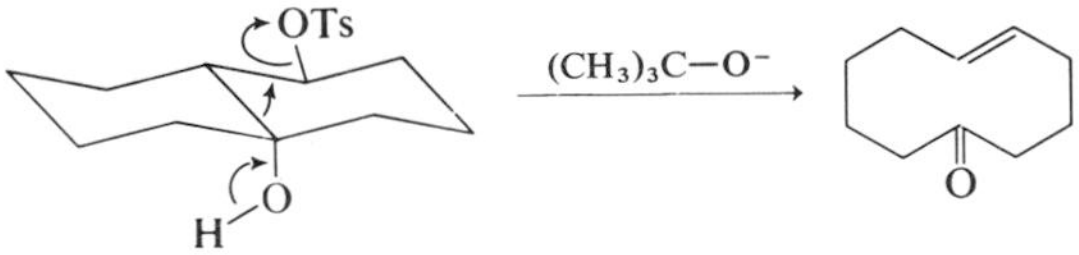

(*v*) *Electrophilic addition to olefins*. The stereoelectronic requirement for *trans* addition (p. 179) results in the formation of di-axial products which, if ring-flipping is possible, can convert predominantly into the di-equatorial products, e.g.

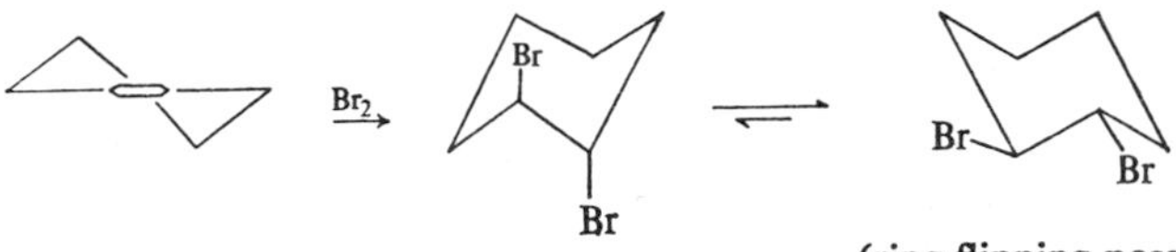

(ring-flipping possible)

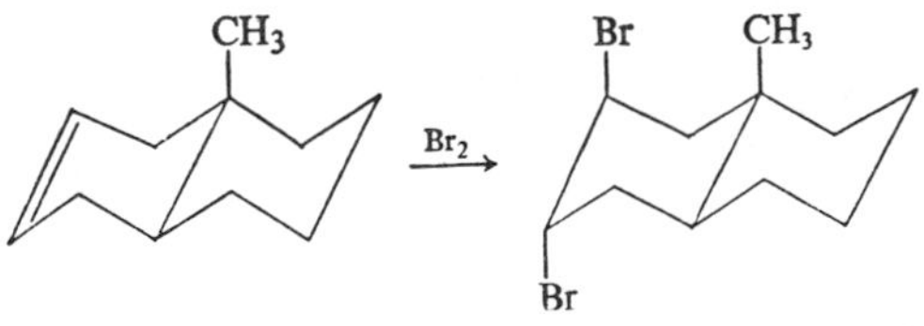

(ring-flipping not possible)

(*vi*) *Intramolecular reactions*. The stereoelectronic requirement—the *anti* conformation—can lead to different behaviour by diastereoisomers.

For example, treatment of *trans*-2-aminocyclohexanol with nitrous acid gives
only cyclopentanecarboxaldehyde, by rearrangement:

However, the *cis* isomer gives a mixture of the ring-contracted aldehyde
and cyclohexanone; in this case, alternative *anti* conformations are available
for rearrangement, one involving ring-contraction and the other, hydride
migration:

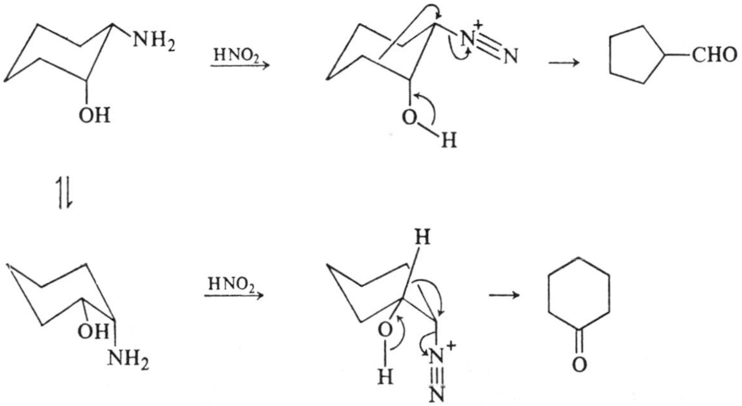

A slightly different situation holds for *cis*- and *trans*-2-chlorocyclohexanol.
The oxyanion formed by the *trans* isomer with base is suitably sited to displace
chloride ion intramolecularly by back-side attack; reaction occurs rapidly to
give the epoxide. The *cis* isomer cannot achieve the necessary conformation for
displacement and instead a slower reaction occurs in which a hydride ion migrates to give cyclohexanone:

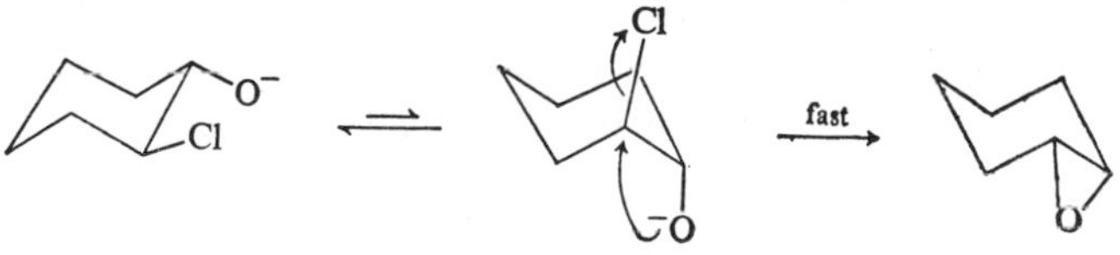

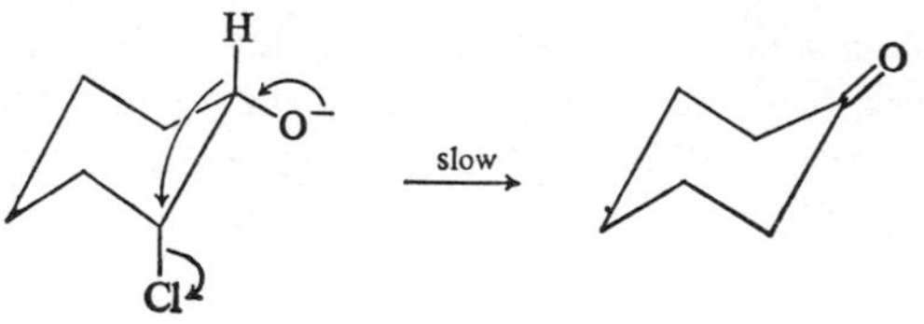

Just as epoxides are formed from diaxial halohydrins as in the example above, so also they yield diaxial products on ring-opening: a nucleophile attacks from the opposite side to the quasi-equatorial bond from carbon to oxygen, e.g.

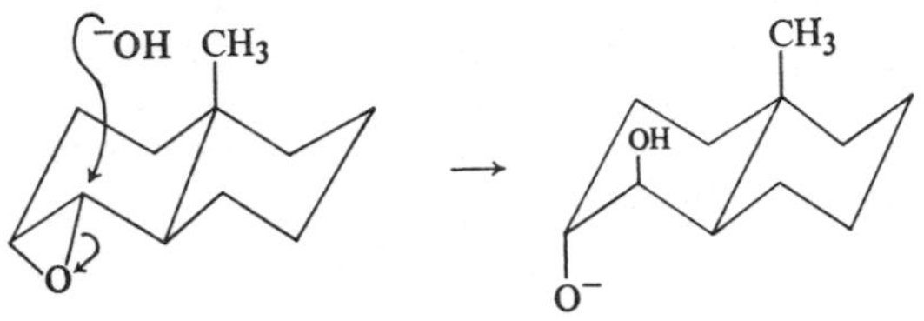

(vii) *Cyclohexanone and related compounds.* Cyclohexanone reacts with cyanide ion, to give the cyanohydrin, and with borohydride ion, to give cyclohexanol, faster than cyclopentanone. This is because cyclohexanone possesses conformational strain, owing to the eclipsing of its oxygen atom by two adjacent hydrogens, which is relieved in passage to saturated derivatives, whereas the conformational strain in the five-membered system is increased in passage to saturated derivatives.

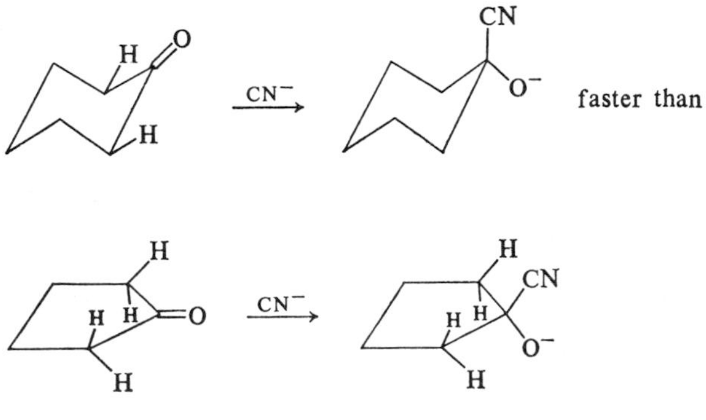

Conversely, cyclohexyl derivatives are less reactive in S_N1 and S_N2 reactions than their cyclopentyl analogues, for passage to the transition state in the former case is accompanied by an increase in steric strain and in the latter by a decrease, e.g.

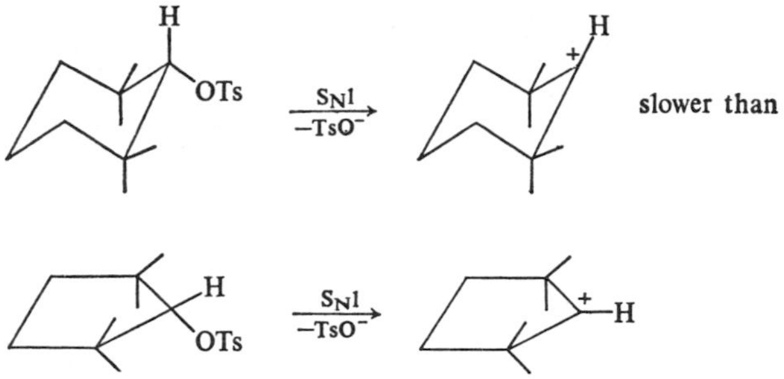
H
OTs
SN1
−TsO⁻
H
+
slower than
H
OTs
SN1
−TsO⁻
+
H

In general, if steric strains are greater in an sp^2-hybridized system than in the corresponding sp^3-hybridized system, reactions of the $sp^2 \rightarrow sp^3$ type, such as additions to carbonyl groups, are facilitated and those of $sp^3 \rightarrow sp^2$ type, such as solvolyses, are retarded; and the converse holds if steric strains are smaller in the sp^2- than in the sp^3-hybridized system.

Further Reading

ELIEL, E. L., *Stereochemistry of Carbon Compounds,* McGraw-Hill (New York 1962).

GUNSTONE, F. D., *Guidebook to Stereochemistry*, Longmans (London and New York 1975).

NEWMAN, M. S., *Steric Effects in Organic Chemistry,* Wiley (New York and London 1956).

WHITHAM, G. H., *Alicyclic Chemistry,* Oldbourne (London 1963).

Problems

1. In what stereoisomeric forms would you expect the following compounds to exist?

(*a*) 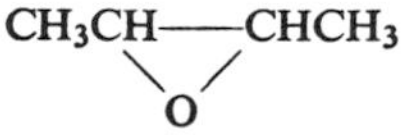
CH₃CH—CHCH₃
O

(*b*) $(CH_3)_2C{=}C{=}C{=}C(CH_3)_2$

(*c*) $(CH_3)(C_2H_5)C{=}C{=}C(CH_3)(C_2H_5)$

(*d*) $PhCH(OH){-}C{\equiv}C{-}CH(OH)CH_3$

(*e*) 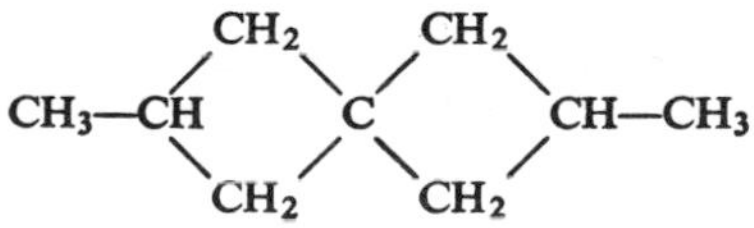
CH₂
CH₂
CH₃—CH
C
CH—CH₃
CH₂
CH₂

(*f*) 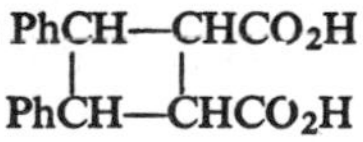
PhCH—CHCO₂H
PhCH—CHCO₂H

(g)

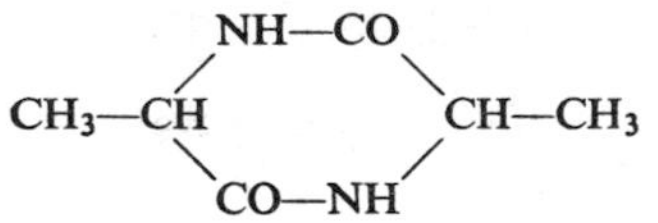

$$CH_3-CH \begin{matrix} NH-CO \\ \\ CO-NH \end{matrix} CH-CH_3$$

(h)

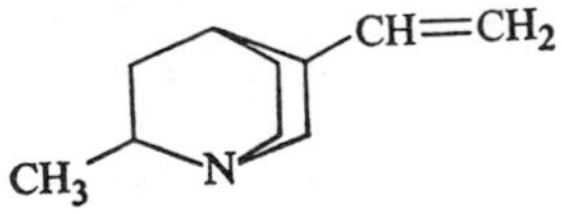

(i) Perhydrophenanthrene

(j) Cyclo-octene

(k) 1-Bromo-2-chlorocyclohexane

(l) 1, 2-Dichlorocyclohexane

(m)

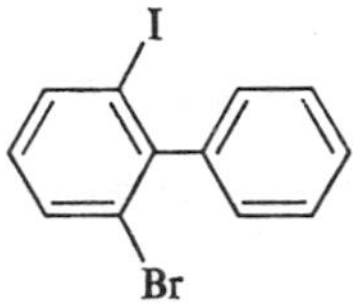

(n)

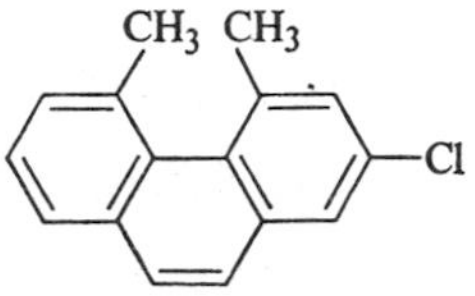

2. Draw Newman projection diagrams of the following structures, which are shown as Fischer projections. Which compound will be optically inactive?

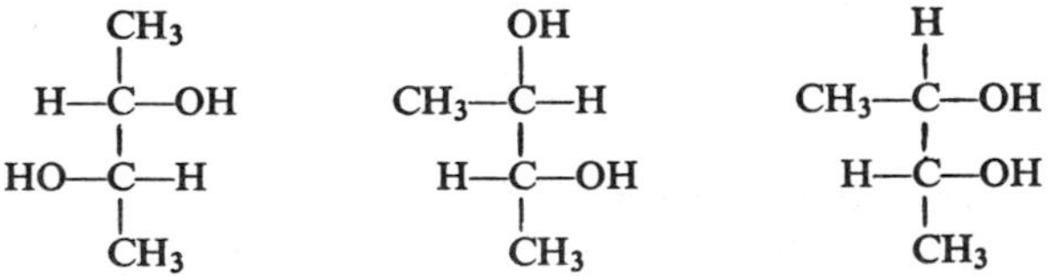

$$\begin{matrix} CH_3 \\ | \\ H-C-OH \\ | \\ HO-C-H \\ | \\ CH_3 \end{matrix} \qquad \begin{matrix} OH \\ | \\ CH_3-C-H \\ | \\ H-C-OH \\ | \\ CH_3 \end{matrix} \qquad \begin{matrix} H \\ | \\ CH_3-C-OH \\ | \\ H-C-OH \\ | \\ CH_3 \end{matrix}$$

3. Which of the following pairs of diastereoisomers are epimers, and which of the pairs of epimers could be readily interconverted?

(a)

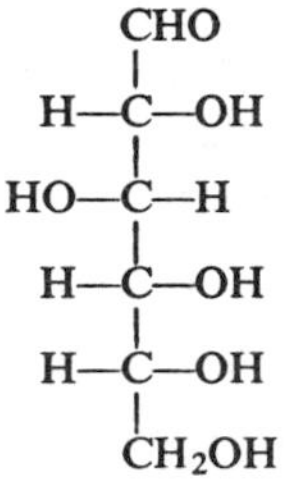

$$\begin{matrix} CHO \\ | \\ H-C-OH \\ | \\ HO-C-H \\ | \\ H-C-OH \\ | \\ H-C-OH \\ | \\ CH_2OH \end{matrix} \qquad and \qquad \begin{matrix} CHO \\ | \\ HO-C-H \\ | \\ HO-C-H \\ | \\ H-C-OH \\ | \\ H-C-OH \\ | \\ CH_2OH \end{matrix}$$

(b)

$$\begin{matrix} CHO \\ | \\ H-C-OH \\ | \\ H-C-OH \\ | \\ H-C-OH \\ | \\ CH_2OH \end{matrix} \qquad and \qquad \begin{matrix} CHO \\ | \\ HO-C-H \\ | \\ H-C-OH \\ | \\ HO-C-H \\ | \\ CH_2OH \end{matrix}$$

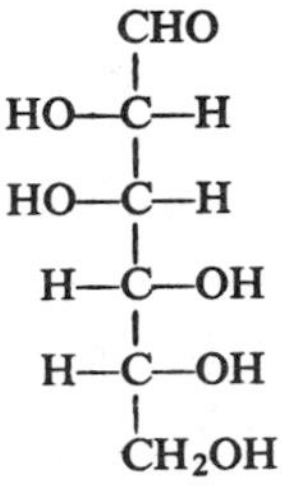

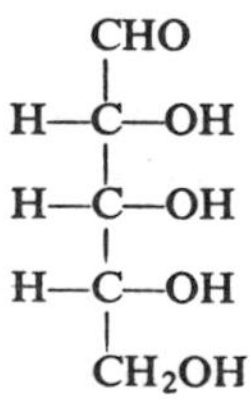

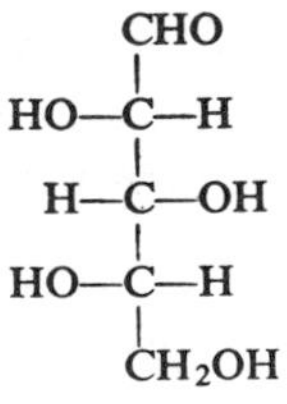

(c)

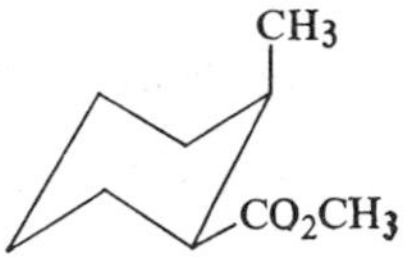

and

(d)

and

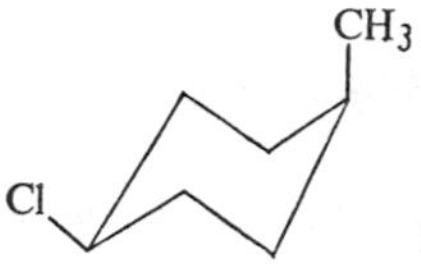

4. What products would you expect from the following reactions?

    (*i*) The dehydration of 1,1,3-triphenyl-3-*p*-chlorophenylpropan-2-ol.

    (*ii*) The treatment of (I) with (*a*) OH⁻ and (*b*) HBr.

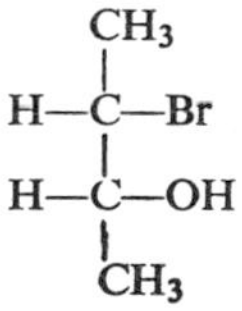

    (*iii*) The elimination of bromine from *meso*-1,2-dibromo-1,2-diphenylethane by treatment with iodide ion.

    (*iv*) The treatment of $CH_3CHXCH_2CH_3$ with base, (*a*) when X = Cl, (*b*) when X = $\overset{+}{N}Me_3$.

5. Account for the following:

    (*i*) When an optically active sample of 2-iodobutane is treated with radioactive iodide ion in solution, the initial rate of racemization is twice the initial rate of uptake of radioactivity.

    (*ii*) *R*-α-Bromopropionic acid gives *S*-lactic acid with very concentrated alkali, but *R*-lactic acid with dilute alkali.

    (*iii*) *R*-2-Butyl acetate is hydrolyzed in basic conditions to give *R*-2-butyl alcohol; the alcohol, with hydrochloric acid, gives a mixture of *R*- and *S*-2-butyl chloride.

    (*iv*) The base-induced elimination of hydrogen chloride occurs more rapidly from *cis*- than from *trans*-1-chloro-4-methylcyclohexane.

Part II

Introduction to Part II

Whereas Part I was concerned with the principles of organic reactions, Part II is concerned with their practice; the structure of each Chapter, with the exception of the last, is designed to answer the question, 'How may one go about constructing a particular type of bond in a given environment?'. Experimental details are not included, but many of the general reactions described are illustrated by examples taken from *Organic Syntheses,* Collective Volumes 1–5 or individual volumes 50–55; these describe carefully tested experimental procedures and are referred to by inclusion in the text of the appropriate collective or individual volume number in square brackets.

It will have become apparent from the earlier Chapters that, in a large number of instances, the construction of a bond necessitates the generation, from a relatively stable compound, of a reactive species which is able to form a bond with a second compound. Chapters 6–8, each of which is concerned with the synthesis of aliphatic carbon-carbon bonds, all illustrate this principle. In the first of the three Chapters the method adopted is to bond one of the two carbon atoms to a metal; then, because of the electropositive character of metals, the carbon atom is negatively polarized and able to react with a second carbon atom which is attached to an electronegative group and is therefore positively polarized. In the second of the Chapters the method involves the use of basic media in order to generate a carbanion from a C—H-containing group of a suitable type; this, like the negatively polarized carbon in a C—metal bond, reacts with a carbon atom which is in an electronegative environment. In the last of the three Chapters the opposite approach is adopted; acidic conditions are used to generate a carbonium ion which then reacts with a carbon atom which is in a relatively electron-rich environment.

A different principle is employed in Chapter 9. Here, two molecules are brought together in such a way that two new bonds are formed in a concerted manner, to give a cyclic product; variants of this method include intramolecular cyclization and rearrangement. In most, but not all, cases new carbon-carbon bonds are formed.

The methods for forming aliphatic carbon-nitrogen bonds (Chapter 10) are related respectively to the methods of Chapters 7 and 8 for forming carbon-carbon bonds. One approach, the more widely applicable, employs a species containing a nucleophilic nitrogen atom to form a bond to an electronegatively substituted carbon atom; the other involves the generation of a positively charged nitrogen species which reacts at an electron-rich carbon atom.

The stability of aromatic nuclei necessitates the use of certain different methods for bonding to aromatic carbon compared with those appropriate in the aliphatic series. In general, reagents of both electrophilic and nucleophilic type are employed, the former class being of wider applicability; they are described in Chapters 11 and 12, respectively. Aromatic chemistry is completed by a survey of the reactions of diazonium ions (Chapter 13); amino substituents may be replaced, *via* diazotization, by a wide variety of other groupings. Since the amino group may usually be introduced readily into aromatic rings through nitration followed by reduction, these reactions are of considerable use.

Intramolecular rearrangements (Chapter 14), in which the skeleton of a molecule is transformed by the migration of a group to an adjacent atom, are characteristic of a large number of structural situations. Many, especially those which involve electron-deficient intermediates, may be usefully employed in synthesis; but it is perhaps equally necessary to recognize the need, when planning a synthesis, to guard against the possibility that an intermediate will be formed which will undergo an unwanted rearrangement.

Reagents in which the key atom is phosphorus, sulphur, or boron provide a relatively recent addition to the synthetic chemist's armoury. They are now recognized as of major importance for their versatility and, in several cases, for their specificity. They are grouped together in Chapter 15.

As well as carbonium ions and carbanions, two other classes of highly reactive carbon-containing species provide the basis for a variety of synthetic methods. One comprises electronically excited molecules, which can be formed by the ultraviolet or visible irradiation of unsaturated or aromatic compounds. These take part in a diversity of reactions and find especial use in the construction of complex cyclic compounds, often of a very strained kind (Chapter 16). The other class are free carbon radicals. These can be generated in several ways and are capable of reacting with a number of organic groupings. Although their reactions are less commonly employed in synthesis than those of carbonium ions and carbanions, largely because free radicals are far less selective entities, there is nevertheless a diversity of processes in which they are usefully engaged (Chapter 17).

Almost any multi-stage synthesis involves steps in which the oxidation or the reduction of particular groups is required, and there have been greater recent advances in the control of these than of any other fundamental processes. In particular, highly selective methods have been developed which make it possible to oxidize (or reduce) at a required position in a molecule without affecting other oxidizable (or reducible) groups (Chapters 18 and 19).

The construction of heterocyclic systems (Chapter 20) utilizes only those reactions which have already been discussed. The subject is, however, suitably treated separately for two reasons: first, ring-closure reactions which lead to five- or six-membered rings occur in milder conditions than analogous reactions in acyclic chemistry; secondly, mechanistically complex processes which might, in acyclic chemistry, give correspondingly complex mixtures frequently give particular aromatic heterocycles essentially specifically, the driving force being the development of the associated aromatic stabilization energy.

The final Chapter describes and discusses the syntheses of a number of interesting naturally occurring compounds. These have been chosen, above all, as illustrating the fine control which can now be exercised in complex synthetic operations as a result of our understanding of the electronic and stereoelectronic principles of organic chemistry. In terms of structural sophistication, no synthesis can have been more exacting in its planning and execution than that of chlorophyll; in terms of stereochemical sophistication, none can have been more elegantly conceived than that of epiandrosterone in which, in one stage, six chiral carbon atoms were assembled in the stereospecifically required manner. It is to be hoped that the student will not attempt to *learn* these syntheses; their planning, and their individual stages, are dissected in such a manner that, hopefully, he will both develop some insight into the planning of such synthetic monuments and appreciate their artistic and scientific aspects.

6. Formation of Carbon-Carbon Bonds: Organometallic Reagents

6.1 Principles

All the non-metallic elements to which carbon is commonly bonded are more electronegative than carbon. As a consequence, the carbon atom is positively polarized and, if the atom or group attached to it has a marked capacity for accepting an electron-pair, the carbon is susceptible to attack by nucleophiles, e.g.

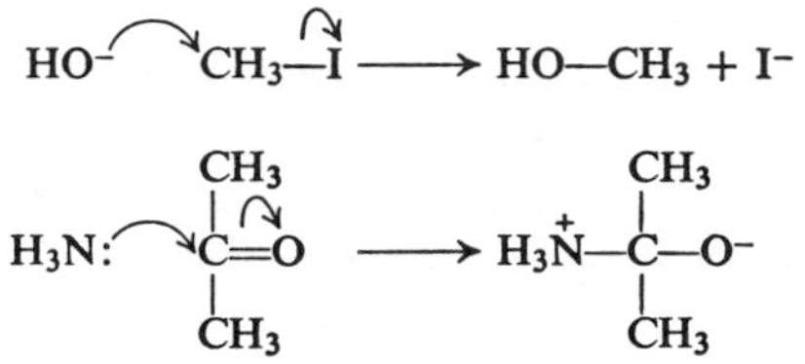

$$HO^- \quad CH_3{-}I \longrightarrow HO{-}CH_3 + I^-$$

$$H_3N: \quad \overset{CH_3}{\underset{CH_3}{C}}{=}O \longrightarrow H_3\overset{+}{N}{-}\overset{CH_3}{\underset{CH_3}{C}}{-}O^-$$

Conversely, carbon bonded to an electropositive element—a metal—is negatively polarized, and because a metal can release a bonding electron-pair to form a cation, the carbon atom is susceptible to attack by electrophiles. The counterparts to the nucleophilic reactions above are:

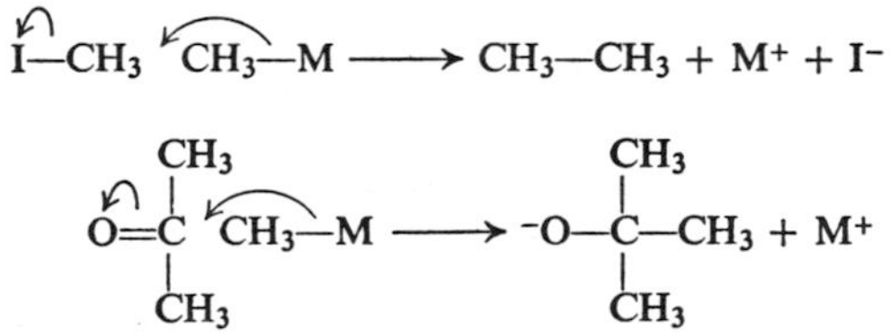

$$I{-}CH_3 \quad CH_3{-}M \longrightarrow CH_3{-}CH_3 + M^+ + I^-$$

$$O{=}\overset{CH_3}{\underset{CH_3}{C}} \quad CH_3{-}M \longrightarrow {}^-O{-}\overset{CH_3}{\underset{CH_3}{C}}{-}CH_3 + M^+$$

where M is a monovalent metal and methyl iodide and acetone are regarded as electrophilic reagents attacking the negatively polarized carbon in $CH_3{-}M$.

Organometallic compounds are enormously versatile reagents.* Their reactivity depends on the nature of the metal atom, there being a steady decrease in reactivity with decrease in the electropositive character of the metal, and by appropriate choice of the metal a wide range of syntheses may be carried out. This, together with the ease with which organometallic compounds can be obtained, gives the reagents extensive applications.

*The term, organometallic compound, is here used in the restricted sense of describing compounds which contain a carbon-metal bond.

(a) PREPARATION

The following are the general methods of preparation of organometallic compounds. They are discussed in more detail in the later sections concerned with individual metals, together with their limitations and the methods of choice in particular instances.

(1) *From metals and organic halides.* The simplest method of preparation is by treatment of the metal with an organic halide in an unreactive solvent, e.g.

$$R—Br + 2\ Li \longrightarrow R—Li + LiBr$$

(2) *Metal-halogen exchange.* The exchange reaction between an organometallic compound and an organic halide, e.g.

$$R—Br + R'—Li \rightleftharpoons R—Li + R'—Br$$

is an equilibrium which favours the organometallic compound in which the metal is attached to the more electronegative carbon.* For example, since aromatic (sp^2) carbon is more electronegative than aliphatic (sp^3) carbon, aryl-metal compounds can be prepared from alkyl-metal compounds, e.g.

$$PhBr + BuLi \rightleftharpoons PhLi + BuBr$$

(3) *Metal-metal exchange.* Organometallic compounds react with metallic salts by interchanging metals. The equilibrium favours formation of the organometallic compound which contains the less electropositive metal, e.g.

$$2\ RLi + CdCl_2 \rightleftharpoons R_2Cd + 2\ LiCl$$

(4) *Metalation of hydrocarbons.* A C—H bond which is significantly acidic (i.e. which gives rise to a reasonably stable carbanion) reacts with an organometallic reagent in which the carbon is less electronegative to give the corresponding C-metal derivative, e.g.

$$CH{\equiv}CH + C_2H_5MgBr \longrightarrow CH{\equiv}C—MgBr + C_2H_6$$

(b) STRUCTURE AND REACTIVITY

The organic compounds of sodium and potassium are salt-like compounds which are insoluble in non-polar solvents. Those of less electropositive elements such as magnesium and zinc are essentially covalent and are soluble in, e.g. ether.

---

*The following is a simple way of understanding this. The carbon-metal bond has considerable ionic character and may be represented as a hybrid of structures such as $CH_3—Li$ (purely covalent) and $CH_3^-\ Li^+$ (purely ionic). The more electronegative the carbon, the more strongly bonded is the compound.

There is, however, no sharp distinction between the two groups and it is possible to classify C—metal bonds according to their percentage ionic character. Values for the commoner bonds are as follows:

Metal	K	Na	Li	Mg	Zn	Cd
% Ionic character	51	47	43	35	18	15

The reactivity of organometallics can be correlated with their ionic character. Compounds of sodium and potassium are by far the most reactive; for example, they are spontaneously inflammable in air, whereas organomagnesium compounds react with oxygen but far less vigorously. Lithium compounds are more reactive than magnesium compounds, e.g. unlike the latter, they react with carboxylate ions (p. 218); and cadmium compounds are less reactive than magnesium compounds, e.g. they do not react with ketones.

6.2 Organomagnesium Compounds (Grignard Reagents)

Organomagnesium compounds, known as Grignard reagents after their discoverer, are the most widely used of the organometallic reagents. Curiously, there is even today some doubt about their structure: although conventionally written as RMgX, where X is chlorine, bromine, or iodine, there is evidence that they exist in solution in equilibrium with the di-organomagnesium and the magnesium halide:

$$2 \text{ RMgX} \rightleftharpoons R_2Mg + MgX_2$$

Grignard reagents are soluble in ethers in which they are customarily prepared and used. The solubility is due to the coordination of magnesium by ether molecules, and a likely formulation for the reagents is

$$
\begin{array}{ccc}
R & X & OR'_2 \\
\diagdown & \diagup \diagdown & \diagup \\
& Mg \quad\quad Mg & \\
\diagup & \diagdown \diagup & \diagdown \\
R & X & OR'_2
\end{array}
$$

(a) PREPARATION

The standard method for the preparation of a Grignard reagent is the reaction of an organic halide in dry diethyl ether with magnesium metal (as turnings, granules or powder). The reagents are not isolated from solution but are used directly for the required synthesis.

$$RX + Mg \longrightarrow RMgX$$

The group R may be alkyl, aryl, or vinyl, although vinyl halides require special conditions. Amongst the halides, the order of reactivity is I > Br > Cl ≫ F;

organomagnesium fluorides have not been prepared. The choice amongst chloride, bromide, and iodide is not critical except in the case of aryl halides, where the chlorides are unreactive. The choice is therefore based on the accessibility of the halide, and in general chlorides are used because they are cheaper to obtain. Since methyl chloride and methyl bromide are gases at room temperature, methyl iodide (b.p. 43°C) is used to make the methyl Grignard, and likewise ethyl bromide is preferred to ethyl chloride for the ethyl Grignard. Grignard reagents will be referred to in this discussion as RMgX, where the nature of X is unspecified.

Alkyl and aryl halides (other than aryl chlorides) normally react readily with magnesium provided that water is rigorously excluded. The magnesium is covered with a solution in ether of about 10% of the halide to be used, the exothermic reaction is normally initiated by stirring, and the remainder of the halide in ethereal solution is then added at such a rate that the ether is maintained at reflux by the heat evolved in the reaction. With some halides it is necessary to add a crystal of iodine or a small quantity of another (preformed) Grignard such as methylmagnesium iodide in order to initiate reaction. Grignard reagents react with oxygen, and it is helpful, although not usually essential, to exclude air.

Vinyl halides do not react under these conditions but do so when tetrahydrofuran is used in place of diethyl ether as solvent. Reaction requires longer times and higher temperatures than with alkyl or aryl halides, e.g.

$$CH_2{=}CHCl \xrightarrow[\text{9 hours at 50°C}]{\text{Mg/THF}} CH_2{=}CHMgCl$$

Grignard reagents are also frequently prepared by the metalation procedure (p. 203), using a preformed Grignard reagent. The method is only suitable when the atom attached to magnesium in the compound to be prepared is markedly more electronegative than that in the preformed Grignard compound so that the equilibrium,

$$RH + R'MgBr \rightleftharpoons RMgBr + R'H$$

lies well to the right. Acetylenic Grignards are normally prepared by this method, e.g.

$$R{-}C{\equiv}CH + C_2H_5MgBr \longrightarrow R{-}C{\equiv}C{-}MgBr + C_2H_6$$

and so are those from other acidic hydrocarbons such as cyclopentadiene,

(b) REACTIVITY

Grignard reagents react with all organic functional groups except tertiary amines, olefinic and aromatic double bonds, acetylenic triple bonds, and ethers, * and they cannot therefore be prepared from compounds containing groupings other than these. Thus, while their reactivity is usefully applied in synthesis, it also provides a limiting factor. In general, the compounds are of more use in the synthesis of small and simple molecules than for polyfunctional molecules, and it is to be noted that they are rarely used in the syntheses of complex naturally occurring compounds such as quinine and cholesterol (Chapter 21) where highly specific reactions are required.

A useful working guide to the mode of reaction of Grignard reagents is that the direction of reaction is such that the magnesium atom is transferred to a more electronegative atom, as in the metalation procedure for their synthesis. All OH- and NH-containing compounds react by replacement of hydrogen,† e.g.

$$ROH + R'MgX \longrightarrow RO{-}MgX + R'H$$

$$R_2NH + R'MgX \longrightarrow R_2N{-}MgX + R'H$$

and in their reactions at carbon centres the magnesium is similarly transferred to oxygen or nitrogen when one of these elements is present, e.g.

$$R_2C{=}O + R'MgX \longrightarrow R_2R'C{-}O{-}MgX$$

One complicating feature of their reactions should be noted: a Grignard derived from an allylic system can react at either of two carbon atoms:

(E$^+$ = an electrophilic centre, such as the carbon atom in a carbonyl compound; p. 209).

The proportion of each possible product depends on the steric environments of the two carbon atoms and the nature of the electrophile. In some cases reaction at the γ-carbon atom is essentially exclusive; e.g. the Grignard reagent from

---

*They react with *strained* cyclic ethers; p. 212.

†The reaction of a Grignard reagent with heavy water (deuterium oxide) may be used to introduce deuterium into organic compounds:

$$RMgX + D_2O \longrightarrow RD + MgX(OD)$$

2-butenyl bromide gives only α-vinylpropionic acid with carbon dioxide (see p. 213):

$$CH_3-CH=CH-CH_2Br \xrightarrow[\substack{1) \ Mg \\ 2) \ CO_2 \\ 3) \ H^+}]{} CH_3-\underset{\underset{CO_2H}{|}}{CH}-CH=CH_2$$

This characteristic is usefully employed in the synthesis of pyrrole and indole derivatives. Pyrrole reacts with Grignard reagents at its NH group to give a N—Mg derivative which reacts with electrophiles at the α-carbon atom, e.g.

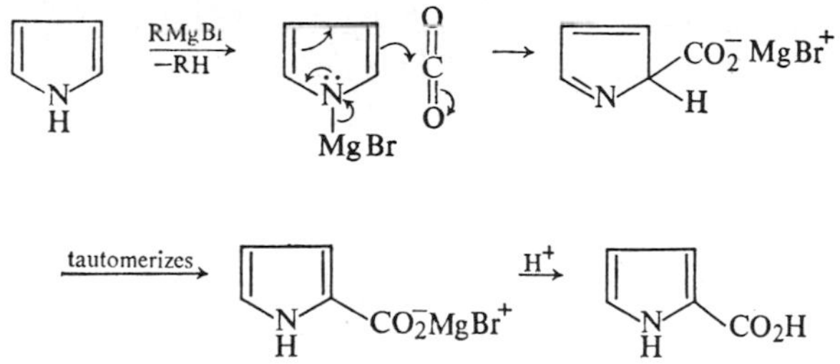

The Grignard reagents from indole behave analogously in that reaction occurs at the β-position, as in the synthesis of the plant-growth hormone, heteroauxin (indole-β-acetic acid).*

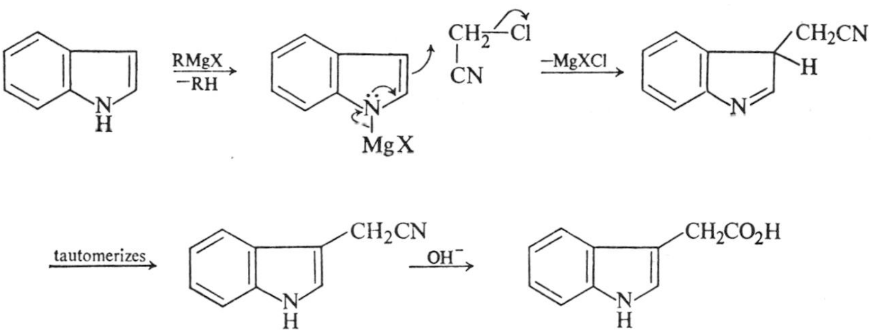

Heteroauxin

(c) FORMATION OF CARBON-CARBON BONDS

The reactions of Grignard reagents at carbon atoms in various environments are classified here with reference to the type of compound which is obtained.

*The reasons why pyrrole and indole react with electrophiles chiefly at their α- and β-positions, respectively, are discussed later (p. 381).

(*i*) *Hydrocarbons.* Grignard reagents react with alkyl halides and related compounds in the S_N2 manner, e.g.

$$XMg\!-\!R \quad CH_3\!-\!X \longrightarrow R\!-\!CH_3 + MgX_2$$

The yields from saturated halides are low, but allylic and benzylic halides (which are more reactive than alkyl halides in S_N2 reactions) react efficiently, e.g.

$$CH_3CH_2MgBr + CH_2\!=\!CH\!-\!CH_2Br \longrightarrow CH_3CH_2CH_2CH\!=\!CH_2 + MgBr_2$$
$$\text{1-Pentene}$$
$$94\%$$

Alkyl compounds containing better leaving groups than the halides, such as alkyl sulphates and sulphonates, react in much higher yields than the alkyl halides. For example, n-propylbenzene is obtained in 70–75% yield from benzyl chloride and diethyl sulphate [1],

$$PhCH_2Cl \xrightarrow[\text{2) C}_2\text{H}_5\text{OSO}_2\text{OC}_2\text{H}_5]{\text{1) Mg}} PhCH_2CH_2CH_3 + MgCl(OSO_2OC_2H_5)$$
$$\text{n-Propylbenzene}$$

n-pentylbenzene is obtained in 50–60% yield from benzyl chloride and butyl toluene-*p*-sulphonate [2],

$$PhCH_2Cl \xrightarrow[\text{2) C}_4\text{H}_9\text{OTs}]{\text{1) Mg}} PhCH_2CH_2CH_2CH_2CH_3 + MgCl(OTs)$$
$$\text{n-Pentylbenzene}$$

and isodurene may be prepared in up to 60% yield from mesityl bromide and dimethyl sulphate [2],

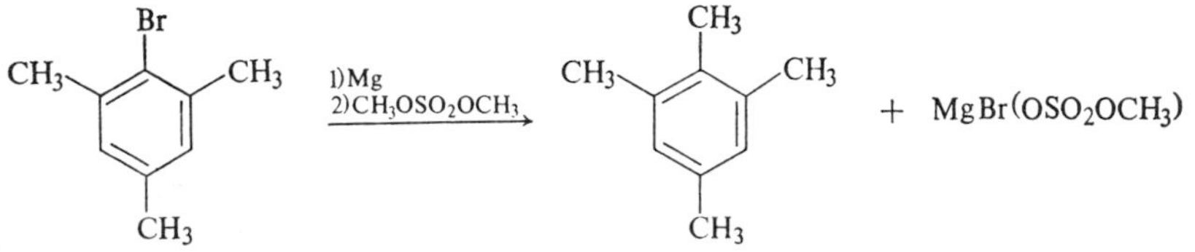

Isodurene

Since Grignard reagents are usually prepared from organic halides, their reactions with halides can be disadvantageous. Fortunately, alkyl and aryl halides react sufficiently slowly for this side-reaction not to be of significance, but allyl halides present a more serious problem. In order to minimize the extent of the S_N2 coupling reaction it is advisable to add the allyl halide in very dilute solution to a large excess of magnesium.

(*ii*) *Alcohols*. Grignard reagents react at the carbonyl groups of aldehydes and ketones to give the magnesium derivatives of alcohols which are converted into alcohols by treatment with acid:

$$\underset{}{XMg-R} \quad \underset{R''}{\overset{R'}{C}}=O \longrightarrow R-\underset{R''}{\overset{R'}{C}}-O-MgX \overset{H^+}{\longrightarrow} R-\underset{R''}{\overset{R'}{C}}-OH$$

Formaldehyde ($R'=R''=H$) gives primary alcohols, other aldehydes give secondary alcohols, and ketones give tertiary alcohols. For example, the Grignard reagent from cyclohexyl chloride reacts with formaldehyde to give cyclohexylcarbinol in 65% yield [1]:

Cyclohexylcarbinol

When the alcohol is sensitive to acid (e.g. tertiary alcohols, which are readily dehydrated) it is advisable to decompose the magnesium salt of the alcohol with aqueous ammonium chloride; basic magnesium salts are precipitated and the alcohol remains in the ethereal layer.

There are two interesting applications of the reaction of the Grignard reagent from ethoxyacetylene with carbonyl groups.* The reaction gives an α-acetylenic alcohol,

$$RR'CO + BrMg-C{\equiv}C-OEt \longrightarrow \underset{OMgBr}{RR'C-C{\equiv}C-OEt} \longrightarrow \underset{OH}{RR'C-C{\equiv}C-OEt}$$

which may be treated in one of two ways. (1) Acid-catalyzed hydration of the acetylenic group (p. 116) gives a β-hydroxy-ester which may readily be dehydrated to give the αβ-unsaturated ester (*cf.* the Reformatsky reaction, p. 221):

*The reagent is prepared from the diethylacetal of chloroacetaldehyde. Treatment with amide ion in liquid ammonia brings about E2 elimination (of hydrogen chloride and ethanol) to give ethoxyacetylene which is converted into the Grignard reagent with ethylmagnesium bromide.

$$CH_2Cl-CH(OEt)_2 \xrightarrow[-HCl, -EtOH]{NH_2^-/NH_3} CH{\equiv}C-OEt \xrightarrow[-C_2H_6]{C_2H_5MgBr} BrMg-C{\equiv}C-OEt$$

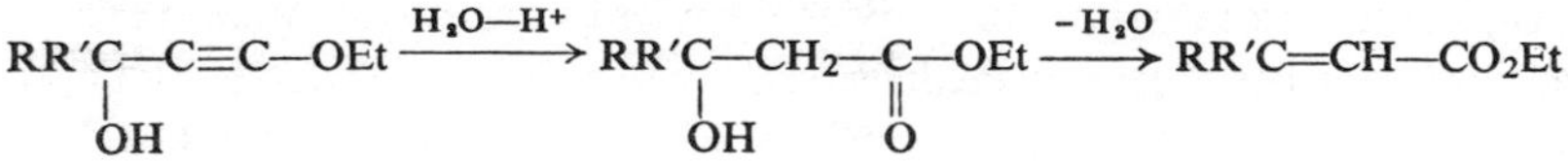

$$RR'C-C{\equiv}C-OEt \xrightarrow{H_2O-H^+} RR'C-CH_2-C-OEt \xrightarrow{-H_2O} RR'C{=}CH-CO_2Et$$
$$\underset{OH}{|} \qquad\qquad \underset{OH}{|}\ \underset{O}{\|}$$

(2) Partial catalytic reduction (p. 619) gives a vinyl ether which is readily hydro-lyzed by mineral acid to a β-hydroxyaldehyde (p. 110); dehydration gives the $\alpha\beta$-unsaturated aldehyde:

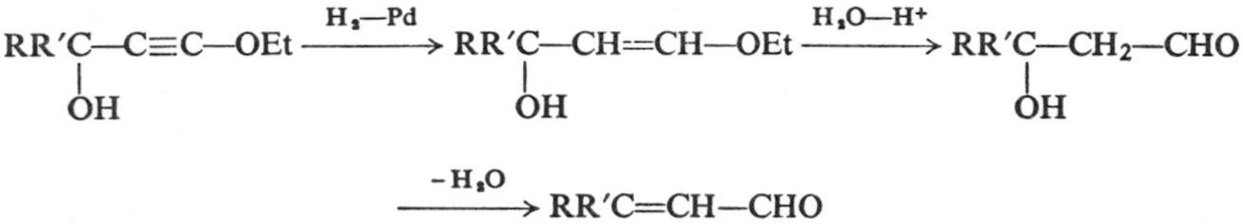

$$RR'C-C{\equiv}C-OEt \xrightarrow{H_2-Pd} RR'C-CH{=}CH-OEt \xrightarrow{H_2O-H^+} RR'C-CH_2-CHO$$
$$\underset{OH}{|} \qquad\qquad\qquad \underset{OH}{|} \qquad\qquad\qquad \underset{OH}{|}$$

$$\xrightarrow{-H_2O} RR'C{=}CH-CHO$$

The latter reaction was applied in a synthesis of Vitamin A (21.1) to bring about the conversion,

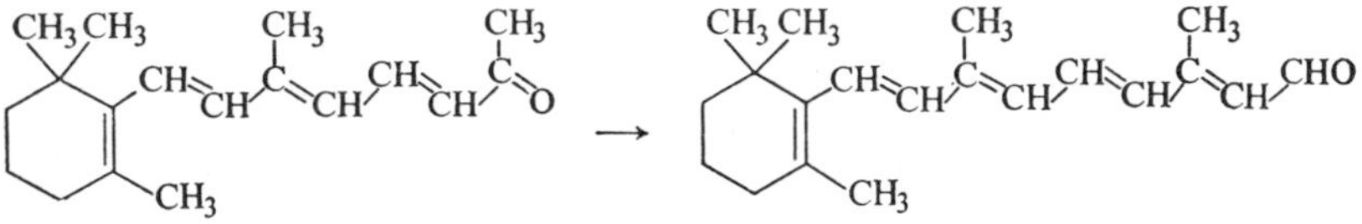

Limitations. The reactions with aldehydes and ketones usually give good yields of alcohols except with ketones which contain bulky groups. For example, di-t-butyl ketone does not give tertiary alcohols with Grignard reagents. If the ketone contains somewhat less bulky groups but the Grignard reagent contains a branched alkyl group, yields are again low; e.g. whereas di-isopropyl ketone reacts with methylmagnesium bromide to give the tertiary alcohol in 95% yield,

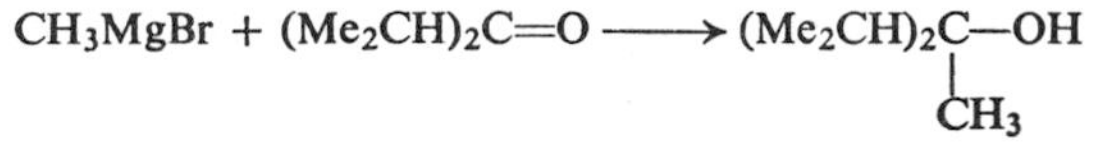

$$CH_3MgBr + (Me_2CH)_2C{=}O \longrightarrow (Me_2CH)_2C-OH$$
$$\underset{CH_3}{|}$$

Methyldi-isopropylcarbinol

it fails to give tertiary alcohols with isopropyl and t-butyl Grignard reagents.

In these cases one or both of two side-reactions take place. If the ketone has at least one hydrogen atom on one of its two α-carbon atoms, *enolization* can occur: the Grignard reagent acts as a base rather than as a nucleophile and abstracts an activated hydrogen, giving the *enolate*. Treatment with acid regenerates the ketone.

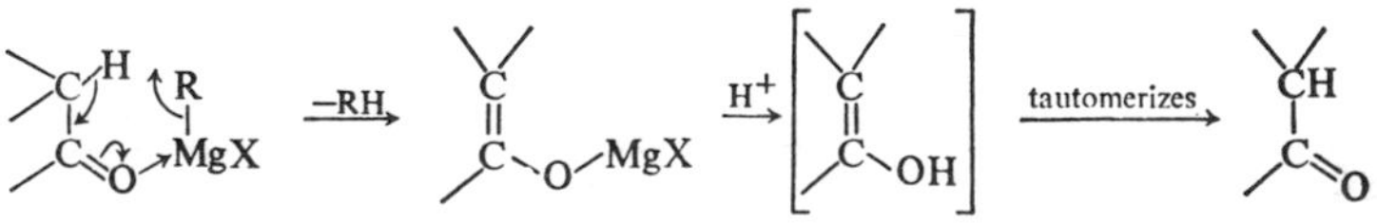

If the Grignard reagent contains at least one hydrogen atom on its β-carbon atom, *reduction* can occur by hydride-ion transfer within a six-membered cyclic transition state (*cf.* Meerwein-Ponndorf-Verley reduction; p. 635).

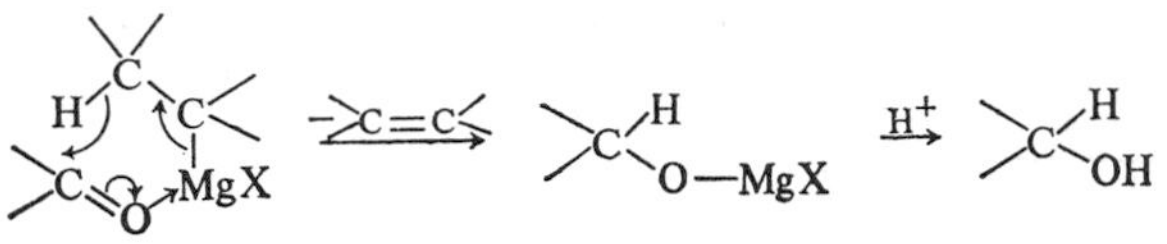

When each of these structural features is present, enolization and reduction are competitive; e.g. di-isopropyl ketone and t-butylmagnesium bromide give 35% of the enolate and 65% of di-isopropylcarbinol.

α,β-Unsaturated carbonyl compounds. Olefinic double bonds do not react with Grignard reagents, but if the olefinic bond is conjugated with a carbonyl group, addition occurs to give the magnesium derivative of an enol (*cf.* p. 253):

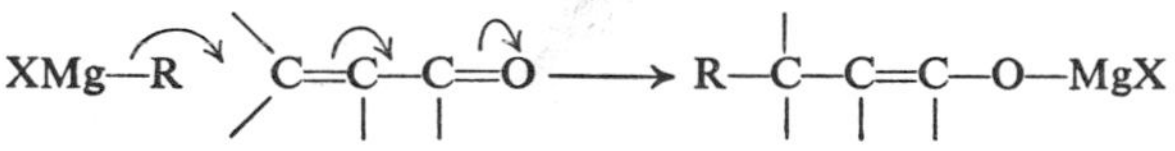

Treatment of the product with acid leads to the enol which rapidly tautomerizes to the more stable carbonyl compound:

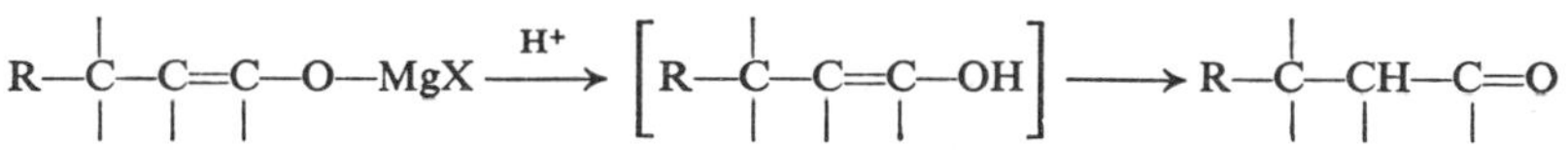

This type of reaction (1,4-addition) competes with addition to the carbonyl group (1,2-addition), and the balance between the two is determined mainly by steric influences. For example, crotonaldehyde reacts with ethylmagnesium bromide entirely at the carbonyl group,

$$CH_3\!-\!CH\!=\!CH\!-\!CH\!=\!O \xrightarrow[\text{2) H}^+]{\text{1) C}_2\text{H}_5\text{MgX}} CH_3\!-\!CH\!=\!CH\!-\!\underset{\underset{\displaystyle C_2H_5}{|}}{CH}\!-\!OH$$

but 3-penten-2-one reacts mainly by 1,4-addition,

$$CH_3\!-\!CH\!=\!CH\!-\!CO\!-\!CH_3 \xrightarrow[\text{2) H}^+]{\text{1) C}_2\text{H}_5\text{MgX}}$$

$$CH_3\!-\!\underset{\underset{\displaystyle C_2H_5}{|}}{CH}\!-\!CH_2\!-\!CO\!-\!CH_3 \;+\; CH_3\!-\!CH\!=\!CH\!-\!\underset{\underset{\displaystyle C_2H_5}{|}}{\overset{\overset{\displaystyle OH}{|}}{C}}\!-\!CH_3$$

$$\qquad\quad 75\% \qquad\qquad\qquad\qquad\qquad 25\%$$

Alternative methods for synthesizing alcohols. Acid chlorides react with Grignard reagents to give ketones which react further to give tertiary alcohols:

$$XMg-R \quad \overset{R'}{\underset{Cl}{C}}=O \rightarrow R-\overset{R'}{\underset{Cl}{\underset{|}{C}}}-O-MgX \xrightarrow{-MgXCl} \overset{R'}{\underset{R}{C}}=O \xrightarrow[2)H^+]{1)RMgX} R'-\overset{R}{\underset{R}{\underset{|}{C}}}-OH$$

Although acid chlorides are more reactive than ketones, it is only rarely possible to isolate the ketone because of the high reactivity of Grignard reagents towards them. However, cadmium alkyls may be successfully employed for the synthesis of ketones from acid chlorides (p. 220).

Esters react analogously to acid chlorides; e.g. phenylmagnesium bromide and ethyl benzoate give triphenylcarbinol in 90% yield [1]:

$$2 \ PhMgBr + PhCO_2Et \xrightarrow{-MgBr(OEt)} Ph_3C-OMgBr \xrightarrow{H^+} Ph_3C-OH$$

Whereas acyclic ethers and cyclic ethers containing essentially strainless rings (e.g. tetrahydrofuran) do not react with Grignard reagents, the strained small-ring cyclic ethers do so, for the strain is relieved during ring-opening. For example, butylmagnesium bromide and ethylene oxide give n-hexyl alcohol in 60% yield [1]:

$$CH_2-CH_2 \xrightarrow[\substack{2) \ H^+}]{1) \ BuMgBr} CH_3CH_2CH_2CH_2CH_2CH_2OH$$
$$\underset{O}{\diagdown \diagup}$$

n-Hexyl alcohol

Trimethylene oxide reacts similarly. These procedures provide a rapid means of extending carbon chains by two and three atoms, respectively.

(*iii*) *Aldehydes.* The reaction of a Grignard reagent with ethyl orthoformate gives an acetal which is converted by mild acid hydrolysis into the aldehyde:

$$XMg-R \quad \overset{EtO}{\underset{EtO}{\diagdown}} CH-OEt \xrightarrow{-MgXOEt} R-CH(OEt)_2 \xrightarrow[-2 \ EtOH]{H_2O-H^+} R-CHO$$

Ethyl orthoformate

For example, n-pentyl bromide can be converted into n-hexaldehyde in up to 50% yield [2].

(*iv*) *Ketones.* Two methods are available. First, Grignard reagents add to the triple bond in nitriles to give magnesium derivatives which are unreactive to further addition and on hydrolysis give ketones *via* the unstable ketimines:

$$RMgX + R'-C\equiv N \longrightarrow \underset{R'}{\overset{R}{>}}C=N-MgX \overset{H^+}{\longrightarrow}$$

$$\underset{R'}{\overset{R}{>}}C=NH \underset{-NH_3}{\overset{H_2O}{\longrightarrow}} \underset{R'}{\overset{R}{>}}C=O$$

Secondly, *NN*-disubstituted amides react with Grignard reagents to give magnesium derivatives* which yield ketones with acid:

$$RMgX + R'-CO-NR''_2 \longrightarrow R'-\underset{R}{\overset{OMgX}{\underset{|}{\overset{|}{C}}}}-NR''_2 \overset{H^+}{\longrightarrow} R-CO-R' + MgX^+ + R''_2NH$$

(*v*) *Carboxylic acids.* Grignard reagents add to carbon dioxide to give salts of carboxylic acids from which the free acids are generated by treatment with mineral acid:

$$XMg-R \underset{\overset{\|}{O}}{\overset{\overset{O}{\|}}{C}} \longrightarrow R-C\underset{O}{\overset{O^-}{<}} MgX^+ \underset{-MgX^+}{\overset{H^+}{\longrightarrow}} R-CO_2H$$

(Grignard reagents, unlike organolithium compounds (p. 218), are not sufficiently reactive to add to the resonance-stabilized carboxylate ion.)

Reaction may be carried out either by pouring an ethereal solution of the Grignard reagent on to solid carbon dioxide ('Dry Ice') or by passing gaseous carbon dioxide from a cylinder into the Grignard solution. For example, using the latter method trimethylacetic acid (pivalic acid) can be obtained in 70% yield from t-butyl chloride [1].

$$Me_3C-Cl \overset{Mg}{\longrightarrow} Me_3C-MgCl \overset{CO_2}{\longrightarrow} Me_3C-CO_2^-MgCl^+ \overset{H^+}{\longrightarrow} Me_3C-CO_2H$$
$$\text{Trimethylacetic acid}$$

(*d*) REACTION AT ELEMENTS OTHER THAN CARBON

Grignard reagents may be used to attach various other elements to carbon. The following types of compounds may be obtained.

---

*The corresponding derivatives from acid chlorides and esters liberate Cl^- and RO^-, respectively, giving ketones, which react further. The difference here results from the high energy of amide ions, R_2N^-, which are consequently not eliminated.

(1) *Hydroperoxides*. The slow addition of a Grignard reagent at low temperatures to ether through which oxygen is bubbling gives the magnesium derivative of a hydroperoxide from which the hydroperoxide is obtained with acid. t-Butyl hydroperoxide can be obtained in 90% yield in this way:

$$Me_3C—MgX \xrightarrow{O_2} Me_3C—O—O—MgX \xrightarrow{H^+} Me_3C—O—OH$$
t-Butyl hydroperoxide

(2) *Alcohols*. When the above reaction is carried out in conditions in which excess of the Grignard reagent is present (e.g. by bubbling oxygen into the Grignard solution), the hydroperoxide derivative reacts with a second molecule of Grignard reagent to give an alcohol:

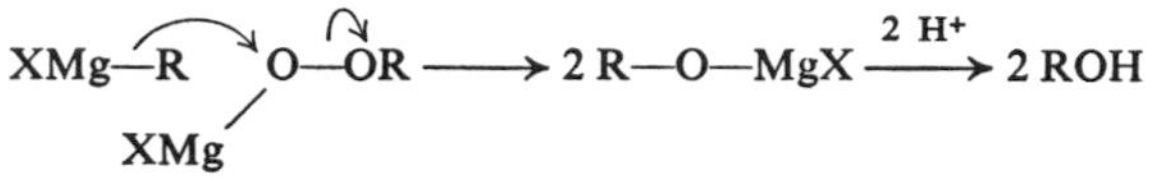

$$XMg—R \quad O—OR \longrightarrow 2\,R—O—MgX \xrightarrow{2\,H^+} 2\,ROH$$
$$XMg$$

(3) *Thiols*. Reaction with sulphur leads to thiols:

$$RMgX + S \longrightarrow R—S—MgX \xrightarrow{H^+} RSH$$

(4) *Sulphinic acids*. Sulphur dioxide reacts analogously to carbon dioxide (p. 213):

$$RMgX + SO_2 \longrightarrow R—S\overset{O^-}{\underset{O}{\diagup\diagdown}} MgX^+ \xrightarrow{H^+} R—\overset{}{\underset{\overset{\|}{O}}{S}}—OH$$
A sulphinic acid

(5) *Iodides*. The reaction with iodine,

$$XMg—R \quad I—I \longrightarrow R—I + MgXI$$

provides a useful method for preparing iodides when the standard methods are unsuccessful. For example, iodides are commonly prepared from chlorides by treatment with sodium iodide in acetone (S_N2 displacement), but this method fails for the highly hindered neopentyl chloride (*cf.* p. 128). However, the Grignard reagent may be formed from neopentyl chloride and it reacts with iodine to give neopentyl iodide in good yield:

$$Me_3C—CH_2Cl \xrightarrow{Mg} Me_3C—CH_2MgCl \xrightarrow[- MgClI]{I_2} Me_3C—CH_2I$$
Neopentyl iodide

(6) *Amines*. The reagent employed is *O*-methylhydroxylamine:

$$XMg\overset{\frown}{-}R \quad NH_2\overset{\frown}{-}OCH_3 \longrightarrow R-NH_2 + MgX(OCH_3)$$

This provides a useful method for t-alkyl amines such as $(CH_3)_3C-NH_2$, for these are not obtainable from S_N2 reactions between t-alkyl halides and ammonia.

(7) *Derivatives of phosphorus, boron, and silicon.*

$$3 \ RMgX + PCl_3 \longrightarrow R_3P + 3 \ MgXCl$$
$$3 \ RMgX + BCl_3 \longrightarrow R_3B + 3 \ MgXCl$$
$$4 \ RMgX + SiCl_4 \longrightarrow R_4Si + 4 \ MgXCl$$

In the last case it is possible to isolate the intermediate silanes, $RSiCl_3$, R_2SiCl_2, and R_3SiCl, by using calculated quantities of the Grignard reagent.

6.3 Organosodium Compounds

Organosodium compounds react in the same way as Grignard reagents but far more vigorously. In addition, they react with ethers by S_N2 displacement and must therefore be prepared in hydrocarbon solvents. Even so, their reaction with the halides from which they are prepared (*Wurtz* coupling),

$$Na\overset{\frown}{-}R \quad \overset{\diagdown}{\underset{\diagup|}{C}}\overset{\frown}{-}X \longrightarrow R-\overset{\diagup}{\underset{|\diagdown}{C}} + NaX$$

is so much more rapid than that of the corresponding Grignard reagents that special techniques are required to obtain them. These difficulties, combined with their spontaneous inflammability in air, have restricted the synthetic value of the compounds, particularly since the advent of organolithium compounds.

6.4 Organolithium Compounds

Organolithium compounds are somewhat less reactive than their sodium analogues but more reactive than Grignard reagents. The fact that they undergo some reactions of which Grignard reagents are incapable gives them their special uses in synthesis.

(a) PREPARATION

Like Grignard reagents, organolithium compounds can often be made by treating an organic halide in ether with lithium metal:

$$RX + 2 \ Li \longrightarrow RLi + LiX$$

Because of the reactivity of organolithiums with oxygen, the reaction is carried out in an atmosphere of dry nitrogen or (better) argon. Organolithium compounds are more reactive than Grignard reagents towards alkyl halides, and it is advantageous to cool the reaction mixture to about $-10°C$ to minimize the extent of the Wurtz coupling reaction,

$$RLi + RX \longrightarrow R—R + LiX$$

However, aryl halides are much less reactive towards nucleophiles, and the synthesis of aryl-lithium compounds can be carried out at the boiling point of the solvent.

Lithium metal does not always react well with aryl and vinyl halides and the corresponding lithium compounds are then conveniently prepared by the metal-halogen exchange reaction using, e.g. preformed butyl-lithium:

$$RBr + BuLi \longrightarrow RLi + BuBr$$

The metalation reaction is suitable for the preparation of the lithium derivatives of comparatively acidic hydrocarbons,* e.g.

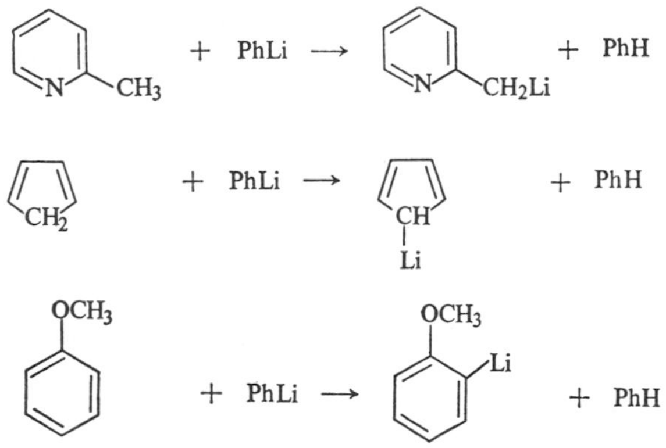

Some of these metalations (e.g. those of α-picoline and anisole) cannot be brought about by Grignard reagents; thus, the preparation of the o-anisyl Grignard reagent would require first the preparation of o-bromoanisole.

(b) REACTIONS
The reactions of organolithium compounds in the main parallel those of Grignard reagents, and attention is directed here to those reactions which either

*The metalation of anisole occurs at the ortho-position because the ortho carbon atoms, being the nearest of the nuclear carbon atoms to the electron-attracting oxygen atom, can tolerate a negative charge the most readily.

only the lithium compounds undergo, or in which they are the more effective reagents.

(1) Lithium compounds are less readily prevented by steric hindrance from reacting at carbonyl groups. For example, whereas isopropylmagnesium bromide does not add to di-isopropyl ketone but brings about enolization and reduction (p. 210), isopropyl-lithium adds successfully to give tri-isopropylcarbinol:

$$Me_2CHLi + (Me_2CH)_2C{=}O \longrightarrow (Me_2CH)_3C{-}O{-}Li \xrightarrow{H^+} (Me_2CH)_3C{-}OH$$

(2) Whereas Grignard reagents often react with $\alpha\beta$-unsaturated ketones predominantly by 1,4-addition (p. 211), lithium compounds react predominantly by 1,2-addition, e.g.

$$Ph{-}CH{=}CH{-}CO{-}Ph \begin{cases} \xrightarrow{PhMgX} Ph_2CH{-}CH_2{-}CO{-}Ph \text{ (mainly)} \\ \\ \xrightarrow{PhLi} Ph{-}CH{=}CH{-}\underset{\underset{OH}{|}}{C}Ph_2 \text{ (mainly)} \end{cases}$$

(3) Organolithium compounds react more efficiently with alkyl halides, and Wurtz coupling reactions can be carried out in good yield. An interesting example is the synthesis of optically active 9,10-dihydro-3:4,5:6-dibenzophenanthrene* from optically active 1,1'-binaphthyl-2,2'-dicarboxylic acid:

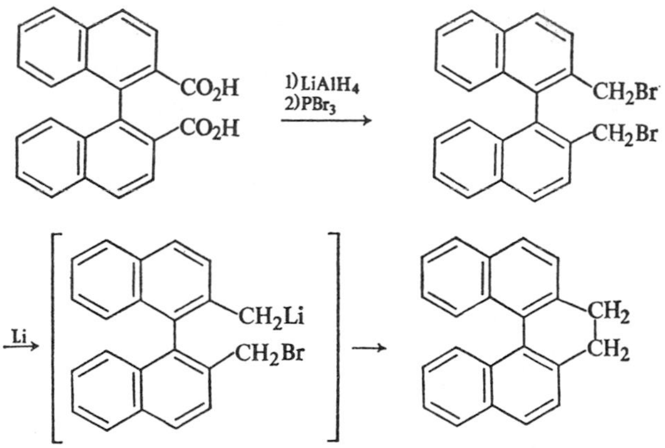

9,10-Dihydro-3:4,5:6-dibenzo-
phenanthrene

*This was the first synthesis (1950) of an optically active compound of this type. Repulsion between the *peri*-hydrogen atoms forces the lower left-hand benzenoid ring to lie somewhat above or somewhat below the upper left-hand ring, giving the two enantiomers. The molecule thus has a spiral shape.

(4) Carbon dioxide, which reacts with Grignard reagents to give carboxylic acids, reacts with organolithium compounds to give ketones:

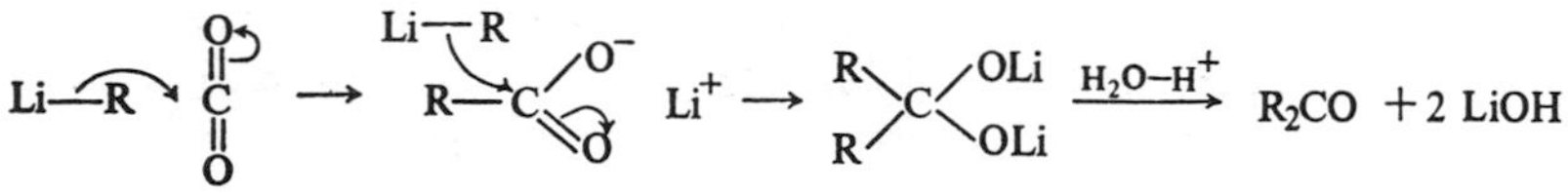

$$Li{-}R \nwarrow \overset{O}{\underset{O}{\overset{\parallel}{C}}} \xrightarrow{Li{-}R} R{-}\overset{O^-}{\underset{O}{C}} \; Li^+ \longrightarrow \underset{R'}{\overset{R}{>}}C\underset{OLi}{\overset{OLi}{<}} \xrightarrow{H_2O-H^+} R_2CO \; + 2 \; LiOH$$

The difference arises from the fact that the lithium compounds are more strongly nucleophilic than Grignard reagents and are able to react with the intermediate resonance-stabilized carboxylate anion.

Similarly, carboxylic acids may be converted into ketones:

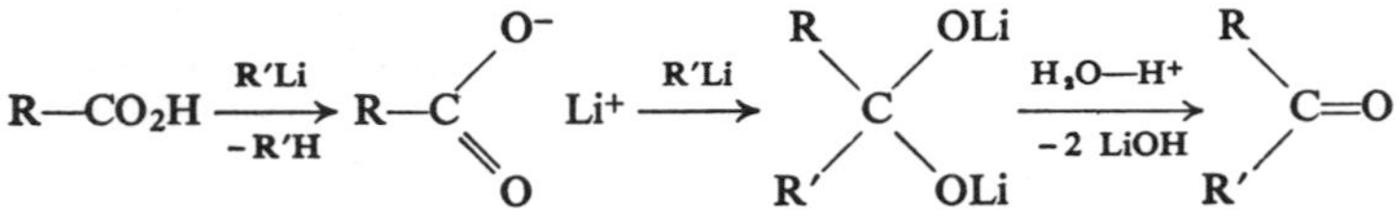

$$R{-}CO_2H \xrightarrow[-R'H]{R'Li} R{-}\overset{O^-}{\underset{O}{C}} \; Li^+ \xrightarrow{R'Li} \underset{R'}{\overset{R}{>}}C\underset{OLi}{\overset{OLi}{<}} \xrightarrow[-2\,LiOH]{H_2O-H^+} \underset{R'}{\overset{R}{>}}C{=}O$$

For example, cyclohexanecarboxylic acid and methyl-lithium give methyl cyclohexyl ketone in up to 94% yield [5]:

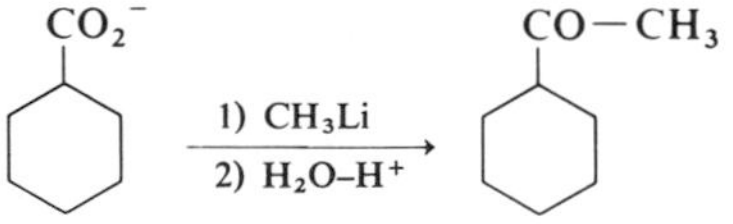

(5) Organolithium compounds, unlike Grignard reagents, react with olefinic double bonds. Simple olefins such as ethylene react only at high pressures (100–500 atmospheres) and give mixtures of compounds of different chain-lengths,

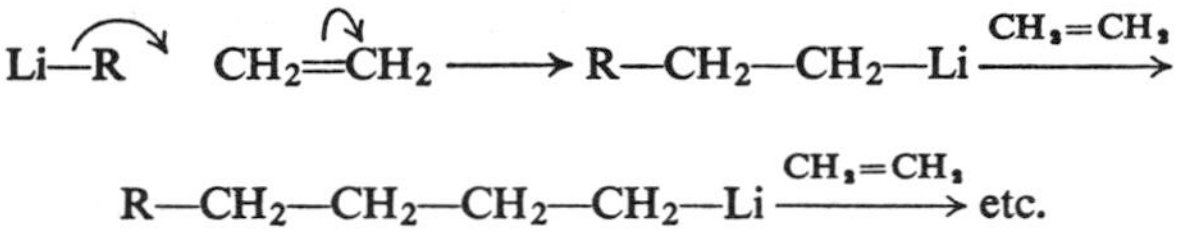

$$Li{-}R \quad CH_2{=}CH_2 \longrightarrow R{-}CH_2{-}CH_2{-}Li \xrightarrow{CH_2=CH_2}$$

$$R{-}CH_2{-}CH_2{-}CH_2{-}CH_2{-}Li \xrightarrow{CH_2=CH_2} etc.$$

but conjugated olefins react at atmospheric pressure. If the first adduct contains lithium attached to a hindered carbon atom, reaction stops at this stage, e.g.

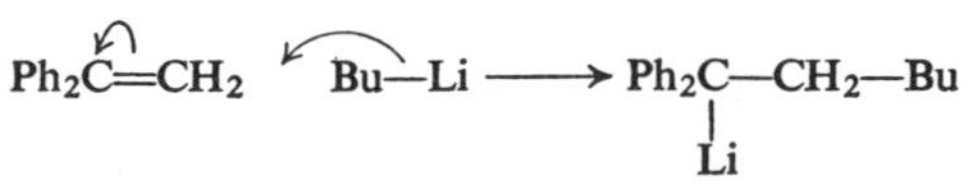

$$Ph_2C{=}CH_2 \quad Bu{-}Li \longrightarrow Ph_2\underset{Li}{\overset{}{C}}{-}CH_2{-}Bu$$

(6) Unlike Grignard reagents, lithium compounds are sufficiently nucleophilic to react at the nuclear carbon atoms of aromatic systems which are activated to nucleophiles (see Chapter 12 for a fuller discussion). For example,

phenyl-lithium reacts with pyridine at 110°C to give an adduct which is decomposed by water to 2-phenylpyridine (40–49%) [2]:

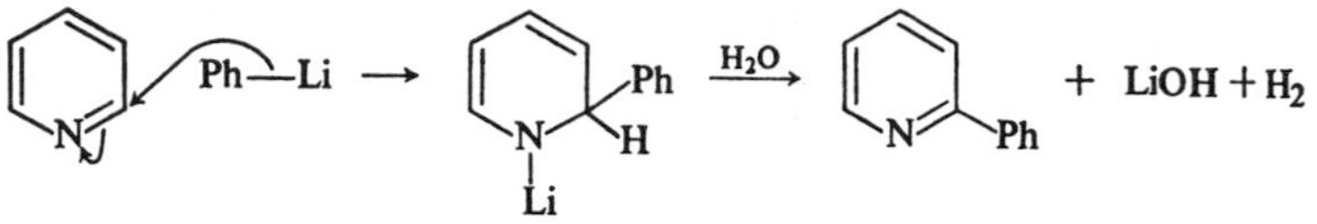

2-Phenylpyridine

6.5 Organocopper Compounds

αβ-Unsaturated carbonyl compounds can react with Grignard reagents by both 1,2- and 1,4-addition (p. 211) and with lithium reagents by 1,2-addition (p. 217). The desirability of having a reagent which reacts specifically in the 1,4-manner has led to the development of organocopper intermediates.

These are of two types: organocopper compounds, RCu, and lithium organocuprates, R_2CuLi, which are formed as follows:

$$RLi + CuI \longrightarrow RCu + LiI$$
$$2\ RLi + CuI \longrightarrow R_2CuLi + LiI$$

Each type reacts solely by 1,4-addition, although the reason is not clear. The overall reaction can be represented:

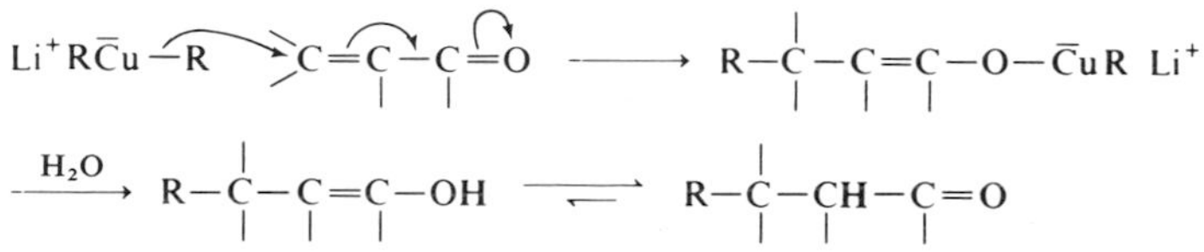

The organocopper compounds also react with alkyl halides, e.g.

$(C_4H_9)_2CuLi$	+	$C_{10}H_{21}Br$	$\longrightarrow$	$C_{14}H_{30}$
Lithium di-n-butylcuprate		1-Bromodecane		Tetradecane (80%)

Further, vinyl, aryl, and acetylenic copper compounds can be made, and the corresponding halides can be used as substrates, so increasing the synthetic utility. The first two of the following illustrate the retention of configuration which characterizes both the formation of a vinylic copper compound and the displacement of halide from a vinylic derivative:

$$CH_3\!\!\diagdown\!\!CH\!\!=\!\!CH\!\!\diagdown\!\!Cl \quad\xrightarrow{\text{Li}}\quad CH_3\!\!\diagdown\!\!CH\!\!=\!\!CH\!\!\diagdown\!\!Li \quad\xrightarrow[\text{Et}_2\text{O}\,(-78°\text{C})]{\text{CuI}}$$

$$(CH_3\!\!\diagdown\!\!CH\!\!=\!\!CH)_2\!\!\diagdown\!\!CuLi \quad\xrightarrow[(\text{Me}_2\text{N})_3\text{PO}\,(-35°\text{C})]{CH_3(CH_2)_6CH_2I}\quad CH_3\!\!\diagdown\!\!CH\!\!=\!\!CH\!\!\diagdown\!\!CH_2(CH_2)_6CH_3$$

90% [55]

$$(C_4H_9)_2CuLi + Ph\!\!\diagdown\!\!CH\!\!=\!\!CH\!\!\diagdown\!\!Br \quad\longrightarrow\quad Ph\!\!\diagdown\!\!CH\!\!=\!\!CH\!\!\diagdown\!\!C_4H_9$$

71%

$$(C_4H_9)_2CuLi + PhI \quad\longrightarrow\quad Ph\!-\!C_4H_9$$

75%

Copper(I) salts catalyze the reactions of diazoalkanes with unsaturated compounds. Olefins give cyclopropanes,

$$\diagup\!\!\!C\!\!=\!\!C\!\!\diagdown + CH_2N_2 \xrightarrow{\text{Cu}^+} \underset{CH_2}{\diagup\!\!\!C\!\!-\!\!C\!\!\diagdown} + N_2$$

and aromatic compounds give ring-expanded products, through addition and Cope rearrangement (p. 314), e.g.

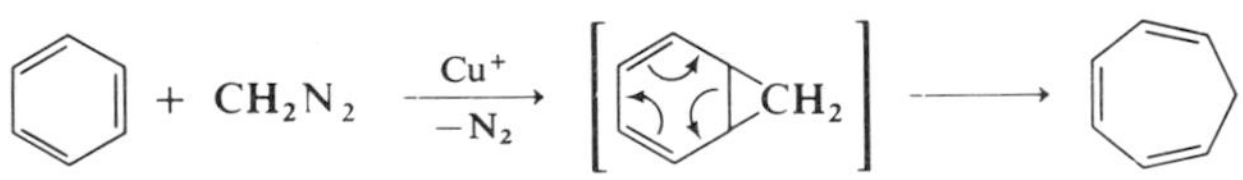

$$\bigcirc + CH_2N_2 \xrightarrow[-N_2]{\text{Cu}^+} \left[\bigcirc\!\!\!CH_2 \right] \longrightarrow \bigcirc$$

It is thought that a copper-carbene complex is involved.

6.6 Organocadmium Compounds

Cadmium alkyls and aryls may be obtained by the metal-metal exchange reaction (p. 203), using either a Grignard reagent or an organolithium compound:

$$2\ RMgX + CdCl_2 \longrightarrow R_2Cd + 2\ MgXCl$$

$$2\ RLi + CdCl_2 \longrightarrow R_2Cd + 2\ LiCl$$

They are far less reactive compounds than both Grignard reagents and organolithium compounds. In particular, they do not react with ketones or esters, although they react with acid chlorides:

$$2\ R'COCl + R_2Cd \longrightarrow 2\ R'COR + CdCl_2$$

This specificity gives organocadmium compounds their main use in synthesis. Although it is possible to obtain ketones from both Grignard reagents and organolithium compounds, it is often necessary to introduce a ketonic function into a molecule which possesses other functions susceptible to attack by magnesium and lithium compounds. The cadmium compound may then be used. A specific example occurs in the Harvard synthesis of cholesterol in which the conversion —COCl → —COCH$_3$ was effected by dimethylcadmium in the presence of an ester group (21.3).

6.7 Organozinc Compounds

Dialkylzinc compounds are obtained from alkyl iodides and a zinc-copper couple,

$$2\ RI + 2\ Zn \longrightarrow R_2Zn + ZnI_2$$

They are less reactive than Grignard reagents and resemble organocadmium compounds; e.g. they react readily with acid chlorides but only very slowly with ketones, so that it is possible to employ them for the synthesis of ketones. However, they offer no advantages over their cadmium analogues in this reaction, and moreover their spontaneous ignition in air makes them difficult to handle. They have therefore been superseded by organocadmium reagents.

There is, however, one reaction involving a zinc alkyl which is of considerable synthetic importance. In the *Reformatsky reaction* an aldehyde or ketone is treated with zinc metal and an α-bromo-ester to give, after hydrolysis, a β-hydroxy-ester:

$$RCOR' + Zn + BrCH_2CO_2Et \longrightarrow R\underset{\underset{OZnBr}{|}}{\overset{\overset{R'}{|}}{C}}\!-CH_2\!-CO_2Et$$

$$\overset{H_2O}{\longrightarrow} R\underset{\underset{OH}{|}}{\overset{\overset{R'}{|}}{C}}\!-CH_2\!-CO_2Et + ZnBr(OH)$$

The reaction is usually carried out in ether, as are Grignard syntheses, but an important practical difference is that in the Reformatsky reaction all the reactants are mixed in the same vessel, whereas in Grignard syntheses it is necessary to form the magnesium compound before the introduction of the carbonyl compound. Clearly, however, the mechanism of reaction is fundamentally similar: the organozinc compound is formed first and reacts at the carbonyl group in the manner of a Grignard reagent. In a typical example, benzaldehyde and ethyl bromoacetate give ethyl β-hydroxydihydrocinnamate in 61–64% yield [3]; dehydration is brought about readily to give ethyl cinnamate:

$$PhCHO + BrCH_2CO_2Et \xrightarrow[\text{2) H}_2\text{O}]{\text{1) Zn}} \underset{\underset{OH}{|}}{PhCH}-CH_2CO_2Et \xrightarrow{-H_2O} PhCH{=}CHCO_2Et$$

Ethyl cinnamate

The reaction is specific to α-bromo-esters and vinylogous compounds such as ethyl γ-bromocrotonate,

$$R_2CO + BrCH_2-CH{=}CH-CO_2Et \xrightarrow[\text{2) H}_2\text{O}]{\text{1) Zn}} \underset{\underset{OH}{|}}{R_2C}-CH_2-CH{=}CH-CO_2Et$$

The reason for this is not clear, although it is possibly relevant that α-bromo-esters and their vinylogues are more reactive in, e.g. S_N2 reactions than other bromo-esters.

An interesting example of the application of the Reformatsky reaction using methyl γ-bromocrotonate occurs in a synthesis of Vitamin A from β-ionone (p. 713), which also illustrates the use of two other organometallic reagents (21.1):

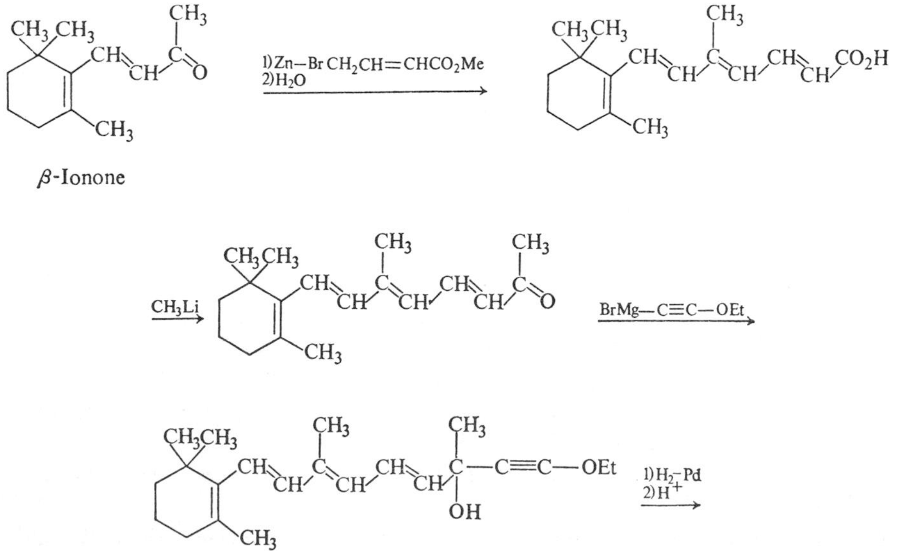

β-Ionone

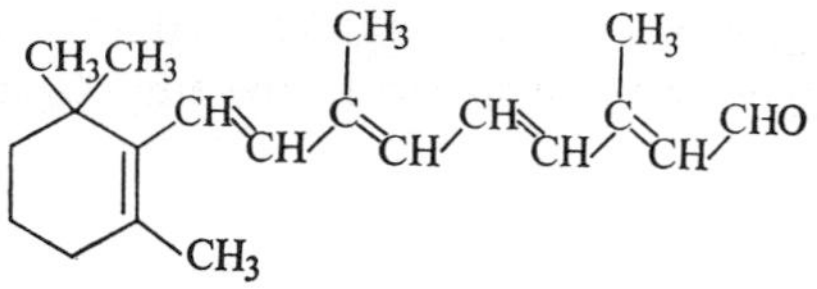

Vitamin A aldehyde

A second useful application of organozinc compounds has recently been developed. An olefin reacts with methylene iodide in the presence of a zinc-copper couple to give a cyclopropane:

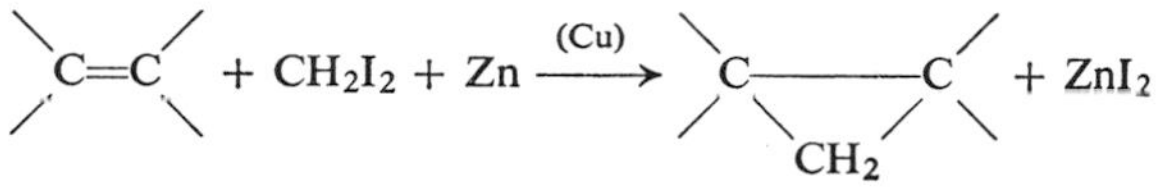

For example, cyclohexene gives 57% of norcarane:

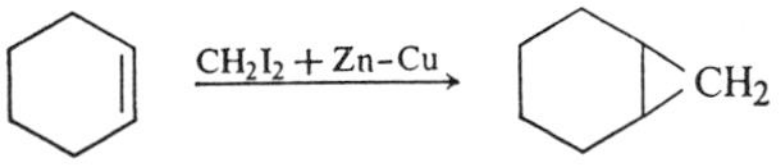

Norcarane

The reaction, which is thought to involve a complexed carbene,

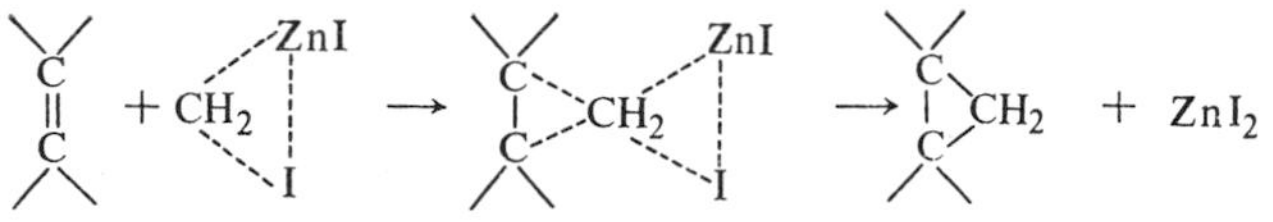

occurs in a stereospecific manner, giving the *cis*-adduct.

Further Reading

JORGENSON, M. J., 'Preparation of ketones from the reaction of organolithium reagents with carboxylic acids,' *Organic Reactions*, 1970, **18**, 1.

KHARASCH, M. S., and REINMUTH, O., *Grignard Reactions of Nonmetallic Substances*, Constable (London 1954).

POSNER, G. H., 'Conjugate addition reactions of organocopper reagents,' *Organic Reactions*, 1972, **19**, 1.

POSNER, G. H., 'Substitution reactions using organocopper reagents,' *Organic Reactions*, 1975, **22**, 253.

RATHKE, M. W., 'The Reformatsky reaction,' *Organic Reactions*, 1975, **22**, 423.

SHIRLEY, D. A., 'The synthesis of ketones from acid halides and organometallic compounds of magnesium, zinc, and cadmium,' *Organic Reactions*, 1954, **8**, 28.

Problems

1. How would you obtain each of the following from phenylmagnesium bromide? Ph_3C—OH; Ph_2C=CH_2; $PhC(CH_3)$=CH_2; $PhCO_2H$; $PhCH_2Ph$; PhCHO; $PhCH_2CH_2OH$; PhD.

2. What products would you expect from the reaction of methylmagnesium iodide with each of the following? CH_3CO_2Et; CH_3CN; $ClCO_2Et$; CH_3CO—O—$COCH_3$; $ClCH_2OCH_3$; CH_2=CH—CH_2Br; $PhCOCH$=CH_2; $(CH_3)_3C$—CO—$C(CH_3)_3$.

3. How would you employ organometallic reagents to make the following?

    (a) $CH_3CHCH_2CH_3$
        |
        OH

    (b) CH_2=CH—CO_2H

    (c) $(CH_3)_2C$=CH—CHO

    (d) $(CH_3)_2C$=CH—CO_2Et

    (e) $PhCH_2CH_3$

    (f) CH_3—C≡C—CH_2OH

    (g) $(CH_3)_3C$—CHO

    (h) $PhCH$=CH—CH=CH—CO_2Et

    (i) $(CH_3)_3C$—CO—CH_3

    (j)

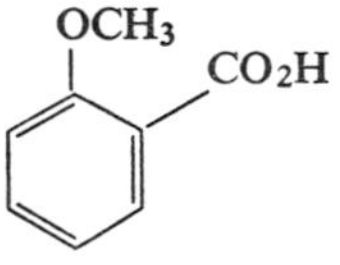

    (k)

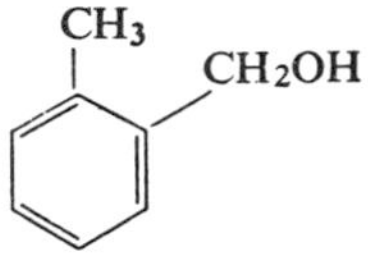

    (l)

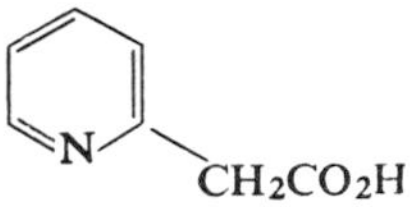

7. Formation of Carbon-Carbon Bonds: Base-catalyzed Condensations

7.1 Principles

The formation of carbon-carbon bonds by base-catalyzed condensations is closely related to their formation from organometallic reagents. In each method negatively polarized carbon reacts with the electrophilic carbon of carbonyl groups, alkyl halides, and related compounds. The difference between the two procedures is that whereas the method of inducing negative polarization in carbon in the latter case is to attach carbon to an electropositive metal, in the former case a base is employed to abstract a proton from a C—H bond to give a *carbanion*.

The scope of base-catalyzed condensations depends on three facts: a wide range of compounds is able to give rise to carbanions; these anions react with electrophilic carbon in a variety of environments; and the basicity of the reagent used to abstract the proton in the generation of the carbanion may be varied widely so as to produce suitable conditions for a particular reaction.

The factors which govern the formation of a carbanion have been briefly discussed (2.6). The structural requirement is that the C—H bond from which proton-abstraction is to occur must be adjacent to one or more groups of $-M$ type which can stabilize the anion, e.g.

$$CH_3-CH{=}O \underset{BH^+}{\overset{B:}{\rightleftharpoons}} \bar{C}H_2-CH{=}O \leftrightarrow CH_2{=}CH-\bar{O}$$

Thus paraffins do not undergo base-catalyzed reactions because the corresponding anions, being of very high energy content, are formed at a negligible rate. The more common of the activating groups, arranged in approximately decreasing power of activation, are:

$$-NO_2 > -COR > -CN > -CO_2R$$

A hydrogen atom attached to carbon which is bonded to one of these groups (an 'α-hydrogen') is described as being *activated*. The presence of two such substituents adjacent to a C—H bond considerably increases the acidity of that bond (2.6), and many useful condensations employ di-activated compounds such as malonic ester, $CH_2(CO_2Et)_2$, and acetoacetic ester, $CH_3COCH_2CO_2Et$.

225

The carbanions, although resonance-stabilized, are nevertheless high-energy species which react with positively polarized carbon. The two classes of reaction of general importance are:

(*a*) with carbonyl-containing compounds:

$$\overset{\diagdown}{\underset{\diagup}{C^-}} \quad \overset{|}{\underset{|}{C}}{=}O \rightleftharpoons -\overset{|}{\underset{|}{C}}-\overset{|}{\underset{|}{C}}-O^-$$

(*b*) with alkyl halides:

$$\overset{\diagdown}{\underset{\diagup}{C^-}} \quad \overset{\diagdown}{\underset{\diagup|}{C}}{-}Hal \rightarrow -\overset{|}{\underset{|}{C}}-\overset{|}{\underset{|}{C}}- + \text{Halide}^-$$

The reactions are similar in principle in that each involves the displacement of an electron-pair from the carbon which is attacked on to an electronegative atom. They differ in that only the former reaction is readily reversible (so that many of the reactions described in the ensuing text can be employed in the reverse sense for the cleavage of C—C bonds).

The commonly used bases, in order of decreasing base-strength (2.6), are:

$$Ph_3C^- > NH_2^- > Me_3CO^- > EtO^- > HO^- > R_3N$$

The choice of base is determined by the reactivity of the carbanion-forming component and is discussed below.

It should be noted that the charge on a carbanion is shared by carbon and a second, more electronegative element (usually oxygen, but in the case of activation by —CN, nitrogen) so that reaction with a carbonyl group or a halide might occur at the more electronegative atom, e.g.

$$\bar{C}H_2{-}CH{=}O \leftrightarrow CH_2{=}CH{-}O^- \quad \overset{CH_3}{\underset{}{CH}}{=}O \longrightarrow CH_2{=}CH{-}O{-}\overset{CH_3}{\underset{}{CH}}{-}O^-$$

$$\xrightarrow[-\,EtO^-]{EtOH} CH_2{=}CH{-}O{-}\overset{CH_3}{\underset{}{CH}}{-}OH$$

$$\bar{C}H_2{-}CH{=}O \leftrightarrow CH_2{=}CH{-}O^- \quad CH_3{-}I \longrightarrow CH_2{=}CH{-}O{-}CH_3 + I^-$$

In fact, reaction at carbon normally dominates in each case, probably because the products are thermodynamically more stable than those derived by reaction through oxygen, this difference in stability being reflected in the transition states for the processes. One exception to this generalization is described later (p. 252).

7.2 Condensations of Carbanions with Aldehydes and Ketones

(*a*) THE ALDOL CONDENSATION

The treatment of acetaldehyde with dilute caustic soda solution gives, initially, aldol:

$$2 \text{ CH}_3\text{CHO} \xrightarrow{\text{OH}^-} \text{CH}_3\text{—CH—CH}_2\text{—CHO}$$
$$\underset{\text{OH}}{|}$$

Aldol

This is the simplest example of a condensation which is common to those aldehydes and ketones in which the carbon adjacent to the carbonyl group bears one or more hydrogen atoms. The mechanism is as follows:

$$\text{CH}_3\text{CHO} + \text{OH}^- \rightleftharpoons \bar{\text{C}}\text{H}_2\text{CHO} + \text{H}_2\text{O}$$

$$\text{CH}_3\text{CHO} + \bar{\text{C}}\text{H}_2\text{CHO} \rightleftharpoons \text{CH}_3\text{CH—CH}_2\text{CHO}$$
$$\underset{\text{O}^-}{|}$$

$$\text{CH}_3\text{CH—CH}_2\text{CHO} + \text{H}_2\text{O} \rightleftharpoons \text{CH}_3\text{CH—CH}_2\text{CHO} + \text{OH}^-$$
$$\underset{\text{O}^-}{|} \qquad\qquad\qquad\qquad \underset{\text{OH}}{|}$$

The kinetics (rate = $k[\text{CH}_3\text{CHO}][\text{OH}^-]$) show that the first step, generation of the carbanion, is rate-determining. In a closely related case, however, the self-condensation of acetone to give diacetone alcohol, the kinetics (rate = $k[\text{CH}_3\text{COCH}_3]^2[\text{OH}^-]$) show that the rate-determining step is the reaction of the carbanion with a second molecule of acetone:

$$\text{CH}_3\text{COCH}_3 + \text{OH}^- \rightleftharpoons \bar{\text{C}}\text{H}_2\text{COCH}_3 + \text{H}_2\text{O}$$

$$\text{CH}_3\text{COCH}_3 + \bar{\text{C}}\text{H}_2\text{COCH}_3 \underset{\xleftarrow{\text{slow}}}{\rightleftharpoons} (\text{CH}_3)_2\text{C—CH}_2\text{COCH}_3$$
$$\underset{\text{O}^-}{|}$$

$$(\text{CH}_3)_2\text{C—CH}_2\text{COCH}_3 + \text{H}_2\text{O} \rightleftharpoons (\text{CH}_3)_2\text{C—CH}_2\text{COCH}_3 + \text{OH}^-$$
$$\underset{\text{O}^-}{|} \qquad\qquad\qquad\qquad\qquad \underset{\text{OH}}{|}$$

Diacetone alcohol

The difference is attributed to the fact that the carbonyl group in acetone is less rapidly attacked by nucleophiles than that in acetaldehyde, partly because the electron-releasing character of methyl compared with hydrogen renders the adduct from acetone (and the preceding transition state) less stable, and partly because the carbonyl group in acetone is more hindered. It is a general feature of aldol condensations that the step involving C—C bond-formation is facilitated by electron-attracting groups on the carbonyl component and retarded by electron-releasing groups.

Other general features of the reaction which must be borne in mind in synthetic applications are the following.

(1) Reaction is readily reversible, and the position of equilibrium is not always favourable to the product. This is true, e.g. for the formation of diacetone alcohol from acetone, and a special experimental procedure is used to obtain a high yield of the product. Barium hydroxide is placed in the thimble of a Soxhlet extractor over a flask of boiling acetone. As acetone comes into contact with the catalyst a small proportion of diacetone alcohol is formed which is then automatically returned to the boiler by a syphon mechanism. Since this product boils over 100°C higher than acetone, it accumulates in the boiler while acetone continues to reflux, condense on to the catalyst, and return to the boiler together with further product. In this way diacetone alcohol may be obtained in 70% yield [1].

(2) Aldols as such are not always isolated from the condensation. For example, acetaldol readily forms a cyclic hemi-acetal:

$$CH_3-CH-CH_2-CHO \xrightarrow{CH_3CHO} \left[\begin{array}{c} CH_2-CHO \\ CH_3-CH \qquad OH \\ O-CH \\ CH_3 \end{array} \right] \longrightarrow \begin{array}{c} OH \\ CH_2-CH \\ CH_3-CH \qquad O \\ O---CH \\ CH_3 \end{array}$$

On being heated, this trimer is converted back into acetaldol which, in hot basic conditions, undergoes dehydration to crotonaldehyde:

$$CH_3CH-CH_2CHO \xrightarrow[-H_2O]{(OH^-)} CH_3CH=CHCHO$$
$$\qquad\quad OH \qquad\qquad\qquad\qquad \text{Crotonaldehyde}$$

Diacetone alcohol is more stable, but dehydration can be brought about by acid or base to give mesityl oxide:

$$(CH_3)_2C-CH_2COCH_3 \xrightarrow[-H_2O]{(H^+)\ or\ (OH^-)} (CH_3)_2C=CHCOCH_3$$
$$\qquad\quad OH \qquad\qquad\qquad\qquad\qquad \text{Mesityl oxide}$$

(3) The use of more concentrated alkali can lead directly to the unsaturated carbonyl product which can then undergo further condensations. For instance, acetaldehyde gives resins as a result of successive condensations of the type,

$$CH_3CHO \xrightarrow[\text{(OH}^-)]{CH_3CHO} CH_3CH{=}CHCHO \xrightarrow[\text{(OH}^-)]{CH_3CHO} CH_3CH{=}CH{-}CH{=}CHCHO \xrightarrow[\text{(OH}^-)]{CH_3CHO} \text{etc.}$$

Unsymmetrical ketones. Whereas an aldehyde has only one carbon atom activated by the carbonyl group from which proton-abstraction may occur, a ketone may have two. For example, methyl ethyl ketone might undergo condensation *via* either of the anions $\overline{C}H_2COCH_2CH_3$ and $CH_3CO\overline{C}HCH_3$, leading to

$$\underset{C_2H_5 \quad OH}{\overset{CH_3}{\diagdown}}C{-}CH_2COCH_2CH_3 \text{ and } \underset{C_2H_5 \quad OH}{\overset{CH_3}{\diagdown}}C{-}\underset{}{\overset{CH_3}{\diagup}}CH\underset{COCH_3}{\diagdown}$$

respectively. In practice, the former product predominates, possibly because there is greater steric hindrance to attack on the carbonyl group of methyl ethyl ketone by the carbanion $CH_3CO\overline{C}HCH_3$ than by $\overline{C}H_2COCH_2CH_3$. In general, the carbanion component from an unsymmetrical ketone is that derived from the less highly alkylated carbon.

Mixed condensations. If each of two aldehydes contains an α-hydrogen atom, aldol condensation can give each of four products: each aldehyde can provide a carbanion and each can act as the carbonyl component. Such condensations are only of synthetic value if a required product can easily be separated from the resulting mixture.

If, however, one of the two components has no α-hydrogen, only two condensation products can be formed. Further, if this component has the more reactive carbonyl group of the two, one product predominates. For example, the action of alkali on a mixture of formaldehyde and acetaldehyde leads, initially, to β-hydroxypropionaldehyde,

$$CH_3CHO \underset{OH^-}{\rightleftharpoons} \overline{C}H_2CHO \underset{CH_2O}{\rightleftharpoons} \underset{O^-}{\overset{}{\underset{|}{CH_2}}}{-}CH_2CHO \underset{H_2O}{\rightleftharpoons} \underset{OH}{\overset{}{\underset{|}{CH_2}}}{-}CH_2CHO$$

β-Hydroxypropionaldehyde

for only acetaldehyde can form a carbanion and the carbonyl group in formaldehyde is the more reactive to nucleophilic addition.

In practice, this condensation does not give β-hydroxypropionaldehyde as the final product. If the reaction is carried out in the gas phase at a high temperature (e.g. using sodium silicate as catalyst, at 300°C), dehydration occurs to give acrolein,

$$\underset{OH}{\overset{}{\underset{|}{CH_2}}}{-}CH_2CHO \xrightarrow[-H_2O]{\text{heat}} CH_2{=}CHCHO$$

Acrolein

whereas at low temperatures further aldol condensations occur until each of the three α-hydrogens of acetaldehyde has been replaced,

$$CH_3CHO + 3 CH_2O \xrightarrow{(OH^-)} HOCH_2-\underset{\underset{CH_2OH}{|}}{\overset{\overset{CH_2OH}{|}}{C}}-CHO$$

This product, like formaldehyde, has no active hydrogen and next undergoes the crossed Cannizzaro reaction characteristic of such aldehydes (p. 634):

$$(CH_2OH)_3C-CHO + HCHO \xrightarrow{(OH^-)} C(CH_2OH)_4 + HCO_2H$$
$$\text{Pentaerythritol}$$

The product, pentaerythritol, can be obtained in about 55% yield in this way using calcium hydroxide as the basic catalyst [1]. This reaction is also carried out industrially, using lime as the catalyst; the pentaerythritol is converted into its tetranitrate ester, which is used as an explosive.

The aldol condensation on a mixture of an aldehyde and a ketone each of which contains active hydrogen may also in principle yield four products, but since the carbonyl group in aldehydes is usually more reactive towards nucleophiles than that in a ketone, only two products are likely to be formed in significant amounts. These are the self-condensation product of the aldehyde and the product derived by reaction of the carbanion from the ketone with the carbonyl of the aldehyde. Moreover, the self-condensation product can be minimized by the slow addition of the aldehyde to a mixture of the ketone and the base. The formation of ψ-ionone from citral, used in a total synthesis of Vitamin A (21.1), illustrates the synthetic utility of this process:

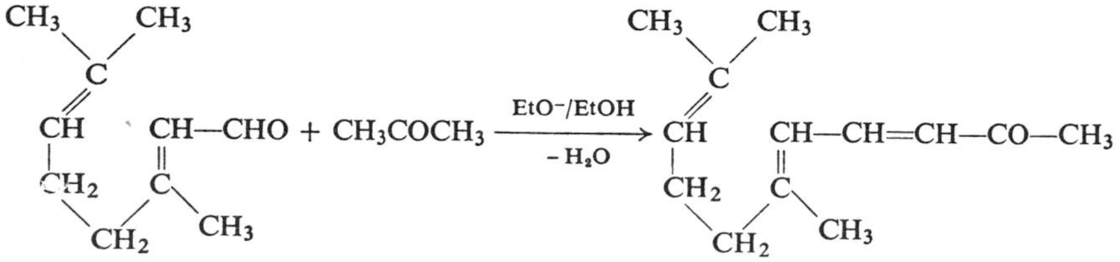

ψ-Ionone
45%

The aldol condensation on a mixture of a ketone and an aldehyde which possesses no active hydrogen usually gives essentially one product. For example, benzalacetone is obtained in about 70% yield from benzaldehyde and excess of acetone in the presence of 10% sodium hydroxide solution [1],

$$PhCHO + CH_3COCH_3 \xrightarrow[-H_2O]{(OH^-)} PhCH=CHCOCH_3$$
Benzalacetone

and, if acetone is not in excess, dibenzalacetone can be obtained in about 80% yield [2]:

$$2 \ PhCHO + CH_3COCH_3 \xrightarrow[-2 \ H_2O]{(OH^-)} PhCH=CH-CO-CH=CHPh$$
Dibenzalacetone

Any ketone which possesses the grouping $-CH_2-CO-CH_2-$ can give an analogous dibenzal derivative, and the reaction with benzaldehyde is usefully employed to reveal the presence of such groupings.

(b) THE CLAISEN REACTION

The base-catalyzed reaction between an ester which contains active hydrogen and an aldehyde which does not is known as the Claisen reaction. For example, ethyl acetate and benzaldehyde react in the presence of sodium ethoxide to give ethyl cinnamate in about 70% yield [1]:

$$PhCHO + CH_3CO_2Et \xrightarrow[-H_2O]{(EtO^-)} PhCH=CHCO_2Et$$
Ethyl cinnamate

The success of the reaction depends on the fact that the carbonyl in an aldehyde is more reactive towards nucleophiles than one in an ester. Aldehydes which contain an α-hydrogen atom are not suitable components in the Claisen reaction because they preferentially undergo self-condensation.

(c) THE PERKIN REACTION

This reaction consists of the condensation of an acid anhydride with an aromatic aldehyde catalyzed by a carboxylate ion. The anhydride provides the carbanion under the influence of the basic carboxylate ion, and attack by this anion on the carbonyl of the aldehyde is followed by dehydration and hydrolysis of the anhydride group, e.g.

$$CH_3CO.O.COCH_3 + CH_3CO_2^- \rightleftharpoons \bar{C}H_2CO.O.COCH_3 + CH_3CO_2H$$

$$PhCHO + \bar{C}H_2CO.O.COCH_3 \rightleftharpoons PhCH-CH_2CO.O.COCH_3$$
$$\qquad\qquad\qquad\qquad\qquad\qquad\qquad\qquad | $$
$$\qquad\qquad\qquad\qquad\qquad\qquad\qquad\qquad O^-$$

$$\xrightarrow[\longleftarrow]{CH_3CO_2H} PhCH-CH_2CO.O.COCH_3 \xrightarrow[\text{2) hydrolysis}]{\text{1) } -H_2O} PhCH=CHCO_2H + CH_3CO_2H$$
$$\qquad\qquad\qquad | $$
$$\qquad\qquad\qquad OH \qquad\qquad\qquad\qquad\qquad\qquad\qquad\qquad \text{Cinnamic acid}$$
$$\qquad\qquad\qquad\qquad\qquad\qquad\qquad\qquad\qquad\qquad\qquad\qquad\qquad 50\%$$

Furylacrylic acid may be made in this way in 65–70% yield [3],

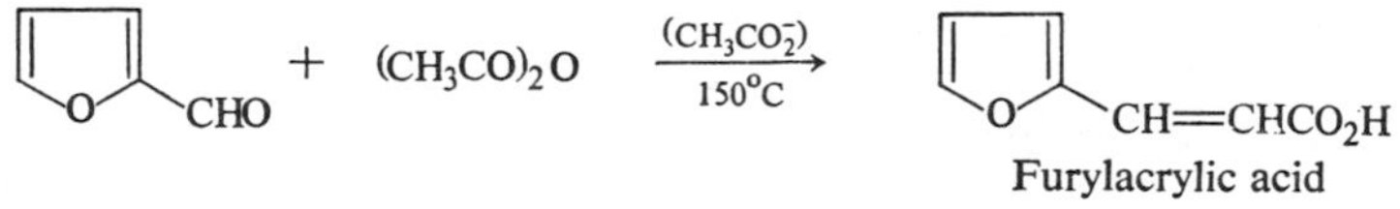

Furylacrylic acid

Electron-attracting groups on the aromatic ring increase the efficiency of the reaction: e.g. p-nitrobenzaldehyde gives an 82% yield of p-nitrocinnamic acid. Conversely, electron-releasing groups reduce the efficiency: e.g. p-dimethyl-aminobenzaldehyde fails to condense. Because of the relatively weak basicity of carboxylate ions, fairly high temperatures are generally required.

Methods based on the Perkin reaction have been used in the synthesis of α-amino-acids. The first of these was Erlenmeyer's azlactone synthesis, in which an acylglycine is condensed with an aromatic aldehyde in the presence of acetic anhydride and sodium acetate. The acylglycine is converted into an azlactone by the dehydration of its enolic tautomer with acetic anhydride (cf. the ready formation of γ-lactones, p. 90), e.g.

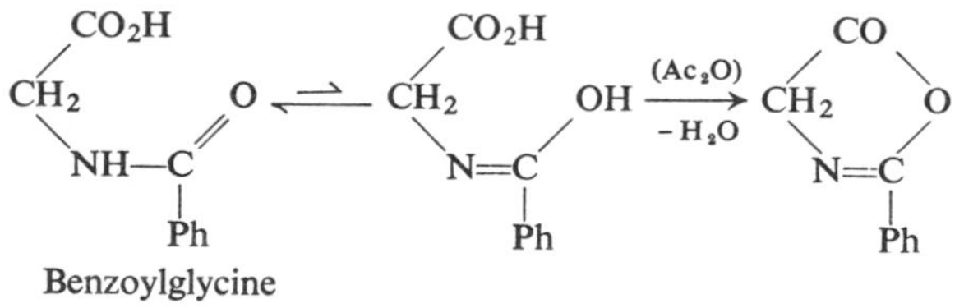

Benzoylglycine

The azlactone, whose methylene group is activated by the carbonyl group, undergoes the Perkin condensation with the aldehyde,

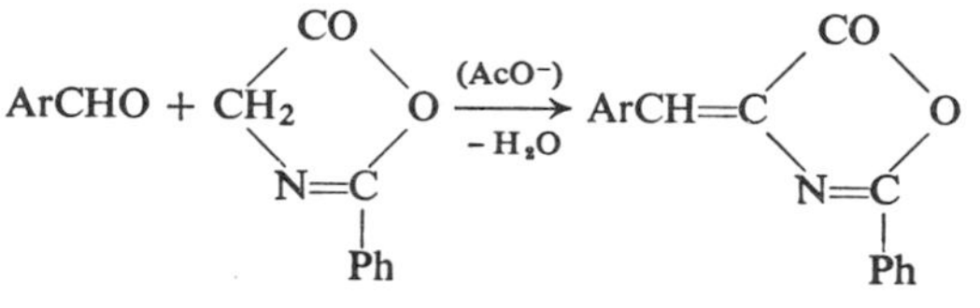

This product is then reduced and hydrolyzed to give the α-amino-acid. For example, $(\pm)$-phenylalanine may be obtained in about 65% yield by treating the azlactone of α-benzoylaminocinnamic acid with red phosphorus and aqueous hydriodic acid [2]:

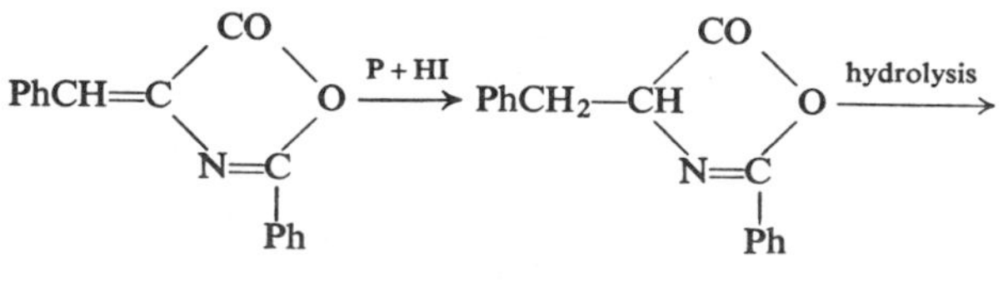

$$\left[\begin{array}{l} \overset{\displaystyle CO_2H}{\underset{\displaystyle N=C}{PhCH_2-CH}} \quad OH \rightleftharpoons \overset{\displaystyle CO_2H}{\underset{\displaystyle NH-CO}{PhCH_2-CH}} \\ \qquad\qquad\quad Ph \qquad\qquad\qquad\qquad Ph \end{array} \right] \longrightarrow \overset{\displaystyle CO_2H}{\underset{\displaystyle NH_2}{PhCH_2-CH}}$$

(±)-Phenylalanine

The Erlenmeyer reaction proceeds much more readily than the Perkin reaction; e.g. the condensation above is complete after about one hour at 100°C. However, if the acylglycine cannot form an azlactone (e.g. benzoyl-N-methylglycine) condensation occurs much less readily.

Variants of the azlactone synthesis in which analogues of azlactones are used are sometimes advantageous. Hydantoin, thiohydantoin, and rhodanine have each been employed as the carbanion-forming component of the condensation.

$$\underset{\text{Hydantoin}}{\begin{array}{c} CO \\ CH_2 \quad NH \\ NH-CO \end{array}} \qquad \underset{\text{Thiohydantoin}}{\begin{array}{c} CO \\ CH_2 \quad NH \\ NH-CS \end{array}} \qquad \underset{\text{Rhodanine}}{\begin{array}{c} CO \\ CH_2 \quad NH \\ S-CS \end{array}}$$

By omitting the reduction step, α-keto-acids may also be obtained from azlactones, e.g.

$$\begin{array}{c} CO \\ PhCH=C \quad O \\ N=C \\ Ph \end{array} \xrightarrow{\text{hydrolysis}} \left[\overset{\displaystyle CO_2H}{\underset{\displaystyle NHCOPh}{PhCH=C}} \right] \longrightarrow$$

$$\left[\overset{\displaystyle CO_2H}{\underset{\displaystyle NH_2}{PhCH=C}} \rightleftharpoons \overset{\displaystyle CO_2H}{\underset{\displaystyle NH}{PhCH_2-C}} \right] \xrightarrow{H_2O} \overset{\displaystyle CO_2H}{\underset{\displaystyle O}{PhCH_2-C}}$$

(d) THE STOBBE CONDENSATION

Ketones normally react with esters in the presence of base by a mechanism in which the carbanion from the ketone displaces alkoxide ion from the carboalkoxy group of the ester (Claisen condensation, p. 238). Dialkyl succinates, however, behave differently from other esters in that the carbanion from the ester adds to the carbonyl group of the ketone:

$$EtO_2CCH_2CH_2CO_2Et \underset{EtOH}{\overset{EtO^-}{\rightleftharpoons}} EtO_2CCH_2-\overset{-}{C}H-CO_2Et$$

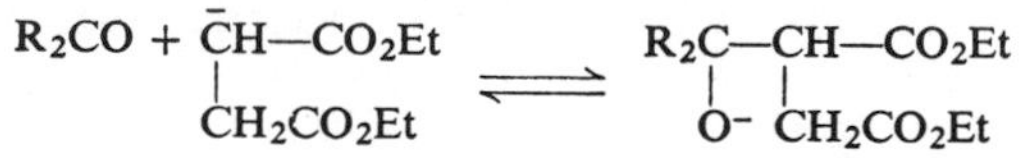

$$R_2CO + \overset{-}{C}H\text{---}CO_2Et \;\rightleftharpoons\; R_2C\text{---}CH\text{---}CO_2Et$$

(with CH_2CO_2Et below the anion carbon, and O^- and CH_2CO_2Et below the product carbons)

The resulting adduct cyclizes to a γ-lactone which then undergoes base-catalyzed ring-opening to give a carboxylate salt:

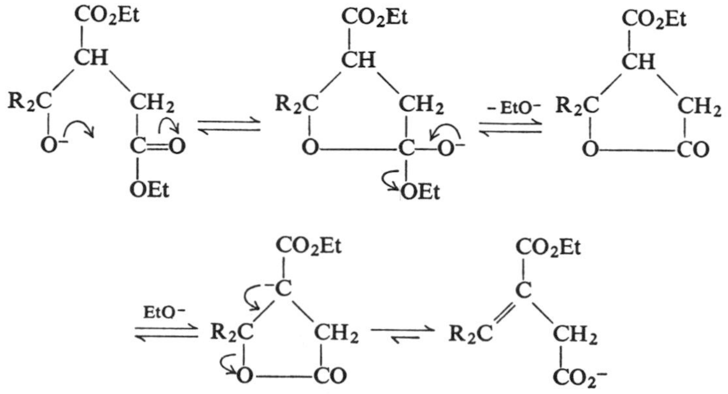

The stability of the carboxylate anion is the basis for the success of the reaction, for it results in the equilibrium being favourable to this product, whereas in the reaction of monobasic esters with ketones, where the corresponding situation does not exist, the equilibria established favour the product of the Claisen condensation.

Aldehydes containing α-hydrogen undergo self-condensation in preference to the Stobbe reaction, but those with no α-hydrogen, e.g. benzaldehyde, react successfully.

The Stobbe condensation leads to the attachment of a three-carbon chain to a ketonic carbon atom, whereas the condensations so far described lead to the attachment of a two-carbon chain.

The procedure is usefully employed in the extension of aromatic rings, e.g.

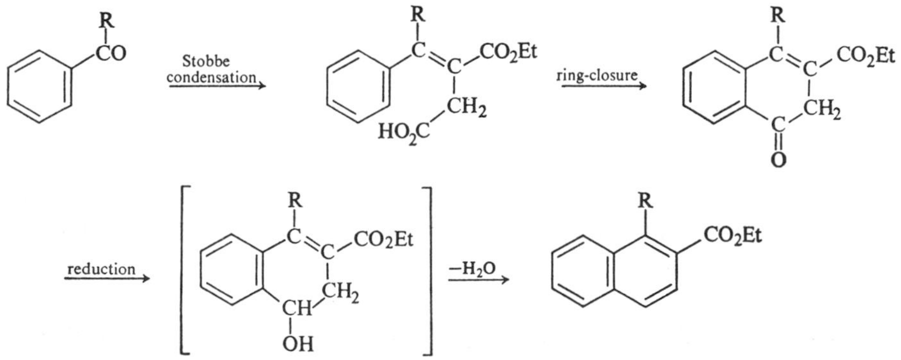

Ring-closure is accomplished by the Friedel-Crafts reaction (11.3) and the final dehydration occurs spontaneously because of the accompanying gain in aromatic stabilization energy.

(e) THE DARZENS REACTION

The base-catalyzed condensation between an α-halo-ester and an aldehyde or ketone gives an epoxy-containing ester (glycidic ester):

$$ClCH_2CO_2Et \xrightleftharpoons{EtO^-} Cl\bar{C}HCO_2Et$$

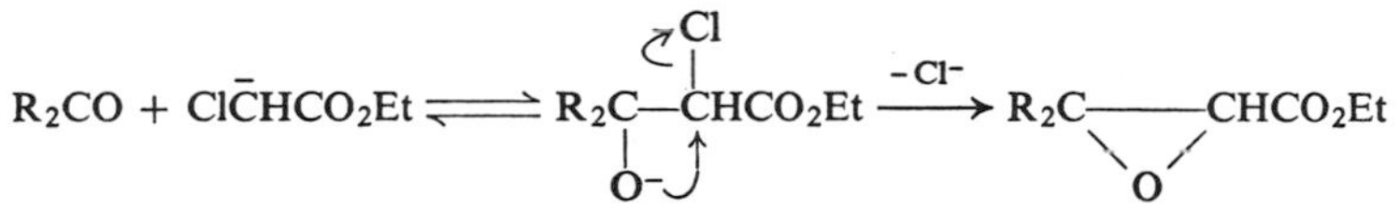

$$R_2CO + Cl\bar{C}HCO_2Et \rightleftharpoons R_2C\underset{\underset{O^-}{|}}{\overset{\overset{Cl}{|}}{-}}CHCO_2Et \xrightarrow{-Cl^-} R_2C\underset{O}{\diagdown\diagup}CHCO_2Et$$

Thus, the difference between this and the Claisen reaction is that in the former the oxyanion formed by the addition of the carbanion to the carbonyl group preferentially displaces halide ion in an intramolecular nucleophilic substitution (*cf*. p. 129), whereas in the latter the oxyanion removes a proton from the solvent.

The glycidic acids obtained by base-catalyzed hydrolysis of the esters readily undergo a decarboxylative rearrangement in the presence of acids:

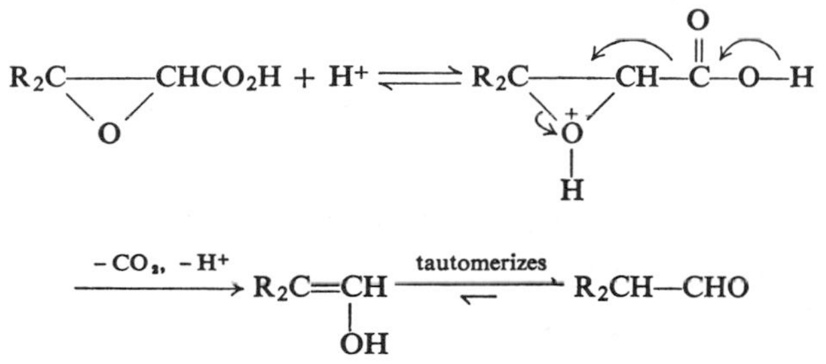

$$R_2C\underset{O}{\diagdown\diagup}CHCO_2H + H^+ \rightleftharpoons R_2C\text{———}CH\text{—}\overset{O}{\overset{\|}{C}}\text{—}O\text{—}H$$

$$\xrightarrow{-CO_2, -H^+} R_2C=\underset{\underset{OH}{|}}{CH} \underset{\xrightarrow{\hspace{1cm}}}{\overset{tautomerizes}{\rightleftharpoons}} R_2CH\text{—}CHO$$

The overall process therefore consists of the addition of *one* carbon atom, as aldehyde, to a carbonyl group,

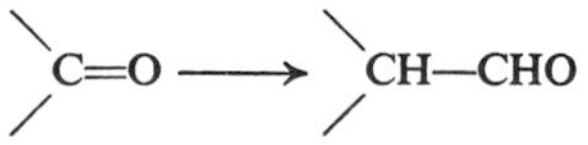

$$\diagdown C=O \longrightarrow \diagdown CH\text{—}CHO$$

A typical example of the synthetic utility of the reaction occurs in a total synthesis of Vitamin A (21.1).

(f) OTHER CONDENSATIONS OF ALDOL TYPE

The reactions of aldehydes and ketones described so far have been with carbanions derived from aldehydes, ketones, esters, and anhydrides. In addition, useful condensations may be carried out using carbanions derived from nitro-compounds, nitriles, and potentially aromatic systems.

(*i*) *Nitro-compounds.* Nitromethane condenses with aldehydes in the presence of base to give β-hydroxy-nitro-compounds. With formaldehyde, for example, each of the three α-hydrogen atoms of nitromethane is successively replaced, giving a trimethylol derivative (*cf.* the reaction of formaldehyde with acetaldehyde, p. 229):

$$CH_3NO_2 \xrightleftharpoons[BH^+]{B:} \bar{C}H_2-\overset{+}{N}\overset{O}{\underset{O^-}{\diagup\!\!\diagdown}} \leftrightarrow CH_2=\overset{+}{N}\overset{O^-}{\underset{O^-}{\diagup\!\!\diagdown}}$$

$$\xrightleftharpoons{CH_2=O} \underset{\underset{O^-}{|}}{CH_2}-CH_2NO_2 \xrightleftharpoons{BH^+} \underset{\underset{OH}{|}}{CH_2}-CH_2NO_2 \xrightarrow[\text{condensations}]{\text{further}} HOCH_2-\underset{\underset{CH_2OH}{|}}{\overset{\overset{CH_2OH}{|}}{C}}-NO_2$$

The product of condensation with aromatic aldehydes is normally the $\alpha\beta$-unsaturated nitro-compound, for dehydration of the initial product occurs readily as a result of the increased conjugation in the product; e.g. benzaldehyde and nitromethane give ω-nitrostyrene in 80% yield [1]:

$$PhCHO + CH_3NO_2 \xrightarrow{(OH^-)} \underset{\underset{OH}{|}}{PhCH}-CH_2NO_2 \xrightarrow{-H_2O} \underset{\omega\text{-Nitrostyrene}}{PhCH=CHNO_2}$$

ω-Nitrostyrene and its derivatives are useful starting materials in the Bischler-Napieralski synthesis of isoquinolines (p. 683) since the necessary β-arylethyl-amines may readily be derived from them by reduction.

(*ii*) *Nitriles.* Nitriles containing α-hydrogen atoms behave analogously to nitro compounds, e.g.

$$PhCH_2CN \xrightleftharpoons{EtO^-} [Ph-\bar{C}H-C{\equiv}N \leftrightarrow Ph-CH=C=\bar{N}]$$

$$\xrightleftharpoons{PhCHO} \underset{\underset{O^-}{|}}{PhCH}-\underset{\diagdown CN}{\overset{\diagup Ph}{CH}} \xrightleftharpoons{EtOH} \underset{\underset{OH}{|}}{PhCH}-\underset{\diagdown CN}{\overset{\diagup Ph}{CH}} \xrightarrow{-H_2O} PhCH=\underset{\diagdown CN}{\overset{\diagup Ph}{C}}$$

(*iii*) *Potentially aromatic systems.* The anion derived by proton-abstraction from cyclopentadiene is an aromatic system containing six π-electrons (p. 59), and the resulting stabilization energy renders cyclopentadiene acidic enough to undergo base-catalyzed condensations, e.g.

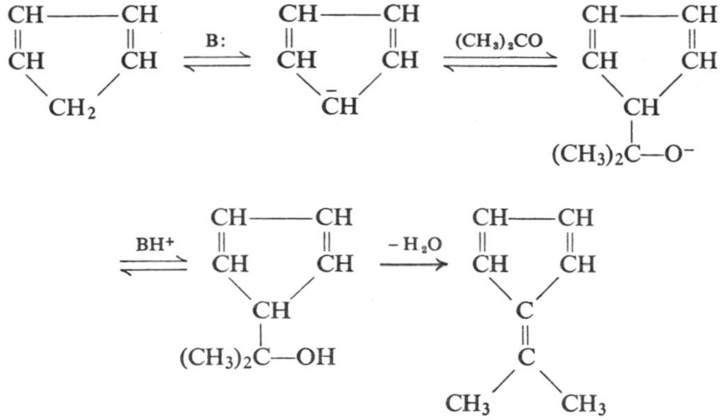

Dimethylfulvene

Related compounds such as indene and fluorene react similarly.

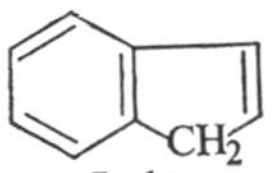

Indene 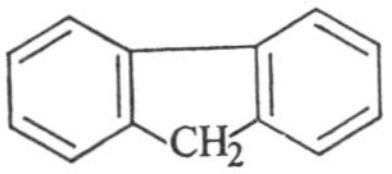   Fluorene

(g) THE KNOEVENAGEL REACTION

The base-catalyzed condensation between an aldehyde or ketone and malonic acid or a related compound is known as the Knoevenagel reaction. It differs from the reactions so far considered in that the methylene group in the carbanion-forming component is activated by *two* groups of $-M$ type (e.g. two carboxyls in malonic acid) and this allows condensation to be effected by weaker bases than those used for aldol condensations. In a typical procedure a solution of malonic acid and the aldehyde in pyridine containing a little piperidine as catalyst is heated under reflux. Condensation is followed by dehydration to give an $\alpha\beta$-unsaturated dibasic acid which, at the reflux temperature of pyridine, is decarboxylated. In this way cinnamic acid can be obtained in 80% yield from benzaldehyde in one stage:

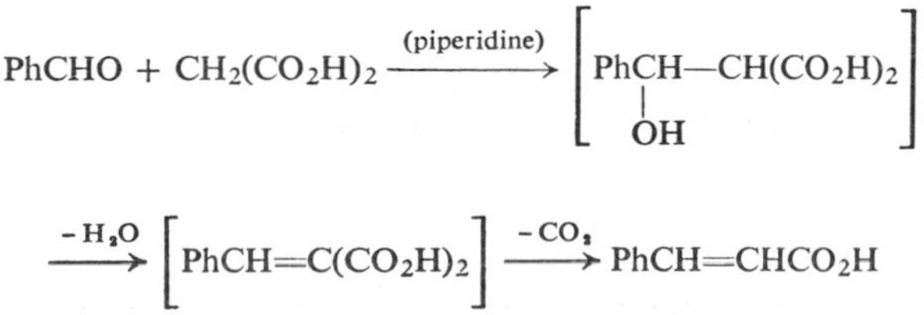

$$PhCHO + CH_2(CO_2H)_2 \xrightarrow{\text{(piperidine)}} \left[\begin{array}{c} PhCH-CH(CO_2H)_2 \\ | \\ OH \end{array} \right]$$

$$\xrightarrow{-H_2O} \left[PhCH=C(CO_2H)_2 \right] \xrightarrow{-CO_2} PhCH=CHCO_2H$$

The use of malonic ester (diethyl malonate) in place of malonic acid leads to $\alpha\beta$-unsaturated dibasic esters, $RCH{=}C(CO_2Et)_2$.

The Knoevenagel reaction is more useful with aromatic than with aliphatic aldehydes. The latter react readily, but the olefinic bond in the products, being activated to nucleophiles by the conjugated carbonyl, usually reacts further (Michael addition, 7.5), e.g.

$$CH_3CHO + CH_2(CO_2Et)_2 \xrightarrow[-H_2O]{(B:)} CH_3CH{=}C(CO_2Et)_2$$

$$\xrightarrow[(B:)]{CH_2(CO_2Et)_2} CH_3CH \begin{array}{c} CH(CO_2Et)_2 \\ \diagup \\ \diagdown \\ CH(CO_2Et)_2 \end{array}$$

The corresponding addition to the olefinic bond of the product from an aromatic aldehyde occurs less readily because of the loss of conjugation to the aromatic system which such addition involves.

Ketones do not undergo the Knoevenagel reaction with malonic acid or its esters, but do so with cyanoacetic acid and its esters which contain more highly activated α-hydrogen.

(h) SUMMARY

It is apparent from the preceding discussion that a variety of types of condensation is often available for the synthesis of a particular compound. For example, cinnamic acid may be prepared from benzaldehyde in each of the following ways:

Claisen reaction (CH_3CO_2Et, EtO^-; two hours at 0–5°C, followed by hydrolysis of ethyl cinnamate): *ca.* 70%.

Perkin reaction (Ac_2O, AcO^-; five hours at 180° C): 55%.

Knoevenagel reaction ($CH_2(CO_2H)_2$, piperidine; one hour at 110° C): 80%.

Reformatsky reaction (CH_2BrCO_2Et, Zn; see p. 221): 50–60%.

The Perkin reaction requires the most vigorous conditions and usually gives the lowest yield. However, it has the advantage over the Claisen and Reformatsky reactions that if a nitro- or halogen-substituent is present in the aromatic aldehyde the yield is increased whereas the other two methods cannot be employed. Nevertheless, the Perkin reaction has generally been superseded by the Knoevenagel reaction.

7.3 Condensations of Carbanions with Esters

(a) THE CLAISEN CONDENSATION

The self-condensation of an ester which contains an α-hydrogen atom is known as the Claisen (ester) condensation. The simplest example is the formation of acetoacetic ester (ethyl acetoacetate) from ethyl acetate, catalyzed by ethoxide ion:

$$CH_3CO_2Et \underset{\longleftarrow}{\overset{EtO^-}{\rightleftharpoons}} \left[\bar{C}H_2-\underset{\underset{O}{\|}}{C}-OEt \leftrightarrow CH_2{=}\underset{\underset{O^-}{|}}{C}-OEt \right]$$

$$\underset{\longleftarrow}{\overset{CH_3CO_2Et}{\rightleftharpoons}} CH_3-\underset{\underset{O^-}{|}}{\overset{\overset{OEt}{|}}{C}}-CH_2-CO_2Et \underset{\longleftarrow}{\overset{-EtO^-}{\rightleftharpoons}} CH_3-\underset{\underset{O}{\|}}{C}-CH_2-CO_2Et$$

$$\underset{\longleftarrow}{\overset{EtO^-}{\rightarrow}} \left[CH_3-\underset{\underset{O}{\|}}{C}-\bar{C}H-\underset{\underset{O}{\|}}{C}-OEt \leftrightarrow CH_3-\underset{\underset{O^-}{|}}{C}{=}CH-\underset{\underset{O}{\|}}{C}-OEt \leftrightarrow CH_3-\underset{\underset{O}{\|}}{C}-CH{=}\underset{\underset{O^-}{|}}{C}-OEt \right]$$

$$\overset{H^+}{\longrightarrow} CH_3COCH_2CO_2Et$$

Acetoacetic ester

The reaction is usually carried out by refluxing very dry ethyl acetate over sodium wire. The sodium reacts with the 2-3% of ethanol which is present in commercial ethyl acetate, generating ethoxide ion which then catalyzes the condensation. The product is obtained as the sodio-derivative of acetoacetic ester from which the free ester is liberated by treatment with acetic acid. Yields of about 30% are obtained [1].

The Claisen condensation differs from the aldol condensation only after the oxyanion has been formed by addition of the carbanion to the carbonyl group: in the Claisen condensation this anion eliminates ethoxide ion to give a β-keto-ester, whereas in the aldol condensation the anion gains a proton to give a β-hydroxy-aldehyde or ketone.

Each step in the Claisen condensation is reversible, and acetoacetic ester can be isolated in significant yield only because it is essentially completely removed from equilibrium by conversion into its anion under the influence of ethoxide ion (i.e. acetoacetic ester is a much stronger acid than ethanol). However, in those cases in which the β-keto-ester product does not contain a C—H bond adjacent to both carbonyl and carbalkoxy groups so that this marked acidic character is absent, the equilibria are unfavourable to condensation when ethoxide ion is the base. For example, ethyl isobutyrate, Me_2CHCO_2Et, fails to give $Me_2CH—CO—CMe_2—CO_2Et$ in these conditions. This problem can be surmounted by using a very much stronger base than ethoxide ion, and the triphenylmethide ion, added as sodium triphenylmethyl, is conveniently employed. This results in the essentially complete removal of the ethanol formed in the condensation:

$$2\ Me_2CHCO_2Et \rightleftharpoons Me_2CH-CO-CMe_2-CO_2Et + EtOH$$

$$EtOH + Ph_3C^- \rightleftharpoons EtO^- + Ph_3CH$$

Other strong bases have been used. One type is a hindered Grignard reagent such as mesitylmagnesium bromide which, though sterically prevented from reacting as a nucleophile with the carbonyl group of the ester, is nevertheless able readily to abstract a proton from an acidic C—H bond,

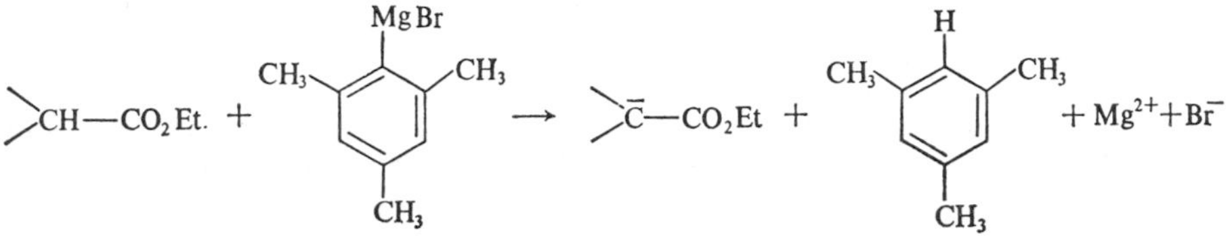

Amide ion, as sodamide, has also been used, but it has the disadvantage that it brings about a side-reaction by reacting with the ester to form an amide.

Mixed condensations. As with aldol condensations, the Claisen condensation on a mixture of two esters each of which contains α-hydrogen may yield four products. In general, therefore, this is an unsatisfactory method for obtaining a particular β-keto-ester. However, if only one of the two esters has α-hydrogen and the other ester has the more reactive of the two carbonyl groups, one product predominates.

This principle was utilized by Woodward and Doering in their synthesis of quinine (21.6). The ethyl ester of *N*-benzoylhomomeroquinene, which contains α-hydrogen, was condensed with ethyl 6-methoxyquinoline-4-carboxylate, which does not contain α-hydrogen but whose carbonyl group is the more reactive of the two because of the presence of the electron-attracting heterocyclic nitrogen:

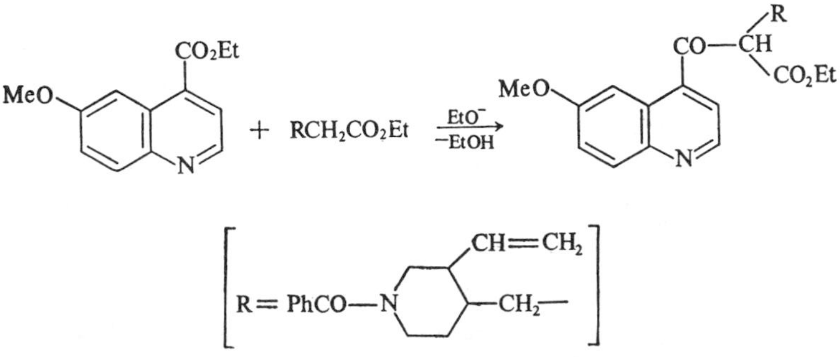

Oxalic ester and formic ester, which possess no α-hydrogen but very reactive carbonyl groups, have been widely applied in mixed Claisen condensations. The

use of oxalic ester leads to compounds which are both β-keto-esters and α-keto-esters. Since α-keto-esters decarbonylate on being heated, the method provides a route to β-dibasic-esters, as in the preparation of diethyl phenylmalonate in 80–85% yield from ethyl phenylacetate [2]:

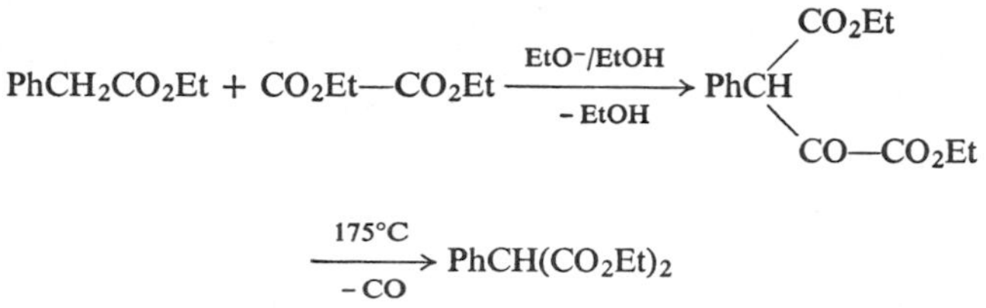

$$PhCH_2CO_2Et + CO_2Et—CO_2Et \xrightarrow[-\ EtOH]{EtO^-/EtOH} PhCH \begin{array}{c} CO_2Et \\ \\ CO—CO_2Et \end{array}$$

$$\xrightarrow[-\ CO]{175°C} PhCH(CO_2Et)_2$$

Since phenyl halides are not reactive towards nucleophiles in normal conditions (p. 426) and do not react with malonic ester in the presence of ethoxide, this provides a convenient route to phenylmalonic acid and related compounds.

The products from condensations with oxalic ester may also be used to prepare α-keto-acids, for hydrolysis of the initially formed dibasic keto-ester gives a dibasic acid which, as a β-keto-acid, is readily decarboxylated by heat:

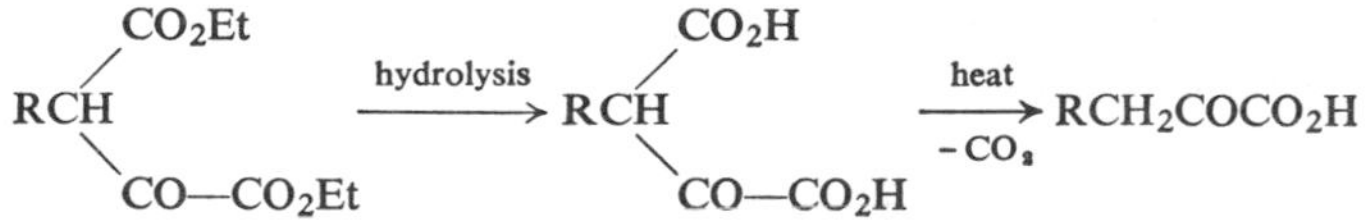

$$RCH \begin{array}{c} CO_2Et \\ \\ CO—CO_2Et \end{array} \xrightarrow{hydrolysis} RCH \begin{array}{c} CO_2H \\ \\ CO—CO_2H \end{array} \xrightarrow[-CO_2]{heat} RCH_2COCO_2H$$

Formic ester reacts in Claisen conditions with esters containing α-hydrogen, giving β-aldehydo-esters:

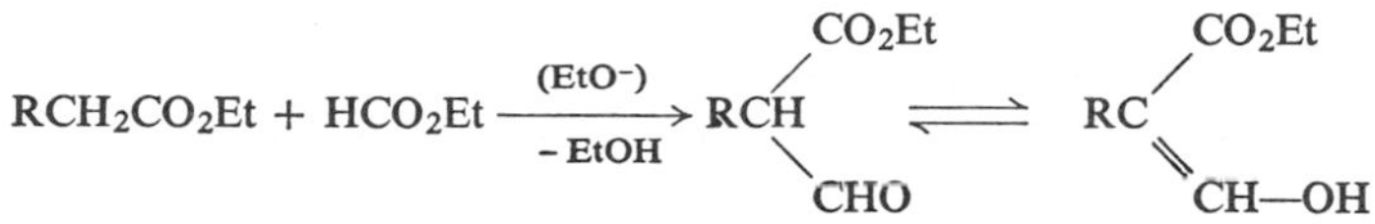

$$RCH_2CO_2Et + HCO_2Et \xrightarrow[-\ EtOH]{(EtO^-)} RCH \begin{array}{c} CO_2Et \\ \\ CHO \end{array} \rightleftharpoons RC \begin{array}{c} CO_2Et \\ \\ CH—OH \end{array}$$

(b) THE DIECKMANN CONDENSATION

The Claisen condensation on the di-esters of C_6 and C_7 dibasic acids occurs intramolecularly to give five- and six-membered cyclic β-keto-esters, respectively. For example, diethyl adipate in toluene reacts with sodium metal to give the sodio-derivative of 2-carboethoxycyclopentanone from which the free ester (75–80%) is liberated with acetic acid [2]:

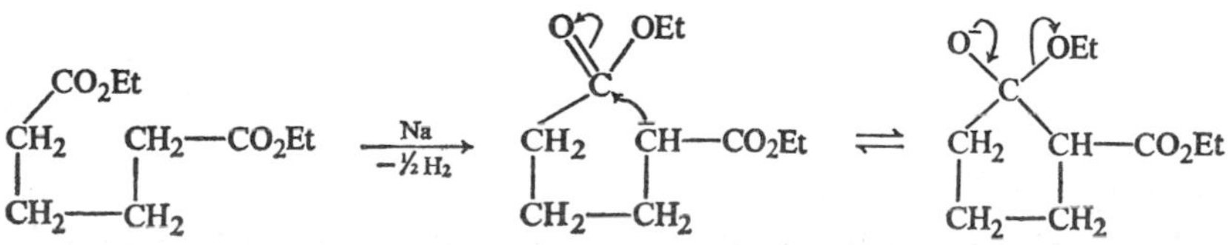

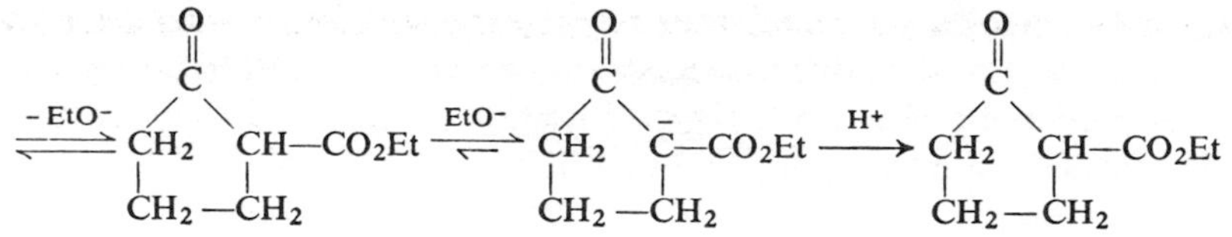

$$\xrightarrow[\longleftarrow]{- \text{EtO}^-} \underset{\text{CH}_2-\text{CH}_2}{\overset{\overset{\displaystyle\text{O}}{\underset{\displaystyle\text{C}}{\|}}}{\text{CH}_2 \quad \text{CH}-\text{CO}_2\text{Et}}} \xrightarrow[\longleftarrow]{\text{EtO}^-} \underset{\text{CH}_2-\text{CH}_2}{\overset{\overset{\displaystyle\text{O}}{\underset{\displaystyle\text{C}}{\|}}}{\text{CH}_2 \quad \overset{-}{\text{C}}-\text{CO}_2\text{Et}}} \xrightarrow{\text{H}^+} \underset{\text{CH}_2-\text{CH}_2}{\overset{\overset{\displaystyle\text{O}}{\underset{\displaystyle\text{C}}{\|}}}{\text{CH}_2 \quad \text{CH}-\text{CO}_2\text{Et}}}$$

2-Carboethoxycyclopentanone

Diethyl pimelate reacts analogously to form 2-carboethoxycyclohexanone in 60% yield.

Diethyl suberate undergoes the Dieckmann condensation (to give the seven-membered 2-carboethoxycycloheptanone) in only very small yield, and longer-chain dibasic esters do not form alicyclic compounds. This is because the probability factor for the intramolecular reaction becomes smaller as the chain-length is increased (less favourable entropy of activation) so that ring-closure competes less effectively with intermolecular condensation.

Di-esters of shorter-chain dibasic acids do not undergo the Dieckmann condensation because of the strain which would result in the small rings formed (higher heat of activation). In the case of diethyl succinate, an intermolecular condensation between two molecules is followed by cyclization to the cyclohexanedione system:

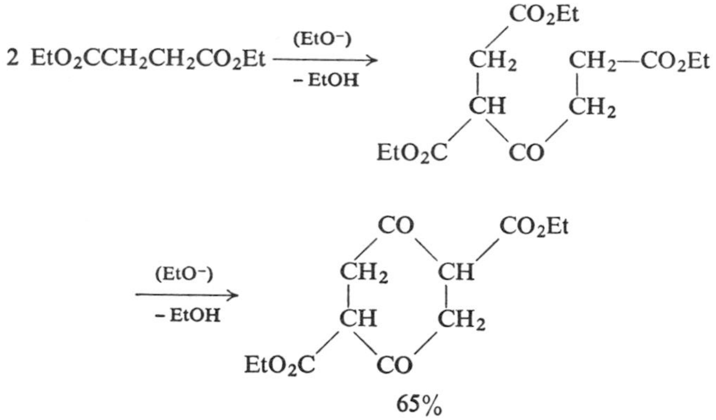

$$2\ \text{EtO}_2\text{CCH}_2\text{CH}_2\text{CO}_2\text{Et} \xrightarrow[-\ \text{EtOH}]{(\text{EtO}^-)}$$

$$\xrightarrow[-\ \text{EtOH}]{(\text{EtO}^-)}$$

65%

Dieckmann cyclization has been used to form five- and six-membered rings in a number of syntheses of natural products, typified as follows:

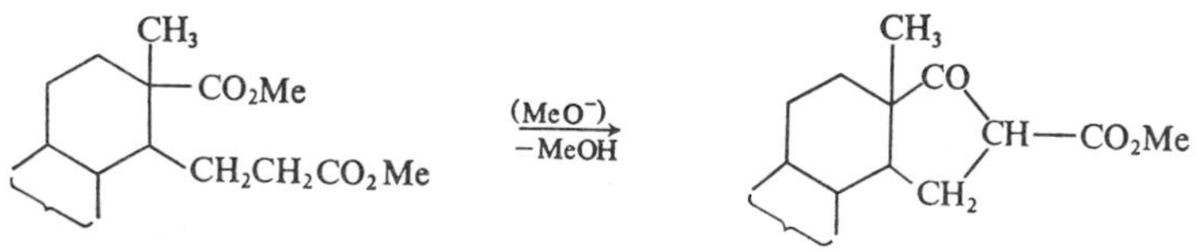

$$\xrightarrow[-\text{MeOH}]{(\text{MeO}^-)}$$

Five- and six-membered cyclic ketones are also formed in good yield, with the liberation of carbon dioxide, when the metal salts of (substituted) adipic or pimelic acids are heated strongly; for example, adipic acid, heated over barium hydroxide at 290°C, gives cyclopentanone in about 75% yield.

(c) THE THORPE REACTION

The cyclization of α,ω-dinitriles in the presence of base is closely analogous to the Dieckmann reaction. The initial product, a β-imino-nitrile, is readily hydrolyzed to a β-keto-nitrile:

$$
\begin{array}{ccc}
\underset{\underset{CH_2-CH_2}{|}}{\overset{\overset{CN}{\diagup}}{CH_2}} \quad \underset{|}{\overset{CH_2-CN}{}} & \xrightarrow{(\text{B:})} & \underset{\underset{CH_2-CH_2}{|}}{\overset{\overset{\overset{NH}{\parallel}}{\overset{C}{\diagup\ \diagdown}}}{CH_2}}\ \ \overset{CH-CN}{\underset{|}{}} \quad \xrightarrow{H_2O} \quad \underset{\underset{CH_2-CH_2}{|}}{\overset{\overset{\overset{O}{\parallel}}{\overset{C}{\diagup\ \diagdown}}}{CH_2}}\ \ \overset{CH-CN}{\underset{|}{}}
\end{array}
$$

As in the Dieckmann condensation, satisfactory yields are normally obtained only for five- and six-membered rings. However, the problem that arises in trying to make larger rings, that intermolecular reaction competes increasingly favourably with intramolecular reaction, has been overcome by employing an experimental technique devised by Ziegler. The dinitrile is added in very dilute solution in benzene to a solution of the basic catalyst in a large quantity of ether. This reduces considerably the probability of the intermolecular reaction, and in this way yields of 95% (7-ring), 88% (8-ring), and 60–80% (rings with 14 or more members) have been obtained; yields of 9–13 membered rings are, however, less than 15%. The basic condensing agent must be soluble in ether and highly hindered so as to minimize its reaction as a nucleophile with the nitrile group; lithium salts of amines, such as Li^+NPhEt^-, are commonly used.

(d) CONDENSATIONS OF ESTERS WITH KETONES

Four products might result from a base-catalyzed condensation on a mixture of an ester and a ketone each of which contains α-hydrogen: each of the two carbanions might react with each of the two carbonyl groups. In practice, one product, that derived from the carbanion of the ketone and the carbonyl of the ester, normally predominates, e.g.

$$CH_3COCH_3 \underset{\overset{}{\rightleftharpoons}}{\overset{EtO^-}{}} \overset{-}{C}H_2-\underset{\underset{O}{\parallel}}{C}-CH_3 \longleftrightarrow CH_2=\underset{\underset{O^-}{|}}{C}-CH_3$$

$$\underset{\overset{}{\rightleftharpoons}}{\overset{CH_3CO_2Et}{}} \quad CH_3-\underset{\underset{O^-}{|}}{\overset{\overset{OEt}{\diagup}}{C}}-CH_2-CO-CH_3 \underset{\overset{}{\rightleftharpoons}}{\overset{-EtO^-}{}} CH_3-\underset{\underset{O}{\parallel}}{C}-CH_2-\underset{\underset{O}{\parallel}}{C}-CH_3$$

$$\xrightarrow[]{EtO^-} \left[CH_3-\underset{O}{\underset{\|}{C}}-\overset{-}{C}H-\underset{O}{\underset{\|}{C}}-CH_3 \leftrightarrow CH_3-\underset{O^-}{\underset{|}{C}}=CH-\underset{O}{\underset{\|}{C}}-CH_3 \leftrightarrow CH_3-\underset{O}{\underset{\|}{C}}-CH=\underset{O^-}{\underset{|}{C}}-CH_3 \right]$$

$$\xrightarrow{H^+} CH_3COCH_2COCH_3$$
Acetylacetone

Two factors are responsible. First, the two Claisen products are converted into their anions by ethoxide ion so that the equilibria are favourable to these products but not to the two aldol products (*cf.* the self-condensation of acetone, p. 228). Secondly, acetone and acetylacetone are more acidic than ethyl acetate and acetoacetic ester, respectively, so that the formation of acetylacetone as its anion is the predominant reaction.

Unfortunately, this procedure for preparing β-diketones is limited in scope to reactions between acetate esters and methyl ketones because of the reversibility of Claisen condensations. Consider the reaction between diethyl ketone and ethyl acetate. The expected β-diketone,

$$CH_3CH_2COCH_2CH_3 + CH_3CO_2Et \xrightarrow[-EtOH]{(EtO^-)} CH_3CH_2CO\overset{\overset{\displaystyle CH_3}{\displaystyle |}}{C}H-COCH_3$$

can yield, in the reverse reactions, not only the initial reactants (cleavage at *a*) but also methyl ethyl ketone and ethyl propionate (cleavage at *b*):

$$CH_3CH_2CO\overset{\overset{\displaystyle CH_3}{\displaystyle |}}{-}\underset{b \quad a}{\overset{|\ |\ |}{C}H}-COCH_3 \begin{cases} \xrightarrow{a} CH_3CH_2COCH_2CH_3 + CH_3CO_2Et \\ \\ \xrightarrow{b} CH_3CH_2CO_2Et + CH_3CH_2COCH_3 \end{cases}$$

Condensations may now occur involving the reactants produced in *b*, giving rise to a complex mixture of products.

Condensation may also be effected on to aromatic esters. For example, ethyl benzoate and acetophenone react in the presence of sodium ethoxide to give the sodio-derivative of dibenzoylmethane; the diketone is liberated (60–70% yield) on acidification [3]:

$$PhCO_2Et + CH_3COPh \xrightarrow{(EtO^-)} \left[PhCOCHCOPh \right]^- \xrightarrow{H^+} PhCOCH_2COPh$$
Dibenzoylmethane

Both oxalic ester and formic ester are usefully employed in condensations with ketones, just as they are with esters (p. 240). It is possible to convert ketones into β-keto-esters by the sequence,

$$-CO-\overset{|}{C}H + CO_2Et-CO_2Et \xrightarrow[-EtOH]{(EtO^-)} -CO-\overset{|}{\underset{|}{C}}-CO-CO_2Et$$

$$\xrightarrow[-CO]{heat} -CO-\overset{|}{\underset{|}{C}}-CO_2Et$$

and this provides a method for increasing the activation of the α-hydrogen in a ketone by introducing a second activating substituent. This enables the α-carbon to be alkylated in the presence of ethoxide ion, whereas ketones themselves are unreactive in these conditions; the carboethoxy group which has been introduced may later be removed, if desired, by hydrolysis to the β-keto-acid followed by heating. In this way a necessary methyl substituent was introduced in the course of a synthesis of the steroid equilenin:

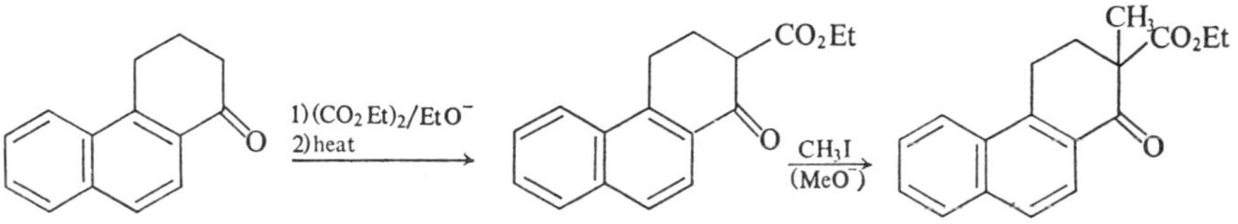

Since it is now possible to alkylate ketones themselves by using stronger bases than ethoxide ion (e.g. t-butoxide and triphenylmethide; p. 248), this use of oxalic ester has been largely superseded.

The condensation of formic ester with ketones gives β-keto-aldehydes,

$$-CO-CH_2- + HCO_2Et \xrightarrow[-EtOH]{(EtO^-)} -CO-\overset{|}{C}H-CHO \rightleftharpoons -CO-\overset{|}{C}=CH-OH$$

which have two important synthetical applications.

(1) The C—H bond between the two carbonyl groups is more strongly activated than that in the original ketone (*cf.* the use of oxalic ester in this respect, above). This property was employed in a synthesis of cholesterol (21.3).

(2) When it is necessary to alkylate a ketone at the less reactive of its two α-methylenic positions, the more reactive position may be blocked by reaction with formic ester followed by treatment with N-methylaniline. Alkylation is then effected at the other α-methylene carbon and the blocking substituent is removed by hydrolysis. This technique has been employed in the introduction of angular methyl groups in steroidal synthesis, e.g.

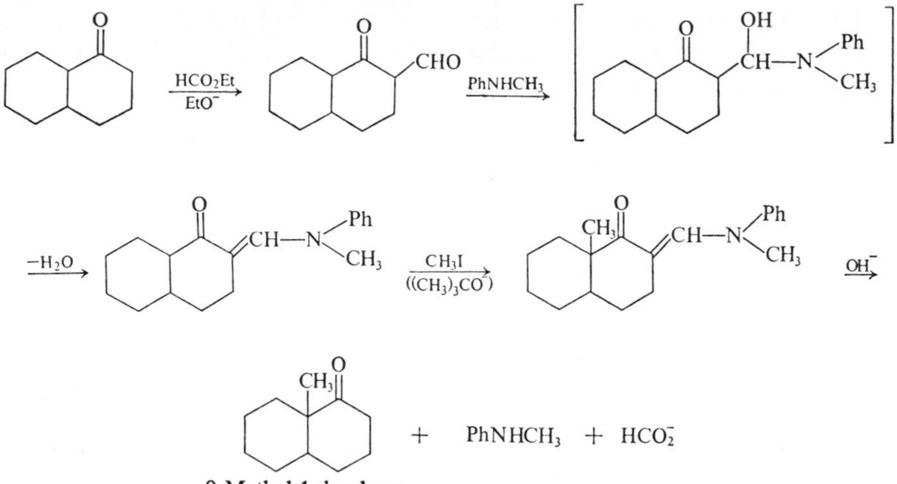

9-Methyl-1-decalone

(e) ENAMINES

Vinylamines (enamines) may be prepared from secondary amines and ketones which contain an α-hydrogen atom:*

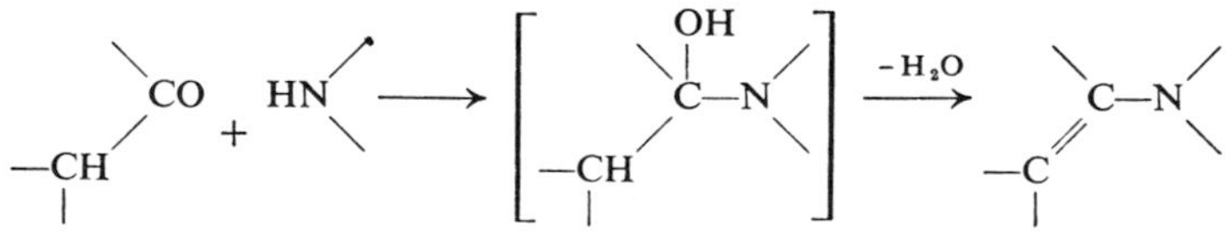

They are closely related to the carbanions derived from esters, etc.:

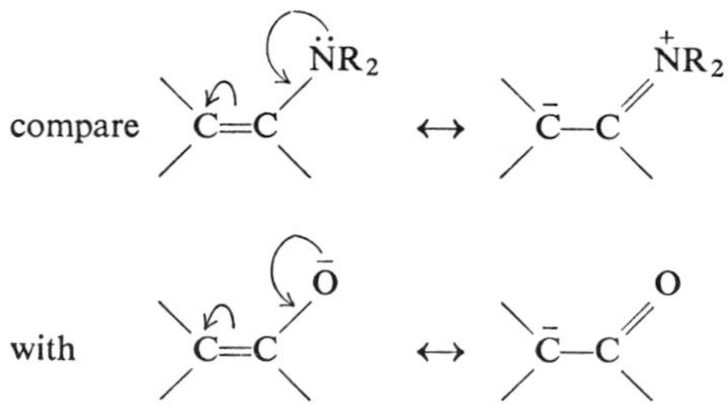

*Those prepared from ammonia or primary amines are unstable with respect to their tautomers,

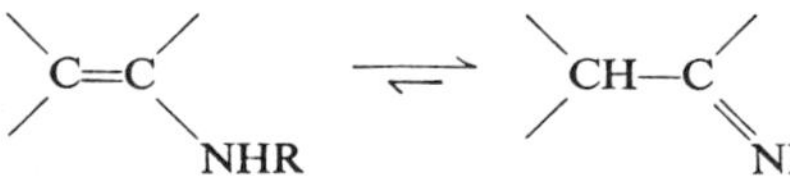

and cannot be isolated (cf. the instability of vinyl alcohols with respect to their carbonyl tautomers).

and behave analogously in their reactions with acyl chlorides, e.g.

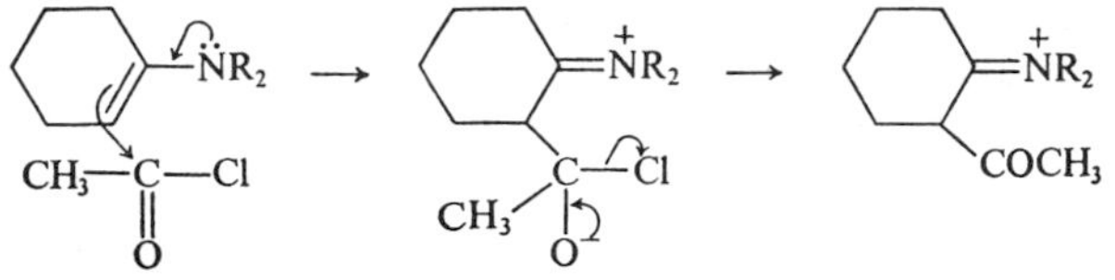

Mild hydrolysis removes the original amine, so that the overall reaction is the formation of a β-diketone:

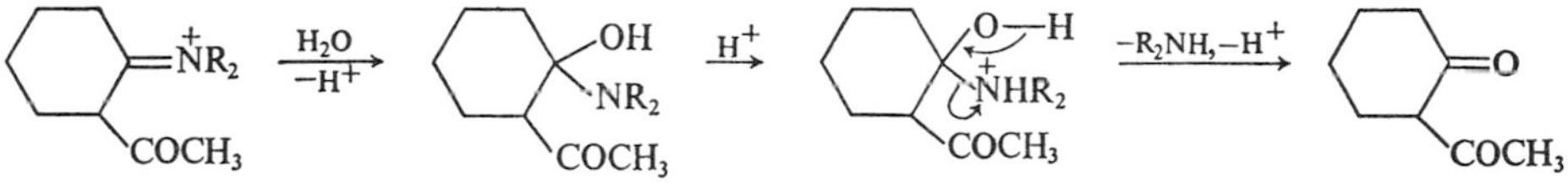

7.4 The Alkylation of Carbanions

In addition to their being able to add to carbonyl groups, carbanions, like other nucleophiles, displace halide ions from alkyl halides with the formation of C—C bonds,

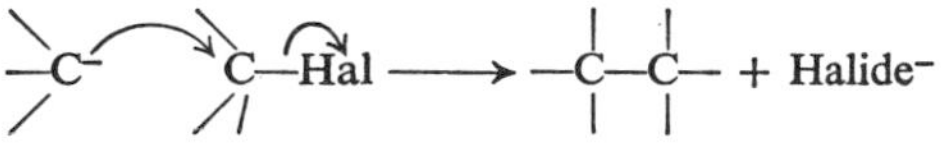

Likewise, toluene-p-sulphonates and methanesulphonates undergo displacement, e.g.

The following discussion centres on alkyl halides, since these have usually been used, but it is now clear that the toluene-p-sulphonate (tosylate) often reacts more efficiently. Moreover, tosylates are readily available from alcohols with toluene-p-sulphonyl chloride (TsCl):

$$ROH + TsCl \longrightarrow ROTs + HCl$$

It is convenient to group these reactions into those in which the carbanion is derived from (a) a monofunctional compound and (b) a bifunctional compound.

(a) ALKYLATION OF MONOFUNCTIONAL COMPOUNDS

Although compounds containing a hydrogen atom activated by one group of $-M$ type react readily with carbonyl groups in the presence of ethoxide ion, their reactions with alkyl halides in these conditions are inefficient. This is because ethoxide ion is a relatively weak base, so that the carbanion is formed only in small concentration and most of the ethoxide ion added remains as such and is able itself to displace on the alkyl halide.

One way of obviating this problem is to use a stronger base so as to generate a larger concentration of carbanions, and if possible also a base which is sterically hindered from reacting as a nucleophile. The t-butoxide ion is frequently used, as in the conversion of cyclohexanone into 2-methylcyclohexanone:

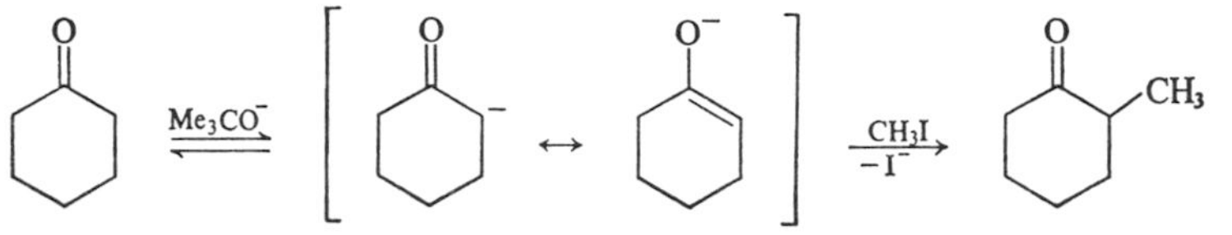

These alkylations are S_N2 processes and show the characteristics of those reactions which have been described previously (p. 126). These are:

(1) Amongst the halides, the order of reactivity is $I > Br > Cl \gg F$.

(2) Primary alkyl groups give higher yields than secondary, and secondary higher than tertiary. Along this series, elimination competes increasingly favourably with substitution, and in general tertiary halides are unsuitable alkylating components.

(3) Vinyl and aryl halides are very unreactive, although aryl halides react if they are substituted with strongly electron-withdrawing groups, such as nitro, in the *ortho* or *para* positions (p. 426).

Unsymmetrical ketones can undergo alkylation at each of two positions (*a* and *b*):

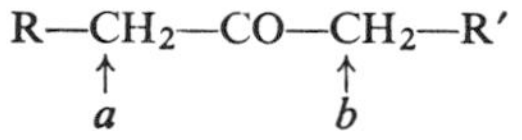

$$R\text{---}CH_2\text{---}CO\text{---}CH_2\text{---}R'$$
$$\phantom{R\text{---}}\overset{\uparrow}{a}\phantom{\text{---}CO\text{---}}\overset{\uparrow}{b}$$

However, those ketones in which one α-carbon is attached to more alkyl groups than the other normally undergo alkylation mainly at the more highly substituted carbon, e.g.

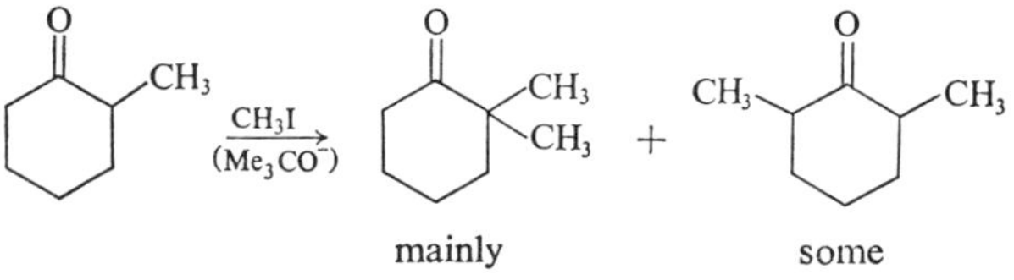

mainly   some

This is the opposite result to the aldol condensation (p. 229), and a possible reason is that the relative ease of attack is here dictated primarily by the relative concentrations of the two carbanions. These are both delocalized species which owe their stability largely to the capacity of oxygen to accommodate the negative charge; that is, the first canonical in each of the following pairs is the more important contributor:

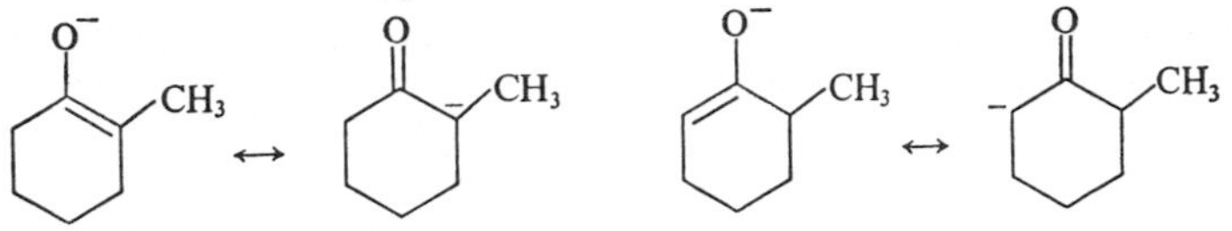

In the first structure the olefinic bond is conjugated to methyl, and the resulting stabilization energy (p. 53) could be responsible for this ion being present in the higher concentration.

In some cases alkylation occurs at the less highly substituted carbon, e.g.,

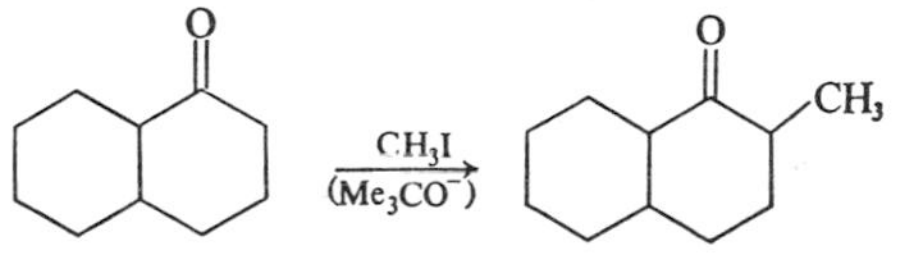

This is possibly the result of steric hindrance to reaction at the more highly substituted angular position. Methylation at this position necessitates a special procedure (p. 245).

(b) ALKYLATION OF BIFUNCTIONAL COMPOUNDS

A C—H bond which is adjacent to two groups of $-M$ type is more acidic than one adjacent to one such group (p. 67) and may be alkylated in milder conditions and usually in better yield. In practice it is often more satisfactory to alkylate such a bifunctional compound and then remove one of the two activating groups than to alkylate the monofunctional analogue. In other respects the characteristics of alkylation of bifunctional compounds parallel those for monofunctional compounds; in particular it should be noted that tertiary halides are unsuitable as alkylating agents.

Two particularly important bifunctional compounds in synthetic procedures are malonic ester and acetoacetic ester.

(i) *Malonic ester.* Malonic ester may be successfully monoalkylated in the presence of ethoxide ion:

$$CH_2(CO_2Et)_2 \underset{EtO^-}{\rightleftharpoons} \bar{C}H(CO_2Et)_2 \xrightarrow{R-Hal} R-CH(CO_2Et)_2 + Halide^-$$

The monoalkylated product still contains an activated α-hydrogen and a second alkyl group may be introduced. However, slightly more vigorous conditions are usually required for the second alkylation, and it is therefore possible to isolate the product of monoalkylation in good yield by using only one mole of the alkyl halide. For example, methyl bromide gives methylmalonic ester (diethyl methylmalonate), without external heating, in about 80% yield [2].

The substituted malonic esters may be hydrolyzed to the acids which, having two carboxyl groups on one carbon atom, are readily decarboxylated on being heated (cf. β-keto-acids, p. 318):

$$RR'(CO_2Et)_2 \xrightarrow{\text{hydrolysis}} RR'C(CO_2H)_2 \xrightarrow[-CO_2]{\text{heat}} RR'CH-CO_2H$$

The ease of all these reactions, together with the ready availability of malonic ester, makes this a convenient synthetical method for aliphatic acids of the types RCH_2CO_2H and $RR'CHCO_2H$ from halides containing the groups R and R'. For example, pelargonic acid can be prepared in about 70% yield from n-heptyl bromide [2]:

$$CH_2(CO_2Et)_2 + C_7H_{15}Br \xrightarrow{\text{(EtO}^-\text{)}} C_7H_{15}-CH(CO_2Et)_2$$

$$\xrightarrow[\text{2) H}^+]{\text{1) KOH}} C_7H_{15}-CH(CO_2H)_2 \xrightarrow[-CO_2]{180°C} C_8H_{17}-CO_2H$$
$$\text{Pelargonic acid}$$

Malonic ester may also be used for the synthesis of three- and four-membered alicyclic compounds from ω-dibromides, e.g.

$$CH_2(CO_2Et)_2 + BrCH_2CH_2CH_2Br \xrightarrow{\text{EtO}^-} [BrCH_2CH_2CH_2-CH(CO_2Et)_2]$$

$$\xrightarrow{\text{EtO}^-} \begin{array}{c} CO_2Et \\ | \\ CH_2-C-CO_2Et \\ | \quad | \\ CH_2-CH_2 \end{array} \xrightarrow[\text{3) heat}]{\substack{\text{1) KOH} \\ \text{2) H}^+}} \begin{array}{c} CH_2-CH-CO_2H \\ | \quad | \\ CH_2-CH_2 \end{array}$$
$$\text{Cyclobutanecarboxylic acid}$$
$$80\%$$

When malonic ester or a monoalkylated derivative is treated with iodine in the presence of a base, condensation occurs to give a tetrabasic ester. A probable mechanism is:

$$RCH(CO_2Et)_2 \underset{\text{EtO}^-}{\rightleftharpoons} R\bar{C}(CO_2Et)_2 \xrightarrow[-I^-]{I_2} \begin{array}{c} RC(CO_2Et)_2 \\ | \\ I \end{array}$$

$$(EtO_2C)_2\bar{C} \quad \begin{array}{c} R \\ | \\ C-I \\ / \quad \backslash \\ CO_2Et \quad CO_2Et \end{array} \xrightarrow{-I^-} (EtO_2C)_2CR-CR(CO_2Et)_2$$

Hydrolysis and decarboxylation give succinic acid or a symmetrical dialkyl derivative:

$$(EtO_2C)_2\overset{\overset{\displaystyle R}{|}}{C}-\overset{\overset{\displaystyle R}{|}}{C}(CO_2Et)_2 \xrightarrow[\text{3) heat}]{\text{1) KOH} \atop \text{2) H}^+} HO_2C-\overset{\overset{\displaystyle R}{|}}{C}H-\overset{\overset{\displaystyle R}{|}}{C}H-CO_2H$$

The reaction of malonic ester and its monoalkyl-derivatives with bromine or chlorine in the presence of base is different from that with iodine: reaction stops at the first stage and the bromo- or chloro-derivative may be isolated (*cf.* the readier reaction of nucleophiles with alkyl iodides than with bromides or chlorides, p. 126):

$$RCH(CO_2Et)_2 \underset{EtO^-}{\rightleftharpoons} R\bar{C}(CO_2Et)_2 \xrightarrow[-Br^-]{Br_2} \underset{Br}{\overset{|}{R}}C(CO_2Et)_2$$

The products are useful intermediates in the synthesis of α-amino-acids (p. 350).

(*ii*) *Acetoacetic ester.* Like malonic ester, acetoacetic ester can be mono- and di-alkylated by alkyl halides in the presence of base.

$$CH_3COCH_2CO_2Et \underset{EtO^-}{\rightleftharpoons} CH_3CO\bar{C}HCO_2Et \xrightarrow[-\text{Halide}^-]{R-Hal} CH_3CO\overset{\overset{\displaystyle R}{|}}{C}HCO_2Et$$

$$\underset{EtO^-}{\rightleftharpoons} CH_3CO\underset{R}{\overset{|}{\bar{C}}}CO_2Et \xrightarrow[-\text{Halide}^-]{R'-Hal} CH_3CO\overset{\overset{\displaystyle R}{|}}{\underset{\underset{\displaystyle R'}{|}}{C}}CO_2Et$$

The products undergo two types of hydrolytic cleavage, depending on the conditions. Dilute acid leads to hydrolysis of the ester group and the resulting β-keto-acid loses carbon dioxide (p. 318):

$$CH_3COCRR'CO_2Et \xrightarrow{H_2O-H^+} CH_3COCRR'CO_2H \xrightarrow[-CO_2]{\text{heat}} CH_3COCHRR'$$

Caustic alkali, on the other hand, induces scission of a carbon-carbon bond in the reverse manner of the Claisen condensation:

$$CH_3-CO-CRR'-CO_2Et \underset{OH^-}{\rightleftharpoons} CH_3-\overset{\overset{\displaystyle OH}{|}}{\underset{\underset{\displaystyle O^-}{|}}{C}}CRR'-CO_2Et$$

$$\rightleftharpoons CH_3CO_2H + \bar{C}RR'-CO_2Et \rightleftharpoons CH_3CO_2^- + RR'CH-CO_2Et$$

The equilibria are favourable to the products as the result of the stability of the acetate ion.

It is therefore possible to make, from acetoacetic ester and the alkyl halides R—Hal and R'—Hal, both methyl ketones of the general type $CH_3COCHRR'$ and acids of the type $CHRR'—CO_2H$. The former is the more useful process, for the compounds obtained in the latter process may be contaminated with by-products derived from the ketonic scission and they are therefore more suitably prepared from malonic ester.

Other β-keto-esters, of the types $RCOCH_2CO_2Et$ and $RCOCHR'CO_2Et$, react analogously.

Like malonic ester, acetoacetic ester and its monoalkyl derivatives undergo a base-catalyzed condensation in the presence of iodine. Hydrolysis and decarboxylation of the products give γ-diketones:

$$CH_3COCHRCO_2Et \underset{EtO^-}{\rightleftharpoons} CH_3CO\bar{C}RCO_2Et \xrightarrow[-I^-]{I_2} CH_3COCRCO_2Et$$
$$\underset{\displaystyle I}{|}$$

$$\xrightarrow[-I^-]{CH_3CO\bar{C}RCO_2Et} CH_3CO\underset{\underset{\displaystyle CO_2Et}{|}}{\overset{\overset{\displaystyle R}{|}}{C}}\underset{\underset{\displaystyle CO_2Et}{|}}{\overset{\overset{\displaystyle R}{|}}{C}}COCH_3 \xrightarrow[\text{2) heat}]{\text{1) } H^+} CH_3CO—\underset{}{\overset{\overset{\displaystyle R}{|}}{CH}}—\underset{}{\overset{\overset{\displaystyle R}{|}}{CH}}—COCH_3$$

Unlike malonic ester, acetoacetic ester does not form a four-membered alicyclic compound with ω-dibromopropane. The first alkylation occurs normally, but the second, involving ring-closure, occurs on oxygen instead of carbon for, although the latter is favoured in intermolecular condensations (p. 226), it is here made less favourable by the fact that a strained four-membered ring would be formed whereas O-alkylation gives a six-membered ring:

$$CH_3COCH_2CO_2Et + BrCH_2CH_2CH_2Br \xrightarrow{EtO^-} CH_3COCHCO_2Et$$
$$\underset{\displaystyle CH_2CH_2CH_2Br}{|}$$

7.5 Addition of Carbanions to Activated Olefins

Although carbanions, in common with other nucleophiles, do not react with simple olefins, they do so if the olefinic double bond is conjugated to a group of $-M$ type. The reason is that the anion formed by addition, and therefore the preceding transition state, is then stabilized sufficiently by the delocalization of the charge on to an electronegative element for addition to occur at a practicable rate, e.g.

$$\overset{-}{C}H(CO_2Et)_2 + PhCH{=}CHCO_2Et \longrightarrow$$

$$PhCH{-}\overset{-}{C}H{-}\overset{\overset{\displaystyle O}{\|}}{C}{-}OEt \leftrightarrow PhCH{-}CH{=}\overset{\overset{\displaystyle O^-}{|}}{C}{-}OEt$$
$$\quad\ \ \underset{\displaystyle CH(CO_2Et)_2}{|} \qquad\qquad\quad\ \underset{\displaystyle CH(CO_2Et)_2}{|}$$

The adduct formed may react with a proton on acidification at either carbon or oxygen,

$$PhCH{-}\overset{-}{C}H{-}\overset{\overset{\displaystyle O}{\|}}{C}{-}OEt \leftrightarrow PhCH{-}CH{=}\overset{\overset{\displaystyle O^-}{|}}{C}{-}OEt$$
$$\quad\ \ \underset{\displaystyle CH(CO_2Et)_2}{|} \qquad\qquad\quad\ \underset{\displaystyle CH(CO_2Et)_2}{|}$$

$$\big|\,\text{H}^+$$

$$PhCH{-}CH_2{-}CO_2Et \qquad PhCH{-}CH{=}\overset{\overset{\displaystyle OH}{|}}{C}{-}OEt$$
$$\quad\ \ \underset{\displaystyle CH(CO_2Et)_2}{|} \qquad\qquad\quad\ \underset{\displaystyle CH(CO_2Et)_2}{|}$$

but since these tautomers equilibrate rapidly in the presence of acids and the keto-tautomer is the more stable, this is the product which is isolated. Thus, the overall reaction between malonic ester and ethyl cinnamate is

$$PhCH{=}CHCO_2Et + CH_2(CO_2Et)_2 \xrightarrow{\text{(EtO}^-)} PhCH{-}CH_2CO_2Et$$
$$\qquad\qquad\qquad\qquad\qquad\qquad\qquad\qquad\ \ \underset{\displaystyle CH(CO_2Et)_2}{|}$$

These additions to activated olefins are usually referred to as *Michael reactions,* a name which was originally applied to those reactions involving the carbanions from acetoacetic and malonic esters. The olefin may be activated by conjugation to carbonyl, carboalkoxy, nitro, and nitrile groups, and the carbanion-forming component may be a bifunctional compound such as malonic ester, or a monofunctional compound such as nitromethane, e.g.

$$CH_3CH{=}CHCO_2Et + CH_3NO_2 \xrightarrow{\text{(EtO}^-)} CH_3CH{-}CH_2CO_2Et$$
$$\qquad\qquad\qquad\qquad\qquad\qquad\qquad\qquad\ \ \underset{\displaystyle CH_2NO_2}{|}$$

Michael addition may follow spontaneously the condensation of an aliphatic aldehyde with malonic ester (p. 238),

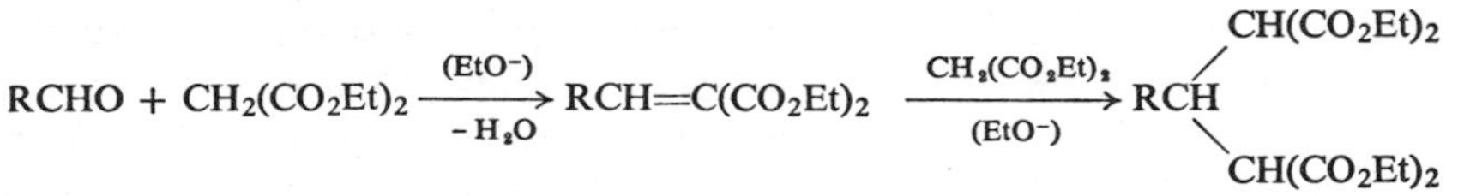

$$RCHO + CH_2(CO_2Et)_2 \xrightarrow[-H_2O]{(EtO^-)} RCH{=}C(CO_2Et)_2 \xrightarrow[(EtO^-)]{CH_2(CO_2Et)_2} RCH\begin{smallmatrix} CH(CO_2Et)_2 \\ \\ CH(CO_2Et)_2 \end{smallmatrix}$$

Hydrolysis and decarboxylation of the products give (substituted) glutaric acids:

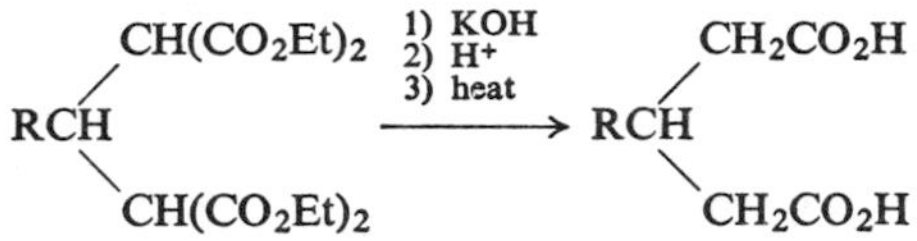

$$RCH\begin{smallmatrix} CH(CO_2Et)_2 \\ \\ CH(CO_2Et)_2 \end{smallmatrix} \xrightarrow[\substack{1)\ KOH \\ 2)\ H^+ \\ 3)\ heat}]{} RCH\begin{smallmatrix} CH_2CO_2H \\ \\ CH_2CO_2H \end{smallmatrix}$$

Michael addition to an $\alpha\beta$-unsaturated ketone may itself be followed by an intramolecular Claisen condensation. For example, dimedone (methone) may be prepared in up to 85% yield from mesityl oxide (p. 228) and malonic ester in the presence of ethoxide ion [2]:

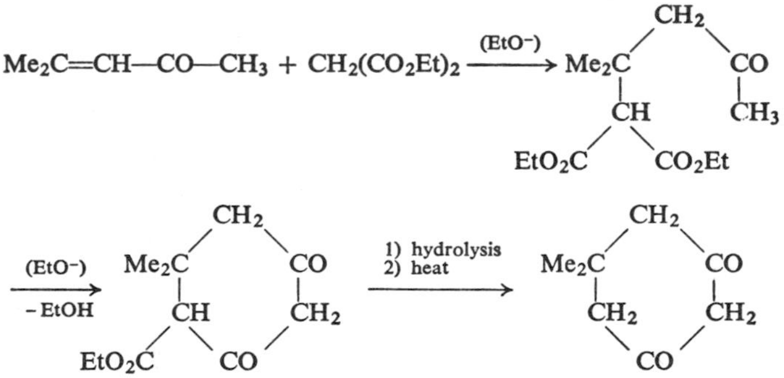

$$Me_2C{=}CH{-}CO{-}CH_3 + CH_2(CO_2Et)_2 \xrightarrow{(EtO^-)}$$

Dimedone, one of whose methylene groups is strongly activated by two adjacent carbonyl groups, is used for the quantitative analysis of formaldehyde with which it readily condenses by aldol condensation followed by Michael addition:

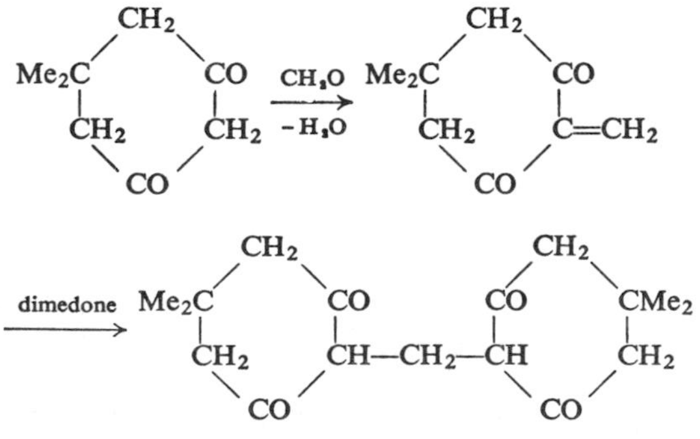

Michael addition has frequently been applied in building up the alicyclic systems of steroids, as in the Harvard sterol synthesis (see also p. 718). Here, a methylene group initially activated by one carbonyl was first further activated by the introduction of a second carbonyl group with formic ester (p. 245) and was then used as the carbanion-forming component of a Michael reaction with ethyl vinyl ketone. Base-catalyzed condensation led to the completion of a new six-membered ring and the removal of the activating formyl group (21.3):

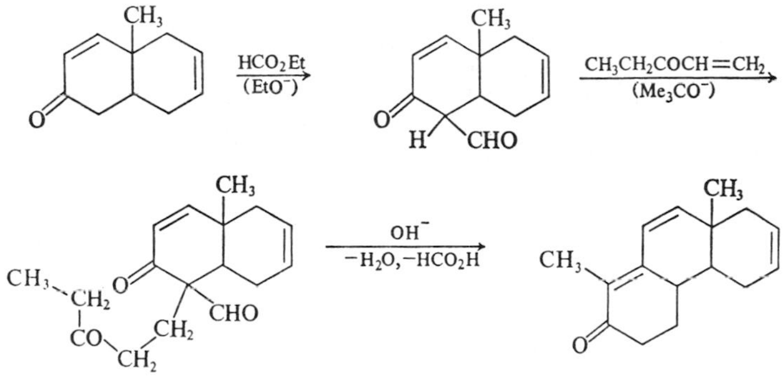

When the olefinic component of a Michael reaction is acrylonitrile, the reaction is usually referred to as *cyanoethylation*. This reaction, too, was used in the Harvard sterol synthesis during the introduction of a six-membered ring (21.3).

7.6 Condensations involving Acetylides

Acetylene and its monosubstituted derivatives are markedly more acidic than olefins and paraffins and are able to take part in base-catalyzed reactions with both carbonyl-containing compounds and alkyl halides. A strong base is necessary, and amide ion in liquid ammonia is commonly used. The typical reactions are:

$$R-C \equiv C-H + NH_2^- \rightleftharpoons R-C \equiv C^- + NH_3$$

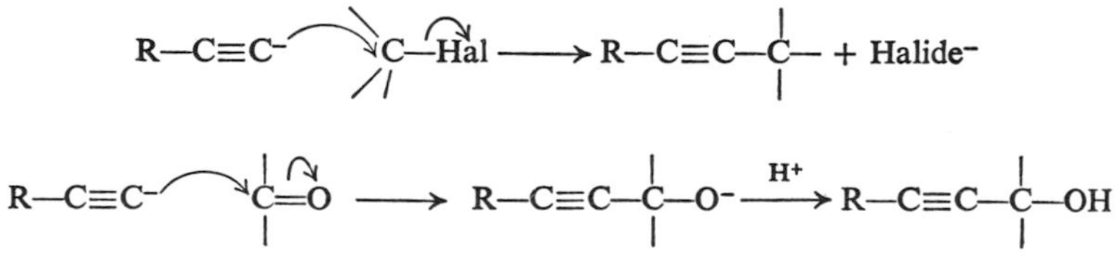

In the usual reaction conditions sodamide is first formed by treating liquid ammonia with sodium metal in the presence of a catalytic quantity of an iron(III) salt. (In the absence of the iron(III) salt the formation of sodamide, $Na + NH_3 \rightarrow Na^+ NH_2 + \frac{1}{2}H_2$, is very slow and the solution contains solvated electrons which are powerfully reducing, p. 617.) The acetylene is then passed into the ammonia solution, the acetylide ion being formed by the acid-base equilibrium above. Pre-formed sodamide or potassium amide may also be used.

An alternative method is to pass the acetylene into liquid ammonia and then to add sodium at such a rate that no blue colour develops for any length of time. This method has the disadvantage that one-third of the acetylene is reduced to the corresponding olefin:

$$3 \ CH{\equiv}CH + 2 \ Na \longrightarrow 2 \ Na^+ \ \bar{C}{\equiv}CH + CH_2{=}CH_2$$

(a) CONDENSATIONS WITH ALKYL HALIDES

The reaction is successful only with halides of the type RCH_2CH_2Hal, i.e. those in which there is no branching at the α- or β-carbon atoms. Good yields are then obtained, as in the formation of n-butylacetylene from acetylene and n-butyl bromide (70–77%) [4]:

$$CH{\equiv}CH \xrightarrow{Na/NH_3} CH{\equiv}C^- \xrightarrow[-Br^-]{CH_3CH_2CH_2CH_2Br} CH{\equiv}C{-}CH_2CH_2CH_2CH_3$$
$$\text{n-Butylacetylene}$$

As usual in nucleophilic displacements on halides, the order of reactivity is iodide > bromide > chloride. Use may be made of this fact when selectivity is required; an example occurs in the synthesis of oleic acid:

$$CH_3{-}(CH_2)_7{-}C{\equiv}CH + I{-}(CH_2)_7{-}Cl \xrightarrow[-I^-]{(NH_2^-)} CH_3{-}(CH_2)_7{-}C{\equiv}C{-}(CH_2)_7{-}Cl$$

$$\xrightarrow[\text{2) hydrolysis}]{\text{1) CN}^-} CH_3{-}(CH_2)_7{-}C{\equiv}C{-}(CH_2)_7{-}CO_2H$$

$$\xrightarrow[\text{(p. 619)}]{\substack{\text{partial} \\ \text{reduction}}} CH_3{-}(CH_2)_7{-}\underset{cis}{CH{=}CH}{-}(CH_2)_7{-}CO_2H$$
$$\text{Oleic acid}$$

(b) CONDENSATIONS WITH CARBONYL GROUPS

Acetylides react with aldehydes and ketones to form α-acetylenic alcohols. For example, acetylene and acetone, with sodamide in liquid ammonia, give 2-methylbut-3-yn-2-ol in about 45% yield [3]:

$$(CH_3)_2CO + CH{\equiv}CH \xrightarrow{(NH_2^-)} (CH_3)_2\underset{O^-}{C}{-}C{\equiv}CH \xrightarrow{H^+} (CH_3)_2\underset{OH}{C}{-}C{\equiv}CH$$

This type of reaction has been widely applied to the synthesis of intermediates in carotenoid and polyene chemistry. An example from the preparation of β-ionone, required for Isler's synthesis of Vitamin A (21.1), is illustrative. 3-Methylpent-2-en-4-yn-1-ol was prepared by treatment of methyl vinyl ketone with acetylide ion in liquid ammonia followed by an acid-catalyzed anionotropic rearrangement for which the thermodynamic driving force is the introduction of conjugation between the olefinic and acetylenic bonds (*cf.* p. 20):

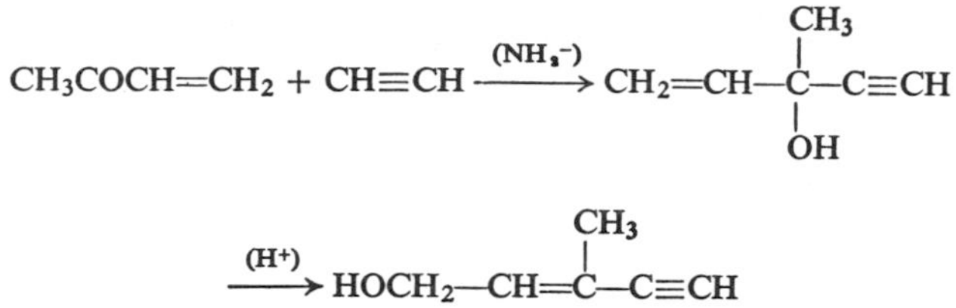

$$CH_3COCH{=}CH_2 + CH{\equiv}CH \xrightarrow{(NH_2{}^-)} CH_2{=}CH{-}\underset{\underset{OH}{|}}{\overset{\overset{CH_3}{|}}{C}}{-}C{\equiv}CH$$

$$\xrightarrow{(H^+)} HOCH_2{-}CH{=}\overset{\overset{CH_3}{|}}{C}{-}C{\equiv}CH$$

Copper(I) acetylides may be made by treating acetylene or its monosubstituted derivatives with an ammoniacal solution of a copper(I) salt. Those from monosubstituted acetylenes are unreactive towards halides and carbonyl groups but that from acetylene itself reacts readily. Since copper(I) acetylide, unlike sodium acetylide, is not decomposed by water, reactions may be conducted in aqueous solution and this makes for both ease of handling and economy. Many commercial products are derived from the reaction between copper(I) acetylide and two molecules of formaldehyde, including tetrahydrofuran, butadiene (for synthetic rubbers), and adipic acid and hexamethylenediamine (the constituents of 6.6-nylon).

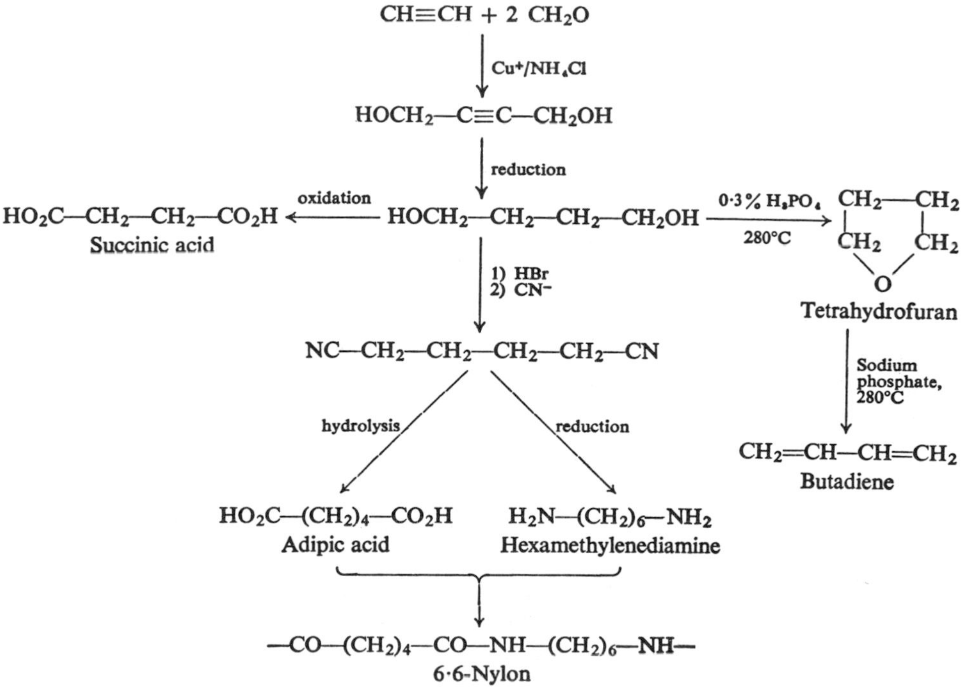

$$CH{\equiv}CH + 2\ CH_2O$$

$$\downarrow {\scriptstyle Cu^+/NH_4Cl}$$

$$HOCH_2{-}C{\equiv}C{-}CH_2OH$$

$$\downarrow {\scriptstyle reduction}$$

HO$_2$C—CH$_2$—CH$_2$—CO$_2$H $\xleftarrow{\text{oxidation}}$ HOCH$_2$—CH$_2$—CH$_2$—CH$_2$OH $\xrightarrow[280°C]{0{\cdot}3\%\ H_3PO_4}$ Tetrahydrofuran
Succinic acid

$$\downarrow {\scriptstyle \begin{array}{l}1)\ HBr\\2)\ CN^-\end{array}}$$

$$NC{-}CH_2{-}CH_2{-}CH_2{-}CH_2{-}CN$$

(hydrolysis / reduction)

Sodium phosphate, 280°C

$$CH_2{=}CH{-}CH{=}CH_2$$
Butadiene

HO$_2$C—(CH$_2$)$_4$—CO$_2$H   H$_2$N—(CH$_2$)$_6$—NH$_2$
Adipic acid   Hexamethylenediamine

$$-CO{-}(CH_2)_4{-}CO{-}NH{-}(CH_2)_6{-}NH{-}$$
6·6-Nylon

7.7 Condensations involving Cyanide

Hydrogen cyanide is isoelectronic with acetylene and, like acetylene, it is a weak acid whose anion may be generated by base and is reactive towards alkyl halides and carbonyl groups. It is often more convenient to introduce the cyanide as cyanide ion (e.g. NaCN) rather than as hydrogen cyanide.

(a) CONDENSATIONS WITH ALKYL HALIDES

Primary and secondary halides undergo nucleophilic displacement to give the corresponding nitriles.

$$N\equiv C \overset{\frown}{} \overset{\diagdown}{\underset{\diagup|}{C}}\text{—}\overset{\frown}{\text{Hal}} \longrightarrow N\equiv C\text{—}\overset{|}{\underset{|}{C}}\text{—} + \text{Halide}^-$$

Tertiary halides do not yield nitriles but undergo elimination to give olefins (*cf.* the reactions of other nucleophiles, including carbanions such as that from malonic ester, p. 248).

These reactions provide a way of extending aliphatic carbon chains by one carbon atom. The following transformations of the nitrile are useful.

(1) To carboxylic acids, by hydrolysis. An important example is the synthesis of malonic acid, and hence malonic ester, in 75–80% yield from chloroacetic acid [2]. Chloroacetic acid is first converted into its sodium salt (or otherwise the addition of cyanide ion would liberate hydrogen cyanide), displacement with cyanide ion gives sodium cyanoacetate, and alkaline hydrolysis yields sodium malonate. The addition of calcium chloride precipitates calcium malonate from which malonic acid is liberated by treatment with hydrochloric acid.

$$\text{ClCH}_2\text{CO}_2\text{H} \xrightarrow{\text{OH}^-} \text{ClCH}_2\text{CO}_2^- \xrightarrow{\text{CN}^-} N\equiv C\text{—CH}_2\text{CO}_2^-$$

$$\xrightarrow{\text{OH}^-} \text{CH}_2(\text{CO}_2^-)_2 \xrightarrow{\text{Ca}^{++}} \text{CH}_2(\text{CO}_2^-)_2\ \text{Ca}^{2+} \xrightarrow{\text{2 HCl}} \text{CH}_2(\text{CO}_2\text{H})_2 + \text{CaCl}_2$$

(2) To amines, by catalytic reduction (—CN → —CH$_2$NH$_2$; p. 641).

(3) To aldehydes, by Stephen reduction (p. 646).

$$\text{R—C}\equiv\text{N} \xrightarrow[\substack{\text{2) H}_2\text{O}}]{\text{1) HCl—SnCl}_2} \text{R—CHO}$$

Aryl halides do not react readily with cyanide ion unless the aromatic nucleus is especially activated towards nucleophilic attack (p. 426). For example, 2,4-dinitrochlorobenzene gives 2,4-dinitrobenzonitrile readily, but to obtain benzonitrile it is necessary to treat bromobenzene with anhydrous copper(I) cyanide at 200°C in the presence of pyridine or quinoline:

$$\text{PhBr} \xrightarrow[\text{200°C}]{\text{CuCN/pyridine}} \text{PhCN}$$

In general, it is more satisfactory to prepare aromatic nitriles through aromatic diazonium salts (p. 438).

(*b*) CONDENSATIONS WITH CARBONYL COMPOUNDS

Hydrogen cyanide adds to aldehydes and ketones to give cyanohydrins:

$$R_2C{=}O + HCN \longrightarrow R_2C{\overset{\displaystyle OH}{\underset{\displaystyle CN}{\big<}}}$$

One of the first mechanistic studies in organic chemistry concerned this reaction. Lapworth discovered in 1903–4 that the reaction is catalyzed by base, and he rationalized this by suggesting that the reactive species is the cyanide ion:

$$HCN + OH^- \rightleftharpoons CN^- + H_2O$$

The complete reaction therefore corresponds to those in which other carbanions react with carbonyl compounds: addition of cyanide ion to the carbonyl group gives an oxyanion which, by abstracting a proton from hydrogen cyanide, generates a further cyanide ion:

$$R_2C{=}O \xrightarrow{\ CN^-\ } R_2C{\overset{\displaystyle O^-}{\underset{\displaystyle CN}{\big<}}} \xrightarrow{\ HCN\ } R_2C{\overset{\displaystyle OH}{\underset{\displaystyle CN}{\big<}}} + CN^-$$

Typical examples of cyanohydrin formation are the preparation of glycolonitrile (*ca.* 76%) from formaldehyde [3],

$$CH_2O + KCN + H_2O \longrightarrow CH_2{\overset{\displaystyle OH}{\underset{\displaystyle CN}{\big<}}} + KOH$$

Glycolonitrile

and of acetone cyanohydrin (77%) from acetone in a similar manner [2].

As usual in addition reactions at carbonyl groups, aldehydes are more reactive than ketones, partly because of the unfavourable inductive effect of the second alkyl group in a ketone and partly because of the increased steric hindrance in addition to ketones. For example, di-isopropyl ketone does not react. In the aromatic series, alkyl aryl ketones react unless the alkyl group is particularly bulky or the aromatic ring has large *ortho*-substituents, but diaryl ketones such as benzophenone are inert. Aromatic aldehydes behave anomalously (see below).

Activated olefins react with cyanide in the manner of the Michael addition. For example, benzalacetophenone adds hydrogen cyanide, in ethanol containing acetic acid, at its olefinic bond [2]:

$$PhCH{=}CHCOPh + HCN \longrightarrow PhCH{-}CH_2COPh$$
$$\underset{CN}{\overset{|}{}}$$

It should be noted that this olefin is reactive enough for addition to occur readily in acidic conditions, in which it is probably helped by protonation of the carbonyl oxygen (p. 113). The product, which contains a carbonyl group, is inert to further addition in these conditions, but if the solution is allowed to become basic a second molecule of hydrogen cyanide is added.

Cyanohydrins are of synthetic value because of their ready conversion by hydrolysis into α-hydroxy-acids or esters. For example, (±)-lactic acid may be obtained from acetaldehyde,

$$CH_3CHO \xrightarrow{HCN} CH_3CH\underset{CN}{\overset{OH}{<}} \xrightarrow{H_2O-H^+} CH_3CH\underset{CO_2H}{\overset{OH}{<}}$$
$$(\pm)\text{-Lactic acid}$$

and methyl α-methacrylate (polymerization of which gives Perspex) may be obtained from acetone cyanohydrin by treatment with sulphuric acid in methanol, which brings about both esterification and dehydration:

$$(CH_3)_2C\underset{CN}{\overset{OH}{<}} \xrightarrow{MeOH-H_2SO_4} CH_2{=}\underset{}{\overset{CH_3}{\underset{}{C}}}{-}CO_2Me$$
$$\text{Methyl } \alpha\text{-methacrylate}$$

The reaction of an aliphatic aldehyde with sodium cyanide in the presence of ammonium chloride gives an α-amino-nitrile (p. 331); hydrolysis gives an α-amino-acid (Strecker synthesis, pp. 331, 552), e.g.

$$CH_3CHO \xrightarrow{NaCN-NH_4Cl} CH_3CH\underset{CN}{\overset{NH_2}{<}} \xrightarrow{H_2O-H^+} CH_3CH\underset{CO_2H}{\overset{NH_2}{<}}$$
$$(\pm)\text{-Alanine}$$

Aromatic aldehydes. These do not form cyanohydrins but instead undergo the *benzoin condensation.* Benzaldehyde and sodium cyanide in ethanol give benzoin itself in about 80% yield [1]:

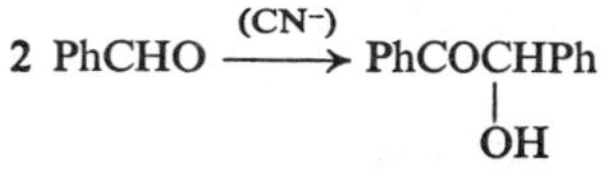

$$2 \text{ PhCHO} \xrightarrow{\text{(CN}^-)} \underset{\underset{\text{OH}}{|}}{\text{PhCOCHPh}}$$

Benzoin

Reaction occurs through the cyanide addition product which, by base-abstraction of a proton from the α-carbon, gives a carbanion; this reacts with a second molecule of the aldehyde, and hydrogen cyanide is then eliminated.

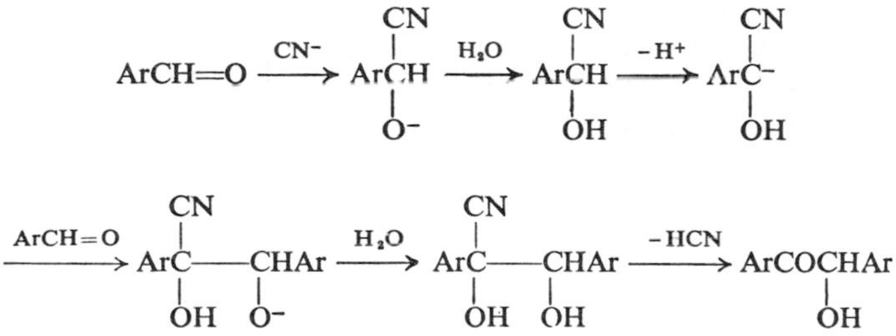

Cyanide ion owes its ability to effect this condensation to two properties: first it is a reactive nucleophile; secondly, the cyanide group, by its capacity to delocalize the negative charge on the carbanion, assists the formation of this species. The difference between aliphatic and aromatic aldehydes may then be ascribed to the fact that the further delocalization of the negative charge over the aromatic ring provides sufficient extra driving force for the reaction to occur.

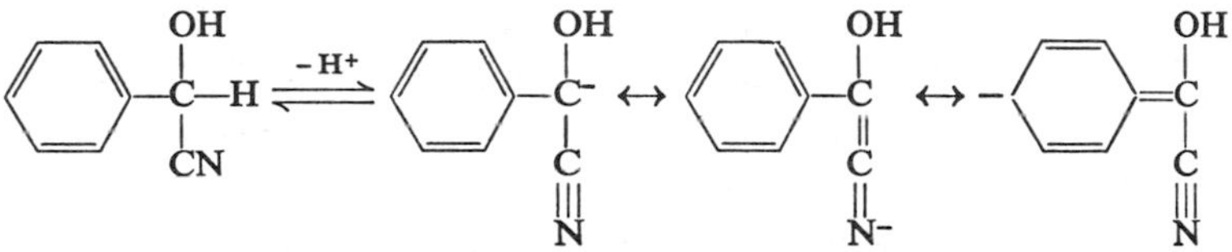

It is interesting to find that Vitamin B_1 (thiamine) brings about similar reactions both in the laboratory and in living organisms. The activity is due to the thiazolium ring whose C_2-hydrogen (shown) is acidic enough to exchange with deuterium in heavy water.

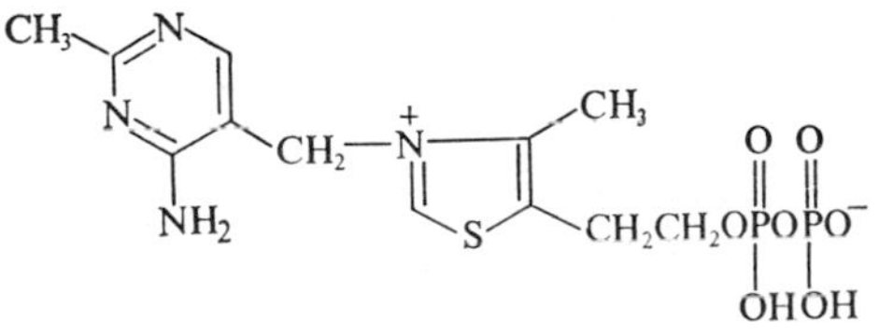

Vitamin B_1

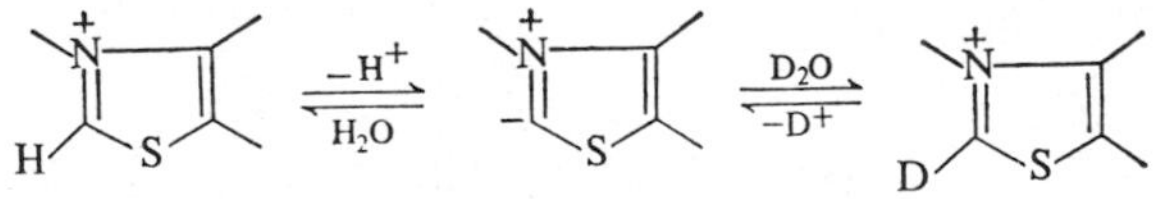

The carbanion intermediate evidently owes its relative stability (compared, say, with that of the anion formed by ionization of the methylene group adjacent to the quaternary nitrogen) to a combination of the presence of an adjacent positive pole and the fact that the charge resides on unsaturated carbon (p. 68). It brings about the benzoin condensation in a manner completely analogous to cyanide ion, to which it bears a close resemblance:

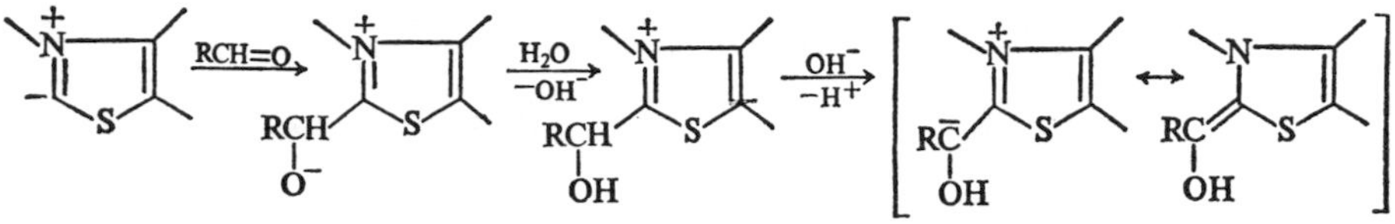

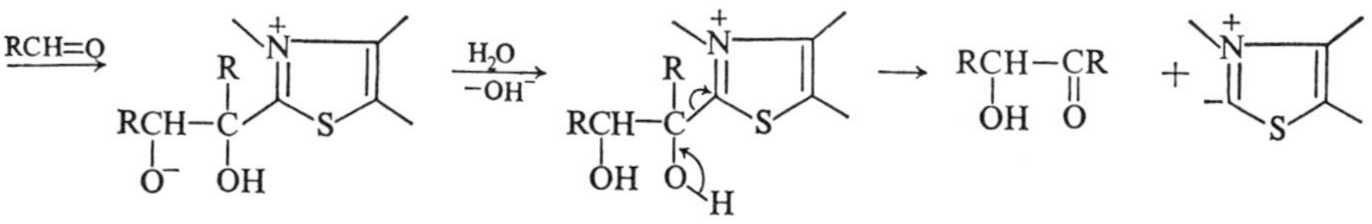

Further Reading

The wide scope and utility of base-catalyzed condensations are demonstrated by the following Chapters in *Organic Reactions*:

'The Perkin and related reactions,' 1942, **1**, 210.

'The acetoacetic ester condensation and certain related reactions,' 1942, **1**, 266.

'Cyanoethylation', 1949, **5**, 79.

'The Darzens glycidic ester condensation,' 1949, **5**, 413.

'The Stobbe condensation,' 1951, **6**, 1.

'The acylation of ketones to form β-diketones or β-keto-aldehydes,' 1954, **8**, 59.

'The Michael reaction,' 1959, **10**, 179.

'The Dieckmann condensation,' 1967, **15**, 1.

'The Knoevenagel condensation,' 1967, **15**, 204.

'The aldol condensation,' 1968, **16**, 1.

Problems

1. How would you synthesize the following compounds?

    (a) PhCH=CHCOPh   (b) $PhCH_2CH_2CO_2H$

(c) PhCOCH$_2$COPh

(d) PhCH(CH$_2$CO$_2$H)$_2$

(e) PhCH$_2$COCO$_2$H

(f) PhCH=CHNO$_2$

(g) (CH$_3$)$_2$CHCOCH$_3$

(h) (CH$_3$)$_2$CHCOCH(CH$_3$)$_2$

(i) CH$_3$COCH$_2$CH$_2$COCH$_3$

(j) CH$_3$C≡CC$_2$H$_5$

(k)
$$\underset{\text{CH}_2=\text{C}-\text{CH}=\text{CH}_2}{\overset{\text{CH}_3}{|}}$$

(l)
$$\underset{\text{C}_2\text{H}_5}{\overset{\text{CH}_3}{\diagdown}}\text{CH}-\text{CO}_2\text{H}$$

(m)
$$\underset{\text{CH}_3\text{CHCO}_2\text{H}}{\overset{\text{CH}_3\text{CHCO}_2\text{H}}{|}}$$

(n) $\underset{}{\overset{\text{CH}_3}{\underset{|}{\text{PhC}}}}$————CHCO$_2$Et (with epoxide O bridge)

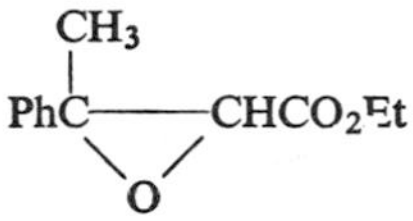

(o)

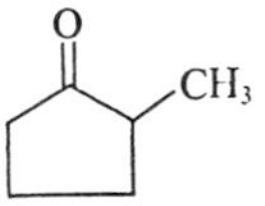

(p)

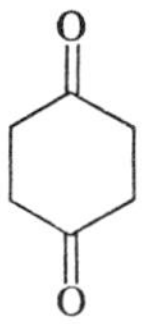

2. What products would you expect from the following reactions?

(a) CH$_3$COCH$_2$CH$_3$ $\xrightarrow{\text{OH}^-}$

(b) CH$_3$COCOCH$_3$ $\xrightarrow{\text{OH}^-}$

(c) CH$_3$COCH$_3$ + EtO$_2$C—CO$_2$Et $\xrightarrow{\text{EtO}^-}$

(d) EtO$_2$CCH$_2$CH$_2$CO$_2$Et $\xrightarrow{\text{EtO}^-}$

(e) PhCOCH$_3$ + CH$_2$O $\xrightarrow{\text{EtO}^-}$

(f) CH$_2$(CO$_2$Et)$_2$ + CH$_2$=CHCN $\xrightarrow{\text{EtO}^-}$

(g) NC—(CH$_2$)$_4$—CN $\xrightarrow{\text{EtO}^-}$

3. Account for the following:

(i) Amongst simple reagents, only cyanide ion catalyzes the self-condensation of benzaldehyde.

(ii) Although diethyl succinate undergoes self-condensation with base, it may be successfully used in the Stobbe condensation to react with aldehydes and ketones.

(*iii*) When ketones containing α-hydrogen react with esters, the product is that derived by displacement of alkoxide ion from the ester by the carbanion from the ketone. However, when the ester is ethyl chloro-acetate, the product is that derived by addition of the carbanion from the ester to the carbonyl group of the ketone.

(*iv*) 2,4,6-Trinitrotoluene reacts with benzaldehyde in the presence of pyridine, but toluene does not.

8. Formation of Aliphatic Carbon-Carbon Bonds: Acid-catalyzed Condensations

8.1 Principles

The principle which is applied in all acid-catalyzed condensations is the generation of an electrophilic species, with the aid of an acid, in the presence of a nucleophile with which the electrophile then reacts.

The electrophile may be obtained either from an alkyl or acyl halide by treatment with a Lewis acid, as in the alkylation of olefins (p. 267), e.g.

$$(CH_3)_3C\!\frown\!Cl \quad AlCl_3 \longrightarrow (CH_3)_3C^+ + AlCl_4^-$$

or, more commonly, by the addition of a proton to a double bond. The double bond may be olefinic, as in the dimerization of olefins (p. 266), e.g.

$$(CH_3)_2C{=}CH_2 + H^+ \longrightarrow (CH_3)_3C^+$$

or a carbonyl group, as in the self-condensations of aldehydes and ketones, e.g.

$$CH_3{-}CH{=}O + H^+ \longrightarrow CH_3{-}CH{=}\overset{+}{O}H \leftrightarrow CH_3{-}\overset{+}{C}H{-}OH$$

The electrophile in the Mannich reaction (8.6) is generated from an aldehyde and an amine in the presence of an acid, e.g.

$$(CH_3)_2\overset{..}{N}H \quad CH_2{=}\overset{+}{O}H \longrightarrow (CH_3)_2\overset{+}{N}H\,{-}CH_2{-}OH \xrightarrow{-H_2O}$$

$$(CH_3)_2\overset{+}{N}{=}CH_2 \leftrightarrow (CH_3)_2N{-}\overset{+}{C}H_2$$

The nucleophile may be an olefin, an enol, or, in some applications of the Mannich reaction, a compound of related type such as indole. A typical example is the reaction of ethylene with t-butyl chloride in the presence of aluminium trichloride,

$$(CH_3)_3C{-}Cl \xrightarrow[-AlCl_4^-]{AlCl_3} (CH_3)_3C^+ \xrightarrow{CH_2{=}CH_2} (CH_3)_3C{-}CH_2{-}CH_2^+$$

$$\xrightarrow[-AlCl_3]{AlCl_4^-} (CH_3)_3C{-}CH_2{-}CH_2{-}Cl$$

The ensuing discussion delineates the various combinations of electrophile and nucleophile which are of value in synthesis.

8.2 The Self-Condensation of Olefins

The treatment of isobutylene with 60% sulphuric acid gives a mixture of 2,4,4-trimethyl-1-pentene and 2,4,4-trimethyl-2-pentene. Reaction occurs by the protonation of one molecule of the olefin to give a carbonium ion which adds to the methylene group (Markovnikov's rule) of a second molecule; the new carbonium ion then eliminates a proton:

$$(CH_3)_2C{=}CH_2 + H^+ \longrightarrow (CH_3)_3C^+ \xrightarrow{\ (CH_3)_2C=CH_2\ }$$

$$(CH_3)_3C{-}CH_2{-}\overset{+}{C}(CH_3)_2 \xrightarrow{\ -H^+\ } (CH_3)_3C{-}CH_2{-}C\!\!\begin{array}{c} {}^{CH_2} \\[-2pt] \diagdown \\ CH_3 \end{array} \;+\; (CH_3)_3C{-}CH{=}C\!\!\begin{array}{c} {}^{CH_3} \\[-2pt] \diagdown \\ CH_3 \end{array}$$

$$\qquad\qquad\qquad\qquad\qquad\quad \text{4 parts} \qquad\qquad\qquad \text{1 part}$$

The conditions must be carefully controlled. When dilute sulphuric acid is used, the first carbonium ion reacts preferentially with water to give t-butyl alcohol,

$$(CH_3)_3C^+ + H_2O \longrightarrow (CH_3)_3C{-}\overset{+}{O}H_2 \xrightarrow{\ -H^+\ } (CH_3)_3C{-}OH$$

and if more concentrated acid is employed, the second carbonium ion reacts with a further molecule of isobutylene,

$$(CH_3)_3C{-}CH_2{-}\overset{+}{C}(CH_3)_2 \xrightarrow{\ (CH_3)_2C=CH_2\ } (CH_3)_3C{-}CH_2{-}C(CH_3)_2{-}CH_2{-}\overset{+}{C}(CH_3)_2$$

and further polymerization can occur. Thus, the concentration of the basic species, water, must be such that the simple hydration reaction is minimized while sufficient water is present to favour removal of a proton from the dimeric carbonium ion over further addition of olefin.

This discussion indicates the difficulties necessarily present in effecting the dimerization of an olefin. Nonetheless, the example cited is of great industrial importance; the mixture of pentenes is reduced catalytically to 2,2,4-trimethyl-pentane (iso-octane), used as a high-octane fuel.

In a modified procedure, iso-octane is obtained directly from isobutylene and concentrated sulphuric acid by carrying out the reaction in the presence of iso-butane. Dimerization occurs as above, but the resulting carbonium ion, instead of eliminating a proton or reacting with more isobutylene, abstracts hydride ion from isobutane:

$$(CH_3)_2C{=}CH_2 \xrightarrow{H^+} (CH_3)_3C^+ \xrightarrow{(CH_3)_2C{=}CH_2} (CH_3)_3C{-}CH_2{-}\overset{+}{C}(CH_3)_2$$

$$(CH_3)_3C{-}CH_2{-}\overset{+}{C}(CH_3)_2 + (CH_3)_3C{-}H \longrightarrow (CH_3)_3C{-}CH_2{-}CH(CH_3)_2 + (CH_3)_3C^+$$

A new t-butyl carbonium ion is produced and adds to isobutylene, so that a chain reaction is propagated.

Di-olefins undergo acid-catalyzed cyclization provided that the stereochemically favoured five- or six-membered rings are formed. The example chosen is the conversion of ψ-ionone into α- and β-ionone, the second of which is required for a synthesis of Vitamin A (21.1):

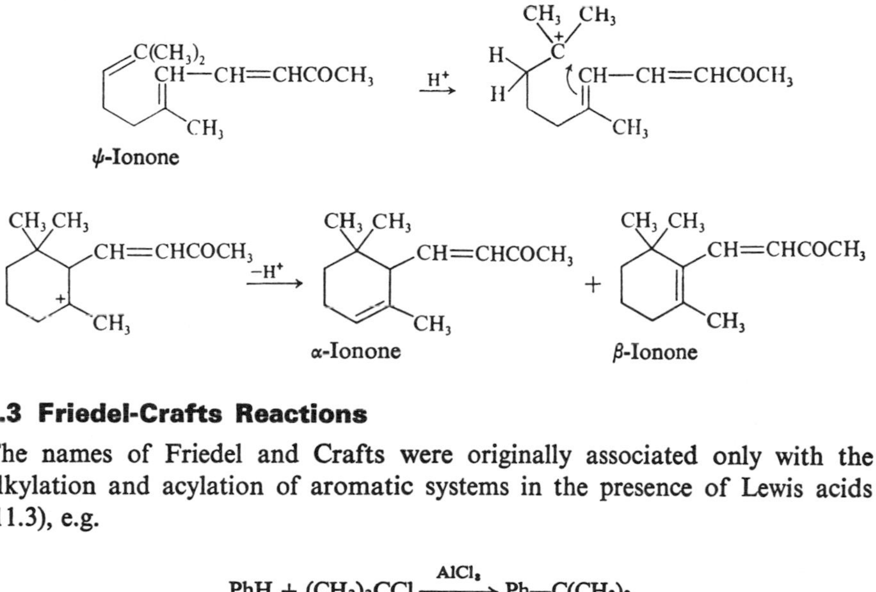

ψ-Ionone

α-Ionone   β-Ionone

8.3 Friedel-Crafts Reactions

The names of Friedel and Crafts were originally associated only with the alkylation and acylation of aromatic systems in the presence of Lewis acids (11.3), e.g.

$$PhH + (CH_3)_3CCl \xrightarrow[-HCl]{AlCl_3} Ph{-}C(CH_3)_3$$

$$PhH + CH_3COCl \xrightarrow[-HCl]{AlCl_3} Ph{-}COCH_3$$

These reactions are fully discussed in Chapter 11. Analogous processes occur with olefins, although they are not so widely applicable.

(a) ALKYLATION

The simplest example of an efficient alkylation is the preparation of neohexyl chloride (75%) from t-butyl chloride and ethylene at about −10°C in the presence of aluminium trichloride:

$$(CH_3)_3C—Cl \quad AlCl_3 \longrightarrow (CH_3)_3C^+ + AlCl_4^-$$

$$(CH_3)_3C^+ + CH_2{=}CH_2 \longrightarrow (CH_3)_3C—CH_2—CH_2^+$$

$$(CH_3)_3C—CH_2—CH_2^+ \quad Cl—\bar{A}lCl_3 \longrightarrow (CH_3)_3C—CH_2—CH_2Cl + AlCl_3$$

Neohexyl chloride

Several side-reactions are normally encountered in alkylation. First, olefins are often isomerized by aluminium trichloride. Secondly, alkyl halides may rearrange: e.g. n-propyl chloride gives isopropyl derivatives (*cf.* p. 387). Finally, the halide produced may react further. These problems do not arise in the example cited because ethylene cannot isomerize, t-butyl halides are not rearranged by Lewis acids, and the primary halide product is far less reactive than the tertiary halide towards aluminium trichloride. It must be emphasized, however, that the synthetic utility of alkylation is restricted.

(*b*) ACYLATION

The acylation of an olefin is brought about by an acid chloride or acid anhydride in the presence of a Lewis acid. The electrophile is probably an acylium ion,

$$R—CO—Cl \quad AlCl_3 \longrightarrow R—\overset{+}{C}{=}O + AlCl_4^-$$

$$R—CO—OCOR \quad AlCl_3 \longrightarrow R—\overset{+}{C}{=}O + AlCl_3(OCOR)^-$$

which adds to the olefinic bond; reaction is completed by the uptake of a nucleophile, e.g.

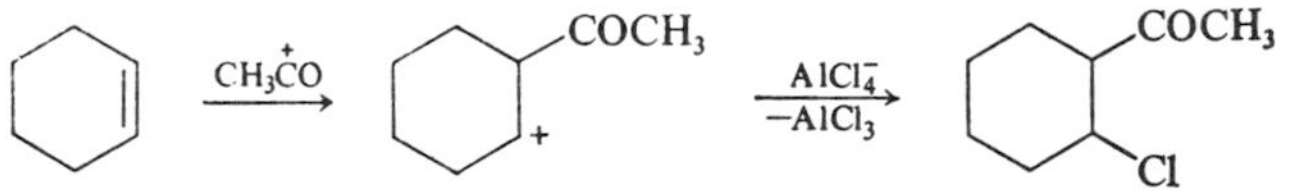

2-Acetylcyclohexyl chloride

The products, β-substituted ketones, readily undergo elimination to give (conjugated) αβ-unsaturated ketones. This can occur spontaneously if the reaction is carried out at elevated temperature, or otherwise can be induced by a weak base. For example, 2-acetylcyclohexyl chloride formed as above gives 1-acetylcyclohexene, in 42% overall yield from cyclohexene, when treated with dimethylamine:

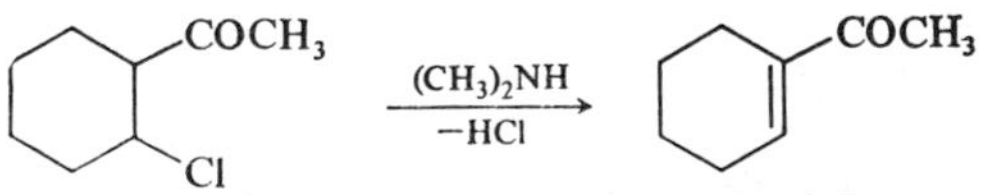

1-Acetylcyclohexene

Acylations suffer from the same disadvantage as alkylations in that the olefins are rearranged by aluminium trichloride, and it is therefore usually more satisfactory to use a less vigorous Lewis acid such as tin(IV) chloride. However, the other side-reactions encountered in alkylations do not apply: the acid halides and anhydrides do not rearrange, and the products are much less reactive than the starting materials. Acylation is therefore a useful synthetic process, as illustrated by the (1955) synthesis of (±)-thioctic acid:*

$$Cl-CO-(CH_2)_4-CO_2Et + CH_2=CH_2 \xrightarrow{AlCl_3} ClCH_2-CH_2-CO-(CH_2)_4-CO_2Et$$

$$\xrightarrow{NaBH_4} ClCH_2-CH_2-CH(OH)-(CH_2)_4-CO_2Et \xrightarrow[\text{2) 2 PhCH}_2\text{SH—KOH}]{\text{1) SOCl}_2}$$

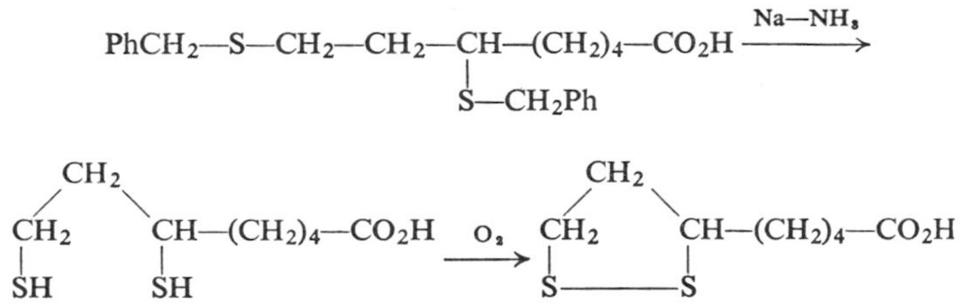

$$PhCH_2-S-CH_2-CH_2-\underset{\underset{S-CH_2Ph}{|}}{CH}-(CH_2)_4-CO_2H \xrightarrow{Na-NH_3}$$

(±)-Thioctic acid

8.4 Prins Reaction

The treatment of an olefin with formaldehyde in the presence of an acid gives a 1,3-diol together with the cyclic acetal derived from this and a second molecule of formaldehyde:

$$CH_2=O + H^+ \rightleftharpoons CH_2=\overset{+}{O}H \leftrightarrow \overset{+}{C}H_2-OH$$

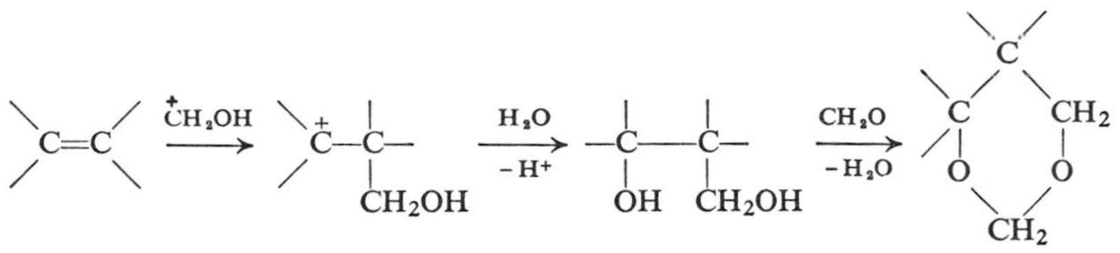

Ethylene itself requires very vigorous conditions for reaction, but alkylated olefins, which are more reactive towards electrophiles, react fairly readily. Olefins of the type RCH=CHR give mainly 1,3-diols, in low yield, whereas those of the types $RCH=CH_2$ and $R_2C=CH_2$ give mainly acetals (1,3-dioxans),

*Thioctic acid (lipoic acid) is a cofactor in the physiological oxidative-decarboxylation of pyruvic acid to acetic acid:

$$CH_3COCO_2H \xrightarrow{\frac{1}{2}O_2} CH_3CO_2H + CO_2$$

often in good yield. For example, 4-phenyl-1,3-dioxan can be isolated in about 80% yield by refluxing a mixture of styrene, 37% formalin, and concentrated sulphuric acid [4]:

$$PhCH{=}CH_2 + 2\ CH_2O \xrightarrow{H^+} \quad \begin{array}{c} Ph \quad CH_2 \\ \diagdown\diagup\diagdown \\ CH \quad CH_2 \\ | \qquad | \\ O \qquad O \\ \diagdown\diagup \\ CH_2 \end{array}$$

8.5 Condensations of Aldehydes and Ketones

(a) SELF-CONDENSATIONS

Aldehydes and ketones which are capable of enolization undergo self-condensation when treated with acids. The acid has two functions: first, it enhances the reactivity of the carbonyl group towards the addition of a nucleophile (p. 113), e.g.

$$CH_3{-}CH{=}O \underset{\longleftarrow}{\overset{H^+}{\rightleftharpoons}} CH_3{-}CH{=}\overset{+}{O}H \leftrightarrow CH_3{-}\overset{+}{C}H{-}OH$$

Secondly, it catalyzes the enolization of the carbonyl compound, e.g.

$$CH_3{-}CH{=}O \overset{H^+}{\rightleftharpoons} \overset{H}{\underset{}{CH_2}}{-}CH{=}\overset{+}{O}H \overset{-H^+}{\rightleftharpoons} CH_2{=}CH{-}OH$$

A molecule of the enol then reacts with a molecule of the (activated) carbonyl compound,

$$\underset{CH{=}CH_2}{\overset{HO:}{}} \quad \underset{CH{=}\overset{+}{O}H}{\overset{CH_3}{|}} \longrightarrow \underset{CH{-}CH_2{-}CH{-}OH}{\overset{HO}{\overset{\|}{}} \overset{CH_3}{|}} \xrightarrow{-H^+} CH_3{-}CH(OH){-}CH_2{-}CHO$$

Acid-catalyzed dehydration normally follows:

$$\underset{OH}{\overset{}{CH_3{-}CH{-}CH_2{-}CHO}} \overset{H^+}{\rightleftharpoons} \underset{\overset{+}{O}H_2}{\overset{H}{CH_3{-}CH{-}CH{-}CHO}} \xrightarrow{-H_2O^+} \underset{\text{Crotonaldehyde}}{CH_3{-}CH{=}CH{-}CHO}$$

A ketone which possesses at least one hydrogen atom on each of its α-carbon atoms can condense further. For example, when acetone is saturated with hydrogen chloride, a mixture of mesityl oxide and phorone is produced:

$$2\ CH_3COCH_3 \xrightarrow[-H_2O]{HCl} \underset{\text{Mesityl oxide}}{(CH_3)_2C{=}CHCOCH_3} \xrightarrow[-H_2O]{CH_3COCH_3} \underset{\text{Phorone}}{(CH_3)_2C{=}CH{-}CO{-}CH{=}C(CH_3)_2}$$

The yields in these acid-catalyzed aldol condensations are usually low; e.g. in the formation of mesityl oxide, phorone is inevitably a byproduct. The base-catalyzed aldol condensation is therefore normally preferred. It should also be noted that some aldehydes are polymerized by acid through oxygen: e.g. acetaldehyde and a little concentrated sulphuric acid give the cyclic trimer, paraldehyde, and some of the cyclic tetramer, metaldehyde:

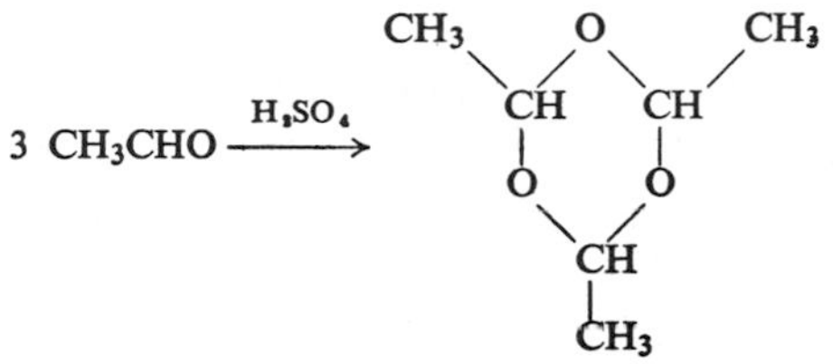

$$3 \text{ CH}_3\text{CHO} \xrightarrow{\text{H}_2\text{SO}_4}$$

Paraldehyde

$$4 \text{ CH}_3\text{CHO} \xrightarrow{\text{H}_2\text{SO}_4}$$

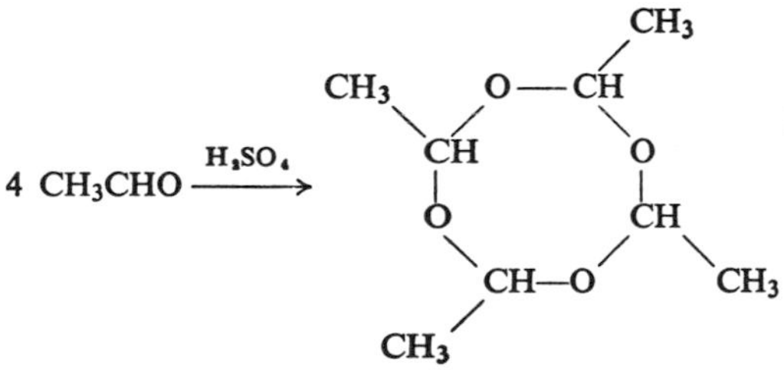

Metaldehyde

Polymerization is reversible and the aldehyde may be recovered by warming with dilute acid. Formaldehyde yields a solid polymer, paraformaldehyde, on evaporation of formalin,

$$n \text{ CH}_2\text{O} \xrightarrow{\text{H}_2\text{O}} \text{HOCH}_2\text{—(O—CH}_2)_{n-1}\text{—OH}$$

and treatment with sulphuric acid gives trioxan (trioxymethylene):

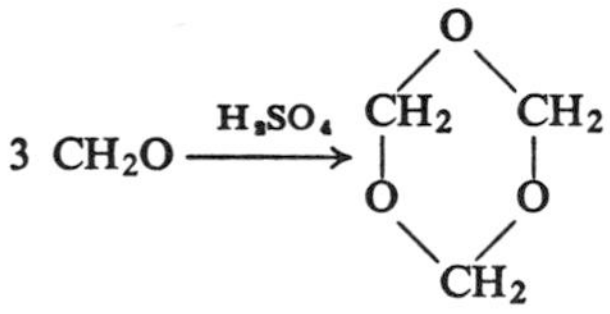

$$3 \text{ CH}_2\text{O} \xrightarrow{\text{H}_2\text{SO}_4}$$

Trioxan

Formaldehyde and acetaldehyde are conveniently stored in the form of para-formaldehyde and paraldehyde, respectively; treatment with dilute acid immediately before use or *in situ* generates the free aldehyde. Ketones do not form analogous oxygen-linked polymers, but treatment of acetone with concentrated sulphuric acid gives mesitylene.

(b) CROSSED CONDENSATIONS

As in base-catalyzed condensations, crossed condensations between two carbonyl compounds each of which can enolize are likely to result in a mixture of four products. Again, however, if only one of the two compounds can enolize and the other has the more reactive carbonyl group, a single product may be formed in good yield.

An example is the condensation between acetophenone and salicylaldehyde, catalyzed by anhydrous hydrogen chloride; only the former compound can enolize and the latter has the more reactive carbonyl group. Condensation is followed by the acid-catalyzed elimination of water to give an oxonium salt which is the parent of the anthocyanidin system:*

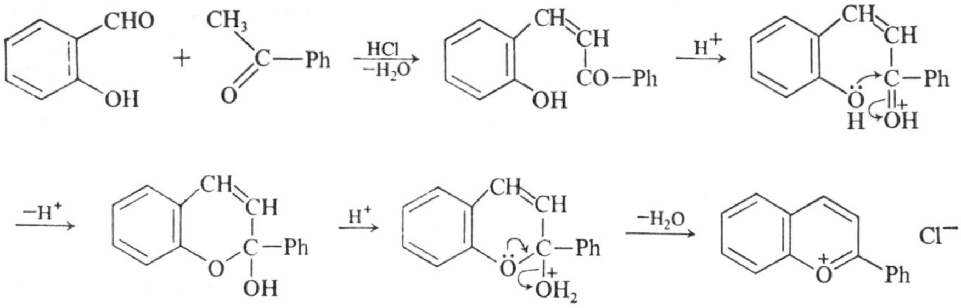

The application of this synthesis to the formation of the naturally occurring anthocyanidins and anthocyanins is described later (p. 690).

(c) CONDENSATIONS BETWEEN KETONES AND ACID CHLORIDES OR ANHYDRIDES

Condensations can also be effected between ketones and compounds with very reactive carbonyl groups such as acid chlorides and anhydrides. Aqueous conditions must be avoided to prevent hydrolysis of the chloride or anhydride, and it is convenient to use a Lewis acid to catalyze the enolization step, e.g.

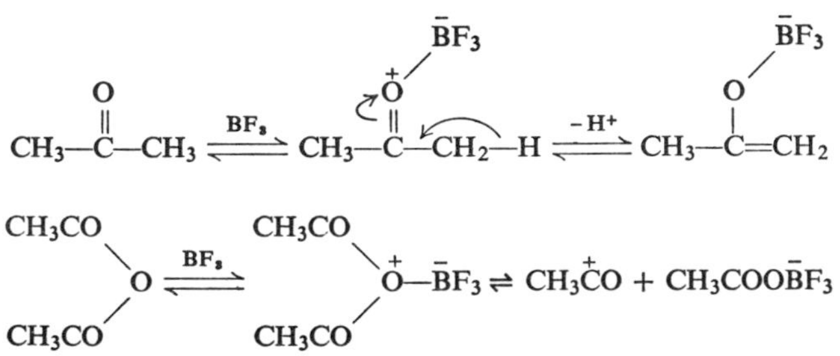

*Oxonium salts normally exist only in solution. The stability of the anthocyanidin cation evidently results from the aromatic character of the six-membered conjugated oxonium ring which contains six π-electrons.

$$\text{CH}_3\text{—}\overset{|}{\underset{}{\text{C}}}\text{=CH}_2 \quad \overset{+}{\text{C}}\text{—CH}_3 \xrightarrow{-\text{BF}_3} \text{CH}_3\text{—CO—CH}_2\text{—CO—CH}_3$$

Acetylacetone

Acetylacetone can be obtained in this way by passing gaseous boron trifluoride into a mixture of acetone and acetic anhydride, adding copper(II) acetate to precipitate the product as the copper(II) derivative of the enol, and reconverting this into acetylacetone with acid [3]. The yield (80%) is considerably higher in this case than in the alternative base-catalyzed process from acetone and ethyl acetate (p. 243).

In general, however, the base-catalyzed method is superior. In particular, only the base-catalyzed procedure is successful for replacing the α-hydrogen atom of a ketone by oxalyl (—COCO$_2$Et), formyl (—CHO), and aromatic acyl (—COAr) groups; and for reactions on aldehydes, which tend to undergo addition at the carbonyl group in the acid-catalyzed process, e.g.

$$\text{CH}_3\text{CH}_2\text{CHO} + (\text{CH}_3\text{CO})_2\text{O} \xrightarrow{\text{BF}_3} \text{CH}_3\text{CH}_2\text{CH(OCOCH}_3)_2$$

Nevertheless, the acid-catalyzed method has one important application in which it complements the base-catalyzed method: a methyl alkyl ketone, $\text{CH}_3\text{COCH}_2\text{R}$, is acylated mainly at the methylene group by the former method but mainly at the methyl group by the latter, e.g.

$$\text{CH}_3\text{COCH}_2\text{R} \begin{cases} \xrightarrow{(\text{CH}_3\text{CO})_2\text{O—BF}_3} & \text{CH}_3\text{COCHR} \\ & \qquad\quad |\\ & \qquad \text{COCH}_3 \\ \xrightarrow{\text{CH}_3\text{CO}_2\text{Et—EtO}^-} & \text{CH}_3\text{COCH}_2\text{COCH}_2\text{R} \end{cases}$$

(*d*) REACTIONS OF α-PICOLINE AND RELATED COMPOUNDS

α-Picoline condenses with aldehydes in the presence of zinc chloride, which catalyzes the conversion of α-picoline into the nitrogen-analogue of an enol, e.g.

α-Picoline

Stilbazole

Similar reactions occur with γ-picoline, 2- and 4-methylquinoline and 1-methylisoquinoline, each of which can undergo the corresponding acid-catalyzed reaction, e.g.

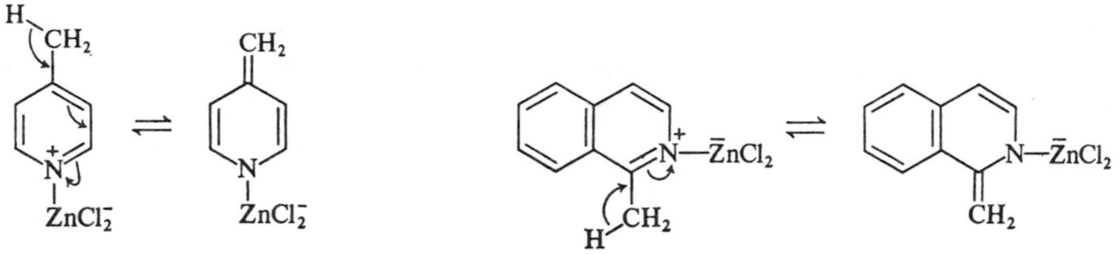

but the other methyl-derivatives of these heterocyclic compounds are either unable to form the analogous methylene derivatives (e.g. β-picoline) or fail to do so because the loss of aromatic stabilization energy on enolization is too great, e.g.

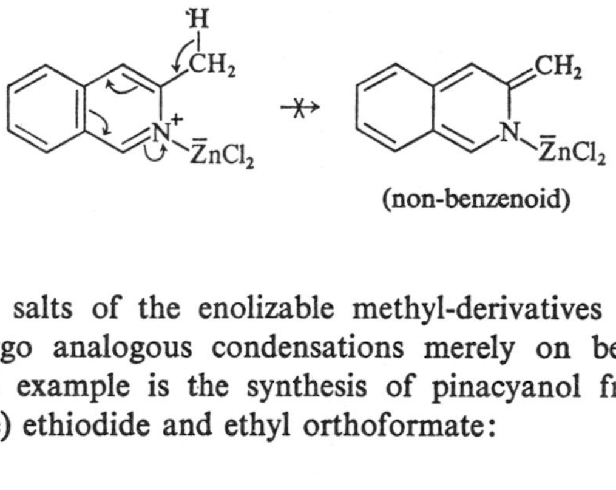

(non-benzenoid)

The quaternary salts of the enolizable methyl-derivatives of heterocyclic compounds undergo analogous condensations merely on being heated. A particularly useful example is the synthesis of pinacyanol from quinaldine (2-methylquinoline) ethiodide and ethyl orthoformate:

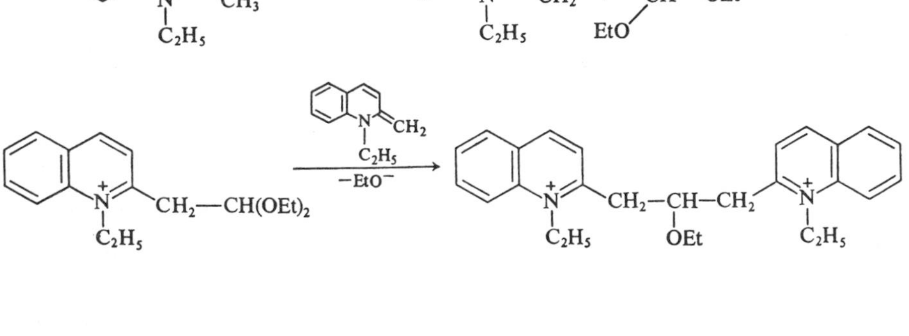

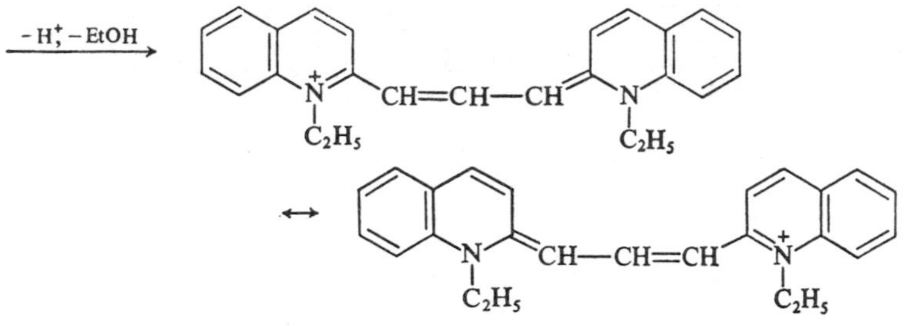

Pinacyanol (cation)

Pinacyanol and related compounds made by analogous routes have highly coloured cations and are used as photographic sensitizers.

8.6 Mannich Reaction

Compounds which are enolic or potentially enolic, and also certain acetylenes, react with a mixture of an aldehyde (usually formaldehyde) and a primary or secondary amine in the presence of an acid to give, after basification, an aminomethyl derivative. For example, by refluxing a mixture of acetone, diethylamine hydrochloride, paraformaldehyde, methanol, and a little concentrated hydrochloric acid, and treating the product with base, 1-diethylamino-3-butanone can be obtained in up to 70% yield [4]:

$$CH_3COCH_3 + CH_2O + (C_2H_5)_2NH_2{}^+Cl^- \xrightarrow{-H_2O}$$

$$CH_3COCH_2CH_2NH(C_2H_5)_2{}^+Cl^- \xrightarrow{OH^-} CH_3COCH_2CH_2N(C_2H_5)_2$$

(a) MECHANISM

The probable mechanism of the reaction can be illustrated with reference to the example cited above. The amine reacts with formaldehyde in the presence of acid to give an adduct which eliminates water to form an electrophile,

$$(C_2H_5)_2\overset{..}{N}H \quad CH_2 = \overset{+}{O}H \longrightarrow (C_2H_5)_2\overset{+}{N}H - CH_2 - OH$$

$$\xrightarrow{-H_2O} (C_2H_5)_2\overset{+}{N} = CH_2 \longleftrightarrow (C_2H_5)_2N - \overset{+}{C}H_2$$

and the acid also catalyzes the conversion of acetone into its enolic tautomer,

$$\underset{CH_3-C-CH_3}{\overset{O}{\parallel}} \overset{H^+}{\rightleftarrows} \underset{CH_3-C-CH_3}{\overset{\overset{+}{O}H}{\parallel}} \overset{-H^+}{\rightleftarrows} \underset{CH_3-C=CH_2}{\overset{OH}{\mid}}$$

The enol then reacts with the electrophile and the resulting adduct tautomerizes to the amine salt,

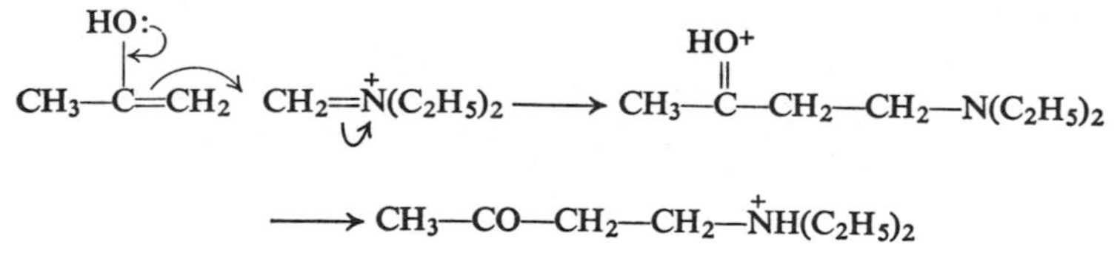

$$CH_3-\overset{\overset{\displaystyle HO:}{\Large\frown}}{C}=CH_2 \quad CH_2=\overset{+}{\underset{\curvearrowleft}{N}}(C_2H_5)_2 \longrightarrow CH_3-\overset{\overset{\displaystyle HO^+}{\|}}{C}-CH_2-CH_2-N(C_2H_5)_2$$

$$\longrightarrow CH_3-CO-CH_2-CH_2-\overset{+}{N}H(C_2H_5)_2$$

(b) STRUCTURAL REQUISITES IN THE REACTANTS

The amine may be primary or secondary. In the former case, the product, a secondary amine, usually reacts further, e.g.

$$RNH_2 + CH_2O + R'COCH_3 \xrightarrow[-H_2O]{H^+} RNHCH_2CH_2COR' \xrightarrow[-H_2O]{CH_2O + R'COCH_3}$$

$$RN(CH_2CH_2COR')_2$$

Ammonia may also be employed, usually resulting in three successive reactions, e.g.

$$NH_3 + 3\ CH_2O + 3\ CH_3COPh \xrightarrow[-3\ H_2O]{H^+} N(CH_2CH_2COPh)_3$$

The aldehyde is commonly formaldehyde, but higher aldehydes have been used successfully and several examples are given later.

The third reactant may be an enol, a related compound, or any compound capable of undergoing enolization in the presence of acid. The more important classes of compound are the following; each is illustrated by the reaction of a typical member with formaldehyde and dimethylamine.

Aldehydes,

$$(CH_3)_2CHCHO \longrightarrow (CH_3)_2\underset{\underset{\displaystyle CH_2N(CH_3)_2}{|}}{C}-CHO$$

Ketones,

$$PhCOCH_3 \longrightarrow PhCOCH_2CH_2N(CH_3)_2$$

β-Dibasic acids, β-cyano-acids, β-keto-acids, etc.,

$$CH_3CH(CO_2H)_2 \longrightarrow CH_3\underset{\underset{\displaystyle CH_2N(CH_3)_2}{|}}{C}(CO_2H)_2$$

α-Picoline and related compounds (see p. 273),

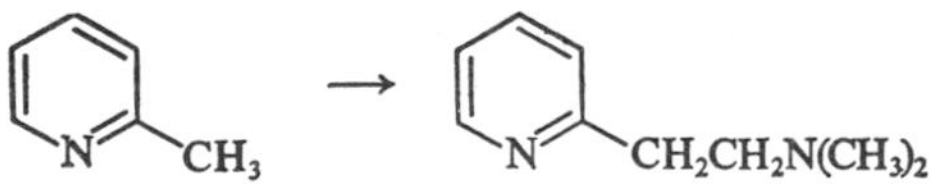

Phenols, *

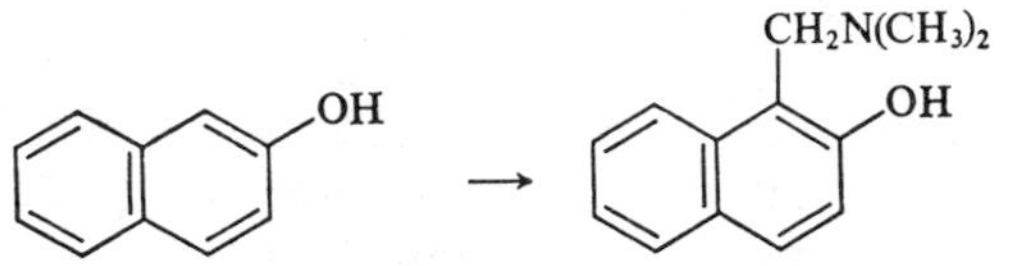

Furan, pyrrole, indole, and derivatives, *

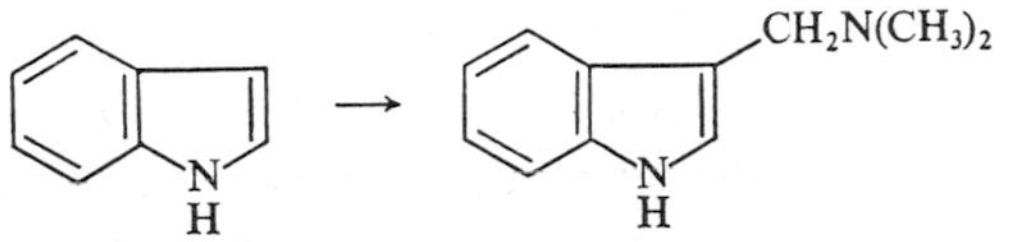

Gramine

(These last-named compounds are sensitive to acids (p. 381) and the Mannich reaction must be carried out in a weak acid, usually acetic acid, at a low temperature.)

Phenylacetylene and some of its nuclear-substituted derivatives, although not enolic, also react, e.g.

$$PhC{\equiv}CH + CH_2O + (CH_3)_2NH \xrightarrow[-H_2O]{H^+} PhC{\equiv}CCH_2N(CH_3)_2$$

A compound which possesses two enolic or potentially enolic hydrogen atoms can undergo a second Mannich reaction. The extent to which this occurs when only one mole of each of the other reactants is employed varies with the structure of the compound: e.g. whereas acetone, formaldehyde, and diethylamine give 1-diethylamino-3-butanone contaminated by only a trace of the *bis*-derivative, phenol reacts at all three of its activated nuclear positions (*cf.* its reaction with bromine, p. 411):

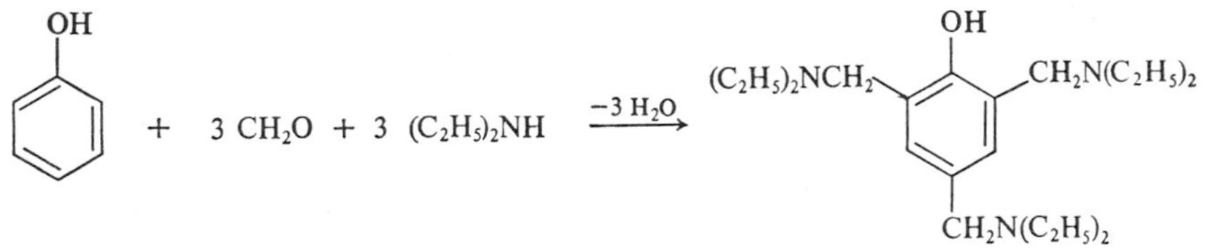

(*c*) MANNICH BASES AS INTERMEDIATES IN SYNTHESIS

The facts that the Mannich reaction is applicable to a wide range of compounds, is usually efficient, and gives products (Mannich bases) which undergo a number

*The mechanisms of reactions of electrophiles with phenols, furan, pyrrole, indole, and related compounds, and the reasons for the preferred positions of reactivity in the nucleus, are discussed in Chapter 11.

of types of transformation, give the process extensive applications in synthesis. The following are the more important uses.

(i) Formation of αβ-unsaturated carbonyl compounds. The Mannich reaction normally gives the hydrochloride of the Mannich base (p. 275). These salts are usually stable at room temperature but those derived from aliphatic compounds eliminate an amine hydrochloride on being heated, e.g.

$$RCOCH_2CH_2NH(CH_3)_2^+Cl^- \longrightarrow RCOCH=CH_2 + (CH_3)_2NH_2^+Cl^-$$

The reaction is similar to the Hofmann elimination of a quaternary ammonium salt (p. 121), but whereas the latter process requires a strong base (hydroxide ion), Mannich base hydrochlorides undergo elimination readily because the double bond which is generated is conjugated with a second unsaturated group.

Eliminations of this type have two uses. First, reduction of the olefinic double bond (e.g. catalytically; p. 611) gives the next highest homologue of the ketone from which the Mannich base was derived:

$$RCOCH_3 \xrightarrow[\text{reaction}]{\text{Mannich}} RCOCH_2CH_2NHR_2^+Cl^- \xrightarrow{\text{heat}} RCOCH=CH_2$$

$$\xrightarrow{\text{reduction}} RCOCH_2CH_3$$

Secondly, the quaternary salts of Mannich bases are latent sources of αβ unsaturated carbonyl compounds required for condensation reactions. Thus, for a base-catalyzed condensation utilizing methyl vinyl ketone, it is better to employ the quaternized Mannich base from acetone, from which the unsaturated ketone is generated *in situ* by the action of base, than to use the free ketone, since this readily polymerizes. The best known example of this application is Robinson's ring-extension, used in building up the ring system of steroids (see also pp. 718, 726):

$$CH_3COCH_2\overset{+}{N}(CH_3)_3 \xrightarrow{EtO^-} CH_3COCH=CH_2$$

(ii) Replacement of the amino group. The Mannich bases derived from aromatic systems such as phenols and indoles are benzylic-type compounds and, as such, are particularly susceptible to nucleophilic displacements (S_N2 reactions). Advantage can be taken of this to replace the amino group by other functions.

For this purpose they are first quaternized, for a tertiary amine is a more labile leaving-group than an amide ion.

(1) An alternative to the synthesis of heteroauxin (indole-β-acetic acid) described previously (p. 207) employs the Mannich base from indole (gramine). Methylation with dimethyl sulphate followed by treatment with cyanide ion and then hydrolysis gives heteroauxin:

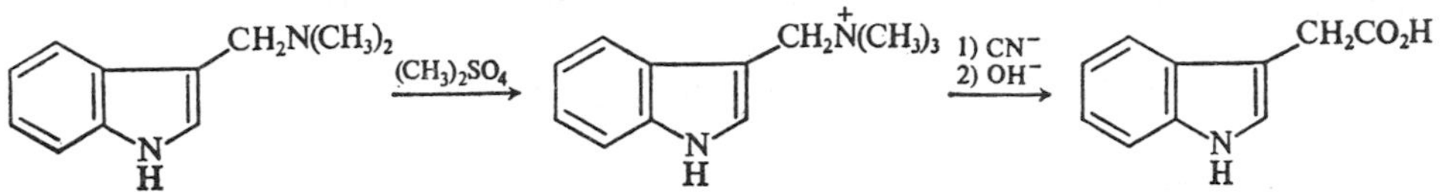

$$\text{(indole-CH}_2\text{N(CH}_3)_2) \xrightarrow{(CH_3)_2SO_4} \text{(indole-CH}_2\overset{+}{N}(CH_3)_3) \xrightarrow[2) \ OH^-]{1) \ CN^-} \text{(indole-CH}_2CO_2H)$$

(2) Gramine is also the starting material in a synthesis of tryptophan, one of the α-amino-acids contained in proteins. The quaternized compound is treated with acetamidomalonic ester (p. 352) in the presence of base; the malonate carbanion displaces on the quaternary salt to give a compound which is readily hydrolyzed and decarboxylated to tryptophan:

$$CH_3CONH-CH(CO_2Et)_2 \underset{EtO^-}{\rightleftharpoons} CH_3CONH-\bar{C}(CO_2Et)_2$$

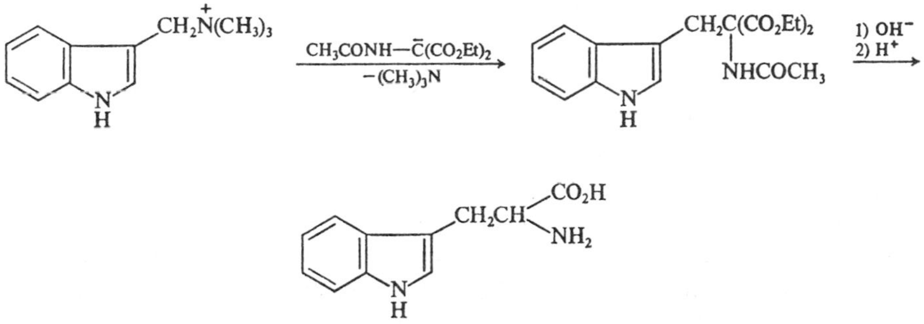

$$\text{(indole-CH}_2\overset{+}{N}(CH_3)_3) \xrightarrow[-(CH_3)_3N]{CH_3CONH-\bar{C}(CO_2Et)_2} \text{(indole-CH}_2C(CO_2Et)_2 \ / \ NHCOCH_3) \xrightarrow[2) \ H^+]{1) \ OH^-}$$

$$\text{(indole-CH}_2CH(CO_2H)(NH_2))$$

Tryptophan

(3) Benzylic systems are also susceptible to hydrogenolysis,

$$ArCH_2X \xrightarrow{2 \ H} ArCH_3 + HX$$

where X is an oxygen or nitrogen function (p. 625). This characteristic was applied in the synthesis of quinine to introduce a methyl group into the 8-position of 7-hydroxyisoquinoline *via* a Mannich reaction (21.6). 7-Hydroxyisoquinoline was converted into a Mannich base with formaldehyde and piperidine (reaction occurring at the 8-position just as β-naphthol reacts preferentially at the 1-position, p. 379). Reduction of the 8-methyl-derivative was brought about by sodium methoxide at a high temperature (hydride-ion transfer):

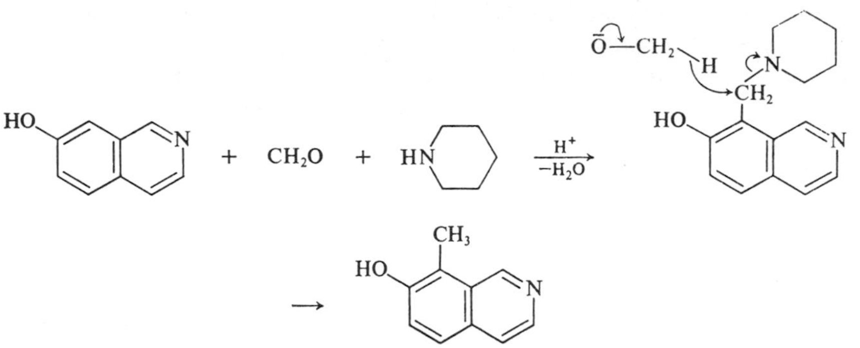

(iii) *The use of aldehydes other than formaldehyde: alkaloid synthesis.* A number of elegant syntheses of alkaloids have been based on the Mannich reaction, perhaps the most outstanding being Robinson's (1917) synthesis of tropinone, required for the synthesis of atropine. A mixture of succindialdehyde, methylamine, and the calcium salt of acetonedicarboxylic acid, on standing for several days at pH 5–7, gave tropinone in 40% yield. *

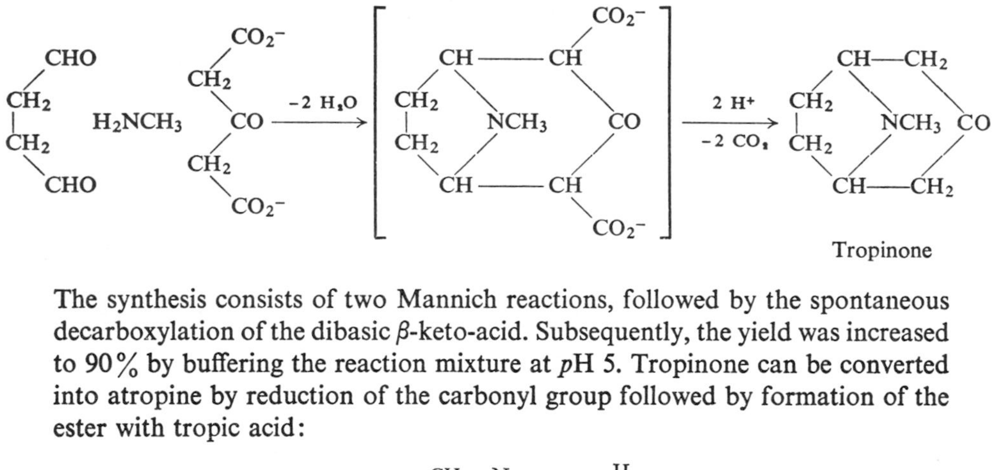

The synthesis consists of two Mannich reactions, followed by the spontaneous decarboxylation of the dibasic β-keto-acid. Subsequently, the yield was increased to 90% by buffering the reaction mixture at pH 5. Tropinone can be converted into atropine by reduction of the carbonyl group followed by formation of the ester with tropic acid:

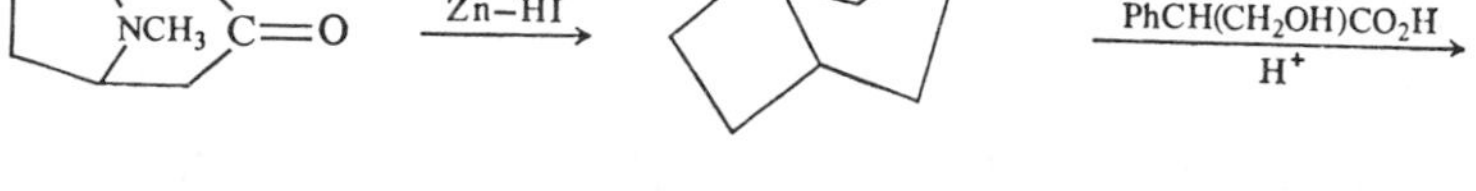

*Robinson's synthesis was inspired by his consideration of the possible ways in which alkaloids such as atropine and cocaine may arise naturally. His successful synthesis of atropine in the mild physiological conditions of pH and temperature was the first example of a probable biosynthetic process achieved in the laboratory. It has since become apparent, by the use of radioactive tracer techniques, that Mannich-type reactions play a central role in the formation in plants of the heterocyclic systems of several groups of alkaloids.

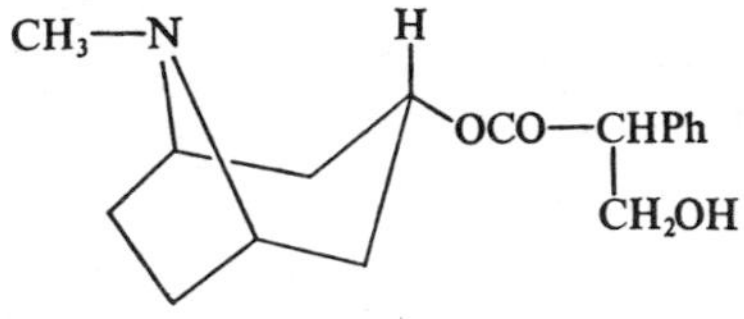

Atropine

This synthesis has been adapted to the syntheses of cocaine (from coca leaves),

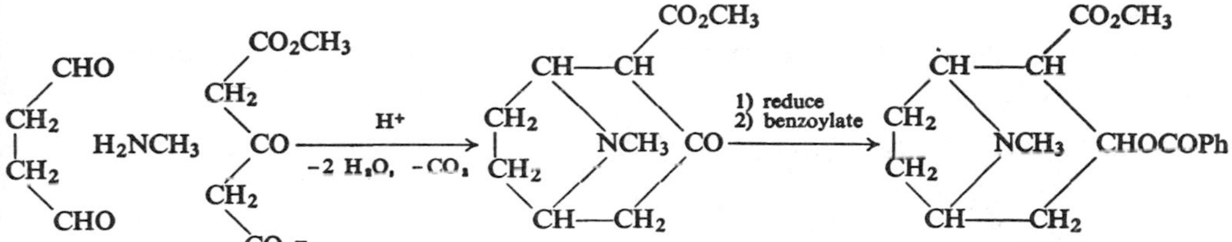

and pseudopelletierine (58–68%) (an alkaloid obtained from the root bark of the pomegranate tree) [4],

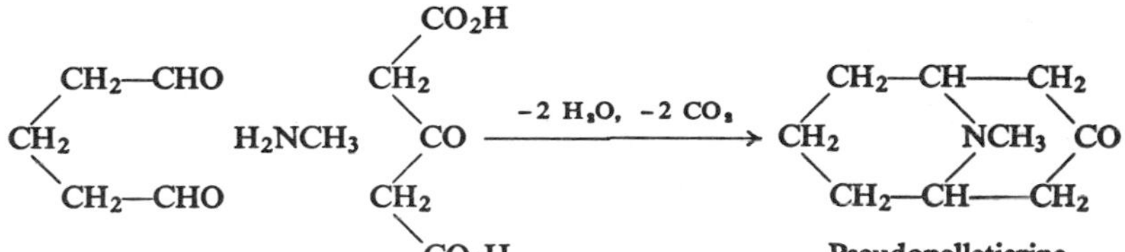

(iv) *Local anaesthetics.* Both atropine and cocaine have certain anaesthetic properties, and an examination of a range of such compounds has indicated that the critical arrangement of atoms which is associated with local anaesthetic action is

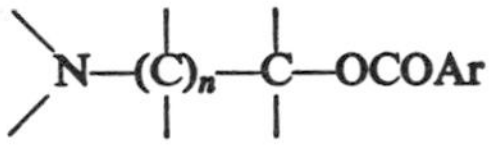

$$\begin{array}{c}\diagdown\\ \diagup\end{array}N-(\overset{|}{\underset{|}{C}})_n-\overset{|}{\underset{|}{C}}-OCOAr$$

i.e. an amino function connected by a chain of carbon atoms to an alcoholic function which is esterified by an aromatic acid. These structures, where $n = 2$, can be obtained by Mannich reactions on ketones followed by reduction of the carbonyl group and esterification of the resulting alcohol, and compounds whose anaesthetic properties are as good as those of cocaine but which are less toxic have been made in this way. Tutocaine is a typical example:

$$CH_3COCH_2CH_3 + CH_2O + (CH_3)_2NH \xrightarrow{-H_2O} CH_3COCHCH_2N(CH_3)_2$$
$$\underset{\displaystyle CH_3}{|}$$

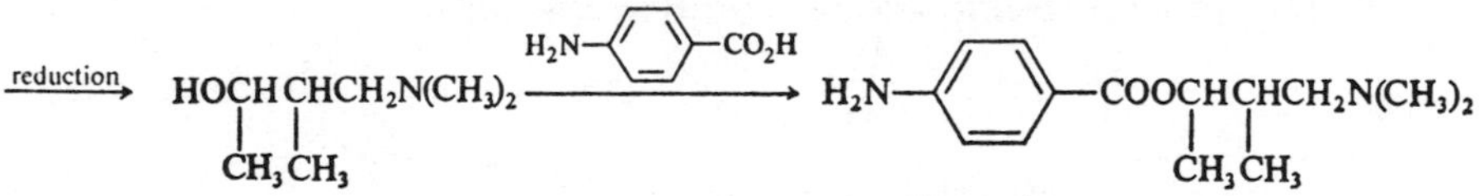

$$\xrightarrow{\text{reduction}} \underset{\underset{CH_3 CH_3}{|\quad|}}{HOCHCHCH_2N(CH_3)_2} \xrightarrow{\;H_2N-\underset{}{\bigcirc}-CO_2H\;} H_2N-\bigcirc-\underset{\underset{CH_3 CH_3}{|\quad|}}{COOCHCHCH_2N(CH_3)_2}$$

Tutocaine

Further Reading
BLICKE, F. F., 'The Mannich reaction', *Organic Reactions*, 1942, **1**, 303.

Problems
1. Write the structures of the products you would expect from the following reactions:

    (a) $CH_3CH{=}CH_2 + (CH_3)_3CCl \xrightarrow{\;AlCl_3\;}$

    (b) $CH_3CH{=}CH_2 + CH_2O \xrightarrow{\;H_2O{-}H^+\;}$

    (c) $CH_3COCH_3 + CH_2O \xrightarrow{\;H^+\;}$

    (d) $2\ CH_3CH_2COCH_3 \xrightarrow{\;BF_3\;}$

    (e) $CH_3CH_2COCH_3 + CH_2O + HN(CH_3)_2 \xrightarrow{\;H^+\;}$

    (f)
$+ PhCOCH_3 \xrightarrow{\;HCl\;}$

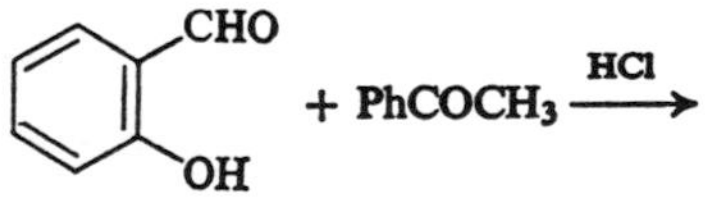

2. Outline synthetic methods for the following:

    (a) $CH_3CH{-}CH(CH_3){-}CH_2OH$ (b) $CH_3COCH(CH_3)CH_2NEt_2$
            $\overset{\displaystyle |}{OH}$

    (c) $CH_3COCHPhCOCH_3$ (d) $CH_3CHCH_2NMe_2$
                                        $\overset{\displaystyle |}{CO_2H}$

    (e) 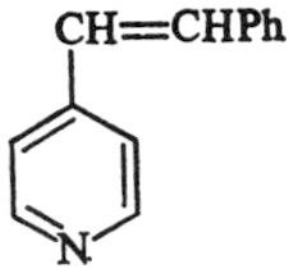
    (f)

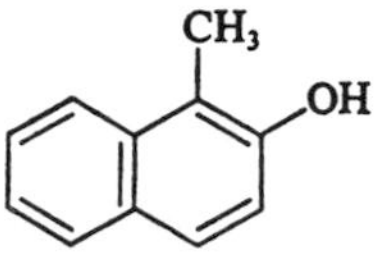

    (g)

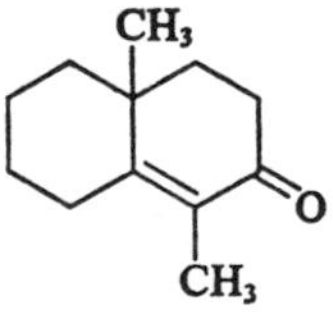

3. Formulate mechanisms for the following reactions:

(i) $(CH_3)_2C$=$CHCH_2CH_2C(CH_3)$=$CHCHO$ $\xrightarrow{H_2O-H^+}$

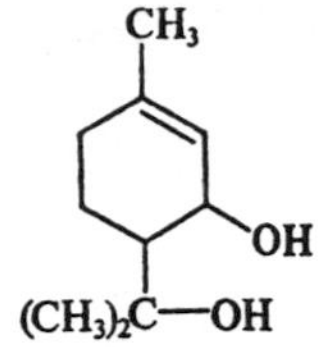

(ii) $(CH_3)_2C$=$CHCH_2CH_2C(CH_3)_2OH$ $\xrightarrow{H^+}$

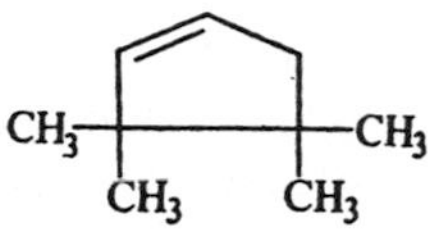

(iii) $(CH_3)_2C$——$CHCH_2CH_2C(CH_3)$=CH_2 $\xrightarrow[2)\ H_2O]{1)\ BF_3}$

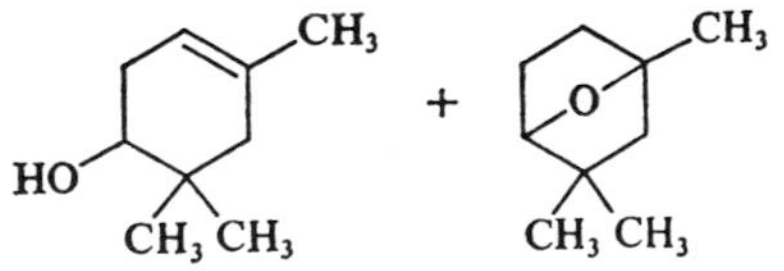

9. Pericyclic Reactions

9.1 Principles

Pericyclic reactions are those in which the bond changes which occur in the conversion of reactants into products do so in a concerted manner within a cyclic array of the participating atomic centres. The reactions fall into a number of groups, which were summarized in Section 4.8. The majority lead to the formation of at least one new C—C bond. One of the most widely used is the Diels-Alder reaction, in the group of cycloadditions, e.g.

The curly-arrow symbolism used above is strictly not appropriate to these reactions, as will become apparent from the discussion of the courses of the reactions. Essentially, the processes involve the interconversion of σ- and π-bonds within cyclic transition states and without the mediation of ions, radicals, or any other intermediates. Nevertheless, curly-arrow representation can still provide a useful check that valency-rules are satisfied.

Other characteristics of the reactions are as follows. (1) They have negative entropies of activation. These are particularly large for cycloadditions, consistent both with the loss of translational entropy when two molecules come together to form the transition state, and also with the high degree of ordering which corresponds to the mutual orientation of several atomic centres (four in the example above); however, the activation enthalpy in these reactions is often small, so that the rates are large even at moderate temperatures. (2) They are usually stereospecific. (3) They are reversible. With cycloadditions, in which two molecules react to give one, the reverse reaction can often be brought about at high temperatures (i.e. as the $T\Delta S$ term becomes increasingly important). With electrocyclic ring-closures, in which one reactant molecule gives one of product, temperature has less effect on the position of equilibrium; the preferred position is determined by enthalpy factors such as the strain in a small ring and the differences in conjugation between reactant and product.

It is only recently (since 1965) that the mechanism and stereochemistry of pericyclic reactions have been satisfactorily rationalized. The theory followed from the recognition that the reactions have orbital-symmetry requirements; and it has proved as successful in explaining why certain reactions which appear reasonable on paper do not occur as in accounting for the courses of those that do occur. The theory has been developed in several forms, of which the frontier-orbital treatment is the one used here.

Frontier orbitals and orbital symmetry. Just as the outer shell of electrons of an atom is regarded as especially significant in determining the chemistry of that atom, so it is reasonable that, for a molecule, it is the highest occupied molecular orbital (HOMO) which is the key to determining reactivity. This is termed the *frontier orbital.*

Since the ground state of almost all molecules has a pair of electrons in the HOMO, bonding interaction between two molecules* cannot involve only the HOMO of each for this would lead to an orbital occupancy greater than two, in contravention of Pauli's principle. The HOMO of one reactant needs therefore to interact with an unoccupied MO of the second. Now, since the bonding interaction between two orbitals increases as the energies of the two become more nearly equal, it is expected that the HOMO of one reactant should interact efficiently with the lowest unoccupied molecular orbital (LUMO) of the second. This is one tenet of the theory.

The second tenet is that simple orbital overlap is not a sufficient condition for reaction. In addition, the two orbitals which are to overlap must have the same phase (or sign); overlap of orbitals of unlike phase results in repulsion.

Consider ethylene. The HOMO is that which results from overlap of carbon $2p$ atomic orbitals of like phase (p. 39); the LUMO is constructed from these two orbitals with opposite phases:†

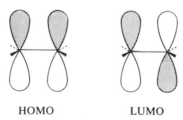

HOMO   LUMO

The HOMO and the LUMO differ in symmetry with respect to reflection in a plane which bisects the olefin perpendicularly to the plane of the molecule: the HOMO is symmetric with respect to this mirror plane and the LUMO is correspondingly antisymmetric.

---

*This discussion also applies when reaction occurs between two centres in the same molecule.
†For clarity, lobes of opposite phase are shown respectively dark and clear.

Consider now the hypothetical cycloaddition of two molecules of ethylene to give cyclobutane:

$$2\ CH_2{=}CH_2 \longrightarrow \begin{array}{c} CH_2{-}CH_2 \\ |\qquad | \\ CH_2{-}CH_2 \end{array}$$

If the molecules were to approach each other with their molecular planes parallel, the LUMO of one molecule and the HOMO of the other would interact as follows:

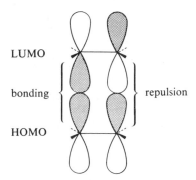

A concerted reaction cannot occur; it is described as *symmetry-forbidden*. Only one of the two new C—C bonds could be formed, and the bonding energy associated with this, when offset against the loss of π-bonding of two ethylene molecules, serves to make this (two-step) process of such high activation energy that it is effectively impossible.

The approach of the molecules towards each other described above is such that one surface of the π-orbital of one molecule interacts with one surface of the π-orbital of the other. The interaction is described as *suprafacial* with respect to each reactant. However, there is in principle an alternative geometry of approach, namely one in which the molecular planes are perpendicular, so that one surface of the π-orbital of one molecule interacts with *both* surfaces of the π-orbital of the other:

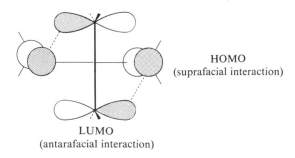

The interaction is described as suprafacial with respect to the former component and *antarafacial* with respect to the latter. In principle, bonding interaction could occur between the two pairs of lobes of like phase, as indicated ($\cdots$), but for this to be significant, the nearer olefinic molecule would need to twist about its original π-bond. The energy required for this is evidently so much greater than that which would be gained through the partial formation of two new C—C bonds at the transition state that the reaction does not occur. That is, the process, though symmetry-allowed, is sterically inaccessible.

Consider next the reaction between an olefin and a 1,3-diene. The π-molecular orbitals of the diene are formed by combination of the carbon $2p$ orbitals:*

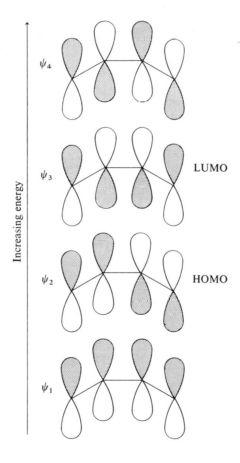

*In constructing π-MOs for dienes and polyenes, it is useful to remember that the lowest-energy MO has no nodes, the next has one, the next two, and so on. In this case, the nodes are at the middle of the central C—C bond (ψ_2), the middles of the two terminal bonds (ψ_3), and the middles of all three bonds (ψ_4).

It is apparent that the symmetries of the HOMO of the diene and the LUMO of ethylene are such that, when the reactants approach each other with their molecular planes parallel, two new C—C bonds can be formed at the same time:*

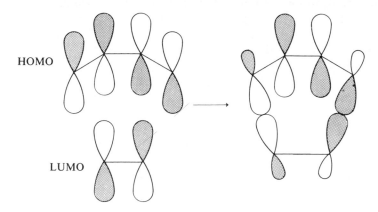

HOMO

LUMO

Consequently, two new C—C bonds are partially formed at the transition state, and when the bonding energy associated with these is set against that lost as two C—C π-bonds are partially destroyed, the activation energy for this concerted reaction is relatively low.

9.2 Cycloadditions

(a) DIELS-ALDER REACTION

The reaction between butadiene and ethylene described above is the simplest example of a general procedure for forming six-membered rings, developed by Diels and Alder, in which a conjugated diene reacts with a compound containing a carbon-carbon double or triple bond (a dienophile). The reactions are described as [4 + 2] cycloadditions, the number denoting the numbers of interacting π-electrons in each component. They are normally brought about by heating the components alone or in an inert solvent, the temperature required depending on the structures of the reactants.

(i) *The dienophile.* Although the Diels-Alder reaction between butadiene and ethylene is successful (in low yield), it is generally unsatisfactory, or fails completely, with other simple olefins or acetylenes. However, when the unsaturated bond of the dienophile is conjugated to a group of $-M$ type, such as carbonyl, nitro, or cyano, reaction occurs under milder conditions and normally gives good yields. It has been suggested that this is because the substituent lowers the energy of the LUMO of the dienophile, thereby bringing it closer in energy to the HOMO of the diene and increasing the bonding interaction in the transition state. Sometimes greater reactivity can be achieved by carrying out the reaction

*The same conclusion follows if the LUMO of the diene and the HOMO of ethylene are considered.

in the presence of a Lewis acid such as BF_3, which complexes with the $-M$ substituent and so serves to withdraw electrons even more strongly; such reactions can occur at room temperature or below. The following reactions of butadiene with various dienophiles are illustrative (the solvent, if any, is given in parentheses):

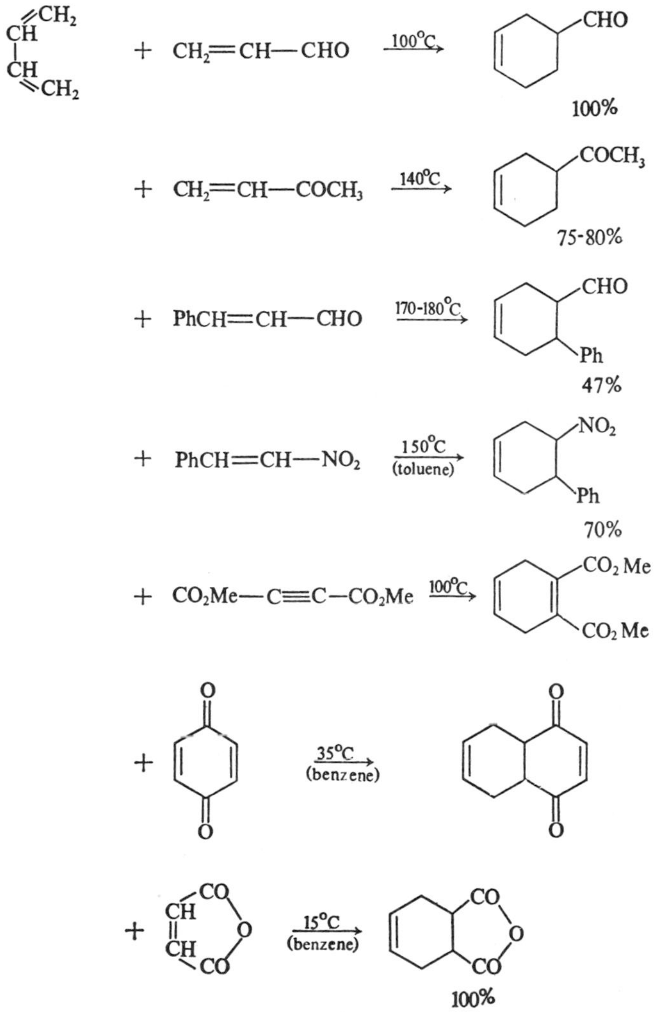

Olefins containing electron-attracting groups of $-I$ type (e.g. chlorine) are in general less satisfactory dienophiles, although some dienes react to give moderate yields, e.g.

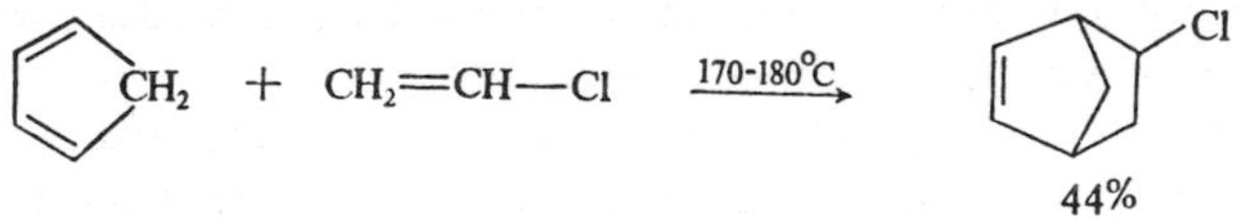

44%

Acetylene itself also reacts with cyclopentadiene to give norbornadiene which rearranges at high temperatures to tropilidene (cycloheptatriene):

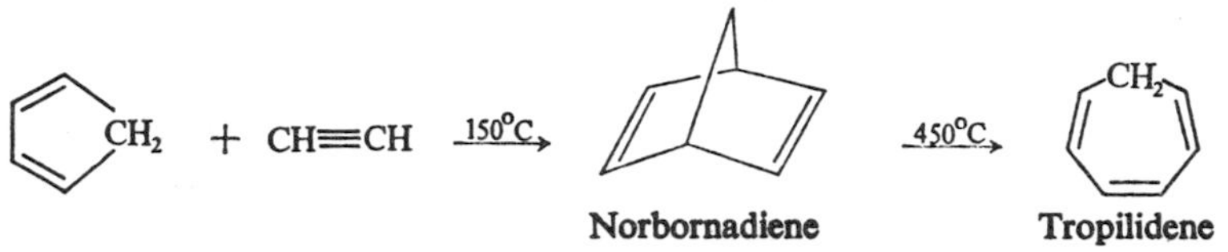

Norbornadiene   Tropilidene

(*ii*) *The diene.* In order for reaction to occur, the diene must be capable of achieving the *s-cis* conformation which the formation of a six-membered ring requires. This is always possible for acyclic dienes, but it can result in wide variations in reactivity. Thus, *cis*-1-substituted butadienes are less reactive than their *trans*-isomers because the substituent in the former suffers steric crowding in the *s-cis* conformation which raises the activation energy:

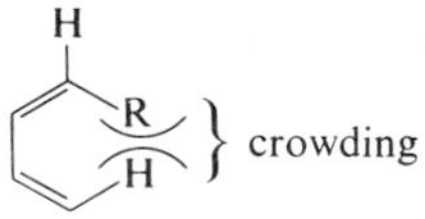

Cyclic dienes react only if they are of *cis* type; for example,

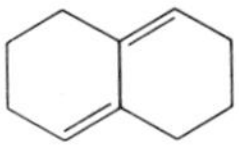

fails to react.

The reactivity of the diene is increased by electron-releasing substituents. It is thought that this is because they raise the energy of the diene's HOMO and so make it more compatible with the dienophile's LUMO.

When both diene and dienophile are substituted, more than one product can be formed, e.g.

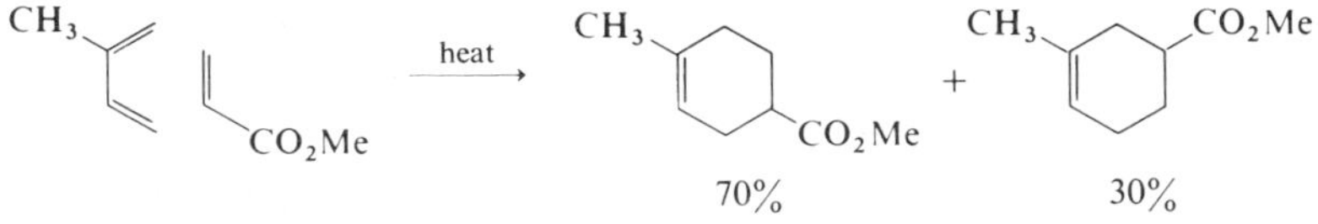

70%   30%

The reason for the preferential formation of a particular product is not clear. It may be that when one substituent is electron-donating and the other is electron-attracting, there is a stabilizing interaction between the two in the transition states which lead to 1,2- or 1,4-substituted products, but not 1,3-, which may be likened to that which occurs across aromatic rings.

Conjugated systems which are formally aromatic can also react. Benzene itself is usually inert, but certain other aromatic compounds react fairly readily. For example, anthracene reacts with maleic anhydride almost quantitatively at 80°C:

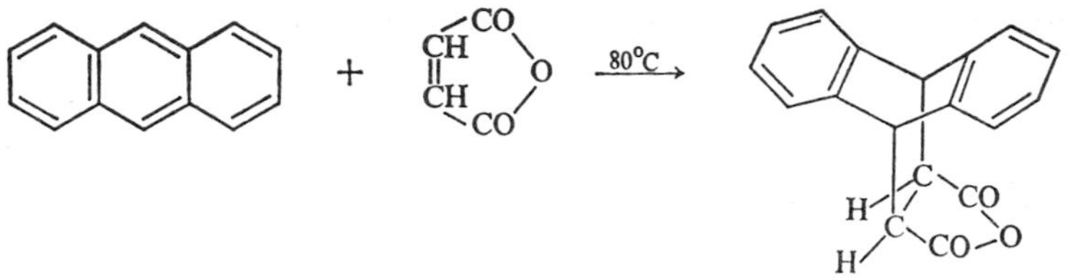

The difference in the behaviour of benzene and anthracene stems from the fact that addition to benzene would give a non-conjugated product, with a corresponding loss of aromatic stabilization energy of 150 kJ mol^{-1}, whereas addition at the 9,10-positions of anthracene gives a dibenzenoid derivative, the loss of aromatic stabilization energy being only $(349 - 2 \times 150)$, i.e. about 50 kJ mol^{-1} (cf. p. 57).

Other aromatic compounds take part in Diels-Alder reactions provided that the loss of stabilization energy is not too great. The best known monocyclic compound which acts as a diene is furan, e.g.

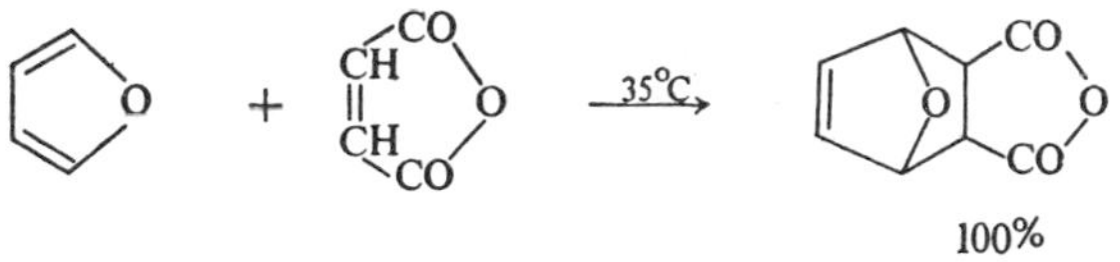

100%

whereas the more stabilized aromatic, thiophen, is inert. Pyrrole reacts with maleic anhydride, but not in the Diels-Alder manner (p. 381).

Diels-Alder reactions on furans have a particular use in synthesis because the resulting bicyclic ethers can be hydrolyzed to diols which are dehydrated to aromatic compounds which might otherwise be difficult to obtain, e.g.

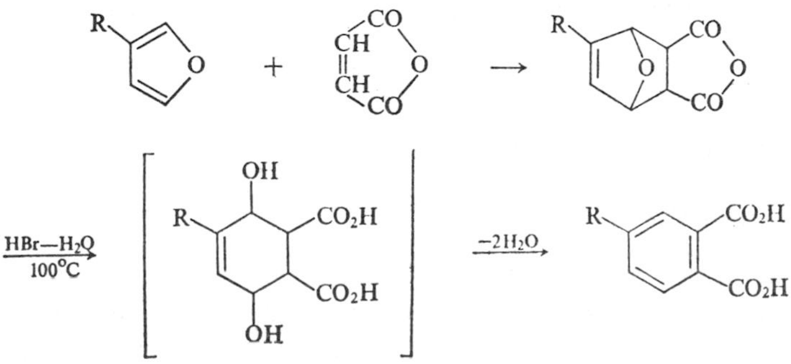

(*iii*) *Stereochemistry*. The cycloadditions are stereospecific. For example, dimethyl maleate and dimethyl fumarate react with butadiene to give, respectively, *cis*- and *trans*-dimethyl cyclohexene-4,5-dicarboxylates:

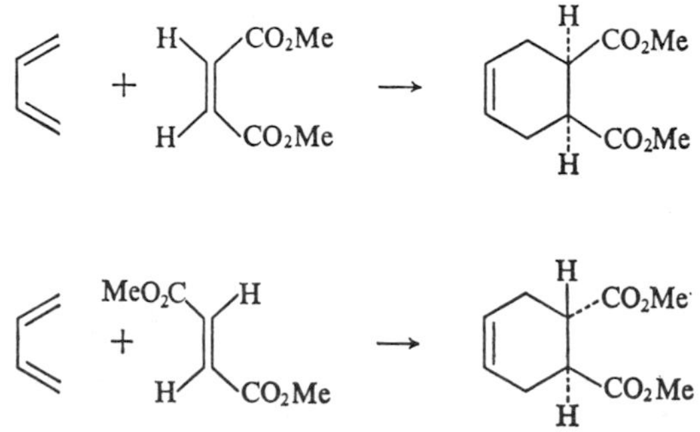

This is a result of the concerted nature of the reaction. The diene and dienophile can only achieve appropriate orbital overlap if they come together with their molecular planes (nearly) parallel; the orbital interaction is suprafacial with respect to both reactants. As the sp^2-hybridized carbon atoms of the dienophile undergo simultaneous rehybridization to sp^3, the relative positions of the substituents remain the same.

A further stereochemical factor applies when the diene is cyclic. For example, in the dimerization of cyclopentadiene, where one molecule acts as diene and the other as dienophile, two orientations in the product are possible, each of which arises from suprafacial-suprafacial interaction of the reactants:

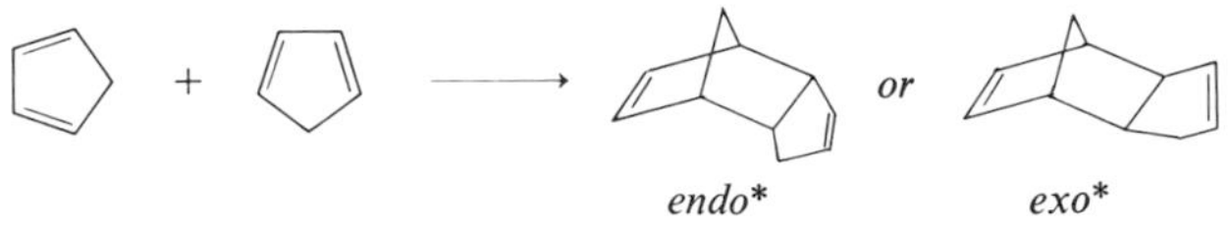

In practice, in this and most other cases the thermodynamically less stable *endo* product predominates. This has been accounted for in terms of a secondary interaction of orbitals in the transition state for *endo*-addition which is sterically impossible for *exo*-addition:

*The term *endo* indicates that the substituent is directed to the *inside* of the boat-shaped cyclohexene ring; *exo* indicates that the substituent is directed to *outside* of the the ring.

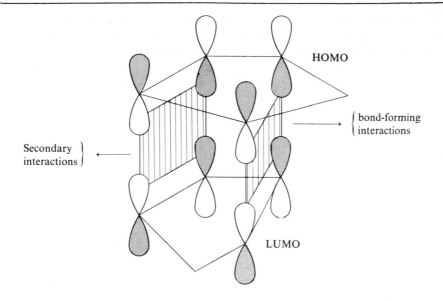

Transition state for
endo addition

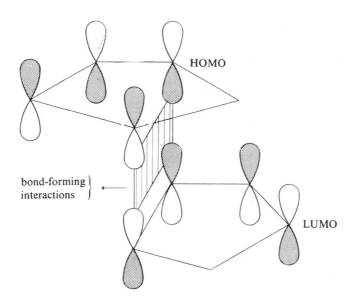

Transition state for
exo addition

This secondary interaction does not lead to bonding in the product, but, because the overlapping orbitals are of like phase, it can stabilize the transition state.

Since, usually, the *exo* product is thermodynamically more stable, it is sometimes possible to form it by the use of conditions under which formation of the kinetically favoured adduct is reversible. The reaction of furan with maleimide is an example:

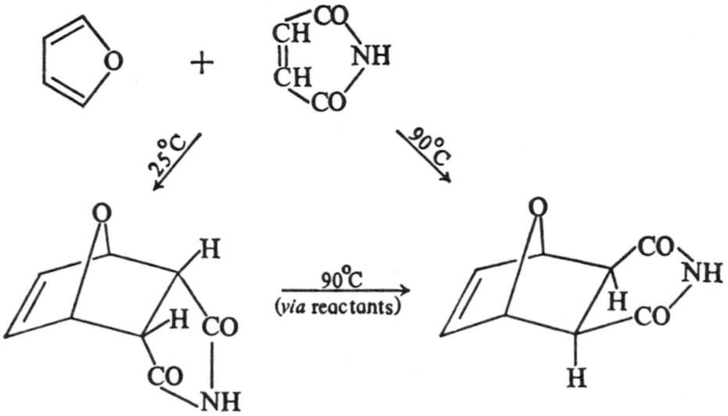

The less stable *endo* isomer is formed the faster and predominates at 25°C, where the reaction is effectively irreversible; at 90°C, this product is in fairly rapid equilibrium with the reactants, and the less rapidly formed but more stable *exo* isomer gradually accumulates.

(e) APPLICATIONS

The Diels-Alder reaction is of very great importance in the synthesis of naturally occurring compounds, partly because of the large number of structures to which it is applicable and partly because of its stereospecificity. By the appropriate choice of starting materials, it is often possible in one step to assemble a product with several groupings in the required stereochemical configuration. For example, the reaction between *trans*-vinylacrylic acid and benzoquinone gives a product with both a *cis* ring-junction and a *trans* relationship between the hydrogens at the ring-junction and the carboxyl group:

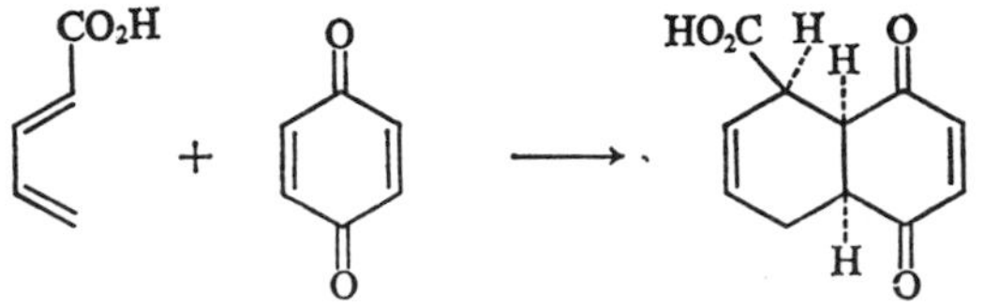

This particular stereochemical arrangement was required for the synthesis of the alkaloid, reserpine (21.7).

Other examples of Diels-Alder reactions in the synthesis of natural products are:

(1) The reaction between butadiene and 2-methoxy-5-methylbenzoquinone, used to establish the C and D rings in a synthesis of cholesterol (21.3). This illustrates the use of the Diels-Alder reaction to introduce an angular methyl group, a recurring characteristic of steroids and related compounds. The reaction gives, as usual, a *cis* ring-junction, but the required (more stable) *trans* compound is readily obtained since the hydrogen atom at the point of fusion of the rings, being adjacent to carbonyl, is enolizable.

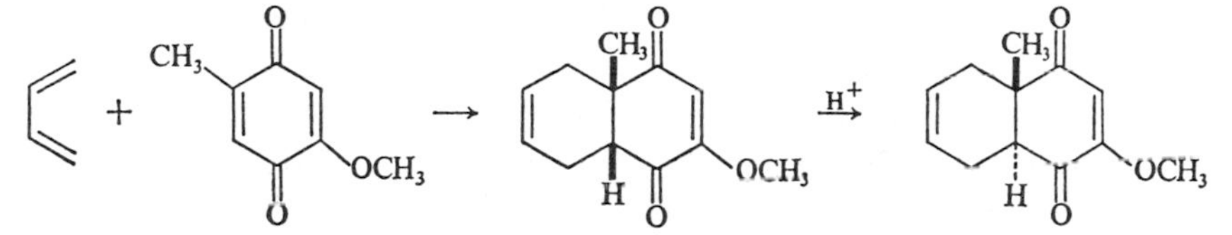

(2) The reaction between furan and dimethyl acetylenedicarboxylate, in a synthesis of cantharidin (from the body fluid of cantharides beetles such as the Spanish fly). Catalytic hydrogenation of the product gave a new dienophile which

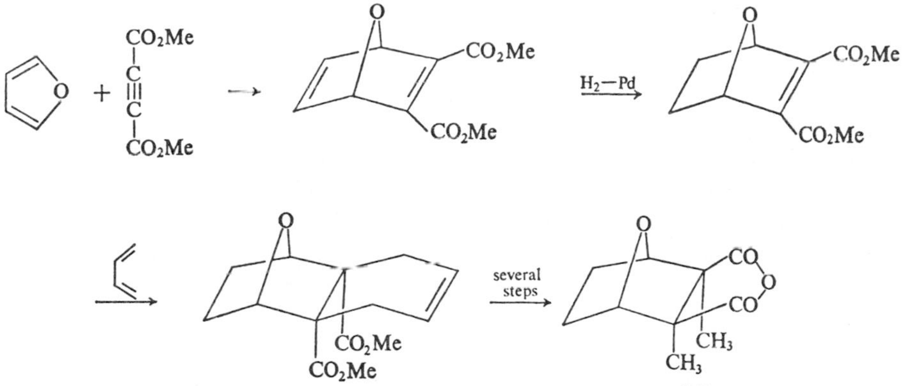

Cantharidin

reacted with butadiene in a second Diels-Alder reaction.* The required stereochemistry having been established, the carbomethoxy groups were converted into methyl and the ring derived from butadiene was oxidized in several steps to the required anhydride.

---

*The stereochemistry of this reaction results from the fact that addition of butadiene is less hindered by the oxide bridge than by the ethylene bridge.

(b) 1,3-DIPOLAR ADDITIONS

A large number of types of five-membered rings may be formed by the reaction of '1,3-dipolar' compounds with unsaturated bonds (dipolarophiles) such as $C=C$, $C\equiv C$, $C=O$, and $C\equiv N$. The 1,3-dipolar components are compounds whose representation requires ionic structures which include ones with charges on atoms bearing a 1,3-relationship, e.g.

diazomethane $CH_2 = \overset{+}{N} = \overset{-}{N} \leftrightarrow \overset{-}{C}H_2 - \overset{+}{N} \equiv N$

benzonitrile oxide $Ph - C \equiv \overset{+}{N} - \overset{-}{O} \leftrightarrow Ph - \overset{-}{C} = \overset{+}{N} = O$

The 1,3-dipole is merely a structural variant of the diene component in the Diels-Alder reaction; in the dipolar compound, four π-electrons are distributed over three atoms instead of the four in a diene. Moreover, the HOMO and LUMO of a 1,3-dipole are of similar symmetry to those of a diene with respect to the two-fold axis and to the mirror plane which bisects the molecule:

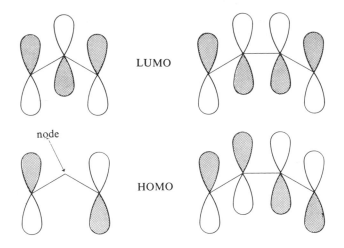

Consequently, concerted cycloaddition to an olefin is symmetry-allowed.

Unlike dienes which are stable, isolable molecules, 1,3-dipoles are often short-lived species which must necessarily be generated *in situ* and trapped as soon as they are formed; even those which are isolable, such as diazoalkanes, $R - \overset{-}{C}H - \overset{+}{N} \equiv N$, and azides, $R - \overset{+}{N} - \overset{-}{N} \equiv \overset{+}{N}$, often tend to be thermally labile materials, best handled in solution and used immediately after synthesis. In general, the stabler dipoles are those where the central atom is other than carbon and possesses a lone pair of electrons. This increases the number of canonical structures which may represent the dipole, e.g.

Diazoalkanes: $\bar{R}\overset{..}{CH}-\overset{..}{N}=\overset{+}{N} \longleftrightarrow RCH=\overset{+}{N}=\bar{N} \longleftrightarrow$

$\quad\quad\quad\quad \bar{R}\overset{+}{CH}-\overset{+}{N}\equiv N \longleftrightarrow R\overset{+}{CH}-\overset{..}{N}=\bar{N} \longleftrightarrow RCH=\overset{..}{N}-\overset{..}{N}:$

Ketocarbenes: $\overset{+}{R}\overset{..}{C}=CR-\bar{O} \longleftrightarrow R\overset{..}{C}-CR=O$

The following reactions illustrate those which are used for preparing 1,3-dipolar compounds *in situ*:

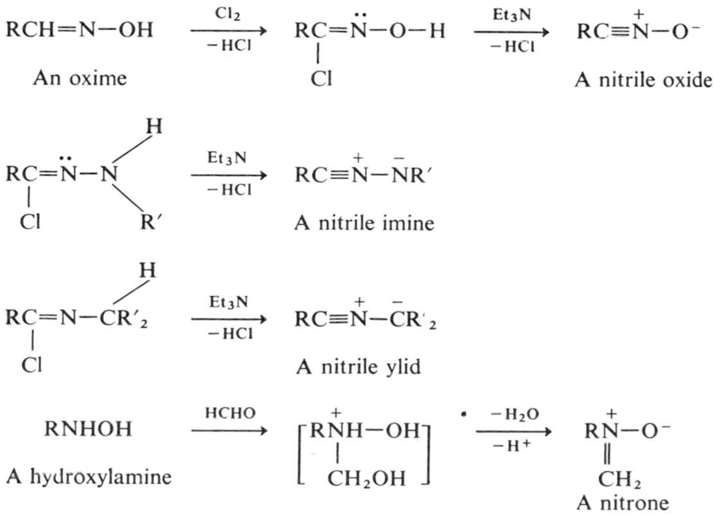

$$RCH=N-OH \xrightarrow[-HCl]{Cl_2} R\overset{..}{C}=\overset{..}{N}-O-H \xrightarrow[-HCl]{Et_3N} RC\equiv\overset{+}{N}-O^-$$

$$\text{An oxime} \quad\quad\quad \underset{Cl}{\big|} \quad\quad\quad \text{A nitrile oxide}$$

$$R\overset{..}{C}=\overset{H}{\underset{Cl}{N}}-N\overset{H}{\underset{R'}{\diagdown}} \xrightarrow[-HCl]{Et_3N} RC\equiv\overset{+}{N}-\bar{N}R'$$

A nitrile imine

$$R\overset{}{C}=\overset{}{\underset{Cl}{N}}-C\overset{H}{\underset{}{R'_2}} \xrightarrow[-HCl]{Et_3N} RC\equiv\overset{+}{N}-\bar{C}R'_2$$

A nitrile ylid

$$RNHOH \xrightarrow{HCHO} \left[R\overset{+}{N}H-OH \atop \underset{CH_2OH}{\big|} \right] \xrightarrow[-H^+]{\bullet -H_2O} R\overset{+}{N}-O^- $$

A hydroxylamine $\quad\quad\quad\quad\quad\quad\quad\quad\quad\quad$ $\underset{CH_2}{\|}$

$\quad\quad\quad\quad\quad\quad\quad\quad\quad\quad\quad\quad\quad\quad\quad\quad$ A nitrone

Applications. A variety of heterocyclic compounds can be constructed by reactions of 1,3-dipolar compounds with olefins and acetylenes. Some examples are:

(*i*) Diazoalkanes yield pyrazole derivatives, e.g.

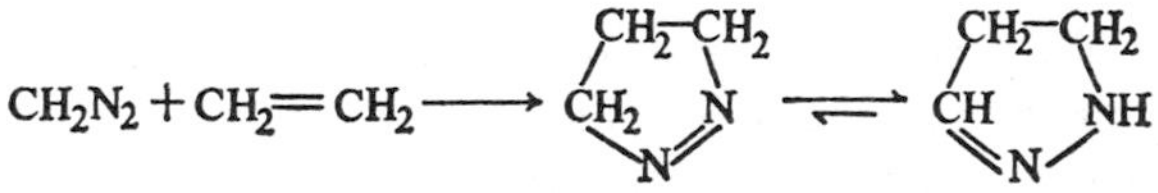

Pyrazoline

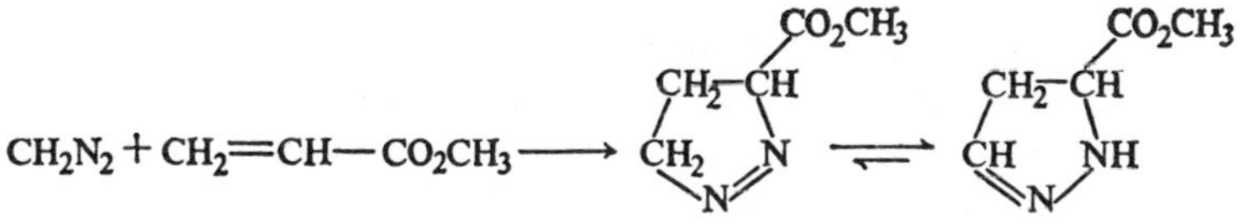

(*ii*) Azides yield triazole derivatives,

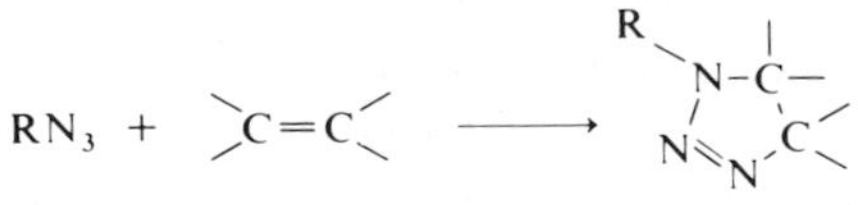

$$RN_3 + \quad \diagup^{}C{=}C^{}\diagdown \quad \longrightarrow$$

(*iii*) Nitrile oxides yield isoxazole derivatives,

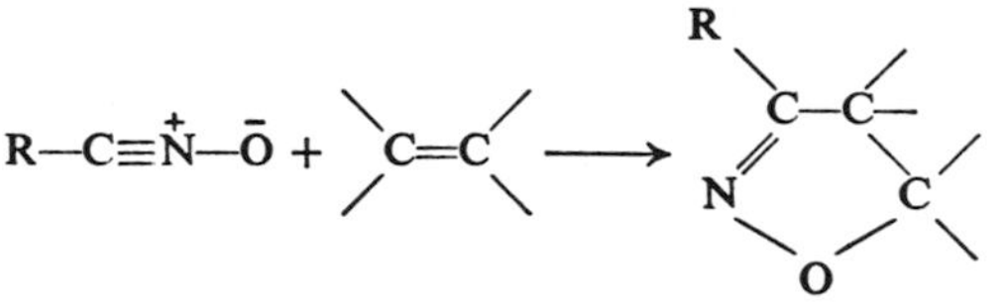

$$R{-}C{\equiv}\overset{+}{N}{-}\overset{-}{O} + \quad \diagdown C{=}C \diagup \quad \longrightarrow$$

An isoxazoline

(*iv*) Nitrile ylids yield pyrrole derivatives,

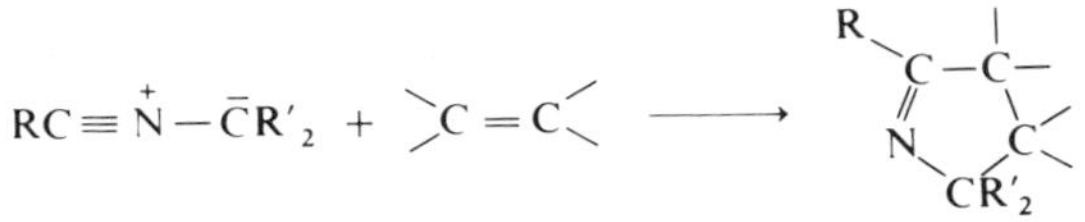

$$RC{\equiv}\overset{+}{N}{-}\overset{-}{C}R'_2 + \quad \diagdown C{=}C \diagup \quad \longrightarrow$$

The dipolarophile may also be a carbonyl compound or a nitrile. In these cases, there are two possible orientations for the addition, e.g.

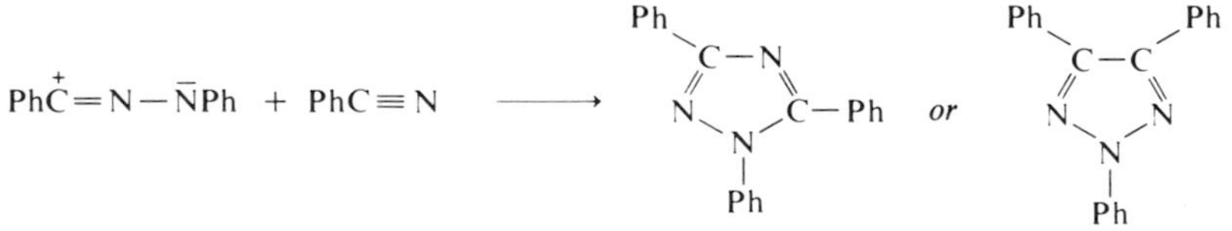

$$Ph\overset{+}{C}{=}N{-}\overset{-}{N}Ph + PhC{\equiv}N \quad \longrightarrow \qquad or$$

One product usually predominates; in the above example it is the first of the two. This and related results can be accounted for by considering the energies of the bonds being formed. For example, in the first product above, two C—N bonds are formed, whereas in the second, one C—C and one N—N bond are formed; the first combination corresponds to the greater bonding energy, and this is reflected in the preceding transition states.

Much less use has so far been made of 1,3-dipolar addition compared with the Diels-Alder reaction in the synthesis of natural products. However, its value is now being recognized. An elegant example, the synthesis of racemic luciduline (an alkaloid whose dextrorotatory enantiomer occurs in *Lycopodium lucidulum*) makes use of both these forms of cycloaddition:

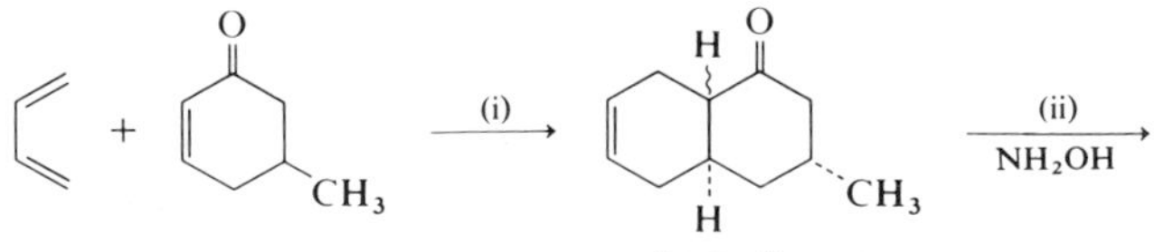

2 *cis*: 3 *trans*

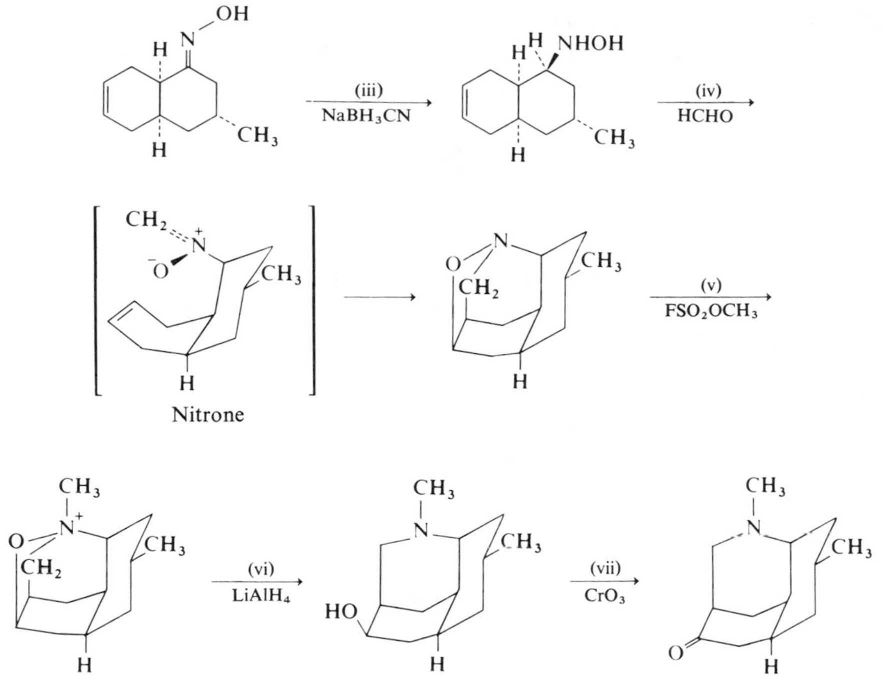

Nitrone

Luciduline

The Diels-Alder reaction—step (i)—was carried out at room temperature in the presence of tin(IV) chloride. This results initially in the formation of a *cis*-ring-junction and also in a *cis*-relationship between the hydrogen atoms at the ring-junction and the methyl substituent (the latter because the suprafacial addition of the diene is to the less hindered face of the dienophile). However, the carbonyl group activates the adjacent bridgehead hydrogen towards acid-catalyzed enolization, and the major product was the (unwanted) *trans*-isomer. It was found that the *cis*-isomer reacted the faster of the two with hydroxylamine in step (ii), and by conducting this reaction at high pH so as to effect rapid *cis-trans* isomerization *via* the enolate ion, it was possible to obtain almost entirely the required *cis*-ring-fused oxime. This was reduced with sodium cyanoborohydride in methanol; this is a hydride-transfer agent (p. 633) which delivered H⁻ from the less hindered side of the oxime, to create a hydroxylamine function stereoselectively as shown. The nitrone, generated in step (iv) with paraformaldehyde in hot toluene, underwent a spontaneous cycloaddition to give a bridged oxazolidine. In step (v) this was methylated with methyl fluorosulphonate, and the resulting salt was reduced in step (vi) with lithium aluminium hydride (p. 633). In the final step, the secondary alcohol was oxidized to carbonyl with chromium(VI) oxide in acetone.

(c) [2 + 2] CYCLOADDITIONS

As shown above, the [2 + 2] cycloaddition of olefins to form cyclobutanes is a symmetry-forbidden process when the geometry of the transition state requires suprafacial-suprafacial interaction of the orbitals.

However, olefins do combine in the [2 + 2] manner photochemically. The simplest interpretation of this is as follows. When an olefin molecule absorbs a quantum of light, an electron is promoted from the HOMO to the LUMO, so that there are now two singly occupied molecular orbitals (SOMO). The highest occupied orbital now therefore has the symmetry of the LUMO, and consequently can interact with the LUMO of a second molecule of the olefin which is in its ground state:

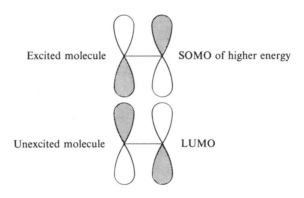

It has been recognized that this description is an oversimplification. Nevertheless, it provides a helpful approach to predicting whether a reaction will proceed photochemically; for example, it is simple to show that concerted [4 + 2] cycloaddition will not occur photochemically, as is found. These photochemical reactions are considered in more detail in Chapter 16.

Cumulenes. Cumulenes—compounds which contain adjacent double bonds, such as allene ($CH_2{=}C{=}CH_2$), keten ($CH_2{=}C{=}O$), and isocyanates ($RN{=}C{=}O$)—participate readily in [2 + 2] cycloadditions which have been proved in a number of cases to be concerted. For example, keten reacts with dienes to give [2 + 2] cycloadducts in a concerted reaction, to the exclusion of the formation of [4 + 2] adducts.

From the earlier analysis of the [2 + 2] cycloaddition of olefins, [2 + 2] addition for ketens is forbidden when the geometry of approach is suprafacial-suprafacial. However, consider the alternative suprafacial-antarafacial interaction. The efficiency of bonding interaction is low unless one of the components undergoes twisting about its original π-bond, and this is the constraint which makes a transition state derived from this array inaccessible for olefins. On the other hand, if the component whose π-orbital interacts in an antarafacial manner is a cumulene, one of the carbons involved is *sp*-hybridized. This reduces the geometric constraint but, more importantly, makes available another orbital whose participation stabilizes the transition state. This further orbital is the LUMO of the adjacent double bond. This is very low-lying, in keten especially, and its participation accounts for the preference by keten of [2 + 2] over [4 + 2] cycloaddition.

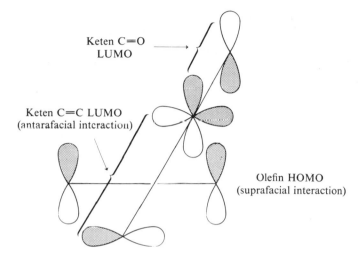

Keten C=O LUMO

Keten C=C LUMO (antarafacial interaction)

Olefin HOMO (suprafacial interaction)

The propensity for [2 + 2] cycloaddition amongst cumulenes causes them readily to dimerize. Substituted ketens form cyclobutanediones,

$$2R_2C=C=O \longrightarrow$$

but keten itself dimerizes in the liquid phase to a β-lactone, the [2 + 2] addition involving the carbonyl group of one component:

$$2CH_2=C=O \longrightarrow$$

This dimerization forms part of the industrial synthesis of acetoacetic ester; the keten is obtained by the pyrolysis of acetone, and the dimer is converted into acetoacetic ester by reaction with ethanol.

Other examples of dimerizations of cumulenes and of [2 + 2] cycloadditions involving two cumulenes or a cumulene and an olefin are the following:

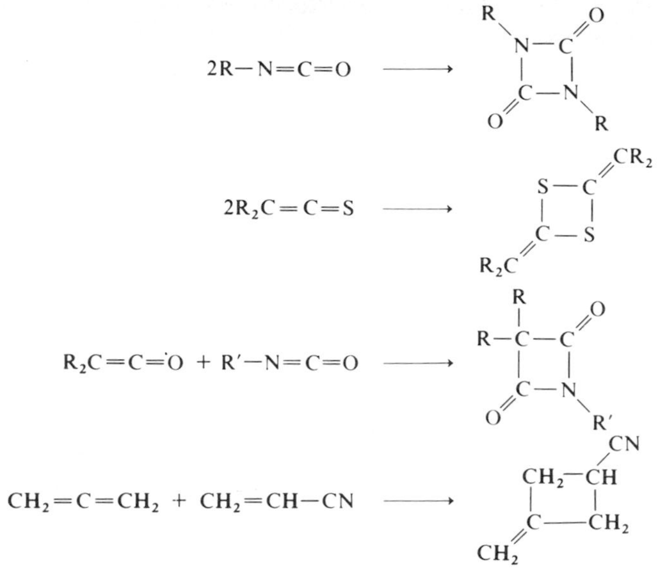

$$2R-N=C=O \longrightarrow$$

$$2R_2C=C=S \longrightarrow$$

$$R_2C=C=O + R'-N=C=O \longrightarrow$$

$$CH_2=C=CH_2 + CH_2=CH-CN \longrightarrow$$

Some of the reactions are equilibria which may be displaced in a synthetically useful sense, as in the formation of an unsymmetrical carbodi-imide from a symmetrical one and an isocyanate:

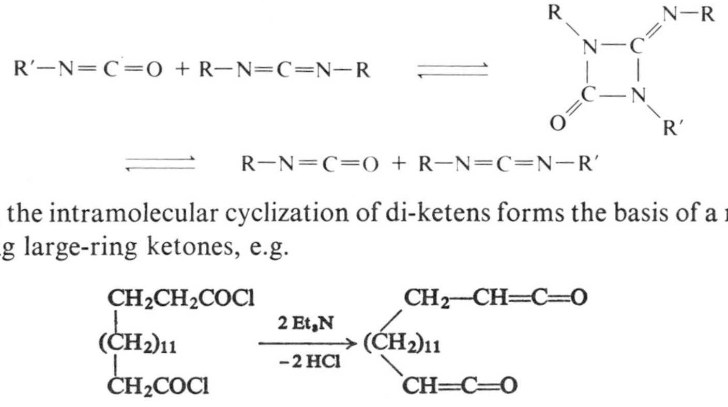

$$R'-N=C=O + R-N=C=N-R \rightleftharpoons$$

$$\rightleftharpoons R-N=C=O + R-N=C=N-R'$$

Finally, the intramolecular cyclization of di-ketens forms the basis of a method for making large-ring ketones, e.g.

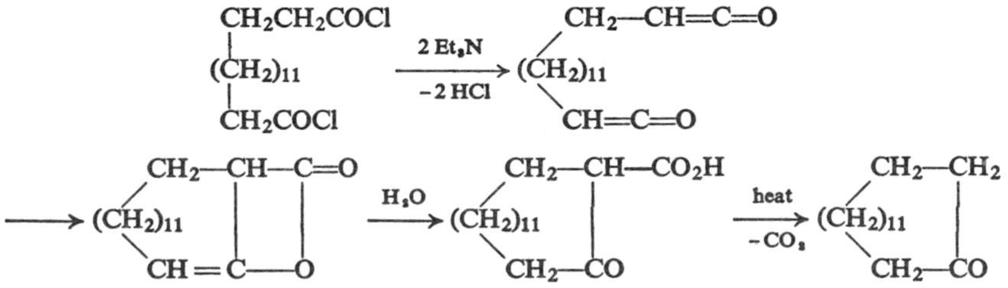

Yields are lower than in the Ziegler process from ω-dinitriles (p. 243), but the starting materials are more readily accessible.

9.3 Electrocyclic Reactions

Electrocyclic reactions are those in which either a ring is formed with the generation of a new σ-bond and the loss of a π-bond or a ring is broken with the opposite consequence, e.g.

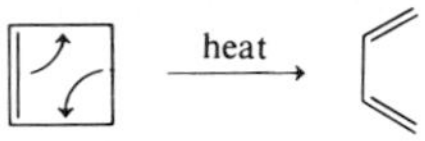

The reactions are stereospecific. For example, *trans*-3,4-dimethylcyclobutene gives solely *trans,trans*-2,4-hexadiene:

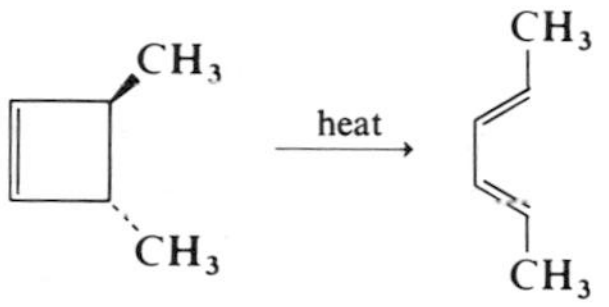

Considerations of orbital symmetry again offer an explanation. In the transition state for ring-opening of a cyclobutene, the HOMO of the σ-bond which is undergoing fission interacts with the LUMO of the olefinic bond. This can only happen if the σ-bond opens in a *conrotatory* manner, as shown (this also follows if the σ-LUMO and the π-HOMO are considered):

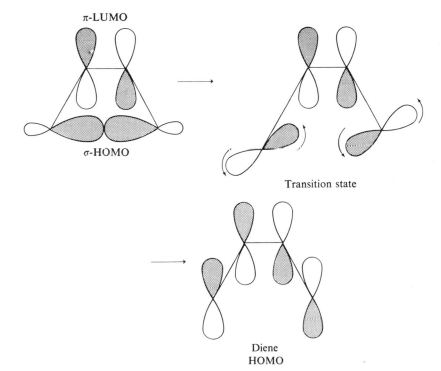

π-LUMO

σ-HOMO

Transition state

Diene
HOMO

Now, conrotatory ring-opening of *trans*-3,4-dimethylcyclobutene could in principle lead to two products, depending upon the sense of the rotation: the *trans,*
trans-dimethyl diene and the *cis,cis*-isomer. However, severe steric crowding
would result from the conrotatory movement which turns both methyl groups
inwards in the formation of the latter compound,

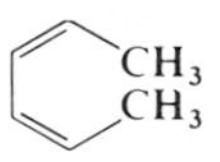

CH₃
CH₃

so raising the activation energy relative to that for the *trans,trans*-isomer.
Consequently, only the *trans,trans*-compound is formed.

For the reverse reaction—the ring-closure—precisely the opposite path is
followed (principle of microscopic reversibility) so that conrotatory motion of
the termini is likewise required.

The theory is readily extended to polyenes. For example, the opening of a
cyclohexa-1,3-diene, and likewise the ring-closure, require *disrotatory* motion:

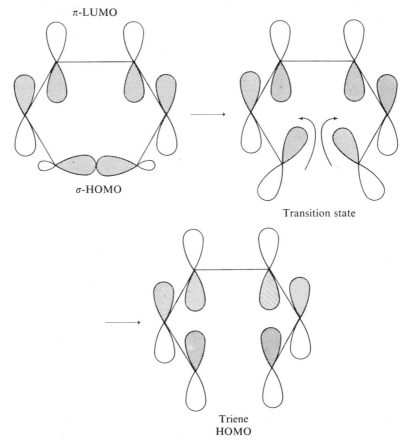

π-LUMO

σ-HOMO

Transition state

Triene
HOMO

It can be understood why, for example, *trans,cis,trans*-2,4,6-octatriene gives specifically *cis*-5,6-dimethylcyclohexadiene:

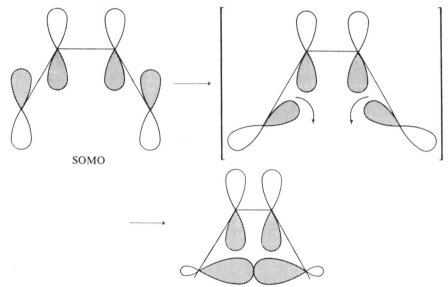

When the ring-closures of dienes and trienes are carried out photochemically, the stereospecificity is the opposite to that of the thermal reaction. If it is assumed that it is the photo-excited state which cyclizes, the symmetry of the SOMO of higher energy governs the reaction and the observed stereochemical course follows, e.g.

SOMO

However, it is recognized that this is an oversimplified treatment. Nevertheless, it does give a rule-of-thumb for the prediction of the course of a photochemical electrocyclic reaction.

The stereochemistry of electrocyclic reactions can be summarized as follows:

Direction of motion of the termini in electrocyclic ring-closure or the reverse ring-opening:

	Rotatory motion	
Number of π-electrons in ring-closure reaction	Thermal reaction	Photochemical reaction
4	con	dis
6	dis	con

Applications. The formation and reactivity of benzocyclobutenes demonstrate the utility of electrocyclic reactions in synthesis. They can be formed from tetrabromo-*o*-xylenes with iodide ion; the *o*-quinodimethane intermediate readily undergoes electrocyclic ring-closure owing to the gain of aromatic stabilization energy, e.g.

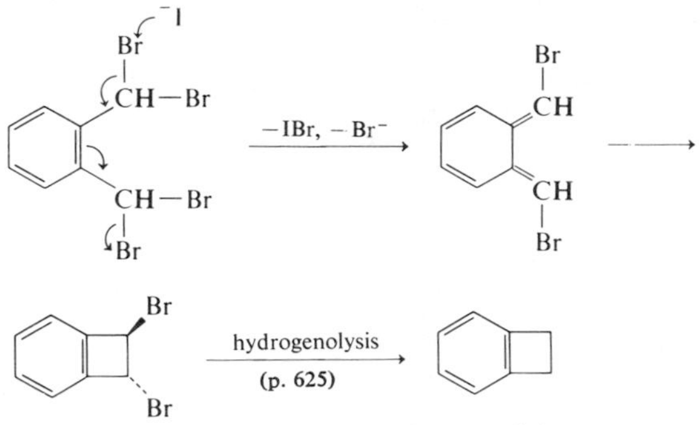

Benzocyclobutene

The ring-opening of a cyclobutene, followed by a cycloaddition, has been used in an elegant steroid synthesis; two of the rings of the product are formed with a predetermined *trans* stereochemistry:

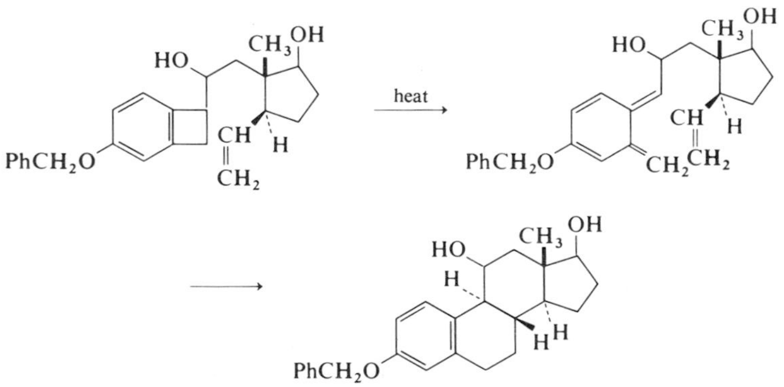

Valence tautomerism. Electrocyclic equilibria that occur at ambient temperature give rise to the phenomenon of valence tautomerism. For example, cyclooctatetraene is in equilibrium with *cis*-bicyclo[4.2.0]octa-2,4,7-triene, which it forms by the symmetry-allowed disrotatory ring-closure:*

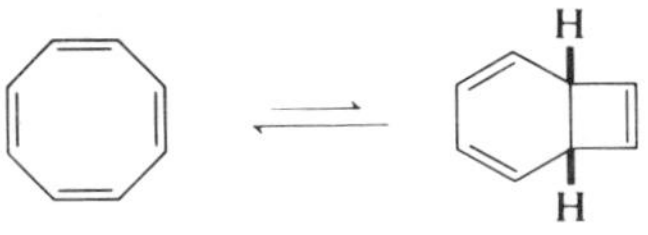

---

*Valence tautomerism can be studied by n.m.r. spectroscopy. If equilibrium occurs slowly with respect to the n.m.r time-scale, the spectrum of each tautomer is observed; if it is fast, a time-averaged spectrum of the two is observed. At intermediate rates, line-broadened spectra are obtained, analysis of which enables the rate of interconversion to be derived.

9.4 Cheletropic Reactions

Cheletropic reactions are ones in which two σ-bonds which terminate at a single atom are made or broken in a concerted reaction, e.g.

3-Sulpholene

Cheletropic addition is related to Diels-Alder addition in that the 2π-electron system in the dienophile of the latter is replaced by an unshared pair of electrons on a single atom in the former.

(*a*) CHELETROPIC REACTIONS OF DIENES
The HOMO of a molecule like sulphur dioxide or carbon monoxide is that which has a lone-pair of electrons in the plane containing the atoms; the LUMO is a *p* orbital perpendicular to this plane:

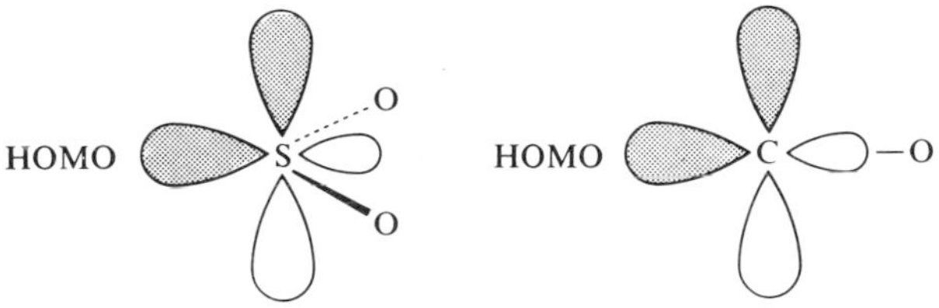

For a symmetry-allowed cycloaddition of sulphur dioxide to a diene, the SO_2 must lie in a plane which bisects the *s-cis* conformation of the diene:

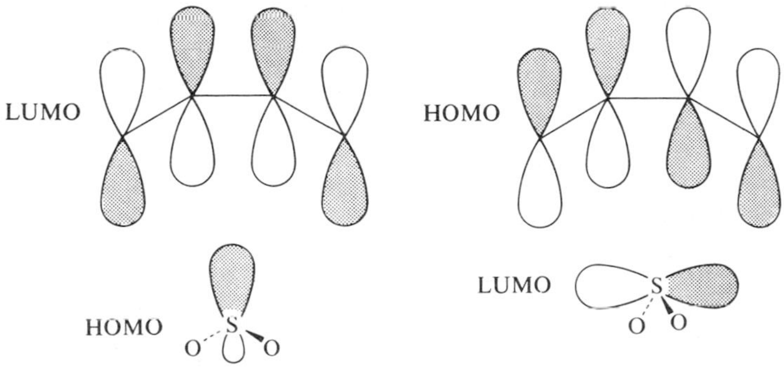

The interaction is suprafacial for both orbitals of the diene and also for the HOMO of SO_2; it is antarafacial for the LUMO of SO_2. In the transition state,

the terminal carbon atoms of the diene must move in the disrotatory manner
so that the HOMO of SO_2 can interact with the LUMO of the diene, or the
LUMO of SO_2 with the HOMO of the diene. The reality of this prediction of
symmetry theory is proved by the fact that *trans,trans*-1,4-disubstituted dienes
give specifically *cis*-substituted 3-sulpholenes, e.g.

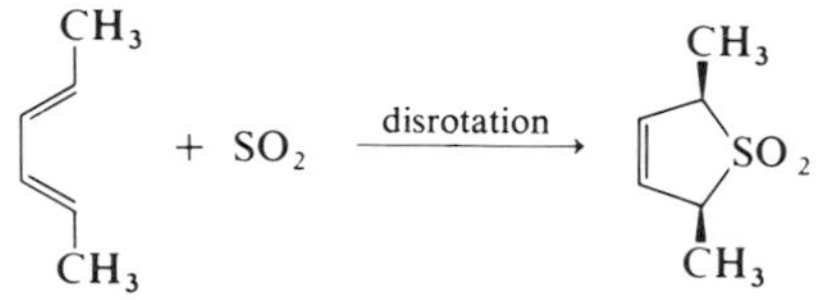

and *cis,trans*-disubstituted dienes give *trans*-substituted 3-sulpholenes.

As with electrocyclic reactions, the opposite stereochemistry is observed if the
reaction is photochemical rather than thermochemical:

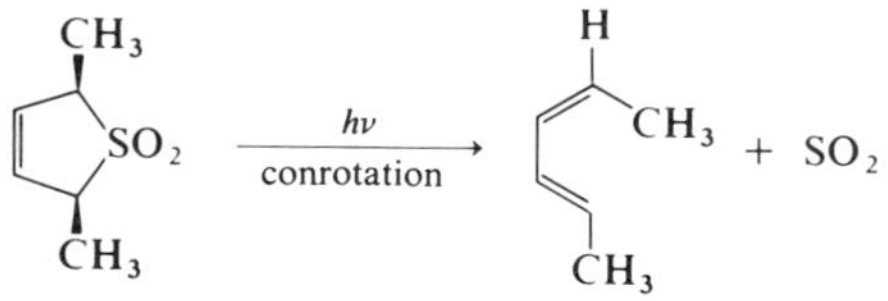

For the photochemical reaction, the rule-of-thumb for predicting the correct
stereochemistry is to assume that the reaction occurs from the higher-energy
SOMO of the photoexcited state. Again, this is likely to be an oversimplification.

Applications. The reaction between butadiene and sulphur dioxide, and the
reverse reaction which occurs at high temperatures, provide a useful method for
'carrying' butadiene, so avoiding the experimental inconvenience of handling
gaseous butadiene at the elevated temperatures which may be needed for reaction. For example, when 3-sulpholene is heated in the presence of a dienophile,
butadiene is released and immediately trapped in a Diels-Alder reaction.

There are few other reactions where a molecule like sulphur dioxide is seques-
tered by a diene, but there are others where the reverse process—extrusion of a,
usually, stable molecule—occurs. For instance, Diels-Alder adducts of cyclo-
pentadienones extrude carbon monoxide, e.g.

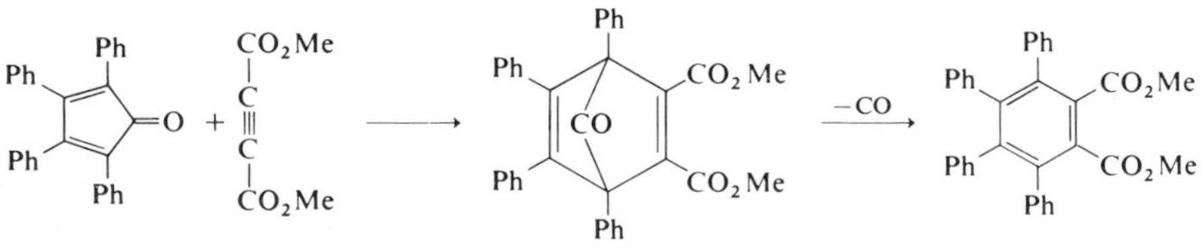

(b) CHELETROPIC REACTIONS OF TRIENES

2,7-Dihydrothiepin dioxides undergo thermolysis with a high degree of stereo-specificity, e.g.

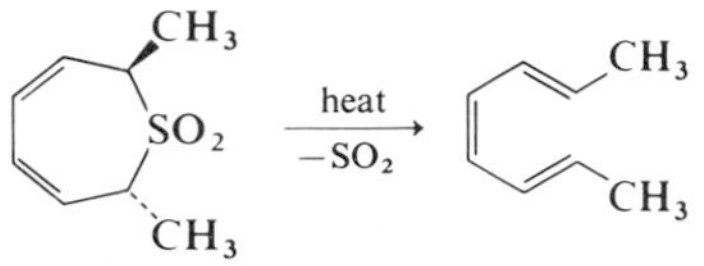

The reaction therefore occurs with a conrotatory motion of the triene termini. If the cheletropic reaction is linear—i.e. if the HOMO of SO_2 reacts supra-facially as it does in the case of dienes—the triene must react antarafacially, which is consistent with the observed conrotation:

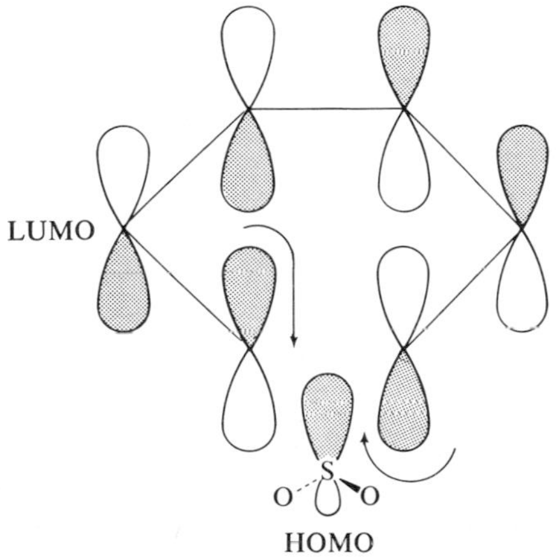

If the HOMO of SO_2 were to react antarafacially, and the LUMO of the triene suprafacially, which would also be symmetry-allowed, the motion of the triene's termini would have to be disrotatory. Since the result of this non-linear chele-tropic reaction is not observed, it follows that the linear process is preferred (presumably for steric reasons).

Cheletropic reactions have as yet found little practical use, other than as examples of the principles of symmetry-controlled reactions!

(c) CHELETROPIC REACTIONS OF SIMPLE OLEFINS

Cheletropic reaction of sulphur dioxide with olefins is symmetry-forbidden if both components interact suprafacially. Antarafacial reaction of a simple olefin is sterically very unlikely (cf. discussion of [2 + 2] cycloaddition of olefins), so that the reaction is likely to involve the SO_2 antarafacially. It is known in the stereospecific extrusion of SO_2 from episulphones, e.g.

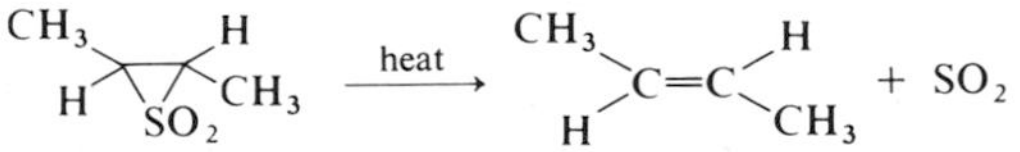

The orbital interaction in the transition state is thus represented by:

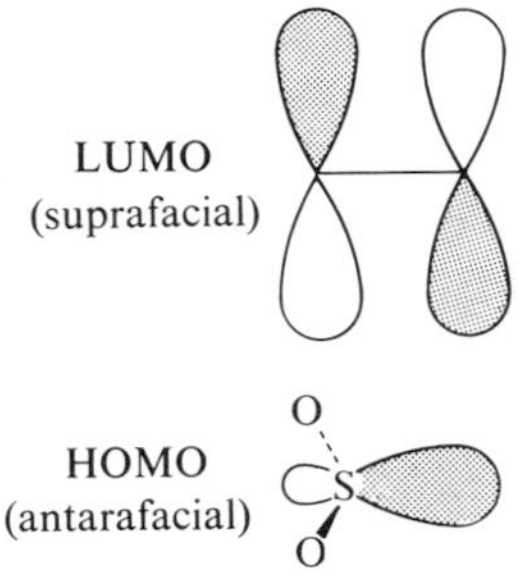

LUMO
(suprafacial)

HOMO
(antarafacial)

An analogous reaction is the stereospecific extrusion of nitrous oxide from nitrosoaziridines, e.g.

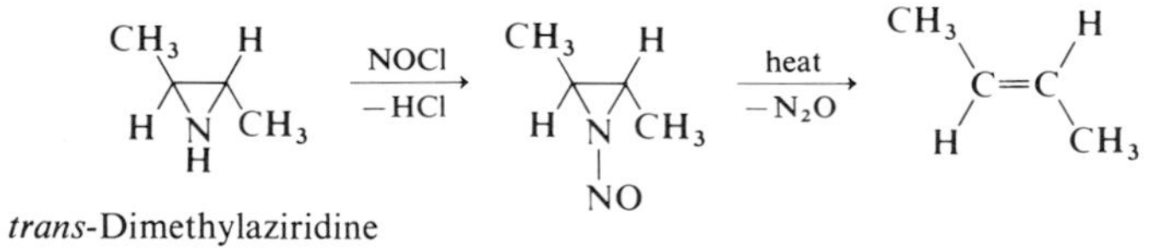

trans-Dimethylaziridine

The reactions of singlet carbenes and nitrenes with olefins represent reactions of like character, e.g.

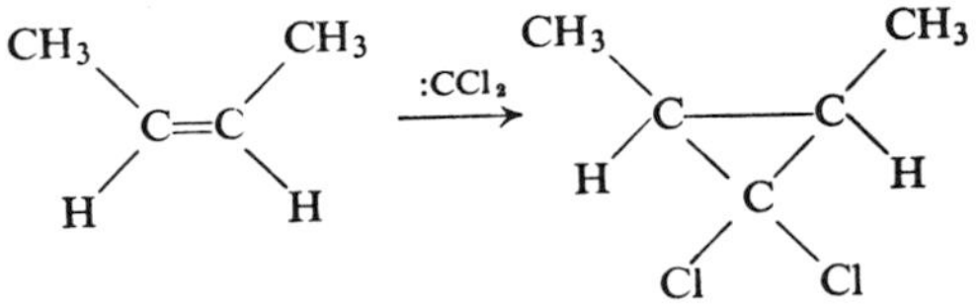

Only singlet carbenes and nitrenes behave in this way; the addition of triplet carbenes or nitrenes to a double bond results in a diradical which is sufficiently long-lived, during the time required for spin-inversion, for stereospecificity to be lost as a result of rotations about bonds (p. 181).

The development of phase-transfer catalysis has greatly increased the synthetic utility of dichlorocarbene. A phase-transfer catalyst is usually a quaternary ammonium salt where the cation contains large alkyl groups which confer upon it solubility in organic solvents (e.g. $C_{16}H_{33}\overset{+}{N}Me_3$). Such a cation will transport hydroxide as its counter-ion into chloroform from aqueous solution. This greatly facilitates the reaction,

$$HO^- + H{-}CCl_3 \rightleftharpoons H_2O + CCl_3{}^-$$

which otherwise could only occur at the interface of the organic and aqueous phases. The trichloromethyl carbanion, $CCl_3{}^-$, is relatively short-lived, decomposing to give dichlorocarbene:

$$CCl_3{}^- \longrightarrow Cl^- + {:}CCl_2$$

Suitable substrates such as olefins, dissolved in the chloroform, then react with the carbene in synthetically useful quantity.

9.5 Sigmatropic Rearrangements

A sigmatropic rearrangement is a pericyclic reaction which involves the migration of an atom or group within a π-electron system; overall, the numbers of π- and σ-bonds remain separately unchanged. The rearrangements can be divided into two classes: (1) those where the group which migrates is bonded through the same atom in both reactant and product; (2) those where the migrating group is bonded through different atoms in the reactant and the product. An example of the former is,

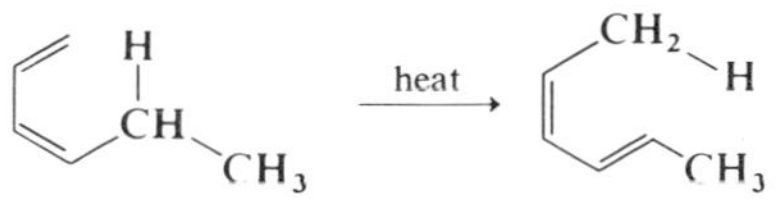

where hydrogen is transferred from the methylene group to the alternative terminus of the diene; the latter is exemplified by the Claisen rearrangement, e.g.

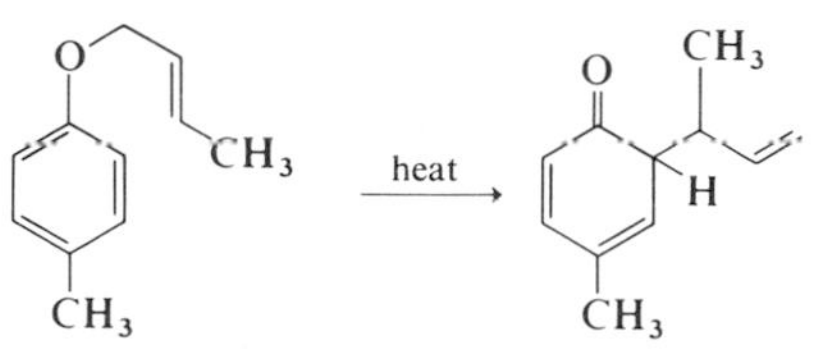

where the allyl group which migrates is bound by different carbon atoms before and after the reaction.

These two reactions illustrate, respectively, a [1,5]-shift and a [3,3]-shift. The figures in parentheses denote the numbers of essential interacting centres in the two groups which are formed by breaking the migrating σ-bond.

(a) [1, j]-SIGMATROPIC REARRANGEMENTS

The occurrence and stereochemistry of sigmatropic rearrangements can be accounted for, like other pericyclic reactions, in terms of the symmetry of frontier orbitals. Consider the [1,5]-shift of hydrogen as in the first example above:

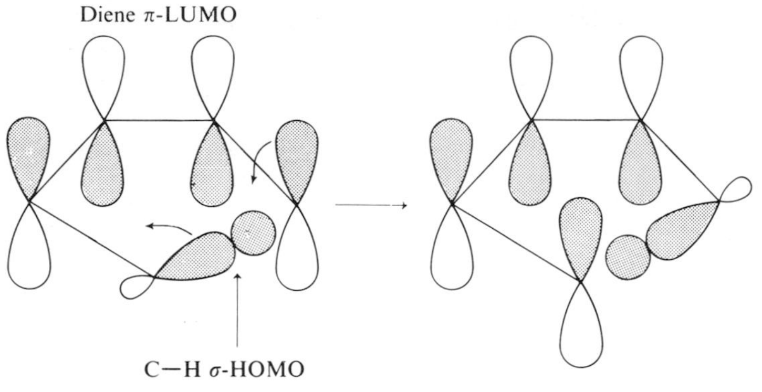

Diene π-LUMO

C—H σ-HOMO

For a maximum of bonding to occur in the transition state when the HOMO of the C—H σ-bond interacts with the LUMO of the diene π-system, the hydrogen is transferred suprafacially. Since this arrangement is easily accessible geometrically, the [1,5]-shift of hydrogen in dienes readily occurs thermally.

Owing to the difference in symmetry between the LUMO of a simple olefin and that of a diene, a similar suprafacial [1,3]-transfer of hydrogen in a substituted olefin is symmetry-forbidden. A [1,3]-hydrogen transfer would be allowed if the π-bond were to interact antarafacially,

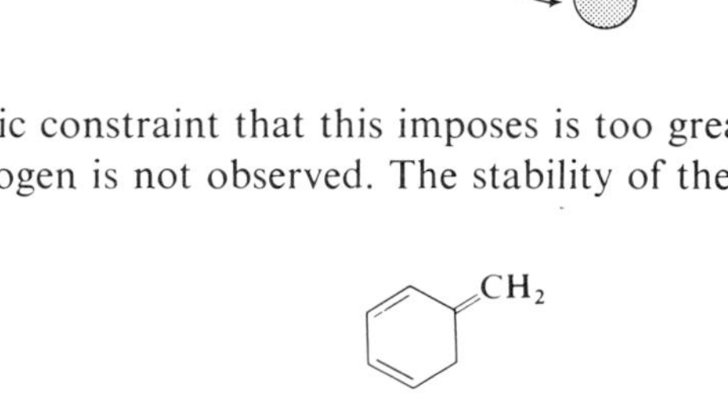

π-LUMO

C—H σ-HOMO

but the geometric constraint that this imposes is too great and concerted [1,3]-transfer of hydrogen is not observed. The stability of the triene,

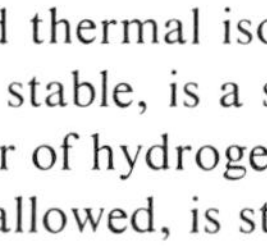

derives from the fact that concerted thermal isomerization to toluene, which is thermodynamically much the more stable, is a symmetry-forbidden process.

On the other hand, a [1,7]-transfer of hydrogen in a triene, which must also be antarafacial if it is to be symmetry-allowed, is sterically feasible in certain cases. The classic example is the thermal interconversion of vitamin D and precalciferol:

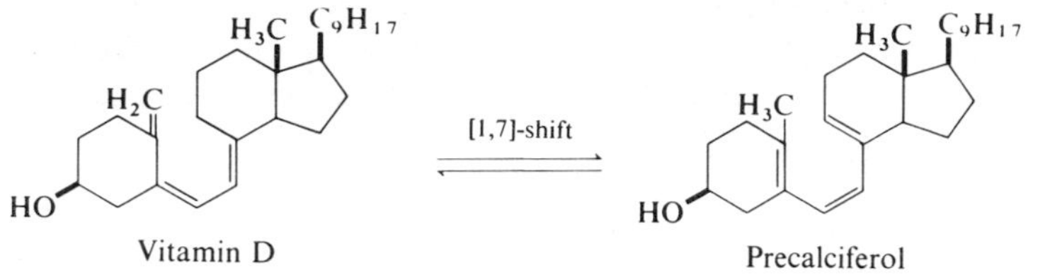

Vitamin D   Precalciferol

In sigmatropic shifts of hydrogen, only the polyene can participate in an antarafacial manner on account of the spherical symmetry of the hydrogen s-orbital. With other functions such as alkyl groups, the migrating group can be the antarafacial component and this expands the possibility for reaction. Consider the [1,3]-shift of an alkyl group in an olefin:

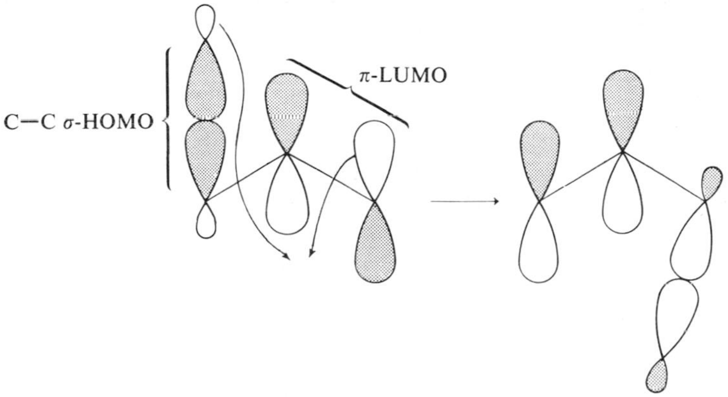

As shown, there is the possibility of an interaction which is antarafacial for the C—C σ-bond and suprafacial for the π-bond. Again, this is likely to require a geometrically difficultly accessible transition-state, particularly for flexible molecules, but it is known in molecules whose rigid framework builds into the ground state of the reactant much of the ordering which would otherwise have to be achieved at the expense of an unfavourable entropy of activation in a flexible analogue. An example is:

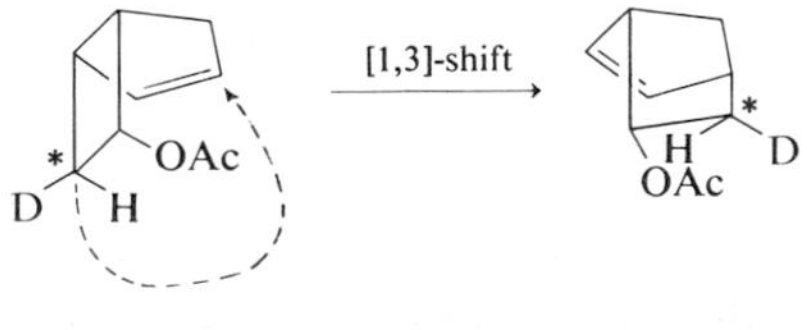

The antarafacial character of the interaction of the alkyl σ-bond is demonstrated by the fact that the configuration of the migrating carbon (marked *) is inverted in the change.

Alkyl migrations are most familiar in carbonium ions and other electron-deficient species. Again, the stereochemistry observed is that predicted by orbital symmetry. An example is the Wagner-Meerwein rearrangement:

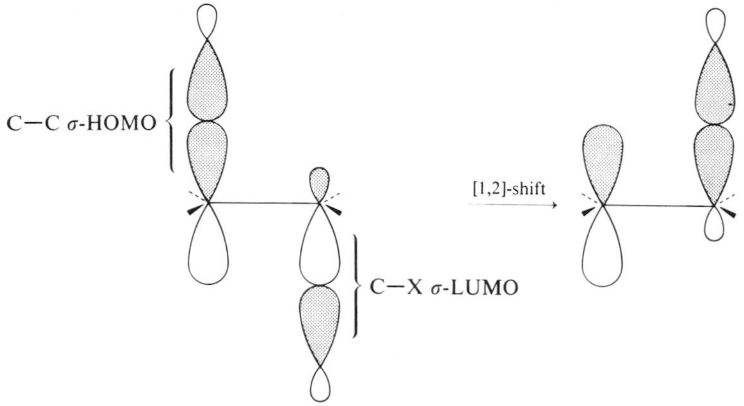

The HOMO of the migrating σ-bond interacts with the LUMO of the electron-deficient centre (or the empty p orbital of a carbonium ion). This can occur in a suprafacial-antarafacial manner, in accord with the observed result that the migrating group moves with retention of configuration whilst the configuration at the migration terminus is inverted. These rearrangements are considered in detail in Chapter 14.

In contrast, the concerted [1,2]-shift in a carbanion is predicted to be symmetry-forbidden.* Such rearrangements do, however, occur, but not by a concerted mechanism (p. 474).

(b) [i, j]-SIGMATROPIC REARRANGEMENTS

The most frequently encountered sigmatropic rearrangements of this category are [3,3]-shifts for, as the notation implies, the transition state involves the stereochemically favourable six-membered ring.

The *Cope rearrangement* is the [3,3]-sigmatropic rearrangement of 1,5-dienes, e.g.

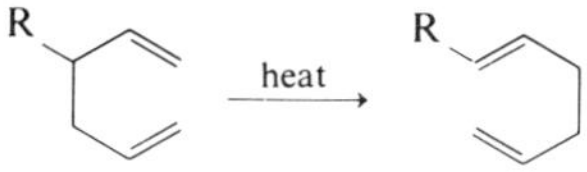

---

*This is usually discussed in terms of a treatment of pericyclic reactions which is an alternative to the frontier-orbital method. The transition state for the [1,2]-shift of a carbanion would contain a cyclic array of 4π-electrons. This is an 'anti-aromatic', high-energy configuration, in contrast to the stable 'aromatic' arrangements in the [1,2]-shift of a carbonium ion (2π-electrons) or a [1,5]- or [3,3]-shift (6π-electrons).

In acyclic systems there is strong evidence that the reaction proceeds through a chair-shaped transition state. For example, *meso*-3,4-dimethyl-1,5-hexadiene rearranges virtually exclusively to *cis,trans*-2,6-octadiene, consistent only with a chair-shaped transition state; a boat-shaped transition state would lead to either the *trans,trans*- or the *cis,cis*-product, depending on conformation:

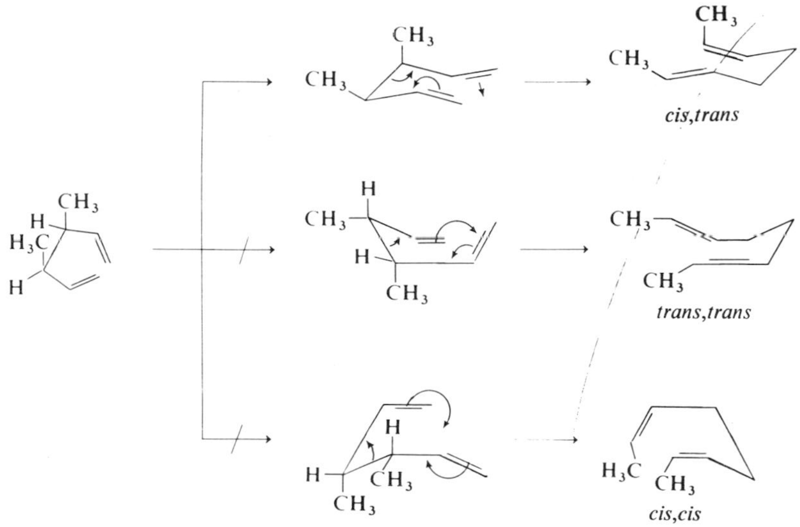

cis,trans

trans,trans

cis,cis

If the structure of the reactant prevents attainment of a chair-shaped transition state, the reaction occurs through a boat-shaped one. The Cope rearrangement of *cis*-1,2-divinylcyclopropane is of this type:

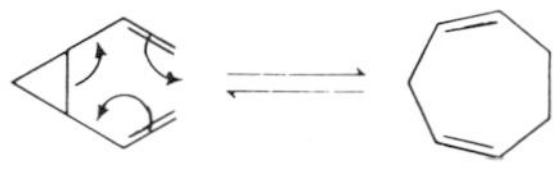

The reaction occurs spontaneously at room temperature, presumably because the strain in the three-membered ring is thereby relieved. Such processes augment the range of the valence-tautomerism phenomenon.

On occasion, the valence tautomerism can be *degenerate*; that is, the product is identical with the reactant. The ultimate example is bullvalene which exists in 10!/3 (i.e. over a million) identical forms:*

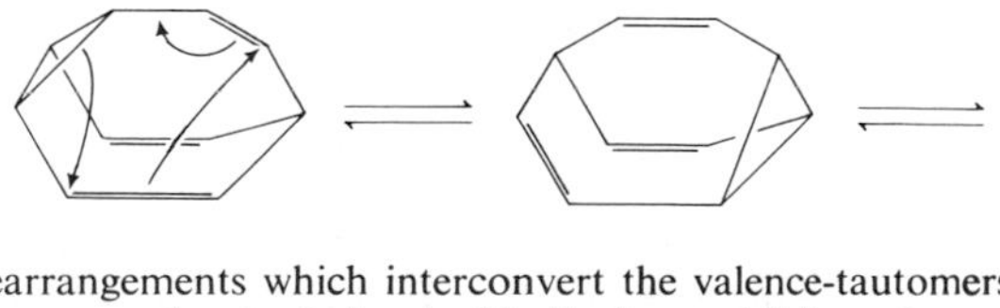

etc.

*The Cope rearrangements which interconvert the valence-tautomers are sufficiently rapid at 120°C that the n.m.r. (proton) signal of bullvalene exhibits a sharp singlet. At −85°C, by contrast, signals from four separate types of proton are detected.

The *Claisen rearrangement* is another example of a [3,3]-sigmatropic rearrangement. The mechanism can be represented as:

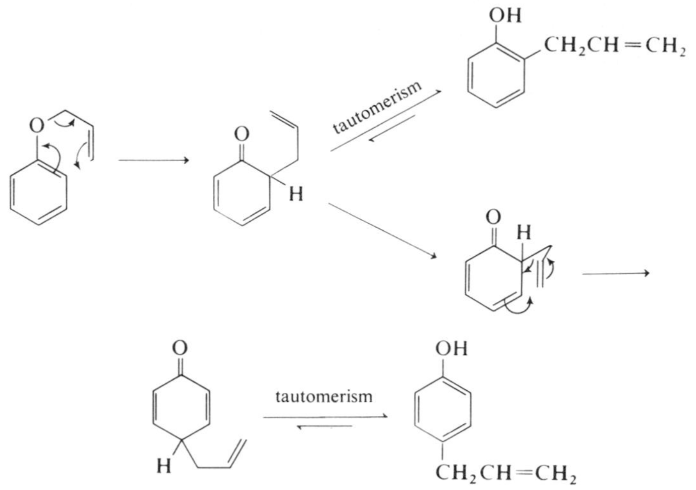

The rearrangement of the aryl allyl ether to the *o*-allylphenol is the first [3,3]-
sigmatropic rearrangement. It may be followed by a further [3,3]-shift, particularly in the absence of a *para*-substituent, to give the *p*-allylphenol.*

The reaction also occurs with aliphatic allyl ethers and, as with the Cope rearrangement, a transition state of chair conformation is preferred; for example,
trans,trans-crotyl propenyl ether rearranges to the *threo*- and not the *erythro*-
2,3-dimethylpentenal:

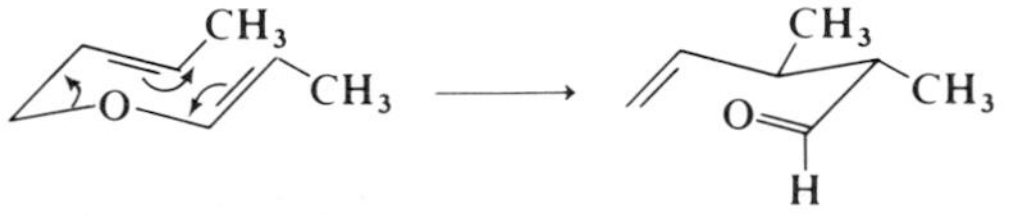

trans,trans-ether   *threo*-aldehyde

The Claisen rearrangement can be valuably employed in synthesis. For
example, guaiacol allyl ether is converted into *o*-eugenol in 80–90% yield [3],

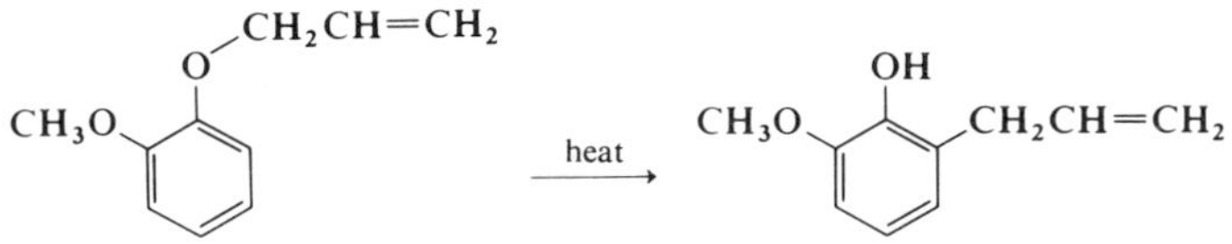

o-Eugenol

*Support for this two-step reaction is provided by the observation that a terminal ^{14}C label
is inverted once in the *ortho*-shift and twice in the *para*-shift. Further, the cyclohexa-2,4-dienone
intermediate may be trapped as a Diels-Alder adduct.

and the isoprene unit is built into the following polyolefinic ketone, in a manner most valuable in natural-product synthesis, by repeated aliphatic Claisen rearrangements:

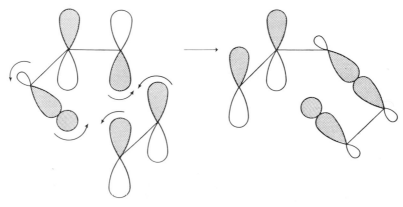

[3,2]-Sigmatropic rearrangements are also well established. The reactants possess a two-atom migrating unit in which one atom has a negative charge; this is equivalent to the two π-electrons of an unsaturated bond in a three-atom unit. An example is the Sommelet rearrangement in which the two-atom unit is a nitrogen ylid, $-\overset{+}{N}R_2-CH_2^{-}$ (p. 474).

9.6 The Ene-reaction and Related Reactions

The ene-reaction is a reaction of an allylic compound with an olefin in a manner which resembles both cycloaddition and a [1,5]-sigmatropic shift of hydrogen, e.g.

It is seen to be a symmetry-allowed process by consideration of the frontier orbitals:

The reaction occurs through a suprafacial interaction of all the participating orbitals, as in the Diels-Alder reaction. The observed stereochemistry is consistent with this: *trans*-2-butene reacts with maleic anhydride to form predominantly the *erythro*-adduct, which implies also a preference for *endo*-addition:

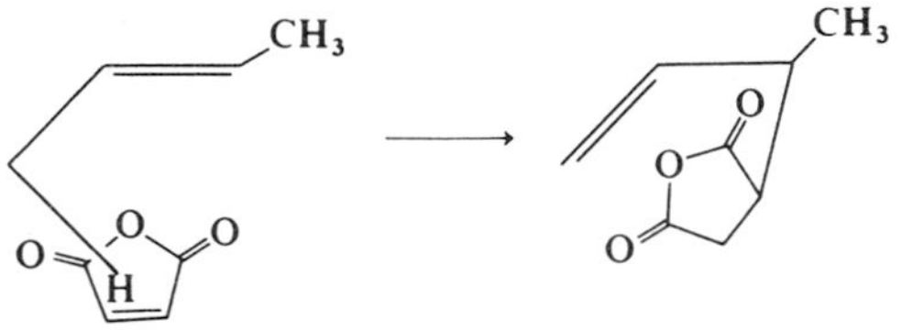

The reaction of allylic alcohols provides a useful synthesis of aldehydes:

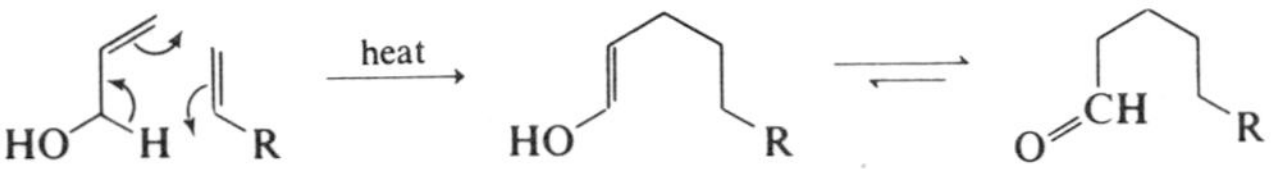

As with other pericyclic reactions, reversal is, of course, symmetry-allowed. The preferred direction depends upon thermodynamic factors; in particular, the *retro-ene* reaction is favoured when the system contains hetero-atoms, especially oxygen, for then the bond energy of the strong carbonyl bond is partially released at the transition state. An example is the pyrolysis of an ester:

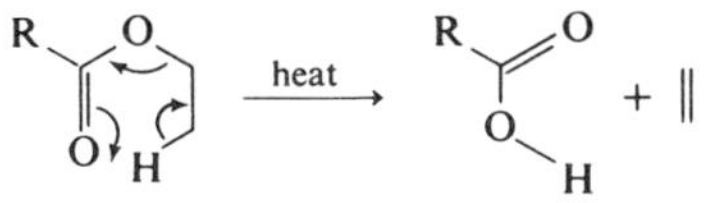

The temperature required for this reaction is high (300–500°C), which restricts its utility. Conditions are much milder, however, for the related *Chugaev reaction* in which a xanthate ester is heated at 100–200°C:

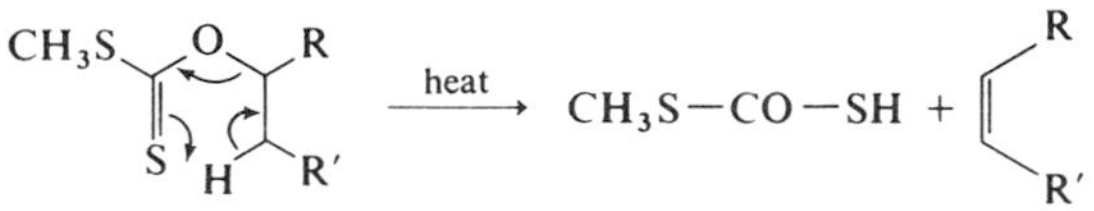

$$CH_3S-CO-SH \longrightarrow CH_3SH + COS$$

The synthetic value of the Chugaev reaction lies in the stereospecificity which its concertedness confers.

The ready decarboxylation of β-keto-acids has been shown to occur analogously, e.g.

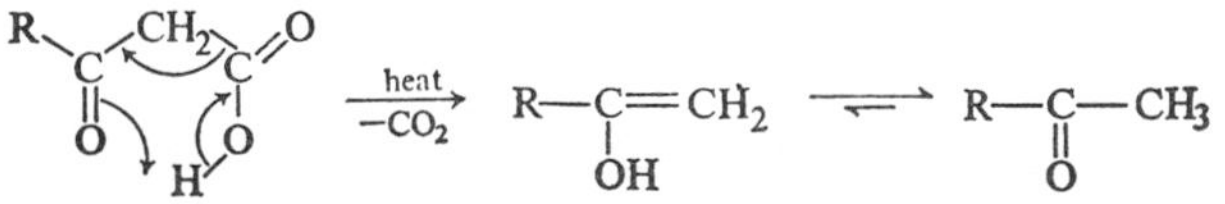

The *Cope reaction* is a related pericyclic reaction of amine oxides which involves a five-membered cyclic transition state. It occurs under mild conditions and is useful in the generation of non-conjugated polyenes, e.g.

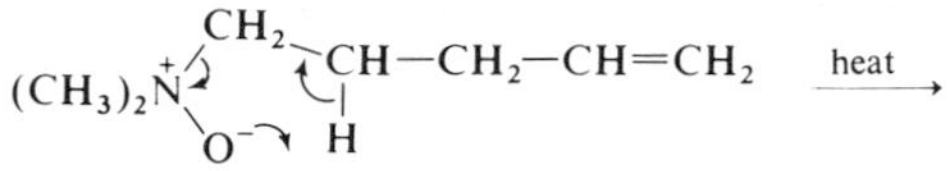

$$(CH_3)_2\overset{+}{N} \quad \overset{CH_2}{\underset{O^- \ H}{\cdots CH-CH_2-CH=CH_2}} \quad \xrightarrow{\text{heat}}$$

$$(CH_3)_2N-OH + CH_2=CH-CH_2-CH=CH_2$$

Sulphoxides react similarly, and applications of the process are described later (p. 495).

Further Reading

GILL, G. B., and WILLIS, M. R., *Pericyclic Reactions*, Chapman and Hall (London 1974).

KIRMSE, W., 'Carbenes,' *Progress in Organic Chemistry*, Vol. 6, Butterworths (London 1964), p. 164.

RHOADS, S. J., and RAULINS, N. R., 'The Claisen and Cope rearrangements,' *Organic Reactions*, 1975, **22**, 1.

ROBERTS, J. D., and SHARTS, C. M., 'Cyclobutane derivatives from thermal cycloaddition reactions,' *Organic Reactions*, 1962, **12**, 1.

WOODWARD, R. B., and HOFFMANN, R., 'The conservation of orbital symmetry,' *Angewandte Chemie* (International Edition), 1969, **8**, 781.

Problems

1. Draw the structure of the chief product of each of the following thermal reactions:

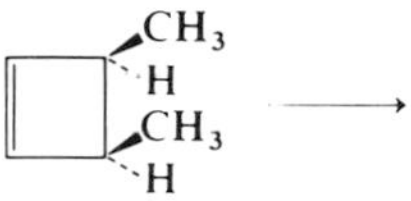

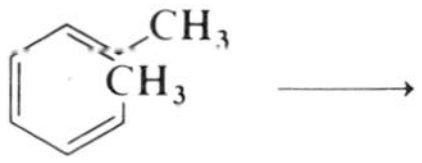

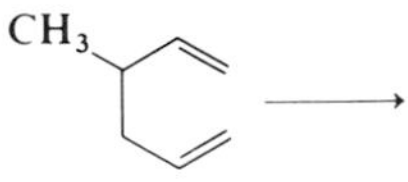

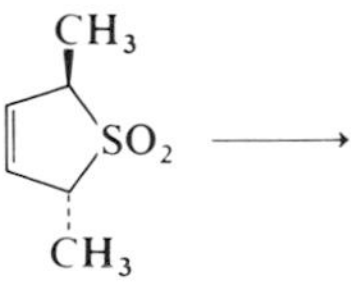

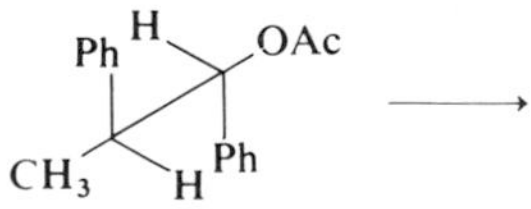

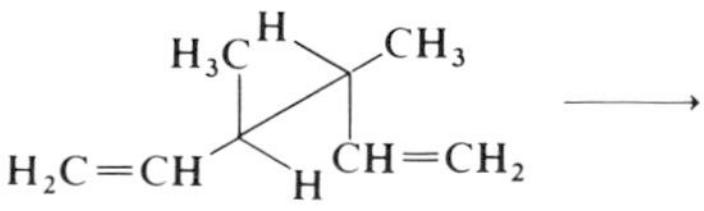

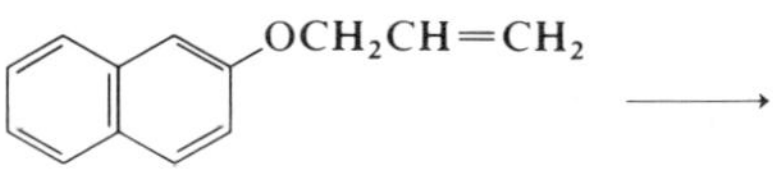

(h) $\quad$ + $MeO_2C-C\equiv C-CO_2Me$ $\longrightarrow$

(i)

$$\begin{array}{c} CH_3 \\ | \\ CH=CH \\ | \\ CH=CH_2 \end{array} \quad + CH_2=CH-CO_2Me \quad \longrightarrow$$

(j) $\quad$ + PhN_3 $\longrightarrow$

(k) $\quad PhCH=CH_2$ + $PhC\equiv\overset{+}{N}-\overset{-}{O}$ $\longrightarrow$

(l) $\quad CH_2=CH-CH_2-O-CH=CH_2$ $\longrightarrow$

2. How would you employ pericyclic reactions in the synthesis of the following?

(a)
$$\begin{array}{c} CHO \\ CH_3 \end{array}$$

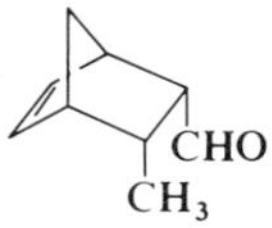

(b) 
$$NC \quad \overset{}{\underset{CN \; NC}{}} \quad CN$$

(c)
$$\begin{array}{c} Ph \quad CH_3 \\ N \\ O \quad CH_3 \end{array}$$

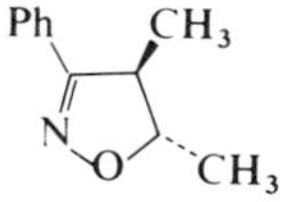

(d)
$$\begin{array}{c} Ph \quad Ph \\ N \\ N \\ H \end{array}$$

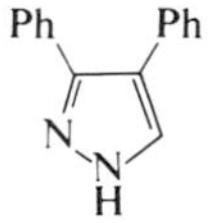

(e) $\quad CH_3NHCOCH_2COCH_3$

(f) $\quad Ph(CH_2)_5CHO$

3. Rationalize the following thermal reactions:

(a)
$$\begin{array}{c} Ph \\ \overset{+}{N}(CH_3)_2 \\ | \\ O^- \end{array} \longrightarrow \quad \begin{array}{c} Ph \end{array} + (CH_3)_2NOH$$

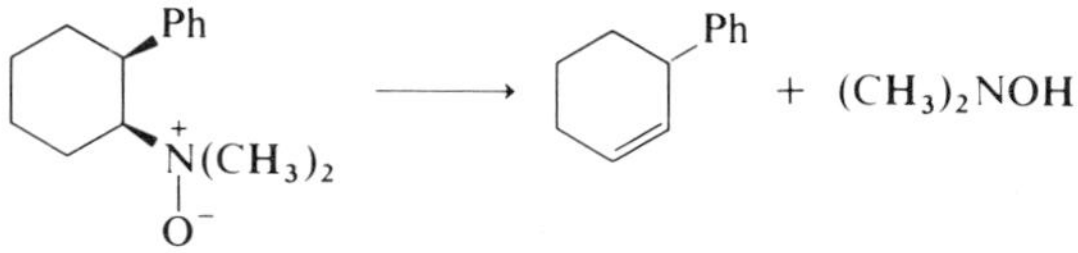

(b)
$$\begin{array}{c} Ph \\ Ph \end{array} \quad + \quad O=\!\!\!\!\!\begin{array}{c} \\ O \\ \end{array}\!\!\!\!\!=O \quad \longrightarrow \quad \begin{array}{c} Ph \quad O \\ O \\ Ph \quad O \end{array}$$

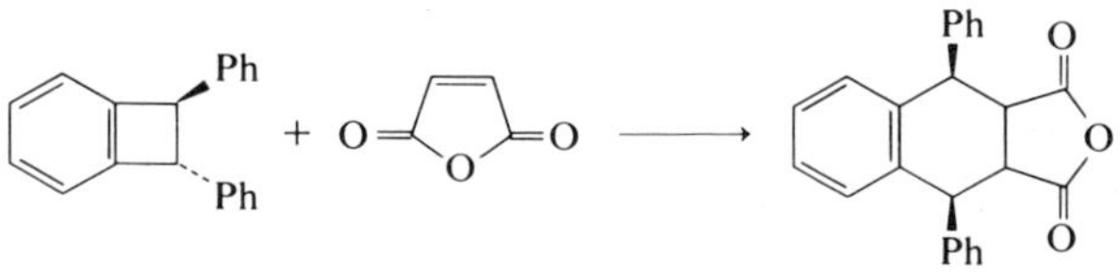

(c)

(d)

10. The Formation of Aliphatic Carbon-Nitrogen Bonds

10.1 Principles

With the exception of a few methods involving free-radical reactions (17.4), the methods for forming bonds between nitrogen and aliphatic carbon fall into two categories. In the first, nucleophilic nitrogen reacts with electrophilic carbon, and in the second, electrophilic nitrogen reacts with nucelophilic carbon. The first category is by far the more important.

(a) NUCLEOPHILIC NITROGEN

A ternary nitrogen atom possesses an unshared pair of electrons and is therefore nucleophilic. Like a carbanion, ternary nitrogen can react both with saturated carbon from which a group can be displaced with the covalent bonding-pair (S_N2 reaction),

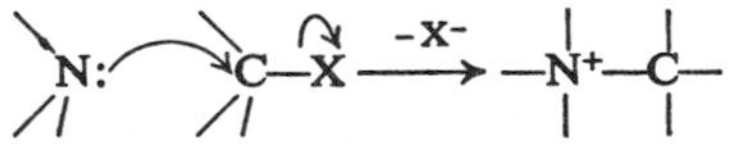

$$\begin{array}{c}\diagdown \\ \diagup \end{array}N: \quad \begin{array}{c}\diagdown \\ \diagup \end{array}C-X \xrightarrow{-X^-} -\overset{+}{N}-\overset{|}{\underset{|}{C}}-$$

and with unsaturated carbon, leading initially to an adduct, e.g.

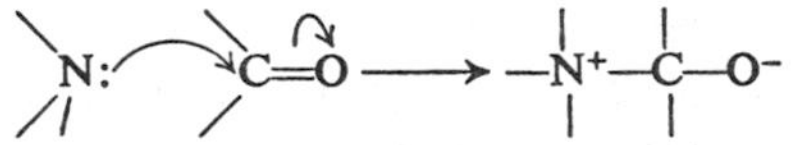

$$\begin{array}{c}\diagdown \\ \diagup \end{array}N: \quad \begin{array}{c}\diagdown \\ \diagup \end{array}C=O \longrightarrow -\overset{+}{N}-\overset{|}{\underset{|}{C}}-O^-$$

The products of these reactions depend on the structures of the reactants. In the S_N2 process, if the nitrogen atom is not bonded to hydrogen, a quaternary salt is formed:

$$R_3N: \quad \begin{array}{c}\diagdown \\ \diagup \end{array}C-X \longrightarrow R_3\overset{+}{N}-\overset{|}{\underset{|}{C}}- + X^-$$

but if one or more hydrogen atoms are bonded to nitrogen, a proton may be lost to a basic species (such as a second molecule of the nitrogen compound), giving a new ternary nitrogen species which can react further, e.g.

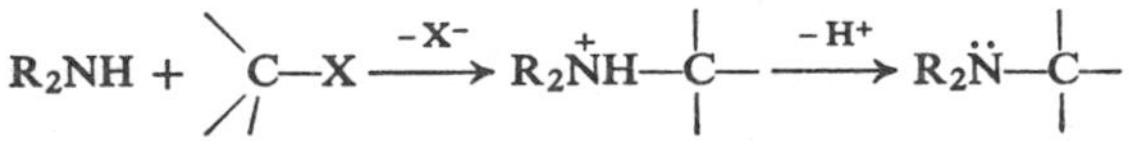

$$R_2NH + \begin{array}{c}\diagdown \\ \diagup \end{array}C-X \xrightarrow{-X^-} R_2\overset{+}{N}H-\overset{|}{\underset{|}{C}}- \xrightarrow{-H^+} R_2\overset{..}{N}-\overset{|}{\underset{|}{C}}-$$

322

Reactions at unsaturated centres may be followed by a tautomeric shift, to give an adduct, e.g.

$$R_2\overset{..}{N}H \quad \overset{\backslash}{\underset{/}{C}}=O \longrightarrow R_2\overset{+}{N}H-\overset{|}{\underset{|}{C}}-O^- \longrightarrow R_2N-\overset{|}{\underset{|}{C}}-OH$$

or, if the carbonyl group is attached to an electronegative leaving-group, by the elimination of a proton and an anion, e.g.

$$R_2\overset{..}{N}H \quad \overset{\backslash}{\underset{\underset{Cl}{|}}{C}}=O \longrightarrow R_2\overset{+}{N}H-\overset{|}{\underset{Cl}{C}}-O^- \xrightarrow{-H^+,-Cl^-} R_2N-\overset{|}{C}=O$$

In the former case, elimination from the adduct commonly ensues in appropriate structural situations, e.g.

$$R\overset{..}{N}H_2 \quad \overset{\backslash}{\underset{/}{C}}=O \longrightarrow R\overset{+}{N}H_2-\overset{|}{\underset{|}{C}}-O^- \longrightarrow RNH-\overset{|}{\underset{|}{C}}-OH \xrightarrow{-H_2O} RN=\overset{/}{\underset{\backslash}{C}}$$

Thus, the overall process may consist of substitution at saturated carbon, addition to unsaturated carbon (sometimes followed by elimination), or substitution at unsaturated carbon. Each of these processes has an exact counterpart in the reactions of carbanions (Chapter 7).

The commonest nitrogen nucleophiles are ammonia and its derivatives. These may be alkyl- or aryl-derivatives (primary, secondary, or tertiary amines), nitrogen-containing derivatives such as hydrazine, NH_2-NH_2, and related compounds (e.g. $PhNH-NH_2$), and oxygen-containing derivatives such as hydroxylamine, $HO-NH_2$.

Nitrogen attached to electron-attracting groups of $-M$ type has very little nucleophilic reactivity. The reason is that these systems are resonance-stabilized, e.g.

$$R-\overset{..}{\underset{\overset{||}{O}}{C}}-\overset{..}{N}H_2 \leftrightarrow R-\overset{}{\underset{\overset{|}{O^-}}{C}}=\overset{+}{N}H_2$$

and the loss of resonance energy which occurs when the nitrogen atom bonds to carbon renders the transition states for the reactions of such high energy-content, relative to the reactants, that reaction is very slow. For example, amides, imides, sulphonamides, and nitramide (NO_2-NH_2) are very weak nucleophiles.

Amide ions (e.g. NH_2^-, present in a solution of sodamide in liquid ammonia) are considerably more nucleophilic than the corresponding amines. This is of particular value in effecting reactions with imides: the grouping —CO—NH—CO— is relatively acidic (p. 66) so that, in basic conditions, the nucleophile (—CO—$\bar{\text{N}}$—CO—) is present and, for example, can react with alkyl halides (pp. 326, 350).

Other nitrogen nucleophiles include azide ion ($\bar{\text{N}}{=}\overset{+}{\text{N}}{=}\bar{\text{N}}$) and nitrite ion ($\bar{\text{O}}$—N=O). The latter is an *ambident* nucleophile, being capable of reacting at either nitrogen or oxygen. Its mode of reaction is determined by the structure of the second reactant and the reaction conditions.

(b) ELECTROPHILIC NITROGEN

Nitrogen is electrophilic both in certain cations such as aromatic diazonium ions, $\text{ArN}_2{}^+$ (13.4), the nitronium ion, $\text{O}{=}\overset{+}{\text{N}}{=}\text{O}$, and the nitrosonium ion, $\overset{+}{\text{N}}{=}\text{O}$, and in neutral molecules such as alkyl nitrites, R—O—N=O, and nitroso compounds, R—N=O. The former class of electrophile is of only limited importance in aliphatic chemistry, although of major importance in aromatic chemistry (11.4).

The nitrogen atom in alkyl nitrites and related compounds is electrophilic by virtue of the fact that the addition of a nucleophile gives a relatively stable oxyanion; reaction is completed by the elimination of an anion, e.g.

$$\text{Nu:} \overset{\frown}{} \text{N}{=}\overset{\frown}{\text{O}} \longrightarrow \text{Nu—N}\overset{\frown}{\text{—O}^-} \xrightarrow{-\text{RO}^-} \text{Nu—N}{=}\text{O}$$
$$\text{RO} \qquad\qquad\qquad \text{RO}$$

This reaction is the analogue of that between a nucleophile and a carboxylic ester (p. 134).

The carbon centres to which electrophilic nitrogen bonds include both carbanions and neutral molecules such as enols which are particularly reactive towards electrophiles.

10.2 Substitution by Nucleophilic Nitrogen at Saturated Carbon

(a) REACTIONS OF AMMONIA AND AMINES

The treatment of a primary alkyl halide with ammonia gives, initially, the conjugate acid of the corresponding primary amine, e.g.

$$\text{H}_3\text{N:} \overset{\frown}{} \text{CH}_3\overset{\frown}{\text{—I}} \longrightarrow \text{H}_3\overset{+}{\text{N}}\text{—CH}_3 + \text{I}^-$$

This then reacts with more ammonia in an acid-base equilibrium,

$$\text{CH}_3\text{—NH}_3{}^+ + \text{NH}_3 \rightleftharpoons \text{CH}_3\text{—NH}_2 + \text{NH}_4{}^+$$

and the primary amine so generated reacts with a second molecule of the halide:

$$CH_3NH_2 + CH_3I \longrightarrow (CH_3)_2\overset{+}{N}H_2 + I^-$$

Further reactions lead to the tertiary amine and the quaternary ammonium salt.

Consequently a mixture of products is usually formed, and the simple alkylation of ammonia is an inefficient process for the preparation of secondary and tertiary amines. However, primary amines can be obtained in good yield by the use of a considerable excess of ammonia, for then the amine formed is a relatively ineffective competitor for reaction with the halide.

Secondary amines can be obtained by treating an alkyl halide with an excess of a primary amine, but this is a wasteful process and it is often more convenient to employ one of the methods for the reductive condensation of a carbonyl compound and a primary amine (p. 651), or to use tertiary aromatic amines as starting materials (p. 430). Tertiary amines are obtained by the alkylation of secondary amines. Quaternary ammonium salts may be made either by the alkylation of tertiary amines or, for quaternary salts of the type R_4N^+, by the treatment of ammonia with a large excess of an alkyl halide.

Halides. As in other S_N2 processes, aryl and vinyl halides are inert towards amines and ammonia. Amongst alkyl halides, primary halides react efficiently but secondary halides usually give a significant yield of an elimination product and tertiary halides give entirely the elimination product (i.e. ammonia and amines react preferentially as bases and bring about E2 elimination; *cf.* p. 128). Tertiary alkyl amines such as $(CH_3)_3C—NH_2$ are usually prepared either by reaction of a Grignard reagent with *O*-methylhydroxylamine (p. 215) or by the *Ritter reaction.* The latter consists of treating a tertiary alcohol, or the corresponding olefin, with concentrated sulphuric acid and a nitrile. A tertiary carbonium ion is formed which is attacked by the nitrile;* the resulting quaternary ion is decomposed by water to give an amide which, in the acidic conditions, is hydrolyzed to the amine, e.g.

$$(CH_3)_3C—OH \xrightarrow[-H_2O]{H^+} (CH_3)_3C^+ \xrightarrow{N\equiv C—R} (CH_3)_3C—\overset{+}{N}\equiv C—R$$

$$\xrightarrow[-H^+]{H_2O} (CH_3)_3C—N=\underset{\underset{OH}{|}}{C}—R \xrightarrow{tautomerizes} (CH_3)_3C—NHCOR$$

$$\xrightarrow{H^+—H_2O} (CH_3)_3C—NH_2 + RCO_2H$$
$$\text{t-Butylamine}$$

*The nitrogen atom of a nitrile is sufficiently nucleophilic to react with a carbonium ion but not so as to react with the less electrophilic carbon atoms of halides, ketones, etc. The greater nucleophilicity of an amine than a nitrile has a parallel in the greater basicity of the former; p. 69.

Gabriel's method. The problem which arises in attempts to make primary amines by the monoalkylation of ammonia (namely, that further alkylation in some degree is unavoidable) is circumvented by a procedure devised by Gabriel. This is based on the fact that phthalimide, having an acidic N—H group (p. 66), reacts with base to give a nitrogen-containing anion which, as a strong nucleophile, displaces on alkyl halides. Hydrolysis of the resulting compound with alkali gives the primary amine.

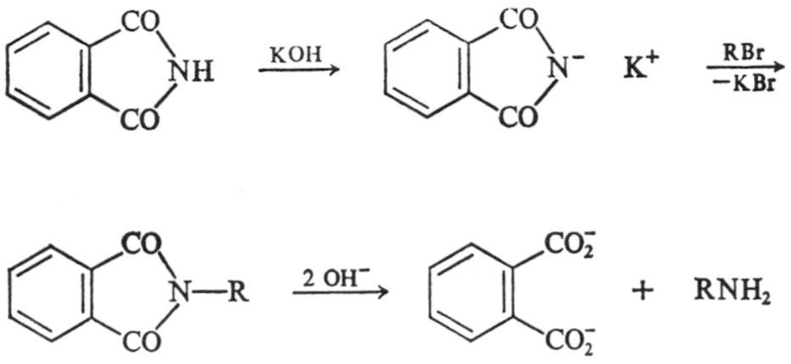

For example, potassium phthalimide and excess of ethylene dibromide react at 180–190°C to give β-phthalimidylethyl bromide in 75% yield [1]; hydrolysis gives β-bromoethylamine.

Other leaving groups. As usual in S_N2 reactions, alcohols are relatively inert to displacement by nucleophilic nitrogen. However, reaction can be brought about in vigorous conditions in the presence of a Lewis acid, which aids the departure of hydroxide ion by coordinating with oxygen. For example, methylamine is made industrially by the reaction of methanol with ammonia at 400°C under pressure, in the presence of alumina:

$$CH_3OH + NH_3 \xrightarrow[400°C]{Al_2O_3} CH_3NH_2 + H_2O$$

Ethers, like alcohols, are relatively inert, but strained cyclic ethers react. For example, ethylene oxide reacts with secondary amines to give β-hydroxy tertiary amines,

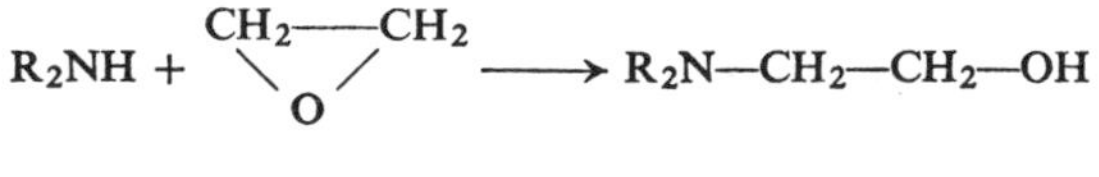

$$R_2NH + \overset{\displaystyle CH_2\!-\!\!-\!\!-\!CH_2}{\underset{\displaystyle O}{\diagdown \quad \diagup}} \longrightarrow R_2N\!-\!CH_2\!-\!CH_2\!-\!OH$$

and ammonia reacts with ethylene oxide to give successively ethanolamine, diethanolamine, and triethanolamine,

$$NH_3 + \overset{CH_2-\!-\!-CH_2}{\underset{O}{\diagdown\diagup}} \longrightarrow H_2NCH_2CH_2OH$$
Ethanolamine

$$\overset{CH_2-\!CH_2}{\underset{O}{\diagdown\diagup}} \longrightarrow HN(CH_2OH)_2 \overset{CH_2-\!CH_2}{\underset{O}{\diagdown\diagup}} \longrightarrow N(CH_2OH)_3$$
Diethanolamine   Triethanolamine

These reactions are analogous to those of small ring cyclic ethers with other nucleophiles, such as hydroxide ion (p. 571) and Grignard reagents (p. 212); in each case reaction is facilitated by the release of ring-strain which occurs in passage to the transition state.

A related example is the reaction of ammonia with β-propiolactone to give β-alanine:

$$H_3N: \overset{CH_2-CH_2}{\underset{O-\!-\!CO}{\diagdown\diagup}} \longrightarrow H_3\overset{+}{N}CH_2CH_2CO_2^-$$
β-Alanine

Here, the relief of ring-strain in the transition state so lowers the free energy of activation as to give this reaction preference over the normal reaction of ammonia with lactones (or esters), i.e. attack at the carbonyl group to give an amide (the base-catalyzed hydrolysis of β-propiolactone occurs in a similar manner).

Intramolecular nucleophilic substitution can occur readily to give three, five-, and six-membered alicyclic rings (but not other ring-sizes; cf. p. 91). For example, distillation of (the zwitterionic) ethanolamine hydrogen sulphate with aqueous caustic soda gives ethyleneimine,

$$H_2NCH_2CH_2OH \xrightarrow{H_2SO_4} H_3\overset{+}{N}CH_2CH_2OSO_2O^- \xrightarrow{NaOH}$$

$$\left[\overset{CH_2-CH_2-OSO_2O^-}{\underset{H_2N:}{\diagdown\diagup}} \right] \xrightarrow{-H^+, \ -SO_4^{2-}} \overset{CH_2-\!-\!-CH_2}{\underset{\underset{H}{N}}{\diagdown\diagup}}$$
Ethyleneimine

and pyrrolidine and piperidine are formed by heating the dihydrochlorides of tetramethylenediamine and pentamethylenediamine, respectively, e.g.

$$\begin{array}{c}CH_2-CH_2 \\ | \qquad | \\ CH_2 \quad CH_2-\overset{+}{N}H_3 \\ \overset{+}{N}H_3\end{array} \ 2\ Cl^- \xrightarrow[-HCl]{heat} \left[\begin{array}{c}CH_2-CH_2 \\ | \qquad | \\ CH_2 \quad CH_2-\overset{+}{N}H_3 \\ \overset{..}{N}H_2\end{array} \right] Cl^- \xrightarrow{-H^+, \ -NH_3, \ -Cl^-} \begin{array}{c}CH_2-\!-\!-CH_2 \\ | \qquad | \\ CH_2 \quad CH_2 \\ \underset{H}{N}\end{array}$$
Pyrrolidine

Carbylamine reaction. Primary amines react with chloroform and alkali-metal hydroxides to give carbylamines (isocyanides). (The unpleasant odour of these compounds has in the past been used as a test for primary amines.) Reaction probably occurs *via* dichlorocarbene (p. 124).

$$CHCl_3 \xrightarrow{OH^-} CCl_3^- \longrightarrow :CCl_2 + Cl^-$$

$$RNH_2 \xrightarrow{:CCl_2} R\overset{+}{N}H_2-\overset{-}{C}Cl_2 \longrightarrow RNH-CHCl_2 \xrightarrow{-HCl} RN=CHCl$$

$$\xrightarrow{-HCl} RN=C: \leftrightarrow R\overset{+}{N}\equiv\overset{-}{C}$$

(*b*) REACTIONS OF OTHER NITROGEN NUCLEOPHILES

(*i*) *Nitrite.* Metal nitrites can react with alkyl halides at both nitrogen and oxygen, giving nitro-compounds and nitrites, respectively.

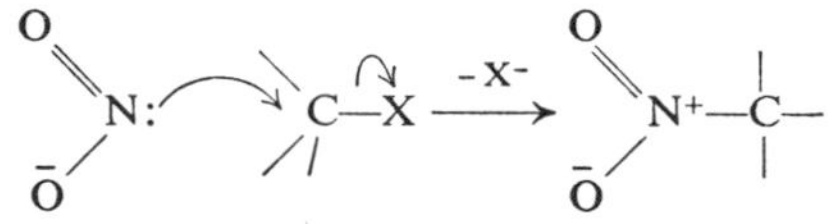

A nitro compound

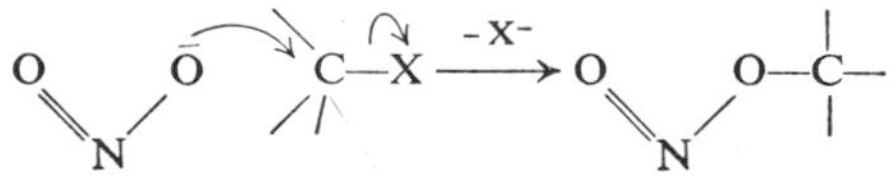

A nitrite

The proportions of the two products are determined by the structures of the reactants and the reaction conditions. First, with a given halide, silver nitrite suspended in ether gives a higher proportion of nitro-compound than alkali-metal nitrites. Secondly, with a given nitrite, the proportion of nitro-compound decreases from primary to secondary to tertiary halides, e.g.

$$CH_3CH_2CH_2CH_2Br \xrightarrow[ether]{AgNO_2} CH_3CH_2CH_2CH_2-NO_2 + CH_3CH_2CH_2CH_2-ONO$$
$$70\% \qquad\qquad 15\%$$

$$(CH_3)_2CHBr \xrightarrow[ether]{AgNO_2} (CH_3)_2CH-NO_2 + (CH_3)_2CH-ONO$$
$$20\% \qquad\qquad 30\%$$

$$(CH_3)_3CCl \xrightarrow[ether]{AgNO_2} (CH_3)_3C-NO_2 + (CH_3)_3C-ONO$$
$$(trace) \qquad\qquad 60\%$$

Thirdly, although alkali-metal nitrites normally give very low yields of nitro-compounds, the use of the polar solvents dimethylformamide and dimethyl sulphoxide markedly increases the yields (to *ca.* 50–60%) from primary and secondary alkyl bromides and iodides. However, tertiary halides undergo elimination in these conditions (e.g. t-butyl bromide gives isobutylene; *cf.* p. 128).

Nitromethane can be obtained conveniently, although not in high yield, by treating sodium chloroacetate with sodium nitrite. The resulting nitro-compound readily undergoes decarboxylation (*cf.* the analogous β-keto-acids) and nitromethane may be isolated by distillation, the nitrite byproduct remaining in solution.

$$Na^+ \ {}^-O_2C-CH_2-Cl \xrightarrow{\ Na^+ \ NO_2^-\ } Na^+ \ {}^-O_2C-CH_2-NO_2 + Na^+ \ {}^-O_2C-CH_2-ONO$$

$$Na^+ \ {}^-O_2C-CH_2-NO_2 \xrightarrow[-CO_2]{heat} \left[{}^-\overset{}{C}H_2-NO_2 \leftrightarrow CH_2=\overset{+}{N}\diagdown^{\overset{\bar{O}}{}}_{\diagdown \bar{O}} \right] Na^+ \xrightarrow{H_2O} CH_3-NO_2$$
$$35\%$$

(*ii*) *Azide ion.* Halides react with azides such as sodium azide to give alkyl azides. The reaction provides a method for preparing primary amines *via* catalytic reduction of the azide, e.g.

$$CH_3-\underset{Br}{\underset{|}{CH}}-CO_2H \xrightarrow[-Br^-]{\bar{N}=\overset{+}{N}=\bar{N}} CH_3-\underset{\underset{N^-}{\overset{N^+}{\underset{\|}{\overset{\|}{N}}}}}{\underset{|}{CH}}-CO_2H \xrightarrow{H_2-Pd} CH_3-\underset{NH_2}{\underset{|}{CH}}-CO_2H$$
$$\alpha\text{-Alanine}$$

(*iii*) *Hydrazine.* Hydrazine reacts with alkyl halides to give alkylhydrazines. The introduction of the first alkyl group increases the nucleophilicity of the alkylated nitrogen (since alkyl groups are electron-releasing relative to hydrogen), so that further alkylation tends to occur, e.g.

$$NH_2-NH_2 \xrightarrow[-HI]{CH_3I} CH_3NH-NH_2 \xrightarrow[-HI]{CH_3I} (CH_3)_2N-NH_2$$
$$NN\text{-Dimethylhydrazine}$$

Monoalkylhydrazines are obtained more satisfactorily from alkylureas (p. 337) by the Hofmann bromination procedure (p. 468):

$$RNHCONH_2 \xrightarrow{Br_2-KOH} RNHNH_2$$

10.3 Addition of Nucleophilic Nitrogen to Unsaturated Carbon

(*a*) REACTION WITH ALDEHYDES AND KETONES

(*i*) *Ammonia.* Ammonia reacts with aldehydes and certain ketones, but the products are usually complex. With aldehydes, the first step is a simple nucleophilic addition,

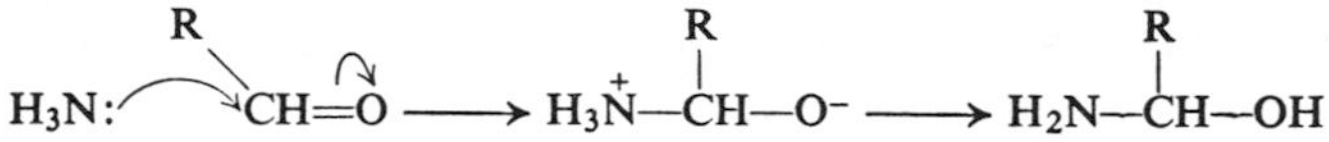

$$H_3N: \quad CH{=}O \longrightarrow H_3\overset{+}{N}{-}\underset{R}{CH}{-}O^- \longrightarrow H_2N{-}\underset{R}{CH}{-}OH$$

but the products, aldehyde-ammonias, are unstable except when the aldehydic carbon is attached to a strongly electron-attracting group. For example, an ethereal solution of acetaldehyde absorbs ammonia and gives a white crystalline precipitate which is probably a polymer of the adduct,

$$CH_3CHO + NH_3 \longrightarrow CH_3CH\overset{NH_2}{\underset{OH}{\big\langle}} \longrightarrow Polymer$$

but it cannot be obtained pure, whereas chloral gives a stable adduct, *

$$Cl_3C{-}CHO + NH_3 \longrightarrow Cl_3C{-}CH\overset{NH_2}{\underset{OH}{\big\langle}}$$

The fate of the aldehyde-ammonia depends on the structure of the aldehyde. Formaldehyde is so reactive towards nucleophiles that the primary addition product reacts further, giving ultimately urotropine (hexamethylenetetramine) whose structure is similar to that of adamantane (p. 174):

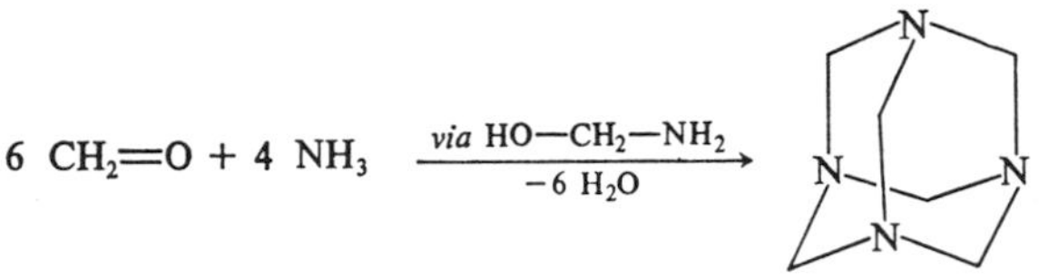

$$6\ CH_2{=}O + 4\ NH_3 \quad \xrightarrow[-6\ H_2O]{via\ HO{-}CH_2{-}NH_2}$$

Urotropine

Many aromatic aldehydes (e.g. benzaldehyde and furfural) give condensation products whose formation probably involves dehydration of the initial adduct

*The stability of the ammonia-adducts of aldehydes parallels roughly the stability of the corresponding hydrates (*gem*-diols); e.g. chloral forms a stable hydrate whereas that of acetaldehyde exists only in solution (p. 113). Likewise, *gem*-diamines, RCH(NH_2)_2, are unstable.

followed by further reaction:

$$\text{ArCH=O} \xrightarrow{\text{NH}_3} \text{ArCH} \begin{smallmatrix} \diagup \text{NH}_2 \\ \diagdown \text{OH} \end{smallmatrix} \xrightarrow{-\text{H}_2\text{O}} \text{ArCH=NH} \xrightarrow{\text{NH}_3}$$

$$\text{ArCH} \begin{smallmatrix} \diagup \text{NH}_2 \\ \diagdown \text{NH}_2 \end{smallmatrix} \xrightarrow{\text{2 ArCHO}} \text{ArCH} \begin{smallmatrix} \diagup \text{NH—CH(OH)Ar} \\ \diagdown \text{NH—CH(OH)Ar} \end{smallmatrix} \xrightarrow{-\text{2 H}_2\text{O}} \text{ArCH} \begin{smallmatrix} \diagup \text{N=CHAr} \\ \diagdown \text{N=CHAr} \end{smallmatrix}$$

Most other aldehydes give polymers of the corresponding imines, but if the reaction is carried out in the presence of a reagent which can react with the imine, the process can yield useful products. Two examples are shown.

(1) *Strecker synthesis.* The reaction between the aldehyde and ammonia is carried out in the presence of cyanide ion. The dehydration product of the aldehyde-ammonia reacts with cyanide to give an α-aminonitrile:

$$\text{RCHO} + \text{NH}_3 \xrightarrow{-\text{H}_2\text{O}} \text{RCH=NH} \xrightarrow{\text{CN}^-} \text{RCH} \begin{smallmatrix} \diagup \bar{\text{N}}\text{H} \\ \diagdown \text{CN} \end{smallmatrix} \xrightarrow[-\text{OH}^-]{\text{H}_2\text{O}} \text{RCH} \begin{smallmatrix} \diagup \text{NH}_2 \\ \diagdown \text{CN} \end{smallmatrix}$$

It is convenient to employ aqueous ammonium chloride and sodium cyanide as the reagents, ammonia being generated *in situ* by hydrolysis. Hydrolysis of the α-aminonitrile gives an α-amino-acid, and the total process constitutes a useful method of synthesis of these compounds.

(2) *Reductive amination.* When the reaction between an aldehyde and ammonia is carried out in the presence of hydrogen and a hydrogenation catalyst such as Raney nickel (p. 612), the imine is reduced as it is formed to a primary amine:

$$\text{RCHO} + \text{NH}_3 \xrightarrow{-\text{H}_2\text{O}} [\text{RCH=NH}] \xrightarrow{\text{H}_2\text{—Ni}} \text{RCH}_2\text{—NH}_2$$

Alternatively, the *Leuckart procedure* may be used. The aldehyde is heated with ammonium formate at about 170°C; the ammonia released from ammonium formate forms the imine and this is reduced by formic acid (hydride-ion transfer; p. 650) with liberation of carbon dioxide:

$$\text{NH}_4^+\text{HCO}_2^- \longrightarrow \text{NH}_3 + \text{HCO}_2\text{H}$$

$$\text{RCH=O} + \text{NH}_3 \xrightarrow{-\text{H}_2\text{O}} \text{RCH=NH} \xrightarrow{-\text{CO}_2} \text{RCH}_2\text{—}\bar{\text{N}}\text{H} \xrightarrow[-\text{OH}^-]{\text{H}_2\text{O}} \text{RCH}_2\text{—NH}_2$$

The scope and limitations of these reductive condensations are described later (p. 650).

Ketones do not give adducts with ammonia. Reaction at the carbonyl group occurs reversibly, but only leads to products if further reaction can follow so as to give a stable product. Three examples are illustrative.

First, acetoacetic ester reacts to give an imine which tautomerizes to a conjugated amine (the nitrogen analogue of the stable enolic form of acetoacetic ester):

$$\text{CH}_3\text{COCH}_2\text{CO}_2\text{Et} \xrightarrow{\text{NH}_3} \underset{\underset{\text{NH}_2}{|}}{\overset{\overset{\text{OH}}{|}}{\text{CH}_3\text{C}}}\text{—CH}_2\text{CO}_2\text{Et} \xrightarrow{-\text{H}_2\text{O}}$$

$$\underset{\underset{\text{NH}}{\|}}{\text{CH}_3\text{C}}\text{—CH}_2\text{CO}_2\text{Et} \rightleftharpoons \underset{\underset{\text{NH}_2}{|}}{\text{CH}_3\text{C}}\text{=CHCO}_2\text{Et}$$

Secondly, acetonylacetone reacts at one carbonyl group to give an adduct whose amino-group is suitably placed to react intramolecularly at the second carbonyl group. Dehydration occurs to give 2,5-dimethylpyrrole.

Thirdly, the imine can be trapped by the two reductive methods described for aldehydes:

$$\text{R}_2\text{C}=\text{O} + \text{NH}_3 \xrightarrow{-\text{H}_2\text{O}} [\text{R}_2\text{C}=\text{NH}] \xrightarrow[\text{or HCO}_2\text{H}]{\text{H}_2\text{—Ni}} \text{R}_2\text{CH—NH}_2$$

Acetone behaves in a different manner, first undergoing self-condensation under the influence of ammonia which here acts as a base rather than as a nucleophile. Two products are isolable. The first, diacetoneamine, is derived by Michael-type addition of ammonia to the condensation product, mesityl oxide:

$$2\ \text{CH}_3\text{COCH}_3 \xrightarrow[-\text{H}_2\text{O}]{\text{NH}_3} (\text{CH}_3)_2\text{C}=\text{CH—CO—CH}_3 \xrightarrow{\text{NH}_3} \underset{\underset{\text{NH}_2}{|}}{(\text{CH}_3)_2\text{C}}\text{—CH}_2\text{—CO—CH}_3$$

$$\text{Diacetoneamine}$$

The second product, triacetoneamine, is derived from a similar reaction on the product of further condensation, phorone:

$$(CH_3)_2C{=}CHCOCH_3 + CH_3COCH_3 \xrightarrow[-H_2O]{NH_3} (CH_3)_2C{=}CHCOCH{=}C(CH_3)_2$$

Phorone

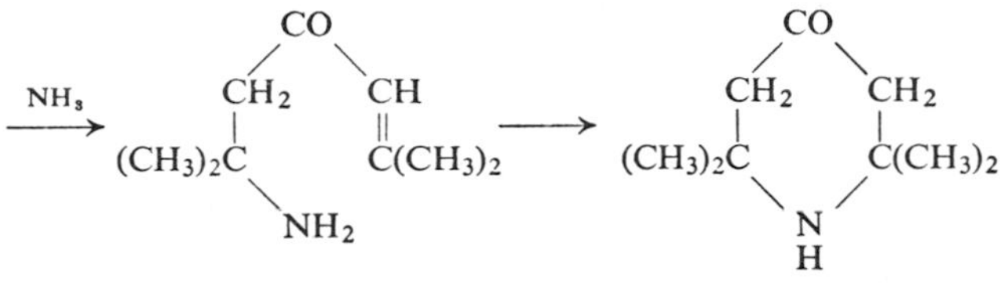

Triacetoneamine

(*ii*) *Primary amines.* As with ammonia, aldehydes and ketones react with primary amines initially by addition. Dehydration then occurs to give an imine:

$$RR'C{=}O + R''NH_2 \longrightarrow RR'C(NHR'')(OH) \xrightarrow{-H_2O} RR'C{=}NR''$$

The imines, except for those from aromatic amines, are unstable and tend to polymerize unless they are trapped in a further addition process. For example, reductive amination, catalytically or by the Leuckart procedure, gives secondary amines,

$$RR'C{=}O + R''NH_2 \longrightarrow [RR'C{=}NR''] \xrightarrow[\text{or } HCO_2H]{H_2\text{—}Ni} RR'CH\text{—}NHR''$$

Imines (Schiff bases) from aromatic amines are usually stable and can be isolated. For example, benzaldehyde and aniline react exothermically to give benzylideneaniline in about 85% yield [1]:

$$PhCHO + H_2NPh \xrightarrow{-H_2O} PhCH{=}NPh$$

Benzylideneaniline

The stability of these imines is evidently associated with the conjugation between the aromatic ring of the amine and the imino double bond.

α-Dicarbonyl compounds react with aromatic *ortho*-diamines by consecutive addition-elimination processes to give quinoxalines:

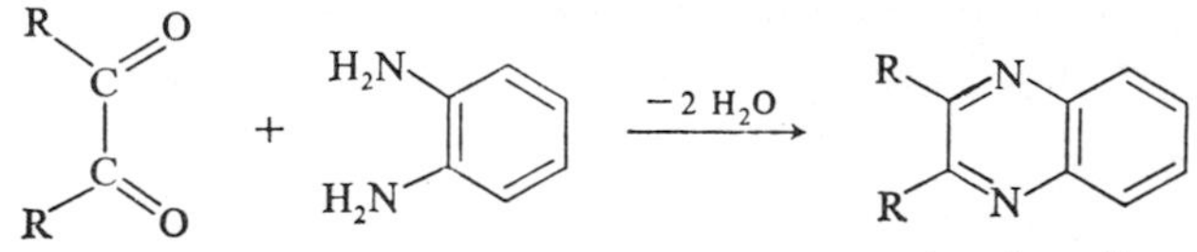

A quinoxaline

Apart from the synthetic value of the reaction, phenanthrenequinone may be used to characterize aromatic *ortho*-diamines and *o*-phenylenediamine may be used to characterize α-dicarbonyl compounds.

(*iii*) *Secondary amines.* Aldehydes and ketones which possess a hydrogen atom attached to an α-carbon atom react with secondary amines to give *enamines*, e.g.

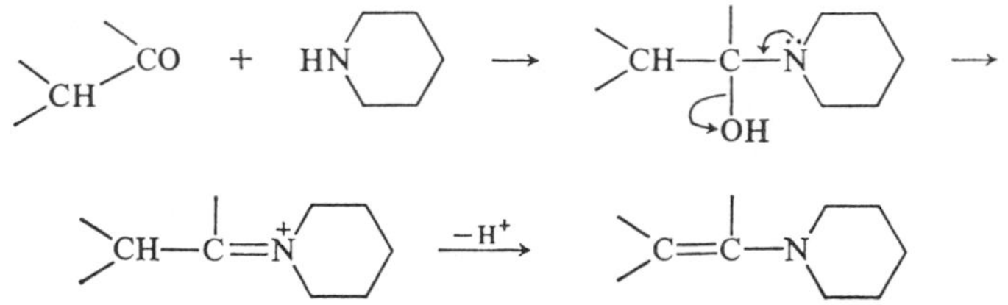

The synthetic uses of enamines are discussed elsewhere (p. 246).

The reductive amination procedures lead to tertiary amines:

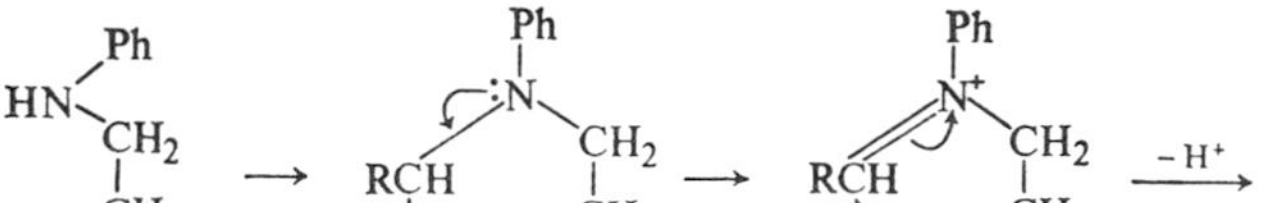

Dianilinoethane reacts with aldehydes by consecutive nucleophilic reactions, the second occurring intramolecularly:

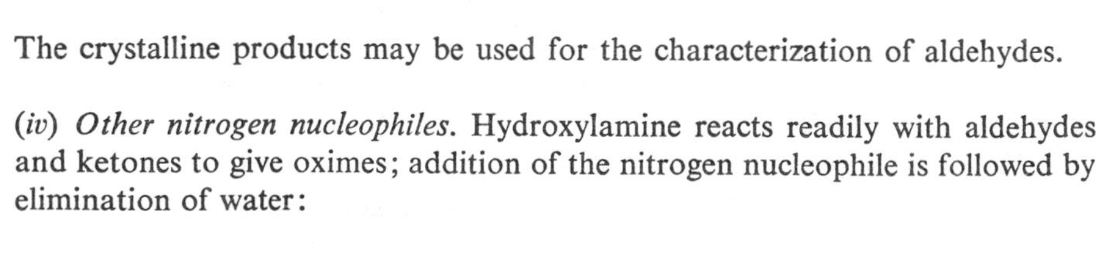

The crystalline products may be used for the characterization of aldehydes.

(*iv*) *Other nitrogen nucleophiles.* Hydroxylamine reacts readily with aldehydes and ketones to give oximes; addition of the nitrogen nucleophile is followed by elimination of water:

$$RR'C{=}O + NH_2{-}OH \longrightarrow RR'C \Big\langle \begin{array}{c} NH{-}OH \\ \\ OH \end{array} \xrightarrow{-H_2O} RR'C{=}N \Big\backslash OH$$

The reaction may be carried out simply by treating a mixture of the carbonyl compound and hydroxylamine hydrochloride with sodium hydroxide; the hydroxylamine reacts as it is released from its salt. In this way, for example, benzophenone oxime may be obtained in over 98% yield [2]. Alternatively, for less reactive carbonyl compounds, the reaction is carried out by adding sodium acetate to the carbonyl compound and hydroxylamine chloride; hydroxylamine is then present in a buffered medium of near-neutral pH. The reason for adopting this procedure is that the reactivity of a carbonyl group towards nucleophiles is increased by the addition of an acid (p. 113) but if the pH is too low hydroxylamine is present almost entirely as its unreactive conjugate acid. As a result of these conflicting requirements the reaction has a pH-optimum of about 5.

Oximes are used both for the characterization of aldehydes and ketones, since they are stable, crystalline compounds, and in synthesis; ketoximes give amines *via* the Beckmann rearrangement (p. 470) and aldoximes can be dehydrated by acetic anhydride to nitriles:

$$\begin{array}{c} R \\ \diagdown \\ H \diagup \end{array} C{=}N \diagdown OH \xrightarrow[-HOAc]{Ac_2O} \begin{array}{c} R \\ \diagdown \\ H \diagup \end{array} C{=}N \diagdown OAc \xrightarrow{-HOAc} R{-}C{\equiv}N$$

Hydrazine reacts with aldehydes and ketones analogously to hydroxylamine except that the first product, a hydrazone, can react with a second molecule of the carbonyl compound to give an azine, so that a mixture of two products is usually obtained:

$$RR'C{=}O \xrightarrow[-H_2O]{NH_2NH_2} RR'C{=}N{-}NH_2 \xrightarrow[-H_2O]{RR'CO} RR'C{=}N{-}N{=}CRR'$$

$$\text{A hydrazone} \qquad\qquad\qquad \text{An azine}$$

In order to obtain derivatives for the characterization of the carbonyl compound it is more satisfactory to use a monosubstituted hydrazine; semicarbazide, phenylhydrazine, and 2,4-dinitrophenylhydrazine are commonly chosen.

$$RR'C{=}O + H_2N{-}NHCONH_2 \xrightarrow{-H_2O} RR'C{=}N{-}NHCONH_2$$

$$\text{A semicarbazone}$$

$$RR'C{=}O + H_2N{-}NHPh \xrightarrow{-H_2O} RR'C{=}N{-}NHPh$$

$$\text{A phenylhydrazone}$$

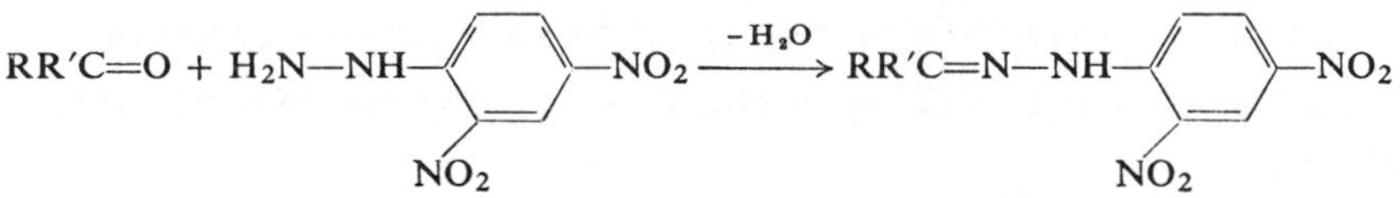

$$RR'C{=}O + H_2N{-}NH{-}\!\!\bigcirc\!\!{-}NO_2 \xrightarrow{-H_2O} RR'C{=}N{-}NH{-}\!\!\bigcirc\!\!{-}NO_2$$

A 2,4-dinitrophenylhydrazone

2,4-Dinitrophenylhydrazine is usually preferred to phenylhydrazine because it gives products which have higher melting points and are more easily isolated. A convenient small-scale method is to add hydrochloric acid dropwise to a mixture of the carbonyl compound and 2,4-dinitrophenylhydrazine in diethyleneglycol dimethyl ether; the derivative precipitates.

β-Dicarbonyl compounds and β-keto-esters react in a more complicated manner with hydroxylamine, hydrazine, and substituted hydrazines. The initial addition-elimination process gives a derivative which is stereochemically suited to intramolecular reaction, e.g.

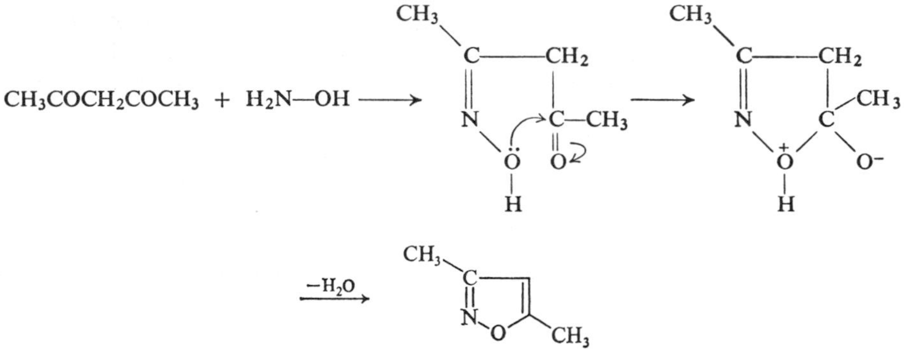

$$CH_3COCH_2COCH_3 + H_2N{-}OH \longrightarrow$$

An isoxazole

$$CH_3COCH_2CO_2Et + H_2N{-}NHPh \longrightarrow$$

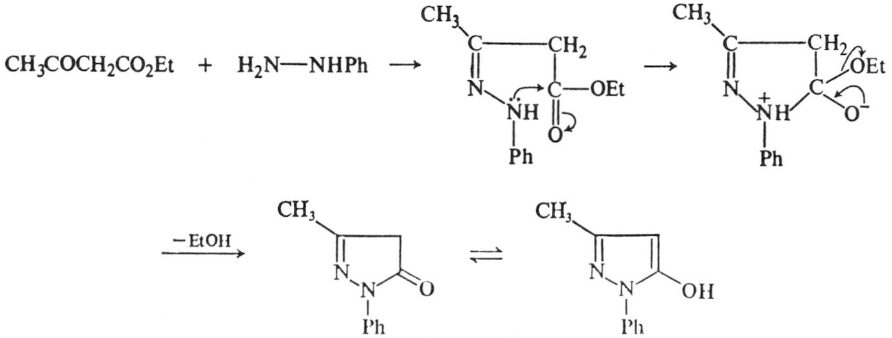

A pyrazolone

These reactions are important in the synthesis of heteroaromatic compounds and are discussed later (p. 670).

(b) REACTION WITH OTHER TYPES OF UNSATURATED CARBON

(i) *Nitriles.* Amidines may be obtained by heating nitriles with ammonium chloride:

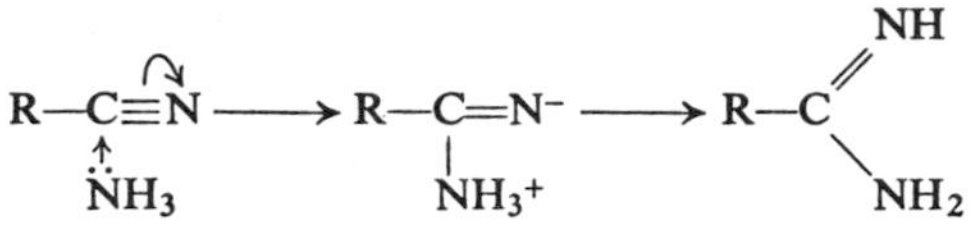

$$R—C≡N \longrightarrow R—C=N^- \longrightarrow R—C \overset{NH}{\underset{NH_2}{<}}$$
$$\underset{\overset{\uparrow}{\ddot{N}H_3}}{} \qquad \underset{NH_3^+}{}$$

An amidine

Likewise, cyanamide gives guanidine:

$$NH_2—CN + NH_3 \longrightarrow (NH_2)_2C=NH$$

Guanidine

(ii) *Cyanates and isocyanates.* The action of heat on ammonium cyanate gives urea, reaction probably occurring *via* ammonia and (the unstable) isocyanic acid:

$$NH_4^+NCO^- \rightleftharpoons NH_3 + H—N=C=O$$

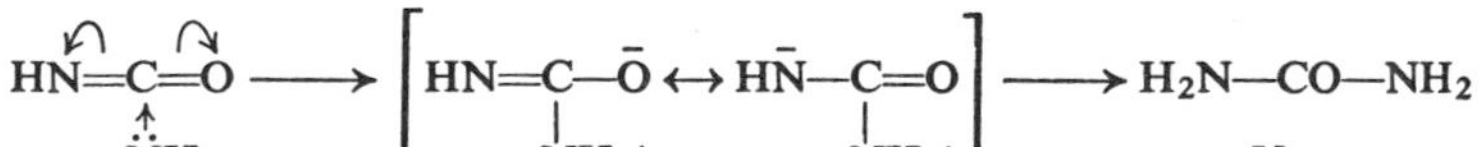

$$HN=C=O \longrightarrow \left[HN=C—\bar{O} \leftrightarrow H\bar{N}—C=O \right] \longrightarrow H_2N—CO—NH_2$$
$$\underset{\overset{\uparrow}{\ddot{N}H_3}}{} \qquad \underset{NH_3^+}{} \qquad \underset{NH_3^+}{}$$

Urea

The principle underlying this reaction can be applied to the synthesis of both mono- and di-substituted ureas. Mono-substituted ureas may be obtained by heating an amine with an alkali cyanate, e.g.

$$CH_3—NH_2 + K^+NCO^- \xrightarrow{H_2O} CH_3—NH—CO—NH_2 + KOH$$

N-Methylurea

$$NH_2—NH_2 + K^+NCO^- \xrightarrow{H_2O} NH_2—NH—CO—NH_2 + KOH$$

Semicarbazide

α-Carboalkoxyamines react further to give hydantoins, e.g.

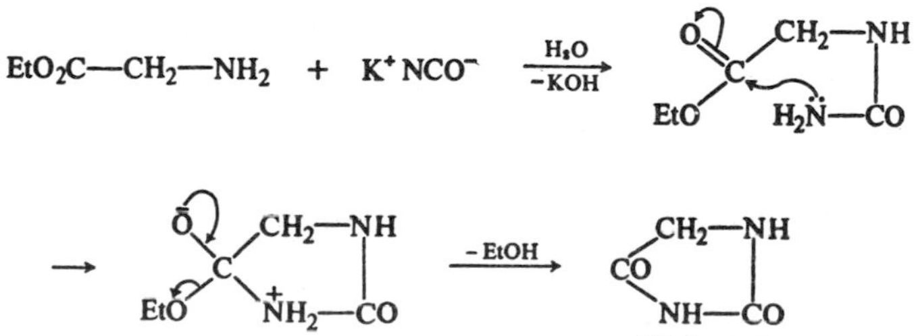

$$EtO_2C—CH_2—NH_2 + K^+NCO^- \xrightarrow[-KOH]{H_2O}$$

Hydantoin

Disubstituted ureas may be prepared by heating organic isocyanates with amines, e.g.

$$PhN{=}C{=}O + CH_3NH_2 \longrightarrow PhNH{-}CO{-}NHCH_3$$
$$\text{\textit{N}-methyl-\textit{N}'-phenylurea}$$

A reaction of this type is used in the formation of polyurethane foams. A di-isocyanate is first polymerized with a dihydric alcohol such as ethylene glycol:

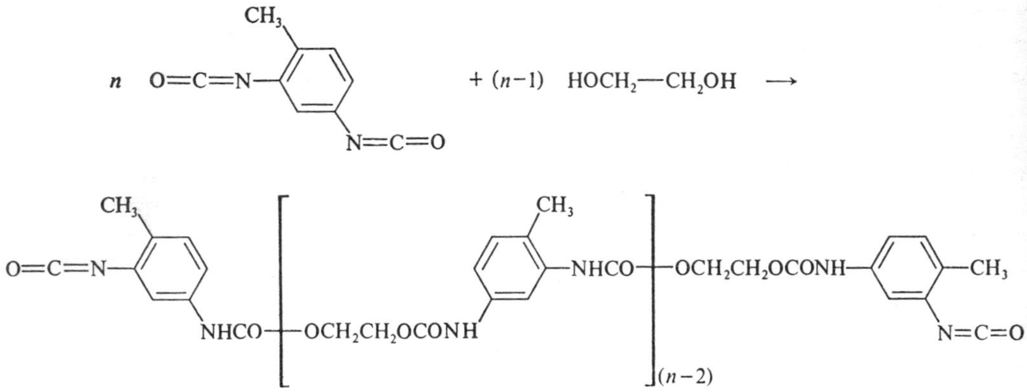

Treatment of the resulting polymer with water converts some of the terminal isocyanate groups into amino-groups,

$$\text{\small$\sim\sim$}Ar{-}N{=}C{=}O + H_2O \longrightarrow [Ar{-}NH{-}CO_2H] \longrightarrow \text{\small$\sim\sim$}Ar{-}NH_2 + CO_2$$

which react with the remaining isocyanate groups to extend the polymeric chain:

$$\text{\small$\sim\sim$}Ar{-}NH_2 + \text{\small$\sim\sim$}Ar'{-}N{=}C{=}O \longrightarrow \text{\small$\sim\sim$}Ar{-}NH{-}CO{-}NH{-}Ar\text{\small$\sim\sim$}$$

The carbon dioxide which is liberated in the first step is embedded in the polymer and causes the characteristic foam.

(*iii*) *Thiocyanates and isothiocyanates.* Thiourea is obtained by heating ammonium thiocyanate at 160–170°C (*cf.* the formation of urea from ammonium cyanate):

$$NH_4^+NCS^- \rightleftharpoons NH_3 + H{-}N{=}C{=}S$$

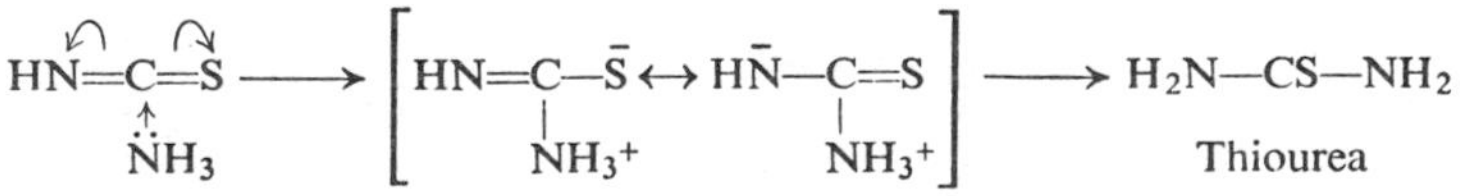

Likewise, organic isothiocyanates react with ammonia to give monosubstituted thioureas; e.g. methyl isothiocyanate gives *N*-methylthiourea in about 75% yield [3]:

$$CH_3—N{=}C{=}S + NH_3 \longrightarrow CH_3—NH—CS—NH_2$$
N-Methylthiourea

Reaction with amines gives disubstituted thioureas,

$$R—N{=}C{=}S + R'—NH_2 \longrightarrow R—NH—CS—NH—R'$$

Symmetrically substituted dithioureas are more easily obtained by treating amines with carbon disulphide in refluxing alcohol. Reaction probably occurs by addition of one molecule of the amine to carbon disulphide to give a dithiocarbamic acid; this loses hydrogen sulphide and the resulting isothiocyanate reacts with a second molecule of the amine, e.g.

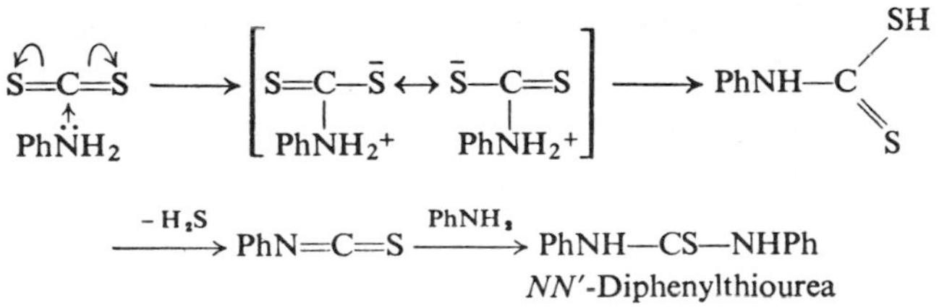

$$\xrightarrow{-H_2S} PhN{=}C{=}S \xrightarrow{PhNH_2} PhNH—CS—NHPh$$
NN'-Diphenylthiourea

The last reaction is reversible, and by treatment with acid to remove the amine, isothiocyanates can be obtained:

$$(PhNH)_2C{=}S \rightleftharpoons Ph—N{=}C{=}S + PhNH_2$$

$$PhNH_2 + H^+ \longrightarrow PhNH_3^+$$

The treatment of disubstituted thioureas with mercuric oxide (to remove hydrogen sulphide) has been used to obtain carbodi-imides, reagents used in peptide synthesis (p. 361):

$$RNHCSNHR \xrightarrow[-H_2S]{HgO} RN{=}C{=}NR$$
A carbodi-imide

10.4 Substitution by Nucleophilic Nitrogen at Unsaturated Carbon

A carbonyl group attached to a group capable of departing with the covalent bonding-pair is susceptible to substitution by nitrogen nucleophiles, e.g.

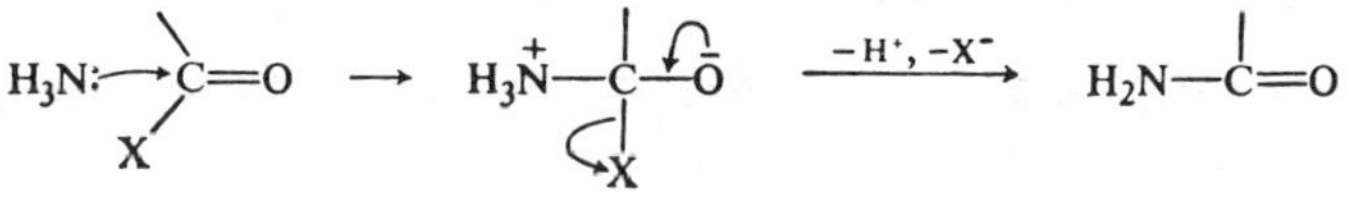

These reactions are analogous to the base-catalyzed hydrolyses of the carbonyl-containing compounds (p. 133). As in hydrolysis, the order of reactivity of compounds containing different leaving-groups, X, is: acid halides > anhydrides > esters > amides; i.e. X = Cl (etc.) > OCOR > OR > NH_2. Amides are effectively inert to displacement (except in suitable intramolecular processes; e.g. formation of succinimide, below). Carboxylic acids react, but vigorous conditions are required because the nitrogen nucleophile, acting as a base, first generates the unreactive carboxylate ion.

(a) REACTIONS OF AMMONIA AND AMINES

Acids react with ammonia only at high temperatures; e.g. benzoic acid and ammonia, heated in a sealed tube at 200°C, give benzamide:

$$PhCO_2H + NH_3 \rightleftharpoons PhCO_2^-NH_4^+ \xrightarrow{-H_2O} PhCONH_2$$

Amides and substituted amides are more conveniently prepared by treating acid chlorides or anhydrides with ammonia or an amine; reaction can be very vigorous, particularly when acid chlorides are used, and care is needed.

$$RCOCl + NH_3 \longrightarrow RCONH_2 + HCl$$

$$(RCO)_2O + NH_3 \longrightarrow RCONH_2 + RCO_2H$$

$$RCOCl + R'NH_2 \longrightarrow RCONHR' + HCl$$

$$RCOCl + R'R''NH \longrightarrow RCONR'R'' + HCl$$

Amides themselves are so weakly nucleophilic that further reaction, e.g. $RCONH_2 + RCOCl \rightarrow RCONHCOR + HCl$, occurs only slowly and the mono-acylated derivative may be isolated in high yield. However, intramolecular displacement can occur if a five- or six-membered ring is thereby formed; e.g. succinimide can be obtained by heating the acyclic diamide of succinic acid:

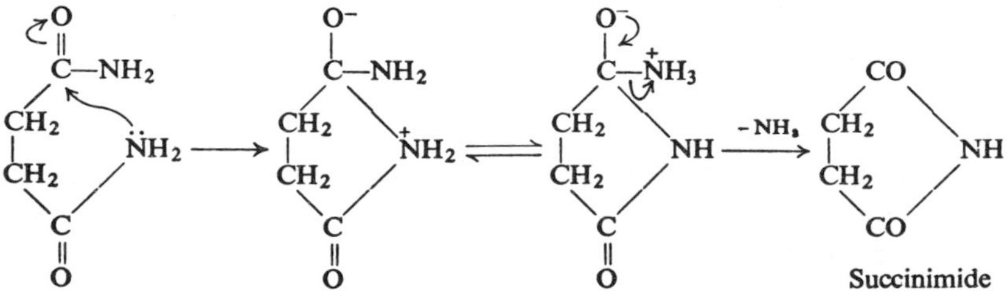

In practice, it is easier to obtain succinimide and related cyclic imides directly from the corresponding anhydrides by treatment with ammonia at a high temperature. For example, by heating phthalic anhydride with aqueous ammonia so that the water is gradually distilled and the temperature rises to about 300°C, phthalimide is obtained in over 95% yield [1]:

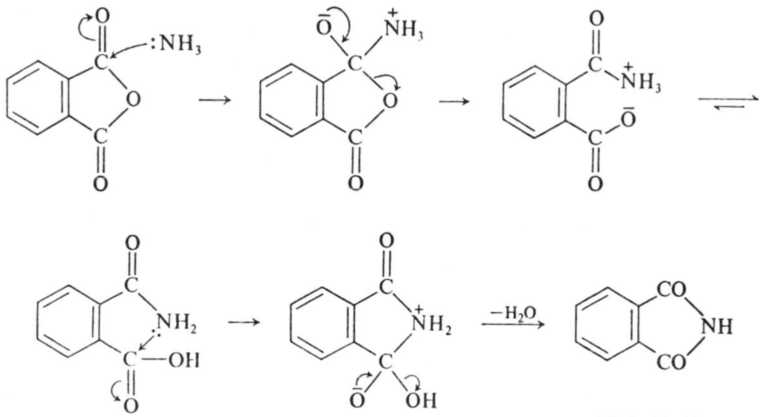

Phthalimide

Ethyl chloroformate and phosgene (carbonyl chloride) react with amines to give urethanes and substituted ureas, respectively:

$$RNH_2 + ClCO_2Et \longrightarrow RNHCO_2Et + HCl$$

$$2\ RNH_2 + COCl_2 \longrightarrow RNHCONHR + 2\ HCl$$

$$2\ R_2NH + COCl_2 \longrightarrow R_2NCONR_2 + 2\ HCl$$

Finally, the grouping, $-C\overset{\textstyle OEt}{\underset{\textstyle NH_2^+}{\diagdown}}$, contained in imino-ether hydrochlorides, reacts analogously to the ester grouping. For example, acetamidine hydrochloride can be obtained in about 85% yield from acetonitrile by the acid-catalyzed addition of ethanol followed by treatment of the resulting imino-ether hydrochloride with ammonia [1]:

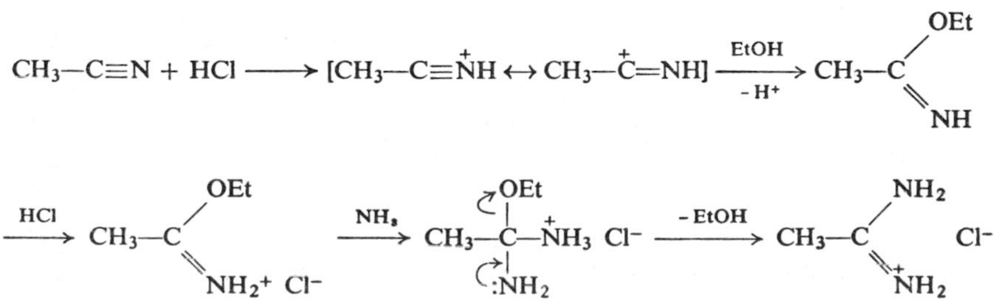

(b) REACTIONS OF OTHER NITROGEN NUCLEOPHILES

Hydrazine reacts with acid chlorides, anhydrides, and esters to give acid hydrazides, e.g.

$$RCOCl + NH_2NH_2 \longrightarrow RCONHNH_2 + HCl$$

Introduction of the acyl group reduces the nucleophilicity of both nitrogens, more strongly affecting that to which it is attached; further acylation requires more vigorous conditions and occurs at the unsubstituted nitrogen atom:

$$RCONHNH_2 + R'COCl \longrightarrow RCONHNHCOR' + HCl$$

The reaction between hydrazine and N-substituted phthalimides,

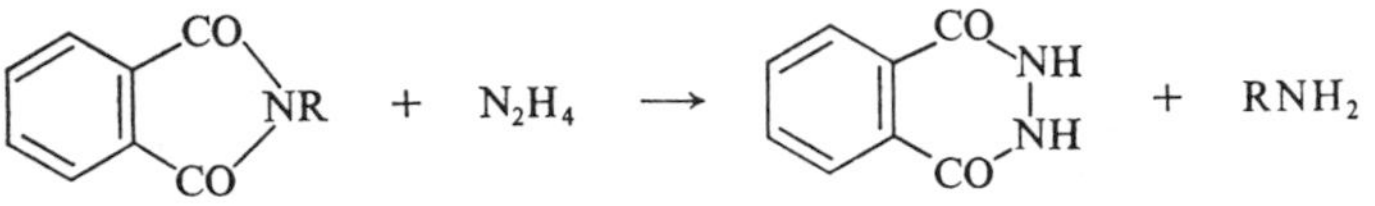

takes place in mild conditions in which, in particular, amide bonds, —CO—NH—, are not hydrolyzed. This has been applied to the generation of terminal amino groups in the synthesis of peptides (p. 358).

Hydroxylamine also reacts with carboxylic acid derivatives giving (tautomeric) hydroxamic acids, e.g.,*

The reverse reaction, hydrolysis of a hydroxamic acid to a carboxylic acid and hydroxylamine, is applied in an industrial process for making carboxylic acids. An aliphatic nitro compound, available through vapour-phase nitration (p. 557), is heated with hydrochloric acid; the acid catalyzes the formation of the *aci*-tautomer, an anionotropic shift *via* the addition and elimination of water, and the reverse of hydroxamic acid formation:

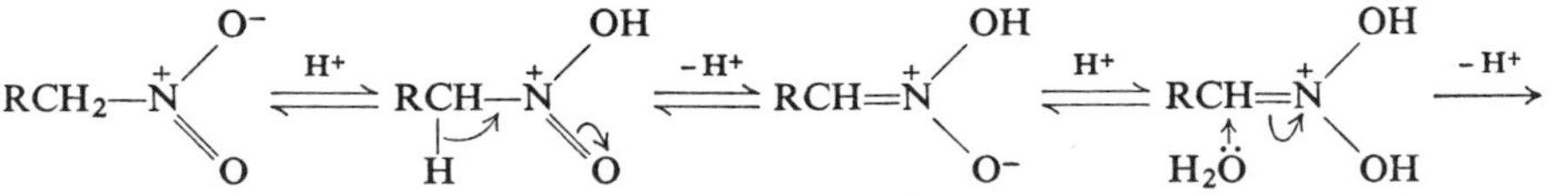

*Hydroxamic acids give deeply coloured complexes with iron(III) ion, and this provides a method for detecting the ester group in qualitative analysis.

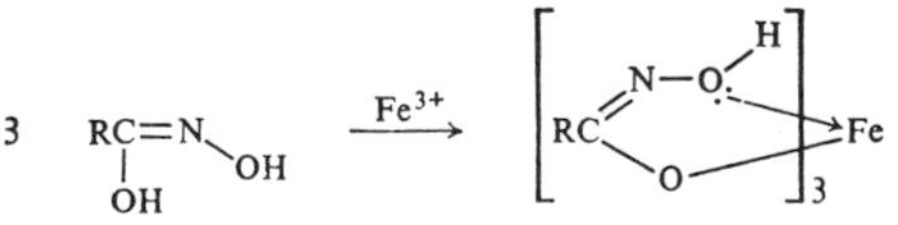

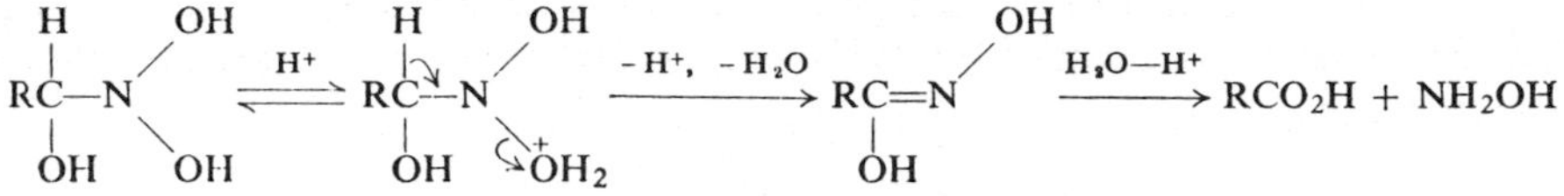

$$\underset{\substack{| \\ \text{OH}}}{\overset{\substack{\text{H} \quad \text{OH} \\ | \quad /}}{\text{RC}-\text{N}}} \overset{\text{H}^+}{\rightleftharpoons} \underset{\substack{| \quad | \\ \text{OH} \quad \overset{+}{\text{OH}_2}}}{\overset{\substack{\text{H} \quad \text{OH} \\ | \quad /}}{\text{RC}-\text{N}}} \xrightarrow{-\text{H}^+, \ -\text{H}_2\text{O}} \underset{\substack{| \\ \text{OH}}}{\overset{\substack{\text{OH} \\ /}}{\text{RC}=\text{N}}} \xrightarrow{\text{H}_2\text{O}-\text{H}^+} \text{RCO}_2\text{H} + \text{NH}_2\text{OH}$$

Acid azides, required for the preparation of amines *via* the Curtius rearrangement (p. 469), may be made by treating acid chlorides with sodium azide:

$$\text{RCOCl} + \text{Na}^+\text{N}_3^- \longrightarrow \text{RCON}_3$$

10.5 Reactions of Electrophilic Nitrogen

Four nitrogen-containing groups may be bonded to aliphatic carbon by procedures involving electrophilic nitrogen. The groups are nitroso (—NO), nitro (—NO$_2$), arylazo (—N=NAr), and arylimino (=NAr); the introduction of the arylazo group is discussed subsequently (13.4).

(a) NITROSATION
The nitroso group may be introduced in one of two ways. First, olefinic carbon which is strongly activated towards electrophiles, such as that in an enol, reacts with nitrous acid or an organic nitrite in an acid-catalyzed process; the electrophile appears to be the nitrosonium ion, NO$^+$:

$$\text{HO}-\text{N}=\text{O} + \text{H}^+ \rightleftharpoons \text{H}_2\overset{+}{\text{O}}-\text{N}=\text{O} \rightleftharpoons \text{H}_2\text{O} + \text{NO}^+$$

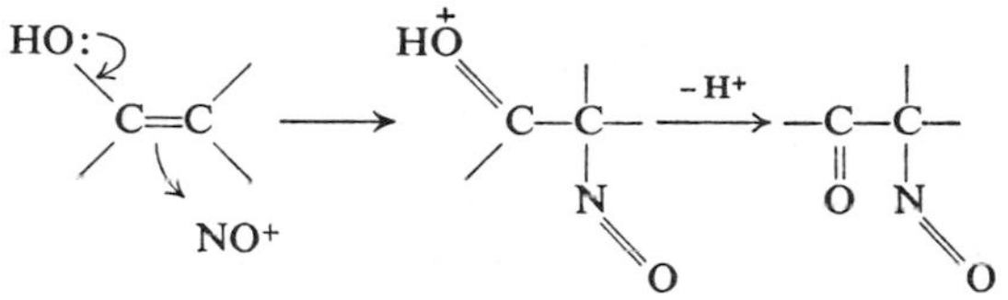

If the product has a hydrogen atom attached to the nitroso-bearing carbon, it tautomerizes to the more stable oxime:

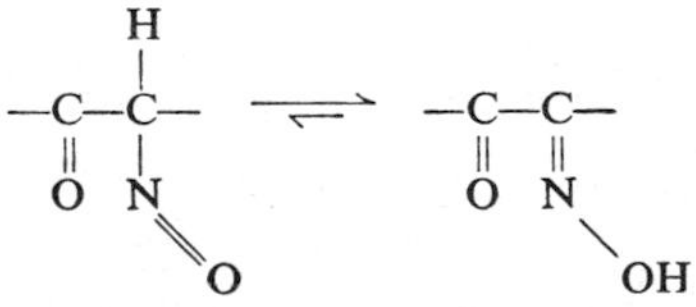

For example, acetoacetic ester, when treated in acetic acid solution with aqueous sodium nitrite, gives the oximino-derivative:

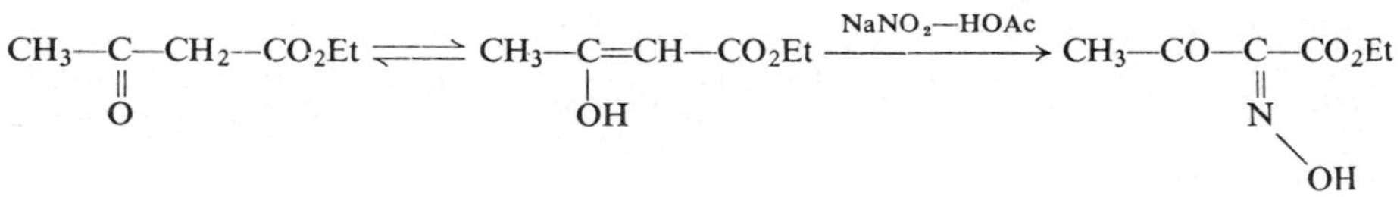

$$CH_3-C-CH_2-CO_2Et \rightleftharpoons CH_3-C=CH-CO_2Et \xrightarrow{NaNO_2-HOAc} CH_3-CO-C-CO_2Et$$

(with $\overset{\parallel}{O}$ under first, $\overset{|}{OH}$ under second, and $\overset{\parallel}{N}\diagdown_{OH}$ under the product)

Compounds which are not significantly enolic but are capable of enolization also react, for the enolization is catalyzed by acid. For example, methyl ethyl ketone reacts with ethyl nitrite* and hydrochloric acid to give biacetyl monoxime in about 70% yield [2]; note that reaction occurs at the methylene and not the methyl group.

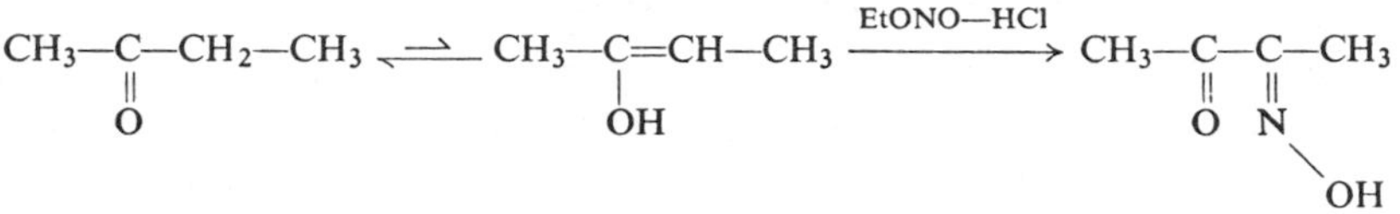

$$CH_3-C-CH_2-CH_3 \rightleftharpoons CH_3-C=CH-CH_3 \xrightarrow{EtONO-HCl} CH_3-C-C-CH_3$$

(with $\overset{\parallel}{O}$ under first, $\overset{|}{OH}$ under second, and $\overset{\parallel}{O}\ \overset{\parallel}{N}\diagdown_{OH}$ under the product)

Likewise, nitromethane gives methyl nitrolic acid *via* its *aci*-tautomer:

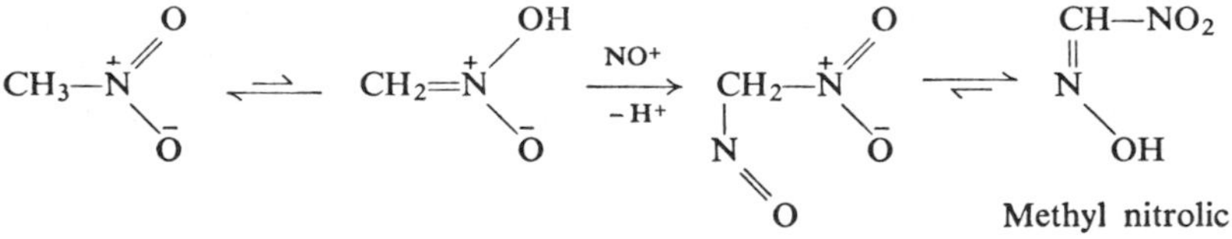

Methyl nitrolic acid

In the second method a compound containing a C—H group adjacent to one or more groups of $-M$ type is treated with base (usually ethoxide ion in ethanol) in the presence of an alkyl nitrite. A carbanion is generated and reacts with the nitrite in a manner analogous to the Claisen condensation (p. 238):

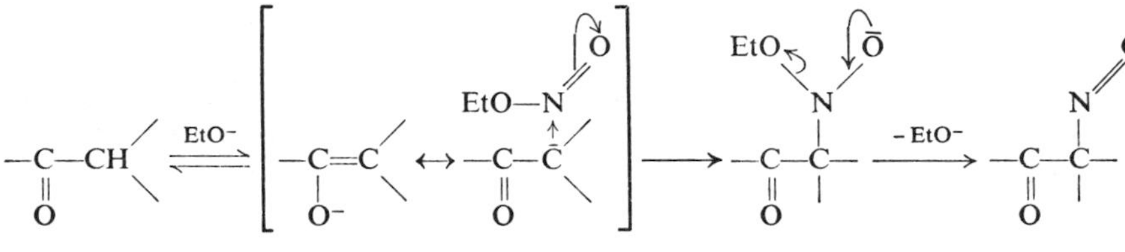

*Ethyl nitrite is prepared by treating an aqueous ethanolic solution of sodium nitrite with sulphuric acid, analogously to the esterification of a carboxylic acid (p. 135):

$$NaNO_2 + H^+ \longrightarrow HO-NO + Na^+$$

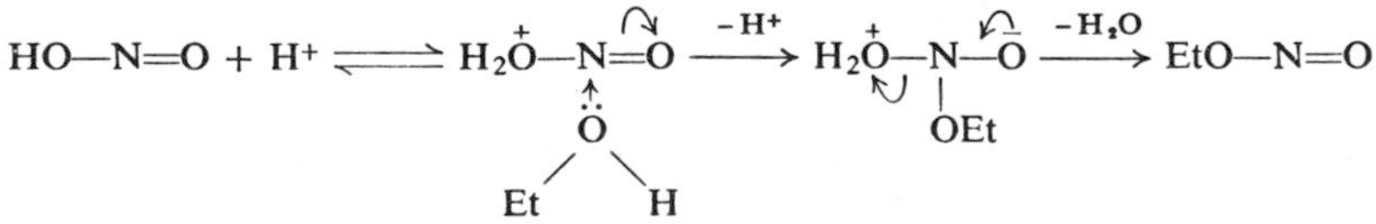

$$HO-N=O + H^+ \rightleftharpoons H_2\overset{+}{O}-N=O \xrightarrow{-H^+} H_2\overset{+}{O}-N-O \xrightarrow{-H_2O} EtO-N=O$$

(with $\overset{\ddot{.}}{O}$ and Et, H below; OEt below)

If the product contains a hydrogen atom on the nitroso-bearing carbon, its acidity is such that it is converted essentially completely into its (delocalized) conjugate base (*cf.* the synthesis of β-keto-esters; p. 239); acidification gives the oxime.

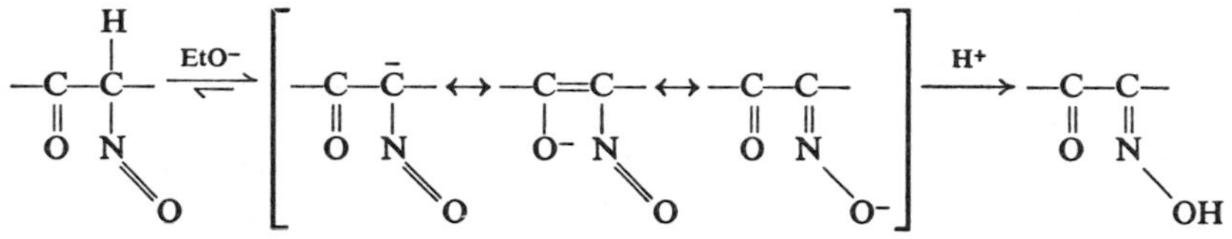

If, however, the nitroso product cannot be removed from the system in this way, a reverse Claisen condensation involving heterolytic carbon-carbon bond-fission can take place:

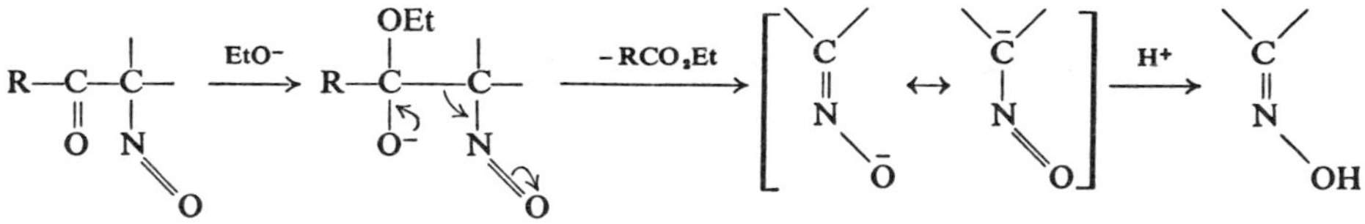

This principle was applied in the synthesis of quinine (21.6) to effect a required ring-opening:

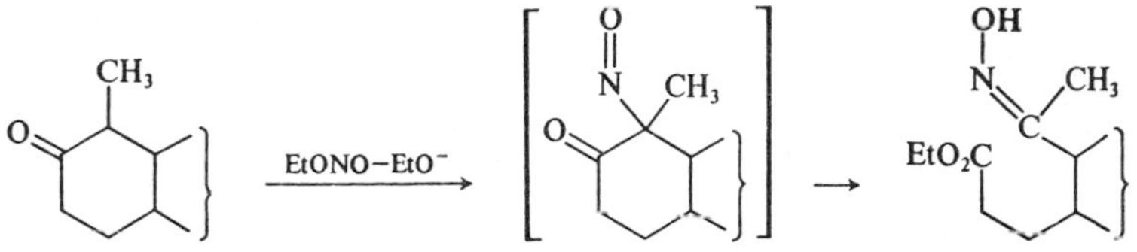

Enols and potentially enolic compounds are also those which are capable of giving carbanions, so that the acid- and base-catalyzed methods are alternative procedures. The following examples include both methods.

$$CH_3COCH_3 \xrightarrow{HNO_2-H^+} CH_3COCH{=}N{-}OH$$
$$69\%$$

$$PhCOCH_2Cl \xrightarrow{BuONO-H^+} \underset{\underset{\underset{OH}{\overset{|}{N}}}{\overset{\parallel}{\underset{}{}}}{PhCOCCl}}$$
$$85\%$$

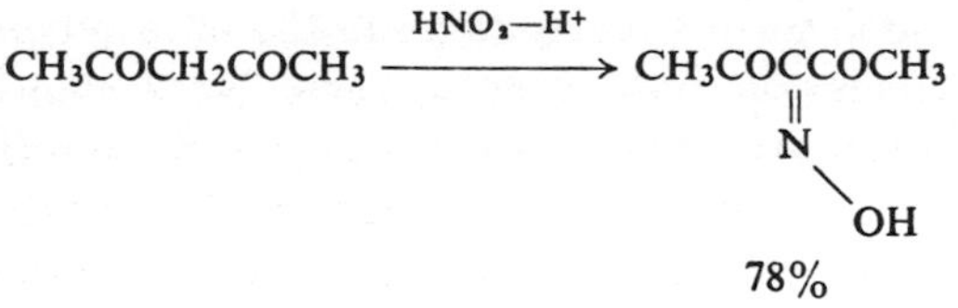

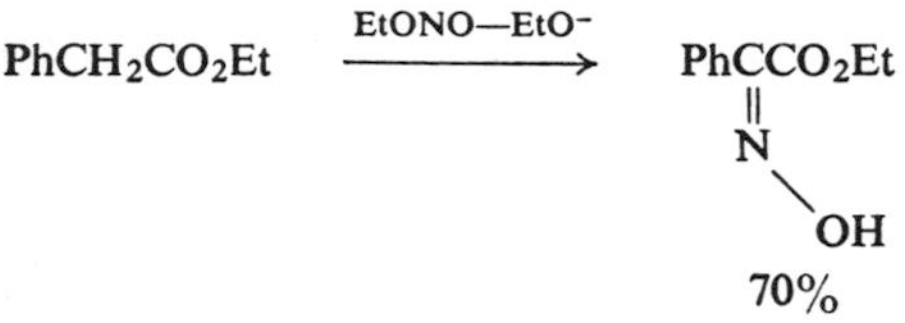

$$CH_3COCH_2COCH_3 \xrightarrow{\quad HNO_2 - H^+ \quad} CH_3COCCOCH_3$$

with N—OH group

78%

$$PhCH_2CO_2Et \xrightarrow{\quad EtONO - EtO^- \quad} PhCCO_2Et$$

with N—OH group

70%

The products of nitrosations have two main synthetic uses. First, reduction leads to amino-derivatives. Those derived from ketones are unstable because they readily undergo self-condensation, but several heterocyclic syntheses are successfully carried out by reducing β-keto-oximes in the presence of compounds with which the products react to form ring systems such as pyrroles, e.g.

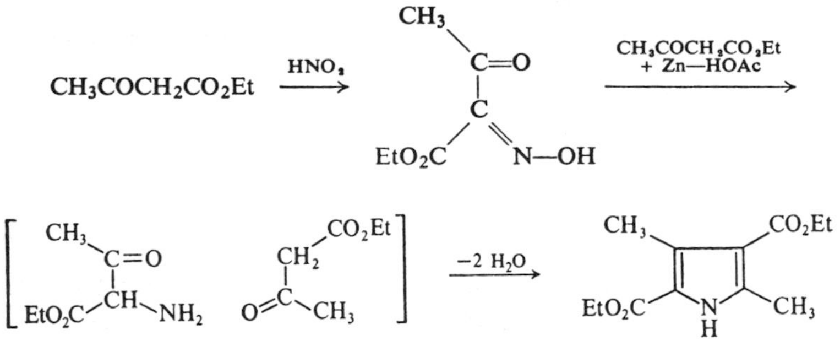

$$CH_3COCH_2CO_2Et \xrightarrow{\quad HNO_2 \quad}$$

Secondly, hydrolysis converts the oxime into a carbonyl group, e.g.

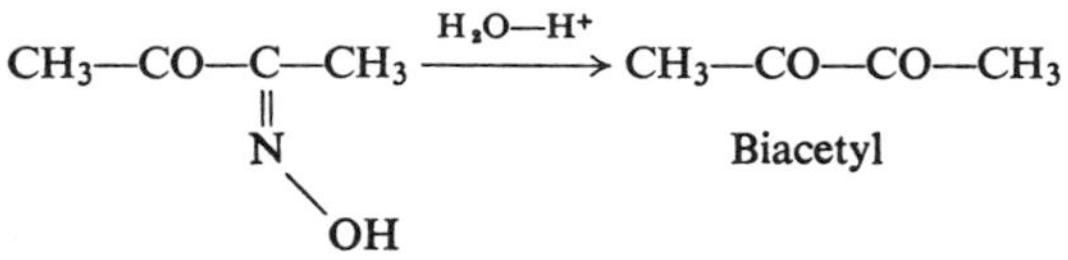

$$CH_3{-}CO{-}C{-}CH_3 \xrightarrow{\quad H_2O - H^+ \quad} CH_3{-}CO{-}CO{-}CH_3$$

with N—OH group   Biacetyl

so that the overall process can be used for the transformation, —CO—CH$_2$— → —CO—CO—.

(b) NITRATION

The nitronium ion, NO_2^+, is generated, analogously to the nitrosonium ion, by treating concentrated nitric acid with a powerful acid such as sulphuric acid. This method is widely used for bonding aromatic carbon to the nitro group (p.

400), but it is not suitable for the nitration of aliphatic systems because of the oxidations and degradations which tend to occur in these very vigorous conditions. However, a base-catalyzed procedure analogous to nitrosation may be used: a carbanion-forming compound is treated with base in the presence of an organic nitrate. For example, benzyl cyanide and methyl nitrate* react in the presence of ethoxide ion to give a nitro compound which, on alkaline hydrolysis followed by acidification, gives phenylnitromethane in 50–55% yield [2]:

$$PhCH_2CN \underset{EtO^-}{\rightleftharpoons} \left[PhCH=C=\bar{N} \overset{CH_3O-N^+=O}{\underset{O^-}{}} \leftrightarrow Ph\bar{C}H-C\equiv N \right] \longrightarrow \overset{CH_3O-N^+-O}{\underset{O^-}{}} Ph\dot{C}H-CN \xrightarrow{-CH_3O^-}$$

$$\underset{PhCH-CN}{\overset{NO_2}{|}} \xrightarrow{OH^-} \underset{PhCH-CO_2^-}{\overset{NO_2}{|}} \xrightarrow{H^+} \left[\underset{PhCH-CO_2H}{\overset{NO_2}{|}} \right] \xrightarrow{-CO_2} PhCH_2NO_2$$

Phenylnitromethane

(c) FORMATION OF IMINES

Carbanion-forming compounds react with aromatic nitroso compounds to give Schiff bases:†

$$\underset{Ar}{\overset{\diagdown}{/}}CH^- \quad N=O \longrightarrow \underset{Ar}{\overset{\diagdown}{/}}CH-N-\bar{O} \xrightarrow{H^+} \underset{Ar}{\overset{\diagdown}{/}}CH-N-OH \xrightarrow{-H_2O} \overset{\diagdown}{/}C=NAr$$

The reaction provides a method for the oxidation of activated methylene groups to carbonyl groups which are released on acidic hydrolysis of the Schiff base. The readily prepared p-nitrosodimethylaniline (p. 405) is usually employed as the nitroso compound. For example, reaction with 2,4-dinitrotoluene, whose methyl group is activated by the *ortho* and *para* nitro groups, gives 2,4-dinitrobenzaldehyde:

---

*Methyl nitrate is prepared from methanol and nitric acid in the presence of concentrated sulphuric acid (*cf.* ethyl nitrite; p. 344). It is not very stable and is used immediately.

†This reaction bears the same relationship to that between a carbanion and an alkyl nitrite as the aldol condensation bears to the Claisen condensation.

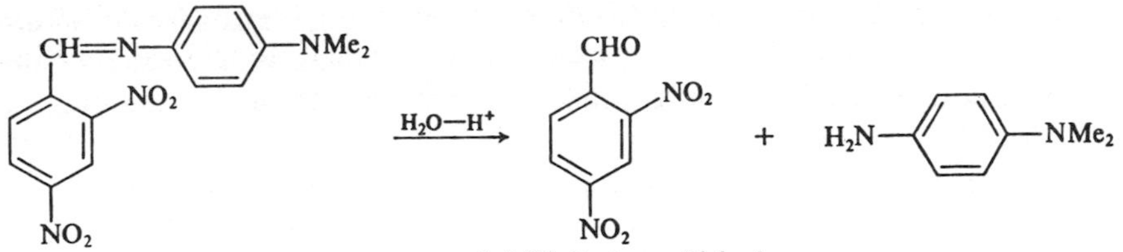

2,4-Dinitrobenzaldehyde

10.6 α-Amino-acids, Peptides, and Proteins

Proteins are naturally occurring long-chain polymers, of universal occurrence in living systems, which are derived from α-amino-acids linked together by amide bonds:

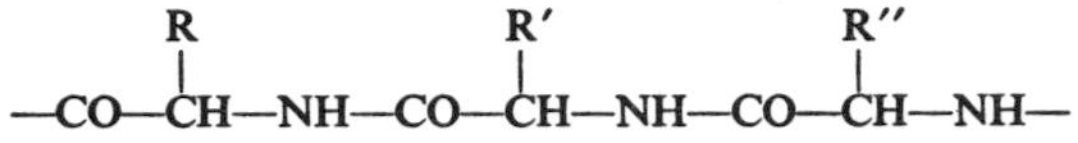

$$
\begin{array}{ccc}
R & R' & R'' \\
| & | & | \\
\end{array}
$$
—CO—CH—NH—CO—CH—NH—CO—CH—NH—

The commonest α-amino-acids, R—$CH(NH_2)CO_2H$, in proteins are: R— = H— (glycine), CH_3— (alanine), $(CH_3)_2CH$— (valine), $(CH_3)_2CHCH_2$— (leucine), $CH_3CH_2CH(CH_3)$— (isoleucine), $HOCH_2$— (serine), $HSCH_2$— (cysteine), $CH_3SCH_2CH_2$— (methionine), HO_2CCH_2— (aspartic acid), $HO_2CCH_2CH_2$— (glutamic acid), H_2N—$(CH_2)_3$—CH_2— (lysine), $PhCH_2$— (phenylalanine), together with

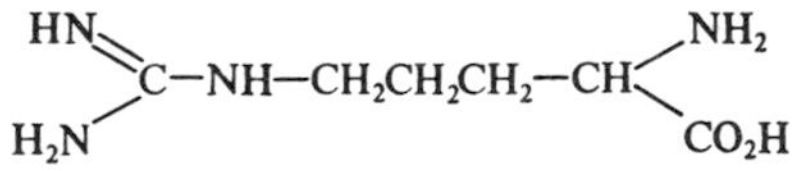

Arginine

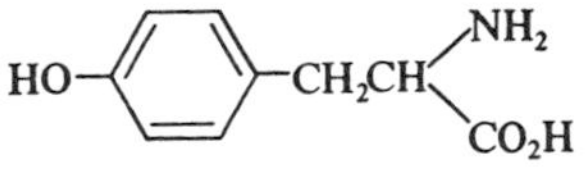

Tyrosine

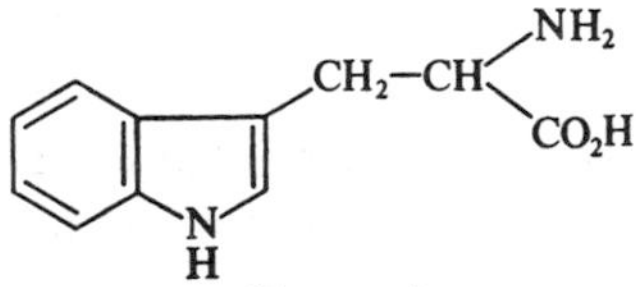

Tryptophan 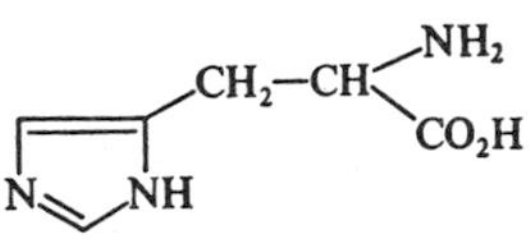   Histidine 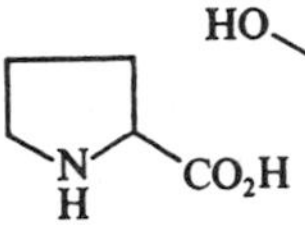   Proline 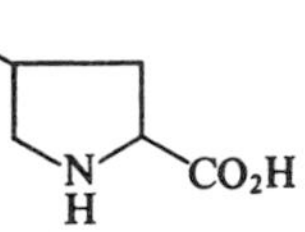   Hydroxyproline

Except in one case (glycine) the carbon atom of the —CHR— fragment in the protein chain is asymmetric and each has the same (L) configuration at this centre.

Peptides, which also occur naturally, are similar polymeric materials but of smaller chain-length. The (arbitrary) distinction between peptides and proteins is that those polymers with molecular weights less than 10,000 are termed peptides and those with higher molecular weights are termed proteins.

The acid-catalyzed hydrolysis of peptides and proteins yields the constituent α-amino-acids, and the syntheses of peptides achieved to date have been based on the reverse of this process. The α-amino-acids are therefore of considerable importance. They are zwitterionic compounds, $\overset{+}{H_3N}$—CHR—CO_2^-, both in the solid state and in solution, although in the latter conditions the zwitterion is in equilibrium with both its conjugate base and its conjugate acid, the position of equilibrium depending on the acidity of the medium:

$$H_2N-CHR-CO_2^- \xrightleftharpoons{-H^+} H_3\overset{+}{N}-CHR-CO_2^- \xrightleftharpoons{+H^+} H_3\overset{+}{N}-CHR-CO_2H$$

The physical properties of α-amino-acids reflect their zwitterionic structure: they are high-melting solids ($>200°C$), insoluble in non-polar solvents, and soluble in water (the solubility decreasing as the group R is made increasingly non-polar). Since they are normally obtained in aqueous solution together with inorganic salts, the isolation of the more soluble members of the group is difficult. They can be obtained in one of three ways.

(i) Copper(II) ion is added to precipitate the copper chelate,

$$2 \; H_3\overset{+}{N}-CHR-CO_2^- \xrightarrow[-2\;H^+]{Cu^{2+}} \begin{array}{ccc} RCH-NH_2 & O-CO \\ | & \diagdown \diagup & | \\ & Cu & \\ | & \diagup \diagdown & | \\ CO-O & H_2N-CHR \end{array}$$

(ii) Hydrochloric acid is added, the solution is evaporated to dryness, and the amino-acid hydrochloride is extracted into alcohol (leaving the inorganic salts as residue). Lead oxide is then added to remove chloride ion, followed by hydrogen sulphide to remove lead ion, and evaporation leaves the amino-acid which can be recrystallized from water or aqueous alcohol.

(iii) The simplest and most modern method is to separate them from inorganic salts on an ion-exchange resin.

Synthetic methods for α-amino-acids give both optical isomers. They are usually resolved by fractional crystallization of the diastereoisomers formed by their *N*-acetyl-derivatives with an optically active base such as brucine or strychnine.

(*a*) THE SYNTHESIS OF α-AMINO-ACIDS
The following are the more general methods for the synthesis of α-amino-acids, illustrated with reference to some of those derived from proteins.

(*i*) *From α-halo-acids.* The simplest method consists of converting a carboxylic acid into its α-bromo-derivative and treating this with ammonia:

$$R-CH_2-CO_2H \longrightarrow \underset{\underset{Br}{|}}{R-CH-CO_2H} \xrightarrow[-HBr]{NH_3} \underset{\underset{NH_2}{|}}{R-CH-CO_2H}$$

The *Hell-Volhard-Zelinsky procedure* is normally employed for the first step. The acid is treated with bromine in the presence of a small quantity of phosphorus, the phosphorus tribromide formed converts the acid into its acid bromide, this undergoes (electrophilic) bromination at the α-position *via* its enolic tautomer, and the resulting α-bromo-acid bromide exchanges with unreacted acid to give α-bromo-acid together with more acid bromide for further bromination:

$$2\ P + 3\ Br_2 \longrightarrow 2\ PBr_3$$

$$R-CH_2-CO_2H \xrightarrow{PBr_3} R-CH_2-COBr \rightleftharpoons R-CH=C\!\!\begin{array}{c}OH\\ \diagup\\ \diagdown\\ Br\end{array} \xrightarrow[-HBr]{Br_2} R-CHBr-COBr$$

$$R-CHBr-COBr + R-CH_2-CO_2H \longrightarrow R-CHBr-CO_2H + R-CH_2-COBr$$

The conversion of the α-bromo-acid into the α-amino-acid may be accomplished with *excess* of ammonia (p. 325), but better yields and purer products are usually obtained by the Gabriel procedure (p. 326).

The use of malonic ester considerably increases the versatility of this general method. First, the appropriate alkyl group can be attached by the standard alkylating procedure (p. 249), and secondly, bromination of the resulting substituted malonic acid occurs readily with bromine alone because these acids, unlike monocarboxylic acids, are significantly enolic. The synthesis of leucine is illustrative:

$$CH_2(CO_2Et)_2 + (CH_3)_2CHCH_2Br \xrightarrow{EtO^-} (CH_3)_2CHCH_2CH(CO_2Et)_2 \xrightarrow{hydrolysis}$$

$$(CH_3)_2CHCH_2CH\!\!\begin{array}{c}CO_2H\\ \diagup\\ \diagdown\\ CO_2H\end{array} \xrightarrow[-HBr]{Br_2} (CH_3)_2CHCH_2\underset{\underset{Br}{|}}{C}(CO_2H)_2 \xrightarrow[-CO_2]{130°C}$$

$$(CH_3)_2CHCH_2\underset{\underset{Br}{|}}{C}HCO_2H \xrightarrow{NH_3} (CH_3)_2CHCH_2CH\!\!\begin{array}{c}CO_2H\\ \diagup\\ \diagdown\\ NH_2\end{array}$$
$$\text{Leucine}$$

This approach may be combined with the Gabriel procedure, as in a synthesis of methionine:

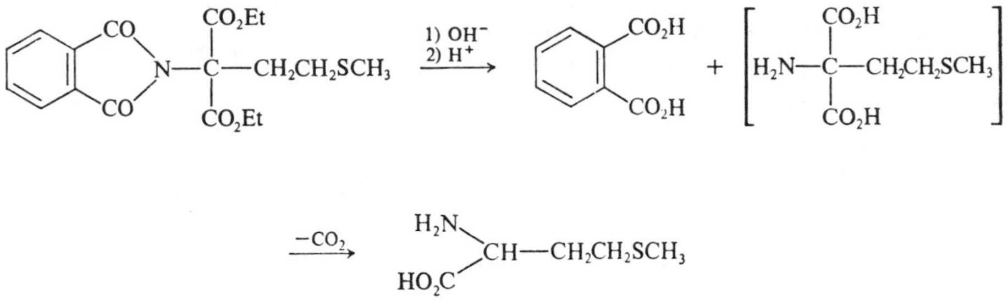

Methionine

Application of the same method to the synthesis of cysteine illustrates the use of a protective group: the —SH group in cysteine is held in the form of its benzyl-derivative until the final stage and is then released by hydrogenolysis (p. 625):

$$PhCH_2SH + CH_2O + HCl \longrightarrow [PhCH_2SCH_2OH] \longrightarrow PhCH_2SCH_2Cl$$

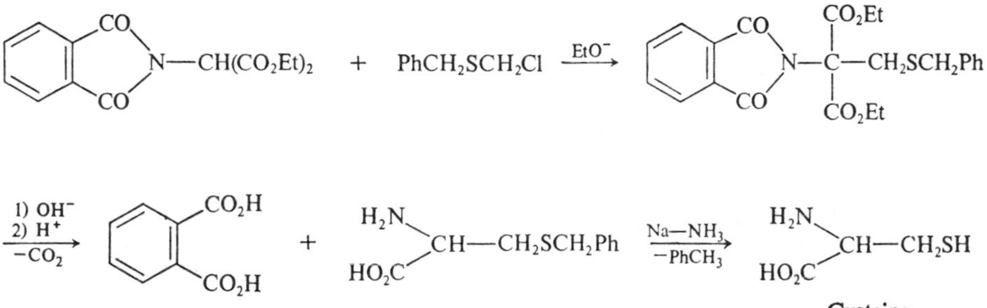

Cysteine

The amino-group may also be introduced *via* nitrosation at an activated C—H group. For example, alkylation of acetoacetic ester with s-butyl bromide followed by treatment with an alkyl nitrite in the presence of base gives a nitroso-derivative which, being incapable of ionizing, undergoes the reverse Claisen condensation (*cf.* p. 345). Reduction of the resulting oxime gives isoleucine:

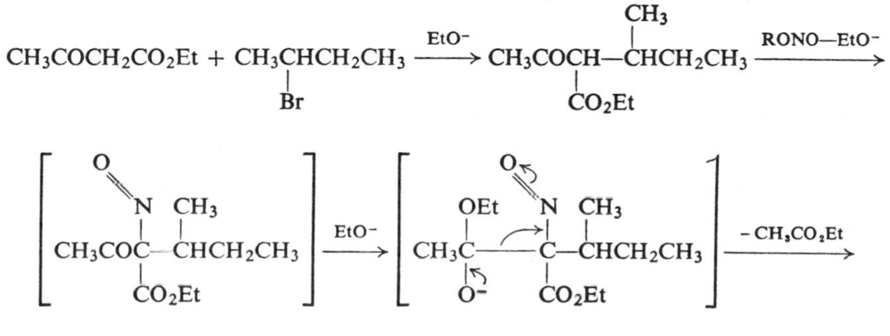

$$\underset{\underset{EtO_2C}{\diagup}}{\overset{\overset{\bar{O}}{\diagdown}}{N}}\underset{C\!-\!CHCH_2CH_3}{\overset{CH_3}{|}} \xrightarrow{H_2O\!-\!H^+} \underset{\underset{HO_2C}{\diagup}}{\overset{\overset{HO}{\diagdown}}{N}}\underset{C\!-\!CHCH_2CH_3}{\overset{CH_3}{|}} \xrightarrow{Zn\!-\!HCl} \underset{\underset{HO_2C}{\diagup}}{\overset{H_2N}{\diagdown}}\underset{CH\!-\!CHCH_2CH_3}{\overset{CH_3}{|}}$$

Isoleucine

A further modification employs acetamidomalonic ester; i.e. the amino-group is introduced in a protected form *before* attachment of the required alkyl residue. Acetamidomalonic ester is obtained by nitrosation of malonic ester (*cf.* p. 343) followed by reduction in the presence of acetic anhydride. An example of its use occurs in a synthesis of glutamic acid:

$$CH_2(CO_2Et)_2 + HNO_2 \xrightarrow{H^+} HO\!-\!N\!=\!C(CO_2Et)_2 \xrightarrow{Zn\!-\!HOAc\!-\!Ac_2O} AcNH\!-\!CH(CO_2Et)_2$$

$$AcNH\!-\!CH(CO_2Et)_2 + ClCH_2\!-\!CH_2\!-\!CO_2Et \xrightarrow{EtO^-}$$

$$\underset{NHAc}{\overset{}{EtO_2C\!-\!CH_2\!-\!CH_2\!-\!C(CO_2Et)_2}} \xrightarrow[\text{2) decarboxylation}]{\text{1) hydrolysis}} HO_2C\!-\!CH_2\!-\!CH_2\!-\!CH\overset{CO_2H}{\underset{NH_2}{\diagup\diagdown}}$$

Glutamic acid

A variation of this method is used in the synthesis of lysine where the alkyl residue is introduced by Michael addition:

$$AcNH\!-\!CH(CO_2Et)_2 + CH_2\!=\!CH\!-\!CHO \xrightarrow{EtO^-} \underset{CO_2Et}{\overset{CO_2Et}{AcNH\!-\!C\!-\!CH_2CH_2CHO}}$$

$$\xrightarrow{HCN} \underset{CO_2Et}{\overset{CO_2Et}{AcNH\!-\!C\!-\!CH_2CH_2CH}}\overset{OH}{\underset{CN}{\diagup\diagdown}} \xrightarrow[\text{2) reduction}]{\text{1) dehydration}}$$

$$\underset{CO_2Et}{\overset{CO_2Et}{AcNH\!-\!C\!-\!CH_2CH_2CH_2CH_2NH_2}} \xrightarrow[\text{2) decarboxylation}]{\text{1) hydrolysis}} \underset{CO_2H}{\overset{}{H_2N\!-\!CH\!-\!CH_2CH_2CH_2CH_2NH_2}}$$

Lysine

(*ii*) *Strecker reaction.* This reaction may be carried out either with potassium cyanide and ammonium chloride (p. 331) or with hydrogen cyanide and ammonia; the latter conditions obviate the need to separate the product from potassium chloride. The intermediate α-aminonitrile is conveniently hydrolyzed with sulphuric acid, sulphate ion then being removed from the solution by the addition of barium carbonate.

Methionine has been synthesized by the Strecker procedure:[*]

$$CH_3SCH_2CH_2CHO \xrightarrow{HCN-NH_3} CH_3SCH_2CH_2\overset{\displaystyle CN}{\underset{\displaystyle NH_2}{CH}} \xrightarrow{H_2O-H^+} CH_3SCH_2CH_2\overset{\displaystyle CO_2H}{\underset{\displaystyle NH_2}{CH}}$$

Methionine

The following synthesis of serine by the Strecker procedure illustrates a method for introducing an alcoholic group in a protected form. The ether linkage is hydrolyzed at the final stage by boiling with hydrobromic acid.

$$C_2H_5OCH_2CHO \xrightarrow{HCN-NH_3} C_2H_5OCH_2\overset{\displaystyle CN}{\underset{\displaystyle NH_2}{CH}} \xrightarrow{HBr} HOCH_2\overset{\displaystyle CO_2H}{\underset{\displaystyle NH_2}{CH}}$$

Serine

(iii) *Curtius method.* Acid azides undergo thermal rearrangement to isocyanates (p. 469). In the presence of water the isocyanate reacts to form an amine *via* the unstable carbamic acid. The required acid azides are readily obtained from malonic ester and its derivatives, e.g.

$$\overset{\displaystyle CO_2Et}{\underset{\displaystyle CO_2Et}{CH_2}} \xrightarrow{KOH} \overset{\displaystyle CO_2^-K^+}{\underset{\displaystyle CO_2Et}{CH_2}} \xrightarrow{N_2H_4} \overset{\displaystyle CO_2^-K^+}{\underset{\displaystyle CONHNH_2}{CH_2}} \xrightarrow[-2\,H_2O]{HNO_2} \overset{\displaystyle CO_2H}{\underset{\displaystyle CON_3}{CH_2}} \xrightarrow[-N_2]{heat}$$

$$\overset{\displaystyle CO_2H}{\underset{\displaystyle N=C=O}{CH_2}} \xrightarrow{H_2O} \left[\overset{\displaystyle CO_2H}{\underset{\displaystyle NHCO_2H}{CH_2}}\right] \xrightarrow{-CO_2} \overset{\displaystyle CO_2H}{\underset{\displaystyle NH_2}{CH_2}}$$

Glycine

[*]β-Thiomethylpropionaldehyde may be prepared as follows:

$$CH_2=CH-CHO \xrightarrow{HCl-EtOH} ClCH_2CH_2CH(OEt)_2 \xrightarrow[-Cl^-]{CH_3S^-}$$

$$CH_3SCH_2CH_2CH(OEt)_2 \xrightarrow{H_2O-H^+} CH_3SCH_2CH_2CHO$$

The addition of hydrogen chloride to acrolein ('anti-Markovnikov'; see p. 108) is followed by conversion of the aldehyde into its diethyl acetal which protects the compound against base-catalyzed condensation during treatment with the methanethiol anion.

(*iv*) *Condensation methods.* The aromatic-containing α-amino-acids are usually prepared by Perkin-type condensations between aromatic aldehydes and the activated methylene groups of hydantoin and related cyclic compounds. The general procedures have been outlined earlier (pp. 232, 233), and are suitable for the following conversions:

$$PhCHO \longrightarrow PhCH_2\overset{\displaystyle CO_2H}{\underset{\displaystyle NH_2}{CH}}$$

Phenylalanine

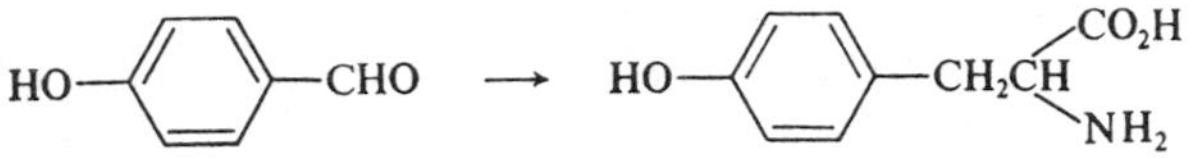

Tyrosine

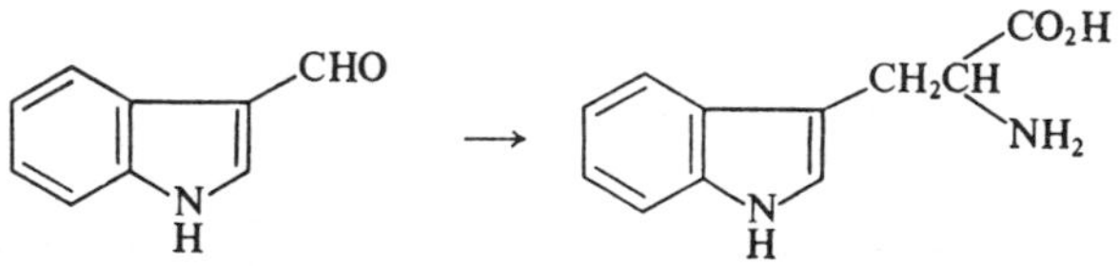

Tryptophan

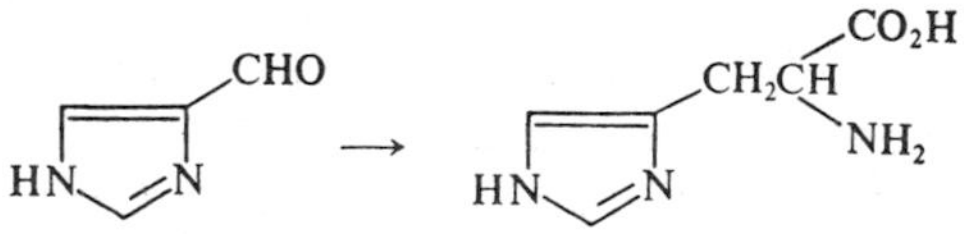

Histidine

(*b*) THE SYNTHESIS OF PEPTIDES

The synthesis of peptides from α-amino-acids presents three main problems. First, a reaction between two α-amino-acids could give each of four products (two by self-condensation and two by crossed condensation). In order to achieve a specific mode of reaction it is necessary to protect the amino-group of one reactant and the carboxyl group of the other so that reaction can only occur in one way. After the peptide link has been formed, the protecting groups are removed. Secondly, it is necessary to activate the carboxyl-group which is to be bonded to amino in order that the peptide-forming step should take place in mild conditions in which undesirable side-reactions do not occur.

Thirdly, it is necessary to prevent racemization of the optically active centres in the peptide units, for the synthesis of a natural peptide requires that all these centres should have the L-configuration. Racemization usually occurs as follows: a basic reagent initiates formation of the oxazolone system, (where X^- is a leaving group), the optically active centre racemizes *via* the enolic form of the oxazolone,

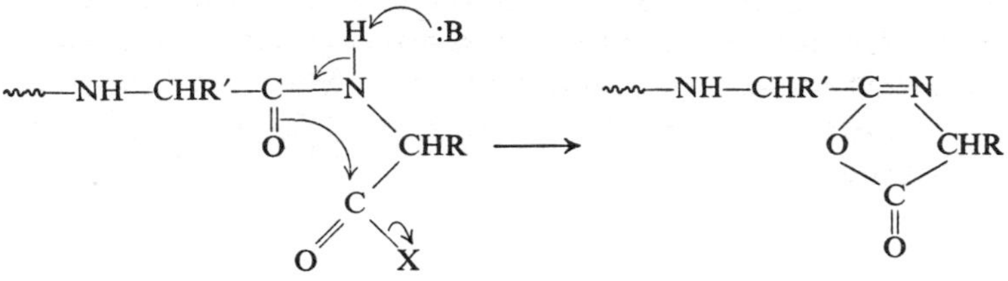

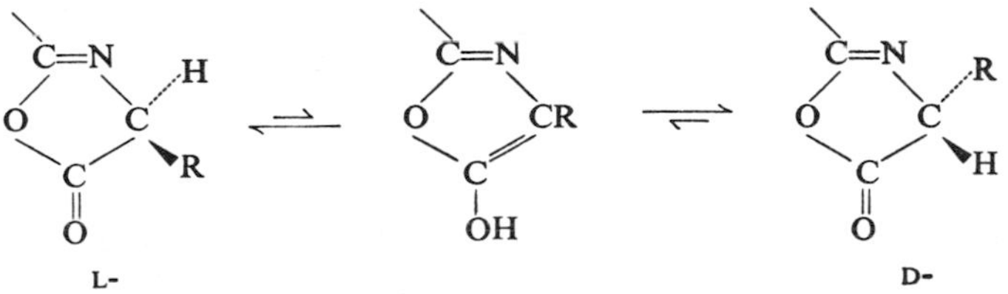

and ring-opening of the DL-mixture of oxazolones (by, for example, reaction with the amino-group of the amino-acid to be bonded) gives DL-peptide.*

Finally, when α-amino-acids which contain other functional groups such as —SH are employed, additional protection is necessary.

These difficulties have led to the development of synthetic methods of great elegance, and several important peptides, including ACTH (β-corticotropin) which contains 39 α-amino-acid units, have been synthesized. However, there is clearly an ultimate limitation to this synthetic approach: to build a peptide containing fifty α-amino-acid units requires at least fifty synthetic steps, and even if 90% yields were achieved at each stage the overall yield would be only $(0.9^{50} \times 100) = 0.5\%$; for 100 units the corresponding figure is 0.003%. Since the smallest protein contains about 100 such units, it is clear that different approaches will be necessary for the synthesis of the larger proteins.

In the synthesis of peptides, the *step-wise* technique is usually adopted. One amino-acid is protected at its carboxyl-end with a group Y and the second is protected at its amino-end with a group Z and activated at its carboxyl group by conversion into a derivative —COX. Reaction between the two forms the peptide bond. The group Z is now removed and a third amino-acid, protected at its amino-end and activated at its carboxyl end, is introduced to form the second peptide bond. Repetition of the procedure gives the required peptide.

---

*The problem of racemization is not encountered when the amino-acid unit is glycine, which is not asymmetric, or proline or hydroxyproline, which cannot form oxazolones of this kind.

$$Z—HN.CHR .CO—X + H_2N.CHR.CO—Y$$

$$\downarrow -HX$$

$$Z—HN.CHR'.CO—NH.CHR.CO—Y$$

$$\downarrow -Z$$

$$Z—HN.CHR''.CO—X + H_2N.CHR'.CO—NH.CHR.CO—Y$$

$$\downarrow -HX$$

$$Z—HN.CHR''.CO— NH.CHR'.CO —NH.CHR.CO—Y$$

$$\downarrow -Z$$

etc.

The chief advantage of the step-wise procedure is that the likelihood of race-mization is minimized because the oxazolone-forming intermediates (i.e. peptide chains terminating in —COX) are not involved. In some syntheses, however, it is desirable to bond two pre-formed peptide units, and racemization is then only avoided by using the acid azide method of activation (p. 360).

(*i*) *Methods of protection.* A practicable method of protection must have the following characteristics: the protective group must be capable of introduction in conditions in which side-reactions, including racemization, do not occur; it must be inert in the conditions in which the peptide link is formed; and it must be removable in conditions which do not affect other bonds, and in particular the peptide bonds.

The carboxyl group is now normally protected by converting it with isobutylene in the presence of sulphuric acid into its t-butyl ester:

$$—CO_2H + (CH_3)_2C{=}CH_2 \xrightarrow{H^+} —CO—O—C(CH_3)_3$$

The protecting group may be removed by mild acid hydrolysis *via* the readily formed t-butyl carbonium ion:

$$—CO—O—C(CH_3)_3 \rightleftharpoons —CO—\overset{+}{\underset{H}{O}}{-}C(CH_3)_3 \xrightarrow{-(CH_3)_3C^+} —CO_2H$$

$$(CH_3)_3C^+ + H_2O \xrightarrow{-H^+} (CH_3)_3C—OH$$

The amino group is commonly protected in organic synthesis by acetylation or benzoylation. Neither method is suitable here because hydrolysis of the amide linkages to the protective groups also cleaves the peptide bonds. Certain acylamino-groups can, however, be cleaved by methods other than basic hydrolysis, in particular, by catalytic hydrogenolysis and mild acid hydrolysis. Two derivatives are widely employed.

(1) *The benzyloxycarbonyl group.* Benzyl chloroformate (from benzyl alcohol and phosgene) reacts with amino-groups to give benzyloxycarbonyl-derivatives:

$$PhCH_2OCOCl + H_2N\text{---}\text{\char`\~\char`\~} \xrightarrow{-HCl} PhCH_2OCONH\text{---}\text{\char`\~\char`\~}$$

The protecting group may be removed by hydrogenolysis,[*] carried out with hydrogen on a palladium catalyst,

$$PhCH_2OCONH\text{---}\text{\char`\~\char`\~} \xrightarrow{2H} PhCH_3 + [HO_2C\text{---}NH\text{---}\text{\char`\~\char`\~}] \xrightarrow{-CO_2} H_2N\text{---}\text{\char`\~\char`\~}$$

but sulphur-containing compounds (i.e. units derived from cysteine and cystine) poison the catalysts. Suitable alternative reducing agents are sodium in liquid ammonia and hydrogen bromide in acetic acid.

(2) *The t-butyloxycarbonyl group.* The group is introduced with t-butyloxycarbonyl azide,[†]

$$(CH_3)_3C\text{---}O\text{---}\underset{\underset{O}{\|}}{C}\text{---}N_3 \quad H_2\ddot{N}\text{---}\text{\char`\~\char`\~} \xrightarrow{-HN_3} (CH_3)_3C\text{---}O\text{---}CO\text{---}NH\text{---}\text{\char`\~\char`\~}$$

and is removed by standing the protected peptide in cold trifluoroacetic acid:

$$(CH_3)_3C\text{---}O\text{---}CO\text{---}NH\text{---}\text{\char`\~\char`\~} \xrightarrow{H^+} (CH_3)_3C\text{---}\overset{+}{\underset{\underset{H}{|}}{O}}\text{---}CO\text{---}NH\text{---}\text{\char`\~\char`\~}$$

$$\xrightarrow{-(CH_3)_3C^+} [HO_2C\text{---}NH\text{---}\text{\char`\~\char`\~}] \xrightarrow{-CO_2} H_2N\text{---}\text{\char`\~\char`\~}$$

The trityl (triphenylmethyl) group is also employed. It may be removed either by catalytic hydrogenation or by dilute acid, as appropriate.

[*]Systems of the types $ArCH_2O\text{---}$ and $ArCH_2N<$ undergo hydrogenolysis (i.e. hydrogenation with bond-fission) in catalytic and other conditions; see p. 625.

$$† Cl\text{---}CO\text{---}O\text{---}Ph \xrightarrow{(CH_3)_3C\text{---}O^-} (CH_3)_3C\text{---}O\text{---}CO\text{---}O\text{---}Ph \xrightarrow[-PhOH]{N_2H_4}$$

$$(CH_3)_3C\text{---}O\text{---}CO\text{---}NHNH_2 \xrightarrow[-2\,H_2O]{HNO_2} (CH_3)_3C\text{---}O\text{---}CO\text{---}N_3$$

$$\text{Ph}_3\text{C—NH—}\rightsquigarrow \xrightarrow{\quad 2\text{H} \quad} \text{Ph}_3\text{CH} + \text{H}_2\text{N—}\rightsquigarrow$$

$$\text{Ph}_3\text{C—NH—}\rightsquigarrow \xrightarrow{\quad \text{H}_2\text{O—H}^+ \quad} \text{Ph}_3\text{C—OH} + \text{H}_2\text{N—}\rightsquigarrow$$

The phthaloyl group has been used, though not in the synthesis of large peptides. It is introduced either by heating the amino-acid with phthalic anhydride or, in milder conditions (room temperature), by treating the amino-acid with N-carboethoxyphthalimide. Its removal can be accomplished in mild conditions by treatment with hydrazine (p. 342), but this has the disadvantage that ester groups which are present in the peptide are converted into acid hydrazides.

(3) *Additional protection.* Glutamic acid occurs in peptides and proteins as both α-glutamyl and γ-glutamyl residues,

$$\begin{array}{cc}
\overset{\displaystyle \text{CH}_2\text{CH}_2\text{CO}_2\text{H}}{\underset{\displaystyle \alpha\text{-glutamyl}}{\text{—NH—CH—CO—}}} &
\overset{\displaystyle \text{CO}_2\text{H}}{\underset{\displaystyle \gamma\text{-glutamyl}}{\text{—NH—CH—CH}_2\text{—CH}_2\text{—CO—}}}
\end{array}$$

and it is therefore necessary to be able to protect each of the two carboxyl groups separately. The γ-carboxyl group can be protected as follows: (1) both carboxyl groups are benzylated; (2) mild acid hydrolysis preferentially regenerates the γ-carboxyl group, * (3) the amino-protecting group Z is introduced; (4) the free acid is esterified with isobutylene; and (5) the α-benzyl group and Z are cleaved by hydrogenolysis:

$$\overset{\displaystyle \text{CH}_2\text{CH}_2\text{CO}_2\text{H}}{\underset{(1)}{\text{H}_2\text{N—CH—CO}_2\text{H} + 2\ \text{PhCH}_2\text{OH} \xrightarrow{\text{H}^+}}} \overset{\displaystyle \text{CH}_2\text{CH}_2\text{CO—O—CH}_2\text{Ph}}{\overset{+}{\text{H}_3\text{N—CH—CO—O—CH}_2\text{Ph}}}$$

$$\xrightarrow[(2)]{\text{HI—HOAc}} \overset{\displaystyle \text{CH}_2\text{CH}_2\text{CO}_2\text{H}}{\overset{+}{\text{H}_3\text{N—CH—CO—O—CH}_2\text{Ph}}} \xrightarrow[4)\ (\text{CH}_3)_2\text{C=CH}_2]{3)\ \text{Introduce Z}}$$

$$\overset{\displaystyle \text{CH}_2\text{CH}_2\text{CO—O—C(CH}_3)_3}{\text{Z—HN—CH—CO—O—CH}_2\text{Ph}} \xrightarrow[(5)]{\text{H}_2\text{—Pd}} \overset{\displaystyle \text{CH}_2\text{CH}_2\text{CO—O—C(CH}_3)_3}{\text{H}_2\text{N—CH—CO}_2\text{H}}$$

*The alkyl-oxygen of the α-carboxyl group is less basic than that of the γ-carboxyl group because it is more affected by the electron-withdrawing positive pole. Consequently acid-catalyzed hydrolysis,

$$\rightsquigarrow\text{—CO—O—CH}_2\text{Ph} \underset{}{\overset{\text{H}^+}{\rightleftharpoons}} \rightsquigarrow\text{—CO—}\overset{+}{\underset{\displaystyle \text{H}}{\text{O}}}\text{—CH}_2\text{Ph} \xrightarrow{-\text{PhCH}_2^+} \rightsquigarrow\text{—CO}_2\text{H}$$

occurs more readily at the γ-ester.

The α-carboxyl group is protected by acid-catalyzed methylation of the γ-carboxyl, t-butylation of the α-carboxyl, and cleavage of the γ-methyl ester with base:*

$$\underset{\overset{|}{H_3\overset{+}{N}}-CH-CO_2H}{\overset{CH_2CH_2CO_2H}{}} + CH_3OH \xrightarrow{H^+} \underset{\overset{|}{H_3\overset{+}{N}}-CH-CO_2H}{\overset{CH_2CH_2CO_2CH_3}{}}$$

$$\xrightarrow[\text{2) }(CH_3)_4C=CH_2]{\text{1) Introduce Z}} \underset{Z-HN-CH-CO_2C(CH_3)_3}{\overset{CH_2CH_2CO_2CH_3}{}} \xrightarrow{OH^-} \underset{H_2N-CH-CO_2C(CH_3)_3}{\overset{CH_2CH_2CO_2H}{}}$$

Lysine contains an amino group in the side-chain, although it is always present in peptides and proteins as the α-lysyl group. The ϵ-amino groups may be selectively protected by adding the protecting agent to lysine's copper chelate, e.g.

The —SH group in cysteine is protected by treatment with benzyl chloride and regenerated by hydrogenolysis,

$$\text{vvv—SH} + PhCH_2Cl \xrightarrow{-HCl} \text{vvv—SCH}_2Ph$$

$$\text{vvv—SCH}_2Ph \xrightarrow{Na-NH_3} \text{vvv—SH} + PhCH_3$$

The hydroxyl group in serine does not usually require protection; when it does, the benzyl ether is employed; removal is by catalytic hydrogenolysis (—OCH_2Ph → —OH + $PhCH_3$). The guanidino-group in arginine can be protected by nitration, or by conducting the synthesis at a pH at which the group is essentially fully protonated (the pK_a of —NH—C(NH$_2$)=NH is $\sim$12, so that at pH 7 only about one molecule in 10^5 is not protonated).

(ii) *Methods of activation.* Carboxylic acids react with amines only under very vigorous conditions (*cf.* ammonia, p. 340). It is consequently necessary to convert the acid into a derivative which is more reactive towards nucleophiles, the

*The readier acid-catalyzed esterification of the γ-carboxyl has the same basis as the readier acid-catalyzed hydrolysis of the γ-ester (preceding footnote). The selective base-catalyzed hydrolysis occurs because the t-butyl ester is more hindered than the methyl ester.

requirement being that the group X in the derivative R—CO—X should be a good leaving-group. Derivatives of several types have been used.

(1) *Acid chlorides* ($X = Cl$). These are exceptionally reactive towards nucleophiles and peptide-bond formation occurs readily, but chloride ion is so good a leaving-group that N-carboxyanhydrides are formed from benzyloxycarbonyl-protected acid chlorides:

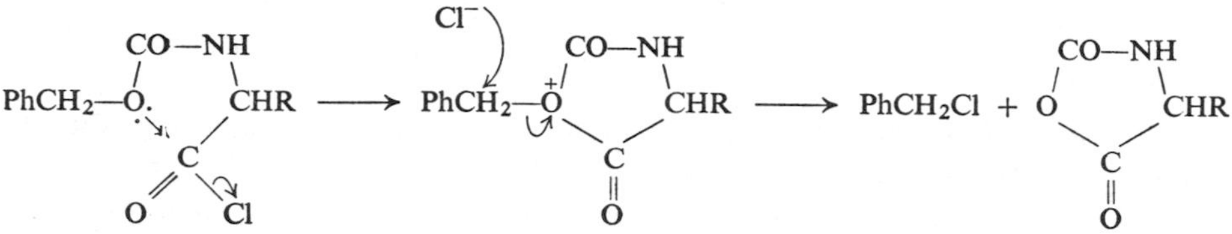

(2) *Acid azides* ($X = N_3$). Conversion of an acid into its azide can be accomplished under relatively mild conditions (esterification, treatment with hydrazine and then with nitrous acid; p. 469). Azides, though not so reactive as chlorides, are sufficiently reactive for peptide-bond formation to occur smoothly without the competing reaction leading to the N-carboxyanhydride.

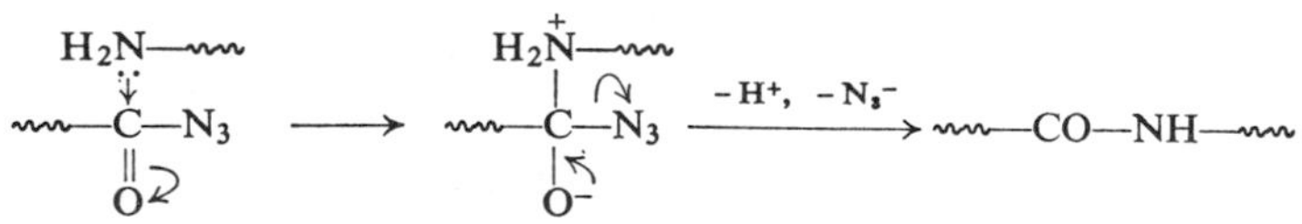

(3) *Mixed anhydrides* ($X = OCOR$). Mixed anhydrides are readily formed by displacement by the nucleophilic carboxylate anion on an acid chloride. A typical example is the reaction with ethyl chloroformate, triethylamine being added to generate the carboxylate ion:

$$\text{~~—CO}_2\text{H} \xrightarrow{\text{Et}_3\text{N}} \text{~~—CO}_2^- \xrightarrow[-\text{Cl}^-]{\text{ClCO}_2\text{Et}} \text{~~—CO—O—CO—OEt}$$

Reaction of an amino group with the resulting mixed anhydride occurs smoothly:

$$\text{~~—CO—O—CO—OEt} \xrightarrow{\text{~~—NH}_2} \text{~~—CO—NH—~~}$$

$$+ [\text{EtO—CO}_2^- + \text{H}^+] \longrightarrow \text{EtOH} + \text{CO}_2$$

(4) *Activated esters* ($X = OR$). Although alkyl esters are not very reactive towards amines, aryl esters, particularly those with electron-attracting substituents, react readily. The reason is that the negative charge in aryloxide anions

is delocalized over the aromatic ring and over *ortho-* and *para-*substituents of $-M$ type, e.g.

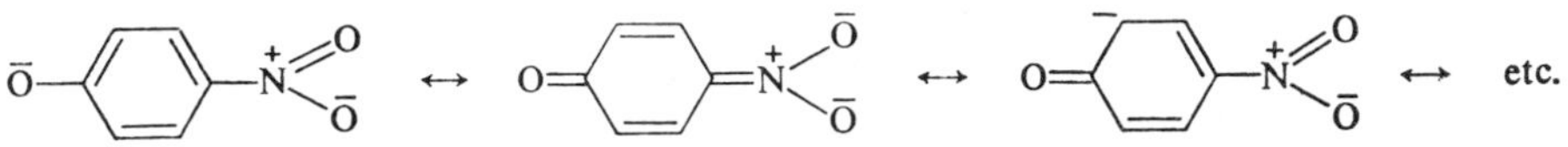

so that ArO^- is a much better leaving-group than RO^-.* As a consequence, the *p*-nitrophenyl group is widely used for activation of carboxyl:

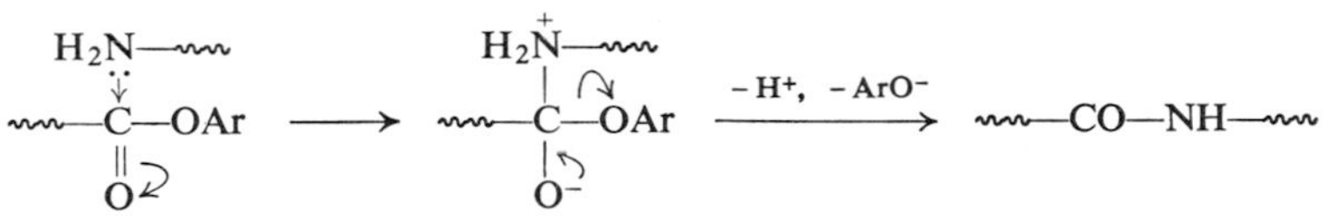

(5) *Carbodi-imide.* Carboxylic acids react readily with amines in the presence of acid and a carbodi-imide (usually dicyclohexylcarbodi-imide, p. 339), a di-substituted urea precipitating from solution. The probable mechanism is:†

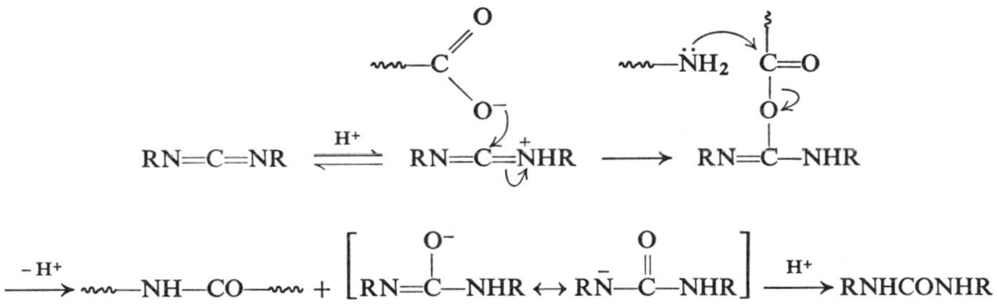

Both the activated ester and the carbodi-imide methods are suitable only for the step-wise synthesis of peptides, for carboxyl-terminating peptides are race-mized by these procedures.

(*iii*) *The synthesis of ACTH (β-corticotropin).* A recent synthesis of the peptide hormone ACTH incorporates many of the preceding methods (R. Schwyzer and P. Sieber, *Nature*, 1963, **199**, 172). The first reactants were phenylalanine, pro-tected at its carboxyl-end as the t-butyl ester, and glutamic acid, protected at its amino-end as the benzyloxycarbonyl derivative and at its *γ*-carboxyl group as

---

*This is also, of course, the basis for the greater acidity of phenol than an alcohol, and of *p*-nitrophenol than phenol; p. 64.

†The two C=N bonds make the central carbon atom very reactive towards nucleophiles, particularly in the presence of an acid. The resulting adduct, containing the grouping —CO—O—C=NR, is analogous to an anhydride and reacts readily with amino-groups.

its t-butyl ester. Condensation was brought about *via* the *p*-nitrophenyl ester and the benzyloxycarbonyl-group was removed by reduction over palladium in acetic acid solution:

$$CO_2C(CH_3)_3$$
$$|$$
$$CH_2$$
$$|$$
$$CH_2 \qquad\qquad CH_2Ph$$
$$| \qquad\qquad\quad |$$
$$PhCH_2OCO{-}NH{-}CH{-}CO{-}OAr + H_2N{-}CH{-}CO_2C(CH_3)_3$$

$$\downarrow \ {-\,ArOH}$$

$$CO_2C(CH_3)_3$$
$$|$$
$$CH_2$$
$$|$$
$$CH_2 \qquad\qquad CH_2Ph$$
$$| \qquad\qquad\quad |$$
$$PhCH_2OCO{-}NH{-}CH{-}CO{-}NH{-}CH{-}CO_2C(CH_3)_3$$

$$\downarrow \ {H_2{-}Pd}$$

$$CO_2C(CH_3)_3$$
$$|$$
$$CH_2$$
$$|$$
$$CH_2 \qquad\qquad CH_2Ph$$
$$| \qquad\qquad\quad |$$
$$H_2N{-}CH{-}CO{-}NH{-}CH{-}CO_2C(CH_3)_3$$

The next amino-acid, protected at its amino-end as the benzyloxycarbonyl-derivative, was introduced in the same way and this process was repeated with the appropriate amino-acids a further 12 times. Condensations were then carried out, successively, with an octapeptide, a hexapeptide, and a decapeptide, using the mixed anhydride, acid azide, and carbodi-imide activation methods. The side-chain amino-groups in the lysine residues were protected as t-butyloxy-carbonyl-derivatives and were therefore retained during the hydrogenative removal of the benzyloxycarbonyl-groups, finally being removed by trifluoro-acetic acid.

(*iv*) *The synthesis of bradykinin.* A recent synthesis of bradykinin (a natural nonapeptide with important physiological properties) illustrates the application of a new technique in step-wise synthesis. The principle is to attach the first amino-acid, through its carboxyl group, to an insoluble but readily filterable solid. The second amino-acid, protected at its amino-end, is introduced and peptide-bond formation is induced by an appropriate method. The peptide formed remains attached to the solid, soluble reagents and byproducts are removed by filtration, the amino-protecting group is removed, and the next amino-acid is introduced. The process is repeated until the desired peptide is obtained and this is then removed from its solid support by a suitable cleavage reaction.

The main advantage of this method is that no purification procedures other than washing are necessary until the final peptide, having been removed from the solid, is purified. The losses normally sustained during the conventional purification methods are thereby avoided, so that yields are high and the individual steps can be carried out in quick succession.

The solid support used in the synthesis of bradykinin was a copolymer of styrene and divinylbenzene. Styrene itself gives a linear polymer, $(CH_2—CHPh)_n$, and the introduction of a small percentage of divinylbenzene yields a polymer with cross-linked chains:

$$—CH_2—CHPh—CH_2—CH—CH_2—CHPh—$$

$$—CH_2—CHPh—CH_2—CH—CH_2—CHPh—$$

The resulting material has a gel structure with good permeability. About 5% of the benzene rings were then chloromethylated (p. 393) and the first amino-acid, arginine, was introduced as t-butyloxycarbonylnitroarginine, its free carboxylate group reacting at the benzyl chloride centres so that the arginine was held to the solid as a benzyl ester (step 1). The t-butyloxycarbonyl group was removed with hydrochloric acid (step 2), triethylamine was added to liberate the amino group (step 3), and the second amino-acid, phenylalanine (protected as its t-butyloxycarbonyl-derivative), was introduced, condensation being brought about by dicyclohexylcarbodi-imide (step 4). The operations in steps 2–4 were repeated with the appropriate amino-acids until the nonapeptide had been formed and the peptide was then detached from the solid by passing hydrogen bromide through a suspension of the solid in trifluoroacetic acid. This acid treatment also removed the final t-butyloxycarbonyl group and the benzyl protecting group of a serine unit, and catalytic hydrogenation removed the nitro protecting groups of two arginine units. Bradykinin was isolated in 68% yield.

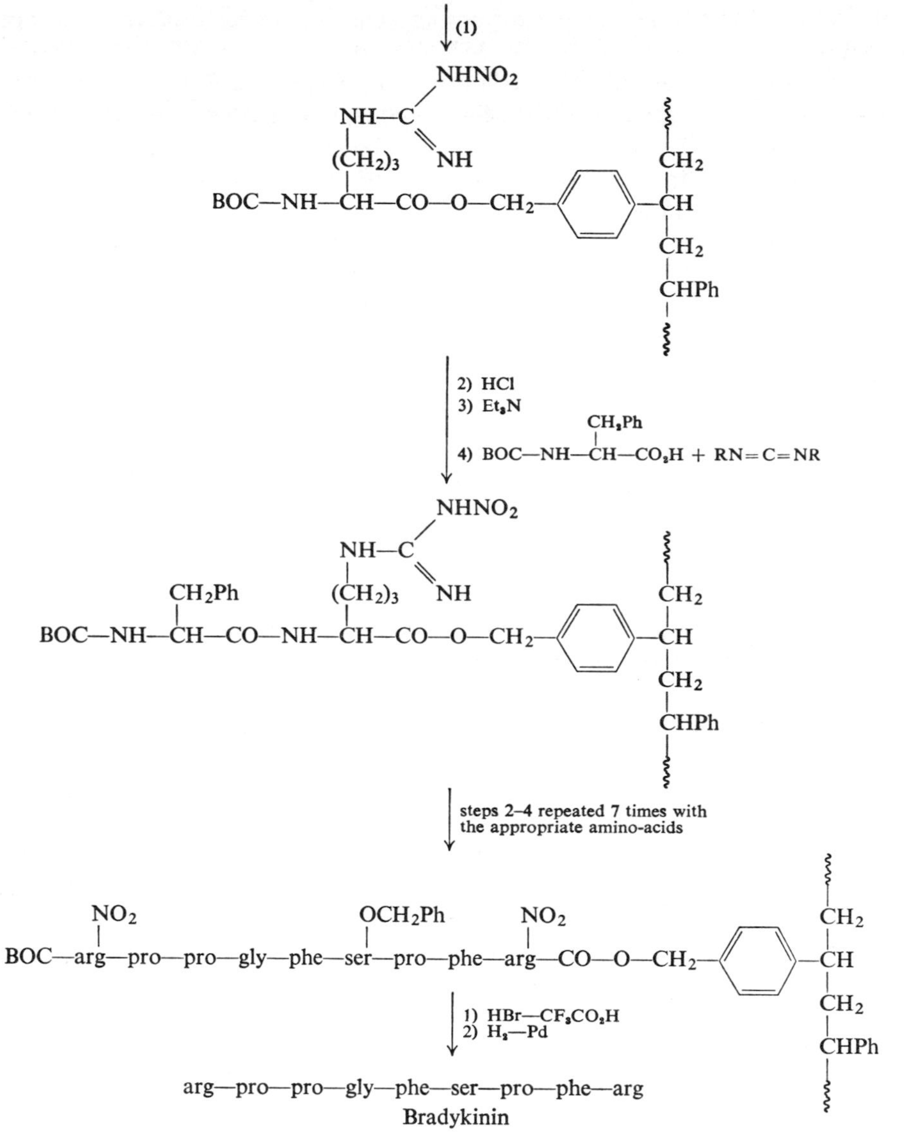

(BOC = t–butyloxycarbonyl; arg = arginine; pro = proline; gly = glycine; phe = phenylalanine; ser = serine)

The solid-phase method has now been developed by automation of the chemical operations. A recent remarkable example of its use is the synthesis of the enzyme ribonuclease. This contains 124 amino-acids, and the synthesis required nearly 12,000 automated operations.

Further Reading

ALBERTSON, N. A., 'Synthesis of peptides with mixed anhydrides,' *Organic Reactions*, 1962, **12**, 157.

EMERSON, W. S., 'The preparation of amines by reductive alkylation,' *Organic Reactions*, 1948, **4**, 174.

KRIMEN, L. I., and COTA, D. J., 'The Ritter reaction,' *Organic Reactions*, 1969, **17**, 213.

MCOMIE, J. F. W., 'Protective groups,' *Advances in Organic Chemistry, Methods and Results*, Vol. 3, Interscience (New York and London 1963), p. 191.

MOORE, M. L., 'The Leuckart reaction,' *Organic Reactions*, 1949, **5**, 301.

Problems

1. What products may be obtained from the reaction of ammonia with each of the following: methyl iodide; acetyl chloride; formaldehyde; acetaldehyde; chloral; acetone; benzaldehyde; ethylene oxide; acetylacetone; acetonylacetone; acetonitrile; phenyl isothiocyanate; cyanamide?

2. Illustrate the types of product which can be obtained from the reactions of acetoacetic ester with compounds containing nucleophilic nitrogen.

3. Outline routes to the following compounds:

    (a) $(CH_3)_3N$

    (b) $(CH_3)_3C-NH_2$

    (c) $PhNHCOCH_3$

    (d) $(C_2H_5)_2NH$

    (e) $PhNHCSNHCH_3$

    (f) CH_3
    $\backslash$
    $CH-NHCH_3$
    $/$
    C_2H_5

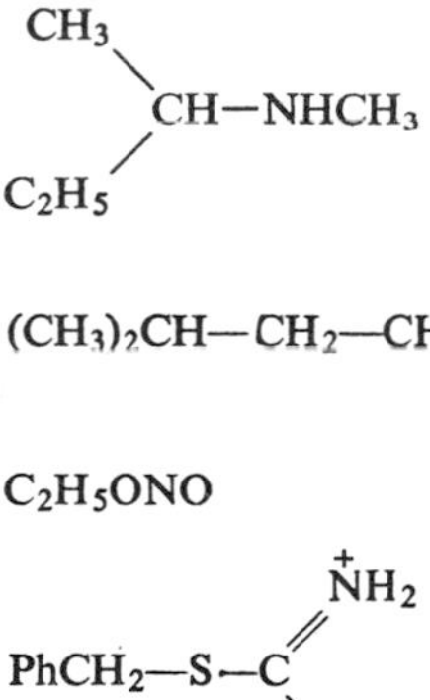

    (g) $HS-CH_2-CH\overset{\displaystyle NH_2}{\underset{\displaystyle CO_2H}{\Big\langle}}$

    (h) $(CH_3)_2CH-CH_2-CH\overset{\displaystyle NH_2}{\underset{\displaystyle CO_2H}{\Big\langle}}$

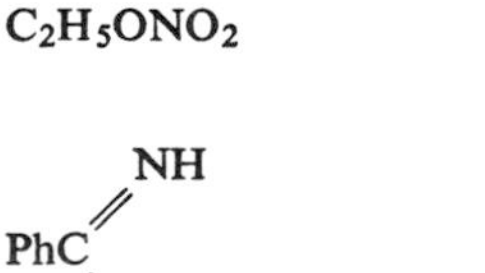

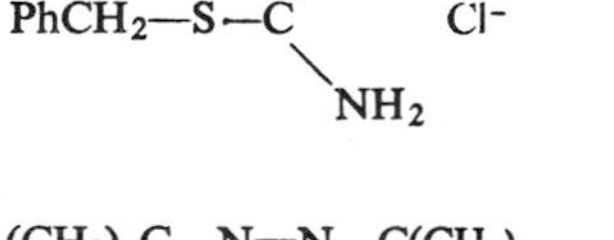

    (i) $C_2H_5NO_2$

    (j) C_2H_5ONO

    (k) $C_2H_5ONO_2$

    (l) $PhCH_2-S-C\overset{\displaystyle \overset{+}{N}H_2}{\underset{\displaystyle NH_2}{\Big\langle}}\quad Cl^-$

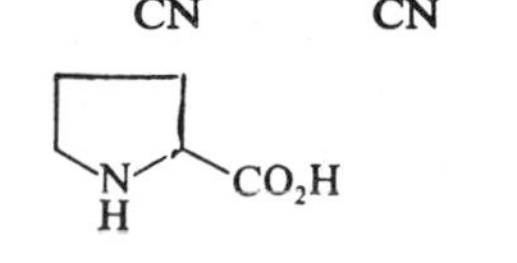

    (m) $PhC\overset{\displaystyle NH}{\underset{\displaystyle NH_2}{\Big\langle}}$

    (n) $(CH_3)_2C-N=N-C(CH_3)_2$
    $||$
    $CNCN$

    (o) —N=C=N—

    (p)

(*q*)

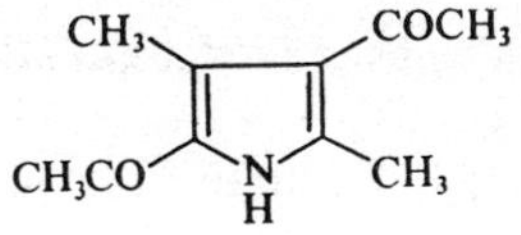
CH3
COCH3
CH3CO
N
H
CH3

(*r*)

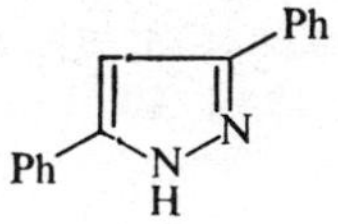
Ph
N
Ph
N
H

(*s*)

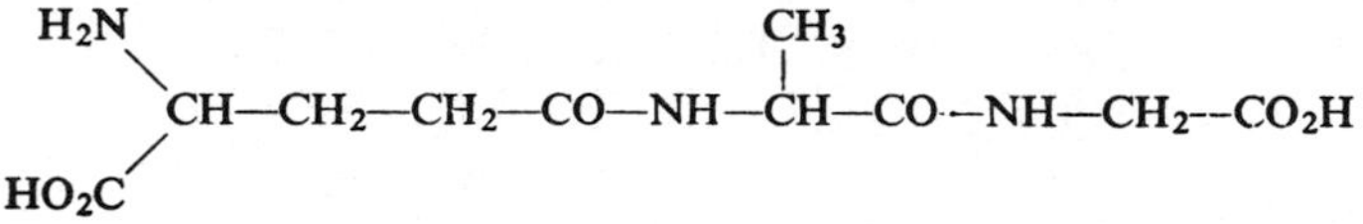
H2N
CH3
CH—CH2—CH2—CO—NH—CH—CO—NH—CH2—CO2H
HO2C

11. Electrophilic Aromatic Substitution

11.1 The Mechanism of Substitution

The substitution of benzene by an electrophilic reagent (E^+) occurs in two stages: the reagent adds to one carbon atom of the nucleus, giving a carbonium ion in which the positive charge is delocalized over three carbon atoms, and a proton is then eliminated from this adduct:

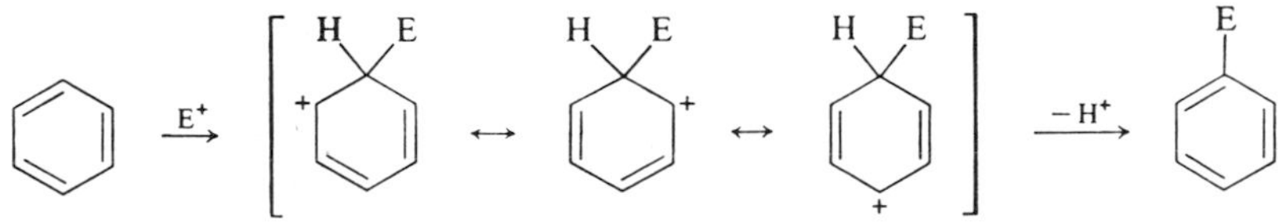

The electrophile may be a charged species, such as the nitronium ion, NO_2^+, which participates in nitration (11.4), and the t-butyl carbonium ion, Me_3C^+, which participates in Friedel-Crafts t-butylation (11.3), or it may be a neutral species which can absorb the pair of electrons provided by the aromatic nucleus. The latter class includes reagents such as sulphur trioxide (in sulphonation, 11.5) which absorb the electron-pair without bond-breakage,

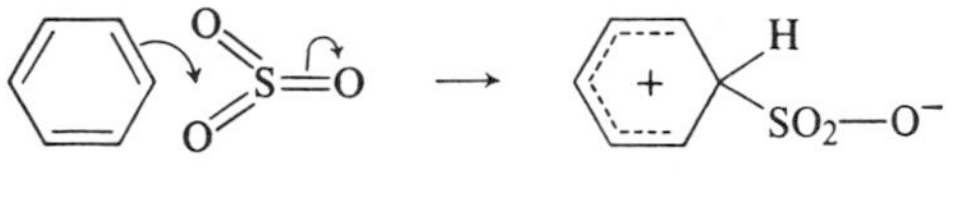

and those such as the halogens in which uptake of the electron-pair leads to bond-breakage and the formation of a stable anion,

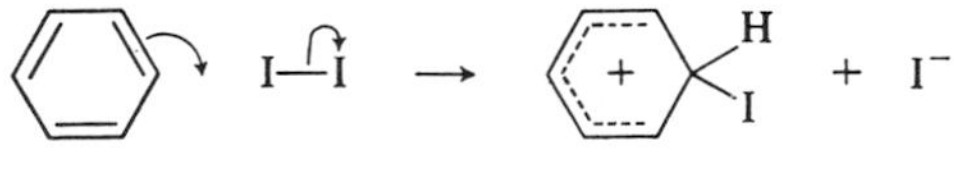

The following are the chief characteristics of these reactions.

(1) The intermediate carbonium-ion adducts are too unstable to be isolated as salts except in special circumstances. For example, benzotrifluoride reacts with nitryl fluoride (NO_2F) and boron trifluoride at low temperatures to give a crystalline product thought to be the salt,

367

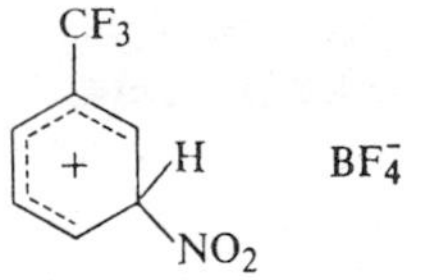

This product is stable only to $-50°C$, above which it decomposes into *m*-nitrobenzotrifluoride, hydrogen fluoride, and boron trifluoride. The relative stability of this adduct is associated with the stability of the fluoborate anion.

(2) In most instances, the first step in the process is rate-determining, e.g. in the nitration and bromination of benzene. There are some reactions in which the second step (loss of the proton) is rate-determining; sulphonation is the best known example.

(3) The reactions are, with few exceptions, irreversible and the products formed are kinetically controlled (p. 92). Two important exceptions are sulphonation (11.5) and Friedel-Crafts alkylation (11.3a); the reversibility of these reactions can lead to the formation of the thermodynamically controlled products in appropriate conditions. Use may be made of this fact in synthesis (p. 388), but in some situations it proves to be disadvantageous (p. 415).

(4) Substitution is subject to electrophilic catalysis. For example, benzene reacts with bromine at a negligible rate but in the presence of iron(III) bromide reaction is comparatively fast. The catalyst acts by aiding the removal of bromide ion:

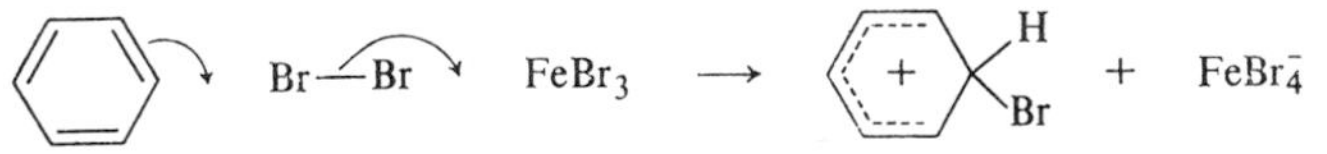

Lewis acids, such as iron(III) bromide in the example above, are customarily used as catalysts, as in halogenation (11.6) and Friedel-Crafts alkylation (11.3a).

(5) An atom or group other than hydrogen may be displaced from the aromatic ring. For example, phenyltrimethylsilane gives benzene with acids:

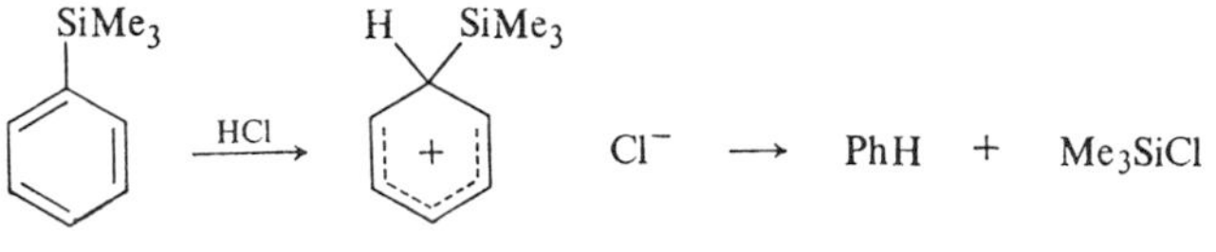

and the treatment of salicylic acid with bromine gives 2,4,6-tribromophenol:

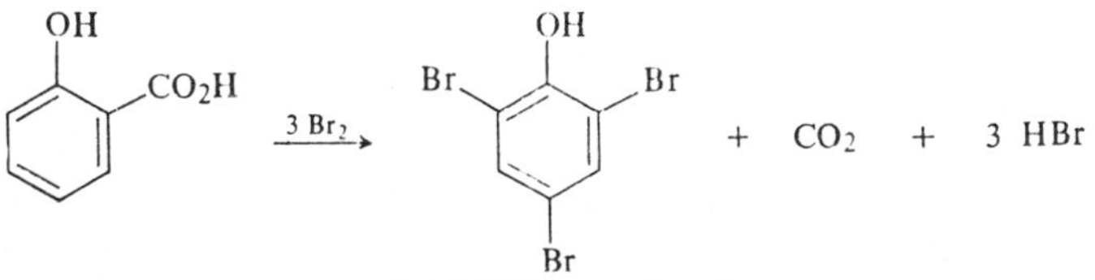

2,4,6-Tribromophenol

Comparison with the reactions between olefins and electrophiles. In their reactions with electrophilic reagents, aromatic compounds resemble olefins in that the first step of the process consists of the addition of the electrophile to sp^2-hybridized carbon with the formation of a carbonium ion. There are, however, two *general* differences, important exceptions to which are discussed below.

First, whereas the carbonium-ion adduct from an olefin and an electrophile normally reacts with a nucleophile by addition (p. 107), that from an aromatic compound reacts by elimination. The difference arises from the fact that in the latter case elimination regenerates the aromatic system and liberates the associated stabilization energy. In fact, the addition of one mole of hydrogen to benzene is endothermic, whereas the reduction of ethylene is exothermic by about 140 kJ mol^{-1}. Thus aromatic compounds are characterized *in general* by their undergoing substitution, just as olefins are characterized by their undergoing addition.

Secondly, aromatic compounds react less rapidly than olefins with a given electrophile: e.g. whereas benzene is hardly affected by bromine, ethylene reacts instantly. This is because the formation of the intermediate carbonium ion is accompanied, in the addition to benzene, by the loss of the aromatic stabilization energy; although this loss is offset to some extent by the delocalization energy in the resulting ion (p. 367), it leads to a more endothermic reaction than in the case of ethylene.

Although the causes of the two principal differences between aromatic compounds and olefins are revealed in the above discussion, two qualifications are necessary. First, although benzene and its simple derivatives are relatively inert to addition, compounds in which two or more benzene rings are fused together are often quite susceptible to addition. For example, anthracene reacts with bromine to give, initially, 9,10-dibromo-9,10-dihydroanthracene,

We have discussed previously the principles which underlie the relative ease of addition across the 9,10-positions of anthracene (p. 57).

Secondly, some substituted benzenes react very rapidly with electrophiles: e.g. acetanilide reacts rapidly with bromine to give mainly *p*-bromoacetanilide, and the reaction of dimethylaniline with bromine is even faster (its activation energy is close to zero). This is because certain substituents are able to stabilize the carbonium-ion intermediate, and hence the preceding transition state, thereby lowering the activation energy; in the case of the bromination of acetanilide, the stabilization is represented by the contribution of the canonical structure,

$$\overset{+}{N}HCOCH_3$$

which symbolizes the delocalization of the positive charge over nitrogen as well as over three of the nuclear carbon atoms.

11.2 Directive and Rate-controlling Factors

The role played by nitrogen in the bromination of acetanilide described above illustrates one principle of great importance in electrophilic aromatic substitution: the rate of reaction is strongly dependent upon the nature of the substituent(s) in the aromatic nucleus. Further, the relative ease of substitution at different positions in an aromatic compound is also determined by the nature of the substituent(s): in the case of acetanilide, the order of reactivity at the nuclear carbons is *para* > *ortho* ≫ *meta*. Both the directive effects and the rate-controlling effects of substituents are of great importance in synthetic applications of electrophilic aromatic substitutions. The effects are conveniently discussed under four headings: (*a*) monosubstituted benzenes, (*b*) di- and poly-substituted benzenes, (*c*) bi- and poly-cyclic hydrocarbons, and (*d*) hetero-aromatic compounds.

(*a*) MONOSUBSTITUTED BENZENES

In the first step of its reaction with an electrophile, the benzene ring provides two electrons to form a new covalent bond with the electrophile, yielding a carbonium ion. It is possible to represent this ion as a hybrid of three canonical structures (p. 367), but it is not possible to represent adequately the structure of the preceding transition state. At the transition state, the electron-pair which ultimately forms the new covalent bond has been partly transferred from the aromatic ring to the electrophile. The residual aromatic system therefore bears a fraction of the unit positive charge which it bears in the intermediate, so that a useful working representation of the transition state is as follows:

Transition state   Intermediate

Thus, any factor which stabilizes the intermediate in a particular case also stabilizes the transition state, and in practice it is convenient to employ the intermediate as a model for the transition state (*cf.* Hammond's postulate, p. 85).

In a reaction on a monosubstituted benzene, the intermediates for *ortho*, *meta*, and *para* substitution are, respectively,

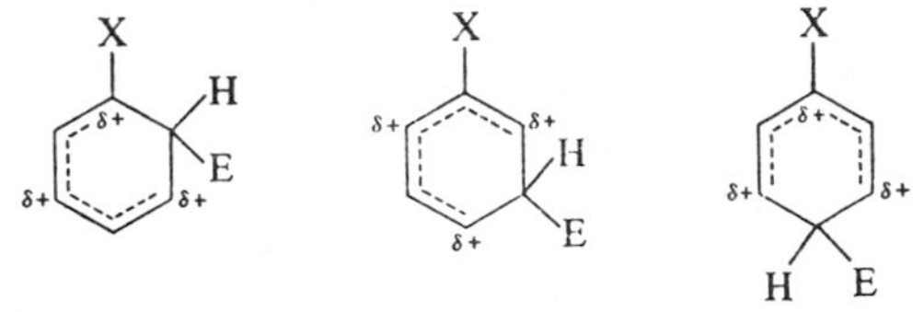

The effect on the stabilities of these ions of groups, X, of different polar types will now be considered.

If X is an electron-releasing substituent of $+I$ type (e.g. CH_3), each of the three ions is stabilized relative to that formed by benzene. Reaction occurs more rapidly at each position than at any one position in benzene and the substituent is said to be *activating*. In the case of reaction at the *ortho* or *para* positions, the substituent is adjacent to a carbon atom which bears a portion of the positive charge in the transition state; the appropriate contributing structures are

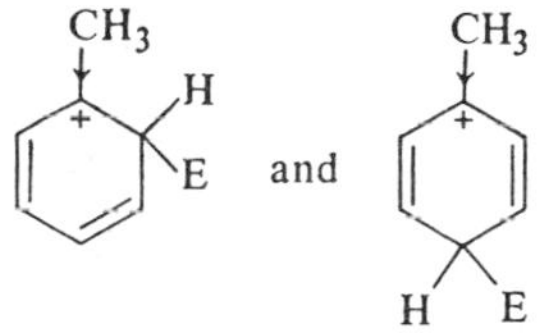

When reaction occurs at the *meta* position, however, the substituent is further removed from a positively polarized carbon,

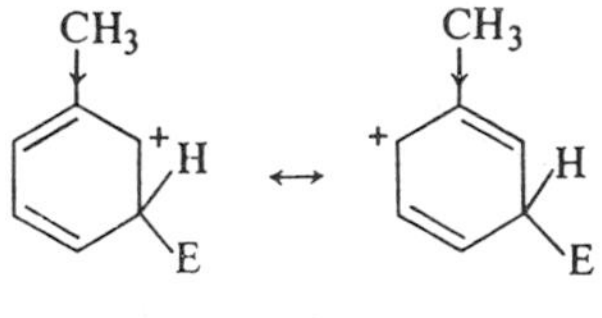

and its stabilizing influence is smaller. Consequently, the *ortho* and *para* positions are more strongly activated than the *meta* position and the substituent is said to be *ortho*, *para-directing*.

If X is an electron-attracting substituent of $-I$ type (e.g. NMe_3^+), the converse applies: each of the three intermediate ions is destabilized relative to that formed by benzene, the effect being least for reaction at the *meta* position. Such substituents are therefore *deactivating* and *meta-directing*.

Some substituents also have conjugative, or mesomeric ($+M$ or $-M$), effects. Those of $+I$, $+M$ type (e.g. $-O^-$) are activating and *ortho*, *para*-directing; the transition state for *para* (and *ortho*) substitution is stabilized not only by inductive electron-release but also by further delocalization involving the substituent,

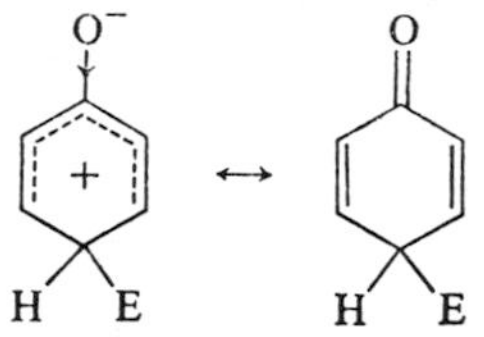

and the transition state for *meta* substitution is stabilized by the inductive effect but not appreciably by the mesomeric effect: the contributing structure,

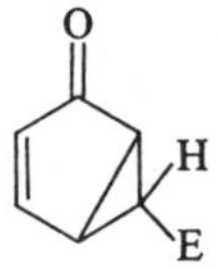

is of very high energy-content and of little significance.

Substituents of $-I$, $-M$ type (e.g. NO_2) are deactivating and *meta*-directing; the $-M$ effect serves to increase the deactivation of the ring due to the $-I$ effect.

Some substituents have opposed inductive $(-I)$ and mesomeric $(+M)$ effects, e.g. amino and hydroxyl groups and halogen atoms. Reaction at the *para* (or *ortho*) position then leads to an intermediate in which the positive charge is delocalized onto the substituent,

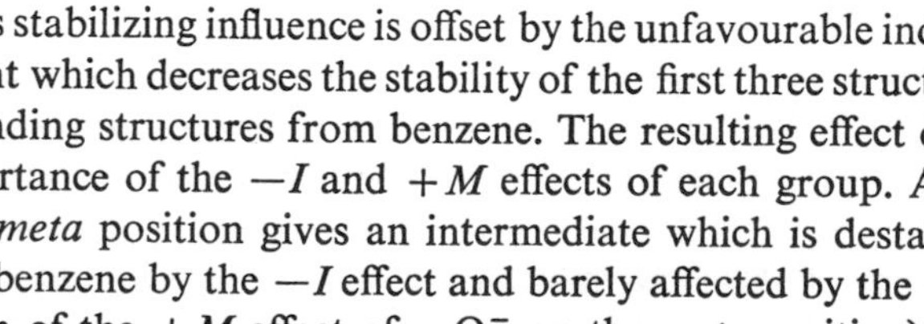

However, this stabilizing influence is offset by the unfavourable inductive effect of the substituent which decreases the stability of the first three structures relative to the corresponding structures from benzene. The resulting effect depends on the relative importance of the $-I$ and $+M$ effects of each group. Addition of the reagent at a *meta* position gives an intermediate which is destabilized relative to that from benzene by the $-I$ effect and barely affected by the $+M$ effect (see the discussion of the $+M$ effect of $-O^-$ on the *meta* position). Consequently the *meta* position is deactivated, and always less reactive than the *para* (or *ortho*) position, whether or not this is itself deactivated.

The similarity in the mode of operation of the polar effect on the *ortho* and *para* positions leads to the expectation that the ratio of the reactivities at these positions should be 2:1, there being two *ortho* positions and one *para* position. In practice, the ratio is normally different from this, for two main reasons. First, more subtle electronic effects than those apparent from the above discussion seem to be involved; and secondly, steric hindrance between substituent and reagent can lead to *ortho:para*-ratios which are considerably less than 2:1.

Individual groups of compounds behave as follows.

(*i*) *Alkylbenzenes.* The nitration of toluene by nitric acid in acetic anhydride at 0°C gives *o*-, *m*- and *p*-nitrotoluene in the ratios (expressed as percentages) 61·5:1·5:37. The total reactivity of toluene compared with benzene is 27, and combination of this result with the isomer distribution of the nitrotoluenes leads to the following data for the relative reactivities of each nuclear position in toluene compared with one position in benzene: *o*, 50; *m*, 1·3; *p*, 60. (This measure of the reactivity of a particular nuclear carbon in a given reaction is referred to as the *partial rate factor*.)

This result illustrates the directive and activating behaviour of a substituent, CH_3, of $+I$ type. Other alkyl-benzenes behave similarly in giving predominantly the *ortho* and *para* derivatives, as illustrated by the following partial rate factors for the nitration of four compounds in acetic anhydride and the chlorination of two in acetic acid.

Nitration:

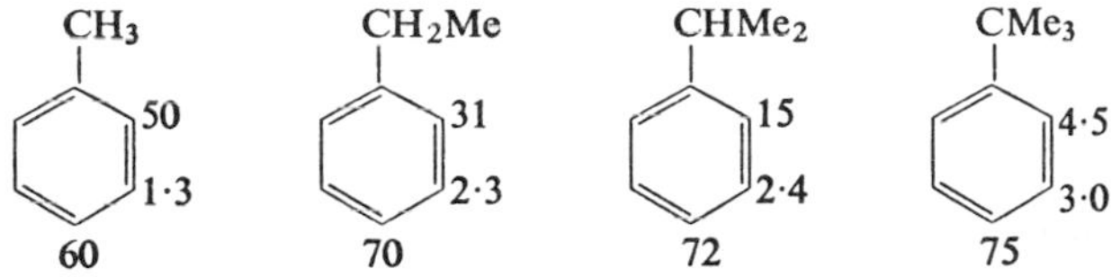

CH_3	CH_2Me	$CHMe_2$	CMe_3
50 (ortho), 1·3 (meta), 60 (para)	31 (ortho), 2·3 (meta), 70 (para)	15 (ortho), 2·4 (meta), 72 (para)	4·5 (ortho), 3·0 (meta), 75 (para)

Chlorination:

CH_3	CMe_3
620 (ortho), 5 (meta), 820 (para)	60 (ortho), 6 (meta), 400 (para)

Three points should be noted. First, the *ortho:para* ratio falls off sharply as the size of the alkyl group is increased, as a result of steric hindrance to *ortho*-substitution. This imposes limitations in synthesis: thus, whereas it is possible to obtain *o*-nitrotoluene in about 60% yield, the yield of *o*-nitro-t-butylbenzene is less than 10%.

Secondly, the absolute values of the partial rate factors for reaction on a given compound depend on the nature of the reagent. This results from variation in the *selectivity* of the reagent. A reagent of such reactivity that it reacted on every collision with an aromatic compound would not discriminate between the three nuclear positions of toluene, or between toluene and benzene: the partial rate factors would all be unity. As the reactivity of the reagent decreases, there is an increasing demand for an electron-pair to be supplied at the relevant nuclear carbon in order that the activation energy barrier can be surmounted. The ease of supply of this pair is determined by the polar character and the position of the substituent, so that the greater the demand by the reagent, the greater will be the differential effects of substituents. Thus the biggest differences in selectivity between different positions occur with the least reactive electrophiles; in the above case, chlorine is less reactive and more selective than the nitrating agent.

Thirdly, although the reactivities at the *para* positions of toluene and t-butyl-benzene are in the expected order in nitration, since the t-butyl group has a stronger inductive effect than the methyl group (p. 49), the opposite order obtains in chlorination and in many other reactions. One reason suggested for this is that the methyl group is able to delocalize the charge at the transition state by a hyperconjugative interaction (p. 52), represented by the contribution of the structure

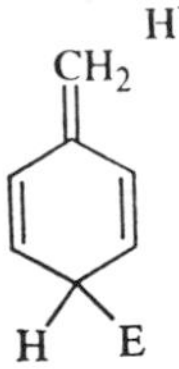

This phenomenon, of theoretical interest, is not of significance in preparative chemistry.

In summary, the greater reactivity of alkylbenzenes than benzene enables electrophilic substitutions to be carried out in slightly milder conditions than are necessary for benzene, and the directive properties of the alkyl groups are such that *para* derivatives and, in suitable cases, *ortho* derivatives, can usually be obtained.

(*ii*) *Benzenes substituted with electron-attracting groups.* These include compounds in which the substituent is a positive pole, such as $PhNMe_3^+$, and those in which it is a neutral unsaturated group of strong dipolar character in which the positive end of the dipole is attached to the benzene ring, e.g.

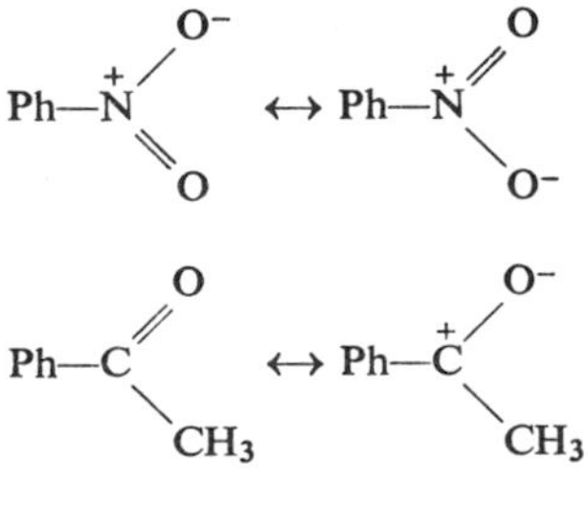

$$Ph—C{\equiv}N \quad \leftrightarrow \quad Ph—\overset{+}{C}{=}N^-$$

The substituents are strongly deactivating and *meta*-directing: e.g. nitro-benzene undergoes nitration with a mixture of concentrated nitric and sulphuric acids at only about a hundred-thousandth the rate of benzene, and *m*-dinitro-benzene accounts for about 93 % of the dinitrobenzenes formed.

Because of their low nuclear reactivities, these compounds require much more vigorous conditions than benzene and in many cases reaction cannot be brought about (e.g. nitrobenzene does not undergo Friedel-Crafts reactions; 11.3).

(*iii*) *Benzenes substituted with groups of* −*I*, +*M type.* These are all *ortho, para*-directing, but the ease of reaction (activation or deactivation) varies over a wide range, depending on the substituent: e.g. in the same conditions, the relative rates of chlorination at the *para* positions of *NN*-dimethylaniline and bromo-benzene are about 10^{20}:1. Consequently, as will be described later, the reaction conditions for the successful substitution of a particular compound in this group vary widely.

(1) *Amino and hydroxyl substituents and related groups.* For these substituents, the conjugative effect (+*M*) dominates over the inductive effect (−*I*). That is, the contribution made to the transition state by structures of the type

far more than outweighs the deactivating influence due to the inductive effect of nitrogen or oxygen when substitution occurs at the *ortho* or *para* positions. These positions are consequently strongly activated.

The amino group is more strongly activating than hydroxyl, for it has the larger +*M* and the smaller −*I* effect; quaternary cationic nitrogen is more stable than ternary cationic oxygen. However, the —O⁻ substituent is more strongly activating than both —NH₂ and —OH because, first, its +*M* effect is greater and, secondly, it possesses a +*I* effect. Electrophilic reactions are normally conducted in acidic media in which phenoxide ions are present in negligible concentration compared with phenols, but for reaction of a phenol with the electrophilic aromatic diazonium ions it is necessary to use basic conditions in order to make use of the strong activating influence of —O⁻ (p. 446).

N-Alkyl-derivatives of aniline and *O*-alkyl-derivatives of phenol are of re-activity comparable with aniline and phenol themselves but the acyl-derivatives are much less reactive. This is because the unshared electron-pair on nitrogen or oxygen is already delocalized within the substituent,

$$\underset{\text{R}}{\text{Ph}-\ddot{\text{N}}\text{H}-\text{C}=\text{O}} \leftrightarrow \underset{\text{R}}{\text{Ph}-\overset{+}{\text{N}}\text{H}=\text{C}-\text{O}^-}$$

$$\underset{\text{R}}{\text{Ph}-\ddot{\text{O}}-\text{C}=\text{O}} \leftrightarrow \underset{\text{R}}{\text{Ph}-\overset{+}{\text{O}}=\text{C}-\text{O}^-}$$

and is not so readily available for π-orbital overlap in the electron-deficient transition state.

The smaller activating effect of acetamido compared with amino is usefully applied in synthesis: e.g. whereas aniline is so reactive towards bromine that it gives 2,4,6-tribromoaniline essentially instantaneously, acetanilide reacts more slowly and the monobromo-derivatives (principally the *para*, together with some of the *ortho* compound) may be isolated. Further, since aromatic amines are very susceptible to oxidation, it is normally necessary to carry out substitutions on their acyl derivatives and then remove the acyl group by hydrolysis.

(2) *Halogen substituents.* Fluorine has a weaker $+M$ and a stronger $-I$ effect than oxygen, and for *para*-substitution the effects approximately nullify each other so that reaction occurs about as readily as at one carbon in benzene. Both the $+M$ and the $-I$ effects of the halogens fall in the order, $F > Cl > Br > I$ (p. 52), and the resultant effect leads to the order, $F \sim H > Cl \sim Br \sim I$; in many substitutions, chloro-, bromo-, and iodobenzene are about one-tenth as reactive at their *para* positions as benzene is at any one position. The *ortho*-positions of all four halobenzenes are less reactive than the *para* positions, and the *meta* positions are strongly deactivated.

In summary, substitutions of the halobenzenes require conditions comparable in vigour with those for benzene and yield mainly the *para*-derivatives.

(3) *Biphenyls and styrenes.* Biphenyl is activated in the *ortho* and *para* positions and weakly deactivated in the *meta* position. The latter result follows from the weak $-I$ effect of sp^2-hybridized carbon (p. 68) and the former from the ability of the phenyl substituent to delocalize the positive charge on the transition states for *ortho* and *para* substitution, e.g. that for *para* substitution is represented as the hybrid,

Consequently, the compound is *ortho, para*-directing and of reactivity comparable with toluene.

The introduction of an electron-attracting substituent into biphenyl reduces the ease of reaction and causes substitution to occur at the *ortho* and *para* positions of the unsubstituted ring. Conversely, an electron-releasing substituent increases the reactivity and causes reaction to occur in the substituted ring.

Styrene and its derivatives behave similarly. For example, cinnamic acid is nitrated predominantly in the *ortho* and *para* positions and reacts at about one-tenth the rate of benzene. In effect, the —CH=CH—CO₂H group is similar to a halogen atom: the inductive effect of the group (—*I*) reduces the ease of reaction, but the deactivation at the *ortho* and *para* positions is somewhat offset by the extra delocalization of the positive charge onto the aliphatic side-chain:

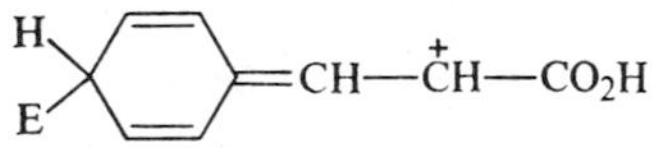

Thus, the pair of π-electrons in the olefinic group of cinnamic acid behaves analogously to a pair of *p*-electrons on a halogen substituent.

(iv) Summary. The following Table summarizes the directive and rate-controlling effects of the more common substituents in electrophilic aromatic substitution.

Substituent	Polar character	Directive effect	Rate-controlling effect
O⁻	+I, +M	o, p	Very powerfully activating
NH₂, NHR, NR₂ OH, OR	−I, +M	o, p	Powerfully activating
NHCOR, OCOR	−I, +M	o, p	Activating
Ph	−I, +M	o, p	Moderately activating
CH₃ and other alkyl groups	+I	o, p	Moderately activating
(H)			—
F	−I, +M	o, p	Comparable with benzene
Cl, Br, I CH=CH—CO₂H (etc.)	−I, +M	o, p	Weakly deactivating
CO₂R, CO₂H, CHO, COR, CN, NO₂, SO₂OH	−I, −M	m	Strongly deactivating
NH₃⁺, NR₃⁺	−I	m	Strongly deactivating

(b) DI- AND POLY-SUBSTITUTED BENZENES

To a first approximation, two or more substituents affect the ease of reaction at a particular position in benzene independently of each other. * For example, the introduction of successive methyl groups into benzene increases the reactivity and pentamethylbenzene is the most reactive compound in the series.

Use can be made of this principle in synthesis. For instance, the acetamido group activates *ortho* positions and slightly deactivates *meta* positions, whereas the methyl group activates *ortho* less strongly than acetamido but also slightly activates *meta* positions. Consequently, in *p*-acetamidotoluene,

$$CH_3-\!\!\underset{b\quad a}{\bigcirc}\!\!-NHCOCH_3$$

position *a*, which is *ortho* to acetamido and *meta* to methyl, is more reactive than position *b*, which is *meta* to acetamido and *ortho* to methyl. Substitution therefore occurs predominantly at *a*, and *meta*-substituted toluenes, which are obtained in only very low yield by the direct substitution of toluene, may in this way be obtained by removal of the acetamido group (e.g. p. 440).

(c) BI- AND POLY-CYCLIC HYDROCARBONS

The delocalization of the positive charge in the transition states of electrophilic substitutions is increased by the fusion of two or more benzene rings and the polycyclic hydrocarbons are therefore all more reactive than benzene. Directive effects are also introduced.

Substitution in naphthalene is illustrative. The transition states for reaction at the 1- and 2-positions may be represented as follows:

1-substitution:

*A linear free-energy relationship holds approximately. That is, if the introduction of each of two substituents separately alters the free energy of activation for reaction at a particular position by amounts x and y, the presence of both substituents alters the free energy of activation by $(x + y)$. Since $\log k \propto \Delta G^{\ddagger}$, the partial rate factor for substitution at a particular position in the disubstituted compound is equal to the *product* of the partial rate factors for the appropriate positions of the two monosubstituted compounds.

2-substitution:

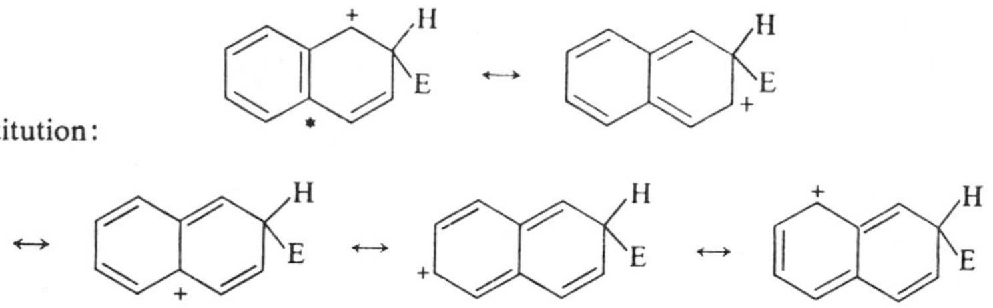

In each of the two transition states, the positive charge is more extensively delocalized than in reaction on benzene, leading to lower activation energies. Further, the three starred structures are of benzenoid type and therefore of lower energy content than the remainder, in which the benzenoid nature of the second ring has been interrupted. Since there are two such low-energy contributors for 1-substitution as compared with one for 2-substitution, it is understandable that the 1-position should be the more reactive. In a typical reaction, nitration, the partial rate factors for 1- and 2-substitution are 470 and 50, respectively.

There are, however, two conditions in which substitution at the 2-position predominates. One applies when the reaction is thermodynamically controlled, as in sulphonation at high temperatures (p. 408), for the 1-derivative, in which there is steric repulsion between the substituent and the *peri*-hydrogen atom,

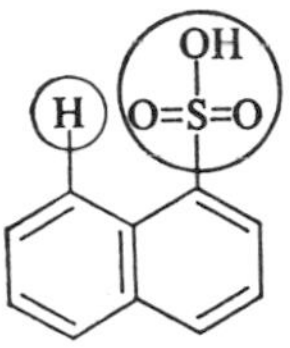

is thermodynamically the less stable. The other applies when the reaction is kinetically controlled but the reagent is particularly bulky, as in Friedel-Crafts acetylation in nitrobenzene (p. 392), for reaction at the 1-position is then markedly hindered sterically by the *peri*-hydrogen.

The presence of an electron-attracting group in naphthalene reduces the reactivity and causes substitution to occur in the unsubstituted ring, mainly at the 5- and 8-positions (i.e. the two 1-positions of that ring). An electron-releasing group activates the molecule further and reaction occurs in the substituted ring. If the group is in the 1-position, substitution occurs at the 2- and 4-positions (i.e. *ortho* and *para* to the electron-releasing group), but a 2-substituent directs almost entirely to the 1-position, although the 3-position is also an *ortho* position. The reason is that the stabilization of the transition state which is provided by the substituent is more effective when the appropriate canonical structure is benzenoid (1-substitution) than when it is not (3-substitution), e.g.

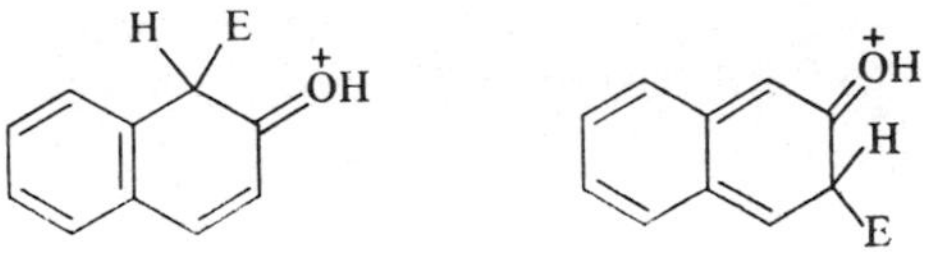

The reactivities of polycyclic hydrocarbons follow from the principles described for naphthalene. Examples, including data for partial rate factors in nitration at the most reactive position, are tabulated.

Anthracene

Phenanthrene

Chrysene

Pyrene

Perylene

3:4-Benzopyrene

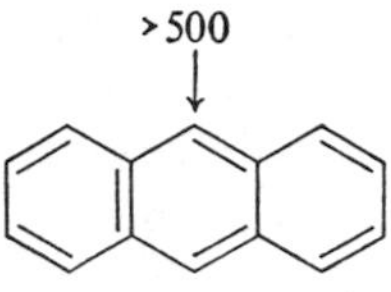

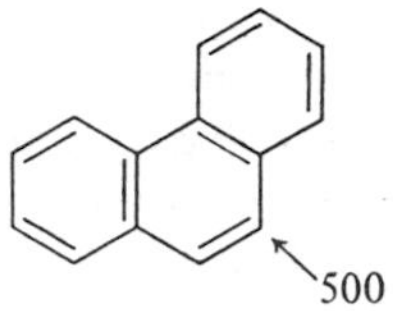

(*d*) HETEROAROMATIC COMPOUNDS

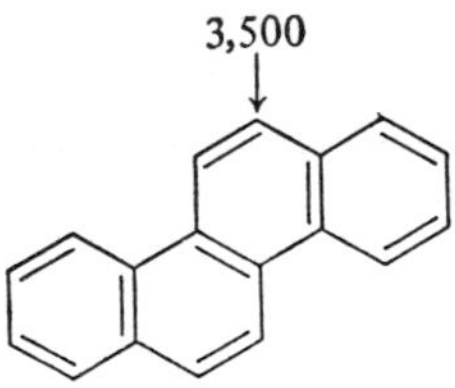

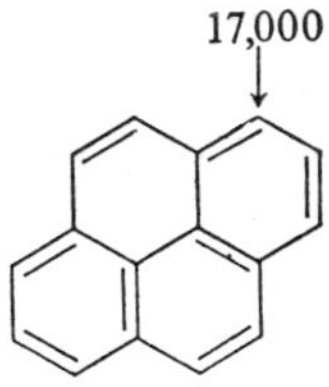

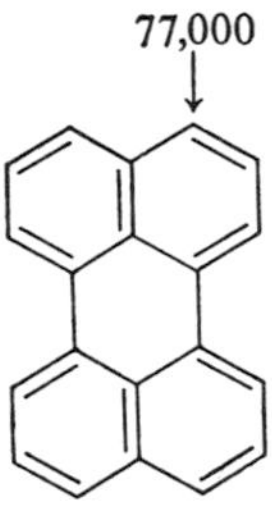

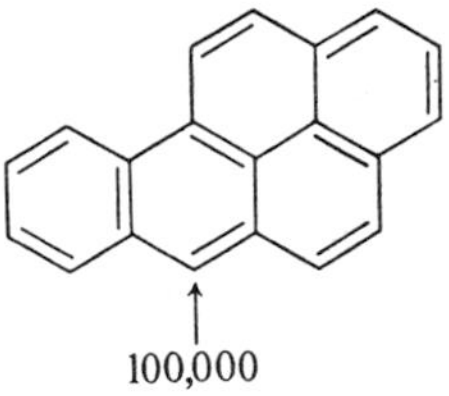

(*i*) *Five-membered rings.* The principles which govern the electrophilic substitutions of this group of heteroaromatic compounds will be illustrated by reference to pyrrole.

Pyrrole is highly reactive at both the 2- and 3-positions. The reason is that the transition state for substitution at each position is strongly stabilized by the accommodation of the positive charge by nitrogen,

2-substitution:

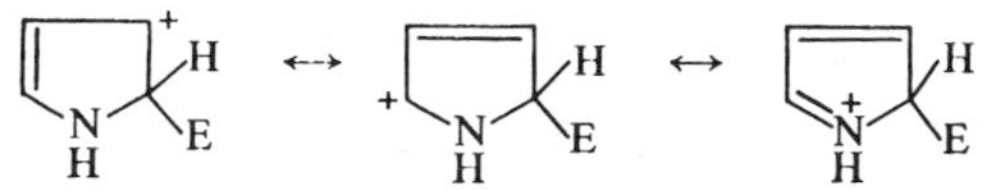

3-substitution:

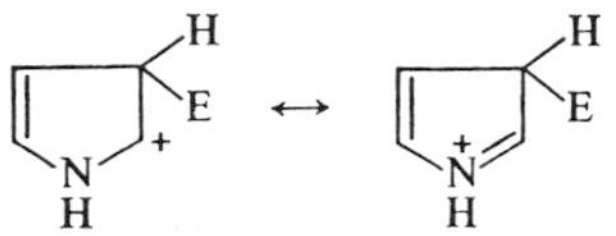

in just the way that aniline owes its reactivity to the exocyclic nitrogen (p. 375). 2-Substitution predominates because the positive charge in the transition state is delocalized over a total of three atoms, compared with two for 3-substitution.

The similarity of pyrrole and aniline is particularly apparent in their reactions with bromine: each reacts at all its activated carbon atoms, pyrrole giving tetra-bromopyrrole and aniline giving 2,4,6-tribromoaniline. In fact, pyrrole is even more strongly activated than aniline and should perhaps be compared with the phenoxide ion: each undergoes the Reimer-Tiemann reaction, unlike other benzenoid compounds (p. 398). In addition, pyrrole undergoes Friedel-Crafts acylation in the absence of a catalyst (p. 389) and the Hoesch reaction, characteristic of polyhydric phenols (p. 396).

One difficulty in dealing with pyrrole is that it is readily converted by acid into the trimer,

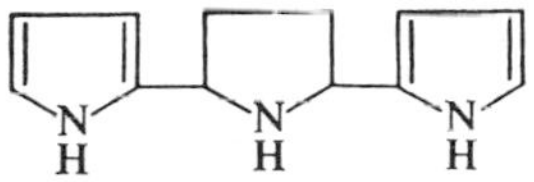

Accordingly, special conditions need to be used to carry out reactions normally involving acids (e.g. sulphonation, p. 407). Pyrroles substituted with electron-attracting groups are less prone to polymerization and can be more conveniently handled.

Furan and thiophen are also activated towards electrophiles and react predominantly at the 2-position. The underlying theory is similar to that for pyrrole, namely, that the heteroatom is able to delocalize the positive charge on the transition state. Since oxygen accommodates a positive charge less readily than nitrogen, furan is less reactive than pyrrole, just as phenol is less reactive than aniline. The $+M$ effect of sulphur is smaller than that of oxygen because the overlap of the differently sized p-orbitals of carbon and sulphur is less than in the case of carbon and oxygen, so that understandably thiophen is less reactive than furan.

A revealing trend of reactivities is shown by the behaviour of these three heterocyclic compounds towards maleic anhydride. Pyrrole is sufficiently reactive towards electrophiles to take part, as a nucleophile, in a Michael addition (p. 115):

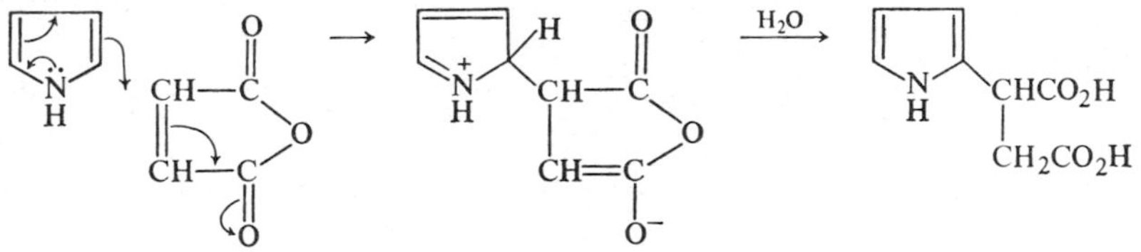

Furan is less reactive towards electrophiles than pyrrole and instead undergoes the Diels-Alder reaction (p. 291), differing in this respect from benzene because less aromatic stabilization energy is lost on 1,4-addition. Thiophen, which is less reactive both to electrophiles and as a conjugated diene, does not react.

The 2:3-benzo-derivatives of pyrrole, furan, and thiophen are activated and react in the hetero-ring. Substitution occurs mainly at the 3-position as a result of the fact that the stabilizing influence of the hetero-atom on the transition state is more effective when the appropriate canonical structure is benzenoid (3-substitution) than when the benzenoid system is disrupted (2-substitution), as illustrated for indole.

3-substitution   2-substitution

(ii) Six-membered rings. The principles governing the reactivity of these compounds are illustrated by reference to pyridine. The transition states for substitution at the 3- and 4-positions can be represented as the hybrids,

3-substitution:

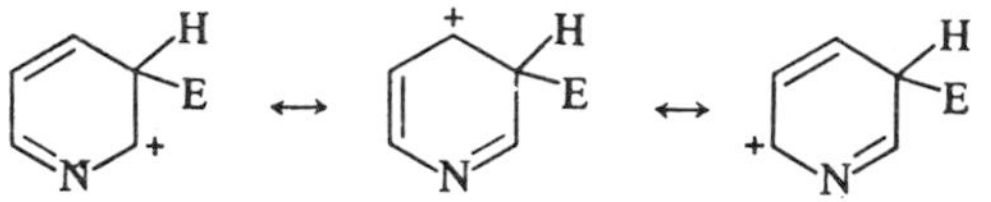

4-substitution:

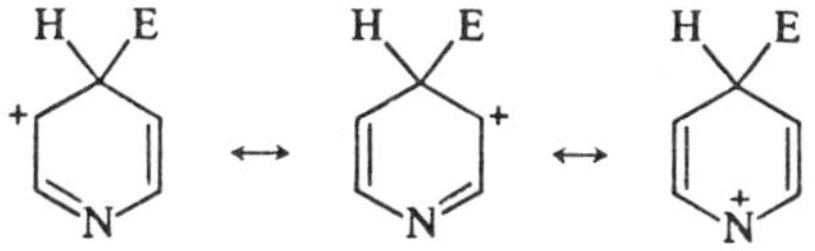

In each, the positive charge is less well accommodated than in reactions on benzene because nitrogen is more electronegative than carbon. Hence both the 3- and 4-positions are deactivated, the latter the more strongly because of the high energy of the contributing structure which contains divalent positive nitrogen.* The 2-position resembles the 4-position, as reference to the appropriate canonical structures will show.

Many electrophilic substitutions are conducted in acidic media in which pyridine is present almost entirely as its conjugate acid, and this is even less reactive than pyridine itself. Very vigorous conditions are required to bring about reaction: e.g. nitration requires 100% sulphuric acid with sodium and potassium nitrates at 300°C, and even then the yield of 3-nitropyridine is only a few percent.

Quinoline and isoquinoline are also deactivated, though less so than pyridine, and reaction normally occurs in the homocyclic ring at the 5- and 8-positions (*cf.* the behaviour of naphthalene containing an electron-attracting substituent, p. 379). There are, however, many exceptions.

Quinoline   Isoquinoline

Pyridazine, pyrimidine, and pyrazine (p. 659), containing two heterocyclic nitrogen atoms, are more strongly deactivated than pyridine and electrophilic substitutions are unsuccessful. Derivatives are obtained either *via* nucleophilic substitution (12.2) or by synthesizing a compound which contains a powerfully activating substituent such as hydroxyl or amino and with which electrophiles react (see, e.g. p. 697).

11.3 Formation of Carbon-Carbon Bonds

A carbon atom which is bonded to an electronegative atom or group, X, is positively polarized and therefore electrophilic, although not normally sufficiently so for it to react with aromatic compounds other than those of exceptionally high reactivity. However, its electrophilicity may be enhanced by the addition of a species which can accept electrons from X, and reaction then occurs with less reactive aromatic compounds.

This principle may be applied to the formation of bonds between aromatic and aliphatic carbon atoms in a variety of ways, some of which are of wider applicability than others. The most commonly used processes are Friedel-Crafts alkylation and acylation.

---

*Careful distinction should be made between the ability of (*a*) divalent nitrogen, and (*b*) quaternary nitrogen, to accommodate a positive charge. The latter system, in which nitrogen possesses an octet, is relatively stable, and this stability is responsible for the activating effect of nitrogen in aniline and pyrrole.

(a) FRIEDEL-CRAFTS ALKYLATION

Aliphatic compounds which take part in alkylations are halides, alcohols, esters, ethers, olefins, aldehydes, and ketones. Reactions of the first four classes of compound are normally catalyzed by Lewis acids and those of the last three (the unsaturated compounds) by proton acids.

Halides are the reagents of most frequent choice in alkylation, and aluminium trichloride is usually employed as the catalyst. The reaction of benzene with primary and secondary halides occurs by the S_N2 mechanism in which nucleophilic attack by benzene on the aliphatic carbon atom is aided by the removal of halide by the Lewis acid:*

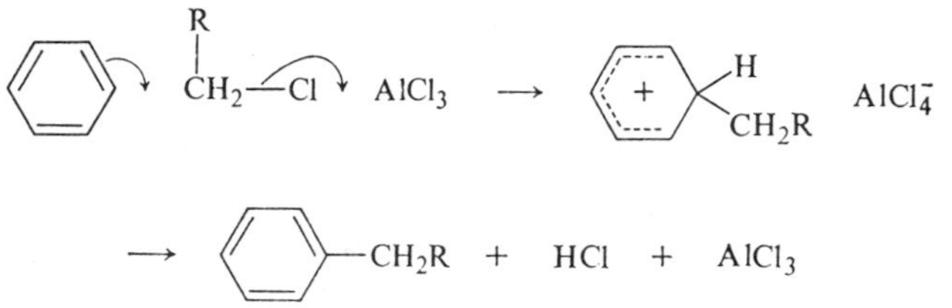

Tertiary halides differ in that the S_N1 mechanism occurs because tertiary carbonium ions are formed comparatively readily. For example, benzene, t-butyl chloride, and iron(III) chloride give t-butylbenzene in 80% yield *via* the t-butyl cation:

$$Me_3C\!-\!Cl \quad FeCl_3 \longrightarrow Me_3C^+ + FeCl_4^-$$

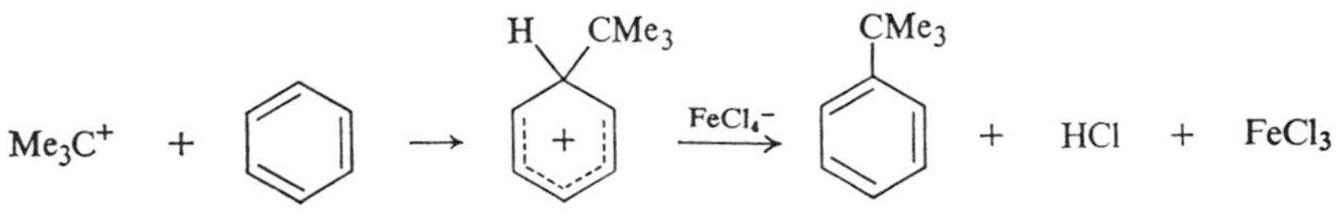

t-Butylbenzene

Benzyl halides are very reactive as alkylating agents, just as they are in other nucleophilic displacements, but vinyl and aryl halides are inert. The use of di- and poly-halides leads to successive alkylations: benzene reacts with methylene dichloride in the presence of aluminium trichloride to give diphenylmethane,

$$PhH + CH_2Cl_2 \xrightarrow[-HCl]{(AlCl_3)} [PhCH_2Cl] \xrightarrow[-HCl]{PhH\ (AlCl_3)} PhCH_2Ph$$

Diphenylmethane

with ethylene dichloride to give bibenzyl,

---

*These reactions may be regarded alternatively as nucleophilic substitutions at aliphatic carbon or electrophilic substitutions at aromatic carbon.

$$2 \text{ PhH} + \text{ClCH}_2\text{CH}_2\text{Cl} \xrightarrow[-2\text{HCl}]{(\text{AlCl}_3)} \underset{\text{Bibenzyl}}{\text{PhCH}_2\text{CH}_2\text{Ph}}$$

and with carbon tetrachloride to give triphenylmethyl chloride (steric hindrance evidently preventing the displacement of the fourth chlorine),

$$3 \text{ PhH} + \text{CCl}_4 \xrightarrow[-3\text{HCl}]{(\text{AlCl}_3)} \underset{\substack{\text{Triphenylmethyl} \\ \text{chloride}}}{\text{Ph}_3\text{CCl}}$$

Alcohols, esters, and ethers react analogously to halides.

Olefins react *via* the carbonium ions which they form with protonic acids:

$$R-CH{=}CH_2 + H^+ \longrightarrow R-\overset{+}{C}H-CH_3$$

$$PhH + R-\overset{+}{C}H-CH_3 \xrightarrow{-H^+} Ph-CH\overset{\displaystyle R}{\underset{\displaystyle CH_3}{\big<}}$$

For example, the addition of cyclohexene to benzene in concentrated sulphuric acid at 5–10°C gives cyclohexylbenzene in about 65% yield [2]:

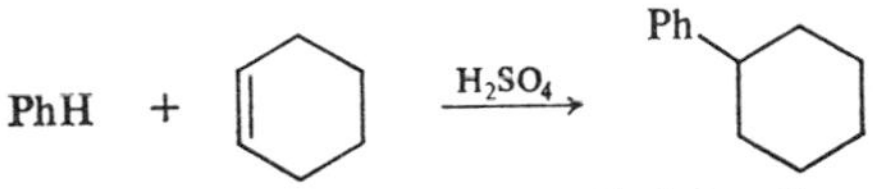

Cyclohexylbenzene

Styrene (required for the production of polystyrene, styrene-butadiene copolymer rubbers, etc.) is made industrially by the ethylation of benzene on an aluminium trichloride catalyst followed by dehydrogenation (p. 585):

$$PhH + CH_2{=}CH_2 \xrightarrow{\text{AlCl}_3} PhCH_2-CH_3 \xrightarrow[600°C]{\text{ZnO}} PhCH{=}CH_2$$

The Bogert-Cook synthesis of phenanthrene and its derivatives makes use of intramolecular alkylation by an olefin, e.g.

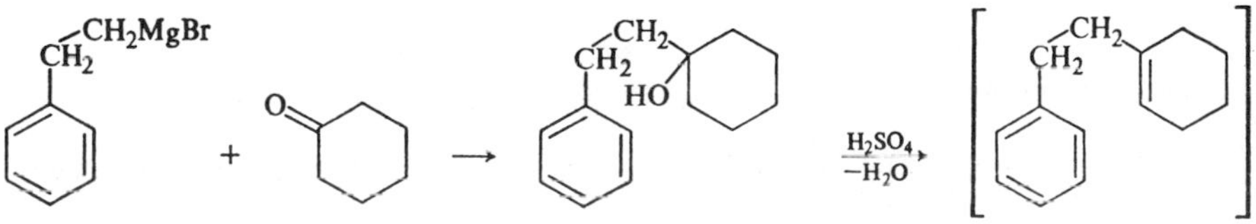

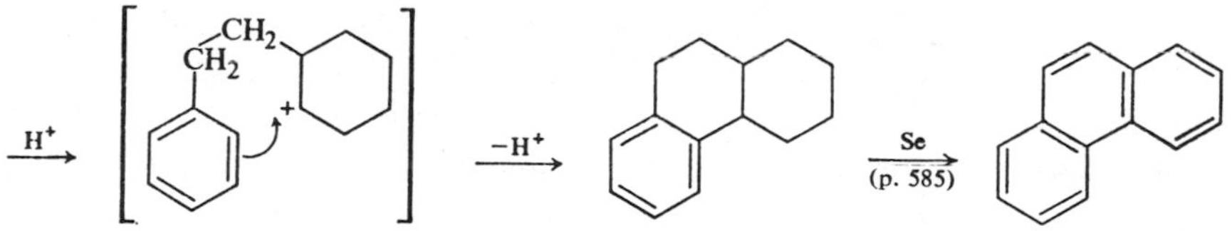

Phenanthrene

Aldehydes and ketones can also act as alkylating agents in the presence of proton acids but this process does not compete favourably with acid-catalyzed self-condensation of the carbonyl compound (p. 270) except when alkylation is intramolecular and stereochemically favoured. For example, 4-methyl-2-quinolone can be obtained by acid-treatment of the amido-ketone formed from aniline and acetoacetic ester (p. 681):

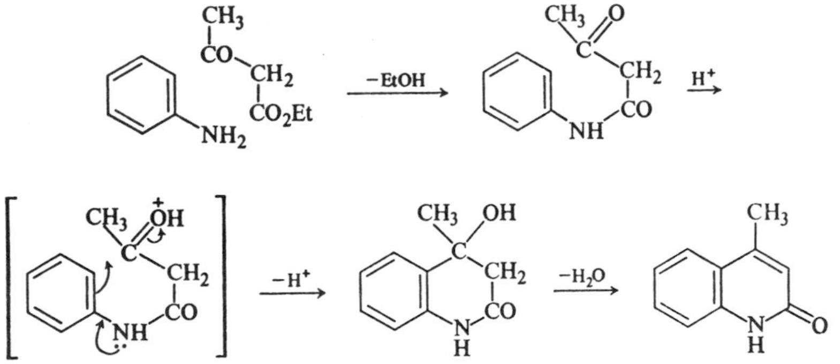

4-Methyl-2-quinolone

The Skraup synthesis of quinoline also involves acid-catalyzed intramolecular alkylation (p. 680).*

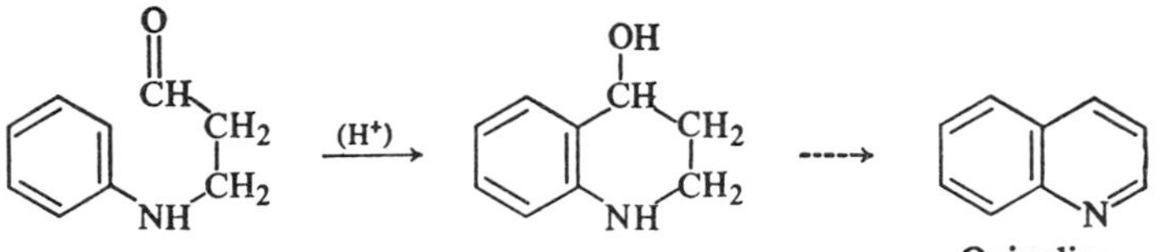

Quinoline

Reactivity of the aromatic compound. Aromatic compounds whose nuclear reactivity is comparable with or greater than that of benzene can be successfully alkylated but strongly deactivated compounds do not react. Thus chlorobenzene reacts but nitrobenzene is inert. Phenols do not react satisfactorily because they react with the Lewis acids at oxygen ($ArOH + AlCl_3 \rightarrow ArOAlCl_2 + HCl$) and

*These reactions may also be regarded as acid-catalyzed nucleophilic additions to carbonyl groups.

the resulting compound is usually only sparingly soluble in the reaction medium so that it reacts slowly. Conversion of the phenol into its methyl ether leads to successful alkylations providing that a low temperature is used to minimize acid-catalyzed cleavage of the ether group. Aromatic amines complex strongly with Lewis acids and are not suitable for alkylations.

The Lewis acids used as catalysts differ in activity, the order for the commoner compounds in alkylations with halides being: $AlCl_3 > SbCl_5 > FeCl_3 > SnCl_4 > ZnCl_2$.

The catalyst is chosen with reference to the reactivities of the aromatic compound and the alkylating agent; in general, tertiary halides require milder catalysts than primary halides. It is advisable to employ the least active Lewis acid consistent with the occurrence of alkylation at a practicable rate, for the more active catalysts tend to induce the isomerizations described below.

Alkylations with alcohols and ethers may be catalyzed by either a Lewis acid or a proton acid. Amongst Lewis acids, boron trifluoride is the reagent of choice because of its strong tendency to complex with oxygen. The proton acids used are hydrogen fluoride, concentrated sulphuric acid, and phosphoric acid; sulphuric acid is often unsuitable because of its capacity for forming sulphonated byproducts.

The reactions of olefins, aldehydes, and ketones are normally proton-catalyzed.

Problems attendant upon alkylation. Three problems are encountered in alkylation, each of which considerably reduces the general scope of the process.

(1) Since alkyl groups are activating in electrophilic substitutions, the product of alkylation is more reactive than the starting material and further alkylation inevitably occurs. For example, the methylation of benzene with methyl chloride in the presence of aluminium chloride gives a mixture containing toluene, the xylenes, the tri- and tetra-methylbenzenes, pentamethylbenzene, and hexamethylbenzene. Careful control of the molar ratio of the reactants and the reaction conditions can yield a particular polyalkylbenzene as the predominant product: for example, a mixture of the tetramethylbenzenes may be obtained by fractionation and durene (1,2,4,5-tetramethylbenzene) may then be isolated by freezing. In general, in order to obtain a monoalkylated product minimally contaminated by polyalkylated products, an excess of the aromatic compound should be used.

(2) Many alkyl groups rearrange during alkylation: e.g. benzene and n-propyl halides give mixtures of n-propylbenzene and isopropylbenzene. The extent to which isomerization competes with direct alkylation depends on the structure of the alkyl halide, the nature of the catalyst, and the reactivity of the aromatic compound.

First, isomerization is especially common with primary halides, as in the formation of isopropylbenzene from an n-propyl halide, and is fairly common with secondary halides. These rearrangements are understandable in that primary carbonium ions are less stable than secondary, and secondary less stable than tertiary. However, in some instances tertiary halides rearrange during alkylation: e.g. benzene and t-pentyl chloride in the presence of aluminium trichloride give mainly 2-methyl-3-phenylbutane by rearrangement of the t-pentyl cation:

$$CH_3-CH_2-\overset{\overset{\displaystyle CH_3}{|}}{\underset{\underset{\displaystyle CH_3}{|}}{C}}-Cl \quad AlCl_3 \longrightarrow CH_3-\overset{\overset{\displaystyle H}{|}}{CH}-\overset{\overset{\displaystyle CH_3}{|}}{\underset{\underset{\displaystyle CH_3}{|}}{C^+}}$$

$$\longrightarrow CH_3-\overset{+}{CH}-CH(CH_3)_2 \xrightarrow[-H^+]{PhH} Ph-\overset{\displaystyle CH_3}{\underset{\displaystyle CH(CH_3)_2}{CH}}$$

2-Methyl-3-phenylbutane

A probable explanation is that, although the tertiary ion is the more stable of the two, it is also the less reactive, so that the faster reaction of the secondary ion dominates, even though this ion is present in smaller concentration. t-Butyl halides do not undergo rearrangement during alkylation, probably because this would involve formation of the highly energetic primary carbonium ion.

Secondly, the extent of isomerization is reduced by using a less powerful Lewis acid catalyst. For example, the rearrangement of the t-pentyl group above does not occur when iron(III) chloride is the catalyst; t-pentylbenzene is then the product.*

Thirdly, isomerization becomes less significant as the reactivity of the aromatic compound is increased.

(3) Alkylation is reversible, so that reaction is thermodynamically controlled. For example, a monosubstituted benzene usually gives mainly the *meta*-alkyl derivative, since this is thermodynamically the most stable. This principle may be usefully applied: e.g. benzene and excess of ethyl bromide react in the presence of aluminium trichloride to give 1,3,5-triethylbenzene in 87% yield, further substitution probably being sterically impeded.

$$\text{C}_6\text{H}_6 + 3 \text{ C}_2\text{H}_5\text{Br} \xrightarrow[-3HBr]{AlBr_3}$$

1,3,5-Triethylbenzene

Just as tertiary alkyl groups are the most readily introduced during alkylation,

*This is probably because the weaker catalyst, while polarizing the C-halogen bond sufficiently for alkylation to occur, does not completely break the bond to form the alkyl cation through which isomerization occurs.

so they are the most readily removed by the reverse reaction, departing as the relatively stable tertiary carbonium ions.

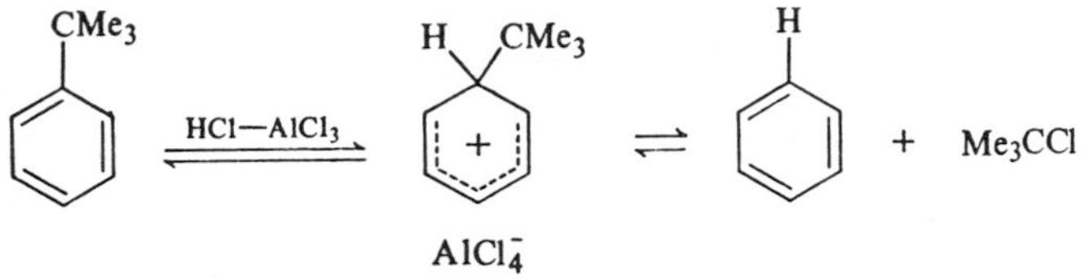

CMe$_3$ HCl—AlCl$_3$ H CMe$_3$ (+)   AlCl$_4^-$ ⇌ H   +   Me$_3$CCl

This enables the t-butyl group to be used to protect the most reactive position in a compound in order to effect reaction elsewhere; the t-butyl group is subsequently removed by the addition of an excess of benzene which draws the equilibrium in the desired direction. For example, the Friedel-Crafts acylation of toluene gives largely the *para*-acyl derivative (p. 391). The following scheme enables the *ortho* isomer to be obtained:

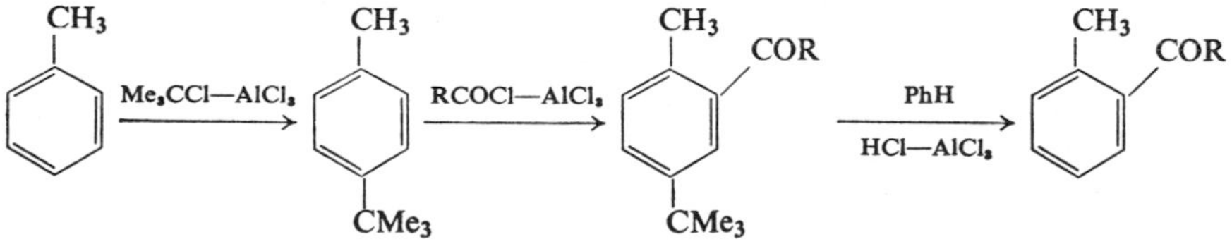

CH$_3$ Me$_3$CCl—AlCl$_3$ →   CH$_3$ / CMe$_3$ RCOCl—AlCl$_3$ →   CH$_3$ COR / CMe$_3$ PhH / HCl—AlCl$_3$ →   CH$_3$ COR

The less easily accessible 1,2,3-trialkylbenzenes may also be prepared by application of this principle. t-Butylation of a *m*-dialkylbenzene gives the 5-t-butyl derivative (thermodynamic control), the desired alkyl group is then introduced (reaction occurring at the 2- rather than the 4-position because of the hindrance due to the t-butyl group), and the t-butyl group is removed by reaction with more of the starting *m*-dialkylbenzene.

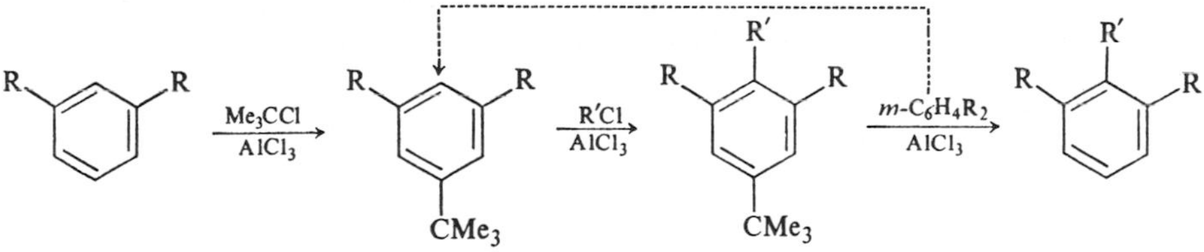

R   R   Me$_3$CCl / AlCl$_3$ →   R   R / CMe$_3$ R'Cl / AlCl$_3$ →   R' / R   R / CMe$_3$ *m*-C$_6$H$_4$R$_2$ / AlCl$_3$ →   R' / R   R

(b) FRIEDEL-CRAFTS ACYLATION

The acylation of aromatic rings may be brought about by an acid chloride or anhydride in the presence of a Lewis acid or, in some circumstances (p. 393), by a carboxylic acid in the presence of a proton acid.

The Lewis acid-catalyzed methods apparently occur by each of two mechanisms, depending on the reaction conditions, although the overall reaction is the same in each case. In one path, the acyl chloride or anhydride is converted into an acylium cation which reacts with the aromatic:

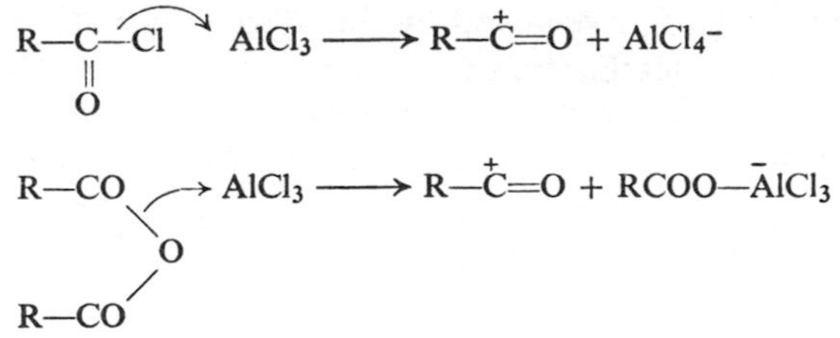

$$R-\underset{\overset{\|}{O}}{C}-Cl \quad AlCl_3 \longrightarrow R-\overset{+}{C}=O + AlCl_4^-$$

$$\underset{R-CO}{\overset{R-CO}{\diagdown}} O \longrightarrow AlCl_3 \longrightarrow R-\overset{+}{C}=O + RCOO-\bar{A}lCl_3$$

$$ArH + RCO^+ \longrightarrow Ar-COR + H^+$$

In the second path, the acylating agent is polarized through oxygen, the reactivity of the carbonyl carbon to nucleophilic displacement thereby being enhanced:

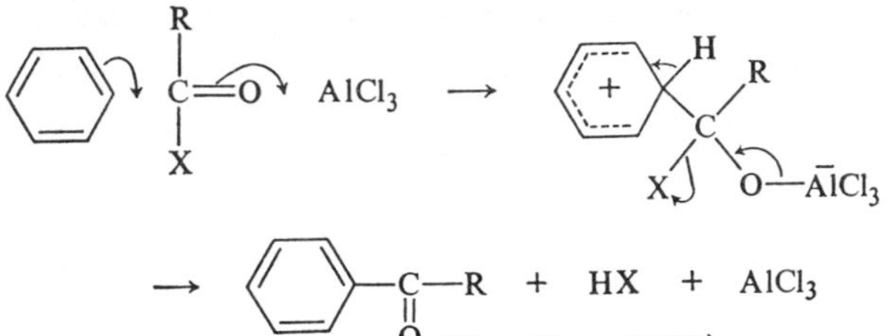

$$\longrightarrow \quad \text{(ring)}-\underset{\overset{\|}{O}}{C}-R \quad + \quad HX \quad + \quad AlCl_3$$

$$(X = Cl \text{ or } OCOR)$$

In a typical procedure, acetic anhydride is added to a refluxing solution of bromobenzene in carbon disulphide containing suspended aluminium trichloride. After removal of the solvent and decomposition of the resulting complex with hydrochloric acid, p-bromoacetophenone is obtained in about 75% yield [1].

The aromatic compounds which may be acylated are in general the same as those which may be alkylated. Benzene and compounds of comparable or greater reactivity undergo the reaction, but deactivated molecules such as benzaldehyde, benzonitrile, and nitrobenzene are inert; nitrobenzene is a common solvent for acylation. The difficulty encountered in the alkylation of phenols (p. 386) applies also in acylation and may be surmounted by acylating the methyl ether at a low temperature. An alternative in some instances is to employ concentrated sulphuric acid as the catalyst, as in the preparation of phenolphthalein from phenol and phthalic anhydride:

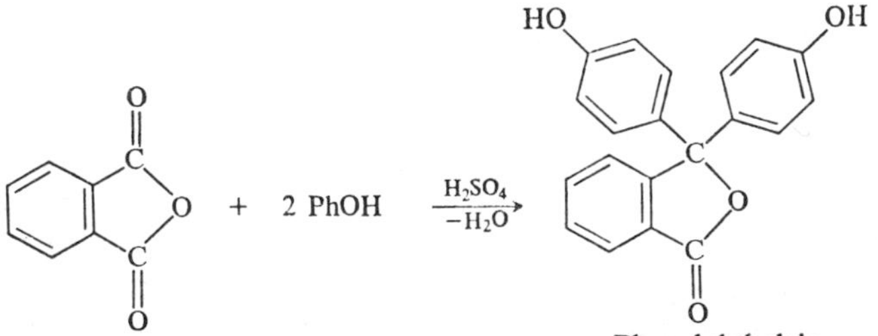

Phenolphthalein

The order of activity of Lewis acids in the benzoylation of toluene and chlorobenzene is: $SbCl_5 > FeCl_3 > AlCl_3 > SnCl_4$. As with alkylation, it is often wise to employ the weakest reagent consistent with successful reaction in order to minimize side-reactions; e.g. thiophen is polymerized during acetylation by acetyl chloride in the presence of aluminium trichloride but gives 2-acetylthiophen in about 80% yield when tin(IV) chloride is used [2].

$$\text{(thiophene)} + CH_3COCl \xrightarrow[-HCl]{SnCl_4} \text{(2-acetylthiophene)} S\text{---}COCH_3$$

2-Acetylthiophen

There are a number of important differences between acylation and alkylation.

(1) Whereas alkylation does not require stoicheiometric quantities of the Lewis acid since this is regenerated in the last stage of the reaction, acylation requires greater than mole quantities because the ketone which is formed complexes with the Lewis acid.

(2) Since acyl groups deactivate aromatic nuclei towards electrophilic substitution, the products of acylation are less reactive than the starting materials and the mono-acylated product is easy to isolate. This makes acylation a more useful procedure than alkylation, and alkyl-derivatives are often more satisfactorily obtained by acylation followed by reduction of carbonyl to methylene than by direct alkylation.

(3) A further advantage of acylation over alkylation is that the isomerizations and disproportionations which are characteristic of the latter process do not occur in the former. There is, however, one limitation: attempted acylation with derivatives of tertiary acids may lead to alkylation. For example, pivaloyl chloride reacts with benzene in the presence of aluminium trichloride to give mainly t-butylbenzene, liberating carbon monoxide:

$$Me_3C\text{---}COCl \xrightarrow{AlCl_3} Me_3C\text{---}\overset{+}{C}O \xrightarrow{-CO} Me_3C^+ \xrightarrow[-H^+]{PhH} Ph\text{---}CMe_3$$

The driving force for the decarbonylation no doubt resides in the relative stability of tertiary carbonium ions.

However, more reactive aromatic compounds can react with the acylium ion before decarbonylation occurs: e.g. anisole in the same conditions gives mainly p-methoxypivalophenone, $p\text{-}CH_3O\text{---}C_6H_4\text{---}COCMe_3$.

(4) The complex of the acylating agent and Lewis acid is evidently very bulky, for *ortho,para*-directing monosubstituted benzenes give very little of the *ortho* product. For example, toluene, which gives nearly 60% of the *ortho*-derivative on nitration, gives hardly any *o*-methylacetophenone on acetylation; the *para*-isomer may be obtained in over 85% yield. When nitrobenzene is used as the

solvent, steric hindrance to *ortho*-substitution is even more marked, possibly because a solvent molecule takes part in the acylating complex; e.g. whereas the acetylation of naphthalene in carbon disulphide solution gives mainly the expected 1-acetylnaphthalene, reaction in nitrobenzene gives predominantly 2-acetylnaphthalene, providing a convenient route to 2-naphthoic acid by oxidation with hypochlorite (p. 598):

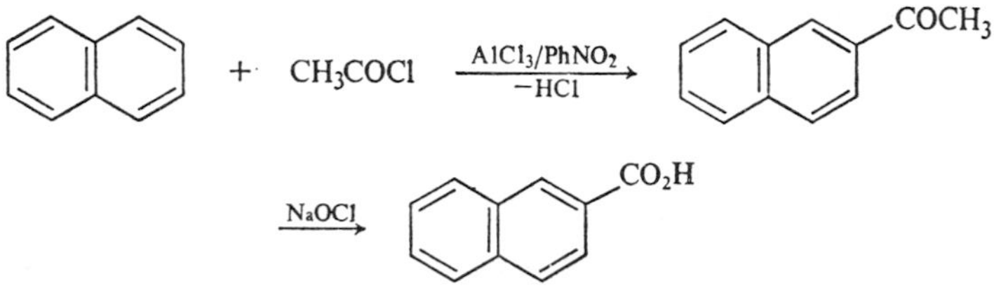

2-Naphthoic acid

Cyclizations. Intramolecular Friedel-Crafts acylations are of particular value in building up cyclic systems; dibasic acid anhydrides are much used in these reactions.

For example, benzene and succinic anhydride in the presence of aluminium trichloride give β-benzoylpropionic acid in 80% yield [2], and reduction of carbonyl to methylene, conversion of the acid group to the acid chloride, and cyclization with aluminium trichloride give α-tetralone:

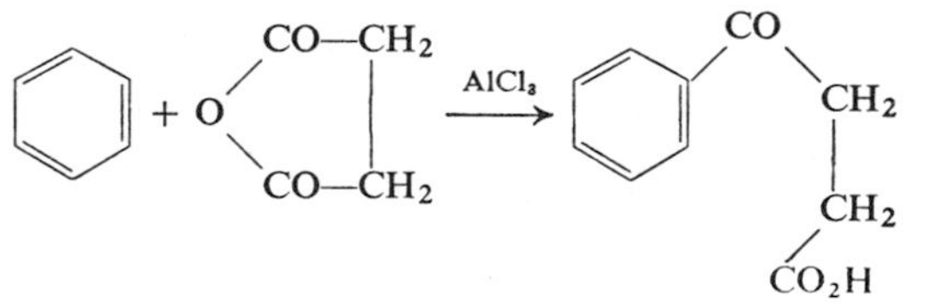

β-Benzoylpropionic acid

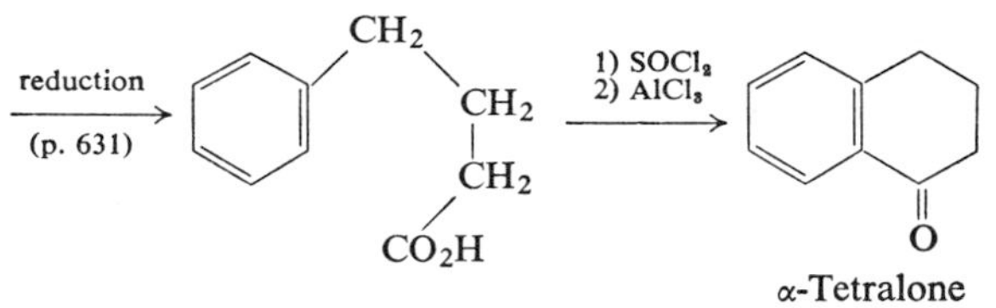

α-Tetralone

1-Alkylnaphthalenes may then be obtained by a Grignard reaction followed by dehydration and dehydrogenation:

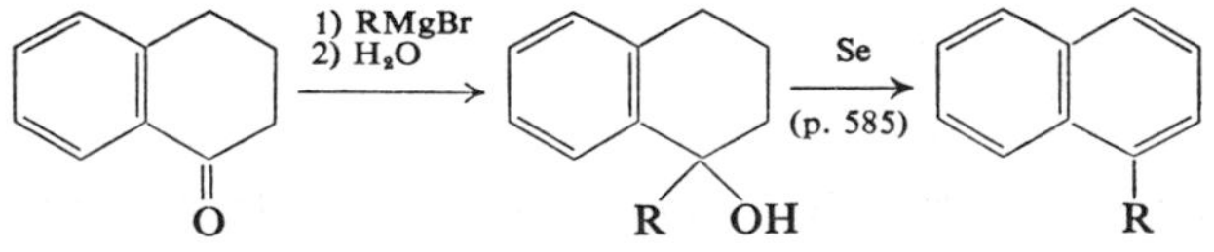

Many variants of this method are possible. By starting with naphthalene, phenanthrene and some of its derivatives may be obtained. By using phthalic anhydride, two new aromatic rings may be built on, as in the synthesis of 1:2-benzanthracene:

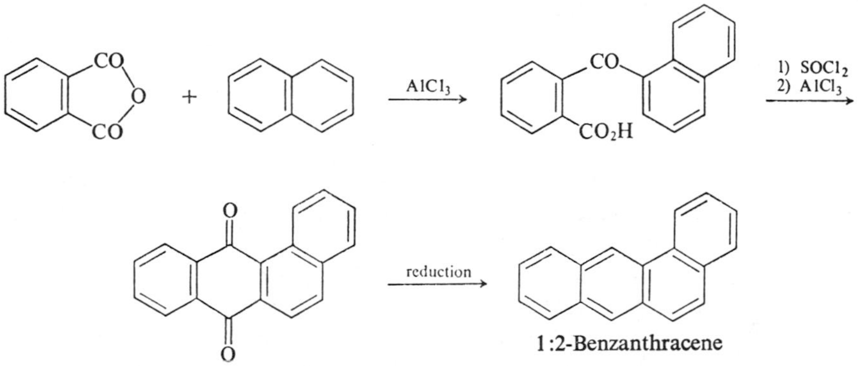

1:2-Benzanthracene

The cyclization step may be carried out either by treating the acid chloride with aluminium trichloride as above or, more conveniently, by treating the acid itself with liquid hydrogen fluoride, concentrated sulphuric acid, or polyphosphoric acid. Sulphuric acid is the least satisfactory reagent since it can lead to sulphonated byproducts.

(c) CHLOROMETHYLATION

The chloromethyl group, —CH$_2$Cl, can be introduced into aromatic compounds by treatment with formaldehyde and hydrogen chloride in the presence of an acid. For example, benzyl chloride may be obtained in about 80% yield by passing hydrogen chloride into a suspension of paraformaldehyde and zinc chloride in benzene; the acid first liberates formaldehyde from paraformaldehyde and then takes part in the condensation.

$$PhH + CH_2O + HCl \xrightarrow{(ZnCl_2)} PhCH_2Cl + H_2O$$

Fluoromethylation, bromomethylation, and iodomethylation may be carried out with the appropriate halogen acid.

Studies of the mechanism indicate that the electrophilic entity is the hydroxymethyl cation. This reacts to give an alcoholic product which, in the presence of hydrogen chloride, is converted into the chloromethyl product.

$$CH_2{=}O + H^+ \longrightarrow [\, CH_2{=}\overset{+}{O}H \leftrightarrow \overset{+}{C}H_2{-}OH\,] \xrightarrow[-H^+]{PhH} PhCH_2OH \xrightarrow[-H_2O]{HCl} PhCH_2Cl$$

Chloromethylation, unlike Friedel-Crafts reactions, is successful even with quite strongly deactivated nuclei such as that of nitrobenzene, although *m*-dinitrobenzene and pyridine are inert.

Two complications can occur in chloromethylation. First, the chloromethyl product can alkylate another molecule of the aromatic compound in the presence of the acid catalyst, e.g.

$$PhH \xrightarrow[ZnCl_2]{CH_2O-HCl} PhCH_2Cl \xrightarrow[ZnCl_2]{PhH} PhCH_2Ph$$

This secondary reaction is of particular significance when the aromatic compound is strongly activated, and for this reason chloromethylation is not a suitable procedure for phenols and anilines.

Secondly, the chloromethyl group is activating, although less so than methyl because the chlorine substituent in the methyl group reduces the $+I$ effect of that group. It is usually difficult to avoid the occurrence of some further chloromethylation, although this is not nearly so important a problem as it is in Friedel-Crafts alkylation.

The reaction conditions may be varied widely. Anhydrous hydrogen chloride may be replaced by the concentrated aqueous acid; formaldehyde may be introduced as paraformaldehyde or methylal ($CH_2(OCH_3)_2$); and zinc chloride may be replaced by sulphuric acid or phosphoric acid or omitted altogether in the chloromethylation of very reactive aromatic compounds such as thiophen. In a typical example, a mixture of naphthalene, paraformaldehyde, glacial acetic acid, 85% phosphoric acid, and concentrated hydrochloric acid, heated at 80°C for 6 hours, gives a 75% yield of 1-chloromethylnaphthalene.

The principal value of chloromethylation lies in the ease of displacement of the benzylic chloride by nucleophiles. Conversion into the corresponding alcohols, $ArCH_2OH$, ethers, $ArCH_2OR$, nitriles, $ArCH_2CN$, and amines, $ArCH_2NR_2$, may be accomplished efficiently, and treatment with carbanions, such as that from malonic ester, leads to extension of the aliphatic carbon chain, e.g.

$$ArCH_2Cl \xrightarrow[EtO^-]{CH_2(CO_2Et)_2} ArCH_2CH(CO_2Et)_2 \xrightarrow[2)\ -CO_2]{1)\ hydrolysis} ArCH_2CH_2CO_2H$$

(*d*) GATTERMANN-KOCH FORMYLATION

The formyl group, —CHO, may be introduced into aromatic compounds by treatment with carbon monoxide and hydrogen chloride in the presence of a Lewis acid:

$$ArH + CO \xrightarrow{HCl\ +\ Lewis\ acid} ArCHO$$

The reaction is carried out either under pressure or in the presence of copper(I) chloride.

It was at one time thought that the reaction occurs through the formation of formyl chloride, HCOCl, from carbon monoxide and hydrogen chloride, followed by a Friedel-Crafts acylation catalyzed by the Lewis acid, but formyl chloride has never been obtained and it is now considered probable that the electrophilic species is the formyl cation, $[HC{\equiv}\overset{+}{O} \leftrightarrow \overset{+}{HC}{=}O]$, formed without the mediation of formyl chloride:

$$HCl + CO + AlCl_3 \longrightarrow HCO^+ + AlCl_4^-$$

$$ArH + HCO^+ \xrightarrow{\ -H^+\ } ArCHO$$

The role of copper(I) chloride may be to aid the reaction between carbon monoxide and hydrogen chloride *via* the complex which it forms with carbon monoxide.

Formylation is unsuccessful with aromatic compounds of lower nuclear reactivity than the halobenzenes; nitrobenzene may be used as solvent. It is also unsuccessful with amines, phenols, and phenol ethers, because of the formation of complexes with the Lewis acid. One drawback in the application of the reaction to polyalkylated benzenes is that rearrangements and disproportionations occur: e.g. *p*-xylene gives 2,4-dimethylbenzaldchyde.

2,4-Dimethylbenzaldehyde

In a typical procedure, carbon monoxide and hydrogen chloride are passed into toluene which contains suspended aluminium trichloride and copper(I) chloride. *p*-Tolualdehyde is formed as a complex with the Lewis acid and is isolated in 50% yield by treatment with ice and distillation in steam [2].

(*e*) GATTERMANN FORMYLATION
This is an alternative to the Gattermann-Koch reaction, employing hydrogen cyanide instead of carbon monoxide. The initial product is an iminium hydrochloride which is converted into the aldehyde with mineral acid.

$$ArH + HCN + HCl \xrightarrow{\text{Lewis acid}} ArCH{=}NH_2^+Cl^- \xrightarrow{H_2O-H^+} ArCHO + NH_4Cl$$

The mechanism of the process is not clear; the ionic intermediate, $[HC{\equiv}\overset{+}{N}H$ $\leftrightarrow H\overset{+}{C}{=}NH]$, analogous to the formyl cation, may be the electrophilic entity.

The reaction is unsuccessful with deactivated compounds such as nitrobenzene, and compounds of moderate reactivity such as benzene and the halobenzenes give only low yields. Yields from more reactive compounds are considerably higher: e.g. anthracene gives the 9-aldehyde in 60% yield.

Unlike the Gattermann-Koch reaction, Gattermann formylation is successful with phenols and phenol ethers: e.g. p-anisaldehyde is formed from anisole almost quantitatively in the presence of aluminium trichloride. More reactive nuclei still can be formylated in the presence of the weaker Lewis acid, zinc chloride, and furan reacts even in the absence of a catalyst to give furfuraldehyde.

To avoid the use of hydrogen cyanide, it is convenient to use zinc cyanide from which hydrogen cyanide is generated *in situ* by reaction with hydrogen chloride. For example, mesitaldehyde is obtained in over 75% yield by passing hydrogen chloride into a solution of mesitylene in tetrachloroethane in the presence of zinc cyanide, adding aluminium trichloride, and decomposing the resulting iminium hydrochloride with hydrochloric acid [2].

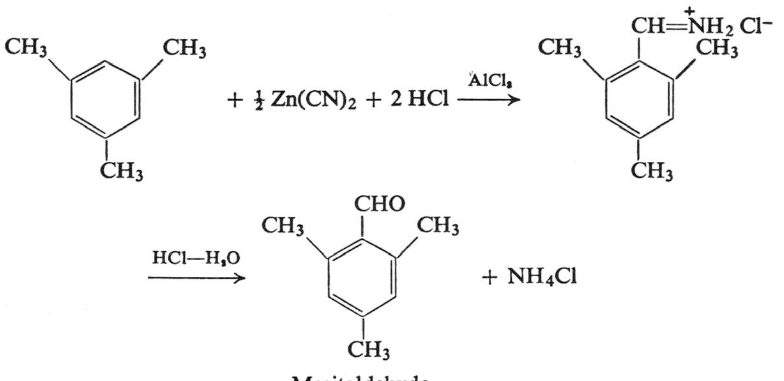

Mesitaldehyde

(f) HOESCH ACYLATION

This reaction is an adaptation of Gattermann formylation: the use of an aliphatic nitrile in place of hydrogen cyanide leads to an acyl-derivative of the aromatic compound.

$$ArH + RCN + HCl \xrightarrow{\text{Lewis acid}} Ar{-}\underset{\underset{NH_2{}^+\ Cl^-}{\|}}{C}{-}R \xrightarrow{\text{H}_2\text{O---H}^+} ArCOR + NH_4Cl$$

The reaction occurs only with the most highly activated aromatic compounds such as di- and polyhydric phenols. Monohydric phenols react mainly at oxygen to give imido-esters,

$$\text{ArOH} + \text{RCN} + \text{HCl} \xrightarrow{\text{Lewis acid}} \underset{\underset{\text{NH}_2^+ \text{ Cl}^-}{\|}}{\text{ArO—C—R}}$$

but the combination of two or three hydroxyl groups *meta* to each other so increases the reactivity of the nuclear positions *ortho* or *para* to hydroxyl that nuclear acylation occurs. For example, phloroacetophenone may be obtained in 80% yield by passing hydrogen chloride into a cooled solution of phloroglucinol and acetonitrile in ether containing suspended zinc chloride and then hydrolyzing the resulting precipitate of the ketimine hydrochloride by boiling in aqueous solution [2].

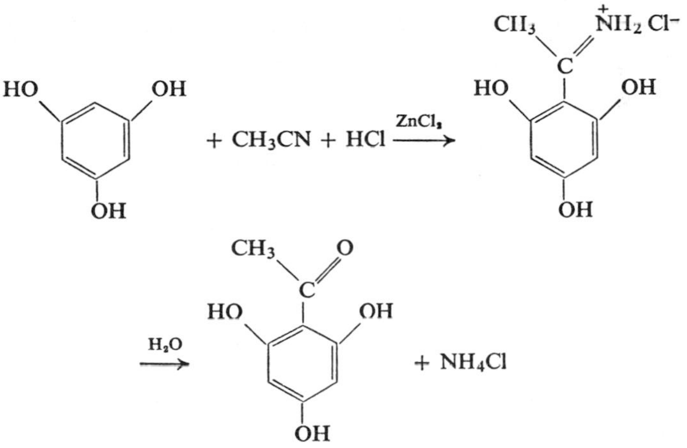

Phloroacetophenone

(g) VILSMEYER FORMYLATION

N-Formylamines, from secondary amines and formic acid, formylate aromatic compounds in the presence of phosphorus oxychloride. The mechanism is thought to be as follows:

$$\text{R}_2\text{NH} + \text{HCO}_2\text{H} \xrightarrow[-\text{H}_2\text{O}]{\text{heat}} \text{R}_2\text{N—CHO}$$

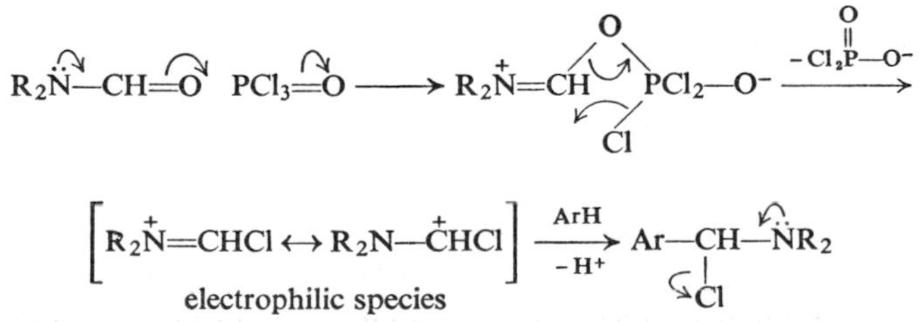

$$\longrightarrow Ar\!-\!CH\!=\!\overset{+}{N}R_2Cl^- \xrightarrow{\ H_2O\ } ArCHO + R_2NH_2{}^+ Cl^-$$

Only the most reactive aromatic compounds are formylated: e.g. benzene and naphthalene are unreactive, but anthracene gives the 9-aldehyde (84%), *NN*-dimethylaniline gives *p*-dimethylaminobenzaldehyde (80%), and thiophen gives the 2-aldehyde (70%). The method is particularly effective for compounds such as pyrroles which are not formylated by other procedures: e.g. pyrrole gives the 2-aldehyde in 85% yield and indole gives the 3-aldehyde in 97% yield, each with dimethylformamide [4].

(*h*) REIMER-TIEMANN FORMYLATION

The treatment of phenols with chloroform in basic solution gives aldehydes, e.g.

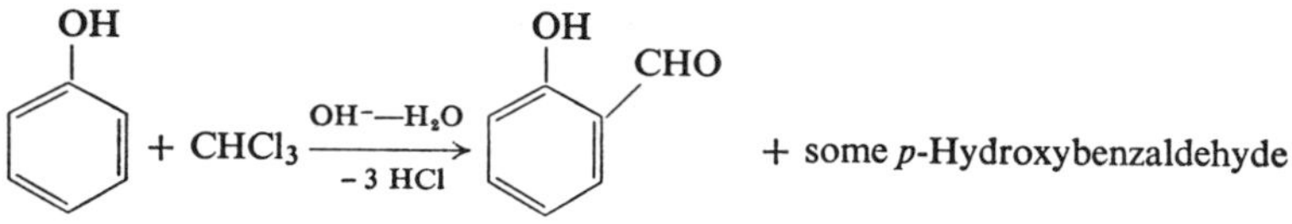

Salicylaldehyde
40%

The reaction occurs through dichlorocarbene, which is generated from chloroform and alkali (p. 124) and, being electrophilic, is attacked by the strongly nucleophilic phenoxide ion. Hydrolysis of the benzal chloride follows, and acidification yields the aldehyde.

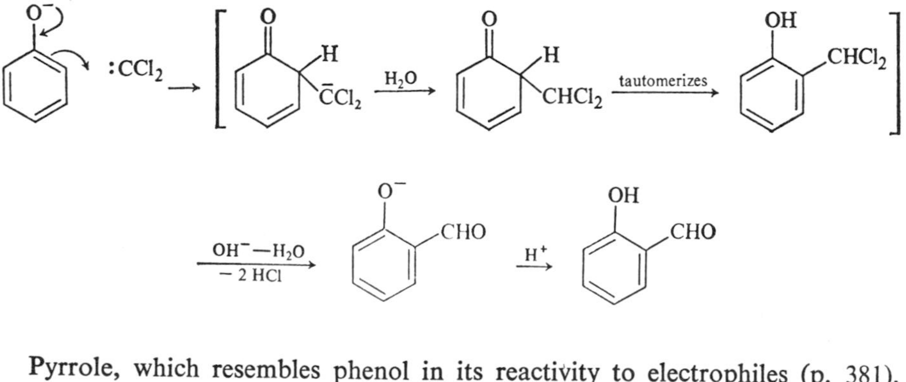

Pyrrole, which resembles phenol in its reactivity to electrophiles (p. 381), undergoes the Reimer-Tiemann reaction, *via* the strongly nucleophilic pyrrolate anion, giving pyrrole-2-aldehyde, but a second product, 3-chloropyridine, is also obtained. Each product derives from the same intermediate:

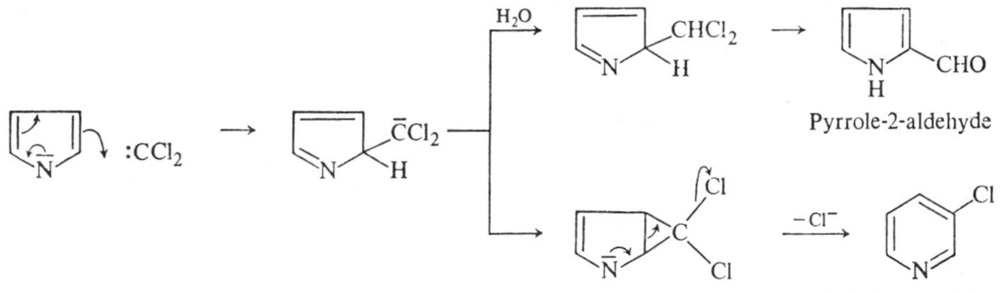

Pyrrole-2-aldehyde

3-Chloropyridine

Phenols with blocked *para* positions give cyclohexadienones in which the two chlorine substituents, being in neopentyl-type environments, are resistant to hydrolysis in the basic conditions (p. 128) and remain intact, e.g.

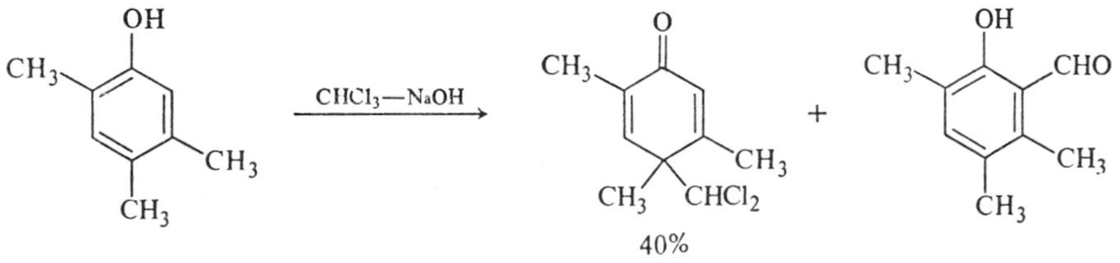

40%

Use may be made of this property for introducing angular methyl groups into decalin derivatives (*cf.* pp. 245 and 724): the angular group is introduced as —CHCl₂ and converted into —CH₃ by hydrogenolysis, e.g.

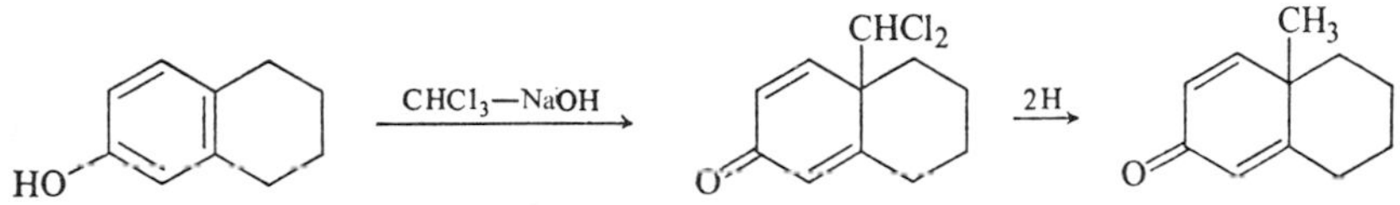

(i) KOLBE-SCHMITT CARBOXYLATION

Phenoxide ions are reactive enough to add to carbon dioxide, which is a relatively weak electrophile; reaction is carried out under pressure at about 100°C. The *ortho* product predominates, probably because of the stabilizing influence of chelation on the transition state for *ortho* substitution, e.g.

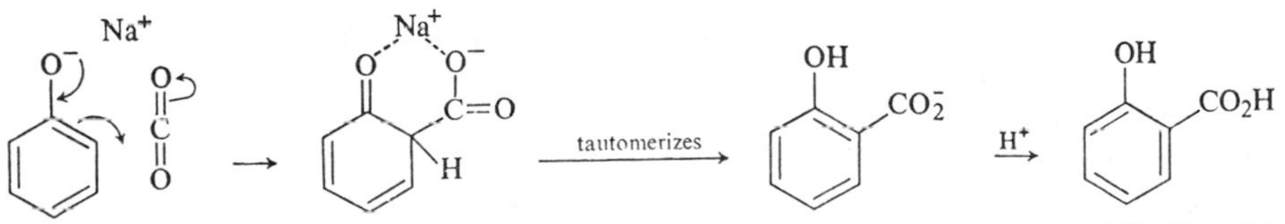

Salicylic acid

Pyrrole undergoes a similar reaction, giving pyrrole-2-carboxylic acid, on being heated with ammonium carbonate at 120°C.

(j) THE MANNICH REACTION

This reaction is suitable for bonding aliphatic carbon to the reactive positions of phenols, pyrroles, and indoles (p. 274). For example, pyrrole, formaldehyde, and dimethylamine give 2-dimethylaminomethylpyrrole,

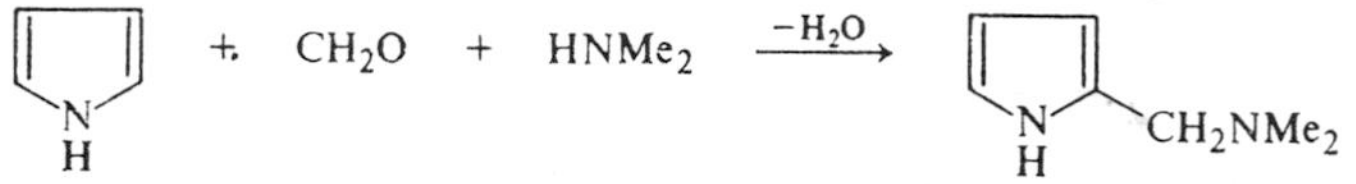

and indole reacts, as usual, at the 3-position to give gramine,

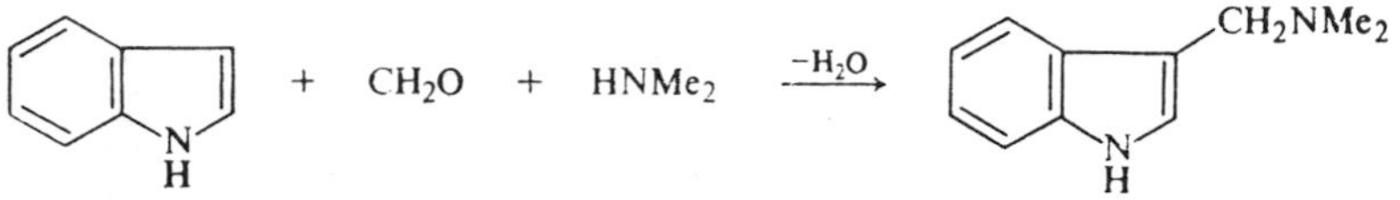

Gramine

Furan resists Mannich conditions, but 2-methylfuran, whose methyl group increases the reactivity of the nucleus to electrophiles, reacts at the 5-position.

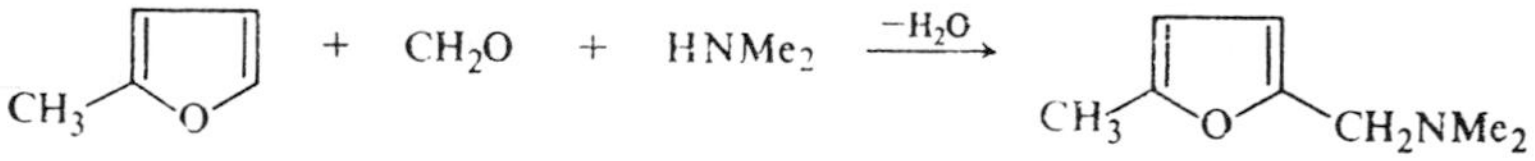

Synthetic uses of the Mannich bases from some of these aromatic compounds have been described previously (p. 279).

11.4 Formation of Carbon-Nitrogen Bonds

(a) NITRATION

By far the most commonly employed method for bonding nitrogen to aromatic systems is by nitration. The reason is that a very wide variety of nitration conditions is available, so that suitable procedures may usually be found for nitrating compounds of both very high and very low nuclear reactivity. The more frequently employed methods, arranged in approximately decreasing order of their vigour, are:

A mixture of concentrated nitric and concentrated sulphuric acid,
Fuming nitric acid in acetic anhydride,
Nitric acid in glacial acetic acid,
Dilute nitric acid.

The mixture of concentrated nitric and sulphuric acids generates the nitronium ion, NO_2^+, and this is the active electrophilic species. Its formation results from protonation of nitric acid by sulphuric acid (i.e. nitric acid acts as a base towards a powerful proton-donor), followed by heterolysis:

$$H_2SO_4 + HO—NO_2 \rightleftharpoons HSO_4^- + H_2\overset{+}{O}—NO_2$$

$$H_2\overset{+}{O}—NO_2 \rightleftharpoons H_2O + NO_2^+$$

The molecule of water formed is essentially completely removed by protonation, so that the equilibria favour the nitronium ion. However, added water can reduce the concentration of nitronium ions and lower the nitrating power of the system.

A solution of nitric acid in acetic anhydride contains a number of species in equilibrium:

$$HO—NO_2 + Ac—OAc \rightleftharpoons HO—Ac + AcO—NO_2$$

$$2\,AcONO_2 \rightleftharpoons AcOAc + O_2N—O—NO_2$$

$$O_2N—O—NO_2 \rightleftharpoons NO_2^+ + NO_3^-$$

It is believed that the nitronium ion is again the principal electrophilic entity.

Dilute nitric acid contains a negligible concentration of the nitronium ion and it acts as a nitrating agent only by virtue of the small concentration of nitrous acid which is normally contained in it. This gives rise to the nitrosonium ion, NO^+, which nitrosates the aromatic ring; the nitroso-compound is oxidized by nitric acid to the nitro-compound which generates more nitrous acid to continue the chain-reaction:

$$ArH + HNO_2 \longrightarrow ArNO + H_2O$$

$$ArNO + HNO_3 \longrightarrow ArNO_2 + HNO_2$$

Consequently, only those compounds (e.g. phenols) which undergo nitrosation can be nitrated with dilute nitric acid.

(*i*) *Benzene.* Benzene is nitrated satisfactorily by a mixture of concentrated nitric and sulphuric acids ('mixed acids'). Since the nitro group deactivates the aromatic nucleus towards electrophiles, it is relatively easy to prevent di- or poly-nitration: reaction at 30–40°C yields nitrobenzene rapidly and in about 95% yield. However, dinitration can be brought about by raising the temperature to 90–100°C.

(*ii*) *Moderately activated nuclei.* Although the alkylbenzenes are more reactive than benzene (e.g. toluene is nitrated at about 25 times the rate of benzene in the

same conditions), nitric acid in sulphuric acid is still the most satisfactory
method. The *o*- and *p*-nitro-derivatives predominate, but the proportion of the
former falls as the alkyl group increases in size (p. 373).

(*iii*) *Strongly activated nuclei.* Naphthalene is reactive enough to be nitrated by
nitric acid in acetic acid, giving mainly 1-nitronaphthalene, and similar conditions give 2-nitrothiophen (70–85%) from thiophen [2] and nitromesitylene
(75%) from mesitylene [2].

Phenol, which is considerably more reactive, is nitrated with dilute nitric acid
and gives comparable amounts of *o*- and *p*-nitrophenol. Aniline is too readily
oxidized for direct nitration to be practicable. The amino group is first protected
by acetylation, acetanilide is nitrated, and the acetyl group is removed by
hydrolysis. *NN*-Dimethylaniline is less easily oxidized than aniline and can be
nitrated with dilute nitric acid.

Furan and pyrrole are polymerized by acidic nitrating systems but nitration
in acetic anhydride at low temperatures is successful; the reagent is acetyl
nitrate. Each compound gives mainly the 2-nitro-derivative, that from furan
being formed *via* an adduct:

$$HNO_3 + Ac_2O \longrightarrow AcONO_2 + HOAc$$

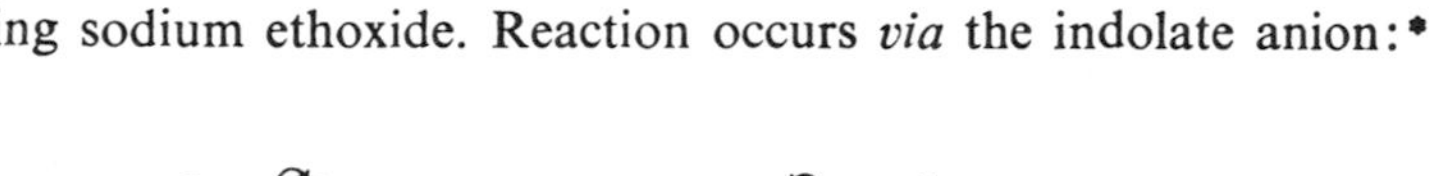

2-Nitrofuran

3-Nitroindole can be obtained by treating indole with ethyl nitrate in ethanol
containing sodium ethoxide. Reaction occurs *via* the indolate anion:[*]

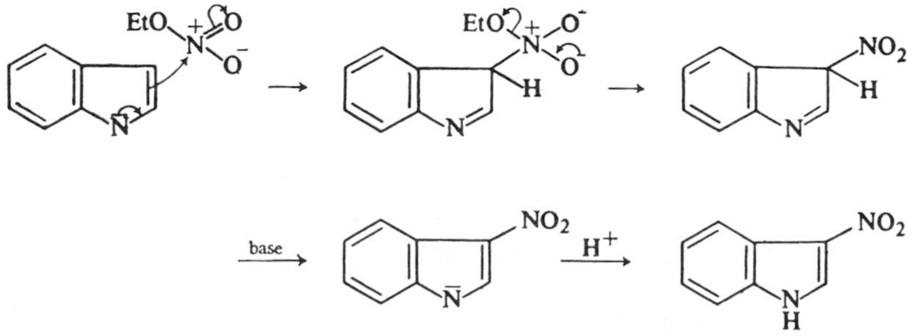

3-Nitroindole

[*]*cf.* the reaction of carbanions with alkyl nitrates, p. 347.

(iv) *Moderately deactivated nuclei.* The halobenzenes, which are about one-tenth as reactive as benzene, are nitrated in the same way as benzene. The *p*-nitro-derivative predominates in each case, and further nitration gives chiefly the 2,4-dinitro-derivatives. Concentrated nitric and sulphuric acids at or below room temperature are also suitable for the nitration of rather less reactive compounds such as benzaldehyde, acetophenone, and methyl benzoate, each of which gives mainly the *m*-nitro-derivative. For example, nitration of methyl benzoate at 5–15°C gives methyl *m*-nitrobenzoate in 80% yield [1].

The nitration of *NN*-dimethylaniline in concentrated sulphuric acid gives the *m*-nitro-derivative in about 60% yield [3]. This is because in the strongly acidic conditions essentially all the amine is present as the *meta*-directing anilinium ion; in less acidic conditions the *p*-nitro-derivative is the main product.

(v) *Strongly deactivated nuclei.* Nitrobenzene is less reactive than the deactivated compounds such as benzaldehyde but the mixed acid procedure is successful at about 90–100°C; *m*-dinitrobenzene is obtained in about 80% yield. *m*-Dinitrobenzene itself is so strongly deactivated that further nitration requires fuming nitric acid and fuming sulphuric acid at 110°C for several days; 1,3,5-trinitrobenzene is formed. This compound is more satisfactorily obtained from toluene: the activating influence of the methyl group allows 2,4,6-trinitrotoluene to be formed with mixed acid, and the methyl group is removed by oxidation and decarboxylation.*

Pyridine is very powerfully deactivated, partly because in the acidic conditions necessary for reaction it is fully protonated (unlike the very weakly basic nitro compounds). Even the use of 100% sulphuric acid and a mixture of sodium and potassium nitrates at 300°C gives only 5% of 3-nitropyridine. 3-Substituted pyridines are better prepared by starting with the naturally occurring nicotinic

*Decarboxylation occurs readily because of the presence of the electron-attracting nitro groups. Reaction probably occurs *via* the carbanion, see p. 414.

acid, converting it into 3-aminopyridine *via* Hofmann bromination (p. 468), and making use of the reactivity of the corresponding diazonium salt (p. 435).

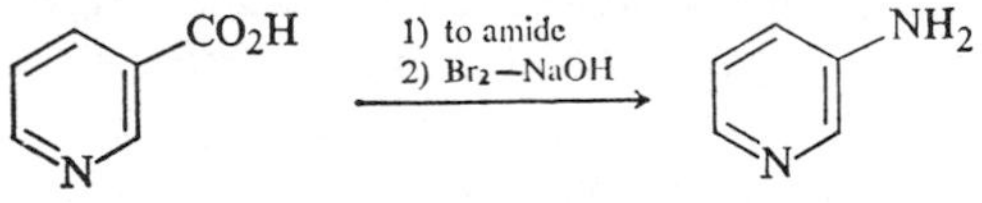

Nicotinic acid   3-Aminopyridine

The properties óf pyridine and related heterocycles are profoundly modified in the corresponding *N*-oxides; for example, pyridine *N*-oxide, obtained from pyridine by oxidation with 30% aqueous hydrogen peroxide in acetic acid at 70–80°C (see p. 602), undergoes nitration with mixed nitric and sulphuric acids at 100°C to give mainly the 4-nitro-derivative. The explanation is that meso-meric electron-release from the oxide oxygen atom stabilizes thé transition state for 4-substitution, as represented by the contribution of the canonical structure *c*:

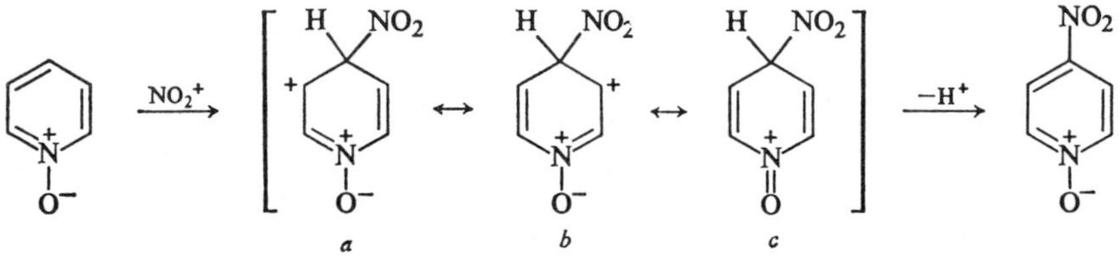

Substitution at the 2-position (but not at the 3-position) should likewise be facilitated, but the activating influence of oxygen at the 2-position is evidently much weaker than at the 4-position, since the 2-nitro-derivative is formed only in very small amount. Since pyridine *N*-oxides are readily deoxygenated (e.g. with phosphorus trichloride), nitration of pyridine *N*-oxide provides a useful route to 4-nitropyridine.

Quinoline and isoquinoline are considerably more reactive than pyridine. The effect of the hetero-atom is not strongly felt in the benzenoid ring and nitration in concentrated sulphuric acid at 0°C gives, with quinoline, comparable quantities of the 5- and 8-nitro-derivatives and, with isoquinoline, almost entirely the 5-nitro-derivative.

(b) NITROSATION

Nitrosation is brought about by treatment of an aromatic compound with sodium nitrite and a strong acid. Nitric acid cannot be used since it oxidizes the nitroso product to a nitro compound (p. 601) and sulphuric acid or hydrochloric acid is normally chosen.

The electrophilic entity is the nitrosonium ion, NO^+, which appears to be generated from nitrous acid in a manner analogous to the formation of the nitronium ion from nitric acid,

$$NaNO_2 + H_2SO_4 \longrightarrow HNO_2 + NaHSO_4$$

$$HO\!-\!NO + H_2SO_4 \rightleftharpoons H_2\overset{+}{O}\!-\!NO + HSO_4^-$$

$$H_2\overset{+}{O}\!-\!NO \rightleftharpoons H_2O + NO^+$$

although this is possibly an oversimplified representation.

Nitrosation is limited to the very reactive nuclei of phenols and tertiary aromatic amines. Primary aromatic amines undergo diazotization (13.1) (but see below) and secondary amines undergo N-nitrosation ($ArNHR + NO^+ \rightarrow ArNR\!-\!N\!=\!O + H^+$). In a typical procedure, β-naphthol in caustic soda solution is treated with sodium nitrite, and concentrated sulphuric acid is then added: 1-nitroso-2-naphthol is obtained in over 90% yield [1].

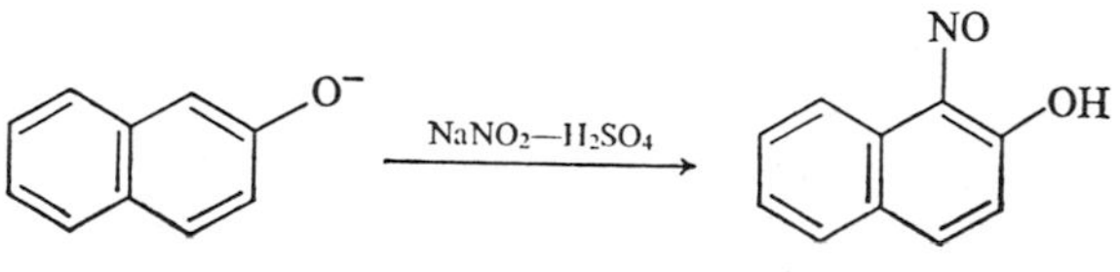

1-Nitroso-2-naphthol

The treatment of NN-dimethylaniline with sodium nitrite and hydrochloric acid gives p-nitrosodimethylaniline, as its hydrochloride, in 85% yield.

The nitroso compounds formed are easily oxidized to nitro compounds (p. 601) and reduced to amines (p. 652). In the latter context, they find a synthetic application when the normal route to an aromatic amine, *via* nitration, cannot be employed because of the disruption of a sensitive compound in nitrating conditions. For example, in one synthesis of adenine (6-aminopurine) the last of three amino groups was introduced *via* nitrosation:*

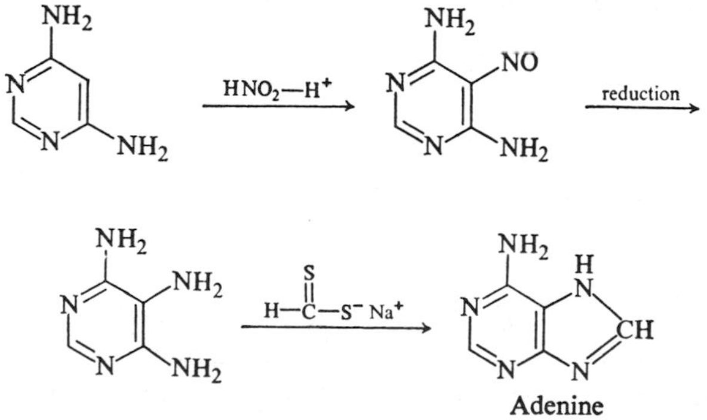

Adenine

*The primary amino groups in this molecule are not diazotized because the unshared electron-pair on each is withdrawn into the ring by the electron-attracting hetero-atoms. The nucleus, however, is sufficiently activated by the *combined* effect of the two amino groups to undergo C-nitrosation.

(c) DIAZONIUM COUPLING

The highly reactive nuclei of phenoxide ions, aromatic amines, pyrrole, etc., are able to react with aromatic diazonium ions, $ArN_2{}^+$. These reactions are discussed subsequently (13.4).

11.5 Formation of Carbon-Sulphur Bonds

(a) SULPHONATION

As in nitration, the conditions for sulphonation may be varied widely and suitable conditions for the reactions of both strongly activated and strongly deactivated nuclei are available.

The electrophilic sulphonating entity is sulphur trioxide, which is present in concentrated sulphuric acid as the result of the equilibrium,

$$2\,H_2SO_4 \rightleftharpoons SO_3 + H_3O^+ + HSO_4{}^-$$

Its concentration is reduced by the addition of water but is greater in fuming sulphuric acid. Sulphur trioxide is also 'carried' by pyridine in the form of 1-protopyridinium sulphonate (from pyridine and sulphur trioxide),

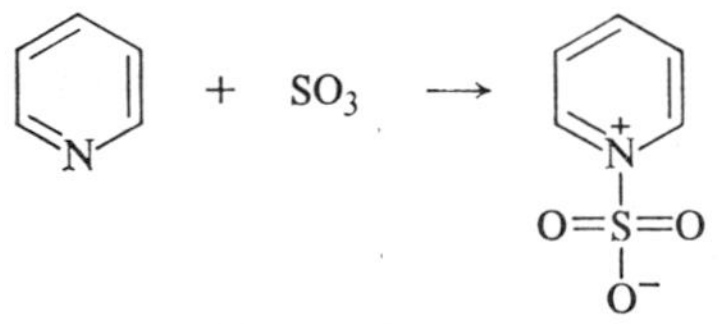

and this compound is a suitable sulphonating agent for compounds which are unstable to acids.

The mechanism of sulphonation is essentially the same as that of other electrophilic substitutions:

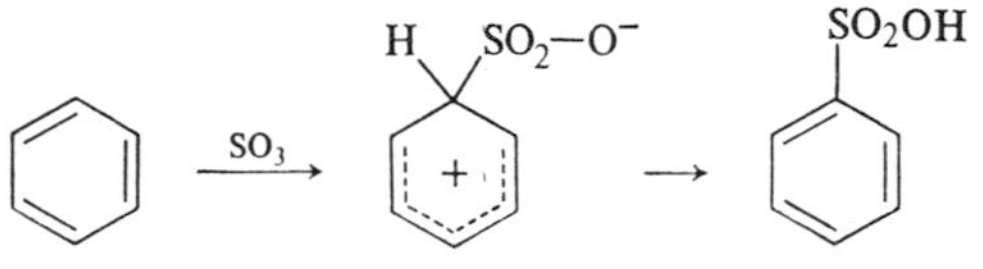

However, unlike most substitutions, sulphonation is readily reversible: the sulphonic acid group is eliminated by heating the compound with dilute sulphuric acid. Applications of the reverse process are described below (p. 407).

Benzene is normally sulphonated with 5–20% oleum, and these conditions are suitable also for the weakly deactivated halobenzenes. For more reactive compounds, such as the alkylbenzenes and the polycyclic hydrocarbons, sulphuric acid is satisfactory; e.g. in these conditions toluene gives the three toluenesulphonic acids in the approximate proportions: *ortho*, 10: *meta*, 1: *para*, 20.

Pyrrole, furan, and indole, which are decomposed by acids, are successfully sulphonated by 1-protopyridinium sulphonate. Reaction occurs at the 2-position in each case (unexpectedly, in the case of indole). The more stable thiophen may be sulphonated with 95% sulphuric acid, but a higher yield (about 90% of thiophen-2-sulphonic acid) is obtained with 1-protopyridinium sulphonate.

Aniline reacts with sulphuric acid to form a salt, but on strong heating this rearranges, through phenylsulphamic acid, to sulphanilic acid.

Aniline bisulphate   Sulphanilic acid

Phenol readily undergoes disulphonation to give the 2,4-disulphonic acid.

Strongly deactivated nuclei can be sulphonated with oleum at high temperatures. Benzenesulphonic acid gives the *m*-disulphonic acid with 20–40% oleum at 200°C, nitrobenzene gives *m*-nitrobenzenesulphonic acid similarly, and even pyridine gives a high yield (70%) of pyridine-3-sulphonic acid when sulphonated with olcum in the presence of mercury(II) sulphate at 230°C. As usual, quinoline and isoquinoline are more reactive than pyridine and give, respectively, mainly the 8-sulphonic acid and the 5-sulphonic acid with sulphuric acid at 220°C.

Applications of the reverse of sulphonation. The reversibility of sulphonation makes possible (1) the use of the sulphonic acid group for *protection*, and (2) the formation, in appropriate conditions, of the thermodynamically most stable product.

(1) The use as a protective group is illustrated by the synthesis of *o*-nitroaniline. Acetanilide is sulphonated to give almost exclusively the *p*-sulphonic acid; nitration then occurs *ortho* to the acetamido group (and is accompanied by the hydrolysis of the acetyl group); and the sulphonic acid group is finally eliminated by treatment with dilute sulphuric acid.

o-Nitroaniline

(2) The principles of the competition between kinetic and thermodynamic control in sulphonation have been discussed earlier (3.7*d*). Sulphonation at low temperatures gives the kinetically controlled products, e.g. mainly toluene-*p*-sulphonic acid from toluene and naphthalene-α-sulphonic acid from naphthalene, whereas at about 160°C the more stable toluene-*m*-sulphonic acid and naphthalene-β-sulphonic acid predominate. This is of particular value in enabling β-naphthyl derivatives to be obtained through conversion of the β-sulphonic acid into β-naphthol (p. 428) and thence into β-naphthylamine (p. 433), diazotization of which enables other groups to be introduced.

(*b*) CHLOROSULPHONATION

Aromatic compounds which are neither powerfully deactivated nor unstable to acids react with chlorosulphonic acid; for example, benzene at 20—25°C gives benzenesulphonyl chloride in 75% yield [1]:

$$PhH + 2\ ClSO_2OH \longrightarrow PhSO_2Cl + H_2SO_4 + HCl$$

An important example occurs in the synthesis of sulphanilamide (the first of the Sulpha drugs).

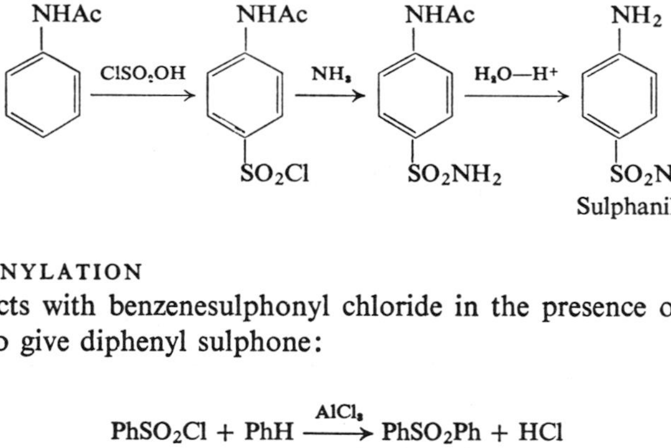

(*c*) SULPHONYLATION

Benzene reacts with benzenesulphonyl chloride in the presence of aluminium trichloride to give diphenyl sulphone:

$$PhSO_2Cl + PhH \xrightarrow{AlCl_3} PhSO_2Ph + HCl$$

Diphenyl sulphone

The reaction, which is analogous to Friedel-Crafts acylation, has the same limitations as chlorosulphonation.

11.6 Formation of Carbon-Halogen Bonds

(*a*) CHLORINATION

Three general procedures are available for chlorination, the choice of method being based on the reactivity of the aromatic compound.

The mildest conditions involve the use of molecular chlorine, usually in solution in acetic acid or a non-polar solvent such as carbon tetrachloride. The chlorine may be introduced as the gaseous element or may be generated *in situ* from an *N*-chloroamide and hydrogen chloride,

$$R_2N\text{—}Cl + HCl \longrightarrow R_2NH + Cl_2$$

Since hydrogen chloride is formed as a result of aromatic chlorination, this procedure for generating chlorine is continuous once started and the concentration of chlorine in solution can be maintained at the desired level according to the amount of hydrogen chloride added initially.

Chlorination in these conditions occurs by the normal S_E2 process:

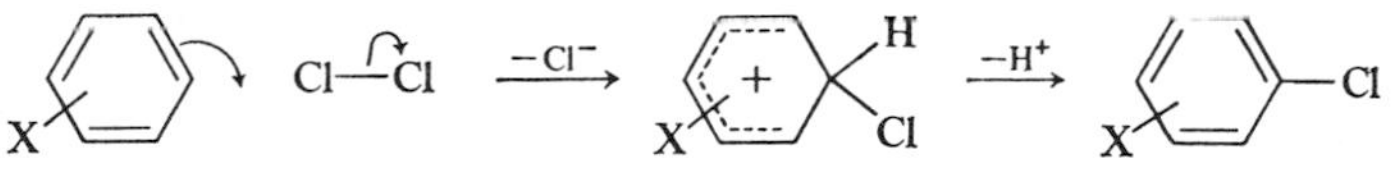

The reactivity of molecular chlorine is increased by the addition of a Lewis acid. Aluminium trichloride and iron(III) chloride are commonly used; they are usually introduced as aluminium amalgam and iron filings, respectively, from which the metal chlorides are generated *in situ*. The function of the Lewis acid is to withdraw electrons from the chlorine molecule, thereby increasing its electrophilic character. It should be emphasized that this polarizing influence does not lead to complete ionization (e.g. $Cl_2 + AlCl_3 \nrightarrow Cl^+ + AlCl_4^-$); i.e. the reaction should be represented as:

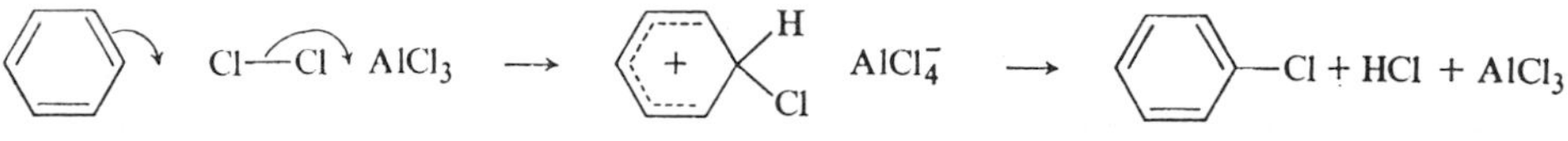

Still more reactive conditions are obtained by using an acidified solution of hypochlorous acid. It has been suggested that the effective electrophile is the chlorinium ion, Cl^+,

$$HO\text{—}Cl + H^+ \rightleftharpoons H_2\overset{+}{O}\text{—}Cl \rightleftharpoons H_2O + Cl^+$$

but this may be an oversimplification of the mechanism. A similarly reactive form of chlorine is derived from a solution of chlorine in carbon tetrachloride in the presence of silver sulphate.

(*b*) BROMINATION

The methods for bromination are closely related to those for chlorination: molecular bromine, added as such or generated *in situ* from an *N*-bromoamide and acid, may be used; increased electrophilic power may be obtained by using bromine in the presence of a Lewis acid such as iron(III) bromide (i.e. addition of iron filings); and a 'positive' brominating entity is present in acidified hypo-bromous acid and can also be obtained from bromine and silver sulphate in concentrated sulphuric acid.

Bromine itself can catalyze molecular bromination by virtue of its ability to form tribromide ion:

$$ArH \quad Br—Br \quad Br—Br \longrightarrow \overset{+}{A}rHBr + Br—\bar{B}r—Br$$

A second-order term in bromine therefore occurs in the kinetics of bromination. However, bromine is a much weaker Lewis acid than iron(III) bromide.

Iodine is often used as a catalyst in brominations. It acts by forming iodine bromide which facilitates bromination by removing bromide ion as IBr_2^-:

$$I_2 + Br_2 \rightleftharpoons 2\,IBr$$

$$ArH \quad Br—Br \quad IBr \longrightarrow \overset{+}{A}rHBr + IBr_2^-$$

(c) IODINATION

Iodine is a relatively weak electrophile which alone reacts with only the most strongly activated aromatic compounds such as phenols and amines. However, less reactive compounds can be iodinated in the presence of nitric acid or peracetic acid; for example, benzene and iodine in the presence of nitric acid give iodobenzene in 86% yield [1].

It used to be thought that iodine and benzene react reversibly to give only a small proportion of iodobenzene,

$$PhH + I_2 \rightleftharpoons PhI + HI$$

and that the function of the additive is to oxidize the hydrogen iodide and so displace the equilibrium to the right. However, it is now realized that this cannot be the explanation, since the rate of the back reaction is very slow. Evidence has now been obtained that, in nitric acid, a more reactive iodinating agent than iodine is formed, namely, the protonated form of NO_2I:

$$PhH + I—\overset{+}{\underset{H}{O}}—NO \longrightarrow PhI + HONO + H^+$$

In peracetic acid, it is possible that the iodinating entity is iodine acetate:

$$I_2 + CH_3CO_2OH \rightleftharpoons IOCOCH_3 + HOI$$

(d) FLUORINATION

Fluorine is far too reactive a compound to be used as a halogenating agent. Aryl fluorides are prepared via aromatic diazonium ions (p. 438).

(e) APPLICATIONS IN SYNTHESIS

Highly activated nuclei react readily with molecular halogens. The introduction of one halogen substituent has a deactivating effect on further reaction, but even so it is not always possible to stop halogenation occurring until all the strongly activated positions have been substituted.

For example, aniline and phenol give the 2,4,6-tribromo-derivatives essentially instantaneously when treated with dilute aqueous solutions of bromine. Pyrrole, each of whose four nuclear positions is comparable in reactivity with the *ortho* and *para* positions of aniline, gives tetrabromo- and tetraiodopyrrole with bromine and iodine, respectively, and with chlorine gives a pentachloro-derivative which is probably

The chlorination of phenol gives ultimately 2,4,6-trichlorophenol but can be controlled to give mainly 2,4-dichlorophenol; the derivative, 2,4-dichloro-phenoxyacetic acid, is manufactured as a selective weed-killer.

Furan is polymerized at room temperature by the acid liberated during halo-genation, but at $-40°C$ chlorination is reasonably efficient; the 2-chloro and 2,5-dichloro products predominate. Thiophen can also be chlorinated at low temperatures and at $-30°C$ gives a mixture of products containing mainly the 2-chloro- and 2,5-dichloro-derivatives. Iodination in the presence of mercury(II) oxide gives 2-iodothiophen.

Compounds of moderately high nuclear reactivity give monohalo-derivatives with the molecular halogens. For example, bromine in carbon tetrachloride reacts with mesitylene to give bromomesitylene (80%), with naphthalene to give 1-bromonaphthalene (75%), and with acetanilide to give *p*-bromoacetanilide (80%).

Compounds of the order of reactivity of benzene require a catalyst for halo-genation. For example, benzene and chlorine give chlorobenzene in the presence of aluminium amalgam, and benzene and bromine give bromobenzene in the presence of iron filings, aluminium amalgam, or iodine.

These conditions are also suitable for quite strongly deactivated compounds, although high temperatures are needed in some cases. For example, nitrobenzene gives a 60% yield of *m*-bromonitrobenzene when treated with bromine in the presence of iron filings at 135–145°C [1] and even pyridine is brominated at 300°C to give the 3-bromo- and 3,5-dibromo-derivatives. *

---

*At about 500°C, 2-bromopyridine and 2,6-dibromopyridine are the main products. In these conditions reaction is homolytic: the bromine *atom* is the reagent.

Strongly deactivated compounds can often be successfully halogenated with a source of 'positive' halogen. For example, *m*-dinitrobenzene reacts with bromine in sulphuric acid in the presence of silver sulphate,

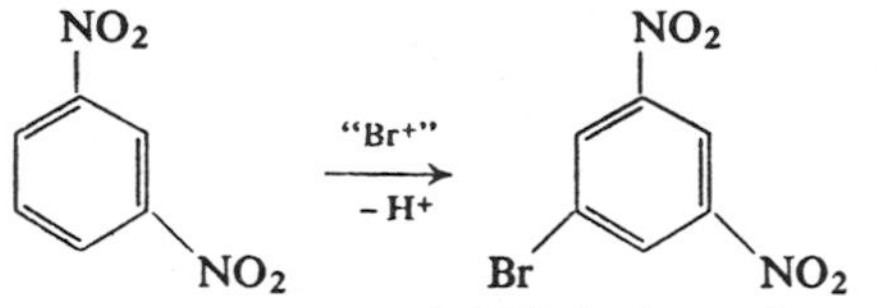

3,5-Dinitrobromobenzene

and benzoic acid and iodine in similar conditions give *m*-iodobenzoic acid in 75% yield.

11.7 Other Reactions

(*a*) HYDROXYLATION

Hydrogen peroxide and peracids in acidic media bring about electrophilic hydroxylation. Proton acids probably act by hydrogen-bonding to one oxygen atom of the peroxy-compound, so increasing the tendency for heterolysis of the O—O bond under the influence of the aromatic compound,

$$\longrightarrow \overset{+}{\text{Ar}}\text{H(OH)} + \text{ROH} + H_2O \longrightarrow \text{ArOH} + \text{ROH} + H_3O^+$$

(R = H or Acyl)

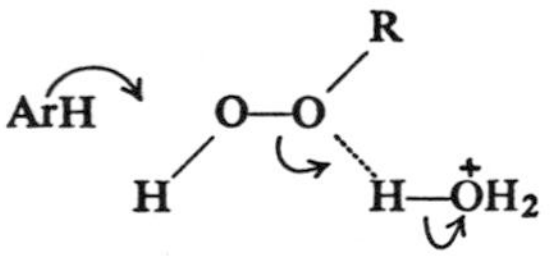

and Lewis acids such as boron trifluoride act by coordination to oxygen,

$$\longrightarrow \overset{+}{\text{Ar}}\text{H(OH)} + \text{RO—}\bar{B}F_3 \longrightarrow \text{ArOH} + \text{ROH} + BF_3$$

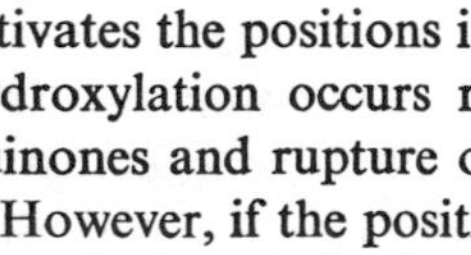

In general, electrophilic hydroxylation is not a satisfactory method for preparing phenols because the hydroxyl group which is introduced powerfully activates the positions in the nucleus which are *ortho* and *para* to it and further hydroxylation occurs readily. The dihydroxy products are then oxidized to quinones and rupture of the aromatic ring may follow.

However, if the positions which are activated by the entering hydroxyl group are substituted, further reaction is less likely to occur. In these circumstances reasonable yields of phenols may be obtained. For example, mesitylene can be converted into mesitol in 88% yield (based on the mesitylene which has reacted) by treatment with trifluoroperacetic acid in the presence of boron trifluoride.

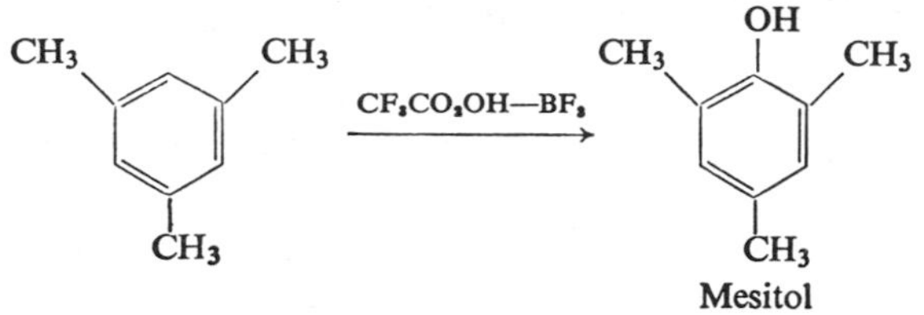

Mesitol

(b) METALATION

Salts of divalent mercury bring about the mercuration of aromatic compounds, * e.g.

$$PhH + Hg(NO_3)_2 \longrightarrow PhHgNO_3 + HNO_3$$

Yields are low unless air and water are excluded and mercury(II) oxide is added to remove nitric acid and prevent the reverse reaction.

The mercuri-derivatives have useful applications. For example, o-iodophenol is conveniently synthesized by treating phenol with mercury(II) acetate, converting the o-acetoxymercuri-derivative which is formed into the o-chloromercuri-derivative, and treating this with iodine.

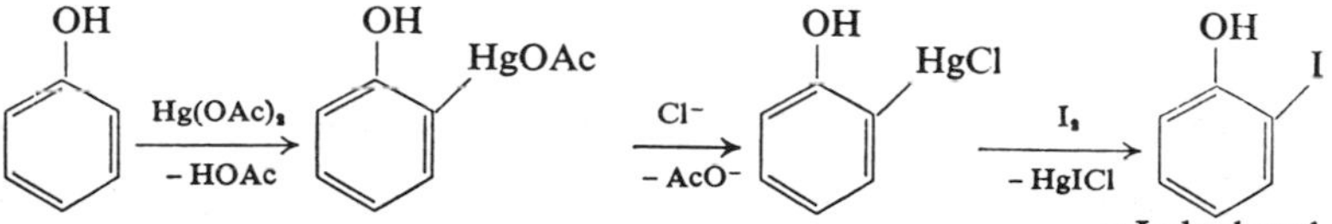

o-Iodophenol

2-Bromo- and 2-iodofuran may be obtained in a similar way from furan.

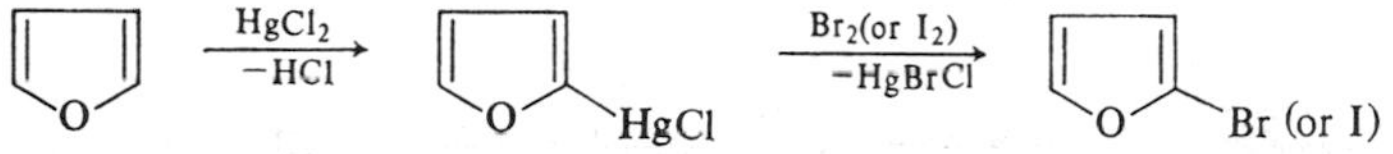

Thallium(III) salts have similar applications. For example, p-xylene and thallium tris(trifluoroacetate) give the thalliated derivative which, with potassium iodide at 0°C, forms the iodo-compound in 80–84% overall yield [55]:

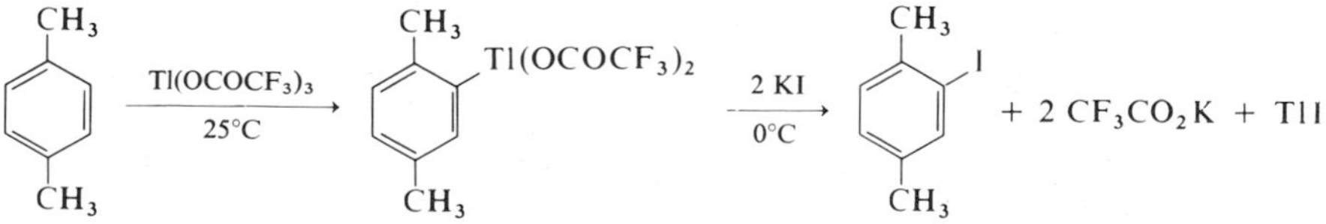

*Electrophilic metalations by metal salts should not be confused with the metalation of aromatic compounds with metal, p. 216.

The thalliation of monosubstituted benzenes with *ortho-*,*para*-directing substituents takes place almost entirely at the *para*-position, probably because of
the bulk of the thallium reagent, and this provides a good method for introducing a *p*-iodo-group.

(*c*) THE DISPLACEMENT OF GROUPS OTHER THAN HYDROGEN
A large number of electrophilic substitutions are known in which atoms and
groups other than hydrogen are displaced from the aromatic ring. Few of these
have synthetic value, but many can compete as unwanted reactions when
attempts are being made to effect other substitutions; particular attention is
drawn in the following to these side-reactions.

(*i*) *Decarboxylation*. Powerfully activating substituents lead to the ready displacement of a carboxyl group, as carbon dioxide, from aromatic compounds. *
For example, pyrrole- and furan-carboxylic acids decarboxylate on being heated;
the reaction can be regarded as involving an internal electrophilic substitution
by hydrogen, e.g.

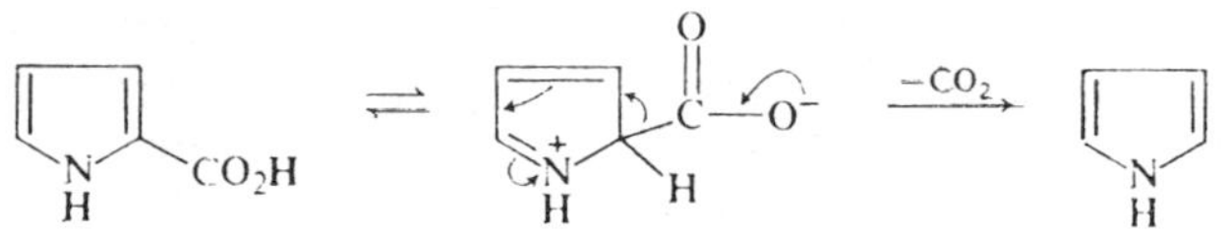

The acid-catalyzed decarboxylation of phenolic acids is similar, e.g.

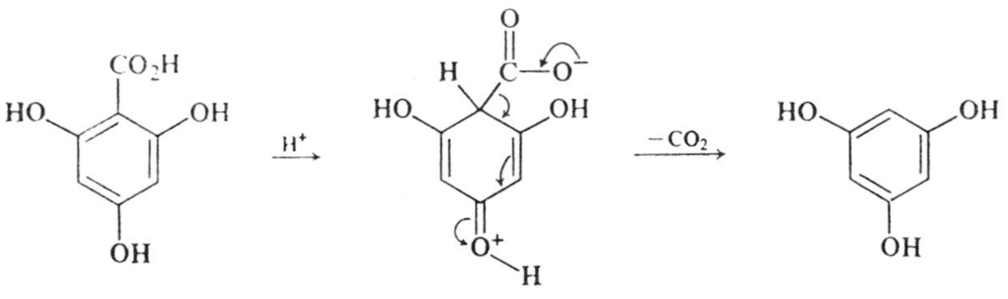

*These decarboxylations are entirely different from those which occur in aromatic car-
boxylic acids containing strongly electron-attracting groups. The latter involve the formation
of *carbanions* which derive moderate stability from the presence of the electron-attracting
group(s). Typical examples are:

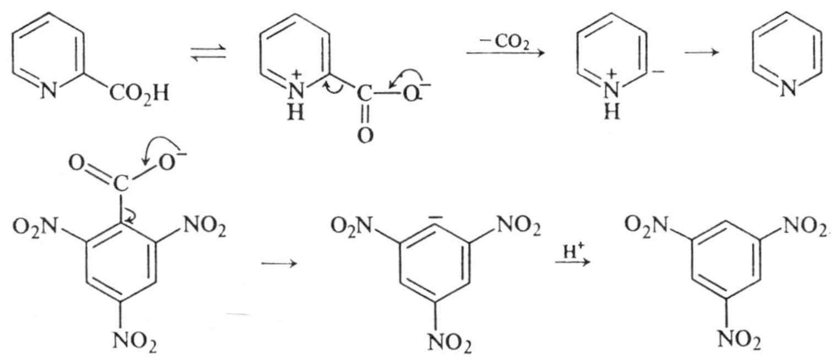

Electrophiles other than the proton can displace carboxyl groups from activated nuclei. For example, salicylic acid and bromine give 2,4,6-tribromophenol.

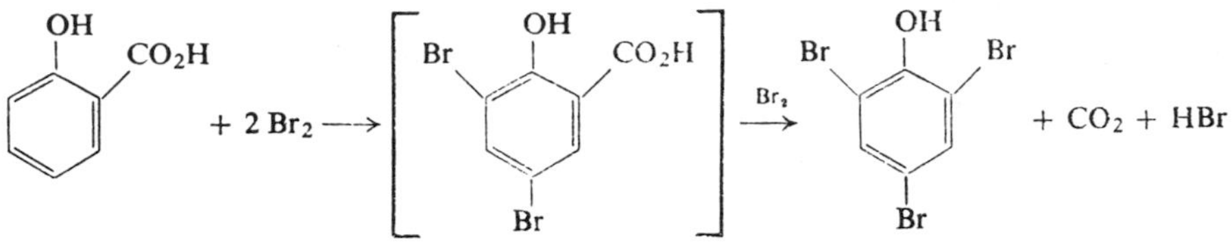

$$\text{(salicylic acid)} + 2\,Br_2 \longrightarrow [\ \cdots\ ] \xrightarrow{Br_2} \text{(2,4,6-tribromophenol)} + CO_2 + HBr$$

(*ii*) *Desulphonation.* The sulphonic acid group is readily displaced from aromatic rings by acids (p. 406) and can also be displaced from strongly activated positions (e.g. *ortho* or *para* to hydroxyl or amino) by halogens, e.g.

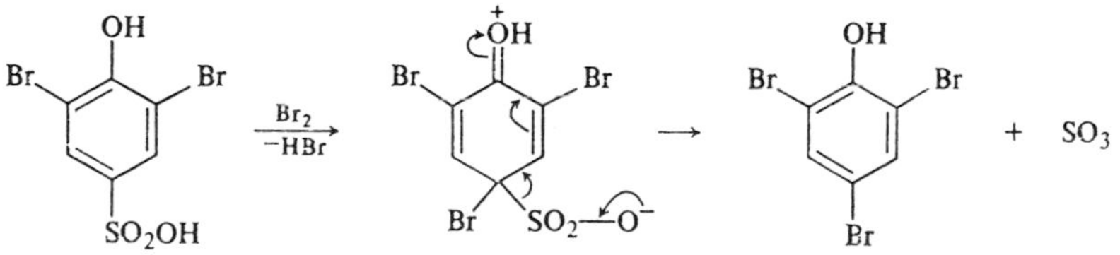

$$\xrightarrow[-HBr]{Br_2} \cdots \longrightarrow \cdots + SO_3$$

and in nitrating conditions, e.g.

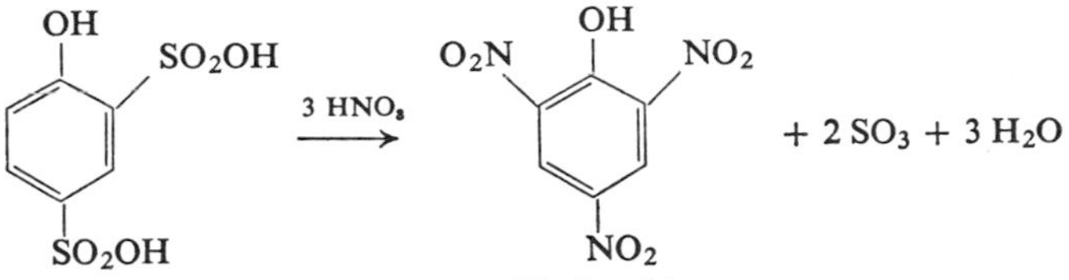

$$\xrightarrow{3\ HNO_3} \text{(Picric acid)} + 2\,SO_3 + 3\,H_2O$$

Picric acid

This last reaction provides a useful method for preparing picric acid, for attempts to trinitrate phenol itself lead to extensive oxidation; phenol-2,4-disulphonic acid is readily obtained by sulphonating phenol (p. 407).

(*iii*) *Dealkylation.* Tertiary alkyl groups are displaced from aromatic rings not only by acids (the reverse of Friedel-Crafts alkylation; p. 388) but also by the halogens. Reaction is especially favourable if the alkyl group is *ortho* or *para* to a strongly activating substituent, but occurs to some extent even in an unactivated situation: e.g. the chlorination and bromination of t-butylbenzene lead to some chlorobenzene and bromobenzene, together with the t-butyl-halobenzenes.

This reaction, which can be a disadvantage in halogenations, is apparently limited to tertiary alkyl groups, no doubt because of the comparative stability of tertiary carbonium ions as leaving groups, e.g.

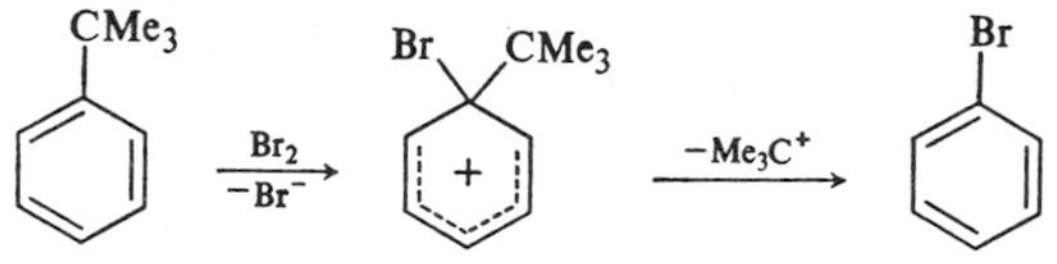

Nitration can also lead to dealkylation: e.g. the nitration of *p*-di-isopropyl-benzene gives comparable amounts of the deprotonated and dealkylated products.

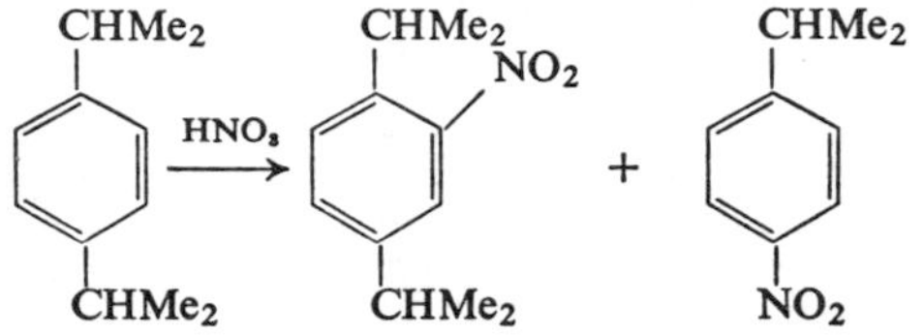

This side-reaction is only significant when highly branched alkyl groups are involved. In some cases it can be the dominant reaction: e.g. the nitration of 1,2,4,5-tetra-isopropylbenzene gives only the tri-isopropylnitro product, evidently because of the steric hindrance to normal nitration.

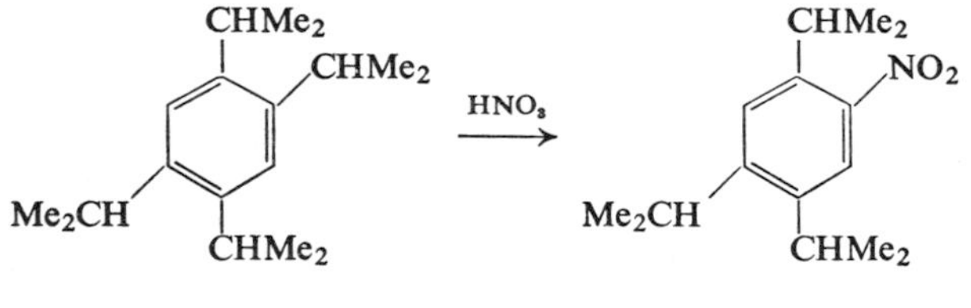

(*iv*) *Dehalogenation.* The reverse of halogenation only occurs when the halogen atom is adjacent to two very large substituents. For example, 2,4,6-tri-t-butyl-bromobenzene is debrominated by strong acid. Evidently the reaction is facili-tated by the release of steric strain between the halogen and the *ortho* sub-stituents in passage from the eclipsed reactant to the staggered intermediate:

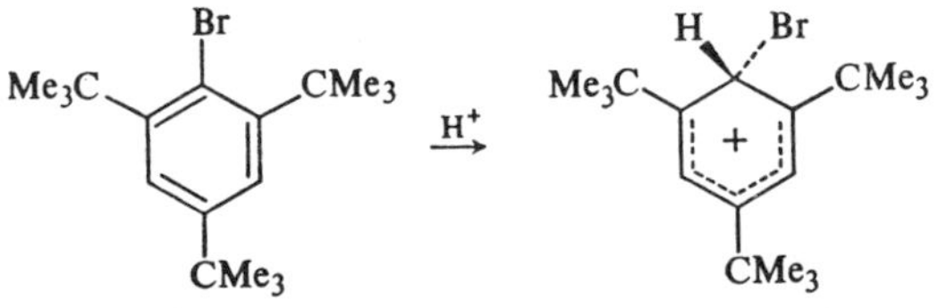

Dehalogenation as a side-reaction is more serious in nitration. For example, p-iodoanisole and nitric acid give p-nitroanisole:

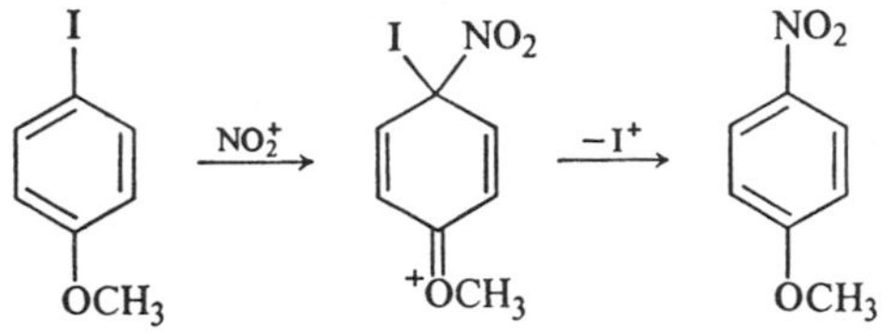

Reaction is here facilitated by the powerful activating effect of p-methoxyl, but deiodination occurs to some extent even in less activated environments: e.g. the nitration of iodobenzene gives a small amount of nitrobenzene. Debromination and dechlorination occur, but less readily.

11.8 The Preparation of Derivatives with Specific Orientations

Unless only one position in an aromatic compound is available for reaction, as in a symmetrically para-substituted benzene, electrophilic reactions invariably lead to mixtures of two or more products. This is usually a disadvantage in synthesis, as it is usually required to obtain only one product in as high a yield as possible. Although it is not possible to write down a series of principles reference to which will indicate an efficient route to every possible aromatic derivative, some generalizations are relevant. These are given here, together with a summary of some of the special techniques mentioned earlier in the text, with respect to substitution in benzenoid rings.

(1) The first consideration in devising a route to a disubstituted benzene, C_6H_4XY, is the directing character of each substituent. If X is meta-directing and Y is ortho, para-directing, it should be possible to obtain as main products both the meta isomer, by starting with PhX, and the ortho and para isomers, by starting with PhY. For example, the chlorination of nitrobenzene in the presence of iron filings gives mainly m-chloronitrobenzene whereas the nitration of chlorobenzene in sulphuric acid gives about 30% of o-chloronitrobenzene and 60% of p-chloronitrobenzene.

(2) This basic approach fails in many cases, a particularly common circumstance being that the meta-directing compound is too deactivated to undergo the necessary reaction: e.g. o- and p-nitrotoluene can be obtained by nitrating toluene, but m-nitrotoluene cannot be obtained by alkylating nitrobenzene. An indirect approach has then to be adopted, and one commonly used is based on the versatility of the amino group, as in the synthesis of m-bromotoluene (p. 378).

(3) In some cases the nature of a substituent may be modified to achieve the appropriate orientation. For example, both m- and p-nitro-n-propylbenzene may

be made from propiophenone: nitration, followed by selective reduction of the carbonyl group (p. 631), gives the *m*-nitro compound,

m-Nitro-n-propylbenzene

and reduction to n-propylbenzene followed by nitration gives about 50% of the *p*-nitro compound.

(4) The main product of a substitution is more easily isolated and purified if it is a solid than if it is a liquid. Fractional distillation of a liquid (on a laboratory scale) is usually inefficient because the boiling points of *ortho, meta,* and *para* isomers are very close together, whereas the purification of a solid by recrystallisation is almost always practicable. It is therefore sometimes more satisfactory to rearrange a synthesis so that the aromatic substitution provides a solid which, after purification, can be transformed into the desired product. For example, although *m*-chloronitrobenzene can be obtained as the main product of the direct chlorination of nitrobenzene, it is more easily obtained in a pure state by the nitration of nitrobenzene, which gives a high yield of the solid *m*-dinitro-derivative, followed by selective reduction of one nitro group (p. 649), diazotization, and Sandmeyer chlorination (p. 438).

(5) It is more difficult to isolate a particular product from the substitution of *ortho, para*-directing compounds than from a *meta*-directing compound. Whereas the latter usually give almost entirely (*ca.* 90%) the *meta*-derivative, the former often give comparable quantities of the *ortho* and *para* products. This not only limits the efficiency of the method for a particular derivative, since yields are likely to be less than 50%, but also makes purification difficult even if both products are solids, and losses result. In general, the *para* product is easier to obtain than the *ortho* isomer because it is normally higher melting and therefore less soluble and can be obtained by recrystallization.

In a few instances a particular structural feature gives rise to a specific difference in physical properties of *ortho* and *para* isomers which can be employed for their separation. For example, the nitration of phenol with dilute nitric acid gives comparable amounts of *o-* and *p*-nitrophenol. The *ortho* isomer is a chelated compound (p. 54) whereas the *para* isomer is highly associated through intermolecular hydrogen-bonding. This results in the former being much more volatile and it may be cleanly separated from the latter by distillation in steam.

(6) Alteration in the reaction conditions can change the proportions of products, as in the sulphonation of naphthalene (p. 408).

(7) A more general and more specific method for modifying the directing character of a compound is to block the most reactive position with a group which can be readily removed after the appropriate substitution has been accomplished. Examples of the use of the t-butyl and sulphonic acid groups have been quoted (pp. 389, 407).

(8) The amino group is particularly widely employed in aromatic substitutions by virtue both of its ease of conversion to other groupings *via* diazotization (Chapter 13) and of its activating power coupled with the ease of its removal from the aromatic nucleus (p. 439). Both mild activation, through the acetyl derivative of the amino group, and powerful activation (e.g. the synthesis of 1,3,5-tribromobenzene, p. 440) can be achieved. The methyl group also has some applications as an activating group which is removable from the nucleus, as in the synthesis of 1,3,5-trinitrobenzene (p. 403).

Further Reading

NORMAN, R. O. C., and TAYLOR, R., *Electrophilic Substitution in Benzenoid Compounds*, Elsevier (London 1965).
OLAH, G. A., *Friedel-Crafts and Related Reactions*, Vols. 1–5, Interscience (New York and London 1963–1965).

Problems

1. What do you expect to be the chief product(s) of the mono-nitration of the following compounds?

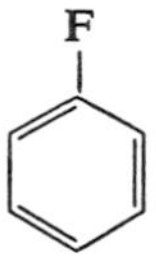

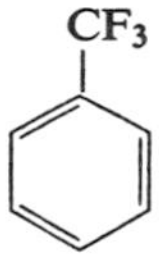

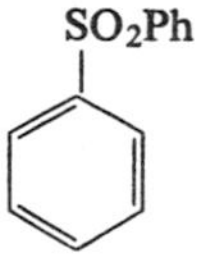

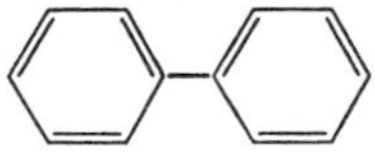

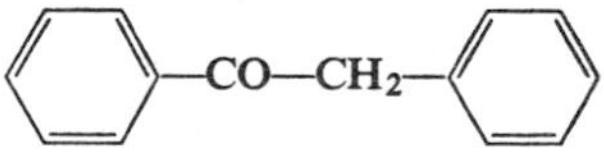

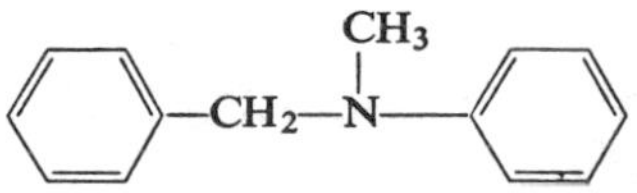

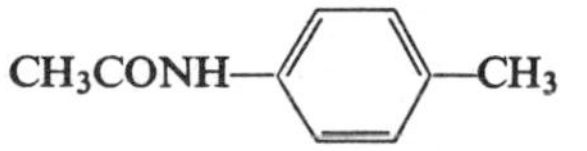

(h) [benzene ring]—CH=CH—CO$_2$CH$_3$

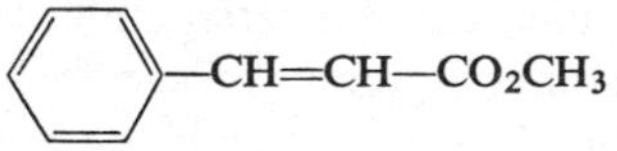

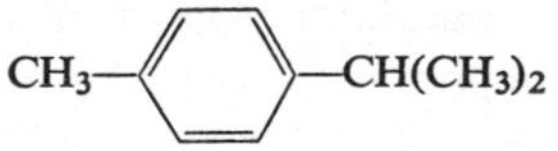

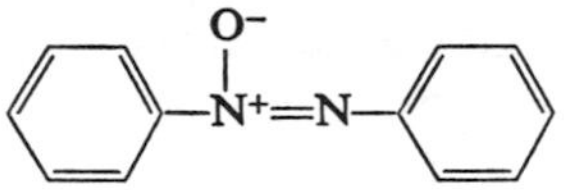

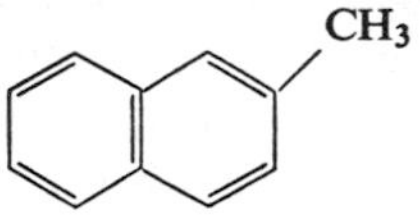

(i) CH$_3$—[benzene ring]—CH(CH$_3$)$_2$

(j) [benzene ring]—N$^+$(O$^-$)=N—[benzene ring]

(k) [naphthalene ring]—CH$_3$

(l) [naphthalene ring with NO$_2$]

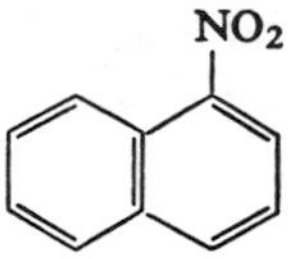

2. Arrange the following in order of decreasing reactivity towards an electrophilic reagent:

$$Ph—Cl \qquad Ph—CH_3 \quad Ph—OCH_3 \quad Ph—N(CH_3)_2$$

$$Ph—\overset{+}{N}(CH_3)_3 \quad Ph—NO_2 \quad Ph—CO_2Et \quad Ph—OCOCH_3$$

3. Account for the following observations:

   (i) Iodine is a catalyst for aromatic bromination.

   (ii) The product of the sulphonation of naphthalene depends on the temperature of the reaction.

   (iii) 2,6-Dimethylacetanilide is nitrated at the 3-position.

   (iv) Pyrrole is more reactive to electrophiles at the 2- than at the 3-position, whereas the opposite holds for indole.

   (v) Nitration of dimethylaniline gives mainly the m-nitro-derivative when concentrated nitric and sulphuric acids are used but mainly the o- and p-nitro-derivatives in less acidic conditions.

   (vi) Pyridine-1-oxide is more reactive than pyridine in nitration, and gives mainly the 4-nitro derivative.

4. Outline methods for the synthesis of the following:

(a) [benzene ring with CH$_3$ and Br]

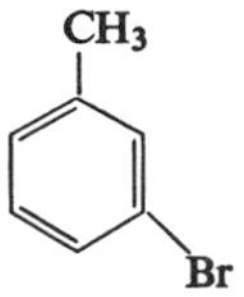

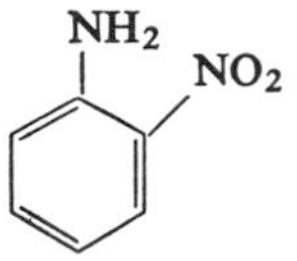

(b) [benzene ring with NH$_2$ and NO$_2$]

(c) [benzene ring with NH$_2$ and Br]

(d) [benzene ring with NH$_2$ and SO$_2$NH$_2$]

(e) [benzene ring with NH$_2$ and CH$_2$CH$_2$CH$_3$]

(f) [benzene ring with NH$_2$ and CH$_2$CH$_2$CH$_3$]

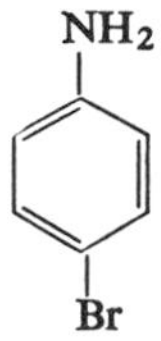

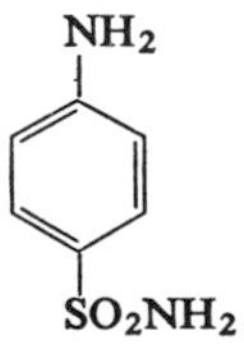

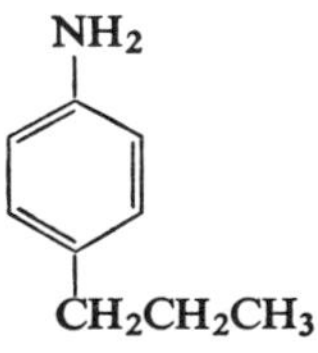

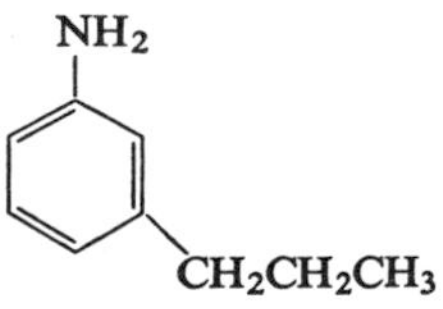

(*g*)
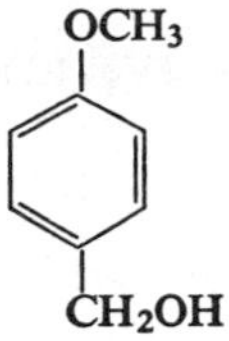
OCH3
CH2OH

(*h*)
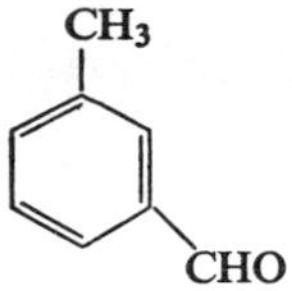
CH3
CHO

(*i*)
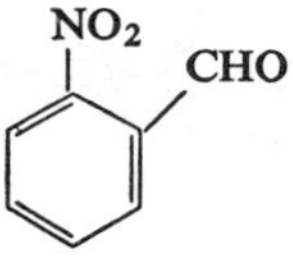
NO2
CHO

(*j*)
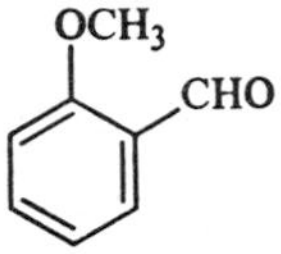
OCH3
CHO

(*k*)
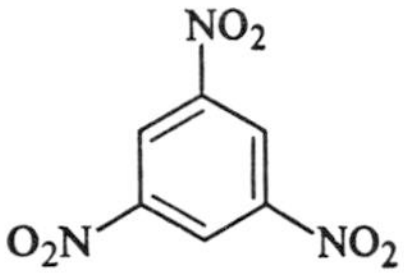
NO2
O2N
NO2

(*l*)
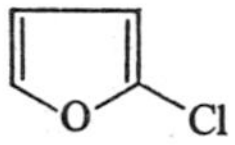
O
Cl

(*m*)
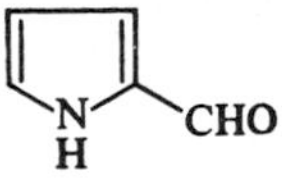
N
H
CHO

(*n*)
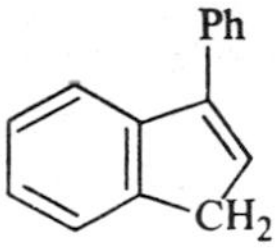
Ph
CH2

(*o*)
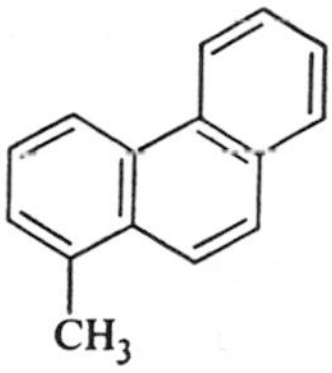
CH3

12. Nucleophilic Aromatic Substitution

12.1 Principles

Although benzene is moderately reactive towards electrophiles, it is inert to nucleophiles. In both respects, therefore, it resembles ethylene and its alkyl-derivatives. The analogies between the two systems may be extended, for just as the attachment of a group of $-M$ type to an olefinic bond activates that bond to nucleophiles, so the attachment of such a substituent to the benzene ring activates the ring to nucleophiles. For example, o- and p-nitrophenol are formed by heating nitrobenzene with powdered potassium hydroxide:

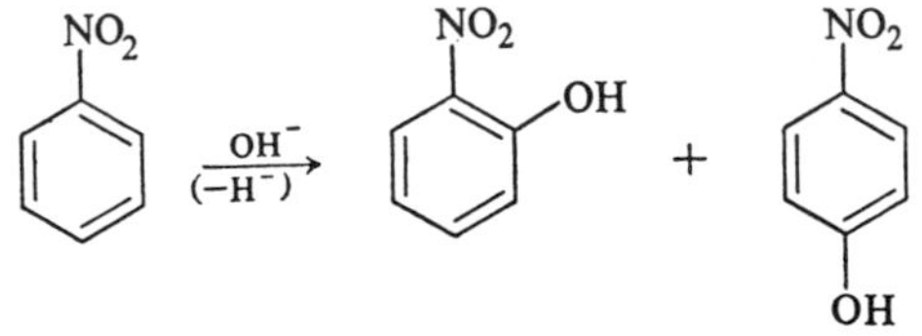

The mechanism of this, and most other, nucleophilic aromatic substitutions is similar to that of electrophilic aromatic substitutions except that an anionic rather than a cationic intermediate is involved. The nucleophile adds to the aromatic ring to give a delocalized anion from which a hydride ion is eliminated:

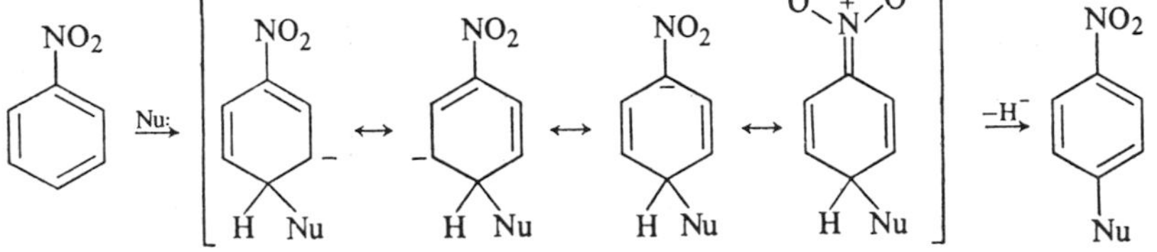

It is apparent that the reactivity towards nucleophiles of nitrobenzene compared with benzene stems from the ability of the nitro-group to stabilize the anionic intermediate, and hence the preceding transition state, by accommodating the negative charge. Further, this delocalization can occur only when the reagent adds to the *ortho* or *para* position; addition to the *meta* position gives an adduct stabilized by the inductive effect of the nitro-group, but not the mesomeric effect, and moreover the inductive effect is relayed through one more carbon atom than is the case for *ortho* or *para* substitution (*cf.* p. 371). Consequently, towards nucleophiles, electron-withdrawing substituents are activating and *ortho, para*-orienting, and conversely electron-releasing groups are deactivating and *meta*-directing. These principles are the reverse of those which apply to electrophilic substitution.

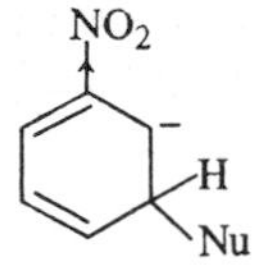

The example of nucleophilic substitution given above involves the displacement of hydride ion. This is a relatively high-energy species, and the reaction would be unfavourable but for the fact that an oxidizing agent, nitrobenzene itself, is present which removes the hydride ion. In doing so, a reduction product, azoxybenzene (p. 647), is formed, so that the yield of nitrophenols based on nitrobenzene is low.

It is therefore more satisfactory to carry out nucleophilic substitutions, where possible, on aromatic compounds from which a more stable anion than hydride can be displaced. Typical leaving groups are halide ions, as in the reaction of *p*-nitrochlorobenzene with hydroxide ion.

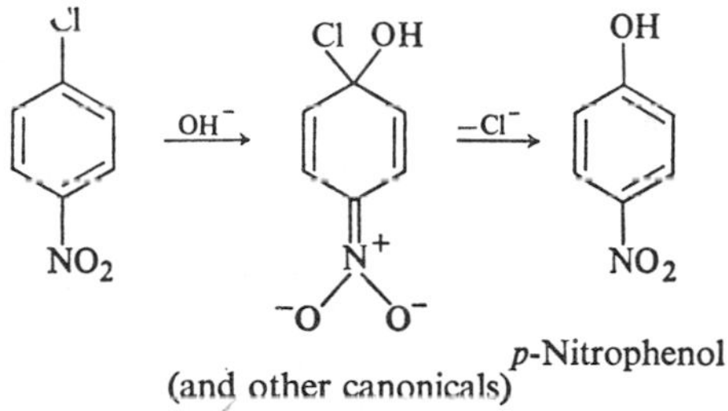

p-Nitrophenol

(and other canonicals)

Three other mechanisms occur, though less widely, in nucleophilic aromatic substitution. The first is an S_N1 process, so far recognized only for reactions of aromatic diazonium ions. When aqueous benzenediazonium chloride is warmed, phenol is formed *via* the phenyl cation. Nuclear-substituted derivatives behave similarly.

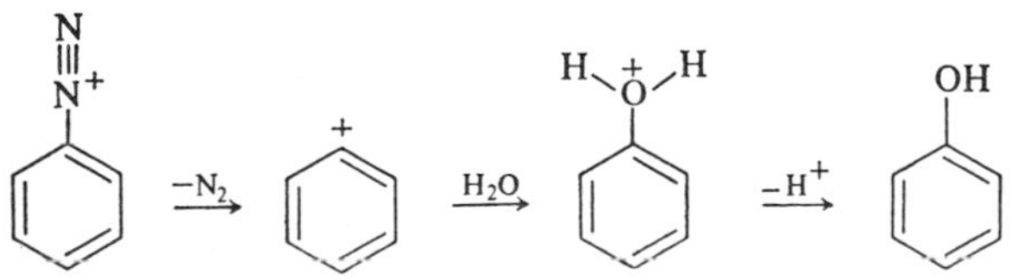

The driving-force for this reaction resides in the strength of the bonding in the nitrogen molecule which makes it a particularly good leaving group.

The two remaining mechanisms, which have closely defined structural require-ments, are described later (12.4 and 12.5).

Substitution in heteroaromatic compounds. The six-membered nitrogen-con-taining heteroaromatic compounds are activated towards nucleophiles. This is because the negative charge on the adduct formed by addition of the nucleophile to positions *ortho* or *para* to nitrogen is stabilized by delocalization onto the electronegative nitrogen atom:

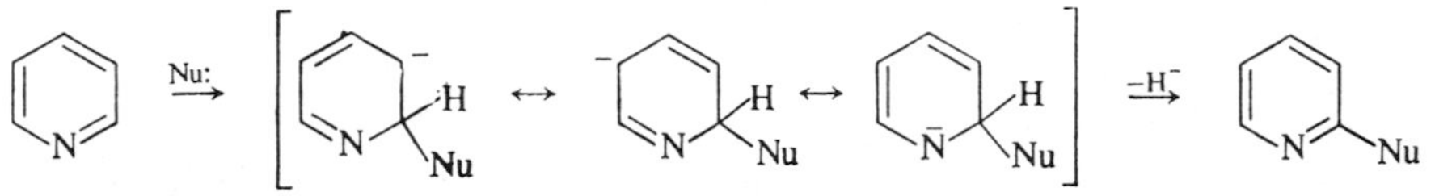

Consequently pyridine resembles nitrobenzene in undergoing nucleophilic dis-placement at the 2- and 4-positions, just as it resembles nitrobenzene in its inert-ness towards electrophiles (p. 382). Similarly, anions such as halide ions can be readily displaced from the 2- and 4-positions of pyridine.

Pyrimidine, as expected, is even more strongly activated than pyridine, sub-stitution being directed to the 2-, 4-, and 6-positions. Quinoline likewise reacts at the 2- and 4-positions, but in isoquinoline, in which both the 1- and 3-positions are *ortho* to the heterocyclic nitrogen, the 1-position is far more reactive than the 3-position.

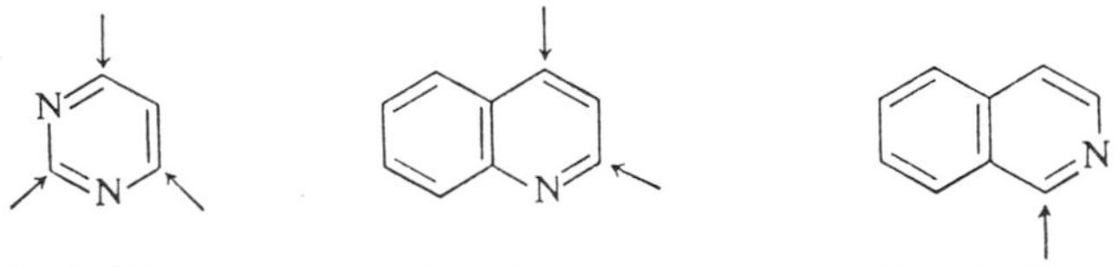

| Pyrimidine | Quinoline | Isoquinoline |

The explanation of the last fact becomes apparent when the structures of the adducts formed at the 1- and 3-positions of isoquinoline are examined. The activating influence in each case is ascribed mainly to the canonical structures in which nitrogen accommodates the negative charge: that for addition to the 1-position is benzenoid and is therefore of lower energy than that for addition to the 3-position, which is not (*cf.* the discussion of why β-naphthol directs electrophiles to the 1-position, p. 379).

1-position:

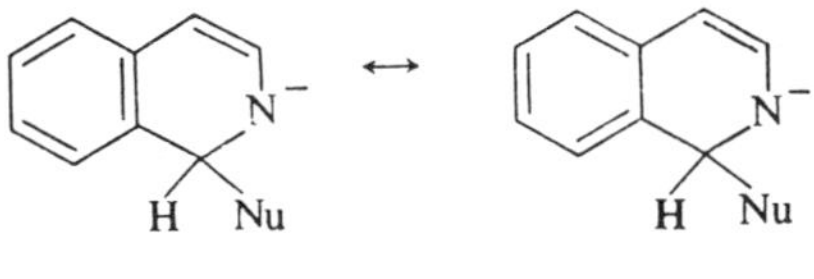

3-position:

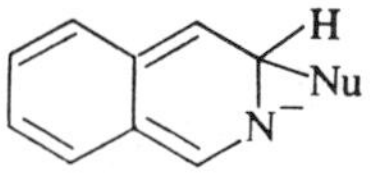

12.2 Displacement of Hydride Ion

Benzene is not attacked by any nucleophile, and nitrobenzene reacts only with the most reactive nucleophiles, such as amide or substituted amide ions,

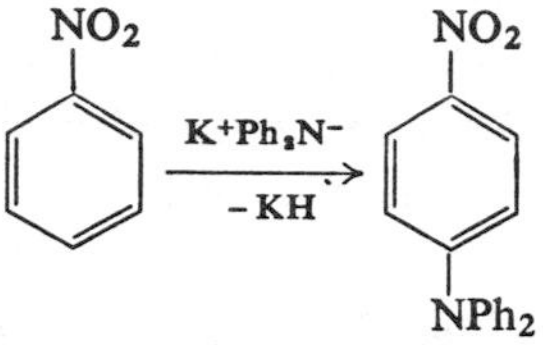

Diphenyl-*p*-nitrophenylamine

or less reactive nucleophiles such as hydroxide ion in very vigorous conditions (p. 422).

m-Dinitrobenzene is more strongly activated than nitrobenzene and reacts, for example, with cyanide ion:

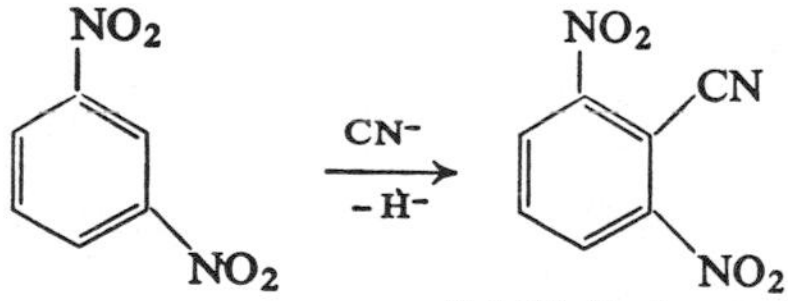

2,6-Dinitrobenzonitrile

Pyridine in general resembles nitrobenzene save that organometallic reagents, which react with the nitro-group of nitrobenzene, substitute in the nucleus of pyridine. The following are typical reactions, the predominant product being the 2-derivative.

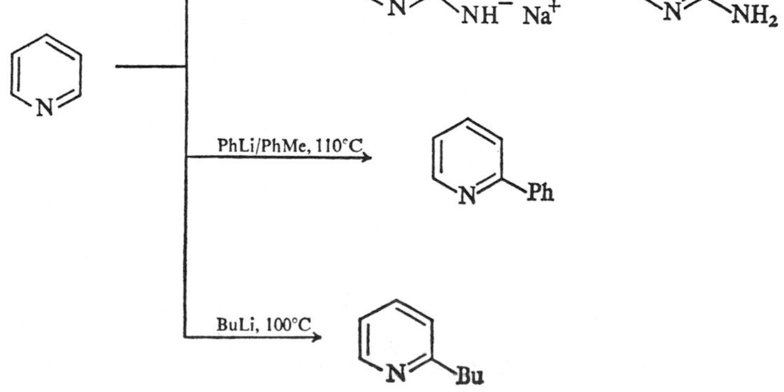

Similar reactions occur on quinoline, mainly at the 2-position, and on iso-quinoline, at the 1-position.

12.3 Displacement of Other Anions

(a) HALIDES

The four halobenzenes are, like the vinyl halides, very inert to nucleophiles in normal conditions. They do not react with methoxide ion in methanol or with boiling alcoholic silver nitrate, conditions in which alkyl halides react readily. Reaction can, however, be brought about in each of two ways. First, very vigorous conditions may be employed, as in the formation of phenol by heating chlorobenzene with 10% caustic soda solution under pressure at 350°C; in these conditions reaction occurs mainly via the benzyne intermediate (12.4). *

$$PhCl \xrightarrow[-Cl^-]{OH^-} [PhOH] \xrightarrow{OH^-} PhO^- \xrightarrow{H^+} PhOH$$

Secondly, it has recently been found that reactions of halobenzenes with alkoxide ions occur many powers of ten faster in dimethyl sulphoxide than in hydroxylic media. For example, bromobenzene and t-butoxide ion give phenyl t-butyl ether in about 45% yield [5]:

$$PhBr \xrightarrow[-Br^-]{Me_3CO^-/DMSO} Ph—O—CMe_3$$

It is argued that this is because alkoxide ions are effectively stabilized by hydrogen-bonding in hydroxylic solvents so that their reactivity is greater in a non-hydroxylic solvent.

The introduction of substituents of −M type into positions ortho or para with respect to the halogen atom considerably increases the ease of nucleophilic substitution. The activating effect increases with the number of such substituents present, so that a typical order of reactivity of chlorobenzenes is

$$p\text{-}NO_2 < 2,4\text{-dinitro} < 2,4,6\text{-trinitro}$$

In fact, 2,4,6-trinitrochlorobenzene is hydrolyzed by dilute alkali at room temperature.

These are S_N2 reactions, being first-order with respect to the halide and to the nucleophile, and in this sense the reactions of aromatic halides are analogous to those of aliphatic halides. Two differences should be noted. First, displacement on an aromatic halide cannot lead to inversion of configuration; the reagent must approach the aromatic carbon from the same side as the halide. Secondly,

---

*This method and the route via cumene hydroperoxide (p. 473) account for most of the industrially produced phenol. Diphenyl ether is formed as a byproduct (ca. 20%) by reaction of phenoxide ion with chlorobenzene and is marketed as a heat-transfer agent.

amongst the aromatic halides the usual order of reactivity is fluoride > chloride ~ bromide ~ iodide, whereas in the aliphatic series fluorides are the least reactive.

The reason for the latter difference is that the rate-determining step for most aromatic substitutions is the addition of the nucleophile to the ring:

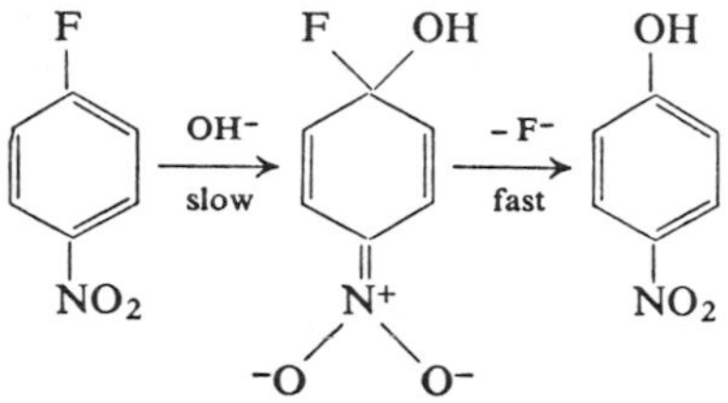

Fluorine, being the most electronegative of the halogens, is best able to stabilize the adduct and hence the preceding transition state. In S_N2 reactions on aliphatic halides, however, the rate-determining step involves the severance of the C-halogen bond and C—F, being the strongest of the four, is the hardest to break.

Nucleophiles other than hydroxide ion react with activated aromatic halides. Typical examples are the formation of aromatic amines, e.g.

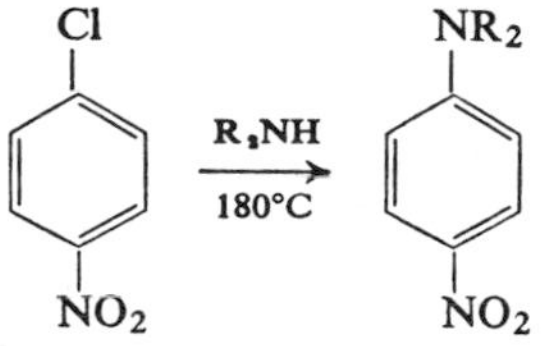

and 2,4-dinitrophenylhydrazine, from 2,4-dinitrochlorobenzene (itself obtained by the dinitration of chlorobenzene), which is used for the characterization of carbonyl compounds.

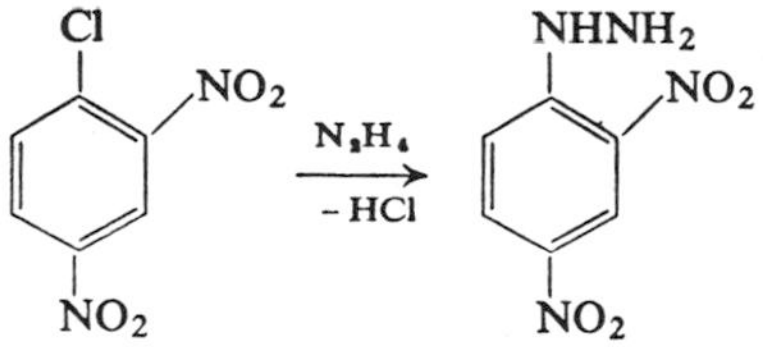

2,4-Dinitrofluorobenzene has a special use in 'marking' the terminal amino group(s) in a protein or polypeptide and was first used by Sanger in his elucidation of the structure of insulin. The nucleophilic amino group displaces fluoride, forming a secondary amine, and hydrolysis of the peptide links then leaves the terminal amino-acid bonded to the 2,4-dinitrophenyl group. Isolation and identification of this derivative indicate the nature of the terminal amino-acid.

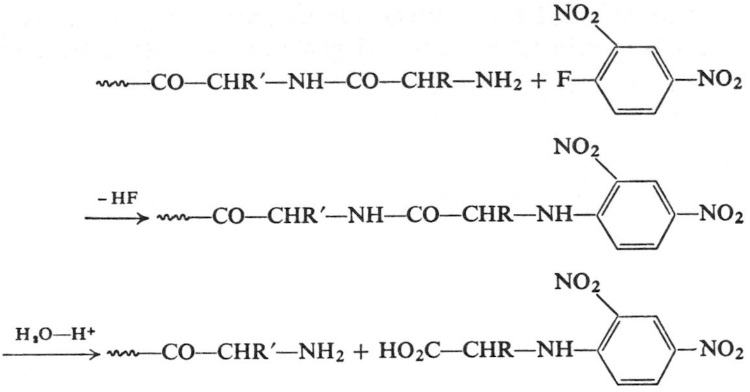

$$\text{\textasciitilde\textasciitilde\textasciitilde—CO—CHR}'\text{—NH—CO—CHR—NH}_2 + F-\underset{}{\bigcirc}(NO_2)(NO_2)$$

$$\xrightarrow{-HF} \text{\textasciitilde\textasciitilde\textasciitilde—CO—CHR}'\text{—NH—CO—CHR—NH}-\underset{}{\bigcirc}(NO_2)(NO_2)$$

$$\xrightarrow{H_2O—H^+} \text{\textasciitilde\textasciitilde\textasciitilde—CO—CHR}'\text{—NH}_2 + HO_2C—CHR—NH-\underset{}{\bigcirc}(NO_2)(NO_2)$$

(b) OXYANIONS

Although phenyl ethers are stable in basic conditions, the introduction of $-M$ substituents in the *ortho* and *para* positions induces hydrolysis, e.g.

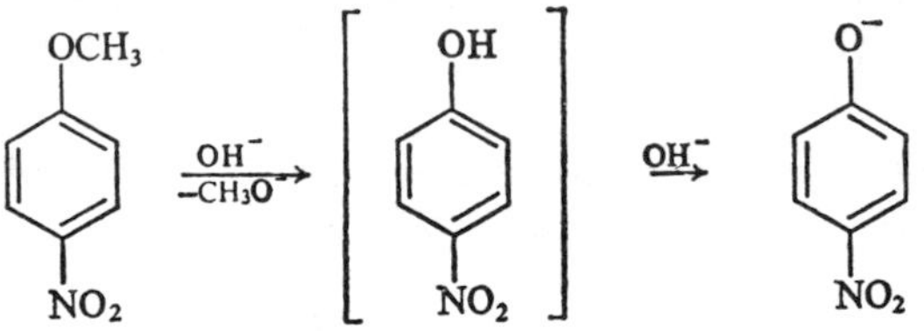

In this respect aromatic ethers resemble esters rather than aliphatic ethers. Indeed, the mechanism of reaction of aromatic ethers is entirely analogous to that of esters in that in each the addition of hydroxide ion is made possible by the presence of an electronegative atom which can accommodate the negative charge.

Compare

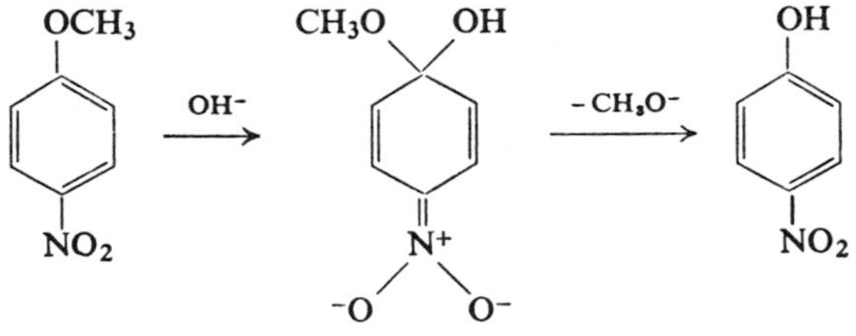

with

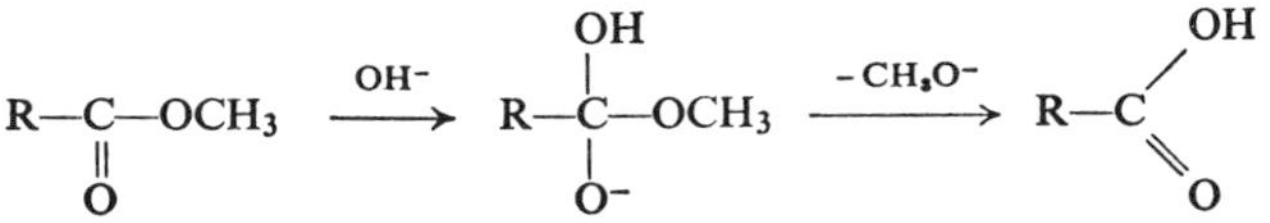

$$\underset{\underset{O}{\parallel}}{R—C—OCH_3} \xrightarrow{OH^-} \underset{\underset{O^-}{|}}{R—C—OCH_3} \xrightarrow{-CH_3O^-} R—C\underset{\diagdown O}{\overset{\diagup OH}{}}$$

(c) SULPHITE ION

The fusion of aromatic sulphonates with caustic alkali at high temperatures gives phenols by displacement of sulphite ion. For example, sodium *p*-toluenesulphonate and a mixture of caustic soda with a little potash (to aid fusion) give *p*-cresol in 65–70% yield at 250–300°C [1]. (This method still accounts for about 15% of the industrially produced phenol.)

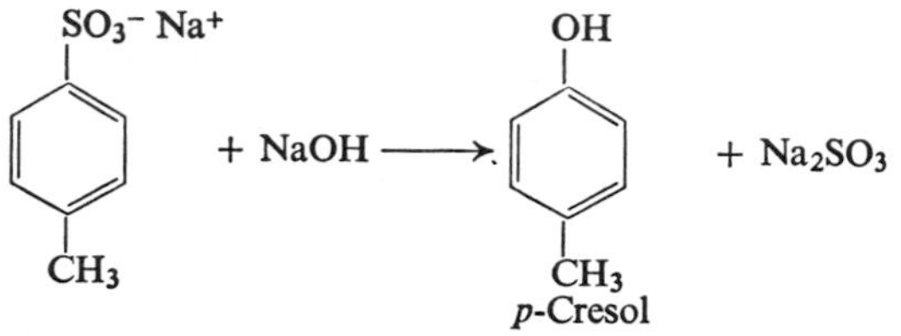

$$SO_3^- \ Na^+ \quad + \ NaOH \longrightarrow \quad OH \quad + \ Na_2SO_3$$

$$CH_3 \qquad\qquad\qquad CH_3$$
$$\textit{p-Cresol}$$

Because of the ready availability of aromatic sulphonic acids by direct sulphonation (11.5), this procedure is useful for the preparation of phenols. For example, β-naphthol can be obtained by the high-temperature sulphonation of naphthalene (p. 408) followed by fusion with hydroxide, and this provides a method for β-substituted naphthalenes *via* the Bucherer reaction (12.5).

There are, however, limitations to the method. At the high temperatures necessary for reaction, halogen substituents are themselves labile, so that halophenols cannot be prepared in this way. Again, a substituent of −*M* type which is *meta* to the sulphonic acid group activates the *ortho* and *para* positions to hydride displacement, so that, for example, *m*-nitrophenol cannot be obtained from *m*-nitrobenzenesulphonic acid. In these instances the phenols are best obtained *via* the corresponding diazonium salt (13.3).

The sulphonate group may also be displaced by cyanide ion, e.g.

$$PhSO_3^-K^+ \xrightarrow{\text{KCN}} PhCN + K_2SO_3$$

but this method is inefficient in practice because the cyanide and sulphonate mixture remains unfused even at very high temperatures and yields are low.

(*d*) NITROGEN ANIONS

The nitrite ion may be displaced readily from aromatic nitro-compounds if other nitro groups are present to activate the nucleus, e.g.

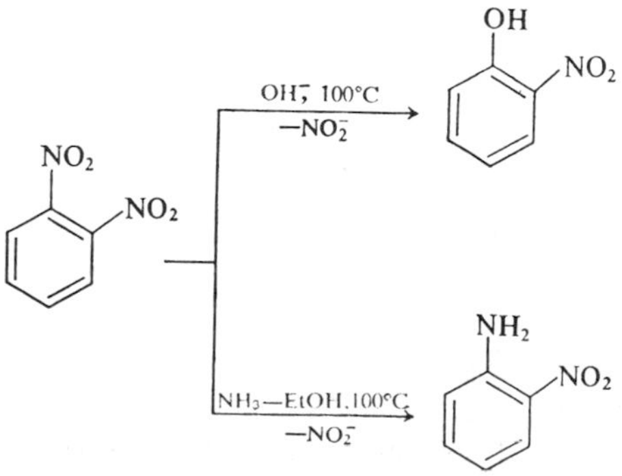

o-Nitrophenol

o-Nitroaniline

When 1,3,5-trinitrobenzene is refluxed in methanol containing methoxide ion, 3,5-dinitroanisole is formed in 70% yield [1]:

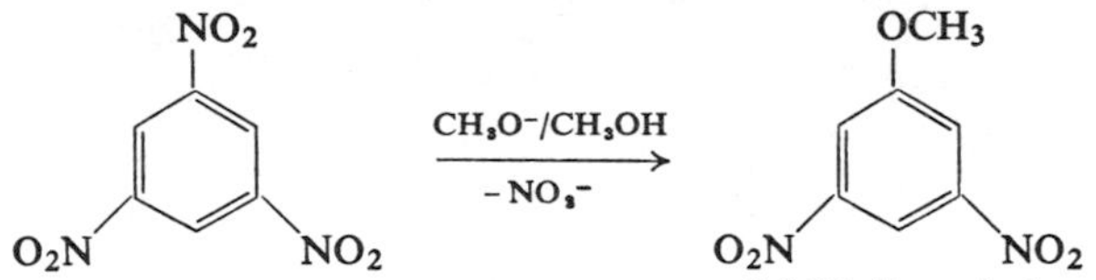

It should be noted that the three positions which are *ortho* and *para* to the nitro groups are more strongly activated to the addition of the nucleophile than a position containing a nitro group which is *meta* to the remaining nitro groups. The course taken in this reaction is evidently determined by the greater tendency of a nitro group to leave as nitrite ion than of hydrogen to leave as hydride.

Amino substituents may be displaced as amide ions. A useful synthetic method for secondary amines consists of converting aniline into a tertiary amine, nitrosating (p. 405), and heating the nitroso-derivative with alkali. The secondary amines produced cannot be contaminated by primary and tertiary amines, as they are when prepared from alkyl halides (p. 325).

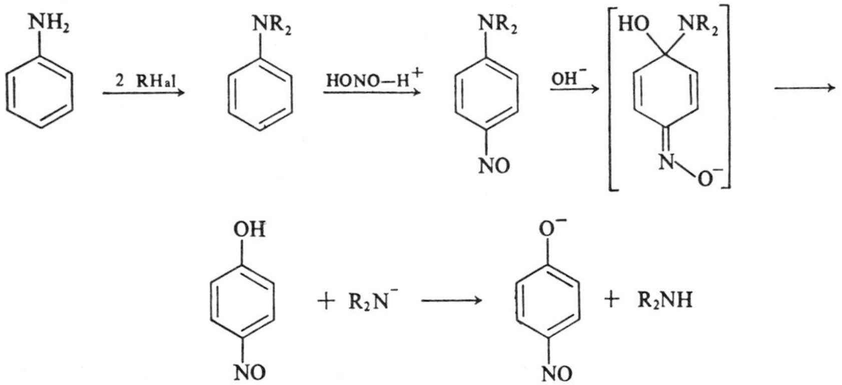

12.4 Substitution *via* Benzynes

(a) FORMATION OF BENZYNES

Treatment of chlorobenzene with sodamide in liquid ammonia at $-33°C$ gives aniline. A mechanistic study has shown that this reaction is entirely different from those nucleophilic substitutions so far described. For example, chlorobenzene labelled with ^{14}C at the position bearing chlorine gives an equimolar mixture of unrearranged and rearranged products, and this, together with other evidence, points to the mediation of a symmetrical species, benzyne, which is formed from chlorobenzene by an E2-type elimination and reacts with ammonia to give aniline.

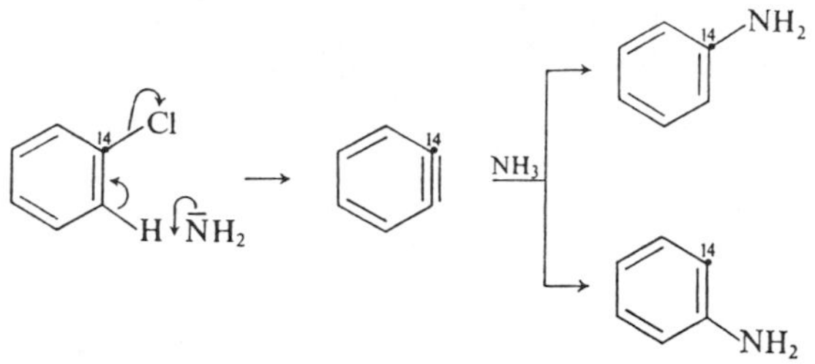

The ease of this reaction is due to the strong basicity of amide ion. Hydroxide ion, which is a much weaker base, reacts with chlorobenzene at about 340°C to give phenol, *via* benzyne. Bromobenzene and iodobenzene react similarly to chlorobenzene, but fluorobenzene does not give benzyne directly with base, differing from the other halobenzenes because of the greater strength of the C—F bond.

Two other methods for generating benzynes are known. First, *o*-halo-substituted organometallic compounds readily lose metal halide to give benzynes, e.g.

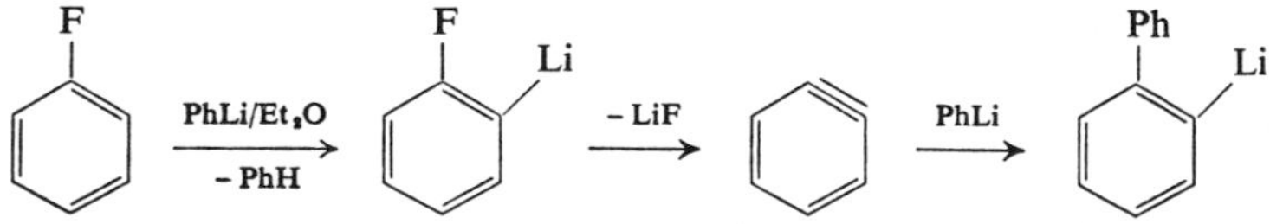

The resulting aryl-metal compound may be usefully employed in synthesis, e.g.

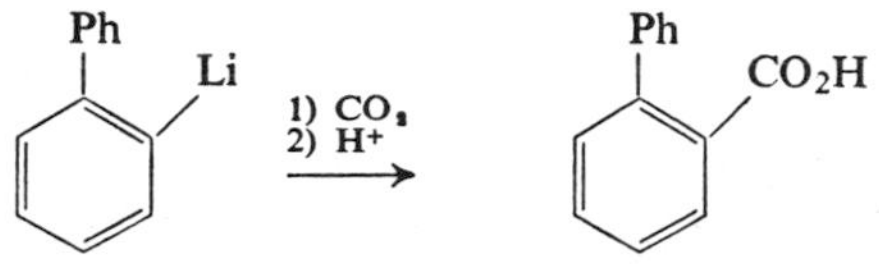

Biphenyl-2-carboxylic acid

Secondly, *ortho*-disubstituted benzenes from which two stable molecules can be formed by elimination give benzynes on thermolysis, e.g.

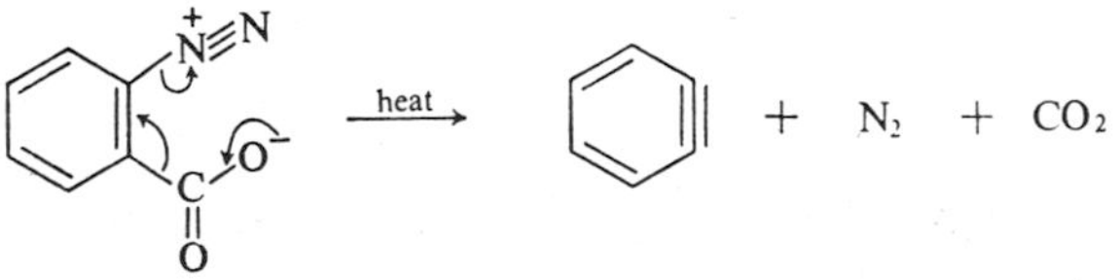

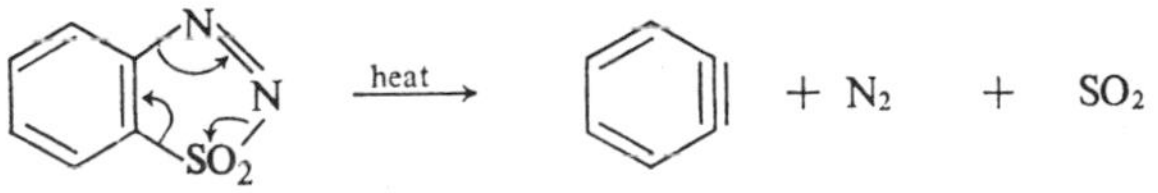

(b) REACTIONS OF BENZYNES

Benzynes may be regarded as highly strained acetylenes. They are too unstable to be isolated and react with any nucleophile which is present: e.g. when generated in liquid ammonia they react to give amines and when formed from organometallic compounds they react with a second molecule of the organometallic compound.

Another nucleophilic species may be specially introduced into the system. For example, the addition of malonic ester to liquid ammonia containing amide ion gives the corresponding carbanion which reacts with benzyne as it is generated.

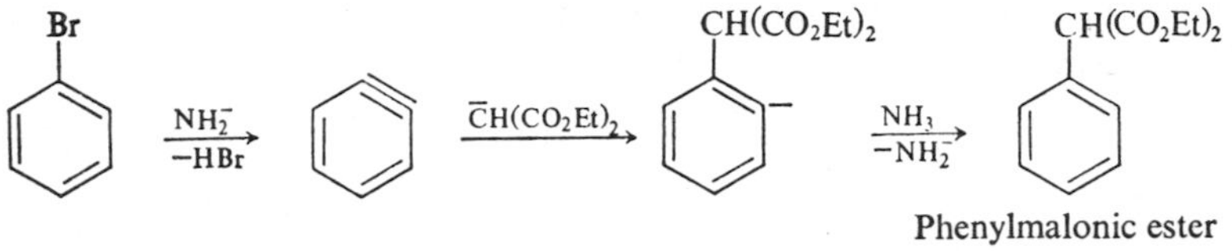

Phenylmalonic ester

One disadvantage attendant upon the use of benzynes in synthesis is that the nucleophile may react at either end of the triple bond of the benzyne, giving a mixture of two products if the benzyne is monosubstituted. For example, *p*-chlorotoluene with hydroxide ion at 340°C gives an approximately equimolar mixture of *m*- and *p*-cresol:

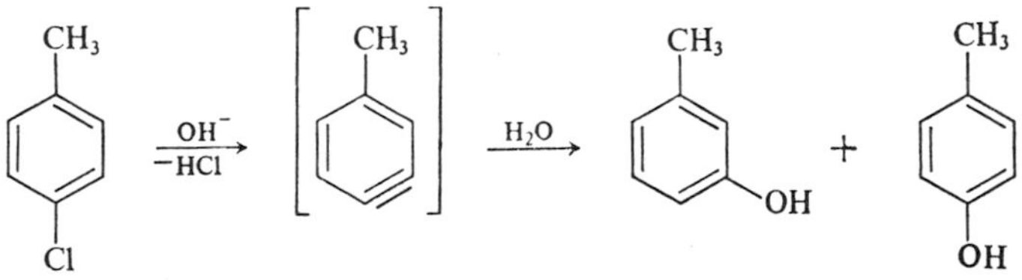

Fortunately, directive effects operate in certain cases: e.g. *m*-aminoanisole is the exclusive product from *o*-chloroanisole with sodamide in liquid ammonia. The direction of addition is such that the more stable of the two possible anions is formed.

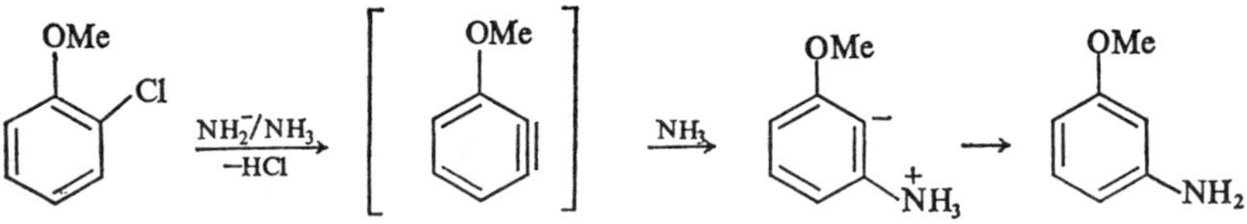

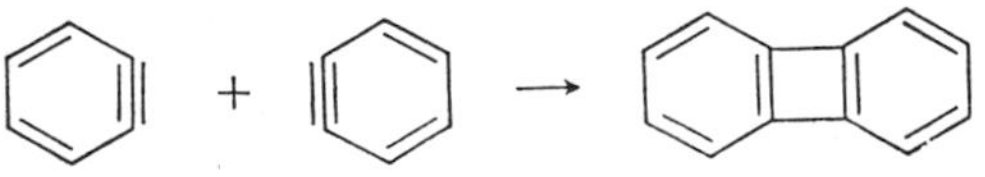

In the absence of nucleophiles, benzyne dimerizes to give biphenylene, and in the presence of dienes it reacts as a dienophile to give Diels-Alder adducts, e.g.

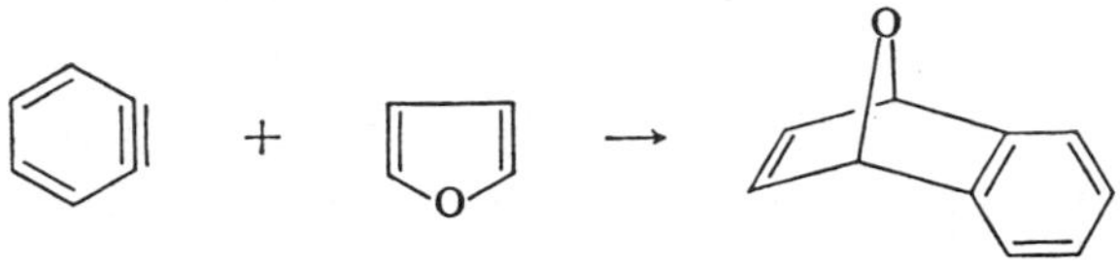

12.5 Bucherer Reaction

Certain phenols react with aqueous ammonium sulphite to give aromatic amines. Reaction occurs through the bisulphite adduct of the keto-tautomer of the phenol, e.g.

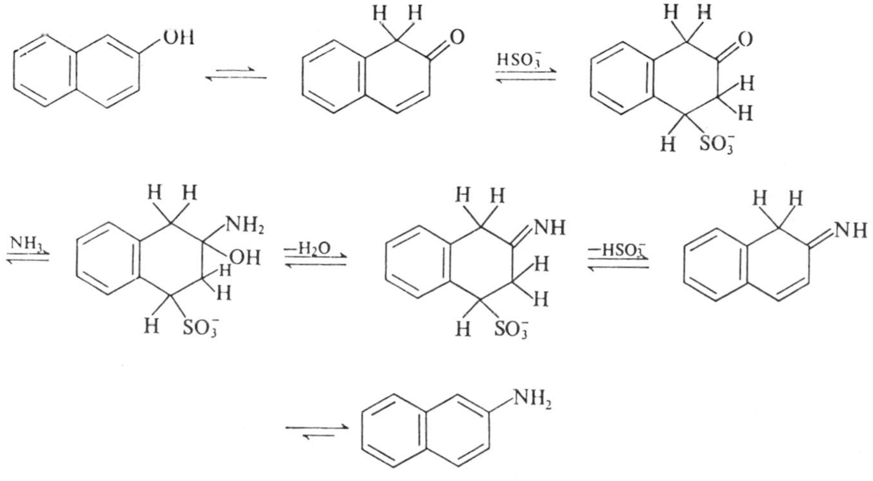

The reaction, named after Bucherer, is limited to those phenols which have a tendency to ketonize, such as α- and β-naphthol and resorcinol (p. 19). In the example above, β-naphthylamine may be obtained in 95% yield by carrying out the reaction at 150°C under pressure, and since β-naphthol may be readily obtained from naphthalene by high-temperature sulphonation (p. 408) followed by alkali-fusion (p. 429), this provides a method for obtaining β-substituted naphthalenes *via* diazotization of β-naphthylamine (p. 436).

The Bucherer reaction is reversible, and by careful control of the relative amounts of ammonia and water it may be used to convert amines into phenols.

Further Reading

BUNNETT, J. F., 'Mechanism and reactivity in aromatic nucleophilic substitution,' *Quarterly Reviews*, 1958, **12**, 1.

DRAKE, N. L., 'The Bucherer reaction,' *Organic Reactions*, 1942, **1**, 105.

HEANEY, H., 'The benzyne and related intermediates,' *Chemical Reviews*, 1962, **62**, 81.

Problems

1. Write the following compounds in order of their decreasing reactivity towards hydroxide ion: chlorobenzene; *m*-nitrochlorobenzene; *p*-nitrochlorobenzene; 2,4-dinitrochlorobenzene.

2. Account for the following:

    (*i*) Whereas aliphatic fluorides are less easily hydrolyzed than the corresponding chlorides, 2,4-dinitrofluorobenzene is more rapidly hydrolyzed than 2,4-dinitrochlorobenzene.

    (*ii*) Isoquinoline is more reactive towards nucleophiles at its 1-position than at its 3-position.

    (*iii*) The treatment of *o*-bromoanisole with sodamide in liquid ammonia gives mainly *m*-aminoanisole.

3. Give methods for carrying out the following conversions:

    (*i*) Naphthalene into β-naphthol.

    (*ii*) Chlorobenzene into 2,4-dinitrophenylhydrazine.

    (*iii*) Chlorobenzene into *p*-chlorophenol.

    (*iv*) Fluorobenzene into *o*-butylbenzoic acid.

    (*v*) Anisole into *p*-nitrophenetole (*p*-EtO—C_6H_4—NO_2).

    (*vi*) Pyridine into 2-aminopyridine.

    (*vii*) Isoquinoline into 1-phenylisoquinoline.

13. Aromatic Diazonium Salts

13.1 The Formation of Diazonium Ions

Primary amines react with nitrous acid in acidic solution to give diazonium ions, $R-\overset{+}{N}\equiv N$. The mechanism of reaction may be summarized as follows:

$$HO-NO + H^+ \rightleftharpoons H_2\overset{+}{O}-NO \rightleftharpoons H_2O + NO^+$$

$$R-NH_2 \xrightarrow{NO^+} R-\overset{+}{N}H_2-N=O \xrightarrow{-H^+} R-NH-N=O \xrightarrow{H^+}$$

$$R-NH-N=\overset{+}{O}H \xrightarrow{-H^+} R-N=N-OH \xrightarrow{H^+} R-\overset{..}{N}=\overset{+}{N}-\overset{-}{O}H_2 \xrightarrow{-H_2O} R-\overset{+}{N}\equiv N$$

The electrophilic nitrosonium ion reacts with the nucleophilic nitrogen of the amine, a series of prototropic shifts occurs, and finally water is eliminated.

This reaction scheme is common to both aliphatic and aromatic amines. However, the two groups differ in that aliphatic diazonium ions are too unstable to be isolated in the form of salts, decomposing into carbonium ions and nitrogen, whereas aromatic diazonium ions are moderately stable in aqueous solution at low temperatures when present with anions of low nucleophilic power and may also, in suitable cases, be isolated as solids. The difference in behaviour may be related to the fact that singly bonded carbon more readily tolerates a positive charge than unsaturated carbon (p. 104).

The diazotization of primary aromatic amines is normally carried out by adding an aqueous solution of sodium nitrite to a solution (or suspension) of the amine hydrochloride in an excess of hydrochloric acid which is cooled by an ice-bath. The rate of addition is controlled so that the temperature of the reaction remains below about 5°C, and addition is continued until the solution just contains an excess of nitrous acid.

Amines which are substituted in the nucleus with electron-attracting groups are less easy to diazotize because the nucleophilicity of the amino-nitrogen is reduced by the partial withdrawal of the unshared electron-pair into the nucleus, e.g.

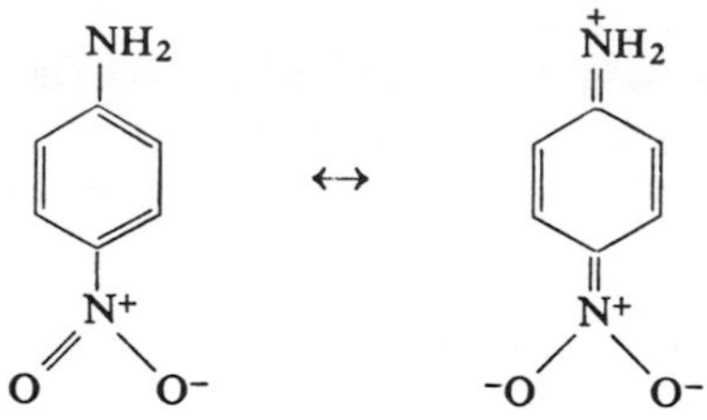

Acetic acid is usually a suitable reaction medium in these instances, and even 2,4,6-trinitroaniline may then be diazotized.

Most reactions employing diazonium salts may be conducted in solution and it is not normally necessary to isolate the solid salt. If necessary, however, two procedures may be employed. First, an aqueous solution of the diazonium salt, usually the chloride or sulphate, is prepared and treated with an alkali fluoroborate; the insoluble diazonium fluoroborate, $ArN_2^+BF_4^-$, is precipitated. Secondly, the amine hydrochloride is treated with an organic nitrite and acetic acid in ether; the diazonium salt, being insoluble in ether, is precipitated.

13.2 The Reactions of Diazonium Ions

Four types of reaction may be distinguished.

(a) REACTIONS OF NUCLEOPHILES AT NITROGEN

Although diazonium ions coexist in solution with stable anions (i.e. those from strong acids) such as chloride, less stable anions (i.e. those from weaker acids) react to give covalent diazo-compounds, e.g.

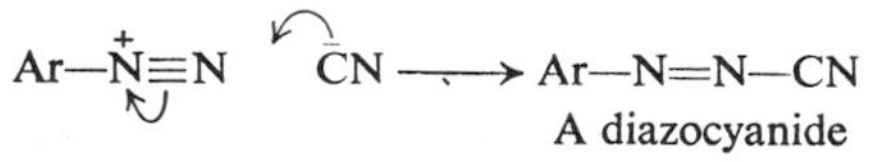

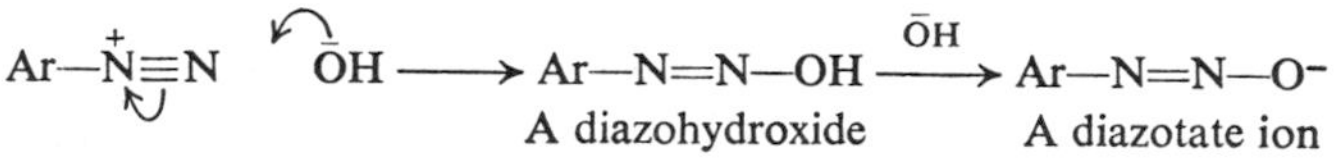

A diazocyanide

A diazohydroxide   A diazotate ion

(Note that these products exist in *cis* and *trans* forms.) Many of these diazo-compounds are unstable and decompose in solution to give aryl radicals (p. 441).

An important group of nucleophiles which react with diazonium ions consists of aromatic nuclei containing powerfully electron-releasing substituents such as the —O⁻ group in a phenoxide ion (13.4):

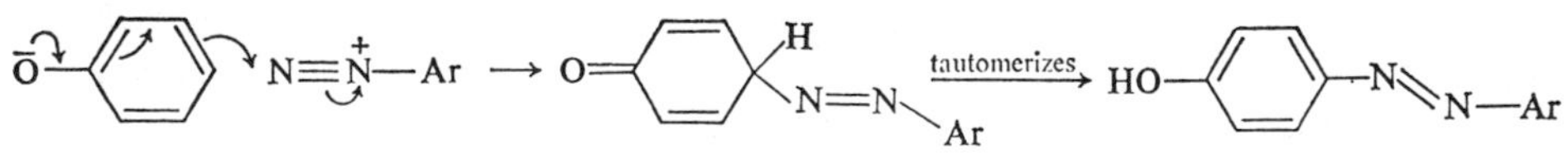

(b) THE S_N1 REACTION
Diazonium ions decompose on warming into aryl cations and nitrogen. The aryl cation is highly reactive and relatively unselective, being attacked rapidly by any nucleophile in its vicinity. Reaction in aqueous solution therefore leads to the formation of phenols:

$$ArN_2^+ \xrightarrow{-N_2} Ar^+ \xrightarrow{H_2O} Ar\overset{+}{-}OH_2 \xrightarrow{-H^+} ArOH$$

The formation of aromatic fluorides by the Schiemann procedure (p. 438) also occurs by the S_N1 mechanism.

(c) THE S_N2 REACTION
The more powerful nucleophiles displace nitrogen by the S_N2 mechanism described previously.

(d) ONE-ELECTRON REDUCTIONS
Diazonium ions are subject to one-electron reduction with the formation of an aryl radical and nitrogen.

$$ArN_2^+ \xrightarrow{e} Ar\cdot + N_2$$

Copper(I) ion is frequently used as the one-electron reducing agent. The aryl radical is highly reactive and is capable of abstracting a ligand from the transition metal ion (Sandmeyer reaction, p. 438) or a hydrogen atom from a covalent bond (p. 439).

The synthetic applications of aromatic diazonium salts are conveniently divided into those in which nitrogen is eliminated and those in which it is retained.

13.3 Reactions in which Nitrogen is Eliminated

(a) REPLACEMENT BY HYDROXYL
When a diazonium salt is warmed in water the corresponding phenol is formed by the S_N1 mechanism. The reaction is normally carried out in acidified solution in order to preserve the phenol in its unionized form, for otherwise a further reaction may occur between diazonium salt and phenoxide ion.

This procedure for phenols is more elaborate and often gives lower yields than that in which the sulphonic acid is fused with alkali (12.3c): starting from an aromatic hydrocarbon, the former process requires nitration, reduction of nitro to amino, diazotization, and hydrolysis, whereas the latter requires only sulphonation and alkali-fusion. However, the diazonium method can be used in circumstances in which the sulphonate method fails: for example, m-nitrophenol, which cannot be obtained from m-nitrobenzenesulphonic acid (p. 429), may be prepared from m-nitroaniline in over 80% yield [1].

(b) REPLACEMENT BY HALOGENS

The procedures used differ according to the halogen to be introduced.

Aromatic fluorides are prepared by the *Schiemann reaction*. An aqueous solution of the diazonium salt is treated with fluoroboric acid, precipitating the diazonium fluoroborate. This is dried and then heated gently until decomposition begins, after which reaction occurs spontaneously. In this way fluorobenzene may be obtained from aniline in about 55% yield [2]. The reaction involves the S_N1 mechanism:

$$ArN_2^+BF_4^- \xrightarrow[-N_2]{heat} Ar^+ \quad F\!-\!\bar{B}F_3 \longrightarrow ArF + BF_3$$

Aromatic chlorides and bromides may be obtained in an analogous manner from diazonium tetrachloroborates and tetrabromoborates, but since these compounds often decompose violently when heated it is in general safer to employ the *Sandmeyer reaction*. The procedure for chlorides consists of adding a cold aqueous solution of the diazonium chloride to a solution of copper(I) chloride in hydrochloric acid; a sparingly soluble complex separates which decomposes to the aryl chloride on being heated. The reaction involves one-electron reduction of the diazonium ion followed by ligand-transfer:

$$ArN_2^+ + CuCl_2^- \longrightarrow Ar\cdot + N_2 + CuCl_2$$

$$Ar\cdot \quad Cl\!-\!Cu\!-\!Cl \longrightarrow ArCl + CuCl$$

Aryl bromides may be prepared similarly from diazonium bisulphates and copper(I) bromide in hydrobromic acid. Typical conversions from the amines are between 70 and 80% for *p*-chlorotoluene and *p*-bromotoluene [1].

Aromatic iodides may be prepared without using a copper(I) salt. An aqueous solution of a diazonium salt is treated with potassium iodide and warmed, and the iodide is usually formed in good yield. For example, aniline gives about 70% of iodobenzene [2].

(c) REPLACEMENT BY CYANO

Aromatic nitriles are obtained in generally good yield by a Sandmeyer procedure using copper(I) cyanide in aqueous potassium cyanide solution. For example, *p*-toluidine gives 64–70% of *p*-tolunitrile [1]:

$$CH_3\!-\!\!\bigcirc\!\!-\!NH_2 \xrightarrow{HONO-H^+} CH_3\!-\!\!\bigcirc\!\!-\!N_2^+ \xrightarrow{CuCN} CH_3\!-\!\!\bigcirc\!\!-\!CN$$

(d) REPLACEMENT BY NITRO

Two general procedures are available for converting a diazonium salt into a nitro-compound.

(1) A neutral solution of the diazonium salt is treated with sodium cobaltinitrite and the resulting diazonium cobaltinitrite is decomposed by aqueous sodium nitrite in the presence of copper(I) oxide and copper(II) sulphate.

$$3\ ArN_2^+ + Co(NO_2)_6^{3-} \longrightarrow (ArN_2^+)_3Co(NO_2)_6^{3-} \xrightarrow[Cu_2O-CuSO_4]{NaNO_2} 3\ ArNO_2 + 3\ N_2$$

(2) A suspension of the diazonium fluoroborate in water is added to aqueous sodium nitrite solution in which copper powder is suspended. For example, p-dinitrobenzene may be obtained from p-nitroaniline in about 75% yield [2].

$$ArN_2^+Cl^- + HBF_4 \longrightarrow ArN_2^+BF_4^- + HCl$$

$$ArN_2^+BF_4^- + NaNO_2 \xrightarrow{(Cu)} ArNO_2 + N_2 + NaBF_4$$

(e) REPLACEMENT BY HYDROGEN
Two general methods may be used for the conversion $ArN_2^+ \to ArH$. In the first, the diazonium solution is warmed with ethanol:

$$ArN_2^+Cl^- + C_2H_5OH \longrightarrow ArH + N_2 + HCl + CH_3CHO$$

The reaction almost certainly involves the aryl radical.
 Yields are often low because of the competitive nucleophilic displacement,

$$ArN_2^+ \xrightarrow{-N_2} Ar^+ \xrightarrow[-H^+]{EtOH} ArOEt$$

although when electron-attracting groups are present in the aromatic nucleus the S_N1 heterolysis occurs less readily and the reduction pathway competes more effectively.
 Because of this disadvantage, the second method is now more commonly chosen. The reducing agent is hypophosphorous acid and reaction occurs at room temperature, when the competing S_N1 reaction is much slower. Copper(I) salts catalyze the reduction and a chain-mechanism operates:

$$\text{initiation: } ArN_2^+ + Cu^+ \longrightarrow Ar\cdot + N_2 + Cu^{2+}$$

$$\text{propagation: } \begin{cases} Ar\cdot + H{-}\overset{\overset{\displaystyle H}{|}}{\underset{\underset{\displaystyle OH}{|}}{P}}{=}O \longrightarrow ArH + \cdot\overset{\overset{\displaystyle H}{|}}{\underset{\underset{\displaystyle OH}{|}}{P}}{=}O \\[3em] ArN_2^+ + O{=}\overset{\displaystyle \cdot}{P}H(OH) \xrightarrow{H_2O} Ar\cdot + N_2 + H_3PO_3 + H^+ \end{cases}$$

The ease of removal of an amino group from an aromatic ring by diazotization and reduction is of considerable value in synthesis, for the amino group and its derivatives are powerfully activating substituents with well defined orienting properties in electrophilic reactions. The group may be introduced in order to promote the necessary reactivity and may later be removed. For example, 1,3,5-tribromobenzene may be made by the bromination of aniline followed by removal of the amino group:

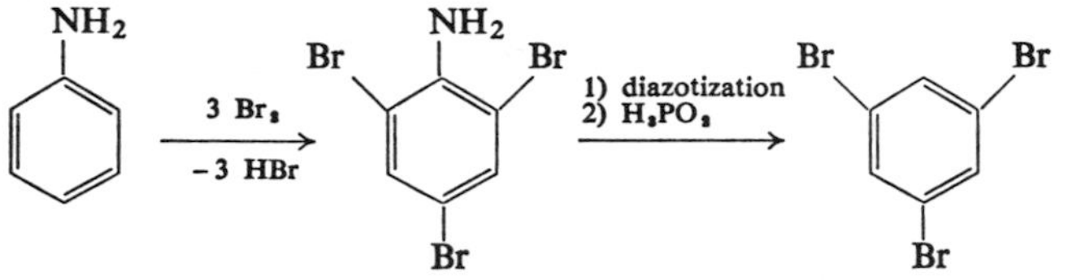

1,3,5-Tribromobenzene

A more elaborate pathway is employed to make *m*-bromotoluene. The nitration of toluene gives the *p*-nitro-derivative in about 40% yield. Reduction of nitro to amino and the acetylation of amino are nearly quantitative reactions, so that *p*-acetamidotoluene is easily available. Since the acetamido group is more strongly activating than the methyl group despite the steric hindrance to reaction at its *ortho*-position, bromination occurs largely *meta* to methyl. Hydrolysis and deamination give *m*-bromotoluene.

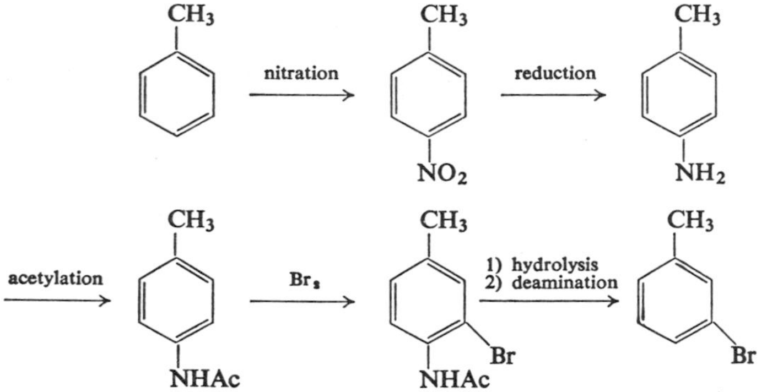

m-Bromotoluene

Despite the number of steps involved, this process is considerably more efficient than the direct bromination of toluene which gives less than 1% of the *meta*-derivative. Similar procedures may be used for preparing other *m*-haloalkylbenzenes (except fluoro compounds) and *m*-nitroalkylbenzenes.

(*f*) REPLACEMENT BY ALIPHATIC CARBON
Olefins in which the double bond is conjugated to a group which is capable of stabilizing a free radical react with diazonium salts in the presence of copper(II)

ion. Reaction is initiated by a trace of copper(I) ion and involves the copper(I)-copper(II) redox system, e.g.

$$PhN_2^+ + Cu^+ \longrightarrow Ph \cdot + N_2 + Cu^{2+}$$

$$Ph \cdot + CH_2{=}CH{-}C{\equiv}N \longrightarrow [Ph{-}CH_2{-}\overset{\cdot}{C}H{-}C{\equiv}N \leftrightarrow Ph{-}CH_2{-}CH{=}C{=}\overset{\cdot}{N}]$$

$$\xrightarrow{\;Cu^{2+},Cl^-\;} Ph{-}CH_2{-}\underset{\underset{Cl}{|}}{CH}{-}C{\equiv}N + Cu^+$$

The radical-adduct in some instances reacts by loss of hydrogen rather than by the uptake of halogen, e.g.

$$PhN_2^+ + CH_2{=}CHPh \xrightarrow{(Cu^{2+})} PhCH{=}CHPh + N_2 + H^+$$

The yields in these processes (the *Meerwein reaction*) are often low, but the method can provide a short route to a compound which is not otherwise readily accessible. Yields are improved by the presence of electron-releasing groups in the diazonium ion and electron-attracting groups in the olefin.

(g) REPLACEMENT BY AROMATIC CARBON
A number of procedures lead to the arylation of aromatic carbon by diazonium salts.

(*i*) *Gomberg reaction.* A two-phase liquid system consisting of an aqueous solution of the diazonium salt and an aromatic liquid, or a solution of an aromatic solid in an inert solvent, is treated with aqueous caustic soda. The covalent diazohydroxide is formed ($ArN_2^+Cl^- + NaOH \rightarrow Ar{-}N{=}N{-}OH + NaCl$) and gives aryl radicals by a complex mechanism similar to that for reaction of diazoacetates, below; these radicals react with the aromatic liquid. Yields are never high and are usually less than 40%, e.g.

$$Br{-}\langle\bigcirc\rangle{-}N_2^+ + PhH \xrightarrow{OH^-} Br{-}\langle\bigcirc\rangle{-}\langle\bigcirc\rangle$$

p-Bromobiphenyl
35%

When the aromatic compound is unsymmetrical, arylation gives a mixture of products. For example, benzenediazonium chloride reacts with nitrobenzene in Gomberg conditions to give a mixture containing all three nitrobiphenyls. The reaction is therefore more valuable when benzene itself or a symmetrical *para*-disubstituted benzene is involved. The mechanism of homolytic arylation, and the factors governing the orientation of the substitution, are discussed subsequently (17.3c).

(ii) Diazoacetates. Aromatic diazoacetates decompose to aryl radicals which bring about the homolytic arylation of aromatic compounds. The mechanism is complex:

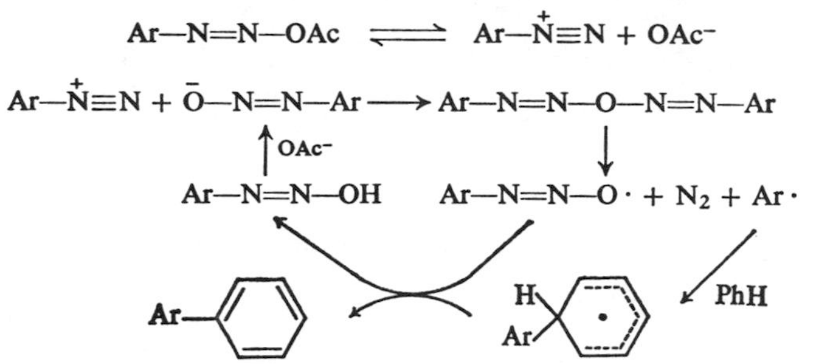

$$Ar-N=N-OAc \; \rightleftharpoons \; Ar-\overset{+}{N}{\equiv}N + OAc^-$$

$$Ar-\overset{+}{N}{\equiv}N + \bar{O}-N=N-Ar \longrightarrow Ar-N=N-O-N=N-Ar$$

Ar—N=N—OH   Ar—N=N—O· + N$_2$ + Ar·

It is usually more convenient to use an *N*-nitrosoacylarylamine rather than a diazoacetate as the starting material; the former rearranges to the latter on being heated:

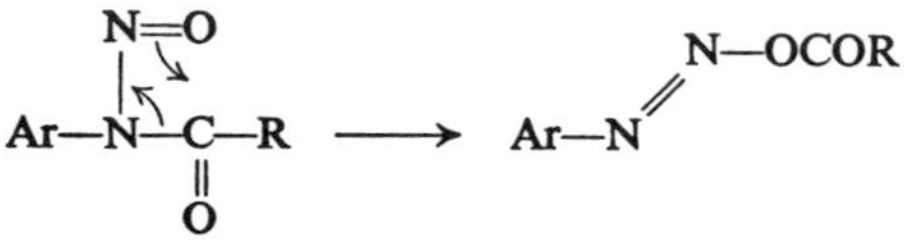

Ar—N—C—R ⟶ Ar—N

The use of either the Gomberg reaction or the diazoacetate method for the formation of biphenyls is normally less satisfactory than the aroyl peroxide method (17.3c). However, when the appropriate peroxide is difficult to obtain the other two methods provide practicable alternatives.

(iii) Pschorr reaction. A convenient synthesis of phenanthrene consists of condensing *o*-nitrobenzaldehyde with phenylacetic acid in the presence of acetic anhydride (Perkin reaction, p. 231), reducing the nitro-group to amino, diazotizing, treating the diazonium salt with copper powder, and finally decarboxylating.

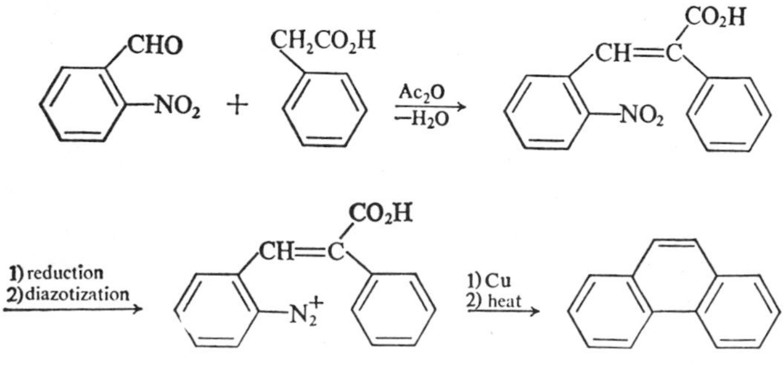

Phenanthrene

This procedure, originally applied by Pschorr to the synthesis of phenanthrene and its derivatives, has been extended to forming other aromatic systems; e.g. fluorenone may be obtained from 2-aminobenzophenone.

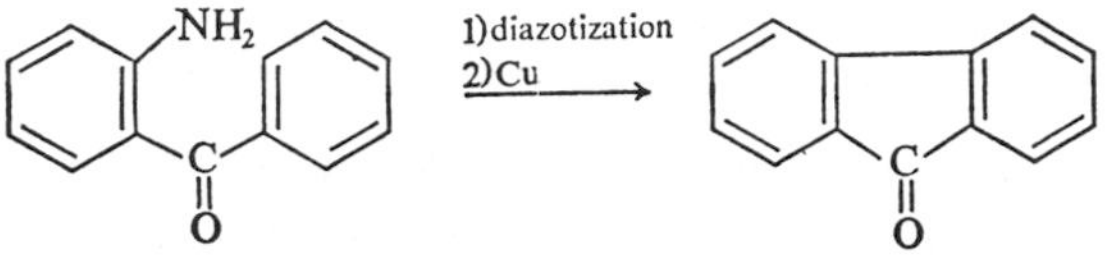

Fluorenone

Reaction in the presence of copper occurs *via* the aryl radical, but it is also possible to carry out the ring-closure in acid-catalyzed conditions in which the diazonium ion undergoes S_N1 heterolysis to an aryl cation which then effects cyclization as an electrophile.

(*iv*) *Reduction by copper*(I) *ammonium ion.* The addition of copper(I) ammonium hydroxide (obtained by treating copper(II) sulphate in ammonia with hydroxylamine) to diazotized anthranilic acid gives diphenic acid in about 90% yield:

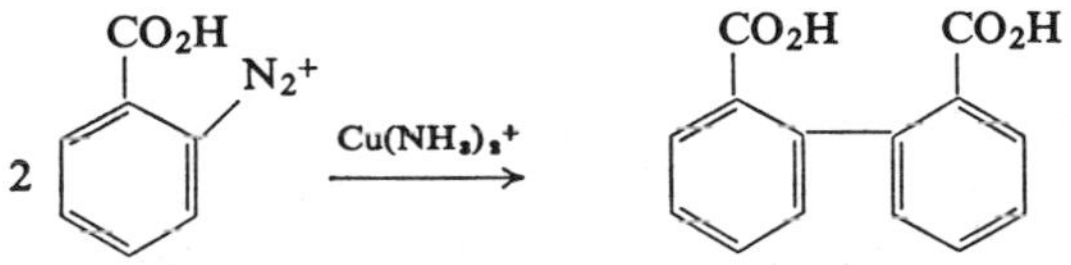

Diphenic acid

Reaction probably occurs by one-electron reduction of the diazonium ion followed by dimerization of the resulting aryl radicals. Nucleophilic substitutions of the Sandmeyer type which occur in the presence of copper(II) ion may here be inhibited because the copper(II) ion formed is bound as copper(II)-ammonium ion so that the ligand-transfer mechanism (p. 438) cannot operate.

Yields of biaryls are rarely as high as that for diphenic acid. Diazotized *o*-nitroaniline does not couple, and diazotized *m*-nitroaniline gives 3,3'-dinitrobiphenyl in 45% yield.

(*h*) REPLACEMENT BY OTHER SPECIES
Like iodide ion, other anions of strong nucleophilic power react directly with diazonium salts. Typical anions are azide ion, sulphydryl ion, disulphide ion, and ethyl xanthate ion. For example, diazotized anthranilic acid and sodium disulphide give dithiosalicylic acid [2] (Ar = *o*-carboxyphenyl),

$$2 \ ArN_2^+Cl^- + Na_2S_2 \longrightarrow ArSSAr + 2N_2 + 2NaCl$$

and diazotized *m*-toluidine and potassium ethyl xanthate give *m*-tolyl ethyl xanthate, hydrolysis of which gives *m*-thiocresol in about 70% yield [3] (Ar = *m*-tolyl).

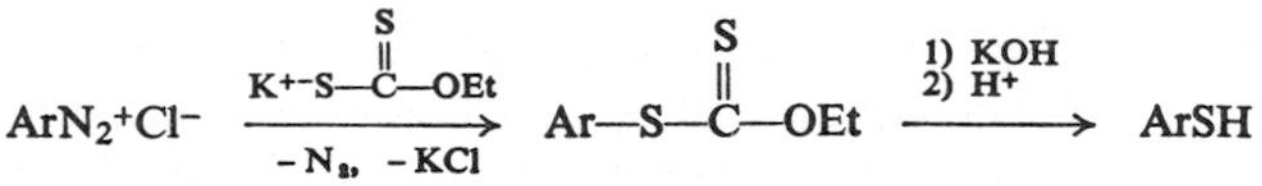

$$ArN_2^+Cl^- \xrightarrow[-N_2, \ -KCl]{K^+ -S-\overset{\overset{S}{\|}}{C}-OEt} Ar-S-\overset{\overset{S}{\|}}{C}-OEt \xrightarrow{\substack{1) \ KOH \\ 2) \ H^+}} ArSH$$

13.4 Reactions in which Nitrogen is Retained

The reaction of a nucleophile with the terminal nitrogen of a diazonium ion gives a covalent azo-compound. In many cases the products are very unstable and readily lose nitrogen (e.g. diazohydroxides, p. 441), but two synthetic procedures lead to retention of the nitrogen atoms.

(a) REDUCTION TO ARYLHYDRAZINES

Aromatic diazonium ions are reduced by sodium sulphite to arylhydrazines. The probable mechanism consists of the reaction of the diazonium ion with a sulphite anion to give a covalent azo-sulphite which, having a double bond conjugated to an electron-accepting group, adds a second nucleophilic sulphite ion (*cf.* Michael addition, p. 253). Hydrolysis yields the hydrazine.

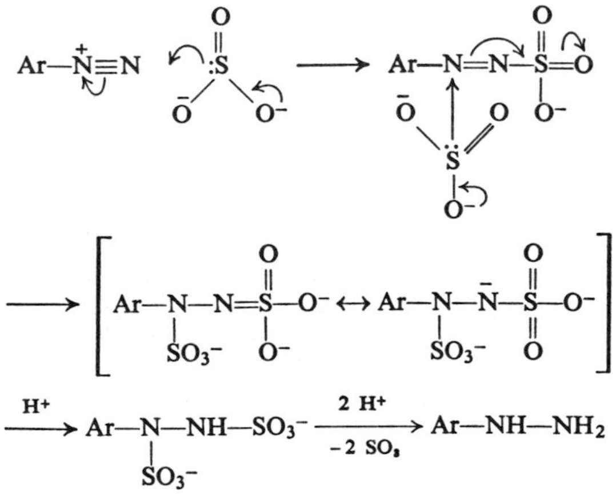

In this way phenylhydrazine may be obtained from aniline in over 80% yield [1].

Reduction to arylhydrazines is also brought about by tin(II) chloride and by electrolysis.

(b) COUPLING REACTIONS

The reactions of diazonium ions with aromatic nuclei are known as coupling reactions. Diazonium ions are weakly electrophilic and react, with some exceptions (p. 448), only with those aromatic compounds which are very powerfully activated towards electrophiles (*cf.* the nitrosonium ion, p. 405). These compounds include amines, phenols, and heterocyclic systems such as pyrrole.

(*i*) *Amines.* Tertiary aromatic amines react with diazonium ions almost exclusively at the *para* position. The mechanism is that common to other electrophilic aromatic substitutions, i.e. the diazonium ion adds to the aromatic compound and a proton is then eliminated:

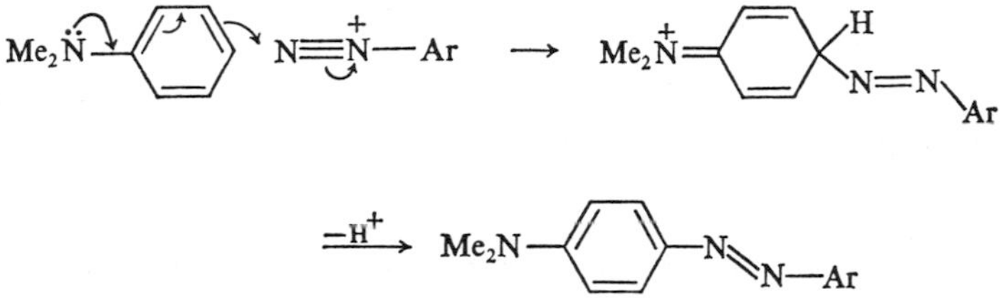

Careful control of the *p*H of the medium is necessary. In strongly alkaline conditions the diazonium ion is converted fairly rapidly into the covalent diazohydroxide, whereas in strongly acidic conditions the amine is converted largely into its unreactive conjugate acid. A *p*H in the region 4–10 is satisfactory, and a sodium acetate buffer is commonly used to maintain suitable conditions.

Primary and secondary aromatic amines usually react preferentially with diazonium ions at their nitrogen atoms, just as they do with the nitrosonium ion (pp. 405 and 435). For example, when aniline is only partially diazotized in hydrochloric acid, the addition of sodium acetate liberates the remaining aniline from its conjugate acid and coupling then occurs to give diazoaminobenzene in about 70% yield [2]:

$$PhNH_3^+ + AcO^- \longrightarrow PhNH_2 + HOAc$$

Diazoaminobenzene

These *N*-coupled products can be rearranged to *C*-coupled products by treatment with mineral acid: e.g. diazoaminobenzene gives *p*-aminoazobenzene. Reaction occurs by uncoupling followed by *C*-coupling:

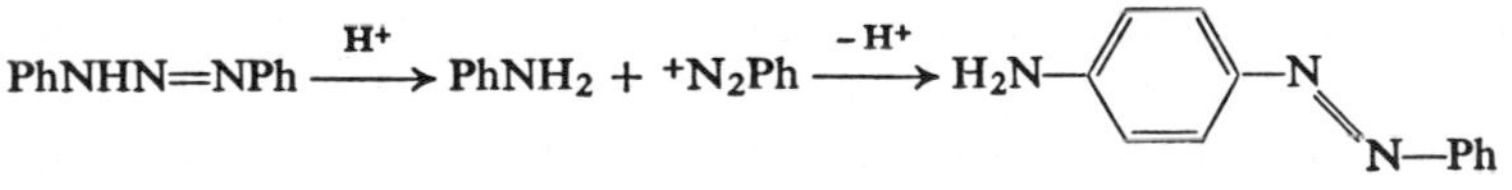

Direct *C*-coupling to primary and secondary amines occurs in two circumstances: first, when the diazonium ion is particularly reactive (e.g. *p*-nitrobenzenediazonium ion); and secondly, in aqueous formic acid solution, which is acidic enough for the *N*-coupled product to rearrange. When the structure and stereochemistry of the diazonium ion are suitable, *intra*molecular *N*-coupling can occur; for example, the diazotization of *o*-phenylenediamine leads directly to benzotriazole in over 75% yield [3]:

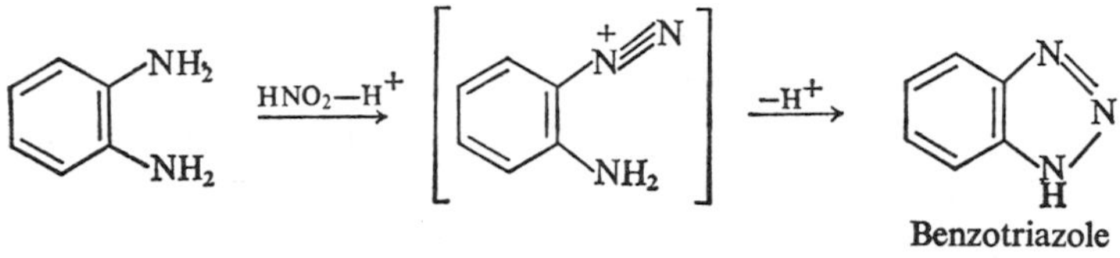

Benzotriazole

(*ii*) *Phenols.* The coupling of diazonium ions with phenols occurs *via* the phenoxide ions which are considerably more strongly activated to electrophiles than phenols themselves (p. 375). It is therefore necessary to carry out the reaction in alkaline solution, and precautions must be taken to prevent the conversion of the diazonium ion into the diazohydroxide. The optimum *p*H for coupling is about 9–10, but it is normally satisfactory to add an acidic solution of the aqueous diazonium salt to a solution of the phenol in sufficient alkali to neutralize the acids formed and to maintain suitable alkalinity; coupling occurs rapidly enough to prevent significant destruction of the diazonium ion.

The mechanism of coupling with phenoxide ions is analogous to that with amines and reaction is similarly directed mainly to the *para* position, e.g.

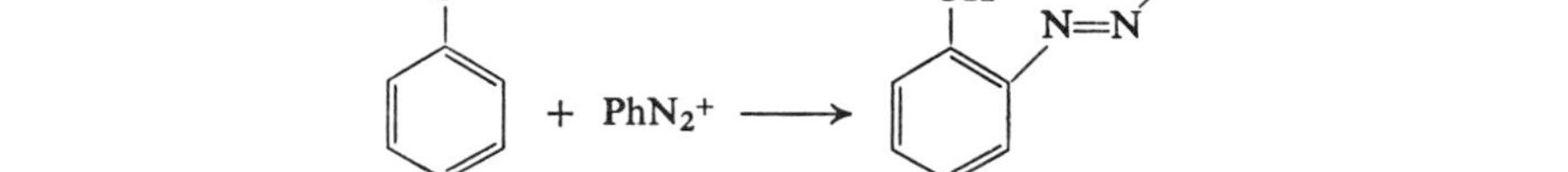

When the *para* position is already substituted, *ortho*-coupling occurs, e.g.

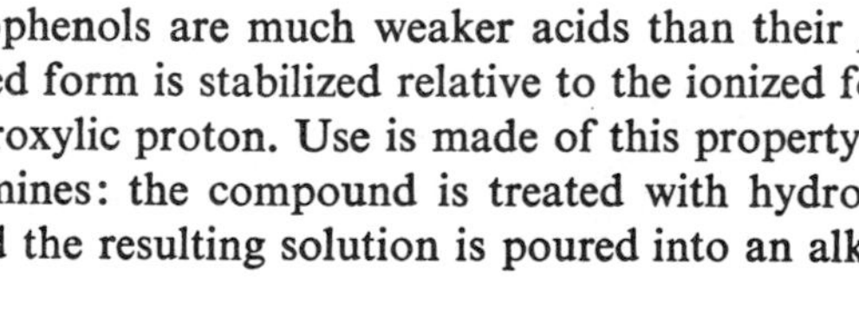

These *ortho*-azophenols are much weaker acids than their *para* isomers because the unionized form is stabilized relative to the ionized form by chelation involving the hydroxylic proton. Use is made of this property in detecting primary aromatic amines: the compound is treated with hydrochloric acid and sodium nitrite and the resulting solution is poured into an alkaline solution of

β-naphthol. Primary aromatic amines form diazonium ions which couple in the α-position of the β-naphtholate (p. 379) to give red azophenols which are insoluble in the alkaline medium.

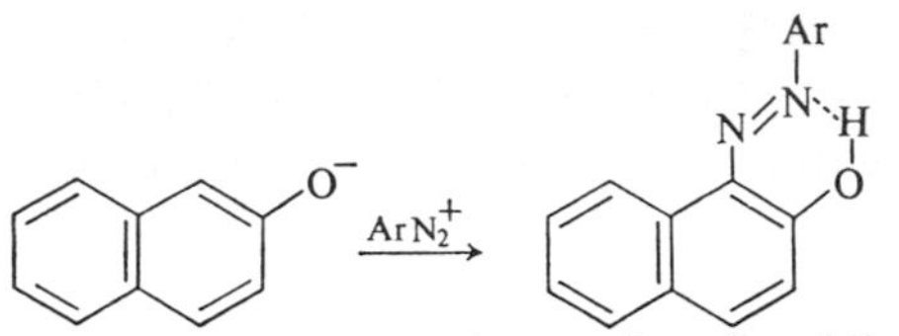

An α-arylazo-β-naphthol

(*iii*) *Enols.* Although olefins do not react with diazonium ions, enols, in which the hydroxylic oxygen activates the double bond to electrophiles (p. 108), do so. Reaction occurs more readily in alkaline conditions since the enolate anion is more reactive than the enol.

A typical example is the reaction of acetoacetic ester with benzenediazonium chloride, in a sodium acetate buffer, which gives ethyl α,β-dioxobutyrate α-phenylhydrazone in 98% yield. Reaction differs from those of phenols in that the initial azo-compound undergoes a tautomeric shift to give the arylhydrazone.

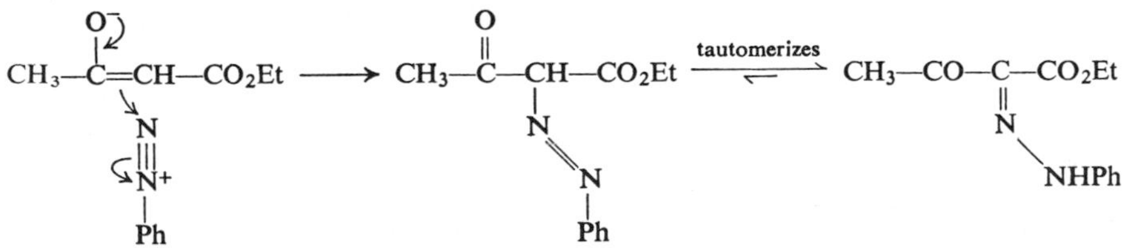

When the carbon atom at which coupling occurs does not possess a hydrogen atom so that the final prototropic shift cannot occur, the initial product readily undergoes C—C bond-fission to give an arylhydrazone (*Japp-Klingemann reaction*) e.g.

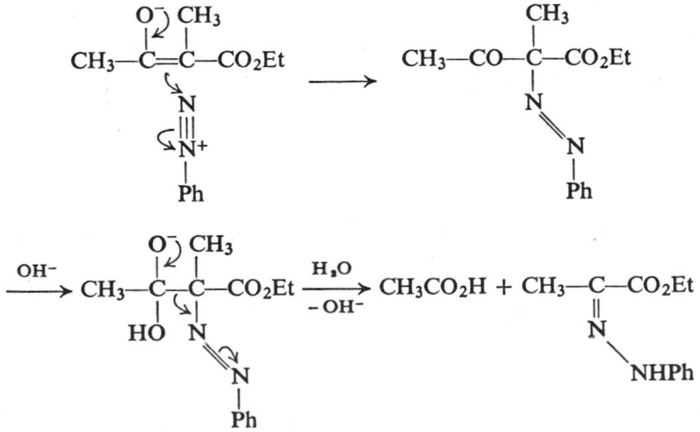

As expected for a reaction involving an enol, carbonyl-containing compounds react both in the normal way and in the Japp-Klingemann manner with decreasing ease as they become decreasingly enolic; for example, acetophenone and its derivatives do not couple with diazonium ions. As usual, however, intramolecular reactions occur more readily than the intermolecular analogues if the stereochemistry is suitable: e.g. diazotization of *o*-aminoacetophenone leads to 4-hydroxycinnoline:

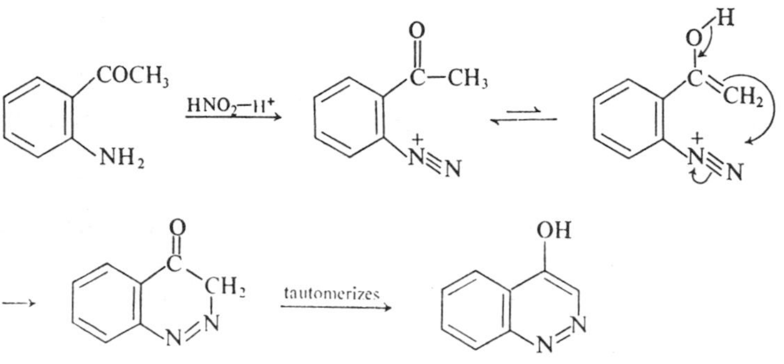

4-Hydroxycinnoline

(*iv*) *Pyrroles.* Pyrrole, whose reactivity towards electrophiles is comparable with that of phenol (p. 381), couples with diazonium salts mainly in the α-position:

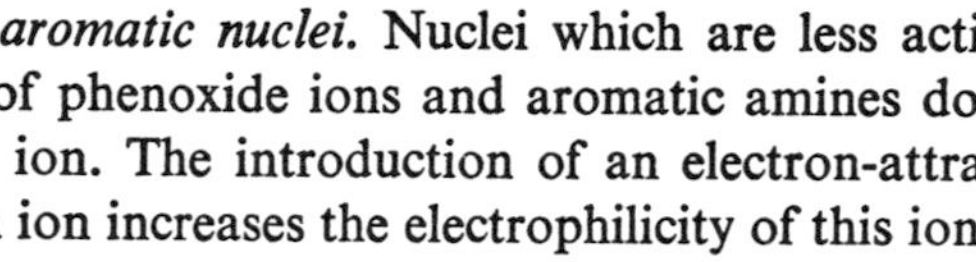

If both α-positions are substituted, coupling occurs at a β-position.

(*v*) *Less activated aromatic nuclei.* Nuclei which are less activated to electrophiles than those of phenoxide ions and aromatic amines do not couple with benzenediazonium ion. The introduction of an electron-attracting substituent into the diazonium ion increases the electrophilicity of this ion and less reactive nuclei then couple. For example, 2,4-dinitrobenzenediazonium ion is reactive enough to couple with anisole, and 2,4,6-trinitrobenzenediazonium ion couples even with mesitylene:

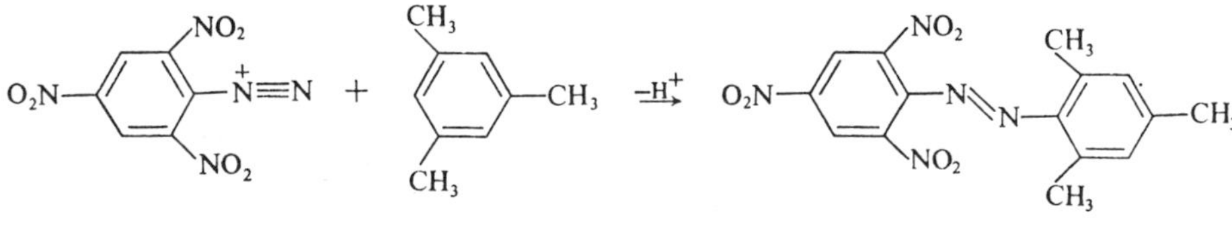

13.5 The Synthetic Value of Diazo-coupling

(a) DYE-STUFFS

Aromatic azo-compounds are all strongly coloured (e.g., azobenzene is orange-red, λ_{max} 448 nm) and many of those prepared by the coupling reaction are used as dye-stuffs. Three classes of dye may be distinguished.

First, neutral azo-compounds are used as *azoic combination* (or *ingrain*) dyes, that is, they are formed *in situ* in the fibre. An example is Para Red, from β-naphthol and *p*-nitrobenzenediazonium ion:

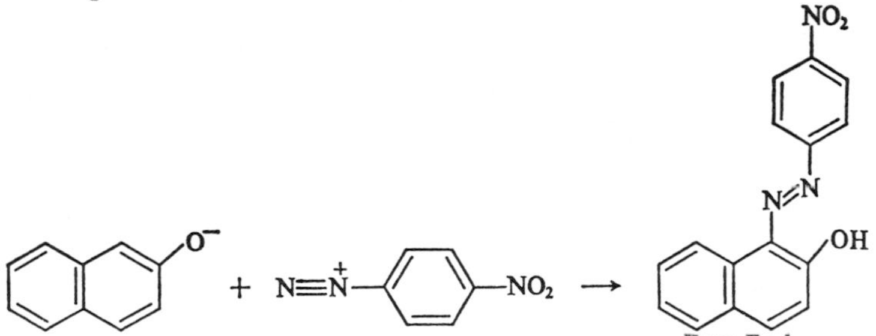

Para Red

Secondly, azo-compounds which contain either a sulphonic acid group or an amino group are adsorbed directly onto the fibre from aqueous solution. Examples are Orange II, an acidic dye, and Bismarck Brown R, a basic dye:

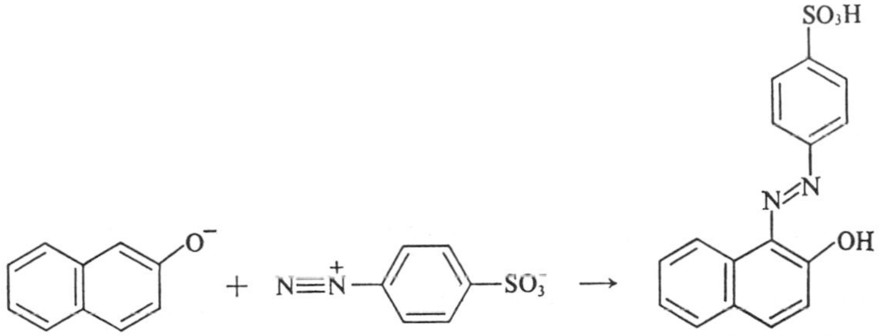

Orange II

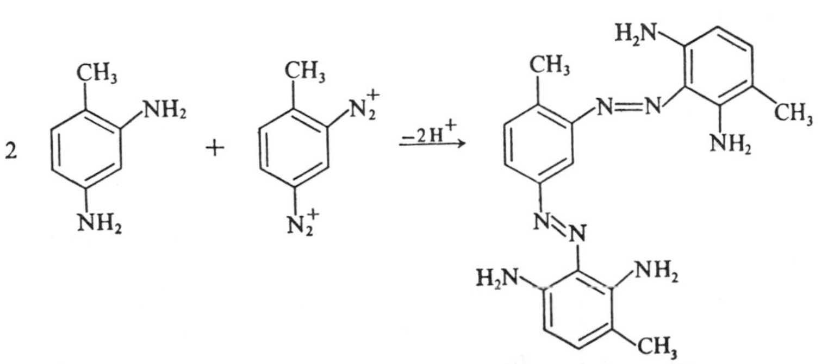

Bismarck Brown R

Thirdly, azo-compounds which contain groups capable of chelation with a metal ion are used as *mordant* dyes: the metal ion is adsorbed on the fibre and binds the dye-stuff. An example is Alizarin Yellow R, which forms chelates by means of its phenolic and carboxyl groups. Aluminium and chromium oxides are commonly used as mordants.

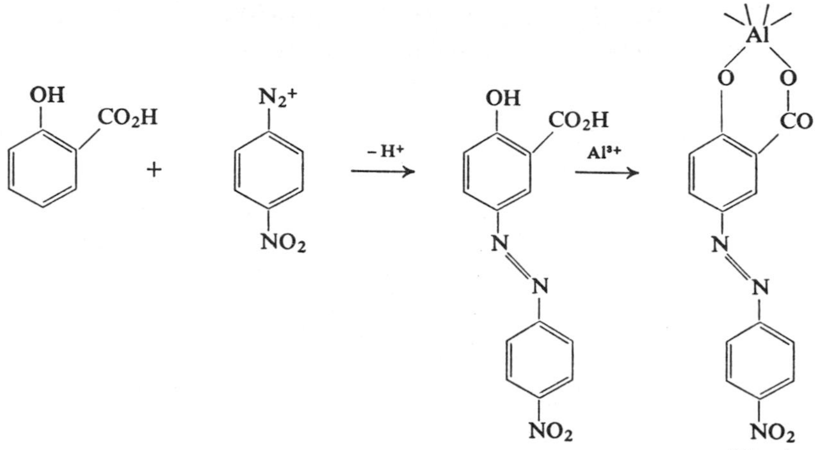

Alizarin
Yellow R

(b) INDICATORS

Azo-compounds which contain both an acidic and a basic group may be used as indicators since the colours of the conjugate base and the conjugate acid are invariably different. Examples are Methyl Red, made from diazotized anthranilic acid and dimethylaniline in about 64% yield [1] and Methyl Orange, from diazotized sulphanilic acid and dimethylaniline.

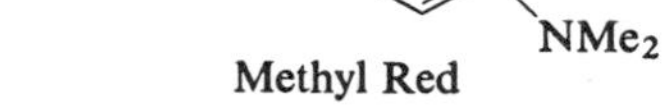

Methyl Red 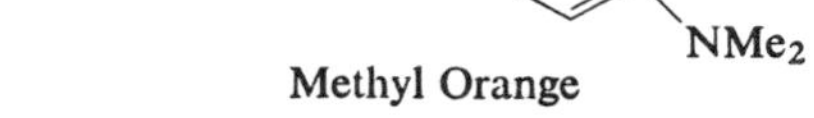   Methyl Orange

(c) SYNTHESIS OF AMINES

Azo-compounds are susceptible to hydrogenolysis (reductive cleavage, p. 625), giving amines. Sodium dithionite is usually employed as the reducing agent, but catalytic methods may be used. For example, 4-amino-1-naphthol is obtained in about 70% yield from α-naphthol by coupling with benzenediazonium chloride followed by reduction with dithionite [1]:

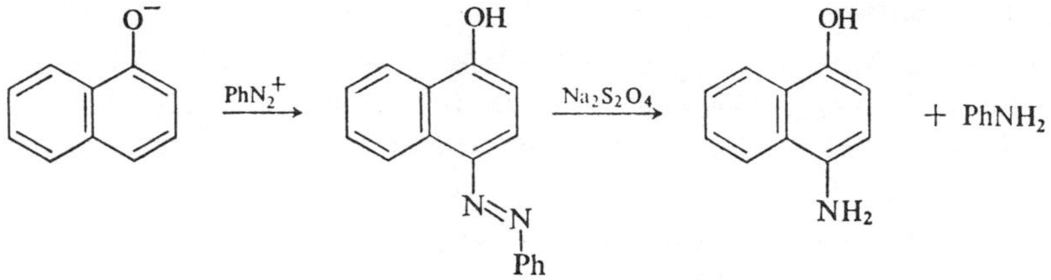

4-Amino-1-naphthol

Aliphatic amines may be obtained similarly by coupling to an enol followed by reduction.

This method for preparing amines has the advantage of occurring in mild conditions and it is therefore of value in some cases as an alternative to the normal method for making aromatic amines, by nitration followed by reduction, which involves vigorous and strongly oxidizing conditions. An interesting example occurs in a synthesis of adenine (p. 699).

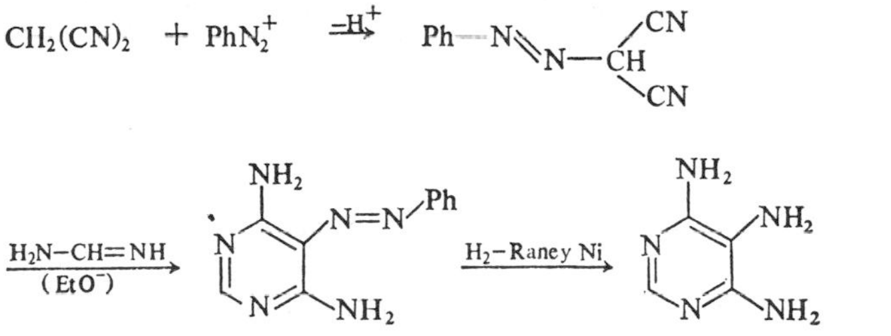

4,5,6-Triaminopyrimidine

(d) SYNTHESIS OF QUINONES

The *ortho* and *para* diamines and aminophenols are readily oxidized to quinones. Both classes of compound are available by diazo-coupling followed by reduction, providing a useful route to quinones, e.g.

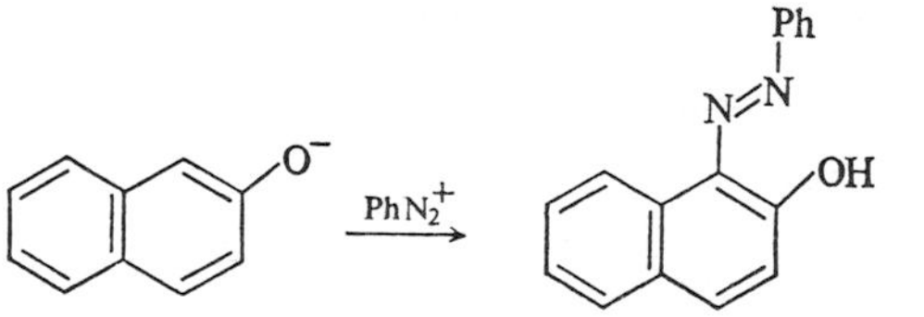

1-Phenylazo-2-naphthol

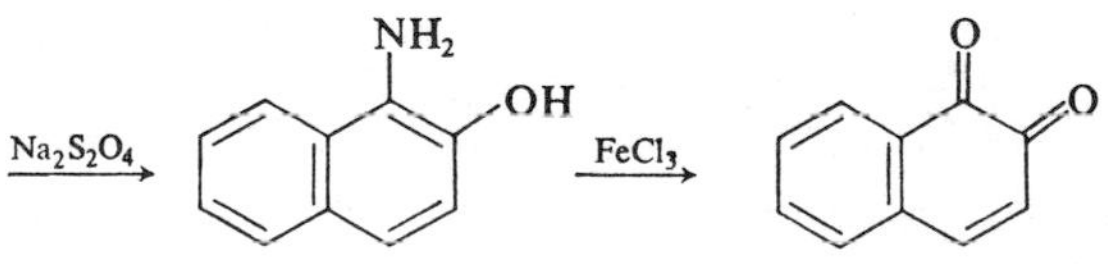

1,2-Naphthoquinone

Further Reading

DETAR, D. F., 'The Pschorr synthesis and related diazonium ring-closure reactions,' *Organic Reactions*, 1957, **9**, 409.

KORNBLUM, N., 'Replacement of the aromatic primary amino group by hydrogen,' *Organic Reactions*, 1944, **2**, 262.

PHILLIPS, R. R., 'The Japp-Klingemann reaction,' *Organic Reactions*, 1959, **10**, 143.

RIDD, J. H., 'Nitrosation, diazotization, and deamination,' *Quarterly Reviews*, 1961, **15**, 418.

ROE, A., 'Preparation of aromatic fluorine compounds from diazonium fluoborates: the Schiemann reaction,' *Organic Reactions*, 1949, **5**, 193.

RONDESTVEDT, C. S., 'Arylation of unsaturated compounds by diazonium salts,' *Organic Reactions*, 1960, **11**, 189.

Problems

1. Outline routes to the following compounds, starting with any monosubstituted benzene:

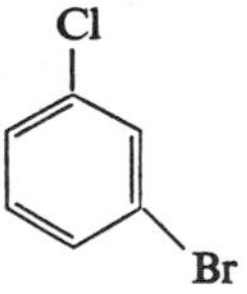

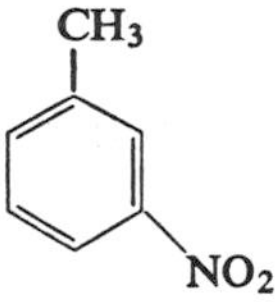

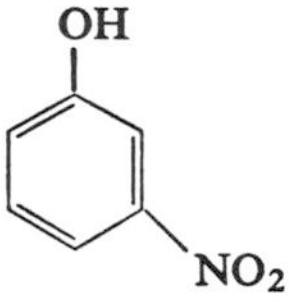

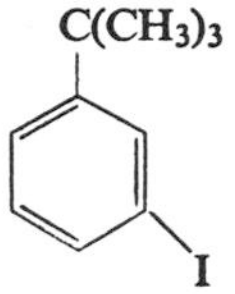

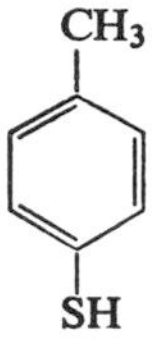

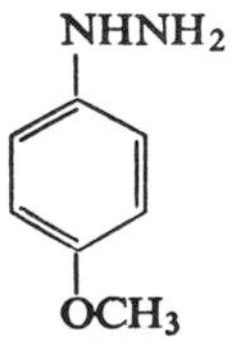

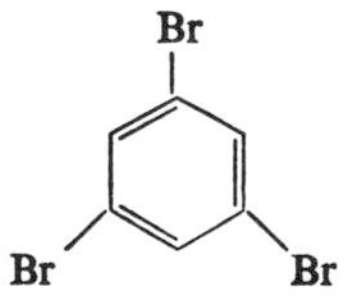

(m), (n), 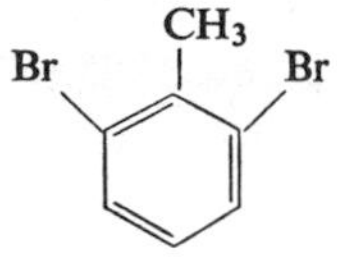 (o)

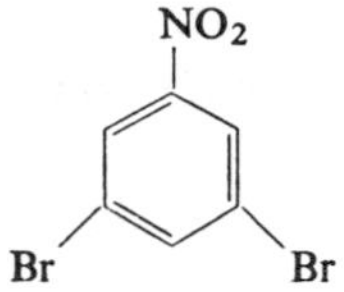

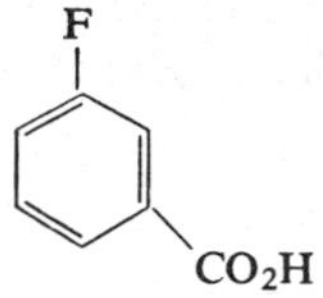

(p), 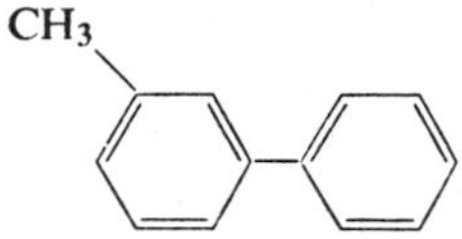 (q)

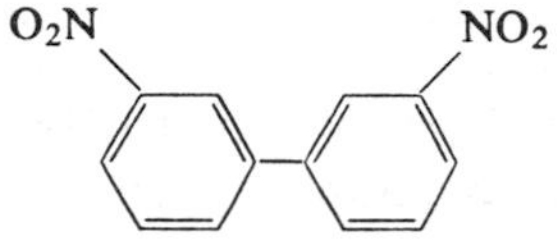

2. How would you obtain the following compounds?

(a) CH_3—CO—C—CO—CH_3
       $\overset{\|}{N}$
       NHPh

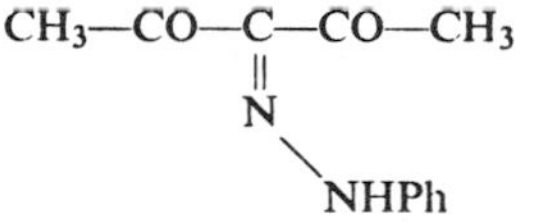

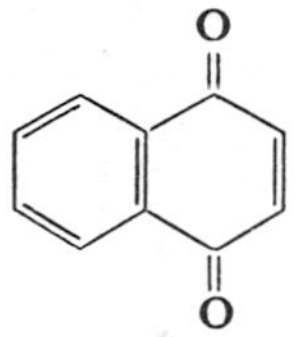

(b)

(c) Me_2N—◯—N≈N—◯—SO_2OH

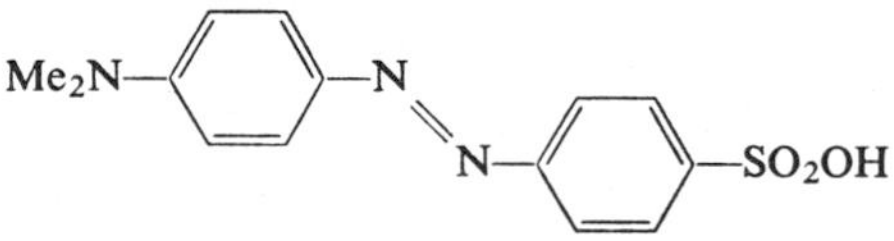

(d)

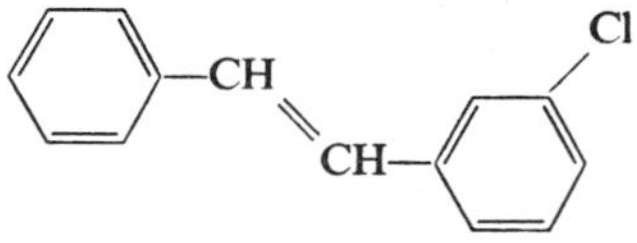

3. Account for the following observations:

(i) Although NN-dimethylaniline couples with benzenediazonium chloride, its 2,6-dimethyl derivative does not.

(ii) 2,4-Dinitrobenzenediazonium chloride couples with anisole, although benzenediazonium chloride does not.

(iii) When p-chloroaniline is diazotized with sodium nitrite and hydrobromic acid, the resulting diazonium salt solution couples with NN-dimethylaniline to give largely 4-bromo-4'-dimethylaminoazobenzene.

14. Molecular Rearrangements

14.1 Types of Rearrangement

Although in the majority of reactions of organic compounds the basic skeleton of the molecule remains intact, there are many in which a skeletal rearrangement occurs. Some of these processes are of value in synthesis, and the main emphasis in this Chapter will be on this aspect of rearrangements. In addition, however, rearrangements can be a disadvantage if they occur during the course of operations designed only to effect other changes of functionality, so that it is important to be familiar with the structural features and reaction conditions which lead to rearrangement.

Rearrangements are of two types: intramolecular processes, in which the group which migrates does not become completely detached from the system in which rearrangement is occurring; and intermolecular processes, in which the migrating group is first detached and later re-attached at another site. The latter group may be regarded as elimination-addition processes and, except in the aromatic series, will not be discussed here; examples include the prototropic and anionotropic shifts which have been discussed earlier, e.g.

$$CH_3-\underset{\underset{OH}{|}}{CH}-CH=CH_2 \underset{\rightleftharpoons}{\overset{H^+}{}} CH_3-CH=CH-\underset{\underset{OH}{|}}{CH_2}$$

Intramolecular rearrangements are conveniently subdivided into those which occur in electron-deficient systems and those which occur in electron-rich systems. Rearrangements of free radicals are also known and are discussed separately (p. 539); they have little synthetic value. Rearrangements within aromatic nuclei are grouped together in the final section.

14.2 Rearrangement to Electron-deficient Carbon

These reactions are classified according to the nature of the group which migrates.

(a) CARBON MIGRATION

(i) *Wagner-Meerwein rearrangement*. One of the simplest systems within which carbon migrates, with its bonding-pair, to an electron-deficient carbon atom is the neopentyl carbonium ion:

$$CH_3-\underset{\underset{CH_3}{|}}{\overset{\overset{CH_3}{|}}{C}}-CH_2^+ \longrightarrow CH_3-\underset{\underset{CH_3}{|}}{\overset{\overset{CH_3}{|}}{\overset{+}{C}}}-CH_2$$

All reactions which give rise to this ion give products derived from the re-arranged ion, the t-pentyl carbonium ion. For example, the solvolysis of neo-pentyl bromide in ethanol gives a mixture of trimethylethylene and ethyl t-amyl ether:

$$CH_3-\underset{\underset{CH_3}{|}}{\overset{\overset{CH_3}{|}}{C}}-CH_2Br \longrightarrow CH_3-\underset{\underset{CH_3}{|}}{\overset{\overset{CH_3}{|}}{C}}-CH_2^+ \longrightarrow CH_3-\underset{\underset{CH_3}{|}}{\overset{\overset{CH_3}{|}}{\overset{+}{C}}}-CH_2 \longrightarrow \begin{cases} \underset{CH_3}{\overset{CH_3}{\diagdown}}C=C\underset{H}{\overset{CH_3}{\diagup}} \\ (CH_3)_2C-CH_2CH_3 \\ \qquad\quad | \\ \qquad\quad OEt \end{cases}$$

The driving force for the rearrangement resides in the greater stability of a tertiary than a primary carbonium ion. Rearrangement is especially favourable under these circumstances, and less so when a secondary carbonium ion is formed.

In alicyclic systems, the relief of strain can provide a powerful driving force for rearrangement. For example, the addition of hydrogen chloride to α-pinene gives the rearranged product, bornyl chloride; the strained four-membered ring in the carbonium ion expands to the less strained five-membered analogue, despite the fact that the former contains a tertiary and the latter a secondary carbonium ion.

α-Pinene   Bornyl chloride

The other principal features of these migrations are as follows.

1. The carbonium ion may be generated in a variety of ways.

(1) From a halide, by using a strongly ionizing solvent or by adding a Lewis acid such as silver ion or mercury(II) chloride, which aids carbonium-ion forma-tion by abstracting the halide:

$$Me_3C-CH_2\overset{\frown}{-Cl} \quad Ag^+ \longrightarrow Me_3C-\overset{+}{C}H_2 + AgCl$$

(2) From an alcohol, by treatment with acid to promote heterolysis:

$$Me_3C-CH_2-OH \underset{}{\overset{H^+}{\rightleftharpoons}} Me_3C-CH_2-\overset{+}{O}H_2{}^+ \xrightarrow{-H_2O} Me_3C-\overset{+}{C}H_2$$

(3) From an alcohol, by conversion into a derivative which provides a stable leaving group such as the toluene-*p*-sulphonate:

$$Me_3C-CH_2-OSO_2C_6H_4Me \longrightarrow Me_3C-\overset{+}{C}H_2 + \overset{-}{O}SO_2C_6H_4Me$$

(4) From an amine, by treatment with nitrous acid; reaction occurs *via* the aliphatic diazonium ion from which molecular nitrogen is rapidly lost:

$$Me_3C-CH_2-NH_2 \xrightarrow{HNO_2-H^+} Me_3C-CH_2-\overset{+}{N}\equiv N \xrightarrow{-N_2} Me_3C-\overset{+}{C}H_2$$

(5) From an olefin, by protonation, e.g.

$$Me_3C-CH=CH_2 \xrightarrow{H^+} Me_3C-\overset{+}{C}H-CH_3$$

2. Hydrogen can also migrate in these systems. For example, reactions which occur through the isobutyl carbonium ion yield mainly products derived from the t-butyl carbonium ion:

$$\overset{\displaystyle H}{\underset{\displaystyle}{Me_2C}}-\overset{+}{C}H_2 \longrightarrow Me_2\overset{+}{C}-CH_3$$

A typical example of a hydride shift occurs in the reaction of a primary aliphatic amine with nitrous acid; e.g. n-propylamine gives isopropanol, together with propylene, and only a trace of n-propanol:

$$CH_3-CH_2-CH_2-NH_2 \xrightarrow{HNO_2-H^+} CH_3-CH_2-CH_2-\overset{+}{N}\equiv N$$

$$\xrightarrow{-N_2} CH_3-CH_2-\overset{+}{C}H_2 \longrightarrow CH_3-\overset{+}{C}H-CH_3 \xrightarrow[-H^+]{H_2O} \underset{\displaystyle OH}{CH_3-CH-CH_3} + CH_3-CH=CH_2$$

The occurrence of rearrangement in the conversion of an amine into an alcohol with nitrous acid renders this process of little synthetic value. It is necessary instead to choose a reaction for making the alcohol which has S_N2 character so that a carbonium-ion intermediate is not formed (e.g. hydrolysis of the corresponding halide).

3. Aryl groups have a far greater migratory aptitude than alkyl groups or hydrogen. For example, neophyl chloride undergoes solvolysis with rearrange-

ment many thousands of times faster than neopentyl chloride in the same conditions. This is ascribed to the fact that, whereas the rate-determining step in the reaction of neopentyl chloride is the formation of the high-energy primary carbonium ion, that in the reaction of neophyl chloride is the formation of a lower-energy bridged phenonium ion:

Neophyl chloride

The aryl group is said to provide *anchimeric assistance* to the reaction (p. 129). *

The phenonium ion is similar in structure to the intermediate in an electrophilic aromatic substitution and understandably, therefore, electron-releasing groups in the aromatic ring (e.g. p-OCH$_3$) give rise to greater rates of migration and electron-attracting groups (e.g. p-Cl) to lower rates.

It should be noted, however, that the enormously greater tendency for rearrangement of aryl groups than of alkyl groups or hydrogen does not apply to deaminations. For example, the treatment of 3-phenyl-2-butylamine with nitrous acid in acetic acid gives comparable amounts of the acetates derived from the migration of phenyl, methyl, and hydrogen,

whereas the solvolysis of the corresponding tosylate in acetic acid gives only the product derived from the migration of phenyl, CH$_3$CH(OAc)—CHMePh. This is because the tendency for nitrogen to depart from the diazonium ion is so great that anchimeric assistance is not required in order to help to 'push off' the leaving group.

4. The rearrangement is stereospecific: the migrating group approaches the electron-deficient carbon atom from the direction opposite to that in which the departing group is moving, just as in the S$_N$2 reaction. Inversion of configuration therefore occurs at the electron-deficient carbon.

This stereospecific requirement of the reaction, together with the *trans*-stereospecificity of eliminations (p. 176), has significant consequences in alicyclic

*If the rate-determining step was the formation of the primary carbonium ion, PhC(CH$_3$)$_2$— CH$_2$$^+$, reaction would occur less rapidly than on neopentyl chloride because the phenyl group is electron-attracting relative to methyl.

chemistry. For example, the treatment of cyclohexanol in acidic dehydrating conditions results in elimination; reaction occurs on the conformational isomer in which hydrogen is *trans* to the axial hydroxyl:

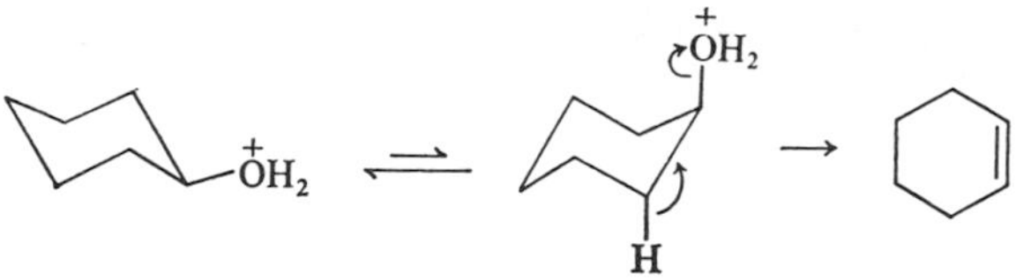

However, in the *trans*-decalin derivative below, the hydroxyl group is held in the equatorial position because the ring system cannot flap (p. 172). In this situation there is no hydrogen *trans* to the hydroxyl, but two of the ring carbon atoms are in the appropriate *trans* position for rearrangement, and in the presence of acid ring-contraction takes place:*

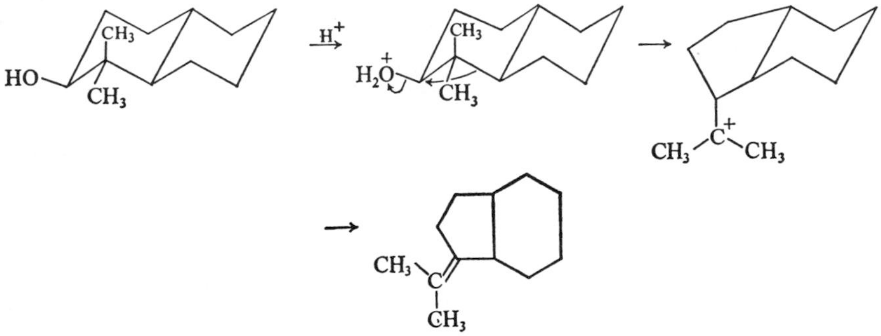

Particular attention must be paid to the possibility of rearrangements such as this in transformations involving steroids and terpenes which contain decalin systems.

5. Rearrangements in bicyclic systems are particularly common, as in the conversion of camphene hydrochloride into isobornyl chloride, catalyzed by Lewis acids. Note that the final uptake of chloride ion also occurs in the *trans* manner, illustrating another general feature of these rearrangements.

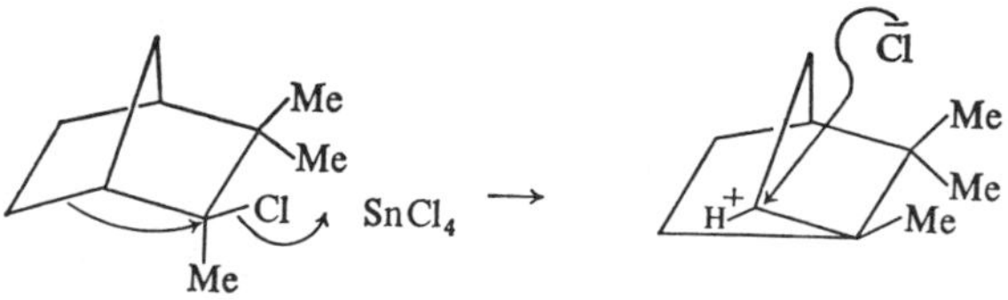

Camphene hydrochloride

*Of the two carbon atoms which might migrate, that one does so which leaves behind the more stable carbonium ion.

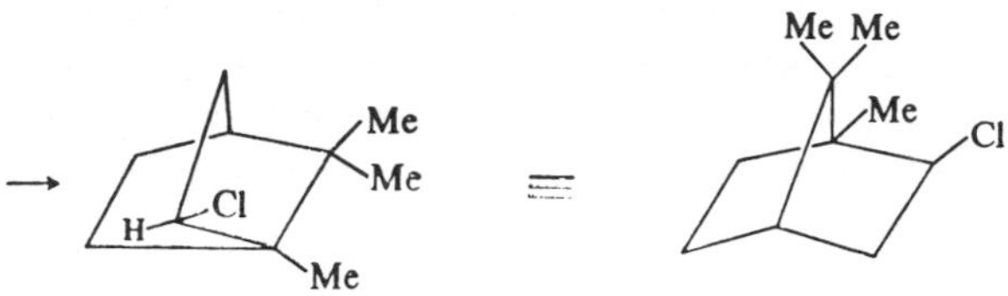

Me Me

Isobornyl chloride

6. Two or more rearrangements may occur successively, leading to extensive skeletal alteration. For example, the initial carbonium ion formed by treatment of diethylcyclobutylcarbinol with acid rearranges by ring-expansion to a secondary carbonium ion containing a five-membered ring (relief of strain) and this then undergoes a further shift to give a tertiary carbonium ion from which the olefinic product is derived:

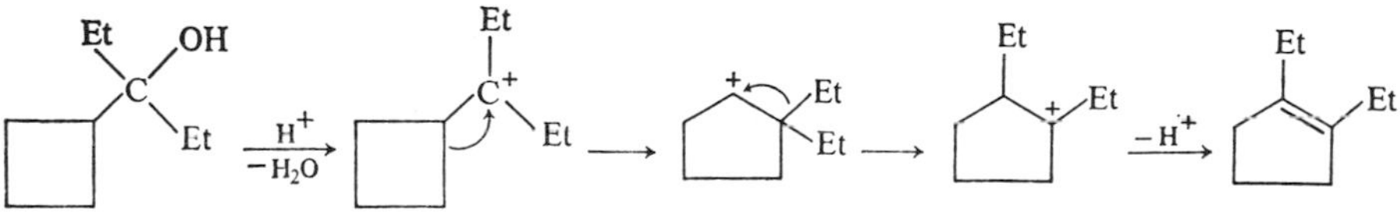

1,2-Diethyl-
cyclopentene

(ii) Pinacol rearrangement. The treatment of 1,2-diols (pinacols) with acid leads to rearrangement, e.g.

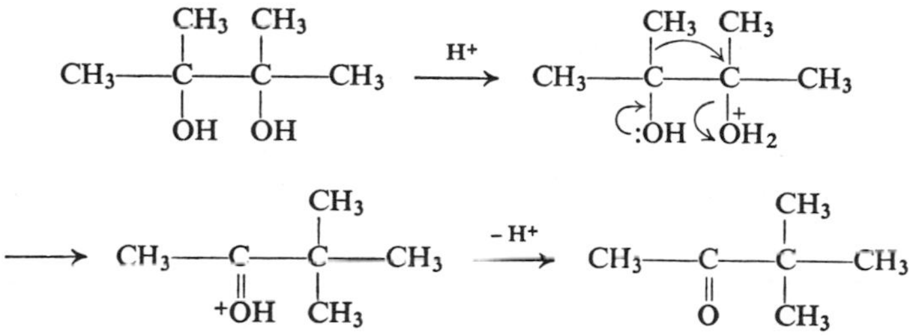

Pinacolone

The pinacol rearrangement, although fundamentally similar to the Wagner-Meerwein rearrangement, differs in that the rearranged ion, the conjugate acid of a ketone, is more stable than the rearranged carbonium ions formed in the Wagner-Meerwein reaction. Consequently, the driving force for the rearrangement of pinacols is much greater: whereas alcohols, other than those with the special structural features discussed above, can usually be dehydrated by acids without the occurrence of rearrangement, pinacols normally rearrange in preference to undergoing simple dehydration.

Pinacols are readily obtained by the one-electron reduction of carbonyl-containing compounds (p. 637) so that the pinacol rearrangement is synthetically useful. For example, pinacolone (methyl t-butyl ketone) can be obtained in about 70% yield by the distillation of a mixture of pinacol hydrate and sulphuric acid [1]. However, the distillation of a mixture of pinacol and aqueous hydrobromic acid gives mainly the dehydration product, 2,3-dimethyl-1,3-butadiene [3]:

$$CH_3\!-\!\underset{\underset{OH}{|}}{\overset{\overset{CH_3}{|}}{C}}\!-\!\underset{\underset{OH}{|}}{\overset{\overset{CH_3}{|}}{C}}\!-\!CH_3 \xrightarrow[-2\,H_2O]{HBr} CH_2\!\!=\!\!\underset{}{\overset{\overset{CH_3}{|}}{C}}\!-\!\underset{}{\overset{\overset{CH_3}{|}}{C}}\!\!=\!\!CH_2$$

2,3-Dimethyl-1,3-butadiene

The characteristics of the Wagner-Meerwein rearrangement apply also to the pinacol rearrangement: 1. Alkyl and aryl groups and hydrogen migrate. 2. The migratory aptitude of an aryl group is much greater than that of alkyl or hydrogen, and amongst aryl groups the migratory aptitude increases as the aromatic nucleus is made increasingly electron-rich (e.g. *p*-chlorophenyl < phenyl < *p*-tolyl < *p*-methoxyphenyl), e.g.

$$Ph\!-\!\underset{\underset{OH}{|}}{\overset{\overset{Tol}{|}}{C}}\!-\!\underset{\underset{OH}{|}}{\overset{\overset{Tol}{|}}{C}}\!-\!Ph \xrightarrow{H_2SO_4} Ph\!-\!\underset{\underset{O}{\|}}{\overset{\overset{}{}}{C}}\!-\!\underset{\underset{Ph}{|}}{\overset{\overset{Tol}{|}}{C}}\!-\!Tol + Tol\!-\!\underset{\underset{O}{\|}}{\overset{}{C}}\!-\!\underset{\underset{Ph}{|}}{\overset{\overset{Ph}{|}}{C}}\!-\!Tol$$

94%　　　　　　　　　6%

(Tol = *p*-tolyl)

3. The reaction occurs in a *trans* manner.

There is a further factor which does not apply to the Wagner-Meerwein reaction: of the two possible hydroxyl groups which are available as leaving groups, that one departs which leaves behind the more stable carbonium ion. This factor takes precedence over the migratory-aptitude factor: e.g. rearrangement of 1,1-dimethyl-2,2-diphenyl glycol leads to 3,3-diphenyl-2-butanone, by migration of methyl rather than phenyl:

$$Ph_2C\!-\!C(CH_3)_2 \xrightarrow[-H_2O]{H^+} Ph_2\overset{+}{C}\!-\!\underset{\underset{:OH}{}}{\overset{\overset{CH_3}{|}}{C}}\!-\!CH_3 \xrightarrow{-H^+} Ph_2C\!-\!\underset{\underset{O}{\|}}{\overset{\overset{CH_3}{|}}{C}}\!-\!CH_3$$

$$\underset{OH\ OH}{}$$

3,3-Diphenyl-2-butanone

The requirement that the migrating group be *trans* to the leaving group has important consequences in alicyclic systems. For example, *cis*-1,2-dimethylcyclohexane-1,2-diol undergoes a methyl shift to give 2,2-dimethylcyclohexanone, whereas the *trans*-isomer undergoes ring-contraction to give a cyclopentane derivative:

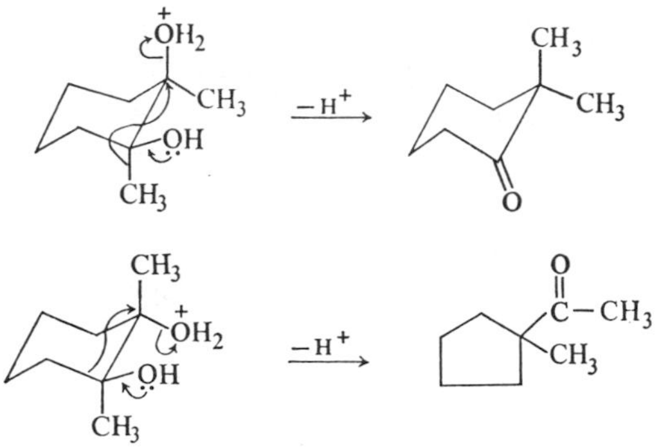

In addition to 1,2-diols, β-halohydrins undergo rearrangement in the presence of Lewis acids, and β-aminoalcohols undergo rearrangement, *via* the diazonium ion, on treatment with nitrous acid.

A synthesis of cycloheptanone illustrates the use of the latter process. The appropriate β-amino-alcohol is obtained from cyclohexanone by base-catalyzed condensation with nitromethane (p. 236) followed by reduction of the nitro group (p. 647), and treatment with sodium nitrite in acetic acid then gives cyclo-heptanone in 40% overall yield [4], which is considerably better than can be achieved from a readily available compound *via* a ring-closure reaction.

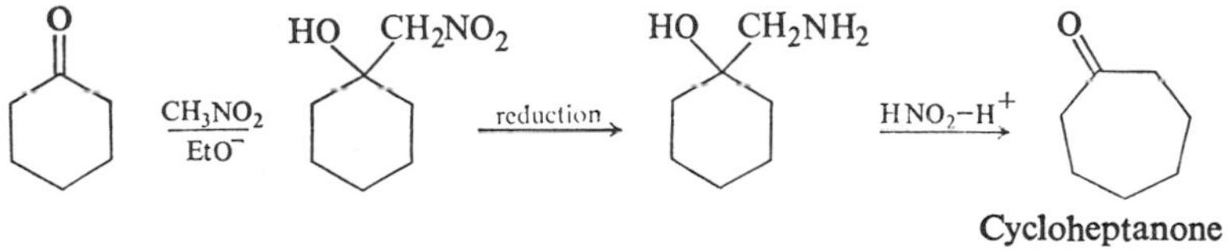

Cycloheptanone

(*iii*) *Benzilic acid rearrangement.* α-Diketones undergo a rearrangement when treated with hydroxide ion, giving α-hydroxy-acids. The best known example is the conversion of benzil into benzilic acid; the migration step is of the 'push-pull' type described previously, and the driving force for the reaction lies in the re-moval of the product by ionization of the carboxyl group.

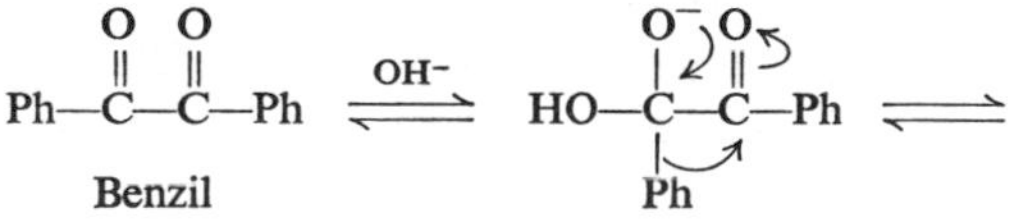

Benzil

$$\underset{\underset{Ph}{|}}{HO-\overset{\overset{O}{||}}{C}-\overset{\overset{O^-}{|}}{C}-Ph} \ \underset{}{\overset{H_2O}{\rightleftharpoons}} \ \underset{\underset{Ph}{|}}{HO-\overset{\overset{O}{||}}{C}-\overset{\overset{OH}{|}}{C}-Ph} \ \overset{OH^-}{\rightleftharpoons} \ Ph_2C\overset{\diagup OH}{\underset{\diagdown CO_2^-}{}}$$

Benzilic acid

Ketones which contain α-C—H bonds usually undergo base-catalyzed condensations in preference to rearrangement, although ketipic acid gives citric acid:

$$HO_2C-CH_2-\overset{\overset{O}{||}}{C}-\overset{\overset{O}{||}}{C}-CH_2-CO_2H \ \overset{1) \ OH^-}{\underset{2) \ H^+}{\longrightarrow}} \ HO_2C-CH_2-\underset{\underset{CO_2H}{|}}{\overset{\overset{OH}{|}}{C}}-CH_2-CO_2H$$

Ketipic acid   Citric acid

Aromatic α-diketones are normally prepared by the oxidation of the α-hydroxy-ketones obtained by the benzoin condensation (p. 260), and benzilic acids may be synthesized directly from benzoins by combining the oxidation and rearrangement reactions. For example, the treatment of benzoin itself with sodium bromate and sodium hydroxide gives benzilic acid in up to 90% yield [1].

(*iv*) *Rearrangements involving diazomethane.* Diazomethane takes part in two types of reaction which lead, as the result of rearrangement, to the insertion of a methylene group into a chain of carbon atoms. In each, it acts first as a carbon nucleophile ($CH_2{=}\overset{+}{N}{=}\overset{-}{N} \leftrightarrow \overset{-}{CH_2}{-}\overset{+}{N}{\equiv}N$), giving a derivative from which nitrogen is readily lost.

(1) Aldehydes and ketones are converted into the next highest homologue:

$$R_2C{=}O \ \overset{CH_2N_2}{\longrightarrow} \ R-\underset{\underset{R}{|}}{\overset{\overset{O^-}{|}}{C}}-CH_2-\overset{+}{N}{\equiv}N \ \overset{-N_2}{\longrightarrow} \ R-\overset{\overset{O}{||}}{C}-CH_2-R$$

The migration step is similar to that in the pinacol reaction: the movement of R may be visualized as being brought about by the combination of the 'pull' from nitrogen and the 'push' from the oxyanion.

Two disadvantages attend the use of this procedure in synthesis. First, unsymmetrical ketones give a mixture of two products, and secondly, an epoxide is formed as a byproduct and in some cases as the main product.

$$R-\underset{\underset{R}{|}}{\overset{\overset{O^-}{|}}{C}}-CH_2-\overset{+}{N}{\equiv}N \ \overset{-N_2}{\longrightarrow} \ R-\underset{\underset{R}{|}}{\overset{\diagup O \diagdown}{C}}\!\!-\!\!CH_2$$

Nevertheless, the reaction in some cases gives practicable yields, in one step, of difficultly accessible compounds: e.g. cyclohexanone gives a 33–36% yield of cycloheptanone [4]. The diazomethane may be generated *in situ* by treating *N*-methyl-*N*-nitrosotoluene-*p*-sulphonamide with base. *

A special example is the reaction of diazomethane with keten to give cyclopropanone:

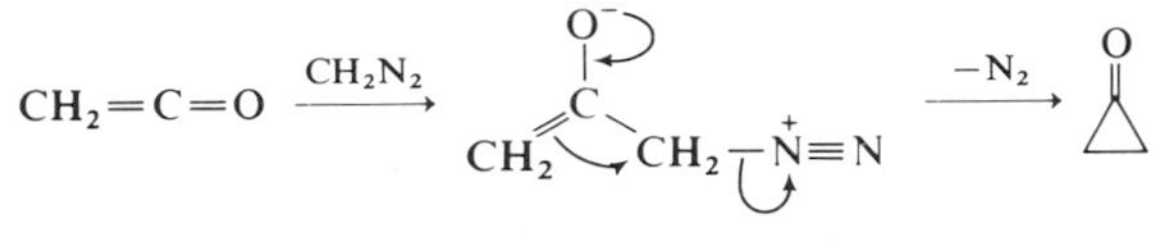

The reaction must be carried out at very low temperatures because of the high reactivity of cyclopropanone. An excess of keten must also be used, for otherwise the cyclopropanone reacts with diazomethane to form cyclobutanone.

(2) The reaction of diazomethane with an acid chloride gives a diazoketone† which, on being heated in the presence of silver oxide, undergoes the *Wolff rearrangement* to give a keten:

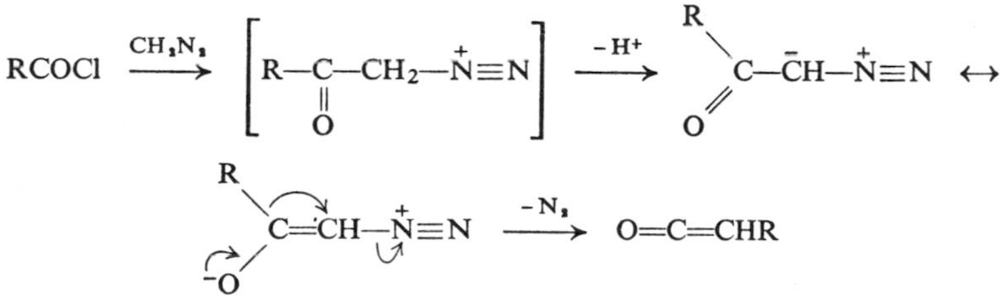

When the rearrangement is carried out in the presence of water or an alcohol, the keten is converted directly into an acid or ester:

$$R-CH=C=O + H_2O \longrightarrow RCH_2CO_2H$$

$$R-CH=C=O + R'OH \longrightarrow RCH_2CO_2R'$$

The overall process (*Arndt-Eistert* synthesis) provides a method for the conversion of an acid RCO_2H into the homologue RCH_2CO_2H in three stages. Total yields are normally good (*ca.* 50–80%).

*A probable reaction path is:

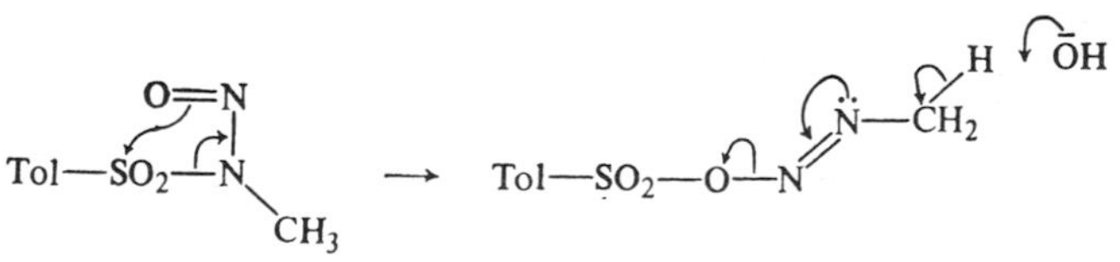

$$\longrightarrow \ Tol-SO_2O^- + CH_2N_2 + H_2O$$

†The acid chloride is added to an excess of diazomethane to minimize the formation of the alternative product, $RCOCH_2Cl$.

(v) *Rearrangement of paraffins.* Saturated carbon chains undergo skeletal re-arrangements when treated with a Lewis acid in the presence of a catalytic quantity of an organic halide. The products are equilibrium mixtures of all the possible isomeric compounds, e.g.

$$CH_3CH_2CH_2CH_3 \underset{150°C}{\overset{AlCl_3(RCl)}{\rightleftharpoons}} (CH_3)_3CH$$

$$ca.\ 20\% \qquad\qquad ca.\ 80\%$$

The rearrangement occurs by way of a carbonium ion, formed from the halide, which abstracts hydride ion from the paraffinic chain:

$$R\!\!-\!\!Cl + AlCl_3 \rightleftharpoons R^+ + AlCl_4^-$$

$$CH_3\!\!-\!\!\underset{\underset{R^+}{\overset{|}{H}}}{\overset{}{CH}}\!\!-\!\!CH_2\!\!-\!\!CH_3 \underset{}{\overset{-RH}{\rightleftharpoons}} CH_3\!\!-\!\!\overset{+}{CH}\!\!-\!\!\overset{\overset{\displaystyle CH_3}{|}}{CH_2}$$

$$\rightleftharpoons CH_3\!\!-\!\!\overset{\overset{\displaystyle CH_3}{|}}{CH}\!\!-\!\!\overset{+}{CH_2} \underset{}{\overset{RH}{\rightleftharpoons}} CH_3\!\!-\!\!\overset{\overset{\displaystyle CH_3}{|}}{CH}\!\!-\!\!CH_3$$

It is to be noted that the predominant product may be derived from the least stable carbonium ion: in the above example, isobutane is formed following the rearrangement of a secondary to a primary carbonium ion. This is the converse of the direction of rearrangement in the reactions previously discussed. The underlying basis is that whereas the latter reactions are kinetically controlled (the carbonium ion formed by rearrangement reacts almost immediately with a nucleophile), the former are thermodynamically controlled. Virtually no nucleo-phile is present, for only a trace of organic halide is added so that the concentra-tion of nucleophilic $AlCl_4^-$ is negligible, and the reactions are freely reversible. Thus the relative proportions of the paraffins alter gradually until they reach values determined by the relative free energies of the compounds.

Since the relative energies of isomeric acyclic hydrocarbons differ little, a complex mixture of products is likely to be obtained by this process. Amongst alicyclic compounds, however, there are greater differences in relative free energies because of the occurrence of strain in ring systems. For example, the isomerization of methylcyclopentane gives cyclohexane but no ethyl- or dimethyl-cyclobutane.

12.5%   87.5%   (25°C)

A particularly novel application of this rearrangement process has recently been described [5]. Catalytic reduction of the readily available dicyclopentadiene (p. 292) gives *endo*-tetrahydrodicyclopentadiene which, on treatment with aluminium trichloride at 150–180°C, gives adamantane (15%), which is the most stable of the saturated hydrocarbons of molecular formula $C_{10}H_{16}$.*

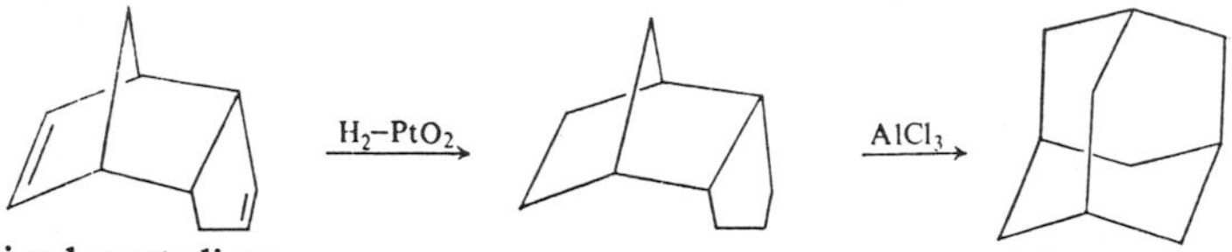

Dicyclopentadiene $\xrightarrow{\text{H}_2\text{–PtO}_2}$ $\xrightarrow{\text{AlCl}_3}$ Adamantane

(*b*) HALOGEN, OXYGEN, SULPHUR, AND NITROGEN MIGRATION

An atom, X, with an unshared pair of electrons, in the system X—C—C—Y, can assist the heterolysis of the C—Y bond in the same way as a phenyl group:

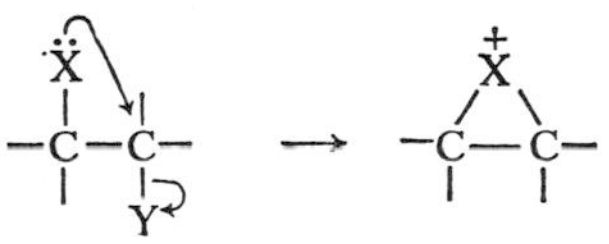

In a symmetrical case, such as the solvolysis of Et—S—CH_2CH_2Cl, no rearrangement occurs because nucleophilic attack at either carbon atom of the bridged ion leads to the same product as would be formed in the absence of neighbouring-group participation. In unsymmetrical cases, however, nucleophilic attack at the less highly substituted carbon of the bridged ion predominates (*cf.* the opening of epoxides; p. 571), and a rearranged skeleton can result:

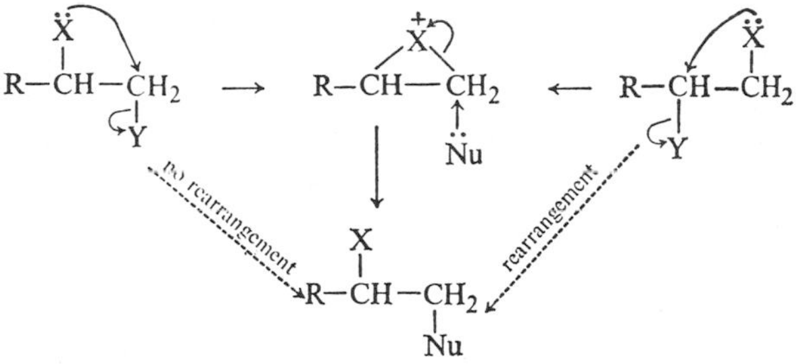

The following are typical examples:.

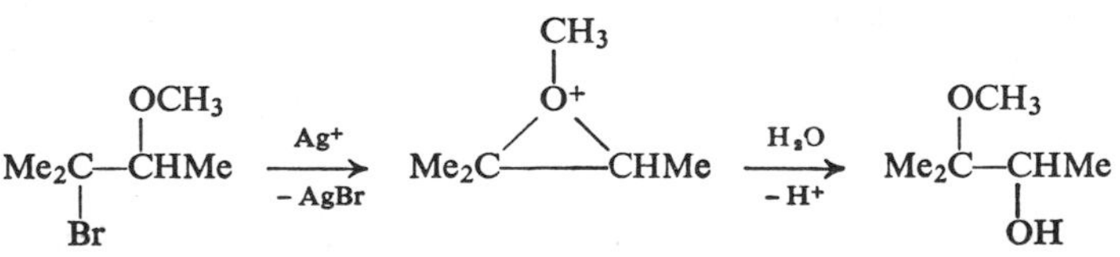

*Adamantane contains four fused cyclohexane rings, each in the strain-free chair conformation, p. 174.

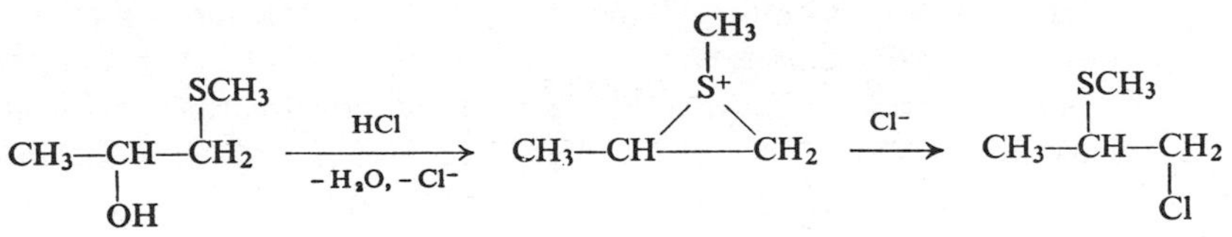

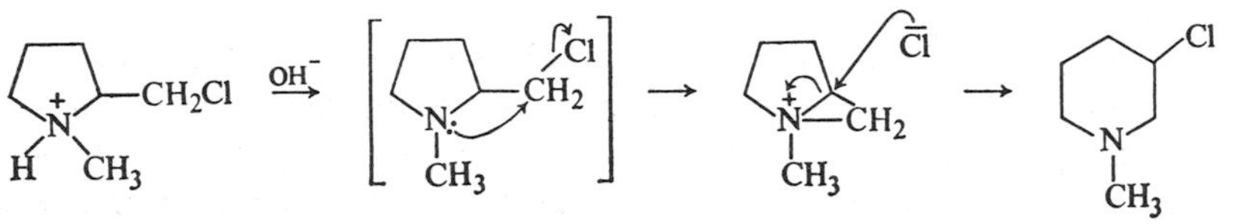

The bridged cation may be generated by protonation of an unsaturated bond, as in the *Rupe rearrangement* of α-acetylenic alcohols, e.g.

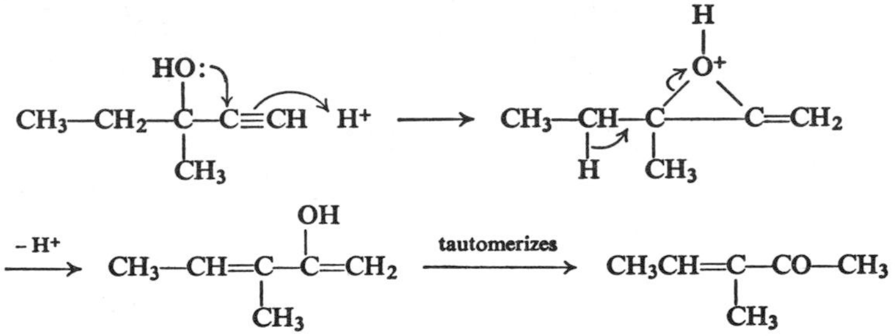

3-Methyl-3-penten-2-one

A neighbouring acetoxy group assists solvolysis by forming a *five*-membered acetoxonium ion:

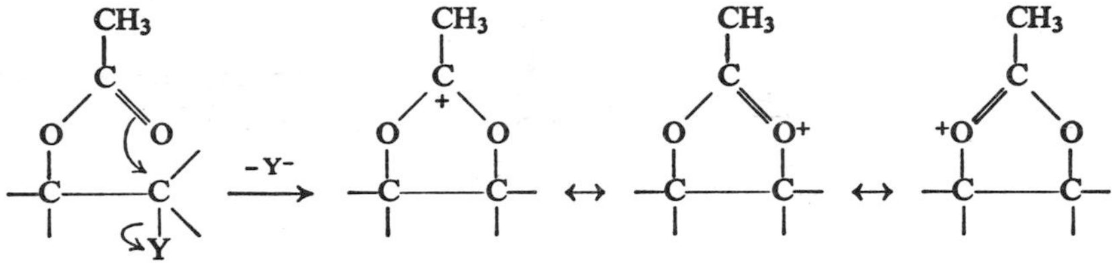

The cyclic ion is then opened by reaction with a nucleophile. Reaction with water occurs at the acetoxy-carbon atom to yield ultimately a *cis*-hydroxy-acetate, whereas reaction with acetate ion occurs at alkyl carbon in the S_N2 manner to yield a *trans*-diacetate:

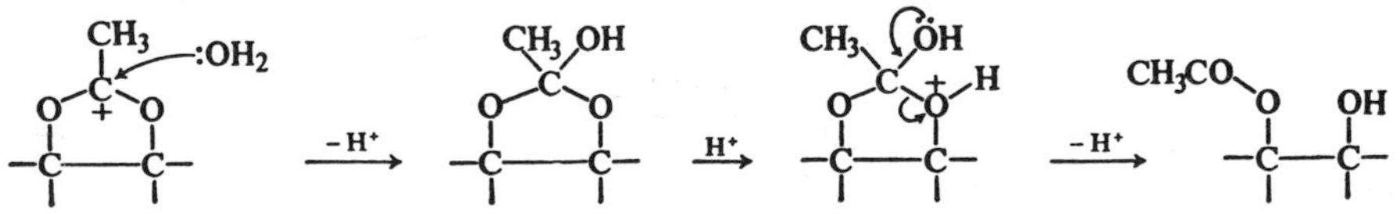

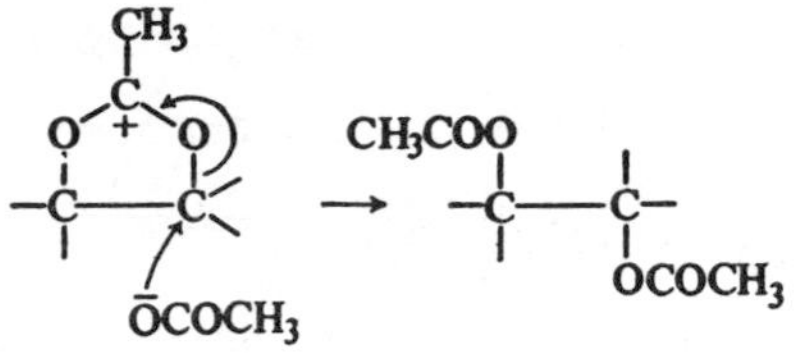

Procedures for making *cis*- and *trans*-diols are based on the principle of acetoxonium-ion participation. Treatment of an olefin with iodine and silver acetate gives the *trans*-iodo-acetate, by normal *trans*-addition to a double bond; iodide ion is then eliminated under the influence of the neighbouring acetoxy group and silver ion; and the acetoxonium ion reacts to give *cis*- and *trans*-derivatives, which are hydrolyzed to the corresponding diols, in 'wet' and 'dry' conditions, respectively, e.g.

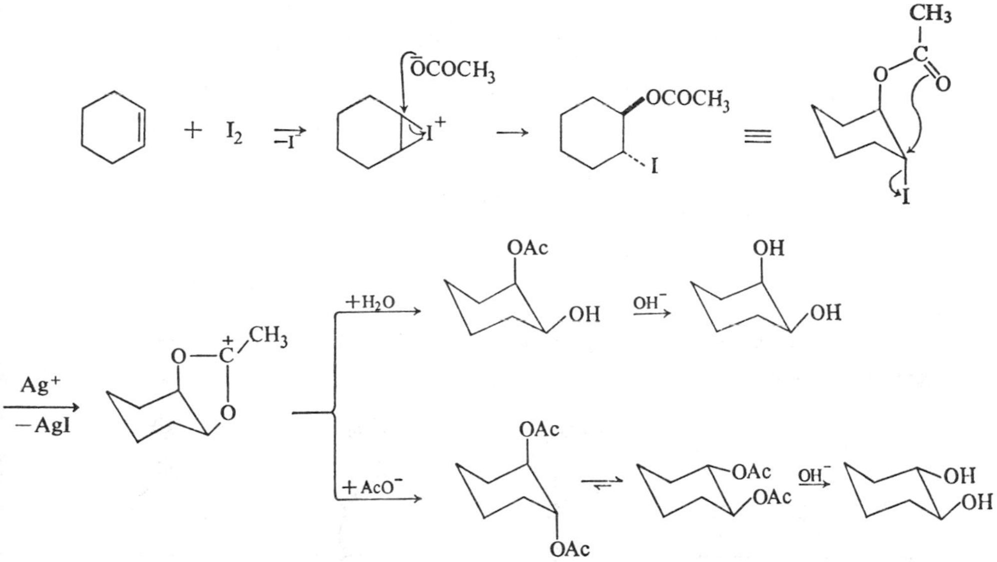

14.3 Rearrangement to Electron-deficient Nitrogen

(*a*) THE HOFMANN, CURTIUS, SCHMIDT, AND LOSSEN REARRANGEMENTS
There is a group of closely related rearrangements in which carbon migrates from carbon to nitrogen. They may be formulated generally as,

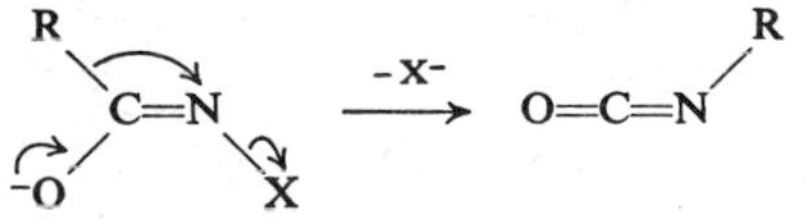

where R is an alkyl or aryl group and —X is a leaving group which may be —Br (*Hofmann rearrangement*), $-\overset{+}{N}\equiv N$ (*Curtius and Schmidt rearrangements*), and —OCOR (*Lossen rearrangement*). In each case, if the alkyl carbon which migrates is asymmetric, it retains its configuration.

(*i*) *Hofmann rearrangement.* A carboxylic acid amide is treated with sodium hypobromite, or bromine in alkali. The *N*-bromo-amide is first formed and reacts with the base to give its conjugate base within which rearrangement occurs. The isocyanate which is produced may be isolated in anhydrous conditions, but reaction is normally carried out in aqueous or alcoholic solution in which the isocyanate is converted into an amine or a urethane, respectively.

$$RCONH_2 + Br_2 \longrightarrow RCONHBr + HBr$$

$$RCONHBr \underset{}{\overset{OH^-}{\rightleftharpoons}} \left[R-\underset{\underset{O}{\|}}{C}-\overset{-}{N}-Br \leftrightarrow R-C=N-Br \atop \quad\quad\quad\;\; O^- \right]$$

$$\xrightarrow{- Br^-} O=C=N \overset{R}{\diagup} \quad \left\{ \begin{array}{l} \xrightarrow{H_2O} RNH_2 + CO_2 \\ \\ \xrightarrow{R'OH} RNHCOR' \end{array} \right.$$

This rearrangement provides an efficient route for making both aliphatic and aromatic primary amines. For example, β-alanine can be obtained in about 45% yield by treating succinimide with bromine and aqueous caustic potash [2]; reaction occurs through the half-amide of succinic acid.

$$\begin{array}{c} CH_2-CO \\ | \quad\quad\;\; \diagdown \\ | \quad\quad\;\;\;\; NH \\ | \quad\quad\; \diagup \\ CH_2-CO \end{array} \xrightarrow{OH^-} \left[\begin{array}{l} CH_2CONH_2 \\ | \\ CH_2CO_2^- \end{array} \right] \xrightarrow{Br_2-KOH} \begin{array}{l} CH_2NH_2 \\ | \\ CH_2CO_2^- \\ \textit{β-Alanine} \end{array}$$

Anthranilic acid (*ca.* 85%) may be obtained in a similar way from phthalimide:

Anthranilic acid

A particularly useful example of the reaction is the preparation of β-amino-pyridine (65–70%) [4] from nicotinamide (available from natural sources), for this cannot be obtained in good yield *via* the nitration of pyridine (p. 403).

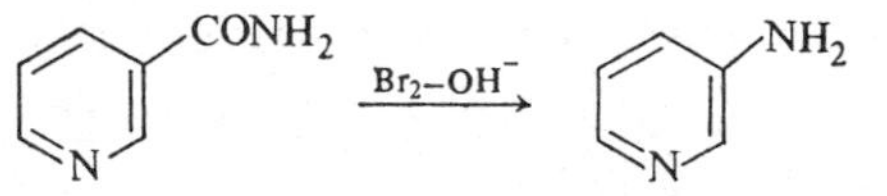

β-Aminopyridine

(*ii*) *Curtius rearrangement.* Acid azides* decompose on being heated to give isocyanates:

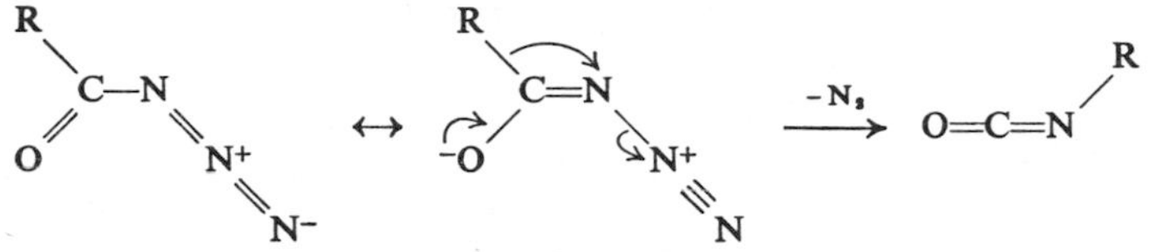

The isocyanate may be isolated by carrying out the reaction in an aprotic solvent such as chloroform, but it is customary to use an alcoholic solvent with which the isocyanate reacts to form a urethane. Acid hydrolysis gives the corresponding amine.

$$R-N=C=O \xrightarrow{R'OH} R-NH-CO_2R' \xrightarrow{H^+} [R-NH-CO_2H] \xrightarrow{-CO_2} RNH_2$$

The Curtius rearrangement has been applied to the synthesis of α-aminoacids, e.g.

$$CH_2(CO_2Et)_2 \xrightarrow{OH^-} CO_2Et-CH_2-CO_2^- \xrightarrow{N_2H_4} NH_2NHCO-CH_2-CO_2^-$$

$$\xrightarrow{HNO_2} N_3-CO-CH_2-CO_2H \xrightarrow{HCl-EtOH} NH_2-CH_2-CO_2H$$
Glycine

(*iii*) *Schmidt reaction.* Carboxylic acids react with hydrazoic acid in the presence of concentrated sulphuric acid to give isocyanates directly. Reaction occurs through the acid azide, but in the strongly acid conditions this is present as its conjugate acid from which nitrogen is lost without heating.

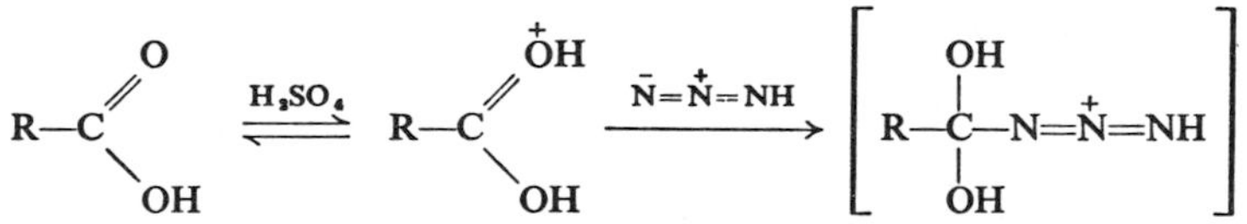

*Acid azides are readily prepared from acid chlorides by treatment with sodium azide,

$$RCOCl \xrightarrow[-NaCl]{NaN_3} RCON_3$$

and from esters by treatment with hydrazine followed by nitrous acid,

$$RCO_2Et \xrightarrow[-EtOH]{N_2H_4} RCONHNH_2 \xrightarrow[-2H_2O]{HNO_2} RCON_3$$

$$\xrightarrow{-H_2O} \left[\begin{array}{c} R{-}C{-}NH \\ \underset{O}{\|} \qquad N^+ \\ \qquad \underset{N}{\lll} \end{array} \right] \xrightarrow{-N_2} R\overset{+}{N}HCO \xrightarrow{H_2O} R\overset{+}{N}H_3 + CO_2$$

(*iv*) *Lossen rearrangement.* This differs from the Hofmann rearrangement only in that the leaving group is a carboxylate anion rather than bromide ion. The starting material is the ester of a hydroxamic acid.

$$RCONHOH \longrightarrow RCONHOCOR' \xrightarrow{OH^-} \begin{array}{c} R \\ \diagdown \\ C{-}\overset{-}{N} \\ O\diagup \qquad \diagdown OCOR' \end{array} \longleftrightarrow$$

$$\begin{array}{c} R \\ \diagdown \\ C{=}N \\ \overset{-}{O}\diagup \qquad \diagdown OCOR' \end{array} \xrightarrow{-R'CO_2^-} O{=}C{=}N \diagup \overset{R}{\ } \xrightarrow{H_2O} RNH_2 + CO_2$$

Of these four related processes the Lossen rearrangement is the least useful in the synthesis of amines because hydroxamic acids are not readily available. The Schmidt reaction is the most direct method but is only applicable if the acid does not contain groups which are sensitive to concentrated sulphuric acid. The Curtius rearrangement involves the mildest conditions but requires the preparation of the azide. The Hofmann rearrangement is convenient providing that other functional groups in the molecule do not react with bromine and alkali.

(*b*) THE BECKMANN REARRANGEMENT

Oximes undergo a rearrangement in acidic conditions to give substituted amides:

$$\begin{array}{c} R \\ \diagdown \\ C{=}N \\ R'\diagup \qquad \diagdown OH \end{array} \underset{}{\rightleftharpoons} \xrightarrow{H^+} \begin{array}{c} R \\ \diagdown \\ C{=}N \\ R'\diagup \qquad \overset{+}{OH_2} \end{array} \xrightarrow{-H_2O} R'{-}C{\equiv}\overset{+}{N}{-}R$$

$$\xrightarrow[-H^+]{H_2O} R'{-}\underset{OH}{C}{=}N{-}R \rightleftharpoons R'{-}\underset{O}{\overset{\|}{C}}{-}NH{-}R$$

The Beckmann rearrangement is stereospecific: the group *trans* to the leaving group migrates. Thus, acetophenone oxime, which has the stereochemistry shown, gives only acetanilide.

$$\begin{array}{c} Ph \qquad CH_3 \\ \diagdown \diagup \\ C \\ \| \\ N \\ \diagdown OH \end{array} \xrightarrow[\text{rearrangement}]{\text{Beckmann}} PhNHCOCH_3$$

Acetanilide

As in other intramolecular rearrangements, if the carbon atom which migrates is asymmetric, it retains its configuration during the reaction.

The rearrangement is also induced by reagents other than proton acids. Boron trifluoride removes the hydroxyl group as HO—$\overline{B}F_3$, toluene-*p*-sulphonyl chloride forms the oxime tosylate which eliminates the stable tosylate anion,

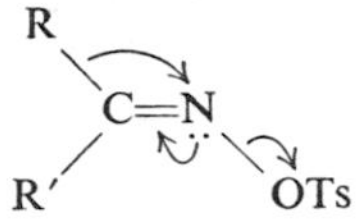

and phosphorus pentachloride, normally used in ether, induces rearrangement by providing a phosphate as leaving group,

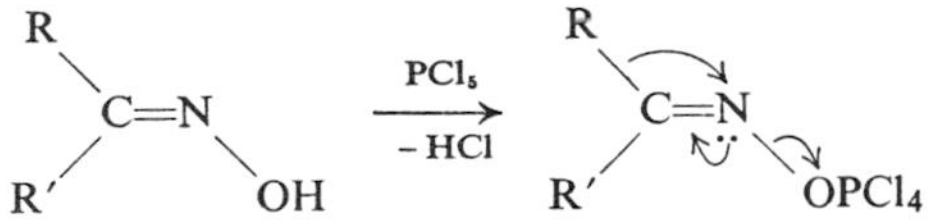

An interesting application of the rearrangement is the synthesis of ε-capro-lactam (70%) from cyclohexanone oxime and concentrated sulphuric acid [2], a ring-expansion of analogous type to the formation of cycloheptanone from cyclohexanone (p. 461). ε-Caprolactam gives a polymer of the Nylon group when heated:

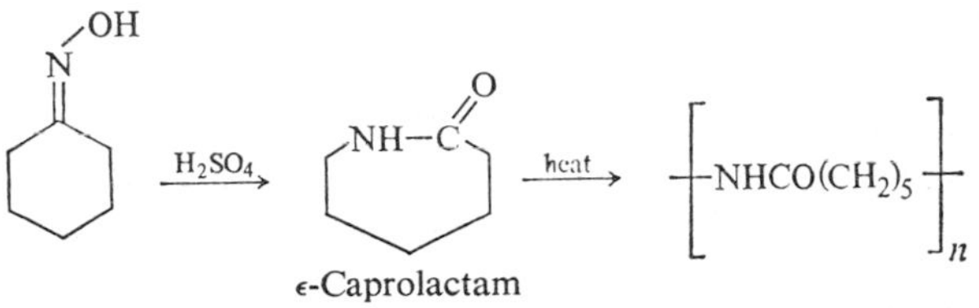

ε-Caprolactam

14.4 Rearrangement to Electron-deficient Oxygen

The most general rearrangement of this type is the *Baeyer-Villiger reaction* in which ketones are converted into esters, and cyclic ketones into lactones, by treatment with a peracid. The mechanism is closely related to that of the pinacol rearrangement: nucleophilic attack by the peracid on the carbonyl group gives an intermediate which rearranges with the expulsion of the anion of an acid.

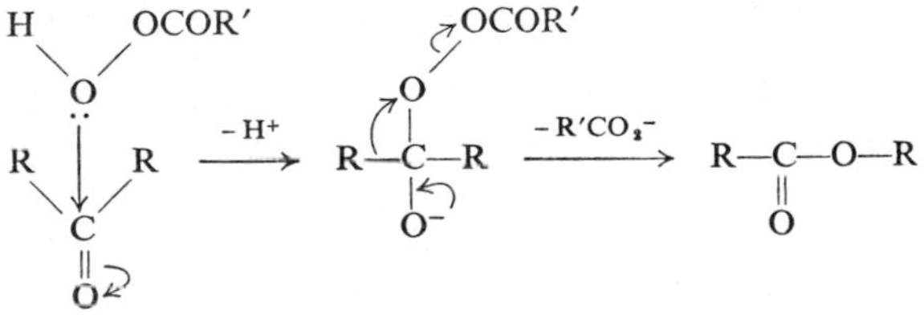

Acids catalyze the reaction by facilitating both the addition to carbonyl and the expulsion of the carboxylate.

A number of peracids, including peracetic, monoperphthalic, monopersulphuric, and trifluoroperacetic acid, have been successfully employed in the reaction. Trifluoroperacetic acid is the most reactive of the peracids, probably because the trifluoroacetate ion, derived from a strong acid, is a very good leaving group; it is necessary to buffer the solution for otherwise transesterification occurs to give the trifluoroacetate ester.

In an unsymmetrical ketone, that group migrates which is the better able to supply electrons, as in the Wagner-Meerwein and related rearrangements. Thus, amongst alkyl groups, the ease of migration is, tertiary > secondary > primary > methyl; e.g. pinacolone gives t-butyl acetate.

$$\text{Me}_3\text{C}\!-\!\text{CO}\!-\!\text{CH}_3 \xrightarrow{\text{peracid}} \text{Me}_3\text{C}\!-\!\text{O}\!-\!\text{CO}\!-\!\text{CH}_3$$
t-Butyl acetate

Amongst aryl groups, the order is, p-methoxyphenyl > p-tolyl > phenyl > p-chlorophenyl, etc., (cf. p. 457). Aryl groups migrate in preference to primary alkyl groups, e.g.

$$\text{Ph}\!-\!\text{CO}\!-\!\text{CH}_3 \xrightarrow{\text{peracid}} \text{Ph}\!-\!\text{O}\!-\!\text{CO}\!-\!\text{CH}_3$$
Phenyl acetate

but only if they contain electron-releasing substituents do they migrate in preference to secondary and tertiary alkyl groups.

Cyclic ketones undergo ring-expansion with peracids. For example, cyclohexanone gives ε-caprolactone (cf. the formation from cyclohexanone of cycloheptanone, p. 461, and ε-caprolactam, p. 471):

ε-Caprolactone
70%

The lactone is hydrolyzed under the reaction conditions. In an aqueous medium, ε-hydroxycaproic acid is formed and undergoes condensation polymerization. Polymerization is prevented by carrying out the reaction in ethanol; ethyl ε-hydroxycaproate is then obtained.

The acid-catalyzed rearrangement of tertiary hydroperoxides is similar to the Baeyer-Villiger reaction. The product from the migration is a hemi-acetal which is hydrolyzed in the reaction conditions (p. 110):

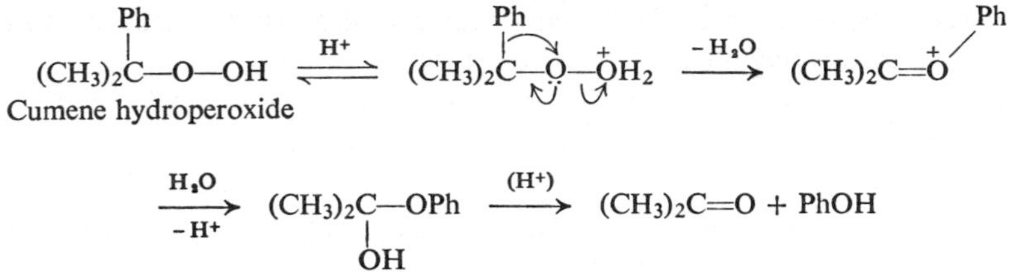

$$
\underset{\text{Cumene hydroperoxide}}{(CH_3)_2\overset{\overset{\text{Ph}}{|}}{C}-O-OH} \quad \underset{}{\overset{H^+}{\rightleftharpoons}} \quad (CH_3)_2\overset{\overset{\text{Ph}}{|}}{C}\overset{+}{\underset{}{O}}-\overset{+}{O}H_2 \quad \xrightarrow{-H_2O} \quad (CH_3)_2C=\overset{+}{O}\overset{\text{Ph}}{\diagup}
$$

$$
\xrightarrow[-H^+]{H_2O} \quad (CH_3)_2\underset{\overset{|}{O}H}{C}-OPh \quad \xrightarrow{(H^+)} \quad (CH_3)_2C=O + PhOH
$$

The above example is of industrial importance, for cumene is cheaply available from the Friedel-Crafts alkylation of benzene with propylene and, like other tertiary hydrocarbons, it readily forms the hydroperoxide by autoxidation (p. 563).

Dakin reaction. Benzaldehydes containing *ortho-* and *para-*hydroxyl groups are converted into catechols and quinols, respectively, by alkaline hydrogen peroxide. For example, catechol itself may be obtained in 70% yield from salicyl-aldehyde [1].

The mechanism is similar to that of the Baeyer-Villiger reaction:*

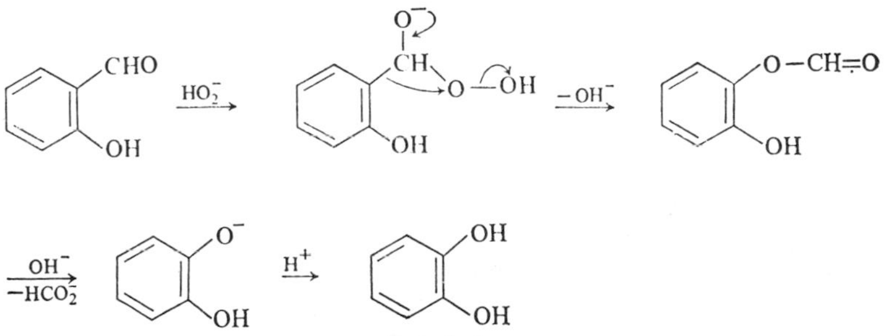

Catechol

14.5 Rearrangement to Electron-rich Carbon

This group of rearrangements has been less extensively studied, and is of less synthetic importance, than rearrangements to electron-deficient carbon. The known examples are of the type,

$$
\overset{\overset{\text{R}}{|}}{X}-\overset{}{\underset{}{C}} \quad \longrightarrow \quad \overset{}{\underset{}{X}}-\overset{\overset{\text{R}}{|}}{C}
$$

$$
(X=N^+, S^+, O)
$$

in which the group R migrates from X to C.

*The hydroxide ion is not a good leaving group and the function of the aromatic hydroxyl substituent may be to provide powerful anchimeric assistance (*o*-hydroxyphenyl ~ *p*-hydroxyphenyl ≫ phenyl) to effect heterolysis of the O—O bond.

(*i*) *Stevens rearrangement.* Quaternary ammonium ions which contain β-hydrogen atoms undergo E2 (Hofmann) elimination with base, e.g.

$$\text{HO}^- \quad\quad \underset{\overset{|}{\overset{+}{\text{N}}\text{Me}_3}}{\text{CH}_2{-}\text{CH}_2} \quad\longrightarrow\quad H_2O + CH_2{=}CH_2 + Me_3N$$

If, however, none of the alkyl groups possesses a β-hydrogen atom but one has a β-carbonyl group, an α-hydrogen is removed by base to give an *ylid* (a species in which adjacent atoms bear formal opposite charges). The role of the carbonyl group is to assist formation of the ylid by stabilizing the negative charge, e.g.

$$\underset{\overset{|}{\text{CH}_2\text{Ph}}}{\overset{+}{\text{Me}_2\text{N}}}{-}\text{CH}_2{-}\overset{\overset{\displaystyle O}{\|}}{C}{-}\text{Ph} \;\overset{OH^-}{\rightleftharpoons}\; \left[\underset{\overset{|}{\text{CH}_2\text{Ph}}}{\overset{+}{\text{Me}_2\text{N}}}{-}\overset{-}{\text{CH}}{-}\overset{\overset{\displaystyle O}{\|}}{C}{-}\text{Ph} \;\leftrightarrow\; \underset{\overset{|}{\text{CH}_2\text{Ph}}}{\overset{+}{\text{Me}_2\text{N}}}{-}\text{CH}{=}\overset{\overset{\displaystyle O^-}{|}}{C}{-}\text{Ph} \right]$$

Rearrangement then occurs:

$$\underset{\text{Me}_2\overset{+}{\text{N}}{-}\overset{-}{\text{CH}}{-}\text{COPh}}{\overset{\text{PhCH}_2}{\overset{|}{}}} \quad\longrightarrow\quad \underset{\text{Me}_2\text{N}{-}\text{CH}{-}\text{COPh}}{\overset{\text{PhCH}_2}{\overset{|}{}}}$$

Sulphonium salts behave analogously (p. 492).

It used to be thought that these rearrangements occurred in a concerted manner, represented by:

$$\underset{R_2N{-}\overset{-}{\text{C}}HR''}{\overset{R'CH_2}{}} \quad\longrightarrow\quad \underset{R_2N{-}CHR''}{\overset{\overset{R'CH_2}{|}}{}}$$

However, the recognition that this would not be an allowed reaction according to orbital-symmetry rules (p. 314) prompted further study which has shown that reaction probably occurs *via* formation and combination of radical-pairs:

$$\underset{R_2\overset{+}{N}{-}\overset{-}{\text{C}}HR''}{\overset{\overset{R'CH_2}{|}}{}} \;\longrightarrow\; \left[\underset{R_2\overset{+}{N}{-}\overset{-}{\text{C}}HR''}{\overset{R'\overset{\cdot}{C}H_2}{}} \;\longleftrightarrow\; \underset{R_2N{-}\overset{\cdot}{\text{C}}HR''}{\overset{R'\overset{\cdot}{C}H_2}{}} \right] \;\longrightarrow\; \underset{R_2N{-}CHR''}{\overset{\overset{R'CH_2}{|}}{}}$$

In the absence of a β-carbonyl group the α-hydrogen is too weakly acidic for rearrangement to be induced by hydroxide ion. A stronger base, such as amide ion in liquid ammonia, is effective, but the rearrangement takes a different course: instead of the [1,2]-shift, a [3,2]-sigmatropic rearrangement (*Sommelet rearrangement*) occurs:*

---

*The benzylic protons are more acidic than the methyl protons because the negative charge of the ylid is delocalized over the benzene ring. However, formation of this ylid is not followed by rearrangement.

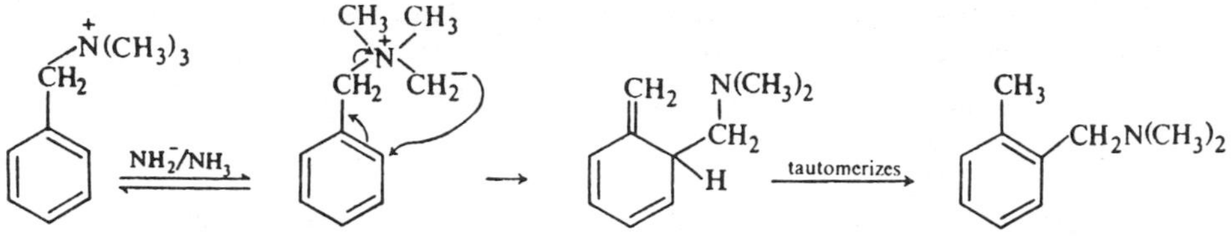

High yields can be obtained by this procedure; in the example above the yield is over 90% [4].

(*ii*) *Wittig rearrangement.* Benzyl and allyl ethers undergo a base-catalyzed rearrangement analogous to the Stevens rearrangement. A benzylic or allylic carbanion* is generated by the action of a powerful base such as amide ion or phenyl-lithium and migration of carbon then leads to the more stable oxy-anion, e.g.

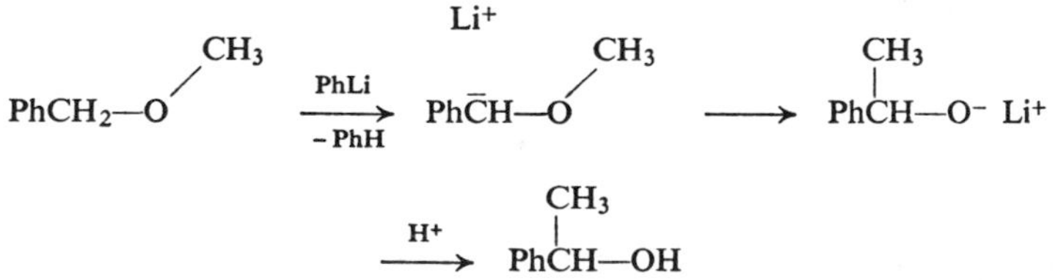

$$PhCH_2\!-\!O\diagup^{CH_3} \xrightarrow[-\,PhH]{PhLi} Ph\bar{C}H\!-\!O\diagup^{CH_3} \xrightarrow{\quad} \underset{\text{Li}^+}{Ph\bar{C}H\!-\!O^-\ Li^+}$$

$$\xrightarrow{H^+} \underset{CH_3}{\overset{\displaystyle|}{PhCH}\!-\!OH}$$

(*iii*) *Favorskii rearrangement.* α-Halo-ketones react with base to give carbanions which rearrange to esters *via* cyclopropanones:

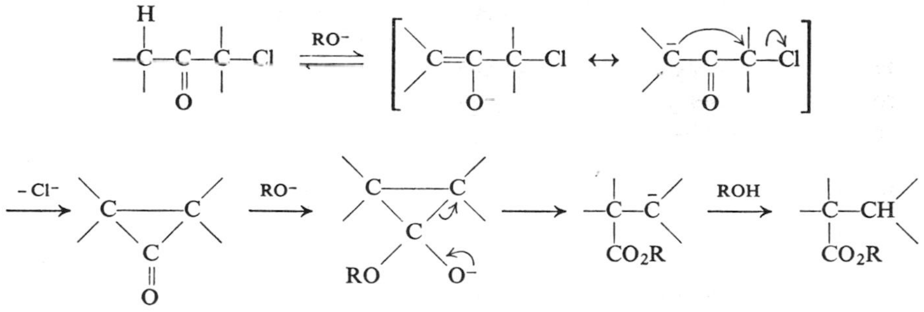

The rearrangement can be employed for bringing about ring-contraction in cyclic systems; e.g. 2-chlorocyclohexanone and methoxide ion give methyl cyclopentanecarboxylate in 60% yield [4]:

*The negative charge is delocalized over the benzylic or allylic system. Alkyl ethers are not sufficiently acidic to react.

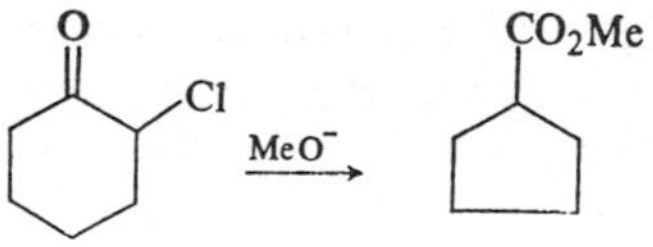

An interesting example is the conversion of the cyclobutyl into the cyclopropyl system, for this must apparently involve the highly strained bicyclobutane system:

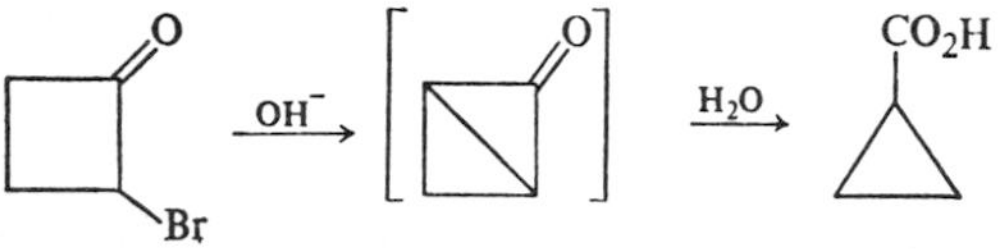

14.6 Aromatic Rearrangements

A number of rearrangements occur in aromatic compounds of the type:

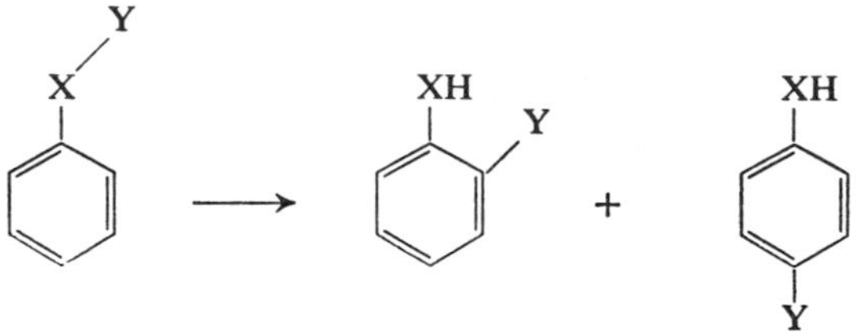

The element X is most commonly nitrogen and in some cases oxygen. Both intermolecular and intramolecular migrations are known.

(*a*) INTERMOLECULAR MIGRATION FROM NITROGEN TO CARBON

Several derivatives of aniline undergo rearrangement on treatment with acid. In the following examples, the conjugate acid of the amine eliminates an electro-philic species which then reacts at the activated *ortho* and *para* positions of the amine.

(*i*) *N-Haloanilides*. For example, *N*-chloroacetanilide and hydrochloric acid give a mixture of *o*- and *p*-chloroacetanilde in the same proportions as in the direct chlorination of acetanilide:

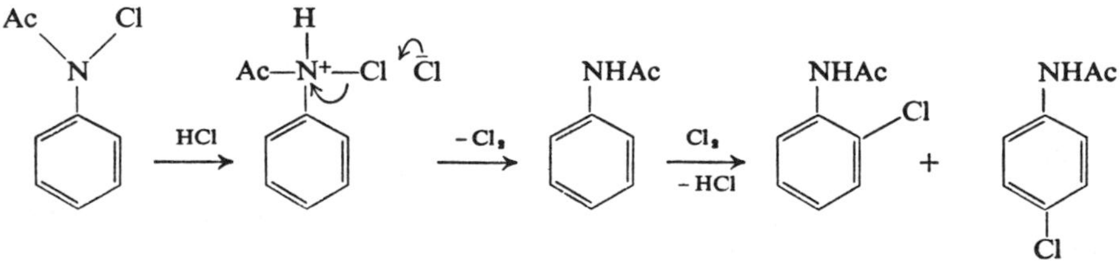

(*ii*) *N-Alkyl-N-nitrosoanilines.* The nitrosonium ion is released from the conjugate acid of the amine and nitrosates the nuclear carbon atoms, giving mainly the *p*-nitroso product:

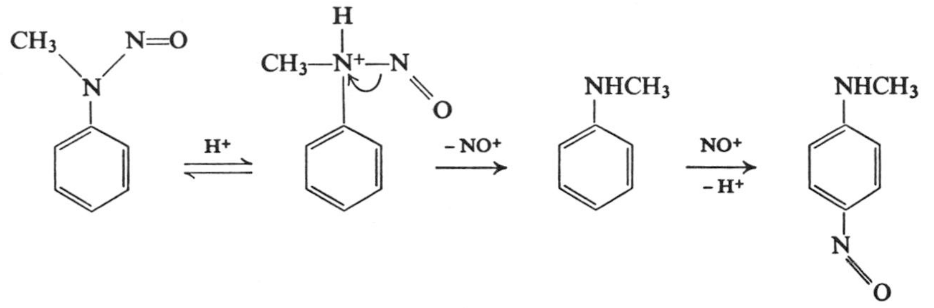

(*iii*) *N-Arylazoanilines.* The aryldiazonium cation is formed and couples, essentially completely, at the *para* carbon atom (p. 445):

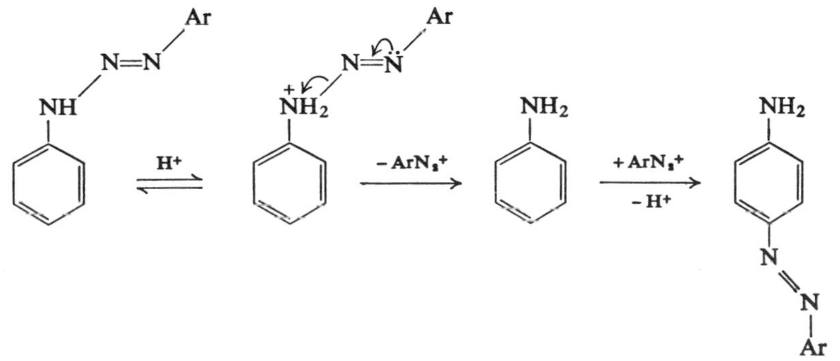

(*iv*) *N-Alkylanilines.* The mechanism of rearrangement of the salts of these amines is the same as those above, although higher temperatures (250–300°C) are required, e.g.

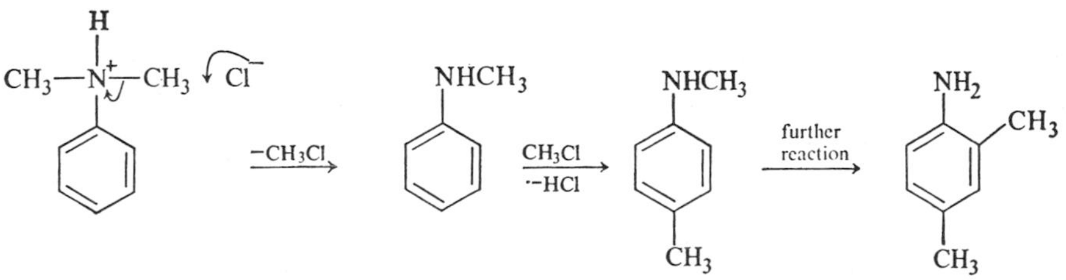

(*v*) *N-Arylhydroxylamines.* The rearrangement of arylhydroxylamines to aminophenols is mechanistically different: the conjugate acid of the hydroxylamine undergoes *nucleophilic* attack by the solvent, e.g.

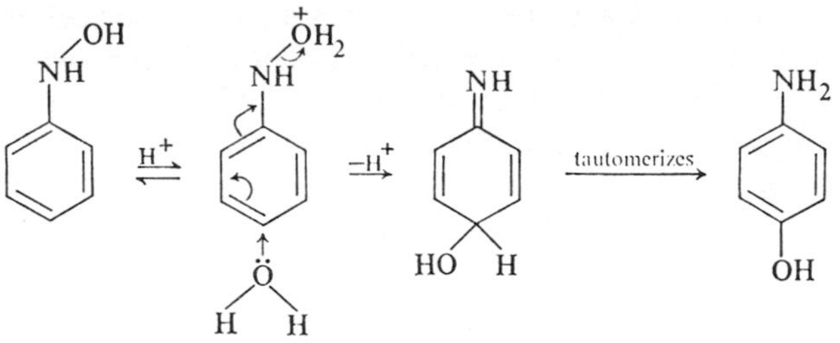

When an alcohol is used as solvent, the corresponding *p*-alkoxy compound is formed.

This rearrangement is useful in synthesis, especially since the hydroxylamine, normally made by reduction of the corresponding nitro compound, need not be isolated. Thus, the electrolytic reduction of nitro compounds in the presence of sulphuric acid leads directly to the aminophenol; e.g. *o*-chloronitrobenzene gives 2-chloro-4-hydroxyaniline [4]:

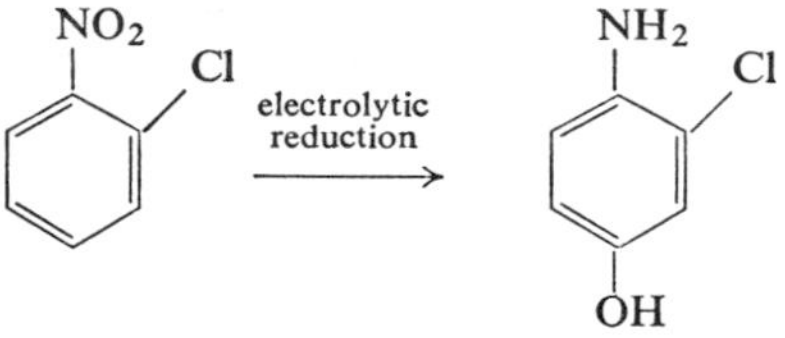

(*b*) INTERMOLECULAR MIGRATION FROM OXYGEN TO CARBON

The only common rearrangement of this type is the *Fries rearrangement* in which aryl esters are treated with Lewis acids to give *ortho* and *para* hydroxy-ketones. The complex between the ester and the Lewis acid eliminates an acylium ion which substitutes at the *ortho* and *para* positions, as in Friedel-Crafts acylation (p. 389), e.g.

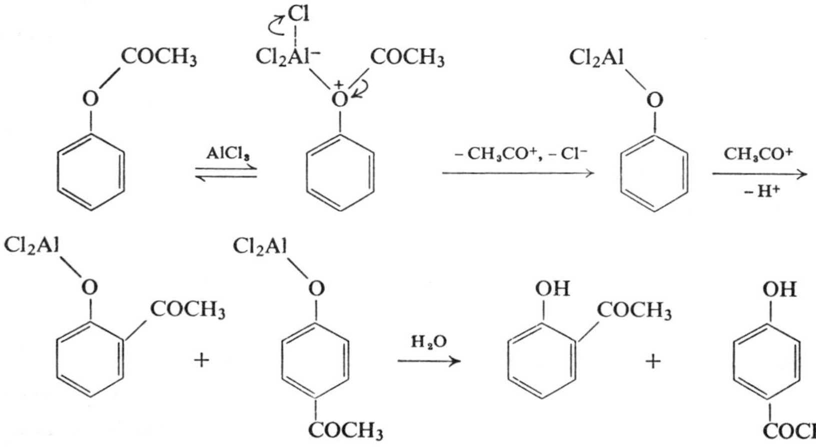

For example, the treatment of phenyl propionate with aluminium chloride at 130°C gives about 35% of o-propiophenol and 45–50% of the *para* isomer [2].

In general, low temperatures favour formation of the *para*-substituted product and high temperatures favour the *ortho*-substituted product. This appears to be because the *para*-derivative is formed the faster (kinetic control) but the *ortho*-derivative is thermodynamically the more stable by virtue of possessing a chelate system:

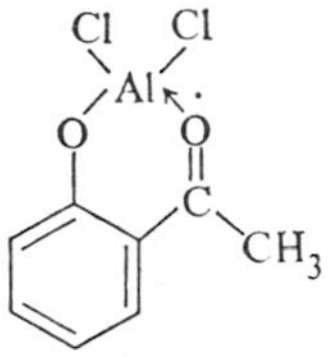

(c) INTRAMOLECULAR MIGRATION FROM NITROGEN TO CARBON
The mechanisms of these reactions are not fully understood, although it is known in each case that the migrating group does not become detached from the aromatic system during rearrangement. Thus, in the third example, all attempts to obtain 'crossed' benzidines by reaction on a mixture of two hydrazo-benzenes have been unsuccessful.

(i) *Phenylnitramines.* These compounds rearrange on being heated with acid, giving mainly the o-nitro-derivative, e.g.

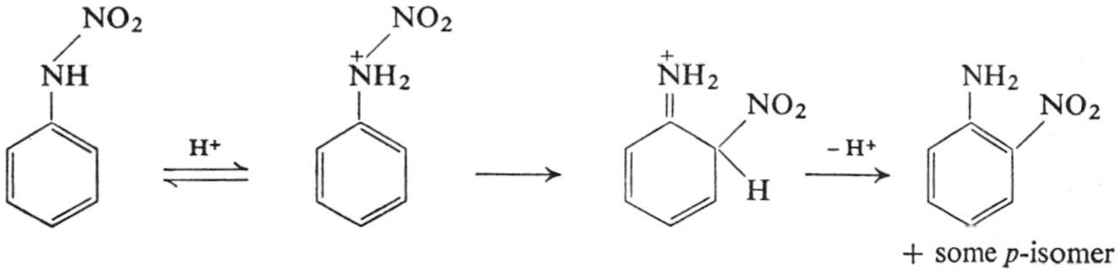

+ some *p*-isomer

(ii) *Phenylsulphamic acids.* These compounds rearrange on being heated, giving mainly the o-sulphonic acid derivatives which, at higher temperatures, are converted into the p-sulphonic acids, e.g.

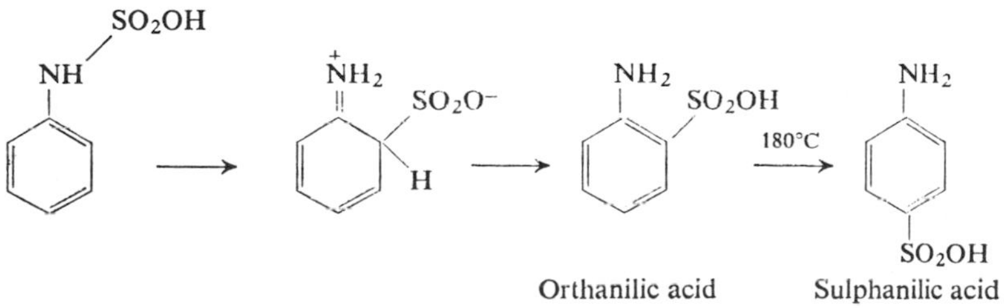

Orthanilic acid   Sulphanilic acid

(*iii*) *Hydrazobenzenes*. Hydrazobenzenes give benzidines on treatment with acid, e.g.

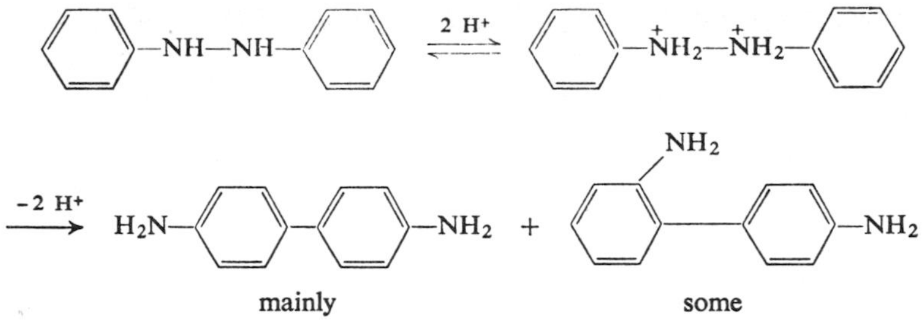

mainly   some

Since hydrazobenzenes are readily obtained by the reduction of aromatic nitro compounds (p. 647), this rearrangement gives access to 4,4'-disubstituted biphenyls.

(*d*) INTRAMOLECULAR MIGRATION FROM OXYGEN TO CARBON
The Claisen rearrangement of aryl allyl ethers to allylphenols is a [3,3]-sigmatropic reaction which was described earlier (p. 316). An example is:

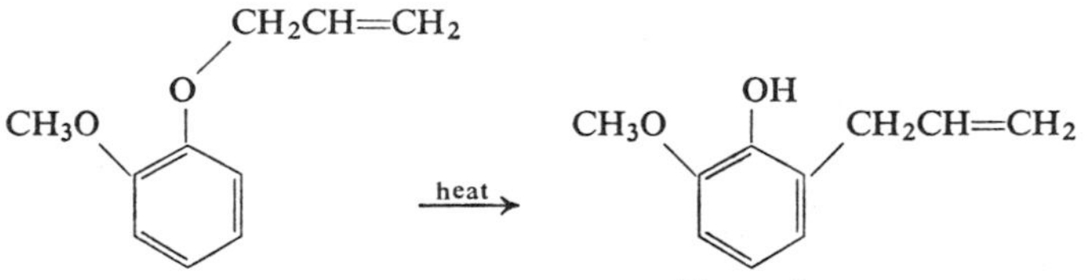

o-Eugenol

Further Reading

BACHMANN, W. E., and STRUVE, W. S., 'The Arndt-Eistert synthesis,' *Organic Reactions*, 1942, **1**, 38.

CAPON, B., 'Neighbouring-group participation,' *Quarterly Reviews*, 1964, **18**, 45.

DONARUMA, L. G., and HELDT, W. Z., 'The Beckmann rearrangement,' *Organic Reactions*, 1960, **11**, 1.

HASSALL, C. H., 'The Baeyer-Villiger oxidation of aldehydes and ketones,' *Organic Reactions*, 1957, **9**, 73.

KENDE, A. S., 'The Favorskii rearrangement of haloketones,' *Organic Reactions*, 1960, **11**, 261.

PINE, S. H., 'The Base-promoted rearrangements of quaternary ammonium salts,' *Organic Reactions*, 1970, **18**, 403.

SMITH, P. A. S., and BAER, D. R., 'The Demjanov and Tiffeneau-Demjanov ring expansions,' *Organic Reactions*, 1960, **11**, 157.

SMITH, P. A. S., 'The Curtius reaction,' *Organic Reactions*, 1946, **3**, 337.
WALLIS, E. S. and LANE, J. F., 'The Hofmann reaction,' *Organic Reactions*, 1946, **3**, 267.
WOLFF, H., 'The Schmidt reaction,' *Organic Reactions*, 1946, **3**, 307.

Problems

1. What products would you expect from the following reactions?

(a) $(CH_3)_3C\!-\!CH\!=\!CH_2 + HCl \longrightarrow$

(b)

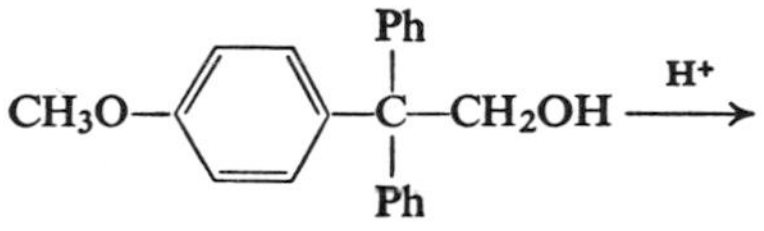

$$CH_3O\!-\!\!\!\bigcirc\!\!\!-\!\!\underset{\underset{Ph}{|}}{\overset{\overset{Ph}{|}}{C}}\!-\!CH_2OH \xrightarrow{\ H^+\ }$$

(c)

$$Ph\!-\!\underset{\underset{OH}{|}}{\overset{\overset{Ar}{|}}{C}}\!-\!\underset{\underset{OH}{|}}{\overset{\overset{Ar}{|}}{C}}\!-\!Ph \xrightarrow{\ H^+\ }$$

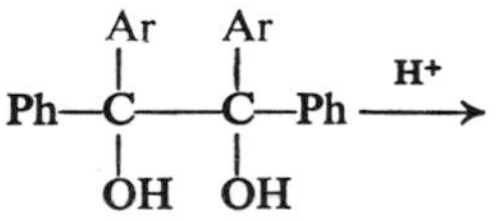

(Ar = *p*-methoxyphenyl)

(d)

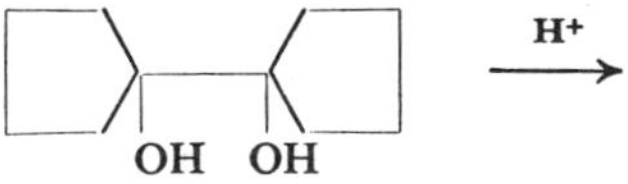

$\xrightarrow{\ H^+\ }$

(e)

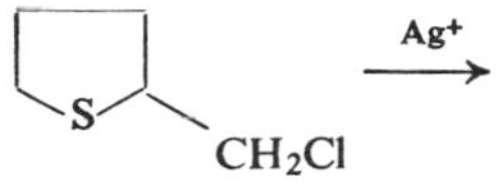

$\xrightarrow{\ Ag^+\ }$

(f) $Ph_2\underset{\underset{OH}{|}}{C}\!-\!C\!\equiv\!C\!-\!Ph \xrightarrow{\ H^+\ }$

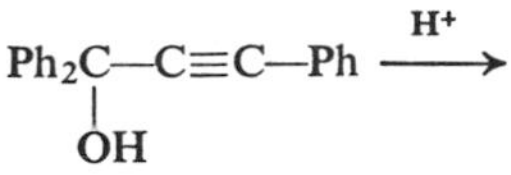

(g)

$\xrightarrow{\ H^+\ }$

(h)

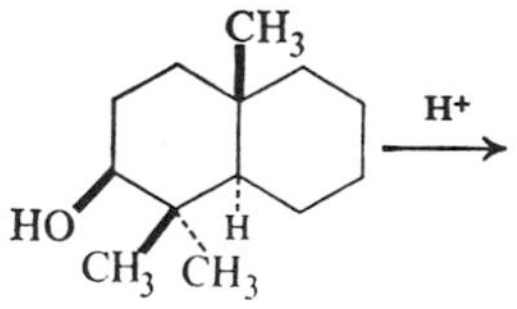

$\xrightarrow{\ H^+\ }$

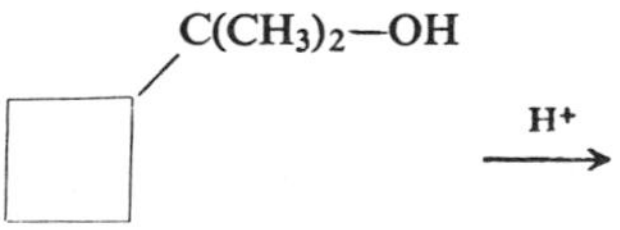

(i) $CHO\!-\!CHO \xrightarrow{\ OH^-\ }$

(j)

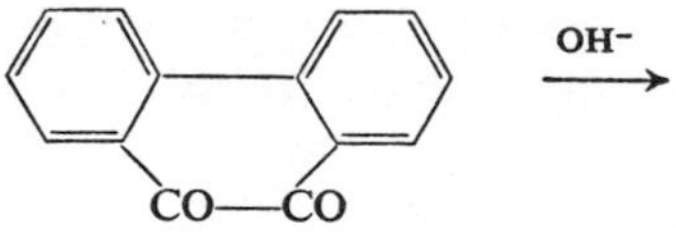

$\xrightarrow{\text{OH}^-}$

(k)

$\xrightarrow{\text{PCl}_5}$

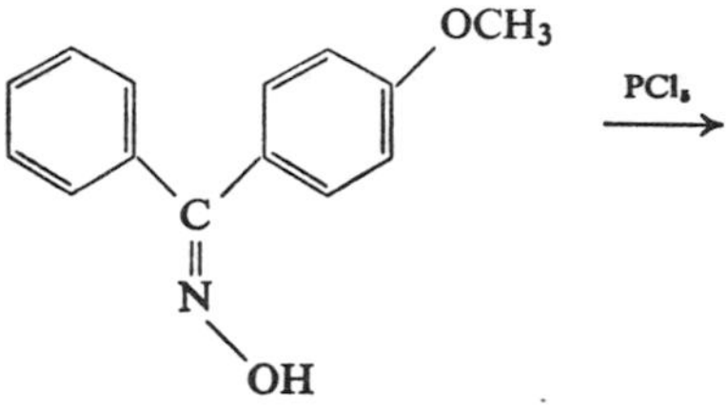

(l) $PhCOCH_3 + CH_2N_2 \longrightarrow$

2. Account for the following:

(i) *cis*- and *trans*-1,2-dimethylcyclohexane-1,2-diol give different products on treatment with concentrated sulphuric acid.

(ii) The dehydration of methyl-t-butylcarbinol gives tetramethylethylene.

(iii) $^{14}CH_3{-}CHPh{-}CH(NH_2){-}CH_3 \xrightarrow{\text{HNO}_2} {}^{14}CH_3{-}CHPh{-}CH(OH){-}CH_3$
                                                                        50%

+ $CH_3{-}CHPh{-}CH(OH){-}^{14}CH_3$
                              50%

(iv) In the pinacol rearrangements of $PhMeC(OH){-}C(OH)PhMe$ and $Ph_2C(OH){-}C(OH)Me_2$, a phenyl group migrates in the former case but a methyl group migrates in the latter.

3. Outline routes to the following from readily available materials:

(a) $(CH_3)_3C{-}CH_2{-}CO_2H$ (b) $(CH_3)_3C{-}NH_2$

(c)

(d)

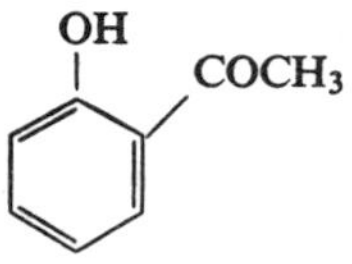

(e)

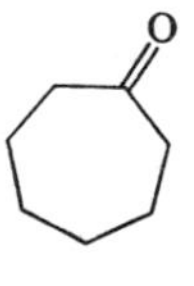

(f)

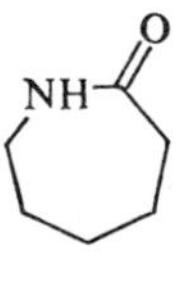

(g)

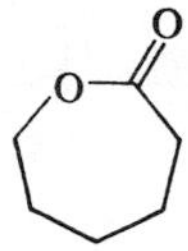

(h) CO_2CH_3

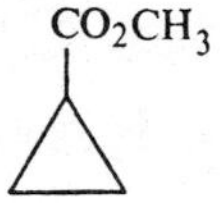

(i)

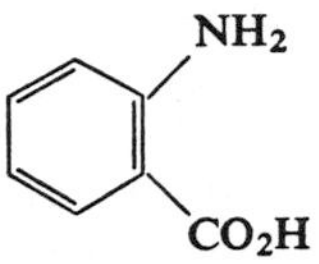

(j)

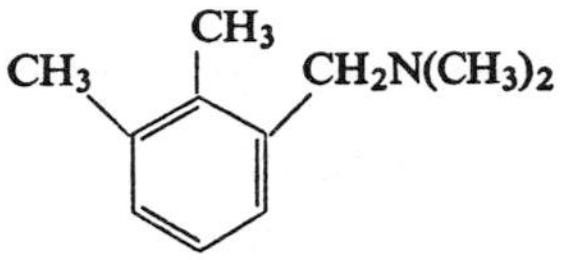

15. Reagents Containing Phosphorus, Sulphur, or Boron

15.1 Introduction

Compared with most of the synthetic methods discussed so far, those based on reagents which contain phosphorus, sulphur, or boron have been introduced relatively recently, almost all within the last 25 years. There is still very active research towards further developments.

Phosphorus-containing reagents owe their usefulness to three characteristics of phosphorus chemistry: the ease with which phosphorus(III) is converted into phosphorus(V); the relatively strong bonds formed by phosphorus to oxygen and to sulphur; and the availability of $3d$ orbitals for bonding. In each of these respects phosphorus differs from nitrogen.

For example, hydroxide ion reacts with a phosphonium salt at phosphorus, to form a P—O bond, and gives ultimately a phosphine oxide, for example:

$$\overset{+}{Me_3P}CH_2Ph + OH^- \longrightarrow Me_3PCH_2Ph \overset{OH^-}{\longrightarrow}$$
$$\underset{OH}{|}$$

$$Me_3P\overset{\frown}{-}CH_2Ph \longrightarrow Me_3P{=}O + PhCH_2^-$$
$$\underset{O^-}{\overset{|}{}}$$

$$(PhCH_2^- + H_2O \longrightarrow PhCH_3 + OH^-)$$

In contrast, reaction with a quaternary ammonium salt occurs at carbon (S_N2 reaction),

$$HO^-\overset{\frown}{}CH_2\overset{\frown}{-}\overset{+}{N}Me_3 \longrightarrow PhCH_2OH + NMe_3$$
$$\underset{Ph}{|}$$

or, if the quaternary ion has a β-hydrogen atom, by the E2 process (p. 121).

The strong affinity of phosphorus for oxygen is shown, too, in the *Arbuzov reaction* of triethyl phosphite with an alkyl bromide:

$$(EtO)_3P\overset{\frown}{:}\,\overset{\frown}{CH_2{-}Br} \longrightarrow (EtO)_3\overset{+}{P}{-}CH_2R + Br^-$$
$$\underset{R}{|}$$

$$(EtO)_2\overset{+}{P}{-}CH_2R \longrightarrow (EtO)_2P{-}CH_2R + CH_3CH_2Br$$
$$\underset{\underset{Br\,\overset{\frown}{}CH_2CH_3}{|}}{\overset{|}{O}}\qquad\qquad\underset{O}{\overset{\parallel}{}}$$

Sulphur-containing reagents also owe their usefulness in part to the capacity of sulphur to utilise $3d$ orbitals for bonding and to occur in valence states higher than 2. However, although the S—O bond is fairly strong, the tendency for its formation does not dominate sulphur chemistry in the way that P—O bond formation dominates phosphorus chemistry. This generates some useful differences between the two groups of reagents.

Two main factors underlie the usefulness of boron-containing reagents. The first is the ability of diborane and of organic boranes to add to olefinic bonds, for example,

$$6 \text{ R}-\text{CH}=\text{CH}_2 + \text{B}_2\text{H}_6 \longrightarrow 2 \text{ (R}-\text{CH}_2-\text{CH}_2)_3\text{B}$$

The second is the case of displacement of the boron atom from these adducts by a variety of reagents, often by a reaction which makes use of the ability of the boron atom to accept an electron-pair.

15.2 Phosphorus-containing Reagents

(a) REACTIONS OF PHOSPHORUS YLIDS

Phosphorus ylids are prepared by quaternizing a tervalent phosphorus compound with an alkyl halide and treating the salt with butyl-lithium, phenyl-lithium, or sodium hydride. The powerfully basic reagent abstracts a proton from the carbon atom adjacent to quaternary phosphorus, e.g.

$$\text{R}_2\text{CHBr} + \text{PPh}_3 \longrightarrow \text{R}_2\text{CH}-\overset{+}{\text{P}}\text{Ph}_3 \text{ Br}^- \xrightarrow{\text{PhLi}} \text{R}_2\overset{-}{\text{C}}-\overset{+}{\text{P}}\text{Ph}_3 + \text{PhH} + \text{LiBr}$$

These compounds, known as alkylidenephosphoranes, are appropriately represented as hybrids of two structures,

$$\text{R}_2\overset{-}{\text{C}}-\overset{+}{\text{P}}\text{Ph}_3 \longleftrightarrow \text{R}_2\text{C}=\text{PPh}_3$$

This is consistent with the much more ready formation of phosphorus ylids compared with their nitrogen analogues (p. 474); that is, bonding within the phosphorus ylid is increased by incorporation of a $3d$ orbital, as represented by the structure $\text{R}_2\text{C}=\text{PPh}_3$, whereas d orbitals of low enough energy are not available to nitrogen.

The importance of alkylidenephosphoranes lies in their reactions with aldehydes and ketones to form olefins:

$$\text{R}'_2\text{C}=\text{O} + \text{R}_2\overset{-}{\text{C}}-\overset{+}{\text{P}}\text{R}''_3 \longrightarrow \text{R}'_2\text{C}=\text{CR}_2 + \text{R}''_3\text{P}=\text{O}$$

The reaction, known after *Wittig*, is thought to occur in two stages: the phosphorane adds as a carbon nucleophile to the carbonyl group, and the resulting intermediate reacts *via* a cyclic transition state to form the products:

$$R'_2C=PR''_3 \longleftrightarrow R'_2\bar{C}-\overset{+}{P}R''_3 \quad \overset{R_2C=O}{\longrightarrow} \quad \underset{R'_2C-\overset{+}{P}R''_3}{R_2C-O^-}$$

$$\longrightarrow \left[\begin{matrix} R_2C\text{---}O \\ R'_2C\text{---}PR''_3 \end{matrix} \right] \longrightarrow R_2C=CR'_2 + R''_3P=O$$

Simple phosphoranes are very reactive and are unstable in the presence of air or moisture. They are therefore prepared in a scrupulously dry solvent (usually tetrahydrofuran) under nitrogen or argon, and the carbonyl compound is added as soon as the phosphorane has been formed.

More stable phosphoranes are obtained when a $-M$ substituent is adjacent to the anionic carbon, for example,

$$Ph_3\overset{+}{P}-\overset{-}{C}H-C\equiv N \longleftrightarrow Ph_3\overset{+}{P}-CH=C=N^-$$

These are usually isolable, crystalline compounds. However, although they react with aldehydes, they fail to do so with ketones, and for this purpose a modification has been introduced in which triphenylphosphine is replaced by triethyl phosphite (*Wadsworth-Emmons reaction*), for example,

$$NC-CH_2Br + P(OEt)_3 \longrightarrow NC-CH_2-\overset{+}{P}(OEt)_3Br^- \overset{PhLi}{\longrightarrow}$$

$$NC-\overset{-}{C}H-\overset{+}{P}(OEt)_3 \overset{R_2C=O}{\longrightarrow} NC-CH=CR_2 + (EtO)_3P=O$$

The importance of the Wittig reaction for the synthesis of olefins stems from two properties. First, it is specific for the conversion $\overset{\diagdown}{\diagup}C=O \rightarrow \overset{\diagdown}{\diagup}C=C\overset{\diagup}{\diagdown}$. For example, the obvious alternative for the conversion of an aldehyde or ketone into an olefin involves the use of a Grignard reagent and can give a mixture of products, e.g.

mainly   some

In contrast, the methylenephosphorane from triphenylphosphine and methyl bromide followed by butyl-lithium reacts with cyclohexanone to give methylenecyclohexane as the only olefinic product, in up to 40% overall yield [5]:

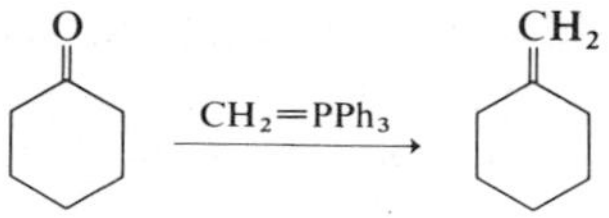

Secondly, the Wittig reaction can be carried out on aldehydes or ketones which contain a variety of other functional groups (e.g. hydroxyl, ester, acetylene) since these are unaffected. This is again in contrast to the use of Grignard reagents.

The reaction has been used in recent years in a number of syntheses of naturally occurring compounds. It is especially suitable for making rather sensitive compounds, such as polyunsaturated ones. An example is an alternative route to Vitamin A (21.1):

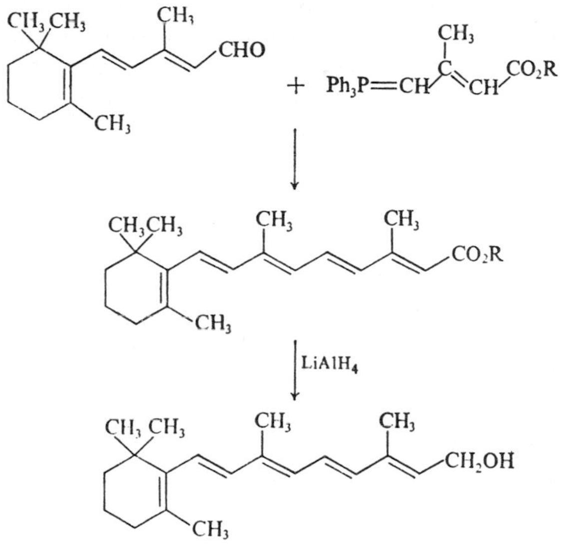

Vitamin A

Developments of the Wittig reaction include the following:

The use of $Ph_3P=CH-OCH_3$ as the phosphorane, to introduce an aldehydic group, e.g.

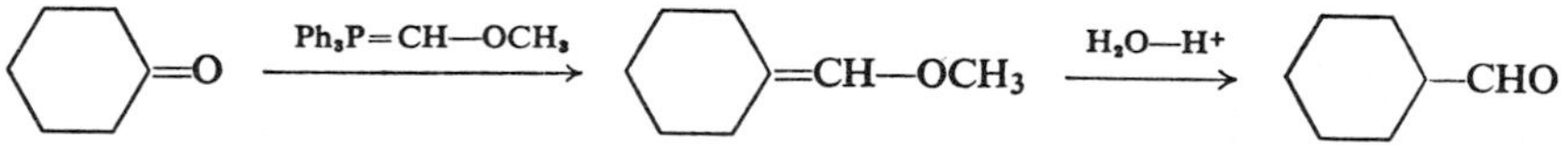

Cyclohexanecarboxaldehyde

The use of bifunctional Wittig reagents, e.g.

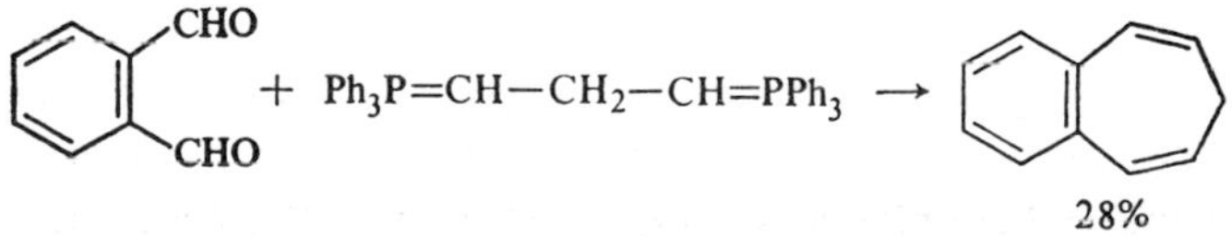

28%

The formation of carbodi-imides (used in the formation of the CO—NH bond in peptide synthesis; p. 361):

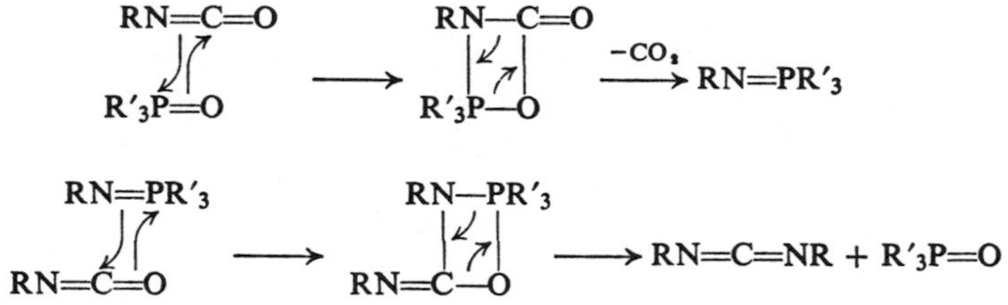

$$RN{=}C{=}NR + R'_3P{=}O$$

A carbodi-imide

(b) REDUCTIVE CYCLIZATION OF NITRO-COMPOUNDS

When an aromatic nitro-compound is heated with triethyl phosphite in an inert solvent such as t-butylbenzene, the oxygen atoms of the nitro group are transferred to phosphorus and the nitrogen atom is inserted into a double bond or aromatic ring which is stereochemically suited for the reaction, for example:

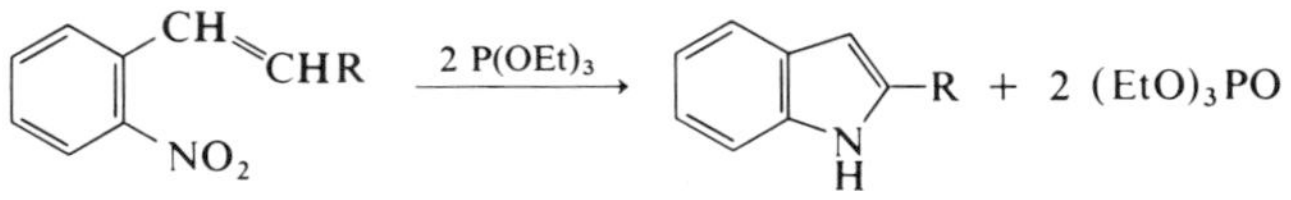

The process probably involves a nitrene as the reactive intermediate, and is a further example of phosphorus's strong affinity for oxygen:

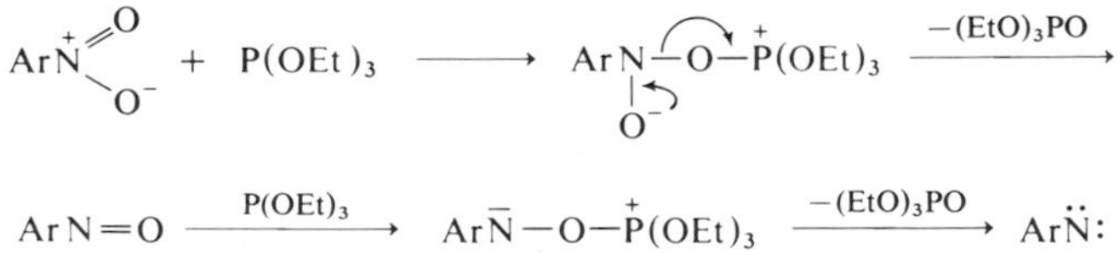

The reaction is capable of wide variation, for example:

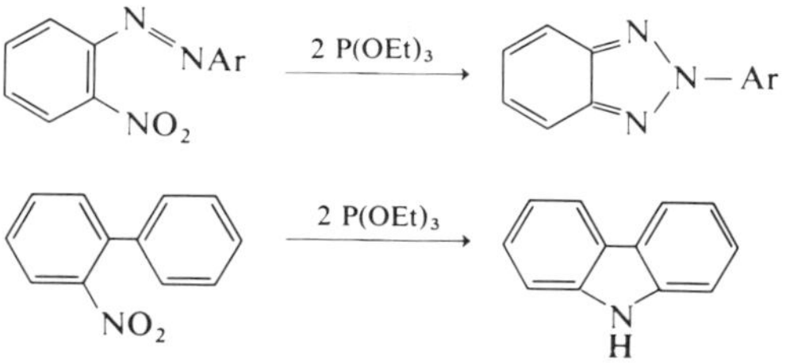

(c) SYNTHESIS OF OLEFINS FROM 1,2-DIOLS

The affinity of phosphorus for sulphur is exploited in a synthesis of olefins from 1,2-diols, via thionocarbonate intermediates:

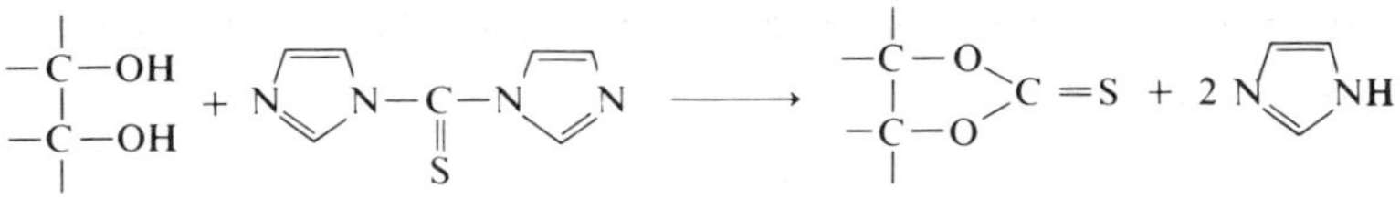

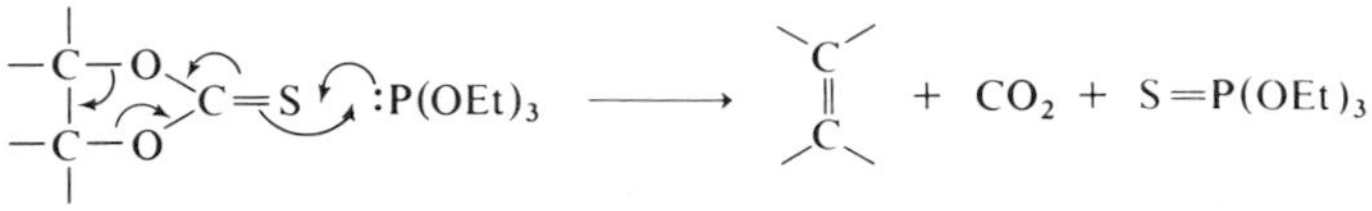

The especial value of the reaction lies in its stereospecificity. For example, the highly strained *trans*-cyclo-octene has been made in 75% yield from the *cis*-isomer by conversion into the *trans*-diol (p. 571) and then elimination; the essence of the procedure is that, in order to form the thionocarbonate, the diol has to adopt a conformation which ensures that the *trans*-olefin is formed:

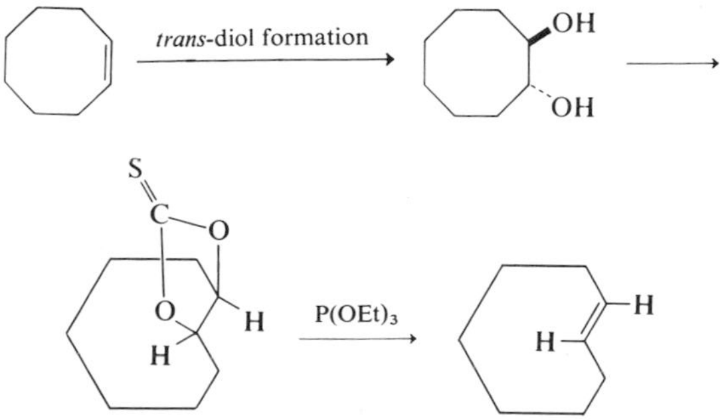

15.3 Sulphur-containing Reagents

(*a*) REACTIONS OF SULPHUR YLIDS

Sulphur ylids are prepared, analogously to phosphorus ylids, from sulphonium salts and an exceptionally strong base. The latter is usually the 'dimsyl' anion, $CH_3{-}SO{-}CH_2^-$, formed from dimethyl sulphoxide and sodium hydride:

$$CH_3{-}\underset{\underset{O}{\|}}{S}{-}CH_3 + NaH \xrightarrow{-H_2} \left[CH_3{-}\underset{\underset{O}{\|}}{S}{-}CH_2^- \longleftrightarrow CH_3{-}\underset{\underset{O^-}{|}}{S}{=}CH_2\right]Na^+$$

Ylid formation then occurs as follows, e.g.

$$(CH_3)_3S^+ + CH_3{-}SO{-}CH_2^- \rightleftharpoons \overset{-}{C}H_2{-}\overset{+}{S}(CH_3)_2 + CH_3{-}SO{-}CH_3$$

Sulphoxonium salts can also be used, e.g.

$$(CH_3)_3\overset{+}{S}{=}O + CH_3{-}SO{-}CH_2^- \rightleftharpoons \overset{-}{C}H_2{-}\underset{\underset{O}{\|}}{\overset{+}{S}}(CH_3)_2 + CH_3{-}SO{-}CH_3$$

The ylids are thought to be stabilized by bonding which involves a sulphur $3d$ orbital, as represented by

$$\overset{-}{C}H_2-\overset{+}{S}(CH_3)_2 \quad\longleftrightarrow\quad CH_2{=}S(CH_3)_2$$

$$\underset{\underset{O}{\|}}{\overset{-}{C}H_2-\overset{+}{S}(CH_3)_2} \quad\longleftrightarrow\quad \underset{\underset{O}{\|}}{CH_2{=}S(CH_3)_2}$$

(*i*) *Reactions with carbonyl compounds.* As with phosphorus ylids, the first stage of the reaction of a sulphur ylid with an aldehyde or ketone consists of nucleophilic addition:

$$\underset{\overset{+}{C}H_2-\overset{+}{S}(CH_3)_2}{R_2C{=}O} \quad\longrightarrow\quad \underset{CH_2-\overset{+}{S}(CH_3)_2}{R_2C-O^-}$$

However, whereas the corresponding adduct from a phosphorus ylid forms an olefin and the phosphine oxide (p. 485), the sulphur-containing adduct undergoes intramolecular nucleophilic substitution, to give an epoxide:

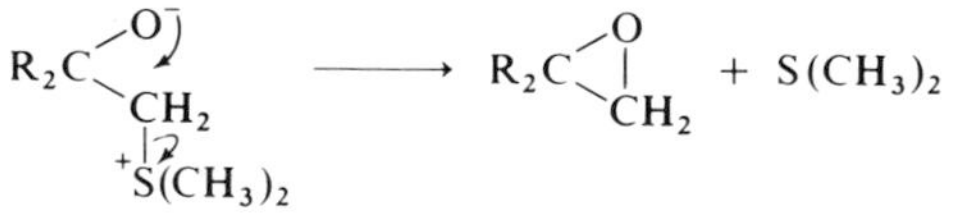

$$R_2C\underset{\underset{\overset{+}{S}(CH_3)_2}{|}}{\overset{O}{\diagdown}CH_2} \quad\longrightarrow\quad R_2C\underset{CH_2}{\overset{O}{\diagup|}} + S(CH_3)_2$$

This illustrates the relatively weaker affinity of sulphur for oxygen compared with that of phosphorus.

Reaction may be brought about with an ylid derived from either a sulphonium salt or a sulphoxonium salt. Yields are generally good; for example, cyclohexanone and the ylid from trimethylsulphoxonium iodide give 70% of the epoxide [5]:

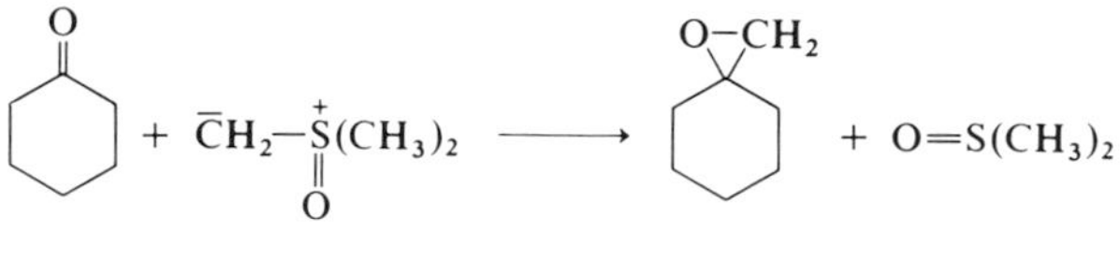

Since epoxides undergo rearrangement to carbonyl compounds when treated with Lewis acids (p. 570), the overall process can be used for the conversion $\diagdown{C}{=}O \rightarrow \diagdown{C}H{-}CHO$; for example,

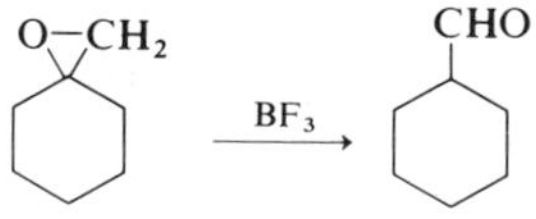

A particularly useful sulphur ylid is the cyclopropyl-substituted one whose synthesis is shown:

$$I(CH_2)_3Cl + Ph_2S \xrightarrow{AgBF_4} Ph_2\overset{+}{S}(CH_2)_3Cl + AgI + BF_4^-$$

$$Ph_2\overset{+}{S}(CH_2)_3Cl \xrightarrow{NaH} \left[Ph_2\overset{+}{S}-\overset{-}{C}H-CH_2-CH_2Cl\right] \xrightarrow{-Cl^-} Ph_2\overset{+}{S}-CH\overset{\diagup CH_2}{\diagdown CH_2}$$

$$Ph_2\overset{+}{S}-CH\overset{\diagup CH_2}{\diagdown CH_2} \xrightarrow{CH_3SOCH_2^-} Ph_2\overset{+}{S}-\overset{-}{C}\overset{\diagup CH_2}{\diagdown CH_2}$$

It is not necessary to make the ylid separately; the precursor sulphonium salt is treated with potassium hydroxide in the presence of a carbonyl compound, and as the ylid is formed it reacts with the carbonyl compound:

$$Ph_2\overset{+}{S}-CH\overset{\diagup CH_2}{\diagdown CH_2} \underset{\xleftarrow{\hspace{1cm}}}{\overset{KOH}{\xrightarrow{\hspace{1cm}}}} Ph_2\overset{+}{S}-\overset{-}{C}\overset{\diagup CH_2}{\diagdown CH_2} \xrightarrow{R_2C=O} R_2C\overset{\diagup O}{\diagdown}\underset{\underset{CH_2}{|}}{C}\diagdown_{CH_2} + Ph_2S$$

The resulting epoxides—oxaspiropentanes—are particularly strained and readily subject to ring-opening reactions. For example, treatment with an acid gives cyclobutanone derivatives:

$$R_2C\overset{\diagup O}{\diagdown}\underset{\underset{CH_2}{|}}{C}\diagdown_{CH_2} \xrightarrow{H^+} R_2C\overset{\overset{\overset{H}{|}}{\diagup}\overset{+}{O}}{\diagdown}\underset{\underset{CH_2}{|}}{C}\diagdown_{CH_2} \longrightarrow$$

$$\left[R_2\overset{+}{C}\overset{O\diagdown^H}{\diagup}\underset{CH_2-CH_2}{\diagdown}C\right] \xrightarrow{-H^+} \underset{CH_2-CH_2}{R_2C-C=O}$$

These in turn can undergo useful transformations, such as ring-expansion to γ-lactones on treatment with alkaline hydrogen peroxide:*

$$\underset{CH_2-CH_2}{R_2C-C=O} \xrightarrow{H_2O_2-OH^-} \underset{CH_2-CH_2}{R_2C\overset{O}{\diagdown}C\overset{\diagup O}{\diagdown}}$$

*This is an example of the Baeyer-Villiger reaction (p. 471). However, the Baeyer-Villiger reaction usually requires a peracid; in this case, the milder reagent no doubt succeeds because of the relief of ring-strain during reaction.

(*ii*) *Rearrangements.* Sulphur ylids rearrange in the same way as nitrogen ylids (p. 474), for example:

When one of the substituents on the sulphur atom is allylic, an alternative pericyclic reaction can occur, represented by:

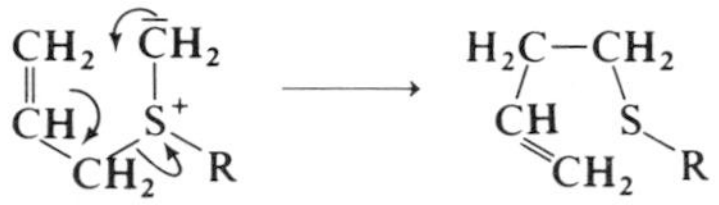

This can be categorized as a [3,2]-sigmatropic rearrangement. It can be usefully employed in synthesis because, first, a sulphur substituent can be readily introduced by making use of the nucleophilic property of sulphides and, secondly, a sulphur group can be removed after the rearrangement. In the following example, the [3,2]-shift occurs in preference to the [1,2]-shift, and the removal of sulphur at the end is made possible by the fact that allylic systems are susceptible to hydrogenolysis (p. 627):

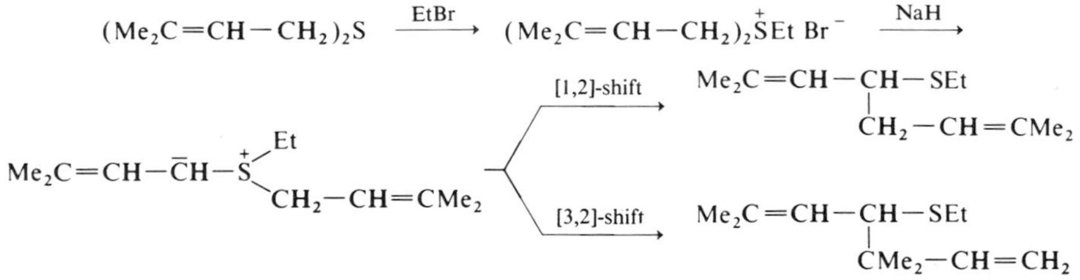

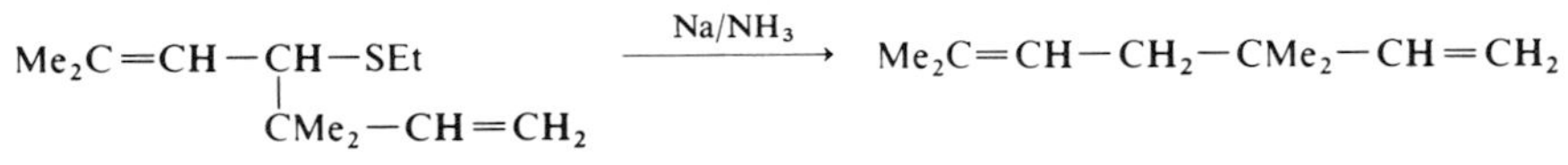

Aza- and oxa-sulphonium salts also give ylids which can rearrange, and special use can be made of this in the aromatic series. The crucial reactions are of the types,

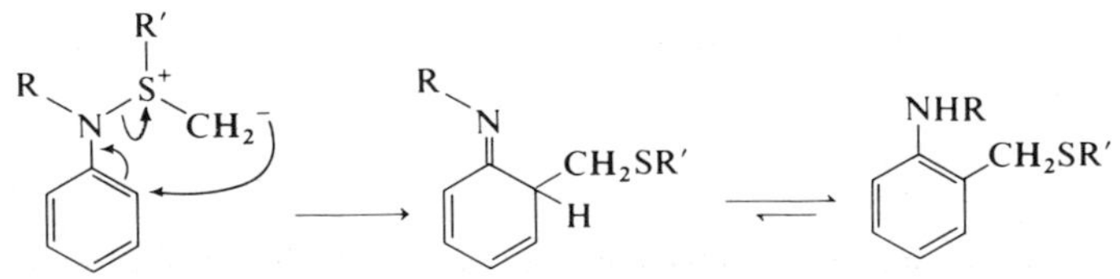

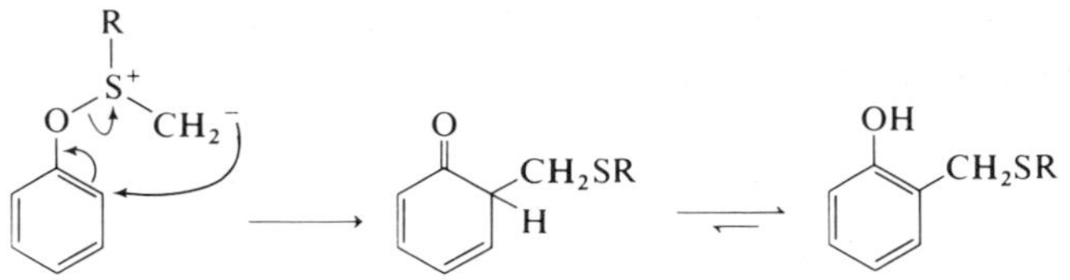

These [3,2]-sigmatropic rearrangements are analogues of the Sommelet re-arrangement (p. 474).

In the nitrogen series, treatment of an aniline with t-butyl hypochlorite gives the N-chloro-compound which, with an organic sulphide, gives the azasulphonium salt, e.g.

$$PhNH_2 \xrightarrow{(CH_3)_3C-OCl} PhNHCl \xrightarrow{(CH_3)_2S} PhNH-\overset{+}{S}(CH_3)_2 \ Cl^-$$

The ylid is generated with sodium methoxide and spontaneously rearranges; reduction over Raney nickel then yields an o-alkylaniline, e.g.

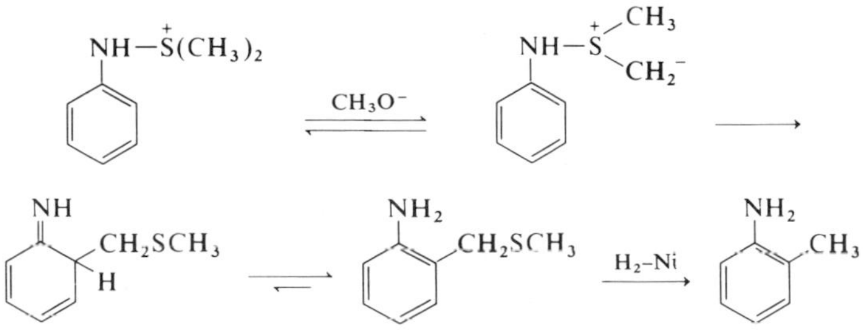

A simple variant enables an o-butyl group to be introduced, e.g.

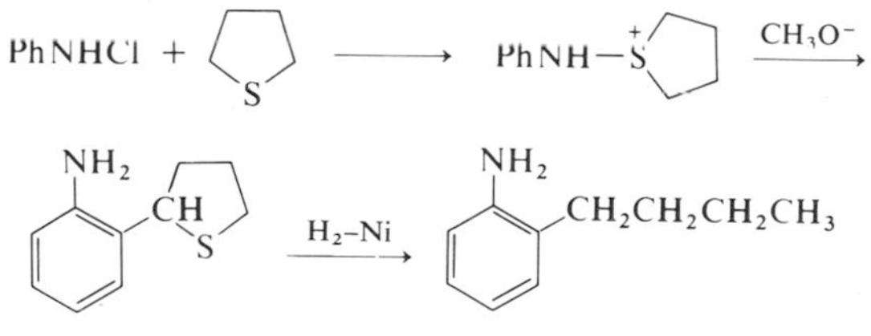

A further variant is to use a β-keto-sulphide. After rearrangement of the ylid,* intramolecular nucleophilic displacement on the carbonyl group occurs spontaneously to form the indole ring, and the sulphur substituent can be removed by reduction, e.g.

---

*The −M effect of the carbonyl group stabilizes the ylid formed by loss of a proton from the adjacent carbon atom and allows the use of a weaker base than methoxide ion.

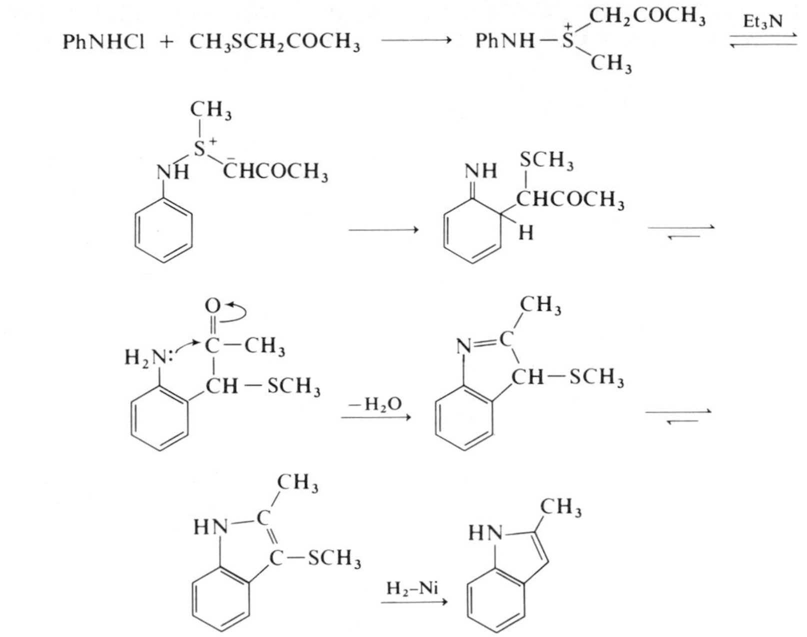

$$PhNHCl + CH_3SCH_2COCH_3 \longrightarrow PhNH\overset{+}{-}\underset{\overset{|}{CH_3}}{\overset{\nearrow CH_2COCH_3}{S}} \quad \underset{\overset{\longrightarrow}{\longleftarrow}}{Et_3N}$$

In the oxygen series, *o*-alkylation of phenols can be effected in a similar manner to that of anilines, e.g.

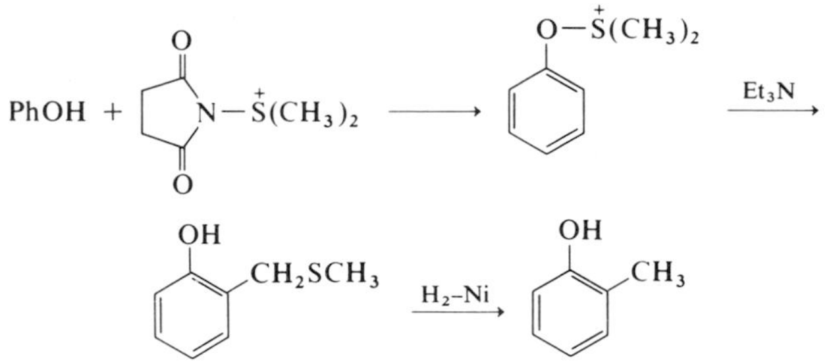

(b) REACTIONS OF THE DIMSYL ANION

Dimethyl sulphoxide is very weakly acidic; a proton can be removed, to a significant extent, only by the strongest of bases. Sodium hydride is commonly used; the removal of hydrogen enables the reaction to go to completion:

$$CH_3-\underset{\overset{\|}{O}}{S}-CH_3 + NaH \xrightarrow{\;-H_2\;} \left[CH_3-\underset{\overset{\|}{O}}{S}-\overset{-}{C}H_2 \longleftrightarrow CH_3-\underset{\overset{|}{O^-}}{S}{=}CH_2 \right] Na^+$$

The delocalized anion above is usually called the dimsyl anion. It is very reactive towards electrophilic carbon, and its use in synthesis is based on this property together with the ease of removal of the sulphur substituent either by reduction or thermally.* For example,

$$CH_3(CH_2)_{10}CH_2Br + CH_3SOCH_2^- \xrightarrow{-Br^-}$$

$$CH_3(CH_2)_{10}CH_2CH_2SOCH_3 \xrightarrow[-CH_3SOH]{heat} CH_3(CH_2)_{10}CH=CH_2$$

$$58\%$$

$$PhCO_2Et + CH_3SOCH_2^- \xrightarrow{-EtO^-} PhCOCH_2SOCH_3 \xrightarrow{Zn/HOAc} PhCOCH_3$$

Since compounds of the type $RCOCH_2SOCH_3$ contain strongly activated methylene groups, the reaction of the dimsyl anion with an ester can be used as the basis of the synthesis of more complex ketones, as for example from $PhCOCH_2SOCH_3$:

$$PhCOCH_2SOCH_3 \xrightarrow[\substack{2) \ 2 \ CH_3I}]{1) \ NaH} PhCOC(CH_3)(CH_3)SOCH_3 \xrightarrow{Zn-HOAc} PhCOCH(CH_3)_2$$

(where the central carbon bears two CH_3 groups: $PhCOCSOCH_3$ with CH_3 above and CH_3 below)

$$PhCOCH_2SOCH_3 \xrightarrow[\substack{2) \ BrCH_2CO_2Et}]{1) \ NaH} PhCOCHSOCH_3 \xrightarrow{Zn-HOAc} PhCOCH_2CH_2CO_2Et$$

(with CH_2CO_2Et substituent below the central carbon)

A synthesis of ninhydrin is based on the reaction of the dimsyl anion with diethyl phthalate:

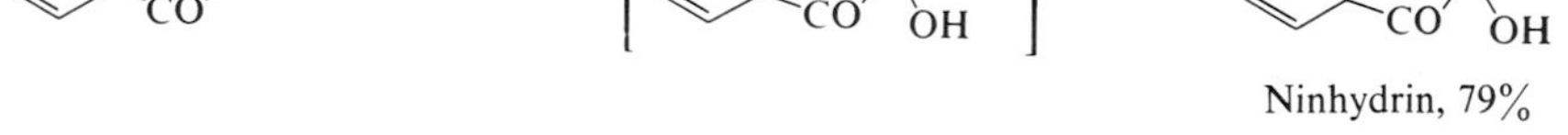

Ninhydrin, 79%

(c) SULPHOXIDE ELIMINATION

A route to sulphoxides which is an alternative to that involving the dimsyl anion is to treat a compound containing an activated C—H bond with diphenyl

*The thermal reaction is a syn-elimination (cf. the Chugaev reaction; p. 318):

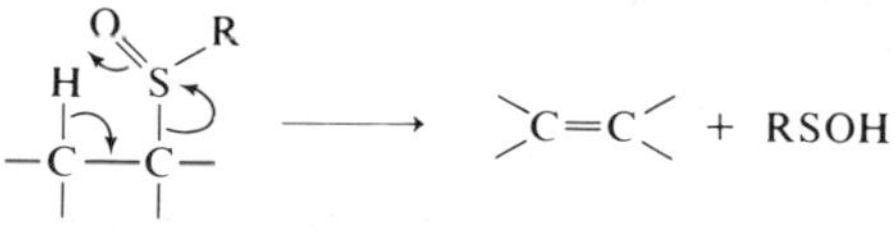

$$\underset{-C-C-}{\overset{H \quad S-R}{\overset{O}{\diagdown}}} \longrightarrow \ >C=C< \ + \ RSOH$$

or dimethyl disulphide in the presence of base, followed by oxidation of the resulting sulphide (p. 605), e.g.

$$RCH_2CH_2COR' \underset{}{\overset{base}{\rightleftharpoons}} RCH_2\bar{C}H-COR'$$

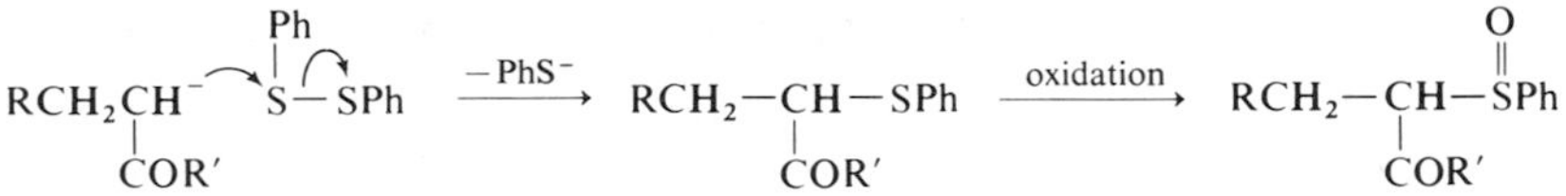

Since sulphoxides undergo elimination on heating,

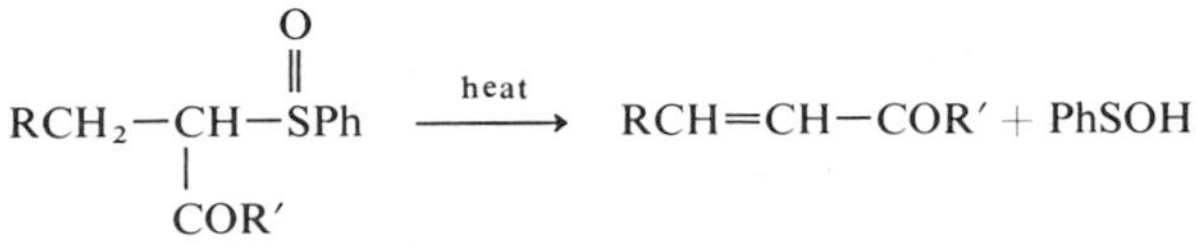

this provides a method for introducing a double bond next to a group of $-M$ type, for example:

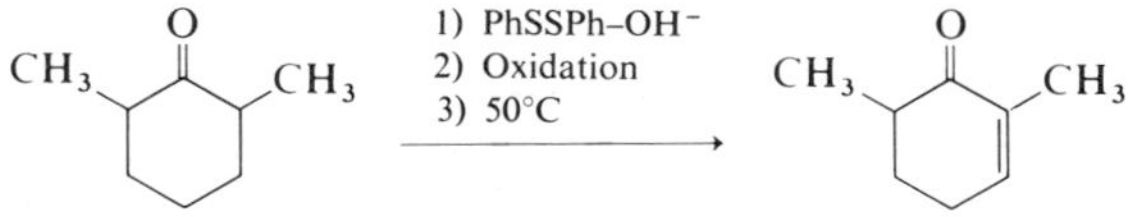

100%

If dimethyl disulphide is used, the final step requires a higher temperature (110°C).

(*d*) USE OF DITHIOACETALS: 'REVERSED POLARITY' OF CARBONYL
     COMPOUNDS

The characteristic of the carbonyl group is its susceptibility to attack by nucleophiles, e.g.

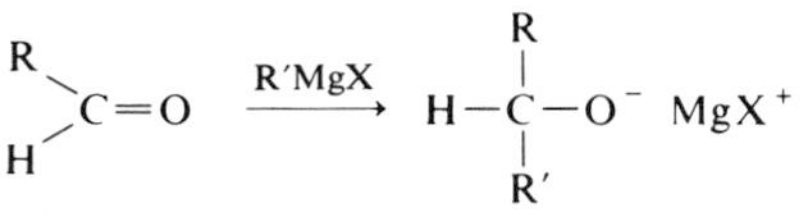

Its versatility in synthesis would be increased if it were possible to render the group itself nucleophilic, i.e. to generate RCO^-. This is not practicable, but an alternative approach is available: to attach groups X and Y to the carbonyl carbon which stabilize an adjacent negative charge, carry out reaction with the corresponding carbanion, and then replace X and Y by oxygen:

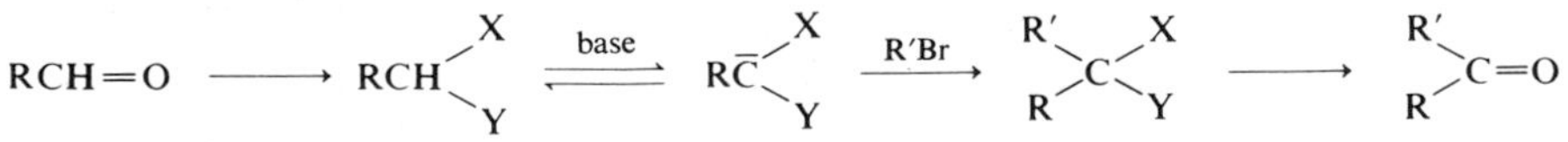

It has been found that the most satisfactory forms of X and Y are RS-groups: they can be readily introduced by an acid-catalyzed reaction between the aldehyde and a thiol, e.g.

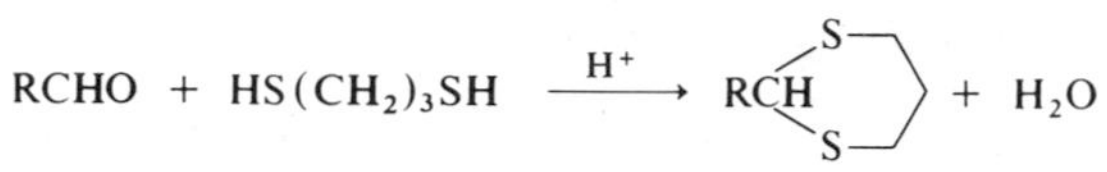

$$RCHO + HS(CH_2)_3SH \xrightarrow{\text{H}^+} RCH\underset{S}{\overset{S}{\langle}} \quad + H_2O$$

the sulphur atoms stabilize carbanions (p. 485), so making possible alkylation,

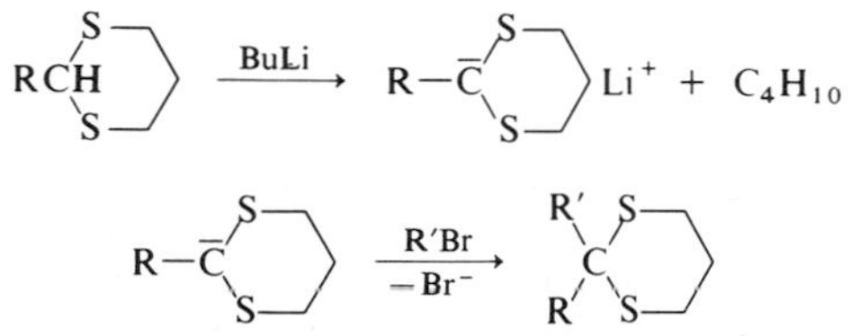

$$RCH\underset{S}{\overset{S}{\langle}} \xrightarrow{\text{BuLi}} R-\bar{C}\underset{S}{\overset{S}{\langle}} Li^+ + C_4H_{10}$$

$$R-\bar{C}\underset{S}{\overset{S}{\langle}} \xrightarrow[-\text{Br}^-]{\text{R'Br}} \underset{R'}{\overset{R'}{}}C\underset{S}{\overset{S}{\langle}}$$

and the dithioketal can be hydrolytically cleaved in the presence of mercury(II) ion,

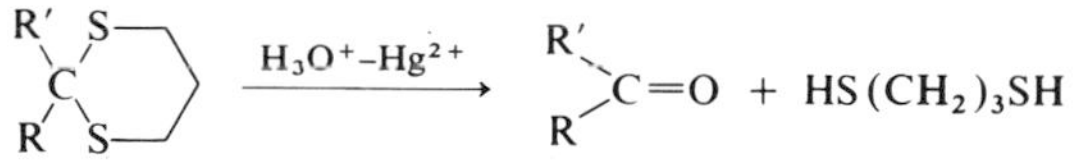

$$\underset{R}{\overset{R'}{}}C\underset{S}{\overset{S}{\langle}} \xrightarrow{\text{H}_3\text{O}^+-\text{Hg}^{2+}} \underset{R}{\overset{R'}{}}C=O + HS(CH_2)_3SH$$

The overall process corresponds to the conversion, $RCHO \rightarrow RR'CO$.

The use of the 1,3-dithian derivative of formaldehyde enables two alkyl groups to be introduced sequentially,

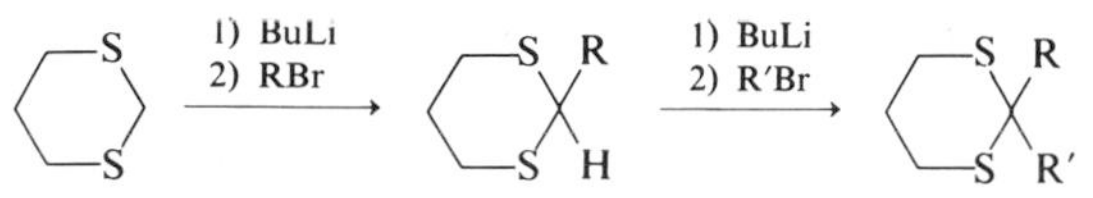

and, if the appropriate dihalide is used, 3- to 7-membered cyclic ketones can be obtained, e.g.

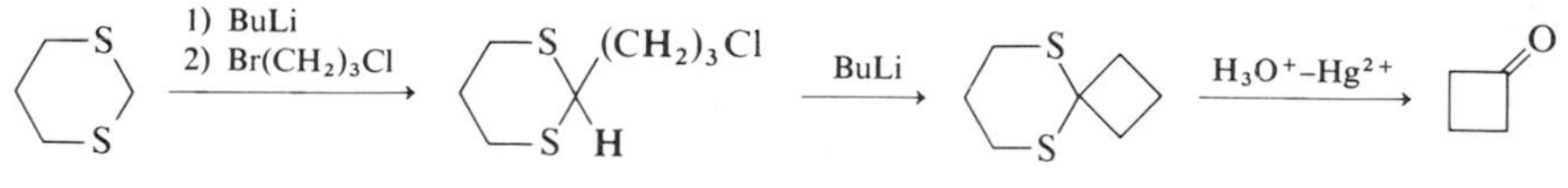

As usual in alkylations of carbanions, primary and secondary aliphatic halides react successfully but aromatic halides are unreactive and tertiary halides undergo elimination. The halide can be replaced by another source of electrophilic carbon in the form of an aldehyde, ketone, epoxide, or aromatic nitrile, e.g.

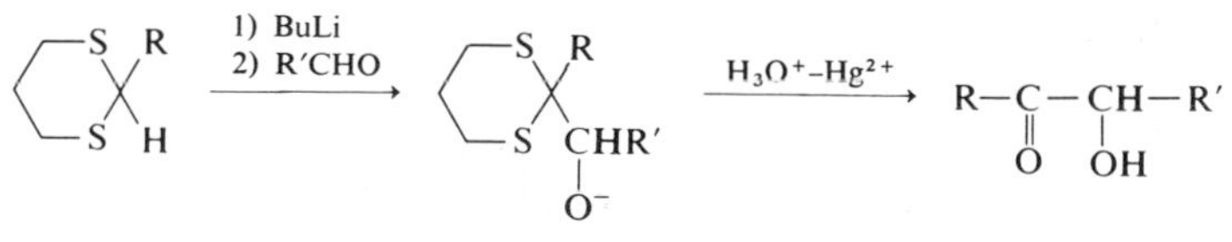

15.4 Boron-containing Reagents

Diborane adds readily to olefins and acetylenes to give organoboranes which have numerous synthetic uses.

The diborane is usually generated *in situ* by the reaction of sodium borohydride with boron trifluoride, introduced as its ether complex, $Et_2\overset{+}{O}-\overset{-}{B}F_3$:

$$3\ NaBH_4 + 4\ BF_3 \longrightarrow 3\ NaBF_4 + 2\ B_2H_6$$

However, when the borohydride would react with a functional centre in the olefin or acetylene, gaseous diborane is passed into the solution. The normal solvents are ether, tetrahydrofuran, and diethylene glycol dimethyl ether ('diglyme'); in these ethereal solvents the diborane forms complexes such as $Et_2\overset{+}{O}-\overset{-}{B}H_3$ and can be regarded as 'BH_3'. The standard procedure is to mix the unsaturated compound and the borohydride in diglyme and then to add boron trifluoride etherate slowly at room temperature. After the rapid reaction, the next appropriate reagent is introduced.

(*a*) FORMATION OF ORGANOBORANES
If the olefin is either monosubstituted or disubstituted with groups which are not bulky, trialkylboranes are formed, e.g.

$$3\ CH_3CH=CH_2 \xrightarrow{\ B_2H_6\ } (CH_3CH_2CH_2)_3B$$

$$3\ CH_3CH=CHCH_3 \xrightarrow{\ B_2H_6\ } \begin{matrix} CH_3CH_2 \\ \\ CH_3 \end{matrix}\!\!\diagdown\!\!CH)_3-B$$

However, with sterically hindered olefins such as trisubstituted ones, reaction can be controlled to give dialkylboranes, e.g.

$$2\ (CH_3)_2C=CHCH_3 \xrightarrow{\ B_2H_6\ } \begin{matrix}(CH_3)_2CH\\ \diagdown \\ CH_3\end{matrix}\!\!CH-BH-CH\!\!\begin{matrix}CH(CH_3)_2\\ \diagup \\ CH_3\end{matrix}$$

or even monoalkylboranes, e.g.

$$(CH_3)_2C=C(CH_3)_2 \xrightarrow{\ B_2H_6\ } (CH_3)_2CH-\underset{\underset{BH_2}{|}}{C}(CH_3)_2$$

With mono-, 1,1-di-, or tri-substituted olefins, the boron atom reacts predominantly at the less substituted carbon atom, as shown above for propylene and trimethylethylene. For example, the preference for attack by boron at the two olefinic carbon atoms of 1-butene is

$$CH_3CH_2CH{=}CH_2$$
$$\uparrow \quad \uparrow$$
$$6\% \quad 94\%$$

With 1,2-disubstituted olefins, the selectivity is less marked, e.g. with 1-methyl-2-t-butylethylene,

$$CH_3CH{=}CHC(CH_3)_3$$
$$\uparrow \quad \uparrow$$
$$58\% \quad 42\%$$

However, greater regioselectivity can be achieved by using a sterically hindered organoborane, such as the dialkylborane from trimethylethylene, shown above, which is known as disiamylborane, $(Sia)_2BH$; for example, the boron atom in this reagent reacts almost exclusively at the methyl-substituted carbon in 1-methyl-2-t-butylethylene. Another useful reagent in this respect is the cyclic dialkylborane known as 9-BBN (9-borabicyclo[3.3.1]nonane) formed by cyclo-octa-1,5-diene:

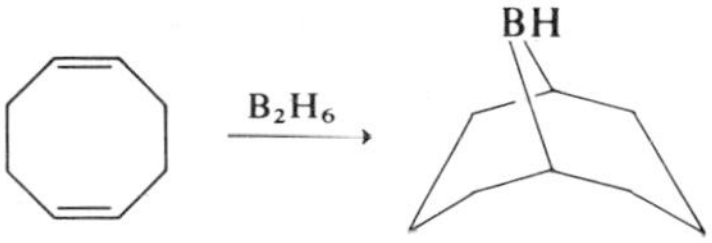

It is thought that boronation occurs by addition of the boron atom in an electrophilic manner to one carbon atom, followed by transfer of hydrogen to the other carbon atom in a reaction which is rapid enough to ensure that the *cis*-adduct is formed:*

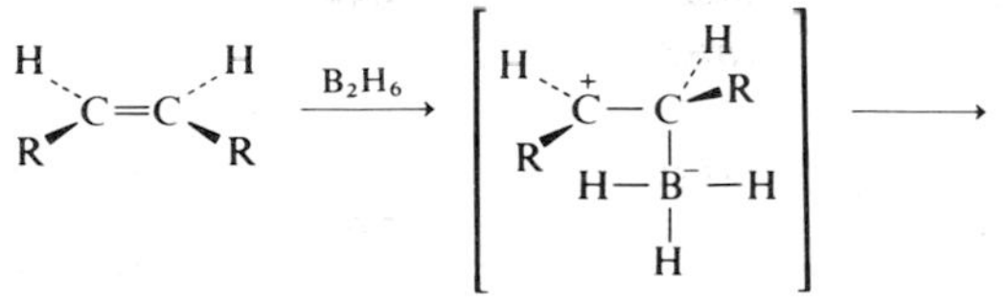

*Note that a concerted cycloaddition is symmetry-forbidden. Moreover, there is evidence for the development of carbonium-ion character during the reaction, for the relative amounts of products (I) and (II) decreases as the electron-attracting capacity of X increases (i.e. as the aryl-conjugated carbonium ion which would lead to (I) becomes increasingly less stable):

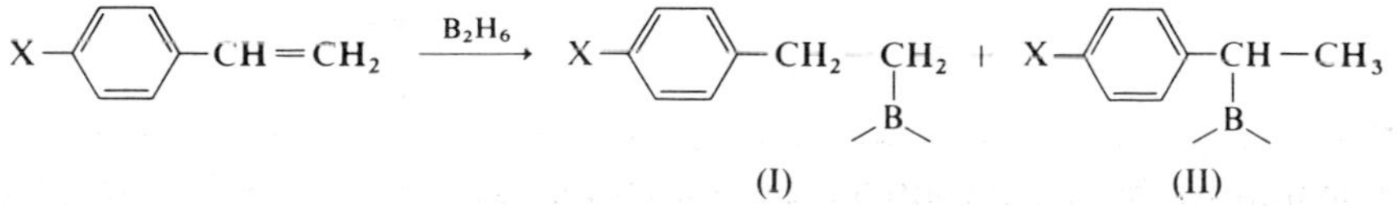

Acetylenes also form organoboranes. With monosubstituted acetylenes, reaction usually gives the diboronated compound,

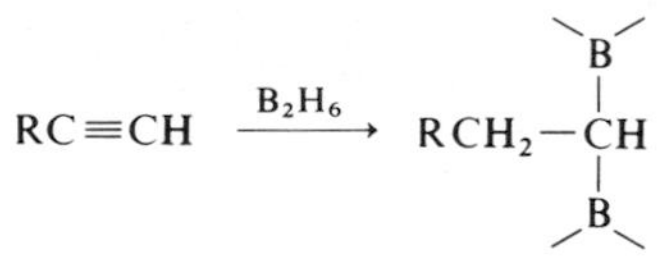

$$RC\equiv CH \xrightarrow{\;B_2H_6\;} RCH_2-\overset{\displaystyle \overset{B}{|}}{\underset{\displaystyle \underset{B}{|}}{CH}}$$

but the monosubstituted derivative is formed with disiamylborane. Disubstituted acetylenes give monoboronated products in a stereospecifically *cis* manner:

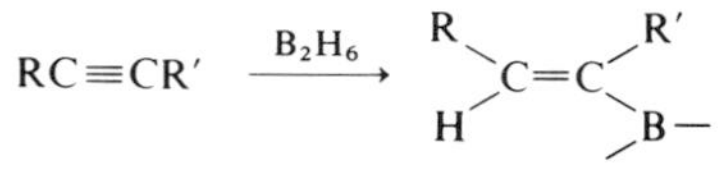

$$RC\equiv CR' \xrightarrow{\;B_2H_6\;} \underset{H}{\overset{R}{\big\diagup}}C=C\underset{B-}{\overset{R'}{\big\diagdown}}$$

The formation of alkanes and alkenes from organoboranes is discussed later (pp. 615, 619). The other synthetic uses of organoboranes are as follows.

(b) FORMATION OF ALCOHOLS

When an organoborane is treated with alkaline hydrogen peroxide at 20–30°C, an alcohol is formed, e.g.

$$(RCH_2CH_2)_3B + 3\,H_2O_2 + NaOH \longrightarrow 3\,RCH_2CH_2OH + NaH_2BO_3 + H_2O$$

Since a monosubstituted olefin, $RCH=CH_2$, yields a boron derivative $(RCH_2CH_2)_3B$ and this in turn yields an alcohol RCH_2CH_2OH, this method is complementary to the conversion $RCH=CH_2 \rightarrow RCH(OH)CH_3$ which can be accomplished by hydration with sulphuric acid (p. 109) or *via* epoxidation followed by reduction with lithium aluminium hydride (p. 570). Yields are usually excellent; for example, 1-hexene gives n-hexanol *via* trihexylborane in 98% yield.

The reaction results in stereospecific *cis*-addition to the double bond, e.g.

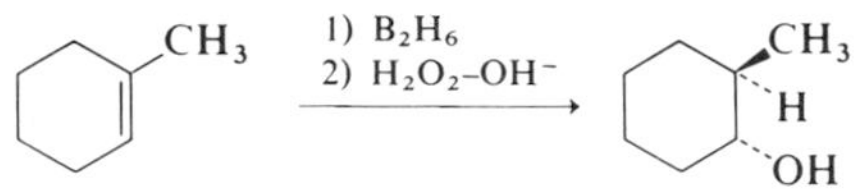

Further, when the directions of approach to the double bond are sterically different, the oxygen atom is introduced at the less hindered side; e.g. the product from norbornene consists of at least 90% of the *exo*-isomer,

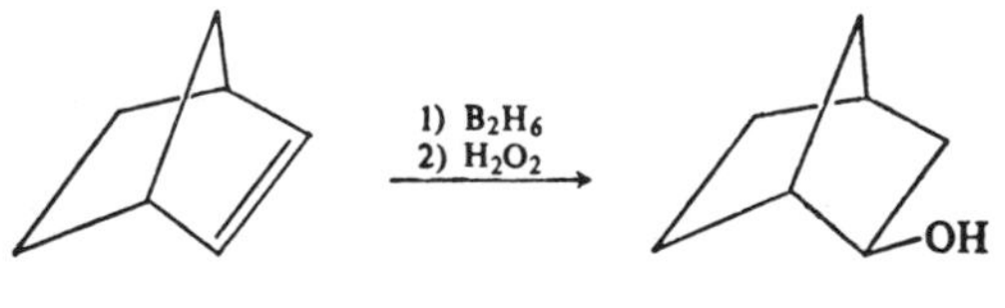

and the reaction of α-pinene with diborane, followed by oxidation of the dialkyl-borane, gives the alcohol shown in 85% yield [52]:

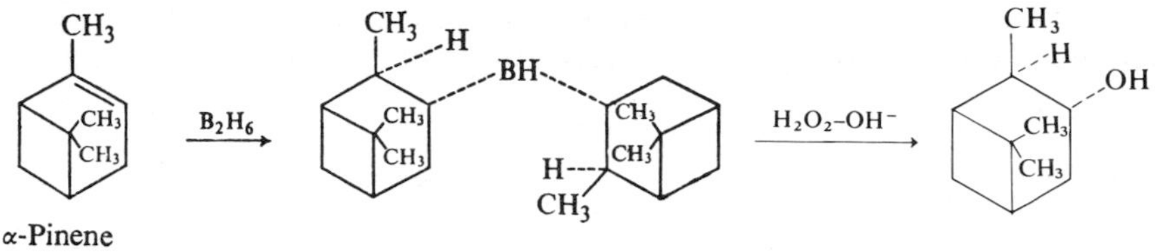

α-Pinene

These facts are consistent with the theory that, first, the boronation occurs in a stereospecifically *cis* manner as described above, with diborane approaching from the less hindered side, and secondly, oxidation occurs through a 1,2-shift which preserves the stereochemistry of the carbon atom which becomes attached to oxygen:

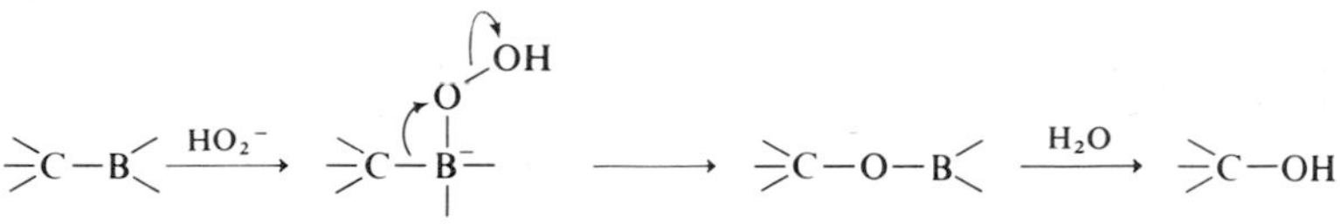

The boron atom is converted eventually into the borate anion, $O=B(OH)_2^-$.

The use of an organoborane derived from an optically active compound enables optically active products to be synthesized from inactive reactants with considerable specificity. An example is the use of the dialkylborane from α-pinene shown above ($\equiv R_2BH$):

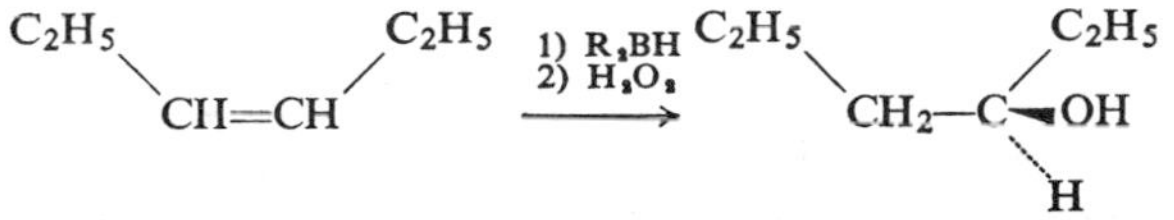

91% optical purity

The selective hydration of a diene can be accomplished efficiently with the use of disiamylborane, provided that the two double bonds differ significantly in the steric hindrance they offer to the boronating agent, e.g.

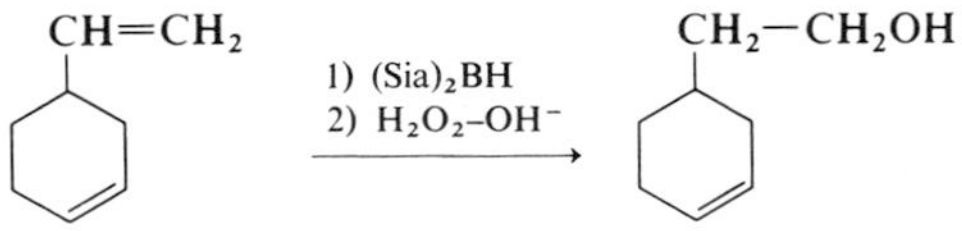

(*c*) FORMATION OF CARBONYL COMPOUNDS
This can be achieved with acid dichromate, e.g.

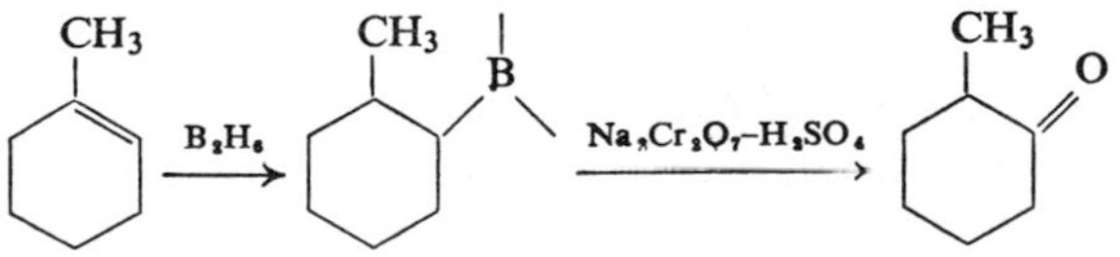

2-Methylcyclo-
hexanone (87%)

Alternatively, ketones can be made from boranes with diazoketones; for example, boronation of 1-hexene followed by treatment with diazoacetophenone gives phenyl heptyl ketone in 75–80% yield [53]:

$$3 \ C_4H_9CH{=}CH_2 \xrightarrow{B_2H_6} (C_5H_{11}CH_2)_3B \xrightarrow[H_2O]{PhCOCHN_2} C_7H_{15}{-}CO{-}Ph$$

It is thought that the mechanism is as follows:

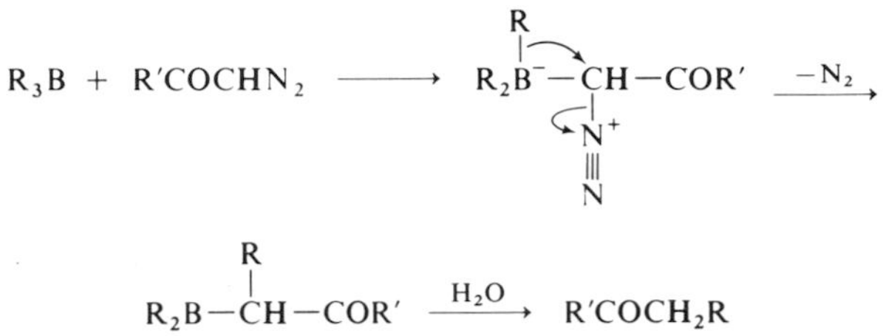

$$R_3B + R'COCHN_2 \longrightarrow R_2\overset{\scriptstyle R}{B}{-}\underset{\underset{\underset{N}{\overset{|||}{}}}{\overset{|}{N^+}}}{CH}{-}COR' \xrightarrow{-N_2}$$

$$R_2B{-}\overset{R}{\underset{|}{CH}}{-}COR' \xrightarrow{H_2O} R'COCH_2R$$

(d) FORMATION OF PRIMARY AMINES
Boranes react with chloramine and with hydroxylamine-O-sulphonic acid to give primary amines:

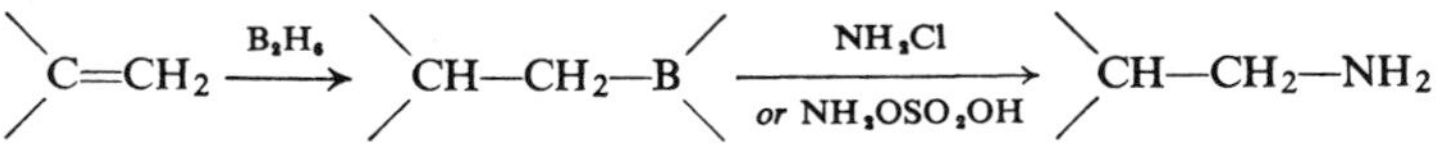

$$\text{>}C{=}CH_2 \xrightarrow{B_2H_6} \text{>}CH{-}CH_2{-}B\text{<} \xrightarrow[or\ NH_2OSO_3OH]{NH_2Cl} \text{>}CH{-}CH_2{-}NH_2$$

(e) FORMATION OF CYCLOPROPANES
Cyclopropanes can be formed from allylic halides *via* the reaction of β-halo-boranes with base. Diborane itself is not satisfactory as the boronating reagent since it is relatively unselective between the two olefinic carbon atoms of the olefinic bond; for example, with allyl chloride it gives 60% of the β- and 40% of the α-haloborane; the latter undergoes elimination to give propylene. However, the hindered borane 9-BBN from cyclo-octa-1,5-diene (p. 499) and disiamyl-borane react with much greater selectivity, e.g.

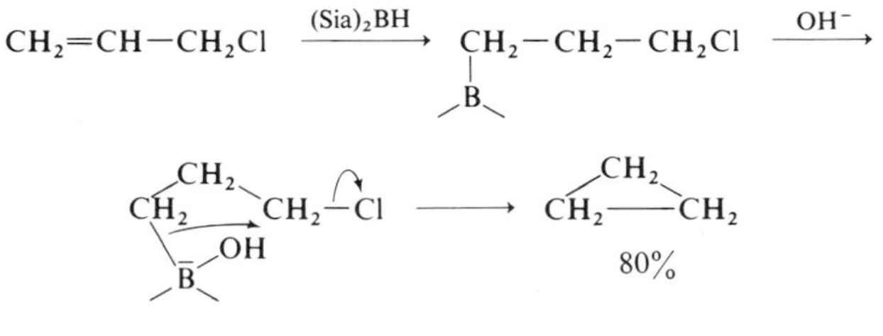

$$CH_2{=}CH{-}CH_2Cl \xrightarrow{(Sia)_2BH} \underset{B}{CH_2{-}CH_2{-}CH_2Cl} \xrightarrow{OH^-}$$

80%

(f) FORMATION OF ORGANOMETALLIC COMPOUNDS
Reaction with metal salts gives organometallic compounds, e.g.

$$2\ (C_2H_5)_3B + 3\ HgCl_2 \xrightarrow[80°C]{NaOH} 3\ (C_2H_5)_2Hg \quad (95\%)$$

(g) CARBONYLATION OF ORGANOBORANES

The reaction of organoboranes with carbon monoxide at *ca.* 100°C followed by alkaline hydrogen peroxide gives tertiary alcohols:

$$R_3B \xrightarrow[\text{2) } H_2O_2-OH^-]{\text{1) CO}} R_3C-OH$$

The mechanism is thought to be as follows:

$$R_3B + C{=}O \longrightarrow R_3\bar{B}-\overset{+}{C}{=}O$$

$$R_2\overset{+}{B}-\overset{-}{C}{=}O \longrightarrow R_2B-\underset{R}{C}{=}O$$
$$\underset{R}{|}$$

$$R-B-\overset{O}{\overset{||}{C}}-R \longrightarrow R-B-\overset{O}{\overset{/\backslash}{C}}-R$$

$$\underset{R}{B-CR_2} \longrightarrow O{=}B-CR_3 \xrightarrow{H_2O_2-OH^-} R_3C-OH + H_3BO_3$$

This scheme is consistent with the fact that two of the suggested intermediates —those formed after, respectively, two and one rearrangement steps—can be trapped by the addition of other reagents, and these conditions can also be applied usefully in synthesis. Thus, first, when carbonylation is carried out in the presence of an equimolar amount of water, the third group R is not transferred and the product formed after oxidation is a ketone. This is in accord with hydrolysis of the three-membered cyclic intermediate:

$$R-B-\overset{O}{\overset{/\backslash}{C}}-R \xrightarrow{H_2O} R-B-\overset{HO}{\underset{|}{C}}\overset{OH}{\underset{|}{R_2}} \xrightarrow{H_2O_2-OH^-} \left[R_2C\overset{OH}{\underset{OH}{\diagdown}}\right] \xrightarrow{-H_2O} R_2CO$$

Secondly, when the carbonylation is carried out in the presence of a reducing agent such as lithium trimethoxyaluminium hydride, the product formed after oxidation is an aldehyde or, if the oxidant is replaced by aqueous acid, an alcohol. It is thought that in this case the intermediate formed by the rearrangement of the first R group is intercepted:

$$R_2B-\underset{O}{\overset{||}{C}}-R \xrightarrow{LiAlH(OCH_3)_3} R_2B-\underset{O-\bar{Al}(OCH_3)_3}{\overset{|}{CH}}-R \begin{cases} \xrightarrow{H_2O_2-OH^-} RCHO \\ \xrightarrow{H_3O^+} RCH_2OH \end{cases}$$

Refinements have been introduced into both the ketone- and the aldehyde- or alcohol-forming reactions which improve their synthetic usefulness. For the former, the use of the hindered monoalkylborane from tetramethylethylene, known as thexylborane, enables different alkyl groups to be introduced by reactions with successive olefins, to give a compound of the type

$$(CH_3)_2CH-C(CH_3)_2$$
$$\underset{R\quad\quad R'}{\overset{|}{\underset{\diagdown B \diagup}{}}}$$

This reacts with carbon monoxide and then hydrogen peroxide in the presence of water to yield the mixed ketone RCOR'; that is, R and R' migrate in preference to the thexyl group. An example is

$$(CH_3)_2CH-\underset{\underset{Thexylborane}{BH_2}}{\overset{|}{C}}(CH_3)_2 \xrightarrow{(CH_3)_2C=CH_2} (CH_3)_2CH-\underset{\underset{H\quad CH_2-CH(CH_3)_2}{B}}{\overset{|}{C}}(CH_3)_2 \xrightarrow{CH_2=CH-CH_2CO_2Et}$$

$$(CH_3)_2CH-\underset{\underset{EtO_2CCH_2CH_2CH_2\quad CH_2CH(CH_3)_2}{B}}{\overset{|}{C}}(CH_3)_2 \xrightarrow[\text{2) } H_2O_2-H_2O]{\text{1) CO}} \underset{84\%}{(CH_3)_2CHCH_2-CO-(CH_2)_3CO_2Et}$$

A further variant is to use a diene for reaction with thexylborane; for example,

$$CH_2=CH-(CH_2)_2-CH=CH_2 \xrightarrow[\substack{\text{1) Thex } BH_2 \\ \text{2) CO} \\ \text{3) } H_2O_2-H_2O}]{} \underset{65\%}{\text{[cycloheptanone]} =O}$$

In the reaction leading to an aldehyde or alcohol, only one of the three alkyl groups in a trialkylborane is converted into the required derivative; this is a disadvantage if the olefin required for the trialkylborane is relatively rare or hard to obtain. The problem has been overcome by boronating an olefin with the monoalkylborane, 9-BBN, from cyclo-octa-1,5-diene (p. 499); reaction with carbon monoxide in the presence of lithium trimethoxylaluminium hydride followed by oxidation then results in the required aldehyde, e.g.

$$\underset{\text{[cyclohexene with CH=CH}_2]}{CH=CH_2} \xrightarrow{\text{9-BBN}} \underset{\text{[cyclohexene with CH}_2-CH_2-B]}{CH_2-CH_2-B\diagup} \xrightarrow[\text{2) } H_2O_2-OH^-]{\text{1) CO-LiAlH(OCH}_3)_3} \underset{\text{[cyclohexene with CH}_2-CH_2-CHO]}{CH_2-CH_2-CHO}$$

(*h*) ISOMERIZATION OF ORGANOBORANES
On heating at about 150°C, organoboranes are isomerized to a mixture in which

the major component has boron attached to the terminal carbon atom of the alkyl group, e.g.

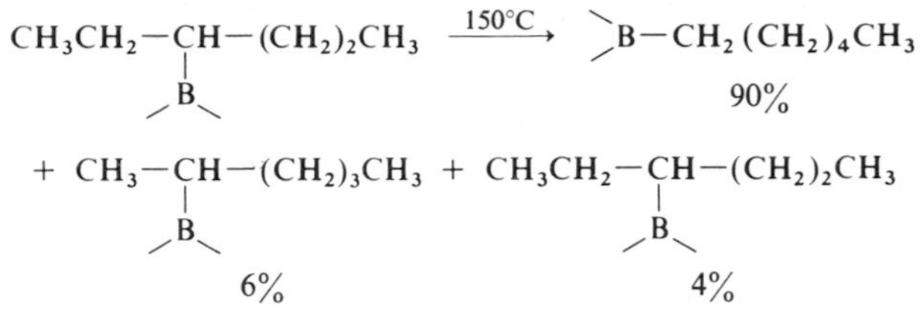

$$CH_3CH_2-CH-(CH_2)_2CH_3 \xrightarrow{150°C} \;>B-CH_2(CH_2)_4CH_3$$
$$\underset{B}{|} \qquad\qquad\qquad 90\%$$

$$+\; CH_3-CH-(CH_2)_3CH_3 \;+\; CH_3CH_2-CH-(CH_2)_2CH_3$$
$$\underset{B}{|} \qquad\qquad\qquad\qquad \underset{B}{|}$$
$$6\% \qquad\qquad\qquad\qquad 4\%$$

The process is catalyzed by diborane, and occurs by sequential elimination and addition steps. It provides a method by which a readily available internal olefin can be converted into, for example, a primary alcohol, e.g.

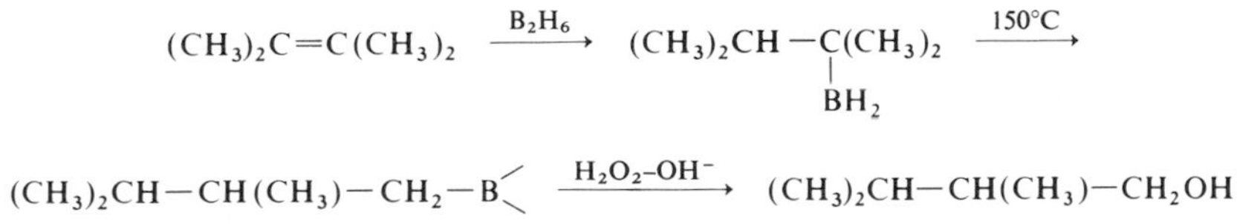

$$(CH_3)_2C=C(CH_3)_2 \xrightarrow{B_2H_6} (CH_3)_2CH-\underset{\underset{BH_2}{|}}{C}(CH_3)_2 \xrightarrow{150°C}$$

$$(CH_3)_2CH-CH(CH_3)-CH_2-B\!\!< \xrightarrow{H_2O_2-OH^-} (CH_3)_2CH-CH(CH_3)-CH_2OH$$

Further Reading

BROWN, H. C., *Hydroboration*, Benjamin (New York 1962).

CADOGAN, J. I. G., and MACKIE, R. K., 'Tervalent phosphorus compounds in organic synthesis,' *Chemical Society Reviews*, 1974, **3**, 87.

MAERCKER, A., 'The Wittig reaction,' *Organic Reactions*, 1965, **14**, 270.

TROST, B. M., 'New alkylation methods,' *Accounts of Chemical Research*, 1974, **7**, 85.

Problems

1. How would you employ phosphorus-, sulphur-, or boron-containing reagents in the synthesis of the following:

   (a) $Ph_2C=CH_2$

   (b) $PhCH_2CH_2OH$

   (c) $Ph_2CH-CHO$

   (d) $Ph(CH_2)_3COPh$

   (e) $PhCOCH\underset{\diagdown C_2H_5}{\overset{\diagup CH_3}{}}$

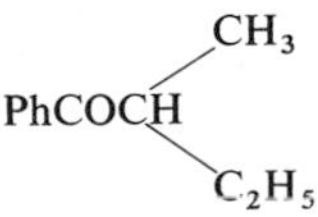

   (f) $PhC(CH_3)=CH-CH_2Ph$

   (g) $PhCH_2CH_2COCH_2CH_2Ph$

   (h) ![structure]

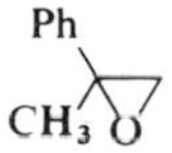

(i)

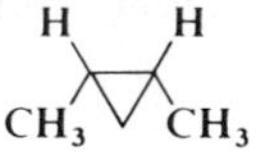

(j) CH₃— (structure with Ph and O)

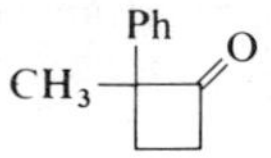

(k)

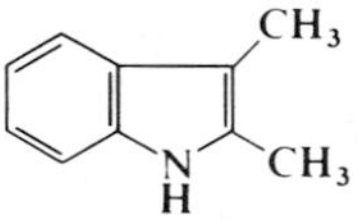

(l)

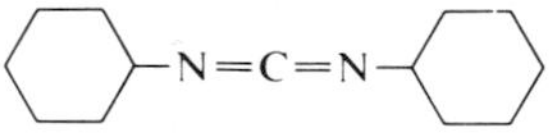

2. Rationalize the following reactions:

(a) $(CH_3)_2\overset{+}{S}-CH_2-C\underset{CH_3}{\overset{CH_2}{\diagdown}}$ $\xrightarrow{EtO^-}$ $CH_3SCH_2CH_2-C\underset{CH_3}{\overset{CH_2}{\diagdown}}$

(b) $(EtO)_2\overset{O}{\overset{\|}{P}}CH_2\overset{+}{S}(CH_3)_2$ $\xrightarrow[\text{2) RCHO}]{\text{1) Base}}$ $RCH=CH\overset{+}{S}(CH_3)_2$

(c) $\xrightarrow[\text{2) H}_2\text{–Ni}]{\text{1) Base}}$

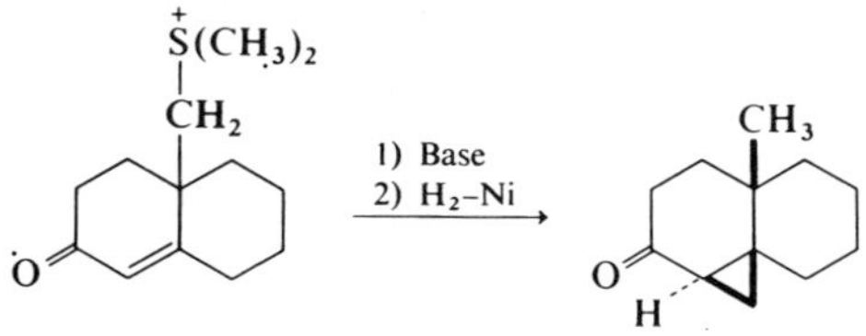

(d) $\xrightarrow[\text{3) H}_2\text{O}_2\text{–OH}^-]{\substack{\text{1) Thex BH}_2 \\ \text{2) CH}_3\text{CO}_2\text{H (1 mol)}}}$

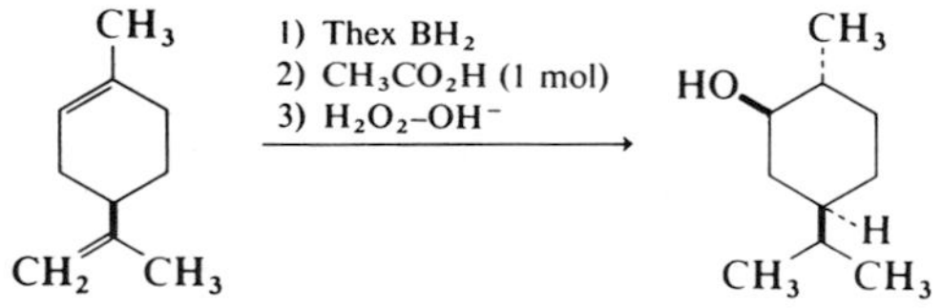

(e) $CH_2=CH-CH_2-S-CH=CH_2$ $\xrightarrow[\substack{\text{3) Heat} \\ \text{4) H}_2\text{O–Hg}^{2+}}]{\substack{\text{1) BuLi} \\ \text{2) RBr}}}$ $RCH=CH-CH_2CH_2CHO$

(f) $\xrightarrow[\text{2) OH}^-]{\text{1) B}_2\text{H}_6}$

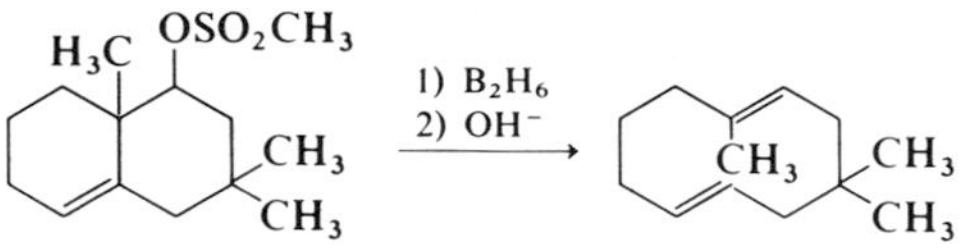

16. Photochemical Reactions

16.1 Principles

The electronic excitations of compounds require quanta of energy which correspond to wavelengths of electromagnetic radiation lying in the ultraviolet or, less frequently, the visible region of the spectrum. When a molecule absorbs a photon of the appropriate energy, one of its electrons is raised to an orbital of higher energy. The resulting molecule may then take part in a reaction while in this, or an alternative, excited state. The reaction so induced may be either intramolecular—for example, rearrangement or dissociation—or intermolecular—for example, addition; alternatively, the excited molecule may transfer its energy to another molecule which in turn undergoes reaction. These variants open up a wide range of synthetic possibilities.

The first requirement, then, is that the reactant should absorb light at a wavelength which is experimentally accessible. In practice, the ultraviolet region below 200 nm is inconvenient because the radiation is absorbed by air, and vacuum techniques are required; moreover, normal solvents absorb in the range up to 220 nm. The useful spectrum therefore begins at 220 nm. For the ultraviolet region, a low-pressure mercury arc lamp has its strongest emission line at 254 nm, which is a wavelength at which aromatic molecules absorb strongly. Broad-band emission is in general more useful; it is obtained with a higher pressure mercury lamp, which also gives more intense radiation, and wavelengths which might bring about unwanted reactions can to some extent be removed by appropriate filters. For the visible region, a tungsten lamp or the more powerful xenon arc are suitable. Saturated organic compounds absorb only at wavelengths well below 200 nm, but many unsaturated ones absorb above 230 nm, so that photochemical reactions are essentially limited to those in which at least one of the reactants is unsaturated or aromatic.

Consider a compound containing a C=C group. Light of the appropriate wavelength—in the region 180–200 nm for compounds with isolated double bonds, but of longer wavelength for conjugated systems—causes a π-electron to be excited to the lowest antibonding π-orbital (i.e. a π^* orbital);† the transition is described as $\pi \rightarrow \pi^*$. The new state of the molecule is described as the first excited singlet, S_1 (since the promoted electron and its original partner still have opposite spins); this particular excited singlet is of (π, π^*) type. For compounds

---

†That is, the excitation in this case is from the HOMO to the LUMO. However, the terminology of pericyclic reactions is generally not appropriate to photochemical reactions.

containing a carbonyl group, two types of excitation are commonly effected: one, which requires the more energy and therefore corresponds to shorter wavelengths (e.g. $\lambda_{max.}$ 187 nm for acetone), is a $\pi \rightarrow \pi^*$ transition. The other, which corresponds to longer wavelengths (e.g. $\lambda_{max.}$ 270 nm for acetone), is an excitation of one of the $2p$ non-bonding electrons on the oxygen atom into an antibonding π-orbital; it is described as an $n \rightarrow \pi^*$ transition and the excited state, S_1, is of (n, π^*) type.

Diagram A shows a typical situation.‡ Initially, almost all molecules are in the lowest vibrational level of the ground state, S_0. The excitation occurs so

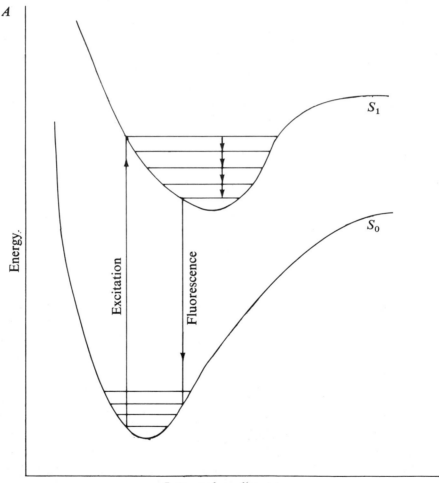

Internuclear distance

‡The x-axis is labelled internuclear distance for simplicity. The diagram is strictly appropriate only for a diatomic molecule; for others, a full description would require a multi-dimensional representation.

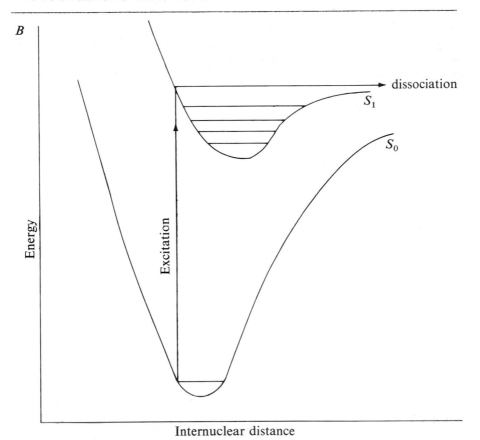

Internuclear distance

rapidly that the nuclei do not have time to alter their relative positions; it is therefore represented by a vertical line. Consequently, the S_1 state can be reached at a point which corresponds to an excess of vibrational energy. However, this energy is rapidly dissipated through molecular collisions, and thereafter one of five processes takes place. (a) The molecule returns to its ground state either by emitting radiation (*fluorescence*) or by giving up its energy as heat (*internal conversion*); as shown, the fluorescent light is of slightly lower energy (and so longer wavelength) than the absorbed light. (b) A chemical reaction occurs. (c) The energy of excitation is transferred to another molecule, which is thereby raised to an excited singlet level. (d) One of the unpaired electrons in the excited molecule undergoes an inversion of its spin, leading to a lower-energy state described as a triplet, T_1, which contains two unpaired electrons of parallel spin; this is described as *intersystem crossing*. (e) In the special case which can be of the type shown in diagram *B*, excitation leads to a point in S_1 which is *above* that of the energy curve on the right-hand side; consequently, the excited molecule, instead of vibrating, dissociates.

The triplet state may return to the ground state, with a further spin inversion, either by emitting radiation (*phosphorescence*) or by giving up energy as heat; it may take part in a chemical reaction; it may transfer its energy to another molecule which is thereby raised to the triplet state; or it may dissociate.

These processes are summarized as follows:

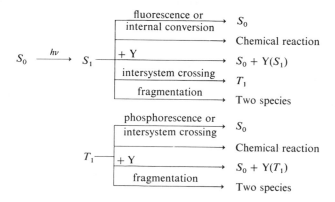

The radiative life-times of excited singlets (i.e. the time-lag before return to the ground state) are very short (10^{-6}–10^{-9} seconds), so that chemical reactions occurring *via* these species, despite their high energy, are comparatively uncommon. Conversion into the triplet state occurs with an efficiency which varies widely with the structure of the molecule; for benzophenone, for example, it is a highly efficient process whereas for olefins it is not. The life-times of triplets are relatively long (> *ca.* 10^{-4} seconds for species containing relatively light atoms) so that the probability of the triplet taking part in a chemical reaction is much higher than for a singlet.

The chemical reactions which occur through excited states comprise the major part of this Chapter. This has been a very active field of study during the last decade; nonetheless, despite the emergence of several principles and helpful generalizations, it is not always possible to predict with assurance whether a reaction will occur, for example, through an excited singlet or triplet, or what its precise course will be.

There are two key features of reactions which occur through excited states which underlie their special synthetic importance. First, since the excited state usually has a large excess of energy compared with the ground state,* it is often possible to effect reactions which, insofar as the *ground*-state reactant is concerned, are thermodynamically unfavourable. Secondly, reaction can usually be carried out, if required, at low temperature, so that the product can be formed 'cold'. For these reasons, it is often possible to make, for example, highly strained ring systems; energy is pumped in as light to overcome the activation-

*For example, light of wavelength 300 nm is equivalent to energy of *ca.* 400 kJ mol^{-1}.

energy barrier in their formation, and they are produced in conditions which allow them to survive.

The reactions of excited states, including both singlets and triplets, can be classified under the following headings: reduction, addition, rearrangement, oxidation, aromatic substitution, and fragmentation.

16.2 Photoreduction

Carbonyl compounds can be converted into their pinacolic dimers by irradiation in the presence of a hydrogen-donating compound. Reaction occurs through the triplet state of the carbonyl compound which abstracts a hydrogen atom from the second reactant; the resulting radicals then combine:

$$R_2C{=}O \xrightarrow{\ h\nu\ } \underset{\text{Triplet}}{R_2\dot{C}{-}\dot{O}} \xrightarrow{\ R'H\ } R_2\dot{C}{-}OH + R'\cdot$$

$$2\ R_2\dot{C}{-}OH \longrightarrow \begin{array}{c} R_2C{-}OH \\ | \\ R_2C{-}OH \end{array}$$

The reaction is usually most efficient when the hydrogen donor is itself the alcoholic reduction product of the carbonyl compound, for then only one dimeric product can be formed, e.g.

$$Ph_2C{=}O \xrightarrow{\ h\nu\ } Ph_2\dot{C}{-}\dot{O} \xrightarrow{\ Ph_2CH{-}OH\ } 2\ Ph_2\dot{C}{-}OH \longrightarrow \begin{array}{c} Ph_2C{-}OH \\ | \\ Ph_2C{-}OH \end{array}$$

Reaction occurs in good yield only when the triplet is of the (n, π^*) type; for example, it is efficient for benzophenone and acetone, where this is so, but not for p-phenylbenzophenone whose lowest triplet state is of (π, π^*) type.

o-Alkyl-substituted aromatic ketones, even though giving (n, π^*) triplets, are not reduced in this way. In preference, intramolecular hydrogen-atom abstraction occurs to give the enol tautomer:

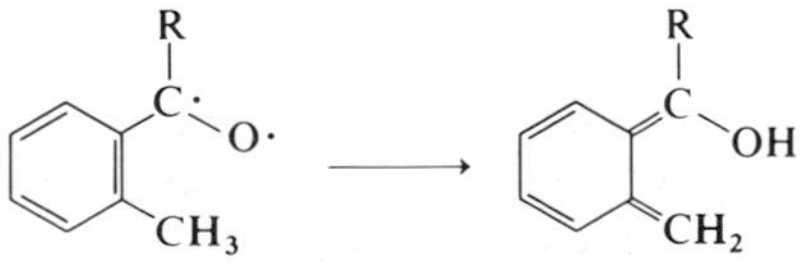

Triplet

The enol then rearranges back to the more stable ketone, though it can be trapped as a Diels-Alder adduct with, e.g. dimethyl acetylenedicarboxylate:

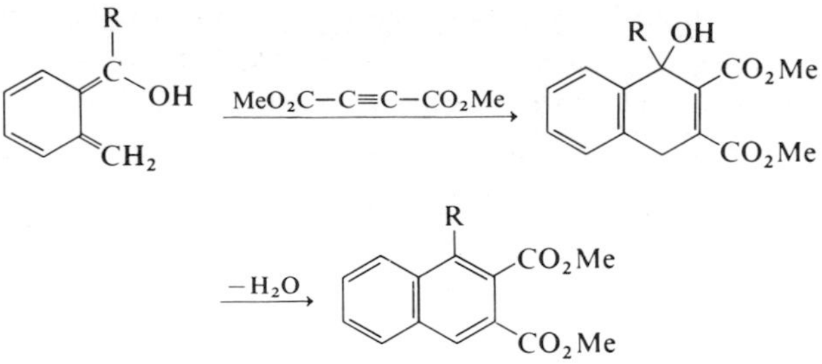

16.3 Photo-addition

Photo-addition is the formation of a $1 : 1$ adduct by reaction of an excited state of one molecule with the ground state of another. The molecule in the ground state is commonly an olefin; the reactant which is excited can be a carbonyl compound, quinone, aromatic compound, or another molecule of the same olefin. The majority of the reactions lead to the formation of a ring.

(a) PHOTO-ADDITION OF OLEFINS TO CARBONYL COMPOUNDS

This photo-addition—the *Paterno-Büchi reaction*—normally occurs by reaction of the triplet state of the carbonyl compound with the ground state of the olefin. As in photo-reduction, it is more efficient when the triplet is of (n, π^*) rather than (π, π^*) type. Typical examples, which also show the marked variations in the yields of products, are the addition of benzophenone to propylene and isobutylene:

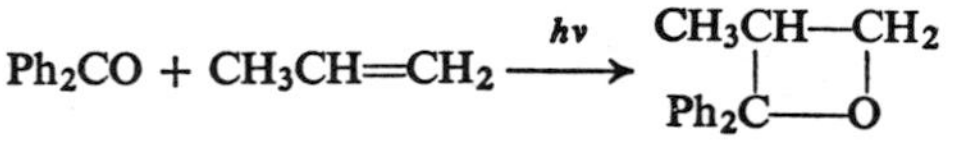

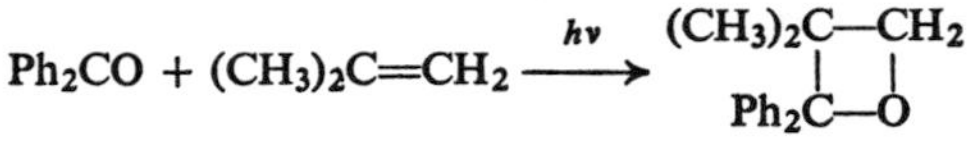

The ring is formed in two stages. The excited carbonyl compound (triplet) first adds through its oxygen atom to the olefin so as to give the more stable of the two possible diradicals.* A spin-inversion then occurs and the second bond is formed.

*The energy-content of alkyl radicals decreases in the order primary > secondary > tertiary; p. 538.

$$\text{Ph}_2\text{C}{=}\text{O} \xrightarrow{hv} \text{Ph}_2\text{C}{\overset{\downarrow}{=}}\text{O}\,\uparrow \xrightarrow{\text{spin-inversion}} \text{Ph}_2\text{C}{\overset{\uparrow}{=}}\text{O}\,\uparrow$$

$$\text{Ph}_2\text{C}{\overset{\uparrow}{=}}\text{O}\,\uparrow + (\text{CH}_3)_2\text{C}{=}\text{CH}_2 \begin{cases} \nearrow \begin{array}{c} (\text{CH}_3)_2\overset{\uparrow}{\text{C}}-\text{CH}_2 \\ | \quad | \\ \text{Ph}_2\overset{\uparrow}{\text{C}}-\text{O} \end{array} \xrightarrow{\text{spin-inversion}} \begin{array}{c} (\text{CH}_3)_2\text{C}-\text{CH}_2 \\ | \quad | \\ \text{Ph}_2\text{C}-\text{O} \end{array} \\ \overset{\times}{} \\ \begin{array}{c} \dot{\text{C}}\text{H}_2-\text{C}(\text{CH}_3)_2 \\ | \\ \text{Ph}_2\dot{\text{C}}-\text{O} \end{array} \end{cases}$$

The reaction is not stereospecific, evidently because the time-lag before the final spin-inversion is more than enough for rotation to occur about single bonds, e.g.

$$\text{RR'CO} + \text{R''CH}{=}\text{CH}_2 \xrightarrow{hv} \begin{array}{c} \text{H} \underset{\text{R''}}{\overset{\uparrow}{\diagdown}}\text{C}-\text{CH}_2 \\ \text{R'} \diagdown \quad | \\ \text{R} \diagup \overset{\uparrow}{\text{C}}{\frown}\text{O} \end{array}$$

$$\xrightarrow{\text{spin-inversion}} \begin{array}{c} \text{H}\diagdown \\ \text{R''}\diagup\text{C}-\text{CH}_2 \\ \text{R'}\diagdown \quad | \\ \text{R}\diagup\text{C}-\text{O} \end{array} + \begin{array}{c} \text{H}\diagdown \\ \text{R''}\diagup\text{C}-\text{CH}_2 \\ \text{R}\diagdown \quad | \\ \text{R'}\diagup\text{C}-\text{O} \end{array}$$

The triplet state of an α,β-unsaturated carbonyl compound is a delocalized species which can be represented as a hybrid,

$$\overset{\diagup}{\diagdown}\text{C}{=}\text{C}-\dot{\text{C}}-\dot{\text{O}} \longleftrightarrow \overset{\diagup}{\diagdown}\dot{\text{C}}-\text{C}{=}\text{C}-\dot{\text{O}} \longleftrightarrow \overset{\diagup}{\diagdown}\dot{\text{C}}-\dot{\text{C}}-\text{C}{=}\text{O}$$

As a consequence, addition to an olefin can occur at C=C as well as at C=O, and in fact the former path is normally followed. Both structural and stereochemical isomers are usually formed, e.g.

There is one general circumstance in which photo-addition of olefins to carbonyl compounds fails: namely, when the energy difference between the triplet and ground state of the carbonyl compound is greater than that between the corresponding states of the olefin. In that event, the excited carbonyl compound can transfer its energy to the olefin, so returning to its ground state and giving triplet-state olefin. This is then followed by olefin dimerization. For example, the energy of the triplet from benzophenone is less than that from norbornene, so that photoaddition occurs:

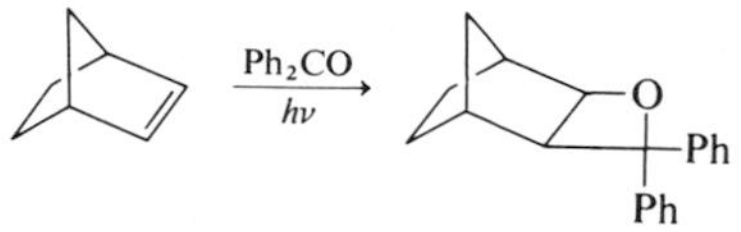

In contrast, the energy of the triplet from acetophenone is greater than that from norbornene, so that irradiation of acetophenone in the presence of norbornene yields mainly norbornene dimers. Acetophenone is here described as a *sensitizer* (p. 515).

(*b*) PHOTO-ADDITION OF OLEFINS TO AROMATIC COMPOUNDS
Both olefins and acetylenes undergo photochemical addition to benzene. The normal mode of reaction is cycloaddition, for example:

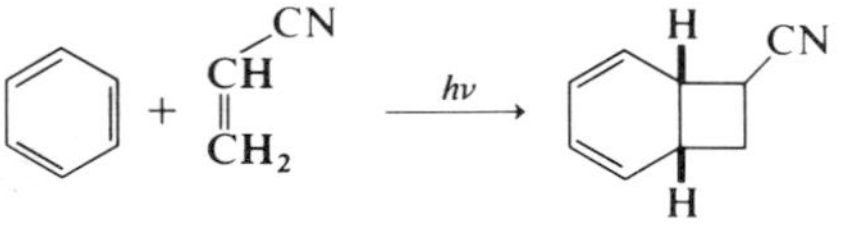

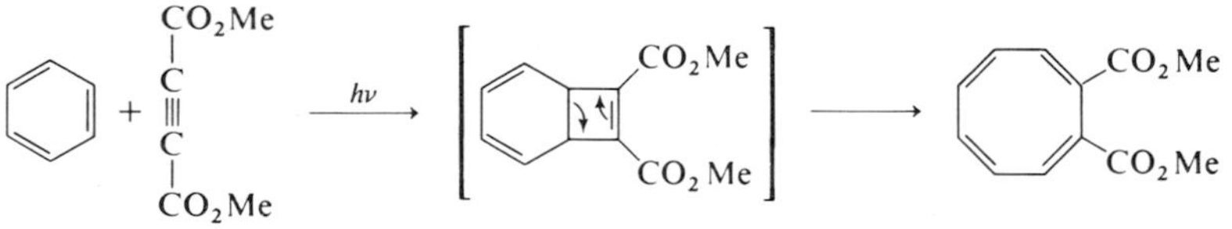

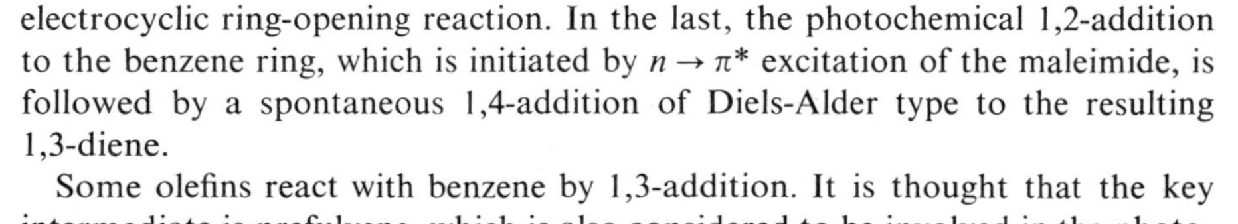

In the second example, the very strained cyclobutene undergoes a spontaneous electrocyclic ring-opening reaction. In the last, the photochemical 1,2-addition to the benzene ring, which is initiated by $n \rightarrow \pi^*$ excitation of the maleimide, is followed by a spontaneous 1,4-addition of Diels–Alder type to the resulting 1,3-diene.

Some olefins react with benzene by 1,3-addition. It is thought that the key intermediate is prefulvene, which is also considered to be involved in the photo-isomerization of benzene (p. 520) and is believed to arise from a high vibrational level of the first excited singlet of benzene, e.g.

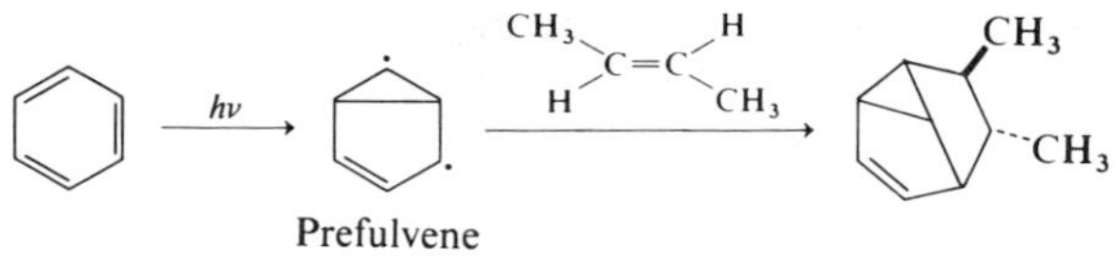

Prefulvene

1,4-Addition to benzene is also known; it occurs when a mixture of benzene and a primary or secondary amine is irradiated, and involves the T_1 state of benzene, e.g.

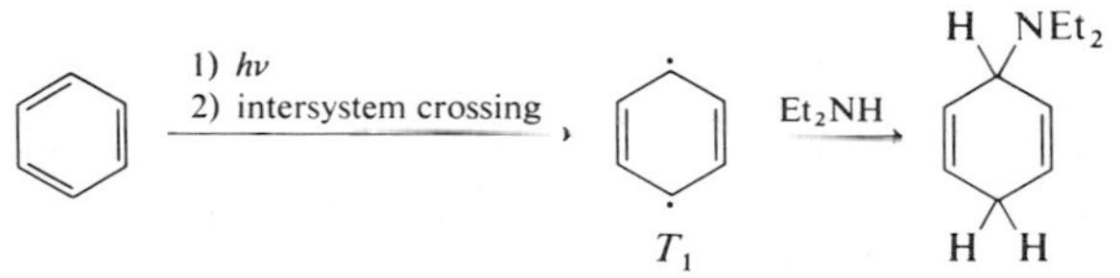

(c) PHOTODIMERIZATION OF OLEFINS, CONJUGATED DIENES, AND AROMATIC COMPOUNDS

A special case of photo-addition is the formation of a 1 : 1 adduct by reaction of an excited state of one molecule of a reactant with another molecule of the same reactant in its ground state. This process—photodimerization—occurs with both olefins and aromatic compounds. Depending on the structure of the reactant and the conditions of excitation, the excited state may be a singlet or a triplet.

Since non-conjugated olefins absorb only in the difficultly accessible region below 200 nm, their unsensitized photodimerizations are comparatively little used. Reaction in this case occurs through the excited singlet, S_1, and is concerted. Reference to the earlier discussion of such reactions (p. 300) shows that the photochemical [2 + 2] cycloaddition is symmetry-allowed, e.g.

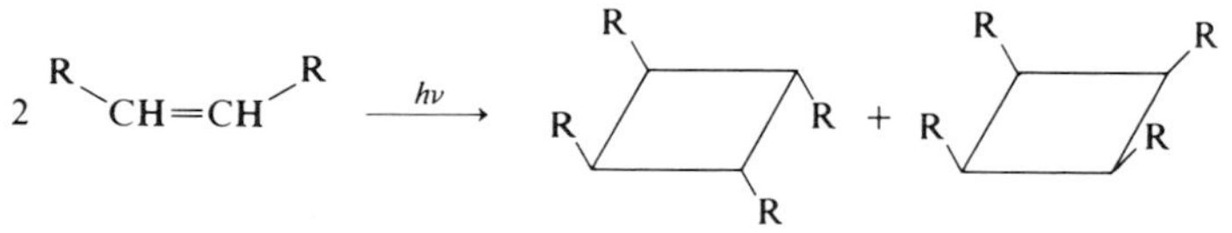

Because of the experimental difficulty, olefin photodimerization is better accomplished through the triplet state, this being generated with a photosensitizer. For example, acetophenone has $\lambda_{max.}$ 270 nm and its S_1 state undergoes intersystem crossing to T_1 efficiently; interaction of this species with the olefin then gives singlet sensitizer and triplet olefin:

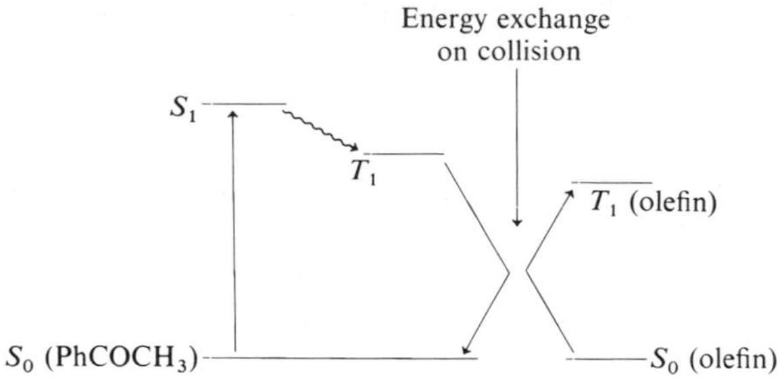

The molecule in its triplet state then adds through one carbon atom to a carbon atom of a second molecule of olefin and, after a spin-inversion, formation of the ring is completed.

Conjugated dienes do not undergo photodimerization through the excited singlet but instead undergo photocyclization (p. 517). However, photodimerization can be effected through triplet sensitization. Mixtures of products are often formed, as in the dimerization of cyclopentadiene in the presence of benzophenone:

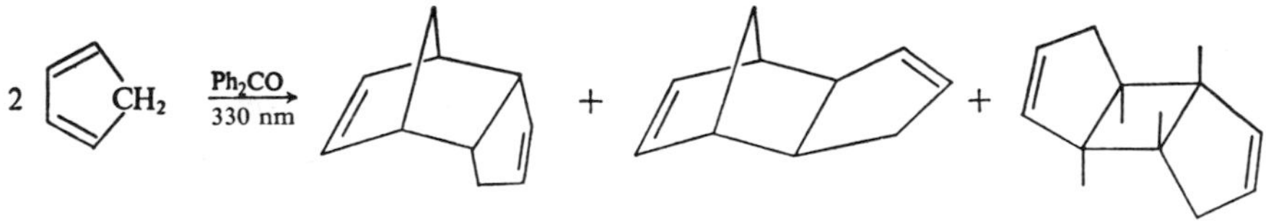

Here, the first two products represent 1,2-1,4 addition and the third represents a 1,2-1,2 addition. The three are formed in approximately equal proportions, whereas the concerted thermal dimerization of cyclopentadiene gives entirely *endo*-dicyclopentadiene (p. 292).

Benzene does not form a photodimer, but polycyclic aromatic compounds can do so. For example, anthracene dimerizes through its excited singlet (π, π^*) state:

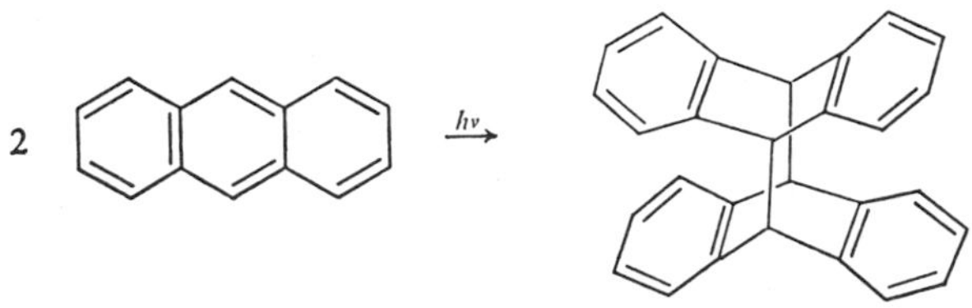

16.4 Photorearrangement

Light can effect rearrangements which lead either to structural isomers (where groups or atoms in reactant and product occupy entirely different positions) or to valence-bond isomers (where groups or atoms essentially maintain their relative positions but the bonding framework changes). These reactions provide the commonest general methods, and certainly often the shortest, for synthesizing highly strained compounds.

(*a*) CIS-TRANS ISOMERIZATION
This provides the simplest case of light-induced structural isomerism. Examples are:

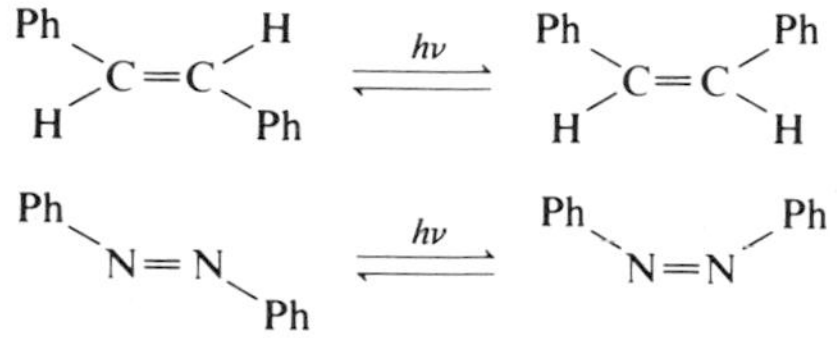

Isomerization can occur because the π-bond which normally prevents it is lost in passage to the excited state. Here, the two sets of substituents tend to occupy mutually perpendicular planes, e.g.

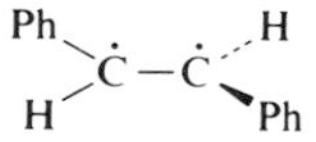

so as to minimize the repulsive forces between them. Consequently, when the molecule returns to the ground state, it can do so by twisting in either of two directions, to give both *cis*- and *trans*-isomers.

The usefulness of the reaction resides in the fact that the final mixture is usually richer in the less stable *cis* form; for example, *trans*-stilbene gives the *cis*- and *trans*-isomers in relative amounts of *ca.* 10 : 1. This is because, at most wavelengths that are used, the *trans*-isomer has the greater molar decadic absorbance; that is, it is the more frequently of the two returned to the excited state to form each of the two isomers.

(*b*) INTRAMOLECULAR PHOTOCYCLIZATION
Many dienes and polyenes are converted photochemically into cyclic isomers.

In those cases where the excited singlet is involved, reaction is of the concerted, electrocyclic type and is stereospecific; its course can be rationalized by the orbital-symmetry principles outlined earlier.

For example, irradiation of *trans,trans*-1,4-dimethyl-1,3-butadiene gives entirely *cis*-3,4-dimethylcyclobutene:

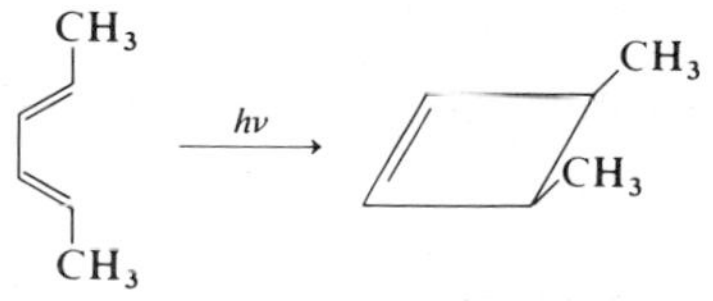

Inspection of the higher energy SOMO of the excited singlet shows that ring-closure should occur in the disrotatory manner, accounting for the stereochemistry of the product:

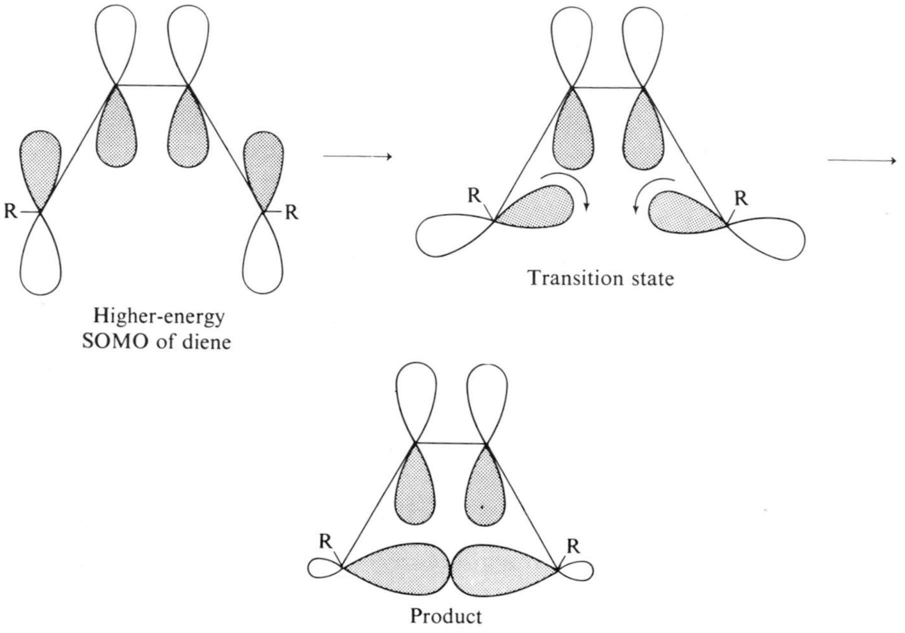

Higher-energy
SOMO of diene

Transition state

Product

It is recognized, however, that this is an oversimplified description of the process (p. 305).

Likewise, the formation of *trans*-5,6-dimethylcyclohexadiene from the octatriene shown is correlated with the conrotatory ring-closure required by the higher energy SOMO of the excited singlet:

$$CH_3 \xrightarrow{h\nu} CH_3 \quad CH_3$$

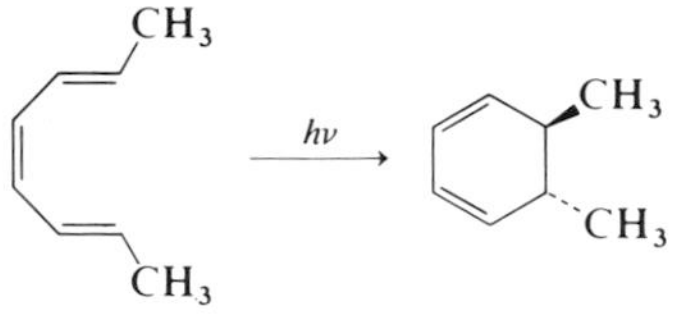

These reactions are reversible, and an early example of their synthetic utility was the electrocyclic ring-opening which occurs on the irradiation of ergosterol, to give precalciferol:

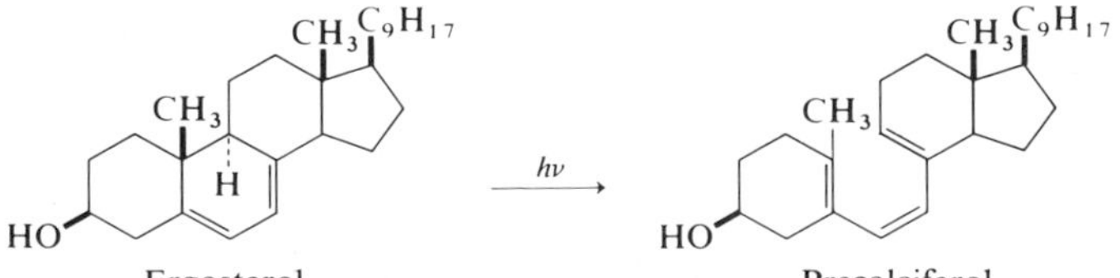

Ergosterol $\xrightarrow{h\nu}$ Precalciferol

More recently, attention has been directed towards forming highly strained ring compounds by photocyclization. For example, bicyclo[2.1.0]pent-2-ene can be obtained by irradiating cyclopentadiene [55],

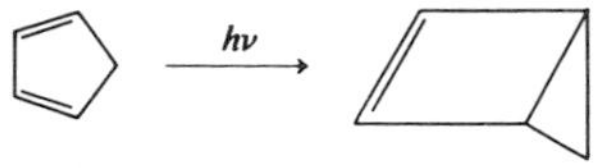

Triplet-sensitized reactions can also be used. For example, the valence-bond isomerization of norbornadiene to give quadricyclane can be effected in 70–80 % yield with acetophenone as sensitizer [51],

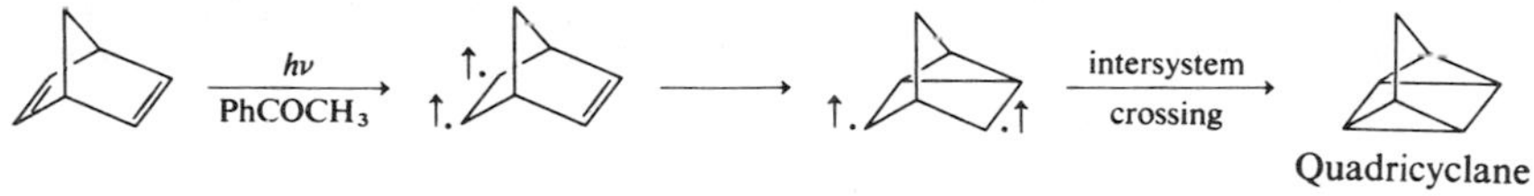

and barrelene likewise gives semibullvalene:

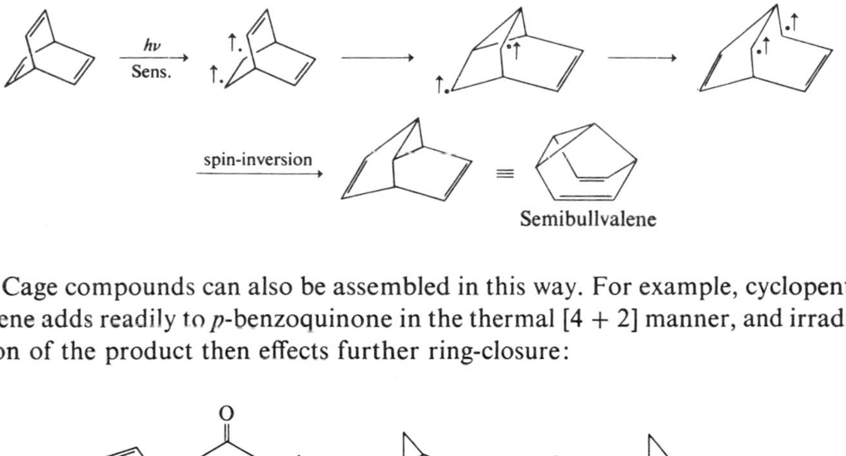

Cage compounds can also be assembled in this way. For example, cyclopentadiene adds readily to p-benzoquinone in the thermal [4 + 2] manner, and irradiation of the product then effects further ring-closure:

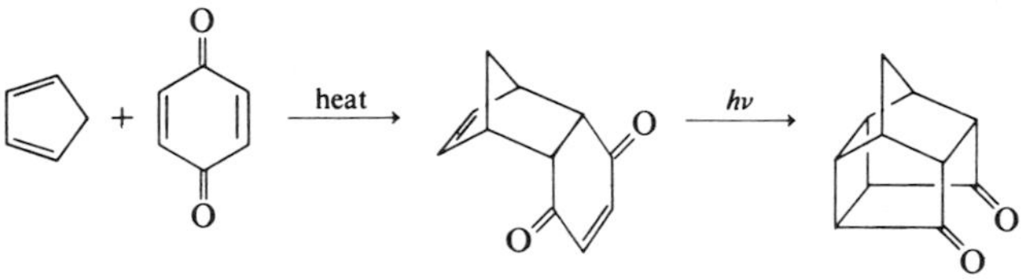

Again, irradiation of dicyclopentadiene yields a cage-like compound:

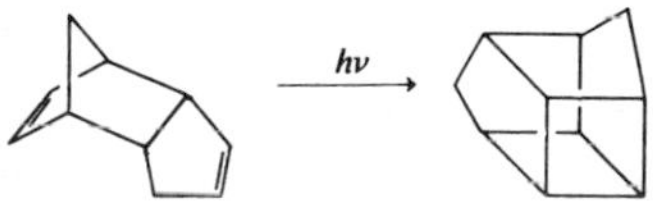

A synthesis of cubane is based on the latter reaction:

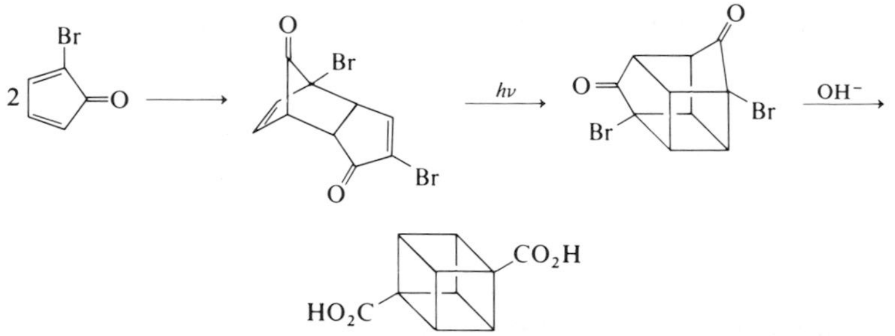

The dimerization of 2-bromocyclopentadienone occurs spontaneously. The irradiation step is best carried out on a ketal derivative of the dimer from which the keto group is subsequently regenerated by hydrolysis. The successive base-induced ring-contractions are reactions of Favorskii type (p. 475). The cubanedicarboxylic acid has been converted into cubane itself by heating the t-butyl perester (from the acid chloride with t-butyl hydroperoxide) in cumene; homolysis of the peroxide bond is followed by decarboxylation and uptake of a hydrogen atom from the solvent:

$$RCO_2H \longrightarrow RCO{-}O{-}O{-}C(CH_3)_3 \xrightarrow{\text{heat}} RCO{-}O\cdot + \cdot OC(CH_3)_3$$

$$RCO_2\cdot \xrightarrow{-CO_2} R\cdot \xrightarrow{PhCHMe_2} RH + Ph\dot{C}Me_2$$

(*c*) PHOTOISOMERIZATION OF BENZENOID COMPOUNDS

The irradiation of benzene itself brings about a low conversion into fulvene and benzvalene; the latter slowly reverts to benzene. Prefulvene is thought to be an intermediate:

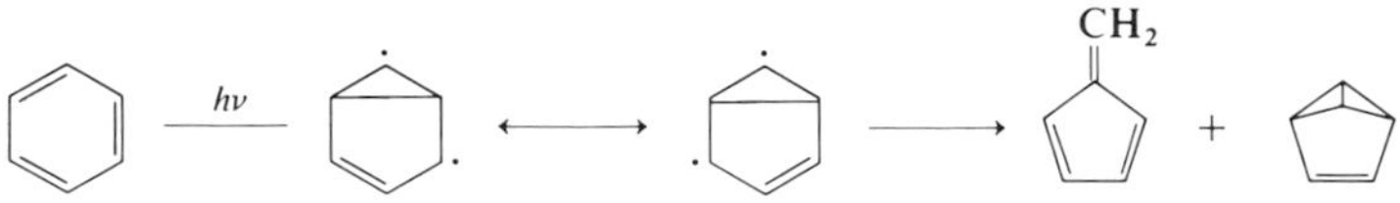

Prefulvene Fulvene Benzvalene

Substituted benzenes give other strained systems. For example, a Dewar benzene can be obtained from 1,2,4-tri-t-butylbenzene,

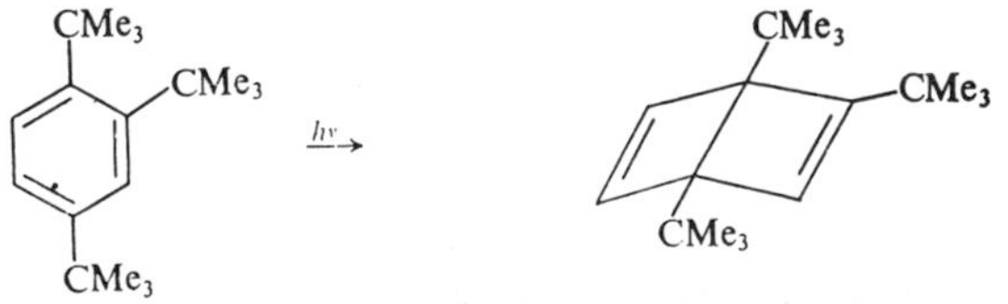

A 'Dewar' benzene

and a prismane from 1,3,5-tri-t-butylbenzene:

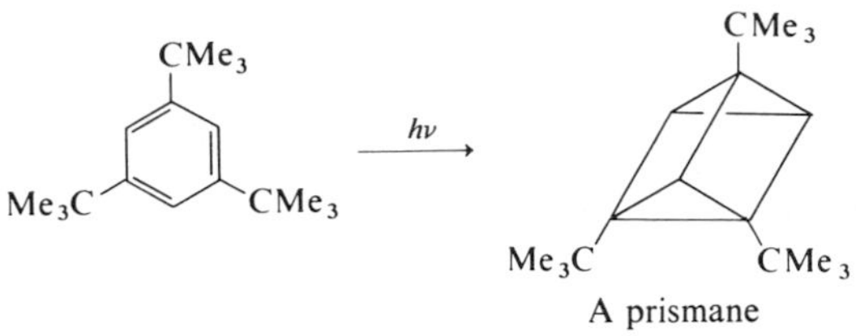

A prismane

In these cases, the products are thought to arise from a 'bent' triplet state, represented for simplicity for benzene:

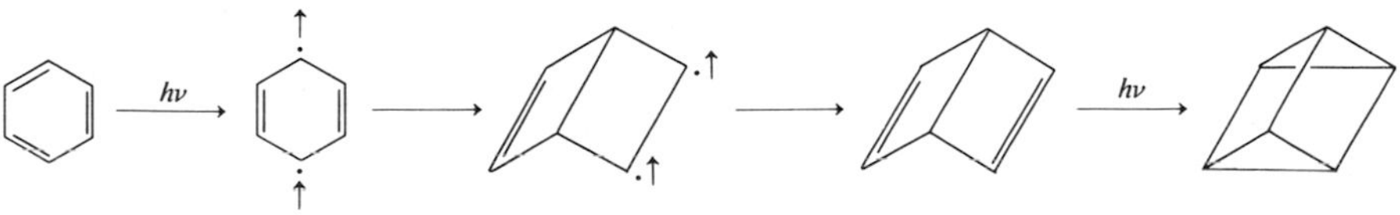

Dewar benzene itself has not been made directly from benzene, but it has been prepared *via* a disrotatory electrocyclic ring-closure:

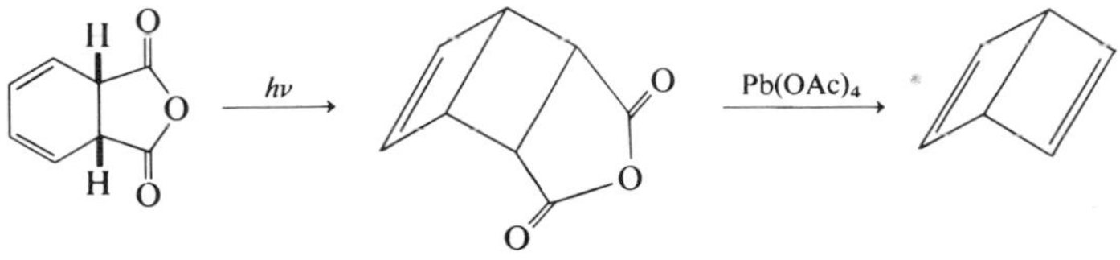

On irradiation, Dewar benzene gives prismane.

Benzvalene, Dewar benzene, prismane, and their derivatives are all thermally labile to varying extents, ultimately reverting to benzenoid compounds; for example, Dewar benzene itself has a half-life of about 2 days at room temperature.*

Some of these species are also thought to be intermediates in photochemical reactions which lead to benzenoid isomers. For example, the irradiation of *o*-xylene gives mixtures containing the *meta*- and *para*-isomers, probably as follows:

---

*The existence of Dewar benzene even for this time is remarkable, considering the strain inherent in the compound on the one hand, and the aromatic stabilization energy to be gained from its conversion into benzene on the other. The significant fact is undoubtedly that the thermal concerted conversion into benzene is symmetry-forbidden.

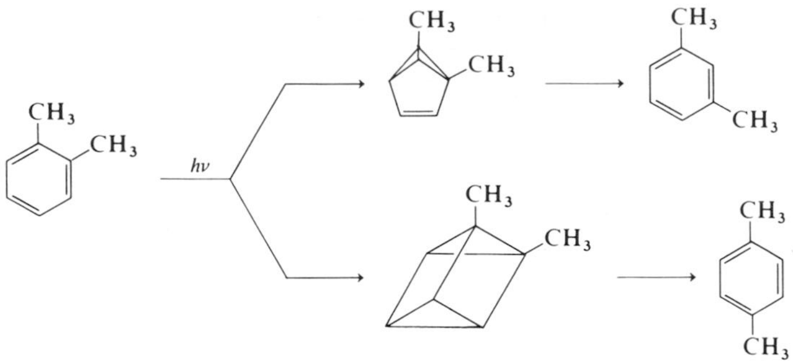

16.5 Photo-oxidation

(a) FORMATION OF PEROXY-COMPOUNDS

Certain types of peroxy-compound can be formed by irradiating the parent organic compound in the presence of oxygen and a sensitizer. Reaction occurs by the excitation of the sensitizer to its triplet state, and thereafter in one of two ways: either the triplet abstracts a hydrogen atom from the substrate to form a radical which then reacts with oxygen, or it interacts with the oxygen molecule so as to activate it.

The best-known example of the first type of reaction is the oxidation of a secondary alcohol to give a hydroxy-hydroperoxide, with benzophenone as the sensitizer. The triplet benzophenone reacts with the alcohol to give a carbon radical (cf. photoreduction, p. 511) which adds oxygen:

$$Ph_2\dot{C}-\dot{O} + R_2CH-OH \longrightarrow Ph_2\dot{C}-OH + R_2\dot{C}-OH$$
Triplet

$$R_2\dot{C}-OH + O_2 \longrightarrow R_2\overset{\overset{\displaystyle O-O\cdot}{|}}{C}-OH$$

A chain reaction is now propagated:

$$R_2\overset{\overset{\displaystyle O-O\cdot}{|}}{C}-OH + R_2CH-OH \longrightarrow R_2\overset{\overset{\displaystyle O-OH}{|}}{C}-OH + R_2\dot{C}-OH$$

$$R_2\dot{C}-OH + O_2 \longrightarrow R_2\overset{\overset{\displaystyle O-O\cdot}{|}}{C}-OH$$

The usefulness of the reaction lies in the fact that hydroxy-hydroperoxides readily eliminate hydrogen peroxide to form carbonyl compounds:

$$R_2C\overset{\displaystyle OH}{\underset{\displaystyle O_2H}{<}} \longrightarrow R_2CO + H_2O_2$$

In the second type of sensitized oxidation, the sensitizer in its triplet state interacts with the triplet oxygen molecule—i.e. the ground state of oxygen*— to give singlet oxygen—i.e. an *excited* oxygen molecule—while the sensitizer returns to its ground state, S_0. Singlet oxygen then interacts directly with the organic substrate. The common sensitizers are dyes such as fluorescein and Rose Bengal (a halogenated fluorescein) .

Three types of oxidation can be brought about in this way. First, conjugated dienes yield cyclic peroxides in a reaction of Diels-Alder type. Examples are:

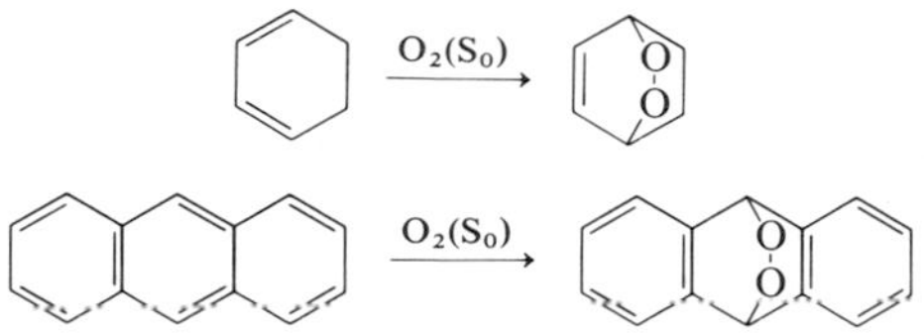

The usefulness of the reaction lies in the reducibility of the peroxides to diols:

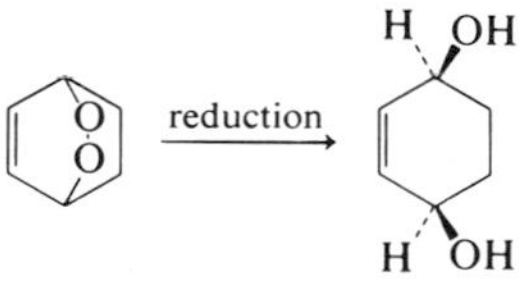

Secondly, olefins with an allylic hydrogen atom form hydroperoxides, probably as follows (*cf.* the ene-reaction, p. 317):

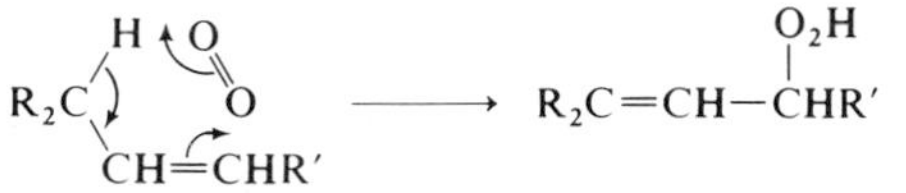

The products can be reduced to allylic alcohols.

Thirdly, olefins with no allylic hydrogen form dioxetans which, on warming, yield carbonyl compounds:

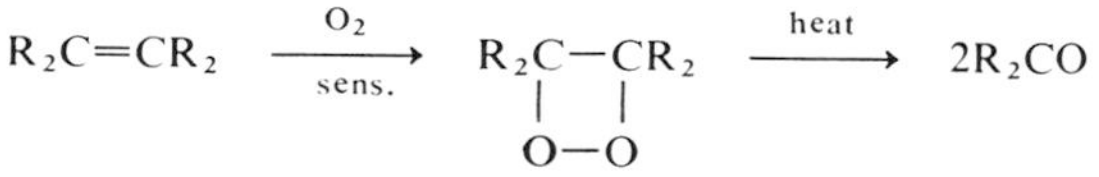

(*b*) OXIDATIVE COUPLING OF AROMATIC COMPOUNDS

Although the irradiation of *cis-* or *trans*-stilbene in the absence of oxygen gives simply an equilibrium mixture of the two (p. 517), in the presence of oxygen phenanthrene is formed. It has been inferred that *cis*-stilbene photocyclizes

*That the ground state of O_2 is a triplet can be readily understood by constructing molecular orbitals as for F_2 on p. 37. Its highest occupied orbitals are the π^*2p; each contains one electron, and the two electrons have parallel spins. The lowest singlet state is only 92.4 kJ mol^{-1} higher in energy.

reversibly to give a small proportion of dihydrophenanthrene which, in the presence of oxygen, is oxidized irreversibly to phenanthrene:

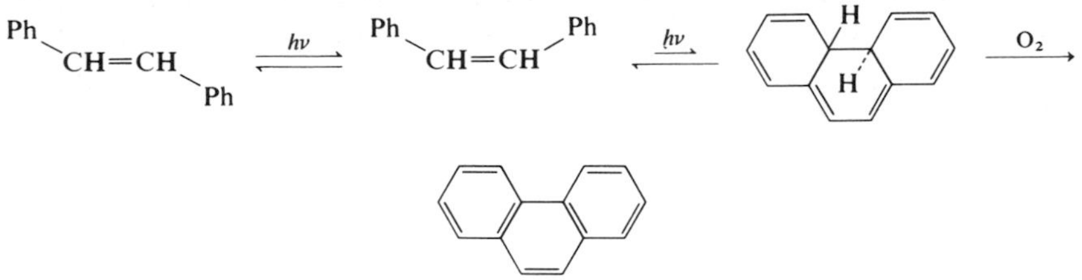

Compounds related to stilbene react analogously, providing a useful method for obtaining polycyclic systems, e.g.

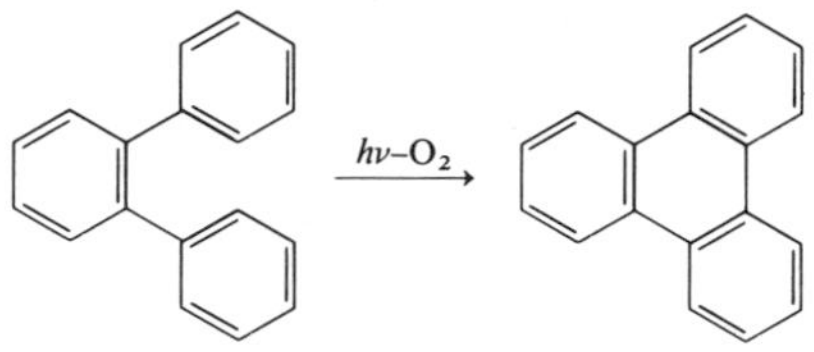

16.6 Photochemical Aromatic Substitution

The distribution of a charge in a species in an excited state can be wholly different from that in the ground state, and especial use has been made of this difference in controlling the positional selectivity in nucleophilic aromatic substitutions.

For example, when 3,4-dimethoxynitrobenzene is heated with hydroxide ion, it is the 4-methoxy substituent which is replaced:

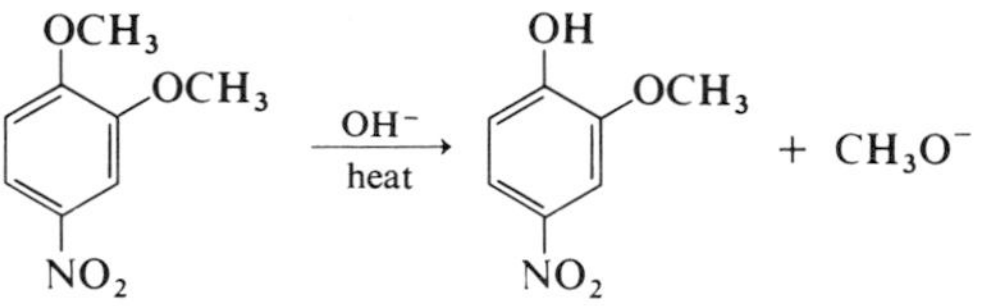

In contrast, when the reaction is carried out at room temperature under ultraviolet irradiation, it is the 3-methoxy substituent which is displaced:

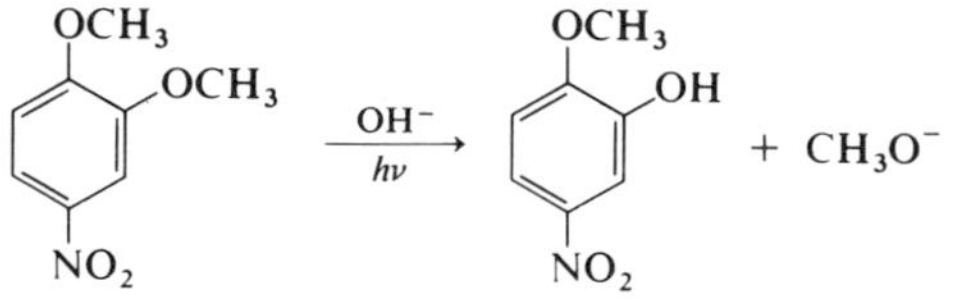

The difference can be related to the difference in distribution of electronic charge in the ground and excited states. In the former, the nitro group, through its $-M$ effect, renders the *ortho-* and *para-*positions positive compared with the *meta-*position; but in the latter, the *ortho-* and *meta-*positions are rendered positive compared with the *para-*position. With an electron-releasing substituent, the converse obtains: in the ground state, the *ortho-* and *para-*positions are negatively charged compared with the *meta-*, whereas in the excited state the *ortho-* and *meta-*positions are negatively charged relative to the *para-*position.

Other examples which illustrate these effects in photoactivated nucleophilic substitution are:

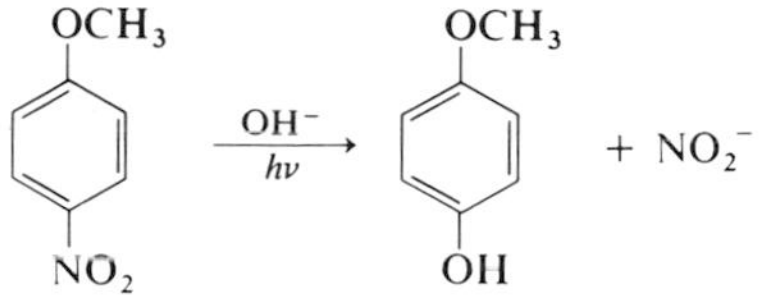

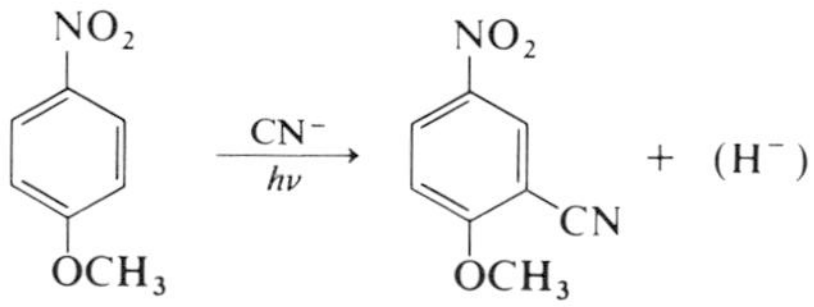

$$\text{(p-nitroanisole)} \xrightarrow[h\nu]{OH^-} \text{(4-methoxyphenol)} + NO_2^-$$

$$\text{(p-nitroanisole)} \xrightarrow[h\nu]{CN^-} \text{(2-cyano-4-nitroanisole)} + (H^-)$$

Compared with the first example, the thermal reaction causes displacement of the methoxy group; and, compared with the second, it leads to cyanation *ortho* to the nitro group.

However, there are also reactions which occur on irradiation but not thermally and yet show the same orienting influences as the thermal process would be expected to display, e.g.

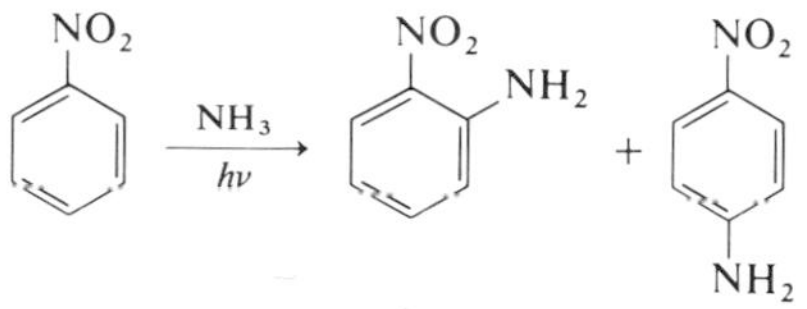

$$\text{(nitrobenzene)} \xrightarrow[h\nu]{NH_3} \text{(o-nitroaniline)} + \text{(p-nitroaniline)}$$

The reason is not clear.

There are so far hardly any examples of photo-activated electrophilic substitutions in which the substituents have the opposite effects to those in thermal reactions. However, in electrophilic deuteriation with CH_3CO_2D, methyl and methoxyl groups direct mainly *meta* and *ortho*, and nitro directs mainly *para*, e.g.

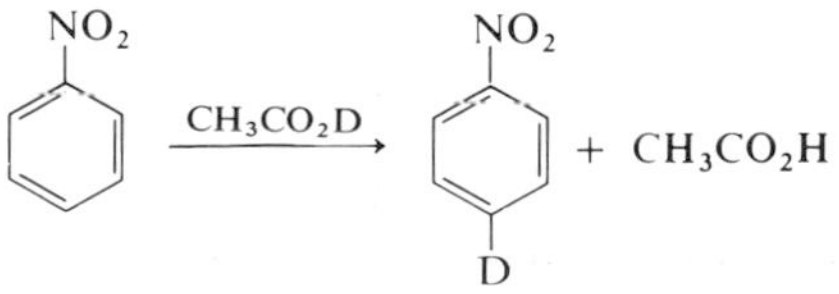

$$\text{(nitrobenzene)} \xrightarrow{CH_3CO_2D} \text{(p-deuterionitrobenzene)} + CH_3CO_2H$$

16.7 Photochemical Fragmentation

In order for light to bring about the fragmentation of a compound, two conditions must be met: the wavelength of the light must correspond both to an energy greater than that of the bond to be cleaved, and to an electronic excitation of the molecule concerned. This effectively eliminates saturated organic compounds since, although they have weaker bonds than correspond to typical, accessible ultraviolet wavelengths, they do not absorb in this region. For example, ethane is unaffected by light of wavelength 300 nm even though this is equivalent to a larger energy (*ca.* 400 kJ mol^{-1}) than the C—C dissociation energy (347 kJ mol^{-1}); and methyl iodide is essentially unaffected by light whereas iodobenzene, having an aromatic chromophore, dissociates into phenyl radicals and iodine atoms.

In general, then, fragmentation is characteristic of certain unsaturated organic compounds together with inorganic compounds such as the halogens. In many instances, the same fragmentation can be brought about by heat, and these are considered in the following Chapter; examples are the fragmentation of peroxides, e.g.

$$PhCO-O-O-COPh \xrightarrow{h\nu} 2\ PhCO_2\cdot$$

and nitrites,

$$RO-N=O \xrightarrow{h\nu} RO\cdot + NO$$

In this section attention is restricted to carbonyl compounds—where the photochemistry is not paralleled thermally—and to aliphatic diazo-compounds.

(a) PHOTOLYSIS OF CARBONYL COMPOUNDS

The $n \to \pi^*$ excitation of aldehydes and ketones can bring about two types of fragmentation.

First, fragmentation can occur of the C—C bond adjacent to carbonyl:

$$RCOR' \xrightarrow{h\nu} R\dot{C}O + R'\cdot$$

This is termed the *Norrish Type I* process, and is followed by one or more of three reactions: disproportionation to give a keten:

$$R_2CH-CO-R' \xrightarrow{h\nu} R_2CH-\dot{C}O + R'\cdot \longrightarrow R_2C=C=O + R'H$$

decarbonylation and then dimerization and/or disproportionation, e.g.

$$RCH_2CH_2-CO-CH_2CH_2R \xrightarrow{h\nu} RCH_2CH_2-\dot{C}O + RCH_2\dot{C}H_2$$

$$RCH_2CH_2-\dot{C}O \longrightarrow RCH_2\dot{C}H_2 + CO$$

$$2\ RCH_2\dot{C}H_2 \left\{ \begin{array}{l} \longrightarrow RCH_2CH_2CH_2CH_2R \\ \\ \longrightarrow RCH=CH_2 + RCH_2CH_3 \end{array} \right.$$

or intermolecular hydrogen-atom abstraction by the acyl radical,

$$RCH_2CH_2-CO-R' \xrightarrow{h\nu} RCH_2\dot{C}H_2 + R'\dot{C}O$$

$$R'\dot{C}O + RCH_2\dot{C}H_2 \longrightarrow R'CHO + RCH=CH_2$$

Secondly, if the carbonyl compound contains a γ-CH group, intramolecular hydrogen-atom abstraction can occur:

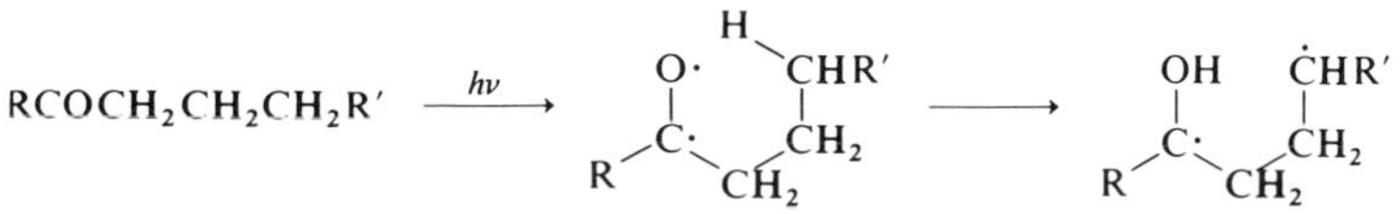

This—the *Norrish Type II* process—is followed by fragmentation,

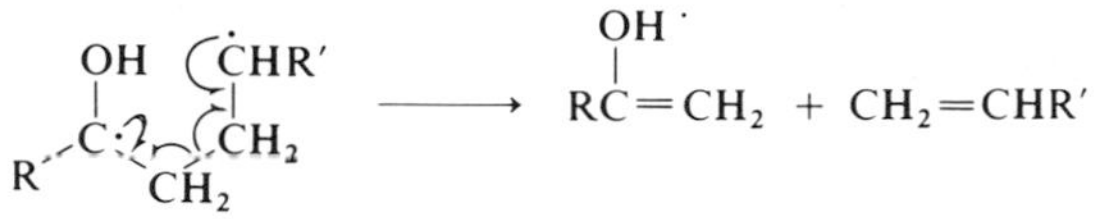

or by ring-closure,

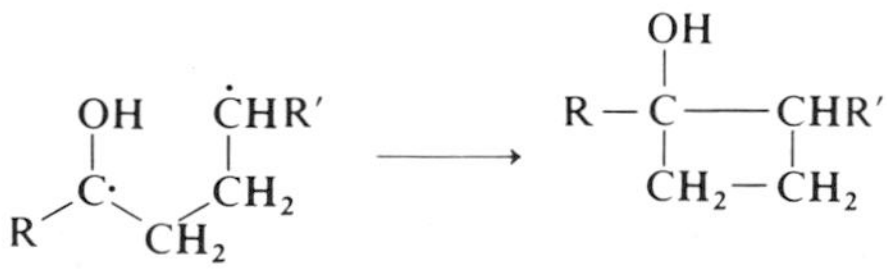

The Type I process occurs more readily in the vapour phase than in solution; under the latter conditions, it is thought that the two radicals usually recombine within the solvent cage in which they are formed, and the process is usually significant only when the alkyl radical formed is of the relatively stable tertiary or benzylic type.

Which of the two types of product is formed by the Type II process depends markedly on the substitution at the carbon atom adjacent to carbonyl; for example, in the following case,

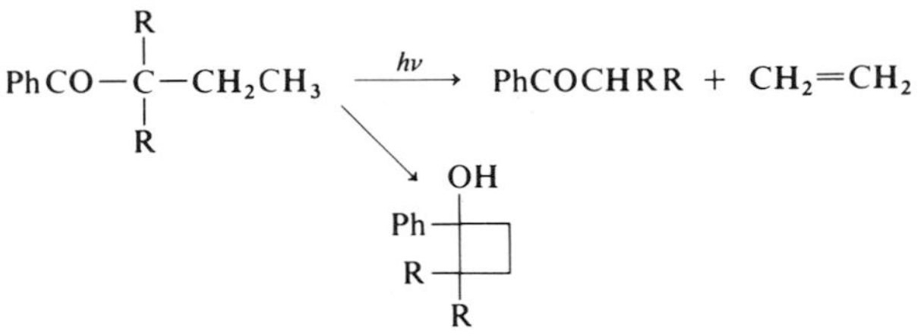

$$PhCO-\underset{\underset{R}{|}}{\overset{\overset{R}{|}}{C}}-CH_2CH_3 \xrightarrow{\ h\nu\ } PhCOCHRR + CH_2{=}CH_2$$

the extent of cyclization relative to fragmentation is 10% when $R = H$ and 89% when $R = CH_3$.

The photochemistry of the smaller cyclic ketones is dominated by the Type I process, followed by intramolecular reactions:

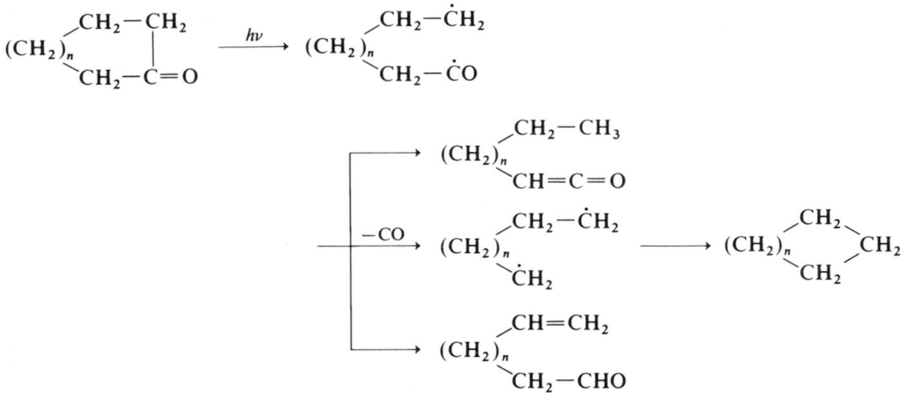

$$(CH_2)_n\underset{\diagdown CH_2-\underset{}{C}=O}{\overset{\diagup CH_2-CH_2}{|}} \xrightarrow{\ h\nu\ } (CH_2)_n\underset{\diagdown CH_2-\dot{C}O}{\overset{\diagup CH_2-\dot{C}H_2}{}}$$

However, in larger rings, cyclization related to the Type II process can occur, e.g.

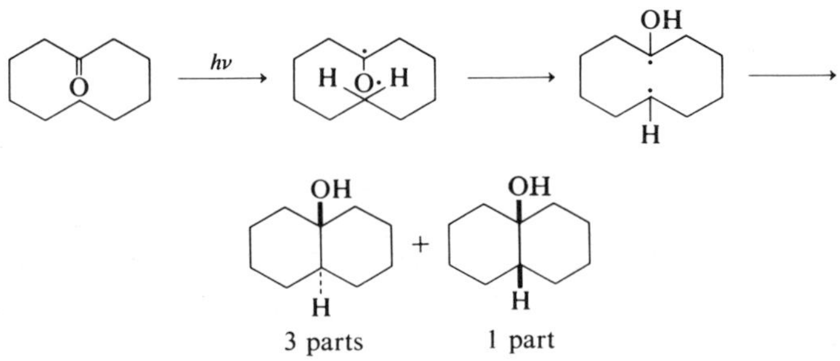

3 parts   1 part

(b) PHOTOLYSIS OF COMPOUNDS CONTAINING THE $\overset{+}{=}N\overset{-}{=}N$ GROUP

(i) *Diazoalkanes*. The irradiation of diazomethane in ether gives methylene in its singlet state:

$$CH_2N_2 \xrightarrow{h\nu} :CH_2 + N_2$$

Methylene is highly reactive, inserting into both unsaturated and saturated bonds (p. 125); its synthetic usefulness is therefore limited. However, suitably substituted methylenes—carbenes—are relatively stable as a result of delocalization, e.g.

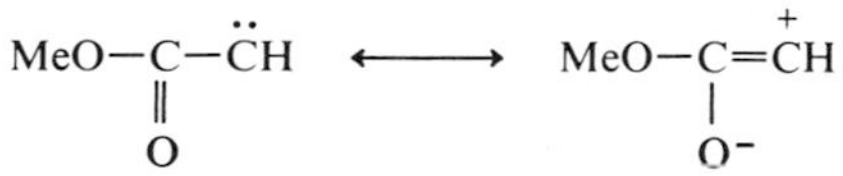

$$MeO-\overset{\cdot\cdot}{C}-\overset{\cdot\cdot}{C}H \quad\longleftrightarrow\quad MeO-\overset{+}{C}=\overset{}{C}H$$

and are consequently much more selective; for example, whereas methylene inserts into both primary and tertiary C—H bonds in 2,3-dimethylbutane, carbomethoxymethylene reacts almost solely at the tertiary position:

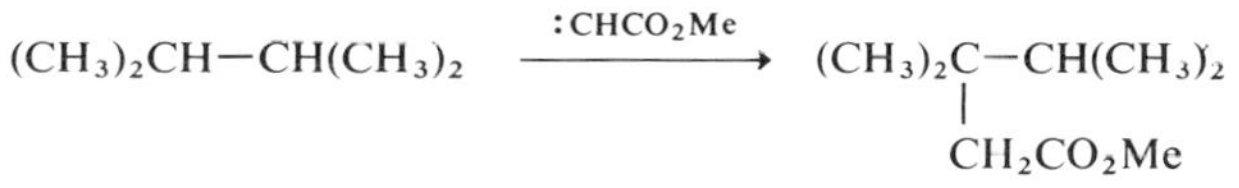

$$(CH_3)_2CH-CH(CH_3)_2 \xrightarrow{:CHCO_2Me} (CH_3)_2\overset{|}{C}-CH(CH_3)_2$$
$$\qquad\qquad\qquad\qquad\qquad\qquad CH_2CO_2Me$$

The most useful reaction of carbenes is with olefins to form cyclopropane derivatives. When the carbene is generated in its singlet state, by direct photolysis, its insertion into the double bond is essentially stereospecific; for example, *cis*-2-butene and dicarbomethoxymethylene give almost entirely the *cis*-dimethyl-cyclopropane:

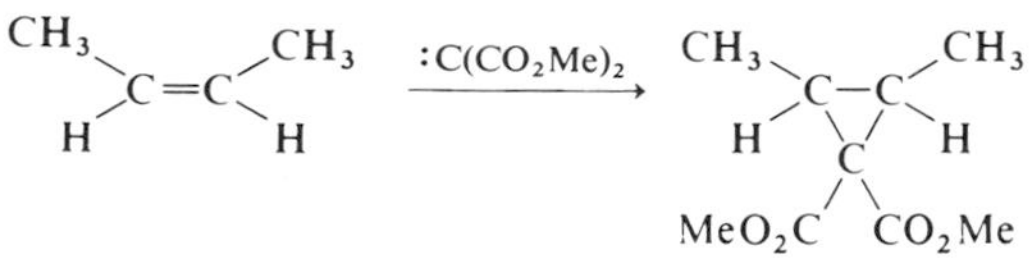

However, when the carbene is generated by photolysis in the presence of a triplet sensitizer, it is formed in its triplet state and the stereospecificity is lost; for example, *cis*-2-butene gives largely the *trans*-dimethylcyclopropane. This is because there is a time-lag between formation of the two C—C bonds during which spin-inversion occurs, and this allows rotation about the original C—C bond:

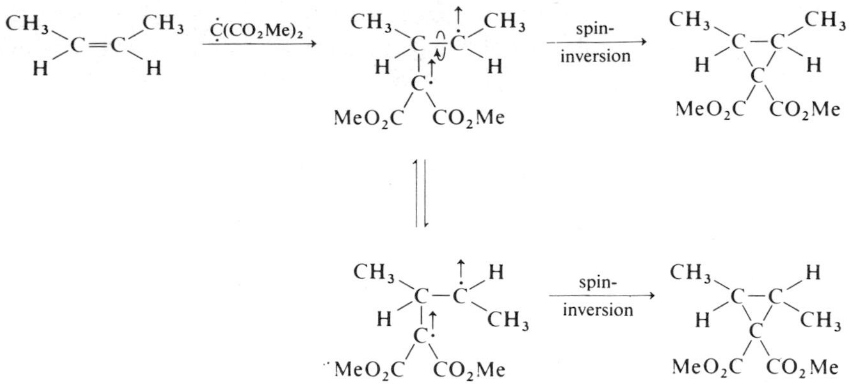

Carbenes also react with aromatic compounds. For example, irradiation of diazomethane in benzene gives about 30% of cycloheptatriene; thus, addition across C=C is followed by a spontaneous Cope rearrangement:

However, this is not nearly so efficient a reaction as that in the presence of copper(I) ion (p. 220).

The carbenes from α-diazoketones have the special property that they undergo rearrangement faster than they are trapped by, for example, olefins. Rearrangement gives a keten which can be trapped by an olefin, and this forms the basis of a method for synthesizing four-membered rings, e.g.

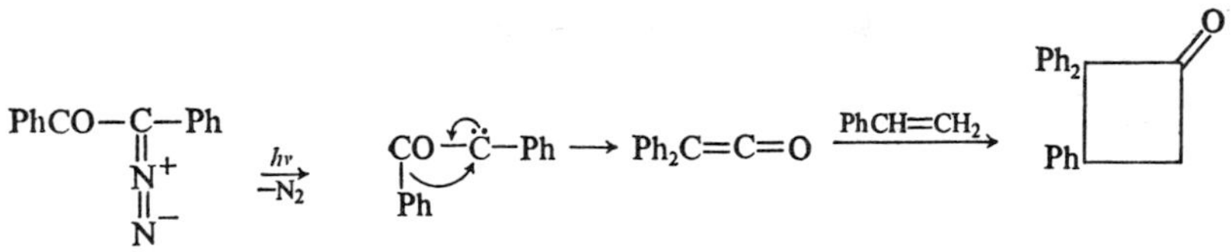

A variant of this reaction is the ring-contraction of carbenes from cyclic α-diazo-ketones, e.g.

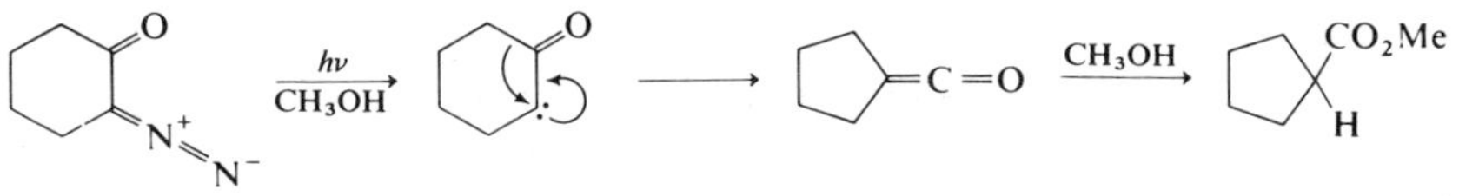

Some highly strained small-ring compounds have been made in this way, e.g.

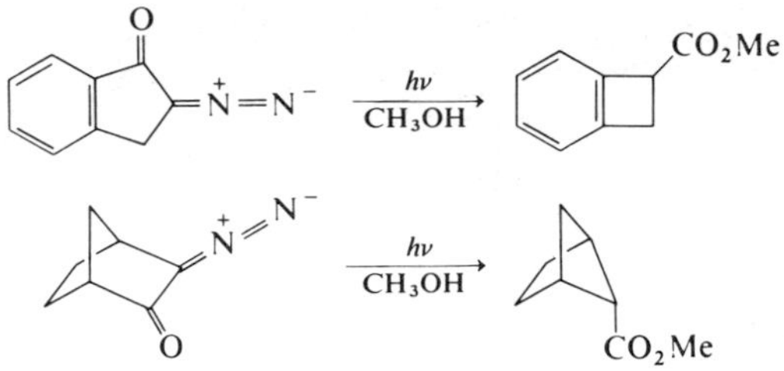

(*ii*) *Alkyl azides*. The photolysis of an azide gives a nitrene,

$$RN=\overset{+}{N}=\overset{-}{N} \quad \xrightarrow{h\nu} \quad R\ddot{N}: + N_2$$

which can undergo a variety of intra- and inter-molecular reactions: intermolecular hydrogen-atom abstraction,

$$R\ddot{N}: + R'H \quad \longrightarrow \quad R\dot{N}H + R'\cdot$$

$$R\dot{N}H + R'H \quad \longrightarrow \quad RNH_2 + R'\cdot$$

intramolecular abstraction of δ-H,

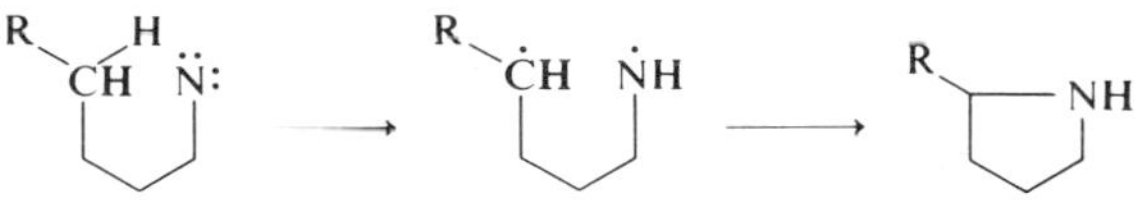

[1,2]-hydrogen shift,

$$R_2CH-\ddot{N}: \quad \longrightarrow \quad R_2C=NH$$

addition to an olefin,

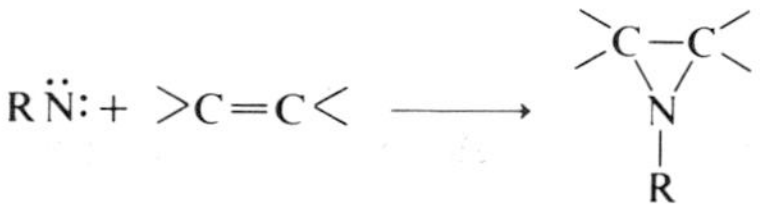

and coupling,

$$2\ R\ddot{N}: \quad \longrightarrow \quad RN=NR$$

Synthetically, the most useful process is that involving abstraction of δ-H followed by cyclization, for this enables functionality to be introduced at an unactivated CH group which is stereochemically suitably placed (*cf.* p. 559).

Vinyl and allyl azides give nitrenes which undergo intramolecular addition to give highly strained compounds, e.g.

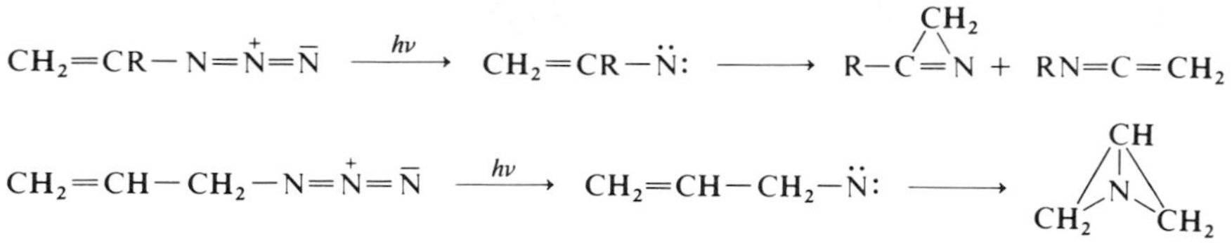

$$CH_2=CR-N=\overset{+}{N}=\overline{\underline{N}} \quad \xrightarrow{h\nu} \quad CH_2=CR-\overset{..}{\underline{N}}: \quad \longrightarrow \quad R-\overset{\overset{\displaystyle CH_2}{\diagup\backslash}}{C=N} + RN=C=CH_2$$

$$CH_2=CH-CH_2-N=\overset{+}{N}=\overline{\underline{N}} \quad \xrightarrow{h\nu} \quad CH_2=CH-CH_2-\overset{..}{\underline{N}}: \quad \longrightarrow \quad \underset{CH_2 \quad CH_2}{\overset{\overset{\displaystyle CH}{\diagup\backslash}}{\underset{\diagdown N \diagup}{}}}$$

Further Reading

COXON, J. M., and HALTON, B., *Organic Photochemistry*, Cambridge University Press (1974).

CUNDALL, R. B., and GILBERT, A., *Photochemistry*, Thomas Nelson (1970).

WELLS, C. H J., *Introduction to Molecular Photochemistry*, Chapman and Hall (London 1972)

Problems

1. Draw the structure of the chief product from ultraviolet irradiation of each of the following:

(a) [structure: cyclobutane ring with CH₃, H, CH₃, H substituents]

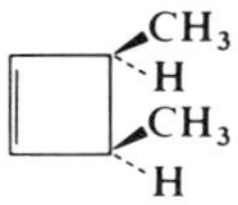

(b) [structure: bicyclic alkene with H H]

(c) [structure: bicyclic diene with H, H]

(d) [structure: ring with CH₃, SO₂, CH₃]

(e) [structure: seven-membered ring with H, OCH₃]

(f) [structure: cyclopentanone with $\overset{+}{N}=\overline{\underline{N}}$ and O]

(g) $\underset{HO_2C}{\overset{H}{\diagdown}}C=C\overset{CO_2H}{\underset{H}{\diagup}}$

(h) [structure: cyclopentene]

(i) [structure: o-xylene, benzene ring with CH₃, CH₃]

(j) [structure: benzene ring with OCH₃ and NO₂] + NH_3

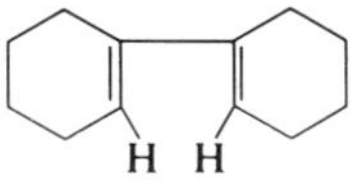

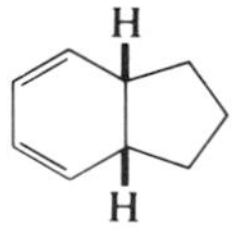

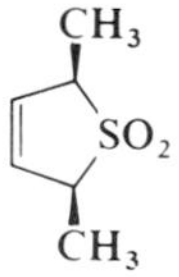

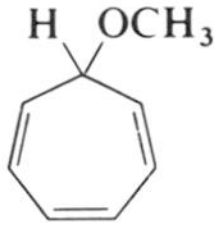

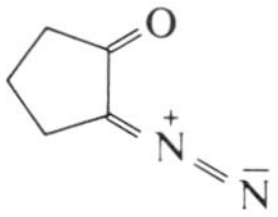

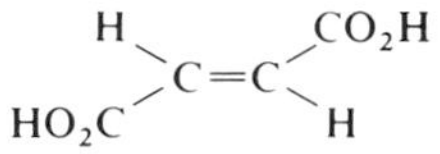

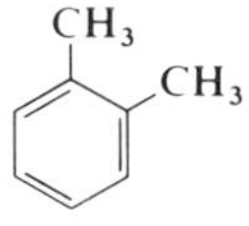

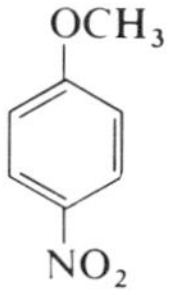

(k) [structure: norbornene] + $PhCOCH_3$

(l) $PhCOCH_2CH_2CH_3$

(m) $Ph_2CO + PhCH=CH_2$

2. Rationalize the following reactions:

(i)

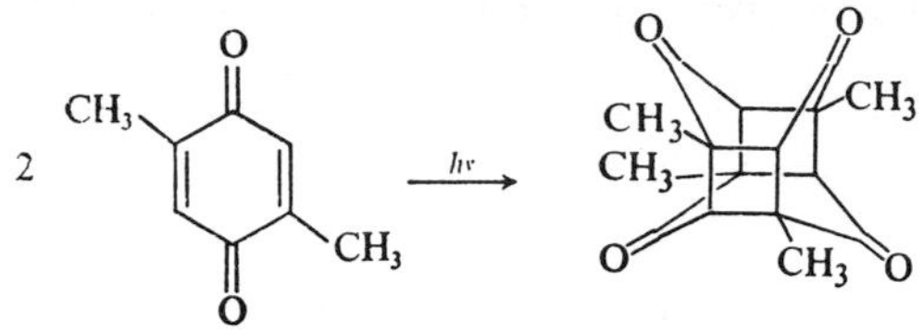

(ii)

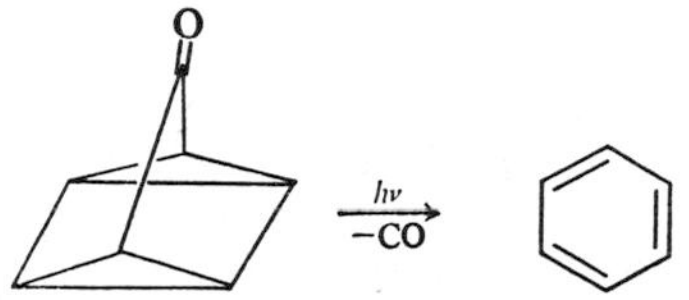

(iii)

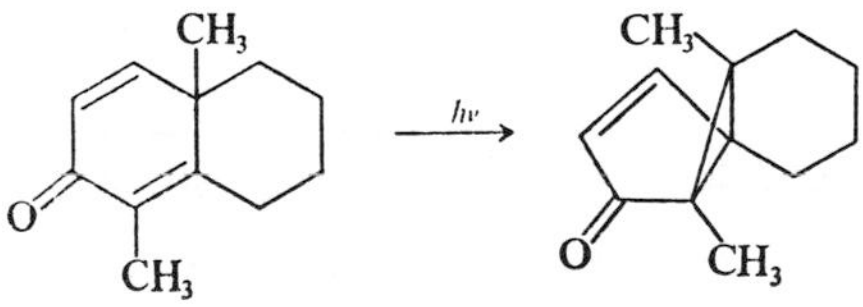

(iv) $CH_3CO(CH_2)_3CH=CH_2$ $\xrightarrow{h\nu}$

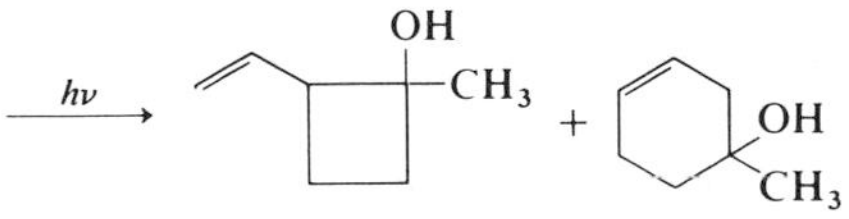

(v)

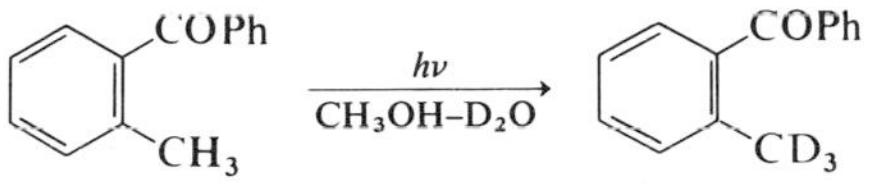

17. Free-radical Reactions

17.1 Principles

A few organic free radicals are relatively stable. For example, the triphenyl-methyl radical, $Ph_3C\cdot$, lives indefinitely in benzene solution in equilibrium with a dimer:

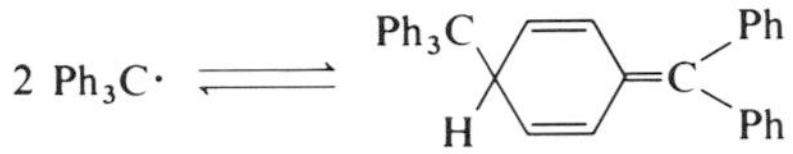

$$2\ Ph_3C\cdot \rightleftharpoons$$

Its stability is accounted for in part by the delocalization of the unpaired electron and in part by the steric forces inhibiting its reactions, for the conversion of the planar tervalent carbon atom into a tetravalent atom involves the compression of the three bulky phenyl groups. Indeed, dimerization gives a compound in which the aromatic stabilization energy of one ring has been sacrificed to retain one of two tervalent carbon atoms, in preference to yielding the fully aromatic but more compressed dimer, Ph_3C-CPh_3.

Even more stable is galvinoxyl (named after Galvin M. Coppinger, who first prepared it), which is a crystalline solid,

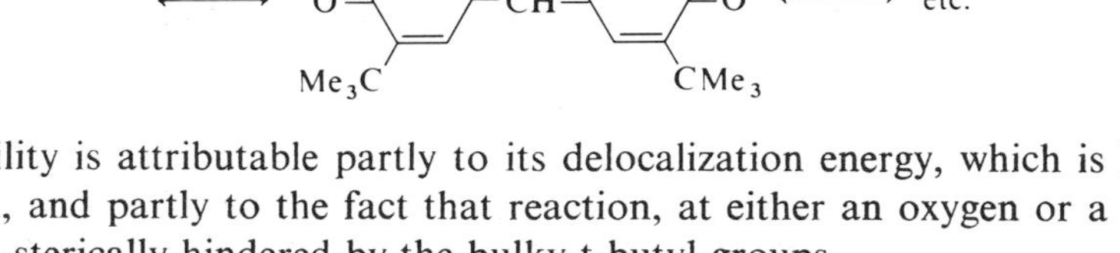

Its stability is attributable partly to its delocalization energy, which is lost on reaction, and partly to the fact that reaction, at either an oxygen or a carbon atom, is sterically hindered by the bulky t-butyl groups.

In the absence of significant stabilizing influences of a steric and/or electronic type, free radicals are very unstable, living for only a fraction of a second before reacting with another species. Nevertheless, they are intermediates in a wide variety of reactions, of which many have important applications in synthesis.

534

(a) THE GENERATION OF FREE RADICALS

Three general methods are used to form free radicals.

(i) *Thermal generation.* The principles have already been described (p. 97) and will be summarized briefly.

Two types of compound undergo dissociation into free radicals at moderate temperatures: (1) compounds which possess an intrinsically weak bond, such as dialkyl peroxides ($D_{O-O} = 155$ kJ mol^{-1}), chlorine ($D_{Cl-Cl} = 238$ kJ mol^{-1}), and bromine ($D_{Br-Br} = 188$ kJ mol^{-1}); (2) compounds which, on fragmentation, form especially strongly bonded products, such as azobisisobutyronitrile, which yields N_2,

$$(CH_3)_2C-N=N-C(CH_3)_2 \longrightarrow 2 (CH_3)_2C\cdot + N_2$$
$$\underset{CN}{|} \qquad \underset{CN}{|} \qquad\qquad\qquad \underset{CN}{|}$$

and peroxalates, which yield CO_2, e.g.

$$(CH_3)_3C-O-O-\underset{\underset{O}{\|}}{C}-\underset{\underset{O}{\|}}{C}-O-O-C(CH_3)_3 \longrightarrow 2 (CH_3)_3C-O\cdot + 2 CO_2$$

In each case, the presence of a substituent which stabilizes a resulting radical increases the rate of decomposition. For example, azomethane is stable up to 200°C whereas azobisisobutyronitrile decomposes with a half-life of 5 minutes at 100°C, to give a delocalized radical,

$$(CH_3)_2\overset{\cdot}{C}-C\equiv N \longleftrightarrow (CH_3)_2C=N=\overset{\cdot}{N}$$

Again, $t_{\frac{1}{2}}$ for dibenzoyl peroxide (30 minutes at 100°C) is far less than for di-t-butyl peroxide (200 hours at 100°C) since the former gives a delocalized radical,

$$PhCO-O-O-COPh \longrightarrow 2 \left[Ph-C\overset{\displaystyle\diagup O}{\diagdown O\cdot} \longleftrightarrow Ph-C\overset{\displaystyle\diagup O\cdot}{\diagdown O} \right]$$

which subsequently decarboxylates to give the phenyl radical:

$$PhCO_2\cdot \longrightarrow Ph\cdot + CO_2$$

(ii) *Photochemical generation.* The principles have been described (p. 526). The method is suitable, for example, for generating chlorine or bromine atoms from the corresponding halogen molecules, alkoxy radicals from alkyl nitrites or hypochlorites,

$$RO-N=O \xrightarrow{h\nu} RO\cdot + NO$$

$$RO-Cl \xrightarrow{h\nu} RO\cdot + Cl\cdot$$

and nitrogen cation-radicals from protonated *N*-chloroamines:

$$\overset{+}{R_2NH}\!-\!Cl \xrightarrow{\;h\nu\;} \overset{+\cdot}{R_2NH} + Cl\cdot$$

(*iii*) *Redox generation.* Some covalent bonds may be broken by the acceptance of an electron from a species of powerful one-electron-donating type, others may be broken by the donation of an electron to a powerful one-electron acceptor.

In the first group, the donor may be a strongly electropositive element such as sodium, as in the acyloin condensation (p. 557),

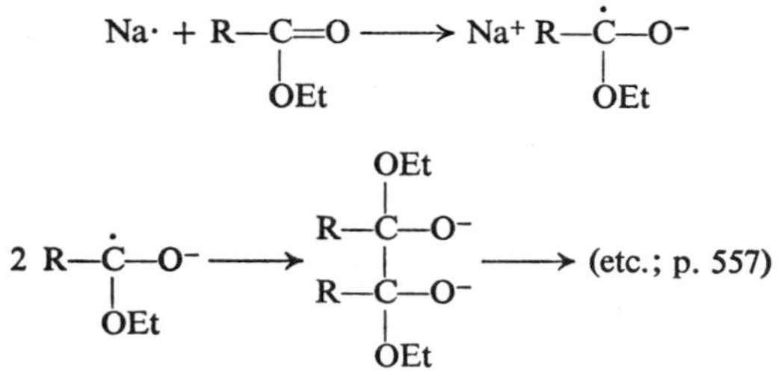

$$Na\cdot + R\!-\!\underset{\underset{OEt}{|}}{C}\!\!=\!\!O \longrightarrow Na^+\, R\!-\!\underset{\underset{OEt}{|}}{\overset{\cdot}{C}}\!-\!O^-$$

$$2\; R\!-\!\underset{\underset{OEt}{|}}{\overset{\cdot}{C}}\!-\!O^- \longrightarrow \begin{array}{c} OEt \\ | \\ R\!-\!C\!-\!O^- \\ | \\ R\!-\!C\!-\!O^- \\ | \\ OEt \end{array} \longrightarrow \text{(etc.; p. 557)}$$

and in other metal-reducing systems (p. 616), or a low-valence transition-metal ion, as in *Fenton's reaction* in which hydroxyl radicals are formed,

$$Fe^{2+} + HO\!-\!OH \longrightarrow Fe^{3+} + HO\cdot + OH^-$$

In the second group, the acceptor is usually a transition-metal ion in a high-valence state, such as hexacyanoferrate(III) ion in the oxidation of certain phenols, e.g.

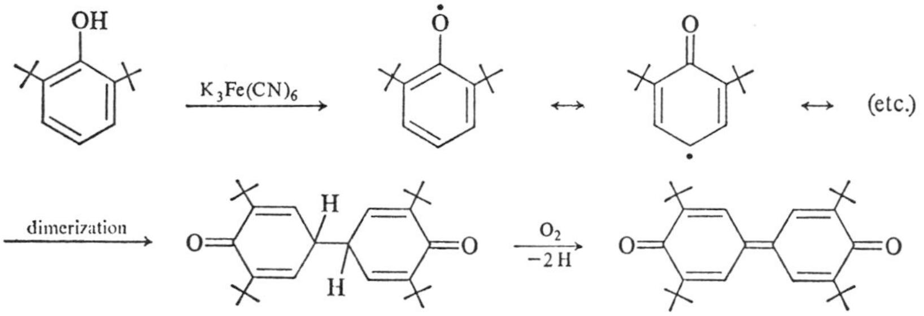

Those redox reactions which lead only to the formation of C—H and C—O bonds are described later under Reduction and Oxidation, respectively.

(*b*) THE REACTIONS OF FREE RADICALS

Radicals normally react in one of three ways: by abstracting an atom from a saturated bond; by adding to an unsaturated bond; and by reacting with other radicals, leading to either combination or disproportionation. A fourth class of reaction, rearrangement, is relatively rare. There is no analogue of the S_N2 reaction; that is, one radical does not displace another from saturated carbon.

(*i*) *Abstraction.* Free radicals react with saturated organic compounds by abstracting an atom, usually hydrogen, from carbon:

$$R\cdot \; + \; H-\overset{\diagup}{\underset{\diagdown}{C}} \longrightarrow R-H + \; \cdot\overset{\diagup}{\underset{\diagdown}{C}}$$

The selectivity of a free radical towards C—H bonds of different types is determined principally by two factors: bond dissociation energies and polar effects. First, the rate of abstraction increases as the bond dissociation energy decreases. The bond dissociation energies for C—H bonds .in four simple paraffins are as follows:

Bond:	H—CH$_3$	H—CH$_2$CH$_3$	H—CH(CH$_3$)$_2$	H—C(CH$_3$)$_3$
Bond dissociation energy (kJ mol^{-1})	426	401	385	372

Consequently, the order of reactivity is, isobutane > propane > ethane > methane, and in general the order is, tertiary C—H > secondary C—H > primary C—H*. Allylic and benzylic C—H bonds (*ca.* 322 kJ mol^{-1}) are significantly weaker than those in saturated systems because the unpaired electron in the resulting radicals is delocalized:

$$CH_2{=}CH{-}\overset{\cdot}{C}H_2 \longleftrightarrow \overset{\cdot}{C}H_2{-}CH{=}CH_2$$

As a result, compounds containing these systems not only react readily with free radicals but also react selectively: e.g. ethylbenzene reacts with bromine atoms essentially entirely at its methylene group:

$$PhCH_2CH_3 \; \underset{-HBr}{\overset{Br\cdot}{\longrightarrow}} \; \begin{cases} \longrightarrow Ph\overset{\cdot}{C}HCH_3 \\ \\ \xrightarrow{\;\;\times\;\;} PhCH_2\overset{\cdot}{C}H_2 \end{cases}$$

Halogenated compounds undergo abstraction of a halogen atom. For example, bromotrichloromethane reacts by loss of the bromine atom (p. 549),

*Expressed differently, the stability of radicals falls in the order, tertiary > secondary > primary. This order is possibly attributable to hyperconjugative stabilization (p. 538).

C—Br rather than C—Cl cleavage occurring because the former bond is the weaker.

Secondly, polar factors are operative in many radical reactions. For example, the relative ease of abstraction by chlorine atoms from the C—H bonds in butyl chloride is:

$$CH_3—CH_2—CH_2—CH_2Cl$$

$$\uparrow \qquad \uparrow \qquad \uparrow \qquad \uparrow$$
$$1{\cdot}5 \qquad 6 \qquad 3 \qquad 1$$

This results from the fact that the chlorine atom is strongly electronegative and preferentially reacts at C—H bonds of relatively high electron-density. The effect of the chloro-substituent is to polarize the carbon atoms so that the electron density is relatively low at the carbon to which chlorine is attached, somewhat greater at the next carbon, and greater still at the third, the order of reactivities reflecting the fall-off in the inductive effect of chlorine. The lower reactivity of the methyl group than the adjacent methylene reflects the bond dissociation energy factor discussed above.

(*ii*) *Addition.* Free radicals add to the common unsaturated groupings. The most important of the unsaturated groups in free-radical synthesis is the C=C bond, addition to which is markedly selective. In particular, addition to an olefin $CH_2{=}CHX$ occurs almost exclusively at the methylene group, irrespective of the nature of X:

$$R{\cdot} \quad + \quad CH_2{=}CHX \quad \begin{cases} \longrightarrow RCH_2—CHX \\ \\ \longrightarrow \dot{C}H_2—CHXR \end{cases}$$

At least two factors are responsible. First, steric hindrance between the radical and X slows reaction at the substituted carbon atom. Secondly, X can stabilize the radical $RCH_2—\dot{C}HX$ relative to the alternative $\dot{C}H_2—CHXR,$* and this can be reflected in the preceding transition state.

*If X = alkyl, stabilization may be the result of (hyperconjugative) delocalization, e.g.

$$RCH_2—\dot{C}H—CH_3 \longleftrightarrow RCH_2—CH{=}CH_2 \overset{\cdot H}{}$$

If X is an unsaturated substituent or a substituent with one or more pairs of *p*-electrons, stabilization results from (conventional) delocalization, e.g.

$$RCH_2—\dot{C}H—CH{=}O \longleftrightarrow RCH_2—CH{=}CH—\dot{O}$$

$$RCH_2—\dot{C}H—Cl \longleftrightarrow RCH_2—\bar{C}H—\overset{+\cdot}{C}l$$

Acetylenes react similarly to olefins, monosubstituted acetylenes reacting preferentially at the unsubstituted carbon. Addition to carbonyl groups is also known, but radicals normally react with carbonyl-containing compounds by abstracting hydrogen from a saturated carbon atom or from —CHO in an aldehyde. The lower reactivity in addition of C=O than C=C probably results from the fact that more energy is required to convert C=O into C—O (*ca.* 350 kJ mol^{-1}) than to convert C=C into C—C (*ca.* 260 kJ mol^{-1}).

(*iii*) *Combination and disproportionation.* Two free radicals can combine by dimerization, e.g.

$$2 \; Br\cdot \longrightarrow Br_2$$

$$2 \; CH_3\cdot \longrightarrow CH_3{-}CH_3$$

$$2 \; CH_3CH_2CH_2\cdot \longrightarrow CH_3{-}(CH_2)_4{-}CH_3$$

by the closely related reaction in which different radicals bond, e.g.

$$CH_3CH_2\cdot + Br\cdot \longrightarrow CH_3CH_2{-}Br$$

and by disproportionation, e.g.

$$2 \; CH_3CH_2\cdot \longrightarrow CH_3CH_3 + CH_2{=}CH_2$$

These reactions are mostly very rapid, some having negligible activation energies.

(*iv*) *Rearrangement.* Free radicals, unlike carbonium ions, seldom rearrange. Thus, whereas the neopentyl cation, $(CH_3)_3C{-}CH_2{}^+$, rearranges to the t-amyl cation, $(CH_3)_2\overset{+}{C}{-}CH_2CH_3$, the neopentyl radical does not rearrange to the t-amyl radical. However, the phenyl group migrates in certain circumstances, e.g.

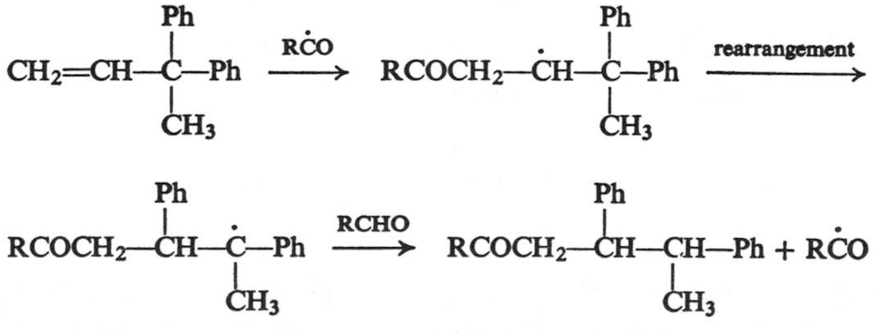

Rearrangements can also occur when they result in the relief of strain in a cyclic system, as in the radical-catalyzed addition of carbon tetrachloride (p. 549) to β-pinene:

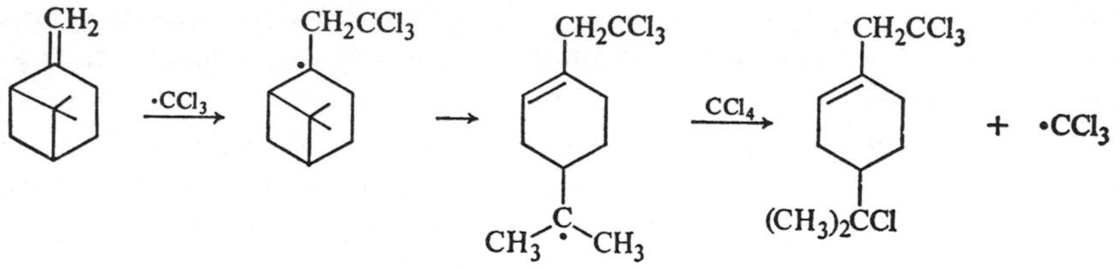

(c) THE CHARACTERISTICS OF FREE-RADICAL REACTIONS

Free-radical reactions may be divided into two classes. In the first, the product results from the combination of two radicals, as in the Kolbe synthesis (p. 552),

$$RCO_2^- \xrightarrow{-e} RCO_2 \cdot \xrightarrow{-CO_2} R\cdot$$

$$2\ R\cdot \longrightarrow R{-}R$$

In the second class, the product results from the reaction of a radical with a molecule, as in the photochemical chlorination of methane (p. 541):

$$Cl_2 \longrightarrow 2\ Cl\cdot$$

$$Cl\cdot + CH_4 \longrightarrow HCl + CH_3\cdot$$

$$CH_3\cdot + Cl_2 \longrightarrow CH_3Cl + Cl\cdot$$

The fundamental difference between the two types of reaction is that reactions in the latter class are *chain reactions*: that is, the step in which the product is formed results in the production of a new free radical which can bring about further reaction. Consequently, were it not for *chain-terminating* reactions, one radical could effect the complete conversion of reactants into products. In practice, chain-terminating processes inevitably occur: in the example above of the chlorination of methane both methyl radicals and chlorine atoms are destroyed by the unions,

$$CH_3\cdot + CH_3\cdot \longrightarrow C_2H_6$$

$$CH_3\cdot + Cl\cdot \longrightarrow CH_3Cl$$

$$Cl\cdot\ + Cl\cdot \longrightarrow Cl_2$$

Nevertheless, these terminating steps, although having large rate-constants, are of relatively low frequency because the concentration of radicals is low, so that the probability of two of them colliding is small compared with the probability of a radical colliding with a molecule of a reactant. Hence the generation of a comparatively small concentration of radicals can suffice for extensive reaction and such reactions are sometimes described as being *radical-catalyzed* (although the use of the term catalyzed is misleading because the catalyst, e.g. a peroxide, is consumed).

Radical-catalyzed reactions are susceptible to *inhibitors*: certain compounds, present in low concentration, which are very reactive towards radicals can react with the radicals as they are generated to give inactive products. Inhibitors may be stable free radicals such as nitric oxide, which combines with organic radicals to give nitroso compounds (R· + NO → R—NO), or, more commonly in organic synthesis, compounds which react with organic radicals to generate radicals of such stability that they do not perpetuate the chain. An example of the latter type is quinol which reacts to give a relatively stable (delocalized) semi-quinone radical:

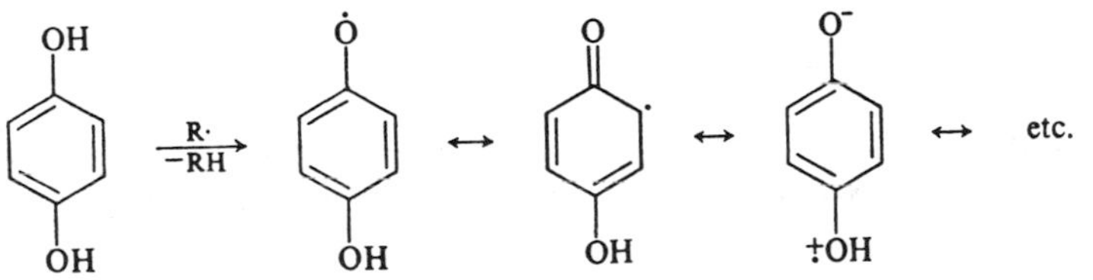

The chain-propagating reaction cannot begin until the inhibitor has been consumed. Since it frequently happens that alternative, ionic reactions can occur between the reactants which are unaffected by the presence of an inhibitor, it is necessary to rid the reactants of impurities which might act as inhibitors in order to achieve maximum efficiency in a radical-catalyzed process. On the other hand, some organic compounds are so susceptible to radical-catalyzed polymerization (p. 549) that it is necessary to store them in the presence of an inhibitor which can remove stray radicals generated by the action of light or oxygen.

17.2 Formation of Carbon-Halogen Bonds

(*a*) SUBSTITUTION IN SATURATED COMPOUNDS
(*i*) *Energetics.* Radical-catalyzed chlorination and bromination can be carried out both in the gas phase and in solution. For a typical chlorination, that of methane, the energetics are as follows:

$$\Delta H \text{ (kJ mol}^{-1}\text{)}$$

$$Cl\cdot + CH_4 \rightarrow HCl + CH_3\cdot \qquad\qquad -2$$

$$CH_3\cdot + Cl_2 \rightarrow CH_3Cl + Cl\cdot \qquad\qquad -101$$

That is, both propagating steps are exothermic. They are rapid reactions, so that the chain reaction competes very effectively with the terminating steps and the chains are long. In light-induced reactions, the quantum yield is high (i.e. one photon can lead to the conversion of many molecules of reactants) and in radical-initiated reactions only a small quantity of initiator is necessary.

In the corresponding bromination, however, the first of the propagating steps is endothermic:

$$\Delta H \text{ (kJ mol}^{-1})$$

$$Br\cdot + CH_4 \rightarrow HBr + CH_3\cdot \qquad\qquad +62$$

$$CH_3\cdot + Br_2 \rightarrow CH_3Br + Br\cdot \qquad\qquad -92$$

Its activation energy is necessarily at least as great as 62 kJ mol^{-1} so that reaction is much slower than in the case of chlorine. As a result, the terminating processes compete more effectively with the propagating step and the chains are short.

The corresponding reaction with iodine atoms, $I\cdot + CH_4 \rightarrow HI + CH_3\cdot$, is so strongly endothermic (129 kJ mol^{-1}) that the reaction is ineffective; indeed, alkyl iodides can be reduced by hydrogen iodide to paraffins and iodine. For fluorine, both propagating steps are strongly exothermic; reaction is violent and is accompanied by the fragmentation of alkyl groups. It is therefore usually better to introduce fluorine by indirect means, although several fluorinations have been conducted successfully in the gas phase in the presence of nitrogen as a diluent and with metal packing to dissipate the heat.

(*ii*) *Applications.* Radical-catalyzed chlorination and bromination occur readily on paraffins and substituted paraffins, both in the gas phase and in solution. Both thermal and photochemical generation of the halogen atoms are employed, and, for chlorinations, sulphuryl chloride in the presence of an initiator such as dibenzoyl peroxide is also used.

$$\text{Initiator} \longrightarrow R\cdot$$

$$SO_2Cl_2 + R\cdot \longrightarrow RCl + \cdot SO_2Cl$$

$$\text{propagation} \begin{cases} \cdot SO_2Cl \longrightarrow SO_2 + Cl\cdot \\ Cl\cdot + RH \longrightarrow HCl + R\cdot \\ R\cdot + SO_2Cl_2 \longrightarrow RCl + \cdot SO_2Cl \end{cases}$$

Although both the bond dissociation energy factor and the polar factor provide some selectivity in chlorinations, a mixture of products is normally obtained. For example, the chlorination of isobutane in the gas phase at 100°C gives comparable quantities of isobutyl chloride and t-butyl chloride,

$$(CH_3)_3CH \xrightarrow[-HCl]{Cl_2} (CH_3)_2CHCH_2Cl + (CH_3)_3CCl$$
$$\qquad\qquad\qquad\qquad\text{Isobutyl chloride} \quad\; \text{t-Butyl chloride}$$

and at 300°C 2-methylbutane gives the following products:

$$(CH_3)_2CHCH_2CH_3 \xrightarrow[- HCl]{Cl_2}$$

$$\underset{\underset{33\cdot5\%}{ClCH_2}}{\overset{CH_3}{\diagdown}}CHCH_2CH_3 + \underset{\underset{22\%}{Cl}}{(CH_3)_2CCH_2CH_3}$$

$$+ \underset{\underset{28\%}{Cl}}{(CH_3)_2CHCHCH_3} + \underset{16\cdot5\%}{(CH_3)_2CHCH_2CH_2Cl}$$

In addition, further chlorination occurs: e.g. methane gives a mixture of methyl chloride, methylene chloride, chloroform, and carbon tetrachloride. Nevertheless, these reactions are of great importance industrially because the products can be separated by efficient fractional distillation. Their use on the laboratory scale is limited, except when only one monochlorinated product can be formed, for this is usually readily separated from any unreacted starting material (lower boiling) and any dichlorinated product (higher boiling); e.g. cyclohexyl chloride can be isolated in 89% yield by treating cyclohexane with sulphuryl chloride in the presence of dibenzoyl peroxide.

$$\text{cyclohexane} + SO_2Cl_2 \xrightarrow{\text{peroxide}} \text{Cl-cyclohexane} + SO_2 + HCl$$

Cyclohexyl
chloride

Bromine atoms are significantly more selective than chlorine atoms in their reactions at primary, secondary, and tertiary C—H.* For example, the gas-phase bromination of isobutane gives t-butyl bromide essentially exclusively:

$$(CH_3)_3CH + Br_2 \longrightarrow (CH_3)_3C\text{—}Br + HBr$$
t-Butyl bromide

(b) SUBSTITUTION IN ALLYLIC AND BENZYLIC COMPOUNDS
Allylic and benzylic compounds are both more reactive than saturated compounds in radical-catalyzed halogenation and react selectively (p. 537). It should

---

*This can be understood by reference to Hammond's postulate (p. 85). Whereas abstraction by Cl· from C—H is exothermic, abstraction by Br· is endothermic. Consequently, the transition state in the latter reaction more closely resembles the product (alkyl radical) (i.e. bond-breaking has made more progress), so that the factors which determine the relative stabilities of the radicals (tertiary > secondary > primary) are of greater significance in determining the relative stabilities of the transition states. The relative reactivities of tertiary and secondary C—H are about 2 : 1 towards Cl· and about 30 : 1 towards Br·.

be noted that allylic compounds of the type $RCH=CH—CH_2R'$ give mixtures of two monohalogenated products because the allylic radical can react at each of two carbon atoms:

$$RCH=CH—CH_2R' \xrightarrow[-HX]{X·} [RCH=CH—\overset{·}{C}HR' \leftrightarrow \overset{·}{R}CH—CH=CHR']$$

$$\xrightarrow[-X·]{X_2} RCH=CH—\underset{X}{CHR'} + RCH—\underset{X}{CH=CHR'}$$

Benzylic compounds, on the other hand, give only one product because the possible isomers, being non-aromatic, are of much higher energy-content.

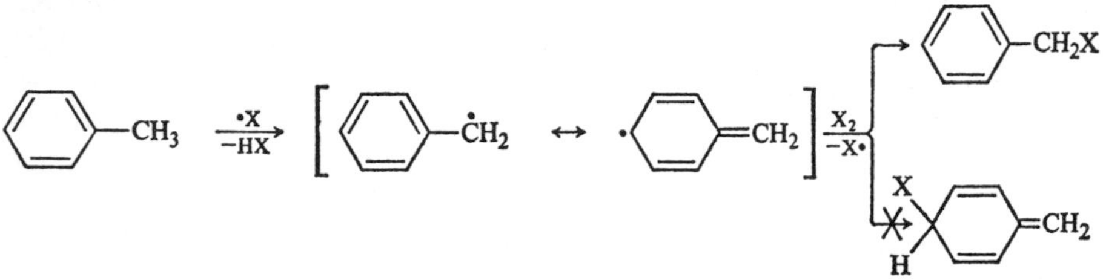

In addition to the thermal and photochemical* methods for the chlorination and bromination of saturated compounds, N-bromosuccinimide is used for allylic and benzylic bromination. It is prepared by treating succinimide in alkaline solution with bromine,

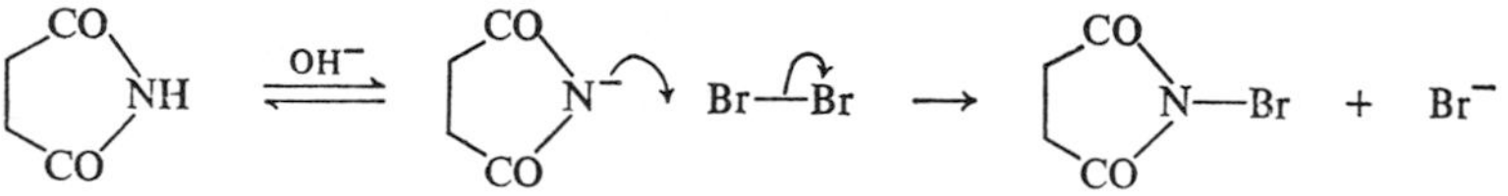

N-Bromosuccinimide

and reacts, usually in the presence of a peroxide as initiator, by a complex chain mechanism; a trace of hydrogen bromide is apparently necessary for initiation:

*Benzenoid compounds can also react by addition with chlorine atoms. For example, the irradiation of benzene and chlorine gives a mixture of stereoisomeric benzene hexachlorides, one of which (the γ–isomer) has powerful insecticidal activity (Gammexane):

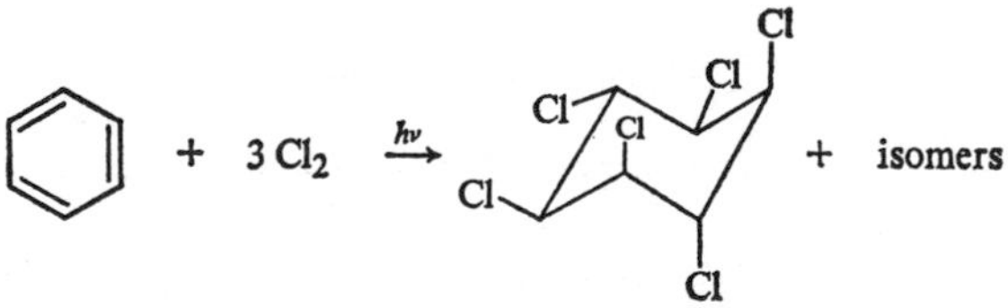

γ-Benzene hexachloride (10%)

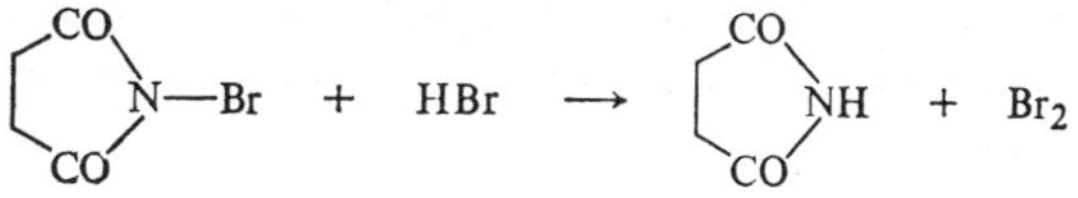

$$\underset{\text{CO}}{\overset{\text{CO}}{\diagdown}}\text{N—Br} \ + \ \text{HBr} \ \longrightarrow \ \underset{\text{CO}}{\overset{\text{CO}}{\diagdown}}\text{NH} \ + \ \text{Br}_2$$

$$\text{Br}_2 \xrightarrow{\ \text{R}\cdot\ } \text{Br}\cdot \ + \ \text{RBr}$$

$$\text{Br}\cdot \ + \ \text{—CH}_2\text{—CH=CH—} \ \longrightarrow \ \text{HBr} \ + \ \text{—}\overset{\cdot}{\text{C}}\text{H—CH=CH—}$$

$$\text{—}\overset{\cdot}{\text{C}}\text{H—CH=CH—} \ + \ \text{Br}_2 \ \longrightarrow \ \underset{\overset{|}{\text{Br}}}{\text{—CH—CH=CH}} \ + \ \text{Br}\cdot$$

Examples of substitution in allylic and benzylic compounds are:

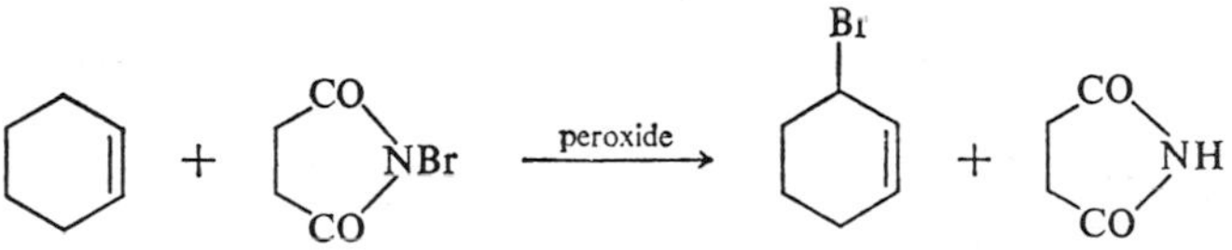

3-Bromocyclohexene

$$\text{PhCH}_3 + \text{SO}_2\text{Cl}_2 \xrightarrow{\ \text{peroxide}\ } \text{PhCH}_2\text{Cl} + \text{SO}_2 + \text{HCl}$$
$$80\%$$

$$\text{PhCH}_3 + \text{Br}_2 \xrightarrow{\ \text{peroxide}\ } \text{PhCH}_2\text{Br} + \text{HBr}$$
$$98\%$$

The bromination of allylic compounds is frequently used in the conversion of olefins into conjugated dienes, which are obtained from the allylic bromides by base-catalyzed elimination. For example, 3-bromocyclohexene gives 1,3-cyclohexadiene:

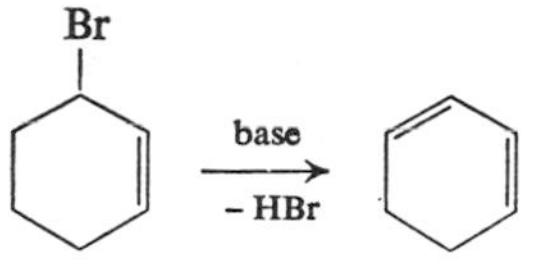

1,3-Cyclohexadiene

(c) ADDITION TO OLEFINIC AND ACETYLENIC COMPOUNDS
Hydrogen bromide adds to C=C and C≡C bonds in radical-catalyzed reactions, illustrated for propylene:

$$\text{Initiator} \longrightarrow \text{R}\cdot$$

$$\text{R}\cdot + \text{HBr} \longrightarrow \text{RH} + \text{Br}\cdot$$

$$\text{propagation} \begin{cases} \text{Br}\cdot + \text{CH}_3\text{—CH=CH}_2 \longrightarrow \text{CH}_3\text{—}\overset{\cdot}{\text{C}}\text{H—CH}_2\text{Br} \\ \text{CH}_3\text{—}\overset{\cdot}{\text{C}}\text{H—CH}_2\text{Br} + \text{HBr} \longrightarrow \text{CH}_3\text{—CH}_2\text{—CH}_2\text{Br} + \text{Br}\cdot \end{cases}$$

Yields are usually good, as in the preparation of n-propyl bromide (87%) and in the following examples:

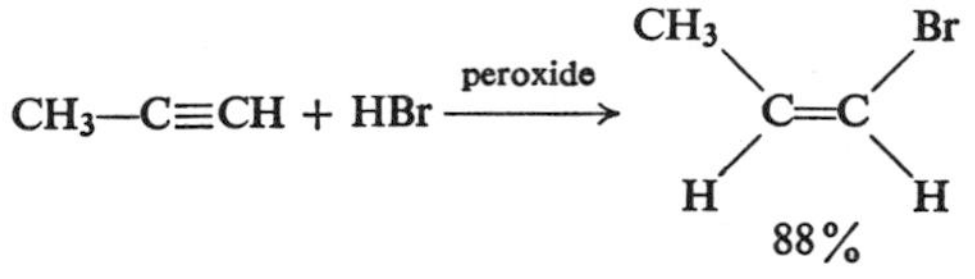

$$CH_3-C\equiv CH + HBr \xrightarrow{\text{peroxide}}$$

with product drawn as:

$$\underset{88\%}{\overset{CH_3}{\underset{H}{>}}C=C\overset{Br}{\underset{H}{<}}}$$

$$ClCH=CH_2 + HBr \xrightarrow{\text{peroxide}} ClCH_2-CH_2Br$$
$$80\%$$

The bromine atom always adds to the terminal carbon atom (p. 538), so that the process is complementary to the *ionic* addition of hydrogen bromide to alkyl- and halo-substituted olefins in which hydrogen adds to the terminal carbon (Markovnikov's rule; p. 107). The ionic and radical-catalyzed processes are competitive and the latter usually occurs the more rapidly. In order to ensure that the ionic addition shall occur when required, it is necessary to free the olefin from the peroxidic impurities formed when olefins are exposed to air (autoxidation; p. 562) which can initiate the radical reaction. The reaction of methylacetylene (above) illustrates another feature of the process, namely, its *trans*-stereospecificity.

The other hydrogen halides do not react at C=C bonds in this way, although hydrogen chloride is partially effective. The reason is apparent from inspection of the energetics of the two propagating steps which, for reaction on ethylene, are as follows:

X in HX	$X\cdot + CH_2=CH_2$ $\to XCH_2-\overset{\cdot}{C}H_2$	$XCH_2-\overset{\cdot}{C}H_2 + HX$ $\to XCH_2-CH_3 + X\cdot$
	ΔH (kJ mol^{-1})	
F	-209	$+159$
Cl	-101	$+27$
Br	-42	-37
I	$+12$	-104

Only for hydrogen bromide are both these steps exothermic. The reaction of hydrogen chloride is slow because of the endothermicity of the reaction between the carbon radical and hydrogen chloride, and hydrogen iodide fails to add because the reaction of iodine atoms with ethylene is endothermic.

(d) BROMO-DECARBOXYLATION (HUNSDIECKER REACTION)

Treatment of the silver salt of a carboxylic acid with bromine in refluxing carbon tetrachloride gives a bromo-compound with elimination of carbon dioxide:

$$RCO_2Ag + Br_2 \longrightarrow RBr + AgBr + CO_2$$

The silver salt is prepared by treating an aqueous solution of the sodium salt of the acid with silver nitrate. Yields are variable but in many cases are high; e.g. that of neopentyl bromide is 62%:

$$(CH_3)_3CCH_2CO_2Ag \xrightarrow[-\,AgBr,\,-\,CO_2]{Br_2} \underset{\substack{\text{Neopentyl}\\ \text{bromide}}}{(CH_3)_3CCH_2Br}$$

Reaction occurs by formation of the acyl hypobromite and its subsequent homolysis; the acyloxy radical loses carbon dioxide and the resulting radical abstracts bromine from a second molecule of the hypobromite:

$$RCO_2Ag + Br_2 \longrightarrow RCO_2Br + AgBr$$

$$RCO_2Br \longrightarrow RCO_2\cdot + Br\cdot$$

$$RCO_2\cdot \xrightarrow{-\,CO_2} R\cdot \xrightarrow{RCO_2Br} RBr + RCO_2\cdot$$

Mercury(II) salts react similarly. For example, treatment of cyclopropane-carboxylic acid with mercury(II) oxide and bromine gives bromocyclopropane in about 45% yield [5]:

$$2\ \triangleright\!\!-CO_2H + HgO + 2\ Br_2 \longrightarrow 2\ \triangleright\!\!-Br + 2\ CO_2 + HgBr_2 + H_2O$$

(e) IODO-DECARBOXYLATION

Iodides may be prepared by a process analogous to the Hunsdiecker reaction. The carboxylic acid is heated in solution, under illumination, with lead tetraacetate and iodine. An exchange equilibrium between lead-bonded acetate and carboxylate groups is established and the acyl hypoiodite is formed and photolyzed:

$$4\ RCO_2H + Pb(OAc)_4 \rightleftharpoons Pb(O_2CR)_4 + 4\ HOAc$$

$$Pb(O_2CR)_4 + I_2 \longrightarrow Pb(O_2CR)_2 + 2\ RCO_2I$$

$$RCO_2I \xrightarrow{h\nu} RCO_2\cdot + I\cdot$$

$$RCO_2\cdot \xrightarrow{-\,CO_2} R\cdot \xrightarrow{RCO_2I} RI + RCO_2\cdot$$

The yields so far reported are in some cases excellent: e.g. n-hexanoic acid gives n-pentyl iodide quantitatively.

17.3 Formation of Carbon-Carbon Bonds

(a) ADDITION TO OLEFINIC COMPOUNDS

A large group of synthetically useful radical-catalyzed reactions is based on the addition of aliphatic carbon radicals to olefinic bonds. The reaction of bromoform with 1-butene is illustrative:

$$PhCOO—OCOPh \xrightarrow{heat} 2\ PhCO_2\cdot$$

$$PhCO_2\cdot + CHBr_3 \longrightarrow PhCO_2H + \cdot CBr_3$$

$$propagation \begin{cases} CH_3CH_2CH{=}CH_2 + \cdot CBr_3 \longrightarrow CH_3CH_2\overset{\cdot}{C}H—CH_2CBr_3 \\ CH_3CH_2\overset{\cdot}{C}H—CH_2CBr_3 + CHBr_3 \longrightarrow CH_3CH_2CH_2CH_2CBr_3 + \cdot CBr_3 \\ \qquad\qquad\qquad\qquad\qquad\qquad\qquad\quad \text{1,1,1-Tribromopentane} \end{cases}$$

The general scope of the reactions is shown by the following examples:

Aldehydes: addition of $R\overset{\cdot}{C}O$, e.g.

$$CH_3CHO + C_6H_{13}—CH{=}CH_2 \xrightarrow{peroxide} C_6H_{13}—CH_2—CH_2—CO—CH_3$$
$$\text{2-Decanone (64\%)}$$

Ketones: addition of $RCO—\overset{\cdot}{C}H—R'$, e.g.

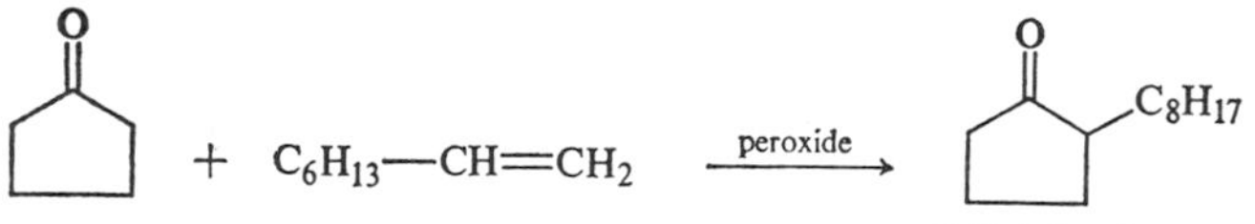

2-Octylcyclo-
pentanone (57%)

Alcohols: addition of $R—\overset{\cdot}{C}(OH)—R'$, e.g.

$$C_2H_5OH + C_6H_{13}—CH{=}CH_2 \xrightarrow{peroxide} C_6H_{13}—CH_2—CH_2—\underset{\underset{OH}{|}}{CH}—CH_3$$
$$\text{2-Decanol (28\%)}$$

Amines: addition of $R—\overset{\cdot}{C}(NHR')—R''$, e.g.

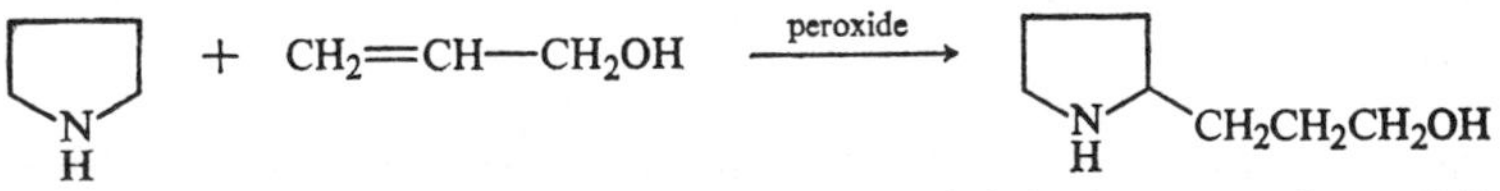

2-(γ-hydroxypropyl)pyrrolidine

Alkyl perhalides react *via* abstraction of a halogen atom to give β-halo-compounds, e.g.

$$CBrCl_3 + CH_3CH_2CH{=}CH_2 \xrightarrow{\text{peroxide}} CH_3CH_2CH{-}CH_2CCl_3$$
$$\underset{Br}{|}$$

3-Bromo-1,1,1-trichloro-
pentane (43%)

$$CCl_4 + (CH_3)_2C{=}CH_2 \xrightarrow{\text{peroxide}} (CH_3)_2C{-}CH_2{-}CCl_3$$
$$\underset{Cl}{|}$$

1,1,1,3-Tetrachloro-3-
methylbutane (70%)

(b) THE POLYMERIZATION OF OLEFINS

Many olefins are polymerized by free-radical initiators. The reaction of a mono-substituted olefin occurs as follows:*

$$\text{Initiator} \longrightarrow R\cdot$$

propagation
$$\begin{cases} R\cdot + CH_2{=}CHX \longrightarrow RCH_2{-}\dot{C}HX \\ RCH_2{-}\dot{C}HX + CH_2{=}CHX \longrightarrow RCH_2{-}CHX{-}CH_2{-}\dot{C}HX \end{cases}$$

termination $\quad R{-}(-CH_2{-}CHX{-})_m CH_2{-}\dot{C}HX + R{-}(-CH_2{-}CHX{-})_n CH_2{-}\dot{C}HX$

combination
$$\longrightarrow R{-}(-CH_2{-}CHX{-})_m CH_2{-}CHX{-}CHX{-}CH_2{-}(-CHX{-}CH_2{-})_n R$$

disproportionation
$$\longrightarrow R{-}(-CH_2{-}CHX{-})_m CH{=}CHX + R{-}(-CH_2{-}CHX{-})_n CH_2{-}CH_2X$$

Initiation is normally effected by a peroxide or an azo-compound and the chains grow until two meet and the reaction terminates by dimerization or disproportionation. As the concentration of initiator is increased the number of growing-chains increases and consequently termination becomes more probable relative to polymerization; hence the average molecular weight of the polymer can be determined by adjusting the concentration of initiator. The molecular weight can also be controlled by the addition of *chain-transfer* agents, i.e. compounds which react with the growing-chain to terminate it and at the same time themselves give radicals which initiate new chains. Thiols are often chosen for the purpose.

---

*Regardless of the nature of X, radicals add to the unsubstituted carbon of the olefin in accord with the principles described earlier, p. 538.

$$R-(-CH_2-CHX-)_m CH_2-\overset{\bullet}{C}HX + RSH \longrightarrow R-(-CH_2-CHX-)_m CH_2-CH_2X + RS\cdot$$

$$RS\cdot + CH_2=CHX \longrightarrow RSCH_2-\overset{\bullet}{C}HX \xrightarrow{\ CH_2=CHX\ } \text{etc.}$$

Polymerization is subject to inhibition (p. 541) and it is advisable to add a small quantity of an inhibitor to the more readily polymerized olefins (e.g. styrene) to prevent polymerization during storage.

Free-radical polymerization is of immense industrial importance. Some of the widely used monomers are vinyl chloride (for polyvinyl chloride, PVC), styrene (for polystyrene), and methyl methacrylate ($CH_2=C(CH_3)CO_2Me$) (for Perspex). The polymerization of a mixture of two monomers (copolymerization) can also give polymers with useful properties: e.g. butadiene is copolymerized with both acrylonitrile and styrene for the manufacture of synthetic rubbers.

Ethylene and propylene are not polymerized in this way. One method for ethylene involves polymerization at 200°C and 1,500 atmospheres pressure in the presence of a trace of oxygen. A second, more recent, method employs Ziegler's catalysts, of which the commonest consists of titanium tetrachloride and an aluminium trialkyl (e.g. $AlEt_3$); the details of the polymerization mechanism are not fully understood. The advantages of this method are that it can be conducted at atmospheric pressure and that the resulting polyethylene has a smaller degree of chain-branching and a higher density. Propylene is also polymerized by the Ziegler technique. In this case, polymerization is *stereoselective*; the methyl groups attached to the chain are ordered uniformly on one side of the chain:

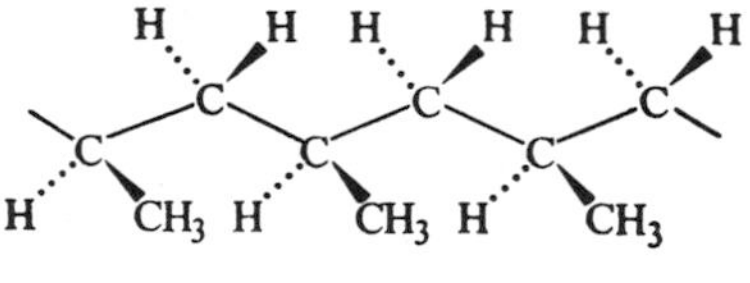

The resulting polymer is described as *isotactic* (as opposed to the randomly ordered *atactic* polymers) and is of higher density and better suited for making films and moulded articles than the atactic polymer.

(c) HOMOLYTIC AROMATIC SUBSTITUTION

Both alkyl and aryl radicals substitute in aromatic nuclei by an addition-elimination sequence:

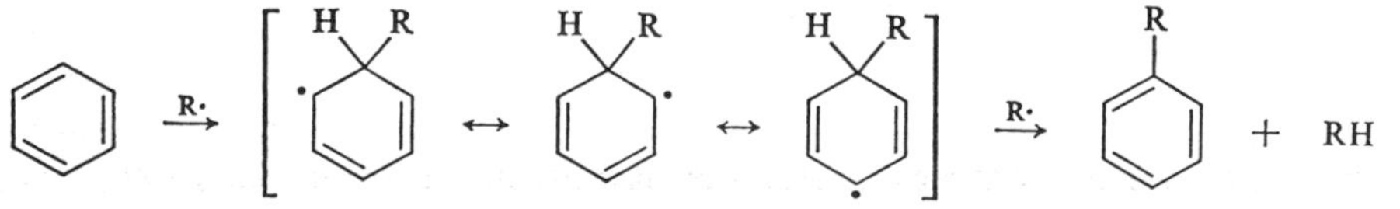

Alkylation can be effected by heating the aromatic compound with a diacyl peroxide or a lead tetra-carboxylate, e.g.

$$CH_3COO-OCOCH_3 \longrightarrow 2\ CH_3CO_2\cdot \longrightarrow 2\ CH_3\cdot + 2\ CO_2$$

$$Pb(OCOCH_3)_4 \longrightarrow Pb(OCOCH_3)_2 + 2\ CH_3\cdot + 2\ CO_2$$

However, the yields are low, at least partly because the alkylbenzenes are very reactive towards free radicals at their benzylic carbon atoms: e.g. the methylation of benzene gives toluene which reacts readily with alkyl radicals to give benzyl radicals and thence bibenzyl. Since alkylated aromatic compounds can normally be obtained through Friedel-Crafts reactions (11.3), homolytic alkylation is of little synthetic utility.

Arylation is normally brought about in one of three ways: by the Gomberg reaction (p. 441), using *N*-nitrosoacylarylamines (p. 442), and using diacyl peroxides such as dibenzoyl peroxide,

$$2\ PhCOO-OCOPh \longrightarrow 2\ PhCO_2\cdot$$

$$PhCO_2\cdot \longrightarrow Ph\cdot + CO_2$$

$$Ph\cdot + ArH \longrightarrow Ar \overset{\cdot}{\underset{Ph}{\diagup}}^{H} \xrightarrow{PhCO_2\cdot} Ar-Ph + PhCO_2H$$

Yields are rarely above 50% and dimeric and polymeric products are formed by processes such as

However, the addition of nitrobenzene and related oxidants increases the yield of the biaryl from peroxide reactions; for example, biphenyl can be obtained in about 80% yield from dibenzoyl peroxide and benzene. The mechanism of action of the nitro compound is complex; effectively, it and species derived from it, such as diphenyl nitroxide, $Ph_2N-O\cdot$, oxidize the intermediate cyclohexadienyl radical and so divert it from dimerization.

Free radicals are not nearly so selective between different positions in aromatic compounds as they are between the two carbon atoms of an olefin $CH_2=CHX$. Reaction tends to occur preferentially at the *ortho*-position of monosubstituted benzenes unless the substituent is particularly large (e.g.

t-butyl), and it is understandable that *ortho*- should predominate over *meta*-substitution since the former reaction leads to a more effectively stabilized radical, e.g.

This cannot be the sole governing factor, however, for then *para*-substitution should always predominate over *meta*-substitution since similar conjugating influences are present for *para*- and *ortho*-substitution; in practice, more than twice as much of the *meta*- as the *para*-derivative is formed in some cases. Some typical data for phenylation are shown; the total spread in the relative reactivities of monosubstituted benzenes is less than a factor of ten, in marked contrast to the spread in electrophilic and nucleophilic reactions where powerful polar effects operate.

X in PhX	*o*	*m*	*p* (%)	Reactivity of PhX relative to benzene
NO_2	62	10	28	3·0
CH_3	67	19	14	1·2
F	55	31	14	1·1
$C(CH_3)_3$	24	49	27	0·6

Despite the fact that homolytic arylation usually gives a mixture of products in low yields, it is nevertheless the only readily applicable method for obtaining many biphenyls, for Friedel-Crafts reactions are unsuccessful and electrophilic substitution in biphenyl is specific only for certain derivatives. For example, although 4-nitrobiphenyl can be obtained in about 50% yield by the nitration of biphenyl, the 3-nitro-derivative is formed in negligible yield and is best prepared from the reaction between benzene and di-*m*-nitrobenzoyl peroxide or *m*-nitrobenzenediazonium chloride in Gomberg conditions. (The alternative route, reaction between phenyl radicals and nitrobenzene, would give all three nitrobiphenyls.)

(d) DIMERIZATION OF ALKYL RADICALS: THE KOLBE ELECTROLYTIC REACTION
Electrolysis of the alkali-metal salts of aliphatic carboxylic acids results in the liberation of alkyl radicals at the anode and their subsequent dimerization:

$$RCO_2^- \xrightarrow{-e} RCO_2 \cdot \longrightarrow R \cdot + CO_2$$

$$2 \, R \cdot \longrightarrow R—R$$

The reaction is usually carried out by dissolving the acid in methanol containing enough sodium methoxide to neutralize about 2% of the acid, and electrolyzing between platinum-foil electrodes. The sodium liberated at the cathode reacts with the solvent to generate further carboxylate anion until the acid has been consumed and the solution is just alkaline. Yields are variable but are sometimes nearly quantitative, e.g.

$$2 \ CH_3CO_2H \longrightarrow CH_3\!-\!CH_3 \ (93\%)$$

$$2 \ CH_3\!-\!(CH_2)_{12}\!-\!CO_2H \longrightarrow CH_3\!-\!(CH_2)_{24}\!-\!CH_3 \ (60\%)$$

Electrolysis of the half-esters of dibasic acids gives long-chain dibasic esters, e.g.

$$2 \ EtO_2C\!-\!(CH_2)_n\!-\!CO_2H \longrightarrow EtO_2C\!-\!(CH_2)_{2n}\!-\!CO_2Et$$
$$n = 2 : 76\%$$
$$n = 8 : 70\% \ [5]$$

Partial hydrolysis and further electrolysis enable still longer chains to be built up.

The electrolysis of a mixture of two acids gives a mixture of three products but nevertheless it sometimes provides a rapid and convenient method for obtaining a compound which is otherwise difficultly accessible. For example, L-tuberculostearic acid has been made by a crossed-Kolbe reaction between octanoic acid and the (optically active) half-ester of β-methylglutaric acid:

$$CH_3\!-\!(CH_2)_6\!-\!CO_2H + HO_2C\!-\!CH_2\!-\!CH(CH_3)\!-\!CH_2\!-\!CO_2CH_3$$

$$\longrightarrow CH_3\!-\!(CH_2)_7\!-\!CH(CH_3)\!-\!CH_2\!-\!CO_2CH_3$$
Methyl L-tuberculostearate

(*e*) DIMERIZATION OF ARYL RADICALS: THE ULLMANN REACTION
Symmetrical biphenyls may be prepared by heating aryl halides (other than fluorides) with copper powder or copper bronze. For example, *o*-nitrochlorobenzene, when heated to 215–225°C with a specially prepared copper bronze in the presence of sand to moderate the reaction, gives 2,2'-dinitrobiphenyl in 50–60% yield [3].

2,2'-Dinitrobiphenyl

The reaction, which probably involves aryl radicals, is more efficient when the aromatic nucleus contains nitro-substituents, as in the example above.

(f) THE COUPLING OF ACETYLENES

Acetylene and monosubstituted acetylenes react with copper(II) acetate in pyridine solution to give diacetylenes. The reaction probably involves one-electron oxidation of the acetylide anion by copper(II) ion followed by dimerization of the resulting acetylide radicals:

$$RC\equiv CH \overset{-H^+}{\rightleftharpoons} RC\equiv C^- \overset{Cu^{2+}}{\longrightarrow} RC\equiv C\cdot + Cu^+$$

$$2\ RC\equiv C\cdot \longrightarrow RC\equiv C-C\equiv CR$$

For example, phenylacetylene gives 1,4-diphenylbutadiyne in 70–80% yield [5]:

$$2\ PhC\equiv CH \xrightarrow[\text{pyridine}]{Cu(OAc)_2} PhC\equiv C-C\equiv CPh$$

An alternative method for carrying out the reaction is to treat the acetylene with an aqueous mixture of copper(I) chloride and ammonium chloride in air. In these acidic conditions, removal of the proton may be facilitated by complexing between the acetylenic bond and copper(I) ion,

$$\underset{\underset{Cu^+}{\downarrow}}{RC\equiv CH} \rightleftharpoons \underset{\underset{Cu^+}{\downarrow}}{RC\equiv C^-} + H^+$$

Some of the copper(I) ion is oxidized by air to copper(II) ion which in turn oxidizes the acetylide ion to the radical, returning to the copper(I) state. Yields can be high, although for acetylene itself the yield is less than 10%, the main product being vinylacetylene ($CH_2=CH-C\equiv CH$).*

The following are examples of the many applications of the coupling reaction.

(i) Extension of carbon chains.

The coupling reaction on 3-hydroxy-1-butyne (from acetaldehyde and acetylene with amide ion in liquid ammonia, p. 256) gives a diacetylene used in a synthesis of β-carotene (p. 716):

$$2\ CH_3CH(OH)-C\equiv CH \xrightarrow{Cu^+-O_2} CH_3CH(OH)-C\equiv C-C\equiv C-CH(OH)CH_3$$

Phellogenic acid has been synthesized from undecylenic acid (obtained from castor oil):

*Diacetylene may be prepared from butyne-1,4-diol (p. 257):

$$HOCH_2-C\equiv C-CH_2OH \xrightarrow{SOCl_2} ClCH_2-C\equiv C-CH_2Cl \xrightarrow[-2\ HCl]{2\ OH^-} HC\equiv C-C\equiv CH$$

$$\left[\underset{ClCH-C\equiv C-CH_2-Cl}{HO^{\frown}H} \longrightarrow \underset{Cl-CH=C=C-CH}{H^{\frown}OH} \longrightarrow CH\equiv C-C\equiv CH \right]$$

$$\text{HO}_2\text{C}-(\text{CH}_2)_8-\text{CH}=\text{CH}_2 \xrightarrow[\text{2) NaNH}_2]{\text{1) Br}_2} \text{HO}_2\text{C}-(\text{CH}_2)_8-\text{C}\equiv\text{CH}$$

$$\xrightarrow{\text{Cu}^+-\text{O}_2} \text{HO}_2\text{C}-(\text{CH}_2)_8-\text{C}\equiv\text{C}-\text{C}\equiv\text{C}-(\text{CH}_2)_8-\text{CO}_2\text{H}$$

$$\xrightarrow{\text{H}_2-\text{Pd}} \text{HO}_2\text{C}-(\text{CH}_2)_{20}-\text{CO}_2\text{H}$$

Phellogenic acid

Most of the poly-ynes of the type $R-(C\equiv C)_n-R$, where $R = CH_3$, $C(CH_3)_3$, and Ph, and $n = 2$–10 have been synthesized, e.g.

$$\text{CH}\equiv\text{C}-\text{C}\equiv\text{CH} + \text{CH}_3\text{I} \xrightarrow{\text{NH}_2^-} \text{CH}_3-\text{C}\equiv\text{C}-\text{C}\equiv\text{CH} \xrightarrow{\text{Cu}^+-\text{O}_2} \text{CH}_3 \ (\text{C}\equiv\text{C})_4-\text{CH}_3$$

(ii) *Synthesis of cyclic compounds.* α,ω-Diacetylenes can couple both inter-molecularly and, if the resulting ring is not highly strained, intramolecularly; the latter reaction is favoured by high dilutions. For example, the macrocyclic lactone, exaltolide, which is partly responsible for the sweet odour of the angelica root and is used in scents, has been synthesized in high yield by this means:

$$\text{HC}\equiv\text{C}-\text{CH}_2\text{CH}_2-\text{OH} + \text{ClCO}-(\text{CH}_2)_8-\text{C}\equiv\text{CH} \xrightarrow{-\text{HCl}}$$

$$\text{HC}\equiv\text{C}-\text{CH}_2\text{CH}_2-\text{O}-\text{CO}-(\text{CH}_2)_8-\text{C}\equiv\text{CH} \xrightarrow{\text{Cu}^+-\text{O}_2}$$

Exaltolide

[18]Annulene* has been synthesized from 1,5-hexadi-yne by coupling to give a mixture of cyclic trimer, tetramer, pentamer, *etc.*, treating the trimer with base to effect conversion into the fully conjugated system *via* prototropic shifts, and partially hydrogenating the remaining triple bonds (p. 619) [54]:

*[18]Annulene was the first compound to be isolated which is cyclic, conjugated, contains $(4n + 2)$ π-electrons where n is greater than 1, and in which the inside hydrogen atoms are sufficiently far apart for the molecule to adopt a planar or near-planar configuration. It should therefore be aromatic (p. 59), and the experimental evidence indicates that this is so. In particular, the n.m.r. spectrum shows 6 protons at unusually high field ($\delta -3.0$ ppm) and 12 at low field (δ 9.3 ppm), indicating that the molecule sustains an aromatic ring-current such that the induced magnetic field causes shielding of the 6 inward-pointing protons and deshielding of the 12 outward-pointing ones.

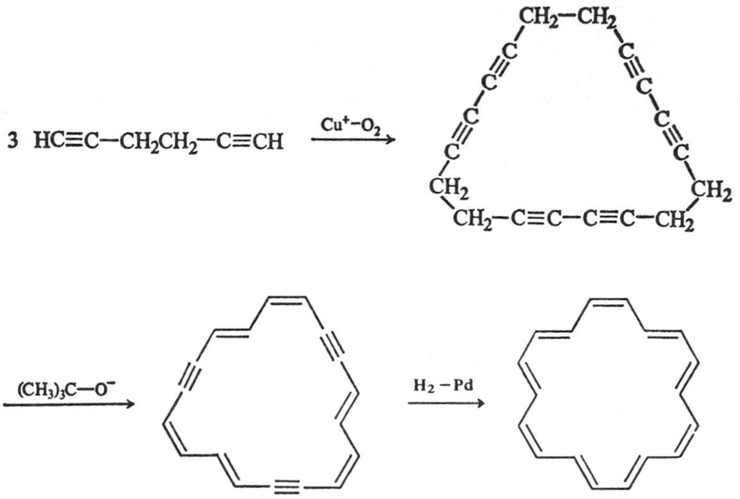

$$3 \ HC\equiv C-CH_2CH_2-C\equiv CH \xrightarrow{\ Cu^+-O_2\ }$$

[18]Annulene

(*iii*) *Synthesis of unsymmetrical di- and poly-ynes.* A coupling reaction between a mixture of two monosubstituted acetylenes normally gives a mixture of three products. Although this procedure has been used to obtain several unsymmetrical di- and poly-ynes in low yield, it is more satisfactory to employ a related coupling process which can give only one product. This procedure (*Chadiot-Chodkiewicz coupling*) consists of treating a monosubstituted acetylene with a 1-bromoacetylene in the presence of copper(I) ion:

$$RC\equiv CH + BrC\equiv CR' \xrightarrow{\ Cu^+\ } RC\equiv C-C\equiv CR' + HBr$$

The copper(I) ion acts catalytically, probably *via* the complexes which it forms with acetylenic triple bonds.

Examples, the second of which gives an essential oil constituent, are:

$$HC\equiv C-C\equiv CH + BrC\equiv CPh \xrightarrow[-HBr]{\ Cu^+\ } H-(C\equiv C)_3-Ph$$

$$HOCH_2-\underset{trans}{CH=CH}-C\equiv CH + BrC\equiv C-C\equiv CCH_3 \xrightarrow[-HBr]{\ Cu^+\ }$$

$$HOCH_2-\underset{trans}{CH=CH}-(C\equiv C)_3-CH_3$$

Dehydromatricarianol

(g) THE ACYLOIN SYNTHESIS

The treatment of aliphatic esters with molten sodium in hot xylene (an inert, fairly high-boiling solvent) gives the disodium derivatives of acyloins from which the acyloin is liberated with acid. The reaction is preferably carried out under nitrogen because acyloins and their anions are readily oxidized. For example, ethyl n-butyrate gives butyroin in 65–70% yield [2].

$$2 \text{ CH}_3\text{CH}_2\text{CH}_2\text{CO}_2\text{Et} \xrightarrow[\text{2) H}^+]{\text{1) Na}} \text{CH}_3\text{CH}_2\text{CH}_2\text{CH(OH)COCH}_2\text{CH}_2\text{CH}_3$$

Butyroin

Reaction is initiated by electron-transfer to the carbonyl group of the ester; the resulting radicals dimerize; alkoxide groups are eliminated; and further electron-transfers give the disodium derivative of the acyloin:

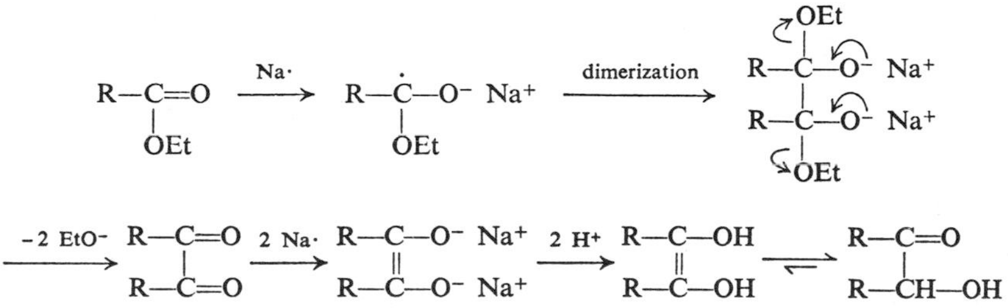

The acyloin reaction is particularly valuable for the synthesis of large rings from ω-dibasic esters. Even rings with 10–13 members (cf. p. 243) are formed in fair to good yield, and for these the method is superior to those of Ziegler (p. 243) and Blomquist (p. 302). (The intramolecular reaction is possibly facilitated by the bringing together of the polar ends of the molecule on the surface of the sodium.) Typical yields are:

Ring-size	9	10	12	13	14	15	16	17	18	20
Yield (%)	9	45	76	67	79	77	84	85	96	96

17.4 Formation of Carbon-Nitrogen Bonds

(a) NITRATION AT SATURATED CARBON

Saturated carbon can be nitrated by nitric acid in both the gas and the liquid phases. The mechanism of reaction is not certain but is probably as follows:

$$\text{initiation} \quad \text{HO-NO}_2 \longrightarrow \text{HO} \cdot + \text{NO}_2$$

$$\text{propagation} \begin{cases} \text{R-H} + \text{NO}_2 \longrightarrow \text{R} \cdot + \text{HONO} \\ \text{R} \cdot + \text{NO}_2 \longrightarrow \text{R-NO}_2 \\ \text{HONO} + \text{HNO}_3 \longrightarrow 2 \text{ NO}_2 + \text{H}_2\text{O} \end{cases}$$

The fragmentation of alkyl groups commonly occurs, but one or more products can often be isolated in good yield and the gas-phase reaction at about 450°C is of considerable industrial importance. Some examples are:

$$CH_3CH_3 + HNO_3 \xrightarrow{450°C} \underset{80-90\%}{C_2H_5-NO_2} + \underset{10-20\%}{CH_3-NO_2}$$

$$(CH_3)_3CH + HNO_3 \xrightarrow{450°C} \underset{65\%}{(CH_3)_2CH-CH_2-NO_2} + \underset{7\%}{(CH_3)_3C-NO_2} + \underset{20\%}{(CH_3)_2CH-NO_2}$$

$$PhCH_3 + HNO_3 \xrightarrow{\text{(liquid)}} \underset{55\%}{PhCH_2-NO_2}$$

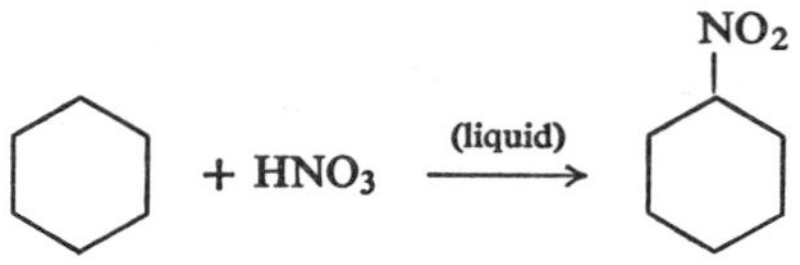

$$+ HNO_3 \xrightarrow{\text{(liquid)}}$$

44%

(b) ADDITION TO OLEFINIC AND ACETYLENIC COMPOUNDS

Dinitrogen tetroxide adds to C=C and C≡C to give a mixture of dinitro-compounds, nitroalcohols, and nitronitrates:

$$O_2N-NO_2 \rightleftharpoons 2\ NO_2{}^*$$

$$RCH=CH_2 + NO_2 \longrightarrow R\overset{\cdot}{C}H-CH_2NO_2$$

$$R\overset{\cdot}{C}H-CH_2NO_2 + O_2N-NO_2 \longrightarrow$$

$$RCH-CH_2NO_2$$
$$\quad\quad\quad\ \ |$$
$$\quad\quad\quad NO_2$$

$$RCH-CH_2NO_2$$
$$\quad\quad\quad\ \ |$$
$$\quad\quad\quad ONO$$

$$RCH-CH_2NO_2$$
$$\quad\quad\quad\ |$$
$$\quad\quad\quad ONO$$

hydrolysis →

$$RCH-CH_2NO_2$$
$$\quad\quad\quad\ |$$
$$\quad\quad\quad OH$$

oxidation →

$$RCH-CH_2NO_2$$
$$\quad\quad\quad\ |$$
$$\quad\quad\quad ONO_2$$

*N_2O_4 is 20% dissociated at 27°C.

Yields of particular products tend to be low, e.g.

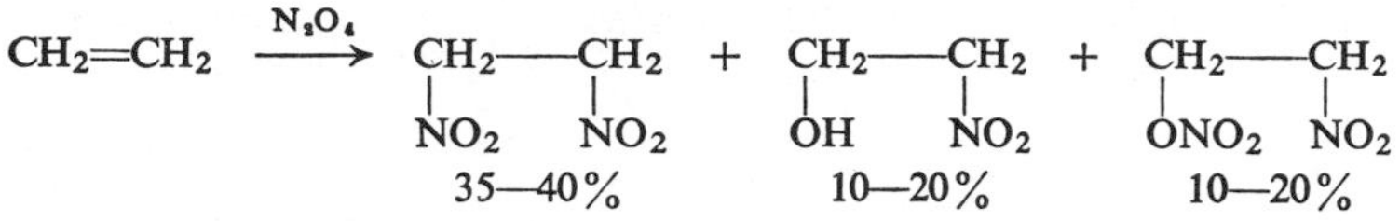

$$CH_2{=}CH_2 \xrightarrow{N_2O_4} \underset{\underset{35-40\%}{\overset{|}{NO_2}\;\overset{|}{NO_2}}}{CH_2{-}CH_2} + \underset{\underset{10-20\%}{\overset{|}{OH}\;\overset{|}{NO_2}}}{CH_2{-}CH_2} + \underset{\underset{10-20\%}{\overset{|}{ONO_2}\;\overset{|}{NO_2}}}{CH_2{-}CH_2}$$

Nitryl and nitrosyl halides add in an analogous manner; the products from nitrosyl halides are often the β-halonitro-compounds which result from oxidation of nitroso-compounds.

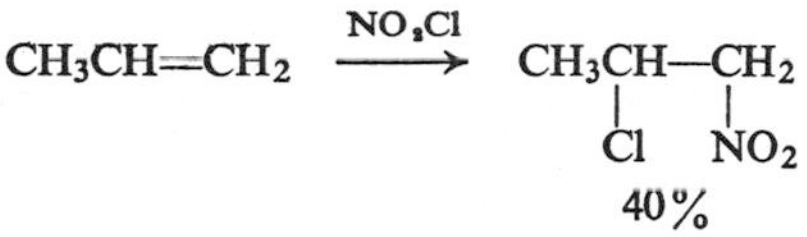

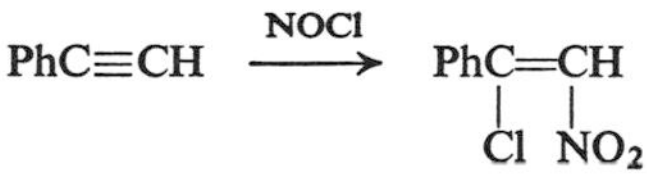

$$CH_3CH{=}CH_2 \xrightarrow{NO_2Cl} \underset{\underset{40\%}{\overset{|}{Cl}\;\overset{|}{NO_2}}}{CH_3CH{-}CH_2}$$

$$PhC{\equiv}CH \xrightarrow{NOCl} \underset{\overset{|}{Cl}\;\overset{|}{NO_2}}{PhC{=}CH}$$

(c) THE HOFMANN-LÖFFLER-FREYTAG REACTION

Irradiation of N-chlorodibutylamine in 85% sulphuric acid, followed by basification, gives N-butylpyrrolidine:

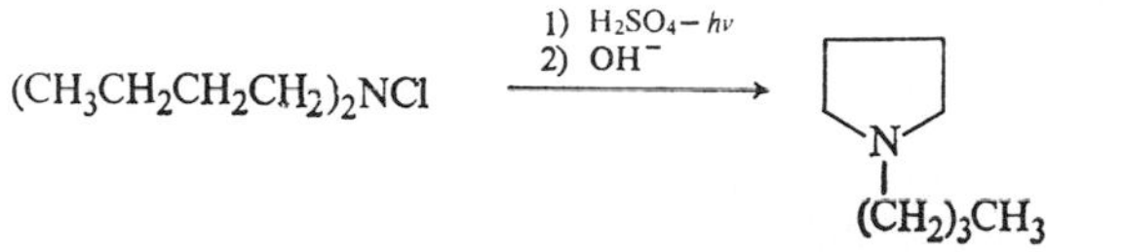

$$(CH_3CH_2CH_2CH_2)_2NCl \xrightarrow[\text{2) OH}^-]{\text{1) H}_2SO_4-h\nu}$$

N-Butylpyrrolidine

This is a general reaction of those N-chloroamines which possess a δ-CH group. Photolysis of the protonated amine gives a nitrogen cation-radical which abstracts a hydrogen atom from the δ-carbon *via* a six-membered cyclic transition state. The resulting alkyl radical abstracts a chlorine atom from more of the chloroamine, and when the amino-group is released with base it displaces intramolecularly on the halide:

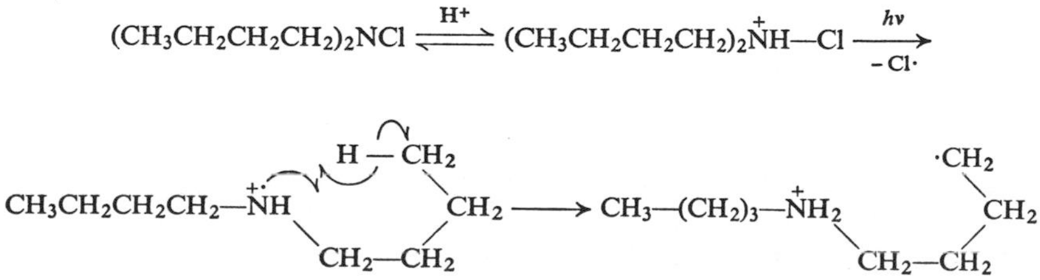

$$(CH_3CH_2CH_2CH_2)_2NCl \rightleftharpoons^{H^+} (CH_3CH_2CH_2CH_2)_2\overset{+}{N}H{-}Cl \xrightarrow[-Cl\cdot]{h\nu}$$

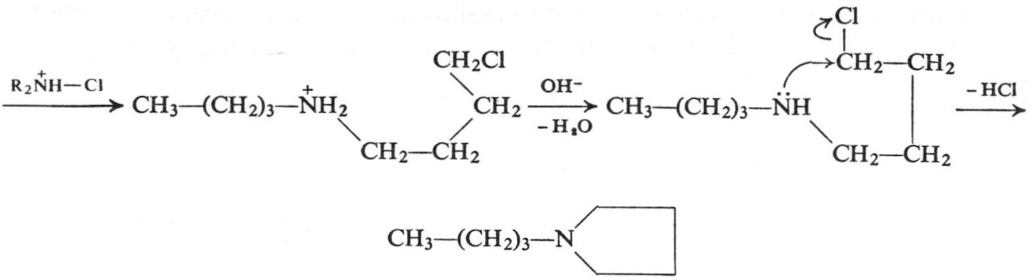

$$R_2\overset{+}{N}H-Cl \longrightarrow CH_3-(CH_2)_3-\overset{+}{N}H_2 \quad \overset{CH_2Cl}{\underset{CH_2-CH_2}{\diagdown CH_2}} \quad \overset{OH^-}{\underset{-H_2O}{\longrightarrow}} \quad CH_3-(CH_2)_3-\overset{\cdot\cdot}{N}H \quad \overset{\overset{Cl}{\diagup}}{\underset{CH_2-CH_2}{\diagdown CH_2-CH_2}} \quad \overset{-HCl}{\longrightarrow}$$

$$CH_3-(CH_2)_3-N\diagdown$$

This provides a method for introducing functionality at an otherwise un-reactive aliphatic carbon atom, and the reaction has been used in the steroidal series to modify angular methyl groups, as in a synthesis of dihydroconessine:

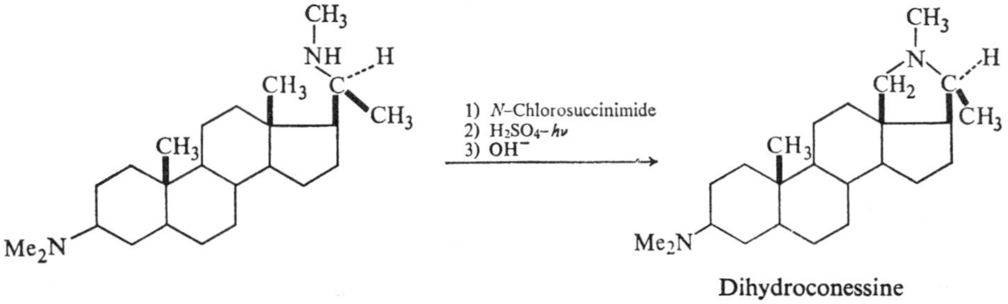

1) *N*-Chlorosuccinimide
2) H_2SO_4-*hν*
3) OH^-

Dihydroconessine

(d) PHOTOLYSIS OF NITRITES: THE BARTON REACTION

The irradiation of organic nitrites may also be used to introduce functionality at an otherwise unreactive aliphatic carbon. The oxy-radical produced by photolysis abstracts hydrogen from a δ-CH bond and the resulting alkyl radical combines with the nitric oxide liberated in the photolysis to give a nitroso compound and hence, when primary and secondary CH groups are involved, the tautomeric oxime.

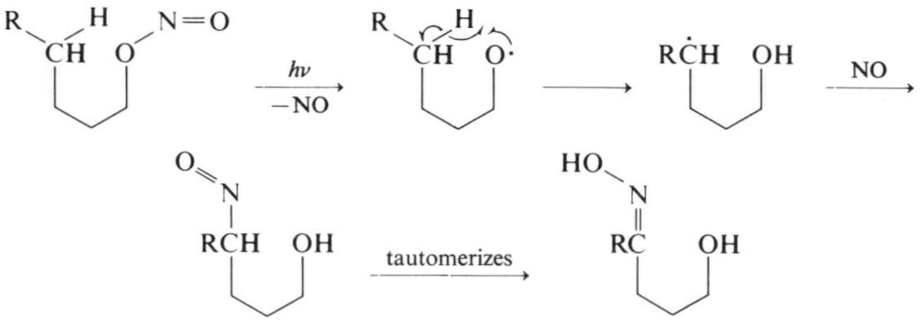

This method has been used in a synthesis of aldosterone 21-acetate from corticosterone acetate. The nitrite ester of the 11β-hydroxyl group was formed from the alcohol with nitrosyl chloride and, after photolysis, the oxime was hydrolyzed in mild conditions (nitrous acid). (Note that aldosterone 21-acetate exists as a hemi-acetal; *cf.* the ring-chain tautomerism of glucose, p. 187.)

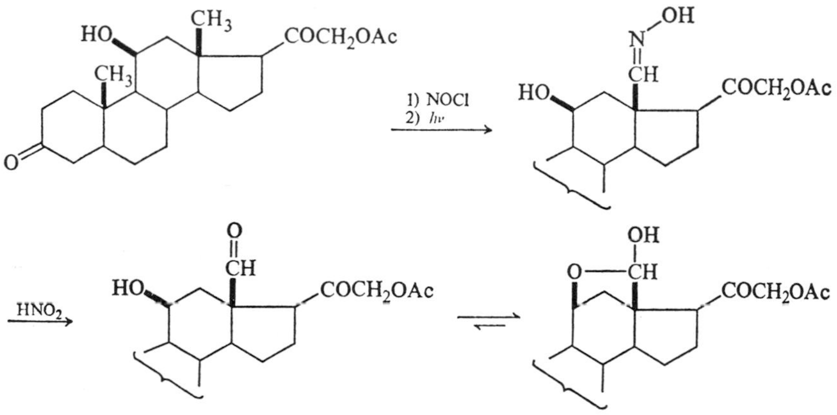

Aldosterone 21-acetate

A closely related method is of interest, although it does not involve the generation of a C—N bond. The photolysis of tertiary hypochlorites (readily prepared from the corresponding alcohols with chlorine in alkali or with chlorine monoxide) gives alkoxy radicals which can abstract hydrogen from δ-CH bonds; the resulting alkyl radical abstracts chlorine from a second molecule of the hypochlorite, giving a δ-chloro-alcohol which may be cyclized with base, e.g. *

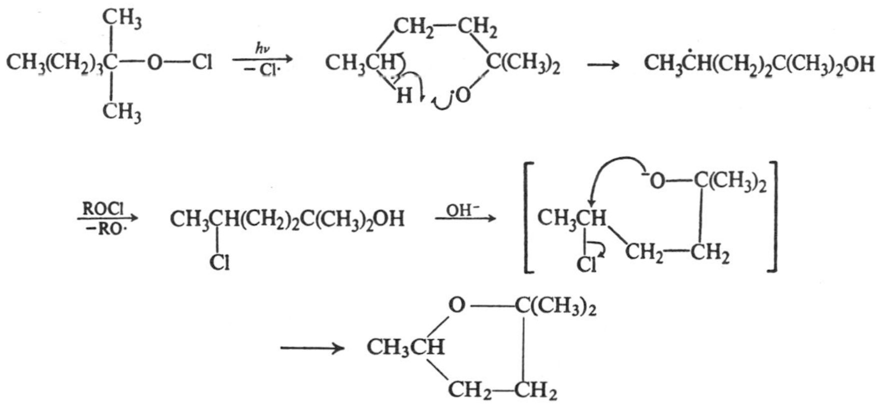

*The fragmentation products, n–butyl chloride and acetone, are also formed in 13% yield; see below.

This method has also been used in the steroid series for introducing functionality at angular methyl groups,

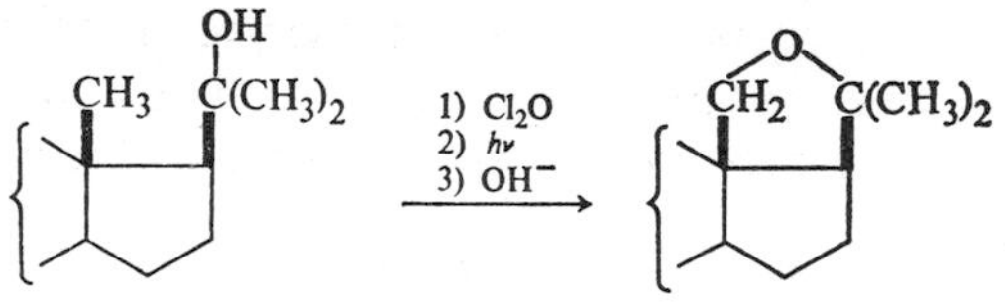

It should be emphasized that in both the Barton and the hypochlorite reactions, intramolecular hydrogen-abstraction has to compete with other reactions of alkoxy radicals. In particular, these radicals undergo fragmentation: for example, the t-butoxy radical gives acetone and the methyl radical. The reaction,

$$(CH_3)_2CH{-}\underset{C_2H_5}{\overset{CH_3}{\underset{|}{\overset{|}{C}}}}{-}OCl \xrightarrow{h\nu} \begin{cases} CH_3COC_2H_5 + (CH_3)_2CHCl & 95\% \\ CH_3COCH(CH_3)_2 + C_2H_5Cl & 4\% \\ CH_3CH_2COCH(CH_3)_2 + CH_3Cl & 0{\cdot}5\% \end{cases}$$

indicates that the ease of fragmentation of a particular group increases with the stability of the radical formed. However, in the examples involving steroidal systems cited above the intramolecular reaction competes favourably with fragmentation, for not only is the transition state of the necessary six-membered type but also the angular methyl group and oxy-radical are suitably placed with respect to each other.

17.5 Formation of Carbon-Oxygen Bonds

C—H bonds in a wide variety of environments are oxidized on standing in air to hydroperoxide groups. Reaction is apparently initiated by the appearance of stray radicals produced, for example, by sun-light photolysis, and thereafter a chain process operates:

$$\geq C{-}H + R\cdot \longrightarrow \geq C\cdot + RH$$

$$\geq C\cdot + O_2 \longrightarrow \geq C{-}O{-}O\cdot$$

$$\geq C{-}O{-}O\cdot + \geq C{-}H \longrightarrow \geq C{-}O{-}OH + \geq C\cdot$$

The rates of these reactions (*autoxidations*) vary markedly with structure: for example, paraffins react at a negligible rate at room temperature whereas allylic compounds react at a significant rate. Autoxidation is catalyzed by the usual initiators and retarded by inhibitors, and it is advisable to store the more readily autoxidized compounds in the presence of an inhibitor.

In the paraffin series the order of reactivity is, tertiary > secondary > primary C—H, as usual in free-radical reactions. For example, isobutane can be converted in good yield into t-butyl hydroperoxide in the presence of an initiator:

$$(CH_3)_3CH + O_2 \longrightarrow (CH_3)_3C—O_2H$$
t-Butyl
hydroperoxide

Allylic compounds owe their greater reactivity to the greater stability of allyl than alkyl radicals. Unsymmetrical compounds give mixtures of products,

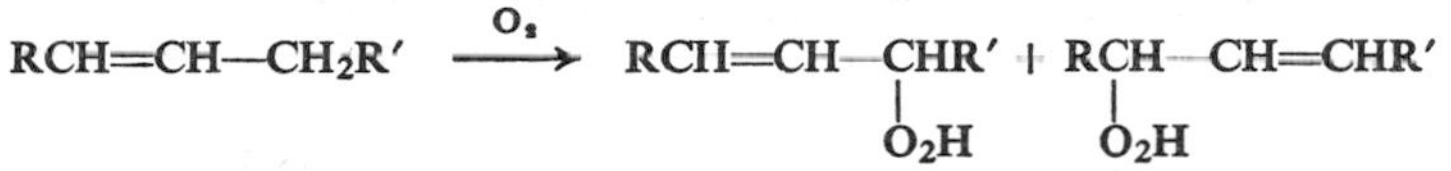

$$RCH{=}CH—CH_2R' \xrightarrow{O_2} RCH{=}CH—\underset{\underset{O_2H}{|}}{CHR'} + RCH—CH{=}CHR'$$
$$\underset{O_2H}{|}$$

Those allylic compounds which, as olefins, are readily polymerized, react differently, for the hydroperoxy radical initiates polymerization:

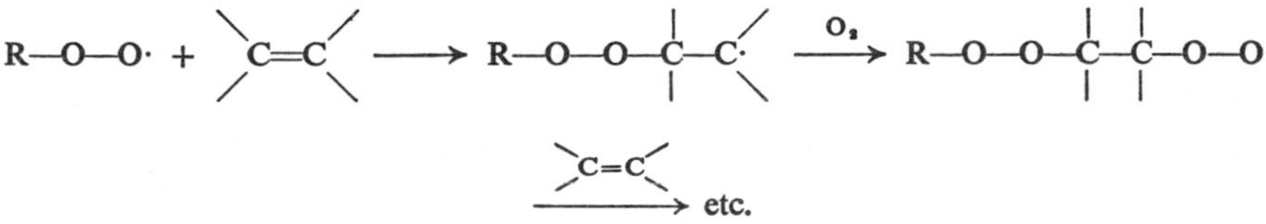

$$\xrightarrow{} \text{etc.}$$

Polymerization of a complex type is also responsible for the formation of a hard skin on drying oils, which are allylic compounds.

Ethers are particularly prone to autoxidation: e.g. tetrahydrofuran gives the α-hydroperoxide

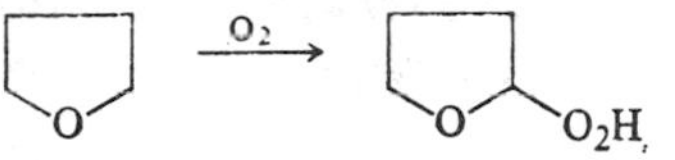

Since hydroperoxides can explode on heating it is essential to remove them from ethers (e.g. by reduction with aqueous iron(II) sulphate) before using the ethers as solvents for reactions which require heat.

Aldehydes also autoxidize readily but the initial product, a peracid, reacts with more of the aldehyde to give the carboxylic acid. For example, benzaldehyde gives benzoic acid on standing in air:

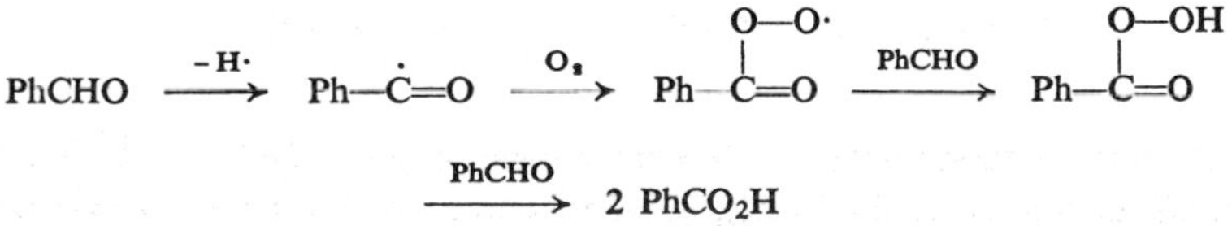

$$PhCHO \xrightarrow{-H·} Ph—\overset{·}{C}{=}O \xrightarrow{O_2} Ph—\overset{\overset{O—O·}{|}}{C}{=}O \xrightarrow{PhCHO} Ph—\overset{\overset{O—OH}{|}}{C}{=}O$$

$$\xrightarrow{PhCHO} 2\ PhCO_2H$$

Hydroperoxides are not themselves of much synthetic value but they are employed as intermediates in certain reactions. Thus, cumene is converted industrially into phenol and acetone *via* cumene hydroperoxide (p. 472),

$$PhCH(CH_3)_2 \xrightarrow{O_2} PhC(CH_3)_2\text{---}O_2H \xrightarrow{H^+} PhOH + (CH_3)_2CO$$

and tetralin can be converted into α-tetralone by E2 elimination on its hydroperoxide,

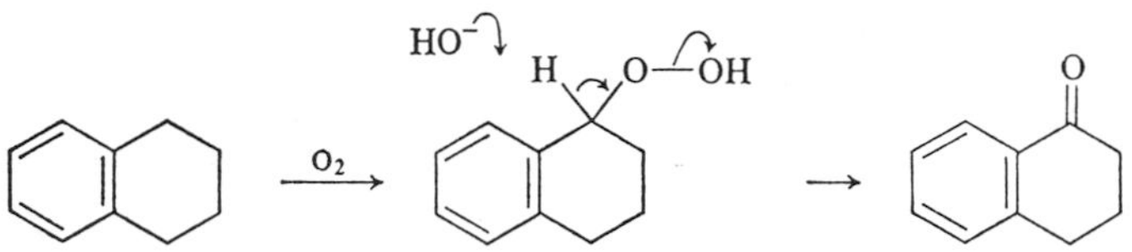

17.6 Formation of Bonds to Other Elements

Elements such as sulphur, phosphorus, and silicon can be bonded to olefinic and acetylenic compounds by radical-catalyzed reactions which have the characteristics of those already described. Typical examples are:

$$H_2S + PhCH\text{=}CH_2 \longrightarrow PhCH_2CH_2SH + (PhCH_2CH_2)_2S$$
$$\qquad\qquad\qquad\qquad\qquad 61\% \qquad\qquad 20\%$$

$$C_2H_5SH + C_6H_{13}CH\text{=}CH_2 \longrightarrow C_8H_{17}\text{---}S\text{---}C_2H_5$$
$$\qquad\qquad\qquad\qquad\qquad\qquad 75\%$$

$$HSCH_2CO_2H + CH_2\text{=}CH\text{---}CN \longrightarrow HO_2CCH_2SCH_2CH_2CN$$
(A mercapto-acid) $\qquad\qquad\qquad\qquad 74\%$

$$CH_3COSH + CH_2\text{=}CHCl \longrightarrow CH_3COSCH_2CH_2Cl$$
(A thiol-acid)

$$PH_3 + 3\ CH_3CH_2CH\text{=}CH_2 \longrightarrow (CH_3CH_2CH_2CH_2)_3P$$
$$\qquad\qquad\qquad\qquad\qquad\qquad 67\%$$

(The intermediate compounds, butylphosphine and dibutylphosphine, may be obtained by using smaller proportions of 1-butene)

$$H_2P(\!\!\begin{array}{c}O\\\|\\\end{array}\!\!)OH + 2\ C_4H_9CH\text{=}CH_2 \longrightarrow (C_6H_{13})_2P(\!\!\begin{array}{c}O\\\|\\\end{array}\!\!)OH$$
$$\qquad\qquad\qquad\qquad\qquad\qquad\qquad 9\%$$

$$Cl_3SiH + (CH_3)_2C\text{=}CH_2 \longrightarrow (CH_3)_2CHCH_2SiCl_3$$
$$\qquad\qquad\qquad\qquad\qquad 95\%$$

Further Reading

BLOOMFIELD, J. J., OWSLEY, D. C., and NELKE, J. M., 'The acyloin condensation,' *Organic Reactions*, 1976, **23**, 259.

EGLINTON, G., and McRAE, W., 'The coupling of acetylenic compounds,' *Advances in Organic Chemistry, Methods and Results*, Vol. 4, Interscience (New York and London 1963), p. 225.

FANTA, P. E., 'The Ullmann synthesis of biaryls,' *Chemical Reviews*, 1964, **64**, 613.

NONHEBEL, D. C., and WALTON, J. C., *Free Radical Chemistry*, Cambridge University Press (1974).

PRYOR, W. A., *Free Radicals*, McGraw-Hill (New York 1966).

SOSNOVSKY, G., *Free Radical Reactions in Preparative Organic Chemistry*, Macmillan (London 1964).

WEEDON, B. C. L., 'The Kolbe electrolytic synthesis,' *Advances in Organic Chemistry, Methods and Results*, Vol. 1, Interscience (New York and London 1960), p. 1.

Problems

1. Account for the following:

   (*i*) The orientation in the addition of hydrogen bromide to allyl bromide depends on whether or not the reactants are contaminated with peroxide impurities.

   (*ii*) The radical-catalyzed chlorination of the optically active amyl chloride, CH_3—CH_2—$CH(CH_3)$—CH_2Cl, gives mainly racemic 1,2-dichloro-2-methylbutane.

   (*iii*) The perester $PhCH{=}CHCH_2$—CO—O—O—CMe_3 decomposes several thousand times faster than the perester CH_3—CO—O—O—CMe_3 at the same temperature.

   (*iv*) The radical-catalyzed chlorination, $ArCH_3 \rightarrow ArCH_2Cl$, occurs faster when Ar = phenyl than when Ar = *p*-nitrophenyl.

   (*v*) The aldehyde Ph—$C(CH_3)_2$—CH_2—CHO undergoes a radical-catalyzed decarbonylation to give a mixture of Ph—$C(CH_3)_2$—CH_3 and Ph—CH_2—$CH(CH_3)_2$. The proportion of the latter product decreases as the concentration of the reactant is increased.

   (*vi*) When *p*-cresol is oxidized by potassium hexacyanoferrate(III), the compound (I) is one of the products.

   (*vii*) When the compound (II) (formed from cyclohexanone and hydrogen peroxide) is treated with iron(II) sulphate, 1,12-dodecanedioic acid is formed.

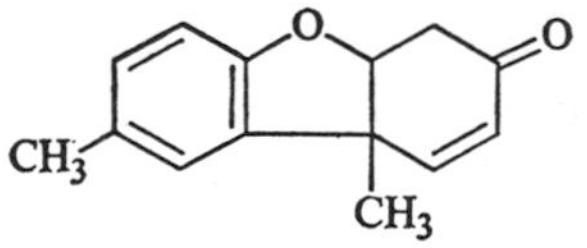

(I) 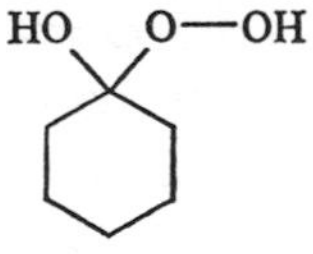   (II)

2. Summarize the free-radical reactions which may be used for:

   (*i*) the extension of carbon chains;

   (*ii*) the introduction of functionality at unactivated methyl groups;

   (*iii*) the formation of medium- and large-sized rings.

   In the last case, compare these methods with those which do not involve radical reactions.

3. What products would you expect from the following reactions?

   (*i*) $(CH_3)_3CH + Br_2 \xrightarrow{\text{light}}$

   (*ii*) $PhCH_3 + N\text{–bromosuccinimide} \xrightarrow{\text{peroxide}}$

   (*iii*) $C_4H_9\text{—}CH\text{=}CH_2 + N\text{–bromosuccinimide} \xrightarrow{\text{peroxide}}$

   (*iv*) $CH_2(CO_2Et)_2 + C_4H_9\text{—}CH\text{=}CH_2 \xrightarrow{\text{peroxide}}$

   (*v*) $CH_2\text{=}CH\text{—}(CH_2)_3\text{—}CH\text{=}CH_2 + CHCl_3 \xrightarrow{\text{peroxide}}$

4. How would you employ radical reactions in the synthesis of the following compounds?

   (*a*) $CH_3CH_2CH_2CBr_3$ (*b*) $(CH_3)_2CH\text{—}CO\text{—}CH(OH)\text{—}CH(CH_3)_2$

   (*c*) $(CH_3)_3C\text{—}CH_2Br$ (*d*) $CH_3\text{—}(CH_2)_4\text{—}CO\text{—}CH_3$

   (*e*) $Ph\text{—}S\text{—}CH_2CH_2CO_2CH_3$ (*f*) $Ph\text{—}(CH_2)_6\text{—}Ph$

   (*g*)   (*h*)

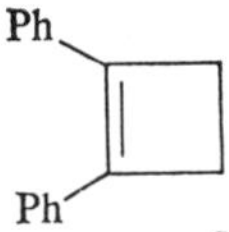

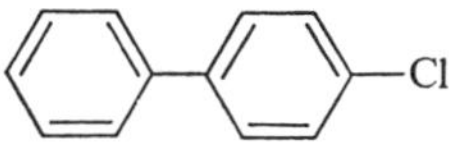

   (*i*) 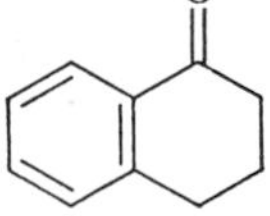   (*j*)

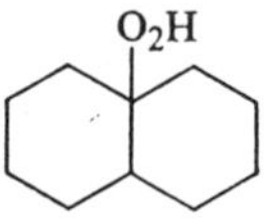

18. Oxidation

18.1 Introduction

The scope of the term oxidation has been discussed earlier (4.9). The processes described in this chapter are those which lead to the incorporation of oxygen or the removal of hydrogen from a molecule; those which result in the partial removal of electrons (e.g. the bromination of methane) are included only where they are employed *en route* to the introduction of oxygen or the removal of hydrogen, as in the conversion of toluene into benzaldehyde *via* benzal chloride (p. 580).

The account is not intended to be comprehensive but is limited to those processes which have proved to be synthetically valuable. Thus, the mechanisms of oxidations by such reagents as cobalt(III) and cerium(IV) ions have been widely studied but are not included because the reagents have not proved to be widely applicable in synthesis.

The different techniques used in industrial and laboratory processes are stressed. In the laboratory, reactions are normally carried out at comparatively low temperatures and at atmospheric pressure, and with group-selective and/or stereoselective reagents which are often expensive. Industrial reactions are more often conducted in the gas phase, at high temperatures and pressures, and over catalysts, followed by efficient fractionation of the mixture of products, but the first essential is cheapness. Thus, the use of osmium tetroxide (£15, or about $25, per gram), which is justified on a small scale in the laboratory, is not an industrial proposition. The natural industrial oxidant is air: a large number of processes involving air and a catalyst are used, many of which are not fully understood or as yet employed in the laboratory. For example, there is a recently developed continuous process for converting ethylene into acetaldehyde, in the presence of palladium(II) and copper(II) ions and air, which may be represented as follows:

$$CH_2{=}CH_2 + Pd^{2+} \longrightarrow \underset{\substack{\downarrow \\ Pd^{2+} \\ \pi-\text{complex}}}{CH_2{=}CH_2} \xrightarrow[-H^+]{H_2O} \underset{Pd^+}{CH_2{-}CH{-}\overset{H}{O}{-}H} \xrightarrow{-H^+} CH_3CHO + Pd$$

$$Pd + 2\,Cu^{2+} \longrightarrow Pd^{2+} + 2\,Cu^+$$

$$Cu^+ \xrightarrow{\text{air}} Cu^{2+}$$

Bacterial oxidation is also employed industrially. For example, *acetobacter suboxydans* is the best reagent for the specific oxidation of the C_2-carbon of sorbitol in the production of vitamin C (L-ascorbic acid) from glucose:

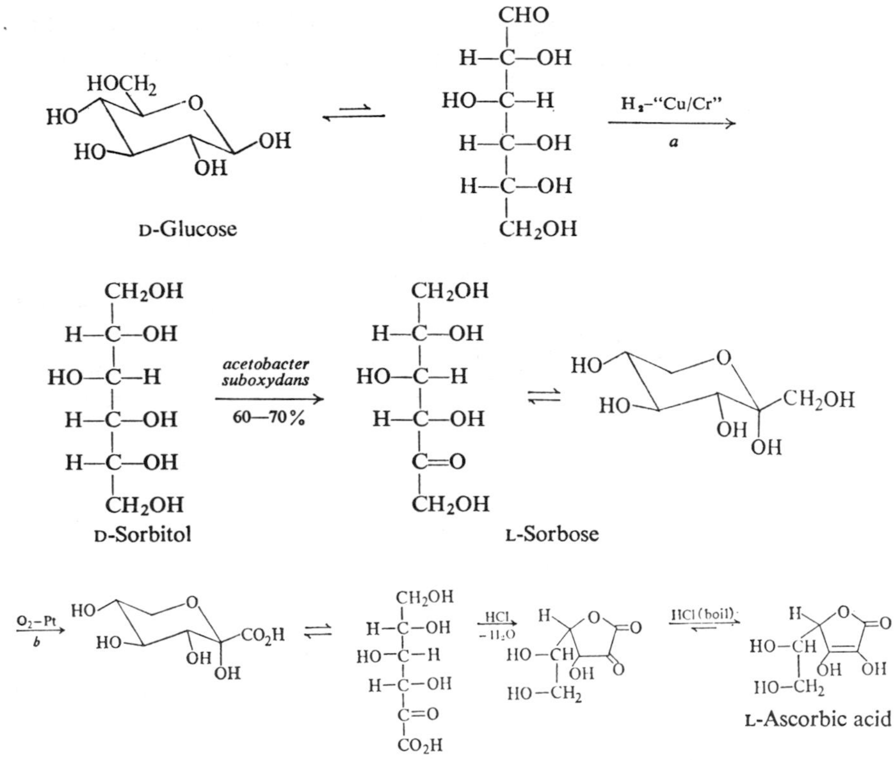

D-Glucose

CHO
|
H—C—OH
|
HO—C—H
|
H—C—OH
|
H—C—OH
|
CH₂OH

$\xrightarrow{\text{H}_2-\text{"Cu/Cr"}}$ *a*

CH₂OH
|
H—C—OH
|
HO—C—H
|
H—C—OH
|
H—C—OH
|
CH₂OH
D-Sorbitol

$\xrightarrow[60-70\%]{\textit{acetobacter suboxydans}}$

CH₂OH
|
H—C—OH
|
HO—C—H
|
H—C—OH
|
C=O
|
CH₂OH

L-Sorbose

$\xrightarrow[b]{O_2-Pt}$

CH₂OH
|
H—C—OH
|
HO—C—H
|
H—C—OH
|
C=O
|
CO₂H

$\xrightarrow[-H_2O]{HCl}$

$\xrightleftharpoons{HCl\,(boil)}$

L-Ascorbic acid

a Reduction over copper chromite; p. 612.
b p. 590.

Both bacterial and enzymatic methods may eventually prove to be powerful tools for selective oxidation in the laboratory.

This chapter is arranged according to the type of system to be oxidized, so that whereas the appropriate method for a particular oxidation can readily be found, the properties of any one oxidizing agent are not collected together.

18.2 Hydrocarbons

(a) OLEFINIC DOUBLE BONDS

(i) *Epoxidation.* Olefinic double bonds react with peracids to give epoxides. The process involves electrophilic attack on the olefin and a simplified representation is:

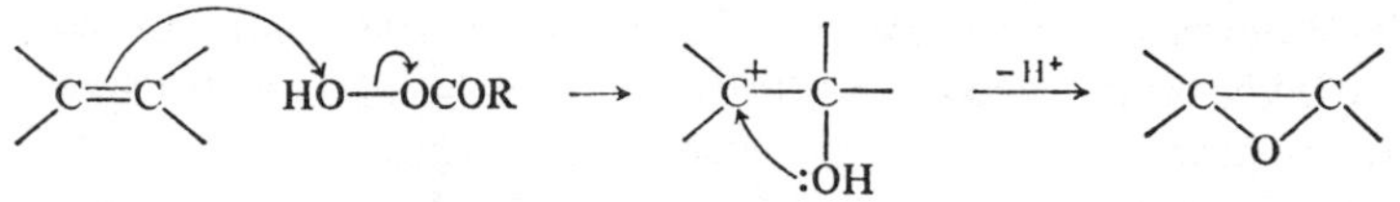

Consequently, electron-releasing groups in the olefin and electron-attracting groups in the peracid facilitate the reaction. In fact, the two steps are closely synchronous, so that a truer representation is:

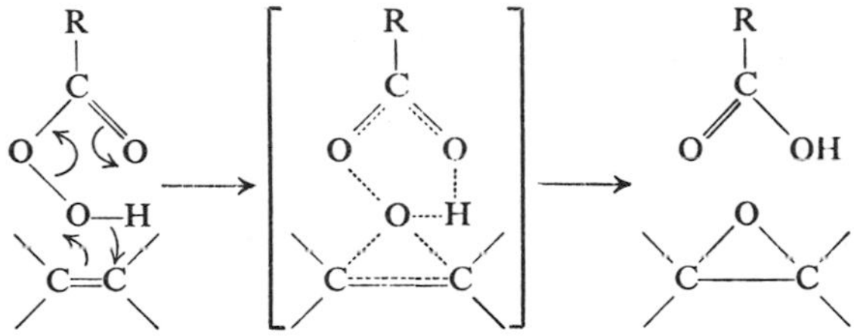

Epoxide-formation is stereospecific, as expected from the above mechanism; thus, *cis*-2-butene gives only the *cis* product:

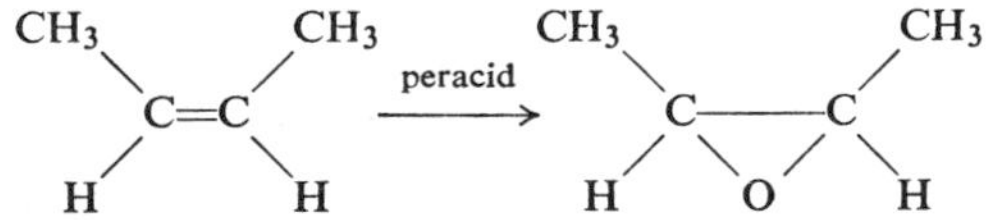

In reactions with cyclic olefins, the approach of the peracid is predominantly from the less hindered side, e.g.

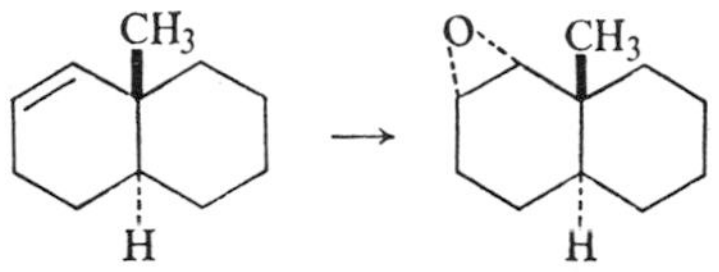

Peracids may be prepared by oxidation of the corresponding carboxylic acid or its anhydride with hydrogen peroxide in water or methanesulphonic acid. It is often more convenient, however, to prepare the peracid *in situ* by using a mixture of the carboxylic acid and hydrogen peroxide as the epoxidizing agent; the addition of a mineral acid increases the rate of attainment of the equilibrium:

$$R-CO_2H + H_2O_2 \underset{}{\overset{H^+}{\rightleftharpoons}} R-CO_2-OH + H_2O$$

Typical peracids used are performic, peracetic, *m*-chloroperbenzoic, mono-perphthalic, trifluoroperacetic, and Caro's acid (H_2SO_5). Of the organic acids, peracetic is the least reactive and trifluoroperacetic (containing the strongly electron-withdrawing $-CF_3$ group) is the most reactive. *m*-Chloroperbenzoic acid is now commercially available, gives good yields, and is often the reagent of choice.

When performic or trifluoroperacetic acid is used, the carboxylic acid produced is sufficiently acidic to bring about ring-opening of the epoxide with the production of a diol monoester. Reaction occurs by S_N2-displacement on the protonated epoxide, the nucleophile attacking the less alkylated carbon,* e.g.

$$CH_3-CH=CH_2 \xrightarrow[-HCO_2H]{HCO_3OH} CH_3-CH\overset{O}{\overbrace{\quad}}CH_2 \xrightarrow{HCO_2H}$$

$$CH_3-CH\overset{\overset{H}{\overset{|+}{O}}}{\overbrace{\quad}}CH_2 \longrightarrow CH_3-\underset{\underset{OCHO}{|}}{\overset{\overset{OH}{|}}{CH}}-CH_2$$
$$\underset{\underset{CHO}{O}}{\uparrow}$$

This can also happen with peracetic acid but only at elevated temperatures, acetic acid being a weaker acid than formic acid. Ring-opening may be prevented by conducting the epoxidation in a suitably buffered medium.

Epoxides have four uses in synthesis:

(1) Reduction (lithium aluminium hydride) gives an alcohol (p. 637),

$$R-CH\overset{}{\underset{O}{\overbrace{\quad}}}CH_2 \xrightarrow{LiAlH_4} R-\underset{\underset{OH}{|}}{CH}-CH_3$$

(2) Treatment with a Lewis acid gives a carbonyl compound, e.g.

$$R-CH\underset{O}{\overbrace{\quad}}CH_2 \xrightarrow{BF_3} R-CH\underset{\underset{-BF_3}{\overset{|}{O}}}{\overbrace{\quad}}\overset{H}{\overset{|}{CH}} \longrightarrow R-CH_2-CH\overset{+}{=}\overset{-}{O}-\overset{-}{B}F_3 \xrightarrow{-BF_3} R-CH_2-CHO$$

(3) Treatment with dimethyl sulphoxide gives an α-ketol (*cf.* p. 589):

$$R-CH\underset{O}{\overbrace{\quad}}CH-R' \longrightarrow R-\underset{\underset{O-}{|}}{CH}-\underset{\underset{R'}{|}}{C}H \longrightarrow R-\underset{\underset{OH}{|}}{CH}-CO-R' + S(CH_3)_2$$

*The order of reactivity of alkyl-derivatives in S_N2 reactions is primary > secondary > tertiary (p. 128).

(4) Aqueous acid or base gives a 1,2-diol. The product has the *trans* stereochemistry as a result of the stereospecificity of S_N2-displacements (p. 175):

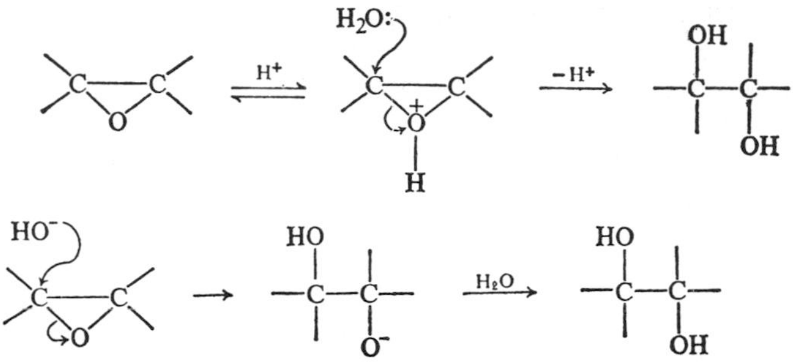

$$\text{epoxide} \underset{H^+}{\rightleftharpoons} \text{protonated epoxide (H)} \xrightarrow{H_2O:} \xrightarrow{-H^+} \begin{array}{c} OH \\ -C-C- \\ OH \end{array}$$

$$\xrightarrow{HO^-} \begin{array}{c} HO \\ -C-C- \\ O^- \end{array} \xrightarrow{H_2O} \begin{array}{c} HO \\ -C-C- \\ OH \end{array}$$

Other nucleophiles also cleave the epoxide ring, reacting predominantly at the less substituted carbon atom, e.g.

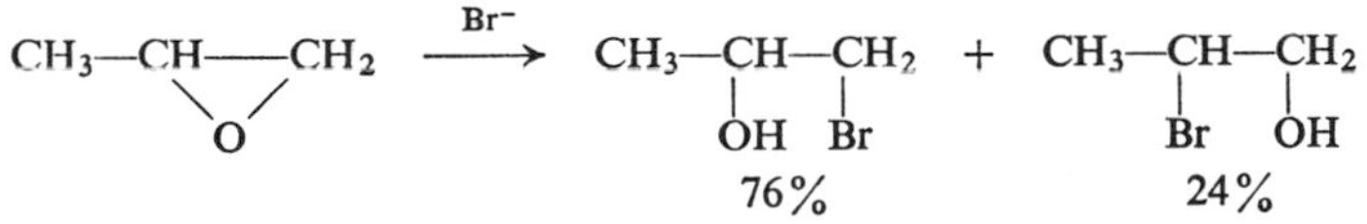

$$CH_3-CH-CH_2 \xrightarrow{Br^-} \underset{76\%}{CH_3-\underset{OH}{CH}-\underset{Br}{CH_2}} + \underset{24\%}{CH_3-\underset{Br}{CH}-\underset{OH}{CH_2}}$$

The ring-opening of epoxides of rigid cyclohexene systems gives *trans*-diaxial products:

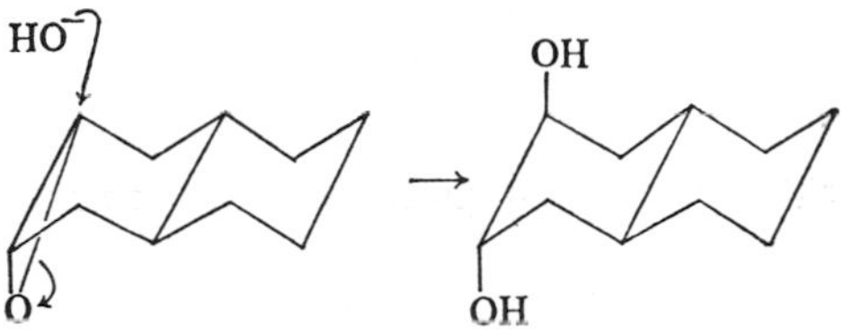

$$\text{epoxide} + HO^- \rightarrow \text{trans-diaxial diol}$$

Industrially, ethylene oxide is made by the oxidation of ethylene with atmospheric oxygen at pressures up to 300 p.s.i. over a silver catalyst at 200–300°C. Its most important use is in the production of ethylene glycol, required, for example, for the manufacture of terylene.

(*ii*) *Diol formation.* As well as the method for obtaining *trans*-1,2-diols *via* epoxides, there are three methods for converting olefins directly into diols.

(1) *Osmium tetroxide.* The addition of an olefin to osmium tetroxide in ether causes the rapid precipitation of a cyclic osmate ester. Pyridine, which complexes the osmium atom in the ester, is often added as a catalyst. The ester is then hydrolyzed, commonly with aqueous sodium sulphite, to give a *cis*-1,2-diol.

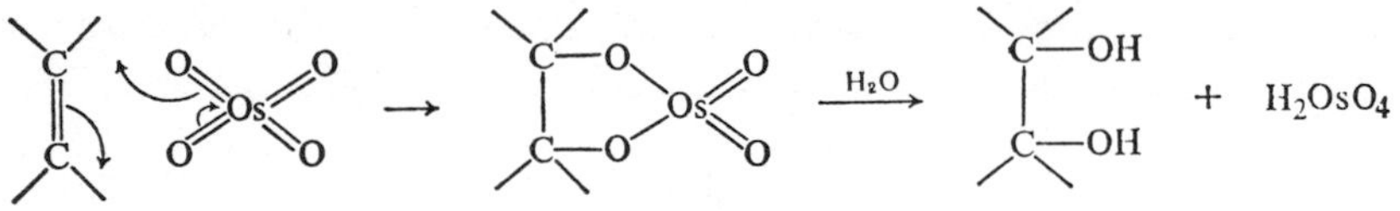

Osmium tetroxide is both highly toxic and expensive, but is a valuable reagent because of its specificity for olefinic bonds* and the ease of its application.

The reagent attacks rigid cyclic systems from the less hindered side, thereby yielding the more stable of the two possible *cis*-diols (compare the iodine-silver acetate method below), e.g.

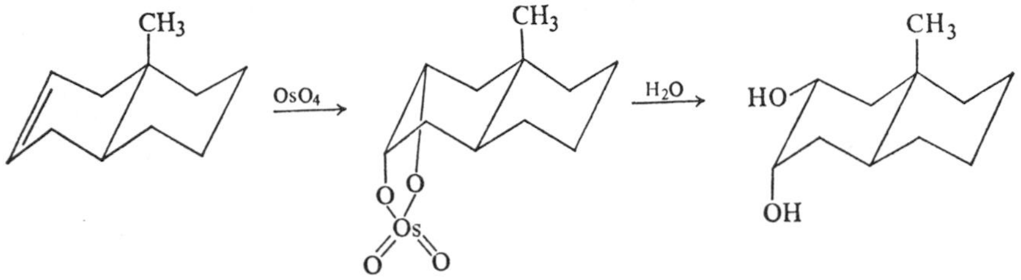

Potassium permanganate acts similarly to osmium tetroxide:

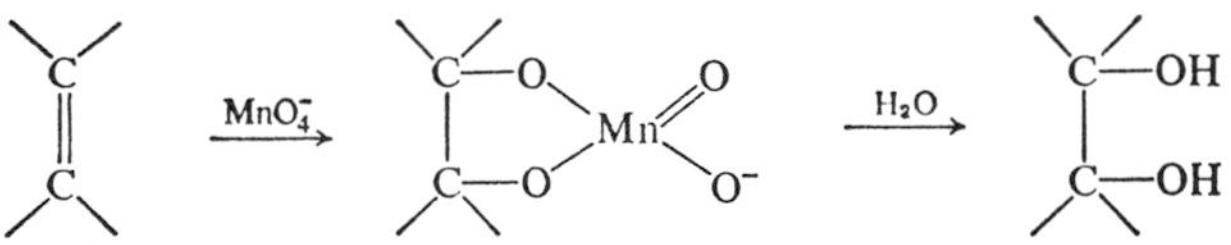

It is normally used in an aqueous solution in which the organic compound is dissolved or suspended; a co-solvent (usually t-butanol or acetic acid) is some-times employed. In order to prevent further oxidation it is necessary to work in alkaline conditions, for otherwise the *cis*-diol is oxidized further with formation of an α-hydroxy-ketone or by cleavage of the C—C bond. In general, perman-ganate is less selective, and therefore less satisfactory, than osmium tetroxide, but it has the advantage of being less hazardous to use and very much cheaper.

(2) *Iodine-silver acetate* ('*wet*'). This method also yields *cis*-diols. The olefin is treated with iodine in aqueous acetic acid in the presence of silver acetate. Iodine reacts with the double bond to give an iodonium ion which undergoes displacement by acetate in the S_N2 manner, giving a *trans*-iodo-acetate. Anchi-meric assistance by the acetate group, together with the powerful bonding capacity of silver ion for iodide, then lead to the formation of a cyclic ace-toxonium ion which in turn reacts with water to give a *cis*-hydroxy-acetate. Final hydrolysis gives the *cis*-diol.

*The very reactive double bonds in certain aromatic compounds, such as the 9,10–bond in phenanthrene, are also oxidized, but very slowly.

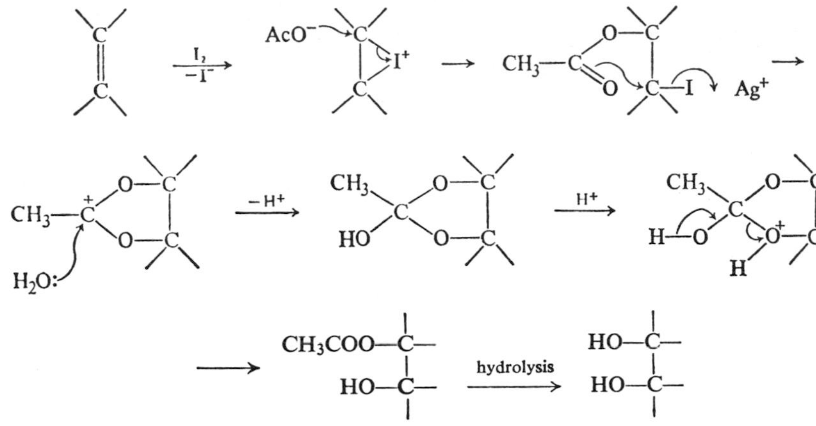

In a rigid cyclic olefin, approach of the iodine from the less hindered side leads ultimately to the *less* stable isomeric diol, e.g.

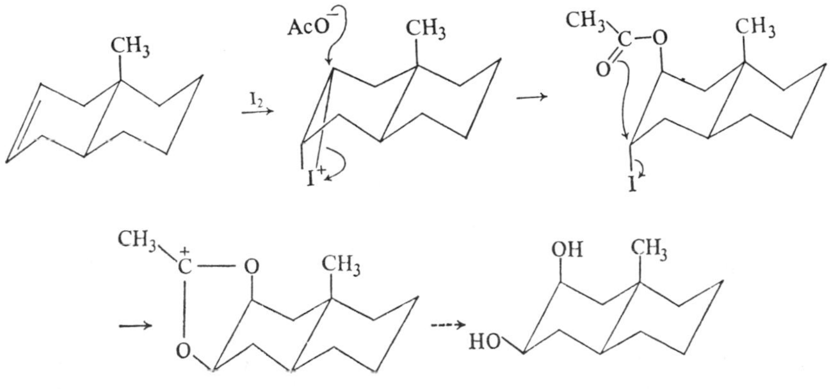

(3) *Iodine-silver acetate* ('*dry*'). The same reaction carried out in the absence of water leads to a *trans*-1,2-diacetate, from which the *trans*-diol is obtained by hydrolysis (*Prévost reaction*):

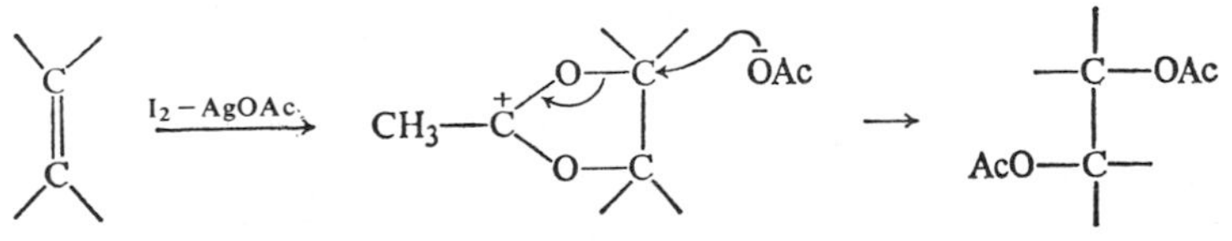

(*iii*) *Cleavage of the double bond.* (1) *Ozonolysis.* Ozone reacts as an electrophile with olefins, forming a primary ozonide of uncertain structure which rearranges through a zwitterionic intermediate to an isolable ozonide:

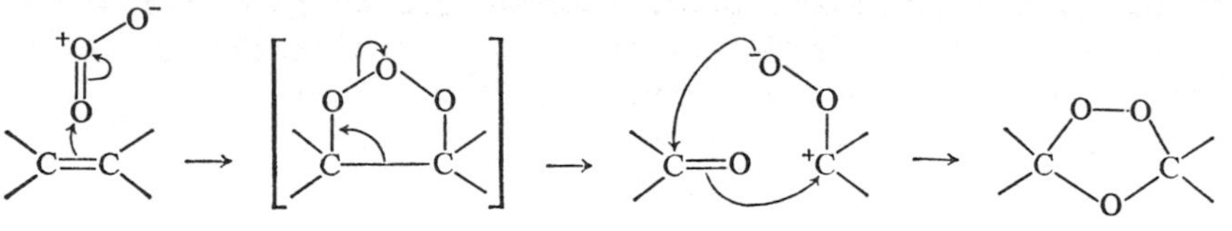

primary   zwitterionic   isolable ozonide
ozonide   intermediate

Ozonides are dangerously explosive, however, and are usually converted at once into products. Direct solvolysis gives ketones and/or acids, depending on the structure of the olefin, e.g.

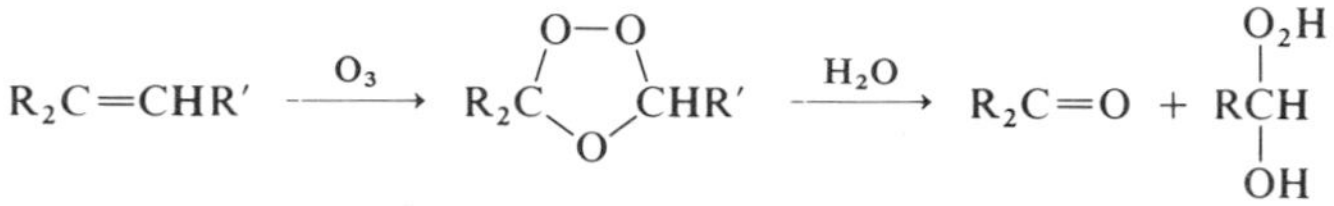

$$R_2C=CHR' \xrightarrow{O_3} R_2C\underset{O}{\overset{O-O}{\diagup\diagdown}}CHR' \xrightarrow{H_2O} R_2C=O + \underset{\underset{OH}{|}}{\overset{\overset{O_2H}{|}}{RCH}}$$

$$\underset{\underset{OH}{|}}{\overset{\overset{O_2H}{|}}{RCH}} \longrightarrow RCO_2H + H_2O$$

It is often more useful to carry out the solvolysis in the presence of a reducing agent. With zinc and acetic acid, hydrogen on a metal catalyst, or dimethyl sulphide, only carbonyl compounds are formed; dimethyl sulphide has the advantage that other reducible groups in the compound are not affected, and its suitability stems from its ability to reduce hydroperoxy compounds:

$$\underset{\underset{OH}{|}}{RCH}\text{—O—O} \quad :S(CH_3)_2 \longrightarrow RCHO + (CH_3)_2SO + OH^-$$

For example,

$$CH_3(CH_2)_5CH=CH_2 \xrightarrow[2)(CH_3)_2S]{1)O_3} \underset{75\%}{CH_3(CH_2)_5CHO + CH_2O}$$

With lithium aluminium hydride, further reduction occurs to give alcohols: e.g. n-amyl alcohol can be obtained in 89% yield from 2-heptene,

$$C_4H_9CH=CHCH_3 \xrightarrow[2) LiAlH_4]{1) O_3} \underset{\text{n-Amyl alcohol}}{C_4H_9CH_2OH}$$

Acetylenic systems are also oxidized by ozone, but they generally react at only about one-thousandth the rate of olefins, just as they are less reactive than olefins towards other electrophiles (p. 104). Selective oxidation of C=C in the presence of $C{\equiv}C$ can therefore be achieved. Reaction with acetylenes gives carboxylic acids, together with small amounts of α-dicarbonyl compounds, probably by a mechanism analogous to that for olefins:

$$R{-}C{\equiv}C{-}R' \xrightarrow{O_3} \left[R{-}\overset{O}{\overset{\|}{C}}{-}\overset{{-}O}{\underset{O}{\overset{|}{\underset{+}{C}}}}{-}R' \right] \longrightarrow R{-}CO{-}O{-}CO{-}R' \xrightarrow{H_2O} RCO_2H + R'CO_2H$$

For example:

$$CH_3{-}(CH_2)_7{-}C{\equiv}C{-}(CH_2)_7{-}CO_2H \xrightarrow[2)\ H_2O]{1)\ O_3} HO_2C{-}(CH_2)_7{-}CO_2H$$

Stearolic acid   Azelaic acid (70 – 80%)

$$+ CH_3{-}(CH_2)_7{-}CO{-}CO{-}(CH_2)_7{-}CO_2H$$

4%

Aromatic rings also undergo ozonolysis (p. 576).

(2) *The Lemieux reagents.* Ozone is unpleasant to handle and is not selective for olefins: e.g. secondary alcohols may be oxidized to ketones and tertiary C—H bonds to alcohols. Ozone has therefore largely been displaced by the Lemieux reagents, which consist of dilute aqueous solutions of sodium periodate with a catalytic quantity of potassium permanganate and of osmium tetroxide, respectively. In each case, the olefin is first oxidized to the *cis*-diol which is then cleaved by the periodate (p. 595) to give aldehydes and/or ketones. The permanganate reagent then oxidizes aldehydic products to carboxylic acids. The low-valent states of manganese and osmium generated during the reaction are re-oxidized by the periodate to their original state, so that only catalytic quantities are required. The reactions are rapid at room temperature and are selective for olefins, e.g.

$$(CH_3)_2C{=}CHCH_3 \xrightarrow{NaIO_4\text{-}KMnO_4} (CH_3)_2CO + CH_3CO_2H$$

$$(CH_3)_2C{=}CHCH_3 \xrightarrow{NaIO_4\text{-}OsO_4} (CH_3)_2CO + CH_3CHO$$

(3) *Chromium*(VI) *oxide.* The cleavage of olefinic bonds with chromic oxide is competitive with oxidation of allylic C—H bonds; e.g. cyclohexene gives a mixture of 3-cyclohexenone and adipic acid:

The use of a partially aqueous medium favours the cleavage process, whereas an anhydrous medium such as glacial acetic acid favours allylic oxidation. In addition, cleavage is promoted by the presence of phenyl substituents, evidently because the first step involves the formation of a carbonium ion which is stabilized by adjacent aromatic rings:

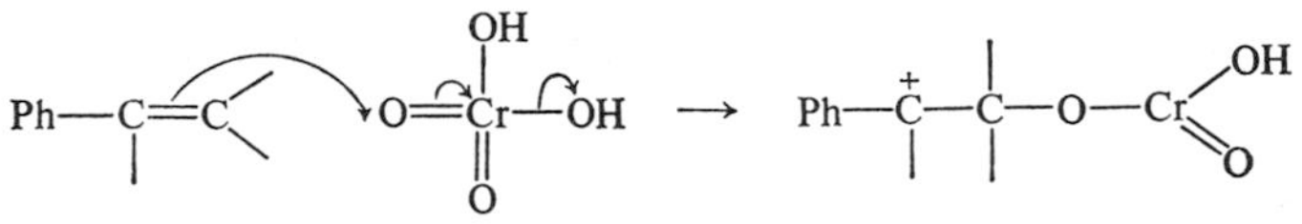

(b) AROMATIC RINGS

The oxidation of unsubstituted aromatic rings, which results in the loss of the associated stabilization energy, requires vigorous conditions. Reaction can result either in the cleavage of the ring or in the formation of quinones.

Ozone effects cleavage, e.g.

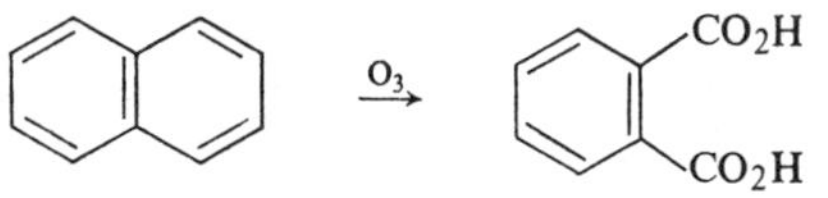

Phthalic acid (88%)

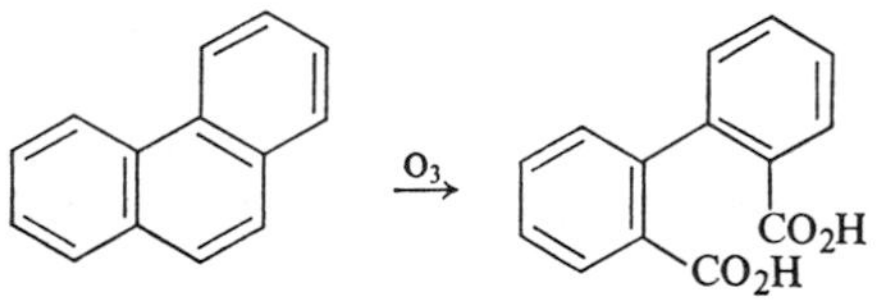

Diphenic acid (65%)

An industrial method, which is far cheaper than ozonolysis, employs aerial oxidation over a vanadium pentoxide catalyst at 400–500°C. For example, maleic anhydride and phthalic anhydride are cheaply available from benzene and naphthalene:

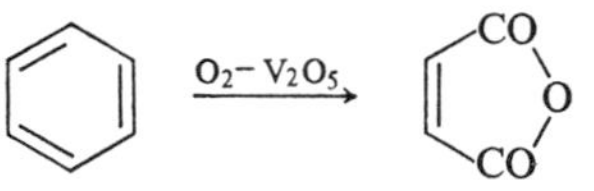

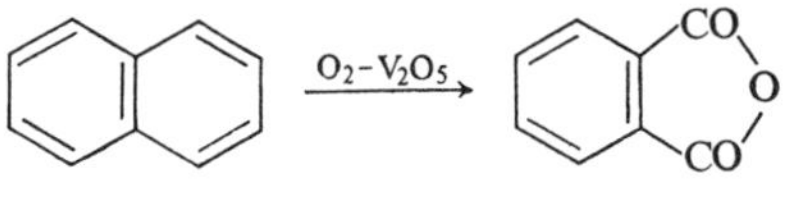

The latter process is particularly important since some of the derived diesters (e.g. dinonyl phthalate) are used as plasticizers for cellulose acetate and polyvinyl chloride. Polyester resins and phthalocyanine dyes are also manufactured from phthalic anhydride.

Chromium(VI) oxide can also be used: for example, quinoline is oxidized to pyridine-2,3-dicarboxylic acid which readily loses the 2-carboxy-substituent on being heated (p. 414), providing an easy route to nicotinic acid:

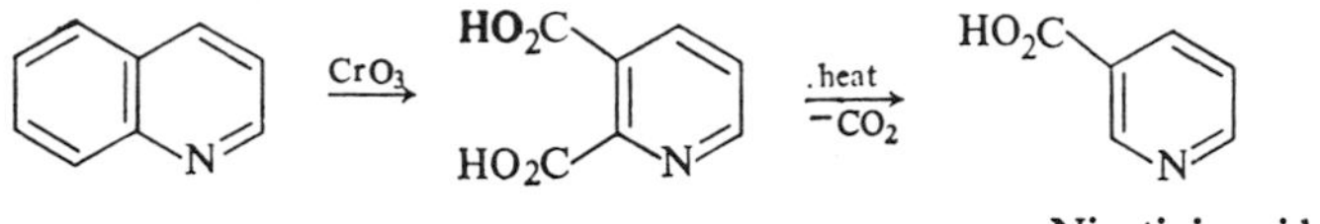

Nicotinic acid

Chromium(VI) oxide does not invariably lead to ring-fission. For example, it reacts with naphthalene in acetic acid solution at room temperature to give 1,4-naphthoquinone in about 20% yield [4]:

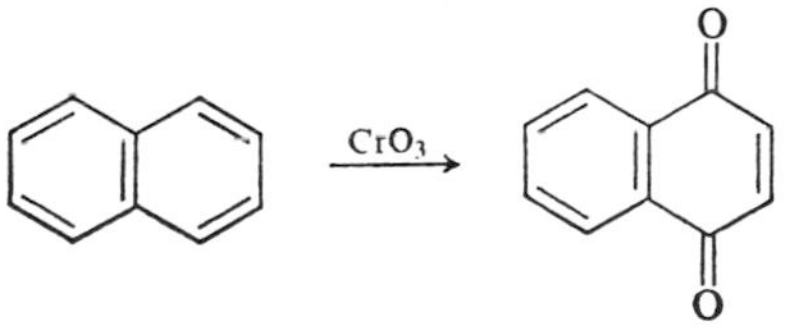

2-Methylnaphthalene gives the corresponding 1,4-quinone similarly, in contrast to toluene, which is oxidized by Cr(VI) compounds preferentially at the methyl group (p. 580).

The aromatic rings of phenols are very susceptible to oxidation by one-electron oxidants, for the removal of a hydrogen atom gives a delocalized aryloxy radical,

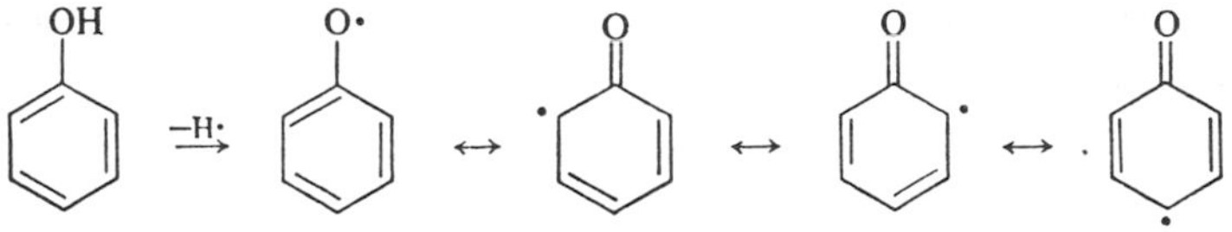

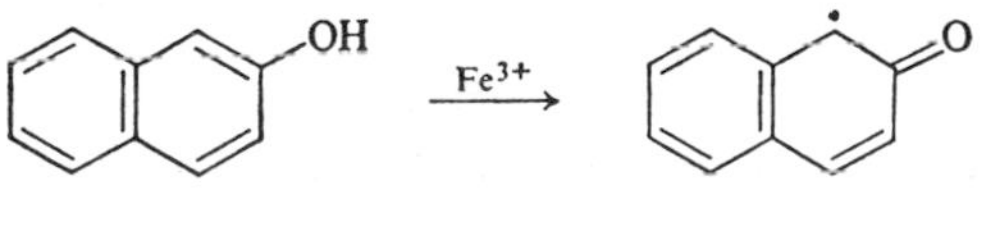

The fate of the radical depends on the structure of the phenol. Diphenols can be formed, as in the oxidation of 2-naphthol by iron(III) ion, further oxidation to quinones can occur, e.g.

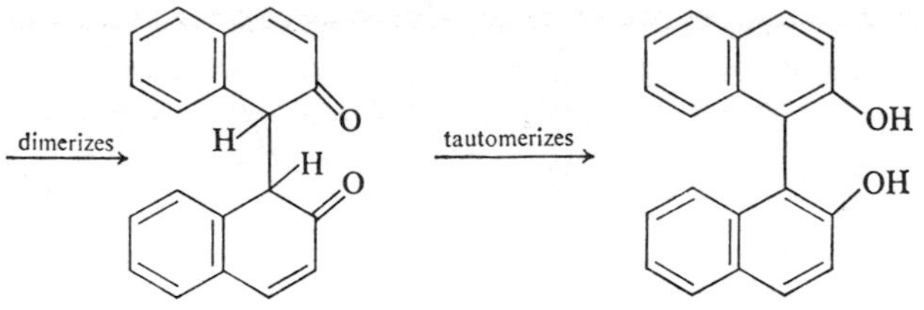

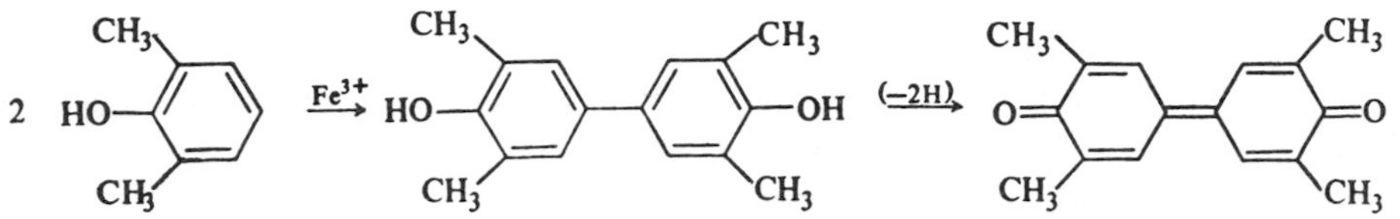

and cyclization to derivatives of dibenzofuran can arise through intramolecular nucleophilic attack on an intermediate quinone. For example, the oxidation of p-cresol by hexacyanoferrate(III) ion gives, in addition to carbon-carbon dimers, Pummerer's ketone:

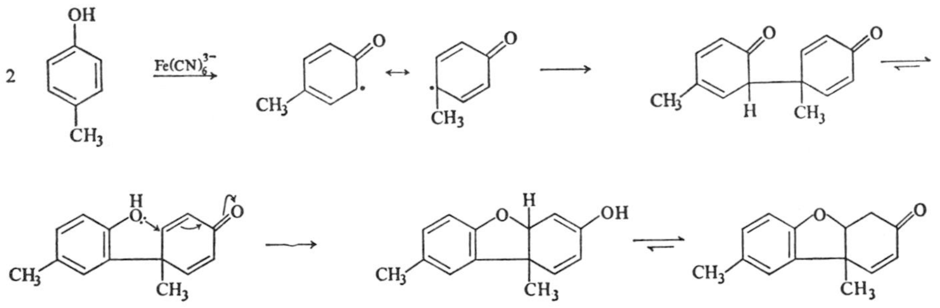

Pummerer's
ketone

A synthesis of usnic acid is based on this mode of coupling. Hexacyano-ferrate(III) oxidation of methylphloroacetophenone gives a dimer which is dehydrated by concentrated sulphuric acid to ($\pm$)-usnic acid:

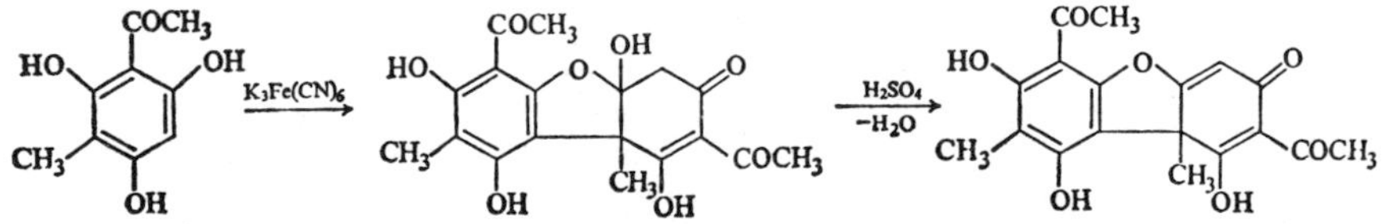

($\pm$)-Usnic acid

Aromatic amines are also sensitive to oxidation and discolour in air. The products are complex and the processes are of little use except in two cases: first, specific methods for oxidizing the amino group rather than the aromatic ring are available (p. 601); and secondly, aniline is oxidized to *p*-benzoquinone in 60% yield by dichromate:

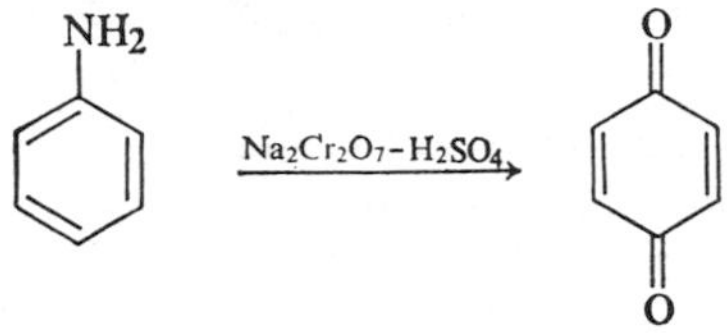

Industrially, aniline is oxidized to *p*-benzoquinone with manganese dioxide and sulphuric acid.

The *ortho* and *para* dihydric phenols are readily oxidized to the corresponding quinones by one-electron oxidants. Reaction occurs through delocalized semi-quinone radicals, e.g.

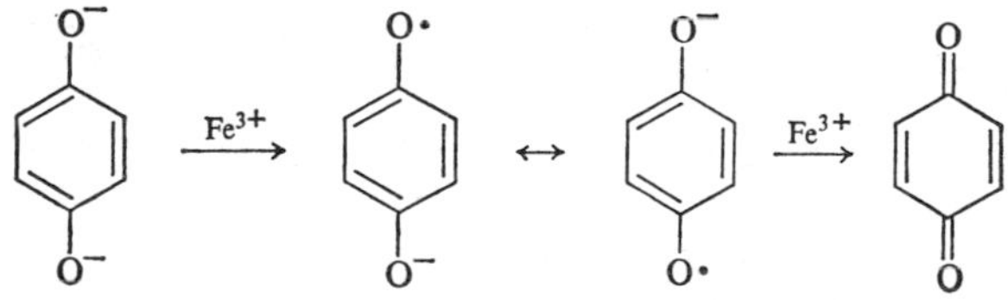

o-Benzoquinone may be prepared by the oxidation of catechol (from salicyl-aldehyde by Dakin's reaction; p. 473) with silver oxide suspended in ether; sodium sulphate is added as a dehydrating agent since the quinone is rapidly attacked by water (addition to the $\alpha\beta$-unsaturated carbonyl system):

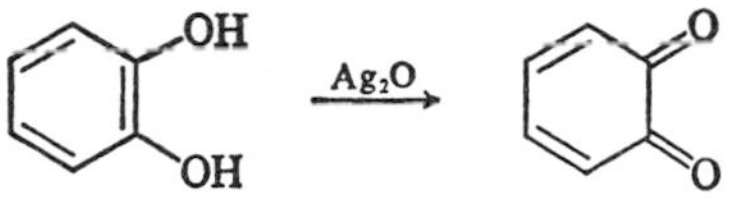

The corresponding aminophenols react similarly. For example, 1-amino-2-naphthol (readily available from 2-naphthol by nitrosation or diazo-coupling followed by reduction; p. 405) is oxidized by iron(III) chloride to 1,2-naphtho-quinone in over 90% yield [2]:

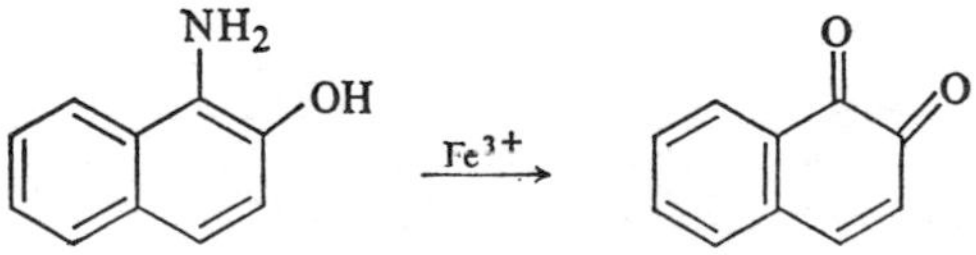

1,2-Naphthoquinone

(c) SATURATED C–H GROUPS

(i) *Allylic and benzylic systems.* The comparative stability of allyl and benzyl radicals, e.g.

$$CH_2{=}CH{-}\dot{C}H_2 \leftrightarrow \dot{C}H_2{-}CH{=}CH_2$$

renders allylic and benzylic systems susceptible to oxidation *via* free-radical reactions. Autoxidation has been described (17.5). An alternative approach is through radical-catalyzed halogenation (p. 543). For example, the chlorination of toluene in the vapour phase or under reflux gives benzyl chloride, benzal chloride, and benzotrichloride, from which benzyl alcohol, benzaldehyde, and benzoic acid, respectively, are available by hydrolysis; the alcohol and aldehyde are produced in this way on an industrial scale. However, the difficulty of separating the halogenated products efficiently in the laboratory necessitates the use of more specific methods.

The first oxidation stage, to give a product at the oxidation level of an alcohol, can be effectively carried out using sulphuryl chloride in the presence of a peroxide; e.g. toluene gives benzyl chloride in 80% yield (p. 545):

$$PhCH_3 + SO_2Cl_2 \xrightarrow{peroxide} PhCH_2Cl \xrightarrow{H_2O} PhCH_2OH$$

Oxidation to the carboxylic acid can be brought about with chromium(VI) oxide, permanganate, or nitric acid. For example, dilute permanganate oxidizes *o*-chlorotoluene to *o*-chlorobenzoic acid in about 65% yield [2],

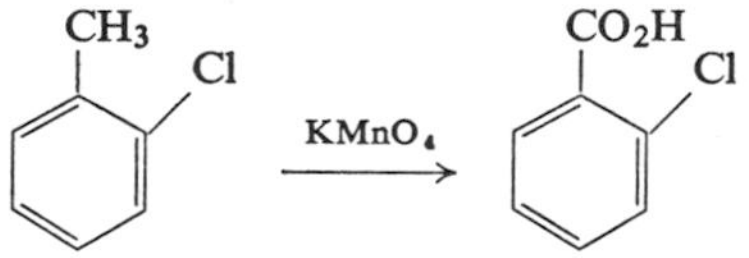

and concentrated nitric acid oxidizes *o*-xylene to *o*-toluic acid in 54% yield [3]:

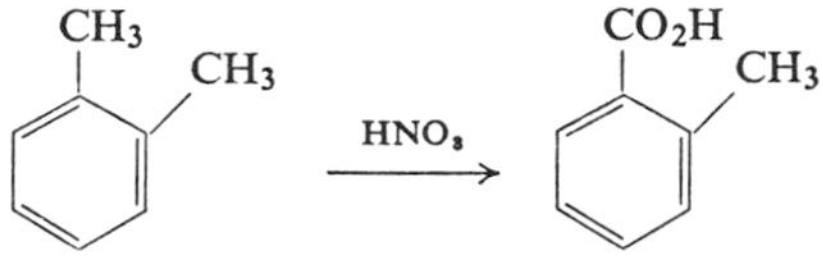

Substituents on the methyl group are also removed in these conditions: e.g. nicotine is oxidized to nicotinic acid by concentrated nitric acid at 70°C:

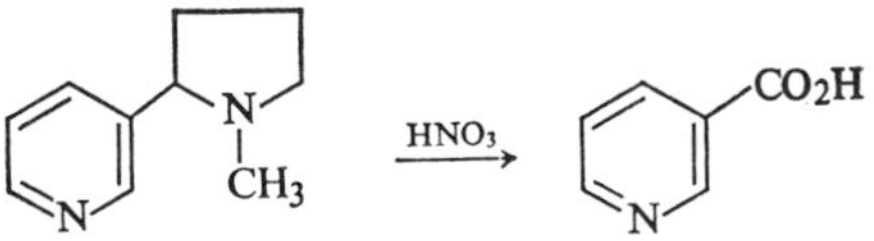

Nicotinic acid

Industrially, benzoic acid is manufactured by the catalytic air oxidation of toluene. Its main use is in the production of sodium benzoate which is employed as a preservative for foodstuffs and pharmaceuticals. Benzoyl chloride is used to prepare dibenzoyl peroxide, used for initiating addition polymerization (and, on the laboratory scale, for other radical-catalyzed reactions) and as a bleaching agent.

$$2PhCOCl + H_2O_2 \xrightarrow{\text{base}} PhCOO\text{---}OCOPh + 2HCl$$

Oxidation to the aldehydic level. Commercially, a particularly important example is the oxidation of propylene to acrolein, CH_2=CH—CHO, over copper(II) oxide at 300–400°C. The product is largely converted into glycerol (required for nitroglycerine), by reduction to allyl alcohol followed by hydroxylation with hydrogen peroxide. A recent development involves the air oxidation of propylene in the presence of ammonia over a catalyst (e.g. bismuth molybdate); acrolein is probably formed first and reacts further:

$$CH_2\text{=}CH\text{---}CHO + \tfrac{1}{2}O_2 + NH_3 \longrightarrow CH_2\text{=}CH\text{---}CN + 2H_2O$$

This process for acrylonitrile has superseded that from acetylene and hydrogen cyanide.

In the laboratory, three methods are available for the selective oxidation of $ArCH_3$ to $ArCHO$.

(1) *With chromyl chloride (Etard reaction).* A solution of chromyl chloride in carbon disulphide is added cautiously to the benzylic compound at 25–45°C and the brown complex which separates is decomposed by water to give the aldehyde and chromic(VI) acid, e.g.

m-Tolualdehyde (75%)

The aldehyde must be removed rapidly by distillation or extraction to prevent its further oxidation.

(2) *With chromium(VI) oxide in acetic anhydride.* Oxidation is carried out with chromium(VI) oxide in a mixture of acetic anhydride, acetic acid, and sulphuric acid at low temperature. As it is formed, the aldehyde is converted into its *gem*-diacetate which is stable to oxidation; this is isolated and reconverted into the aldehyde by acid hydrolysis. For example, *p*-nitrotoluene gives *p*-nitrobenzaldehyde in about 45% overall yield [2]:

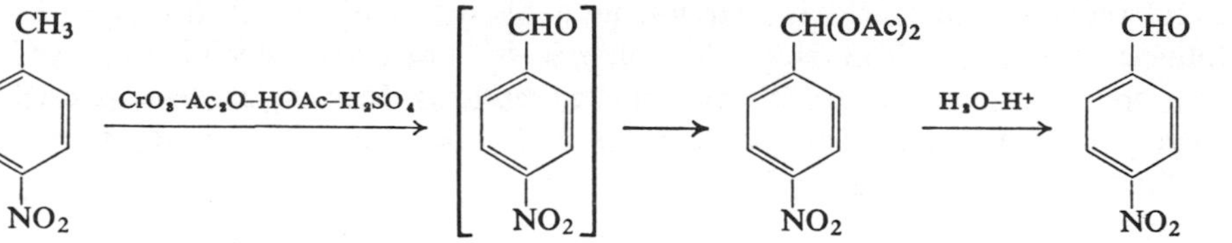

p-Nitrobenzaldehyde

(3) *With p-nitrosodimethylaniline.* The benzylic compound is treated with *p*-nitrosodimethylaniline and the resulting Schiff base is hydrolyzed. The method is applicable only to those compounds whose methyl groups are strongly activated, e.g. 2,4-dinitrotoluene.

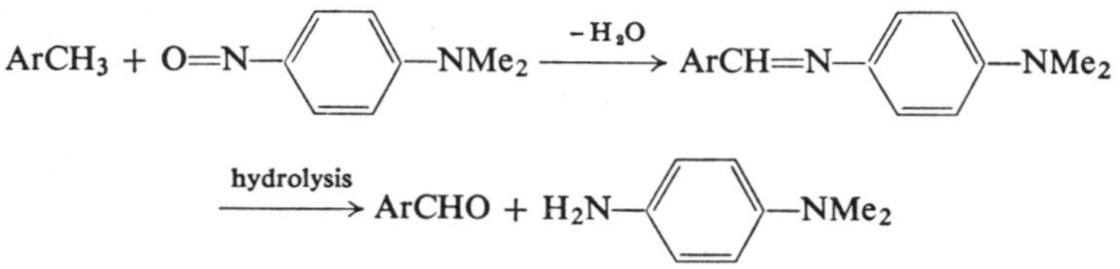

A closely related method was used in the preparation of quininic acid (quinine synthesis; 21.6). The 4-methyl substituent in 4-methyl-6-methoxy-quinoline, activated by the hetero-atom, was condensed with benzaldehyde in the presence of zinc chloride and the resulting benzylidene derivative was oxidized to quininic acid with permanganate:

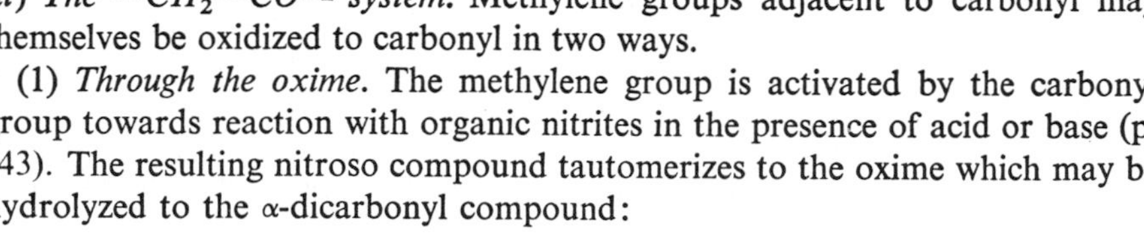

(ii) *The* $-CH_2-CO-$ *system.* Methylene groups adjacent to carbonyl may themselves be oxidized to carbonyl in two ways.

(1) *Through the oxime.* The methylene group is activated by the carbonyl group towards reaction with organic nitrites in the presence of acid or base (p. 343). The resulting nitroso compound tautomerizes to the oxime which may be hydrolyzed to the α-dicarbonyl compound:

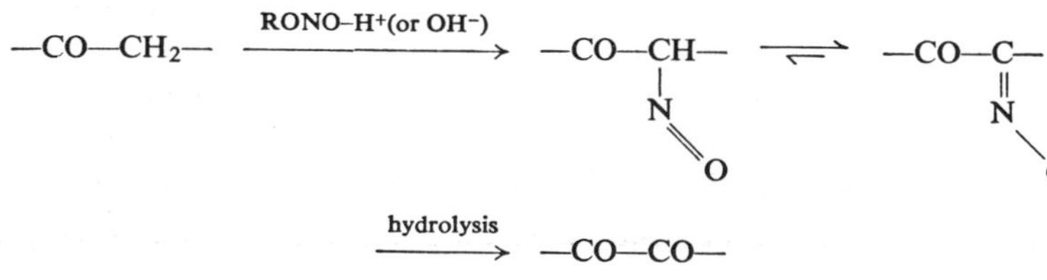

Methylene groups are oxidized in preference to methyl, so that, e.g. methyl ethyl ketone gives biacetyl.

(2) *By selenium dioxide* (*Riley reaction*). The reaction occurs through the enolic form of the carbonyl compound:

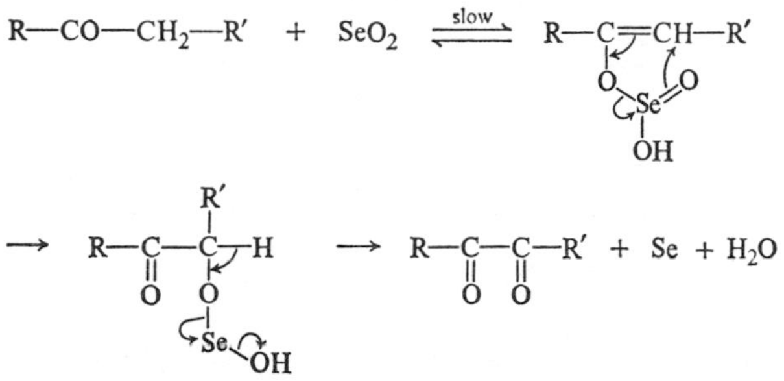

$$R\text{—}CO\text{—}CH_2\text{—}R' \; + \; SeO_2 \; \underset{}{\overset{slow}{\rightleftharpoons}} \; R\text{—}C\text{=}CH\text{—}R'$$

$$\rightarrow \quad R\text{—}C\text{—}C\text{—}H \quad \longrightarrow \quad R\text{—}C\text{—}C\text{—}R' \; + \; Se \; + \; H_2O$$

For example, camphor reacts in refluxing acetic anhydride to give camphorquinone in 95% yield,

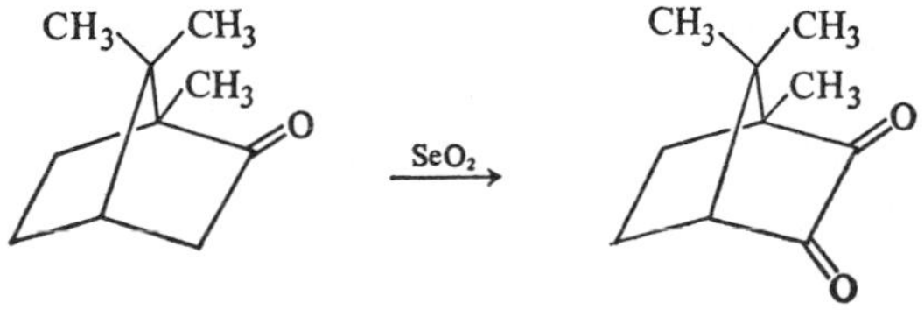

Camphorquinone

and acetophenone reacts in dioxan to give phenylglyoxal in about 70% yield [2]:

$$PhCOCH_3 \; \xrightarrow{SeO_2} \; PhCOCHO$$
Phenylglyoxal

In contrast to the oxime method, selenium dioxide preferentially oxidizes methyl rather than methylene groups. For example, methyl ethyl ketone gives mainly ethylglyoxal, CH_3CH_2COCHO.

Selenium dioxide is of limited scope because it is unselective. Thus, it can bring about the following oxidations:

$$RCH\text{=}CHR' \rightarrow RCOCOR'; \; RCH_2CH\text{=}CHR' \rightarrow RCH(OH)CH\text{=}CHR';$$

$$ArCH_2Ar' \rightarrow ArCOAr'; \; ArCH_2CH_2Ar' \rightarrow ArCH\text{=}CHAr'.$$

Oxidation of $-CH_2-CO-$ and $>CH-CO-$ to $-CH(OH)-CO-$ and $>C(OH)-CO-$ can also be effected. The carbanion is formed in the presence of a strong base such as potassium t-butoxide and reacts with oxygen to give a hydroperoxide. This is reduced to the corresponding alcohol with a trialkyl phosphite (tervalent phosphorus having a very marked affinity for oxygen; 18.6):

$$
\begin{array}{c}
\underset{}{>}\!\text{CH—CO—} \xrightarrow[]{(CH_3)_3CO^-} \underset{}{>}\!\overset{}{\text{C}}\text{—C}{=}\text{O} \ \leftrightarrow\ \underset{}{>}\!\text{C}{=}\text{C—}\bar{\text{O}} \xrightarrow[]{O_2} \overset{\mid}{\underset{\mid}{-}}\!\text{C—CO—}
\end{array}
$$
$$
\text{O—O}^-
$$

$$
\xrightarrow[]{(CH_3)_3COH} \ \underset{\text{O—OH}}{\overset{\mid}{-}\!\text{C—CO—}} \ \xrightarrow[]{(RO)_3P} \ \underset{\text{OH}}{\overset{\mid}{-}\!\text{C—CO—}} \ + \ (RO)_3PO
$$

Finally, the γ-dicarbonyl system undergoes dehydrogenation by autoxidation in the presence of base; reaction occurs through the tautomeric dienediol whose anion donates two electrons successively to oxygen to form first a relatively stable anion-radical and then the enedione (*cf.* p. 579):

$$
\underset{O}{\overset{\parallel}{-}\!\text{C—CH}_2\text{—CH}_2\text{—}}\underset{O}{\overset{\parallel}{\text{C}}\text{—}} \ \rightleftharpoons\ \underset{OH}{\overset{\mid}{-}\!\text{C}{=}\text{CH—CH}{=}}\underset{OH}{\overset{\mid}{\text{C}}\text{—}} \xrightarrow[]{\text{Base}}
$$

$$
\underset{O^-}{\overset{\mid}{-}\!\text{C}{=}\text{CH—CH}{=}}\underset{O^-}{\overset{\mid}{\text{C}}\text{—}} \xrightarrow[(-e)]{O_2} \underset{O\cdot}{\overset{\mid}{-}\!\text{C}{=}\text{CH—CH}{=}}\underset{O^-}{\overset{\mid}{\text{C}}\text{—}} \ \leftrightarrow\ \underset{O^-}{\overset{\mid}{-}\!\text{C}{=}\text{CH—CH}{=}}\underset{O\cdot}{\overset{\mid}{\text{C}}\text{—}}
$$

$$
\xrightarrow[(-e)]{O_2} \underset{O}{\overset{\parallel}{-}\!\text{C—CH}{=}\text{CH—}}\underset{O}{\overset{\parallel}{\text{C}}\text{—}}
$$

(*iii*) *Unactivated C—H.* The selective oxidation of one unactivated C—H group in a molecule possessing alternative centres for attack is attended by the difficulty that the reagents with which C—H bonds react—essentially only free radicals—are relatively unspecific. However, selectivity is obtained in two circumstances. First, a relatively unreactive free radical, such as the bromine atom (p. 543), discriminates quite sharply in favour of tertiary C—H compared with primary or secondary C—H; for example, the gas-phase bromination of isobutane gives t-butyl bromide almost exclusively (p. 543).

The oxidation of tertiary C—H to C—OH can sometimes be achieved directly with alkaline permanganate; reaction occurs with retention of configuration:

$$
\underset{R''}{\overset{R}{\underset{\diagup}{\diagdown}}}\!\!\underset{}{\text{C—H}} \xrightarrow[]{\text{KMnO}_4} \underset{R'''}{\overset{R}{\underset{\diagup}{\diagdown}}}\!\!\underset{}{\text{C—OH}}
$$

Secondly, intramolecular free-radical reactions occur specifically through six-membered cyclic transition states, so that it is possible selectively to oxidize δ-CH bonds, e.g.

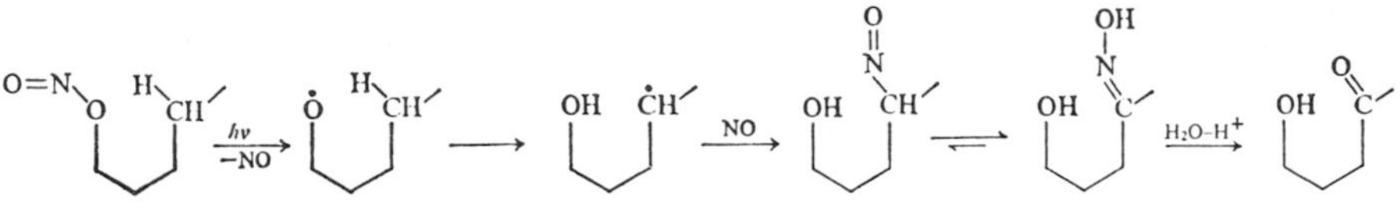

Examples of the application of this principle have been described earlier (pp. 560, 561).

The very vigorous oxidation of hydrocarbon chains by chromium(VI) oxide in concentrated sulphuric acid oxidizes most substances to carbon dioxide and water, but C-methyl groups give mainly acetic acid. This procedure is usefully applied in determining the number of C-methyl groups in a compound of unknown structure (Kuhn-Roth method).

(iv) Aromatization. Alicyclic compounds which are reduction products of aromatic systems can be dehydrogenated to the respective aromatics in several ways.

(1) *With sulphur or selenium.* Reaction occurs with sulphur at about 200°C and with selenium at about 250°C, the hydrogen being removed as hydrogen sulphide or hydrogen selenide. Skeletal rearrangements may occur and carbon atoms may also be removed; in particular, angular methyl groups and *gem*-dialkyl substituents are degraded, e.g.

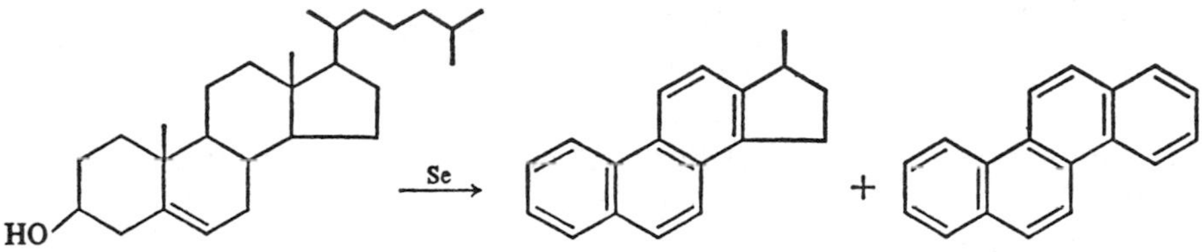

Cholesterol   Diels' hydrocarbon   Chrysene

Selenium is less destructive than sulphur and is usually preferred, but the methods are of more value in degradation than in synthesis. Distillation with zinc dust has essentially the same effect but is more destructive.

(2) *Catalytically.* Alicyclic rings which contain some unsaturation can be dehydrogenated on those catalysts which are successful for hydrogenation (19.2); palladium on charcoal or asbestos is the most commonly used. The conditions are far milder than those using selenium and the procedure is widely applied; e.g. the reduced isoquinolines obtained by the Bischler-Napieralski synthesis are usually oxidized in this way (p. 683):

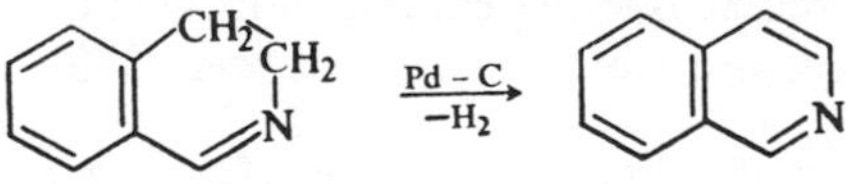

(3) *With quinones.* Partially unsaturated alicyclic rings are oxidized by quinones through hydride-ion transfer, e.g.

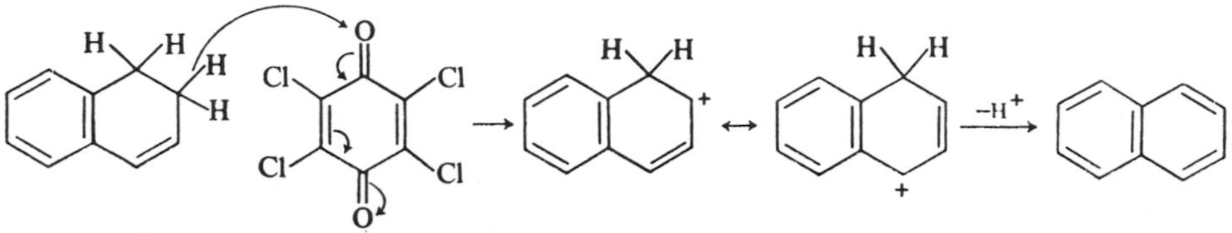

The driving force results from the conversion of both the quinone and the alicyclic system into aromatic compounds. Quinones which contain electron-releasing substituents are stabilized relative to their quinols, e.g.

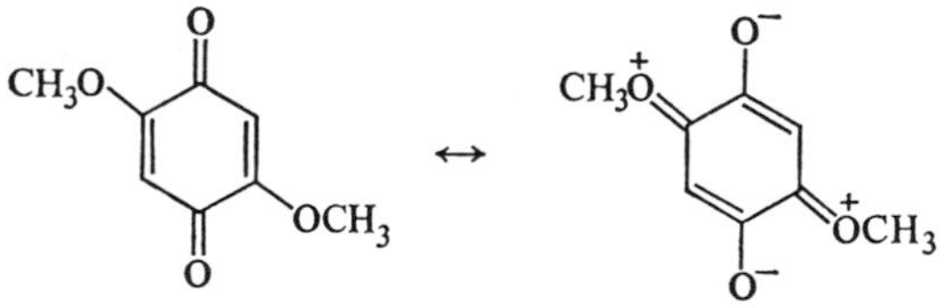

and are less powerful oxidants than those containing electron-attracting substituents. Chloranil* is frequently used, as in a synthesis of *p*-terphenyl,

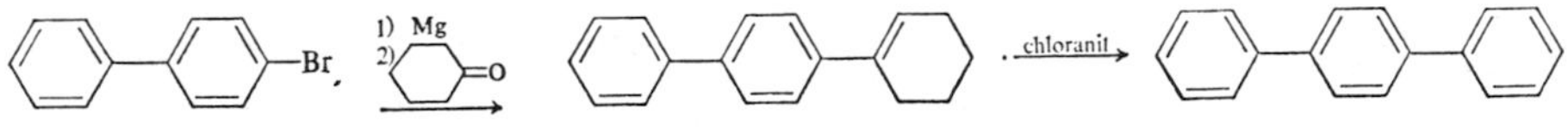

p-Terphenyl

and 2,3-dichloro-5,6-dicyanobenzoquinone is a still more powerful reagent.

*Chloranil is obtained by heating benzoquinone with potassium perchlorate and hydrochloric acid. Nucleophilic attack by chloride ion on the quinone gives a chloroquinol which is oxidized by the perchlorate to the chloroquinone, successive reactions of this type giving chloranil:

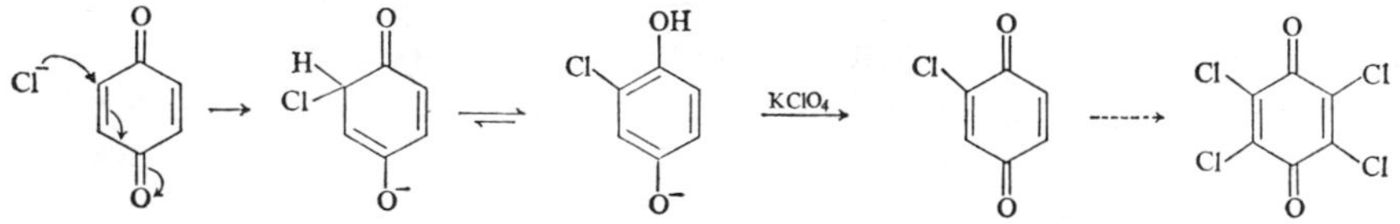

18.3 Systems Containing Oxygen

(a) PRIMARY ALCOHOLS

(i) *To aldehydes.* The oxidation of a primary alcohol to an aldehyde is attended by the difficulties that, first, aldehydes are readily oxidized to acids by many reagents and, secondly, under acidic conditions the aldehyde reacts with unchanged alcohol to give, in an equilibrium mixture, some of the hemi-acetal which is readily oxidized to an ester:

$$RCHO + RCH_2OH \rightleftharpoons RCH{-}OCH_2R \xrightarrow{\text{oxidation}} RCO{-}OCH_2R$$
$$\qquad\qquad\qquad\qquad\qquad | $$
$$\qquad\qquad\qquad\qquad\quad OH$$

Mechanistically, one commonly employed principle is to attach a strongly electrophilic group, X, to the oxygen of the alcohol, so that H^+ and X^- can be eliminated,

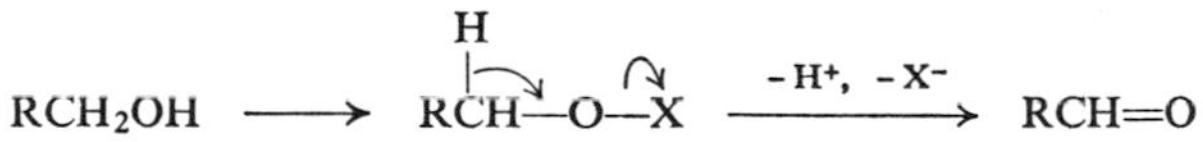

$$RCH_2OH \longrightarrow \overset{H}{\underset{}{RCH{-}O{-}X}} \xrightarrow{-H^+,\ -X^-} RCH{=}O$$

A second approach is to employ a hydride-accepting oxidant,

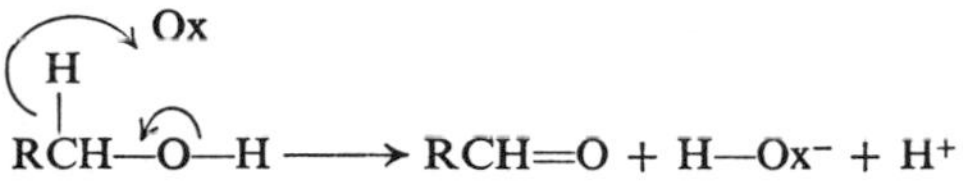

$$\overset{Ox}{\underset{}{\underset{RCH{-}O{-}H}{\overset{H}{|}}}} \longrightarrow RCH{=}O + H{-}Ox^- + H^+$$

(1) *Chromium(VI) oxide.* For the lower-boiling aldehydes, the simplest oxidation procedure is to add an acid solution of potassium dichromate slowly to the alcohol. Use is made of the fact that aldehydes boil at lower temperatures than the corresponding alcohols, so that by maintaining the temperature above the boiling-point of the aldehyde but below that of the alcohol, the aldehyde distils as it is formed; e.g. propionaldehyde can be obtained in about 47% yield [2]:

$$CH_3CH_2CH_2OH \xrightarrow{K_2Cr_2O_7-H_2SO_4} CH_3CH_2CHO$$
$$\text{b.p. } 97°C \qquad\qquad\qquad\qquad \text{b.p. } 48.5°C$$

Reaction occurs through the chromate ester, possibly *via* a cyclic transition state:

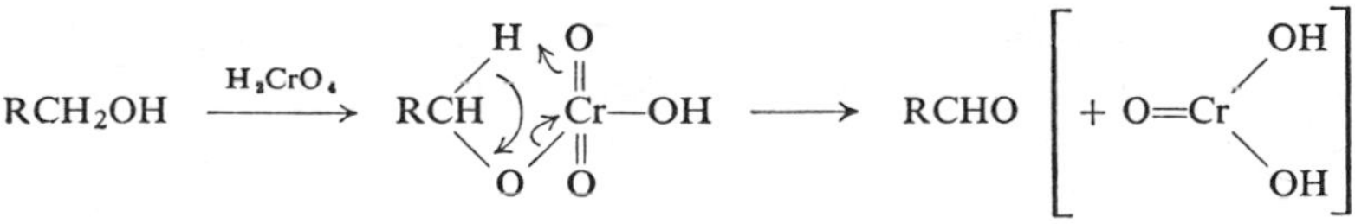

$$RCH_2OH \xrightarrow{H_2CrO_4} \cdots \longrightarrow RCHO \left[+ O{=}Cr{\overset{OH}{\underset{OH}{}}} \right]$$

The chromium(IV) then disproportionates to give chromium(III) and chromium(VI).

The yields of aldehydes can often be increased by use of the complex, $CrO_3 \cdot 2C_5H_5N$, formed by chromium(VI) oxide with pyridine (*cf.* the use of pyridine–sulphur trioxide as a mild sulphonating agent; p. 406). Addition of the chromium(VI) oxide to pyridine is followed either by addition of the alcohol or, better, by isolation of the complex followed by oxidation in methylene chloride; for example, with the latter variant, n-heptanol gives n-heptanal in 70–80% yield [52]:

$$CH_3(CH_2)_5CH_2OH \xrightarrow{CrO_3 \cdot 2C_5H_5N} CH_3(CH_2)_5CHO$$

This method is particularly useful for compounds which contain acid-sensitive groups (e.g. acetals) or other easily oxidized groups (e.g. olefinic bonds).

(2) *Chlorine.* Chlorine appears to oxidize by the acceptance of hydride ion from the alcohol:

$$\overset{\frown}{Cl}-Cl \quad H-\overset{\overset{\textstyle R}{|}}{C}H-O-H \longrightarrow RCHO + 2HCl$$

The hydrogen chloride produced can catalyze chlorination of the C—H bonds adjacent to the carbonyl group, so that the method is not in general suitable for the formation of simple aldehydes. However, a number of important compounds are available through the oxidation of ethanol with chlorine:

$$CH_3CH_2OH \xrightarrow{Cl_2} CH_3CHO \xrightarrow{Cl_2} ClCH_2CHO \xrightarrow{2\,Cl_2} Cl_3C-CHO$$

$$ClCH_2CHO \xrightarrow[H^+]{2\,C_2H_5OH} ClCH_2CH(OC_2H_5)_2 \begin{cases} \xrightarrow{NH_2^-/NH_3} HC{\equiv}C-OC_2H_5 \\ \\ \xrightarrow{NH_3} H_2NCH_2CH(OC_2H_5)_2 \end{cases}$$

$$Cl_3C-CHO \xrightarrow{OH^-} CHCl_3 + HCO_2^-$$

By passing chlorine into ethanol until the specific gravity reaches 1·025, monochloroacetaldehyde is obtained in good yield. Addition of more ethanol to the now acidic solution gives the corresponding acetal from which ethoxyacetylene (a reagent used in terpenoid synthesis; see pp. 711, 712) and the amino-acetal (for the Pomeranz-Fritsch synthesis of isoquinolines; see pp. 684, 730) can be obtained. Further chlorination of monochloroacetaldehyde gives chloral and thence chloroform.

(3) *Uses of dimethyl sulphoxide.* (*i*) The alcohol is oxidized with a mixture of dimethyl sulphoxide and dicyclohexyl carbodi-imide in acidic conditions. It is thought that an adduct of the sulphoxide and the carbodi-imide reacts with the alcohol to give an alkoxysulphonium ion which, *via* the corresponding ylid, yields the carbonyl compound:

$$RN=C=NR + (CH_3)_2\overset{+}{S}-O^- \xrightarrow{H^+} \begin{array}{c} RNH-C=NR \\ | \\ O \\ | \\ (CH_3)_2\overset{+}{S} \end{array} \xrightarrow[-(RNH)_2CO]{R'CH_2OH}$$

$$\begin{array}{c} CH_3 \\ \diagdown \\ \overset{+}{S}-O-CH_2R' \\ \diagup \\ CH_3 \end{array} \xrightarrow{-H^+} R'CHO + (CH_3)_2S$$

(*ii*) *Kornblum's method.* The alcohol toluene-*p*-sulphonate, usually derived from the alkyl iodide and silver toluene-*p*-sulphonate or alternatively from the alcohol and the sulphonyl chloride, is treated with dimethyl sulphoxide in the presence of sodium bicarbonate at 150°C for a few minutes. Yields are usually in the range 60–85%.

$$RCH_2I \xrightarrow{AgOTs} RCH_2OTs \xrightarrow{(CH_3)_2SO} RCHO$$

Reaction probably occurs by an initial S_N2-displacement followed by base-catalyzed elimination on the resulting sulphonium salt:

$$RCH_2{-}OTs \longrightarrow RCH{-}O{-}\overset{+}{S}(CH_3)_2 \longrightarrow RCHO + S(CH_3)_2$$
$$(CH_3)_2\overset{+}{S}{-}O^-$$

Understandably, therefore, benzylic derivatives, which are more reactive than alkyl derivatives towards nucleophiles, react at a lower temperature (100°C), and the very reactive α-bromo-ketones undergo oxidation directly, even though bromide ion is a less good leaving group than toluene-*p*-sulphonate, e.g.

$$Ph{-}CO{-}CH_2Br \xrightarrow{(CH_3)_2SO} Ph{-}CO{-}CHO$$
$$\text{Phenylglyoxal}$$

Amine oxides act similarly to dimethyl sulphoxide, and pyridine *N*-oxide in particular has been used to prepare aldehydes from halides and sulphonates:

$$RCH_2{-}Br \longrightarrow RCH{-}O{-}\overset{+}{N}C_5H_5 \xrightarrow{-H^+} RCHO + NC_5H_5$$

(4) *Catalytic dehydrogenation.* Dehydrogenation over copper or copper chromite occurs at about 300°C. Industrially, silver is now employed as the catalyst for the production of formaldehyde and acetaldehyde. The dehydrogenation step, e.g.

$$CH_3OH \xrightarrow{\text{Ag}} CH_2O + H_2$$

is endothermic ($\Delta H > +100$ kJ mol^{-1}), but in the presence of air a second type of oxidation occurs,

$$CH_3OH + \tfrac{1}{2}O_2 \xrightarrow{\text{Ag}} CH_2O + H_2O$$

This is strongly exothermic ($\Delta H = -150$ kJ mol^{-1}), so that by carefully controlling the air supply, a steady temperature is maintained and the reaction becomes self-supporting. Acetaldehyde is also manufactured from ethylene, as previously described (p. 567).

A recently developed laboratory method is the use of silver carbonate precipitated on Celite. Reaction occurs efficiently under boiling benzene, and other functional groups are unaffected.

(ii) *To carboxylic acids.* Primary alcohols can be oxidized directly to acids by reagents such as chromic(VI) acid, nitric acid, and potassium permanganate, but in each case side-reactions occur and yields are usually not high. Thus, chromic(VI) acid degrades carboxylic acids to smaller molecules, ultimately giving acetic acid (from *C*-methyl groups) and carbon dioxide (see Kuhn-Roth oxidation; p. 585).

A selective method is, however, available: namely, the use of molecular oxygen on a platinum catalyst. For example, pentaerythritol gives trihydroxymethylacetic acid in 50% yield in carefully controlled conditions (sodium bicarbonate buffer, 35°C):

$$C(CH_2OH)_4 \xrightarrow{O_2-Pt} (HOCH_2)_3C-CO_2H$$

An example of the selectivity of this system as between primary and secondary alcohols occurs in a synthesis of ascorbic acid (p. 568):

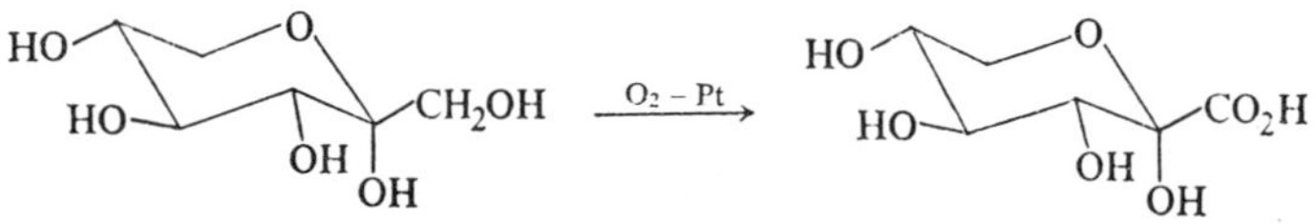

The reagent has the added advantage that double bonds are not attacked, but halides and amines are degraded.

(b) SECONDARY ALCOHOLS

Since ketones are only oxidized (with C—C bond-breakage) under vigorous conditions, the oxidation of secondary alcohols to ketones is not attended by the difficulty which applies to the oxidation of primary alcohols to aldehydes. Three methods are widely employed.

(1) *Chromic oxide.* It is usual to add a solution of chromic oxide in aqueous sulphuric acid from a burette into a cooled solution of the secondary alcohol in acetone. Addition is stopped when a permanent yellow colour, indicating a slight excess of Cr(VI), is obtained. In this way, over-oxidation is prevented and yields of over 90% can be obtained.

Alternative techniques are available. For example, the oxidant power is increased by using acetic acid as the solvent, and milder conditions can be obtained by using chromium(VI) oxide in pyridine (see also primary alcohols).

Since the rate-determining step in this reaction involves the breaking of the C—H bond (p. 587), the more exposed equatorial C—H reacts faster than axial C—H in rigid ring systems so that axial alcohols react faster than equatorial ones, e.g.

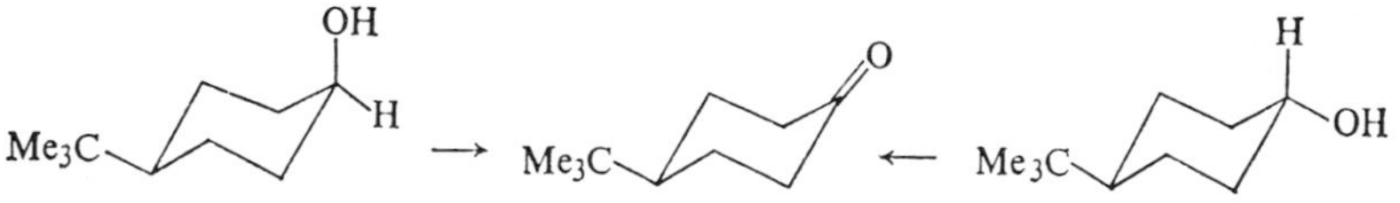

relative rates: 3.2   1

(2) *Oppenauer method.* The secondary alcohol and a ketone are equilibrated with the corresponding ketone and secondary alcohol by heating in the presence of aluminium t-butoxide. By using the added ketone, usually acetone, in large excess, the equilibrium is forced to the right. The reaction, which is the reverse of Meerwein-Ponndorf-Verley reduction (p. 635), involves the transfer of hydrogen within a cyclic complex:

$$3 \ R_2CHOH + Al(OBu^t)_3 \rightleftharpoons (R_2CHO)_3Al + 3 \ Bu^tOH$$

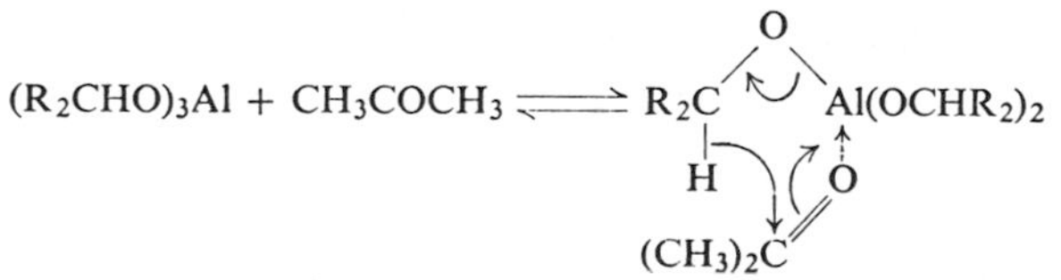

$$\rightleftharpoons R_2C{=}O + (CH_3)_2CH{-}O{-}Al(OCHR_2)_2$$

Benzene or toluene is often added as a co-solvent to raise the temperature of the reaction, or alternatively cyclohexanone is used instead of acetone.

The method is specific for alcohols and is therefore suitable, for example, for compounds containing olefinic bonds or phenolic groups, e.g.

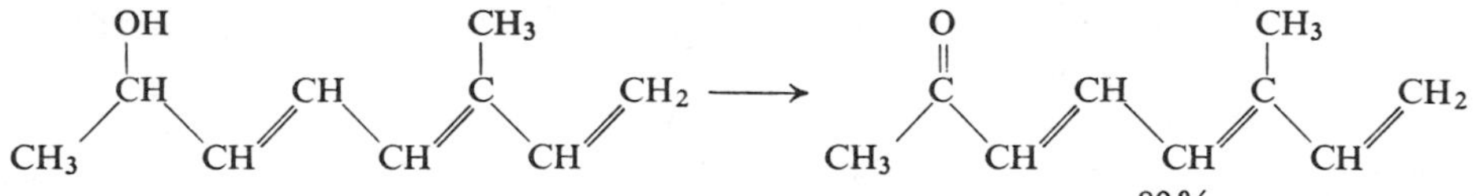

80%

One disadvantage of the method is that the aluminium compounds are basic and can bring about prototropic shifts within the product: e.g. the oxidation of cholesterol is accompanied by migration of the olefinic bond to give the $\alpha\beta$-unsaturated ketone:

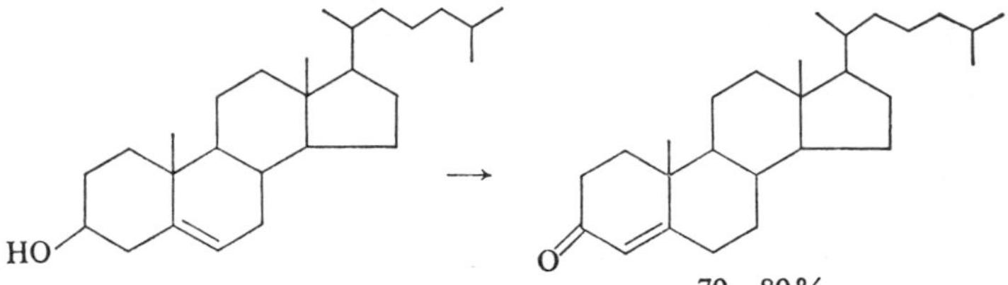

70 – 80%

Aldehydes are not satisfactorily prepared by Oppenauer oxidation because the basic medium induces condensations between the aldehyde and the ketone.

(3) *Catalytic dehydrogenation.* Like primary alcohols, secondary alcohols are dehydrogenated when passed over certain heated catalysts. For example, acetone is obtained industrially by the dehydrogenation of isopropanol over copper or zinc oxide at about 350°C. (Acetone is also produced by the cumene hydroperoxide process; p. 473).

(c) ALLYLIC ALCOHOLS

In addition to the methods described above, allylic alcohols are oxidized to $\alpha\beta$-unsaturated carbonyl compounds on the surface of manganese dioxide suspended in an inert solvent such as methylene chloride. The nature of the manganese dioxide affects the yields: it is best prepared as a non-stoicheiometric compound by reducing permanganate ion with manganese(II) ion in alkaline solution. Yields are then high, e.g.

$$PhCH{=}CH{-}CH_2OH \xrightarrow[\text{15°C}]{\text{MnO}_2/\text{CH}_2\text{Cl}_2} PhCH{=}CH{-}CHO$$
Cinnamaldehyde (76%)

The method is suitable also for benzylic alcohols but not for primary and secondary alcohols which are oxidized only very slowly.

The high-potential quinones (p. 586) are suitable for the oxidation of allylic, benzylic, and propargylic alcohols; reaction occurs through the relatively stable carbonium ions formed by the loss of hydride ion:

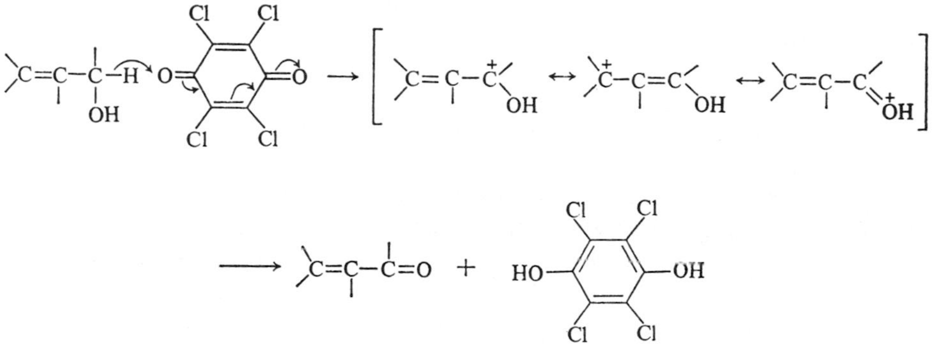

(d) BENZYLIC ALCOHOLS
Several specific processes have been developed.

(1) *Nitrogen dioxide* (*Field's method*). Primary and secondary benzyl alcohols react with dinitrogen tetroxide in chloroform at 0°C, e.g.

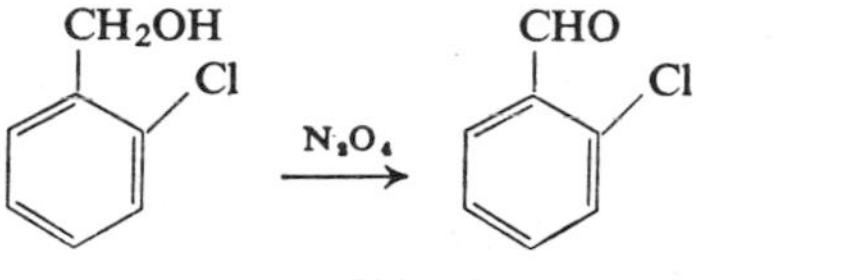

o-Chlorobenzaldehyde (96%)

Reaction apparently occurs through (the radical) nitrogen dioxide, with the formation and decomposition of a hydroxynitro compound:

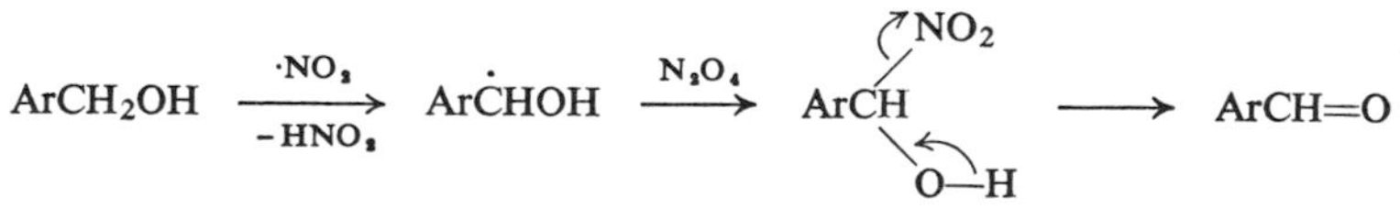

(2) *Hexamethylenetetramine* (*Sommelet reaction*). The halide from the benzyl alcohol is treated with hexamethylenetetramine (p. 330) and the resulting salt is hydrolyzed in the presence of more hexamethylenetetramine, usually with aqueous acetic acid, to the aldehyde, e.g.

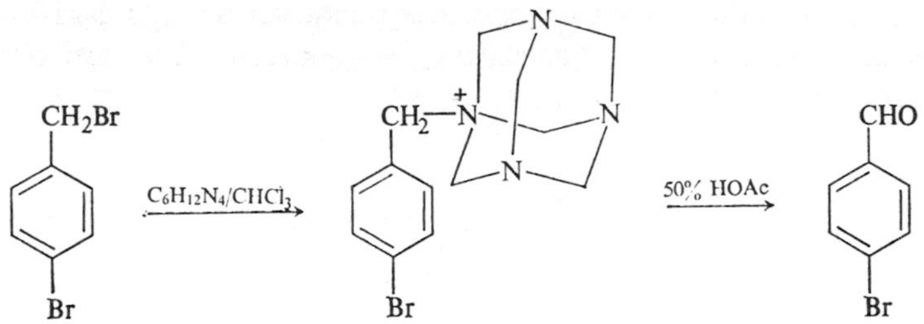

p-Bromobenzaldehyde (70%)

It is not necessary to isolate the intermediate salt.

Electron-withdrawing substituents decrease the yield and *ortho*-substituents hinder the reaction: e.g. neither 2,4-dinitro- nor 2,6-dimethylbenzaldehyde can be prepared in this way.

The reaction involves hydride-ion transfer. At the acidity employed, the quaternary benzyl salt is hydrolyzed to the benzylamine and hexamethylenetetramine itself is hydrolyzed to ammonia and formaldehyde (the reverse of its formation; p. 330). The benzylamine transfers hydride ion to methyleneimine (from formaldehyde and ammonia), giving an imine which is hydrolyzed to the aromatic aldehyde:

$$ArCH_2-(C_6H_{12}N_4)^+ \; Br^- \xrightarrow{H_2O} ArCH_2NH_2$$

$$C_6H_{12}N_4 \xrightarrow{H_2O-H^+} CH_2O + NH_3 \underset{-H_2O}{\rightleftharpoons} CH_2=NH$$

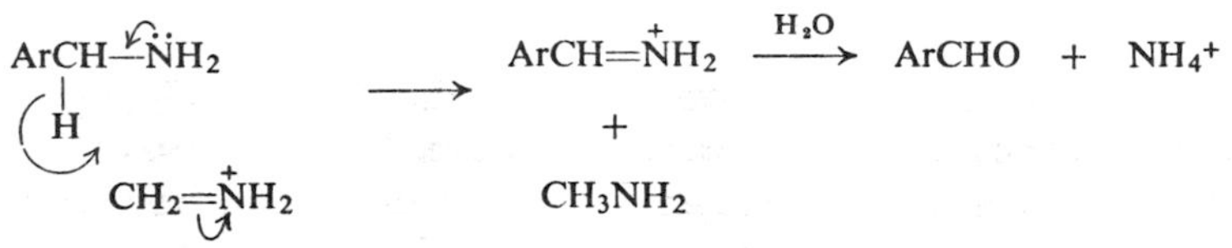

(3) *Duff reaction.* Phenols react with hexamethylenetetramine, usually in acetic acid, to give *ortho*-aldehydes. Reaction occurs by aminomethylation followed by Sommelet reaction:

$$C_6H_{12}N_4 \xrightarrow{H_2O-H^+} CH_2=O + NH_3 \underset{-H_2O}{\rightleftharpoons} CH_2=NH$$

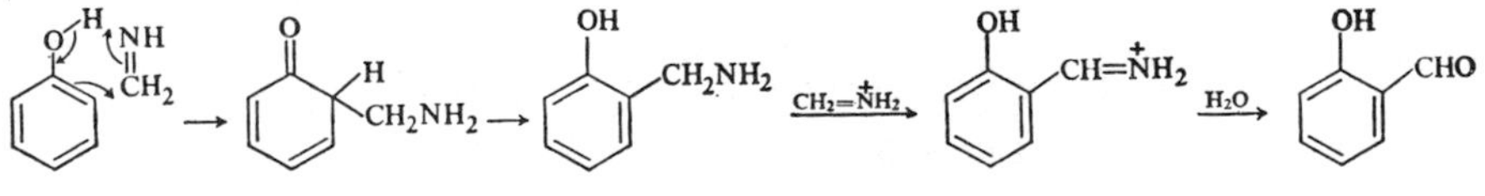

(4) *Kröhnke reaction*. A benzyl halide is converted into its pyridinium salt and thence, with *p*-nitrosodimethylaniline, into a nitrone. Acid hydrolysis gives the aromatic aldehyde.

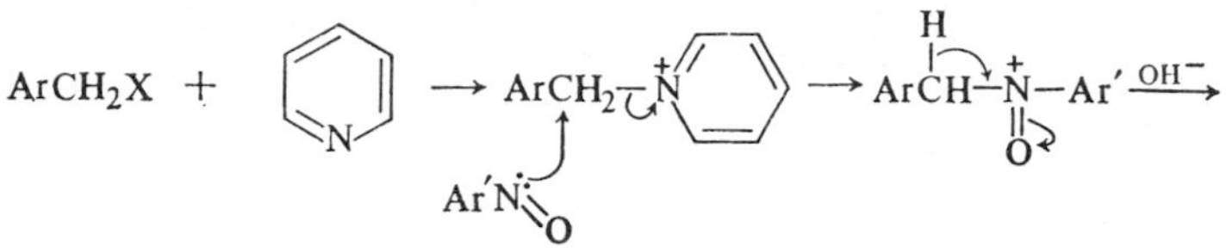

$$\text{ArCH}{=}\overset{+}{\text{N}}{-}\text{Ar}' \quad \xrightarrow{\text{H}_2\text{O-H}^+} \quad \text{ArCHO}$$
$$\underset{\text{O}^-}{|}$$

Reaction occurs under mild conditions so that it is suitable for the preparation of sensitive aldehydes. In addition, it is facilitated by electron-attracting substituents: e.g. 2,4-dinitrobenzaldehyde can be prepared in this way (*cf.* the Sommelet reaction).

(e) 1,2-DIOLS

1,2-Diols (glycols) are cleaved by lead tetra-acetate, phenyliodoso acetate, and periodic acid or sodium metaperiodate:

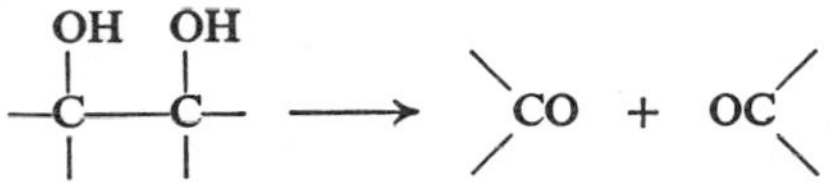

The first two reagents are used in an organic medium, commonly glacial acetic acid, whereas periodate is used in aqueous solution. Phenyliodoso acetate is the least powerful and is not often employed, but all three are entirely specific: the aldehydes formed are not oxidized further. Yields are usually excellent: e.g. di-n-butyl tartrate and lead tetra-acetate give n-butyl glyoxylate in up to 87% yield [4],

$$\text{BuO}_2\text{C}{-}\underset{\underset{\text{OH}}{|}}{\text{CH}}{-}\underset{\underset{\text{OH}}{|}}{\text{CH}}{-}\text{CO}_2\text{Bu} \quad \xrightarrow{\text{Pb(OAc)}_4} \quad 2\,\text{BuO}_2\text{C}{-}\text{CHO}$$

and butane-2,3-diol and sodium periodate give acetaldehyde quantitatively,

$$\text{CH}_3{-}\underset{\underset{\text{OH}}{|}}{\text{CH}}{-}\underset{\underset{\text{OH}}{|}}{\text{CH}}{-}\text{CH}_3 \quad \xrightarrow{\text{NaIO}_4} \quad 2\,\text{CH}_3{-}\text{CHO}$$

Reaction normally occurs by two-electron oxidation [Pb(IV) to Pb(II) and I(VII) to I(V)] within a cyclic intermediate:

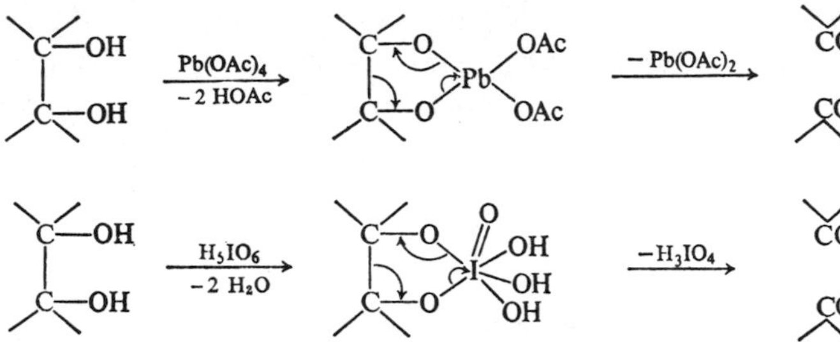

so that *cis*-diols are oxidized faster than *trans*-diols; e.g. *cis*-cyclohexane-1,2-diol reacts about 25 times faster than its *trans* isomer. However, an acyclic path is also followed, e.g.

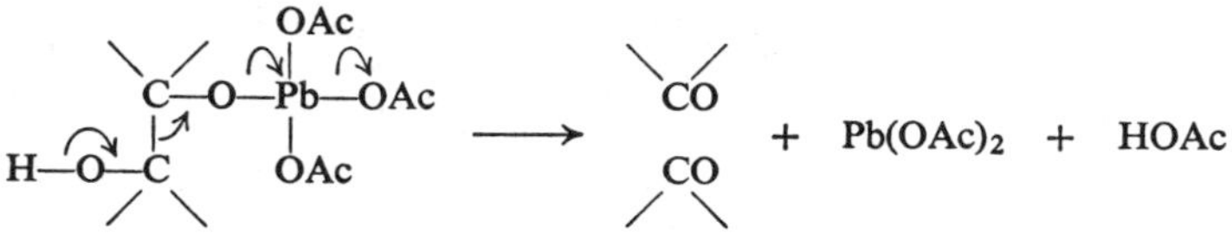

In addition to diols, α-amino-alcohols, α-ketols, and α-dicarbonyl compounds are cleaved:

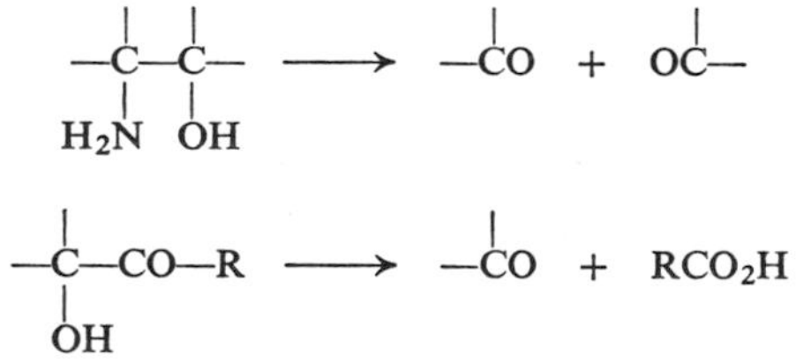

The grouping —CH(OH)—CH(OH)—CH(OH)— is oxidized to a mixture of aldehydes and formic acid,

$$R-CH(OH)-CH(OH)-CH(OH)-R' \xrightarrow{H_5IO_6} R-CH(OH)-CHO + OCH-R'$$

$$R-CH(OH)-CHO \xrightarrow{H_5IO_6} R-CHO + HCO_2H$$

Periodate has proved a valuable degradative agent in carbohydrate chemistry, for it reacts nearly quantitatively. Since the grouping —CH(OH)—CH₂OH is oxidized to formaldehyde and the grouping —CH(OH)—CH(OH)—CH(OH)— gives one mole of formic acid, the estimation of formaldehyde with dimedone

(p. 254) and of formic acid with standard base provides information about the occurrence of these groups, and the total periodate consumed indicates the total number of 1,2-diol groups present.

An interesting example of the synthetic use of periodate, in the oxidation of the system —CO—CH(OH)—CH(OH)—, occurs in the synthesis of reserpine (21.7). Other uses are in the Lemieux reagents (p. 575) and the Grundmann synthesis (p. 643).

Lead tetra-acetate also effects the bisdecarboxylation of succinic acids:

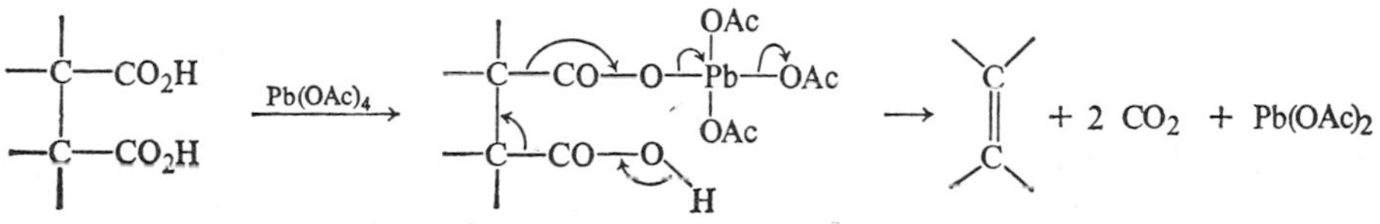

(f) ALDEHYDES

Aldehydes can be oxidized to acids by the vigorous reagents such as chromic(VI) acid and permanganate. This can be satisfactory for compounds which do not possess sensitive groups: e.g. n-heptaldehyde gives n-heptanoic acid in 76–78% yield with permanganate in sulphuric acid at 20°C [2],

$$\text{n-C}_6\text{H}_{13}\text{CHO} \xrightarrow{\text{KMnO}_4-\text{H}_2\text{SO}_4} \text{n-C}_6\text{H}_{13}\text{CO}_2\text{H}$$

n-Heptanoic acid

In most cases, however, milder routes are necessary. One method employs silver oxide: e.g. thiophen-3-aldehyde gives the 3-carboxylic acid nearly quantitatively in 5 minutes at 0°C [4]:

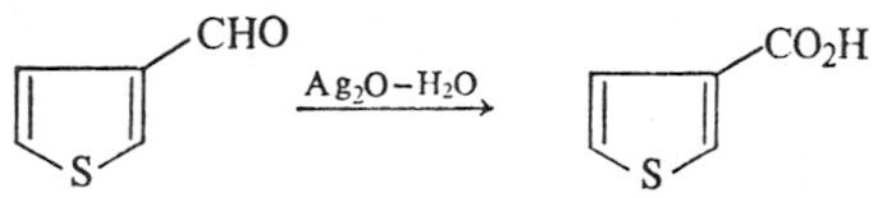

Thiophen-3-carboxylic acid

In general, it is usually more convenient to synthesize acids by routes which do not involve the aldehyde, such as by the carbonation of organometallic compounds or from malonic ester.

Those aldehydes which do not possess α-hydrogen atoms undergo the Cannizzaro reaction with base (p. 634); e.g. benzaldehyde gives benzyl alcohol and benzoate ion. Finally, *ortho* and *para* hydroxybenzaldehydes undergo oxidative rearrangement with alkaline hydrogen peroxide (Dakin reaction; p. 473).

(g) KETONES

The C—CO bond in ketones can be oxidized in three ways.

(1) *By nitric acid or alkaline potassium permanganate.* These powerful conditions give carboxylic acids, reaction occurring through the enol (acid solution) or the enolate anion (basic solution), e.g.

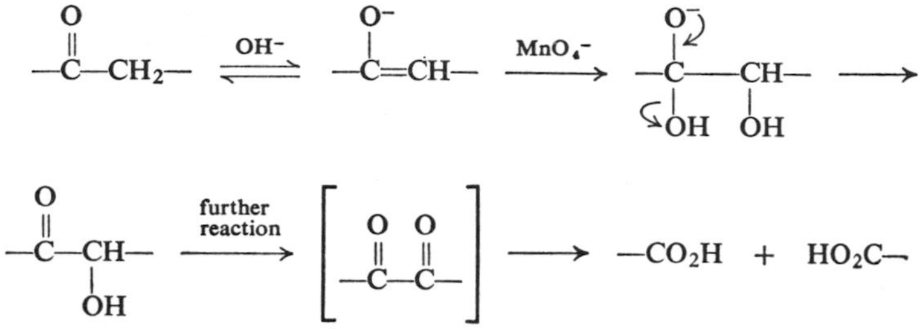

Attack can occur on both sides of the carbonyl group so that, unless the ketone is cyclic, a mixture of products is obtained. With cyclic ketones, however, reasonable yields can be achieved: e.g. cyclohexanol is oxidized by hot 50% nitric acid, *via* cyclohexanone, to give adipic acid in 60% yield [1]:

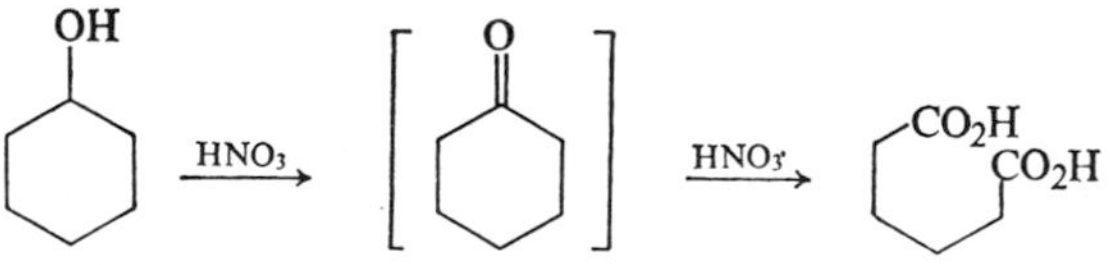

(2) *By the halogens in alkali.* Methyl ketones are oxidized by chlorine, bromine, or iodine in alkaline solution to give acids and the corresponding haloform. Reaction occurs by base-catalyzed halogenation followed by elimination of the conjugate base of the haloform:

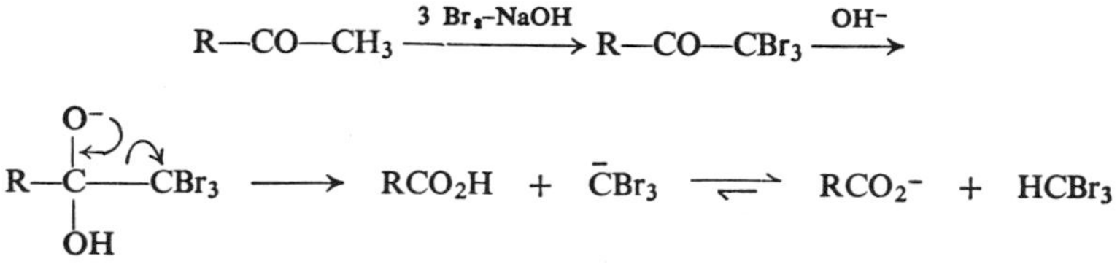

$$R—CO—CH_3 \xrightarrow{\text{3 Br}_2\text{-NaOH}} R—CO—CBr_3 \xrightarrow{OH^-}$$

This provides a particularly useful method for the synthesis of aromatic acids, for the corresponding methyl ketones are often readily available through the Friedel-Crafts reaction (p. 389). For example, the acetylation of naphthalene in nitrobenzene solution gives β-acetylnaphthalene (p. 392) from which β-naphthoic acid can be obtained in 97% yield with chlorine in sodium hydroxide solution at 55°C:

$$\text{(naphthalene)}\text{--COCH}_3 \xrightarrow{\text{Cl}_2\text{-OH}^-} \text{(naphthalene)}\text{--CO}_2\text{H}$$

There are also applications in the aliphatic series. For example, pinacolone, available from acetone by reduction and rearrangement (p. 459), reacts with bromine in sodium hydroxide solution at below 10°C to give trimethylacetic (pivalic) acid in over 70% yield [1]:

$$(CH_3)_3C{-}CO{-}CH_3 \xrightarrow{\text{Br}_2\text{-OH}^-} (CH_3)_3C{-}CO_2H$$
$$\text{Pivalic acid}$$

(3) *By peracids.* Ketones undergo oxidative rearrangement with peracids to give esters or lactones, $RCOR \rightarrow RCO_2R$ (Baeyer-Villiger reaction; p. 471).

(*h*) α-KETOLS

These systems are very readily oxidized to α-dicarbonyl compounds. One-electron oxidants in basic solution are effective, for the carbanion formed by base can donate one electron to the oxidant to give a delocalized radical (*cf.* semiquinone radicals; p. 579); loss of a second electron completes the oxidation.

$$-\overset{|}{\underset{OH}{C}}H{-}\overset{||}{\underset{O}{C}}{-} \xrightleftharpoons{\text{Base}} -\overset{|}{\underset{O^-}{C}}{-}\overset{||}{\underset{O}{C}}{-} \leftrightarrow -\overset{|}{\underset{O^-}{C}}{=}\overset{|}{\underset{O^-}{C}}{-} \xrightarrow{-e}$$

$$-\overset{|}{\underset{O\cdot}{C}}{=}\overset{|}{\underset{O^-}{C}}{-} \leftrightarrow -\overset{|}{\underset{O^-}{C}}{=}\overset{|}{\underset{O\cdot}{C}}{-} \xrightarrow{-e} -CO{-}CO{-}$$

For example, copper sulphate in pyridine at 95°C oxidizes benzoin to benzil in 86% yield [1]:

$$\text{PhCH}{-}\text{COPh} \xrightarrow{\text{CuSO}_4} \text{PhCO}{-}\text{COPh}$$
$$\underset{OH}{|} \qquad\qquad\qquad \text{Benzil}$$

(*i*) α-DICARBONYL COMPOUNDS

Oxidation can be brought about in three ways: peracids in inert solvents give anhydrides, $RCO{-}COR \rightarrow RCO{-}O{-}COR$ (Baeyer-Villiger reaction; p. 471); and both warm hydrogen peroxide in acetic acid and the glycol-cleavage reagents give carboxylic acids, $RCO{-}COR \rightarrow 2\ RCO_2H$.

(*j*) ACIDS: OXIDATIVE DECARBOXYLATION

Carboxylic acids undergo oxidative decarboxylation when heated with lead tetra-acetate in the presence of a catalytic amount of a copper(II) salt:

$$\underset{\displaystyle |}{\overset{\displaystyle |}{>}}CH-\underset{\displaystyle |}{\overset{\displaystyle |}{C}}-CO_2H \xrightarrow[\text{(Cu}^{2+})]{\text{Pb(OAc)}_4} \; >C=C< \; + \; CO_2 \; + \; Pb(OAc)_2 \; + \; 2 \; HOAc$$

Reaction is thought to occur by homolysis of the lead carboxylate, followed by oxidation of the resulting radical by copper(II), e.g.

$$RCH_2CH_2CO-O-Pb(OAc)_3 \longrightarrow RCH_2CH_2CO-O\cdot + \cdot Pb(OAc)_3$$

$$RCH_2CH_2CO-O\cdot \xrightarrow{-CO_2} RCH_2CH_2\cdot$$

$$RCH_2CH_2\cdot + Cu^{2+} \longrightarrow RCH=CH_2 + H^+ + Cu^+$$

The copper(I) ion is then oxidized by lead(IV) to regenerate copper(II).
  vic-Dibasic acids are *bis*-decarboxylated by lead tetra-acetate, e.g.

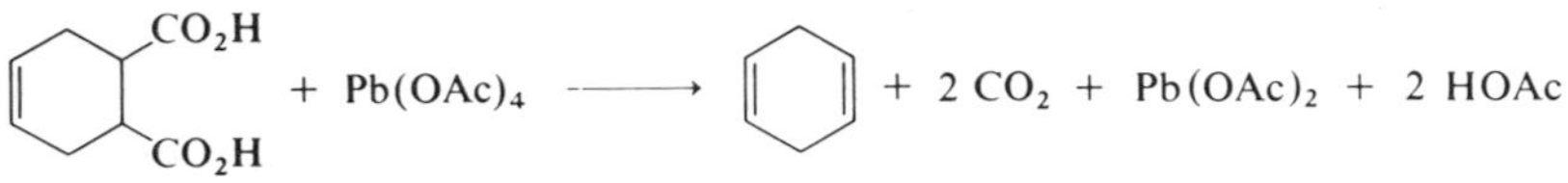

18.4 Systems Containing Nitrogen

(*a*) PRIMARY AMINES
These are very sensitive to oxidation and generally darken on exposure to air, through autoxidation at the surface, to give mixtures of complex products. This is particularly true of aromatic amines: a number of products appear to arise by oxidation to nitroso and nitro compounds followed by condensations; e.g. $ArNH_2 \rightarrow ArNO$; $ArNH_2 + ArNO \rightarrow ArN=NAr$. Repeated condensations give aniline blacks.
  Synthetically useful methods have therefore to be highly selective. The most successful reagents are hydrogen peroxide and peracids.
  (1) *Hydrogen peroxide* converts primary aliphatic amines into aldoximes, e.g.

$$n\text{-}C_3H_7CH_2NH_2 \xrightarrow{H_2O_2} n\text{-}C_3H_7CH=NOH$$

$$\text{n-Butyraldoxime (57\%)}$$

presumably by nucleophilic displacement by the amine on the peroxide:

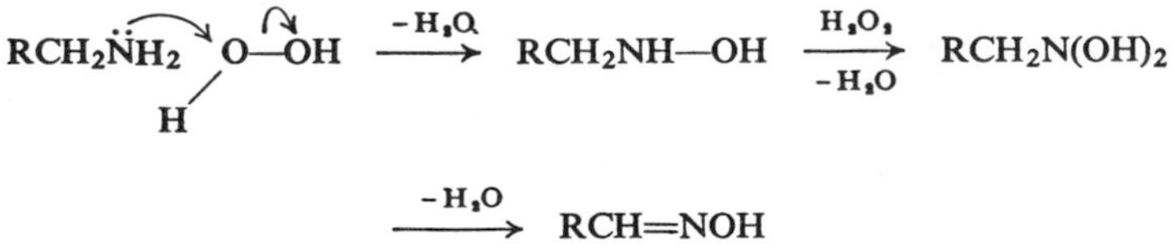

Oxidation of aromatic amines to the corresponding level (nitroso compounds) is generally brought about with perdisulphuric acid, $HO_3S-O-O-SO_3H$ or Caro's acid. For example, the former oxidant converts o-nitroaniline into o-nitrosonitrobenzene in 75% yield:

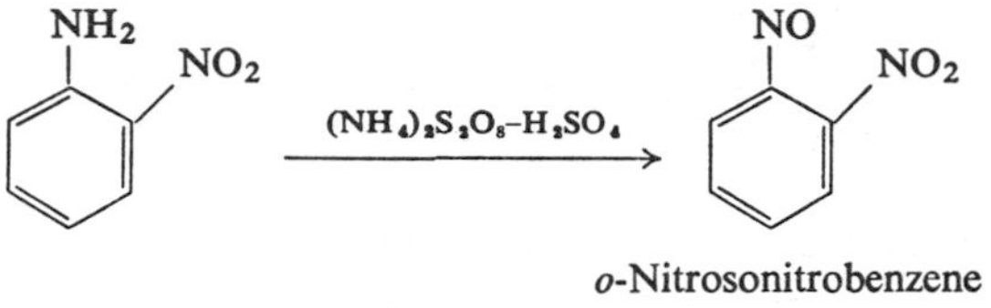

$(NH_4)_2S_2O_8-H_2SO_4$

o-Nitrosonitrobenzene

Aromatic nitroso compounds can also be prepared from the corresponding hydroxylamines by oxidation with dichromate at low temperatures (to prevent further oxidation). For example, phenylhydroxylamine, sodium dichromate, and 50% sulphuric acid at 0°C give nitrosobenzene in about 50% yield [3]:

$$PhNHOH \xrightarrow{\ Na_2Cr_2O_7-H_2SO_4\ } PhNO$$

Since arylhydroxylamines are readily obtained by the reduction of nitro compounds in neutral solution (p. 647), this method is convenient for introducing the nitroso group via nitration.

(2) *Trifluoroperacetic acid*, a more powerful oxidant than hydrogen peroxide, converts primary amines directly into nitro compounds. The yields with aromatic amines are generally high; e.g. o-nitroaniline is oxidized in refluxing methylene chloride to o-dinitrobenzene in 92% yield:

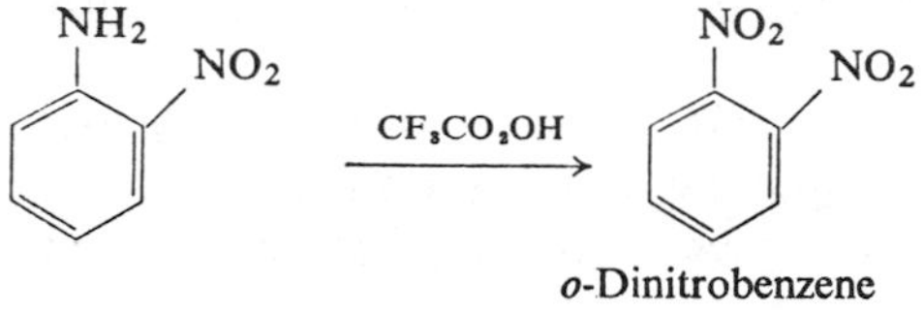

CF_3CO_3OH

o-Dinitrobenzene

With aliphatic amines, however, the yields are low. Other oxidants have been used with moderate success. For example, n-hexylamine is oxidized in 33% yield to the nitro compound by peracetic acid.

Aromatic nitro compounds can also be obtained by the oxidation of nitroso compounds with dichromate or dilute nitric acid, but it is normally more convenient to oxidize the amine.

(b) SECONDARY AMINES

Oxidation with hydrogen peroxide gives hydroxylamines (cf. primary amines):

$$R_2NH \xrightarrow{\ H_2O_2\ } R_2N-OH + H_2O$$

(c) TERTIARY AMINES

Hydrogen peroxide converts tertiary amines into their N-oxide hydrates, by nucleophilic displacement analogous to the reaction of primary amines. The N-oxide is obtained by warming the hydrate *in vacuo*.

$$R_3N \xrightarrow{H_2O_2} R_3\overset{+}{N}-OH \ OH^- \xrightarrow[-H_2O]{warm} R_3\overset{+}{N}-\overset{-}{O}$$

Aromatic amines behave similarly; *e.g.*, pyridine gives pyridine N-oxide.

(d) HYDRAZINES

Hydrazine itself is readily oxidized to nitrogen. Some at least of the oxidants, e.g. copper(II) ion, yield di-imide, $HN=NH$. as an intermediate, and this is employed for the *cis* reduction of olefins (p. 615).

Monosubstituted hydrazines also react with one-electron oxidants such as copper(II) and iron(III) ion to give unstable azo compounds which decompose with loss of nitrogen to hydrocarbons, e.g.

$$PhNH-NH_2 \xrightarrow{Cu^{2+}} [PhN=NH] \longrightarrow PhH + N_2$$

Arylhydrazines are oxidized differently by the two-electron oxidants, chlorine and bromine, giving diazonium salts, e.g.

$$PhNHNH_2 + 2Cl_2 \longrightarrow PhN_2^+ \ Cl^- + 3HCl$$

$$PhNHNH_2 + 3Br_2 \longrightarrow PhN_2^+ \ Br_3^- + 3HBr$$

N,N'-Disubstituted hydrazines give azo compounds readily. For example, azobisisobutyronitrile (a useful reagent for initiating free-radical reactions) can be obtained from acetone, hydrazine, and cyanide (compare the Strecker synthesis of α-amino-acids; p. 352) followed by oxidation with, e.g. mercury(II) oxide:

$$2 \ (CH_3)_2CO + NH_2NH_2 + 2 \ HCN \longrightarrow \underset{\underset{CN}{|}}{(CH_3)_2C}-NH-NH-\underset{\underset{CN}{|}}{C(CH_3)_2}$$

$$\xrightarrow{HgO} \underset{\underset{CN}{|}}{(CH_3)_2C}-N=N-\underset{\underset{CN}{|}}{C(CH_3)_2}$$

Azobisisobutyronitrile

Diarylhydrazines oxidize with exceptional ease since a conjugated system is formed. For example, hydrazobenzene gives azobenzene even on standing in air for some time.

Azo compounds may be further oxidized to azoxy compounds in more vigorous conditions, e.g.

$$PhN{=}NPh \xrightarrow{\text{H}_2\text{O}_2/\text{HOAc}} Ph\overset{+}{N}{=}NPh$$
$$\underset{\quad O^-}{|}$$

Azoxybenzene

(e) HYDRAZONES

Hydrazones are oxidized to diazoalkanes by mercury(II) oxide. Those containing aryl-substituents give moderately stable (conjugated) products which can be isolated: e.g. benzophenone hydrazone gives diphenyldiazomethane nearly quantitatively [3]:

$$Ph_2C{=}N{-}NH_2 + HgO \longrightarrow Ph_2C{=}\overset{+}{N}{=}\overset{-}{N} + Hg + H_2O$$

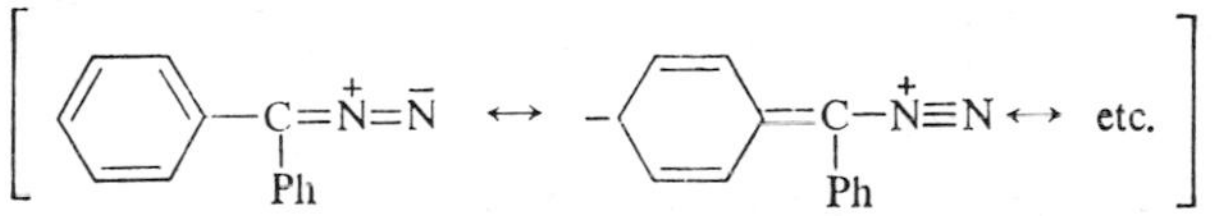

Diazoalkanes containing only saturated groups decompose rapidly to nitrogen and products derived from methylenes ($R_2CN_2 \to R_2C{:} + N_2$; p. 181). Use is made of this in a synthetic procedure for acetylenes: the bishydrazone of an α-diketone is oxidized with mercury(II) oxide and the unstable bisdiazo compound decomposes to the acetylene:

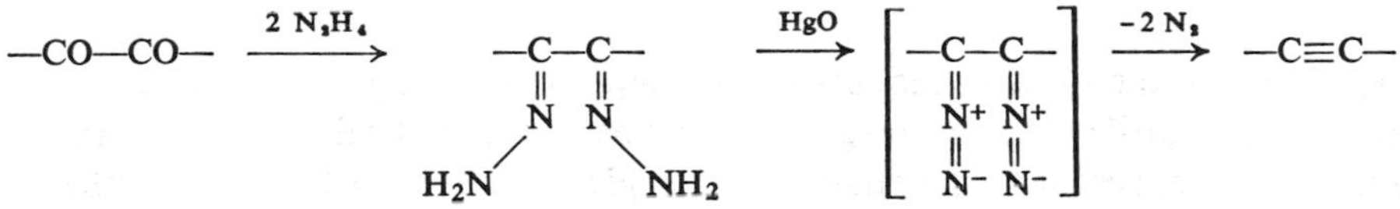

For example, benzil bishydrazone gives diphenylacetylene in about 70% yield [4] and the bishydrazone from cyclo-octane-1,2-dione gives cyclo-octyne in 9% yield, the low yield doubtless resulting from the extreme ring-strain in the product (which is the smallest cyclo-alkyne to have been made).

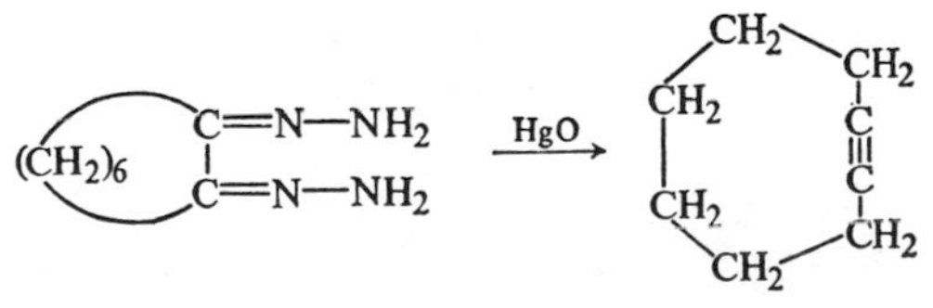

Cyclo-octyne

The monohydrazones of α-diketones give ketens in these conditions: e.g. the treatment of benzil monohydrazone with mercury(II) oxide gives a diazo compound which rearranges on distillation, with loss of nitrogen, to diphenylketen (58%) [3]:

$$
\begin{array}{ccc}
\underset{\substack{\parallel \\ \text{N} \\ \diagdown \\ \text{NH}_2}}{\text{PhC—COPh}} & \xrightarrow{\text{HgO}} & \underset{\substack{\parallel \\ \text{N}^+ \\ \parallel \\ \text{N}^-}}{\text{PhC—COPh}}
\end{array}
\xrightarrow{\text{distil}}
\left[\underset{\substack{.. \\ \curvearrowleft}}{\text{PhC}\!—\!\overset{\text{Ph}}{\underset{|}{\text{C}}}\!=\!\text{O}} \right]
\longrightarrow \underset{\text{Diphenylketen}}{\text{Ph}_2\text{C}\!=\!\text{C}\!=\!\text{O}}
$$

18.5 Systems Containing Sulphur

(a) THIOLS (MERCAPTANS)

Whereas hydroxyl-containing compounds are oxidized at carbon, thiol-containing compounds are oxidized at sulphur, largely because of the comparatively low bond-energy of S—H compared with O—H (p. 65).

(1) *Oxidation to disulphides.* A variety of relatively weak oxidants oxidize thiols to disulphides, such as hydrogen peroxide, iron(III) ion, and iodine:

$$2\ \text{RSH} \xrightarrow{-\text{H}_2} \text{RS—SR}$$

Thus, cysteine is readily oxidized to cystine, and the disulphide ring in thioctic acid is readily formed from the dithiol precursor (p. 624).

(2) *Oxidation to sulphenyl chlorides.* Thiols are oxidized by chlorine, through disulphides, to sulphenyl chlorides, e.g.

$$2\ \text{n-C}_5\text{H}_{11}\text{SH} \xrightarrow{\text{Cl}_2} [\text{n-C}_5\text{H}_{11}\text{S—SC}_5\text{H}_{11}\text{-n}] \xrightarrow{\text{Cl}_2} 2\ \text{n-C}_5\text{H}_{11}\text{SCl}$$

from which derivatives of sulphenic acids may be obtained, e.g.

$$\text{RSCl} + \text{R}'\text{OH} \longrightarrow \underset{\substack{\\ \text{a sulphenic ester}}}{\text{RS—OR}'}$$

$$\text{RSCl} + \text{NH}_3 \longrightarrow \underset{\substack{\\ \text{a sulphenamide}}}{\text{RS—NH}_2}$$

(3) *Oxidation to sulphonic acids.* Vigorous reagents such as nitric acid and permanganate give sulphonic acids, probably through sulphenic and sulphinic acids, which are too easily oxidized to be isolated:

$$
\text{RSH} \longrightarrow \left[\text{RS—OH} \longrightarrow \text{RS}\!\!\underset{\text{OH}}{\overset{\text{O}}{\diagup}} \right] \longrightarrow \underset{\substack{\parallel \\ \text{O}}}{\overset{\substack{\text{O} \\ \parallel}}{\text{RS—OH}}}
$$

Sulphonic acids are also formed by the treatment of lead mercaptides with nitric acid:

$$2\ RSH \xrightarrow{\ Pb(NO_3)_2\ } Pb(SR)_2 \xrightarrow{\ HNO_3\ } 2\ RSO_2OH$$

(b) SULPHIDES

Sulphides can be oxidized to both sulphoxides and sulphones. The former are obtained in high yield (*ca.* 90%) with liquid dinitrogen tetroxide in ethanol cooled with solid carbon dioxide. Further oxidation to the sulphone does not occur, and the product is anhydrous (wet sulphoxides are very difficult to dry).

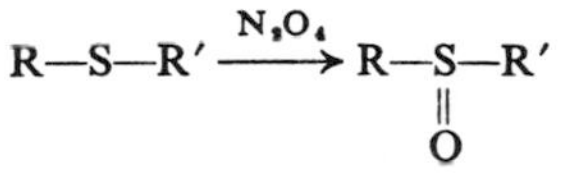

$$R-S-R' \xrightarrow{\ N_2O_4\ } R\overset{\displaystyle}{\underset{\displaystyle \underset{O}{\parallel}}{-}}S{-}R'$$

The oxidation can also be carried out with a slight excess of sodium meta-periodate at 0°C and with hydrogen peroxide, but these methods cause some further oxidation to the sulphone (see below).

Very sensitive sulphides require more delicate treatment. For example, manganese dioxide is the only reagent so far found which oxidizes diallyl sulphide to diallyl sulphoxide in reasonable yield.

Sulphones are normally obtained from sulphides by oxidation with hydrogen peroxide in aqueous or acetic acid solution:

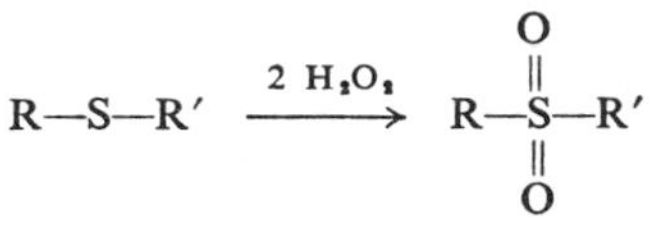

$$R-S-R' \xrightarrow{\ 2\ H_2O_2\ } R\overset{\displaystyle \overset{O}{\parallel}}{\underset{\displaystyle \underset{O}{\parallel}}{-}}S{-}R'$$

18.6 Systems Containing Phosphorus

The characteristics of phosphorus chemistry compared with that of nitrogen are that three-valent phosphorus is readily oxidized to the five-valent state and that P—O bonds are more stable than N—O bonds. Thus, oxidations at phosphorus occur under mild conditions, e.g.

$$R_3P \xrightarrow{\ air\ } R_3P{=}O$$

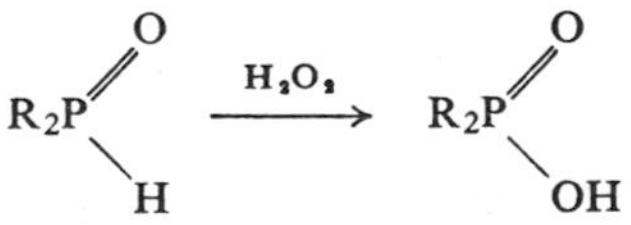

$$R_2P\overset{\displaystyle O}{\underset{\displaystyle H}{\big\langle}} \xrightarrow{\ H_2O_2\ } R_2P\overset{\displaystyle O}{\underset{\displaystyle OH}{\big\langle}}$$

The affinity of phosphorus for combined oxygen is employed in the Wittig reaction (p. 485).

18.7 Systems Containing Iodine

The iodine atom in aryl iodides can be oxidized to both the three- and the five-valent states. For example, iodobenzene reacts with chlorine in dry chloroform to give iodobenzene dichloride in about 90% yield [3]; this may be hydrolyzed to iodosobenzene in about 60% yield [3]; and iodoxybenzene may be obtained in over 90% yield by steam-distilling iodosobenzene to remove the iodobenzene formed by the disproportionation:

$$PhI + Cl_2 \longrightarrow PhICl_2$$
$$\text{Iodobenzene}$$
$$\text{dichloride}$$

$$PhICl_2 + 2\ NaOH \longrightarrow PhIO + 2\ NaCl + H_2O$$
$$\text{Iodosobenzene}$$

$$2\ PhIO \xrightarrow[-\,PhI]{\text{steam distil}} PhIO_2$$
$$\text{Iodoxybenzene}$$

The aliphatic analogues of these higher-valent iodine compounds are unstable.

Further Reading

DJERASSI, C., 'The Oppenauer oxidation,' *Organic Reactions*, 1951, 6, 207.
JACKSON, E. L., 'Periodic acid oxidation,' *Organic Reactions*, 1944, 2, 341.
RABJOHN, N., 'Selenium dioxide oxidation,' *Organic Reactions*, 1949, 5, 331.
STEWART, R., *Oxidation Mechanisms*, W. A. Benjamin (New York 1964).
SWERN, D., 'Epoxidation and hydroxylation of ethylenic compounds with organic peracids,' *Organic Reactions*, 1953, 7, 378.

Problems

1. How would you carry out the following transformations?

   (i) R—CH=CH$_2$ into (a) R—CHO, (b) R—CH$_2$OH, (c) R—CH$_2$—CHO, (d) R—CH(OH)—CHO, (e) R—CH(OH)—CH$_3$.

   (ii) R—CH=CH—CH$_2$OH into (a) R—CH(OH)—CH(OH)—CH$_2$OH, (b) R—CH=CH—CHO.

   (iii) R—CH$_2$—CO—CH$_3$ into (a) R—CH$_2$—CO—CHO, (b) R—CO—CO—CH$_3$.

   (iv) PhCHO into (a) PhCH(OH)—COPh, (b) PhCO—COPh.

   (v) PhCO—COPh into (a) Ph$_2$C=C=O, (b) PhC≡CPh.

(*vi*) PhCH$_2$Br into PhCHO.

(*vii*) PhCOCH$_2$Br into PhCOCHO.

(*viii*) PhNH$_2$ into (*a*) *p*-benzoquinone, (*b*) azoxybenzene, (*c*) *o*-dinitrobenzene.

(*ix*) PhOH into (*a*) *o*-benzoquinone, (*b*) chloranil (tetrachloro-*p*-benzoquinone).

(*x*) PhSH into (*a*) PhS—SPh, (*b*) PhSCl, (*c*) PhSO$_2$OH.

(*xi*) *trans*-2-Butene into (*a*) ($\pm$)-butane-2,3-diol, (*b*) *meso*-butane-2,3-diol.

(*xii*)

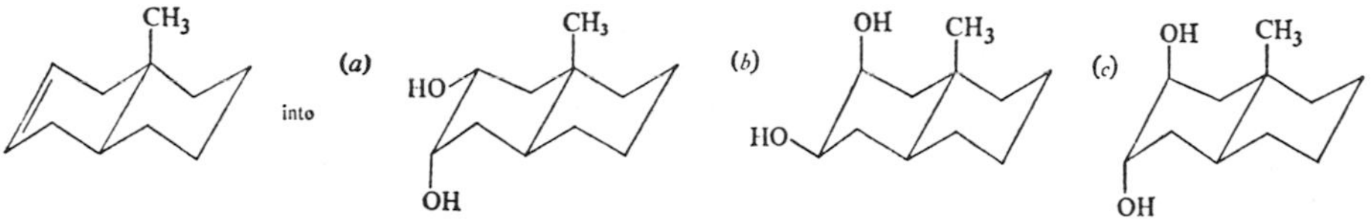

2. Summarize the reagents which may be used to oxidize the methyl group in a compound X—CH$_3$ according to the nature of the grouping X (i.e. X = acetyl, phenyl, etc.).

3. A methylated derivative of D-glucose is thought to have the structure (I). How could an oxidative method be used to provide evidence that the ring is six-membered?

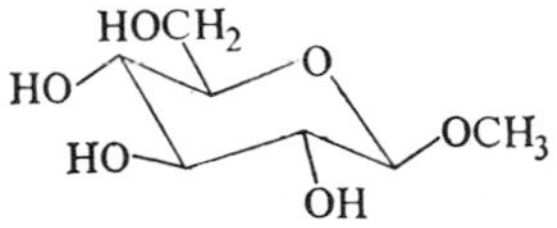

(I)

4. Rationalize the following reactions:

(*i*)

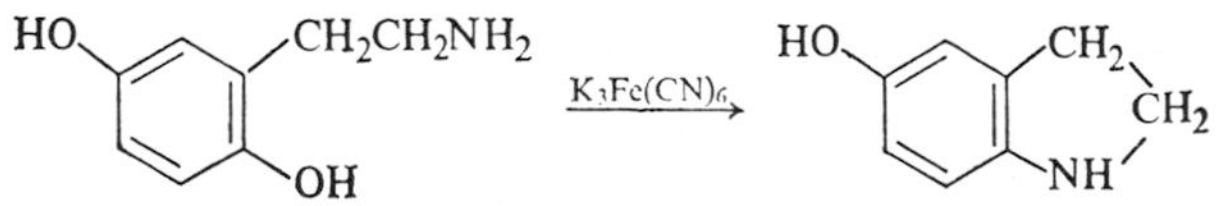

(*ii*)

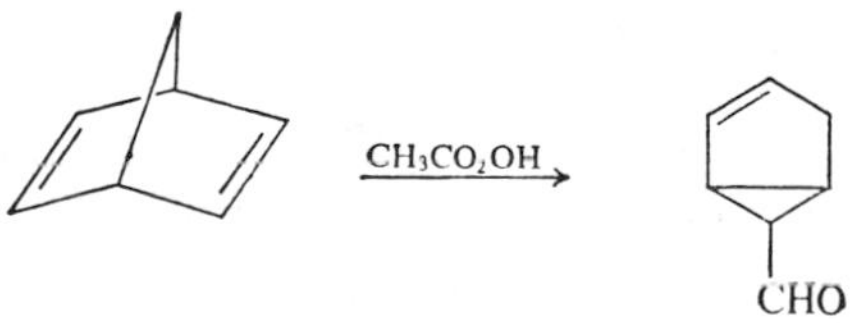

5. Griseofulvin (a fungal metabolite which is an important antibiotic) has been
    made by the following sequence:

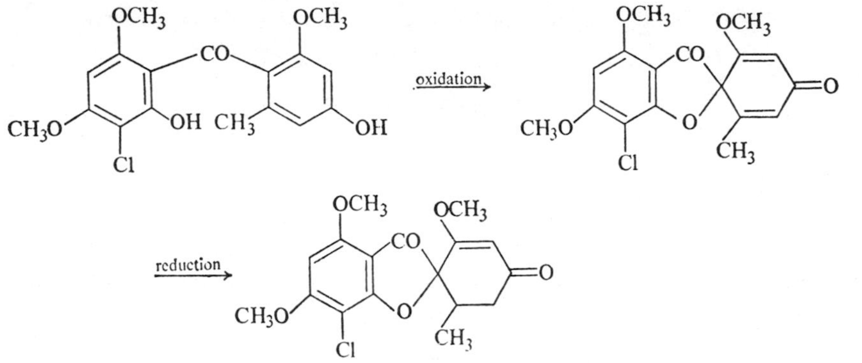

How would you attempt to effect the oxidative step?

19. Reduction

19.1 Introduction

The reductive processes described in this chapter fall into three categories: the removal of oxygen, the addition of hydrogen, and the gain of electrons. The addition of hydrogen may be subdivided into *hydrogenation*, the addition of hydrogen to an unsaturated system, e.g.

$$CH_2{=}CH_2 + H_2 \xrightarrow{\text{catalyst}} CH_3{-}CH_3$$

and *hydrogenolysis*, the addition of hydrogen with concomitant bond-rupture, e.g.

$$ArCH_2{-}NMe_2 + H_2 \xrightarrow{\text{catalyst}} ArCH_3 + HNMe_2$$

Mechanistically, there are three main pathways for reduction:

(1) *By the addition of electrons*, followed either by the uptake of protons, as in the reduction of anisole by sodium in liquid ammonia containing ethanol (p. 623),

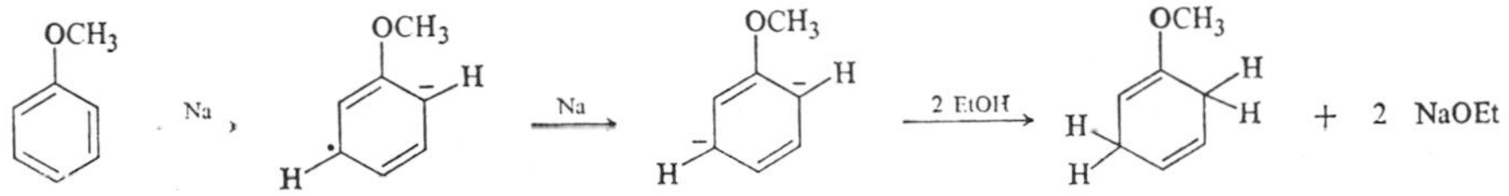

or by coupling, as in the reduction of ketones to pinacols (p. 637),

$$R_2C{=}O \xrightarrow{e} R_2\dot{C}{-}O^- \xrightarrow{\text{dimerizes}} \begin{matrix} R_2C{-}O^- \\ | \\ R_2C{-}O^- \end{matrix} \xrightarrow{2\,H^+} \begin{matrix} R_2C{-}OH \\ | \\ R_2C{-}OH \end{matrix}$$

Electron-transfer reduction can also be brought about electrolytically at the cathode.

(2) *By the transference of hydride ion*, as in the reduction of the carbonyl group by lithium aluminium hydride (p. 633):

$$H_3\bar{Al}-H \quad \diagup C=\ddot{O} \quad \longrightarrow \quad H-\overset{\displaystyle |}{\underset{\displaystyle |}{C}}-O-AlH_3$$

Such transference may also occur intramolecularly, as in Meerwein-Ponndorf-Verley reduction (p. 635).

(3) *By the catalyzed addition of molecular hydrogen*, as in the reduction of olefins on metals (p. 611).

Subsidiary reactions, such as the breaking of C—C bonds, are relatively uncommon in reductive processes, but some are usefully applied. For example, pimelic acid may be obtained in 45% yield from salicylic acid by refluxing for eight hours with sodium in isoamyl alcohol [2]; reaction occurs through a β-keto-acid which undergoes a reverse Claisen condensation:

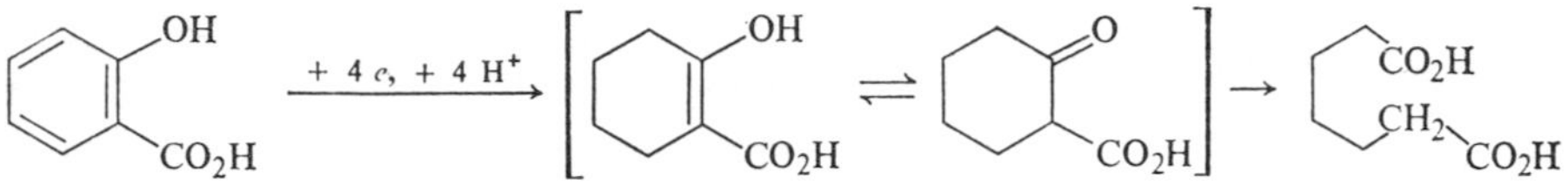

There have been enormous developments in reductive methods during the last twenty years, with respect both to the types of bond which may be reduced and to the selectivity of the processes. The older methods involving electron-transfer, such as sodium and alcohol, and zinc and acetic acid (in which the metal acts as the electron-source and the hydroxylic compound as the proton-donor), are now supplemented by the metal-ammonia and metal-amine systems which have increased the scope of these reductions (as in the above reduction of anisole) and are stereospecific (p. 618). Hydride-transfer agents such as formic acid (Leuckart reduction; p. 650) are supplemented by the complex hydrides, some of which are of remarkable selectivity. Catalytic methods have been improved by procedures for obtaining more active catalysts.

Enzymic systems are as yet of little general importance but are likely to achieve wider use as methods for isolating enzymes are developed. Their particular merit is that many (being themselves optically active) are stereospecific, so that optically active compounds can be obtained from inactive reactants. For example, hydroxyacetone is reduced specifically to R-propylene glycol (50%) by incubation with yeast reductase at 32°C for three days [2]:

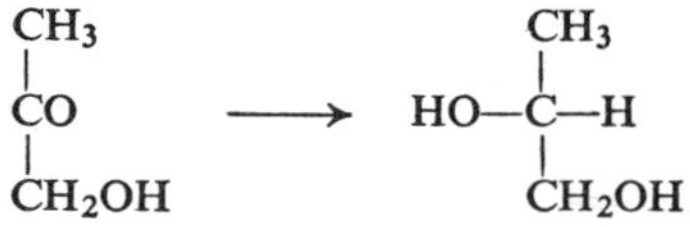

$$
\begin{array}{ccc}
\text{CH}_3 & & \text{CH}_3 \\
| & & | \\
\text{CO} & \longrightarrow & \text{HO}-\text{C}-\text{H} \\
| & & | \\
\text{CH}_2\text{OH} & & \text{CH}_2\text{OH}
\end{array}
$$

This chapter, like that on oxidation, is classified according to the type of group to be reduced. Each major class of reducing agent is described in the context of the system for which it has been most used.

s*

19.2 Hydrocarbons

(*a*) ALKANES

Alkanes can be reduced only by rupturing carbon-carbon bonds. Reducing agents are not normally sufficiently powerful to bring this about, in contrast to oxidizing agents of which the strongest cleave aliphatic chains ultimately to carbon dioxide. There is, however, one circumstance in which reduction can be effected catalytically, namely, with strained cyclic compounds, since C—C cleavage relieves the strain, e.g.

$$\begin{array}{c} CH_2 \\ \diagup \quad \diagdown \\ CH_2 \!\!-\!\!\!-\!\!\!-\!\! CH_2 \end{array} \xrightarrow[120°C]{H_2\text{-Ni}} CH_3CH_2CH_3$$

$$\begin{array}{c} CH_2\!\!-\!\!CH_2 \\ | \qquad | \\ CH_2\!\!-\!\!CH_2 \end{array} \xrightarrow[200°C]{H_2\text{-Ni}} CH_3CH_2CH_2CH_3$$

The nature of the catalyst for the reduction is described in the following section.

(*b*) ALKENES

(*i*) *Catalytic hydrogenation.* Almost all olefins can be saturated in very high yield by treatment with hydrogen and a metal catalyst. The most active catalysts are specially prepared platinum and palladium.

(1) *Adams' catalyst.* Chloroplatinic acid is fused with sodium nitrate to give a brown platinum oxide (PtO_2) which can be stored. When required, it is treated with hydrogen to give a very finely divided black suspension of the metal. This is the reagent usually chosen in the laboratory, reaction being carried out in solvents such as acetic acid, ethyl acetate, and ethanol. About 0·2 grammes of platinum oxide is usually employed per 10 grammes of reactant.

(2) *Palladium.* An active form of palladium can be obtained in a similar way from palladium chloride, but more commonly the palladium chloride is reduced in the presence of a suspension of charcoal or other solid support on which the metal is deposited in a very finely divided state.

A particularly reactive catalyst is obtained by the reduction of platinum oxide *in situ* with sodium borohydride in the presence of carbon. The molecular hydrogen required for hydrogenation is then generated by the addition of acid to the excess of borohydride.

Most olefinic bonds are reduced on these catalysts at temperatures below 100°C and at atmospheric or slightly increased pressure. For example, maleic acid gives succinic acid in 98% yield in 30 minutes on platinum at 20°C and 1 atmosphere [1],

$$
\begin{array}{ccc}
\begin{array}{c}
\text{CO}_2\text{H} \\
\diagup \\
\text{CH} \\
\| \\
\text{CH} \\
\diagdown \\
\text{CO}_2\text{H}
\end{array}
&
\xrightarrow{\text{H}_2\text{-Pt}}
&
\begin{array}{c}
\text{CO}_2\text{H} \\
\diagup \\
\text{CH}_2 \\
| \\
\text{CH}_2 \\
\diagdown \\
\text{CO}_2\text{H}
\end{array}
\end{array}
$$

and this particular reaction is often used for testing the apparatus and assessing the catalyst. The fact that the yields from purely olefinic systems are essentially quantitative has been widely employed in structural investigations: e.g. β-carotene (p. 715) absorbs 11 moles of hydrogen and therefore possesses 11 olefinic bonds; and since the molecular formula of the perhydro product is $C_{40}H_{78}$, it must contain two rings.

(3) *Raney nickel* is a slightly less active catalyst than platinum or palladium. It is prepared by treating a nickel-aluminium alloy with caustic soda,

$$\text{Ni—Al} + \text{NaOH} + \text{H}_2\text{O} \longrightarrow \text{Ni} + \text{Na}^+\text{AlO}_2^- + {}^3/_2\text{H}_2$$

and washing away the sodium aluminate to leave the nickel as a black suspension saturated with hydrogen (*ca.* 50–100 cm^3 per gramme) which is pyrophoric when dry. Most alkenes are hydrogenated over Raney nickel at about 100°C and pressures up to 3 atmospheres. Since the Raney alloy is fairly cheap, it is usually convenient to employ relatively large quantities of the catalyst (*ca.* 2 grammes per 10 grammes) so as to minimize the effects of catalyst-poisons; this is particularly important with sulphur-containing compounds (p. 653).

(4) *'Copper chromite'*. The early catalytic methods developed by Sabatier and Senderens employed metals such as iron, cobalt, nickel, and copper, at temperatures of about 300°C. This approach has been considerably improved by the use of high pressures (up to 300 atmospheres of hydrogen) and more active catalysts. The best, *Adkins' catalyst*, the so-called copper chromite, is made from copper nitrate and sodium dichromate and corresponds to $CuO.CuCr_2O_4$. This is a far cheaper catalyst than palladium or platinum and is therefore more attractive industrially.

(5) *Transfer hydrogenation*. The hydrogen is supplied by a donor such as cyclohexene or hydrazine, e.g.

$$
2 \; \text{C=C} + \bigcirc \xrightarrow[100°C]{\text{Pd}} 2 \; \text{CH—CH} + \bigcirc
$$

The driving force in the case of cyclohexene is the gain in aromatic stabilization energy when benzene is formed; with hydrazine, the strongly bonded N_2 molecule is formed. The advantage of this method is that no special apparatus for handling and measuring gaseous hydrogen is required.

Selectivity. Acetylenic bonds are reduced more readily than olefinic bonds, but other unsaturated groupings, with the exception of nitro groups and acid chlorides, are reduced less readily. Catalytic hydrogenation can therefore be used for the selective reduction of C=C in the presence of aromatic rings and carbonyl groups, whether or not the unsaturated functions are conjugated. For example, benzylideneacetophenone is reduced over platinum at 20°C to phenyl phenethyl ketone in 90% yield [1]:

$$\text{PhCH}\!=\!\text{CHCOPh} \xrightarrow{\text{H}_2\text{-Pt}} \text{PhCH}_2\text{CH}_2\text{COPh}$$

Stereochemistry. Reduction occurs in a stereospecific manner, giving the *cis*-dihydro product. This can be understood in terms of the following crude representation (the detailed course of the reaction is uncertain):

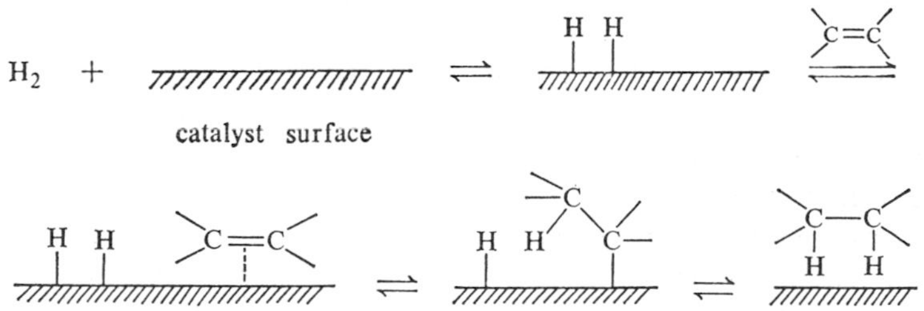

For example, 1,2-dimethylcyclohexene gives *cis*-1,2-dimethylcyclohexane:

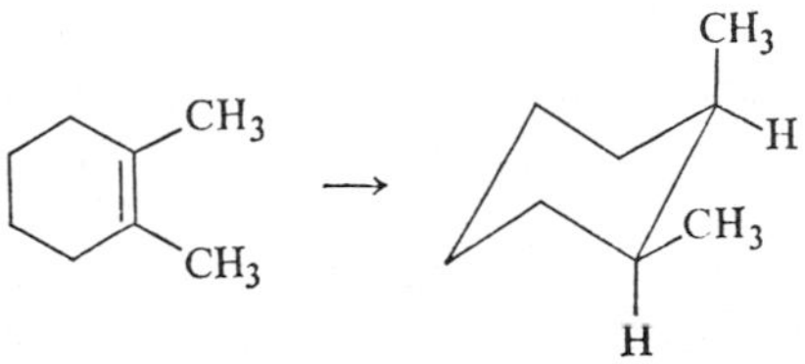

In rigid ring systems, the direction of addition is from the less hindered side. For example, octalins of the type shown below (e.g. cholesterol; X = OH) give mainly *trans* ring-junctions on reduction, since the angular methyl group hinders the fit of the catalyst on the opposite side of the double bond:

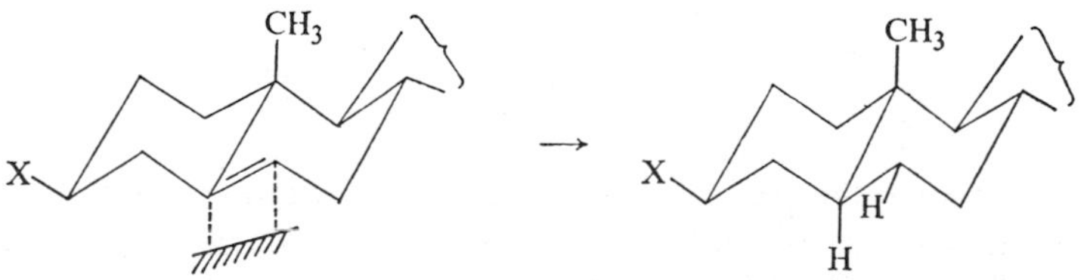

(80% yield for
cholesterol (X=OH) [2])

On the other hand, when the substituent X is axial, the fit to the catalyst is hindered on both sides and reduction gives a mixture of *cis* and *trans* decalins in comparable amounts.

(6) *Homogeneous catalytic hydrogenation.* Recent developments in catalytic hydrogenation have been to replace the metal by a soluble complex of rhodium or ruthenium and so to conduct hydrogenation in homogeneous solution.

The rhodium complex used is $(Ph_3P)_3RhCl$, prepared by heating rhodium chloride, $RhCl_3 \cdot 3H_2O$, with excess of triphenylphosphine in boiling ethanol. It is usually employed in benzene solution, in which olefins and acetylenes are reduced at room temperature and atmospheric pressure.

It is thought to act by exchanging one phosphine ligand for a solvent molecule (S) to give a complex which can then bind two hydrogen atoms on the metal:

$$(Ph_3P)_3RhCl + S \underset{-Ph_3P}{\rightleftharpoons} (Ph_3P)_2Rh(S)Cl \overset{H_2}{\rightleftharpoons} (Ph_3P)_2Rh(S)ClH_2$$

Displacement of the solvent molecule by the olefin is followed by *cis*-transfer of the two hydrogen atoms on to the olefin, and the saturated molecule then leaves the metal centre.

The method has several advantages. First, only olefins and acetylenes are reduced; other common groups such as C=O, C≡N, and NO_2 are unaffected. This makes possible selective reductions such as

$$PhCH{=}CHNO_2 \xrightarrow{H_2 - (Ph_3P)_3RhCl} PhCH_2CH_2NO_2$$

Secondly, mono- and di-substituted olefins are reduced much more rapidly than tri- or tetra-substituted ones. This provides a further degree of selectivity, illustrated by the reduction of carvone to dihydrocarvone—leaving the tri-substituted double bond intact—in over 90% yield [53]:

Thirdly, hydrogenolysis does not occur. For example, benzyl cinnamate gives the dihydro-derivative,

$$PhCH{=}CHCO_2CH_2Ph \xrightarrow{H_2 - (Ph_3P)_3RhCl} PhCH_2CH_2CO_2CH_2Ph$$

whereas hydrogenation on a metal catalyst results also in cleavage of the *O*-benzyl bond to give $PhCH_2CH_2CO_2H$ (*cf.* p. 625).

The major disadvantage of the method is that, because of the strong affinity of the rhodium complex for carbon monoxide, aldehydes are degraded; for example, cinnamaldehyde gives styrene.

The ruthenium complex used is $(Ph_3P)_3RuClH$, which is formed *in situ* from $(Ph_3P)_3RuCl_2$ and molecular hydrogen in the presence of a base such as triethylamine. It is specific for the hydrogenation of monosubstituted olefins and disubstituted acetylenes, which yield *cis*-olefins.

(*ii*) *The di-imide method.* Di-imide is an unstable compound which is obtained by the oxidation of hydrazine with copper(II) ion or by the treatment of sulphonyl hydrazides with base:

$$NH_2-NH_2 \xrightarrow{\ Cu^{2+}\ } HN=NH$$

$$PhSO_2-NH-NH_2 \xrightarrow{\ EtO^-\ } PhSO_2-NH-\overset{..}{N}H \longrightarrow PhSO_2^- + HN=NH$$

In the absence of additives it decomposes to nitrogen and hydrogen, but when generated in the presence of an alkene, rapid *cis*-stereospecific reduction occurs, the driving force being the great stability of the nitrogen molecule compared with the $-N=N-$ system (p. 11):

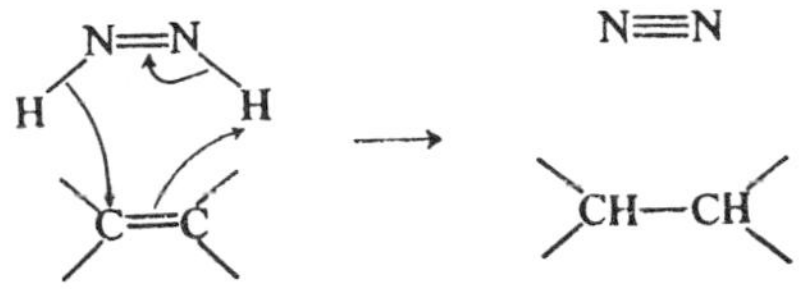

Other homopolar unsaturated bonds, such as those in acetylenes and azo compounds, are also reduced, but carbonyl-containing groups, nitro groups, sulphoxides, and S—S bonds are not affected.

(*iii*) *Hydroboration.* The organoboranes formed by olefins with diborane or other boronating agents (p. 498) are solvolyzed by organic acids (but not by mineral acids) to alkanes:

$$R_3B + 3\ R'CO_2H \longrightarrow 3\ RH + (R'CO_2)_3B$$

It is thought that reaction occurs as follows:

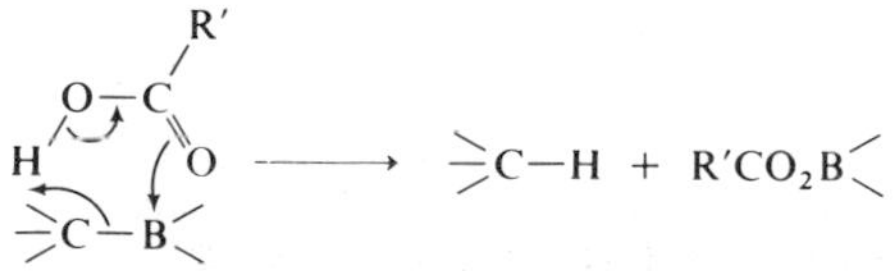

In the case of dienes, it is possible to reduce selectively the less substituted olefin by using a hindered borane such as that from trimethylethylene, $(Sia)_2BH$ (*cf.* p. 499), e.g.

$$
\begin{array}{ccc}
\text{CH}{=}\text{CH}_2 & & \text{CH}_2{-}\text{CH}_3 \\
\text{⬡} & \xrightarrow[\text{2) CH}_3\text{CO}_2\text{H}]{\text{1) (Sia)}_2\text{BH}} & \text{⬡}
\end{array}
$$

Aldehydes, ketones, acids, esters, nitriles, and epoxides are also reduced.

(c) CONJUGATED OLEFINS

(*i*) *Electron-transfer.* In addition to being reducible by catalytic methods, conjugated dienes and polyenes, and compounds containing olefinic bonds conjugated with carbonyl-containing groups, are reducible with electron-transfer agents. The reason is that the uptake of electrons by these systems gives delocalized anions whereas the addition of electrons to olefins does not.

$$-\text{CH}{=}\text{CH}{-}\text{CH}{=}\text{CH}{-} \xrightarrow{\ e\ } -\dot{\text{C}}\text{H}{-}\text{CH}{=}\text{CH}{-}\bar{\text{C}}\text{H}{-} \xrightarrow{\ e\ }$$

$$[-\bar{\text{C}}\text{H}{-}\text{CH}{=}\text{CH}{-}\bar{\text{C}}\text{H}{-} \leftrightarrow -\bar{\text{C}}\text{H}{-}\bar{\text{C}}\text{H}{-}\text{CH}{=}\text{CH}{-} \leftrightarrow -\text{CH}{=}\text{CH}{-}\bar{\text{C}}\text{H}{-}\bar{\text{C}}\text{H}{-}]$$

$$-\text{CH}{=}\text{CH}{-}\underset{|}{\text{C}}{=}\text{O} \xrightarrow{\ e\ } -\dot{\text{C}}\text{H}{-}\text{CH}{=}\underset{|}{\text{C}}{-}\bar{\text{O}} \xrightarrow{\ e\ } -\bar{\text{C}}\text{H}{-}\text{CH}{=}\underset{|}{\text{C}}{-}\bar{\text{O}}$$

The common reducing agents in this category are sodium and an alcohol, sodium amalgam, zinc and acetic acid, and the metal–ammonia and metal–amine systems. The last two groups are discussed more fully below.

The reduction of a conjugated diene occurs by 1,4-addition, evidently because the charges in the dianion tend to adopt the configuration in which they are furthest apart, e.g.

$$\text{PhCH}{=}\text{CH}{-}\text{CH}{=}\text{CH}_2 \xrightarrow{\ 2\ \text{Na}\ } \text{Ph}\bar{\text{C}}\text{H}{-}\text{CH}{=}\text{CH}{-}\bar{\text{C}}\text{H}_2 \xrightarrow{\ 2\ \text{EtOH}\ } \underset{\text{2-Butenylbenzene}}{\text{PhCH}_2{-}\text{CH}{=}\text{CH}{-}\text{CH}_3}$$

An olefinic bond conjugated with an aromatic ring is also reduced, for the aromatic π-system is available to delocalize the charge on the anion, but the 1,2-adduct is formed since formation of the 1,4-adduct would be accompanied by the loss of aromatic stabilization energy. For example, stilbene is reduced to bibenzyl,

$$\text{PhCH}{=}\text{CHPh} \xrightarrow{\ 2\ e,\ 2\ \text{H}^+\ } \underset{\text{Bibenzyl}}{\text{PhCH}_2{-}\text{CH}_2\text{Ph}}$$

The reduction of an $\alpha\beta$-unsaturated carbonyl compound also occurs by 1,4-addition, but the product tautomerizes to the more stable carbonyl-containing form, e.g.

$$PhCH{=}CHCO_2H \xrightarrow{\text{Na–Hg}} \left[Ph\overset{-}{C}H{-}CH{=}C\overset{\overset{\overset{-}{O}}{|}}{\diagdown}_{OH} \right] \xrightarrow{2\,H^+} PhCH_2CH_2CO_2H$$

Hydrocinnamic acid

$$CH_3COCH{=}CH{-}CH{=}CHCOCH_3 \xrightarrow{\text{Zn–HOAc}} \left[\underset{\underset{OH}{|}}{CH_3C}{=}CH{-}CH{=}CH{-}CH{=}\underset{\underset{OH}{|}}{CCH_3} \right]$$

$$\longrightarrow CH_3COCH_2CH{=}CHCH_2COCH_3$$

Metal-ammonia and metal-amine solutions. Strongly electropositive metals dissolve in liquid ammonia and in some amines to give characteristic blue solutions which contain metal cations and solvated electrons. Reduction to give amide ions is slow, but can be accelerated by the addition of catalytic amounts of transition-metal ions, e.g.

$$Na \xrightarrow{NH_3} Na^+ + e^-(\text{solvated}) \xrightarrow{(Fe^{++})} Na^+ + NH_2^- + \tfrac{1}{2}H_2$$

The use of sodamide in liquid ammonia as a powerful catalyst for condensations has already been described (p. 255). The solution before amide formation acts as a powerful source of electrons for the reduction of organic compounds. A proton source is usually added to complete the reaction, for ammonia itself is a very weak proton-donor. The donor should not be a strong enough acid to liberate hydrogen rapidly with sodium; ethanol is commonly used, for it reacts with sodium only very slowly at the boiling point of liquid ammonia. The reducing power of the system increases with the strength of the proton-donor; the stronger this is, the more rapidly is the initial equilibrium displaced to the right:

$$R + 2\,e \rightleftharpoons R^{2-} \xrightarrow{2\,XH} RH_2 + 2\,X^-$$

Thus, as the affinity of R for electrons is reduced, increasingly powerful proton-donors are needed to effect reduction.

The use of ammonia is limited by the low solubility in it of most organic compounds at its boiling-point ($-33°C$). Co-solvents such as ethers can be added, but they depress the solubility of the metals; high mutual solubility is rare. Primary amines such as ethylamine are more powerful solvents, and lithium is the most soluble of the alkali metals in most solvents. Since the reducing power of the system increases with the concentration of both the metal and the reactant, lithium and ethylamine provide about the most vigorous available conditions.

In general, these systems are of less value for the reduction of olefinic bonds than in the aromatic series (p. 621) partly because so many effective alternatives are available for the former and partly because the conditions are often too vigorous; for example, even the isolated double bond in 1-hexene is reduced by sodium in liquid ammonia containing methanol, giving n-hexane in 41% yield, although it is not reduced in the absence of methanol. However, yields in simple cases are high, e.g.

$$CH_2{=}\overset{\overset{\displaystyle CH_3}{|}}{C}{-}CH{=}CH_2 \xrightarrow{\text{Na-NH}_3} (CH_3)_2C{=}CHCH_3$$
2-Methyl-2-butene (98%)

and in certain situations these methods succeed where catalytic methods fail. For example, the C_8—C_9 and especially the C_8—C_{14} double bonds in steroid systems are almost impossible to reduce catalytically but are reduced efficiently with sodium in liquid ammonia, e.g.

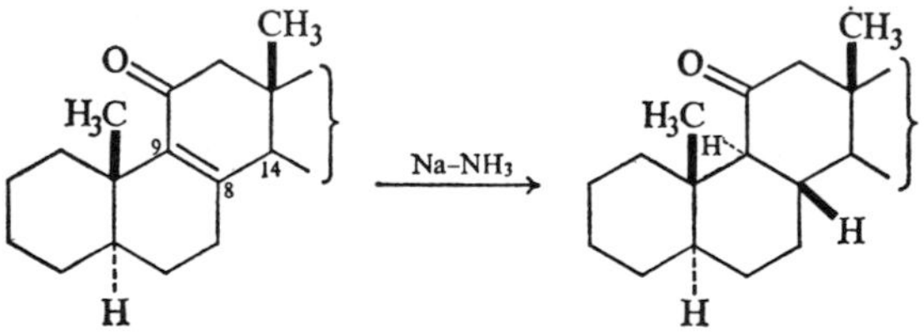

The stereospecificity in this reaction is a general feature of these reductions; the thermodynamically most favourable product is almost invariably formed because the intermediate dianion adopts the most stable configuration before taking up protons,

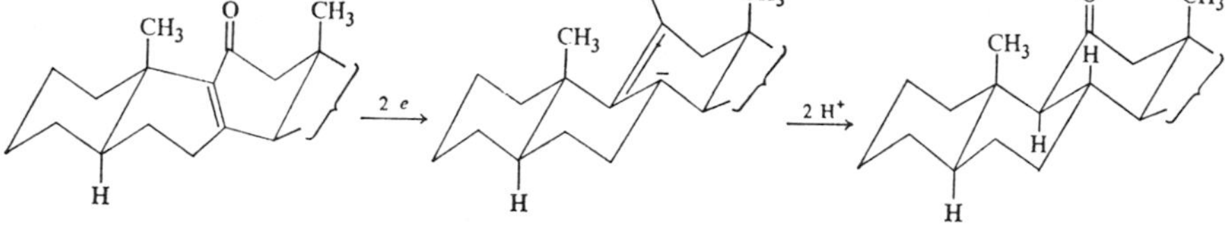

An impressive example of the application of this stereochemical control occurs in the synthesis of epiandrosterone (21.5).

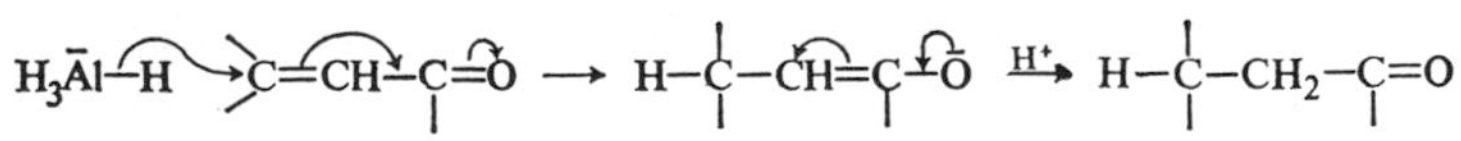

(ii) *Hydride-transfer.* Olefinic bonds which are conjugated to substituents of $-M$ type are susceptible to nucleophiles (p. 115), so that reductions of the type, might be expected with $\alpha\beta$-unsaturated carbonyl compounds. However, the complex metal hydrides usually either react preferentially at the carbonyl-carbon, e.g.

$$CH_3-CH=CH-CHO \xrightarrow{\text{NaBH}_4} CH_3-CH=CH-CH_2OH \quad (85\%)$$

or in some cases, with the more active hydride agents, give mixtures of carbonyl-reduced and fully reduced products. Careful control is sometimes successful, e.g.

$$PhCH=CH-CHO \xrightarrow{\text{LiAlH}_4} \begin{cases} \xrightarrow{0-10°C} PhCH=CH-CH_2OH \\ \xrightarrow{25-35°C} PhCH_2-CH_2-CH_2OH \end{cases}$$

However, if the $-M$ substituent is not itself reducible by the hydride, selective reduction of the olefinic bond can be effected. For example, reduction of an $\alpha\beta$-unsaturated nitro compound by sodium borohydride,

$$-CH=CH-NO_2 \xrightarrow{\text{NaBH}_4} CH_2-CH_2-NO_2$$

was employed in the synthesis of chlorophyll (21.10). The more powerful hydride-donor, lithium aluminium hydride, reduces both the olefinic bond and the nitro group in this situation, and the sequence, $ArCH=CHNO_2 \rightarrow ArCH_2CH_2NH_2$, is employed in the building up of isoquinoline rings from aromatic aldehydes (p. 683).

(d) ALKYNES

(i) Catalytic reduction. Acetylenic bonds can be fully reduced on the catalysts suitable for the reduction of olefins, but this is seldom needed in synthesis. It is much more useful to exploit the fact that triple bonds are reduced to double bonds faster than double bonds themselves are reduced. The technique usually adopted employs palladium on a calcium carbonate support, partially poisoned by lead acetate (Lindlar's catalyst) [46]. The product is exclusively the cis-olefin. The method has been widely applied in the synthesis of terpenoid compounds [see, e.g. vitamin A (21.1) and β-carotene (21.2)].

(ii) Hydroboration. The hydroboration of acetylenes is carried out in the same way as that of olefins. With terminal alkynes, dihydroboration occurs to some extent,

$$R-C\equiv CH \longrightarrow R-CH=CH-B \begin{matrix} / \\ \\ \backslash \end{matrix} \longrightarrow R-CH_2-CH \begin{matrix} \backslash \\ B- \\ / \end{matrix} \begin{matrix} B- \\ / \end{matrix}$$

so that, on hydrolysis, complete reduction occurs. To prevent this, it is best to use the hindered and more selective disiamyl borane (p. 499) which allows quantitative monohydroboration of both mono- and di-substituted acetylenes, e.g.

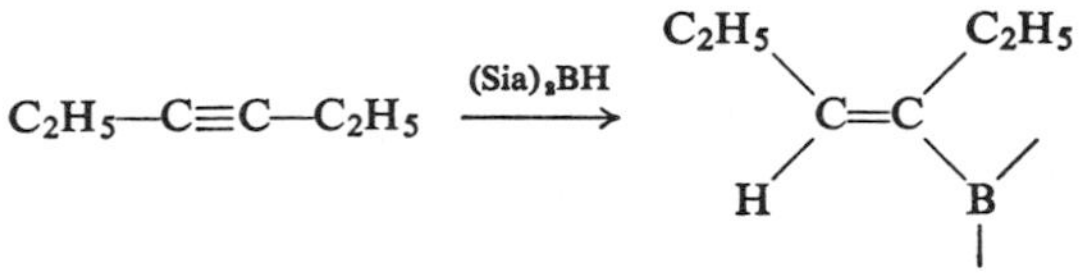

$$C_2H_5-C\equiv C-C_2H_5 \quad \xrightarrow{(Sia)_2BH} \quad \begin{array}{c} C_2H_5 \\ \diagdown \\ \diagup \quad \diagdown \\ H \end{array} C=C \begin{array}{c} C_2H_5 \\ \diagup \\ \diagdown \\ B \\ | \end{array}$$

The products can be hydrolyzed with acetic acid at room temperature to give specifically *cis*-alkenes.

(*iii*) *Electron-transfer.* Electrons are transferred to acetylenes more readily than to olefins (*cf.* the greater reactivity of acetylenes towards nucleophiles). The best reagents are the metal-amine and metal-ammonia systems; no added proton-donor is needed. The reduction is selectively *trans*, since the charges in the dianion optimally assume the configuration in which they are furthest apart,

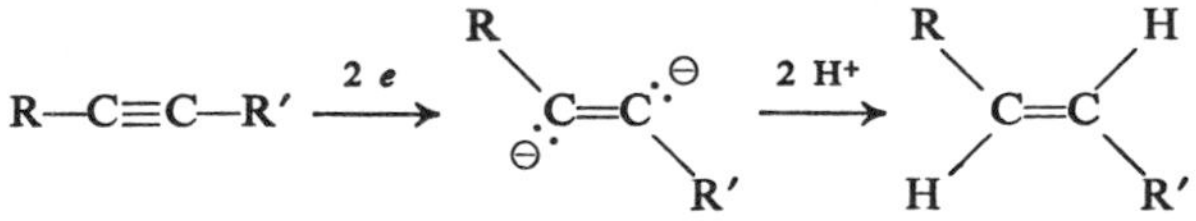

$$R-C\equiv C-R' \quad \xrightarrow{2\ e} \quad \begin{array}{c} R \\ \diagdown \\ \diagup \\ \ominus \end{array} C=C \begin{array}{c} \ominus \\ \diagdown \\ \diagup \\ R' \end{array} \quad \xrightarrow{2\ H^+} \quad \begin{array}{c} R \\ \diagdown \\ \diagup \\ H \end{array} C=C \begin{array}{c} H \\ \diagup \\ \diagdown \\ R' \end{array}$$

so that this method is complementary to the stereospecifically *cis* catalytic methods. For example, both *cis-* and *trans*-cyclononene can be obtained from cyclononyne:

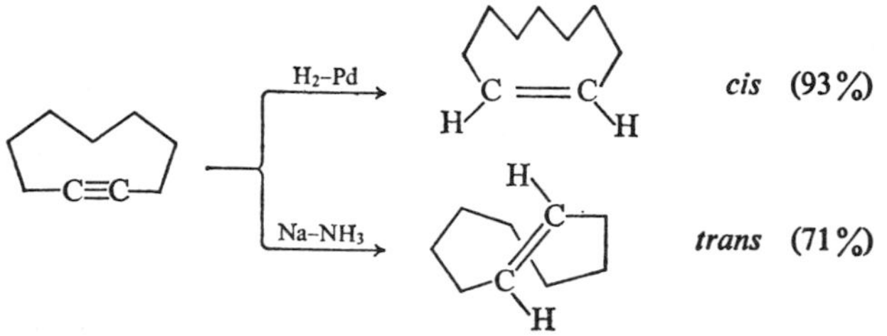

cis (93%)

trans (71%)

(*iv*) *Hydride-transfer.* Di-isobutyl aluminium hydride (p. 642) has recently been found to reduce alkynes to alkenes, but lithium aluminium hydride is effective only when the acetylene has a hydroxyl group in the α-position; the hydroxyl group facilitates reaction through the formation of a complex with aluminium, and the product is the *trans*-olefin.

$$R\text{—}C\equiv C\text{—}\underset{\underset{OH}{|}}{C}R'_2 \quad\xrightarrow{\text{LiAlH}_4}\quad \underset{H}{\overset{R}{\diagdown}}C=C\underset{CR'_2\text{—}OH}{\overset{H}{\diagup}}$$

(e) AROMATIC RINGS

(i) *Catalytic hydrogenation.* Aromatic rings can be reduced on the catalysts suitable for olefins, but more vigorous conditions are required because aromatic stabilization energy is lost in the process. The conditions vary according to the amount of stabilization energy which is sacrificed: e.g. naphthalene is reduced to tetralin more easily than benzene is reduced to cyclohexane (*cf.* p. 57). Typical conditions on Adams' catalyst are 10 hours at 100–150°C and 100–150 atmospheres (i.e. in a steel bomb), compared with 1 hour at 20°C and atmospheric pressure for the reduction of simple olefins.

If there is a danger of the catalyst becoming poisoned, Raney nickel is normally selected because it is cheaper than palladium or platinum and can be used in excess. This is particularly important with sulphur-containing molecules, but it must be remembered that sulphur is removed from the system during reduction (p. 653). Pyridine also poisons catalysts, but its hydrochloride can be reduced efficiently over platinum in ethanol. Ruthenium is gaining use as a catalyst for hydrogenation without concomitant hydrogenolysis, and rhodium is suitable for some of the more resistant aromatic heterocycles.

Reduction in mild conditions gives mainly the *cis* product, but if the *trans* product is the more stable it can be formed in vigorous conditions:* e.g. the reduction of naphthalene in solution over platinum gives mainly *cis*-decalin whereas over copper chromite in the gas phase *trans*-decalin predominates.

Those aromatic compounds which are in equilibrium with significant proportions of aliphatic (olefinic) tautomers can be reduced in conditions appropriate to olefins. For example, resorcinol is reduced, *via* its diketo-tautomer (p. 18), at 50°C and 80 atmospheres over Raney nickel:

Cyclohexane-1,3-dione
('Dihydroresorcinol')

(ii) *Electron-transfer.* Whereas catalytic methods normally reduce aromatic rings to perhydro compounds, electron-transfer reagents can be highly selective.

---

*This is an example of thermodynamic control (p. 92). In vigorous conditions, equilibrium is established with the *cis* product while the slower *trans* reduction gradually removes the reactant to give the more stable product.

A large variety of products can be obtained, depending on the conditions. The reduction of naphthalene is illustrative:

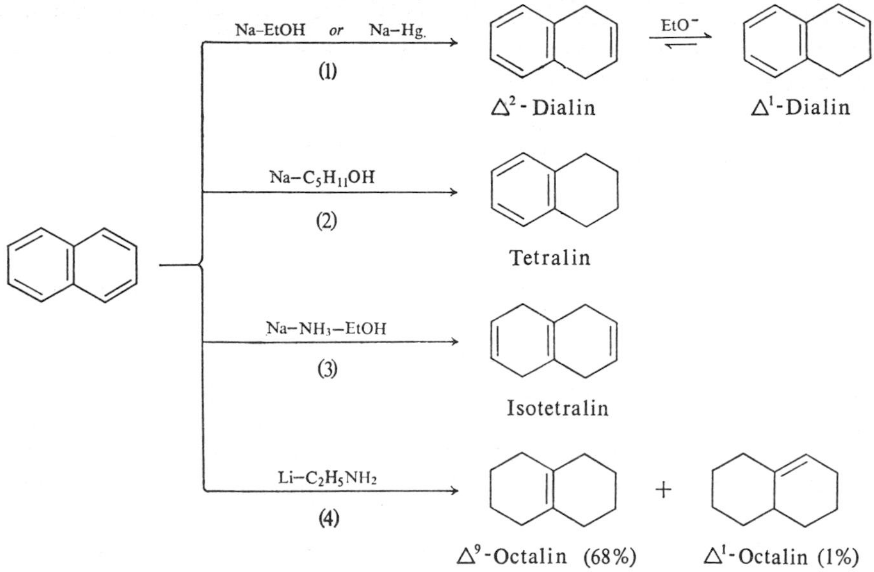

Δ^2-Dialin　　　　Δ^1-Dialin

Tetralin

Isotetralin

Δ^9-Octalin (68%)　　Δ^1-Octalin (1%)

(1) The mildest conditions give a dihydro compound. 1,4-Addition occurs because of the tendency of the charges in the dianion to maintain maximal separation,

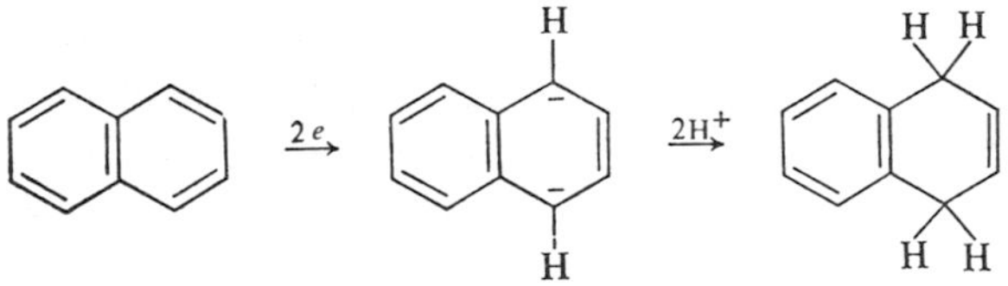

The product is isomerized with base to the more stable (conjugated) Δ^1-dialin, *via* a delocalized anion,

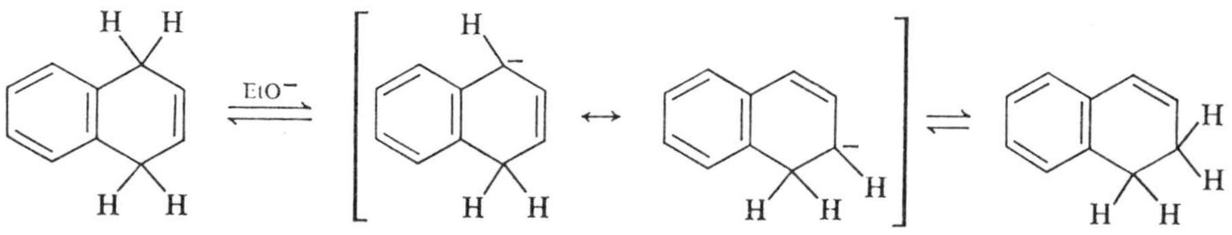

(2) Further reduction occurs in the higher-boiling amyl alcohol, probably through formation of Δ^2-dialin, isomerization as in (1), and reduction of the conjugated Δ^1-dialin.

(3) The use of liquid ammonia with an added proton-donor enables both benzenoid rings to be reduced in the 1,4-manner.

(4) Lithium in ethylamine provides the most vigorous conditions (p. 618): initial reduction as in (3) is followed by isomerization and further reduction.

Mixtures of products are usually obtained unless a strong directive influence is present. For example, the reduction of styrene with lithium in ethylamine gives comparable quantities of 1-ethylcyclohexene and ethylcyclohexane,

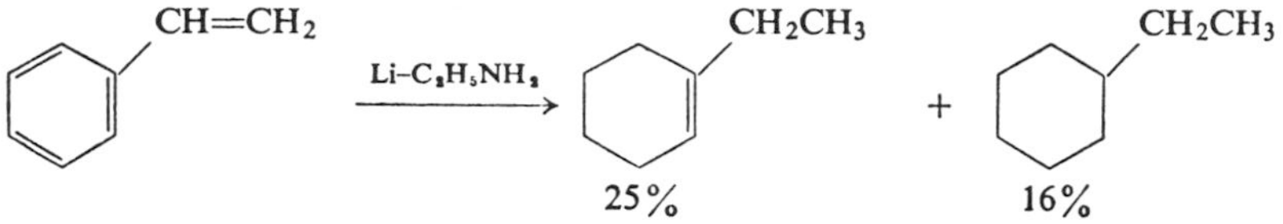

whereas the reduction of phenol gives, after acidification, a 96% yield of cyclo-hexanone:

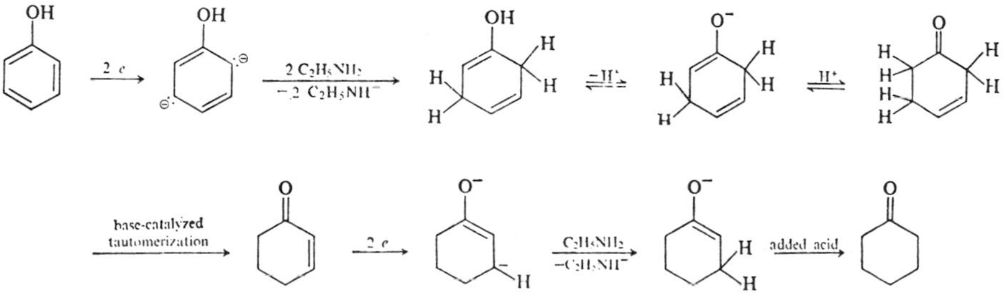

The hydroxyl group dictates the configuration of the charges in the initial dianion as shown above, for in the alternative 1,4-adduct the electron-releasing ($+M$) hydroxyl group would be adjacent to a negatively charged carbon. Two tautomeric shifts give an $\alpha\beta$-unsaturated ketone which is reduced further. The product is the resonance-stabilized enolate anion from cyclohexanone which, being negatively charged, is stable to reduction and is converted by added acid into cyclohexanone.

The same principle governs the mode of reduction of anisole and dimethyl-aniline, each of which contains powerful $+M$ groups:

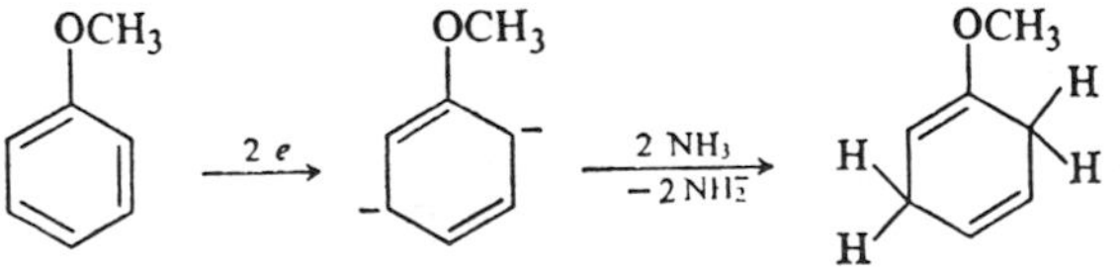

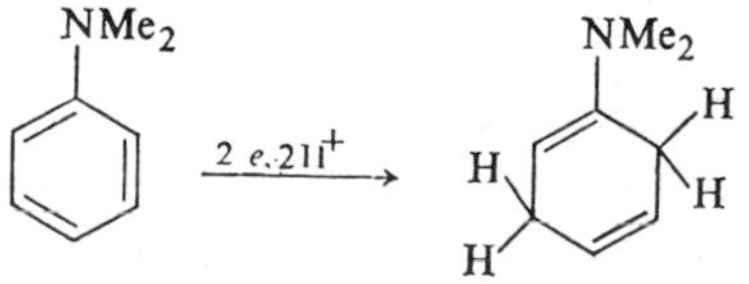

The products, a vinyl ether and a vinyl amine respectively, give cyclohexen-3-one on acidification, e.g.

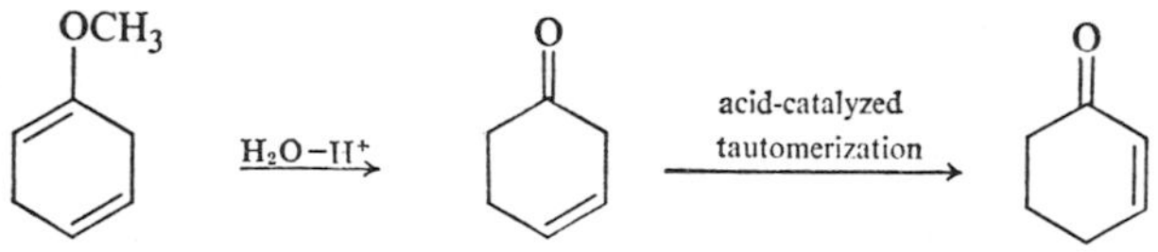

This type of reduction has been employed in a synthesis of ($\pm$)-thioctic acid (see also p. 269):

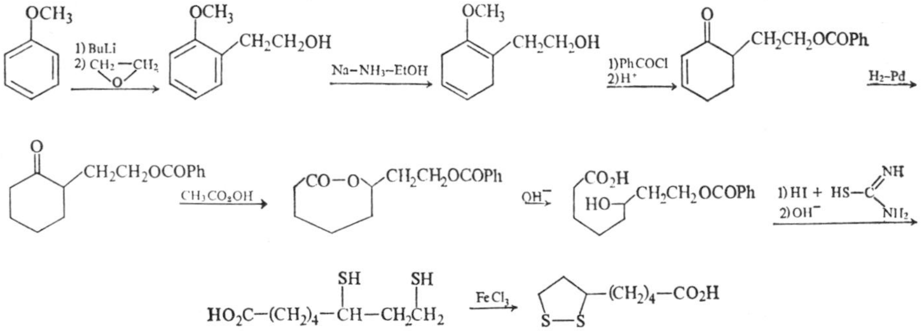

($\pm$)-Thioctic acid

The reductions of α-naphthol and ethyl β-naphthyl ether provide an interesting contrast. The former is reduced by lithium in liquid ammonia containing ethanol in the *unsubstituted* ring in 98% yield [4],

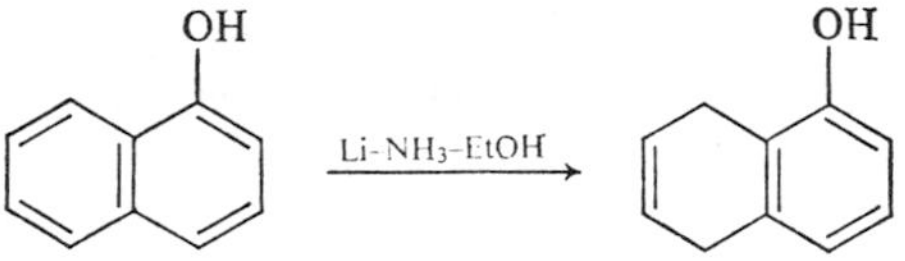

whereas the latter is reduced by sodium and ethanol in the *substituted* ring, to give 2-tetralone in 45% yield after hydrolysis [4]:

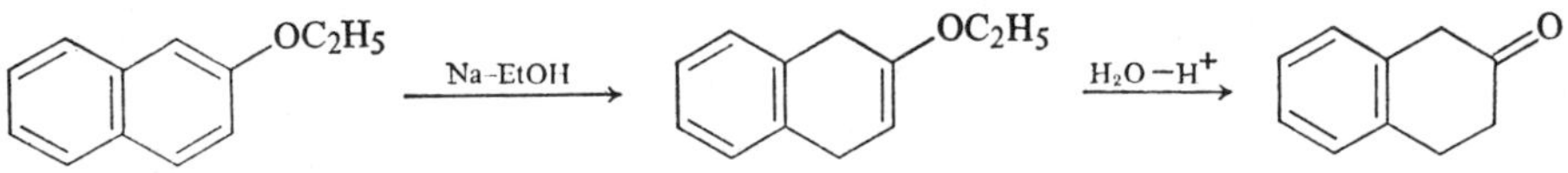

The difference probably arises because the tendency of electrons to add in a 1,4-manner to a six-membered aromatic ring is opposed in the former case by the hydroxyl substituent in one ring so that reaction occurs essentially specifically in the other ring. In the latter case, however, 1,4-addition is not opposed by the electron-releasing group. This principle has been employed in the selective reduction of 1,6-dimethoxynaphthalene in a synthesis of epiandrosterone (21.5).

19.3 Hydrogenolysis

(a) BENZYLIC SYSTEMS

The benzylic group, when attached to OH, OR, OCOR, NR$_2$, SR, or halogen, is particularly susceptible to nucleophilic reducing agents, catalytic reduction, and electron-transfer reagents.

The first method is not commonly employed; one example occurs in the synthesis of quinine in which methoxide ion at a high temperature was used to reduce a Mannich base (21.6):

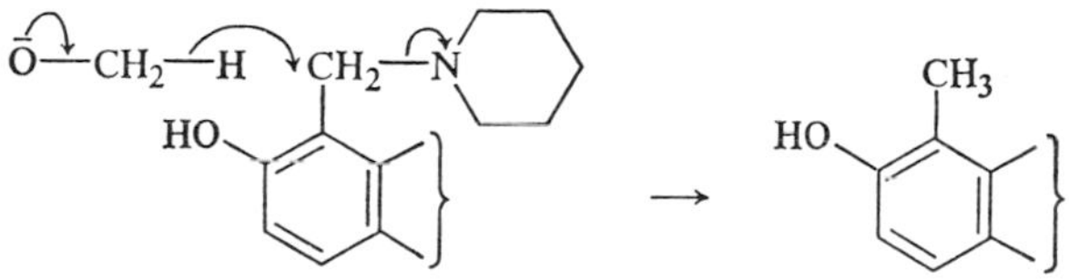

Catalytic reduction is widely employed, as in the following examples:

$$PhCH_2OH \xrightarrow[\text{25°C, 3 atm.}]{\text{H}_2\text{–Pd}} PhCH_3 \quad \text{(quantitative)}$$

$$\underset{\text{PhCHCO}_2\text{H}}{\overset{\text{OH}}{|}} \xrightarrow[\text{25°C, 3·5 atm.}]{\text{H}_2\text{–Pd}} PhCH_2CO_2H \quad (90\%)$$

$$PhCH_2OCOPh \xrightarrow[\text{140°C, 1 atm.}]{\text{H}_2\text{–Pd}} PhCH_3 + PhCO_2H \quad (94\%)$$

$$(PhCH_2)_2NPh \xrightarrow[\text{25°C, 1 atm.}]{\text{H}_2\text{–Pd}} 2\,PhCH_3 + PhNH_2 \quad \text{(quantitative)}$$

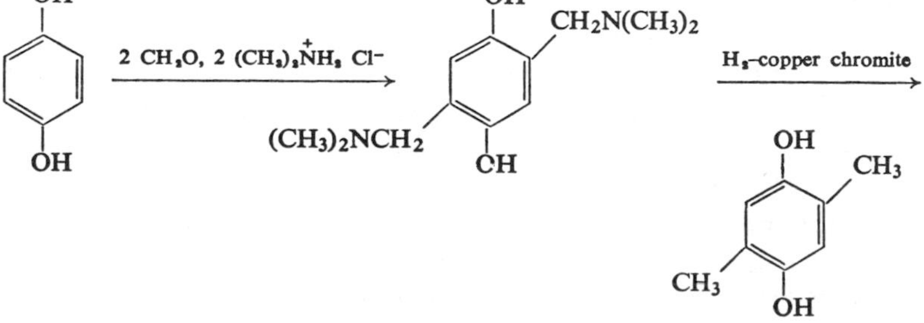

The last example illustrates a method for the introduction of methyl groups into phenols *via* the Mannich reaction (8.6).

Electron-transfer reduction is probably facilitated by the ability of the benzenoid ring to delocalize the charge of an anion, i.e.

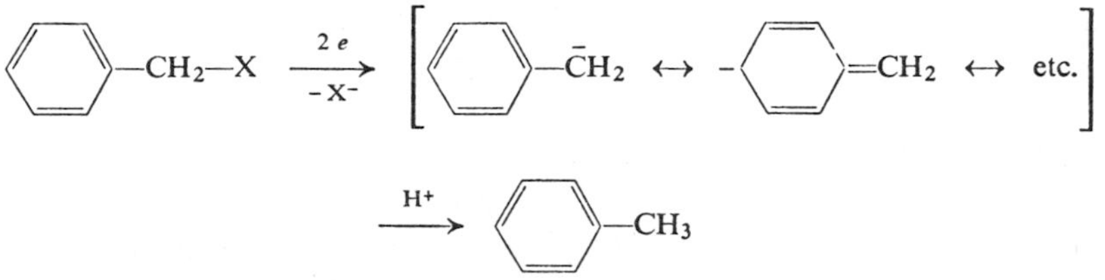

The following are examples:

$$PhCH_2OCONHCH_2CH_2CO_2H \xrightarrow{Na\text{-}NH_3} PhCH_3 + CO_2 + H_2NCH_2CH_2CO_2H$$

β-Alanine (95%)

$$PhCH_2SCH_2CH_2CH(NH_2)CO_2H \xrightarrow{Na\text{-}NH_3} PhCH_3 + HSCH_2CH_2CH(NH_2)CO_2H$$

Cysteine (78%)

$$PhCH{=}CHCH_2\overset{+}{N}(CH_3)_3I^- \xrightarrow{Na\text{-}Hg} PhCH{=}CHCH_3 + N(CH_3)_3$$

1-Phenylpropene

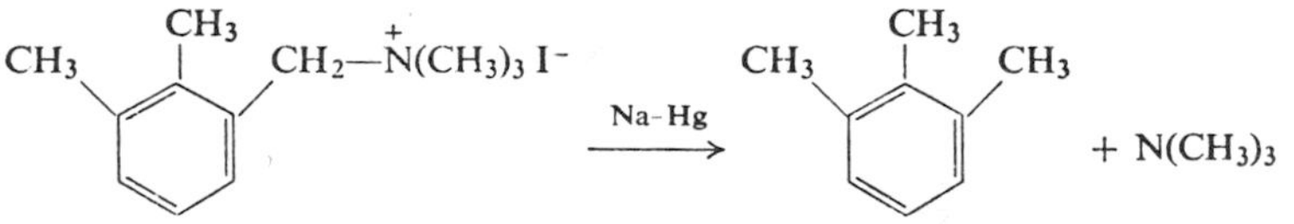

1,2,3-Trimethylbenzene

The first example illustrates the use of the benzyl group as part of a protecting agent in amino-acid chemistry: the benzyloxycarbonyl group is used to protect the amino group in peptide synthesis and is afterwards removed as above or catalytically (p. 357). The second example represents a standard procedure for the protection of thiol groups; the catalytic method is unsuitable for the removal of the benzyl residue because sulphur is present. The third example shows that the activating effect of the benzene ring can be transmitted through conjugated double bonds. The final example illustrates a synthetic method for introducing methyl groups *ortho* to existing methyl groups in aromatic rings; the reactant is obtained from *o*-xylene *via* the Stevens rearrangement (p. 474) and is converted into 1,2,3-trimethylbenzene by sodium amalgam at 80°C in 87% yield [4].

Diphenylmethyl systems are more easily hydrogenolyzed than benzyl systems. For example, benzilic acid is reduced in 95% yield to diphenylacetic acid by treatment with red phosphorus and iodine in refluxing acetic acid [1]:

$$\begin{array}{c} \text{OH} \\ | \\ \text{Ph}_2\text{CCO}_2\text{H} \xrightarrow{\text{red P-I}_2} \text{Ph}_2\text{CHCO}_2\text{H} \\ \text{Diphenylacetic acid} \end{array}$$

Triphenylmethyl (trityl) systems are still more easily reduced. Advantage is taken of this, together with the selectivity of triphenylmethyl chloride for primary alcoholic groups as compared with secondary and tertiary, in protecting primary alcohols (e.g. in the synthesis of adenosine triphosphate; 21.8):

$$\text{Ph}_3\text{C}-\text{Cl} + \text{HOCH}_2-\text{R} \longrightarrow \text{Ph}_3\text{C}-\text{OCH}_2-\text{R} \xrightarrow{\text{H}_2\text{-Pd}} \text{Ph}_3\text{CH} + \text{HOCH}_2-\text{R}$$

Triphenylmethyl systems readily form the trityl carbonium ion and this is particularly susceptible to reduction by hydride-transfer agents. For example, triphenylmethane can be obtained in 70–80% yield by Friedel-Crafts reaction between benzene and carbon tetrachloride in the presence of aluminium trichloride (p. 385) followed by the addition of ether [1]:

$$3\ \text{PhH} + \text{CCl}_4 \xrightarrow[-3\ \text{HCl}]{\text{AlCl}_3} \text{Ph}_3\text{C}-\text{Cl} \xrightarrow[-\text{Cl}^-]{\text{AlCl}_3}$$

$$\text{Ph}_3\overset{+}{\text{C}} \overset{\overset{\text{OC}_2\text{H}_5}{|}}{\text{H}-\text{CH}}_{\underset{\text{CH}_3}{}} \longrightarrow \text{Ph}_3\text{C}-\text{H}\ (+\text{C}_2\text{H}_5\overset{+}{\text{O}}=\text{CHCH}_3 \xrightarrow{\text{Cl}^-} \text{C}_2\text{H}_5\text{Cl} + \text{CH}_3\text{CHO})$$

(b) ALLYLIC SYSTEMS

Allylic systems behave in most reactions analogously to benzylic systems. However, whereas the latter are normally reduced by hydrogenolysis rather than by hydrogenation of the aromatic ring, the former more readily undergo hydrogenation with those reducing agents which react with the olefinic bond. Thus, catalytic reduction is in general unsuitable for the hydrogenolysis of allylic systems.

Fortunately in this context, lithium aluminium hydride and sodium in an amine are suitable reagents since neither reacts with isolated olefinic bonds. For example, allylic ethers are cleaved by lithium aluminium hydride in tetrahydrofuran,

$$\underset{/}{\overset{\backslash}{\text{C}}}=\text{C}-\text{CH}_2-\text{OR} \xrightarrow{\text{LiAlH}_4} \underset{/}{\overset{\backslash}{\text{C}}}=\text{C}-\text{CH}_3 + \text{ROH}$$

and allylic esters are cleaved by sodium in ethylamine,

$$\underset{/}{\overset{\backslash}{C}}{=}C{-}CH_2{-}OCOR \xrightarrow{\text{Na–C}_2\text{H}_5\text{NH}_2} \underset{/}{\overset{\backslash}{C}}{=}C{-}CH_3 + RCO_2H$$

The corresponding alkyl compounds are unaffected in these conditions.

(c) ALKYL SYSTEMS

Alkyl systems are generally much less easy to hydrogenolyze than the corresponding benzyl systems.

(1) *Alcohols* cannot be hydrogenolyzed directly, but can be converted into their toluene-*p*-sulphonates which are reducible with sodium cyanoborohydride or with lithium aluminium hydride (i.e. by taking advantage of the fact that the toluene-*p*-sulphonate anion is so good a leaving group in S_N2 displacements; p. 126):

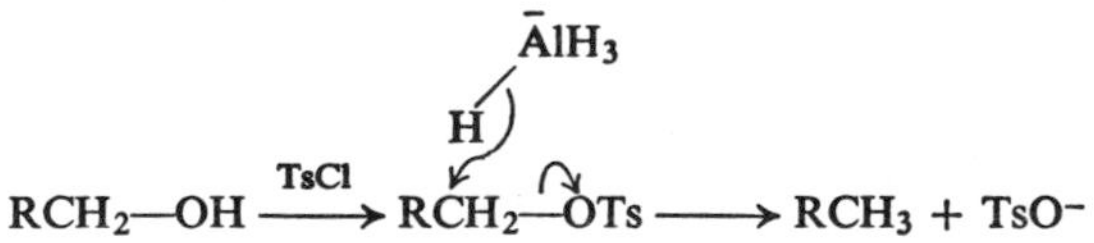

$$RCH_2{-}OH \xrightarrow{\text{TsCl}} RCH_2{-}OTs \longrightarrow RCH_3 + TsO^-$$

The usual alternative is to dehydrate the alcohol and hydrogenate the resulting olefin.

(2) *Halides* are much more reactive than alcohols in S_N2-displacements and are correspondingly more susceptible to hydride-transfer from lithium aluminium hydride or, better, sodium cyanoborohydride. For example, with the latter reagent in hexamethylphosphoramide, 1-iododecane gives decane in about 90% yield [53]:

$$CH_3(CH_2)_8CH_2I \xrightarrow{\text{NaBH}_3\text{CN}} CH_3(CH_2)_8CH_3$$

Electron-transfer agents are also efficient. Since the highly electropositive metals like sodium induce Wurtz coupling, the metal-amine systems are unsuitable, but magnesium amalgam in water and a zinc-copper couple in ethanol can be used, e.g.

$$\text{n-}C_4H_9I \xrightarrow[\text{H}_2\text{O}]{\text{Mg–Hg}} \text{n-}C_4H_{10}\ \text{(nearly quantitative)}$$

Alternatively, magnesium in ether is used to form the Grignard reagent which is subsequently decomposed by water,

$$RX + Mg \longrightarrow RMgX \xrightarrow{\text{H}_2\text{O}} RH$$

A more recently developed electron-transfer agent is chromium(II) ion, usually complexed with ethylenediamine. It is thought to act by first removing the halide to form an alkyl radical,

$$R{-}Br + Cr(II) \longrightarrow R{\cdot} + Cr(III) + Br$$

and then to complete reduction *via* an organochromium species:

$$R{\cdot} + Cr(II) \longrightarrow R{-}Cr(II) \xrightarrow{H^+} RH + Cr(III)$$

This reagent also reduces vinyl halides.

Electron-transfer agents react with 1,2-dihalides in a different manner; both halogen atoms are eliminated and an olefinic bond is introduced. Zinc is normally used as the reducing agent:

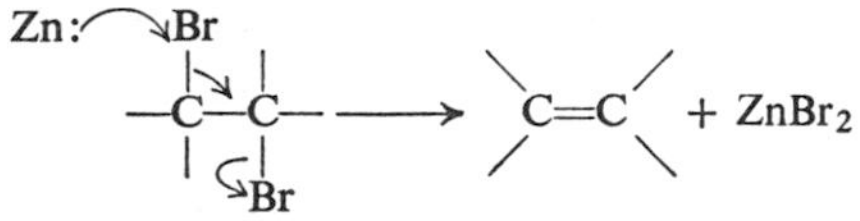

For example, allene may be obtained in 80% yield by the treatment of 2,3-dichloropropylene with zinc in aqueous ethanol:

$$CH_2{=}C{-}CH_2 \xrightarrow[-\,ZnCl_2]{Zn} CH_2{=}C{=}CH_2$$
$$\underset{Cl \quad Cl}{\big| \quad \big|} \qquad\qquad\qquad \underset{Allene}{}$$

Variants of this method, involving other leaving-groups, have been used in the synthesis of cholesterol (p. 717) and of reserpine (p. 733).

(3) *Acetals* and *ketals* are hydrogenolyzed to ethers in high yield by lithium aluminium hydride in the presence of aluminium trichloride:

$$R_2C(OR')_2 \xrightarrow{LiAlH_4-AlCl_3} R_2CH{-}OR' + R'OH$$

(4) *Primary amines* can be reduced by treating the derived toluene-*p*-sulphonamide with a large excess of hydroxylamine-*O*-sulphonic acid in alkaline ethanol. Use is made of the strong leaving group potential of sulphates and sulphinates and the instability of the —N=NH grouping (*cf.* p. 631).

$$RCH_2{-}NH_2 \xrightarrow[-HCl]{TsCl} RCH_2{-}NH{-}Ts \xrightarrow[-H_2SO_4]{NH_2OSO_2OH}$$

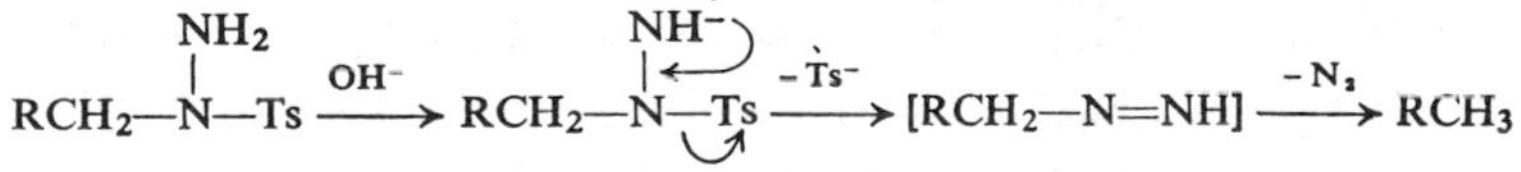

For example, n-hexylamine gives n-hexane in 41% yield.

(d) AROMATIC SYSTEMS

Aryl halides can be reduced by the formation of the Grignard reagent (from bromides and iodides) or the lithium compound (from chlorides) followed by treatment with water, but a recently developed method employing reduction by hydrazine over a palladium catalyst appears more promising. For example, 2-bromonaphthalene is reduced to naphthalene in 95% yield in 5 minutes in boiling ethanol:

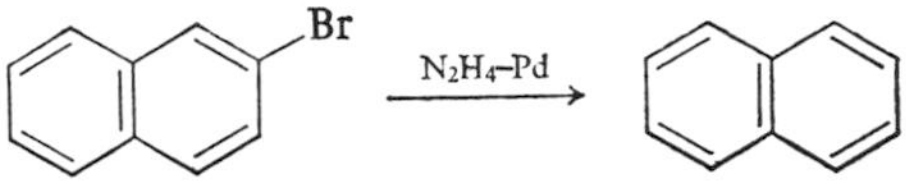

Reduction with chromium(II) ion is also efficient. For example, 1-bromo-naphthalene with chromium(II) complexed with ethylenediamine in aqueous dimethylformamide gives 93–98% of naphthalene [52].

The halogen atoms at the α- and γ-positions of pyridine and its derivatives can be efficiently removed both catalytically by hydrogen and by electron-transfer reducing agents (*cf.* the ease of nucleophilic substitution at these positions; p. 424). Since hydroxyl groups at these positions are convertible into chloro-substituents with phosphorus oxychloride, this provides a method for reducing phenols, as in the Conrad-Limpach synthesis (p. 681), e.g.

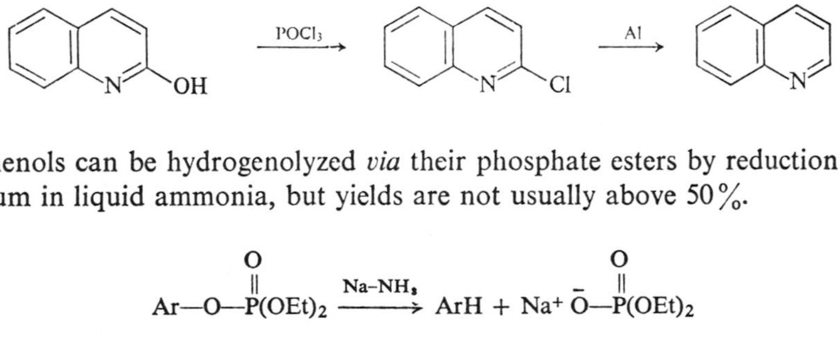

Phenols can be hydrogenolyzed *via* their phosphate esters by reduction with sodium in liquid ammonia, but yields are not usually above 50%.

Aromatic amines cannot be hydrogenolyzed directly; they are deaminated by reduction of the corresponding diazonium ions (p. 439).

19.4 Aldehydes and Ketones

Aldehydes and ketones can be reduced to hydrocarbons, alcohols, and pinacols (1,2-diols or glycols).

(a) TO HYDROCARBONS

Five methods are available. The choice between them is based on the sensitivity of other functional centres in the reactant in the reducing conditions.

(1) *Clemmensen method.* The reducing agent is amalgamated zinc and concentrated hydrochloric acid. The probable mechanism is as follows:

$$R_2C{=}O \; \underset{H^+}{\rightleftharpoons} \; R_2C{-}\overset{+}{O}H \underset{\substack{\uparrow \\ Zn \\ (surface)}}{} \longrightarrow R_2C{-}OH \underset{Zn^+}{} \xrightarrow{3\;Cl^-} R_2C{-}OH \underset{\substack{Cl{-}Zn^{2-}{-}Cl\!:Zn \\ Cl}}{}$$

$$\longrightarrow R_2C{-}OH \underset{Zn^-}{} \underset{-H^+}{\rightleftharpoons} R_2C{-}\overset{+}{O}H_2 \underset{Zn^-}{} \longrightarrow R_2C{=}Zn \xrightarrow{H^+} R_2CH{-}Zn^+ \xrightarrow{H^+} R_2CH_2 + Zn^{2+}$$

The concentrated acid is apparently needed to force the initial protonation; amalgamation of the zinc raises its hydrogen-overvoltage so that hydrogen is not produced. Only halogen acids are effective, probably because, by complexing the initial —Zn$^+$ species, they provide a medium for the reduction of this species by a second atom of zinc.

The method is particularly useful for ketones which contain phenolic or carboxylic groups. For example, the reduction of β-benzoylpropionic acid in toluene gives γ-phenylbutyric acid in 85% yield [2]:

$$PhCOCH_2CH_2CO_2H \xrightarrow{Zn-Hg-HCl} PhCH_2CH_2CH_2CO_2H$$
$$\gamma\text{-Phenylbutyric acid}$$

This type of reduction forms one step in the extension of benzenoid systems *via* Friedel-Crafts acylations (p. 392).

The reagent also reduces the olefinic bond in $\alpha\beta$-unsaturated ketones, acids, and esters; and benzyl halides and alcohols are hydrogenolyzed. Strongly hindered ketones give low yields and sometimes rearrangement products (e.g. $Ph_3C{-}CO{-}Ph \rightarrow Ph_2C{=}CPh_2$).

(2) *Wolff-Kishner method.* The hydrazones of aldehydes and ketones are reduced in vigorously basic conditions with the evolution of nitrogen, probably as follows:

$$R_2C{=}N{-}NH_2 \underset{-H^+}{\rightleftharpoons} [R_2C{=}N{-}\overset{-}{N}H \leftrightarrow R_2\overset{-}{C}{-}N{=}NH] \underset{H^+}{\rightleftharpoons}$$

$$[R_2CH{-}N{=}NH] \longrightarrow R_2CH_2 + N_2$$

The standard procedure is the *Huang-Minlon* modification. The hydrazone is formed by heating the carbonyl compound with hydrazine hydrate and potassium hydroxide in di- or tri-ethylene glycol under a water condenser. After completion of the formation of the hydrazone, the water condenser is removed so that the water liberated in the first reaction is distilled and the temperature rises to about 200°C, so bringing about decomposition of the hydrazone.

A newly developed modification employs potassium t-butoxide as the base and dimethyl sulphoxide as solvent. Alkoxide bases are very much more powerful in this solvent than in water or hydroxylic solvents (p. 89) and reaction occurs at room temperature in high yield. For example, benzophenone gives about 90% of diphenylmethane:

$$\text{PhCOPh} \xrightarrow{\text{N}_2\text{H}_4} \text{Ph}_2\text{C}{=}\text{N}{-}\text{NH}_2 \xrightarrow[25°C]{(CH_3)_3CO^-/(CH_3)_2SO} \text{Ph}_2\text{CH}_2$$

(3) *Mozingo method.* The carbonyl compound is converted with ethylene dithiol in the presence of a Lewis acid into its dithio-acetal or ketal and this is hydrogenolyzed over Raney nickel:

$$\text{R}_2\text{CO} + \text{HS}{-}\text{CH}_2\text{CH}_2{-}\text{SH} \xrightarrow[20°C]{Et_2\overset{+}{O}{-}\bar{B}F_3} \text{R}_2\text{C}\overset{\displaystyle S{-}CH_2}{\underset{\displaystyle S{-}CH_2}{\Big\langle}} \xrightarrow{H_2{-}Ni} \text{R}_2\text{CH}_2$$

Alternatively, the cyclic dithio compound is reduced by hydrogen-transfer from hydrazine at 100–200°C.

The Mozingo reaction is useful for reducing carbonyl compounds which are sensitive to mineral acid and bases, for the Clemmensen and Wolff-Kishner methods are then unsuitable.

(4) *Tosylhydrazone method.* Reaction of the carbonyl compound with toluene-*p*-sulphonylhydrazine gives the tosylhydrazone which is efficiently reduced by sodium borohydride (Ar = *p*-tolyl):

$$\text{R}_2\text{CO} + \text{H}_2\text{N}{-}\text{NH}{-}\text{SO}_2\text{Ar} \xrightarrow{-H_2O} \text{R}_2\text{C}{=}\text{N}{-}\text{NH}{-}\text{SO}_2\text{Ar}$$
$$\xrightarrow{NaBH_4} \text{R}_2\text{CH}_2 + \text{N}_2 + \text{Na}^+\ {}^-\text{O}_2\text{SAr}$$

For example, the keto group in androstan-17β-ol-3-one is reduced in this way in about 75% yield [52].

Reaction probably occurs as follows:

$$\text{R}_2\text{C}{=}\text{N}{-}\text{NH}{-}\text{SO}_2\text{Ar} \xrightarrow{-ArSO_2^-} [\text{R}_2\text{CH}{-}\text{N}{=}\text{NH}] \longrightarrow \text{R}_2\text{CH}_2 + \text{N}_2$$

(5) *Lithium aluminium hydride.* Aromatic ketones are reduced by lithium aluminium hydride in the presence of aluminium trichloride. Reaction occurs by reduction to the alcohol (p. 633) followed by hydrogenolysis of the benzylic system, aided by the Lewis acid:

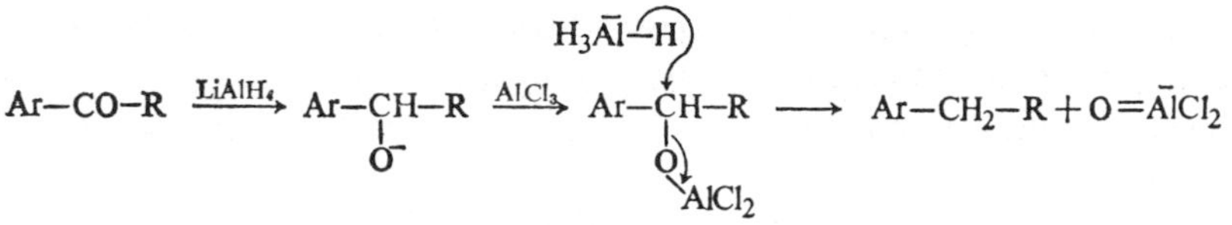

(*b*) TO ALCOHOLS

Carbonyl compounds are reduced to alcohols by a variety of reagents. Of the three general classes of reductive process, catalytic hydrogenation is not normally chosen because it is slow, but both hydride-transfer and electron-transfer reagents are employed.

(*i*) *Hydride transfer.* The alkali-metal hydrides such as sodium hydride are unsuitable reducing agents because of their insolubility in organic solvents and their powerful effects as catalysts for base-catalyzed condensations. The most commonly used hydride reducing agents are lithium aluminium hydride, sodium (or potassium) borohydride, and lithium borohydride.

(1) *Lithium aluminium hydride* is made by treating lithium hydride with aluminium trichloride in ether, and is generally used in very dry ether or tetrahydrofuran. All hydroxyl-, amino-, and thiol-containing compounds liberate hydrogen quantitatively from it, e.g.

$$C_2H_5OH + LiAlH_4 \longrightarrow C_2H_5O\text{—}\bar{A}lH_3\ Li^+ + H_2 \xrightarrow{3\ C_2H_5OH} (C_2H_5O)_4\bar{A}l\ Li^+$$

Each of the four hydrogen atoms in lithium aluminium hydride is available for transfer to carbonyl groups, e.g.

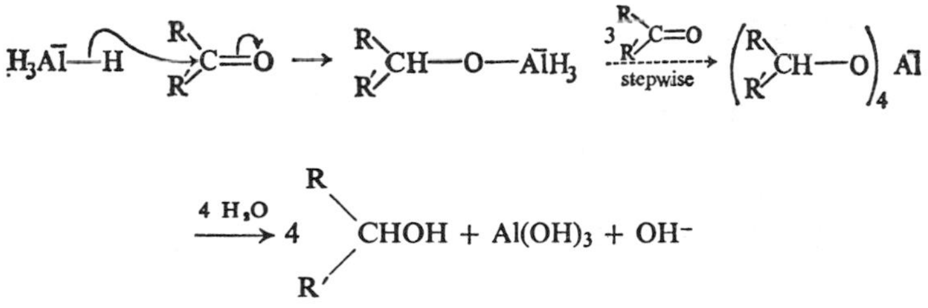

each step occurring less rapidly than the preceding one. (As a result, the replacement of two or three hydrogen atoms of the aluminium hydride anion by alkoxy groups gives less reactive and more selective reducing agents; examples are described later.) Finally, hydrolysis of the aluminium alkoxide gives the alcohol.

(2) *Sodium borohydride* is much less reactive. It can be used in alcoholic solvents and even in water, for it decomposes only enough to make the solution alkaline, after which it is stable.

$$NaBH_4 + 4\,H_2O \longrightarrow NaOH + B(OH)_3 + 4\,H_2$$

(3) *Lithium borohydride* is more reactive than the sodium analogue and reacts with hydroxylic compounds. It is usually employed in solution in tetrahydrofuran or diethylene glycol dimethyl ether (diglyme).

These three reagents differ considerably in their reducing power. Lithium aluminium hydride reduces not only aldehydes and ketones but also acids, acid chlorides, esters, nitriles, imines, and nitro groups, whereas sodium borohydride reduces only aldehydes, ketones, imines, and acid chlorides. Lithium borohydride resembles sodium borohydride except that it also reduces esters and nitriles. None of the reagents normally reduces olefinic, acetylenic, or N=N bonds, although the first reduces acetylenes containing α-hydroxy-substituents (p. 620) and reduces azo compounds in the presence of a Lewis acid (p. 653).

Except in simple cases, therefore, sodium borohydride is the reagent of choice. Typical examples are:

$$CH_3COCH_2COCH_3 \xrightarrow{\text{NaBH}_4} \underset{\underset{OH}{|}}{CH_3CH}CH_2\underset{\underset{OH}{|}}{CHCH_3}\quad (86\%)$$

$$CH_3CH{=}CHCHO \xrightarrow{\text{NaBH}_4} CH_3CH{=}CHCH_2OH\quad (85\%)$$

In rigid ring systems, the stereochemistry of hydride reductions appears to be determined usually by the relative importance of competing influences: steric hindrance to the approach of the reagent, and the stability of the final product. Unless steric effects are particularly severe, the latter factor dominates; for example, 10-methyl-2-decalone gives mainly the more stable equatorial alcohol, although this involves approach of the reagent from the more hindered side:

(4) *Cannizzaro reaction.* Aldehydes which do not have α-CH groups cannot undergo base-catalyzed condensation. Instead they react with bases by disproportionation involving the transfer of hydride ion, e.g.

$$PhCH=O \xrightarrow{OH^-} Ph-\overset{\overset{O\bar{\;}}{|}}{\underset{\underset{OH}{|}}{C}}-H \;\; \overset{Ph}{\underset{H}{C}}=O \longrightarrow PhCO_2H + PhCH_2O^-$$

$$\rightleftharpoons PhCO_2^- + PhCH_2OH$$

Crossed Cannizzaro reactions between one such aldehyde and formaldehyde result in the reduction of the former and the oxidation of the latter, for form-aldehyde is more reactive than other aldehydes towards nucleophiles and rapidly gives a high concentration of the donor anion:

$$CH_2=O \xrightarrow{OH^-} \overset{\overset{O\bar{\;}}{|}}{\underset{\underset{OH}{|}}{C}H}-H \;\; \overset{R}{\underset{H}{C}}=O \longrightarrow HCO_2H + RCH_2O^- \rightleftharpoons HCO_2^- + RCH_2OH$$

This fact can be exploited for reductions. For example, benzaldehyde is reduced by formaldehyde in the presence of potash in refluxing methanol to give 80% of benzyl alcohol [2]. The preparation of pentaerythritol from acetaldehyde and formaldehyde is also dependent on a crossed Cannizzaro reaction (p. 230).

(5) *Meerwein-Ponndorf-Verley reaction.* This is the reverse of Oppenauer oxidation (p. 591): equilibrium is established between the carbonyl group to be reduced and isopropanol on the one hand, and the required alcohol and acetone on the other, in the presence of aluminium isopropoxide. Since acetone is the lowest boiling constituent of the mixture, it can be continuously distilled so that the equilibrium is displaced to the right.

$$R_2C=O \;+\; Al(OCH(CH_3)_2)_3 \rightleftharpoons \overset{\overset{O}{\diagup\!\!\diagup}}{R_2C} \overset{}{Al(OPr^i)_2}$$

$$\xrightarrow[-(CH_3)_2CO]{} R_2CH-O-Al(OPr^i)_2 \xrightarrow[(CH_3)_2CHOH]{} R_2CHOH + Al(OPr^i)_3$$

For example, trichloroacetaldehyde is reduced to trichloroethanol in about 80% yield.

The reaction is specific to aldehydes and ketones; in particular, the olefinic double bond in $\alpha\beta$-unsaturated aldehydes or ketones is not reduced (compare lithium aluminium hydride, p. 619). However, the basic conditions may bring about side-reactions (as in the synthesis of reserpine; 21.7). Very hindered Grignard reagents effect reduction in a similar way (p. 210).

(ii) Electron-transfer reagents. These reagents are less selective than sodium borohydride and the Meerwein-Ponndorf-Verley reagent: e.g. they also reduce the olefinic double bond in $\alpha\beta$-unsaturated carbonyl compounds. Nevertheless, in simple cases they are rapid and efficient: e.g. methyl n-amyl ketone is reduced by sodium in ethanol to 2-heptanol in over 60% yield [2],

$$\text{n-C}_5\text{H}_{11}\text{COCH}_3 \xrightarrow{\text{Na–EtOH}} \text{n-C}_5\text{H}_{11}\text{—CH—CH}_3$$
$$\overset{|}{\text{OH}}$$

2-Heptanol

and n-heptaldehyde is reduced by iron in aqueous acetic acid to n-heptanol in 80% yield.

These reductions are stereoselective. In most cases, the thermodynamically more stable alcohol predominates; for example, 2-methylcyclohexanone gives mainly *trans*-2-methylcyclohexanol. The reason is not fully understood, but one possibility is as follows: the metal transfers one electron to the carbonyl group to form an anion-radical, this is protonated at carbon from the less hindered side, and a second electron is transferred to give the (usually less stable) alkoxide ion. This reacts with more of the ketone by a mechanism similar to that in the Meerwein-Ponndorf-Verley reduction above, giving an equilibrium mixture favourable to the more stable alkoxide, so that the final hydrolysis gives the more stable alcohol.

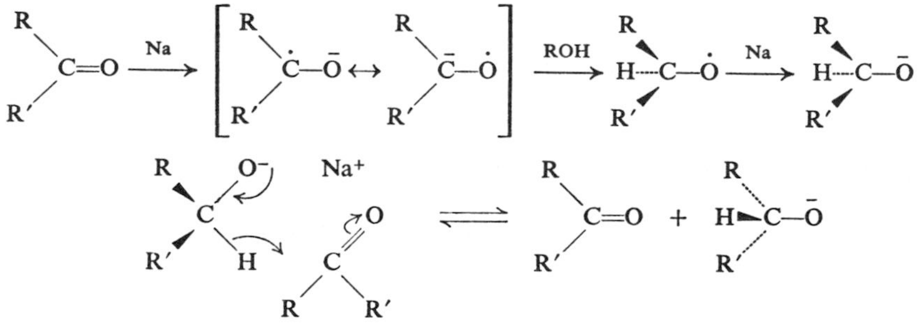

However, if the ketone is very reactive as a result of strain, the rate of the direct reduction may be so much greater than that of the attainment of the final equilibrium that the less stable alcohol is formed predominantly, as in the reduction of camphor:

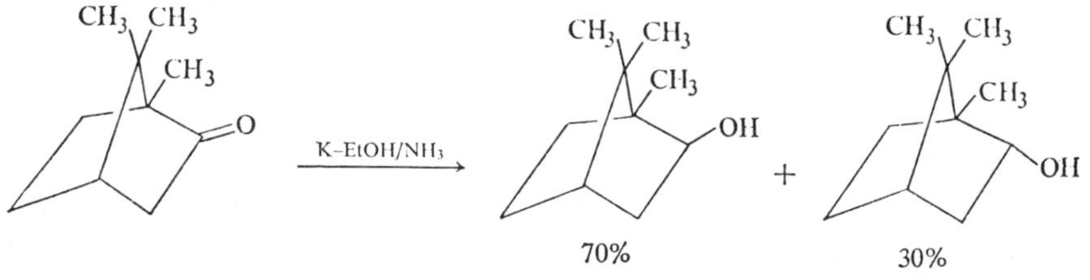

70%   30%

If the ketone contains at the α-position a substituent which is a good leaving-group, the intermediate anion undergoes elimination:

$$-\overset{\overset{O}{\|}}{C}-\underset{\underset{X}{|}}{C}H- \xrightarrow{\text{Na}} -\overset{\overset{O\cdot}{|}}{C}-CH- \xrightarrow{-X^-} -\overset{\overset{O\cdot}{|}}{C}=CH- \xrightarrow{\text{Na}} -\overset{\overset{O^-}{|}}{C}=CH-$$

$$\xrightarrow{\text{ROH}} -\overset{\overset{OH}{|}}{C}=CH- \rightleftharpoons -\overset{\overset{O}{\|}}{C}-CH_2-$$

Zinc is often used as the electron source in these reductions. For example, α-hydroxycyclodecanone gives cyclodecanone in about 75% yield when treated with zinc in a mixture of hydrochloric and acetic acids at 75–80°C [4]:

$$(CH_2)_8 \begin{matrix} C=O \\ | \\ CH-OH \end{matrix} \xrightarrow{\text{Zn-HCl/HOAc}} (CH_2)_8 \begin{matrix} C=O \\ | \\ CH_2 \end{matrix}$$

Cyclodecanone

(c) TO PINACOLS

In the absence of a proton-donor, electropositive metals reduce ketones to pinacols *via* the dimerization of anion-radicals:

$$R_2C{=}O \longrightarrow R_2\overset{\cdot}{C}-\overset{-}{O} \xrightarrow{\text{dimerizes}} \begin{matrix} R_2C-\overset{-}{O} \\ | \\ R_2C-\overset{-}{O} \end{matrix} \xrightarrow{2\ H^+} \begin{matrix} R_2C-OH \\ | \\ R_2C-OH \end{matrix}$$

The standard procedure employs amalgamated magnesium with benzene as solvent; the solid magnesium salt of the pinacol is formed and is hydrolyzed to the pinacol. For example, acetone gives pinacol itself in 45% yield after 2 hours in refluxing benzene [1]:

$$2\ (CH_3)_2CO \xrightarrow{\text{Mg-Hg}} \begin{matrix} (CH_3)_2C-O \\ | \qquad\quad \\ (CH_3)_2C-O \end{matrix}\!Mg \xrightarrow{H_2O} \begin{matrix} (CH_3)_2C-OH \\ | \\ (CH_3)_2C-OH \end{matrix}$$

Pinacol

The reaction is generally ineffective for aldehydes because they are too readily reduced to alcohols. Pinacols may also be formed by photochemical dimerization (p. 511).

19.5 Epoxides

(a) LITHIUM ALUMINIUM HYDRIDE

Epoxides are reduced to alcohols by lithium aluminium hydride. Since epoxides are readily obtained from olefins (p. 568), the overall reaction serves to hydrate the olefin. The procedure is complementary to the hydroboration method (p. 500) since the hydride selectively attacks the less alkylated carbon of the epoxide ring, so giving the more highly alkylated alcohol (p. 570), e.g.

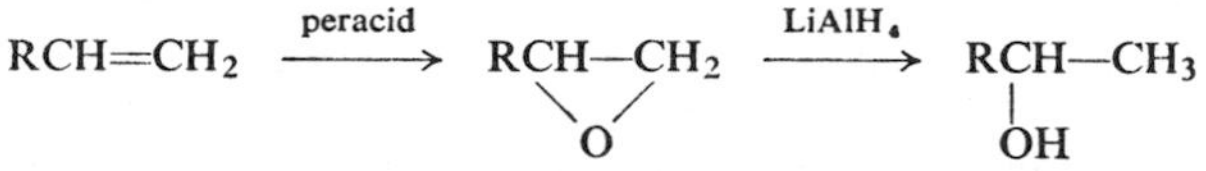

$$RCH=CH_2 \xrightarrow{\text{peracid}} RCH\!-\!CH_2 \xrightarrow{\text{LiAlH}_4} RCH\!-\!CH_3$$

with O bridging the $RCH\!-\!CH_2$ and OH on the $RCH\!-\!CH_3$.

The reaction has the *trans* stereochemistry characteristic of S_N2 reactions. Thus, in a rigid cyclic system the *axial* alcohol is formed, e.g.

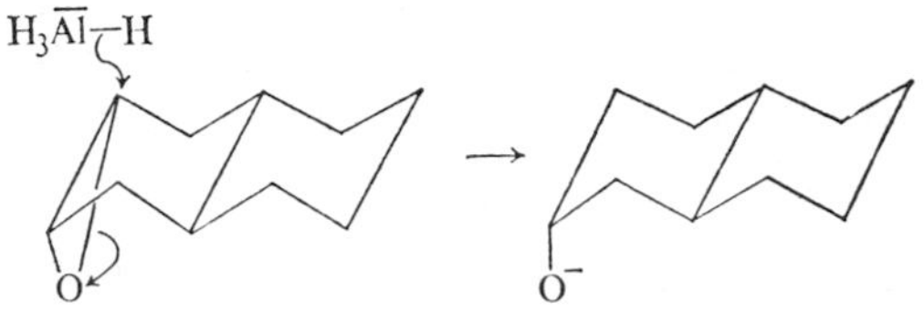

This therefore complements the methods for obtaining the *equatorial* alcohol by reduction of the corresponding ketone with electron-transfer reagents.

The strained ring in four-membered cyclic ethers is also cleaved by lithium aluminium hydride, e.g.

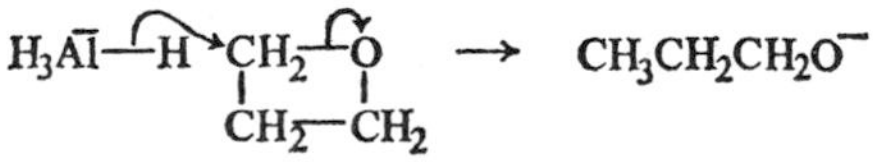

$$\longrightarrow CH_3CH_2CH_2O^-$$

but the near-strainless five-membered ethers are resistant. They are, however, opened in the more vigorous conditions obtained by using lithium aluminium hydride in the presence of aluminium trichloride (*cf.* p. 632); e.g. tetrahydrofuran gives n-butanol:

$+$ LiAlH$_4$ $\xrightarrow{\text{AlCl}_3}$ CH$_3$CH$_2$CH$_2$CH$_2$O$^-$

(b) HYDROBORATION

Epoxides are reduced by diborane to give mainly the *less* substituted alcohol, e.g.

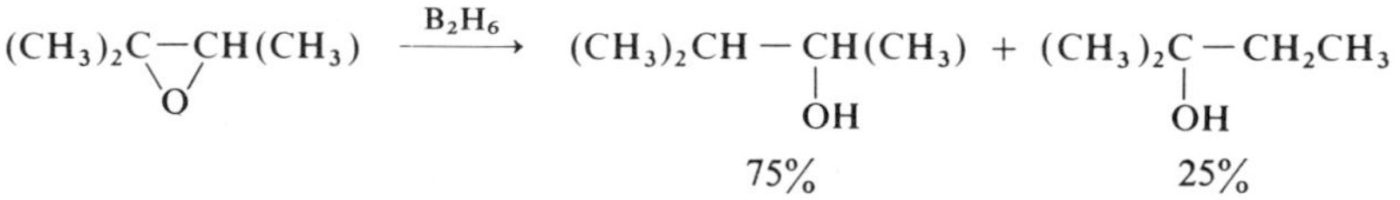

$$(CH_3)_2C\!-\!CH(CH_3) \xrightarrow{B_2H_6} (CH_3)_2CH - CH(CH_3) + (CH_3)_2C - CH_2CH_3$$

with O bridging the left species, OH on the first product, OH on the second product.

$$75\% \qquad\qquad 25\%$$

The method is therefore complementary to the use of lithium aluminium hydride.

With 1-alkylcycloalkene epoxides, the main product is the *cis*-disubstituted alcohol, e.g.

This therefore complements the reaction of 1-alkylcyclohexenes with diborane followed by alkaline hydrogen peroxide, which yields the *trans* product (p. 500).

19.6 Acids and Their Derivatives

Until the advent of the complex metal hydrides (1947), the only methods for reducing acids and related compounds were relatively unselective. For example, esters could be reduced to alcohols over copper chromite in vigorous conditions, although not over other catalysts, e.g.

$$(CH_3)_3C-CO_2Et \xrightarrow[\text{250°C, 220 atm.}]{\text{H}_2\text{-'Cu-Cr'}} (CH_3)_3C-CH_2OH$$

Neopentyl alcohol (88%)

but other reducible groups in the molecule would also react (e.g. C=C, C≡C, C=O).

Many methods, some of unique and others of moderate selectivity, are now available.

(a) REDUCTION TO THE ALCOHOL OR AMINE

Lithium aluminium hydride reduces acids and all their derivatives to the alcoholic or amino level.* The general mechanisms are:

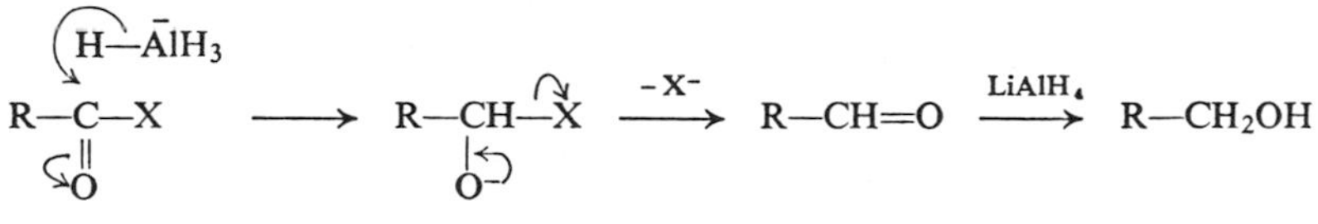

$$\left(\text{H}-\bar{\text{A}}\text{lH}_3\right)$$
$$R-\underset{\substack{\parallel\\O}}{C}-X \longrightarrow R-\underset{\substack{|\\O^-}}{CH}-X \xrightarrow{-X^-} R-CH{=}O \xrightarrow{\text{LiAlH}_4} R-CH_2OH$$

(X = OH, OR, OCOR, halogen)

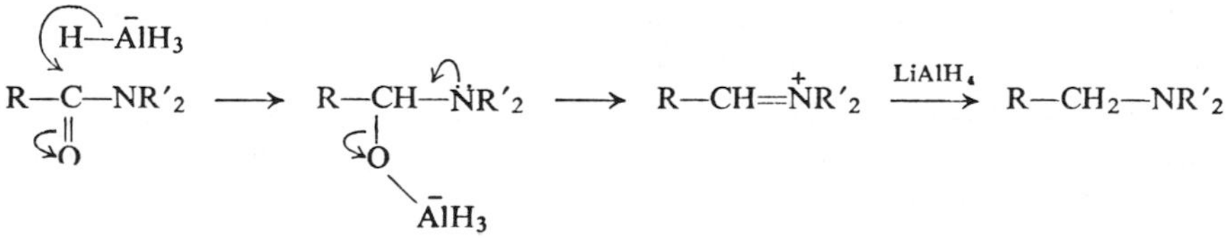

$$\left(\text{H}-\bar{\text{A}}\text{lH}_3\right)$$
$$R-\underset{\substack{\parallel\\O}}{C}-NR'_2 \longrightarrow R-\underset{\substack{|\\O\\ \\ \\ \bar{\text{A}}\text{lH}_3}}{CH}-NR'_2 \longrightarrow R-CH{=}\overset{+}{N}R'_2 \xrightarrow{\text{LiAlH}_4} R-CH_2-NR'_2$$

*Note the difference between the reaction with amides and those with other acid derivatives. It arises because —NR₂ is a less good leaving group than, e.g. —OH or —Cl.

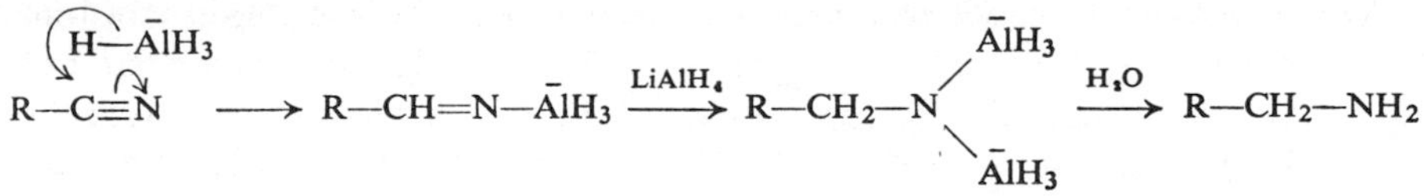

$$R-C\equiv N \xrightarrow{H-\bar{Al}H_3} R-CH=N-\bar{Al}H_3 \xrightarrow{LiAlH_4} R-CH_2-N\underset{\bar{Al}H_3}{\overset{\bar{Al}H_3}{\diagup}} \xrightarrow{H_2O} R-CH_2-NH_2$$

For example:

Acids: $(CH_3)_3C-CO_2H \longrightarrow (CH_3)_3C-CH_2OH$ (92%)

Esters: $Ph-CO_2Et \longrightarrow Ph-CH_2OH$ (90%)

Acid halides: $Cl_2CH-COCl \longrightarrow Cl_2CH-CH_2OH$ (65%) [4]

Anhydrides:

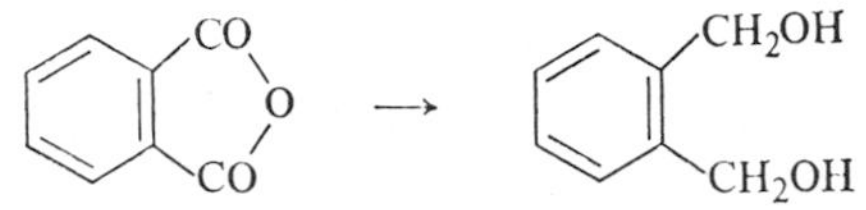

(87%)

Amides: $n\text{-}C_6H_{11}-CONMe_2 \longrightarrow n\text{-}C_6H_{11}-CH_2NMe_2$ (88%) [4]

Nitriles: $Ph-CN \longrightarrow Ph-CH_2NH_2$ (72%)

For acids and nitriles, diborane can be used:

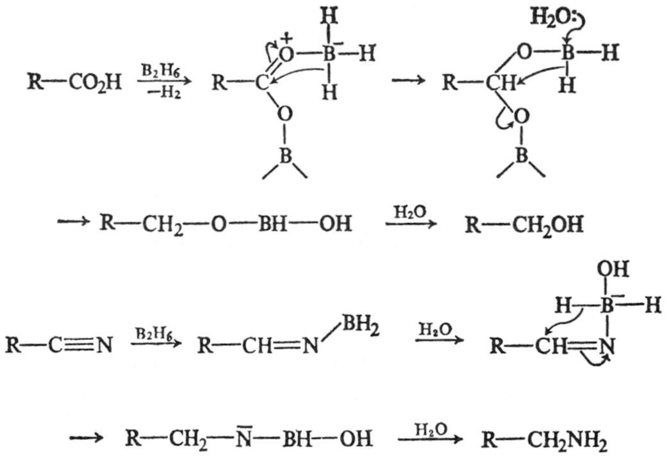

$$\longrightarrow R-CH_2-O-BH-OH \xrightarrow{H_2O} R-CH_2OH$$

$$\longrightarrow R-CH_2-\bar{N}-BH-OH \xrightarrow{H_2O} R-CH_2NH_2$$

This method has the advantage, compared with lithium aluminium hydride, that nitro groups are not reduced; for example, *p*-nitrobenzoic acid gives *p*-nitrobenzyl alcohol in about 80% yield.

Esters can be reduced by the acyloin method (p. 557) and by the *Bouveault-Blanc method* (sodium in refluxing ethanol); e.g. diethyl sebacate gives decamethylenediol in 75% yield:

$$EtO_2C-(CH_2)_8-CO_2Et \xrightarrow{Na-EtOH} HO-(CH_2)_{10}-OH$$
Decamethylenediol

Acid chlorides can be reduced with sodium borohydride in diglyme solution. Acid, ester, amide, C=C and C≡C groups in the molecule are not reduced, but aldehydic and ketonic groups are. Sodium trimethoxyborohydride, $Na[(CH_3O)_3BH]$, also reduces —COCl to —CH_2OH without affecting ester groups.

Nitriles can be reduced catalytically. In the absence of added ammonia, considerable quantities of secondary amines are formed since the primary amine can add to the imino-intermediate (i.e. the half-reduced stage):

$$R—C{\equiv}N \xrightarrow{H_2} R—CH{=}NH \xrightarrow{H_2} R—CH_2—NH_2$$

$$\begin{array}{c} R—CH{=}NH \\ \uparrow \\ R—CH_2—NH_2 \end{array} \longrightarrow \begin{array}{c} R—CH—NH_2 \\ | \\ R—CH_2—NH \end{array} \xrightarrow{-NH_3} \begin{array}{c} R—CH \\ \| \\ R—CH_2—N \end{array} \xrightarrow{H_2} RCH_2NHCH_2R$$

This is circumvented by the addition of ammonia, which assumes the role of nucleophile in place of the primary amine, and yields are then high: e.g. over Raney nickel, benzyl cyanide gives phenethylamine in about 85% yield in liquid ammonia under pressure [3] and sebaconitrile gives decamethylenediamine (79–80%) in ammoniacal ethanol [3]:

$$PhCH_2CN \xrightarrow[120°C,\ 120\ atm.]{H_2–Ni/NH_3} PhCH_2CH_2NH_2$$

Phenethylamine

$$NC—(CH_2)_8—CN \xrightarrow[125°C,\ 100\ atm.]{H_2–Ni/NH_3–EtOH} H_2N—(CH_2)_{10}—NH_2$$

Decamethylenediamine

The reduction of nitriles by sodium in ethanol suffers from the disadvantage that secondary amines are formed, as in the ammonia-free catalytic method.

It should be noted that none of the reagents described above is uniquely selective, but in some cases where two methods are available one complements the other. For instance, diborane is the reagent of choice for acids when ester groups are present but not when C=C or C≡C bonds are, whereas lithium aluminium hydride is suitable in the latter case but not in the former. Benzylic systems are liable to hydrogenolysis, so that reductions of $ArCO_2H$, $ArCO_2R$, $ArCOCl$, $ArCONR_2$, $ArCH_2OCOR$, and $ArCH_2NHCOR$ are likely to give $ArCH_3$.

(b) REDUCTION TO ALDEHYDES
It is often necessary to convert acids and their derivatives into aldehydes, for aldehydes are not readily made by other methods whereas acids are readily available. Selective procedures are required since aldehydes are easily reduced to

alcohols, and a number of older, lengthy procedures (e.g. Grundmann's synthesis) are now being superseded by techniques using modified complex metal hydrides.

(*i*) *Acids.* The only reasonable reducing agent is lithium in ethylamine. Yields are in general low, e.g.

$$\text{n-}C_7H_{15}\text{—}CO_2H \longrightarrow \text{n-}C_7H_{15}\text{—}CHO$$
$$26\%$$

and sometimes zero (e.g. for pivalic acid), but in isolated instances yields of up to 80% have been recorded.

Satisfactory yields can be obtained by reducing the imidazolide of the acid with lithium aluminium hydride, e.g.

$$\text{PhCO}_2H + \underset{}{\text{[imidazole]}}N\text{—CO—N}\underset{}{\text{[imidazole]}} \longrightarrow \text{PhCO—N}\underset{}{\text{[imidazole]}} \xrightarrow{\text{LiAlH}_4} \text{PhCHO}$$
$$77\%$$

(*ii*) *Esters.* (1) *McFadyen and Stevens'* method consists in converting the ester into its hydrazide, treating this with benzenesulphonyl chloride, and hydrolyzing the product with base. The principle of the method is analogous to that of the Wolff-Kishner reduction (p. 631) and to the method for hydrogenolyzing primary aliphatic amines (p. 629):

$$\text{ArCO}_2\text{Et} \xrightarrow[-\text{EtOH}]{N_2H_4} \text{ArCONHNH}_2 \xrightarrow[-\text{HCl}]{\text{PhSO}_2\text{Cl}} \overset{\text{HO}^-}{\underset{H}{\text{ArCO—N—NH—SO}_2\text{Ph}}}$$

$$\xrightarrow[150°C]{Na_2CO_3/(CH_2OH)_2} [\text{ArCO—N}{=}\text{NH}] \longrightarrow \text{ArCHO} + N_2$$

Only aromatic aldehydes can be prepared in this way and yields are often low.

(2) *Di-isobutylaluminium hydride*, prepared by the reduction of isobutylene in the presence of aluminium,

$$(\text{CH}_3)_2\text{C}{=}\text{CH}_2 + 3/2\,H_2 + Al \longrightarrow (\text{Bu}^i)_3\text{Al} \xrightarrow[-(CH_3)_2C=CH_2]{140°C} (\text{Bu}^i)_2\text{AlH}$$

reduces esters in hydrocarbon solvents at −70°C, e.g.

$$\text{n-}C_3H_7\text{—}CO_2\text{Et} \longrightarrow \text{n-}C_3H_7\text{—}CHO \quad (88\%)$$

$$\text{Ph—}CO_2\text{Et} \longrightarrow \text{Ph—CHO} \quad (62\%)$$

Amide, nitrile, and acetylenic groups are also reduced.

(*iii*) *Acid halides*. (1) *Rosenmund's method* consists in reducing the acid chloride with hydrogen on a palladium catalyst supported on barium sulphate or, less commonly, on calcium carbonate or charcoal.

$$R—COCl + H_2 \xrightarrow{Pd} R—CHO + HCl$$

Temperature control is important: the reaction should be conducted at the lowest temperature possible in order to prevent further reduction. Refluxing toluene or xylene (i.e. 110–140°C) is usually suitable, and the reaction can be followed by passing the hydrogen chloride into standard base. It is sometimes helpful partially to poison the catalyst; a commonly used poison is obtained by refluxing sulphur in quinoline.

Yields are normally high, even with sterically hindered compounds, e.g.

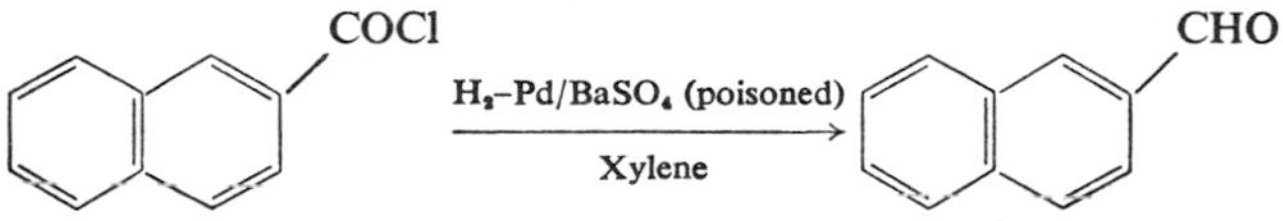

2-Naphthaldehyde
(74–81%) [3]

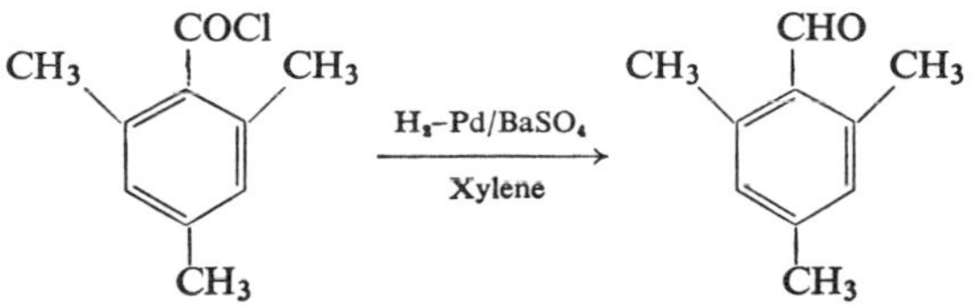

Mesitaldehyde (70–80%) [3]

(2) *Desulphurization* of thiol esters on Raney nickel which has been partially deactivated by boiling with acetone is effective, e.g.

$$CH_3CH_2—COCl \xrightarrow[-HCl]{C_2H_5SH/pyridine} CH_3CH_2—COSC_2H_5 \xrightarrow[-C_2H_5SH]{H_2-Ni} CH_3CH_2—CHO \quad (73\%)$$

(3) *Grundmann's method* is a step-wise procedure, but each step goes in good yield and the overall result is reasonable. The acid chloride is treated with diazomethane to give the diazoketone (p. 463), this is added slowly to acetic acid at 60–70°C and then boiled, the carbonyl group is reduced and the ester group hydrolyzed, and the resulting diol is cleaved with lead tetra-acetate or sodium metaperiodate (p. 595):

$$R—COCl \xrightarrow[-HCl]{CH_2N_2} R—COCHN_2 \xrightarrow[-N_2]{CH_3CO_2H} R—COCH_2OCOCH_3$$

$$\xrightarrow[\text{or } Al(OPr^i)_3]{LiAlH_4} R-CH-CH_2 \xrightarrow[\text{or } NaIO_4]{Pb(OAc)_4} R-CHO + HCHO$$
$$\qquad\qquad\quad |\quad\ |$$
$$\qquad\qquad\ OH\ OH$$

Both aliphatic and aromatic acid chlorides are reduced, but $\alpha\beta$-unsaturated acid chlorides form pyrazolines in the first stage (p. 297):

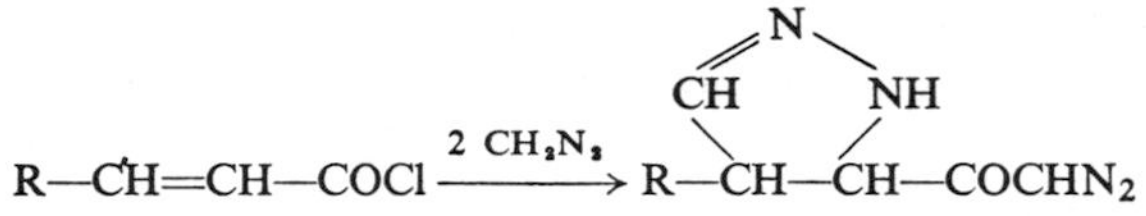

$$R-\overset{.}{C}H=CH-COCl \xrightarrow{2\ CH_2N_2} R-CH-CH-COCHN_2$$

(4) *Lithium tri-t-butoxyaluminium hydride*, $Li[(Bu^tO)_3AlH]$, and *tri-n-butyltin hydride*, $(Bu^n)_3SnH$, have recently been used successfully. The former reagent, prepared from lithium aluminium hydride and t-butanol in ether, acts efficiently in diglyme solution at $-78°C$. Yields are comparable with those from Rosenmund's procedure, and the reagent does not affect nitro, cyano, and ester groups, e.g.

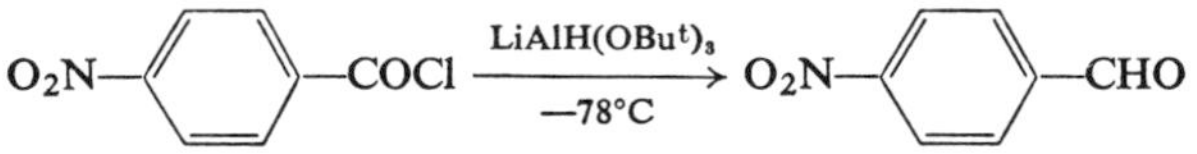

$$O_2N-\langle\ \rangle-COCl \xrightarrow[-78°C]{LiAlH(OBu^t)_3} O_2N-\langle\ \rangle-CHO$$

p-Nitrobenzaldehyde (81%)

(*iv*) *Amides.* (1) The *Sonn-Müller* reaction involves conversion of an anilide or toluidide into the imino-chloride which is reduced to the imine with tin(II) chloride; hydrolysis gives the aldehyde.

$$ArCOCl \xrightarrow[-HCl]{PhNH_2} ArCONHPh \rightleftharpoons Ar\overset{OH}{\overset{|}{C}}=NPh \xrightarrow{PCl_5}$$

$$Ar\overset{Cl}{\overset{|}{C}}=NPh \xrightarrow{SnCl_2} ArCH=NPh \xrightarrow{H_2O-H^+} ArCHO + PhNH_3^+$$

For example, *o*-tolualdehyde can be obtained in up to 70% yield [3]:

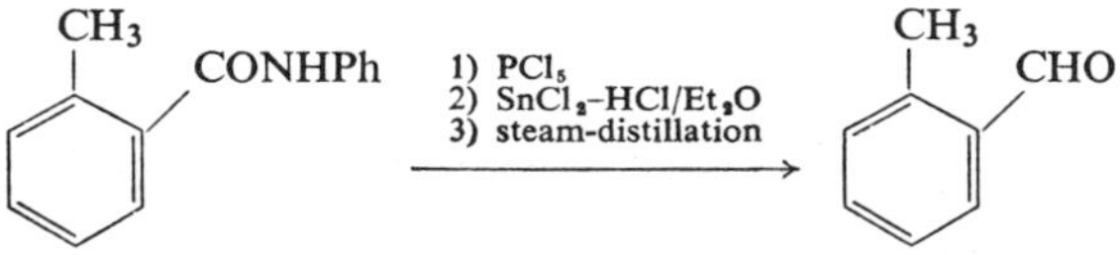

$$\text{(structure CH}_3\text{, CONHPh)} \xrightarrow[\substack{1)\ PCl_5 \\ 2)\ SnCl_2-HCl/Et_2O \\ 3)\ steam-distillation}]{} \text{(structure CH}_3\text{, CHO)}$$

o-Tolualdehyde

This method is not applicable to aliphatic aldehydes because the intermediate imino-chlorides are unstable. However, $\alpha\beta$-unsaturated acid chlorides can usually be reduced so that the procedure complements Grundmann's method.

(2) *The hydrolysis of Reissert compounds* has proved suitable for the preparation of some aromatic aldehydes. The acid chloride is added to quinoline in aqueous potassium cyanide and the resulting Reissert adduct is hydrolyzed with mineral acid:

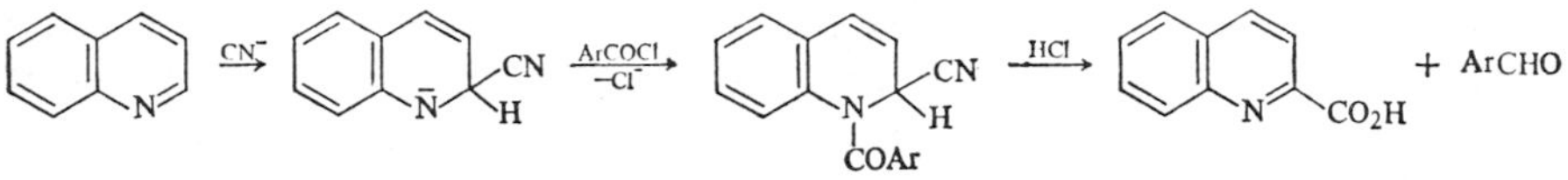

For example, *o*-nitrobenzoyl chloride gives *o*-nitrobenzaldehyde in about 65% yield. The method has also been used for introducing the carboxyl group into quinolines.

(3) *Sodium in liquid ammonia,* containing ammonium acetate as a weak proton-donor, reduces amides to aldehydes:

$$R-\underset{\underset{O}{\|}}{C}-NH_2 \xrightarrow{2e} R-\underset{\underset{O^-}{|}}{\bar{C}}-NH_2 \xrightarrow{NH_4^+OAc^-} R-\underset{\underset{O^-}{|}}{CH}-NH_2 \longrightarrow R-CH=O$$

In the presence of a stronger proton-donor such as ethanol further reduction occurs to give the alcohol.

(4) *Lithium aluminium hydride* reduces disubstituted amides to aldehydes provided that the temperature is kept low:

$$\overset{H-\bar{Al}H_3}{R-\underset{\underset{O}{\|}}{C}-NR'_2} \longrightarrow R-\underset{\underset{O-\bar{Al}H_3}{|}}{CH}-NR'_2 \xrightarrow{H_2O} \left[R-CH\begin{matrix}NR'_2\\OH\end{matrix}\right] \longrightarrow RCHO$$

At higher temperatures the first intermediate undergoes elimination,

$$R-\underset{\underset{O-\bar{Al}H_3}{|}}{CH}-NR'_2 \longrightarrow R-CH=\overset{+}{N}R'_2 + H_3\bar{Al}-O^-$$

and further reduction to the amine ensues (p. 639). The *N*-methylanilide is usually employed, e.g.

$$CH_3-CO-N\begin{matrix}CH_3\\Ph\end{matrix} \xrightarrow{LiAlH_4} CH_3CHO \quad (68\%)$$

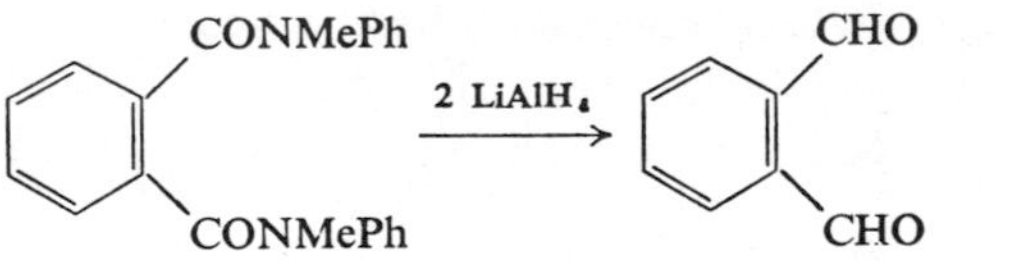

Phthalaldehyde (60%)

(5) *Di-isobutylaluminium hydride* (p. 642) and *lithium diethoxyaluminium hydride*, $Li[(EtO)_2AlH_2]$ (from lithium aluminium hydride with two moles of ethanol), have recently been developed for the conversion $R—CONR'_2 \rightarrow R—CHO$.

(*v*) *Nitriles.* (1) The *Stephen reaction* involves the addition of the nitrile to a suspension of anhydrous tin(II) chloride in ether saturated with hydrogen chloride, followed by hydrolysis. The reduction occurs at room temperature and is equivalent to that in the Sonn-Müller reaction (p. 644). For example, 2-naphthaldehyde can be prepared in about 75% yield [3]:

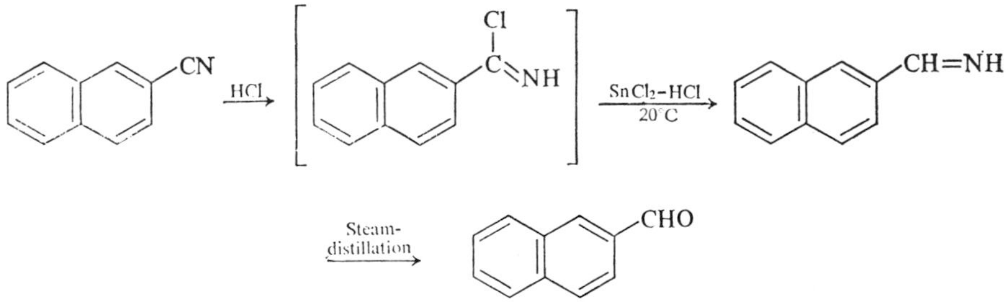

2-Naphthaldehyde

(2) *Hydrogen and Raney nickel* in the presence of semicarbazide and water gives the semicarbazone of the aldehyde from which the aldehyde is liberated by an exchange reaction with formaldehyde. Over-reduction is here prevented by the trapping of the aldehyde.

$$R—C{\equiv}N \xrightarrow{H_2-Ni} [R—CH{=}NH] \xrightarrow{H_2O} [R—CH{=}O]$$

$$\xrightarrow{H_2NNHCONH_2} R—CH{=}NNHCONH_2 \xrightarrow{CH_2=O} R—CHO$$

(3) Both *di-isobutylaluminium hydride* (p. 642) and *lithium triethoxyaluminium hydride*, $Li[(EtO)_3AlH]$ (from lithium aluminium hydride and three moles of ethanol), have recently been used successfully, e.g.

$$CH_3CH_2CH_2CN \xrightarrow[2)H_2O]{1)LiAlH(OEt)_3} CH_3CH_2CH_2CHO$$
n-Butyraldehyde (68%)

The reaction presumably occurs by way of $RCH{=}N—Al(OEt)_3$ (*cf.* p. 640).

19.7 Systems Containing Nitrogen

(a) NITRO COMPOUNDS

The reduction of aliphatic nitro compounds to amines is of little importance because the amino group is easily introduced in a number of other ways (e.g. from the halide and an amine (p. 324)). Lithium aluminium hydride and catalytic reduction on Raney nickel are both suitable methods: e.g. 2-nitrobutane is reduced to 2-aminobutane in 85% yield by the former reagent (compare the mode of reduction of aromatic nitro compounds by $LiAlH_4$; below).

The reduction of aromatic nitro compounds is much more important because the nitro group can be introduced into a wide variety of aromatic systems by nitration whereas the amino group can only be introduced directly into those aromatic compounds which are strongly activated to nucleophiles (p. 425). Several types of reduction product are obtainable, as illustrated for nitrobenzene:

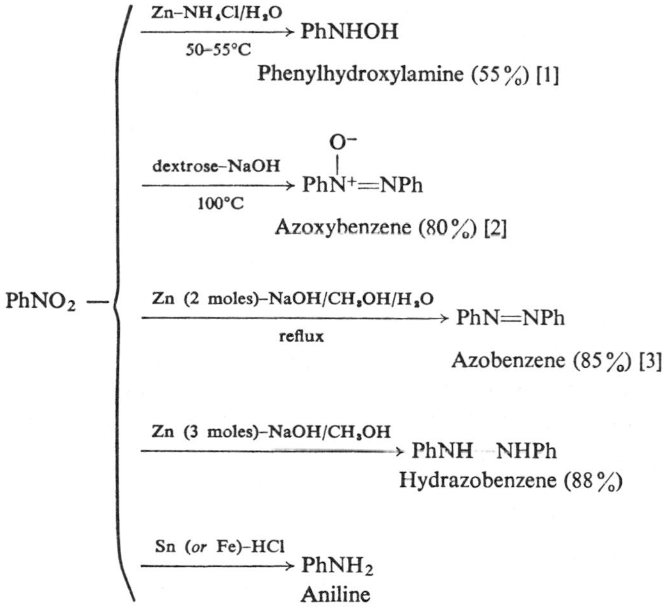

$PhNO_2$

$\xrightarrow[\text{50-55°C}]{\text{Zn-NH}_4\text{Cl/H}_2\text{O}}$ PhNHOH

Phenylhydroxylamine (55%) [1]

$\xrightarrow[\text{100°C}]{\text{dextrose-NaOH}}$ $PhN^+{=}NPh$ with O^- on N^+

Azoxybenzene (80%) [2]

$\xrightarrow[\text{reflux}]{\text{Zn (2 moles)-NaOH/CH}_3\text{OH/H}_2\text{O}}$ PhN=NPh

Azobenzene (85%) [3]

$\xrightarrow{\text{Zn (3 moles)-NaOH/CH}_3\text{OH}}$ PhNH—NHPh

Hydrazobenzene (88%)

$\xrightarrow{\text{Sn (or Fe)-HCl}}$ $PhNH_2$

Aniline

In neutral solution buffered with ammonium chloride, phenylhydroxylamine is the main product. In basic solution, using a weak reducing agent such as dextrose or glucose, the intermediate nitrosobenzene condenses with phenylhydroxylamine to give azoxybenzene, which the more powerful reducing agents convert into azobenzene and hydrazobenzene. Lithium aluminium hydride gives azobenzene quantitatively, and in the presence of Lewis acids bring about further reduction to hydrazobenzene. Metals in acid solution bring about reduction

to aniline; industrially, iron and dilute hydrochloric acid are used, the product is
neutralized with lime, and aniline is isolated by distillation in steam.

The only member of the series which cannot be isolated is nitrosobenzene,
but this is obtainable by the oxidation of phenylhydroxylamine (p. 601).

Controlled electrolytic reduction occurs in exact stages:

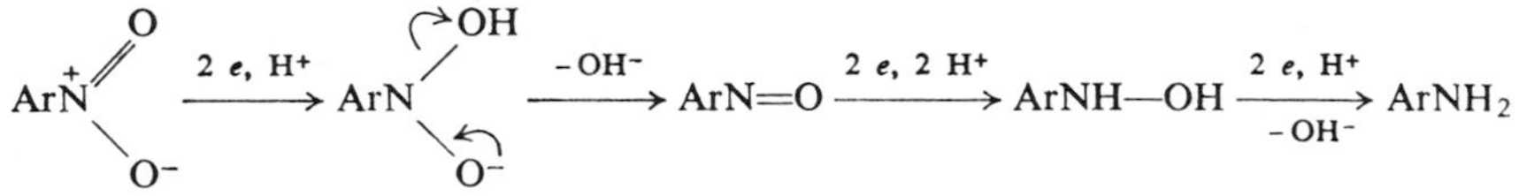

$$
ArN\overset{O}{\underset{O^-}{\diagdown}} \xrightarrow[]{2\ e,\ H^+} ArN\overset{OH}{\underset{O^-}{\diagdown}} \xrightarrow[]{-OH^-} ArN{=}O \xrightarrow[]{2\ e,\ 2\ H^+} ArNH{-}OH \xrightarrow[-OH^-]{2\ e,\ H^+} ArNH_2
$$

It is usually the amino group which is required from the reduction, and the
problem is to find a selective reagent which leaves other reducible groups intact.
Thus, the standard reagent for reducing nitrobenzene to aniline, tin and hydrochloric acid, is unsuitable for many substituted nitro compounds.

Catalytic methods are suitable provided that C=C and C≡C bonds are not
present: e.g.

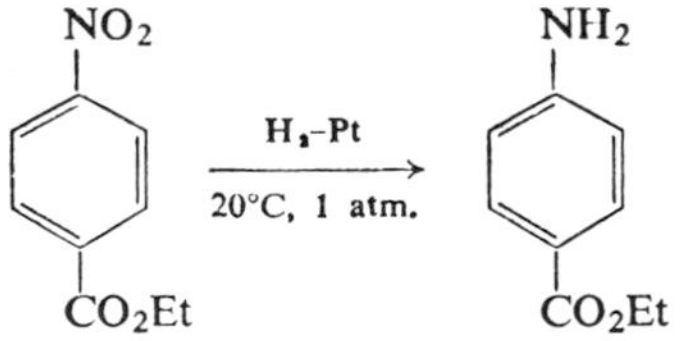

Ethyl p-aminobenzoate
(90–100%) [1]

and transfer-hydrogenation by hydrazine on palladium is also very effective.
Electron-transfer reagents of mild type can be used, e.g.

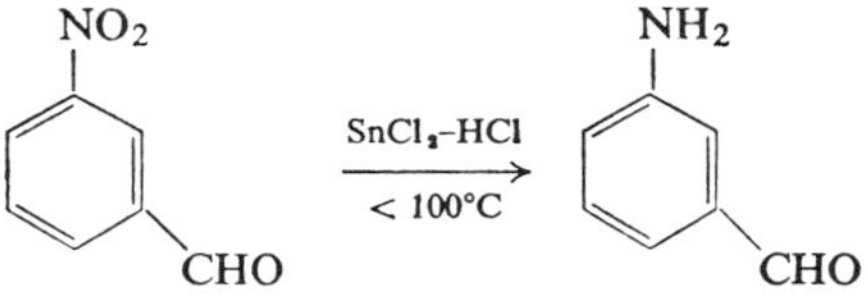

m-Aminobenzaldehyde

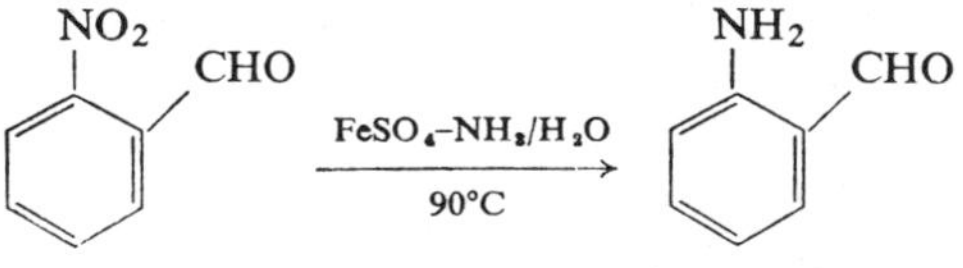

o-Aminobenzaldehyde
(70–75%) [3]

The complement to the last two reactions, reduction of the aldehydic group without reduction of the nitro group, can be effected with sodium borohydride.

It is also possible to reduce one nitro group in the presence of another. The older method employed ammonium sulphide, e.g.

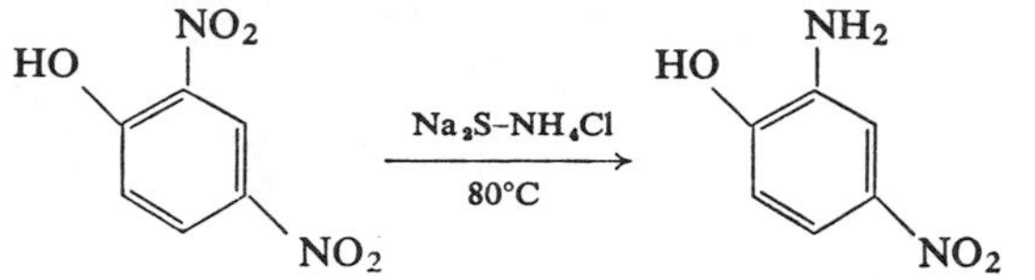

2-Amino-4-nitrophenol
(65%) [3]

but catalytic methods are usually more efficient. For example, *m*-dinitrobenzene gives *m*-nitroaniline almost quantitatively when reduced on palladium with cyclohexene as the hydrogen-donor (p. 612):

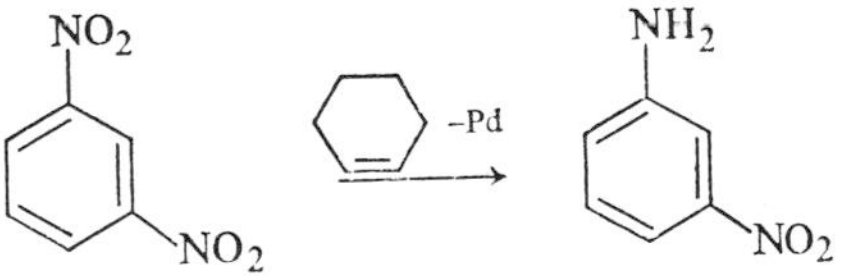

m-Nitroaniline

(b) IMINES
Imines of the type RCH=NR′ and RCH=$\overset{+}{N}$R′$_2$ can be selectively reduced with sodium borohydride:

$$RCH{=}NR' \longrightarrow RCH_2{-}NHR'$$

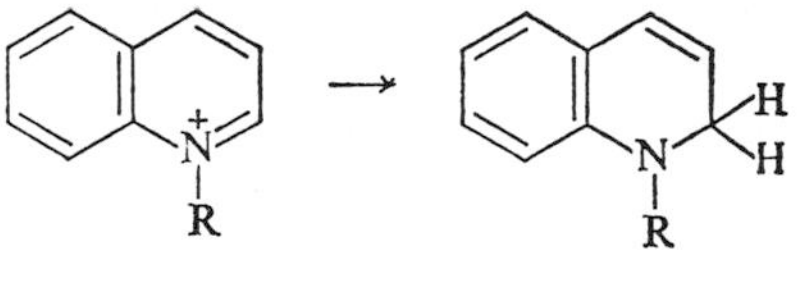

A newer and apparently general reagent is dimethylamine borane, which reduces C=N without reducing acids, esters, or nitro groups. Yields are high, e.g.

$$PhCH{=}NPh \xrightarrow[\text{in } CH_3CO_2H]{(CH_3)_2\overset{+}{N}H-\overset{-}{B}H_3} PhCH_2NHPh + (CH_3)_2NH + B(OCOCH_3)_3$$

84%

Leuckart reactions. When the reaction between an aldehyde or ketone and ammonia or a primary or secondary amine is carried out in the presence of formic acid at fairly high temperatures, the imine or iminium ion which is formed is reduced *in situ* to an amine. Formic acid acts as a hydride-transfer agent:

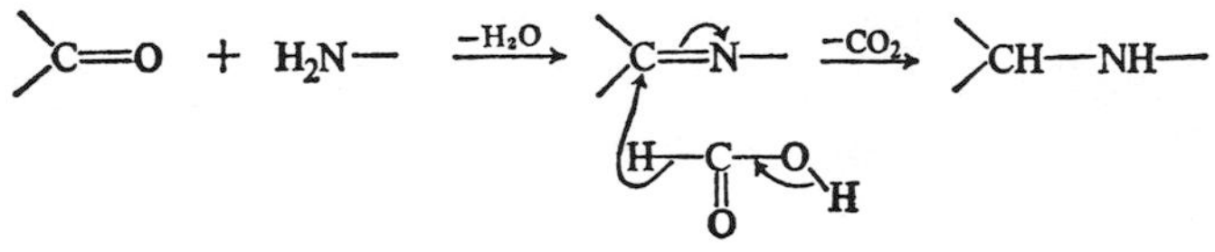

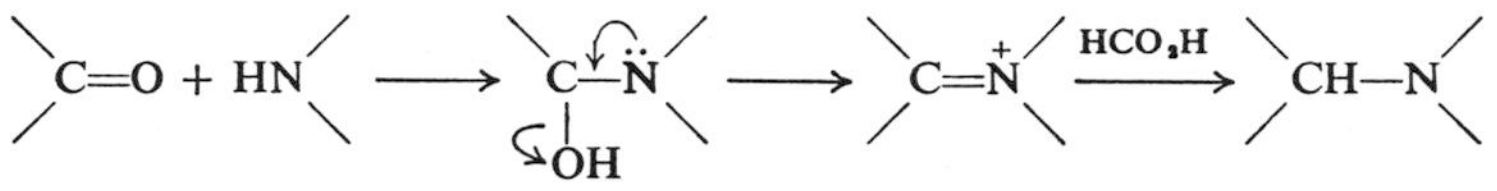

Alternatively, formic acid can be replaced by hydrogen on Raney nickel, or by sodium cyanoborohydride, for example:

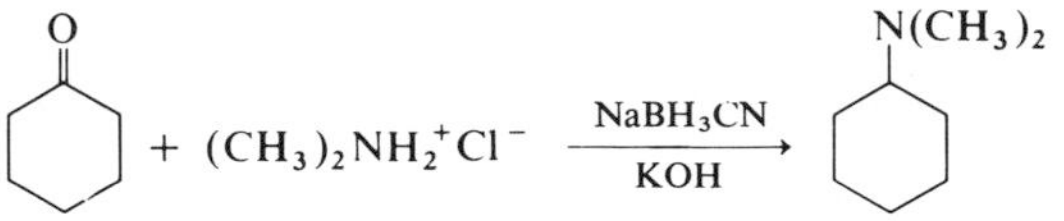

52% [52]

Initially, the reaction with ammonia gives a primary amine, but this may react with the imine to give a secondary amine (*cf.* p. 641):

$$RCHO + NH_3 \longrightarrow [RCH{=}NH] \longrightarrow RCH_2NH_2 \xrightarrow{RCH=NH} (RCH_2)_2NH$$

This is particularly common with the lower aliphatic aldehydes, so that the yields of individual compounds are low, e.g.

$$n\text{-}C_3H_7CHO + NH_3 \xrightarrow{H_2\text{-}Ni} n\text{-}C_4H_9NH_2 + (n\text{-}C_4H_9)_2NH$$
$$\qquad\qquad\qquad\qquad\qquad 32\% \qquad\qquad 12\%$$

but ketones and aromatic aldehydes give primary amines in good yield, evidently because the corresponding imines, like the carbonyl compounds, are less reactive towards nucleophiles. For example,

$$PhCHO + NH_3 \xrightarrow{HCO_2H} PhCH_2NH_2 \quad (85\%)$$

$$(CH_3)_2CO + NH_3 \xrightarrow{HCO_2H} (CH_3)_2CHNH_2 \quad (62\%)$$

Primary amines react to give secondary amines, e.g.

$$CH_3CH_2COCH_3 + CH_3NH_2 \xrightarrow{H_2-Ni} CH_3CH_2\overset{\overset{\displaystyle CH_3}{|}}{C}HNHCH_3 \quad (69\%)$$

(see also the synthesis of riboflavin; p. 696), and it is possible, without a great reduction in yield, to start with the corresponding nitro compound from which the primary amine is generated *in situ*, e.g.

$$CH_3CH_2COCH_3 + CH_3NO_2 \xrightarrow{H_2-Ni} CH_3CH_2\overset{\overset{\displaystyle CH_3}{|}}{C}HNHCH_3 \quad (59\%)$$

Again, the lower aldehydes tend to react further to give tertiary amines; formaldehyde, which is a particularly reactive aldehyde, invariably does so, and this is exploited as a convenient method for the dimethylation of amines. For example, refluxing a mixture of phenethylamine (1 mole), formaldehyde (3 moles), and formic acid (5 moles) gives *N*-dimethylphenethylamine in about 80% yield [3]:

$$PhCH_2CH_2NH_2 + 3 CH_2O + 5 HCO_2H \longrightarrow PhCH_2CH_2N(CH_3)_2$$

Secondary amines can react only to give tertiary amines; the reaction is most efficient when formaldehyde and aliphatic amines are employed, e.g.

$$(C_2H_5)_2NH + CH_2O \xrightarrow{H_2-Ni} (C_2H_5)_2NCH_3 \quad (80\%)$$

(c) OXIMES

Electron-transfer reagents are normally employed, as in the formation of n-heptylamine in 60% yield from heptaldoxime with sodium in ethanol [2],

$$C_6H_{13}-CH=NOH \xrightarrow{Na-EtOH} C_6H_{13}-CH_2-NH_2$$

and in the Knorr synthesis of pyrroles (p. 661), where sodium dithionite is preferably used:

$$R-CO-\overset{\overset{\displaystyle |}{\underset{\underset{\displaystyle OH}{\diagdown}}{N}}}{\underset{\|}{C}}-CO_2Et \xrightarrow{Na_2S_2O_4} R-CO-\overset{\overset{\displaystyle |}{\underset{\underset{\displaystyle NH_2}{|}}{C}}}{\underset{}{C}}H-CO_2Et$$

Catalytic methods are reasonably efficient but tend to produce some of the secondary amine, e.g.

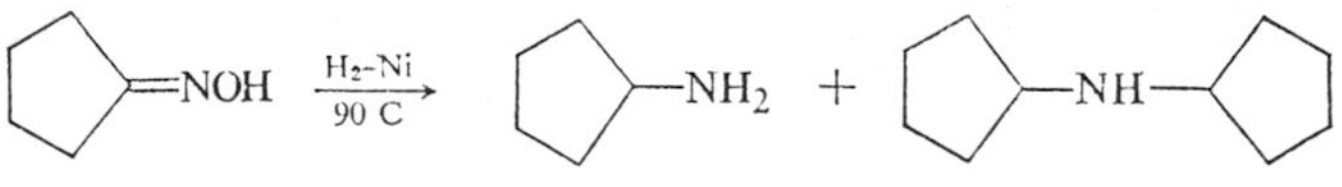

80%   10%

(*d*) NITROSO COMPOUNDS

C-Nitroso compounds can be converted into hydroxylamines by controlled electrolytic reduction (p. 648). Reduction to amines is normally brought about by electron-transfer reagents, e.g.

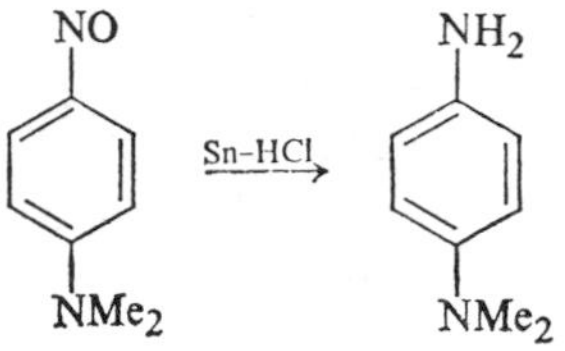

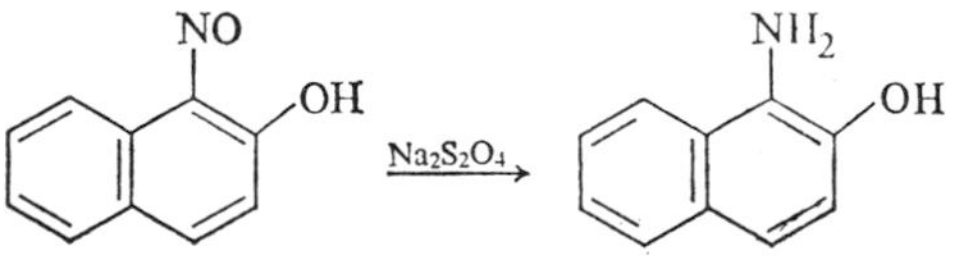

1-Amino-2-naphthol
(70%) [2]

These reactions are commonly used for obtaining amino groups in aromatic nuclei which are strongly activated towards electrophiles, since it is often easier to nitrosate these compounds than to nitrate them (see the synthesis of pteridine; p. 696).

N-Nitroso compounds are reduced by mild electron-transfer reagents to substituted hydrazines: e.g. *N*-nitroso-*N*-methylaniline with zinc in acetic acid gives 55% of *N*-methyl-*N*-phenylhydrazine [2]:

$$\text{PhN—NO} \xrightarrow{\text{Zn-HOAc}} \text{PhN—NH}_2$$
$$\quad\;\;|\qquad\qquad\qquad\;\;|$$
$$\quad\;\text{CH}_3\qquad\qquad\qquad\text{CH}_3$$

Stronger reducing agents rupture the N—N bond, e.g.

$$\text{PhN—NO} \xrightarrow{\text{Sn-HCl (conc.)}} \text{PhNHCH}_3 + \text{NH}_3$$
$$\quad\;\;|$$
$$\quad\;\text{CH}_3$$

(*e*) AZO COMPOUNDS

Azo compounds are reduced to hydrazo compounds by lithium aluminium hydride in the presence of a Lewis acid, as already described (p. 634). A more useful type of reduction is brought about by sodium dithionite which cleaves the N=N bond to give amines.* Since azo compounds can be obtained by diazonium-coupling both to strongly activated aromatic compounds and to activated methylene compounds (13.4), this provides a useful route for the introduction of the amino group, e.g.

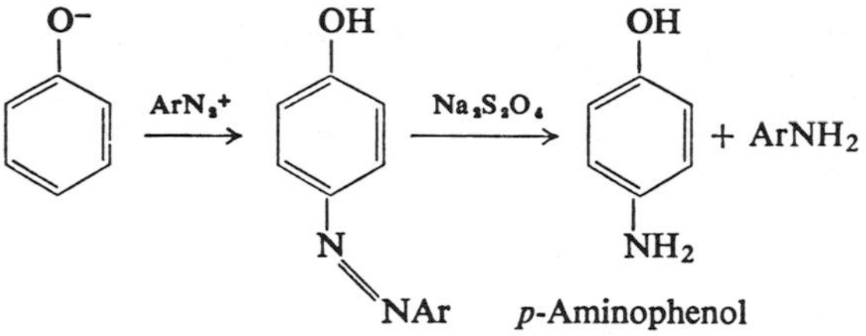

p-Aminophenol

An alternative method, nitrosation followed by reduction, is also available.

The reduction of diazonium salts both to arylhydrazines and to hydrocarbons has been described earlier (pp. 444, 439).

19.8 Systems Containing Sulphur

Catalytic reduction of sulphur-containing compounds removes the sulphur as hydrogen sulphide. Reduction with retention of the sulphur atom requires selective methods; the less active electron-transfer reagents are usually employed.

(*a*) DISULPHIDES

Zinc in refluxing acetic acid is normally used. For example, thiosalicylic acid can be obtained from anthranilic acid in 75–80% overall yield in this way [2]:

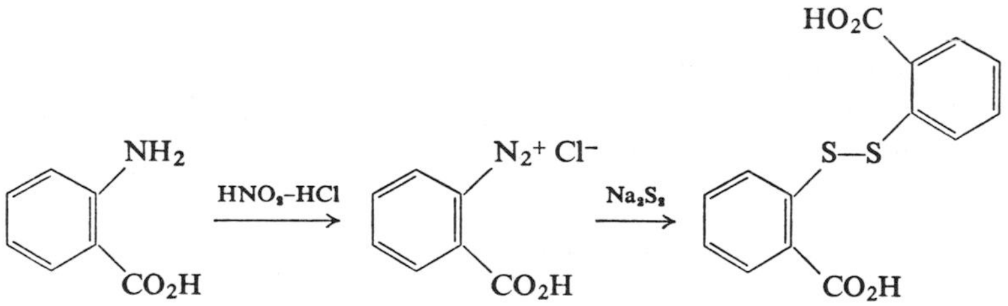

*A probable mechanism is

$$RN{=}NR \xrightarrow{^{-}O{-}S(=O){-}S(=O){-}O^{-}} R\bar{N}{-}\bar{N}R + 2\,SO_2 \xrightarrow{2\,H^+} RNH{-}NHR$$

$$\xrightarrow{S_2O_4^{2-}} R\bar{N}H \quad \bar{N}HR \xrightarrow{2\,H^+} 2\,RNH_2$$

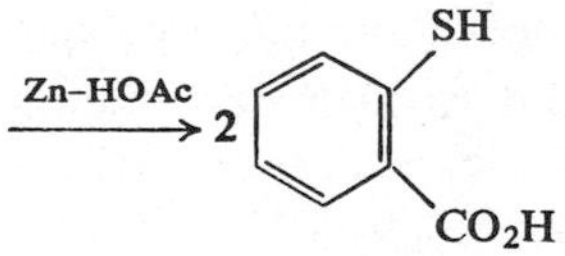

Thiosalicylic acid

(b) SULPHONYL CHLORIDES

Sulphonic acids are not easily reduced, but the readily derived chlorides can be reduced to sulphinic acids and to thiols. There is a resemblance to reduction of the nitro group (p. 647): zinc or tin in mineral acid gives the thiol, whereas in neutral or alkaline solution intermediate products are obtainable. For example, benzenesulphonyl chloride with zinc in sulphuric acid gives thiophenol in 96% yield [1],

$$PhSO_2Cl \xrightarrow{\text{Zn–H}_2\text{SO}_4} PhSH$$

and toluene-p-sulphonyl chloride with zinc in aqueous caustic soda gives sodium toluene-p-sulphinate in 64% yield [1],

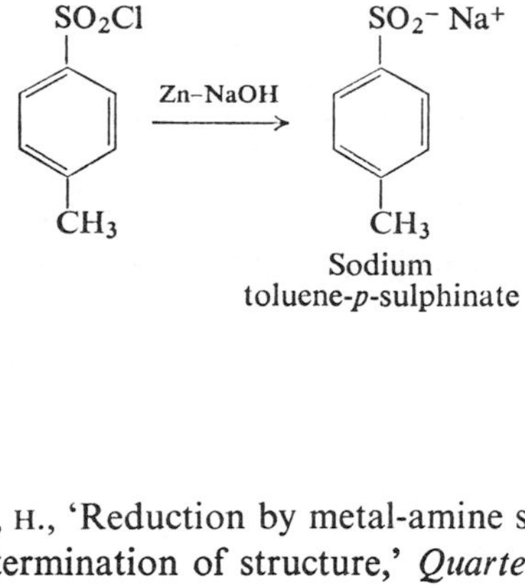

Sodium
toluene-p-sulphinate

Further Reading

BIRCH, A. J., and SMITH, H., 'Reduction by metal-amine solutions: applications in synthesis and determination of structure,' *Quarterly Reviews*, 1958, **12**, 17.

BROWN, H. C., *Hydroboration*, W. A. Benjamin (New York 1962).

BROWN, W. G., 'Reductions by lithium aluminum hydride,' *Organic Reactions*, 1951, **6**, 469.

CAINE, D., 'Reduction and related reactions of $\alpha\beta$-unsaturated compounds with metals in liquid ammonia,' *Organic Reactions*, 1976, **23**, 1.

DENO, N. C., *et al.*, 'The hydride-transfer reaction,' *Chemical Reviews*, 1960, **60**, 7.

EMERSON, W. S., 'The preparation of amines by reductive alkylation,' *Organic Reactions*, 1948, **4**, 174.

GEISSMAN, T. A., 'The Cannizzaro reaction,' *Organic Reactions*, 1944, **2**, 94.

HARTUNG, W. H., and SIMONOFF, R., 'Hydrogenolysis of benzyl groups attached to oxygen, nitrogen, or sulphur,' *Organic Reactions*, 1953, **7**, 263.

JACKMAN, L. M., 'Hydrogenation-dehydrogenation reactions,' *Advances in Organic Chemistry, Methods and Results*, Vol. 2, Interscience (New York and London 1960), p. 329.

TODD, D., 'The Wolff-Kishner reduction,' *Organic Reactions*, 1948, **4**, 378.

VEDEJS, E., 'Clemmensen reduction of ketones in anhydrous organic solvents,' *Organic Reactions*, 1975, **22**, 401.

WILDS, A. L., 'Reduction with aluminum alkoxides (the Meerwein-Ponndorf-Verley reduction),' *Organic Reactions*, 1944, **2**, 178.

Problems

1. How would you carry out the following transformations?

    (*i*) R—CH=CH₂ into (*a*) R—CH₂—CH₂OH, (*b*) R—CH(OH)—CH₃,

    (*c*) R—CH₂—CH₂—NH₂.

    (*ii*) R—C≡CH into (*a*) R—CO—CH₃, (*b*) R—CH₂—CHO.

    (*iii*) R—C≡C—R into the corresponding (*a*) *cis*- and (*b*) *trans*-olefins.

    (*iv*) EtO₂C—CH₂—CH₂—CO₂H into EtO₂C—CH₂—CH₂—CH₂OH.

    (*v*) Ph—CH=CH—CO₂H into (*a*) Ph—CH₂—CH₂—CO₂H, (*b*) Ph—CH=CH—CH₂OH, (*c*) Ph—CH=CH—CHO, (*d*) Ph—CH₂—CH₂—CH₂OH, (*e*) γ-cyclohexyl-propanol.

    (*vi*) *m*-Nitrobenzoyl chloride into *m*-nitrobenzaldehyde.

    (*vii*) *m*-Nitrobenzaldehyde into (*a*) *m*-nitrobenzyl alcohol, (*b*) *m*-aminobenzaldehyde, (*c*) *m*-nitrotoluene.

    (*viii*) 1-Methylcyclohexene into *trans*-2-methylcyclohexanol.

    (*ix*) Diethyl sebacate (EtO₂C—(CH₂)₈—CO₂Et) into (*a*) decamethylenediol, (*b*) cyclodecanone, (*e*) *trans*-cyclodecene.

    (*x*)

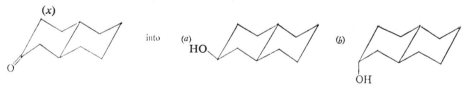

2. Outline a synthesis, which includes a reductive method, of each of the following, from readily available compounds:

    (*a*) Cl₃C—CH₂OH   (*b*) C(CH₃)₄

    (*c*) PhCH₂COPh   (*d*) PhCH₂CH₂N(CH₃)₂

    (*e*)

    (*f*)

(g)

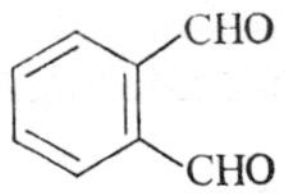

(h)

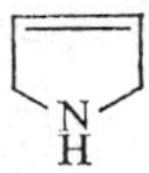

(i)

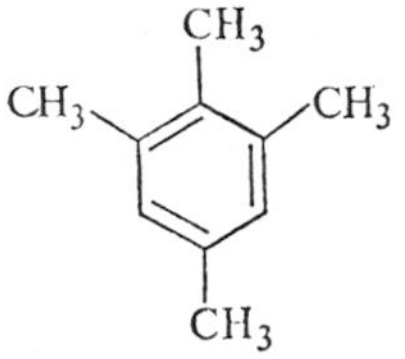

(j)

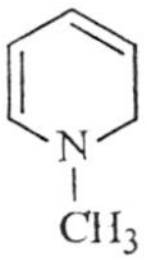

3. How would you attempt to convert acrylonitrile into adiponitrile ($NC—(CH_2)_4—CN$) (which is required for the manufacture of nylon)?

4. An attempt to make the N-dimethyl derivative of the amine (I) by treatment with aqueous formaldehyde in the presence of formic acid gave the compound (II). Discuss.

$(CH_3)_2C=CH—(CH_2)_3—NH_2$

(I)

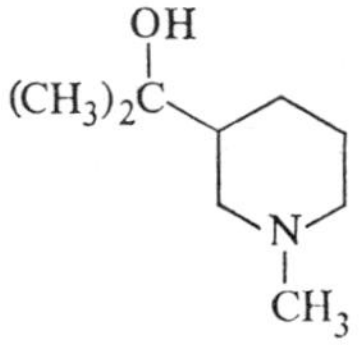

(II)

5. Summarize the uses of hydrazine and its derivatives in effecting the reduction of various types of organic groupings.

6. Annotate the steps in the synthesis of (±)-thioctic acid (p. 624).

20. The Synthesis of Heterocyclic Compounds

20.1 Introduction

Heterocyclic systems are of widespread occurrence in nature, particularly in such natural products as nucleic acids, plant alkaloids, anthocyanins and flavones, and the haem pigments and chlorophyll. In addition, some of the vitamins contain aromatic heterocyclic systems, and proteins contain the imidazole and indole rings (p. 354). In this chapter, the general methods for the synthesis of the more important heterocycles are discussed, and in the following chapter applications of some of these methods in the synthesis of natural products are described.

The reactions employed for making heterocyclic compounds involve only those principles and procedures which have been discussed in earlier chapters. For example, C—N bonds are usually formed by reaction between amino groups and esters, aldehydes, ketones, halides, or activated olefins (Michael-type addition); aliphatic C—C bonds are usually formed by acid- or base-catalyzed condensations involving activated methylene groups and carbonyl groups; and ring-closure on to benzene rings is usually effected by Friedel-Crafts reactions. Two points should, however, be emphasized. First, a reaction giving a strainless or near-strainless ring almost always occurs more readily than the corresponding intermolecular reaction (p. 91). Thus, whereas two molecules of an alcohol are dehydrated to give an acyclic ether only under vigorous conditions, diethanolamine gives morpholine (a useful basic solvent) in comparatively mild acidic conditions:

Morpholine

Secondly, a heterocyclic synthesis often consists of a consecutive series of reactions (see, e.g. discussion of the Skraup quinoline synthesis; p. 96), in some of which one or more steps are thermodynamically unfavourable. In these cases reaction is driven forward by the favourable equilibrium in a subsequent step, which is often the formation of an aromatic ring with the consequent liberation of stabilization energy.

The properties of the heterocyclic compounds whose syntheses are outlined in this chapter will not be described in detail. The non-aromatic compounds have properties closely related to their acyclic analogues: e.g. piperidine behaves like an acyclic secondary amine, being a moderately strong base (pK 11.2; $cf.$ diethylamine, 11.1) and a reactive nucleophile; pyrrolidine is similar. The properties of those aromatic compounds which contain two hetero-atoms can be generally inferred from those of benzenoid and simpler heterocyclic compounds (discussed in chapters 11 and 12). Thus, pyrazole is derived from pyrrole by substitution of —N= for —CH=, just as pyridine is derived from benzene; and the chemistries of pyrazole and pyrrole are related in very much the same way as those of pyridine and benzene. The following are broad outlines of the relevant chemistry.

(1) *Reactions at the nucleus.* Pyridine is much less reactive towards electrophiles than benzene and in the same way the electron-attracting —N= reduces the reactivity of pyrazole and imidazole compared with that of pyrrole. Both compounds readily undergo reactions such as halogenation, but only imidazole reacts (as its conjugate base) with the weaker electrophile, benzenediazonium ion. Pyrazole reacts predominantly at the 4-position, imidazole at the 2-position.

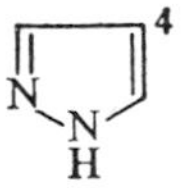

Pyrazole 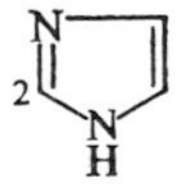   Imidazole

Just as furan and thiophen are less reactive than pyrrole towards electrophiles, so oxazole, isoxazole, thiazole, and isothiazole are less reactive than pyrazole and imidazole, being slightly less reactive than benzene.

The three six-membered diazines are related to pyridine as is pyridine to benzene. Each is even more deactivated than pyridine towards electrophiles, being essentially inert. The reactivity is increased by the incorporation of electron-releasing groups: e.g. a hydroxyl or amino group in the 2-, 4-, or 6-position of pyrimidine enables reaction to occur at the 5-position (see, e.g. p. 696):

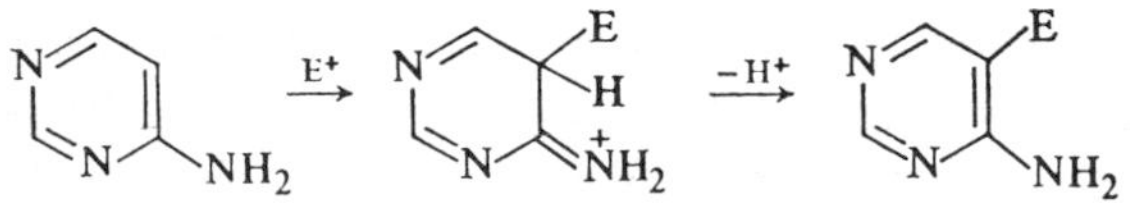

Conversely, the diazines are strongly activated towards nucleophiles, at each carbon in pyridazine and pyrazine and at all save the 5-position in pyrimidine. Methyl substituents in these positions are powerfully activated towards acid- or

base-catalyzed condensation, and halogen atoms are activated towards nucleo-philic displacement (p. 424).

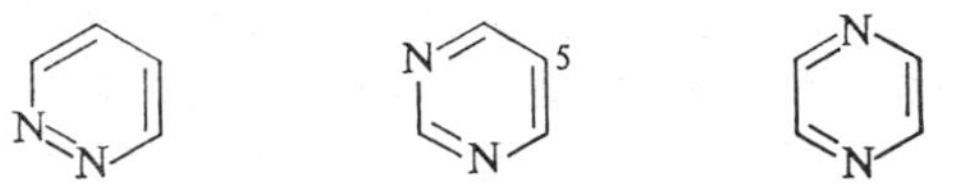

Pyridazine   Pyrimidine   Pyrazine

(2) *Basicity of heterocyclic nitrogen.*

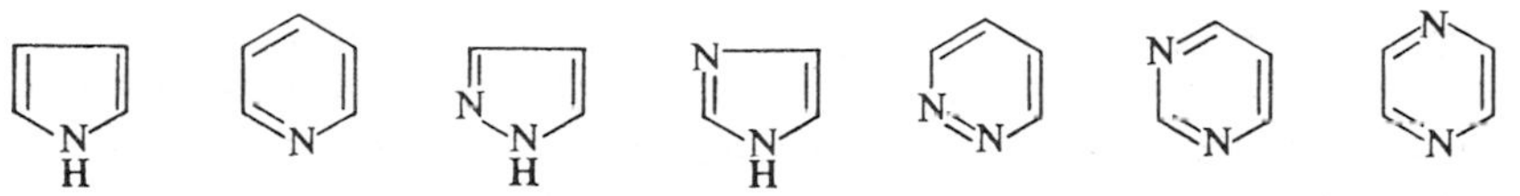

| pK | Ca. 0·4 | 5·2 | 2·5 | 7·2 | 2·1 | 1·1 | 0·6 |

Pyrrole is an extremely weak base because protonation destroys the aromatic system and is accompanied by a significant loss of aromatic stabilization energy (p. 58). This is not so for pyridine, which is nevertheless a weaker base than aliphatic amines because the nitrogen atom is sp^2-hybridized (p. 69). Both pyrazole and imidazole are more strongly basic at the pyridine-like nitrogen than at the pyrrole-like nitrogen, imidazole being a stronger base than pyridine for the same reason that an amidine is a stronger base than an amine; namely, that the conjugate acid is a delocalized cation (p. 69):

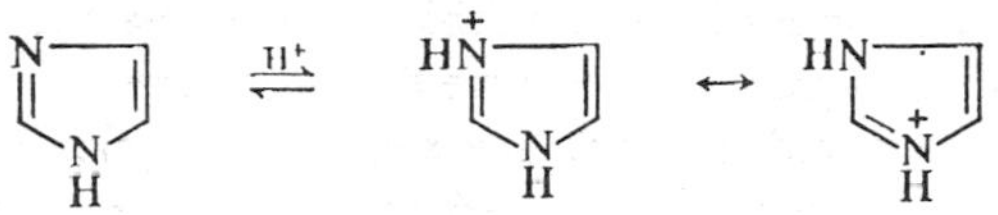

Both pyrazole and imidazole, like pyrrole (p. 69), are weakly acidic.

The weaker basicity of each of the three diazines than pyridine derives from the electron-attracting capacity of nitrogen relative to carbon.

(3) *Tautomeric equilibria.* The phenolic derivatives of the six-membered aromatic nitrogen heterocycles exist in equilibrium with keto tautomers, except for those compounds in which the hydroxyl group is in a β-position. For 2- and 4-hydroxypyridine, equilibrium favours the keto form (K ~ 10^3):

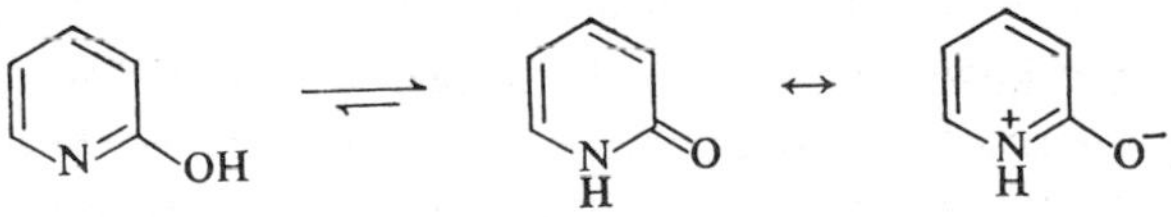

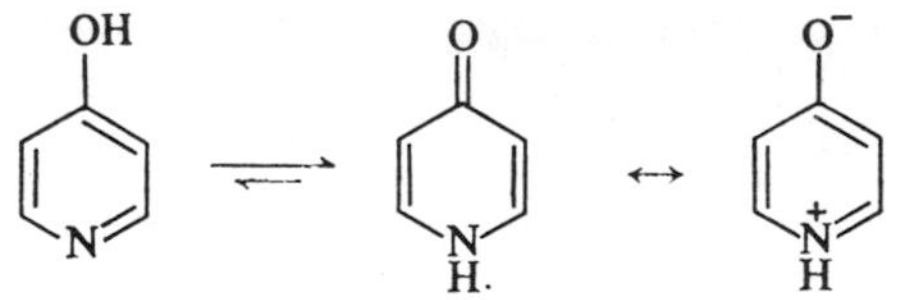

This situation is in strong contrast to that for phenol, where the keto tautomer is undetectable, the reason being that in this case the keto tautomers are themselves aromatic, as shown by the Kekulé representations above, and the resulting stabilization energy serves to make the pyridone the more strongly bonded tautomer.

This is also the case in the 2- and 4-hydroxypyrimidines, e.g.

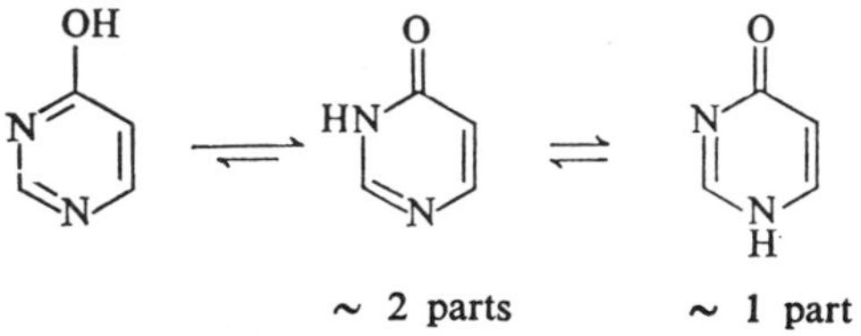

~ 2 parts   ~ 1 part

The same tautomeric possibilities apply to the corresponding amino-derivatives, but equilibrium favours the amino over the imino form, possibly because the involvement of the amino group's unshared pair of electrons in the π-system,

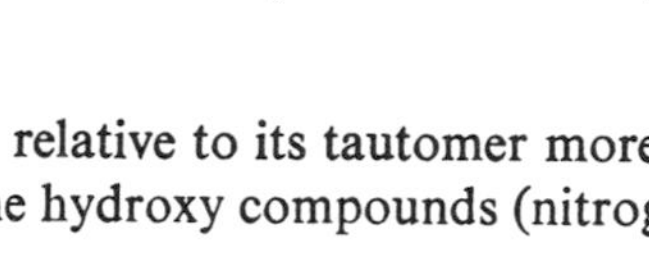

stabilizes the amino form relative to its tautomer more strongly than the corresponding interaction in the hydroxy compounds (nitrogen having a greater $+M$ effect than oxygen).

20.2 Five-membered Rings Containing One Hetero-atom

(a) PYRROLES

Pyrrole itself may be extracted from coal tar and bone oil by distillation, so that its syntheses are relatively unimportant. It has been prepared industrially by passing acetylene and ammonia through a red-hot tube and by passing furan, ammonia, and steam over hot alumina. It has been made in the laboratory by distillation of the ammonium salt of mucic acid (HO_2C—$(CHOH)_4$—CO_2H). It is also produced by the distillation of succinimide over zinc dust:

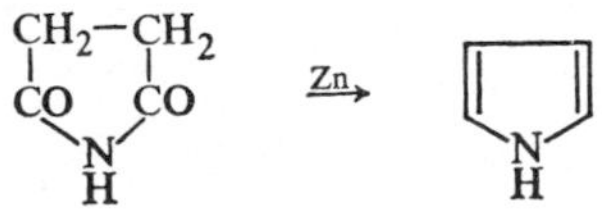

$$\begin{array}{c} CH_2\!-\!CH_2 \\ | \quad\quad | \\ CO \quad CO \\ \diagdown N \diagup \\ | \\ H \end{array} \xrightarrow{\;Zn\;} \quad \boxed{} \begin{array}{c}\\ N \\ | \\ H\end{array}$$

Substituted pyrroles are usually made by combining two aliphatic fragments in one of two ways, represented skeletally as follows:

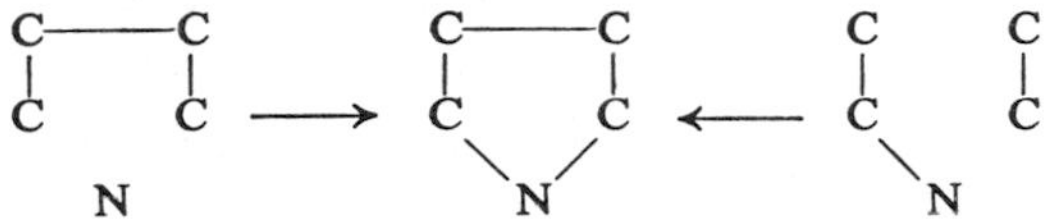

$$\begin{array}{cc} C\!-\!\!-\!\!-C \\ | \quad\quad | \\ C \quad\quad C \\ \\ N \end{array} \longrightarrow \begin{array}{cc} C\!-\!\!-\!C \\ | \quad\quad | \\ C \quad\quad C \\ \diagdown N \diagup \end{array} \longleftarrow \begin{array}{cc} C \quad\quad C \\ | \quad\quad | \\ C \quad\quad C \\ \diagdown N \diagup \end{array}$$

(1) *The Paal-Knorr synthesis.* A 1,4-dicarbonyl compound is treated with ammonia or a primary amine. Successive reactions of the nucleophilic nitrogen at the carbonyl groups are followed by dehydration, which occurs readily because the product is aromatic:

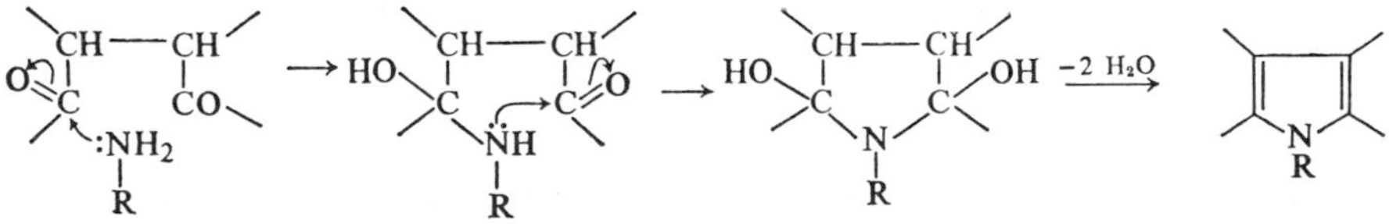

The reaction is limited by the availability of the dicarbonyl compounds. Those which are symmetrical may be obtained by reaction between the appropriate β-keto-ester and iodine (p. 252), and others can be made by treatment of furans with lead tetra-acetate followed by hydrolysis and reduction:

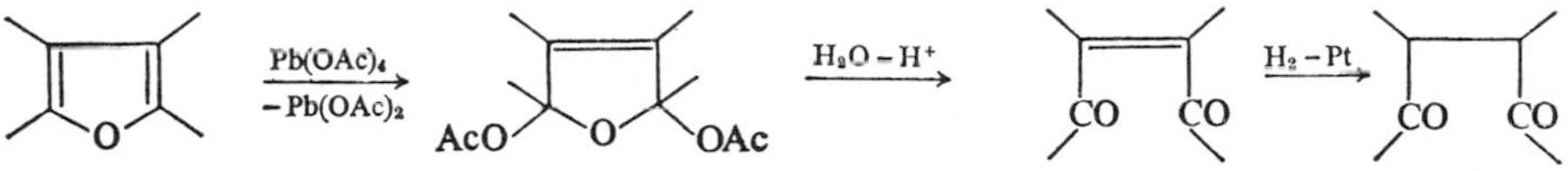

(2) *The Knorr synthesis.* An α-amino-ketone is condensed with a ketone containing an activated α-methylene group:

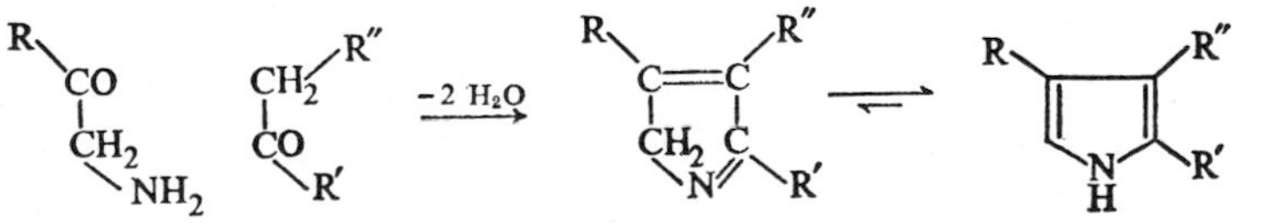

R' should be a methylene-activating group, for otherwise self-condensation of the α-amino-ketone takes place preferentially, giving a dihydropyrazine (p. 695):

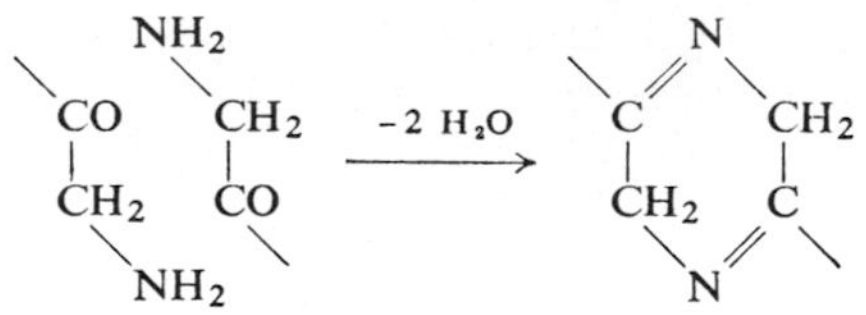

α-Amino-ketones are available from the condensation of β-keto-esters with alkyl nitrites (p. 344) followed by reduction, preferably with sodium dithionite:

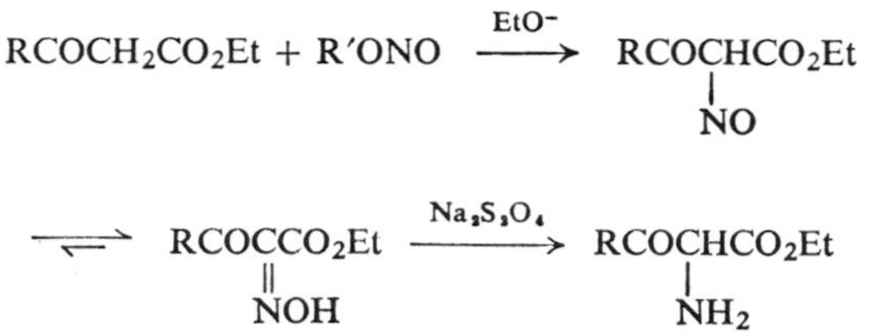

$$RCOCH_2CO_2Et + R'ONO \xrightarrow{EtO^-} RCOCHCO_2Et \atop \underset{NO}{|}$$

$$\rightleftharpoons RCOCCO_2Et \xrightarrow{Na_2S_2O_4} RCOCHCO_2Et \atop \underset{NOH}{\|} \qquad \underset{NH_2}{|}$$

If the ester group is not required in the final product, the benzyl ester is employed, hydrogenolysis and decarboxylation then occurring in the reduction step:

$$RCOCCO_2CH_2Ph \xrightarrow{Na_2S_2O_4} \left[RCOCHCO_2H \atop \underset{NH_2}{|} \right] \xrightarrow{-CO_2} RCOCH_2NH_2$$

(3) *The Hantzsch synthesis.* A β-keto-ester is treated with an α-chloro-ketone in the presence of ammonia:

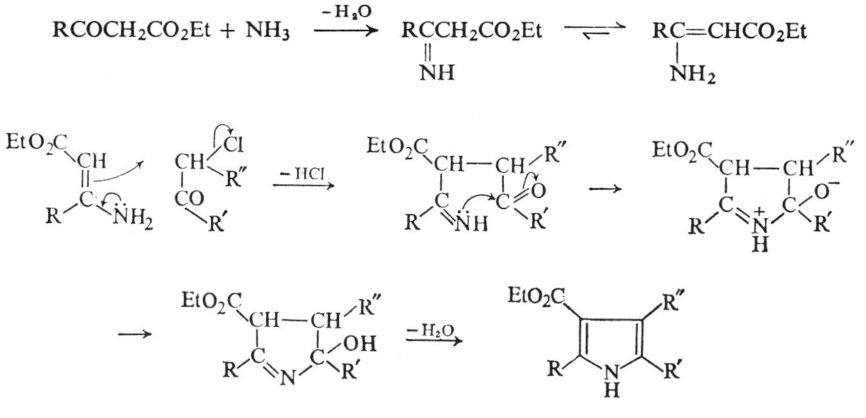

$$RCOCH_2CO_2Et + NH_3 \xrightarrow{-H_2O} RCCH_2CO_2Et \atop \underset{NH}{\|} \rightleftharpoons RC=CHCO_2Et \atop \underset{NH_2}{|}$$

This method is in general not recommended because the chloro-ketone can combine directly with the β-keto-ester to give a furan (p. 668).

Porphyrins. Porphin, which contains four pyrrole units joined across their 2-positions by —CH= fragments,

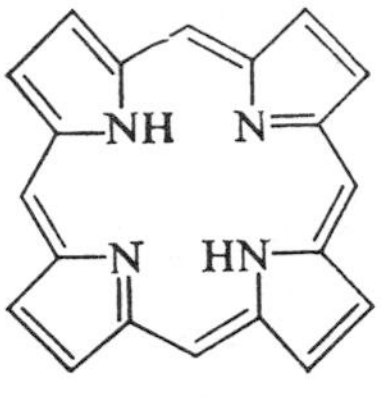

Porphin

is the parent of the porphyrins, a group of naturally occurring compounds which includes haemin (in, e.g. haemoglobin) and chlorophyll. The porphyrin skeleton is usually constructed from four individual pyrroles by first bringing these units together in pairs to give two dipyrrylmethenes and then joining the pairs. The following are three of the methods for synthesizing dipyrrylmethenes.

(1) A pyrrole 2-aldehyde is treated with a second pyrrole, which must possess a free 2-position, in the presence of hydrogen bromide. The acid increases the reactivity of the aldehydic group towards nucleophiles (i.e. its reactivity as an electrophile), and the success of the reaction depends on the great reactivity of pyrroles at the 2-position towards electrophiles. The final dehydration occurs readily because the product is an effectively delocalized cation.*

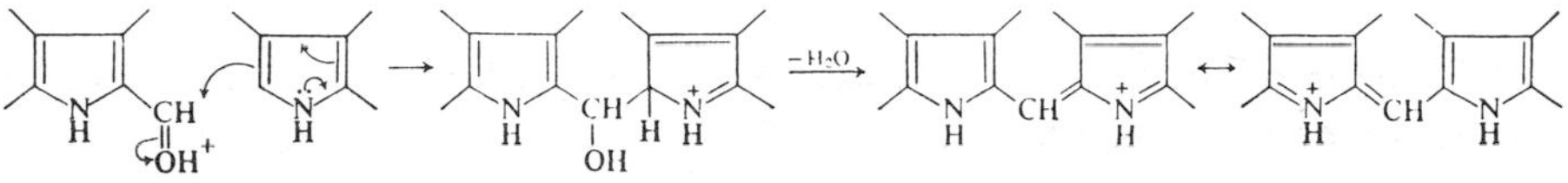

The required aldehydic group is readily introduced into the 2-position by Gattermann formylation (p. 395).

(2) Preparation of the 2-aldehyde as in method (1) may be by-passed if a symmetrical dipyrrylmethene is required. A pyrrole is treated with formic acid in the presence of hydrogen bromide, leading to successive reactions of Friedel-Crafts type:

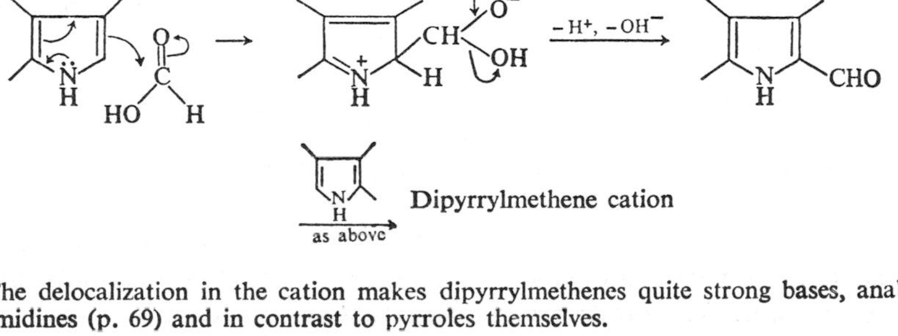

Dipyrrylmethene cation

*The delocalization in the cation makes dipyrrylmethenes quite strong bases, analogous to amidines (p. 69) and in contrast to pyrroles themselves.

(3) A 2-methylpyrrole containing a free 5-position is treated with bromine. A benzylic-type bromide is formed by one of the pyrrole units and this reacts at the 5-position of the second pyrrole in the Friedel-Crafts manner. Nuclear-bromination also occurs.

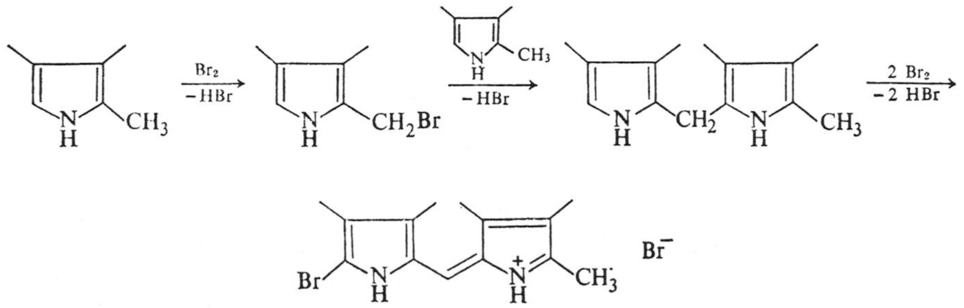

The coupling of two dipyrrylmethenes to give a porphyrin is usually accomplished by heating a 2-methyl-derivative with a 2-bromo-derivative in sulphuric acid at about 220°C. Yields are very low (less than 5%), e.g.

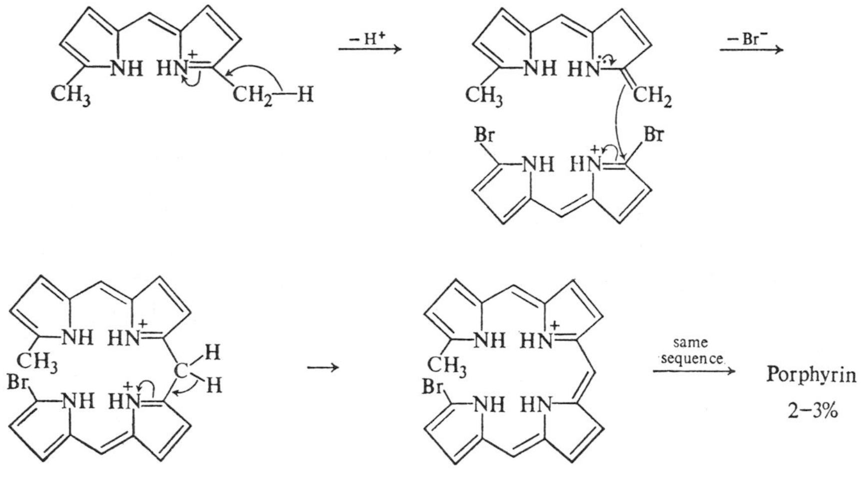

(b) INDOLES

The indole (2:3-benzopyrrole) system occurs in the essential α-amino-acid tryptophan, in the plant alkaloids which owe their biogenetic origin to tryptophan (e.g. reserpine; 21.7), and in indigo and Tyrian purple, two dyestuffs obtained from natural sources. Of the many methods of synthesis which have been developed, four are of particular importance.

(1) *The Fischer synthesis.* The phenylhydrazone of an aldehyde or ketone is heated with an acid catalyst such as boron trifluoride, zinc chloride or polyphosphoric acid. The reaction is analogous to the benzidine rearrangement (p. 480); the fourth step is an example of a [3,3]-sigmatropic rearrangement.

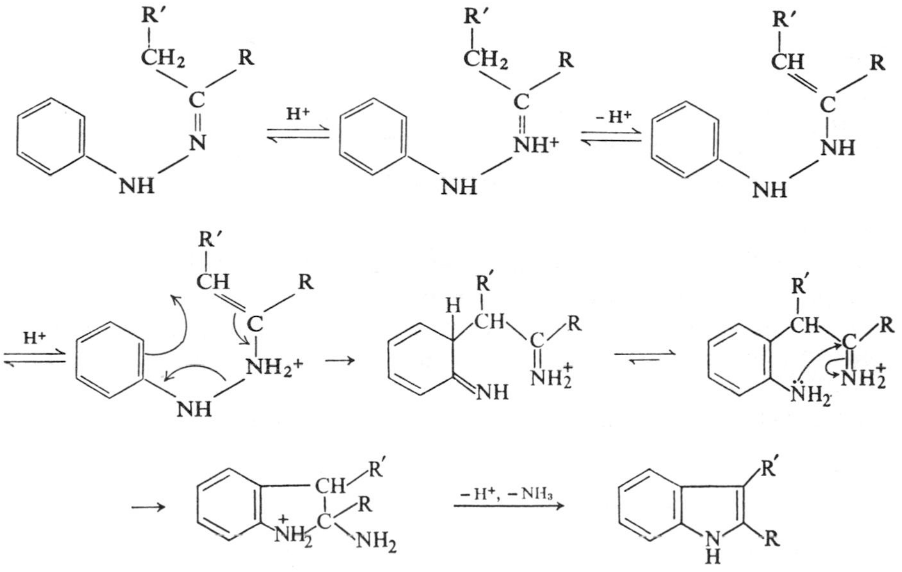

For example, acetophenone phenylhydrazone (R=Ph, R′=H), treated with zinc chloride at 170°C, gives 2-phenylindole in up to 80% yield [3], and yields are usually of this order. However, the reaction fails with acetaldehyde phenylhydrazone (R,R′=H), so that indole itself must be made indirectly. A simple method is to carry out the Fischer reaction on the phenylhydrazone of pyruvic acid (R=CO$_2$H, R′=H) and to decarboxylate the resulting indole-2-carboxylic acid thermally.

(2) *The Madelung synthesis.* An *o*-acylaminotoluene is treated with a base such as potassium t-butoxide. The probable mechanism is as follows:

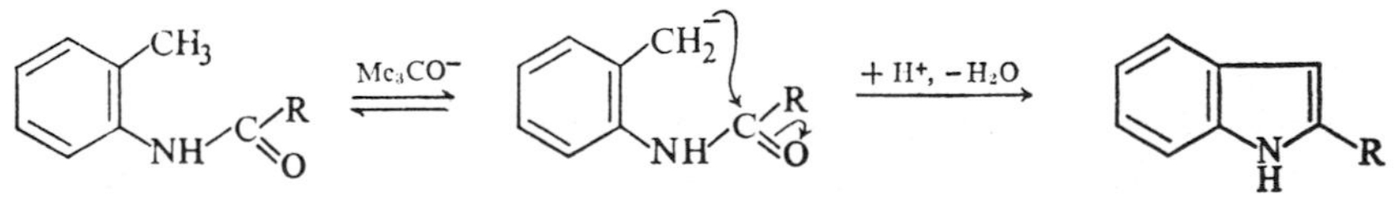

Yields are low, for decarbonylation of the starting material also occurs.

(3) *The Bischler synthesis.* An α-hydroxy- or α-halo-ketone is treated with an arylamine in the presence of acid. The probable mechanism is:

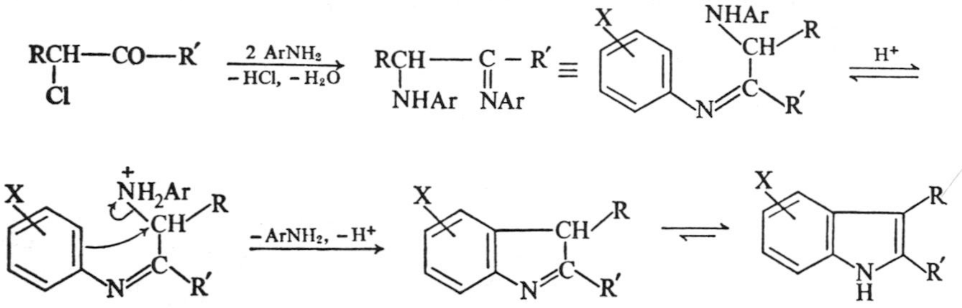

(4) *The Reissert synthesis.* An o-nitrotoluene is condensed with diethyl oxalate in the presence of a base. The nitro group of the resulting α-keto-ester is reduced to amino and cyclization then occurs spontaneously. The 2-carboethoxy-substituent in the indole may be removed, if required, by hydrolysis and thermal decarboxylation.

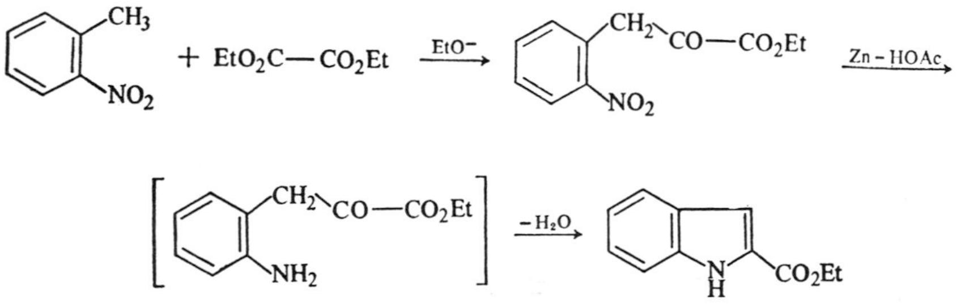

2-Carboethoxyindole

(c) ISOINDOLES

Isoindole itself has only recently been prepared; it is formed by low-pressure sublimation of the *N*-acetoxy compound shown at 50–70°C (*cf.* pyrolysis of esters; p. 318) and is not very stable.

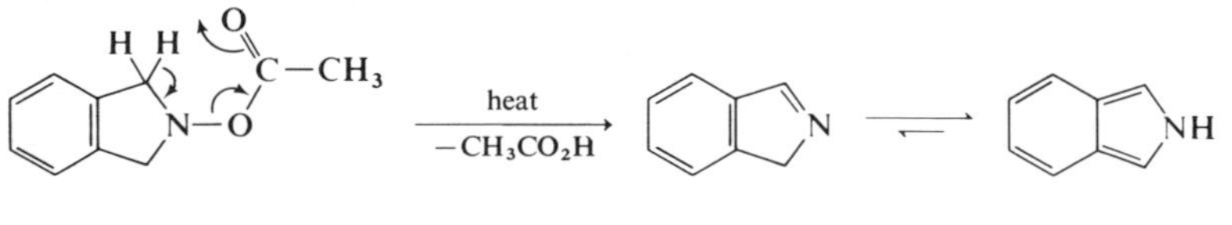

The most important derivatives of isoindole are the metal complexes of the phthalocyanins, used as dyes and pigments. They are prepared by passing ammonia into molten phthalic anhydride (cheaply available by the oxidation of naphthalene; p. 576) in the presence of a metal, e.g.

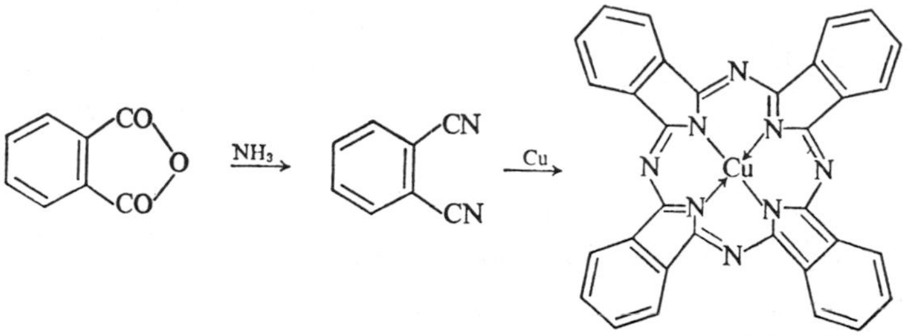

Monastrol Fast Blue

Phthalocyanin itself is obtained by acidification of its metal complexes.

(d) FURANS

Furan itself is obtained industrially by the catalytic decomposition of its 2-aldehyde, furfural, in steam. Furfural is readily available from oat husks or maize cobs (which contain pentoses), by digestion in dilute mineral acid followed by distillation in steam.

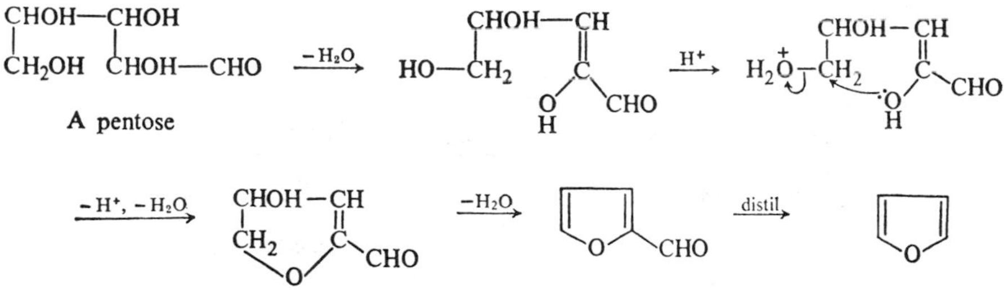

The following are the general methods for synthesizing furans.

(1) *From 1,4-dicarbonyl compounds.* Treatment of 1,4-dicarbonyl compounds with dehydrating agents such as sulphuric acid, zinc chloride, acetic anhydride, and phosphorus pentoxide brings about cyclization and dehydration (*cf.* the Paal-Knorr procedure for pyrroles; p. 661). The reaction is limited by the ease of obtaining the dicarbonyl compounds.

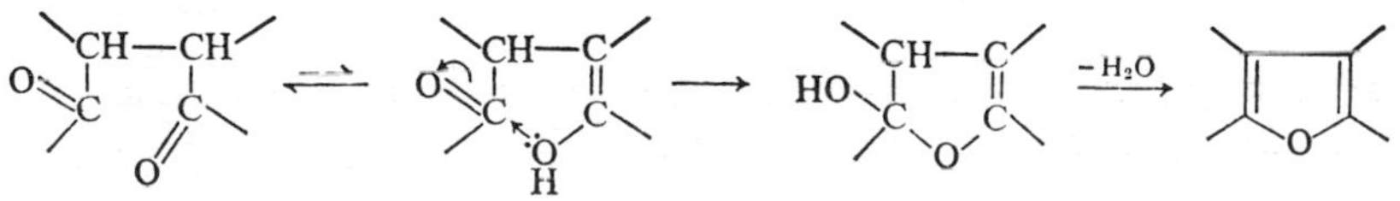

(2) *The Feist-Benary synthesis.* An α-chloro-ketone is condensed with a β-keto-ester in pyridine solution (*cf.* the Hantzsch synthesis of pyrroles; p. 662).

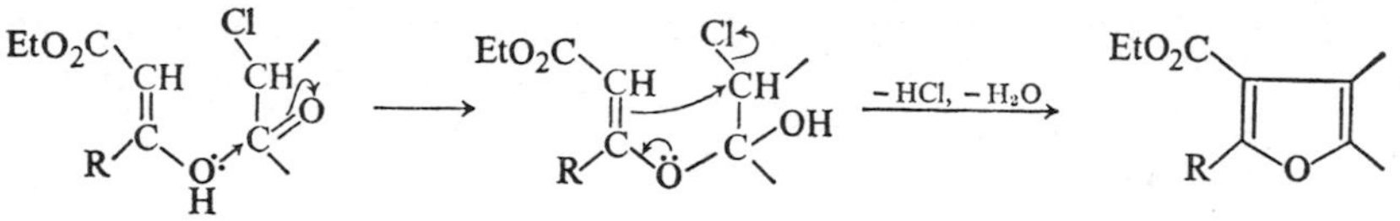

(3) *From 2-pyrones.* 2-Pyrones are prepared by treating malic acid and its derivatives with concentrated sulphuric acid; decarbonylation of the α-hydroxy-acid is followed by cyclization, e.g.

$$HO_2C—CH(OH)—CH_2—CO_2H \xrightarrow[- CO, - H_2O]{H_2SO_4} OCH—CH_2—CO_2H \rightleftharpoons HOCH=CH—CO_2H$$

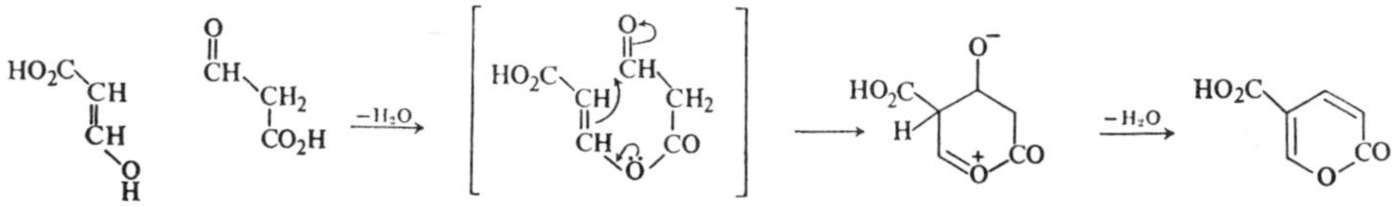

Treatment of the 2-pyrone with bromine and then caustic soda gives the furan, probably as follows:

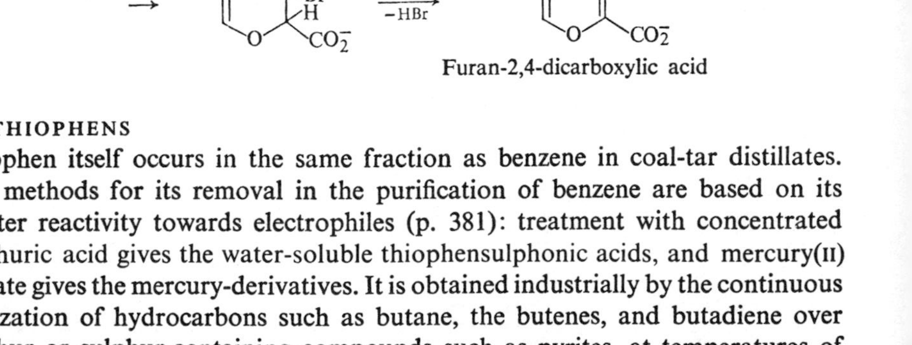

Furan-2,4-dicarboxylic acid

(e) THIOPHENS

Thiophen itself occurs in the same fraction as benzene in coal-tar distillates. The methods for its removal in the purification of benzene are based on its greater reactivity towards electrophiles (p. 381): treatment with concentrated sulphuric acid gives the water-soluble thiophensulphonic acids, and mercury(II) acetate gives the mercury-derivatives. It is obtained industrially by the continuous cyclization of hydrocarbons such as butane, the butenes, and butadiene over sulphur or sulphur-containing compounds such as pyrites, at temperatures of 500–600°C. Furans and pyrroles are converted in moderate yield into their thiophen analogues on treatment with hydrogen sulphide on an alumina catalyst at high temperatures.

The general methods for thiophens are as follows.

(1) *Hinsberg's procedure.* An α-dicarbonyl compound is condensed with a thio-ether in which the sulphur atom is adjacent to two activated methylene groups, such as diethyl thiodiacetate, e.g.

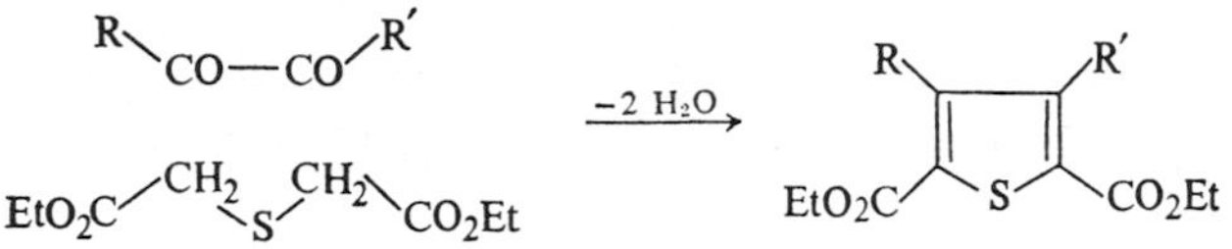

(2) *From 1,4-dicarbonyl compounds.* 1,4-Dicarbonyl compounds react with phosphorus tri- or penta-sulphide, the former usually being the more satisfactory. The reaction is analogous to the Paal-Knorr procedure for pyrroles (p. 661).

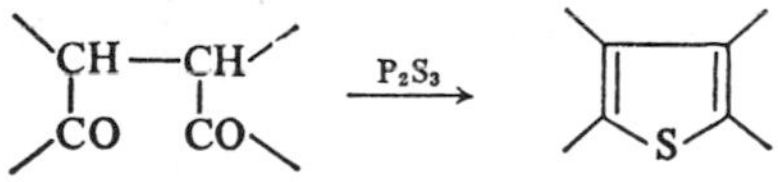

There are many variants of this method, of which the following is one example:

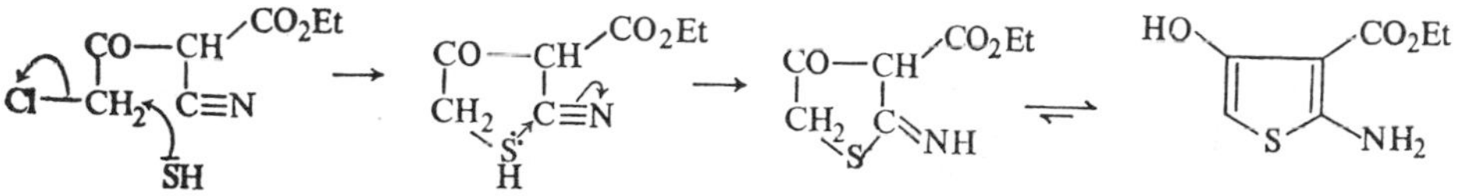

(*f*) REDUCED PYRROLES, FURANS, AND THIOPHENS

These may be made either by the cyclization of an aliphatic compound or, in most cases, by reduction of the corresponding aromatic compound.

(1) Pyrrolidine can be made by reaction between butane-1,4-diol and ammonia or by catalytic reduction. Reduction with zinc and acetic acid, however, results in 1,4-addition of hydrogen:

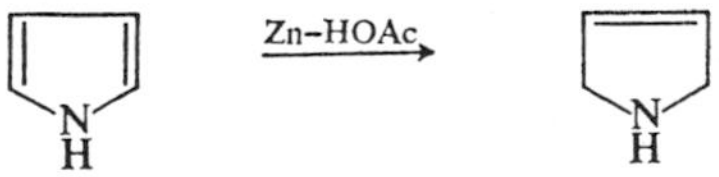

Δ^3-Pyrroline

The pyrrolidine ring is present in the α-amino-acids proline (2-carboxypyrrolidine) and hydroxyproline (2-carboxy-4-hydroxypyrrolidine) which occur in proteins.

(2) The furan ring is fairly resistant to reduction unless it is substituted with electron-withdrawing groups such as carboxyl; furan itself requires Raney nickel at over 80°C for reduction to tetrahydrofuran. Tetrahydrofuran is, however, obtained cheaply by condensation between acetylene and formaldehyde followed by reduction and dehydrative cyclization (p. 257).

(3) Reduction of thiophen cannot be brought about catalytically because the sulphur atom is removed in the process and poisons the catalyst. (The effectiveness of Raney nickel at removing sulphur is the basis of the Mozingo reductive procedure; p. 632). Reduction with sodium in liquid ammonia containing methanol gives a mixture of di- and tetra-hydrothiophens:

$$\text{[structure]} \xrightarrow{\text{Na} - \text{NH}_3/\text{EtOH}} \text{[structure]} + \text{[structure]} + \text{[structure]}$$

Tetrahydrothiophen is best prepared by reaction between 1,4-dichlorobutane and sodium sulphide or between butane-1,4-diol and hydrogen sulphide on alumina at 400°C.

20.3 Five-membered Rings Containing Two Hetero-atoms*

(a) PYRAZOLES

Pyrazoles do not occur naturally, but some of the synthetic derivatives have useful properties (e.g. antipyrine). They are prepared from 1,3-dicarbonyl compounds and hydrazine; e.g. pyrazole itself can be obtained by treating a malondialdehyde diacetal† with hydrazine in the presence of acid:

$$\begin{array}{c} \text{CH}_2\text{—CH(OMe)}_2 \\ | \\ \text{CH(OMe)}_2 \end{array} \xrightarrow{\text{H}_2\text{O} - \text{H}^+} \begin{array}{c} \text{CH}_2\text{—CHO} \\ | \\ \text{CHO} \end{array} \underset{\text{H}_2\text{N}}{\overset{\text{NH}_2}{\diagdown}} \xrightarrow{-2\,\text{H}_2\text{O}} \text{[structure]}$$

Pyrazole

Pyrazolines (partially reduced pyrazoles) are prepared from olefins and diazo-alkanes (p. 297) or from αβ-unsaturated aldehydes or ketones and hydrazine, e.g.

*Nomenclature: oxa, thia, and aza denote the presence of oxygen, sulphur, and nitrogen respectively, and, in compounds containing more than one of these substituents, are used in that order. The aromatic compounds have the suffix –ole; the partially reduced compounds, –oline, and the fully reduced compounds, –olidine. Trivial names are also used (e.g. imidazole for the isomer of pyrazole).

†Malondialdehyde readily undergoes self-condensation and is used in synthesis in the form of its diacetal from which the aldehyde is generated with acid.

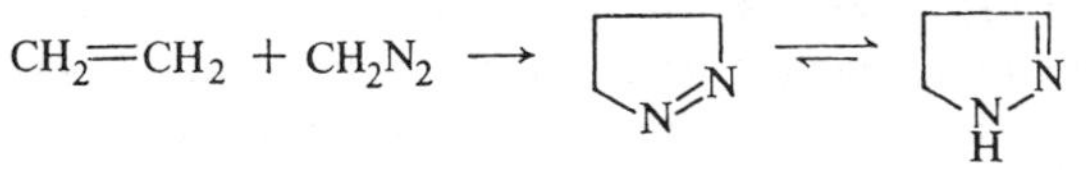

$$CH_2{=}CH_2 + CH_2N_2 \longrightarrow$$

Δ^2-Pyrazoline

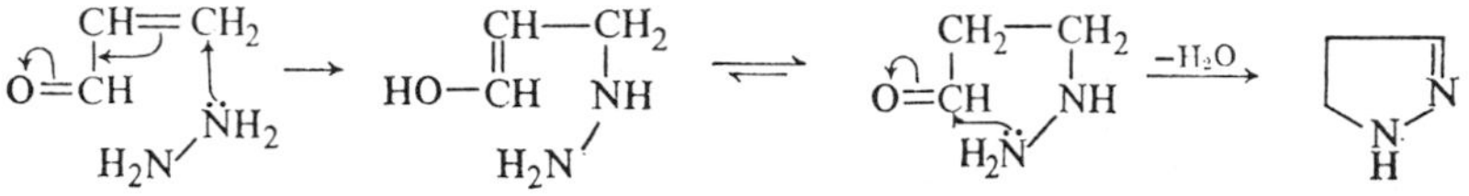

β-Keto-esters give pyrazolones similarly:

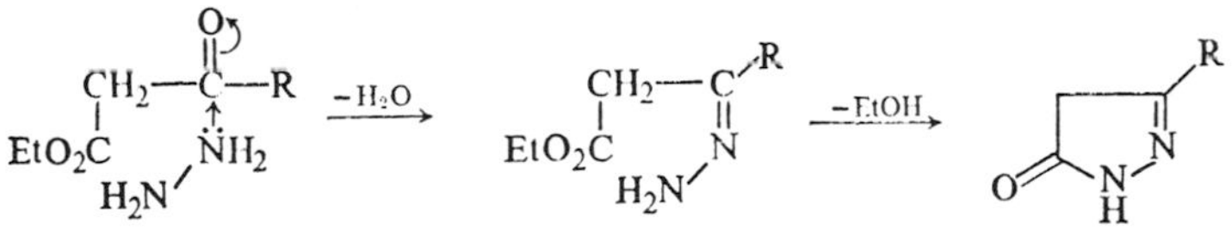

Δ^2-Pyrazolin-5-one

Pyrazolidines (fully reduced pyrazoles) are prepared from 1,3-dihalides and hydrazine, e.g.

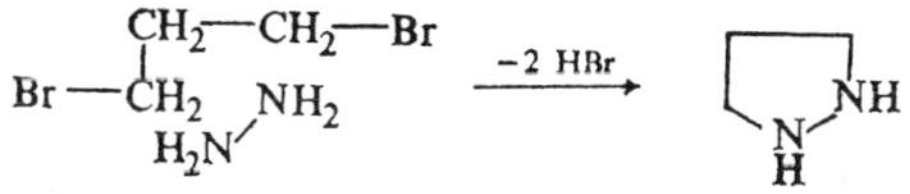

Pyrazolidine

(b) IMIDAZOLES

Imidazoles occur naturally, the best known being histidine, an α-amino-acid contained in proteins (p. 348). Imidazole itself is prepared by treating paraldehyde with bromine in ethylene glycol, giving the cyclic acetal of bromoacetaldehyde, and treating this product with formamide and ammonia:

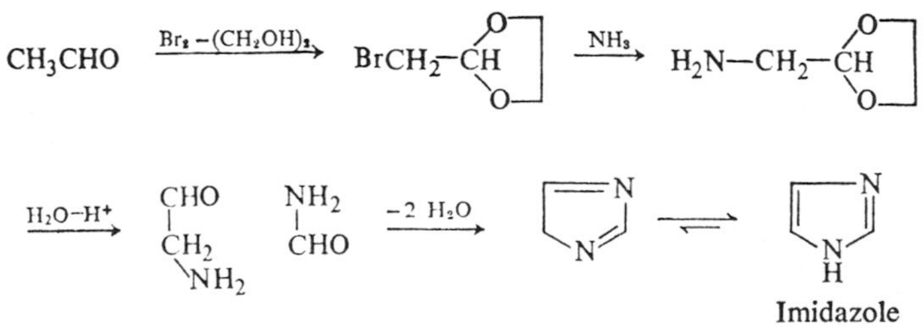

Imidazole

α-Hydroxy-aldehydes and ketones may replace (the intermediate) aminoacetaldehyde in general applications.

In a second general procedure, the hydrochloride of an α-amino-aldehyde or ketone is treated with a hot aqueous solution of potassium thiocyanate, giving a thione from which sulphur is removed with Raney nickel:

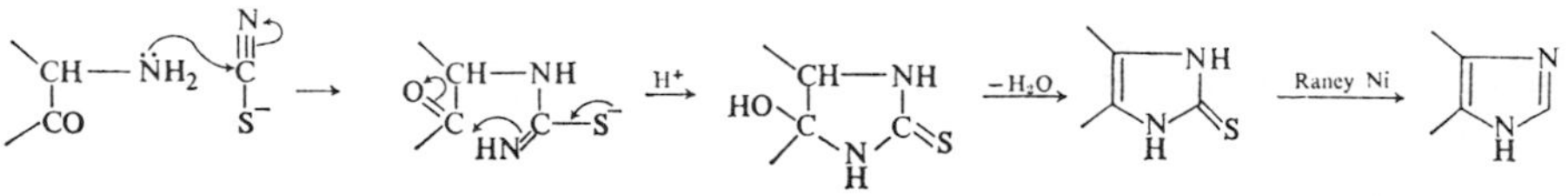

1,2-Diamines react with carboxylic acids and aldehydes or ketones to give, respectively, imidazolines and imidazolidines:

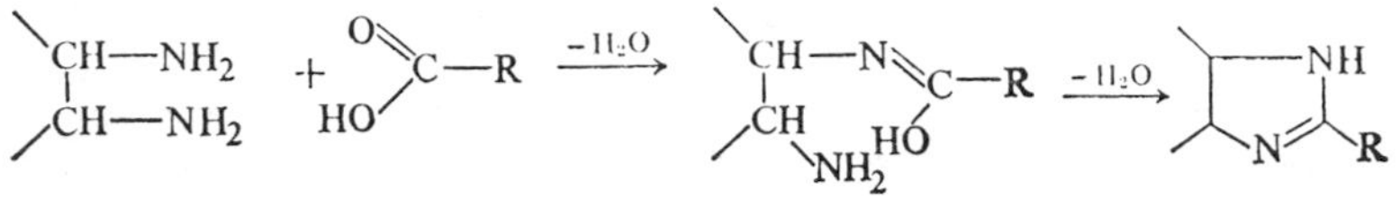

An imidazoline

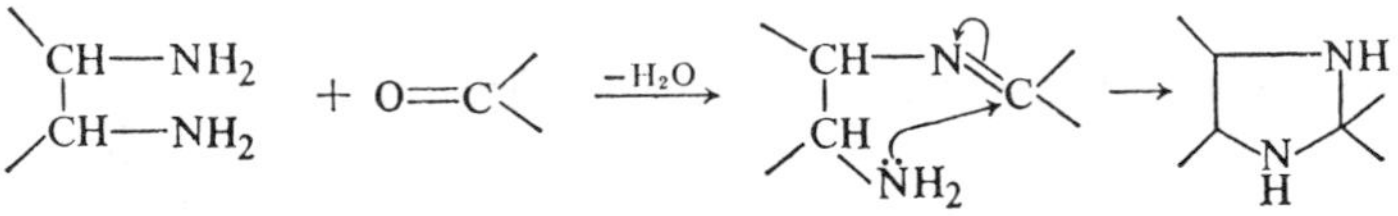

An imidazolidine

(c) ISOXAZOLES

Isoxazoles and their reduction products are prepared by methods analogous to those for pyrazoles and their reduction products, hydrazine being replaced by hydroxylamine. Thus, 1,3-dicarbonyl compounds give isoxazoles,

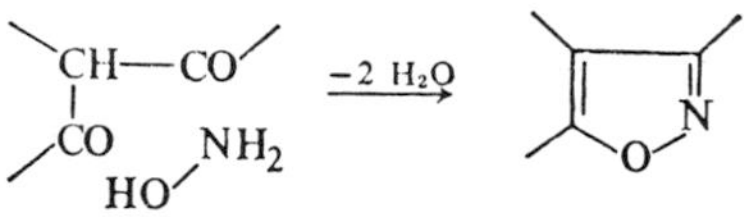

An isoxazole

αβ-unsaturated aldehydes or ketones give isoxazolines,

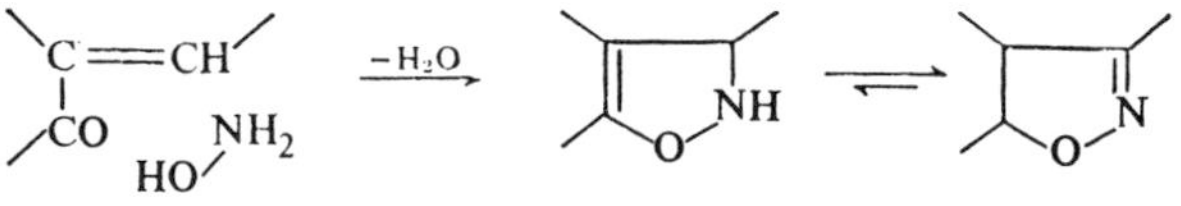

An isoxazoline

β-keto-esters give isoxazolones,

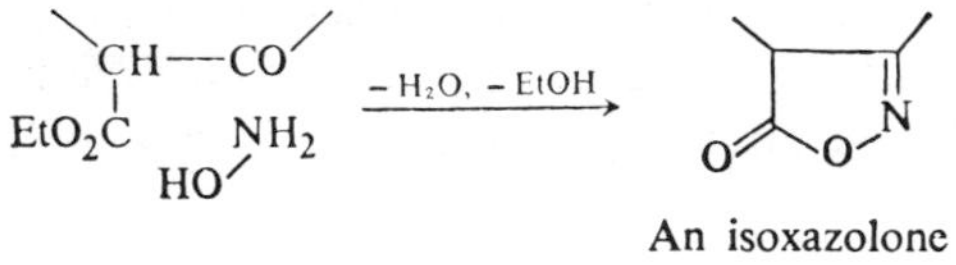

An isoxazolone

and 1,3-dihalides give isoxazolidines,

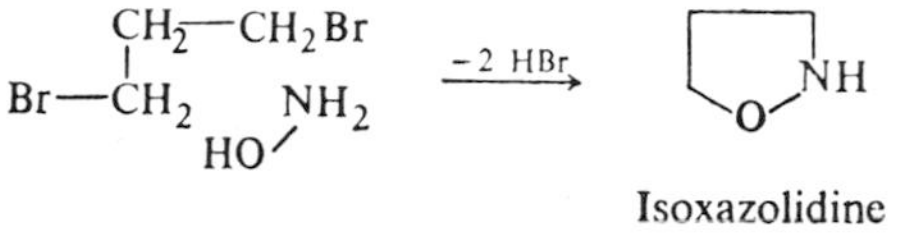

Isoxazolidine

(d) OXAZOLES

The oxazole ring is most simply constructed by reductive acetylation of the α-oximino-derivative of a ketone followed by dehydrative cyclization in the manner in which furans are formed from 1,4-dicarbonyl compounds (p. 667):

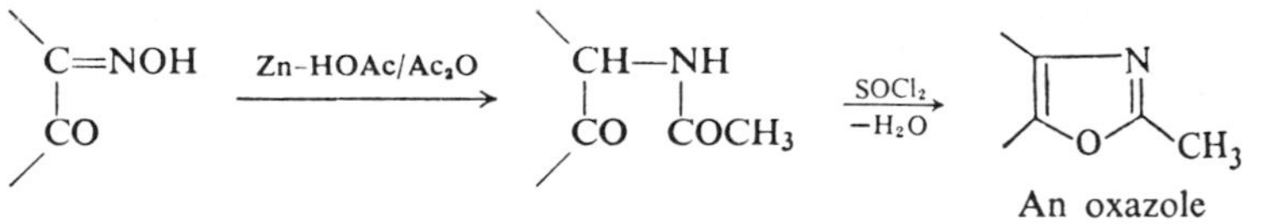

An oxazole

Oxazolines may be made from derivatives of β-chloroethylformamide, e.g.

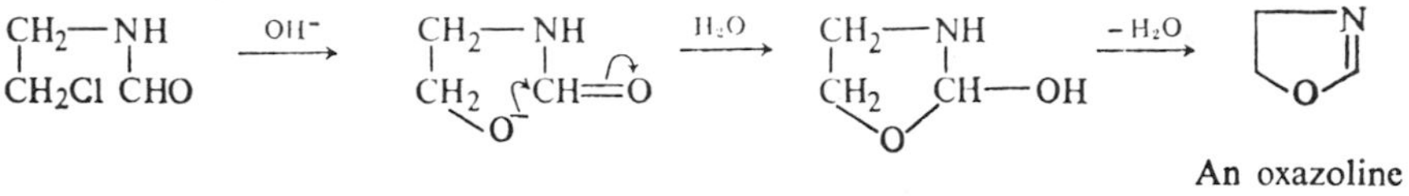

An oxazoline

Their 5-keto-derivatives, the oxazolones or azlactones, prepared by the dehydration of N-acyl-α-amino-acids, are employed in Erlenmeyer's synthesis of α-amino-acids (p. 232). Oxazolidines are made from β-hydroxy-amines and carbonyl compounds, analogously to the preparation of imidazolidines (p. 672).

(e) ISOTHIAZOLES

Isothiazole has only recently been prepared (1959), as follows:—

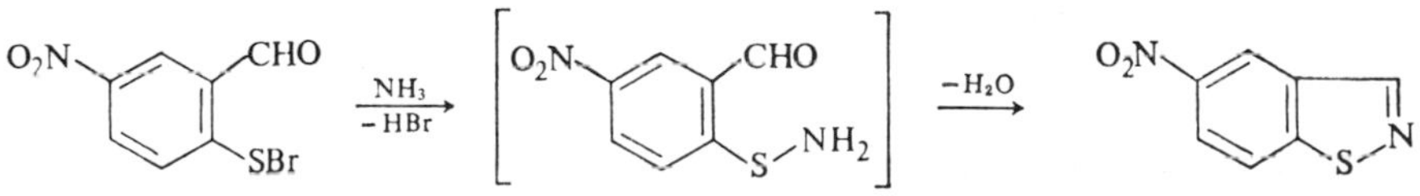

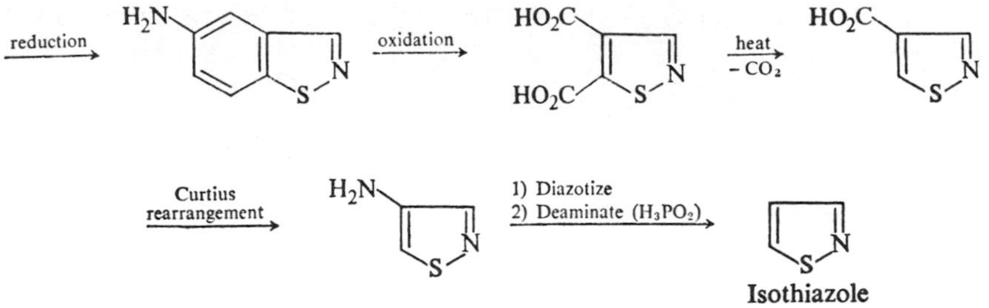

Isothiazole

The sugar-substitute, saccharin, may be regarded as a derivative of isothiazole:

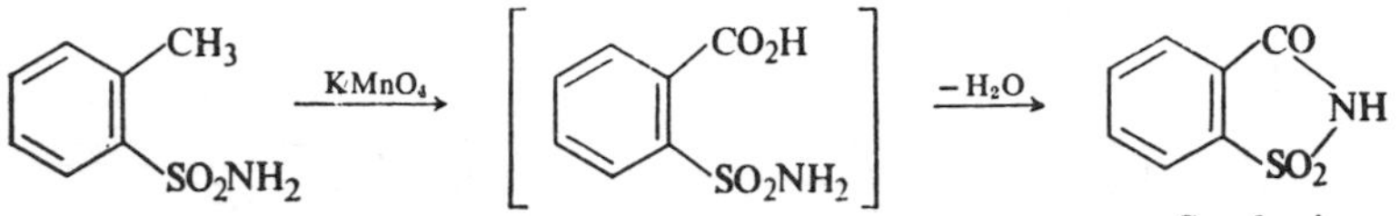

Saccharin

(*f*) THIAZOLES

The thiazole nucleus occurs in vitamin B_1 (thiamine), the penicillins, and the synthetic sulphathiazole (a member of the group of 'Sulpha' drugs).

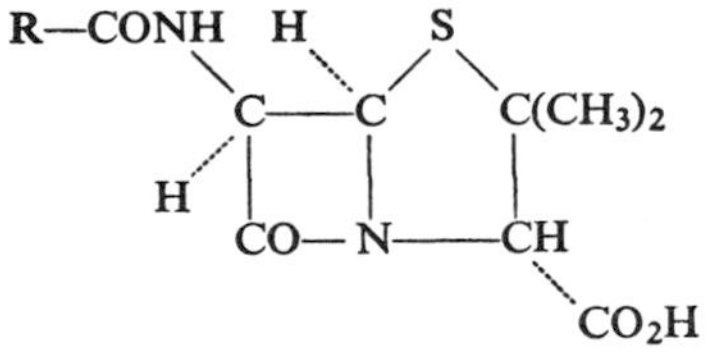

Penicillins

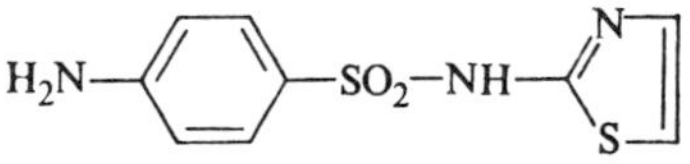

Sulphathiazole

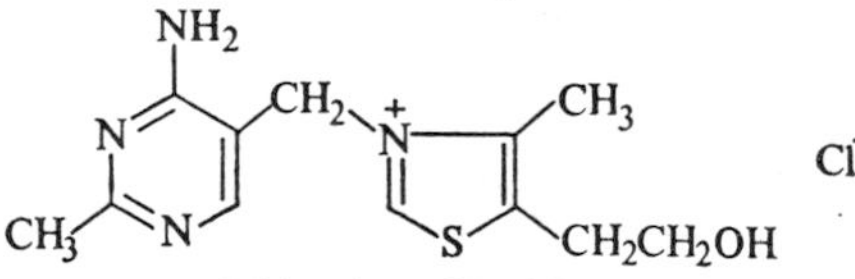

Thiamine chloride

The important physiological properties of these compounds have stimulated interest in thiazole chemistry and three general synthetic routes have been developed.

(1) From thioamides and α-halo-carbonyl compounds:

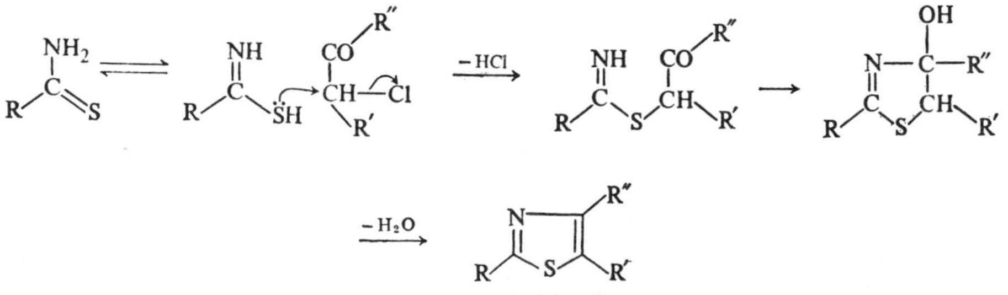

A thiazole

For example, a thiazole required in the synthesis of vitamin B_1 (p. 694) has been prepared in this way:

$$CH_3COCH_2CO_2Et + CH_2\!-\!CH_2 \xrightarrow[]{\text{EtO}^-} \left[\begin{array}{c} CH_3COCHCO_2Et \\ | \\ CH_2CH_2O^- \end{array} \right] \xrightarrow{-\text{EtO}^-} CH_3CO\!\!-\!\!\overset{\displaystyle CO\!-\!O}{\underset{\displaystyle CH_2\!-\!CH_2}{\overset{|}{CH}}} \xrightarrow{SO_2Cl_2}$$

$$CH_3CO\!-\!\underset{Cl}{\overset{}{\underset{|}{C}}}\!\!\overset{CO\!-\!O}{\underset{CH_2\!-\!CH_2}{\overset{|}{}}} \xrightarrow[]{H_2O-H^+} \underset{Cl}{\overset{CH_3}{\underset{}{\overset{|}{C}\!\!\overset{}{}}}}\!\!\begin{array}{c} CO \\ \diagdown CO_2H \\ CH_2CH_2OH \end{array} \xrightarrow[-HCl]{HN=CH-SH} \begin{array}{c} HO \quad CH_3 \\ N \diagup \\ \diagdown CO\!-\!O\!-\!H \\ S \diagdown CH_2CH_2OH \end{array} \xrightarrow{-H_2O,\ -CO_2} \begin{array}{c} N \diagdown \diagup CH_3 \\ \diagup \\ S \diagdown CH_2CH_2OH \end{array}$$

(2) From ammonium dithiocarbamates and α-halo-carbonyl compounds:

$$NH_4^+ \quad \underset{S}{\overset{NH_2}{\underset{\diagdown S}{\overset{|}{C}}}}\!\!\overset{CO}{\underset{R}{\overset{}{\diagup CH\!-\!Cl}}} \longrightarrow \underset{S}{\overset{NH_2}{\underset{\diagdown S}{\overset{|}{C}}}}\!\!\overset{CO}{\underset{R}{\overset{}{\diagup CH}}} \xrightarrow{-H_2O} \underset{S}{\overset{N=C}{\underset{\diagdown S}{\overset{}{C}}}}\!\!\overset{}{\underset{R}{\overset{}{\diagup CH}}} \rightleftharpoons \begin{array}{c} N\diagdown\diagup R' \\ HS \diagup S \diagdown R \end{array}$$

(3) From an α-acylamino-carbonyl compound and phosphorus pentasulphide (analogously to the formation of thiophens from 1,4-dicarbonyl compounds; p. 669):

$$\begin{array}{c} CH_2\!-\!NH \\ | \qquad | \\ R\diagup CO \quad CO\diagdown R' \end{array} \xrightarrow{P_2S_5} \left[\begin{array}{c} CH_2\!-\!\!-NH \\ | \qquad | \\ R\diagup C\diagdown S \quad S\diagup C\diagdown R' \end{array} \right] \rightleftharpoons \left[\begin{array}{c} CH\!-\!\!-N \\ \| \qquad \| \\ R\diagup C\diagdown SH \quad HS\diagup C\diagdown R' \end{array} \right] \xrightarrow{-H_2S} \begin{array}{c} \diagdown\!-\!N \\ R\diagup S\diagdown R' \end{array}$$

Δ^2-Thiazolines have been prepared from β-amino-thiols, e.g.

$$\begin{array}{c} CH_2\!-\!NH_2 \\ | \\ CH_2 \\ \diagdown SH \end{array} \xrightarrow[-EtOH]{HCO_2Et} \begin{array}{c} CH_2\!-\!NH \\ | \qquad | \\ CH_2 \quad CHO \\ \diagdown SH \end{array} \xrightarrow[-H_2O]{P_2O_5} \begin{array}{c} \boxed{\quad\diagdown N} \\ S\diagup \end{array}$$

Δ^2-Thiazoline

and are readily reduced by aluminium amalgam to thiazolidines. Thiazolidine itself is usually prepared from β-aminoethyl mercaptan and formaldehyde (cf. the formation of pyrazolidines; p. 671).

20.4 Six-membered Rings Containing One Hetero-atom

(a) PYRIDINES

The pyridine nucleus occurs in a number of plant alkaloids (e.g. nicotine) and in two members of the vitamin B Group. One of these, pyridoxol, is a component of enzymes responsible for transamination and decarboxylation, and the other, nicotinamide, is a component of enzymes in the electron-transport chain (respiratory enzymes concerned with the utilization of oxygen; see p. 698). Pyridines are also products of the distillation of coal-tar and bone oil.

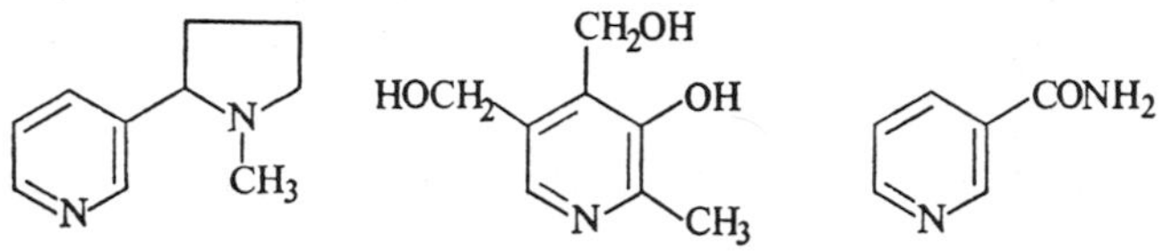

Nicotine   Pyridoxol   Nicotinamide

Whereas the five-membered heterocycles are comparatively reactive towards a wide variety of electrophilic reagents (p. 380), so that substituents may be introduced into accessible pre-formed rings, pyridines are very unreactive towards electrophiles (p. 382) and few nucleophilic reagents are available for introducing substituents. It is therefore usual to synthesize substituted pyridines from aliphatic compounds containing the appropriate groups. A number of general routes is available, the following being the skeletal types from which the ring is constructed:

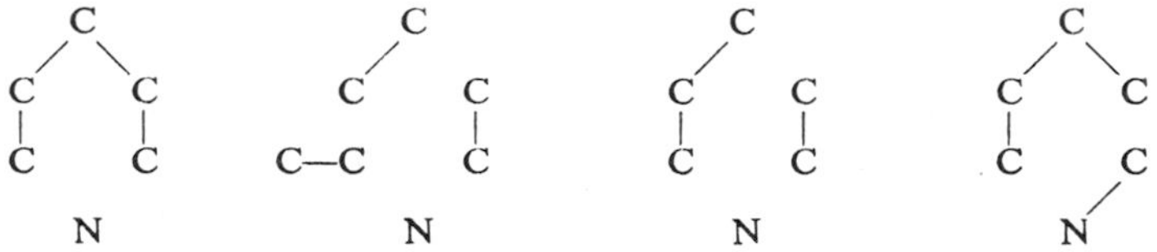

In addition, pyridines can be made from furans and pyrones.

(1) *From five-carbon units and ammonia.* Pyridine itself is formed from glutaconic aldehyde and ammonia:

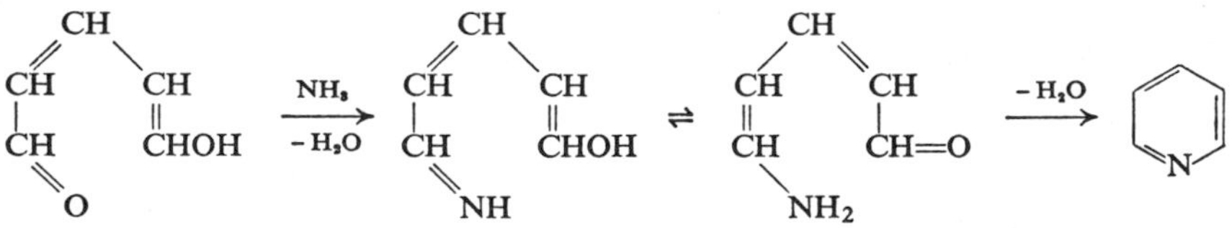

1,5-Dicarbonyl compounds give pyridines with hydroxylamine:

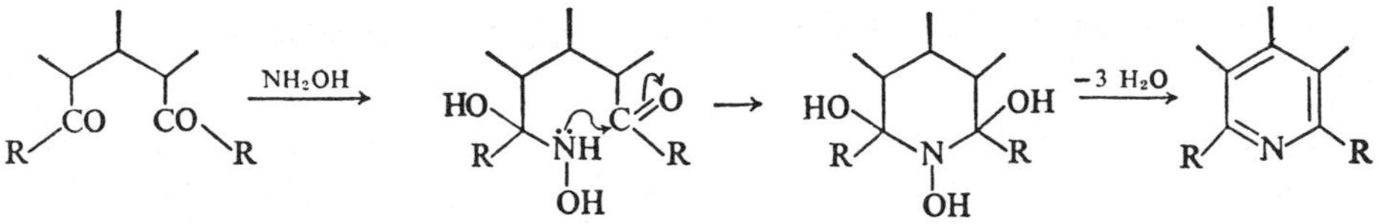

These methods are of limited scope because of the difficulty of obtaining the appropriate starting materials.

(2) *From aldehydes or ketones and ammonia.* Aldehydes and ketones react with ammonia at high temperatures, under pressure, by ammonia-catalyzed aldol condensations together with the incorporation of the nitrogen atom of ammonia, e.g.

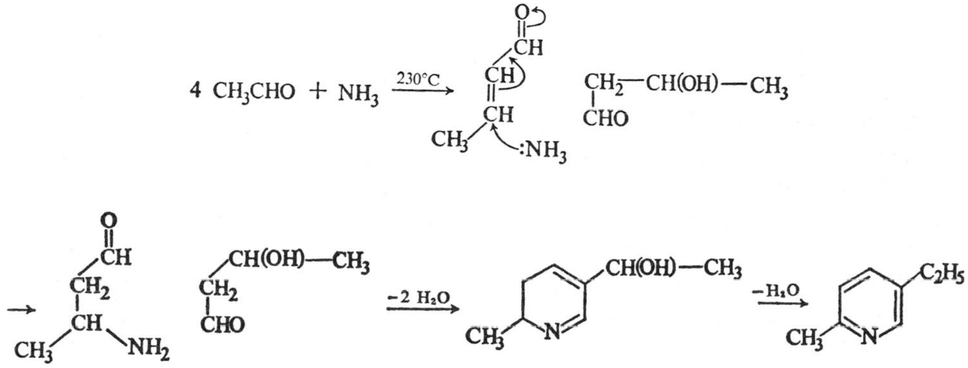

$$4\ CH_3CHO + NH_3 \xrightarrow{230°C}$$

The products are usually mixtures which require separation by chromatography or distillation. For example, the 2-methyl-5-ethylpyridine prepared as above in 53% yield is accompanied by 2-picoline (2-methylpyridine).

(3) *From β-keto-esters, aldehydes, and ammonia: the Hantzsch synthesis.* Two molecules of a β-keto-ester and one of an aldehyde react in the presence of ammonia to give a dihydropyridine which is then dehydrogenated, usually with nitric acid. The aldehyde reacts with one molecule of the β-keto-ester under the influence of ammonia or an added base, and ammonia itself reacts with the second molecule of the β-keto-ester; the two resulting units are then joined by a Michael-type addition followed by ring-closure:

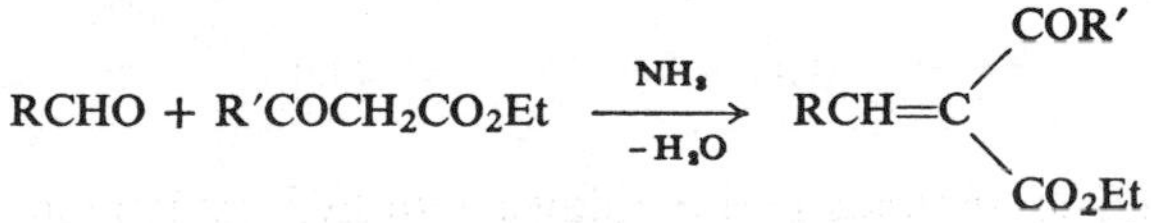

$$RCHO + R'COCH_2CO_2Et \xrightarrow[-H_2O]{NH_3} RCH=C\begin{smallmatrix}COR'\\\\CO_2Et\end{smallmatrix}$$

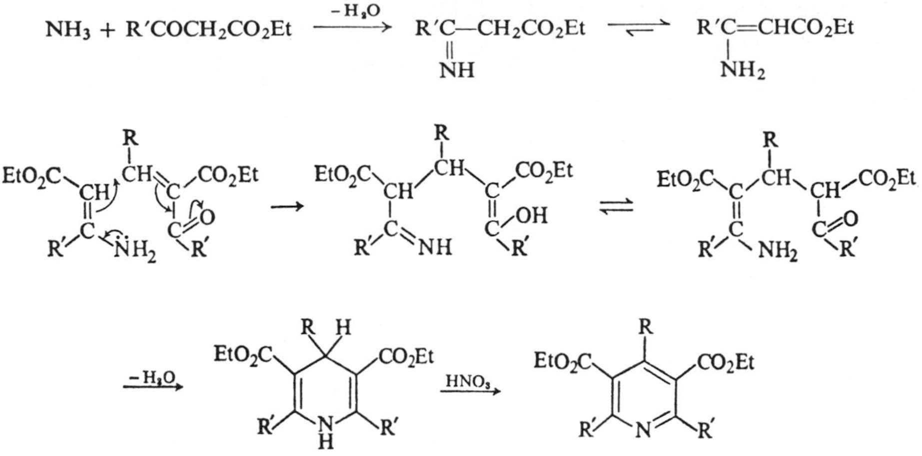

$$NH_3 + R'COCH_2CO_2Et \xrightarrow{-H_2O} \underset{\overset{\|}{NH}}{R'C}-CH_2CO_2Et \rightleftharpoons \underset{\overset{|}{NH_2}}{R'C}=CHCO_2Et$$

For example, acetoacetic ester, formaldehyde, and ammonia, in the presence of diethylamine, give a dihydropyridine which, after oxidation with nitric acid, hydrolysis, and decarboxylation, gives 2,6-lutidine (2,6-dimethylpyridine) in 65% yield [2]:

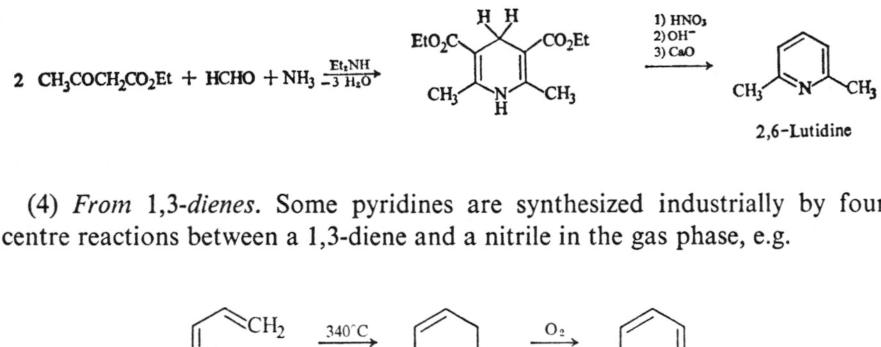

$$2\ CH_3COCH_2CO_2Et + HCHO + NH_3 \xrightarrow[-3\ H_2O]{Et_2NH}$$

2,6-Lutidine

(4) *From* 1,3-*dienes.* Some pyridines are synthesized industrially by four-centre reactions between a 1,3-diene and a nitrile in the gas phase, e.g.

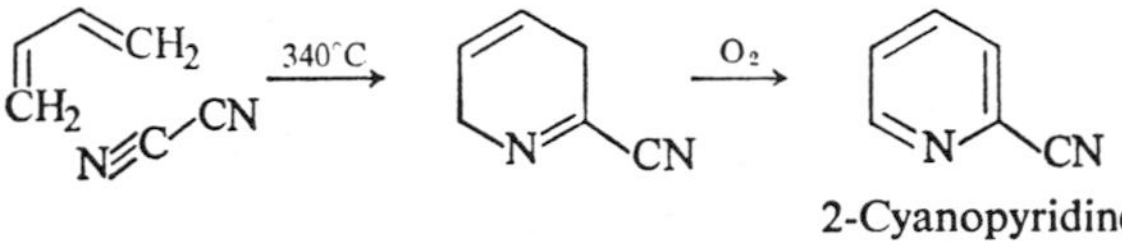

2-Cyanopyridine

(5) *From furans and pyrones.* Furans containing 2-carbonyl substituents react with ammonia to give pyridines, e.g.

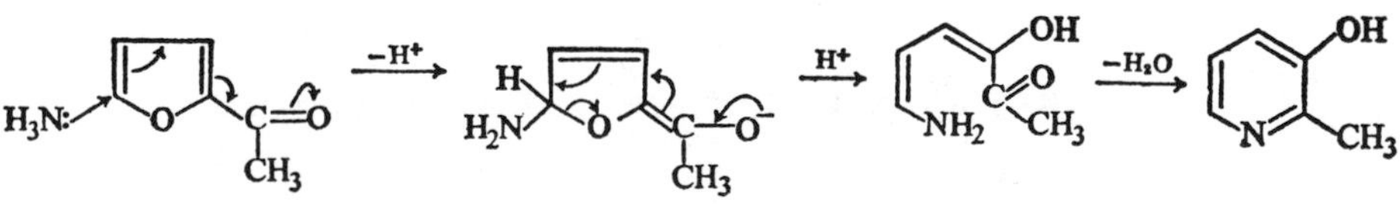

2-Pyrones also give pyridines with ammonia, e.g.

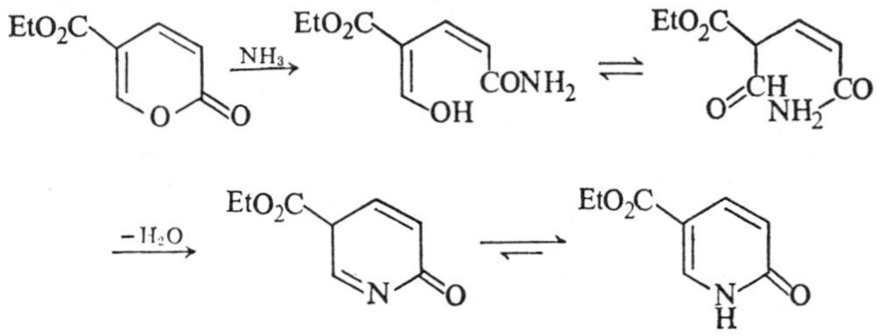

Pyridines are converted into their *N*-oxides with hydrogen peroxide in glacial acetic acid (p. 602) and are catalytically reduced to piperidines over nickel at 200°C. Piperidines are also available by the cyclization of pentamethylene-containing compounds: e.g. pentamethylenediamine dihydrochloride gives piperidine on being heated (p. 327) and pentamethylene dichloride and a primary amine give an *N*-substituted piperidine.

(b) QUINOLINES

The quinoline nucleus is contained in several groups of alkaloids (e.g. quinine; 21.6), in a number of synthetic materials with important physiological properties such as the antimalarial plasmoquin, in the cyanine dyes (p. 274), and in the analytical reagent, oxine (8-hydroxyquinoline).

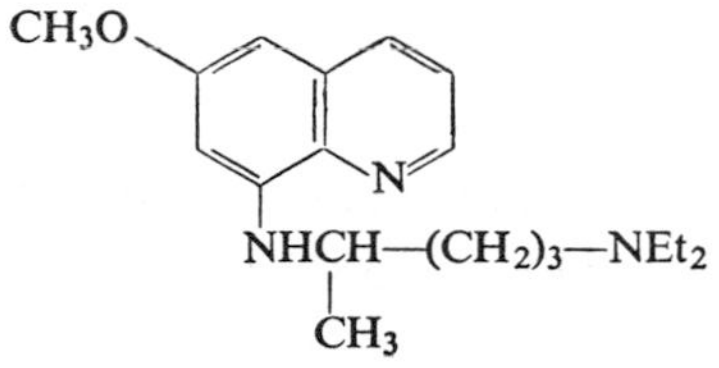

Plasmoquin

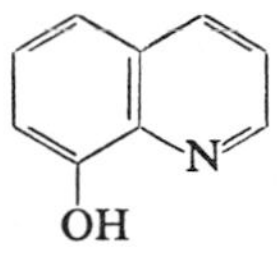

Oxine

The nucleus is usually formed in one of two ways:

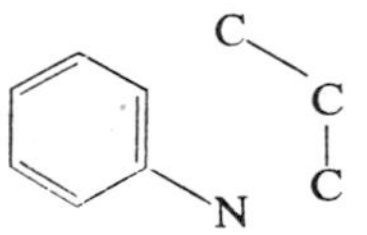

Skraup, Döbner-von Miller, and Conrad-Limpach syntheses

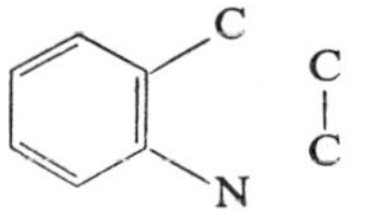

Friedlaender and Pfitzinger syntheses

(1) *The Skraup synthesis.* In the simplest case, aniline, glycerol, nitrobenzene, iron(II) sulphate, and sulphuric acid are mixed and gently heated. A vigorous exothermic reaction occurs which is completed by further heating. After removal of the residual nitrobenzene by distillation in steam, quinoline is liberated from the reaction mixture by basification and is isolated by steam-distillation. The yield of purified product is 84–91% [1].

The reaction involves dehydration of glycerol to acrolein, Michael-type addition of aniline to acrolein, acid-catalyzed cyclization, and dehydrogenation by nitrobenzene. Iron(II) sulphate acts to moderate the reaction. Acrolein itself is not employed because of its tendency to polymerize; the success of the reaction depends on the rapid addition of aniline to the acrolein as it is formed.

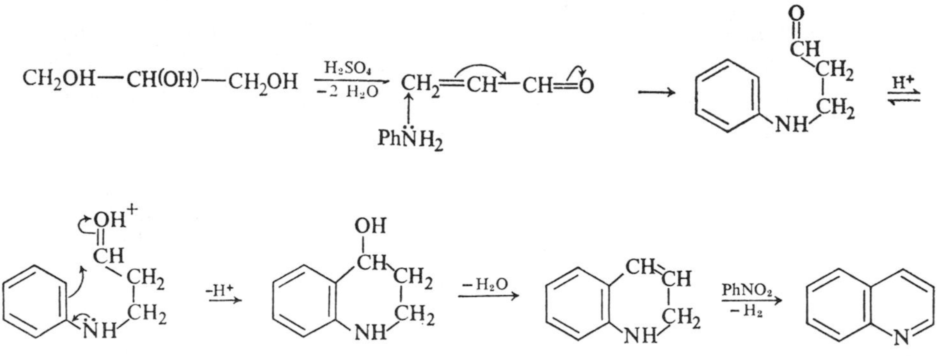

The synthesis is of wide application. The aniline may contain nuclear substituents: *ortho*- and *para*-substituted anilines give 8- and 6-substituted quinolines, respectively, and *meta*-substituted anilines give mixtures of 5- and 7-substituted quinolines, the latter predominating with activating substituents such as methoxyl and the former predominating with deactivating substituents such as nitro. Acrolein (as glycerol) may be replaced by crotonaldehyde, giving 2-methylquinolines, and by methyl vinyl ketone, giving 4-methylquinolines.

(2) *The Döbner-von Miller synthesis.* This is very similar to the Skraup method, the differences being that the three-carbon fragment is formed *in situ* by an acid-catalyzed aldol condensation and that no dehydrogenating agent is added. The oxidative step is thought to be brought about by hydride-transfer to the Schiff base which is formed by reaction between the aniline and the aldehyde or ketone, e.g.

$$2\ CH_3CHO \xrightarrow[-H_2O]{H^+} CH_3-CH=CH-CHO$$

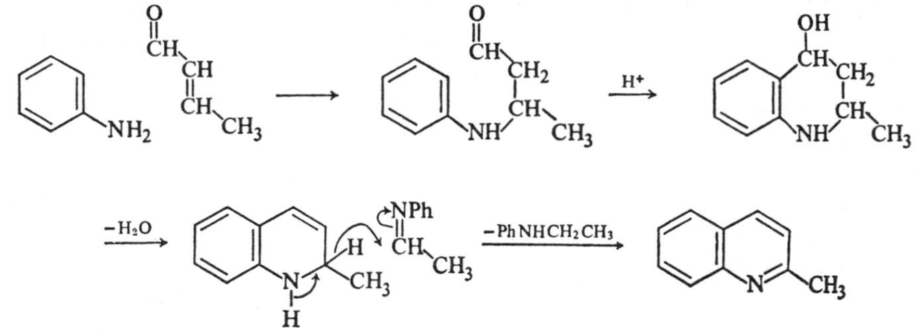

Quinaldine

(3) *The Conrad-Limpach synthesis.* β-Keto-esters react with anilines in each of two ways: at low temperatures reaction occurs at the keto-group to give a Schiff base, and at high temperatures reaction occurs at the ester group to give an amide. The resulting compounds can each be cyclized to quinolines, e.g.

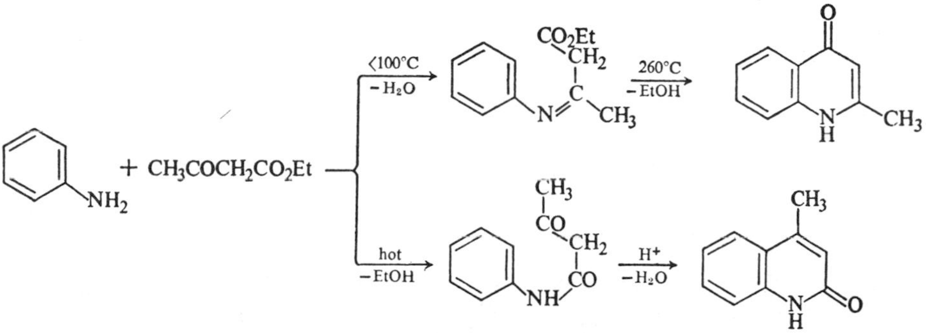

(4) *The Friedlaender synthesis.* An *o*-aminobenzaldehyde is treated with an aldehyde or ketone; formation of the Schiff base is followed by dehydrative cyclization:

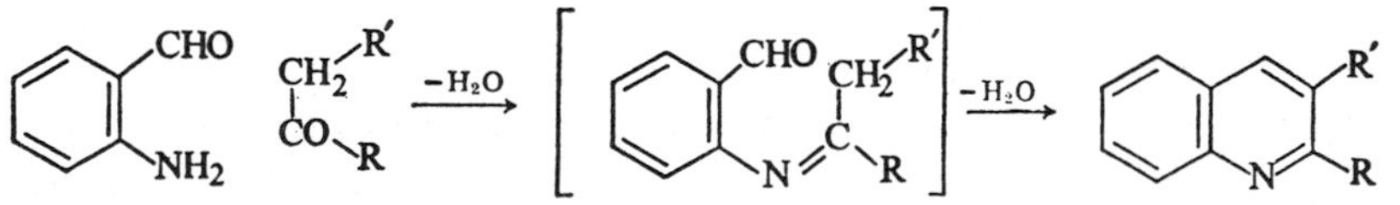

The main problem in this synthesis is that *o*-aminobenzaldehydes are very unstable, readily undergoing self-condensation. One way of overcoming this is to start instead with an *o*-nitrobenzaldehyde: acid- or base-catalyzed condensation gives an intermediate which cyclizes spontaneously on reduction (*cf.* the

Reissert indole synthesis; p. 666), e.g.*

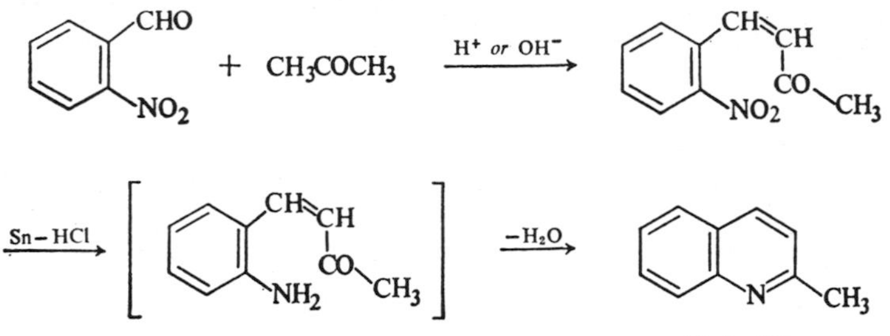

Quinaldine

(5) *The Pfitzinger synthesis.* This modification of Friedlaender's method employs isatin in place of *o*-aminobenzaldehyde. Reaction with base gives *o*-aminobenzoylformate ion which condenses with an aldehyde or ketone to give a 4-carboxyquinoline from which the carboxyl group can be removed thermally:

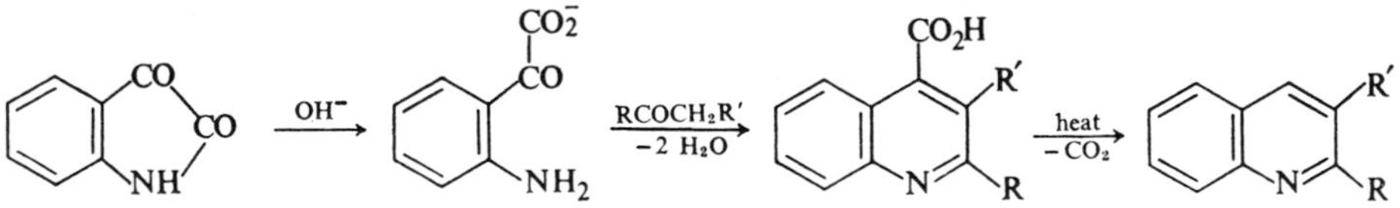

(c) ISOQUINOLINES

The isoquinoline ring occurs in a large number of alkaloids, the synthesis of one of which, papaverine, is outlined below (p. 683). The general synthetic routes to isoquinolines involve the following skeletal types:

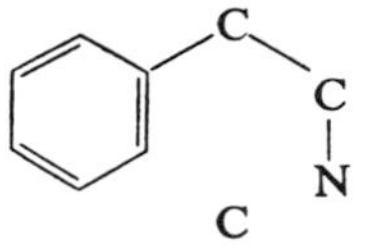

Bischler-Napieralski and
Pictet-Spengler syntheses

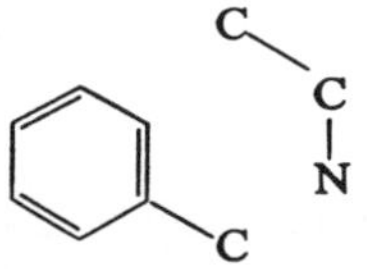

Pomeranz-Fritsch synthesis

*With *hot* alkali, the first-formed aldol product undergoes a series of transformations leading to indigo.

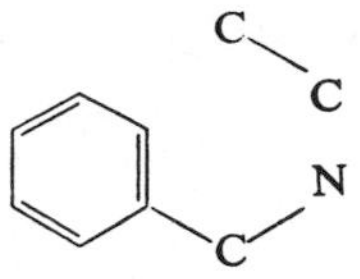

Schlittler-Müller synthesis

(1) *The Bischler-Napieralski synthesis.* A β-arylethylamine* is converted into an amide which is cyclized with an acidic reagent such as phosphorus oxychloride. Dehydrogenation of the product, usually with palladium on charcoal, gives the quinoline. Yields are normally of the order of 90%.

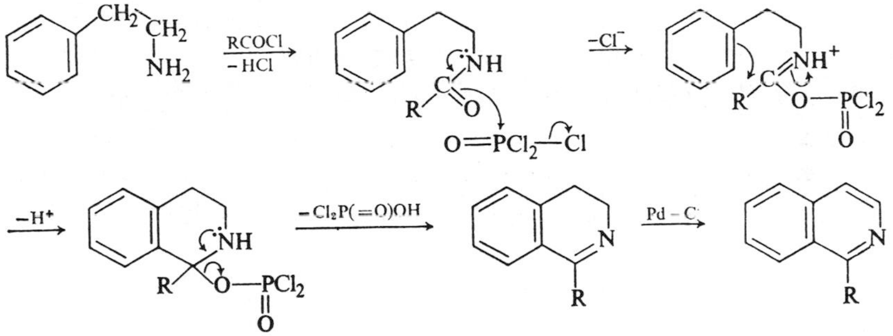

A synthesis of papaverine is based on the Bischler-Napieralski procedure:

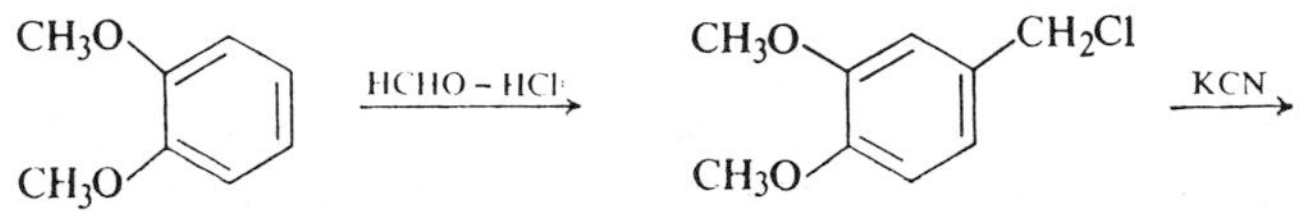

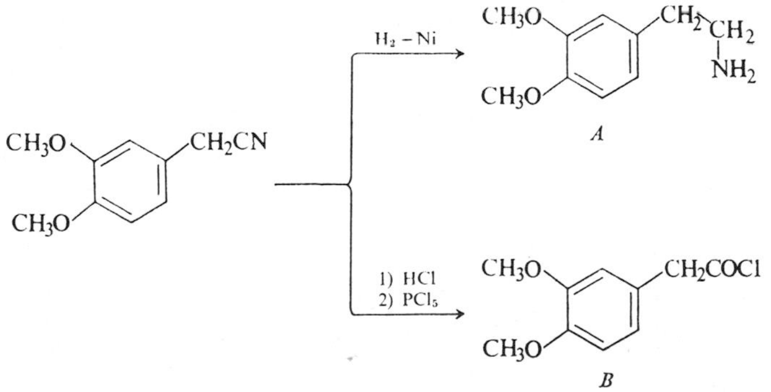

*These compounds are readily obtained from aromatic aldehydes by base-catalyzed condensation with nitromethane followed by reduction: $ArCHO \rightarrow ArCH{=}CHNO_2 \rightarrow ArCH_2CH_2NH_2$.

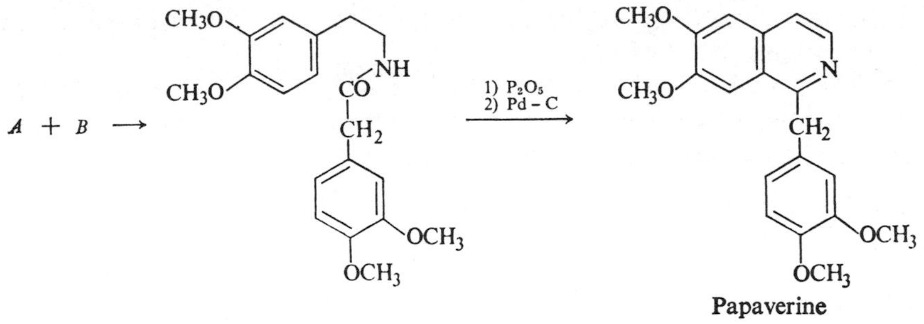

$A + B \longrightarrow$

Papaverine

(2) *The Pictet-Spengler synthesis.* A β-arylethylamine is treated with an alde-hyde in the presence of dilute acid; ring-closure occurs by a reaction of Mannich type and the tetrahydroisoquinoline is dehydrogenated on palladium.

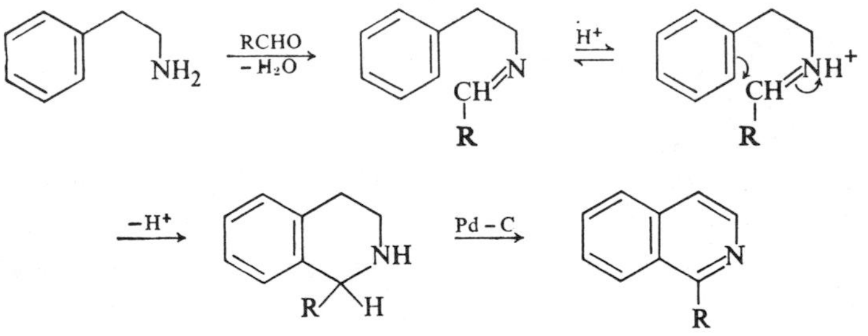

(3) *The Pomeranz-Fritsch synthesis.* The acetal of an α-amino-aldehyde (pre-pared from the corresponding chloro compound with ammonia at 0°C) is treated with an aromatic aldehyde or ketone to give the Schiff base which is then cyclized with acid. (The acetal is employed because of the instability of amino-aldehydes, which undergo self-condensation.)

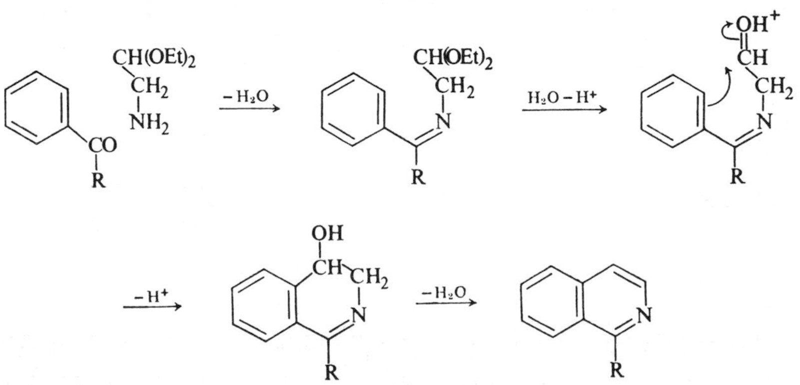

Yields are usually not greater than 50%, being lower from aromatic ketones than from aromatic aldehydes.

(4) *The Schlittler-Müller synthesis.* A basically similar approach is adopted but, by avoiding the inefficient step involving reaction between an aromatic ketone and an amine, yields are improved.

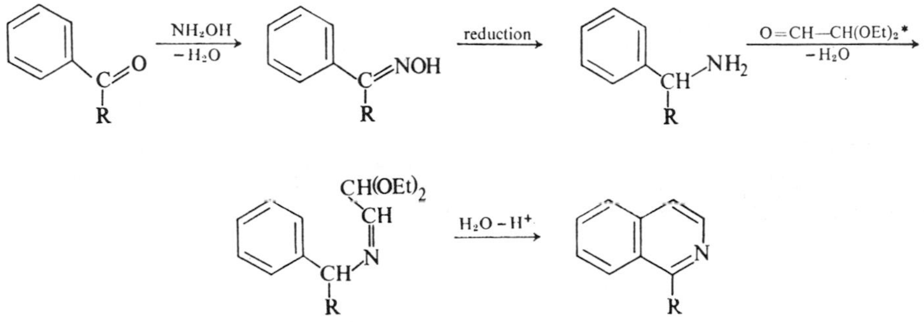

(d) PYRANS, PYRONES, AND PYRYLIUM SALTS

Although there can be no uncharged oxygen analogue of pyridine, six-membered heterocyclic compounds containing oxygen are well known and many occur naturally. The simplest compounds, 2- and 4-pyran, have not themselves been prepared, although simple derivatives can be synthesized, e.g.

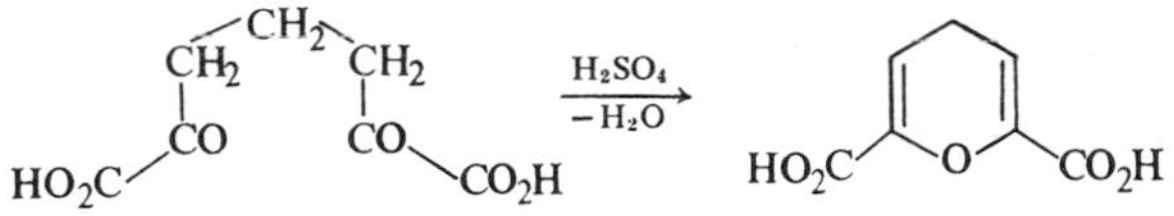

2,6-Dicarboxy-4-pyran

The pyrones are better known. 2-Pyrone can be made by decarboxylation of the 5-carboxy-derivative whose synthesis has been outlined (p. 668), and 4-pyrone may be made *via* the naturally occurring chelidonic acid:

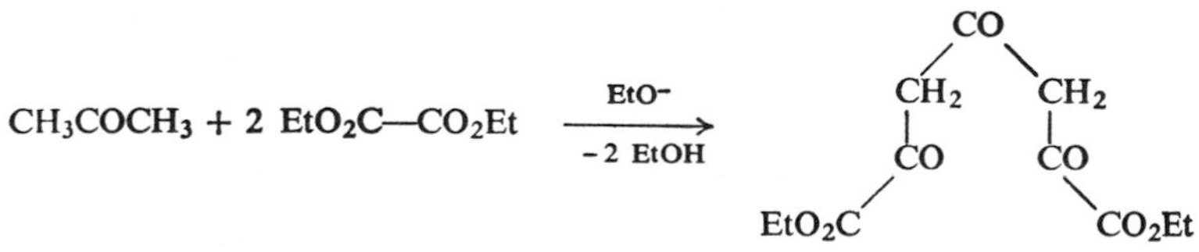

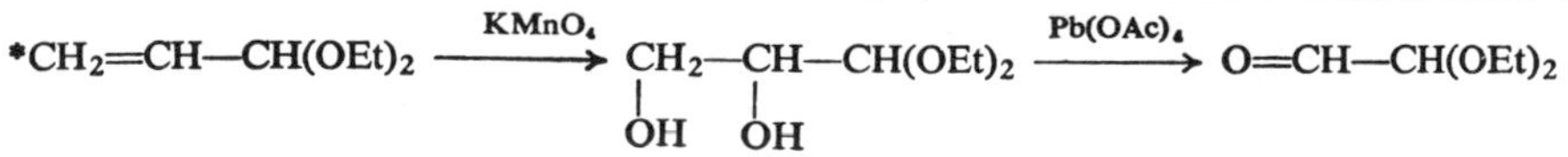

$$\text{CH}_3\text{COCH}_3 + 2\ \text{EtO}_2\text{C}-\text{CO}_2\text{Et} \xrightarrow[-2\ \text{EtOH}]{\text{EtO}^-}$$

$$*\text{CH}_2{=}\text{CH}-\text{CH(OEt)}_2 \xrightarrow{\text{KMnO}_4} \underset{\underset{\text{OH}\quad\text{OH}}{|\qquad|}}{\text{CH}_2-\text{CH}-\text{CH(OEt)}_2} \xrightarrow{\text{Pb(OAc)}_4} \text{O}{=}\text{CH}-\text{CH(OEt)}_2$$

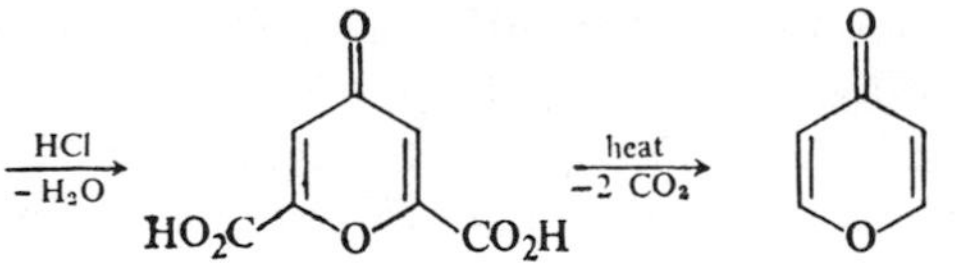

Chelidonic acid   4-Pyrone

$\varDelta^2$-Dihydropyran* can be obtained by the acid-catalyzed rearrangement of tetrahydrofurfuryl alcohol over alumina at 270°C:

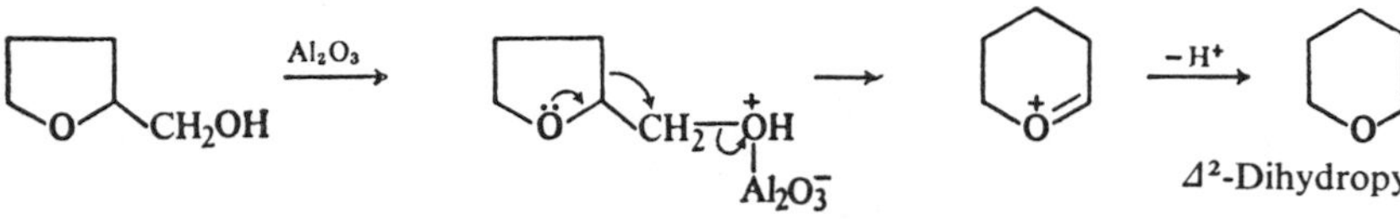

$\varDelta^2$-Dihydropyran

Reduction of the dihydropyran gives the tetrahydro-derivative, used as a solvent. The sugars, in their six-membered cyclic form, are derivatives of tetrahydropyran (pyranoses).

Tetrahydropyran   β-D-Glucose

The formation and transformations of pyrones are controlled by subtle effects which are illustrated by reactions of the dehydroacetic acid system:

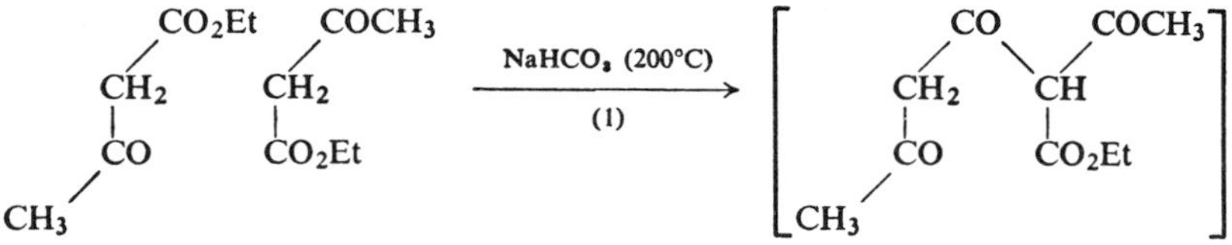

*This compound is used as a protecting agent for alcohols. Reaction with the alcohol in the presence of anhydrous acid gives an acetal from which the alcohol may be regenerated with mineral acid (see p. 114):

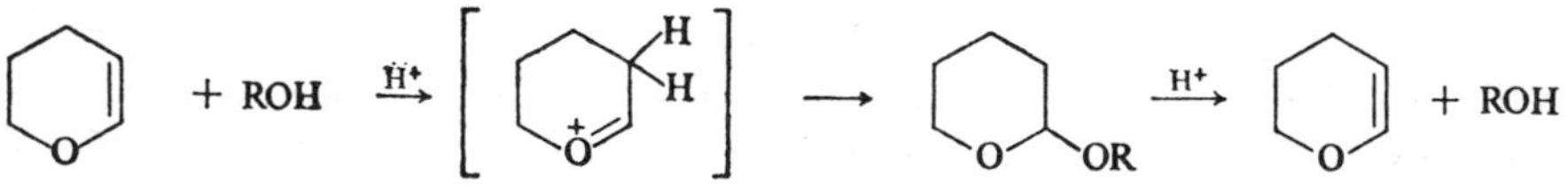

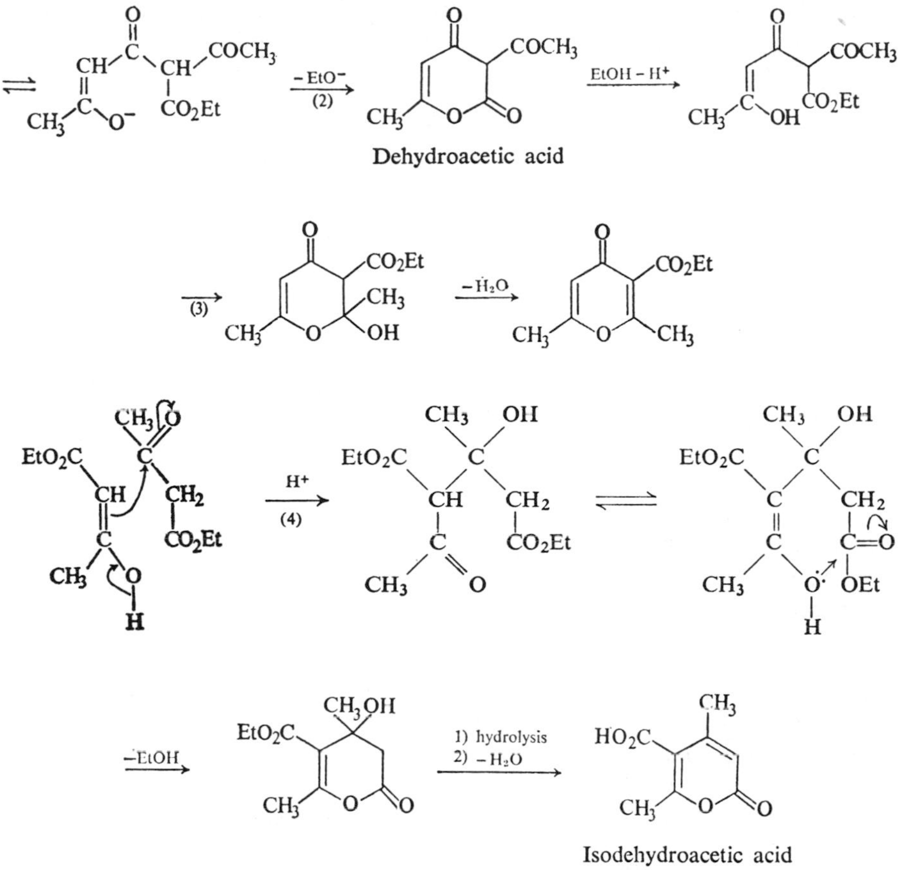

Dehydroacetic acid

Isodehydroacetic acid

(1) The Claisen condensation takes precedence over the aldol condensation (on to the keto group) because the product of the former reaction is removed from equilibrium as a strongly resonance-stabilized anion; *cf.* the condensation between acetone and ethyl acetate, p. 243.

(2) This anion reacts intramolecularly with the carboethoxy group through oxygen rather than through (the more usual) carbon because the latter would lead to a strained four-membered ring; *cf.* the reaction of acetoacetic ester with ω-dibromopropane, p. 252.

(3) Ethanolysis of the lactone is followed by acid-catalyzed formation of a hemi-acetal and dehydration.

(4) Acid-catalyzed self-condensation of acetoacetic ester occurs on the carbonyl group because this is more reactive towards nucleophiles than the ester group and the driving force mentioned in (1) is absent.

Benzopyrones. Two of the three benzopyrones, coumarin and chromone, are notable. Coumarin, which occurs naturally (e.g. in clover), has been synthesized by the Perkin reaction,

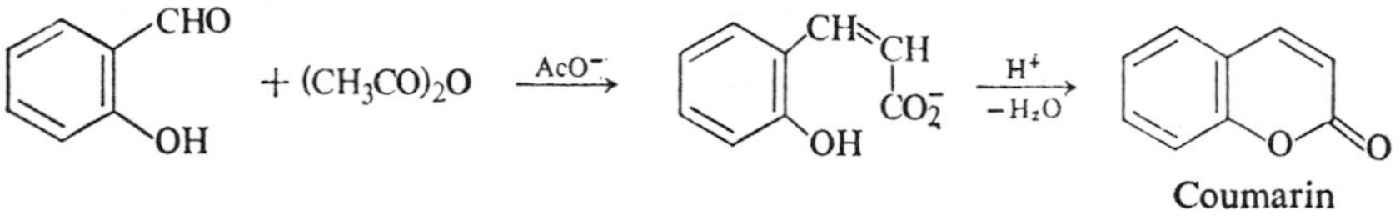

and by the von Pechmann reaction, from phenol, malic acid, and concentrated sulphuric acid,

$$HO_2CCH_2CH(OH)CO_2H \xrightarrow[-CO, \ -H_2O]{H_2SO_4} HO_2C-CH_2-CHO$$

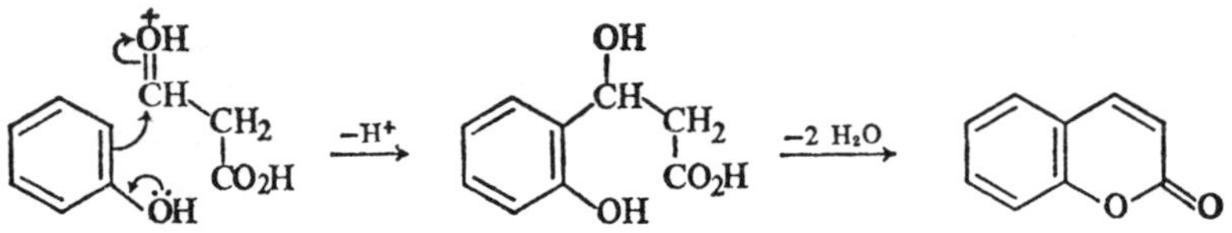

The latter reaction may also be applied to β-keto-esters.

Chromone is the parent member of a group of plant colouring pigments, the flavones (2-arylchromones), which are found both free and as glycosides. Three general synthetic routes are available.

(1) From salicylic acid derivatives:

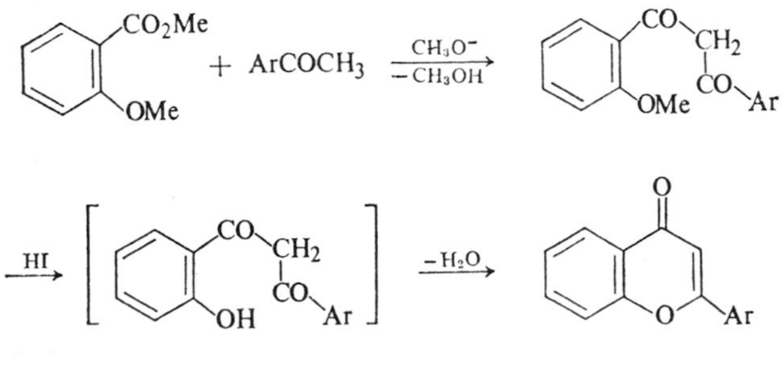

(2) From *o*-hydroxyacetophenones and aromatic acid anhydrides:

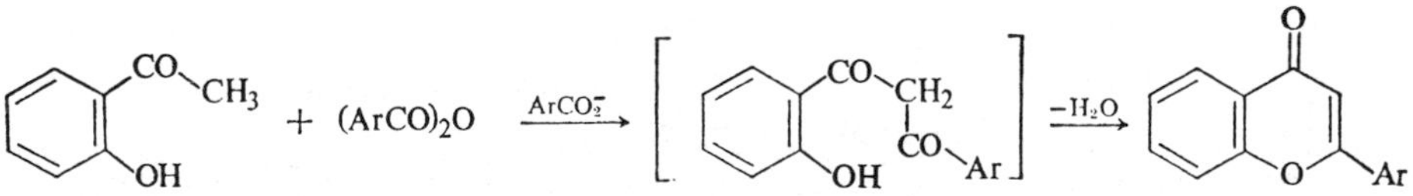

(3) From o-hydroxyacetophenones and aromatic aldehydes. The synthesis of quercetin is illustrative:

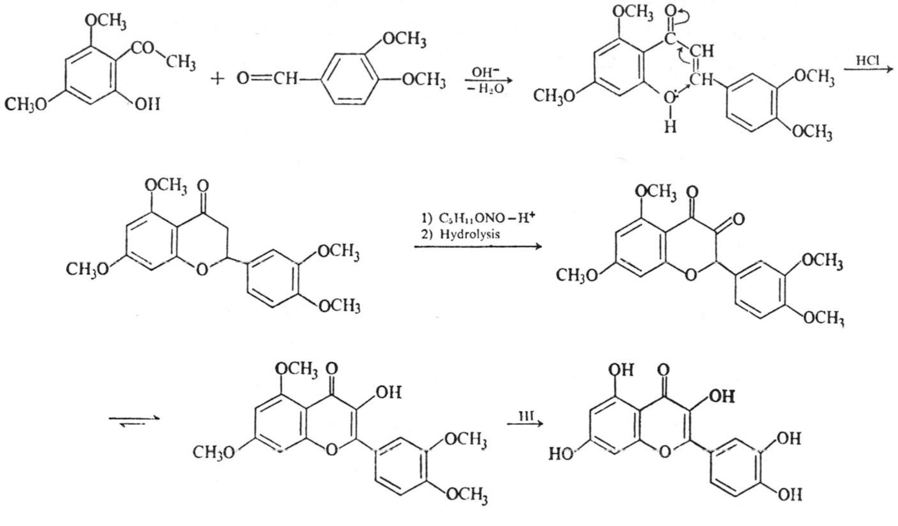

Quercetin

Pyrylium salts. Salts of the type $R_3O^+X^-$ are normally unstable and cannot be isolated unless X^- is an especially stable (non-nucleophilic) anion such as BF_4^-.* However, the oxonium salts derived by oxidation of the pyran nucleus are comparatively stable because the cation possesses aromatic stabilization energy which is lost when an anion bonds covalently to it. The parent member of the series, the pyrylium ion, is formed by acidification of the sodium salt of glutaconic aldehyde at low temperatures:

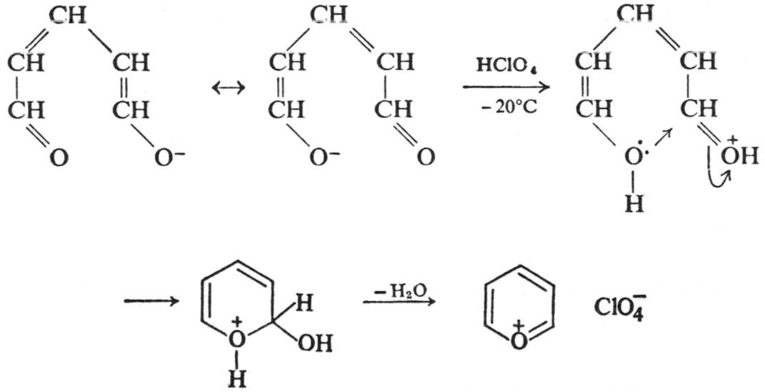

Pyrylium perchlorate

*For example, $(CH_3)_3O^+ BF_4^-$ is a crystalline solid [51] which is a powerful alkylating agent; it can be used to methylate weakly nucleophilic species such as amides.

The glycosides of a number of 2-arylbenzopyrylium salts occur naturally as plant colouring matters, the anthocyanins.

The corresponding aglycons, the anthocyanidins, are synthesized from salicyl-aldehydes and acetophenones by an acid-catalyzed aldol condensation followed by ring-closure in ethyl acetate solution:

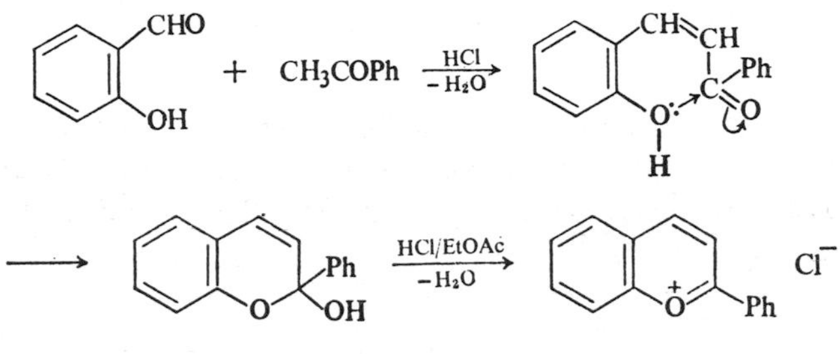

For example, cyanidin chloride has been prepared as follows:

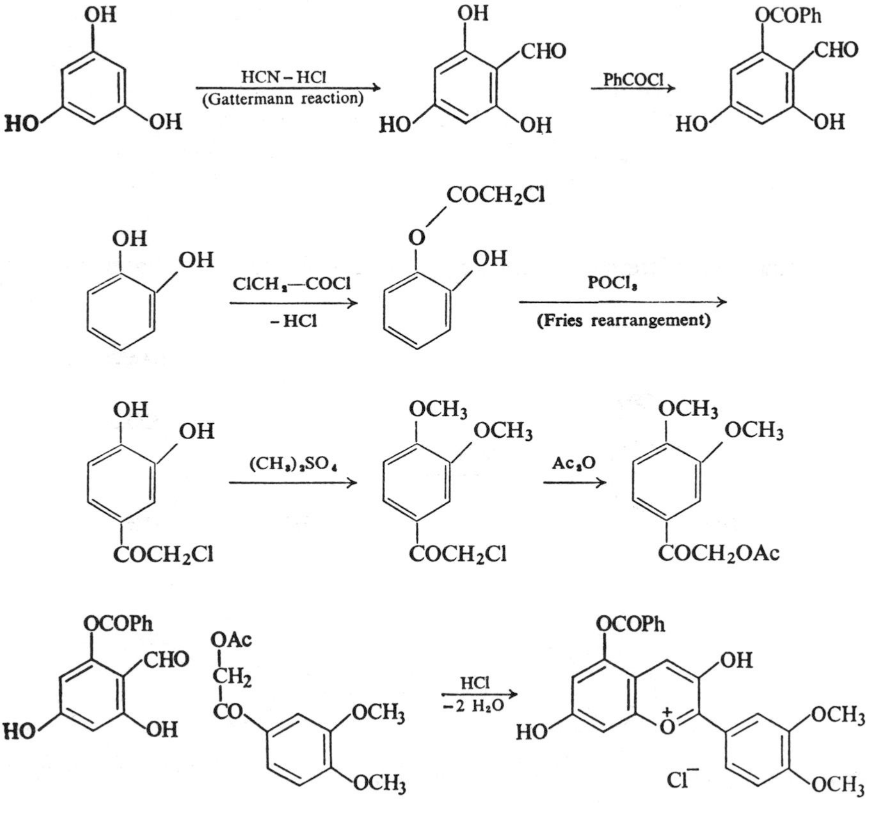

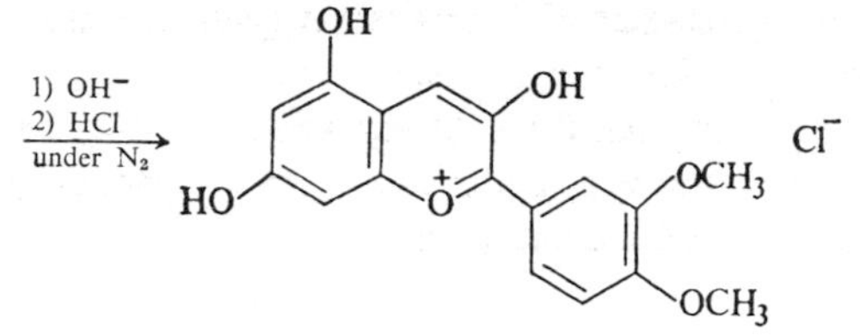

Cyanidin chloride

In the formation of the anthocyanins themselves, it is necessary to introduce the sugar units before the acid-catalyzed condensation. Glucose is introduced by the reaction of tetra-acetyl-α-bromoglucose [Br—$C_6H_7(OAc)_4$] with the appropriate hydroxyl group:

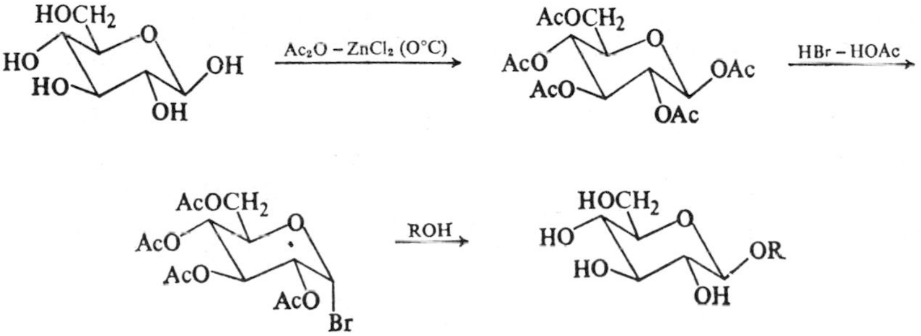

The remainder of the synthesis is an adaptation of that for the anthocyanidins, e.g.

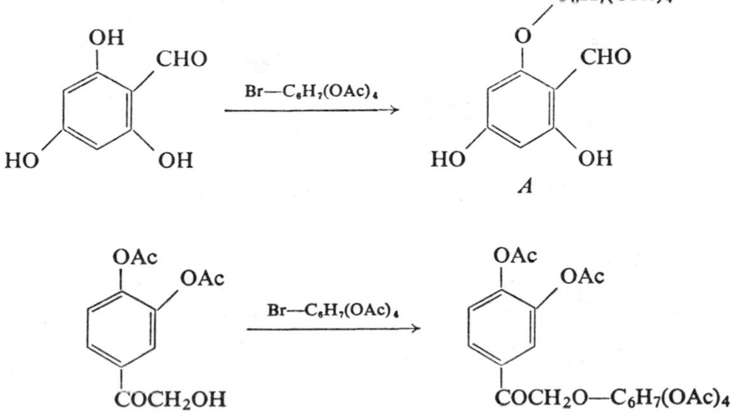

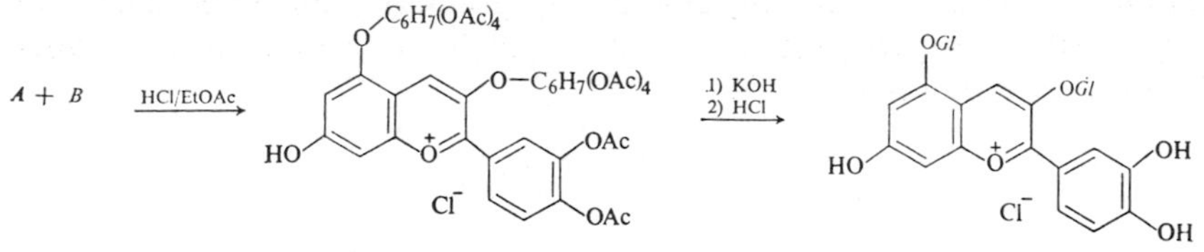

Cyanin chloride
($Gl{\equiv}\beta$-glycosidyl)

20.5 Six-membered Rings Containing Two Hetero-atoms

(a) PYRIMIDINES

Of the three isomeric analogues of benzene which possess two nitrogen atoms (pyridazine, pyrimidine, and pyrazine), pyrimidine and its derivatives are by far the most important. Both the ribonucleic acids and the deoxyribonucleic acids contain cytosine (2-hydroxy-4-aminopyrimidine), the former group contains uracil (2,4-dihydroxypyrimidine), and the latter group contains thymine (2,4-dihydroxy-5-methylpyrimidine). The heterocycles are linked through their 1-nitrogen atoms to ribose and deoxyribose, the hydroxypyrimidines being in their tautomeric -one forms.

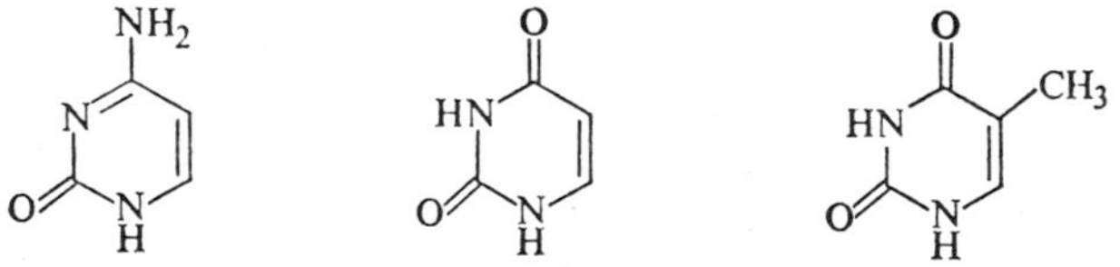

Cytosine   Uracil   Thymine

The full structure of the nucleic acids is as follows:

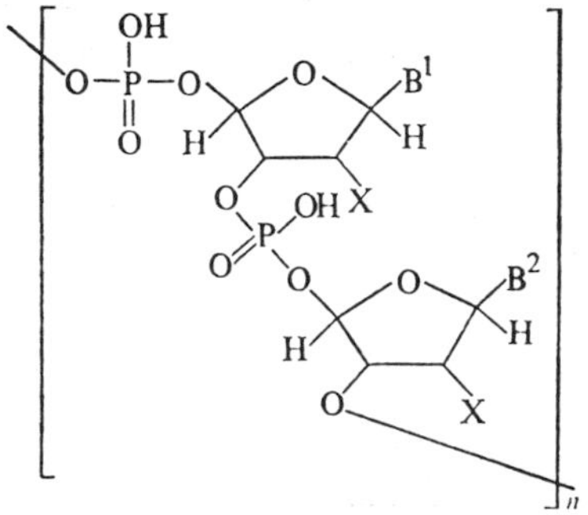

where $X = OH$ (ribonucleic acids) or $X = H$ (deoxyribonucleic acids) and B^1, B^2, etc., are heterocyclic bases which include the three pyrimidines mentioned above and two purines (adenine and guanine; p. 698).

In addition, vitamin B$_1$ (p. 694) contains a pyrimidine ring as do barbituric acid (2,4,6-trihydroxypyrimidine) and several of its derivatives (e.g. veronal) which are used as hypnotics.

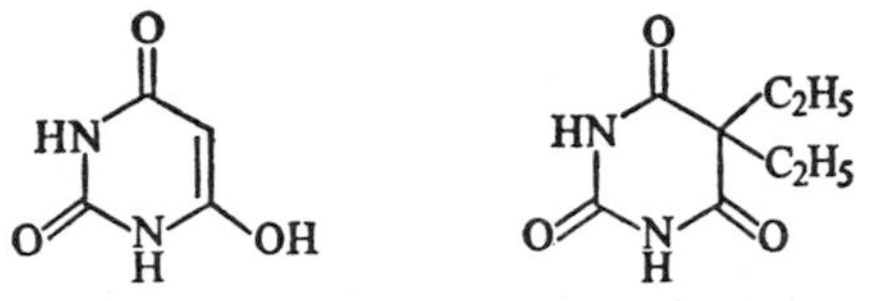

Barbituric acid   Veronal

The pyrimidine nucleus is generally constructed from two three-atom systems:

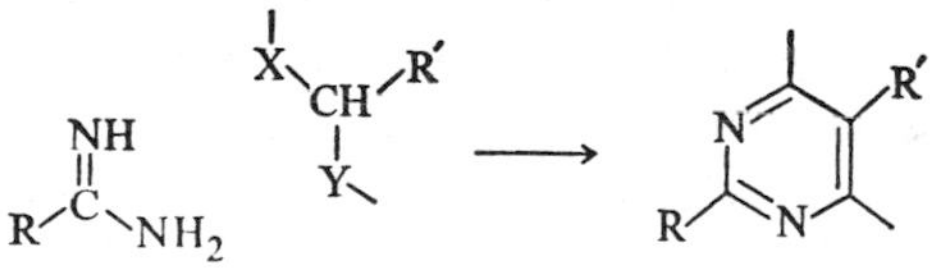

The left-hand component is an amidine derivative in which the choice of R governs the nature of the 2-substituent in the pyrimidine: R may be hydrogen (formamidine), amino (guanidine), hydroxyl (urea, in its tautomeric form), thiol (thiourea), alkyl, aryl, alkoxyl, etc. The choice of R' governs the nature of the 5-substituent (usually hydrogen). X and Y are carbonyl-containing or nitrile groups, the right-hand component being, e.g. a β-dicarbonyl compound, a β-keto-ester, a malonic ester, an α-cyano-ketone or an α-cyano ester. Then, if X = CHO, the 4-position of the resulting pyrimidine is unsubstituted, X = CO$_2$Et gives a 4-hydroxy-derivative, and X = CN gives a 4-amino-derivative; the nature of the 6-substituent is determined likewise by the choice of Y. Reaction between the two fragments is usually brought about in the presence of sodium ethoxide in boiling ethanol. For example, 2,4-diamino-6-hydroxy-pyrimidine (as its dihydrochloride) is obtained in about 80% yield from guanidine and cyanoacetic ester [4],

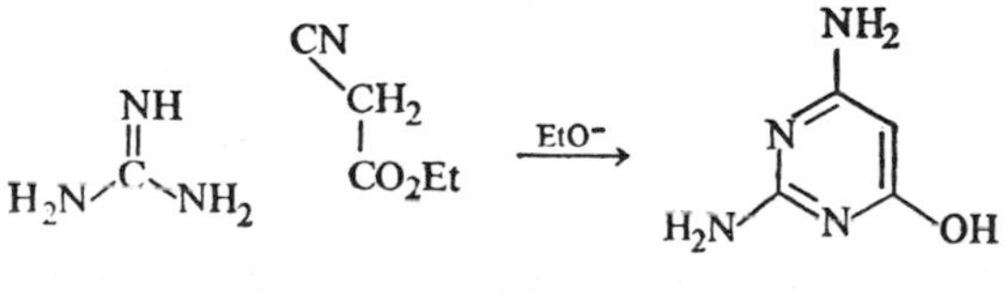

and barbituric acid is obtained in 79% yield from urea and malonic ester [2].

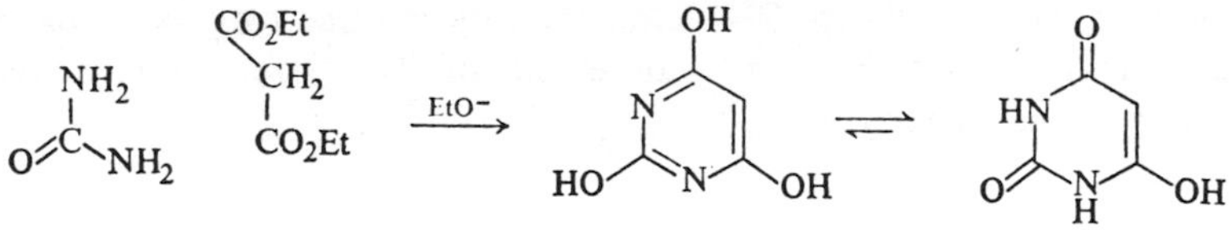

A variant of this approach utilizes Michael-type addition to an $\alpha\beta$-unsaturated ester, e.g.

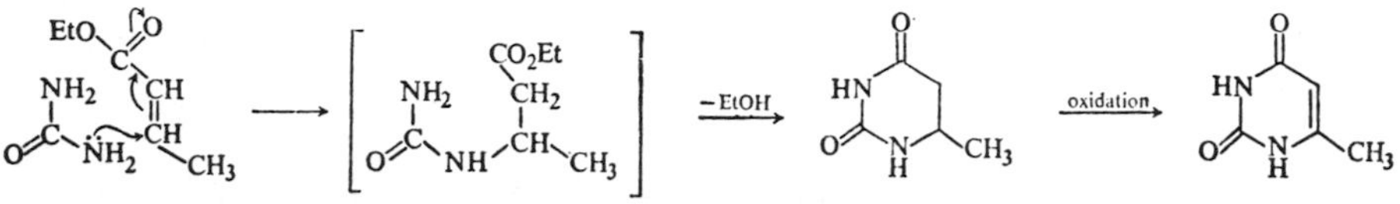

Vitamin B$_1$ (thiamine) is a pyrimidine derivative which has been synthesized as follows (for the synthesis of the thiazole unit, see p. 675):

$$EtO_2C\text{---}CH_2\text{---}CH_2OEt + HCO_2Et \xrightarrow{\ EtO^-\ } EtO_2C\text{---}CH\text{--}CH_2OEt$$
$$\overset{|}{\underset{\displaystyle CHO}{}}$$

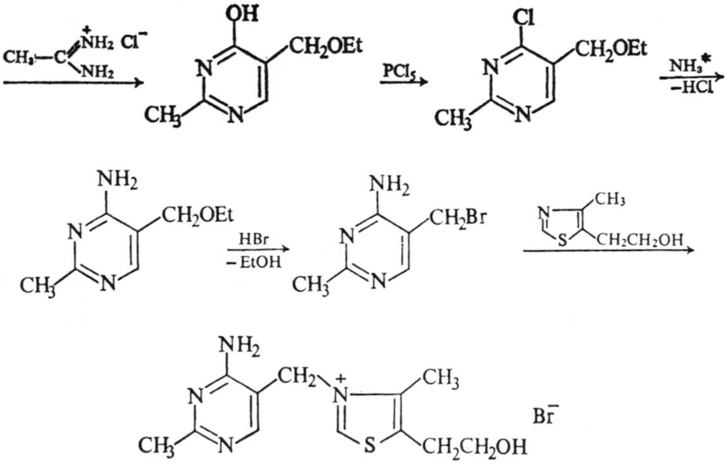

Thiamine bromide

(b) PYRIDAZINES
The ring system is built up from an unsaturated 1,4-dicarbonyl compound and hydrazine, e.g.

*The 2-, 4-, and 6-positions of pyrimidines are strongly activated towards nucleophiles, even more so than those in pyridines; see p. 658.

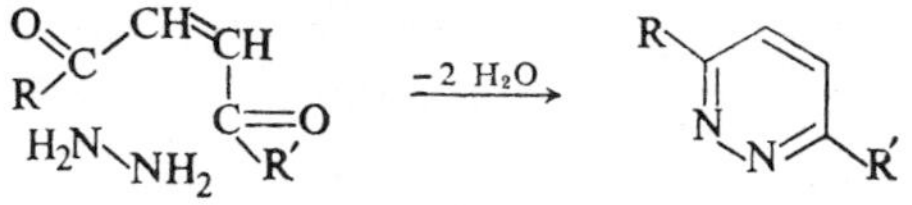

The compounds are of little importance.

(c) PYRAZINES

The pyrazine system occurs naturally (e.g. in aspergillic acid, an antibiotic) and has been incorporated in a number of synthetic drugs, such as sulphapyrazine (a Sulpha drug) and chlorocyclizine (an antihistamine).

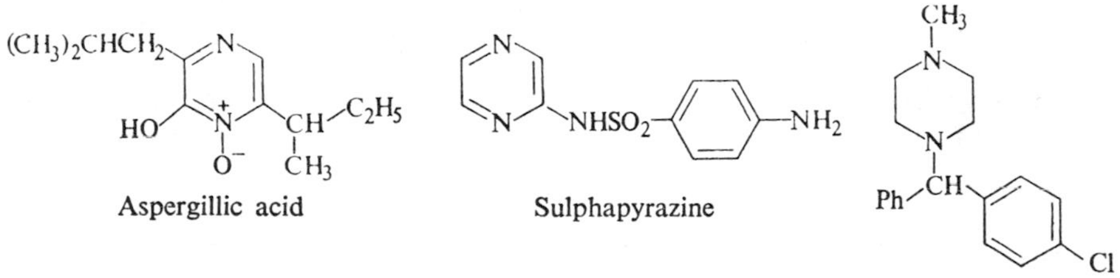

Aspergillic acid   Sulphapyrazine   Chlorocyclizine

The ring is usually constructed by the self-condensation of α-amino-carbonyl compounds followed by aerial oxidation:

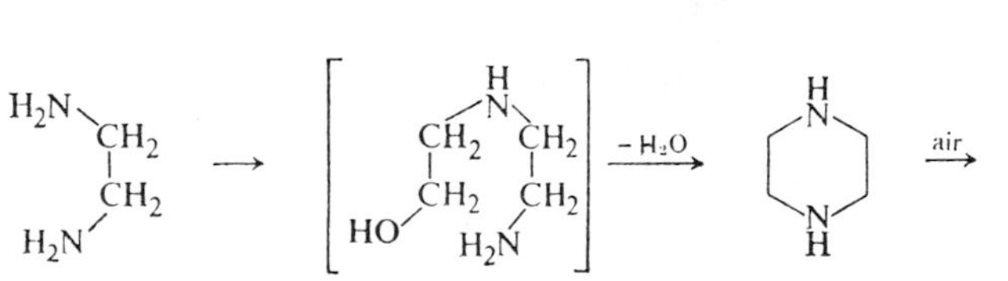

This method is not suitable for pyrazine itself, for aminoacetaldehyde readily polymerizes. A convenient synthesis is from ethylene oxide and ethylenediamine:

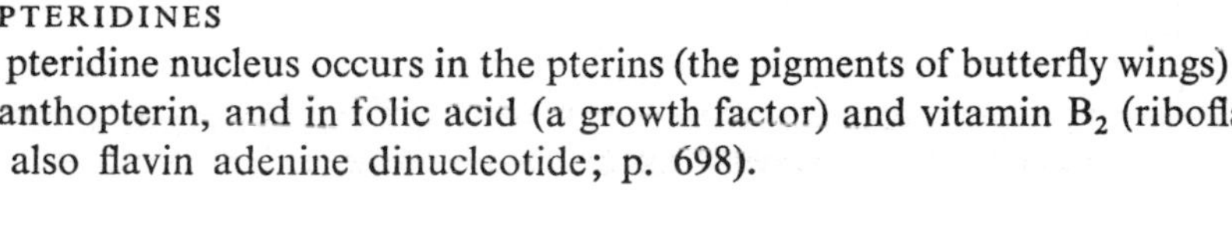

(d) PTERIDINES

The pteridine nucleus occurs in the pterins (the pigments of butterfly wings) such as xanthopterin, and in folic acid (a growth factor) and vitamin B_2 (riboflavin) (see also flavin adenine dinucleotide; p. 698).

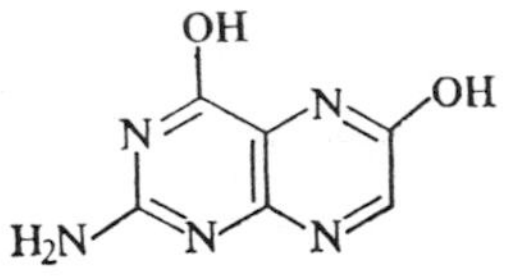

Xanthopterin 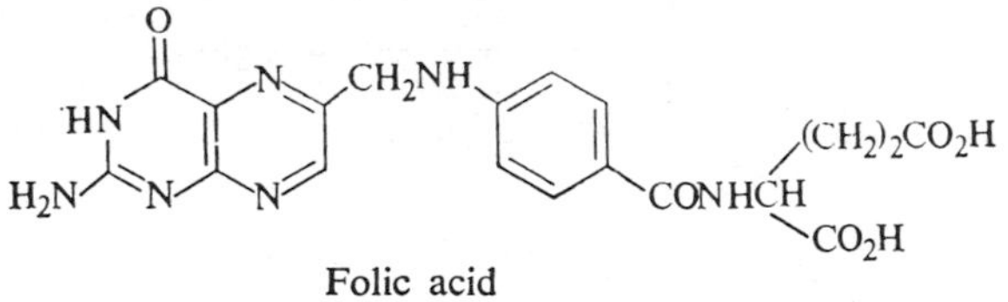   Folic acid

A general synthesis of pteridines, illustrated by that of the parent member, is from a 4,5-diaminopyrimidine and an α-dicarbonyl compound. The former compounds are readily available from 4-aminopyrimidines by nitrosation followed by reduction*.

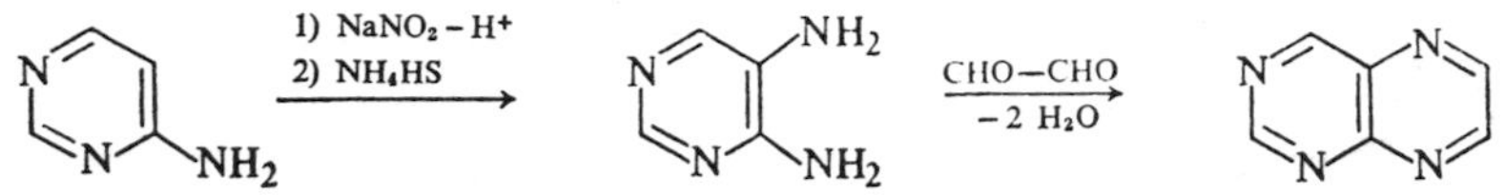

1) NaNO$_2$ – H$^+$
2) NH$_4$HS

CHO—CHO
-2 H$_2$O

Pteridine

A modification of this method gives tetrahydropteridines, as in a synthesis of folic acid:

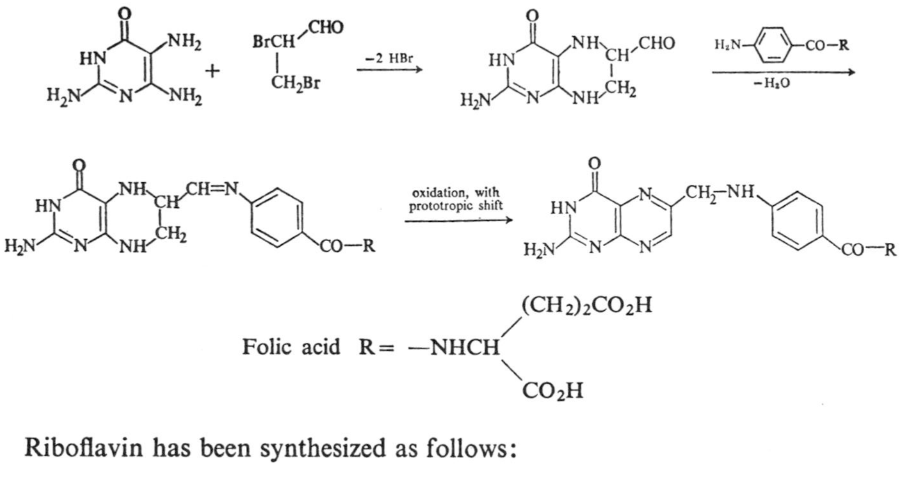

BrCH CHO
CH$_2$Br
-2 HBr

H$_2$N——CO—R
$-$H$_2$O

oxidation, with
prototropic shift

Folic acid   R= —NHCH
(CH$_2$)$_2$CO$_2$H
CO$_2$H

Riboflavin has been synthesized as follows:

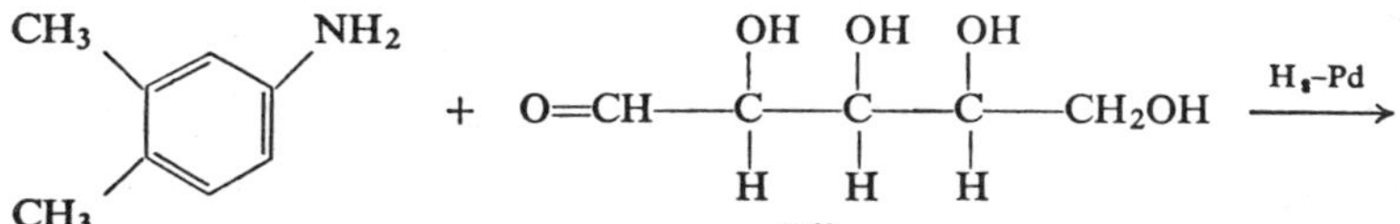

CH$_3$ NH$_2$

CH$_3$

+   O=CH——C——C——C——CH$_2$OH
OH OH OH
H H H

D-Ribose

H$_2$-Pd

*The unshared electron-pair on the amino-nitrogen of a 2- or 4-aminopyrimidine is delocalized over the nucleus and particularly onto the nuclear nitrogens. Diazotization consequently does not occur, but the nucleus is activated towards electrophiles, in this case the nitrosating species.

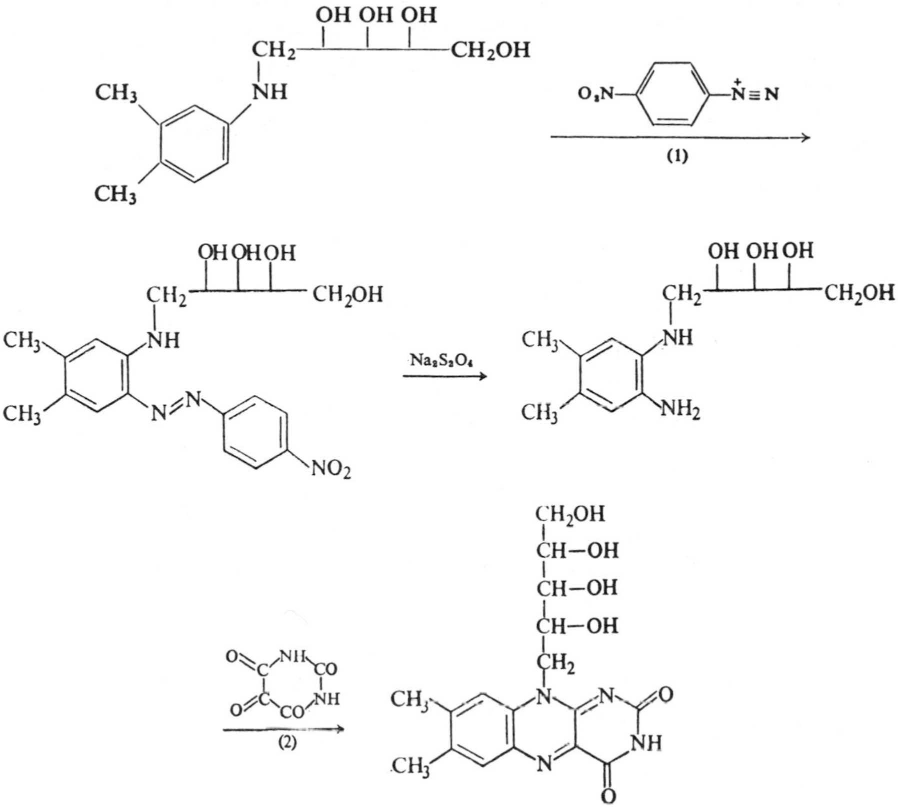

Riboflavin

(1) The amino group activates the *ortho*-position to diazonium coupling; p. 658.

(2) The pyrazine ring could be formed in either of two ways. Only this path enables two molecules of water to be eliminated, and this may provide the necessary driving force. The reagent is alloxan:

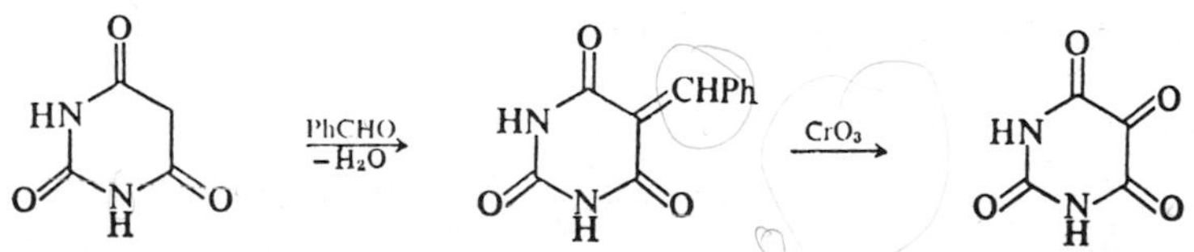

Barbituric acid   Alloxan

(e) PURINES

The purine system occurs widely in nature. Two purines, adenine and guanine, are constituents of the nucleic acids (p. 692); adenine is a component of coenzymes I and II (the so-called phosphopyridine nucleotides), of flavin adenine dinucleotide and of the adenosine phosphates (21.8); caffeine, theophylline, and theobromine are physiologically active constituents of coffee, cocoa, and tea; and uric acid is the end-product of nitrogen metabolism in many animals.

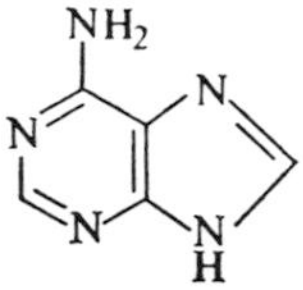

Adenine 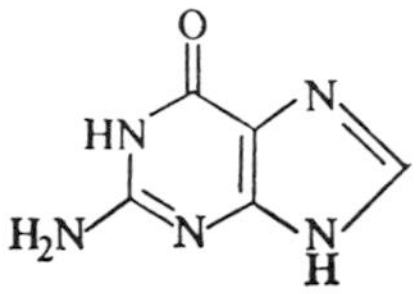   Guanine

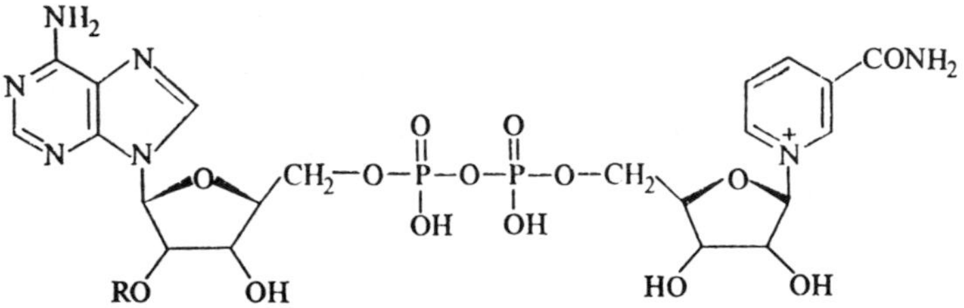

R = H: Coenzyme I (diphosphopyridine nucleotide)
R = P(OH)$_2$: Coenzyme II (triphosphopyridine nucleotide)
    ‖
    O

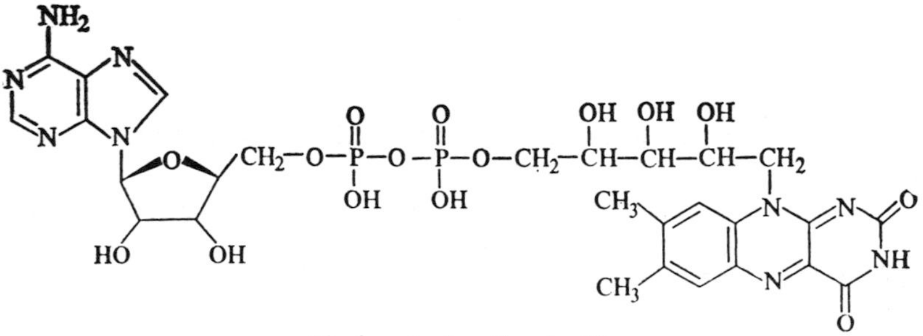

Flavin adenine dinucleotide

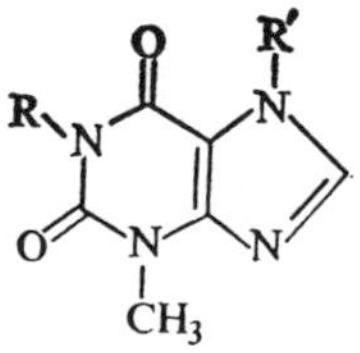

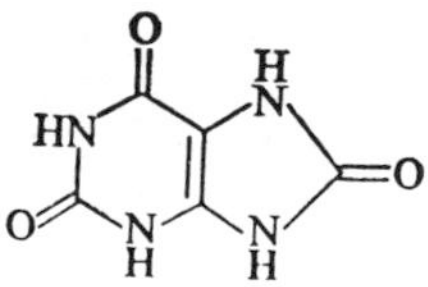

R=R′=CH$_3$: Caffeine   Uric acid
R=CH$_3$, R′=H: Theophylline
R=H, R′=CH$_3$: Theobromine

Purines are usually synthesized by Traube's method in which a 4,5-diamino-pyrimidine is treated with formic acid or, better, sodium dithioformate. 4,5-Diaminopyrimidines are themselves obtained from 4-aminopyrimidines by nitrosation followed by reduction (p. 652) or *via* diazonium coupling of activated methylene compounds (p. 447*) followed by cyclization and reduction. Typical examples are:

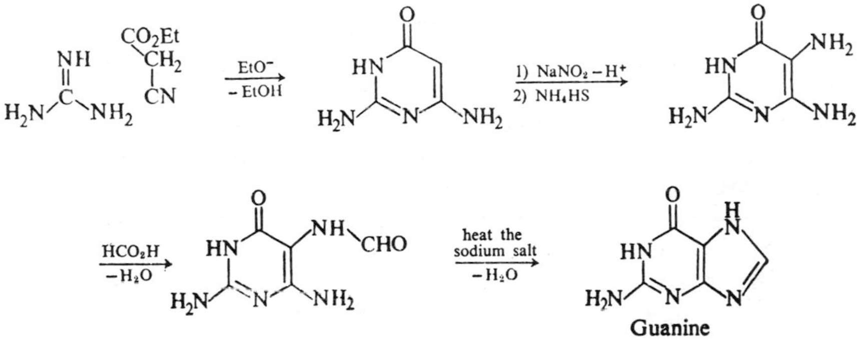

Guanine

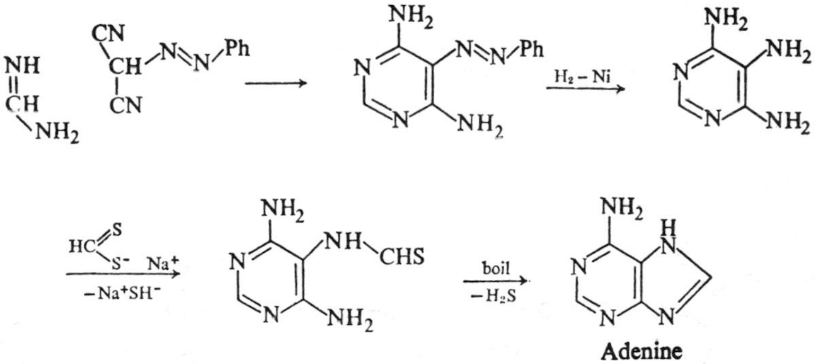

Adenine

8-Hydroxypurines such as uric acid are made by using ethyl chloroformate in place of formic acid:

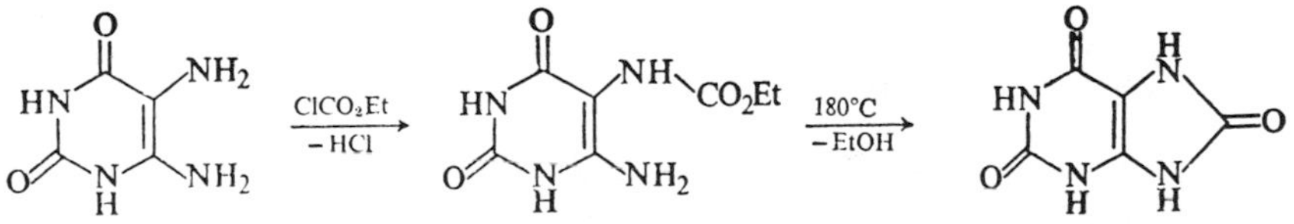

Uric acid

Uric acid is used as the starting material for other purines, e.g.

---

*e.g. $NC-CH_2-CN + Ph-\overset{+}{N}\equiv N \rightarrow (NC)_2CH-N=N-Ph$

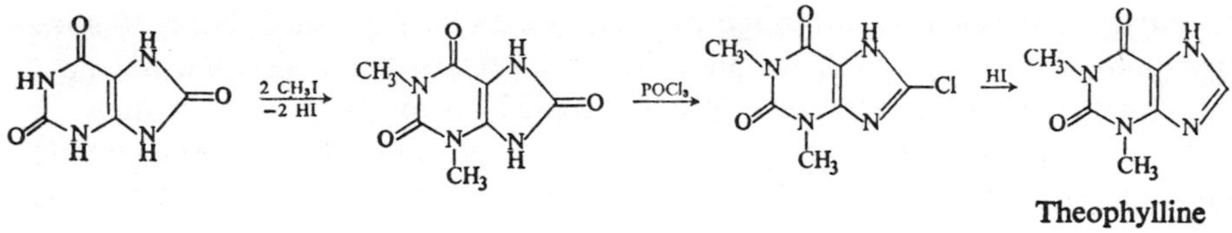

Theophylline

20.6 Six-membered Rings Containing Three Hetero-atoms

Of the three possible analogues of benzene possessing three nitrogen atoms, only 1,3,5-triazine is known. It is obtained by the thermal or base-catalyzed polymerization of formamidine hydrochloride, and is very easily hydrolyzed by acids to formic acid and ammonia. Its best-known derivative, cyanuric chloride, obtained by the vapour-phase polymerization of cyanogen chloride, is now an important intermediate in the dyestuffs industry. Its chlorine atoms are very easily displaced by nucleophiles (*cf.* p. 694), so that a chromophoric system can be attached to the molecule and the whole molecule can then be attached to the cellulose of the fibre by reaction at the latter's hydroxyl groups, e.g.

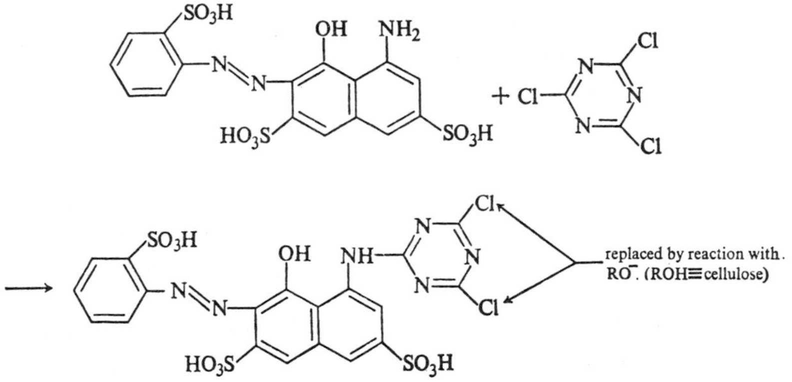

The dyestuffs (Procion dyes) are consequently tightly bound to the fibre and in particular have considerable wet-fastness.

Melamine (2,4,6-triamino-1,3,5-triazine) can be made by treating cyanuric chloride with ammonia,

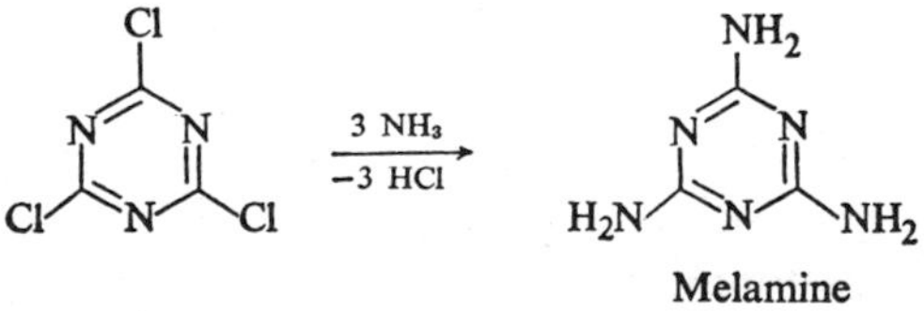

Melamine

but a cheaper industrial method involves the thermal decomposition of urea. The polymers derived from melamine and formaldehyde are used, for example, for plastic vessels and tableware and as electrical insulators.

1,3,5-Trinitrohexahydro-1,3,5-triazine, made by nitrating hexamethylene-tetramine (p. 330), is an important high explosive (Cyclonite, or RDX).

Further Reading

ACHESON, R. M., *An Introduction to the Chemistry of Heterocyclic Compounds*, Interscience (New York and London 1960).

GENSLER, W. J., 'The synthesis of isoquinolines by the Pomeranz-Fritsch reaction,' *Organic Reactions*, 1951, **6**, 191.

MANSKE, R. H. F., and KULKA, M., 'The Skraup synthesis of quinolines,' *Organic Reactions*, 1953, **7**, 59.

WHALEY, W. M., and GOVINDACHARI, T. R., 'The preparation of 3,4-dihydro-isoquinolines and related compounds by the Bischler-Napieralski reaction,' and 'The Pictet-Spengler synthesis of tetrahydroisoquinolines and related compounds,' *Organic Reactions*, 1951, **6**, 74 and 151.

WILEY, R. H., et al., 'The preparation of thiazoles,' *Organic Reactions*, 1951, **6**, 367.

WOLF, D. E., and FOLKERS, K., 'The preparation of thiophenes and tetrahydro-thiophenes,' *Organic Reactions*, 1951, **6**, 410.

Problems

1. Classify the types of heteroaromatic compound which may be synthesized from acetoacetic ester. What structural variant would be introduced in those reactions in which an $\alpha\beta$-unsaturated ester can replace acetoacetic ester?

2. Annotate the steps in the syntheses of (*a*) papaverine (p. 683), (*b*) quercetin (p. 689), (*c*) cyanidin chloride (p. 690).

3. Account for the following:

   (*i*) Pyrrole is both more acidic and less basic than pyrrolidine.

   (*ii*) Pyridine is less basic than piperidine.

   (*iii*) Imidazole is both more acidic than pyrrole and more basic than pyridine.

4. How would you carry out the following transformations?

(*i*)

2-Phenylbenzoxazole

(ii)

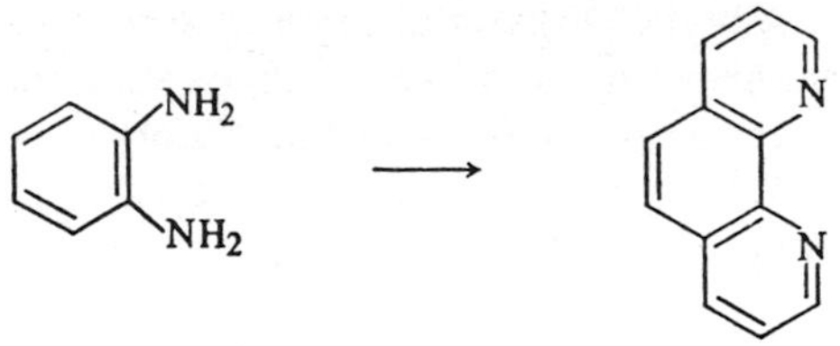

4,5-Diazaphenanthrene

(iii)

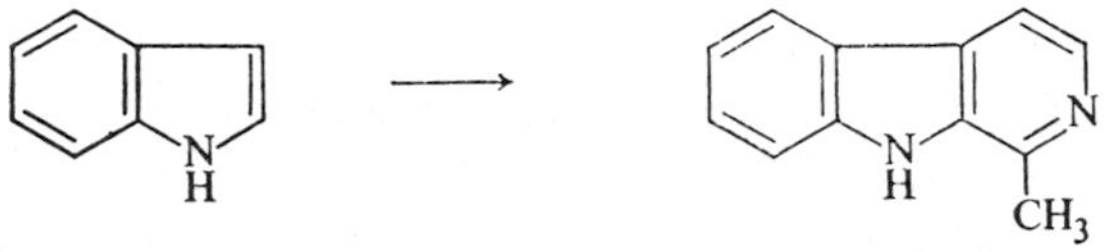

Harman
(a *Harmala* alkaloid)

(iv)

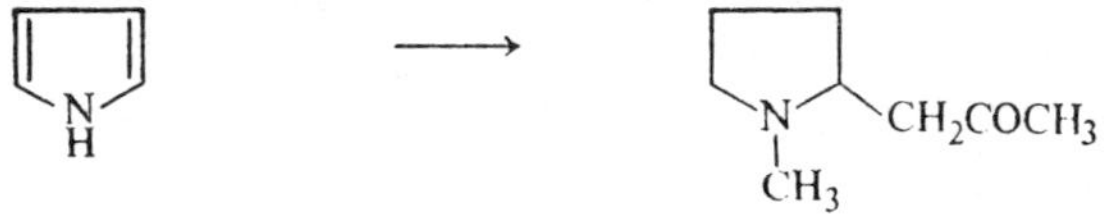

($\pm$)-Hygrine
(a coca alkaloid)

(v)

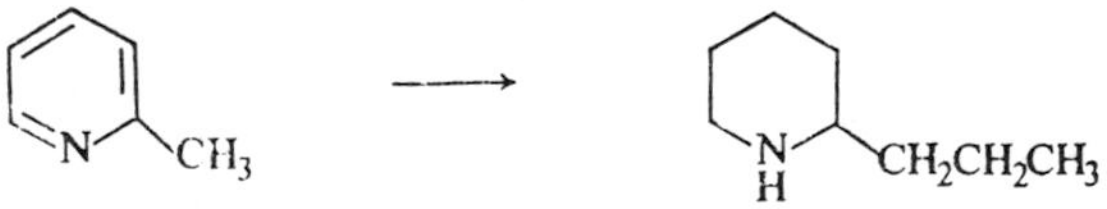

($\pm$)-Coniine
(a hemlock alkaloid)

5. Complete the schemes outlined below by inserting the reagents and the intermediates which have been omitted

   (*i*) Synthesis of Plasmoquin (an antimalarial):

$(C_2H_5)_2NH \longrightarrow (C_2H_5)_2N-CH_2CH_2OH \longrightarrow (C_2H_5)_2N-(CH_2)_3-CH-CH_3$
$\qquad\qquad\qquad\qquad\qquad\qquad\qquad\qquad\qquad\qquad\qquad\qquad\qquad | $
$\qquad\qquad\qquad\qquad\qquad\qquad\qquad\qquad\qquad\qquad\qquad\qquad\qquad Br$

A

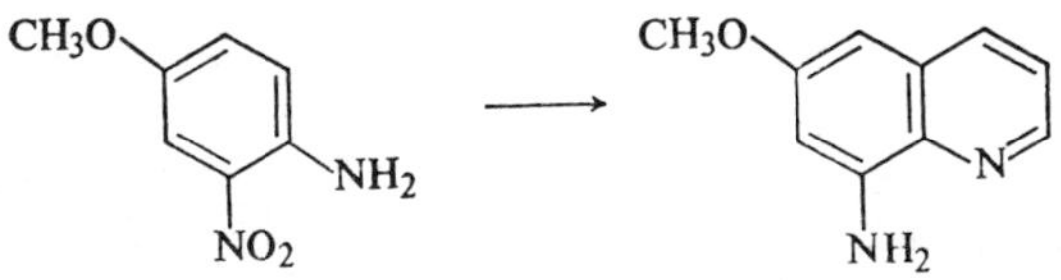

B

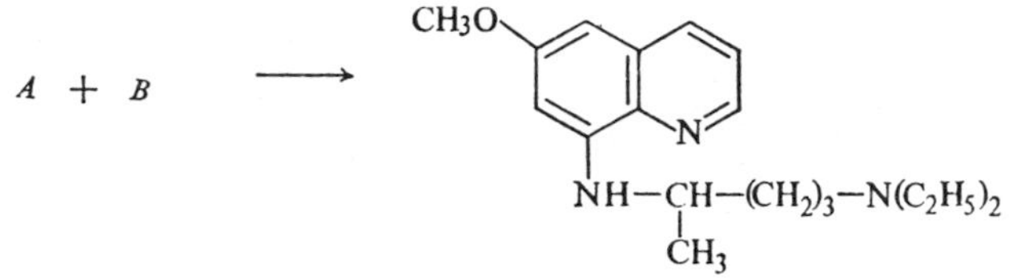

$$A + B \longrightarrow$$

CH₃O — (quinoline structure)

NH—CH—(CH₂)₃—N(C₂H₅)₂
|
CH₃

Plasmoquin

(*ii*) Synthesis of nicotine (a tobacco alkaloid):

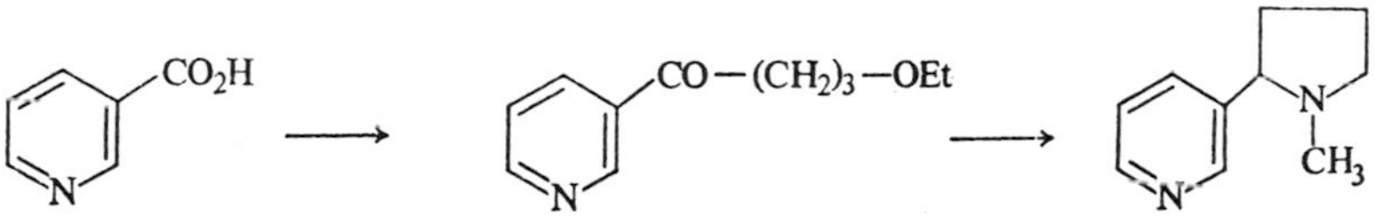

.CO₂H (pyridine) → .CO—(CH₂)₃—OEt (pyridine) → (pyridine with pyrrolidine ring, N—CH₃)

(+)-Nicotine

(*iii*) Synthesis of biotin (vitamin H, a growth substance):

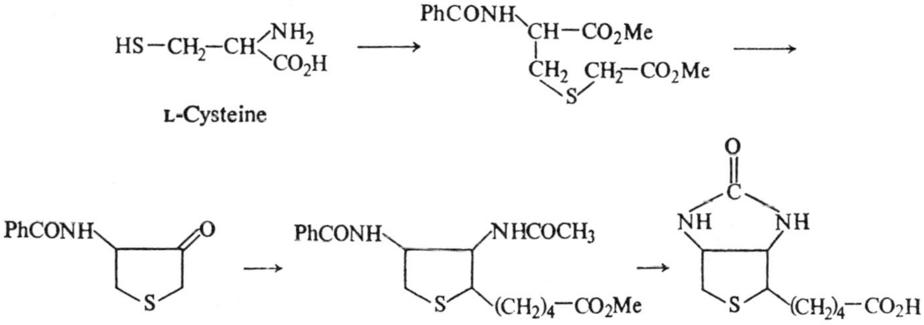

$$HS-CH_2-CH\begin{smallmatrix}NH_2\\CO_2H\end{smallmatrix} \longrightarrow$$

L-Cysteine

PhCONH‚CH—CO₂Me
|
CH₂ CH₂—CO₂Me
\S/ →

PhCONH (thiophene ring) O → PhCONH (thiophene ring) NHCOCH₃ , (CH₂)₄—CO₂Me →

(bicyclic biotin core)
O
‖
C
NH   NH

S   (CH₂)₄—CO₂H

Biotin

(*iv*) Synthesis of pyridoxol (vitamin B₆):

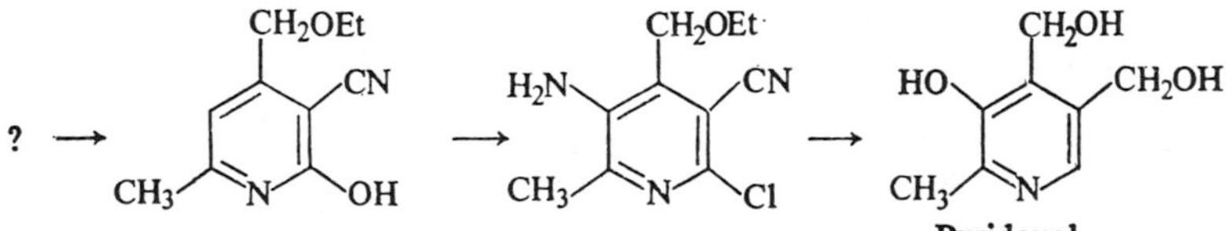

? → (pyridine) CH₂OEt, CN, CH₃, N, OH →
H₂N (pyridine) CH₂OEt, CN, CH₃, N, Cl →
HO (pyridine) CH₂OH, CH₂OH, CH₃, N

Pyridoxol

6. Outline syntheses of the following compounds from readily available materials:

(*a*) (pyrrole: CH₃, C₂H₅, N—H, CH₃)

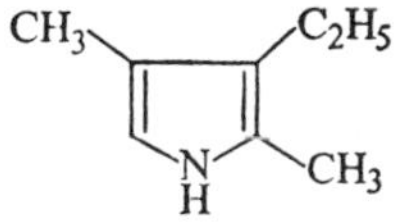

(*b*) (pyrrole: CH₃, N—H, CH₃)

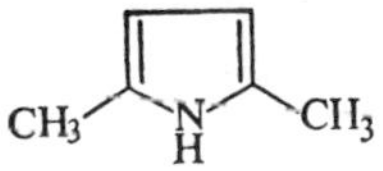

(c)

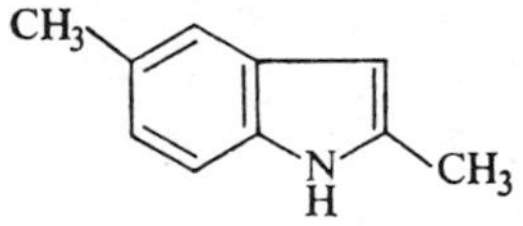

(d)

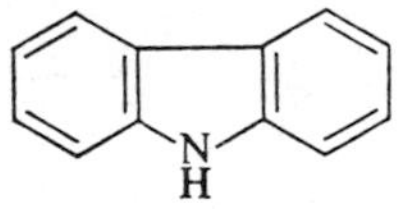

(e)

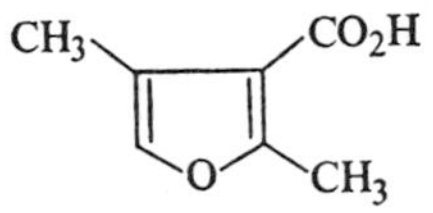

(f)

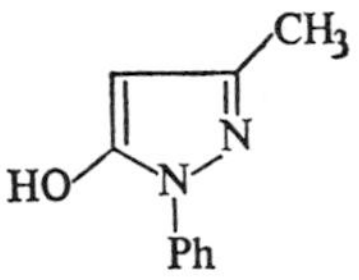

(g)

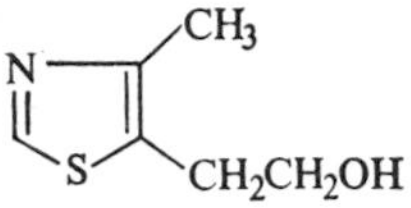

(h)

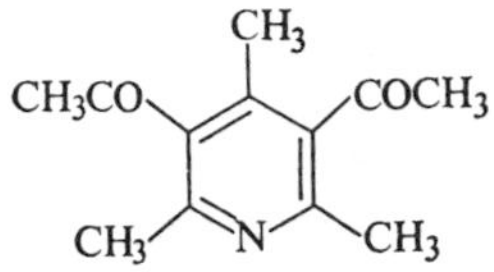

(i)

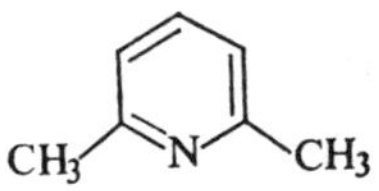

(j)

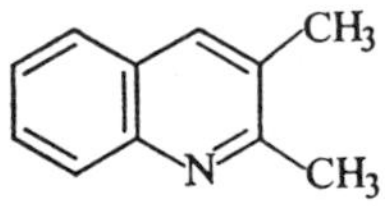

(k)

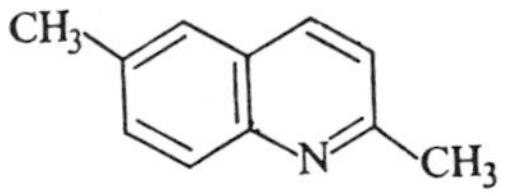

(l)

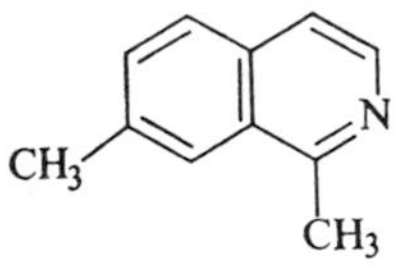

(m)

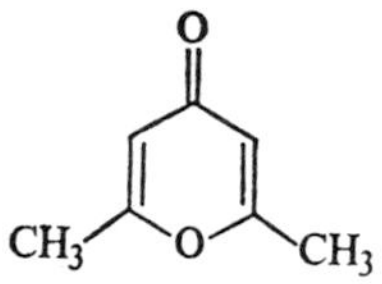

(n)

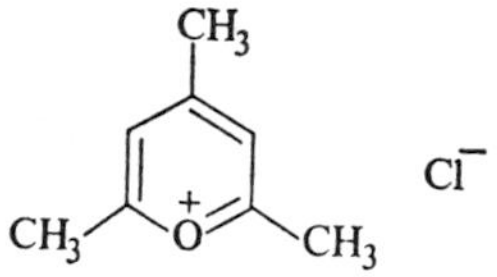

(o)

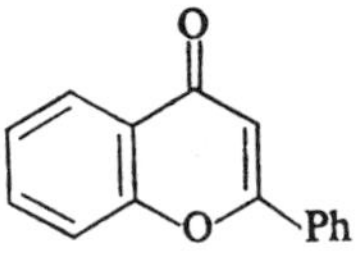

(p)

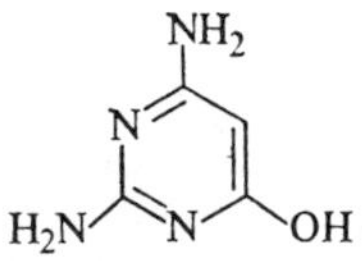

(q)

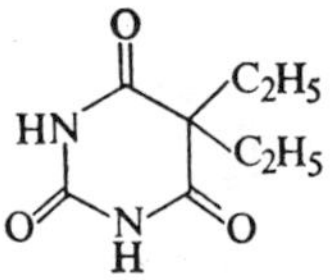

(r)

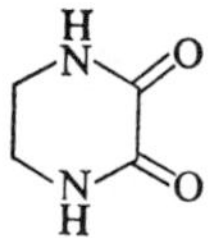

(s)

(t)

21. The Syntheses of Some Naturally Occurring Compounds

The goals of the synthetic organic chemist are of two types: the development of new synthetic methods—for example, a route for making epoxides from olefins; and target synthesis—the multi-stage synthesis of complex molecules, particularly those obtained from natural sources. The majority of this text has been concerned with the former goal; this chapter is devoted to the latter.

Until relatively recently, the total synthesis of a compound represented the essential ultimate proof of the structure of that compound. In general, this is no longer so; physical techniques, especially X-ray crystallography, are capable of supplying proof of structure, usually in a very much shorter time than is required for synthesis. Nonetheless, the synthesis of naturally occurring compounds remains a vigorously pursued field, for two reasons. First, some materials can be obtained from natural sources only in such small quantity or with such difficulty that synthesis can provide cheaper and more abundant material. This is of especial importance if the compound has valuable pharmaceutical properties. Moreover, the total synthesis will usually give precursors of the natural compound from which related structures can be synthesized, and this also can be of value in the pharmaceutical field since the structural variants may be associated with different or more powerful biological properties. Secondly, natural product synthesis is a most fruitful source of new synthetic methods; a particular step in the sequence may lead to the development of a new and general method for its successful completion.

Target synthesis calls for diverse skills in the organic chemist. First, he must be well armed with the amalgam of facts and principles which together constitute chemical knowledge. Beyond this, however, more is needed. He must develop the ability to think 'retrosynthetically'; to work out suitable precursors of the target molecule, then precursors for these, and so on, until he has worked back to available materials. Then he must assess the likelihood of one of the suggested sequences working better than another, for even with the wealth of knowledge of organic reactions at his disposal, he cannot necessarily predict with certainty that a particular reaction in his sequence will be successful; it may be that some other structural feature in the compound will cause an undesired reaction.

In the analytical, or 'retrosynthetic', phase of the planning, it is useful to adopt the habit of mentally disconnecting the target molecule and seeking *synthons*—fragmentary structures of simpler kind which can be related to precursor molecules. To take a simple example, suppose in a target molecule the fragment $\diagdown N-C\diagup$ is present. Synthetic steps dependent upon the chemistry described in Chapter 10 might be appropriate if it is decided to form the C—N bond in creating this fragment. It may be more appropriate, on the other hand, to introduce the N—C unit into the synthesis in a preformed manner. This could be done in various ways: CN^- and $O_2NCH_2^-$ are precursor species which are nucleophilic at carbon, CH_3NH_2 is nucleophilic at nitrogen, imines or isocyanates would be precursor species electrophilic at carbon, and so forth. Which particular precursor is selected will depend on the context of the particular step in the whole synthesis. Clearly accessibility will be one of the criteria upon which selection of precursor molecules is made.

Another factor important in synthetic strategy is the number of synthetic steps which a chosen sequence involves. It is usually desirable to minimize the number; even if individual steps are accomplished with 90% efficiency, after ten such steps the overall yield is only *ca.* 25%. It is often preferable for synthesis to be effected in a *convergent* manner— that is, to assemble a molecule in, say, two separate parts which are then brought together—rather than in a linear manner.*

Some of these principles are exemplified by the following step from R. B. Woodward's synthesis of lysergic acid:

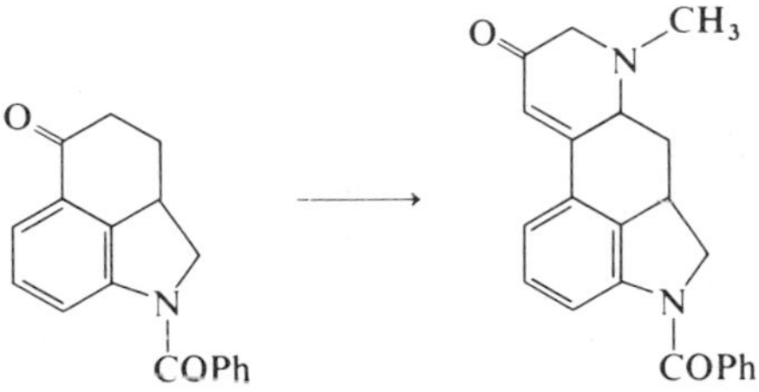

The reaction is an annelation: the tricyclic ketone is extended to give a tetracyclic amino-ketone. The major synthon in the product is immediately recognizable: it is that which is related to the precursor molecule. If the major synthon is disconnected, it is apparent that the fragment

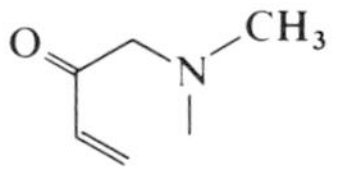

is to be added to the reactant with the loss of ketonic oxygen and an adjacent hydrogen; a carbonyl group has to be transformed into a carbon-carbon double bond and a C—H bond into a C—N bond. In earlier chapters means of bringing

---

*For example, the linear sequence A → B → C → D gives an overall yield of 51% if each step is 80% efficient; in contrast, the scheme V → W, X → Y, W + Y → Z gives an overall yield of 64% with the same number of operations.

about both these conversions have been introduced; now it must be determined how and in which order the two steps should best be done.

What actual molecules are related to the minor synthon? Since the condensation $>C{=}O + -CH_2-CO- \rightarrow >C{=}C(-)-CO-$ can generally be carried out in the presence of base, one such molecule would be of the type

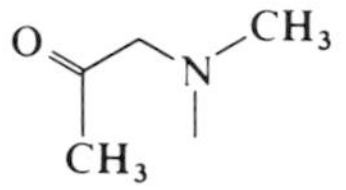

This unit itself can be disconnected to give

$$CH_3COCH_2- \; + \; >N-CH_3$$

and a method of connection should now come to mind: to form the α-bromoketone and treat it with methylamine,

$$CH_3COCH_2- \xrightarrow[(H^+)]{Br_2} CH_3COCHBr- \xrightarrow[-HBr]{CH_3NH_2} CH_3COCH(-)-NHCH_3$$

Moreover, similar treatment of the major synthon can enable the bond between this and the methylamine-unit to be constructed. Therefore suitable precursors to the target molecule appear to be:

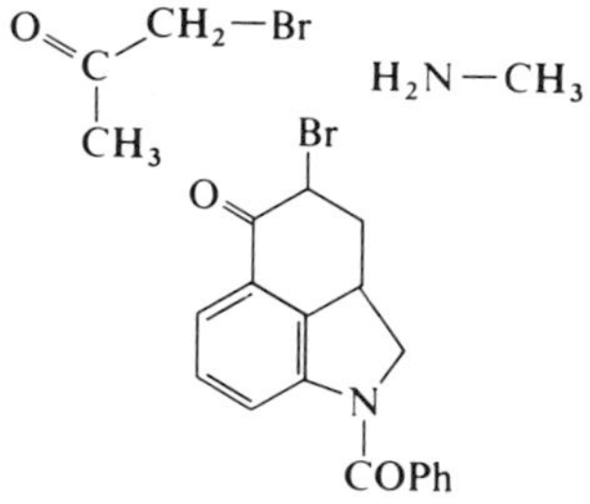

Nevertheless, several problems arise. First, primary amines react with carbonyl compounds to give imines, which may then react further with the amine. It is therefore necessary to protect the adjacent carbonyl group. Now the tricyclic bromo-ketone is the result of several previous synthetic steps, so that it is convenient to minimize the synthetic steps to which it is committed in order to conserve material and effort. Hence it is preferable first to effect reaction of methylamine with protected bromoacetone and then to bring about reaction of the combined product with the tricyclic precursor.

The usual way of protecting a ketonic carbonyl group is as a cyclic ketal. The envisaged reaction now becomes:

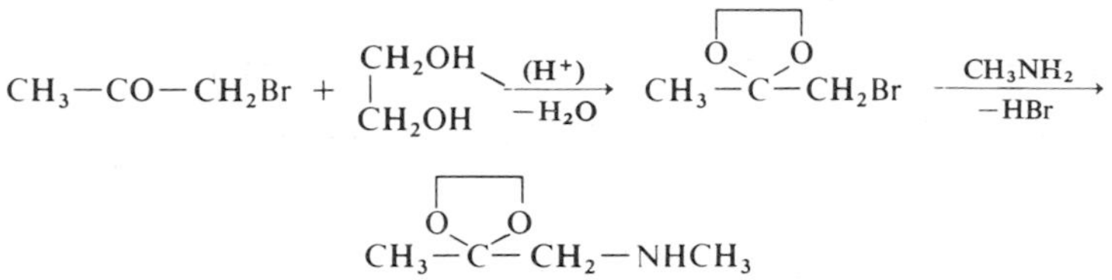

$$CH_3-CO-CH_2Br \; + \; \begin{matrix} CH_2OH \\ | \\ CH_2OH \end{matrix} \xrightarrow[-H_2O]{(H^+)} \; CH_3-\overset{\overset{\displaystyle O \frown O}{|}}{C}-CH_2Br \xrightarrow[-HBr]{CH_3NH_2}$$

$$CH_3-\overset{\overset{\displaystyle O \frown O}{|}}{C}-CH_2-NHCH_3$$

The product is the precursor which relates to the minor synthon, i.e.

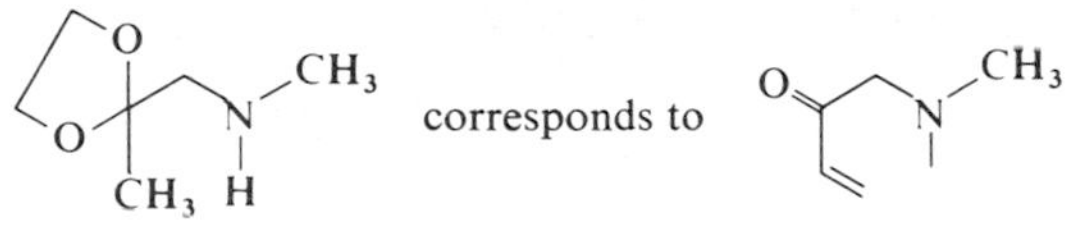

corresponds to

The method of combination of this material with the tricyclic bromoketone has next to be considered. Either the new C=C or the new C—N bond could be formed first. The latter is likely to be the more successful route, as the following argument shows. The first alternative could in principle be effected by removing the protecting group so that the methyl group in the amino-ketone is activated to base-catalyzed condensation by the adjacent carbonyl. However, this condensation would be an intermolecular one involving different ketones, and experience shows that mixtures of products are usually formed in these circumstances; moreover, the condensation of ketones usually yields an equilibrium mixture which is unfavourable to the condensation product (*cf.* the self-condensation of acetone; p. 228), and which could in this case allow reaction of the secondary amino group with carbonyl (to give an enamine) to dominate. In contrast, if the protected amino-ketone is first treated with the tricyclic bromoketone and *then* the protecting group is removed, formation of the new carbon-carbon bond requires *intramolecular* base-catalyzed condensation; in particular, this involves the stereochemically favourable six-membered cyclic transition state which is known usually to lead to efficient condensation. This was the course which was adopted:

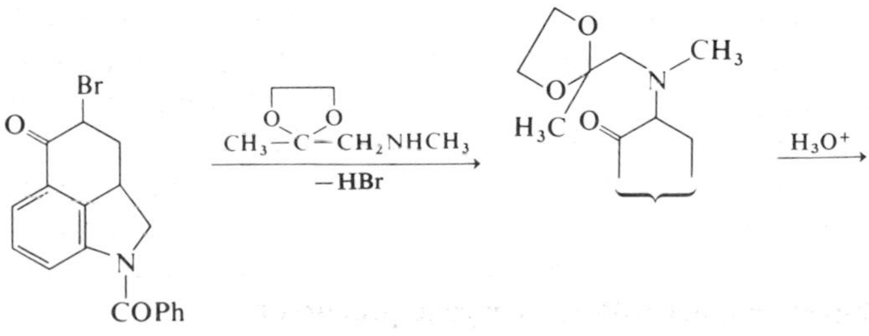

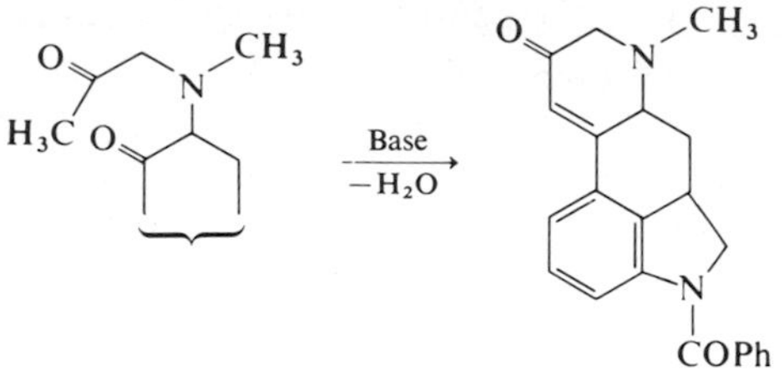

Loss of water from the aldol product is in an unambiguous direction owing to the introduction of conjugation between the aromatic ring and the carbonyl group of the product.

This particular example is simpler than many in having no requirement for stereochemical control. Since the majority of naturally occurring compounds have at least one chiral centre, stereochemical control is vital if conversions are to be efficient. Fortunately, many types of reaction are characterized by their stereospecificity, and this allows *stereochemical control elements* to be built in during synthesis. For example, the product of the Diels-Alder reaction between *p*-benzoquinone and *trans*-vinylacrylic acid,

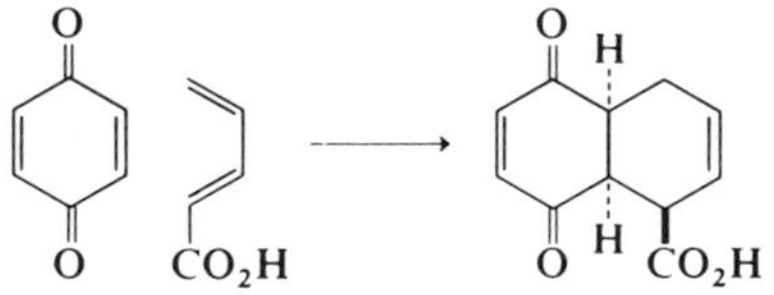

not only has a *cis*-ring-junction and a *trans*-relationship between the hydrogens at the ring-junction and the carboxyl group (*cf.* secondary orbital interaction, p. 294), but also is so constructed that a subsequent reduction of one of the carbonyl groups, and then electrophilic attack on the isolated double bond, both occur at the less hindered underside of the molecule; thus, further chiral centres can be introduced in a selective manner (p. 174). Stereochemical control elements may be related to integral parts of the target molecule, or they may be introduced for a particular purpose and ultimately be eliminated in the manner of protective groupings, as was the case in the above example.

The syntheses which follow have been completed since 1943. The selection has been made on the basis, first, that each compound is of considerable biological importance and secondly, that the synthetic methods illustrate the application of a wide variety of group-selective and stereo-selective reagents, many of which have themselves been recently developed.

21.1 Vitamin A

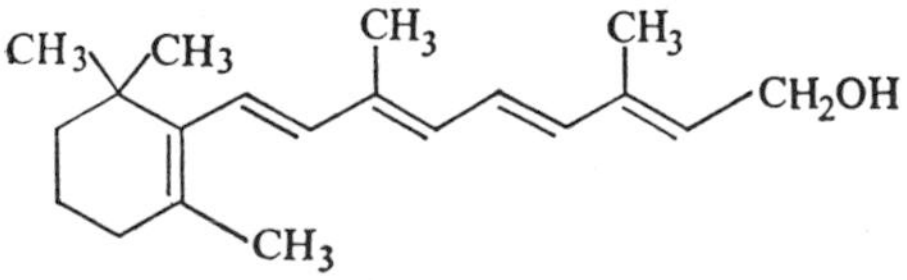

Vitamin A₁   Vitamin A₂

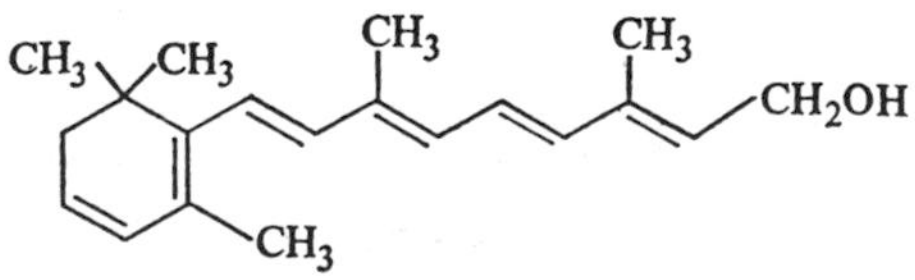

Vitamins A_1 and A_2, given the collective name vitamin A, play a central role in the photochemical processes associated with vision; deficiency leads to xerophthalmia (hardening of the cornea). Vitamin A_1 is found in large amounts in the livers of salt-water fish, and A_2 is found in the livers of fresh-water fish. In addition, a number of naturally occurring compounds containing the A_1 skeleton, such as β-carotene (21.2), are converted *in vivo* into A_1, and it is thought that A_1 is dehydrogenated *in vivo* to A_2.

There are three notable syntheses of vitamin A_1.

(1) O. Isler *et al.* (*Helv. Chim. Acta*, 1947, **30**, 1911). The vitamin was synthesized from β-ionone, a total synthesis of which had previously been completed.

Scheme 1

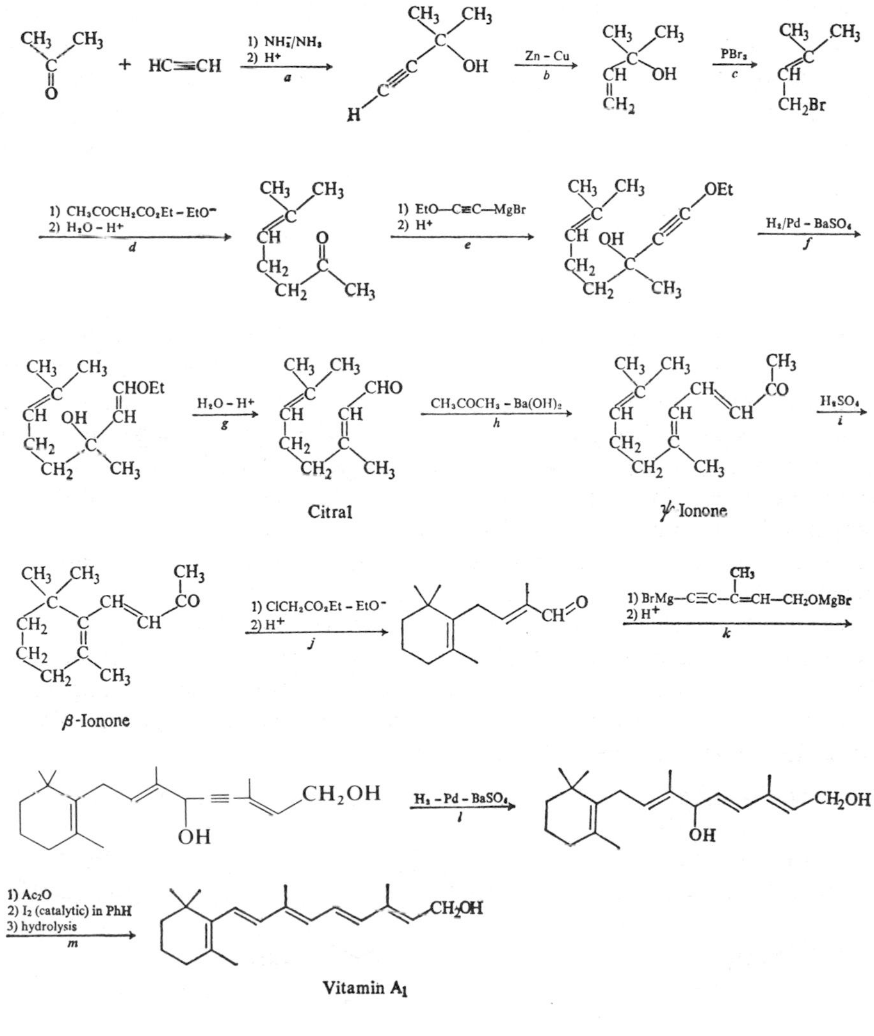

Citral

γ-Ionone

β-Ionone

Vitamin A_1

a   Base-catalyzed ethynylation (p. 256).

b   Reduction by electron-transfer (p. 620).

c   Conversion of the alcohol into the bromide is accompanied by an anion-otropic shift, giving the more stable (trisubstituted) olefinic isomer (p. 53).

d   Base-catalyzed alkylation of acetoacetic ester, followed by ketonic hydrolysis (p. 251).

e   For the preparation of the acetylenic Grignard reagent, see p. 209.

f   Partial reduction on Lindlar's catalyst (p. 619).

g   Hydrolysis of the vinyl ether gives an aldehyde (p. 110) which is readily dehydrated in the acidic conditions to the conjugated unsaturated aldehyde.

h   A base-catalyzed mixed aldol condensation (p. 230).

i   Protonation (Markovnikov's rule), ring-closure, and deprotonation. Some of the non-conjugated isomer (α-ionone) is also formed:

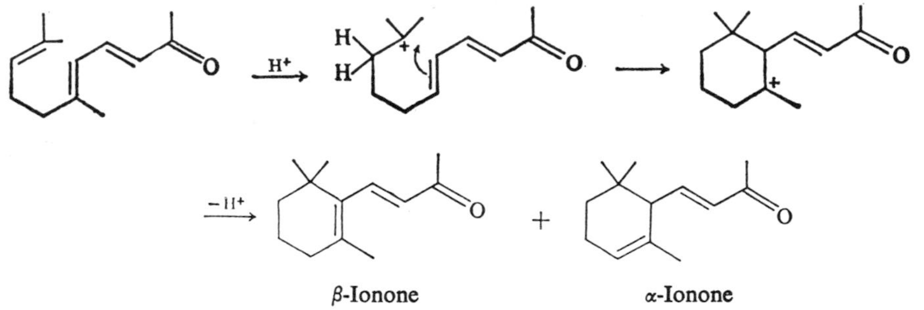

β-Ionone   α-Ionone

j   The Darzens condensation (p. 235). When the glycidic ester is hydrolyzed, a prototropic shift occurs to give the αβ-unsaturated aldehyde:

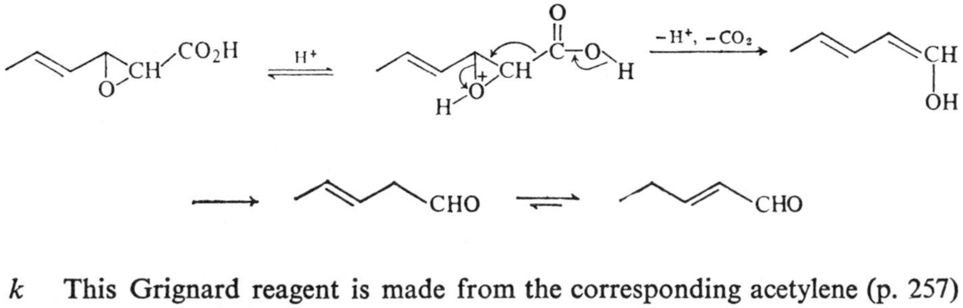

k   This Grignard reagent is made from the corresponding acetylene (p. 257) with ethylmagnesium bromide.

l   Partial reduction, aided in this case by the addition of some quinoline.

m   The primary alcoholic group is protected by acetylation. Dehydration occurs readily (iodine being a weak Lewis acid) to give the fully conjugated system, and the acetyl group is then removed.

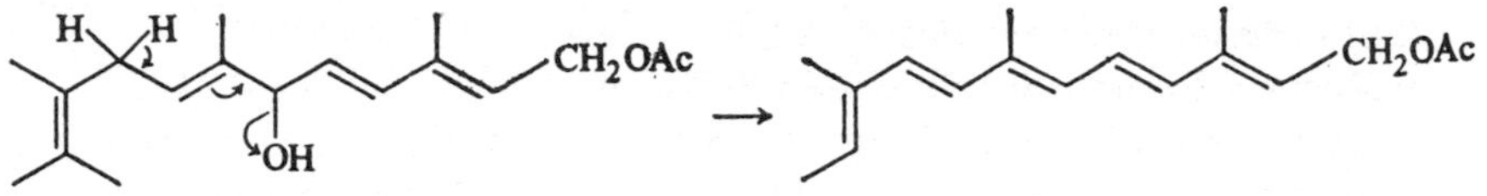

(2) J. F. Arens and D. A. van Dorp (*Rec. Trav. Chim.*, 1949, **68**, 604). The synthesis used as starting material the ketone *A* which is obtainable from natural sources and has been prepared from β-ionone.

Scheme 2

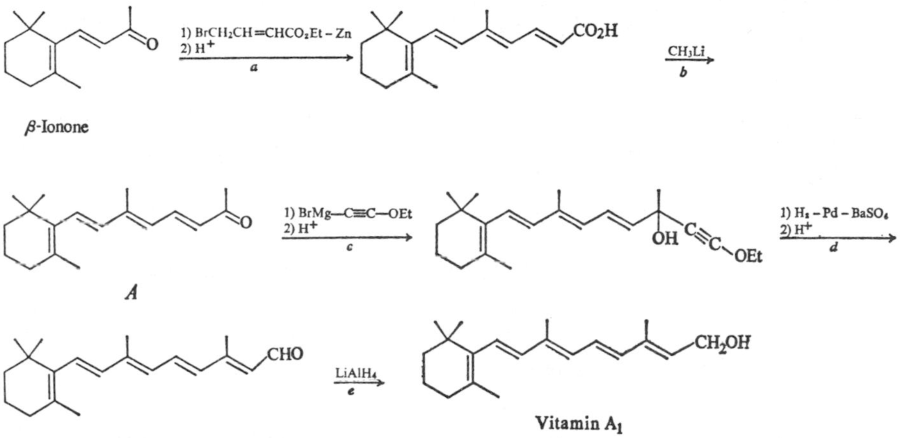

β-Ionone

A

Vitamin A₁

a Reformatsky reaction (p. 221). The product, a β-hydroxyester, is readily dehydrated to give the conjugated system, and the ester is hydrolyzed.

b The acid reacts with methyl-lithium to give the methyl ketone (p. 218).

c, d Introduction of the acetylenic group, partial reduction, hydrolysis of the vinyl ether (3% oxalic acid), and dehydration occur as in the Isler synthesis (steps *e, f, g*).

e Lithium aluminium hydride was found to be more effective than the Meerwein-Ponndorf method.

An alternative route from *A* was also completed in 1949:

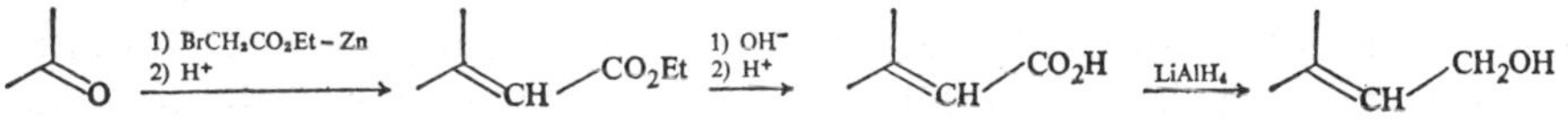

(3) J. Attenburrow *et al.* (*J. Chem. Soc.*, 1952, 1094) synthesized vitamin A_1 by two routes from 2-methylcyclohexanone; the shorter route is described here.

Scheme 3

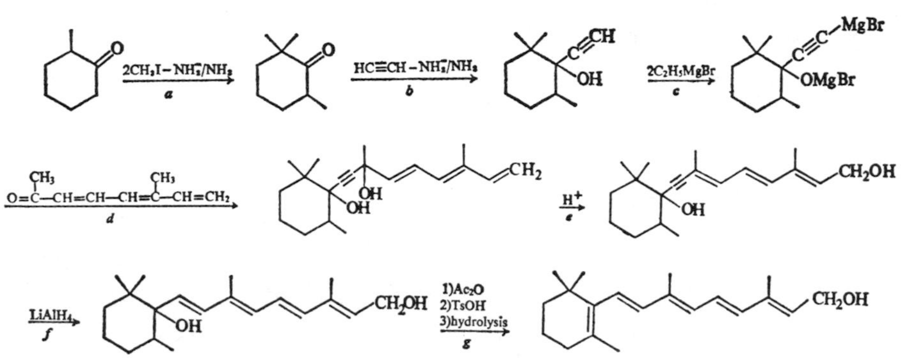

a   Alkylation of a ketone requires a powerful base such as sodamide (p. 248).

b   Ethynylation in liquid ammonia (p. 256).

c   The exchange reaction between a Grignard reagent and an acetylenic hydrogen (p. 205).

d   The required ketone can be prepared by a crossed aldol condensation between crotonaldehyde and acetone, followed by ethynylation, partial reduction, anionotropic rearrangement, and oxidation of the resulting allylic alcohol with manganese dioxide (p. 592):

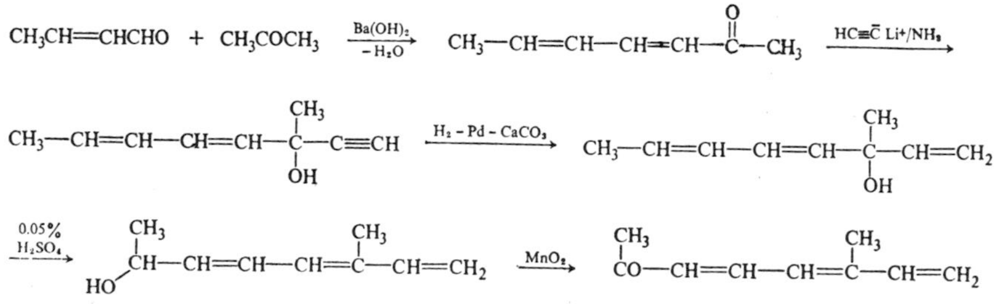

e   The acid-catalyzed anionotropic shift occurs because it leads to an increase in the extent of conjugation within the molecule (p. 20).

f   The partial reduction of the acetylenic bond by lithium aluminium hydride is made possible by the presence of the α-hydroxyl group and gives the *trans* product (p. 620).

g   The primary alcohol is protected by formation of the acetate (the tertiary alcohol being sterically hindered), and the tertiary alcohol is dehydrated with toluene-*p*-sulphonic acid. Hydrolysis gives the vitamin.

Vitamin A_1 has been converted into vitamin A_2 as follows (E. R. H. Jones *et al.*, *J. Chem. Soc.*, 1955, 2765).

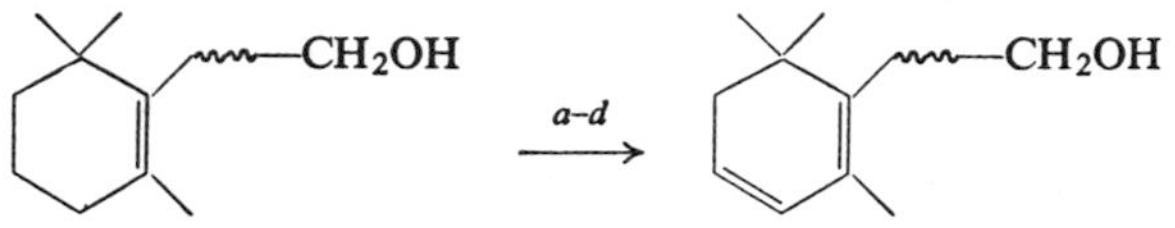

(a) MnO_2/CH_3COCH_3

(b) *N*-bromosuccinimide

(c) *N*-phenylmorpholine

(d) $LiAlH_4$ ($-CHO \rightarrow -CH_2OH$)

a   Mild oxidation of $-CH_2OH$ to $-CHO$, specific for allylic alcohols (p. 592). (This step was necessary because it was found that treatment of the alcohol with *N*-bromosuccinimide as in step *b* gave a mixture of products.)

b   Free-radical bromination of the allylic system (p. 544). The aldehydic group is not affected.

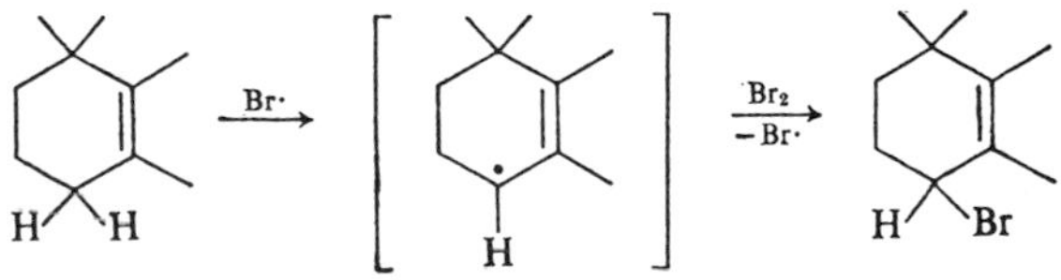

c   Base-catalyzed dehydrobromination. This yielded a mixture of *cis* and *trans* stereoisomers from which the required *trans* compound was separated by formation of the oxime and hydrolysis.

d   Lithium aluminium hydride does not reduce the olefinic bonds of $\alpha\beta$-unsaturated carbonyl compounds in the conditions used ($-30°C$).

21.2 β-Carotene

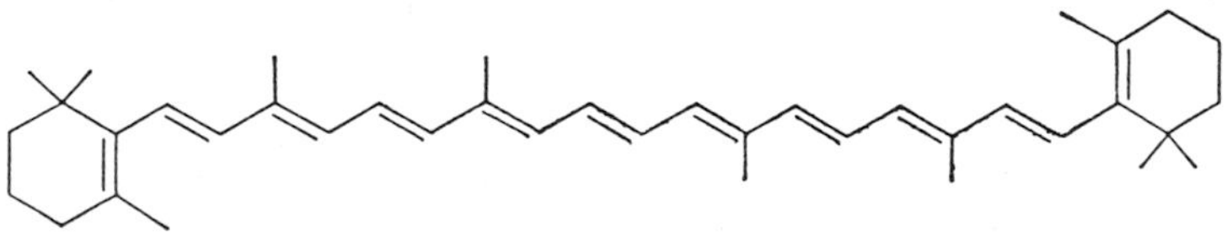

The tetraterpene carotene, first isolated from carrots in 1831, is now known to contain three compounds. β-Carotene, which constitutes 85% of the mixture, is shown above; α- and γ-carotene are isomers of the β-compound. A large number of related compounds (carotenoids) also occur naturally.

The extensive conjugation in the carotenoids is associated with the existence of low-lying electronically excited states; the molecules absorb strongly in the visible region and are mostly yellow or orange. Where chlorophyll is present in association the carotenoids act as photosensitizers.

β-Carotene was first synthesized by P. Karrer and C. H. Eugster (*Helv. Chim. Acta*, 1950, **33**, 1172) who employed an intermediate previously obtained from β-ionone (p. 711),

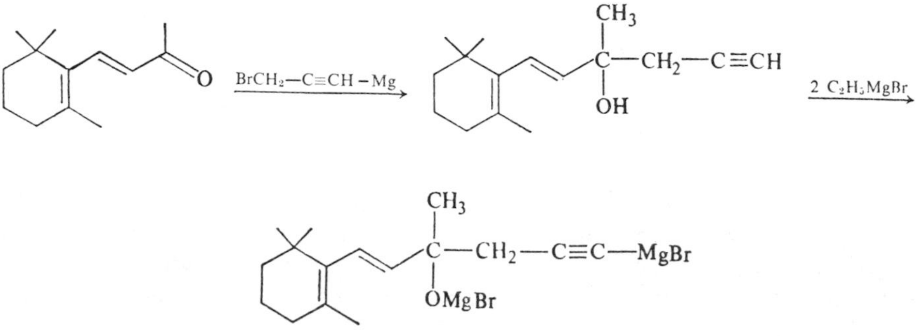

together with oct-4-ene-2,7-dione, which had been prepared as follows:

$$CH_3CHO + CH\equiv CH \xrightarrow[a]{NH_2^-/NH_3} CH_3-\underset{\underset{OH}{|}}{CH}-C\equiv CH \xrightarrow[b]{O_2-CuCl-NH_4Cl}$$

$$CH_3\underset{\underset{OH}{|}}{CH}-C\equiv C-C\equiv C-\underset{\underset{OH}{|}}{CH}CH_3 \xrightarrow[c]{LiAlH_4} CH_3\underset{\underset{OH}{|}}{CH}-CH=CH-CH=CH-\underset{\underset{OH}{|}}{CH}CH_3 \xrightarrow[d]{MnO_2}$$

$$CH_3\underset{\underset{O}{\|}}{C}-CH=CH-CH=CH-\underset{\underset{O}{\|}}{C}CH_3 \xrightarrow[e]{Zn-HOAc} CH_3\underset{\underset{O}{\|}}{C}-CH_2-CH=CH-CH_2-\underset{\underset{O}{\|}}{C}CH_3$$

a   Ethynylation in liquid ammonia (p. 256).
b   Standard procedure for the oxidative coupling of acetylenes (p. 554).
c   Lithium aluminium hydride reduces α-acetylenic alcohols to allylic alcohols (p. 620).
d   Manganese dioxide is a specific oxidant for allylic alcohols (p. 592).
e   1,4-Reduction of a conjugated diene (p. 616).

The synthesis was completed following the combination of these units:

Scheme 4

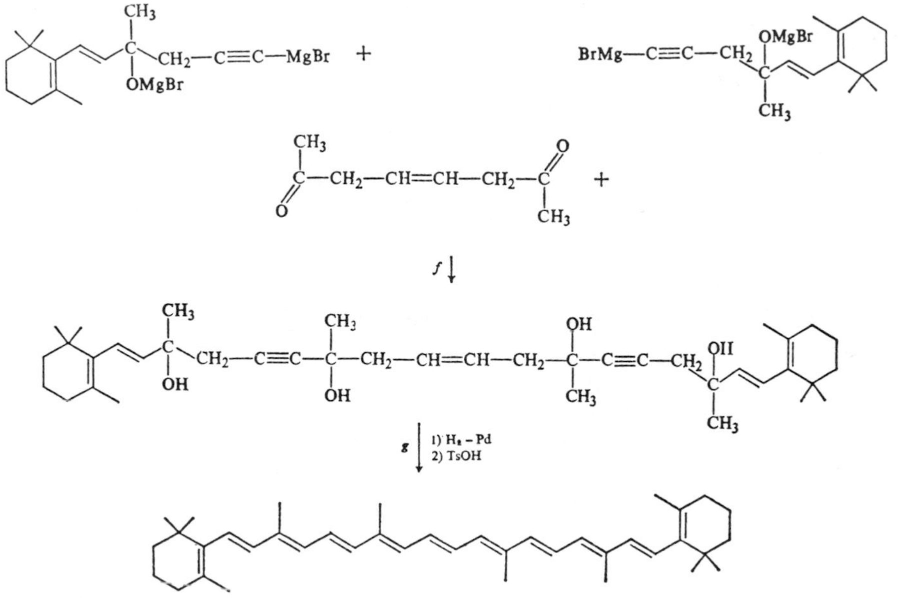

β-Carotene

f   Grignard synthesis, followed by hydrolysis.

g   Partial reduction of the acetylenic bonds (p. 619) and dehydration by toluene-p-sulphonic acid.

Lycopene, a carotenoid found in tomato pigment, differs from β-carotene only in the way ψ-ionone differs from β-ionone (p. 711). It has been made (P. Karrer *et al.*, *Helv. Chim. Acta*, 1950, **33**, 1349) in the same way from ψ-ionone as β-carotene from β-ionone:

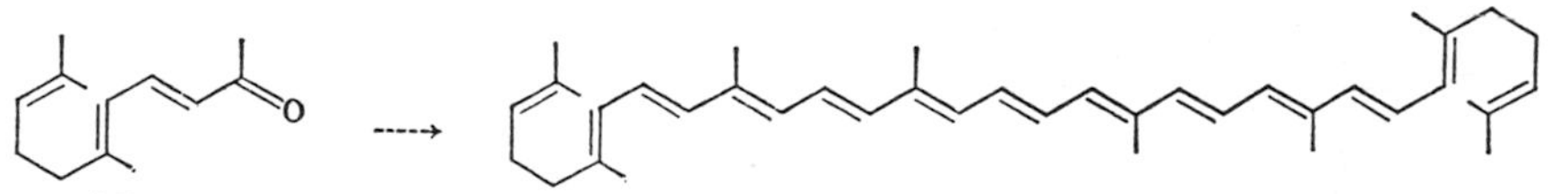

ψ-Ionone   ----→   Lycopene

21.3 Cholesterol

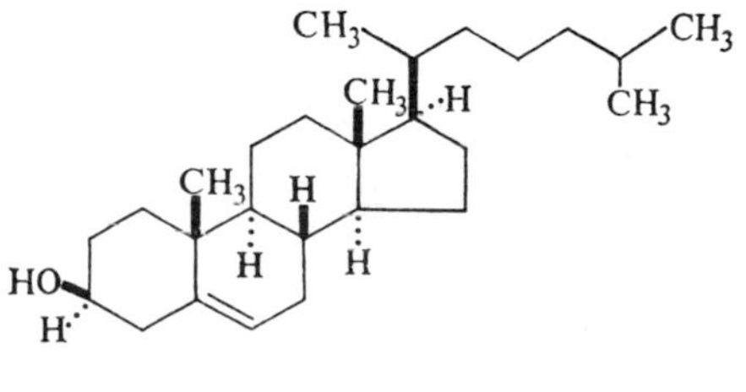

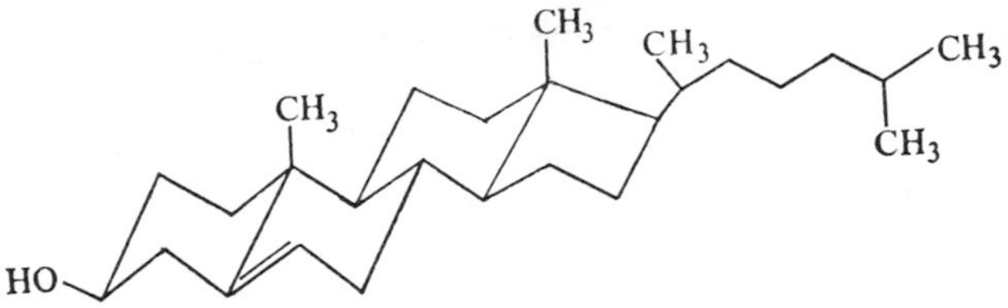

Cholesterol occurs in all tissues of the body, in part as the free alcohol and in part in esters of higher fatty acids. Its biological function is still not fully understood, though one of its roles is almost certainly as precursor to the steroid hormones (see, e.g. oestrone; 21.4).

Two syntheses of cholesterol were reported at about the same time. That by R. B. Woodward *et al.* (*J. Amer. Chem. Soc.*, 1952, **74**, 4223) is the more elegant and is reported here; the product was cholestanol which had previously been converted into cholesterol by the transformations included below. Not all the steps are stereospecific or even stereoselective and several separations are required; in this respect the later synthesis of the related steroid, epiandrosterone (21.5), is more satisfying. Nevertheless, Woodward's synthesis illustrates the fine control that can be imposed in the transformations of polyfunctional compounds by the use of selective reagents.

Scheme 5

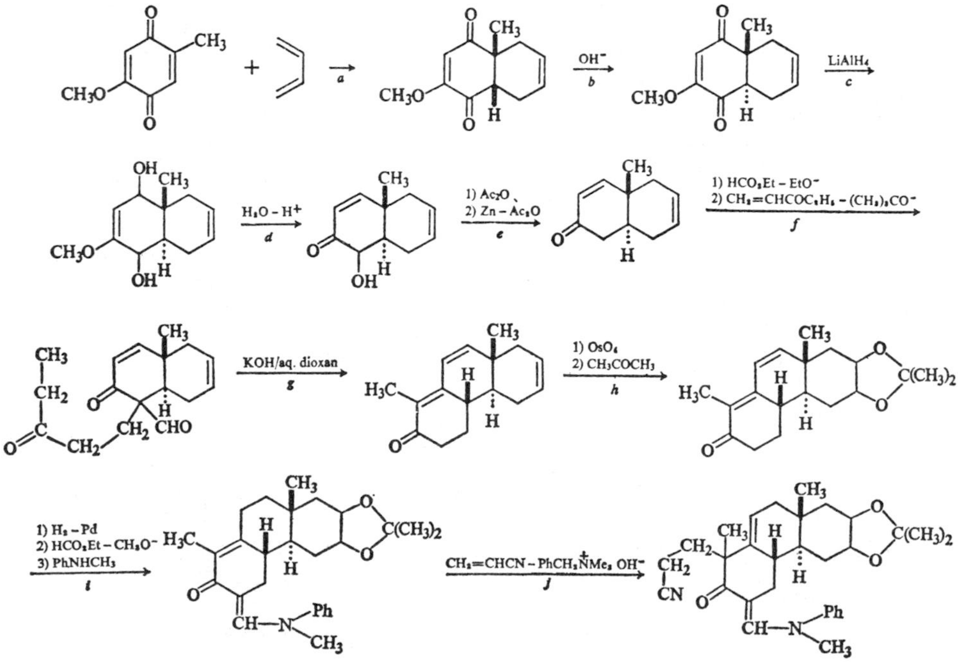

Scheme 5 (cont.)

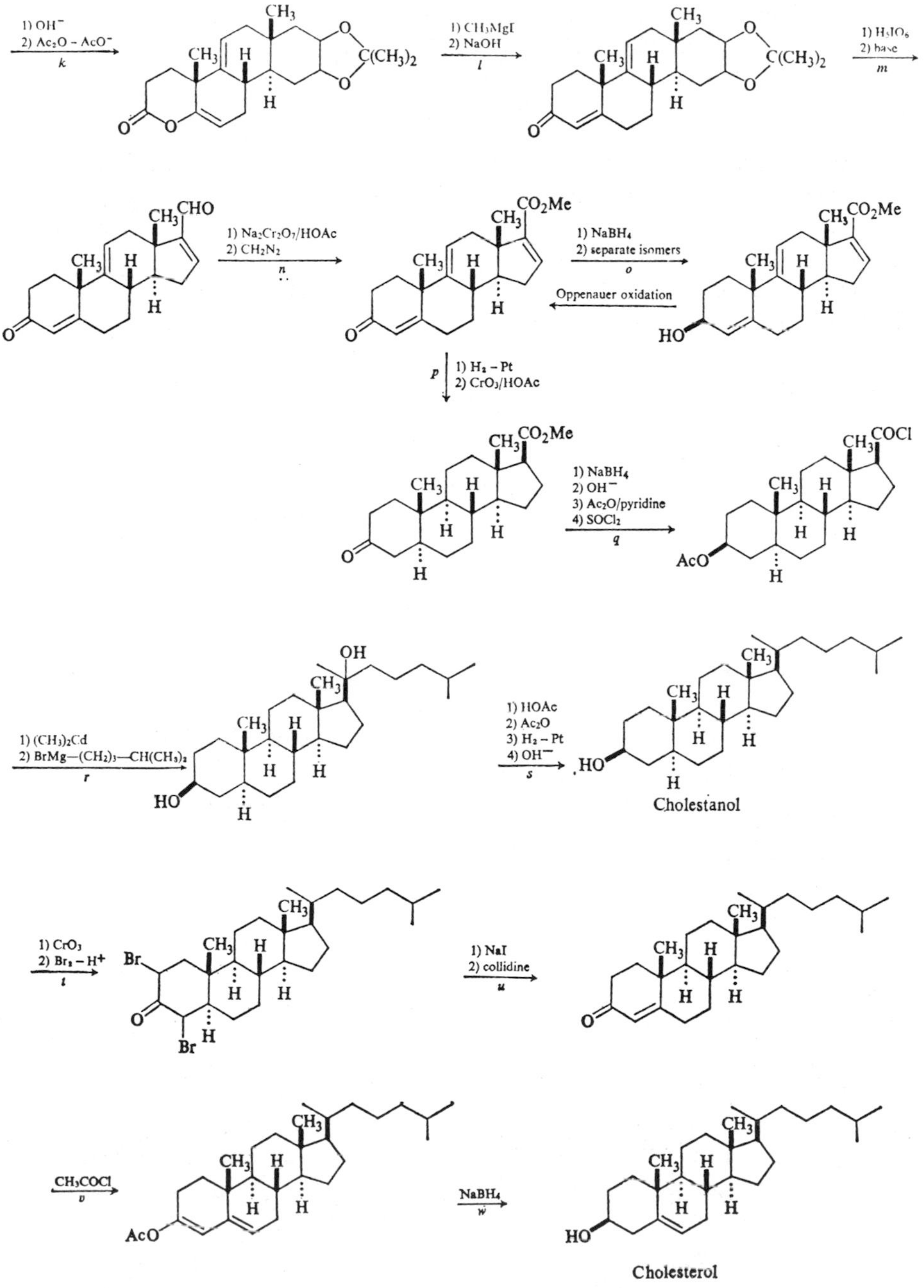

a   A Diels-Alder reaction which gives specifically the *cis* ring-junction (p. 295). The product is a racemic mixture.

b   The isomer with the *trans* ring-junction is the more stable and is obtained almost quantitatively with base, *via* the enolate anion (p. 186).

c   Lithium aluminium hydride reduces the carbonyl groups selectively.

d   The vinyl ether is hydrolyzed with mineral acid (p. 110) and the resulting β-hydroxy-ketone is readily dehydrated in these conditions.

e   The hydroxyl group is acetylated and the acetoxy group is removed by hydrogenolysis. Zinc effects the reaction in a manner comparable with the debromination of 1,2-dibromides:

f   The formyl group is introduced by crossed Claisen condensation (p. 241) in order to increase the activation of the C—H bond for the ensuing Michael addition to ethyl vinyl ketone (p. 253).

g   Treatment of the Michael-adduct with more base causes a reverse Claisen condensation, to eliminate the formyl group, and an intramolecular aldol condensation (p. 227). The ring-closure occurs stereoselectively as shown, for the alternative direction of closure would lead to stronger repulsions between the axial alkyl group at the new angular position and the original angular methyl.

h   The unconjugated olefinic bond requires protection against hydrogenation in the next step. This is achieved by formation of the *cis*-diol with osmium tetroxide (p. 571), reaction occurring more readily than on the stabilized (conjugated) olefinic bonds, followed by conversion into the cyclic ketal (p. 114).

i   It was found possible partially to hydrogenate the diene on a palladium catalyst with benzene as solvent; other conditions gave mixtures of products. The formyl group is then introduced in the standard way (*cf.* step *f*) and converted into an enamine with *N*-methylaniline in order to protect the α-carbon atom against the base-catalyzed condensation in step *j* (p. 245).

j   The abstraction of a proton by base gives a delocalized carbanion able to react with acrylonitrile (Michael addition, specifically cyanoethylation; p. 255) at either of two carbon atoms:

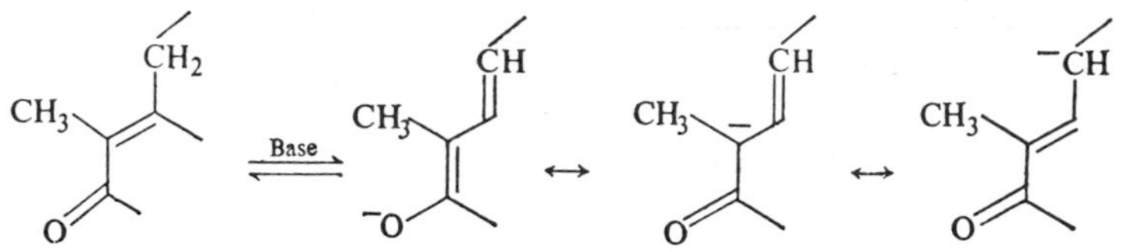

Previous experience with analogous systems suggested that reaction would occur preferentially at the carbon atom α to the carbonyl group rather than at the vinylogous γ-position, and this expectation was borne out in practice.

A new asymmetric centre is created and, unlike step g, reaction is not stereoselective. Separation of the two isomers is achieved in step l.

k   The nitrile is hydrolyzed, the protecting group and then the formyl group are removed, and the acid is lactonized *via* the enol tautomer of the ketone (p. 90):

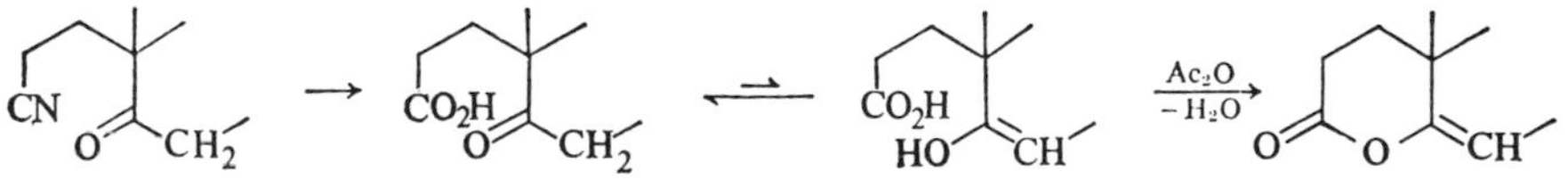

l   Methylmagnesium iodide reacts with the lactone shown faster than with the isomer in which the angular methyl group is equatorial, enabling the compounds formed in step j to be separated. The resulting methyl ketone is cyclized with base (intramolecular aldol condensation; p. 227).

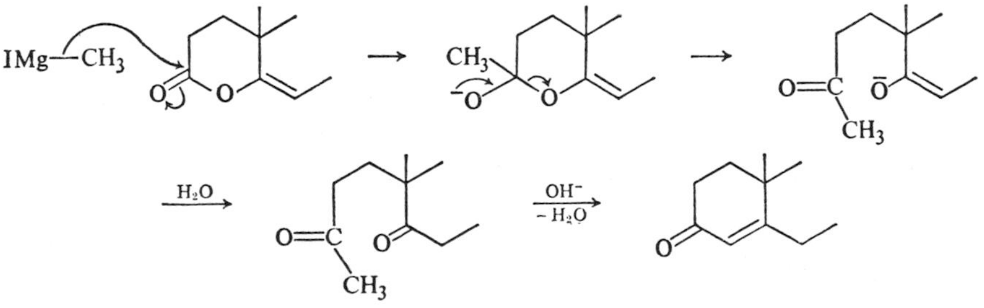

m   Periodic acid first hydrolyzes the cyclic ketal and then cleaves the 1,2-diol (p. 596). An intramolecular aldol condensation is then induced with the weak base, piperidine acetate. Reaction might occur in either or both of two ways:

In practice, hardly any of the second product is formed, probably because the lower methylene group is in the more hindered position and forms the required carbanion less readily.

n   The aldehydic group is oxidized and the carboxylic acid is esterified.

o   The reduction was carried out to give a product to which a successful resolution procedure could be applied. Sodium borohydride reduces the carbonyl group but not the conjugated olefinic bond (p. 634). This gives two products since the hydroxyl group may be α or β*, and the dextrorotatory form of the β-alcohol was separated by preferential precipitation on the addition of digitonin and oxidized to give the required stereoisomer of the ketone.

p   The three olefinic double bonds and the carbonyl group are hydrogenated over platinum in conditions in which the ester group is inert. It was known from previous experience that reduction of the two isolated double bonds would occur in the stereochemically required manner, but reduction of the carbonyl-conjugated double bond in related compounds often gives a mixture of epimers. In fact, the ketone which was isolated after oxidation had the correct stereochemistry; this material had previously been isolated from natural sources and converted into other steroids.

q   Reduction is stereoselective, giving mainly the 3β-hydroxy product (p. 634). After hydrolysis of the ester group, the alcoholic group is protected by acetylation against the reagents used in the following steps to introduce the alkyl side-chain, and then the acid group is activated by conversion into the acid chloride.

r   Dimethylcadmium converts —COCl into —COCH$_3$ without affecting the ester group (p. 220), and the remainder of the alkyl side-chain is introduced by reaction of the new ketone with isohexylmagnesium bromide. The acetyl protecting group is removed during hydrolysis of the Grignard product.

s   The tertiary alcohol is readily dehydrated by hot acetic acid, and, after re-protection of the alcoholic group by acetylation with acetic anhydride, the

---

*The α- and β-substituents are those which project, respectively, below and above the plane of the paper.

new olefinic bond is reduced catalytically. The alcoholic group is then liberated by hydrolysis.

t   The hydroxyl group is oxidized to carbonyl. Acid-catalyzed bromination gives the symmetrical dibromo-ketone (p. 75).

u   Iodide ion, introduced as sodium iodide in acetone, displaces the 2–Br but not the 4–Br substituent. The latter is known to be axial, and is therefore suitably sited for E2 elimination of HBr by the basic collidine. However, elimination of HI across the 1– and 2–positions does not occur, possibly because the iodo-substituent is not in the axial conformation necessary for the *trans*–E2 reaction (its conformation is unknown). Instead, the base induces hydrogenolysis, probably *via* the enolate anion:

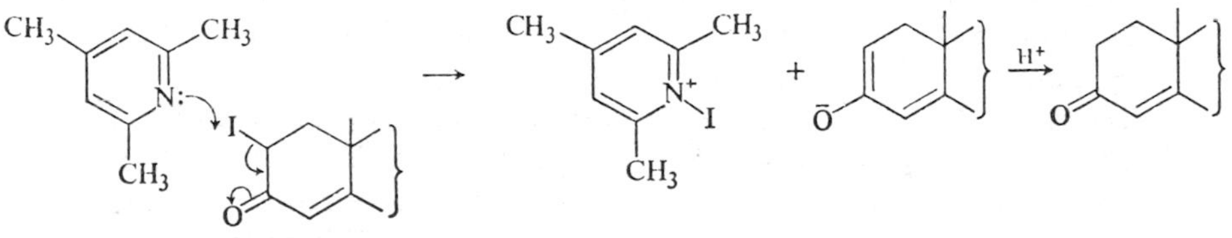

v   Acetyl chloride gives the enol acetate:

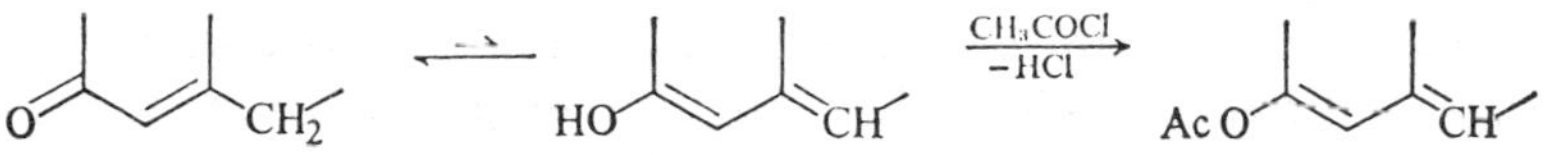

w   In the alkaline conditions, the enol acetate is first hydrolyzed to the enol which tautomerizes to the Δ^5-ketone. This is then reduced to cholesterol before occurrence of the possible tautomeric shift to give the conjugated Δ^4-ketone:

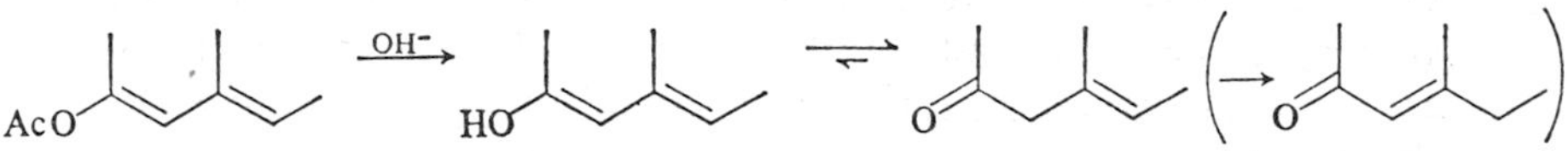

21.4 Oestrone

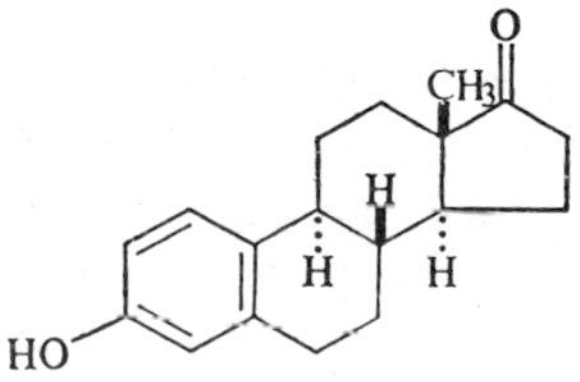

(±)-Oestrone

Oestrone, an aromatic steroid, is a member of the oestrogen family of female sex hormones which are secreted in the anterior lobe of the pituitary gland. It is thought to be built up *in vivo* by a pathway very similar to that involved in the formation of cholesterol, and indeed cholesterol itself may be its precursor.

There are four asymmetric carbon atoms in oestrone, so that there are eight enantiomorphic pairs of stereoisomers, all of which have been synthesized. The earlier syntheses were not entirely stereoselective, but the synthesis by W. S. Johnson, J. Walker, *et al.* (*Proc. Chem. Soc.*, 1958, 114) is stereoselective at each step; it employs the previously synthesized diene *A* as starting material and gives (±)-homomarrianolic acid methyl ether which has been previously converted into oestrone as shown below.

Scheme 6

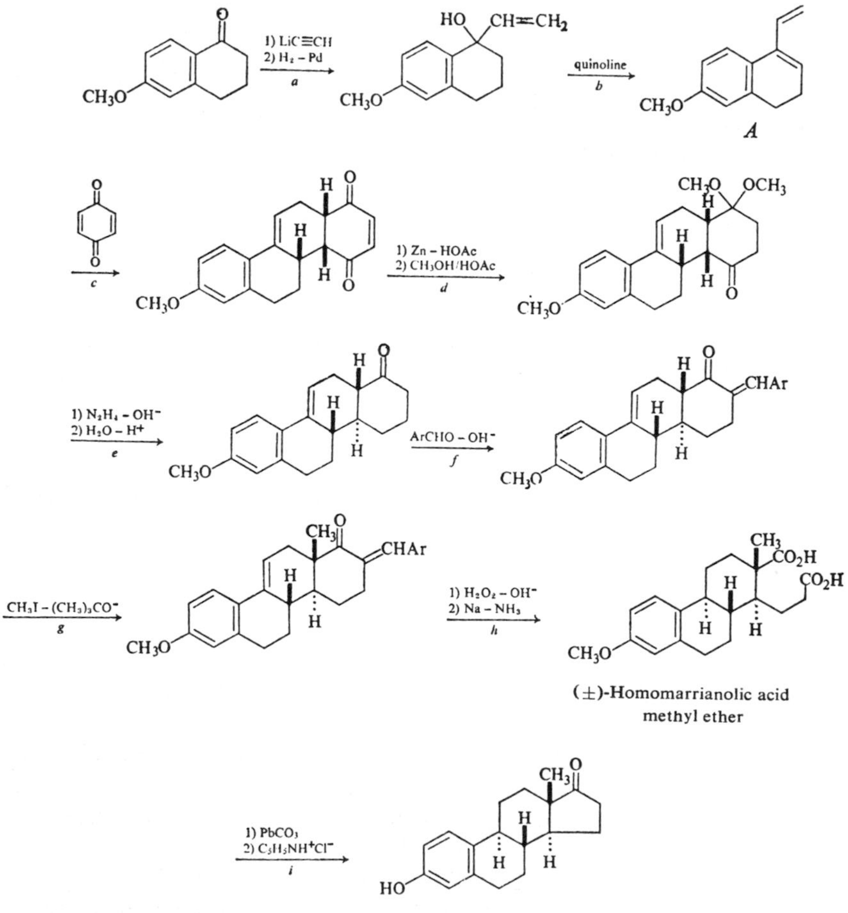

(±)-Homomarrianolic acid
methyl ether

Oestrone

a Ethynylation, followed by partial reduction of the acetylene on Lindlar's catalyst (p. 619).

b Dehydration gives a conjugated diene which polymerizes readily (p. 549). It was found that quinoline effects base-catalyzed elimination and at the same time acts as a polymerization inhibitor. Attempted purification also results in polymerization, and the crude product was used for the next step.

c The Diels-Alder reaction establishes the relative configurations of three asymmetric carbon atoms (p. 292).

d Zinc and acetic acid selectively reduce the carbonyl-conjugated double bond (p. 617). Methanol in the presence of acetic acid selectively forms the ketal of the less hindered carbonyl group.

e Hydrazine and base, in the conditions of Huang-Minlon's modification of the Wolff-Kishner reduction (p. 631), first bring about epimerization *via* the enolate ion (carbonyl-activation) to give the more stable C/D-*trans* ring-junction and then reduce carbonyl to methylene. The ketal is hydrolyzed with aqueous acid.

f Base-catalyzed condensation with furfuraldehyde occurs at the methylene group activated by the carbonyl group, thereby protecting this position against base-catalyzed methylation in the subsequent step (p. 245).

g Alkylation at the angular position, activated by the carbonyl group, occurs in the presence of the strong base, t-butoxide ion (p. 248). It had been observed previously that whereas 2-benzylidene-1-decalone gives 68% of the *cis*- and 23% of the *trans*-methylated product, the related 6,7-dehydro compound gives 56% of the *trans*-compound as the only pure product, probably because in the latter case there is no axial hydrogen at C_6 to hinder the approach of methyl iodide from the direction which leads to the *trans*-derivative.

It was therefore expected that the double bond in ring C would favour formation of the required C/D-*trans* methylated product, and this was borne out in practice.

h The ring is opened by oxidation with alkaline hydrogen peroxide and the olefinic double bond in conjugation with the aromatic ring is reduced by sodium in liquid ammonia (p. 621), giving the (required) more stable *trans*-B/C ring junction.

i The five-membered cyclic ketone is formed from the dibasic acid by heating with lead carbonate at 300°C, and the methoxyl group is hydrolyzed by heating with pyridinium chloride at 210°C. (±)-Oestrone has been resolved to give the natural (−)-isomer *via* the diastereoisomers formed with (+)-menthoxyacetyl chloride.

21.5 Epiandrosterone

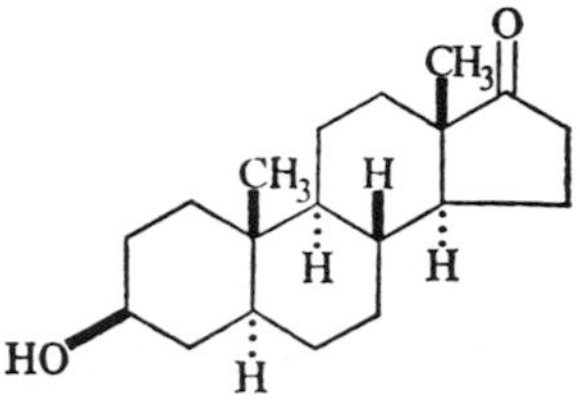

Epiandrosterone is a member of the androgen group of male sex hormones. The related androsterone differs only in that the hydroxyl group at C_3 has the alternative α-configuration. The molecule contains seven asymmetric carbon atoms, so that a satisfactory synthesis requires the application of stereoselective reagents. The approach adopted by W. S. Johnson *et al.* (*J. Amer. Chem. Soc.*, 1956, **78**, 6331) was first to build an unsaturated tetracyclic compound containing only one asymmetric centre and then to reduce the double bonds and effect the required structural modifications by stereoselective processes leading to the required *trans-anti-trans-anti-trans* ring system. Their synthesis is as follows:

Scheme 7

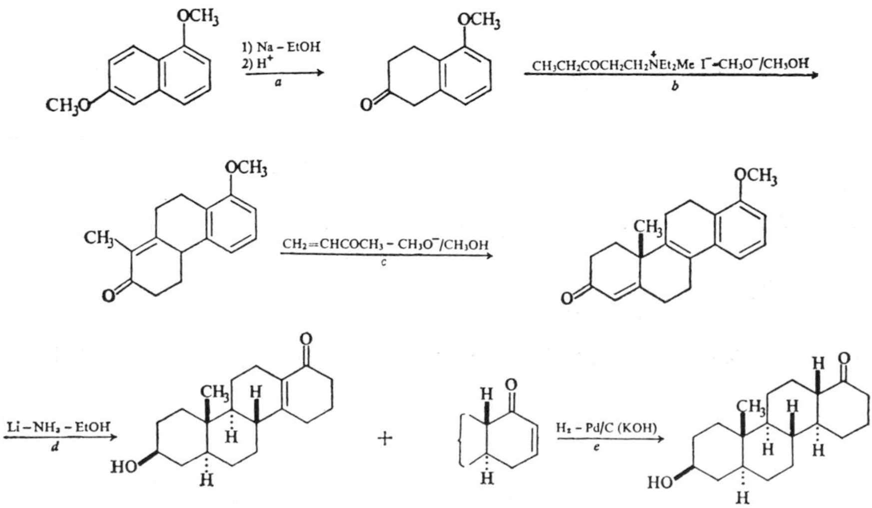

Scheme 7 (*cont.*)

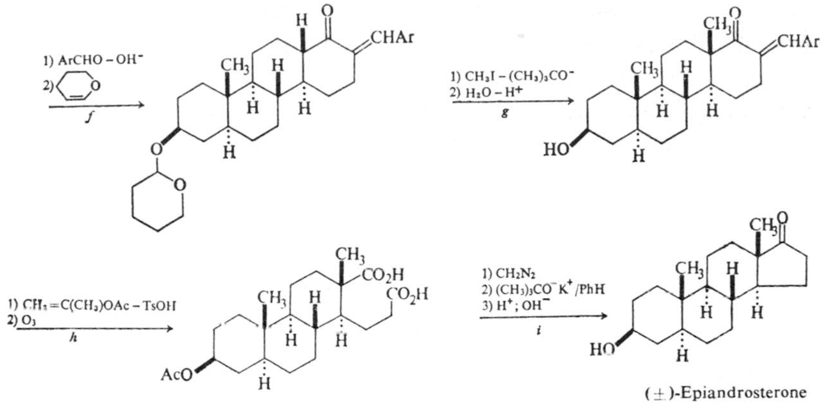

1) ArCHO – OH⁻
2) [cyclohexenone]
→
f

1) CH₃I – (CH₃)₃CO⁻
2) H₂O – H⁺
→
g

1) CH₃=C(CH₃)OAc – TsOH
2) O₃
→
h

1) CH₂N₂
2) (CH₃)₃CO⁻ K⁺ /PhH
3) H⁺; OH⁻
→
i

(±)-Epiandrosterone

a   Reduction of the naphthalene occurs selectively in the benzenoid ring containing the 2-methoxy-substituent, for a reason outlined earlier (p. 624). The resulting 7,8-dihydronaphthalene is an enol ether which is hydrolyzed by acid to the tetralone (p. 110).

b   Robinson's ring-extension (p. 278). The product is a mixture of tautomers from which that shown can be isolated in crystalline form. (In the tautomer which is not shown the olefinic double bond is the bond of fusion between the alicyclic rings and is conjugated with the aromatic ring; see below.) It is not, however, necessary to isolate the tricyclic ketone before carrying out the next step.

c   A second application of Robinson's ring-extension:

MeO⁻

MeO⁻

CH₂=CHCOCH₃

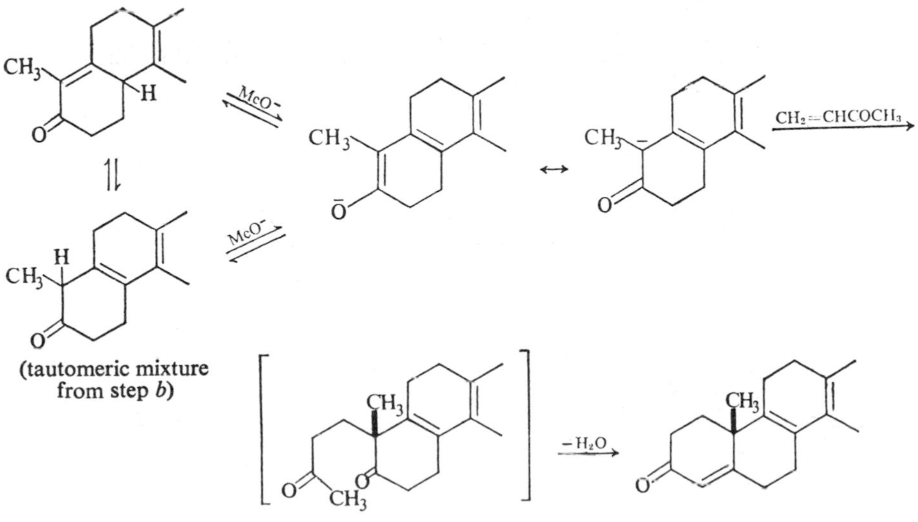

(tautomeric mixture from step *b*)

– H₂O

The methiodide of 1-diethylamino-3-butanone may be employed instead of methyl vinyl ketone.

d   Lithium, liquid ammonia, and ethanol, with dioxan as co-solvent, reduce the olefinic bond conjugated with the aromatic ring and that conjugated with the carbonyl group, each in the *trans* manner (p. 618). The anisole ring is also reduced (p. 623), giving a mixture of two isomers, as shown.

e   The mixture of isomers is reduced catalytically, with *cis*-stereospecificity (p. 613). Stereochemically different products are produced,

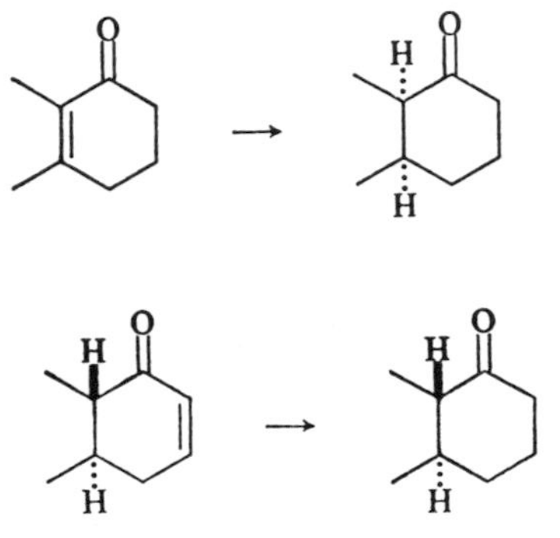

but the inclusion of a base in the medium induces equilibration of the two through the formation of the enolate anion, leading to the preponderance of the more stable *trans* product.

f   The methylene group is protected against methylation in step *g* by formation of the furfurylidene derivative (Ar = 2-furyl) (*cf*. p. 725) and the hydroxyl group is protected by treatment with dihydropyran (p. 686).

g   Methyl iodide in the presence of a strong base gives a mixture of *cis*- and *trans*-methylated products, the former (unwanted isomer) predominating (*cf*. p. 725). Fortunately, the required isomer is readily isolable from the mixture by crystallization and chromatography. Acid work-up removes the pyranyl protecting group.

h   The free hydroxyl group is acetylated with isopropenyl acetate in the presence of toluene-*p*-sulphonic acid, and ring D is cleaved with ozone (p. 573).

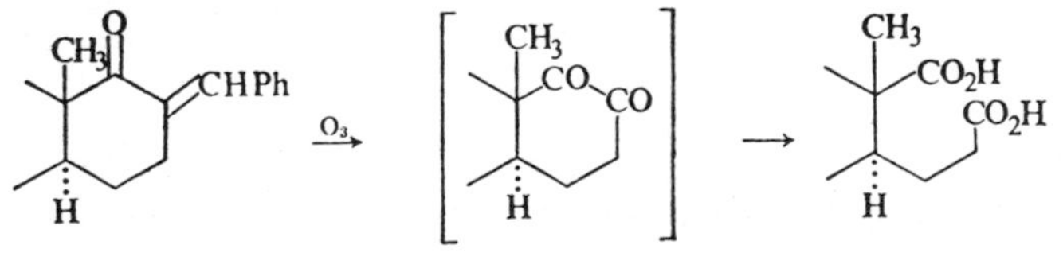

i   Diazomethane methylates the dibasic acid and potassium t-butoxide effects Dieckmann cyclization to give a cyclic β-keto-ester (p. 241). Hydrolysis of the ester followed by decarboxylation, and hydrolysis of the acetate, give (±)-epiandrosterone.

Epiandrosterone has been converted into androsterone by F. Sondheimer *et al.* (*J. Org. Chem.*, 1955, **20**, 542):

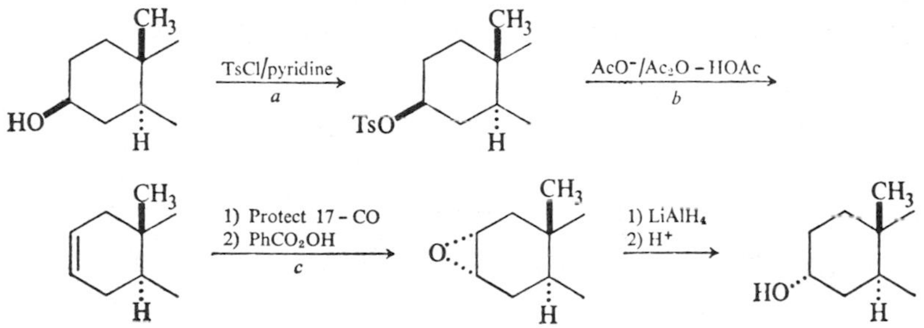

a   The 3β-hydroxyl group is converted into its toluene-*p*-sulphonate.
b   Base-catalyzed elimination.
c   The 17-keto group is protected against Baeyer-Villiger oxidation with the following reagent by conversion into the cyclic ketal with ethylene glycol. Perbenzoic acid induces epoxidation of the olefin from the less hindered side, giving the 2α-3α-oxide.
d   The epoxide ring is opened reductively in the usual *trans*-manner (p. 638).

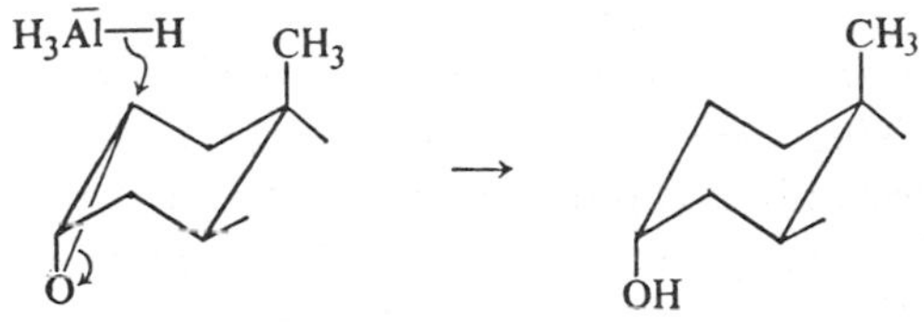

Mineral acid releases the 17-keto group to give androsterone.

21.6 Quinine

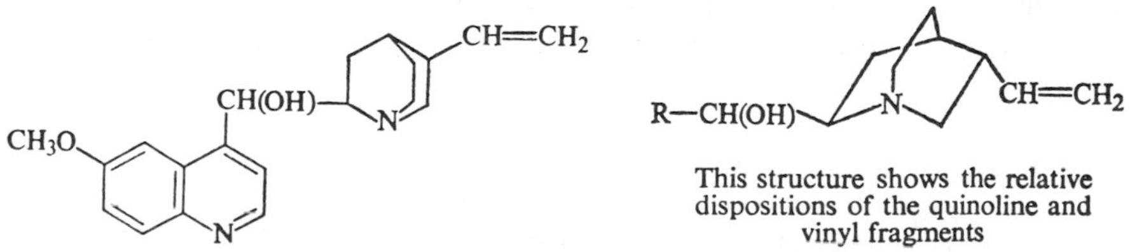

This structure shows the relative dispositions of the quinoline and vinyl fragments

Quinine, which occurs in the bark of *Cinchona* trees, is widely used as an anti-malarial agent. Its synthesis was carried out by R. B. Woodward and W. E.

Doering (*J. Amer. Chem. Soc.*, 1945, **67**, 860) who prepared quinotoxine which had previously been converted into quinine. The total synthesis starts from *m*-hydroxybenzaldehyde, obtainable from benzaldehyde by nitration (p. 403), reduction of the nitro group with tin(II) chloride, diazotization, and treatment with boiling water.

Scheme 8

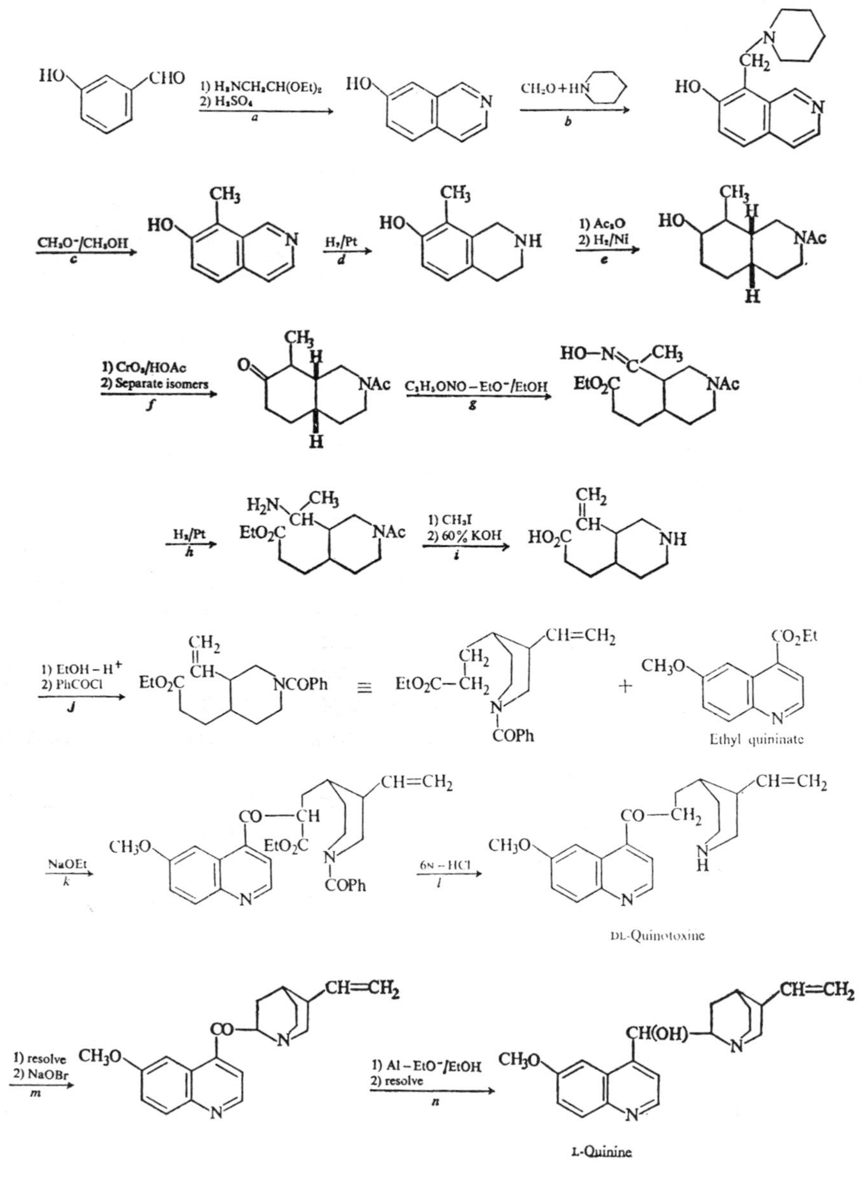

a　The Pomeranz-Fritsch synthesis of isoquinolines (p. 684). The product is accompanied by some of the isomeric 5-hydroxy-isoquinoline (ring-closure *ortho* to the hydroxyl) from which it is freed by fractional crystallization of the mixture of sodium salts, that of the 5-hydroxy compound being more soluble.

b　The Mannich reaction (8.6). The hydroxyl group activates the 8-position towards the electrophilic reagent far more strongly than the 6-position, just as that in β-naphthol directs electrophiles to the 1-position (p. 379).

c　The usual catalytic method for hydrogenolysis was unsuccessful. Methoxide ion at high temperatures acts as a hydride-transfer agent and the benzylic compound is reactive towards nucleophiles:

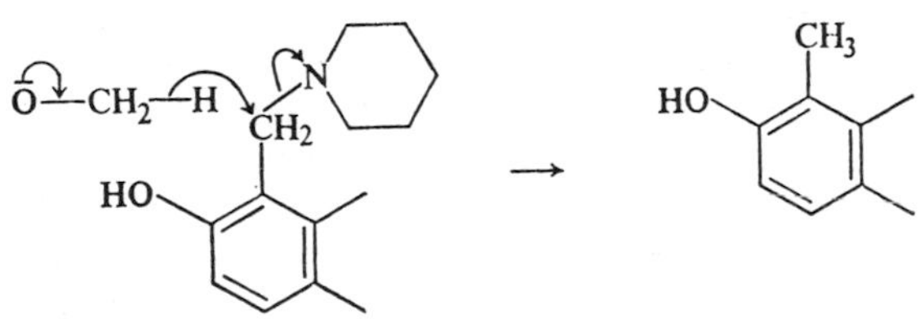

d　A decahydro-derivative with a *cis* ring-junction is required.* Reduction over platinum was found to proceed readily until two moles of hydrogen had been absorbed (reduction of the pyridine ring). Forced hydrogenation reduced the benzenoid ring, but the major product, although having the correct stereochemistry, lacked the hydroxyl group, whereas a minor product, although being a decahydro-derivative, had a *trans* ring-junction.

e　This problem was solved by reducing the *N*-acetylated derivative over nickel at 150°C under pressure. The product was an approximately equimolar mixture of the *cis* and *trans* decahydro-derivatives.

　　The mixture of isomers is oxidized and the isomeric ketones are separated *via* their crystalline hydrates.

g　The base-catalyzed reaction gives a tertiary nitroso compound which is cleaved by a type of reverse-Claisen process (p. 345):

*This is because of the *cis* relationship of the vinyl group to the quinoline ring in quinine which was to be constructed as follows:

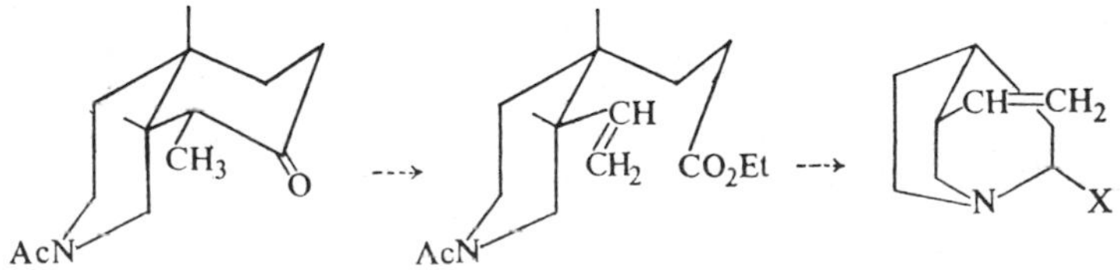

(X = remainder of quinine molecule)

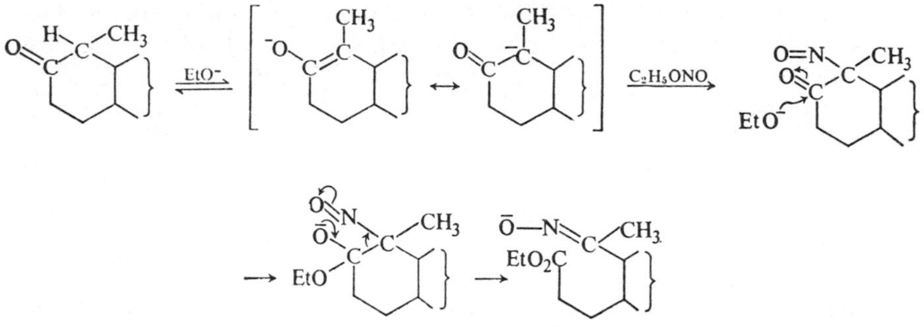

h   A new asymmetric centre is introduced by reduction, but resolution is not necessary since the centre is to be destroyed in the following step. However, the amino-esters require careful handling because heat leads to complex condensations. Reduction is effected at room temperature.

i   Gentle heating with methyl iodide in ethanol over potassium carbonate gives the quaternary salt. The usual procedure for elimination from these salts (heating over silver oxide) gives a stable betaine (zwitterion),

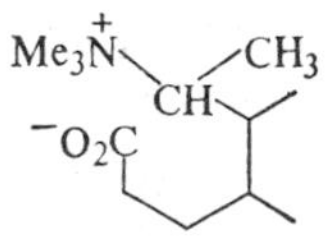

but vigorous conditions (60% potash at 140°C) cause elimination, in the direction expected from Hofmann's rule (p. 121). The amino-acid formed is difficult to isolate and is therefore converted with potassium cyanate into a ureide (*cf.* reaction between ammonia and cyanate; p. 337) which, after recrystallization, is decomposed by acid:

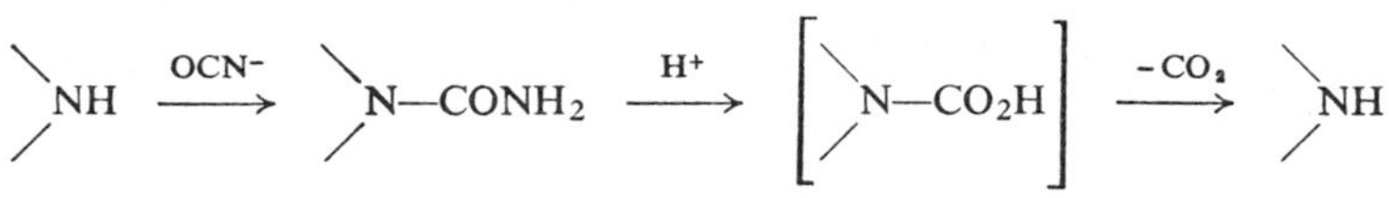

ureide

j   Esterification of the carboxyl group and benzoylation of the amino group, the latter being required for protection during the subsequent Claisen condensation.

k   A crossed Claisen condensation (p. 240). Its success depends on the fact that the quinolinic ester, while lacking an activated C—H bond, has the more reactive carbonyl group because of the presence of the electron-attracting nuclear nitrogen atom. The ethyl quininate can be obtained from *p*-anisidine:

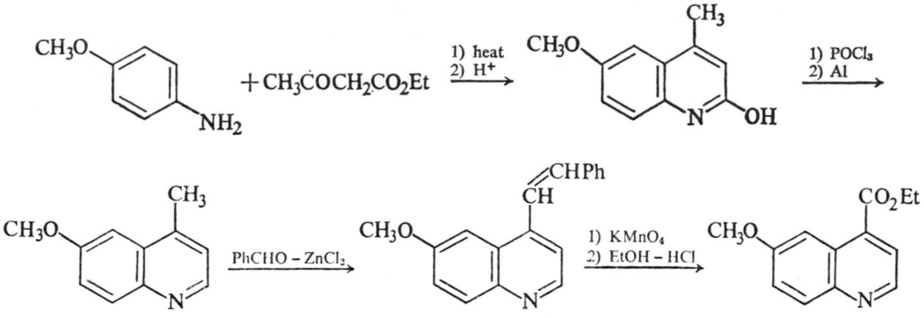

The Conrad-Limpach procedure for quinolines (p. 681) gives a 2-hydroxy-derivative from which hydroxyl is removed by conversion into chloro followed by reduction (p. 630). The activated methyl group is condensed with benzaldehyde in the presence of a Lewis acid (p. 273), and oxidation and esterification give the required product.

l The ester group is hydrolyzed and the resulting β-keto-acid readily decarboxylates (p. 318). At the same time the benzoyl group is removed.

m The quinotoxines were resolved by fractional crystallization of the diastereoisomers formed with D-dibenzoyl tartrate. Ring-closure is brought about with alkaline sodium hypobromite, probably *via* base-catalyzed bromination of the methylene group α to the carbonyl followed by nucleophilic displacement of the bromide by nitrogen:

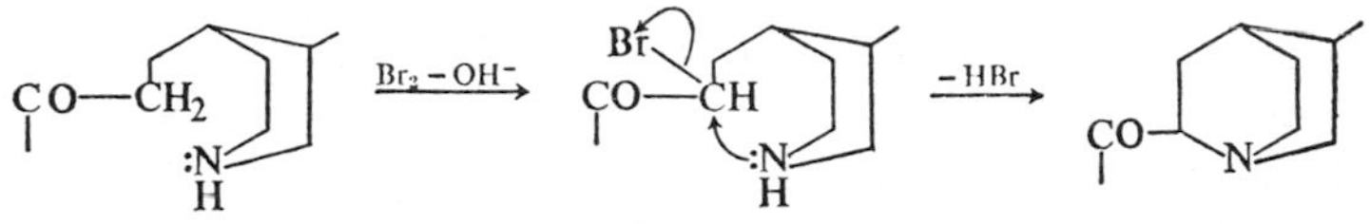

n Reduction with aluminium and ethoxide introduces a new asymmetric carbon atom. Resolution of the products gives the naturally occurring isomer, L-quinine.

21.7 Reserpine

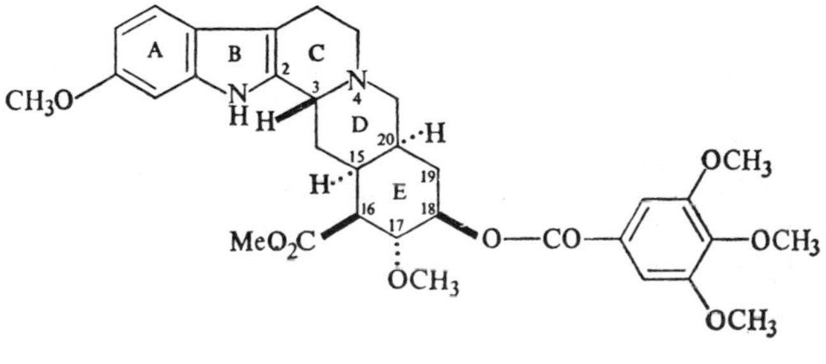

Reserpine is found, with other alkaloids, in the roots of the plant genus *Rauwolfia*. It has useful tranquillizing properties and is used in the treatment of some mental disorders and for the reduction of hypertension. It was synthesized by R. B. Woodward *et al.* (*J. Amer. Chem. Soc.*, 1956, **78**, 2023; *Tetrahedron*, 1958, **2**, 1).

Scheme 9

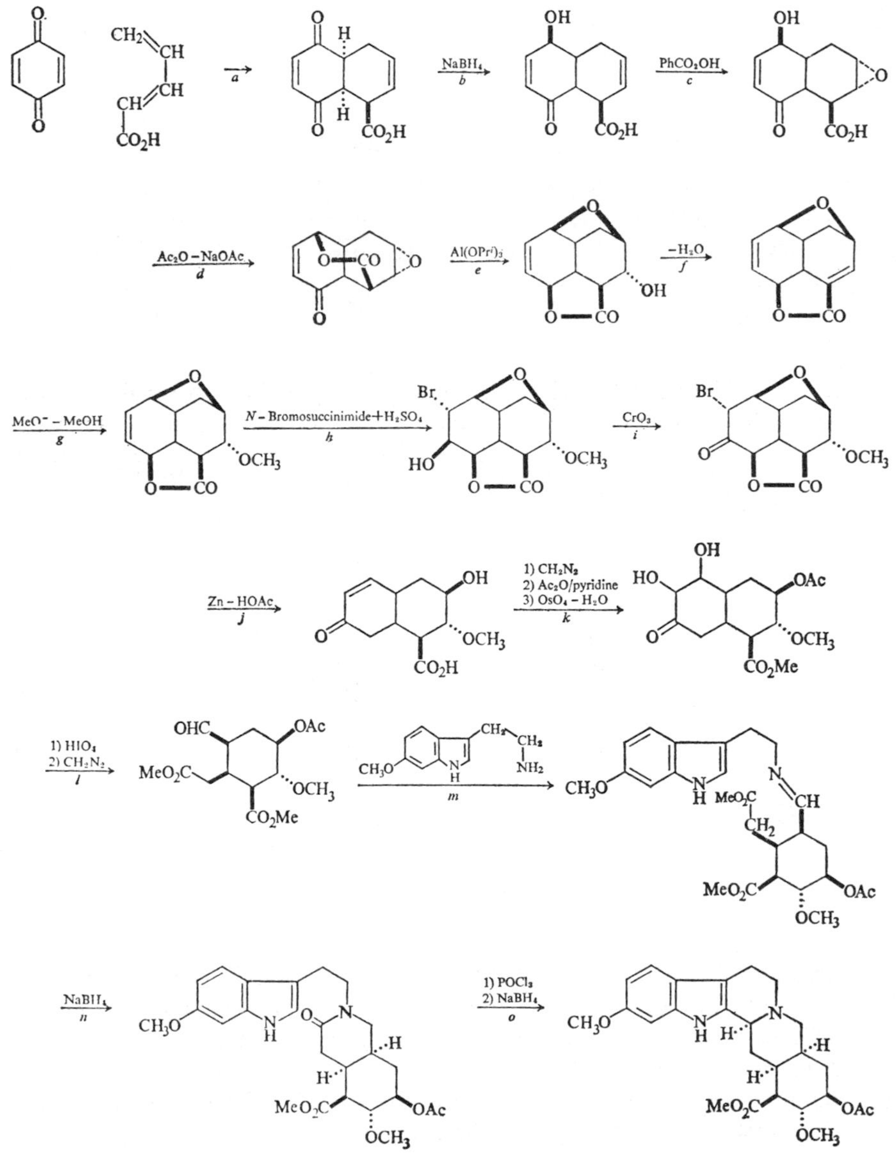

Scheme 9 (cont.)

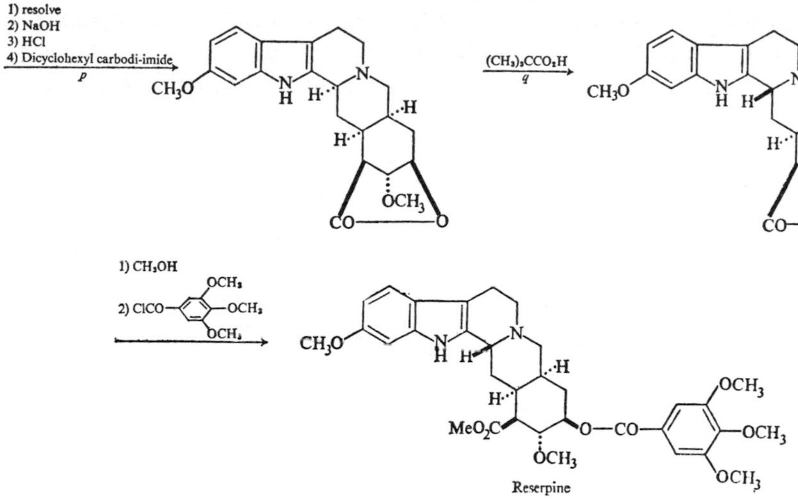

1) resolve
2) NaOH
3) HCl
4) Dicyclohexyl carbodi-imide

p

$(CH_3)_3CCO_2H$

q

1) CH_3OH

2) $ClCO$—(OCH₃ OCH₃ OCH₃ aryl)

Reserpine

a   A Diels-Alder reaction. The stereochemical principles governing this reaction (p. 292) lead to the ring-junction having the *cis* stereochemistry and
the carboxyl group lying on the same side as the rings with respect to the
ring-junction (i.e. reaction is represented as in (*i*) rather than as in (*ii*)).

(*i*)

(*ii*)

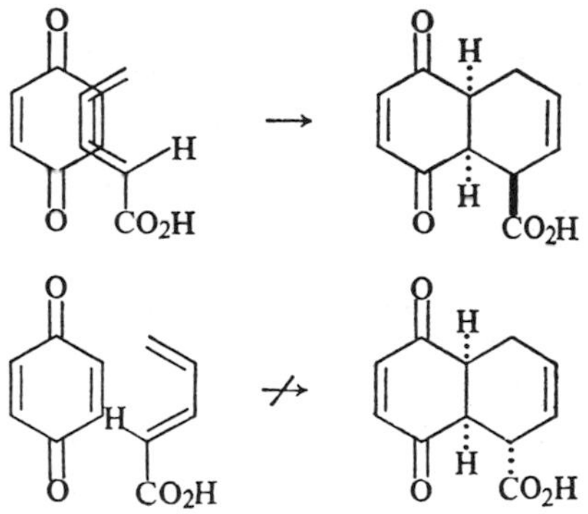

This step fixes the stereochemistry at C_{15}, C_{16} and C_{20} of reserpine. The
remainder of the synthesis concerns the stereochemistry at C_{17}, C_{18} and C_3.
(For clarity, the *cis*-hydrogens at the ring-junction are omitted in the remaining diagrams.)

b   Sodium borohydride reduces the less hindered of the two carbonyl groups.
The nucleophile attacks from the less hindered α side.

c   The peracid reacts more rapidly with the isolated double bond than with the carbonyl-conjugated double bond. Attack is from the less hindered α side. If this epoxide were opened at this stage by reaction with, e.g. acetate ion, the product would be a diaxial diol-derivative having the wrong stereochemistry for ring E in reserpine:

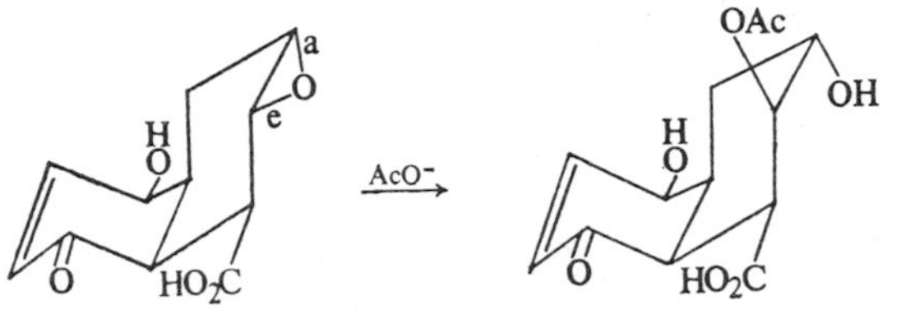

Several transformations are therefore effected to produce the diol-derivative with the required stereochemistry.

d   Almost all the reactant in this step has the conformation in which both the hydroxyl and the carboxyl groups are equatorial, since the conformer in which these groups are axial contains strong steric compressions, but only the latter conformer can undergo dehydration to form a lactone. This reaction therefore has the effect of changing the conformation of the system; in particular, the epoxide becomes axial at C_{17} and equatorial at C_{18}*:

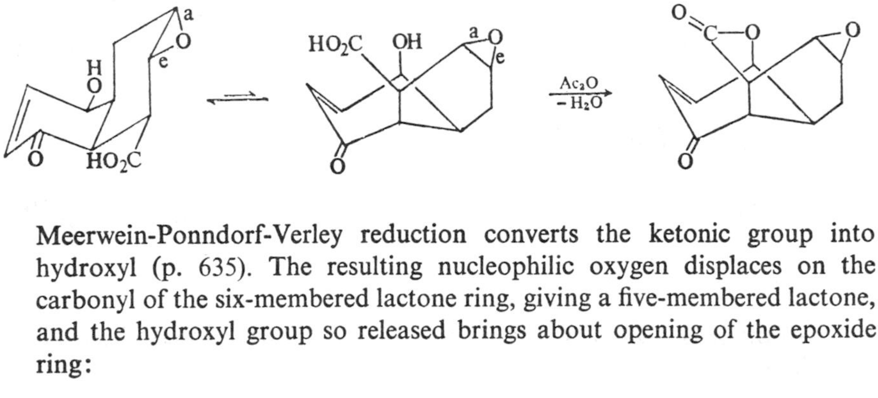

e   Meerwein-Ponndorf-Verley reduction converts the ketonic group into hydroxyl (p. 635). The resulting nucleophilic oxygen displaces on the carbonyl of the six-membered lactone ring, giving a five-membered lactone, and the hydroxyl group so released brings about opening of the epoxide ring:

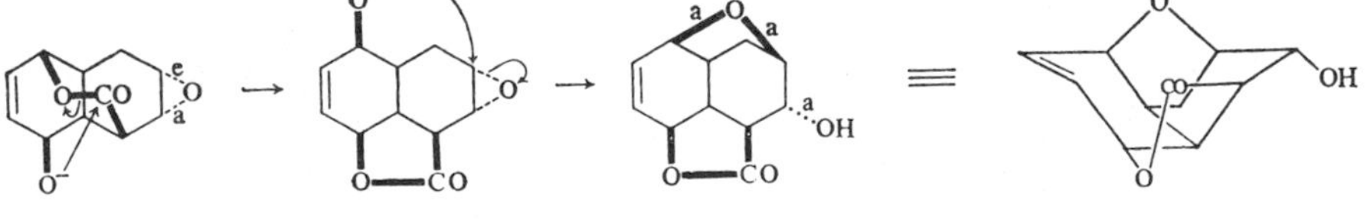

*This transformation is more readily understood with the aid of models.

f   Dehydration, involving elimination of the axial hydroxyl group and formation of an $\alpha\beta$-unsaturated carbonyl-derivative, occurs readily.

g   The olefinic bond is activated towards nucleophiles by the conjugated carbonyl group. Methoxide ion approaches from the less hindered α-side, so that the resulting —OCH_3 substituent is in the axial position at C_{17}. Ring E now has the correct stereochemistry.

h   *N*-bromosuccinimide in acid is an ionic brominating agent. Approach from the α side as usual gives a bromonium ion which is opened by water to give the diaxial bromohydrin:

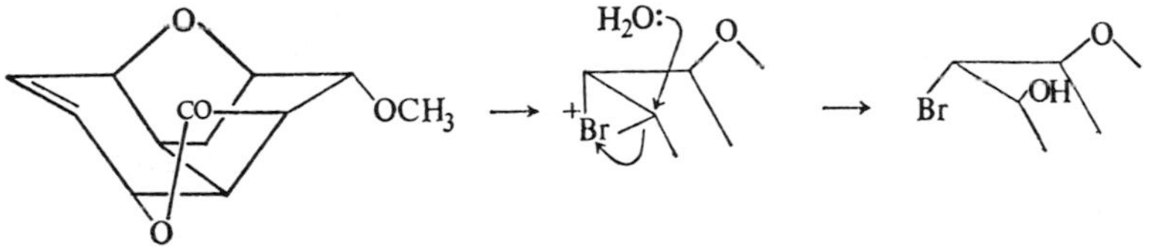

i   Mild oxidation of the secondary alcohol.

j   Zinc in acetic acid brings about the reductive opening of both the lactone and the strained ether:

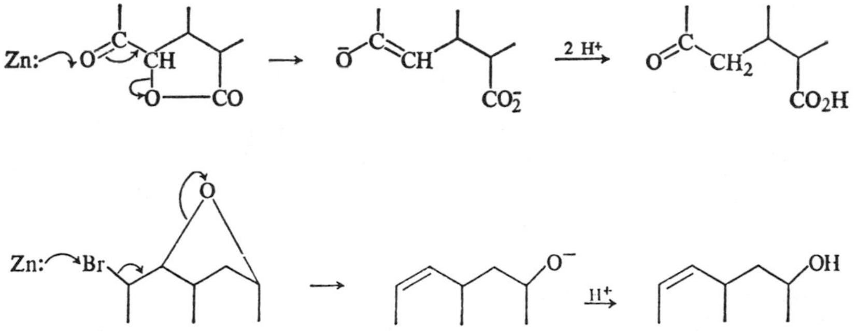

All the substituents in ring E are now able to adopt equatorial conformations:

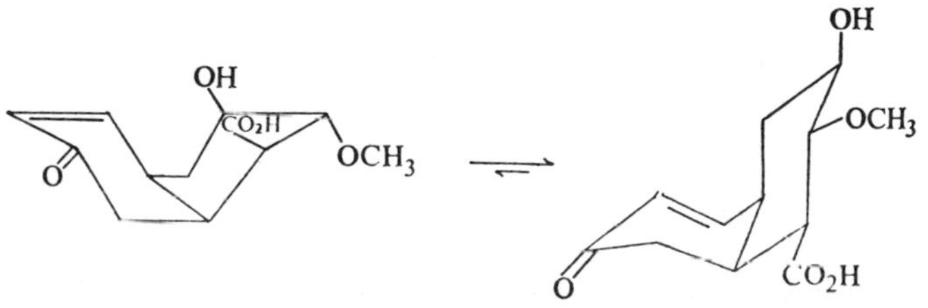

k   The carboxyl group is esterified by diazomethane (which does not react with the alcoholic group), the alcoholic group is acetylated, and the α-glycol is formed with osmium tetroxide (p. 571).

l* Treatment with periodate cleaves ring D and one carbon atom is lost as formic acid (p. 596) (later to be replaced, with ring-closure, by nitrogen in step *n*). The new carboxyl group is then esterified with diazomethane.

m   6-Methoxytryptamine may be prepared as follows:

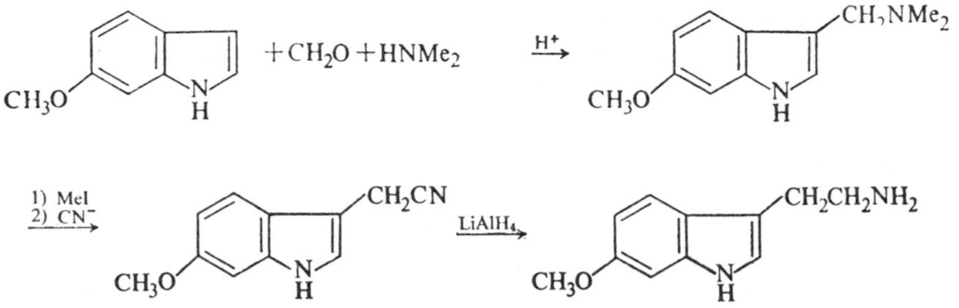

Indoles are very reactive towards electrophiles, particularly at the β-position so that the Mannich reaction can be effected (p. 275). The resulting benzyl-type system is susceptible to nucleophilic displacement by cyanide ion (p. 278), and the final reduction is a standard procedure for nitriles (p. 640).

The tryptamine reacts readily with the aldehydic group in benzene solution.

n   Reduction of the Schiff base gives an amine, and the resulting nucleophilic nitrogen displaces intramolecularly on the ester to close ring D.

o   Phosphorus oxychloride brings about ring-closure as in the Bischler-Napieralskı synthesis of isoquinolines (p. 683), giving an iminium salt which is then reduced with sodium borohydride:

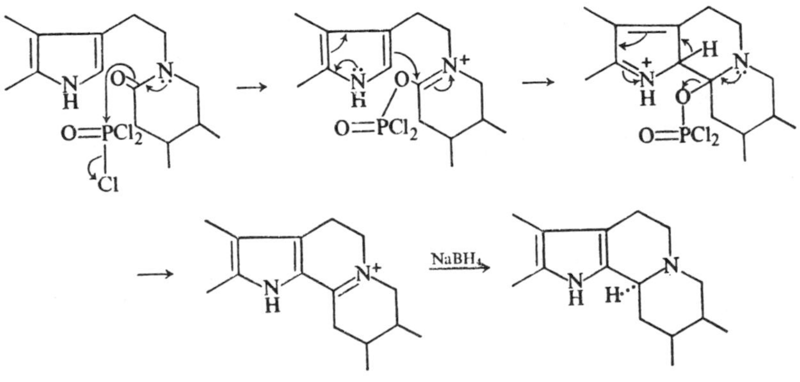

*The intermediates resulting from steps *l–n* were not isolated.

The borohydride attacks from the α side, so that the product has a different stereochemistry at C_3 from that of reserpine. The remaining steps are concerned with correcting this and introducing the aroyl substituent at C_{18}.

p   Treatment with base followed by acid removes the acetyl group (C_{18}) and hydrolyzes the methyl ester (C_{16}). The freed hydroxyl and carboxyl groups are then joined to give a lactone, using the very mild reagent, dicyclohexyl carbodi-imide, which increases the susceptibility of the acid to nucleophilic attack by the alcohol just as it does in the formation of peptide bonds (p. 361). Lactonization occurs through the conformer in which the carboxyl and hydroxyl groups are axial (cf. step d) and in which, therefore, the C_2—C_3 bond is axial rather than equatorial with respect to ring D.

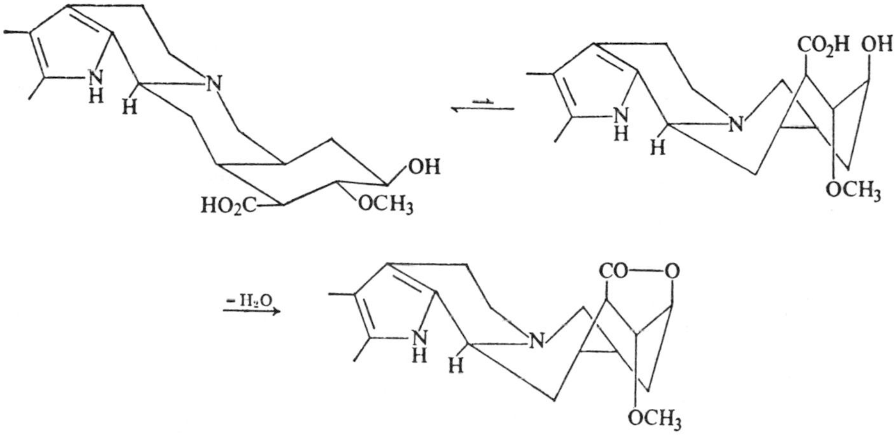

q   The result of the preceding step is that the molecule is now less stable than the stereoisomer obtained by altering the configuration of the hydrogen at C_3.* Equilibration of the two isomers therefore yields almost entirely the product with the stereochemistry required for reserpine at this centre. The equilibrating agent chosen was pivalic acid because of its suitable boiling-point and the fact that it is too weak a nucleophile to open the lactone ring. Interconversion presumably occurs through protonation (activated by the pyrrole nitrogen) and deprotonation:

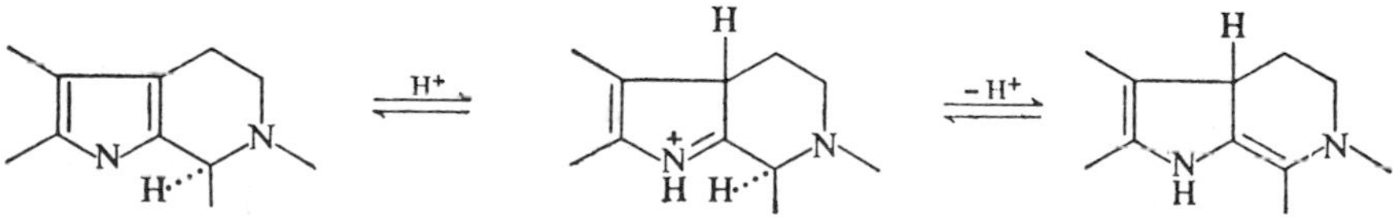

*The steric compressions which are responsible for this can best be seen by constructing molecular models.

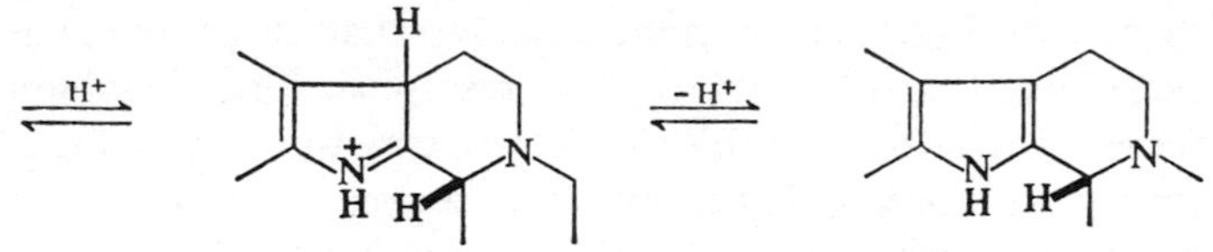

r   The lactone ring is opened with methanol, and the molecule then reverts
to the more stable conformation in which the three substituents on ring E
are equatorial. Finally, the aroyl residue is introduced at C_{18} with 3,4,5-
trimethoxybenzoyl chloride.

21.8 Adenosine Triphosphate

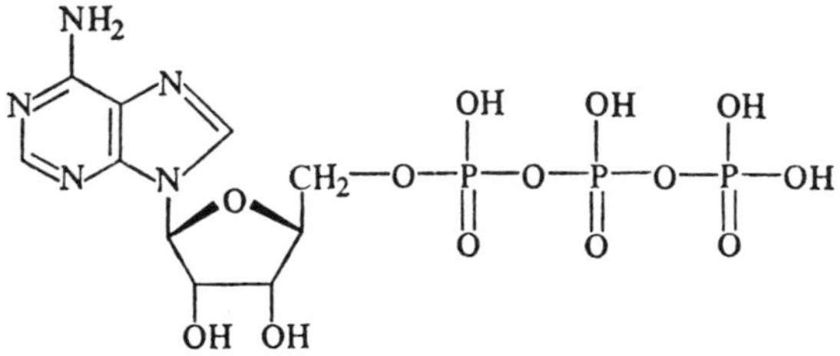

Adenosine occurs naturally in a variety of derived forms each of which performs
important functions.

(1) Adenosine triphosphate, ATP (above), acts in conjunction with its diphosphate, ADP, as a reversible phosphorylating couple and energy store. For
example, it is concerned with the supply of energy for muscular contraction.

(2) Coenzyme A, a derivative of the vitamin pantothenic acid, is involved,
inter alia, in the biological formation and oxidation of fatty acids and in the decarboxylation of α-keto-acids.

(3) *S*-Adenosyl-L-methionine is concerned with the biological transfer of
methyl groups.

(4) Adenosine is a component of the ribonucleic acids (p. 692), di- and triphosphopyridine nucleotides (p. 698), and flavin adenine nucleotide (p. 698).

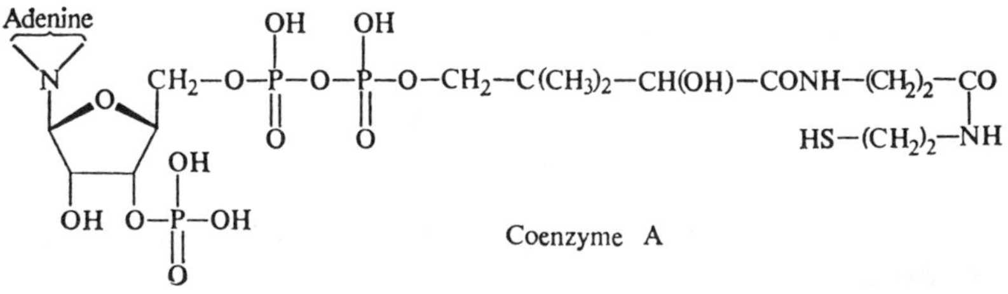

Coenzyme A

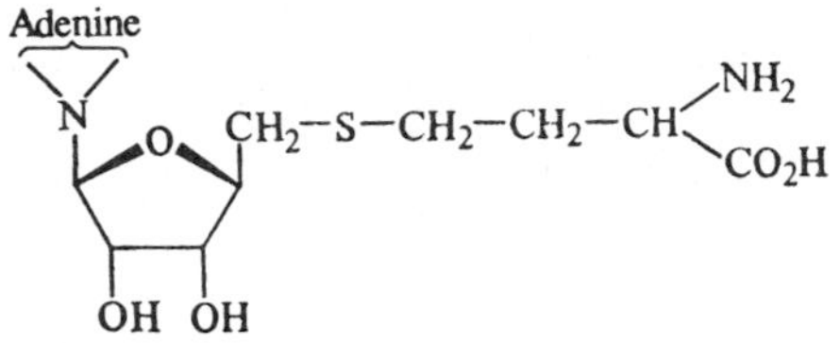

Adenine

S-Adenosyl-L-methionine

The main problems in the synthesis of adenosine and its derivatives involve attaching the sugar, ensuring that the sugar is in the stereochemically correct form, and selectively phosphorylating a sensitive molecule. The first approach was to attach the sugar before closing the five-membered ring of the purine, but it was later found to be far more efficient to attach the sugar to the completed purine. The latter synthesis is described.

(1) *Preparation of the appropriate sugar derivative from* D-*ribose* (A. R. Todd et al., *J. Chem. Soc.*, 1947, 1052).

Scheme 10(a)

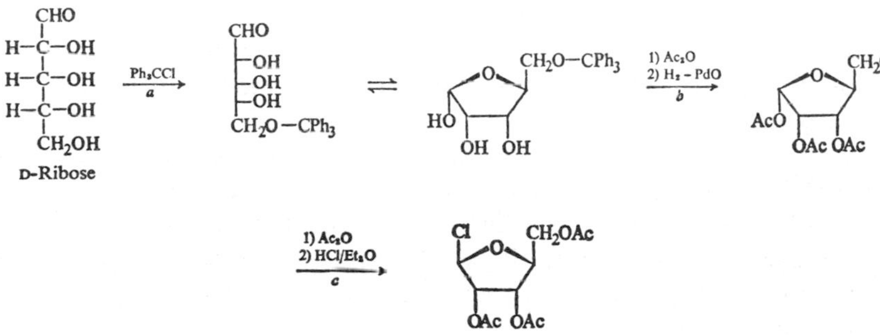

β-Chloro-2, 3, 5-triacetyl-
D-ribofuranose

a   Triphenylmethyl chloride reacts only with the *primary* alcoholic group (p. 627). This ensures that the sugar adopts the furanose ring-system rather than the pyranose system.

b   After acetylation of the remaining hydroxyl groups, the triphenylmethyl group is removed by hydrogenolysis (p. 627).

c   The tetra-acetylated sugar reacts with hydrogen chloride in ether at 0°C to give the β-chloro-derivative (Walden inversion).

(2) *Complete synthesis of adenosine from uric acid* (p. 699). (A. R. Todd *et al.*, *J. Chem. Soc.*, 1948, 967).

Scheme 10(b)

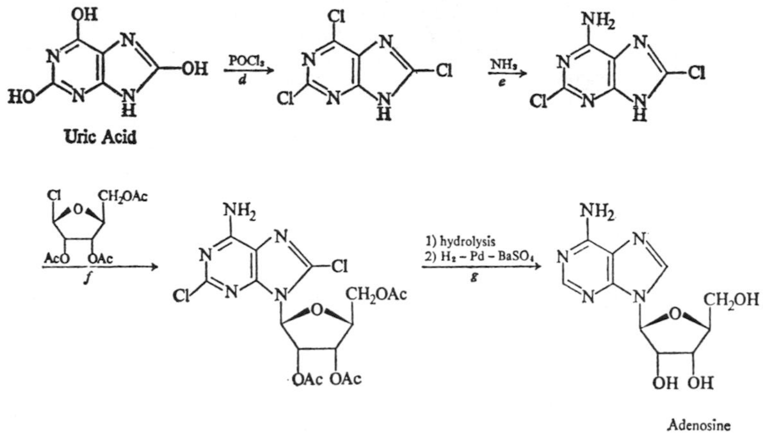

Uric Acid

Adenosine

d   The hydroxyl groups behave differently from those in phenol in being replaced by chlorine with phosphorus oxychloride (p. 630).

e   The chlorine atoms are all activated towards nucleophilic displacement by the heterocyclic nitrogen atoms (p. 426), but careful treatment leads only to the replacement of the 6-substituent.

f   The chloro-furanoside gives the β-furanosidyl-derivative of the purine, presumably as a result of the formation of an acetoxonium ion (p. 466) followed by an S_N2 reaction with inversion:

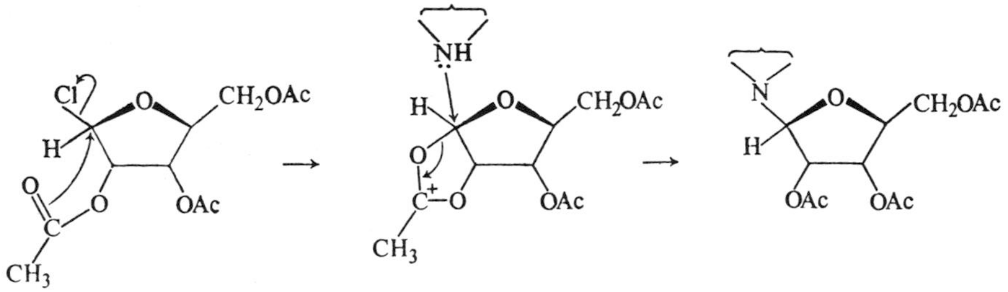

g   Hydrolysis of the acetyl groups, followed by hydrogenolysis of the C—Cl bonds on palladium, gives adenosine.

(3) *Adenosine triphosphate* (A. R. Todd *et al.*, *J. Chem. Soc.*, 1947, 648; 1949, 582). The phosphorylating agent, dibenzyl chlorophosphonate, is prepared by the chlorination of dibenzyl phosphite in carbon tetrachloride:

$$2\ PhCH_2OH + H_3PO_3 \longrightarrow HP(OCH_2Ph)_2 \xrightarrow{\ Cl_2\ } Cl—P(OCH_2Ph)_2$$

$$\overset{\|}{O} \qquad\qquad\qquad \overset{\|}{O}$$

Scheme 11

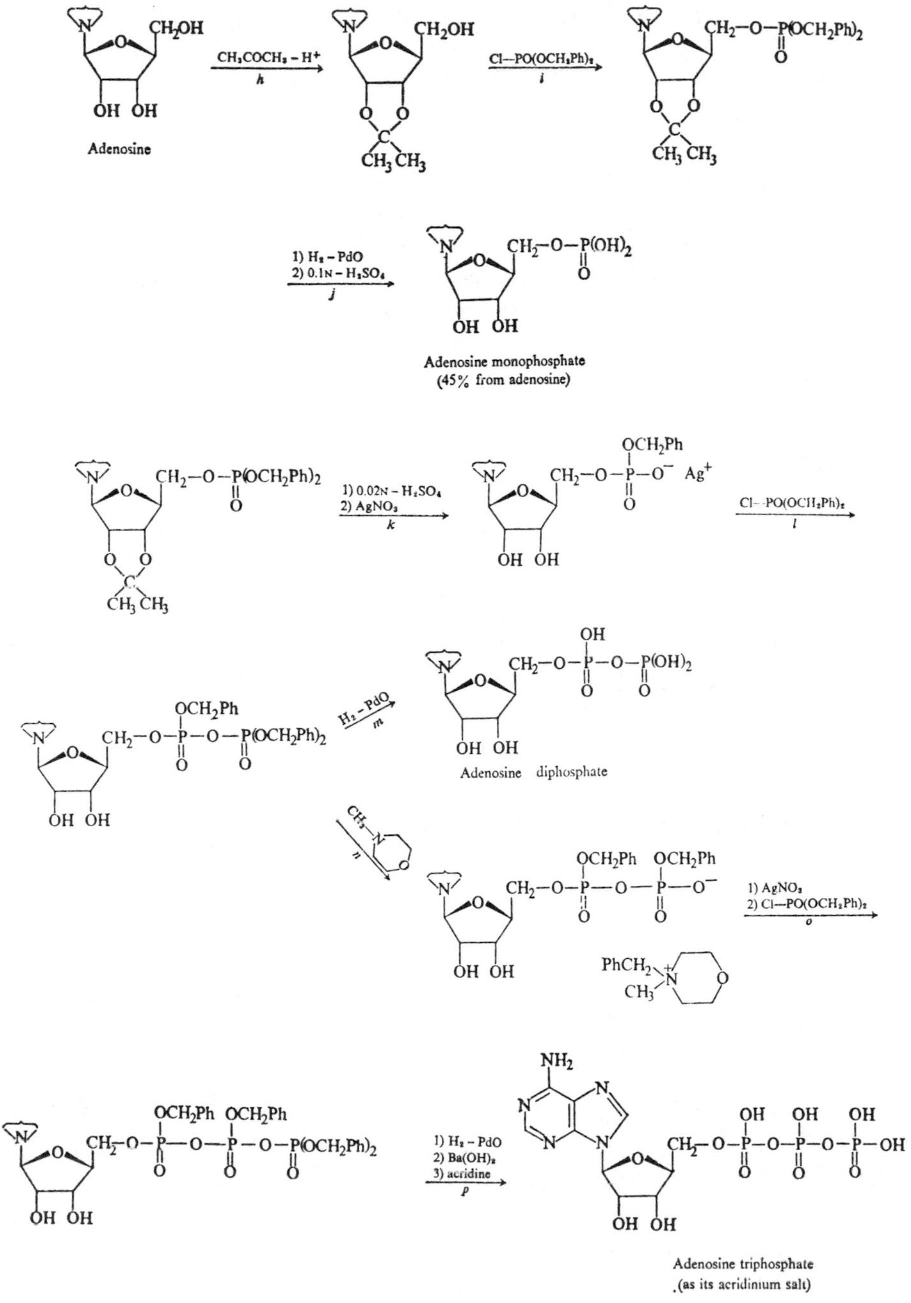
Adenosine
CH₃COCH₃ - H⁺
h
Cl—PO(OCH₂Ph)₂
i
1) H₂ - PdO
2) 0.1N - H₂SO₄
j
Adenosine monophosphate
(45% from adenosine)
1) 0.02N - H₂SO₄
2) AgNO₃
k
Cl—PO(OCH₂Ph)₂
l
H₂ - PdO
m
Adenosine diphosphate
n
1) AgNO₃
2) Cl—PO(OCH₂Ph)₂
o
1) H₂ - PdO
2) Ba(OH)₂
3) acridine
p
Adenosine triphosphate
(as its acridinium salt)

h The 2- and 3-hydroxyl groups of the furanose ring are protected against phosphorylation by formation of the isopropylidene-derivative, the establishment of a cyclic ketal from acetone being selective for *cis*-1,2-diols (p. 154).

i Phosphorylation is effected at a low temperature; pyridine is used as solvent to remove the hydrogen chloride.

j The benzyl groups are removed by catalytic hydrogenolysis (p. 625) and the isopropylidene group is removed with very dilute sulphuric acid. Removal of the acid as barium sulphate allows the monophosphate to crystallize out.

k The dibenzyl phosphate is hydrolyzed under very mild conditions to remove the isopropylidene group and at the same time one of the two benzyl groups. After removal of the acid as barium sulphate, the product is dissolved in alkali and precipitated as its silver salt.

l Phosphorylation in anhydrous acetic acid gives a resinous product which may be converted into both the di- and tri-phosphates.

m Hydrogenolysis, followed by purification of adenosine diphosphate as its acridinium salt.

n It was found that the tribenzyl ester could be selectively debenzylated with *N*-methylmorpholine.

o The product, isolated as its silver salt, is phosphorylated in a mixture of acetonitrile and phenol.

p The four benzyl groups are removed by hydrogenolysis and the product is precipitated as its barium salt, liberated with sulphuric acid, and isolated as its acridinium salt.

21.9 Penicillin

$$R-CONH \quad H \quad S$$
$$C-C \quad C(CH_3)_2$$
$$H \quad | \quad | \quad |$$
$$CO-N-CH$$
$$CO_2H$$

There are at least five naturally occurring penicillins, each based on the same structural nucleus (a β-lactam fused to a thiazolidine). They are produced from the mould *Penicillium notatum*, different strains of which produce different penicillins. The synthesis of penicillin V ($R = PhOCH_2-$) is described below; other penicillins contain $R = CH_3CH_2CH=CHCH_2-$ (F), $PhCH_2-$ (G), and n-heptyl (K). The penicillins owe their importance to their powerful effect on various pathogenic organisms. Another antibiotic, cephalosporin C, obtained from the mould *Cephalosporium* sp., also contains a β-lactam ring, but fused to a 1,3-thiazine ring-system.

The main difficulty in synthesizing the penicillins is to close the β-lactam ring in conditions which do not disrupt the rest of the sensitive molecule. This has been overcome by the carbodi-imide method (also widely used for making peptide bonds; p. 361). Penicillin V was synthesized by J. C. Sheehan and K. R. Henery-Logan (*J. Amer. Chem. Soc.*, 1959, **81**, 3089), a total synthesis being as follows.

(1) (+)-Penicillamine.

Scheme 12

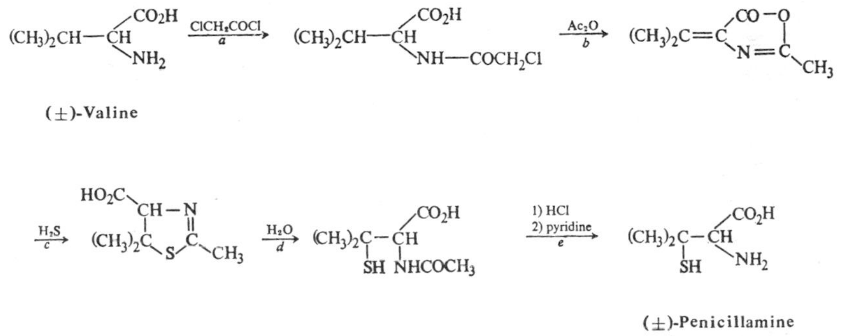

$$(CH_3)_2CH\!-\!\underset{NH_2}{\overset{CO_2H}{CH}} \xrightarrow[a]{ClCH_2COCl} (CH_3)_2CH\!-\!\underset{NH-COCH_2Cl}{\overset{CO_2H}{CH}} \xrightarrow[b]{Ac_2O} (CH_3)_2C\!=\!\underset{N=C}{\overset{CO-O}{C}}\!\underset{CH_3}{\big|}$$

(±)-Valine

$$\xrightarrow[c]{H_2S} (CH_3)_2\underset{S}{\overset{HO_2C\diagdown CH-N}{C}}\underset{C}{\diagup}_{CH_3} \xrightarrow[d]{H_2O} (CH_3)_2C\!-\!\underset{SH\ NHCOCH_3}{\overset{CO_2H}{CH}} \xrightarrow[e]{\substack{1)\ HCl \\ 2)\ pyridine}} (CH_3)_2C\!-\!\underset{SH}{\overset{CO_2H}{CH}}\!\underset{NH_2}{}$$

(±)-Penicillamine

a Acylation. The —COCl group is more reactive towards nucleophiles than —CH₂Cl.

b Acetic anhydride induces dehydrative cyclization, and this is accompanied by the elimination of hydrogen chloride and a prototropic shift:

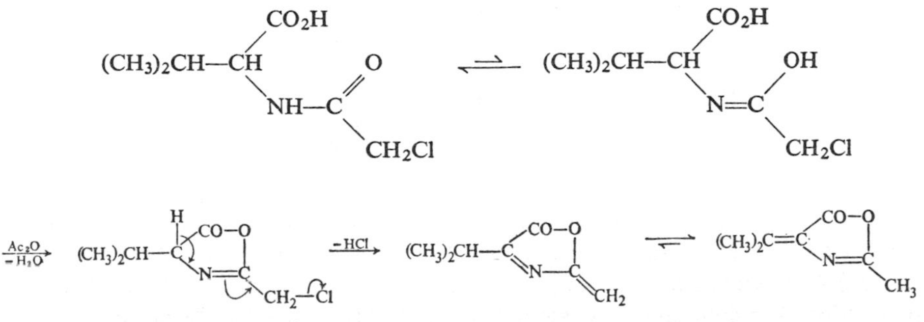

$$(CH_3)_2CH\!-\!\underset{NH-C}{\overset{CO_2H}{CH}}\underset{CH_2Cl}{\overset{O}{\diagup}} \rightleftharpoons (CH_3)_2CH\!-\!\underset{N=C}{\overset{CO_2H}{CH}}\underset{CH_2Cl}{\overset{OH}{\diagup}}$$

$$\xrightarrow[-H_2O]{Ac_2O} (CH_3)_2CH\!-\!\overset{H}{\underset{N=C}{C}}\underset{CH_2-Cl}{\overset{CO-O}{\diagup}} \xrightarrow{-HCl} (CH_3)_2CH\!-\!\underset{N-C}{\overset{CO-O}{C}}\underset{CH_2}{} \rightleftharpoons (CH_3)_2C\!=\!\underset{N=C}{\overset{CO-O}{C}}\underset{CH_3}{}$$

c The azlactone is cleaved by hydrogen sulphide and the resulting thiol cyclizes *via* Michael-type addition to the $\alpha\beta$-unsaturated acid:

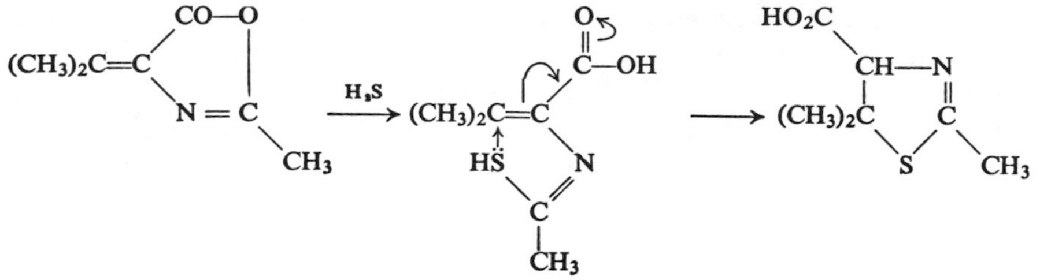

$$(CH_3)_2C=C \overset{CO-O}{\underset{N=C}{\big|}} \quad \overset{H_2S}{\longrightarrow} \quad (CH_3)_2C=C \quad \longrightarrow \quad (CH_3)_2C \overset{HO_2C}{\underset{S}{\overset{CH-N}{\big|}}}$$

d   The thiazoline ring is opened with boiling water.

e   The *N*-acetyl group is removed by acid hydrolysis followed by basification. (±)-Penicillamine may be resolved by formylation (HCO₂H), separation of the diastereoisomers formed with brucine, and hydrolysis (HCl followed by pyridine).

(2) Penicillin V.

Scheme 13

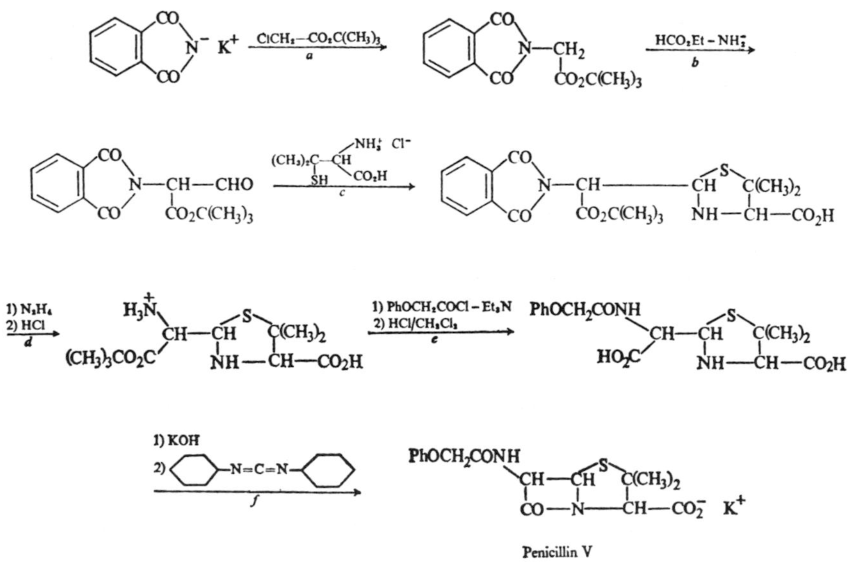

Penicillin V

a   Gabriel's synthesis (p. 326).

b   A crossed Claisen condensation, standard procedure for the formylation of activated methylene groups (p. 245).

c   Condensation with penicillamine hydrochloride occurs at room temperature in sodium acetate buffer.

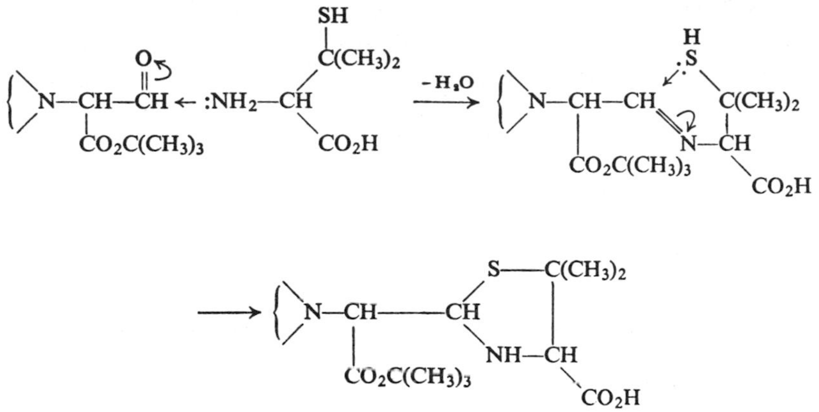

$$
\begin{array}{c}
\text{SH} \\
| \\
\text{C(CH}_3)_2
\end{array}
$$

$\text{N}-\text{CH}-\text{CH} \leftarrow :\text{NH}_2-\text{CH} \quad \xrightarrow{-\text{H}_2\text{O}}$

with $\text{CO}_2\text{C(CH}_3)_3$ and CO_2H substituents, giving the thiazolidine

$$\longrightarrow \quad \text{N}-\text{CH}-\text{CH} \quad \text{with } \text{S}-\text{C(CH}_3)_2, \text{ NH}-\text{CH}, \text{ CO}_2\text{C(CH}_3)_3, \text{ CO}_2\text{H}$$

Yields are higher if (+)- rather than (±)-penicillamine is used, but, as expected, a mixture of diastereoisomers is obtained. There are four possible isomers (due to two new asymmetric carbon atoms), but only two (α and γ) appear to be formed in significant amounts. The major product is the unwanted γ-isomer, but this may be epimerized by boiling in pyridine under hydrogen, giving an equilibrium mixture of 3 parts γ:1 part α. The required α-isomer is isolated and the process repeated on the γ-isomer.

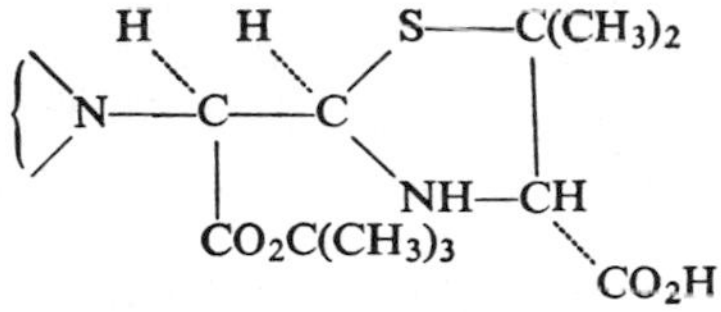

α-isomer structure: $\text{N}-\text{C}-\text{C}$ with H, H, $\text{S}-\text{C(CH}_3)_2$, $\text{NH}-\text{CH}$, $\text{CO}_2\text{C(CH}_3)_3$, CO_2H.

α-isomer

d   The phthalimido group must be removed in conditions which do not cleave the t-butyl ester. This is achieved with hydrazine at 13°C for 3 hours and then at room temperature for one day, followed by treatment with HCl in acetic acid.

e   Acylation of the amino group in the presence of triethylamine is carried out at 25°C so that the thiazolidine ring is not opened. The t-butyl ester is then hydrolyzed by treatment with hydrogen chloride in methylene chloride at 0°C.

f   The potassium salt is cyclized with dicyclohexyl carbodi-imide (p. 361), giving the potassium penicillinate. Penicillin V can be extracted after acidification with phosphoric acid, and it crystallizes from aqueous solution at pH 6·8. It may be purified by counter-current extraction.

21.10 Chlorophyll *

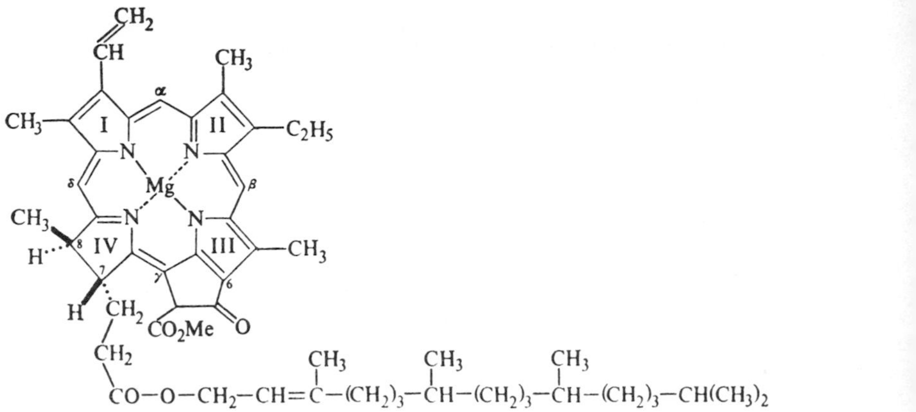

CH2
CH
CH3
α
CH3
I
II
C2H5
N
N
Mg
δ
β
CH3
N
N
H
8
IV
III
CH3
7
γ
6
H
CH2
CO2Me
O
CH2
CH3
CH3
CH3
CO—O—CH2—CH=C—(CH2)3—CH—(CH2)3—CH—(CH2)3—CH(CH3)2

Chlorophyll is a dihydroporphyrin contained in plants and concerned in the photochemical process. There are in fact two chlorophylls of which that shown above is chlorophyll-*a*; chlorophyll-*b* differs in having a formyl group in place of methyl in ring II.

The synthesis described was carried out by R. B. Woodward *et al.* (*J. Amer. Chem. Soc.*, 1960, **82**, 3800; 'The Chemistry of Natural Products,' I.U.P.A.C. Symposium, 1960, p. 383). They based their approach on observations relating to the structure and properties of chlorophyll. In particular, dihydroporphyrins are normally very easily oxidized to porphyrins but chlorophyll is not; the inference is that resistance to removal of the hydrogen atoms at C_7 and C_8 stems from the fact that in the oxidized system the alkyl groups at these positions would eclipse and strongly repel each other; in the reduced form (*trans*) the staggered arrangement reduces the steric strain. In addition, it is apparent from a consideration of bond lengths that the unsaturated ketone ring connecting C_6 and C_γ must be very strained, whereas the ring is opened and closed with ease, the inference being that ring-closure relieves the strain between the C_γ and C_7 substituents. The application of these observations in the synthesis is clear from the discussion below.

The Woodward synthesis was completed with the formation of chlorin e_6 trimethyl ester in step *s*, for this had previously been converted into chlorophyll-*a* in three steps. One of these involves the introduction of phytol which has been synthesized from (+)-citronellol as follows:

* At the time of the synthesis, the absolute configuration of chlorophyll was not known. It has since been determined as the enantiomer of the structure given here. For consistency with the papers by Woodward *et al.* which are referred to, their structures are retained in this section—the relative configurations are of course correct. The absolute configurations at C_7 and C_8 and at the carboxymethyl-bearing carbon are given correctly on the cover of this book.

Scheme 14

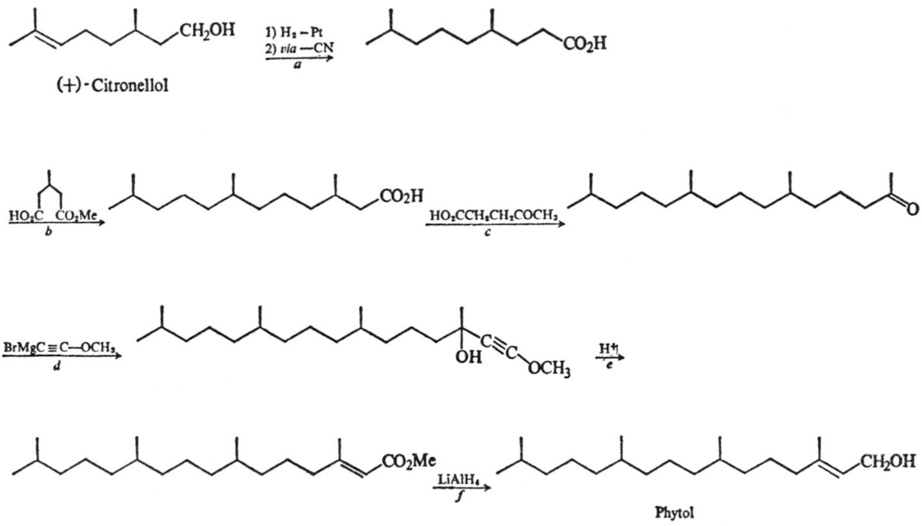

(+)-Citronellol

$$\text{CH}_2\text{OH} \xrightarrow[\text{2) via —CN}]{\text{1) H}_2\text{-Pt}} \quad (a) \quad \text{CO}_2\text{H}$$

$$\xrightarrow[\text{HO}_2\text{C} \quad \text{CO}_2\text{Me}]{(b)} \quad \text{CO}_2\text{H} \xrightarrow[\text{HO}_2\text{CCH}_2\text{CH}_2\text{COCH}_3]{(c)} \quad \text{O}$$

$$\xrightarrow[\text{BrMgC}\equiv\text{C—OCH}_3]{(d)} \quad \underset{\text{OH}}{\overset{\text{C}\equiv\text{C—OCH}_3}{}} \xrightarrow[\text{e}]{\text{H}^+}$$

$$\text{CO}_2\text{Me} \xrightarrow[f]{\text{LiAlH}_4} \quad \text{CH}_2\text{OH}$$

Phytol

a   Reduction of the olefinic bond, followed by ascent of the homologous series through the nitrile (p. 258).

b   Crossed anodic coupling with (+)-methyl hydrogen β-methylglutarate (p. 553).

c   A similar reaction with laevulic acid.

d   Standard procedure using the Grignard reagent from methoxyacetylene (p. 209).

e   Acid-catalyzed rearrangement:

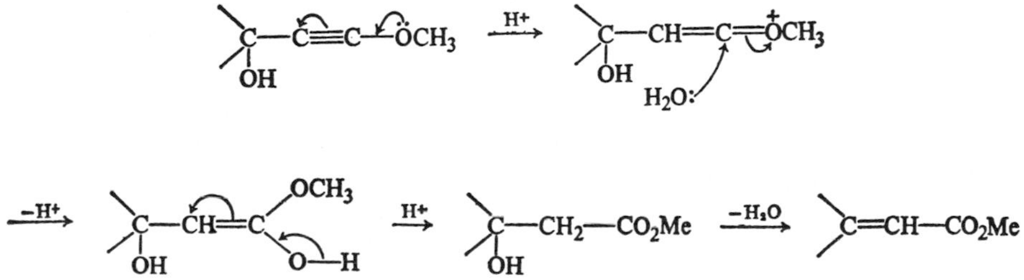

$$\underset{\text{OH}}{\overset{}{\text{C}}}-\text{C}\equiv\text{C}-\text{OCH}_3 \xrightarrow{\text{H}^+} \underset{\text{OH}}{\overset{}{\text{C}}}-\text{CH}=\text{C}=\overset{+}{\text{O}}\text{CH}_3 \quad \underset{\text{H}_2\text{O:}}{}$$

$$\xrightarrow{-\text{H}^+} \underset{\text{OH}}{\overset{}{\text{C}}}-\text{CH}=\text{C}\overset{\text{OCH}_3}{\underset{\text{O—H}}{}} \xrightarrow{\text{H}^+} \underset{\text{OH}}{\overset{}{\text{C}}}-\text{CH}_2-\text{CO}_2\text{Me} \xrightarrow{-\text{H}_2\text{O}} \text{C}=\text{CH}-\text{CO}_2\text{Me}$$

f   The cis and trans isomers are separated after reduction; the (—)-trans-isomer is the naturally occurring compound.

Synthesis of chlorophyll-a.

Scheme 15

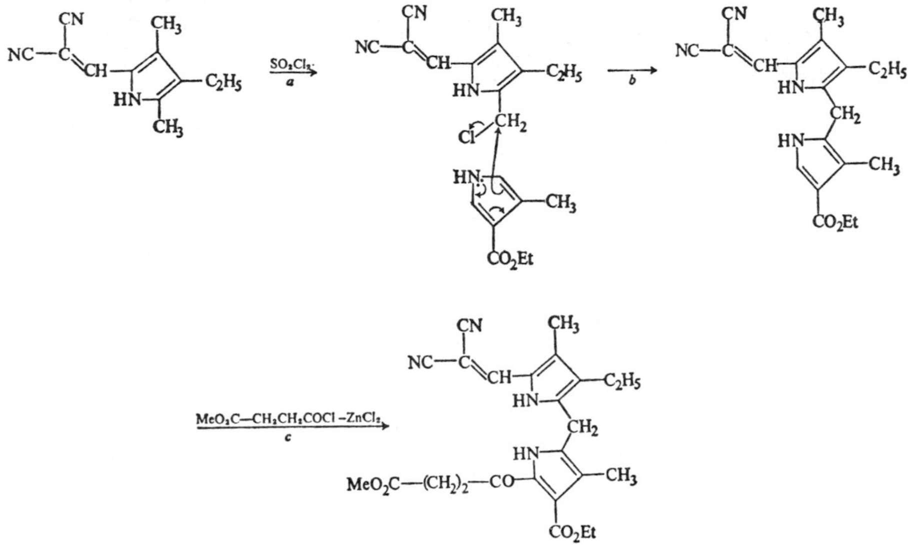

MeO₂C—CH₂CH₂COCl –ZnCl₂,
c

MeO₂C—(CH₂)₂—CO

a The free-radical chlorination of the benzylic-type methyl group (p. 543) occurs rapidly in acetic acid at 55°C; the α- is more activated than the β-position.

b The Friedel-Crafts alkylation takes place readily on the activated pyrrole nucleus (p. 380). Of the two 2-positions available, that adjacent to the methyl-substituent is the more reactive, first, because the electron-releasing methyl group is better able to stabilize the positive charge in the transition state than the electron-attracting estèr group,

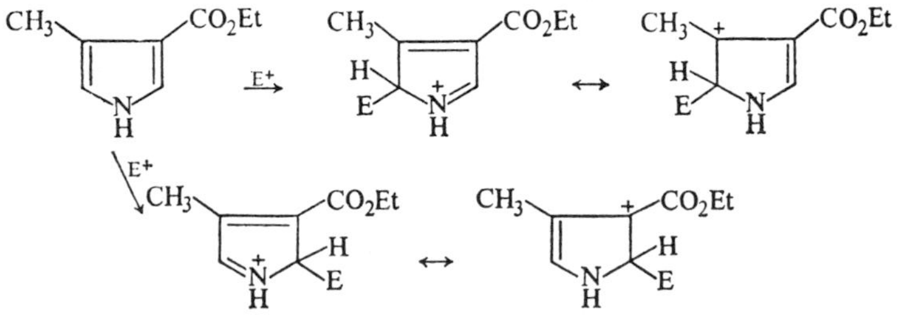

and secondly, because the carboethoxyl group probably provides more hindrance to the reagent than the methyl group.

c This Friedel-Crafts acylation, catalyzed by a mild Lewis acid, takes place at the only unsubstituted nuclear position.

Scheme 15 (cont.)

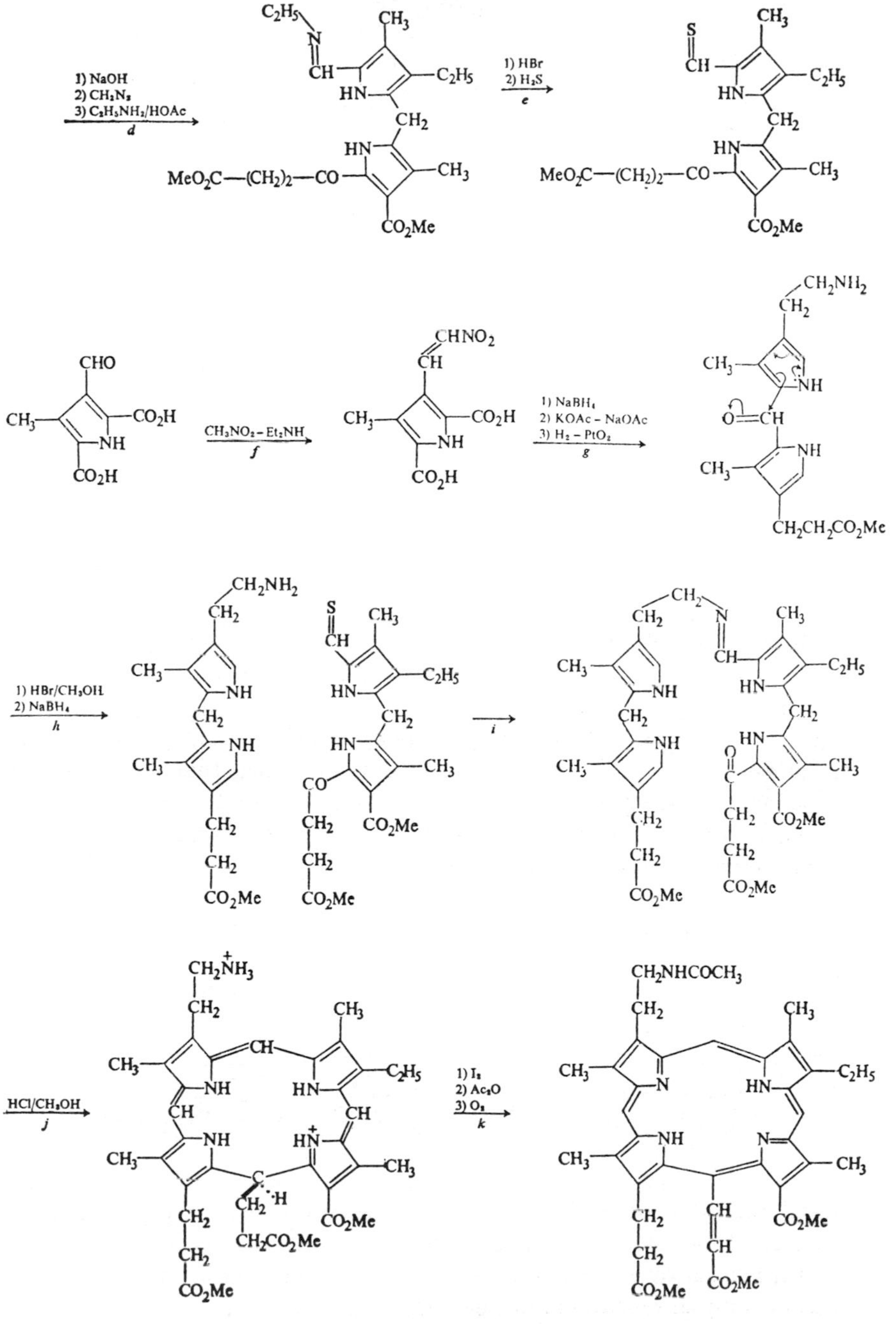

d   The dicyanovinyl group has acted until this point as a protective group for the aldehydic function required at this carbon atom. The group is removed with 33% sodium hydroxide solution by the reverse-aldol reaction (p. 228), use being made of the anion-stabilizing influence of the cyano groups.

$$-CH=C\begin{smallmatrix}C\equiv N\\\\C\equiv N\end{smallmatrix} \quad\overset{OH^-}{\rightleftharpoons}\quad -CH-C\begin{smallmatrix}N^-\\\parallel\\C\\\\CN\end{smallmatrix}$$

$$\rightleftharpoons \quad -CH-CH\begin{smallmatrix}CN\\\\CN\end{smallmatrix} \quad\longrightarrow\quad -CHO + \bar{C}H(CN)_2$$

This procedure also hydrolyzes the ester groups which are then re-esterified with diazomethane. The aldehyde is converted with ethylamine into a Schiff base, preparatory to conversion into the thioaldehyde in the next step.

e   The Schiff base is converted into the thioaldehyde for the following reason. Two dipyrrylmethanes are to be combined (step *j*). Reaction must be specific; that is, the molecules must be oriented as in the representation preceding step *i* and not in the alternative manner. To ensure this, it was hoped to form a Schiff base as in step *i* before coupling the dipyrryl-methenes. The first attempt used an aldehyde group for formation of the Schiff base, but pyrrole aldehydes are unreactive towards nucleophiles such as amines as a result of delocalization,

The reactivity of the carbonyl group is increased by acids, but in this case even the mild acid conditions required (triethylammonium acetate) degraded the dipyrrylmethane. However, the thioaldehyde is much more reactive and proved suitable in this instance.

f   Base-catalyzed condensation involving the activated methyl group of nitro-methane (p. 236).

g   The olefinic bond is activated by the nitro group to nucleophiles and is reduced by borohydride (p. 619). Fusion of the product with a mixture of sodium and potassium acetates at 120°C effects decarboxylation. The nitro group is then reduced catalytically.

h   Hydrogen bromide catalyzes the Friedel-Crafts reaction which joins the

Scheme 15 (cont.)

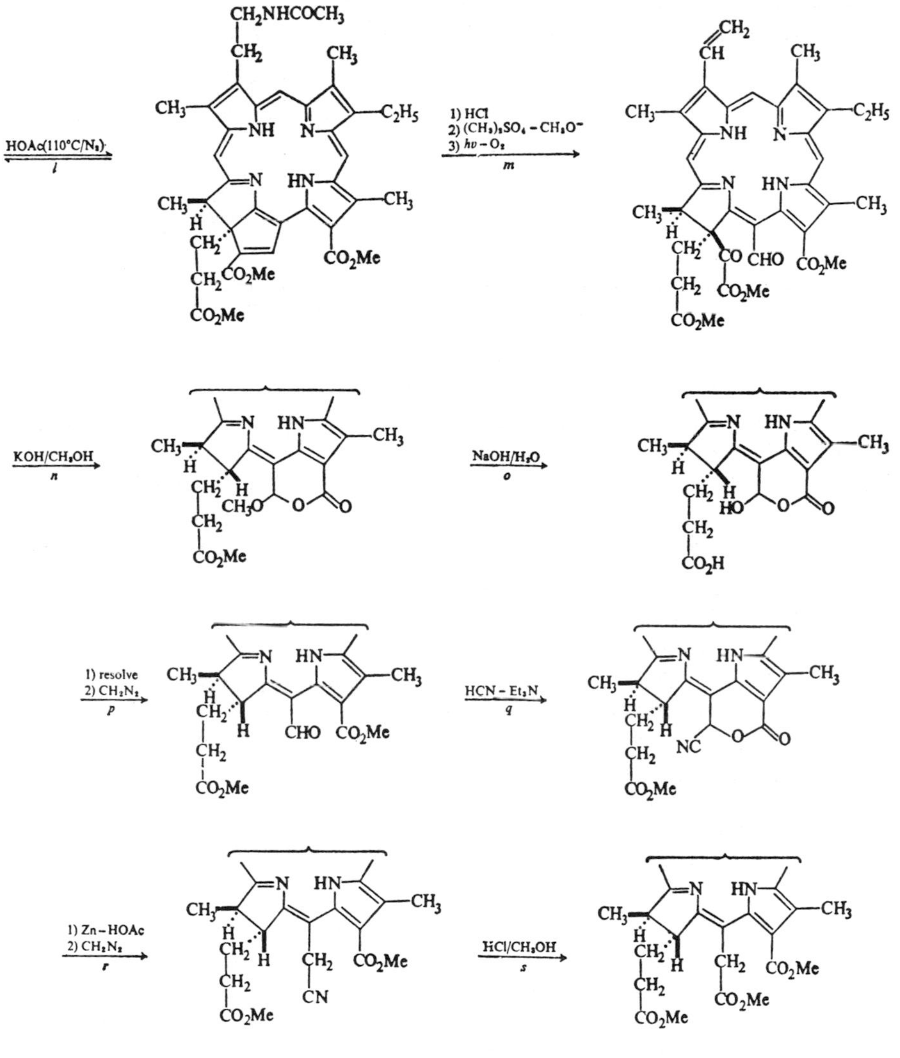

HOAc(110°C/N₂) *l*

1) HCl
2) (CH₃)₂SO₄ – CH₃O⁻
3) hv – O₂
m

KOH/CH₃OH *n*

NaOH/H₂O *o*

1) resolve
2) CH₂N₂
p

HCN – Et₃N *q*

1) Zn – HOAc
2) CH₂N₂
r

HCl/CH₃OH *s*

Chlorin e₆ trimethyl ester

Scheme 15 (*cont.*)

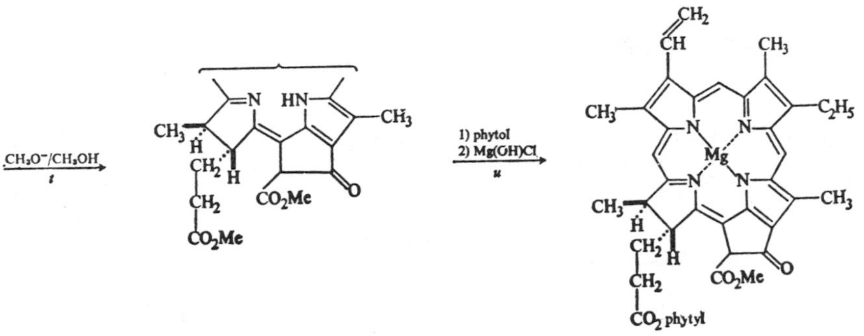

Chlorophyll-*a*

two pyrrole rings; reaction occurs at the 2-position shown rather than at the alternative 2-position as a result of the steric effect. The dipyrrylmethanol is readily reduced by borohydride.

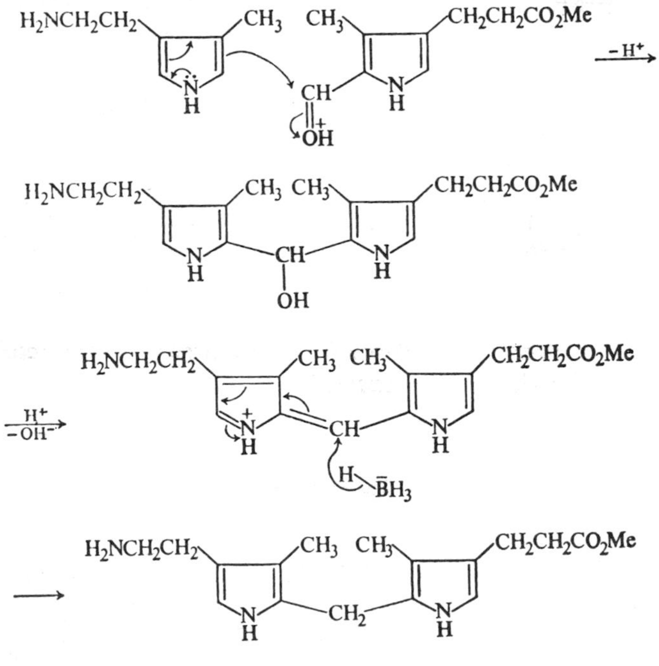

The dipyrrylmethane formed is a labile compound which is extracted into methylene chloride and used immediately in step *i*.

i   Formation of the Schiff base (see *e*).

j 12M-Hydrogen chloride in methanol brings about a series of transformations. The bridges are formed as follows:

(1) Lower bridge:

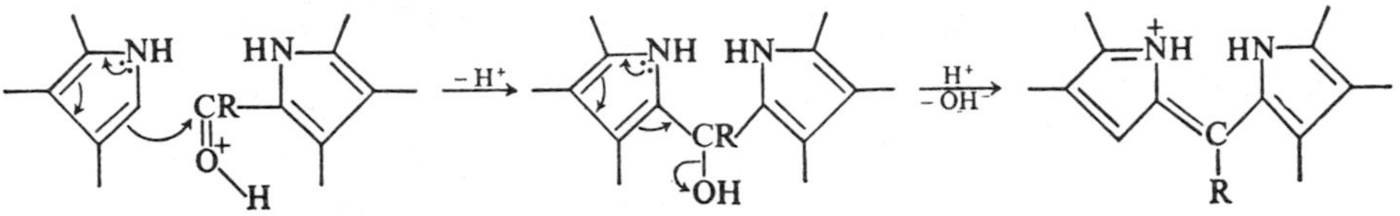

$(R = CH_2CH_2CO_2Me)$

(2) Upper bridge:

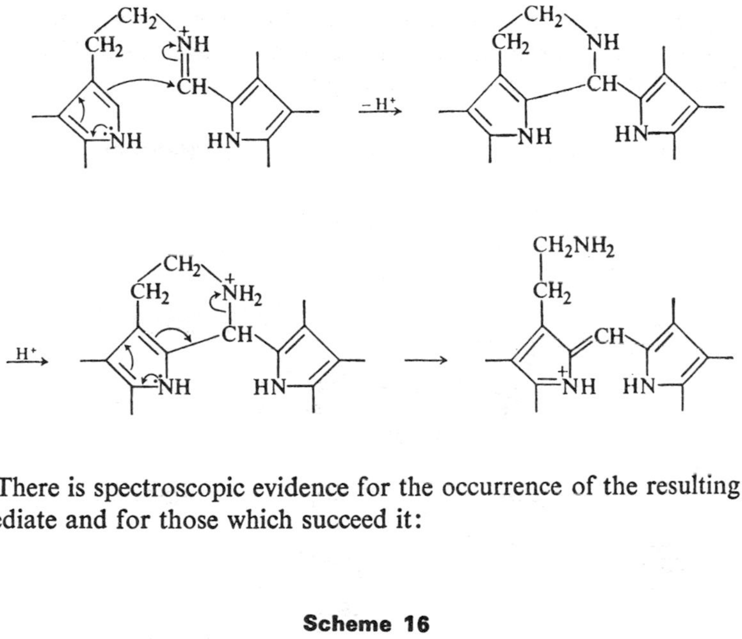

There is spectroscopic evidence for the occurrence of the resulting intermediate and for those which succeed it:

Scheme 16

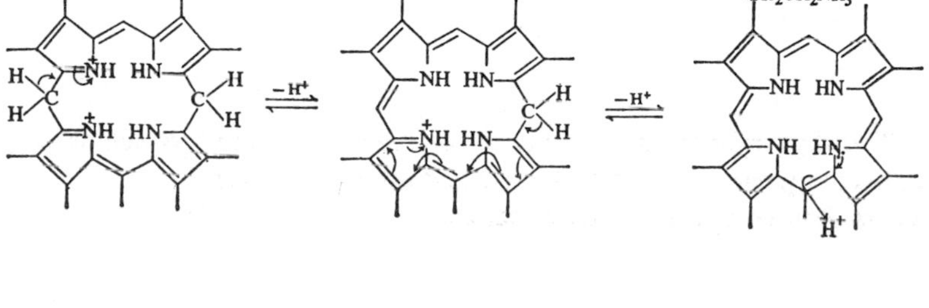

Scheme 16 (*cont.*)

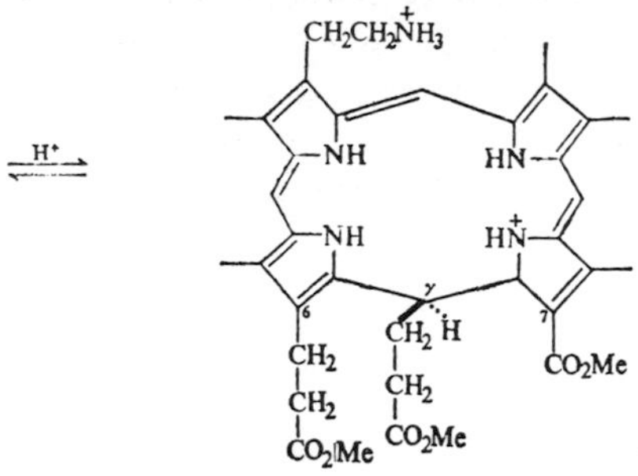

Of the various cations which exist in equilibrium, all save the last contain strong steric compression forces owing to the eclipsing of the three substituents at C_6, C_y and C_7. In the last cation, however, the tetrahedral nature of C_y allows these substituents to assume a staggered relationship to one another, and as a result this is the most stable of the cations. It is not necessary to separate the cations.

k   An excess of iodine is added to the reaction mixture to remove by oxidation two nuclear hydrogen atoms; the amino group is acetylated by the standard procedure (acetic anhydride in pyridine solution), and the C_y-methoxy-carbonylethyl side-chain is dehydrogenated in warm acetic acid in the presence of air.

l   The temperature of the acetic acid solution is raised to 110°C and air is excluded to prevent further oxidation. These conditions establish equilibrium with the dihydroporphyrin (a purpurin), for which $K \sim 2$; the product is isolated and the reactant is recycled to increase the yield.

     The reaction involves electrophilic attack by the activated olefinic bond at the β-position of the pyrrole nucleus:

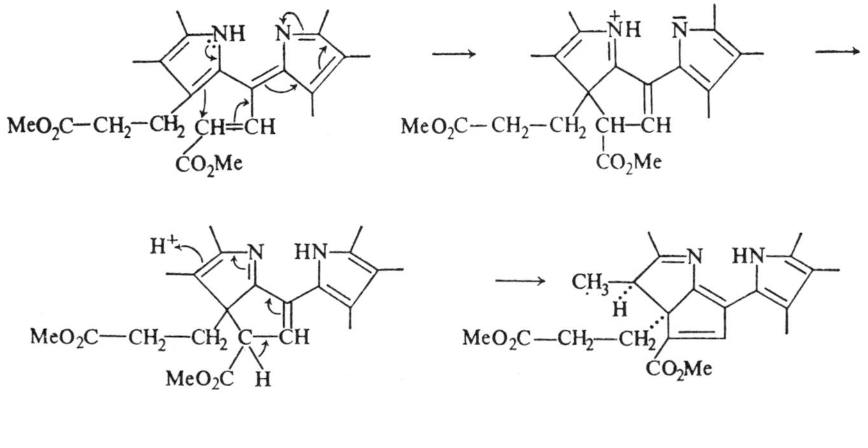

A further important aspect of the reaction is that the uptake of the proton at the last stage is from that direction which results in the methyl and methoxycarbonylethyl groups in ring IV being *trans* to one other, for the alternative product is more severely strained.

m   Hydrogen chloride in methanol removes the protective acetyl group, and the resulting amine is submitted to Hofmann elimination, yielding the required vinyl-substituent. Strong illumination with visible light in the presence of air then oxidatively cleaves the five-membered ring formed in *l*, the driving force for the reaction being derived partly from the release of the ring-strain.

n   The methoxalyl-substituent at C_7 is removed with base. This reaction is no doubt facilitated by the presence of the electron-accepting nitrogen atom in ring II, i.e.

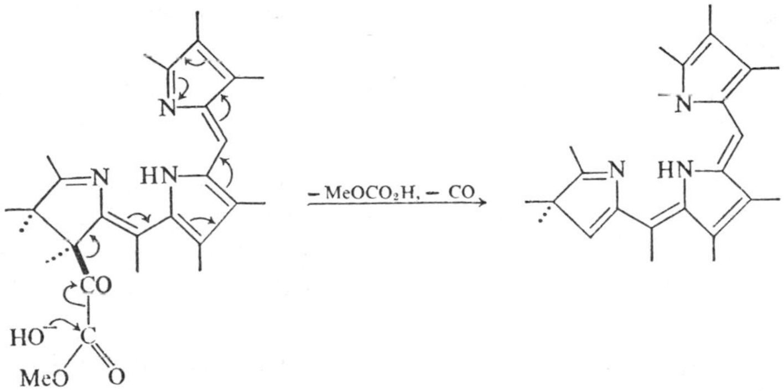

$$- \text{MeOCO}_2\text{H}, - \text{CO}$$

Again, the final protonation occurs so as to give the more stable stereo-isomer (see *l*).

In addition, the base induces lactonization across C_6 and C_γ:

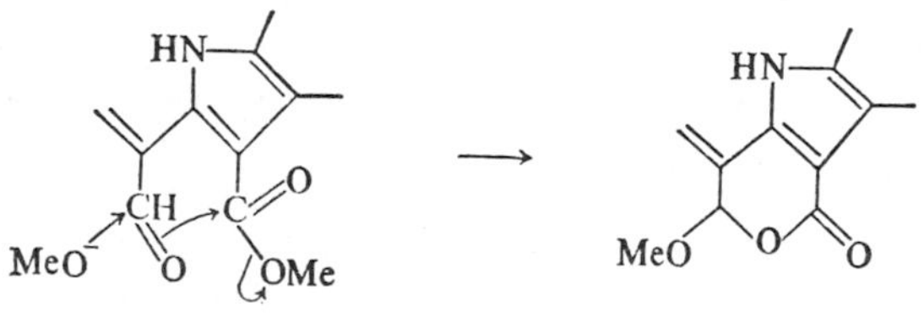

o   Very dilute sodium hydroxide solution hydrolyzes the methyl ester and, by the reverse of the lactonization in step *n* and re-lactonization, gives the hydroxy-containing lactone.

p   The optical isomers are resolved (the diastereoisomers formed with quinine were used). Diazomethane methylates the carboxylic acid and also opens the lactone by reacting with the open-chain tautomer:

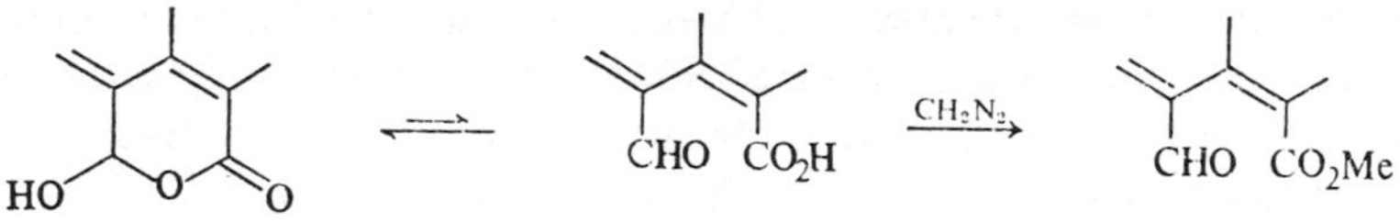

q   Cyanide ion forms a cyano-lactone in the manner of the methoxy-lactoniza-
tion in step n.

r   The lactone is opened reductively and the resulting acid is methylated.

s   Methanolic hydrogen chloride converts the cyano group into carbo-
methoxyl.

t   Dieckmann cyclization to give a five-membered cyclic β-keto-ester (p. 241).

u   The phytyl residue is introduced in place of methyl (alcoholysis) at the less
hindered of the two ester groups and the magnesium atom is inserted.

21.11 Prostaglandins E₂ and F₂α

The prostaglandins are a series of closely related hormones which are derivatives of 'prostanoic acid':

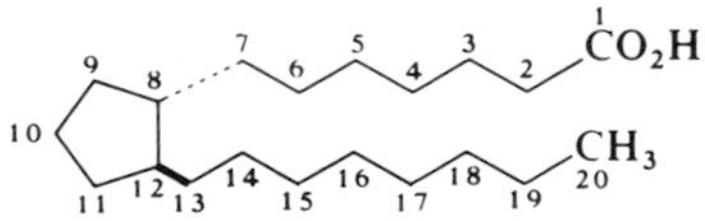

They are present in many mammalian tissues at very low concentrations; one
of the richest known sources is human seminal plasma which contains at least
thirteen different prostaglandins at a total concentration of about 300 $\mu g \, cm^{-3}$.
Prostaglandins have potent effects on various kinds of smooth muscle and they
are of considerable potential medical interest for the control of hypertension,
for gynaecological purposes, and as abortifacients.

Prostaglandins E_2 and $F_{2\alpha}$ are two of the six primary prostaglandins. In
general, the E series of prostaglandins have a β-hydroxy-ketone structure in the
ring and differ in the degree of unsaturation in the side-chains, whereas F series
prostaglandins have a β-dihydroxy function in the ring and likewise differ in the
extent of unsaturation in the side-chains.

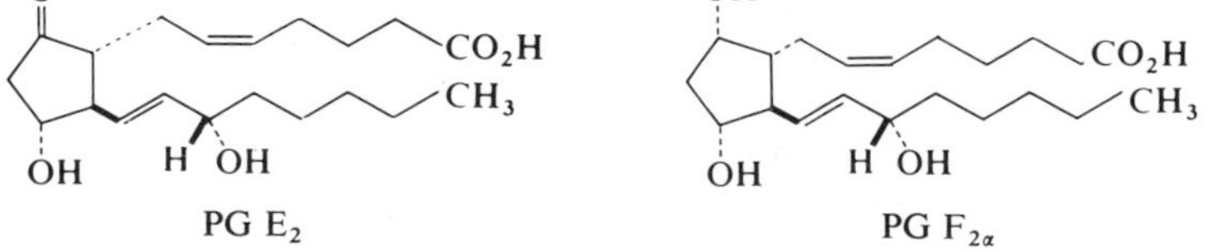

The synthesis of E series prostaglandins poses a particular problem in that the
β-hydroxy-ketone function is especially prone to dehydration (p. 119). (Such
dehydration gives rise to series A prostaglandins, which readily rearrange to

series B prostaglandins which are fully conjugated and have the ring-olefinic bond between carbon atoms 8 and 12.) Further challenges in the syntheses include the control of olefin diastereoisomerism in the side-chains and of the configuration at C-15. The synthesis given below is that of E. J. Corey and co-workers; the evolution of the synthesis from its original form to that of 1975 is discussed (*J. Amer. Chem. Soc.*, 1969, **91**, 5675; *ibid.*, 1971, **93**, 1489; 1491; *ibid.*, 1972, **94**, 8616; *ibid.*, 1975, **97**, 6908).

Scheme 17

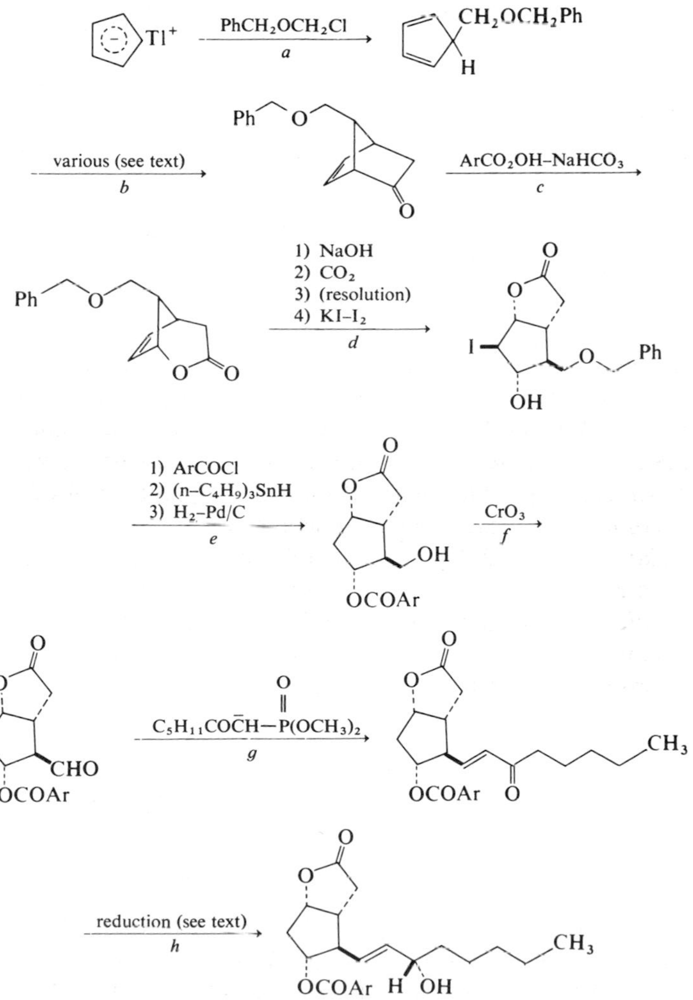

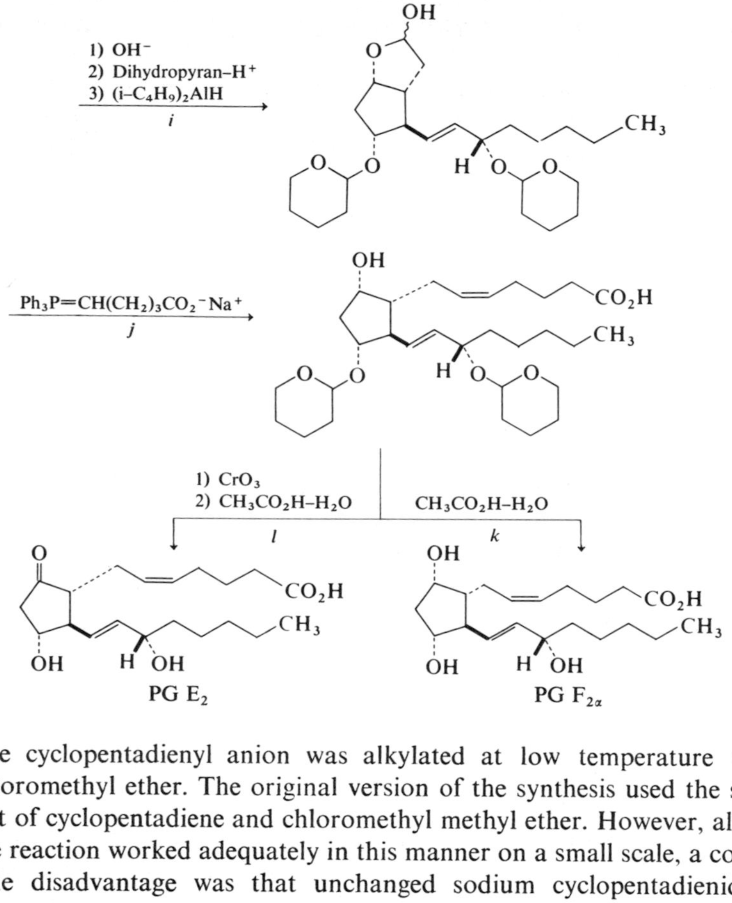

1) OH⁻
2) Dihydropyran–H⁺
3) (i-C₄H₉)₂AlH
i
OH
O
O
O
H
O
O
CH₃
Ph₃P=CH(CH₂)₃CO₂⁻Na⁺
j
OH
CO₂H
CH₃
O
O
H
O
O
1) CrO₃
2) CH₃CO₂H–H₂O
l
CH₃CO₂H–H₂O
k
O
CO₂H
CH₃
OH
H OH
OH
CO₂H
CH₃
OH
H OH

PG E_2 PG $F_{2\alpha}$

a The cyclopentadienyl anion was alkylated at low temperature with a chloromethyl ether. The original version of the synthesis used the sodium salt of cyclopentadiene and chloromethyl methyl ether. However, although the reaction worked adequately in this manner on a small scale, a considerable disadvantage was that unchanged sodium cyclopentadienide was sufficiently basic to catalyze isomerization of the product even at 0°C, and the longer manipulation times necessary for work on a larger scale caused this isomerization to be troublesome:

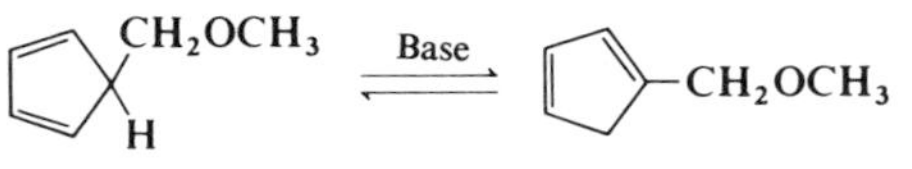

CH₂OCH₃
H
Base
CH₂OCH₃

The problem was overcome by use of the thallium(I) derivative of cyclopentadiene, prepared from the diene with Tl_2SO_4 and KOH in water. An additional advantage was that the thallium(I) cyclopentadienide was capable of manipulation in air. The success deriving from the use of thallium is presumably attributable to the thallium derivative's having more covalent character than the sodium derivative, so that the cyclopentadienyl moiety is correspondingly less basic. Subsequent syntheses used benzyl chloromethyl ether as the alkylating agent, which had the advantage that subsequent debenzylation is more easily accomplished than demethylation (see step *e*).

b   This step was one which underwent the most change during development of the synthesis. Essentially it is a cycloaddition reaction. However, the apparent dienophile in the reaction is keten. As was seen earlier (p. 300), this participates more readily in [2 + 2] than in [4 + 2] cycloadditions, so that an equivalent synthon had to be found. The first dienophile used was 2-chloroacrylonitrile:

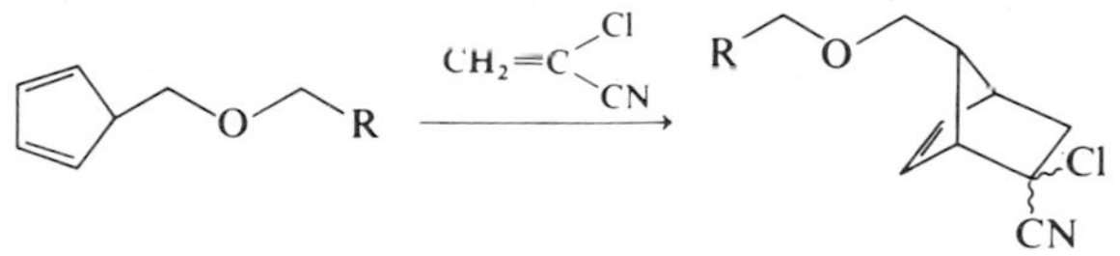

This dienophile was found to be insufficiently reactive in the [4 + 2] mode at temperatures where the isomerization of the precursor (this time by 1,5-sigmatropic rearrangement) did not occur, but the elegant device of catalyzing the Diels-Alder reaction to give it advantage over the competitive pericyclic reaction was successful. The best catalyst was $Cu(BF_4)_2$; it brought about the desired coupling in tetrahydrofuran at $-55°C$ in yields in excess of 80%.

The product of the reaction was a mixture of stereoisomers (both Cl and CN *endo*, and the side-chain mainly *syn* to the double bond) which was efficiently hydrolyzed to the required bicyclic ketone by KOH in aqueous dimethyl sulphoxide.

The second variant in accomplishing step *b* involved a change of dienophile, for 2-chloroacrylonitrile is an unusually hazardous material, undesirable for large-scale work. A preferable alternative was found to be 2-chloroacryloyl chloride. This dienophile is of higher reactivity than its predecessor, so that no catalyst was necessary:

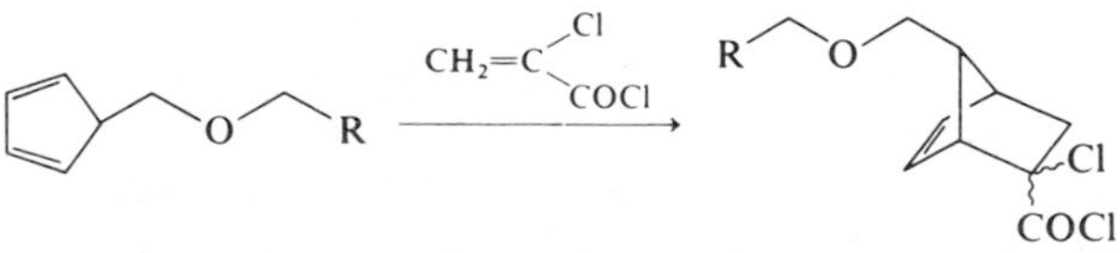

Again the product was a mixture of stereoisomers in respect of the *endo/exo* positions of Cl and COCl, but the side-chain was now exclusively *syn* to the double bond. The conversion into the bicyclic ketone was brought about by Curtius rearrangement (p. 469) and hydrolysis:

$$\xrightarrow{\text{NaN}_3} \quad \left\{ \begin{array}{c} \text{Cl} \\ >\text{C}< \\ \text{CON}_3 \end{array} \right\} \quad \xrightarrow[\text{HOAc–H}_2\text{O}]{\text{heat}} \quad \left[\left\{ \begin{array}{c} \text{Cl} \\ >\text{C}< \\ \text{NH}_2 \end{array} \right\} \right]$$

$$\xrightarrow{-\text{HCl}} \quad \left[\left\{ >\text{C}=\text{NH} \right\} \right] \quad \xrightarrow{\text{H}_2\text{O}}$$

Although this variant of the reaction involves more formal steps than its predecessor, the conversion of cyclopentadiene into the bicyclic ketone was accomplished in greater than 90 % yield, without isolation of intermediates.

The bicyclic ketone contains three chiral centres and the relative stereochemistry at these is governed by the suprafacial character of the Diels-Alder reaction on the less hindered side of the diene. Thus the product is a single diastereoisomer at this stage; it is, however, racemic. The early variants of the synthesis included a resolution step later in the sequence (step d); however, this 'wastes' half the material so far synthesized. The final variant of the synthesis overcame this disadvantage and involved a chiral synthesis of the bicyclic ketone. In order to achieve this, the dienophile for reaction with 5-benzyloxymethylcyclopentadiene was changed again, this time to an acrylate ester:

$$\xrightarrow[]{\quad \text{CH}_2=\text{C} \begin{smallmatrix} \text{H} \\ \text{CO—R}^* \end{smallmatrix} \quad}$$

$$\text{R}^*\!\!-\!\!=\!\!-\text{O}^\cdots$$

The optically pure chiral alcohol R*H was converted into its acrylate ester by reaction with acryloyl chloride in the presence of triethylamine. Diels-Alder reaction of the optically active acrylate in the presence of aluminium chloride as catalyst, in methylene chloride at $-55°$C, gave the bicyclic *endo* ester in 89 % yield. This adduct contains four chiral centres in the acyl moiety. As before, the ring junction diastereoisomerism is determined by the stereochemistry of cycloaddition, whilst the chirality induced in the product depends upon that present in R*. The $(+)$-acrylate gave the stereochemistry required for the synthesis.

Conversion of the bicyclic ester into the bicyclic ketone now required new conditions:

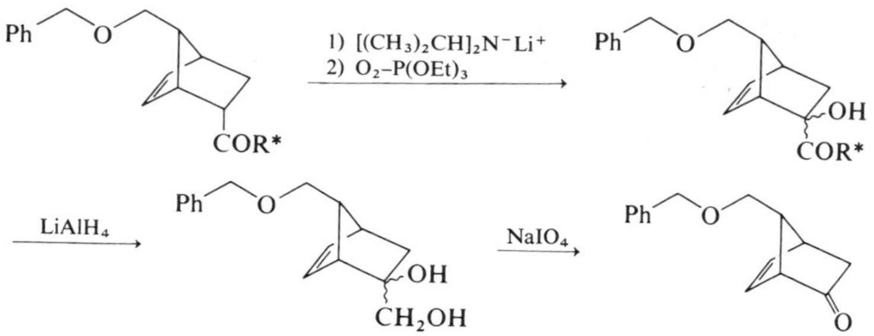

Treatment of the optically active bicyclic ester with lithium di-isopropyl-amide generated its enolate anion which was autoxidized by molecular oxygen. The hydroperoxide so formed was reduced by triethyl phosphite, *in situ*, to the hydroxy-ester. Note that, although the product at this stage was a mixture of diastereoisomers, only one enantiomer of each was present. Reduction of the mixture of hydroxy-esters gave the chiral alcohol, R*H (optically pure and suitable for recycling), and a bicyclic 1,2-diol which was cleaved to give the bicyclic ketone as its required laevorotatory enantiomer.

In this synthesis, therefore, the key prostaglandin intermediate, the bicyclic ketone, was synthesized optically pure without resolution, and the chiral agent used to achieve the asymmetric synthesis was recovered, unaffected, suitable for re-use.

c   Baeyer-Villiger oxidation (p. 471) of the bicyclic ketone gives a single ring-expanded lactone resulting from migration of secondary carbon in preference to the alternative primary carbon. The olefinic double bond avoids epoxidation presumably on account of hindrance of the *exo*-surface by the *syn*-side-chain.

d   The lactone was saponified with aqueous sodium hydroxide and the free acid was obtained by neutralization with carbon dioxide. (In syntheses which had not been chiral a resolution step was then included by crystallization of the (+)-amphetamine salt.) Finally, iodolactonization was achieved by treatment of the acid with iodine in aqueous KI solution (*cf.* Prévost reaction, p. 573):

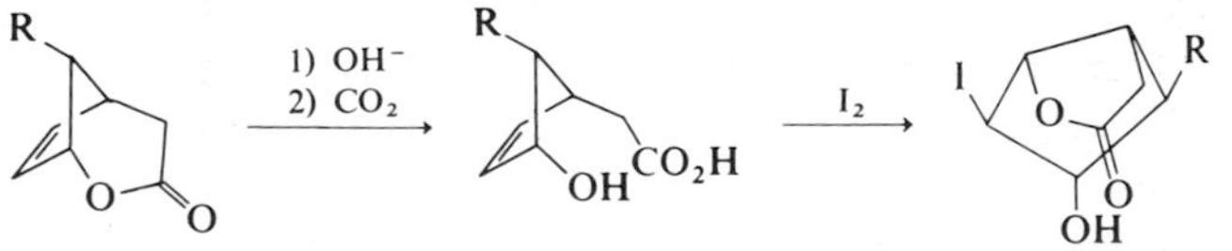

Preference for γ- over δ-lactone formation ensured a unique product now containing five chiral centres.

e   The hydroxy group was protected by acylation. Originally acetylation was used but, later, acylation with 4-phenylbenzoyl chloride was found advantageous in that materials were more consistently crystalline and the aromatic chromophore facilitated location of the products during chromatography. A further important benefit was to accrue later also (step *h*).

Deiodination was effected with tributyltin hydride, the radical-mediated reaction being initiated with azobisisobutyronitrile (p. 535):

$$In\cdot + (C_4H_9)_3Sn\text{—}H \longrightarrow InH + (C_4H_9)_3Sn\cdot$$

$$R\text{—}I + \cdot Sn(C_4H_9)_3 \longrightarrow R\cdot + (C_4H_9)_3SnI$$

$$R\cdot + H\text{—}Sn(C_4H_9)_3 \longrightarrow RH + \cdot Sn(C_4H_9)_3$$

The protecting benzyl group was removed by hydrogenolysis (p. 625) (where methyl was present in place of benzyl, it was removed with BBr_3).

f   The hydroxymethyl group was oxidized to aldehyde with the pyridine complex of chromium(VI) oxide in methylene chloride (p. 587).

g   The aldehyde was coupled with the anion of dimethyl 2-oxoheptyl phosphonate (a variant of the Wadsworth-Emmons reaction; p. 486) to give the *trans*-olefinic bond of the $C_{12}\text{—}C_{20}$ side-chain.

h   Prostaglandins E_2 and $F_{2\alpha}$ have the *S*-configuration at C-15. The evolution of synthetic procedures to build in this configuration stereospecifically is one of the most interesting aspects of this synthesis. The first synthetic sequence used $Zn(BH_4)_2$ as reducing agent. This hydride-transfer reducing agent was preferable to more common alternatives in that the amount of concomitant reduction of the double bond conjugated with the ketone grouping was minimized. However, there was no stereoselectivity in reduction of the carbonyl group, since other chiral centres in the molecule were too distant to influence the course of reaction when the new chiral centre at C-15 was created. In an attempt to introduce stereoselectivity in this reduction, various hindered borohydrides were synthesized and used. The most efficacious combination was one obtained by reduction of the trialkylborane derived from thexylborane (p. 504) and limonene,

and used in tetrahydrofuran at $-120°C$ in the presence of hexamethylphosphoramide as catalytic Lewis base; with this combination the desired α-alcohol predominated over the β-isomer to the extent of 4.5 : 1. It was noted, however, that the $\alpha : \beta$ ratio varied with the nature of the acyl group used to protect the ring-hydroxyl group: the ratio was greater in the desired sense when an elongated rigid function such as 4-phenylbenzoyl was used.

This was rationalized as follows. Molecular models showed that the 4-phenylbenzoyl group could lie alongside the -enone group in the aliphatic side-chain, possibly being weakly bonded to it by π-π interaction,

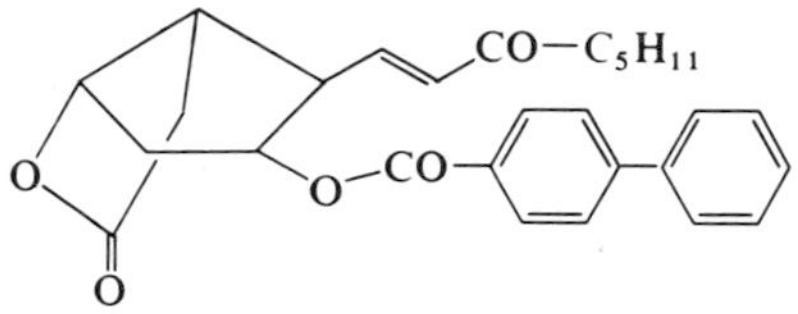

Moreover, the models showed also that the interaction could be more extensive when the -enone function is in the *s-cis* conformation, i.e.

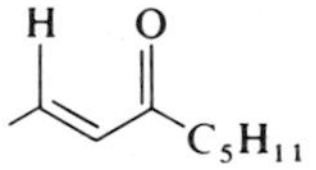

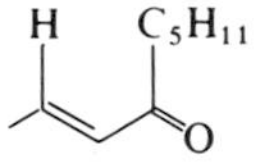

When the -enone is reduced in the *s-cis* conformation from the side opposite to the biphenyl moiety, alcohol of the desired stereochemistry results.

Structural variations were made in the protecting function to improve further the stereoselectivity of reduction. Eventually, by using biphenyl-4-isocyanate in place of the acyl halide as the protecting agent, the resultant urethane could be reduced with a stereoselectivity of 92% α-alcohol and 8% β-alcohol at C-15.

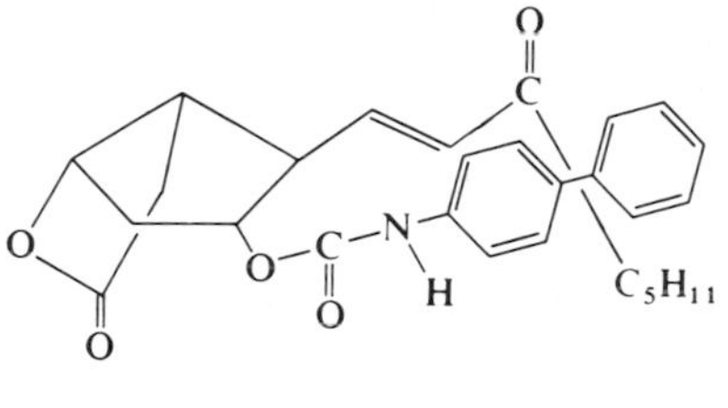

The biphenyl moiety is thus a stereochemical control element as well as having the other advantages listed under step *e*.

i Alkaline hydrolysis removed the protecting group, whether ester or urethane, and the resultant diol was treated with dihydropyran in methylene chloride containing toluene-*p*-sulphonic acid; this protected both hydroxyl groups as 2-tetrahydropyranyl derivatives (p. 686). The lactone ring was then reduced to a hemi-acetal with di-isobutylaluminium hydride in toluene. A hemi-acetal is a masked aldehyde:

j   The C_1—C_8 side-chain was attached through a Wittig reaction. The masked aldehyde, created in the previous step, reacted with the phosphorane resulting from deprotonation of 5-triphenylphosphoniopentanoic acid by the dimsyl anion:

$$Ph_3P + Br(CH_2)_4CO_2H \longrightarrow$$

$$\overset{+}{Ph_3P}-CH_2-(CH_2)_3CO_2H\ Br^- \xrightarrow[\text{CH}_3\text{SOCH}_3]{\text{CH}_3\text{SOCH}_2{}^-\text{Na}^+} Ph_3P{=}CH-(CH_2)_3CO_2{}^-Na^+$$

The noteworthy aspect of this reaction is that the olefin created is stereospecifically *cis*. This is apparently a general property of the Wittig reaction when it is carried out in dimethyl sulphoxide.

The prostanoid material thus formed is the immediate precursor of both prostaglandins E_2 and $F_{2\alpha}$. Treatment with aqueous acetic acid hydrolyzed away the pyranyl protecting groups to give prostaglandin $F_{2\alpha}$ (step *k*), whilst oxidation of the unprotected hydroxyl group by chromium(VI) oxide followed by similar deprotection (step *l*) yielded prostaglandin E_2.

Problems

1. Some of the products of the sequences outlined below have been used in the synthesis of naturally occurring compounds, and others are themselves important natural products. Complete the synthetic schemes in as much detail as you can: insert the formulae of intermediates which have been omitted and state the experimental conditions which you consider suitable for carrying out the individual steps. Comment, where relevant, on the stereochemistry of the intermediates.

(*a*) $CH_2{=}CH-(CH_2)_8-CO_2H \longrightarrow CH_3-(CH_2)_{10}-Br$

$\longrightarrow CH_3-(CH_2)_{10}-C{\equiv}C-(CH_2)_3-Cl \longrightarrow$

$CH_3-(CH_2)_{10}-CH \overset{cis}{=\!=\!=} CH-(CH_2)_4-CO_2H$

(*b*)

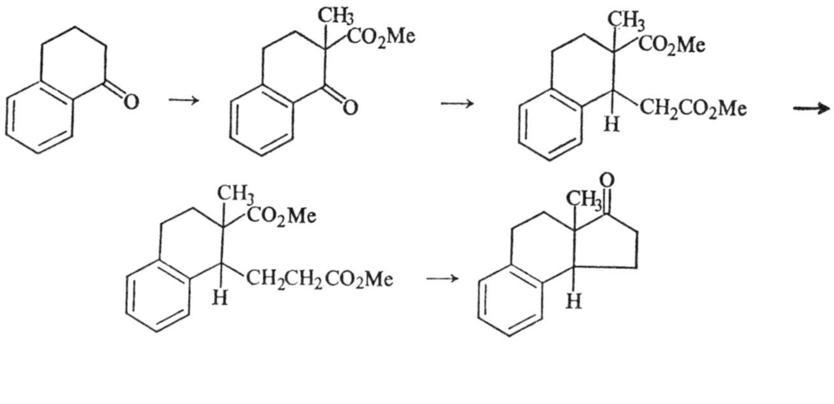

(c)

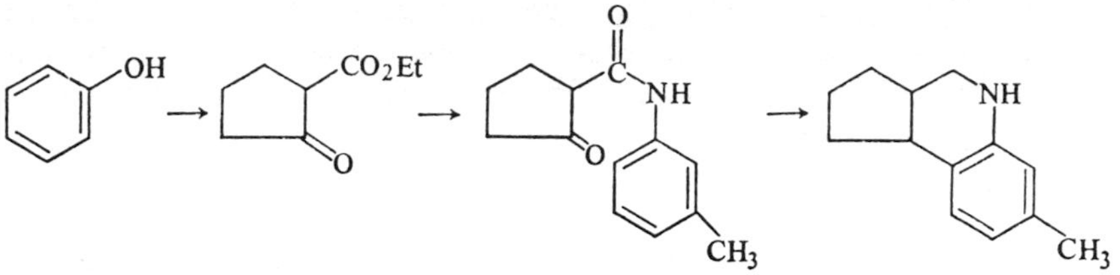
OH
CO_2Et
O
O
C
NH
O
CH_3
NH
CH_3

(d)

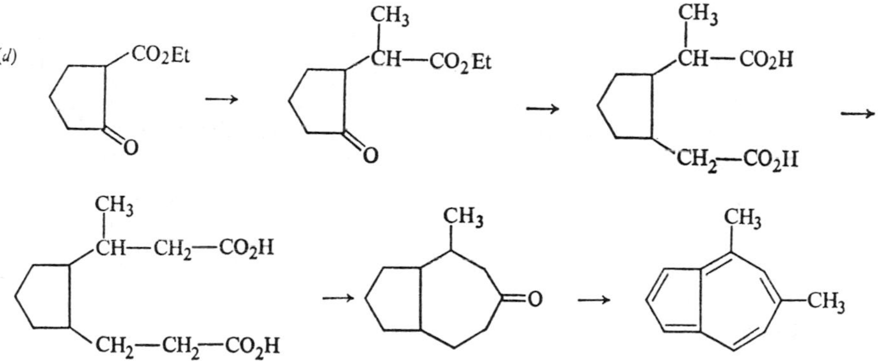
CO_2Et
O
CH_3
CH—CO_2Et
O
CH_3
CH—CO_2H
CH_2—CO_2H
CH_3
CH—CH_2—CO_2H
CH_2—CH_2—CO_2H
CH_3
O
CH_3
CH_3

(e)

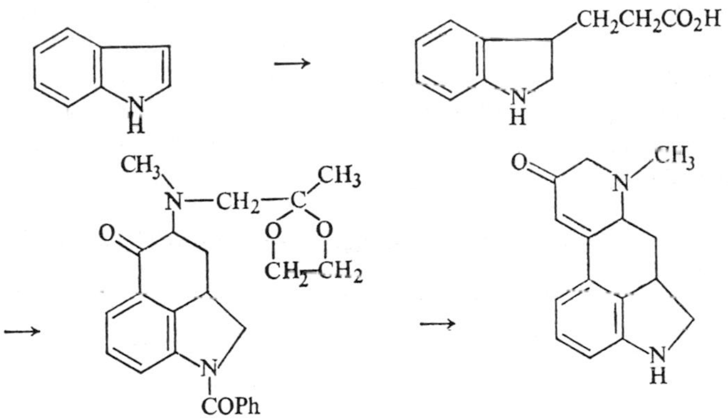
N
H
$CH_2CH_2CO_2H$
N
H
CH_3
N—CH_2—C
CH_3
O
O
CH_2—CH_2
O
N
COPh
O
N
CH_3
N
H

(f) ?→

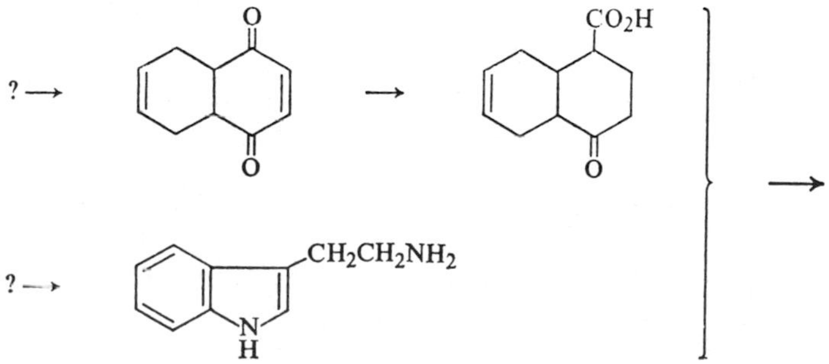
O
O
CO_2H
O
?→
$CH_2CH_2NH_2$
N
H

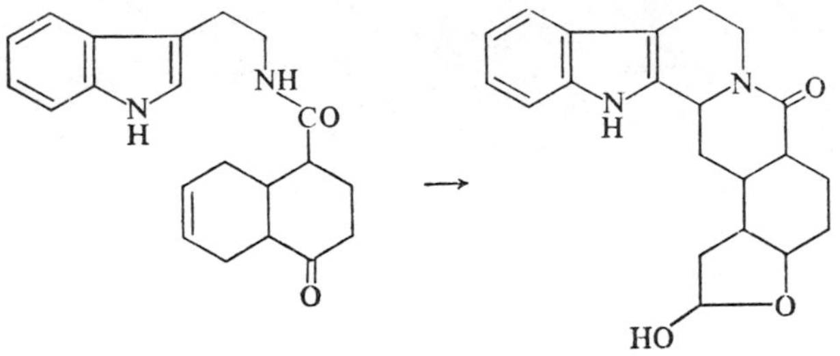

(g) ? →

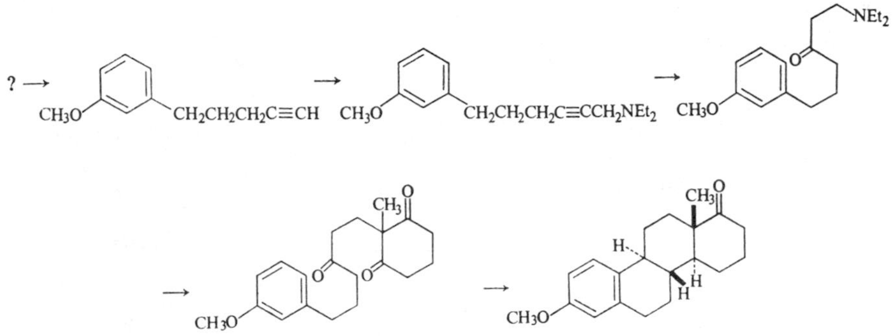

(h) ? →

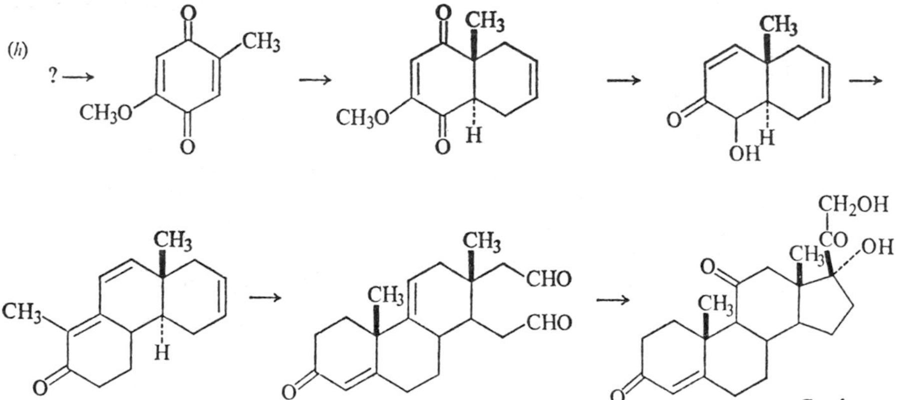

Cortisone

(i) ? →

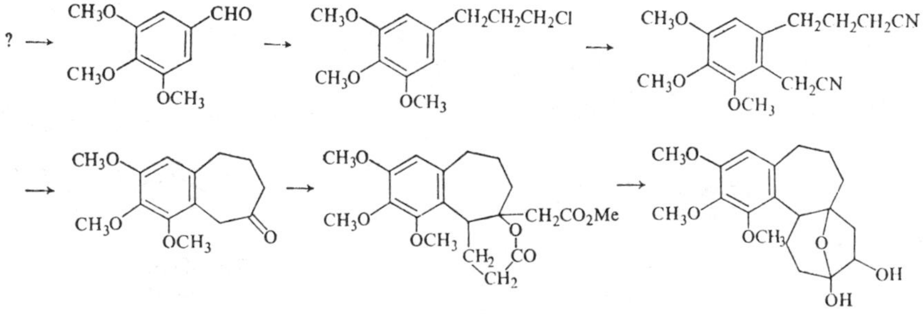

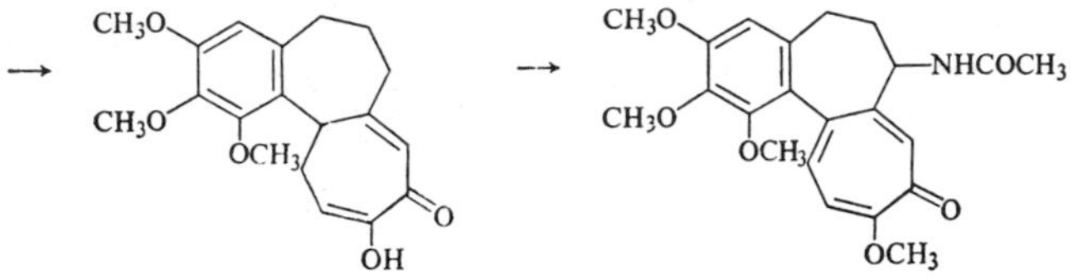

Colchicine

2. Outline methods for the synthesis of the following compounds from readily available materials.

(a)

$$CH_3—(CH_2)_7—CH \overset{cis}{=\!\!=} CH—(CH_2)_7—CO_2H$$

Oleic acid

(b)

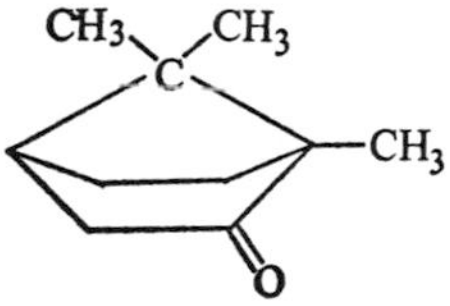

Camphor

(c)

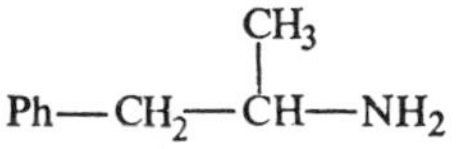

$$Ph—CH_2—\overset{\overset{\displaystyle CH_3}{|}}{CH}—NH_2$$

Benzedrine
(Amphetamine)

(d)

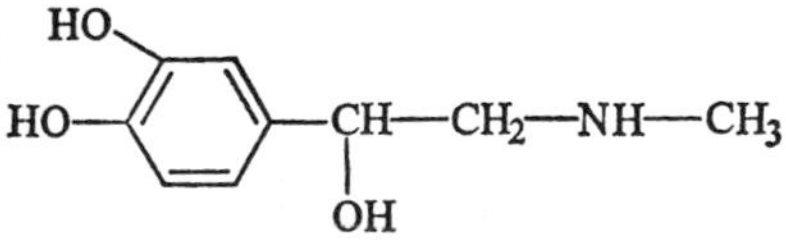

Adrenaline
(Epinephrine)

(e)

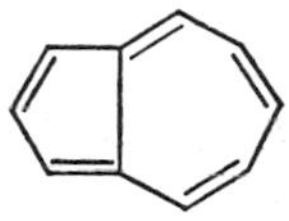

Azulene

(f)

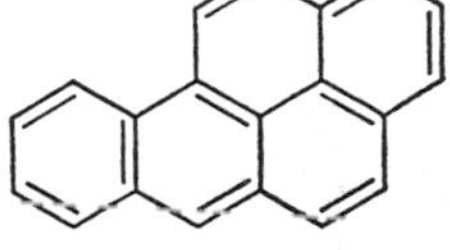

3:4-Benzopyrene

(g)

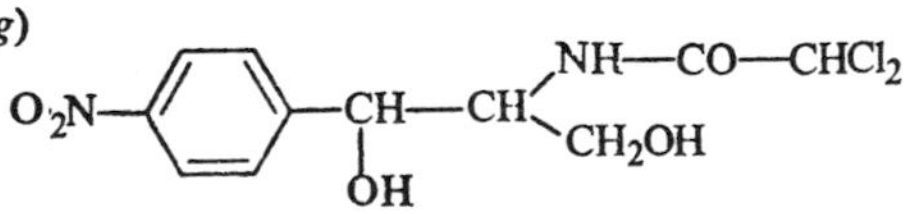

Chloramphenicol
(an antibiotic)

(h)

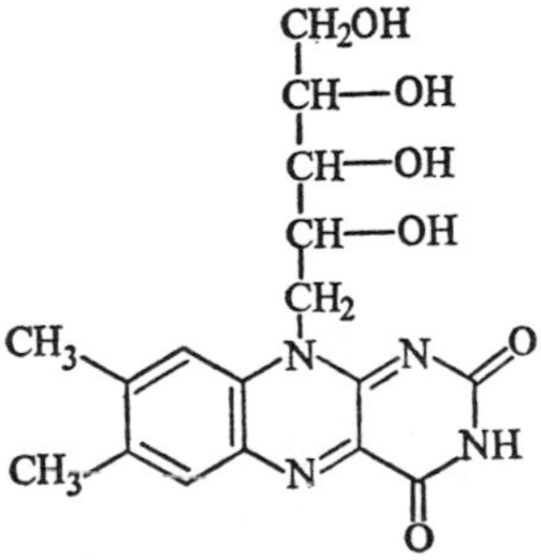

Riboflavin
(Vitamin B$_2$)

Index